Gels, Genes, Grafts and Giants

GELS, GENES, GRAFTS AND GIANTS

Festschrift on the occasion of the 70th birthday of Allan S. Hoffman

Edited by

STUART L. COOPER, THOMAS A. HORBETT,
BUDDY D. RATNER and PATRICK S. STAYTON

VSP
Leiden · Boston

2005

Koninklijke Brill NV incorporates the imprints Brill Academic Publishers, Martinus Nijhoff Publishers and VSP.

ISBN 90 6764 408 0

Printed in the Netherlands by Ridderprint bv, Ridderkerk

Contents

Gels, Genes, Grafts and Giants, p. 1 (2005)

Foreword

In celebration of Allan Hoffman's 70th birthday, a symposium entitled *Gels, Genes, Grafts and Giants* was held in Maui, Hawaii in December of 2002. This symposium, organized through the UWEB NSF Engineering Research Center, was a great success, with many excellent papers presented by scientists from all over the world over the 3 days it was held. At the symposium, Stuart Cooper observed the many fine presentations and took the opportunity to invite the three of us to organize special issues of *Journal of Biomaterials Science Polymer Edition* based on the symposium. After considerable discussion among ourselves and Stuart Cooper, it was decided that we would solicit regular scientific articles with content decided by the authors, although we did suggest that the articles be generally related to the topic of the symposium or the mission of the journal. This approach was consistent with the special issues published in honor of Allan's 60th birthday and with the focused content of the journal.

We sent letters of invitation to all those who made oral or poster presentations at the symposia, but in addition we invited many others who were unable to attend but who worked with Allan in the past or worked in similar research areas. Our goal was to include papers from a cross-section of research areas that Allan had worked in during his career represented in the special issues. The response to our efforts was excellent, as 58 papers were published in 6 Special Issues, and so we wish to thank all of the contributors for their help in celebrating Allan's 70th birthday and advancing the science and technology of polymeric biomaterials! It was finally decided to make these important manuscripts more widely available by publishing a separate book entitled *Gels, Genes Grafts and Giants*.

In the Preface to this volume, following this Foreword, some of the specific contributions of Allan Hoffman are described, so we will not repeat those here. Instead, we wish to say again "Congratulations, Allan!" on your fine career, and congratulations to all the contributors for their advances to the field of polymeric biomaterials — advances that demonstrate creativity, rigorous science and concepts that may improve medical diagnostics and therapies for humans. Further, it is important to remember that many of the papers in this pivotal volume have been inspired by Allan Hoffman's ideas, enthusiasm and encouragement.

Your colleagues

THOMAS HORBETT
BUDDY RATNER
PATRICK STAYTON

January 2005

Gels, Genes, Grafts and Giants, pp. 3–5 (2005)

Preface
Allan Hoffman: A 70th Birthday Celebration

Allan S. Hoffman, Professor of Bioengineering and Chemical Engineering at the University of Washington, was born in Chicago on October 27, 1932. He received BS, MS and ScD degrees from Massachusetts Institute of Technology, completing his doctoral thesis in 1957 under the guidance of E. R. Gilliland, E. W. Merrill and W. H. Stockmayer. His first positions were at MIT, as the Assistant Director of the Oak Ridge Engineering Practice School and then as an instructor. After a few industrial positions, including a stint with Chevron in the Bay Area (California, USA) and Associate Director of Research for Amicon Corporation, Allan returned to MIT, first as Assistant Professor (1958–1960) and then as Associate Professor (1965–1970). In 1970, after a short stay at the Centre d'Etudes Nucleaires in Saclay (France), he came to the University of Washington.

The UW has served for 32 years as a utilitarian platform to build the illustrious career we see today. That is the concise history, but there is so much more to Allan Hoffman that these sparse facts do not even hint at.

Allan Hoffman's contributions to the field of biomaterials have been monumental. His first published paper was in 1955. His first paper on surface technology (radiation grafting) appeared in 1959. By 1969 he had a published a paper on hydrogels for desalination (synthetic hydrogels were first accurately described in 1960; it is this paper that convinced me to come to Seattle to work with Allan). His first papers focusing on biomedical polymers began appearing about 1971. Once Allan had coalesced his vision of polymers in medicine and surgery, it was as if floodgates were opened. A tsunami of hundreds of papers followed, surging through the intellectual landscape and shaping the terrain of biomaterials as we know it today. What has Allan Hoffman given us? Here is a short list (probably incomplete): biomedical hydrogels, mechanics of natural tissue, surface modification, new surface analysis methods, covalent biomolecule immobilization on biomaterials, nonfouling surfaces, drug delivery technology, new concepts in blood compatibility, stimuli-sensitive polymers (polyNIPAM was brought to our community by Hoffman), immunoassays, biomolecule separations, smart bioconjugates and novel gene delivery molecules. Allan Hoffman has profoundly influenced current ideas in biomaterials such that, without his guiding light, many ideas we take for granted would not be with us today and biomaterials would be a different and, most likely, intellectually poorer, endeavor. We see creativity of the highest order in Allan Hoffman.

Allan Hoffman's contributions go much farther than his scholarly activities. With a personality of gold, a love for cultural diversity and willingness to give and contribute, Allan has helped create the collegial, international, supportive biomaterials community we all enjoy today. He has introduced thousands to biomaterials ideas, mentored innumerable young (and some experienced) scientists and befriended hundreds. He has been a world ambassador, not just for biomaterials, but also for the intellectual excitement that lights up science. His creativity, innovation and knack for asking the right question, projected from the podium with the inimitable Hoffman savoir faire, have stimulated us all. How often has he warmed an audience in Japan, China, Italy, Israel, Germany, France, Russia, Spain or Korea with a choice phrase, not badly pronounced, in the local language? He has consumed the cuisine, worn the costumes, sung the songs, danced the dances and downed the local brew of cultures around the globe. And his warmth, concern, great memory and natural curiosity have been rewarded with a lifetime of international friendships.

This symposium honors the 70th birthday and some 50 years of scientific accomplishment of our colleague and friend, Allan S. Hoffman. Hoffman at 70 is not all that different from Hoffman at 30 — only a little slower in the walk, a bit grayer, definitely more savvy, and as creative, energetic and personally warm as ever. Visualize Allan Hoffman as so many have seen him — in an outdoor café in some exotic locale, perhaps overlooking a river. From beneath the table umbrella,

a cloud of smoke wafts — there is Allan, cigar in one hand, calvados in the other, and few close friends, engaged in rich conversation, smiles abound in the glow of a setting sun, the yellow shirt and a shining light — our friend Allan.

Happy Birthday, Allan. Much of the intellectual richness of this symposium can be directly attributed to the inspiration you have provided.

BUDDY D. RATNER
Program Chair

Gels, Genes, Grafts and Giants, pp. 7–12 (2005)

Some personal reflections on the career of Allan Hoffman in honor of his 70th birthday

TOM HORBETT *, BUDDY RATNER † and PAT STAYTON ‡

These special issues were organized by the three of us in honor of Allan Hoffman's 70th birthday, as explained in the Foreword. In the Preface, Buddy Ratner has given a concise and broad overview of the public aspects of Allan Hoffman's career that help to explain the overall impact of his work. As Buddy has described, these contributions can be associated with his creativity and quality research, effectively communicated through his many publications and presentations around the world.

However, as we learn in the progression through our careers, personal interactions with our more experienced colleagues often have profound impacts on us. These less formal, seldom described interactions are important contributions to the development of young investigators, and this was certainly true for the three of us under the tutelage of Allan Hoffman.

So we have decided to further honor Allan for these contributions by describing a few of our personal highlights from our long association with him. Each of us has taken the opportunity to reflect a bit on what Allan has meant to us as a colleague and a scientist, and also about those notable characteristics we see in Allan that perhaps only those working closely with him for a long time really notice.

TOM HORBETT'S REFLECTIONS ON ALLAN HOFFMAN

I was Allan's first post-doc when I joined him in 1970. At that time, Allan and my department chairman and thesis committee member Hans Neurath were both associated with the Batelle center in Seattle, and Neurath recommended me to Allan when he started looking for post-docs. I had finished up my doctoral studies in biochemistry at the University of Washington and was doing a bit more research related to my thesis project, writing up my thesis publications, and applying for

*E-mail: horbett@uweb.engr.washington.edu

†E-mail: ratner@uweb.engr.washington.edu

‡E-mail: stayton@u.washington.edu

post-doc and industrial positions. In that period of searching for my next position, I began to realize that I did not want to continue doing the type of very basic biochemical research I had done in graduate school, and also found that as I had become more familiar with Seattle and the northwest outdoors, I was kind of reluctant to leave. So when Allan asked me to join his lab I became quite interested because biomaterials were actually being used clinically and I was attracted to working in a research area that had more direct relevance to clinical medicine. I remember being impressed by Allan's enthusiasm and positive outlook about his then still new foray into the biomaterials field, and also liked that the work was so different in content and style from what I had been doing. I also really liked the idea of being able to stay in Seattle a little longer. Of course, here I still am, so joining Allan was a pivotal decision in my life.

At the time I joined Allan, he was interested in whether low protein adsorption to the radiation-grafted polyHEMA and polyNVP hydrogels he was making might be a mechanism for their apparently improved blood compatibility. Those materials proved to adsorb less protein than the underlying silicone rubber they were grafted to, and to adsorb it less tightly, but they still adsorbed considerable protein, and were no where near as protein repellant as we would have liked. In addition, these materials also turned out to be more reactive with blood than was originally thought, so that even though they had much less adherent clot when tested in the vena cava ring test, they nonetheless exhibited enhanced platelet turnover in the studies that Hoffman, Ratner, Hanson and Harker conducted in baboons.

Many of us in the biomaterials community gave up on hydrogels after that experience, but I view the current high interest and activity in non-fouling materials as just the latest wave of an approach that Allan and others started in the early 70s. As I noted at the UWEB non-fouling symposium a year or two ago and at the 70th birthday symposium on Maui, I think Allan and others had the right idea but could not make materials that met the theoretical requirements, i.e. it turns out low protein adsorption is in fact a good idea for increased blood compatibility, but you need materials with really low protein adsorption. Hoffman was not totally misguided in spending all that Atomic Energy Commission money making radiation-grafted hydrogels for improved blood compatible biomaterials, he just needed the help of many subsequent investigators to make clear what exactly was needed to achieve low protein uptake. Allan's invitation to work on protein adsorption in his lab has led to my long and continued research in this field, and to my belief that he was right, that we can make better blood compatible materials if we can only make very weakly adsorptive materials.

Another reflection I would like to make has to do more with the style of Allan's research. Any one who knows Allan values his independence in thought, which I always think is a key attribute of any really productive researcher. If you work with Allan, as I did as a post-doc and then for along time after as a colleague, you learn that you must often work independently for considerable periods of time while Allan carries out his many worldwide duties. I am pretty sure this is how he himself functioned in his student days — it was expected that you had to quickly develop your own ideas and approaches to the problem assigned to you, and carry out the studies too. Of course, this is a great way to learn to become a truly independent researcher — when we have to rely on ourselves, we find we can do a lot more than we thought. In my case, Allan really encouraged me to write research proposals largely on my own, very early in my career. Given that free rein, and Allan's generally ready support of my ideas, I developed the necessary knowledge of my own abilities, which gave me confidence to continue despite the typical ups and downs of the funding cycles.

So, I give a lot of credit to Allan for allowing me to learn the true meaning of independent research, embodied in a research proposal. I was also lucky in that I always had Allan and Buddy as collaborators on the proposals I wrote, and the recognized productivity of this association made it easier for us to obtain funding. I never really seriously considered leaving the University of Washington because I never found any other setting that was nearly as good for productive collaborations. So, thanks Allan, for letting me write those early proposals! It had a big impact!

I have learned a lot about other aspects of being a professor from Allan, as he was my mentor in more than just the research aspects of my career. Allan created a nice set of courses in our department, and in recent years I have had the opportunity to teach these courses when Allan could not, and I often followed his excellent lead in regards to both content and style. Just like Allan, I still love transparencies, PowerPoint be damned. And just like Allan, I love guest lecturers when I can get them, but also just like Allan, I worked hard to make the lectures rich in content and insight and perspective, and take our students classroom learning seriously. I also learned a lot from Allan about how to conduct myself in the many meetings any professor has to attend, from student committees to faculty meetings, to departmental strategy meetings, especially to listen carefully to understand the concerns of others before speaking. But, I never learned to be as quiet as he generally is at such meetings. He often limits himself to the few comments really needed and carefully avoids any unnecessary extension of the discussion. Most of all, I have learned from Allan to be my own man in these meetings, to think for myself and say what I think, but try to be as courteous as possible. I learned from Allan that politeness and courtesy and thoughtfulness are a powerful combination in a faculty meeting.

So, Allan, I want to say that the more I reflect on our long association and your long and highly successful career, the more I realize how much I have benefited from it. So, happy 70th again, and congratulations on the wonderful 70th birthday Festschrift issues your colleagues have created in your honor!

BUDDY D. RATNER'S REFLECTIONS ON ALLAN HOFFMAN

In 1970 I was a graduate student in the polymer chemistry PhD at Brooklyn Polytechnic Institute. At that time, most of the world's leaders in polymer science and polymer engineering populated the halls and labs of the revamped commercial building that in the 1920s housed the American Safety Razor factory. There was a seminar by a Professor Allan Hoffman of M.I.T. The topic looked interesting. At the seminar, Prof. Hoffman espoused a philosophy integrating engineering, biology and synthetic polymers. I had never heard this idea expressed so clearly, so persuasively, before. After the seminar there were many questions. I could not get to speak to Hoffman. And then the seminar was over and he was being escorted down the hall by Brooklyn Poly senior professors. I ran after them and asked if I could get a few minutes to talk to Professor Hoffman. I can still remember one of the senior profs glaring back at me — loosely paraphrased he said, 'get lost punk'. I was a bit crestfallen. But, undeterred (the arrogance of youth), when it was time to find a post-doc position, I wrote to Allan at MIT. The letter slowly made its way from MIT to the University of Washington, Allan's new home. I suppose we hit it off in letters (typed on something called a typewriter). He offered me a position.

I arrived in Seattle on May 1, 1972 in a beaten-up, blue, 40 horsepower Renault 10, after a 6000 mile meandering journey across the heartland of the USA. My first stop in Seattle was the University of Washington and Allan Hoffman's office in the faculty row on the second floor of Benson Hall. There was Allan, sitting at his desk and — as I think back, it is almost frightening — he looked, in my mind's eye, almost identical to how he looks today. Did I find a 70-year-old guy sitting in that office in Benson Hall, or am I looking at a 40-year-old guy today? One of the realities of Allan Hoffman is a timeless quality; Peter Pan with a pocket protector?

Shortly thereafter I was introduced to Tom Horbett, another experience I will never forget, but that is another story for another time. Shortly after, we did some scheming on research ideas. Allan pointed me to a laboratory and a desk. He communicated to me a vision of a creative approach to science and handed me some reins to hold. And then, 30 years flew by. Pat Stayton and many others joined us over this period. In fact, looking at the Hoffman research group over 30 years, and the many groups at the University of Washington spawned by that initial Hoffman group, hundreds of faces can be conjured up — these individuals have gone out into the world of science, medicine and industry, and have formed the foundation upon which a world community has been built. And Allan Hoffman's special charisma directly touched every one of these people.

My personal recollections of Allan Hoffman are numerous. I remember Allan patiently coaching me on presentation style in a hotel room in Jerusalem in 1974 before my first international talk. I remember Allan 'bloodying' my draft manuscripts with his ever-sharp red pen, teaching me to write well. I remember a hundred brainstorming sessions, in lab groups, government forums and companies, where Allan consistently led with creative ideas. I remember Allan showing me 'the real' Paris on my first visit to the city of lights, also Istanbul, Kyoto, Capri, New Orleans and who knows how many other cities. In fact, New Orleans has a special story. We were there for a Society For Biomaterials meeting in the 1970s. The first evening after the sessions ended, he took me to Felix's Oyster bar. I swallowed my first raw oyster there, learned to love oysters and now I start every visit to New Orleans with a trip to Felix's. I am not sure I have swallowed every line Allan's fed me, but the oysters were a hit.

Allan Hoffman has taught me about, writing, lecturing, fundsmanship, eating, drinking and savoring this wonderful life that the vibrant biomaterials scientific community creates. Perhaps most importantly, he has communicated to me, by example, lessons in the importance of friendship, mentorship and humanity. I hope I pass on these important lessons to future generations. But for now, I continue to learn at the foot of the master, I appreciate the profound influence Allan has had on my life and career, and I enjoy the ever-mellowing company of Allan Hoffman, my friend for over 30 years.

PAT STAYTON'S REFLECTIONS ON ALLAN HOFFMAN

As the most recent of the UW Biomaterials group to have become associated with Allan, it feels a bit strange at first to be looking backward already. A thousand ideas and projects seem to be swirling around and pulling Allan and I forward together into the future at a dizzying pace. I suppose that is how a decade and more slips away so quickly ... we are lucky indeed to be driven by so much excitement and a feeling that we might help develop something as important as new medicine.

What fun we have had! That is what I remember first. Fun in science and fun in engaging life itself. The two have been inseparable and after all, they really are not so different. I met Allan as a fresh Assistant Professor in 1992, still without a desk in my office, and biomaterials were a foreign vocabulary to me. Allan pulled a couple of lawn chairs out of his car in the university parking lot and we walked over to a beautiful spot on the lawn near his office (one of our many beautiful sunny days in Seattle). In a fraction of a nanosecond, his cigar was lit and we were talking about some type of polymer that he claimed was 'smart'. We were quickly passing a pen back and forth and drawing structures and cartoons on a pad of paper. I had to admit, those polymers did look pretty smart. Soon we had outlined an idea that continues to lead us into new and unexpected directions. Never has a man loved new molecules more than Allan Hoffman, and I still love nothing better than having a free couple of hours with Allan in a room that has an empty blackboard.

What a romantic time that was, just starting a lab, working around the clock, entering a new field. Allan was the best kind of mentor and guide because we were collaborating so closely; I could absorb lessons actively by osmosis. I still marvel at his ability to tell a complicated scientific story so cleanly, providing scientific details while making it comprehensible to the non-specialist. He was also a big help in learning how to work with students from the advisor side, what was fair to expect from them, how to motivate them, and to take a more philosophical approach toward their different talents and natural rhythms of productivity. I learned everything imaginable: local university politics, how to balance the need to publish fast with the need for completeness and how to get upgraded on long airline trips. There is really nothing more valuable to a young professor than having the support of a senior faculty in your department who believes in you, and for that I will be forever indebted. It wasn't just me, Allan is very generous with young faculty and many from around the world have commented to me about his encouragement.

I learned a lot by observation, but believe me, Allan has never been shy about expressing his opinion either. Like the time he started right in with the president of Yale at a parents reception, telling him that he needed to do a better job of keeping the campus safe for Allan's daughter Lisa . . . well, that is another story. There are lots of stories. All of the exciting science is matched to great memories from around the world. Buddy said it perfectly in the Preface, what a wonderful experience to have a simple table outside with Allan in some far away place and time to talk. And some tough times too, away from our profession, the things that cement a friendship. Now we are like an old couple I suppose, grousing at each other at times, but Allan is family.

He loves his real family to death, daughter Lisa and son David, whom I am fortunate to see frequently. But he also exudes a feeling of family in the biomaterials community as well. It is hard to find exactly the right word, but I sort of like graciousness; Allan has a graciousness about him that is rarer these days, always remembering something personal about people, little presents, cards, and that warm handshake at meetings. The great thing about Allan is that the journey continues, for that we can all be thankful of. Take my word for it, though Allan can give a great history or overview, he hates it, he wants to talk about the latest results and exciting progress. What a big, rich life he has had, and continues to have. We can still find him at that outside table, take just a nanosecond to congratulate and thank him, because he wants to hear about your latest discovery.

Drug delivery

Gels, Genes, Grafts and Giants, pp. 15–26 (2005)

Cellular evaluation of insulin transmucosal delivery

JENNIFER E. LÓPEZ [1] and NICHOLAS A. PEPPAS [2,*]

[1] *School of Chemical Engineering, Purdue University, West Lafayette, IN 47907-1283, USA*
[2] *The University of Texas at Austin, Departments of Chemical Engineering, Biomedical Engineering and Pharmaceutics, 1 University Station, CPE 3.466, Mail Code C0400, Austin, TX 78712-0231, USA*

Received 7 August 2003; accepted 30 December 2003

Abstract—P(MAA-g-EG) microparticles have been extensively investigated as carriers for oral delivery of proteins such as insulin. In this study, we investigated the effect of the molecular weight of the PEG tethered chains in the copolymer network and of the microparticle size on the transepithelial electrical resistance (TEER) and insulin epithelial permeability, using monolayers of human intestinal epithelial Caco-2 cells. Two molecular weights of the PEG chains, 400 and 1000, were investigated, as well as three different dry microparticle sizes: 25–90, 90–150 and 150–212 μm. Their effect on the cell monolayer integrity was studied by monitoring TEER as a fraction of time and determining insulin permeability. The presence of insulin-loaded P(MAA-g-EG) microparticles decreases the TEERs value by 50% with respect to the control. This disruption of the cell monolayer was recovered in 3 h after the removal of the polymer microparticles. Within the range of PEG molecular weights studied, there was no significant change of the TEER values. However, decreased microparticle sizes and short PEG chains systems led to higher permeability values. Insulin-loaded P(MAA-g-EG) microparticles enhanced the transport of insulin through the Caco-2 cell monolayers.

Key words: Oral delivery; insulin; Caco-2 cells; transmucosal permeability.

INTRODUCTION

Noninvasive methods of insulin delivery are highly desirable in diabetes treatment, with the oral route clearly being the most convenient [1]. The oral route of administration faces several barriers such as the presence of proteolytic enzymes in the gastrointestinal tract, chemical instability and poor permeability of proteins across biological membranes.

In our laboratory, we have successfully developed microparticulate and nanoparticulate systems based on unique crosslinked copolymers that can be used as carri-

*To whom correspondence should be addressed. Tel.: (1-512) 471-6644. Fax: (1-512) 471-8227. E-mail: peppas@che.utexas.edu ers for oral delivery. These systems consist of hydrogels of poly(methacrylic acid) (PMAA) or poly(acrylic acid) (PAA) grafted with long chains of poly(ethylene glycol (PEG). These networks will be henceforth designated as P(MAA-g-EG) and P(AA-g-EG). These hydrogels exhibit complexation characteristics as well as pH sensitivity. Thus, they can protect the protein from the harsh environment of the stomach and release it in the small intestine when needed [2–4]. These copolymer systems exhibit good mucoadhesive properties due to the presence of the PEG tethers [5]. In addition, we have shown that they can protect and stabilize proteins and inhibit proteolytic enzymes [6].

In previous studies, cultured epithelial cells have been successfully used as tools for the screening of drug permeation and transport [7]. In this study, we used a Caco-2-cell line as an intestinal epithelium model, which can be used as an intestinal model for drug absorption, transport and permeability studies. Caco-2 cells are derived from human colorectal adenocarcinoma. They undergo spontaneous differentiation in culture and form monolayers that exhibit morphological and functional similarities to the small intestinal epithelium [8, 9].

The absorption of large hydrophilic molecules is limited by paracellular and transcellular barriers. Specifically, the paracellular pathway is restricted because of the presence of the tight junctions between cells. Different approaches have been used to induce paracellular transport of large molecules, including proteins, by disrupting the tight junctions. Some of these approaches include the use of surfactants [10], H_2-antagonists [11], polysaccharides such as chitosans [12, 13] and poly(acrylic acid) derivatives [13–16]. However, some of these approaches compromise the integrity of the cell monolayer in order to enhance permeability [17].

Previous studies in our laboratory have shown that of P(MAA-g-EG) microparticles and nanospheres are not cytotoxic [18, 19] when in contact with Caco-2 cell monolayers. We have also proven the ability of P(MAA-g-EG) microparticles to improve the transport of proteins such as insulin [18] and calcitonin [20] through the cell monolayer. Our studies have shown beyond doubt that protein transport is due to the ability of this copolymer system to bind calcium causing disruption of the tight junctions and facilitating the paracellular transport [7].

In our present studies with insulin-loaded P(MAA-g-EG) microparticles we investigated the molecular weight of the PEG tethered chains and the importance of the microparticle size on protein transport. We focused on the effect of PEG molecular weight on transport resistance and permeability because we have previously shown that tethered PEG chains act as mucoadhesion promoters [1, 6]. We wanted to examine if this mucoadhesive process could have an adverse effect on cellular transport. Studies of effect of microparticle size on cellular transport became necessary because of previous studies on the influence of particle transport on bioavailability [1].

The first studies involved the measurement of the transepithelial electrical resistance (TEER) in order to monitor the integrity of the tight junctions in the presence of the protein-loaded microparticles. The second set of studies was done to evaluate the effect of these variables on the insulin permeability through the cell monolayer.

MATERIALS AND METHODS

Materials

Methacrylic acid (MAA) and poly(ethylene glycol) ether monomethacrylate (PEGMA) with PEG molecular weights of 400 and 1000 were purchased from Aldrich (Milwaukee, WI, USA) and used to synthesize P(MAA-g-EG) microparticles. Tetra(ethylene glycol) dimethacrylate (TEGDMA) was purchased from Polysciences (Warrington, PA, USA) and used as the crosslinking agent in the amount of 0.75% mol of the total monomers. Irgacure® 184 was employed as the photoinitiator and obtained from Ciba-Geigy (Hawthorne, NY, USA); it was incorporated in 0.1% (w/w) of the monomer mixture. Bovine insulin was purchased from Sigma (St. Louis, MO, USA).

Preparation of P(MAA-g-EG) microparticles

P(MAA-g-EG) hydrogel films were prepared by free radical solution UV-polymerization of MAA and PEGMA in the presence of TEGDMA. The monomers were mixed in 1 : 1 molar ratio of methacrylic acid-ethylene glycol units. PEGMAs with molecular weights of 400 and 1000 of the PEG graft chains were used in the synthesis. TEGDMA was used as the crosslinking agent and was added in the amount of 0.75 mol% of the total amount of monomers. The photoinitiator, Irgacure 184 was added in the amount of 0.1 wt% of the total amount of monomers. The monomer mixture was diluted with a mixture of 50% (v/v) ethanol and deionized water. Nitrogen was bubbled through the well-mixed solution for 15 min.

This mixture was poured between microscope slides (75 $\times$ 50 $\times$ 1 mm; Fisher, Pittsburgh, PA, USA) separated by Teflon spacers with a thickness of 0.9 mm. The microscope slides were then placed in a nitrogen atmosphere under a UV light source of 11 mW/cm^2 at 365 nm for 30 min. The ensuing polymer films were washed in deionized water for 7 days in order to remove unreacted monomers, crosslinking agent, initiator and sol fraction. The polymer films were then dried in a vacuum oven at 27°C for 3 days, crushed by using a mortar and pestle and then sieved to the desired particle size range (25–90, 90–150, or 150–212 μm).

Insulin loading

Insulin loading was accomplished by equilibrium partitioning of insulin into the P(MAA-g-EG) microparticles. Insulin stock solutions with a concentration of bovine insulin of 0.5 mg/ml and ionic strength of 0.1 M were prepared. In a typical experiment, 7 mg of polymer/ml of insulin stock solution was mixed at a pH of 7.4 for 6 h. Due to the hydrogen bonding decomplexation, the particles swelled, thus absorbing insulin. Then, the system was rendered acidic by adding 0.1 M HCl. The protein-loaded microparticles were vacuum-filtered with a 0.45 μm Millipore filter paper and dried by freeze drying.

Caco-2 cell culture

Caco-2 cells were cultured in 75 cm^2 culturing flasks (VWR Scientific, West Chester, PA, USA) with 10 ml of Dulbecco's Modified Medium, (DMEM, Bio Fluids, Biosource International, Camarillo, CA, USA), at a seeding density for cultivation of 2.5×10^5 cells/flask. Cells were maintained in an incubator at a controlled atmosphere of 37°C, 95% relative humidity and 5% CO_2. The culture medium was replaced with fresh medium every other day for 6 days, until the cells reached 60–80% confluence. At this stage, a passage operation was performed. In this process, cells were detached from the culturing flask by trypsinization and resuspended in fresh culture medium. Then, the cells were counted and transferred with the desired seeding density to a new culturing flask or experimental wells. In these cells studies, cells with passage numbers between 60 and 80 were used.

Transepithelial electrical resistance studies

Caco-2 cell cultures were grown in 96-well plates (4.71 cm^2/well) (Costar, Corning, NY, USA) with DMEM for 21 to 24 days until they achieved a constant transepithelial electrical resistance. The culturing cell density was 2.35×10^5 cells/well. Transwell cells were used for the transport studies. Polycarbonate membranes with pore size of 0.45 μm separated the two half cells. The medium used for these studies was Hanks' balanced salt solution (HBSS) containing Ca^{2+}.

Cell membranes were allowed to equilibrate with the experimental medium, HBSS with Ca^{2+}, for 1 h prior to the initiation of the experiment. The transepithelial electrical resistance (TEER) was monitored using a voltmeter with a chopstick electrode (World Precision Instrument, Sarasota, FL, USA). After the equilibration period, the apical chamber medium was removed, replaced with fresh medium containing the P(MAA-g-EG) microparticles and TEER of the cell monolayer was monitored as a function of time at 37°C.

The reduction in the transepithelial electrical resistance was determined by calculating the degree of TEER and reporting it as a percent of the total: TEER%: $((R_t - R_{\mathrm{w/o}})/(R_{\mathrm{o}} - R_{\mathrm{w/o}})) \times 100\%$, where R_t is the resistance value at time t, R_{o} is the initial resistance value and $R_{\mathrm{w/o}}$ is the resistance value in the absence of the cell monolayer.

Insulin transport studies

Caco-2 cell cultures were grown in 96-well plates with DMEM for 21 to 24 days. DMEM was removed from both the apical and basolateral chamber and then replaced with fresh HBSS with Ca^{2+}. The experimental plates were incubated for

1 h before the experiment. The donor chamber medium was removed and replaced with fresh medium containing the insulin-loaded P(MAA-g-EG) microparticles. Samples were withdrawn from the receiver chamber at different time intervals for 3 h using a micropipette. The amount withdrawn from each chamber depended on the amount of insulin loaded into the P(MAA-g-EG) microparticles. The same percentage (v/v) was taken from both the apical and the basolateral chamber. Once the experiment was completed, the samples were analyzed using a bovine insulin ELISA kit (Alpco Diagnostics, Windham, NH, USA). In all cases triplicates were used.

The apparent permeability coefficient was used to quantify transport in a two-chamber diffusion system. The apparent insulin permeability coefficient, P_{app}, was calculated using $P_{app} = (dQ/dt)/(A \cdot C_O)$, in which dQ/dt (mol/s) was the rate of insulin transport in the receiver chamber, C_O (mol/ml) was the initial insulin concentration in the donor chamber and A (cm^2) was the surface area of the cell monolayer.

RESULTS

As we have indicated in previous studies [1, 6] the presence of PEG grafted chains in the polymer network of the carrier particles contributes to a mucoadhesion promotion mechanism [27–32]. For this purpose, we studied the effect of the molecular weight of the PEG tethered chains on the TEER values of the cell monolayer. As shown in Fig. 1, insulin-loaded P(MAA-g-EG) microparticles in contact with the cell monolayer led to a significant decrease of the TEER, of the ordered 40%. Each datum point represents an average of three wells and one

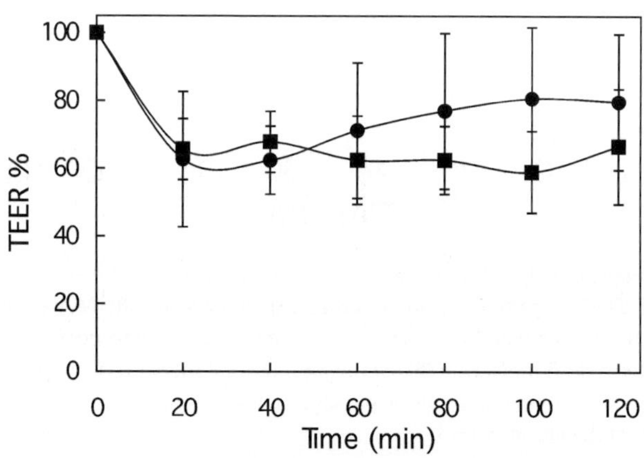

Figure 1. Effect of the molecular weight of the PEG tethered chains on the TEER of the cell monolayer. Insulin-loaded P(MAA-g-EG) microparticles with an initial monomer feed ratio MAA : EG of 1 : 1 and a PEG molecular weight of (●) 400 and (■) 1000 were used in this experiment. 10 mg of microparticles were in contact with the cell monolayers in each well. Caco-2 cell monolayers with a passage number of 62 were used at 37 °C. Error bars represent one standard deviation, $n = 3$.

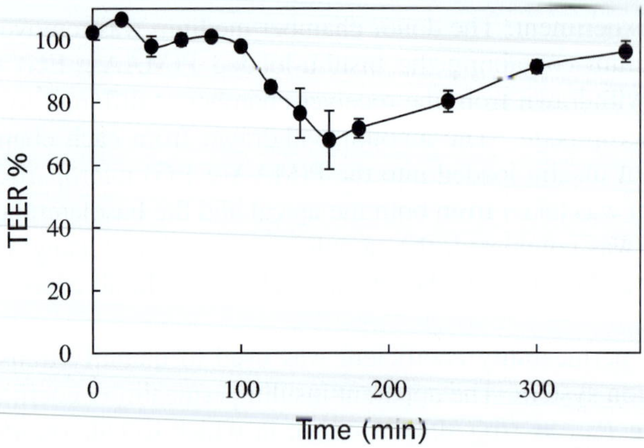

Figure 2. Reversible effect of the reduction of TEER. Insulin-loaded P(MAA-g-EG) microparticles with an initial monomer feed ratio MAA : EG of 1 : 1 and a PEG molecular weight 1000 were used in this experiment. 10 mg of microparticles were in contact with the cell monolayers in each well. Caco-2 cell monolayers with a passage number of 75 were used at 37°C. Error bars represent one standard deviation, $n = 3$.

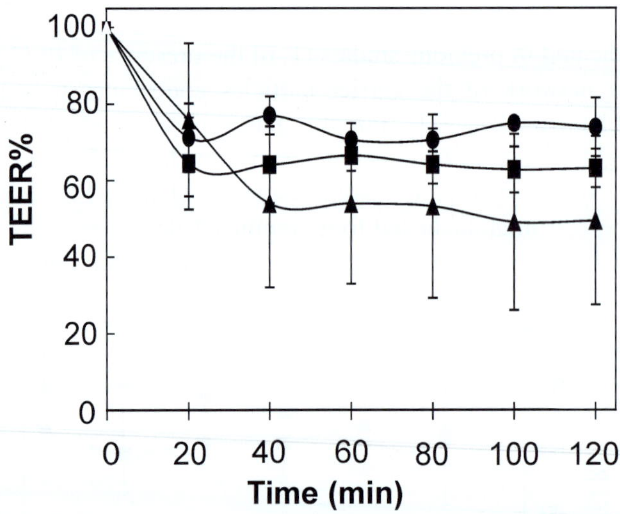

Figure 3. Effect of the amount of P(MAA-g-EG) microparticles on the cell monolayer TEER. Insulin-loaded P(MAA-g-EG) microparticles with an initial monomer feed ratio MAA : EG of 1 : 1 and PEG tethered chains of molecular weight of 1000 were used in this experiment. Different amounts of loaded microparticles, (●) 5 mg/well, (■) 10 mg/well and 20 mg/well, were in contact with the cell monolayer. Caco-2 cell monolayers with a passage number of 62 were used at 37°C. Error bars represent one standard deviation, $n = 3$.

standard deviation. However, the size of the PEG molecular weight did not affect the TEER values measured. Consequently, while the presence of PEG chains contributed to mucoadhesion, as shown by Madsen and Peppas [6], their size did not affect paracellular transport, at least not in the range of molecular weights studied.

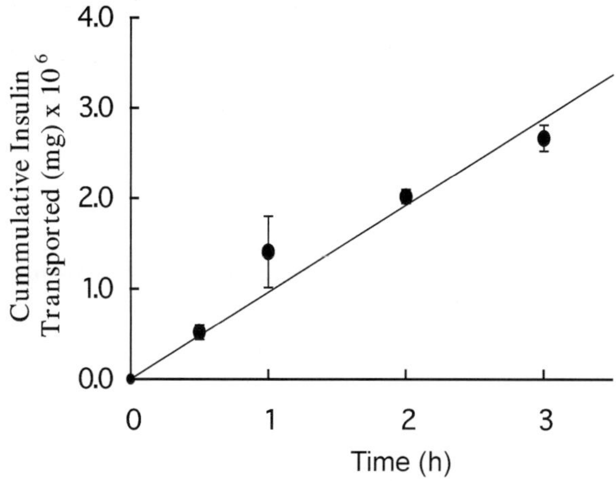

Figure 4. Transport of insulin from apical to basolateral chamber through Caco-2 cell monolayers in the presence of P(MAA-g-EG) microparticles. Microparticles with an initial monomer feed ratio MAA : EG of 1 : 1 and a PEG molecular weight of 1000 were used in this experiment. 20 mg of microparticles with a size of 150–212 μm were in contact with Caco-2 cell monolayers with a passage number of 67 at 37 °C. Error bars represent one standard deviation, $n = 3$.

Figure 2 indicates that the effect of the insulin-loaded microparticles in the TEER reduction was reversible. During the first hour equilibration occurred after DMEM was replaced by the HBSS experimental medium. After this period, 10 mg of loaded microparticles were placed in contact with the monolayer for 3 h. At this point reduction in the TEER value of approx. 35% was observed. Then the microparticles were removed and the experimental medium was replaced with DMEM. Complete and reversible recovery of the TEER was seen within 3 h.

The effect of the amount of particles in contact with the cell monolayer was studied at 5, 10 and 20 mg/well. In all cases, no cytotoxicity was observed when exposed to the Caco-2 cell monolayers. As shown in Fig. 3, the TEER values decreased by up to 50%.

We also studied the difference in insulin transport from the apical to the basolateral side between insulin-loaded P(MAA-g-EG) microparticles and mixtures of insulin-free particles with insulin. In both systems, the initial monomer ratio was MAA : EG 1 : 1, the molecular weight of the PEG tethered chains was 1000 and the microparticle size was 150–212 μm. Figure 4 shows the cumulative amount of insulin transported from the loaded microparticles through the cell monolayer as a function of time. Table 1 shows that the insulin permeability coefficient was much higher for insulin/particle mixtures than for insulin-loaded microparticles.

The cumulative amounts of insulin from the apical to the basolateral as well as from the basolateral to the apical chamber were monitored as a function of time. Insulin-loaded P(MAA-g-EG) microparticles with an initial monomer ratio of MAA : EG 1 : 1, PEG tethered chains with molecular weight of 1000 and microparticle size of 150–212 μm were used in this experiment. The insulin permeability

Table 1.
Insulin permeability coefficient across Caco-2 cell monolayers in contact with two P(MAA-g-EG) microparticle systems

P(MAA-g-EG) microparticle system	Permeability $\times 10^6$ (cm/s)
Mixture of insulin/particles	12.7 ± 1.73
Insulin-loaded particles	4.69 ± 0.03

Transport from apical chamber to basolateral chamber was studied. P(MAA-g-EG) microparticles with an initial monomer feed ratio MAA : EG of 1 : 1 and PEG molecular weight of 1000 were used. 20 mg of microparticles with a size of 150–212 μm were in contact with Caco-2 cell monolayers with a passage number of 75. The experiment was carried out at a constant temperature of 37°C. Error bars represent one standard deviation, $n = 3$.

Table 2.
Effect of transport direction on the permeation of insulin across Caco-2 cell monolayers

Transport	Permeability $\times 10^6$ (cm/s)
Apical to basolateral side	4.69 ± 0.03
Basolateral to apical side	4.94 ± 0.02

Insulin-loaded P(MAA-g-EG) microparticles with an initial monomer feed ratio MAA : EG of 1 : 1 and PEG molecular weight of 1000 were used in this experiment. 20 mg of microparticles with a size of 150–212 μm were in contact with Caco-2 cell monolayers with a passage number of 75. The experiment was carried out at a constant temperature of 37°C. Error bars represent one standard deviation, $n = 3$.

Table 3.
Effect of the molecular weight of the PEG tethered chains of grafted P(MAA-g-EG) microparticles on the permeability coefficient of insulin across Caco-2 cell monolayers

Molecular weight of PEG tethered chains	Permeability $\times 10^6$ (cm/s)
400	6.37 ± 0.02
1000	4.65 ± 0.03

Transport was from the apical chamber to the basolateral chamber. Insulin-loaded P(MAA-g-EG) microparticles with an initial monomer feed ratio MAA : EG of 1 : 1 were used in this experiment. 20 mg of microparticles with a size of 150–212 μm were in contact with Caco-2 cell monolayers with a passage number of 75. The experiment was carried out at a constant temperature of 37°C. Error bars represent one standard deviation, $n = 3$.

coefficients were virtually the same under these experimental conditions as seen in Table 2.

We also studied the effect of the molecular weight of the PEG tethered chains on the insulin transport from the apical to the basolateral chamber. Results from Table 3 show that the insulin permeability values for the P(MAA-g-EG) microparticles containing PEG tethered chains with molecular weight of 400 were higher than for those with molecular weight of 1000.

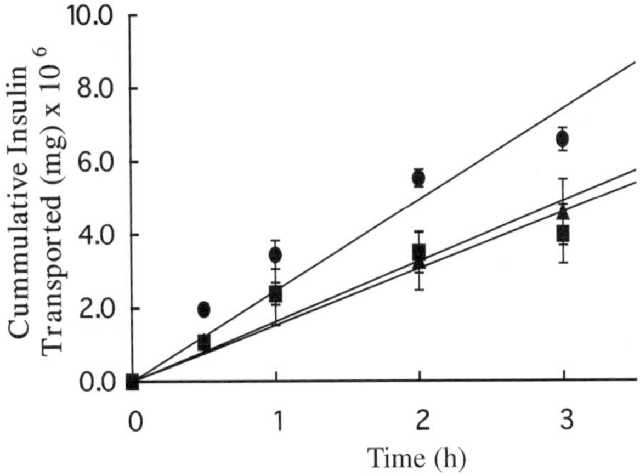

Figure 5. Effect of particle size on insulin transport through Caco-2 cell monolayers. Transport from the apical to the basolateral chamber was studied in this experiment. Loaded P(MAA-g-EG) microparticles with an initial monomer feed ratio MAA : EG of 1 : 1 and a PEG molecular weight of 1000 were used in this experiment. 20 mg of microparticles with a size of (●) 25–90 μm, (■) 90–150 μm and (▲) 150–212 μm were in contact with Caco-2 cell monolayers with a passage number of 72 at 37 °C. Error bars represent one standard deviation, $n = 3$.

Table 4.
Effect of P(MAA-g-EG) carrier particle size on insulin permeability through Caco-2 cell monolayers

P(MAA-g-EG) microparticle size (μm)	Permeability $\times 10^6$ (cm/s)
25–90	3.32 ± 0.03
90–150	2.30 ± 0.02
150–212	2.48 ± 0.01

Transport was from the apical chamber to the basolateral chamber. Loaded P(MAA-g-EG) microparticles with an initial monomer feed ratio MAA : EG of 1 : 1 and a PEG molecular weight of 1000 were used in this experiment. 20 mg of microparticles were in contact with Caco-2 cell monolayers with a passage number of 78, at 37 °C. Error bars represent one standard deviation, $n = 3$.

P(MAA-g-EG) microparticle on the insulin transport through the Caco-2 cell monolayer. Figure 5 shows the cumulative amount of insulin released from P(MAA-g-EG) microparticles of size (25–90, 90–150 and 150–212 μm). There were no statistical differences on the cumulative insulin increase between the various insulin-loaded microparticles studied. However, the 25–90 μm insulin-loaded microparticles showed a statistically higher cumulative insulin increase. Table 4 shows the insulin permeability coefficients from the different P(MAA-g-EG) microparticle sizes.

DISCUSSION

Cell tight junctions are dynamic protein complexes that hold adjacent cells together and impede transepithelial diffusion of solutes between cells. Different approaches have been investigated in order to induce paracellular transport of large molecules, such as proteins, by disrupting the tight junctions [10–16]. As tight junctions are calcium dependent, in the absence of Ca^{2+} these complexes are disrupted. Peppas and Madsen [6] have shown the ability of P(MAA-g-EG) microparticles to bind Ca^{2+}, inhibiting the proteolytic activity of calcium-dependent enzymes such as trypsin. Clearly, Ca^{2+} is bound to P(MAA-g-EG) microparticles because of the presence of negatively charged carboxylic acids on the MAA repeating units at physiological conditions.

The integrity of tight junctions can be monitored by measuring the transepithelial electrical resistance. A decrease in this value indicates a disruption in the tight junctions, which would allow the paracellular transport of the drug. In the present study insulin-loaded P(MAA-g-EG) microparticles were kept in contact with the Caco-2 cell monolayers. Figures 1 and 2 show that the effect of the insulin-loaded microparticles in the TEER reduction is reversible. The first hour shows the equilibration period, after the DMEM media was replaced with the HBSS experimental medium. Torres-Lugo *et al.* [19] previously reported the existence of a second steady state because of an initial drop on the TEER value during the equilibration period. After this period, the microparticles were placed in contact with the monolayer for 3 h. Then the microparticles were removed and the experimental medium was replaced with DMEM, and the TEER values were recovered in a period of 3 h. These results indicate that the insulin-loaded microparticles causes a reversible change in the integrity of the cell monolayer, probably by opening tight junctions and creating conditions of paracellular transport.

The decrease in the TEER value was dependent on the amount of P(MAA-g-EG) microparticles in contact with the Caco-2 cell monolayer. Decreases in TEER of up to 50% are attributed to the ability of the MAA carboxylic acids to bind Ca^{2+} causing the opening of the tight junctions. However, the molecular weight of the PEG tethered chains did not have an effect on the extent to which the TEER values were decreased [33–38].

The insulin released from the polymer carrier and transported across the Caco-2 cell monolayer was monitored as a function of time. The permeability values for the insulin transport from the apical to the basolateral chamber were similar to the ones from the basolateral to the apical chamber and in general agreement with previous studies of our lab [37, 38]. Thus, we believe that active carriers or efflux transporters were not involved in insulin transport through the Caco-2 cell monolayer.

P(MAA-g-EG) microparticles were able to enhance the transport of insulin through the Caco-2 cell monolayers. Clearly, our studies indicate that these systems are a good candidate for oral delivery of insulin. However, for optimal P(MAA-g-EG) characteristics for the oral delivery of insulin, both moderate mucoadhesion by PEG tethers and low hindrance effects them so their transport barriers should be optimized.

Acknowledgements

This work was supported by grant No. EB 000246 from the National Institutes of Health. This work is dedicated to Professor Allan Hoffman on the occasion of his 70th birthday. An academic brother of the senior author of this paper, Allan Hoffman has played a pivotal role in the author's career.

REFERENCES

1. A. L. Lowman, M. Morishita, M. Kajita, T. Nagai and N. A. Peppas, *J. Pharm. Sci.* **88**, 933 (1999).
2. G. P. Carino and E. Mathiowitz, *Adv. Drug Deliv. Rev.* **35**, 249 (1999).
3. C. L. Bell and N. A. Peppas, *J. Control. Rel.* **39**, 201 (1996).
4. A. M. Lowman and N. A. Peppas, *Macromolecules* **30**, 4959 (1997).
5. A. M. Lowman and N. A. Peppas, *Polymer* **41**, 73 (2000).
6. Y. Huang, W. Leobandung, A. Foss and N. A. Peppas, *J. Control. Rel.* **65**, 63 (2000).
7. F. Madsen and N. A. Peppas, *Biomaterials* **20**, 1701 (1999).
8. K. L. Audus, R. L. Bartel, I. J. Hidalgo and R. T. Borchardt, *Pharm. Res.* **7**, 435 (1990).
9. I. J. Hidalgo, T. J. Raub and R. T. Borchardt, *Gastroenterology* **96**, 736 (1989).
10. V. Pade and S. Stavchansky, *Pharm. Res.* **14**, 1210 (1997).
11. M. M. Nerurkar, P. S. Burton and R. T. Borchardt, *Pharm. Res.* **13**, 528 (1996).
12. L. S. L. Gan, S. Yanni and D. R. Thakker, *Pharm. Res.* **15**, 53 (1998).
13. N. G. M. Schipper, K. M. Vårum and P. Artursson, *Pharm. Res.* **13**, 1686 (1996).
14. H. L. Lueßen, C. O. Rentel, A. F. Kotzé, C. M. Lehr, A. G. de Boer, J. C. Verhoef and H. E. Junginger, *J. Control. Rel.* **45**, 15 (1997).
15. M. A. Hurni, A. B. J. Noach, C. M. Blom-Roosemalen, A. G. de Boer, J. F. Nagelkerke and D. D. Breimer, *J. Pharm. Exp. Ther.* **267**, 942 (1993).
16. A. Fasano and S. Uzzau, *J. Clin. Invest.* **99**, 1158 (1997).
17. V. H. L. Lee, A. Yamamoto and V. B. Kompella, *Crit. Rev. Ther. Drug Carrier Syst.* **8**, 91 (1991).
18. E. C. Swenson and W. J. Curatolo, *Adv. Drug Deliv. Rev.* **8**, 39 (1992).
19. A. C. Foss and N. A. Peppas, *Eur. J. Pharm. Biopharm.* (2004) (in press).
20. M. Torres-Lugo and N. A. Peppas, *Biomaterials* **21**, 1191 (2000).
21. N. A. Peppas, P. Bures, W. Leobandung and H. Ichikawa, *Eur. J. Pharm. Biopharm.* **50**, 27–46 (2000).
22. A. M. Lowman, B. A. Cowans and N. A. Peppas, *J. Polym. Sci. Polym. Phys.* **38**, 2823 (2000).
23. N. A. Peppas, Y. Huang, M. Torres-Lugo, J. H. Ward and J. Zhang, *Annu. Rev. Biomed. Eng.* **2**, 9 (2000).
24. M. Morishita, A. M. Lowman, K. Takayama, T. Nagai and N. A. Peppas, *J. Control. Rel.* **81**, 25 (2002).
25. M. Torres-Lugo, M. Garcia, R. Record and N. A. Peppas, *J. Control. Rel.* **80**, 197 (2002).
26. M. Torres-Lugo, M. García, R. Record and N. A. Peppas, *Biotechnol. Progr.* **18**, 612 (2002).
27. D. N. Robinson and N. A. Peppas, *Macromolecules* **35**, 3668 (2002).
28. M. Torres-Lugo and N. A. Peppas, *J. Nanoparticle Res.* **4**, 73 (2002).
29. C. Donini, D. N. Robinson, P. Colombo, F. Giordano and N. A. Peppas, *Int. J. Pharm.* **245**, 83 (2002).

30. B. Kim and N. A. Peppas, *J. Biomater. Sci. Polymer Edn* **13**, 1271 (2002).
31. O. Sipahigil, M. Torres-Lugo and N. A. Peppas, *STP Pharma* **12**, 345 (2002).
32. B. Kim, K. La Flamme and N. A. Peppas, *J. Appl. Polym. Sci.* **89**, 1606 (2003).
33. B. Kim and N. A. Peppas, *Biomed. Microdevices* **5**, 333 (2003).
34. B. Kim and N. A. Peppas, *Int. J. Pharm.* **266**, 29 (2003).
35. H. Ichikawa and N. A. Peppas, *J. Biomed. Mater. Res.* **67**, 609 (2003).
36. A. C. Foss, T. Goto, M. Morishita and N. A. Peppas, *Eur. J. Pharm. Biopharm.* (2004) (in press).
37. P. Artursson, K. Palm and K. Luthman, *Adv. Drug Deliv. Rev.* **46**, 27 (2001).
38. S. Yee, *Pharm. Res.* **14**, 763 (1997).

Gels, Genes, Grafts and Giants, pp. 27–43 (2005)

Hydrogel probe for iontophoresis drug delivery to the eye

ESTHER ELJARRAT-BINSTOCK [1], FREDERIK RAISKUP [2], JOSEPH FRUCHT-PERY [2] and ABRAHAM J. DOMB [1,*]

[1] *Department of Medicinal Chemistry and Natural Products, School of Pharmacy, Faculty of Medicine, The Hebrew University of Jerusalem, Jerusalem 91120, Israel*
[2] *Department of Ophthalmology, Hadassah University Hospital, Jerusalem 91120, Israel*

Received 15 September 2003; accepted 26 November 2003

Abstract—The aim of this study was to evaluate the use of a solid hydrogel loaded with a drug solution as a probe for ejecting drugs to the eye upon application of low current iontophoresis. Hydroxyethyl methacrylate (HEMA), cross-linked with ethylene glycol dimethacrylate (EGDMA), and cross-linked arabinogalactan or dextran were prepared to form solid hydrogels. The hydrogels were examined for their mechanical suitability, absorption of drug solution and *in vitro* release properties when applying an iontophoretic current through the drug-loaded hydrogel into a solid-agar surface. Transconjunctival and transscleral iontophoresis of gentamicin sulfate was studied in healthy rabbits using drug-loaded disposable HEMA hydrogel disc probes. Gentamicin concentrations in different eye segments were assayed using a fluorescence polarization immunoassay. Preliminary corneal toxicity was examined in rabbits using a current intensity of 2.5 and 5.1 mA/cm^2 for 60 and 120 s. The most appropriate hydrogel is composed of HEMA, 2% EGDMA and 75% water. Iontophoresis onto agar gel was found indicative for the evaluation of iontophoretic activity of a hydrogel. Transscleral iontophoretic treatment resulted in high concentrations of drugs in the posterior segments of the eye. Application of iontophoresis onto the rabbit eye caused a reversible swelling of the cornea which lasted a few hours after application. Low current iontophoresis using drug-loaded hydrogel has a potential clinical value in obtaining high drug concentration at posterior segments of the eye.

Key words: Iontophoresis; Hydrogel; Gentamicin; Dexamethasone; Transscleral.

INTRODUCTION

Hydrogels are three-dimensional networks of hydrophilic polymers held together by association bonds as covalent bonds and weaker cohesive forces such as hydrogen and ionic bonds and intermolecular hydrophobic association. These networks

*To whom correspondence should be addressed. Tel.: (972-2) 675-7573. Fax: (972-2) 675-8959. E-mail: adomb@md.huji.ac.il

are able to retain a large quantity of water within their structure. Due to their superior chemical and physical properties, hydrogels have received much attention for preparing biomedical materials, such as contact lenses, wound dressing and drug delivery systems [1]. The monomer composition of a copolymer can be manipulated to influence the permeation and diffusion characteristics of the hydrogel [2].

Hydrogels have been used for transdermal iontophoretic delivery of various drugs [3–6]. Drug-containing hydrogels, rather than drug solutions, facilitate drug handling, minimize skin hydration and allow drug release rate control by changing the characteristics of the hydrogel. Hydrogels have been used for contact lenses while the crosslinked hydroxyethyl methacrylate (HEMA) hydrogel is the most frequently used [1]. No work has been reported yet on iontophoresis to the eye using hydrogels.

Iontophoresis is a non-invasive technique in which a small electric current is used to enhance the penetration of charged molecules into the tissue [7]. Iontophoresis has been used in various fields of medicine, including transdermal administration of local anesthetics [8], testing for cystic fibrosis by transcutaneous delivery of pilocarpine [9], administration of vidarabine to patients with herpes orolabialis [10], fluoride administration to patients with hypersensitive dentin [11] and gentamicin administration for bacterial otitis [12].

Transcorneal or transscleral iontophoresis of various charged drugs has been reported [13–17]. High levels of antibiotics were measured in the cornea and aqueous humor after transcorneal iontophoresis, compared with topical application or subconjunctival and intravenous injection which do not achieve adequate drug levels and involve other complications [13–17]. Moreover, published data confirm that high drug concentrations penetrate also the posterior segments of the eye after transscleral iontophoresis, allowing the treatment of posterior disorders of the eye, such as posterior uveitis and endophthalmitis [15, 18–23]. These studies used iontophoresis of drug solution, which is technically clumsy, may cause mechanical injuries to the cornea and demands sterilization of the solution and cup before each treatment. In order to overcome these limitations, Grossman [15] and Frucht [16, 17] used a drug-loaded agar gel probe. However, this agar-gel is fragile and leaves remains of agar on the eye surface.

The purpose of this study was to prepare and evaluate hydrogel probes suitable for iontophoretic application to the eye and asses the penetration of drugs and toxicity to rabbit eye after transscleral iontophoresis. An attempt to evaluate the usefulness of hydrogel probes by using them for iontophoresis delivery of drugs onto a solid agar gel, resembling the eye surface, is also described. Three drugs were used in this study: gentamicin sulfate, a commonly used antibiotic for treating infections [24], as representative for cationic drug, dexamethasone phosphate and hydrocortisone succinate, common anti-inflammatory drugs for treating eye inflammation, as representatives for anionic drugs (Scheme 1).

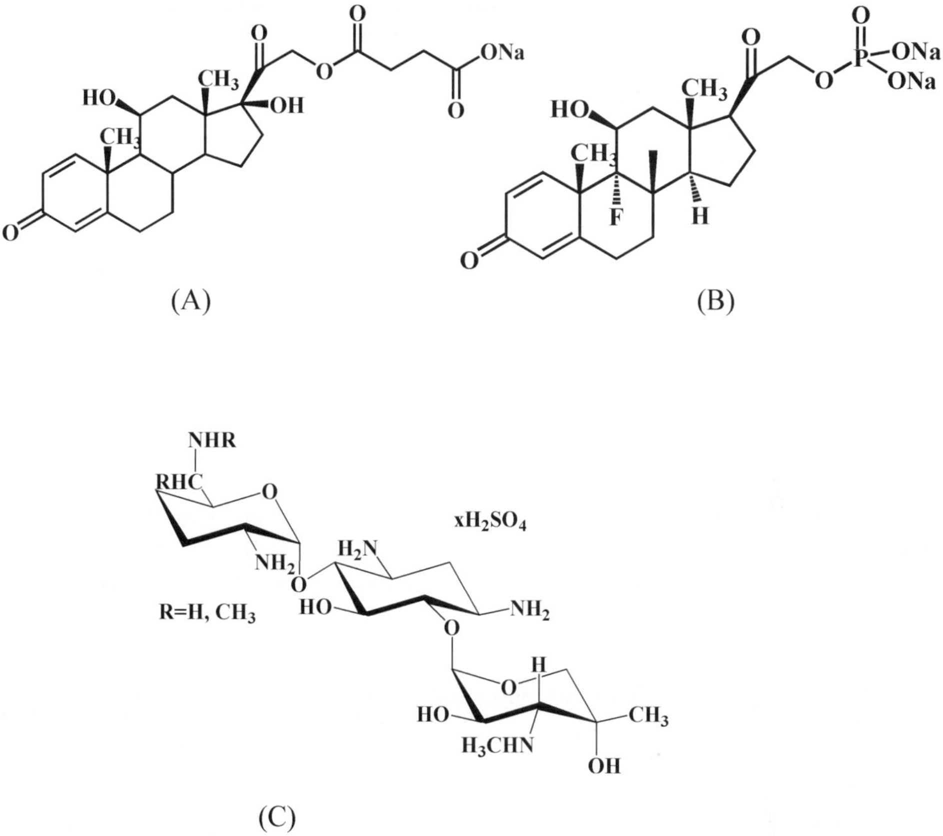

Scheme 1. Schematic structure of hydrocortisone succinate (A), dexamethasone phosphate (B) and gentamicin sulfate (C).

MATERIALS AND METHODS

Gentamicin sulfate was obtained from Sigma (St. Louis, MO, USA). Dexamethasone sodium phosphate was obtained as a gift from Taro Pharmaceutical Industries (Haifa, Israel). All monomers and agents were purchased from Sigma (Rehovot, Israel). All solvents were HPLC grade from BioLab (Jerusalem, Israel).

The iontophoretic device

The iontophoretic device that was applied in this study is a portable Mini Ion device (designed in our laboratory) that can be operated by battery or by an external electric source (see Fig. 1). A drug-loaded hydrogel disk of 5 mm in diameter and 5 mm in height is attached to a cylindrical well of 5 mm in diameter and 3 mm in depth at the edge of the probe. The Mini Ion device applies a variable electrical current that can vary from 100 μA to 1000 μA for pre-set periods from 10 s to 120 s. All materials and agents were purchased from Sigma Israel (Rehovot, Israel).

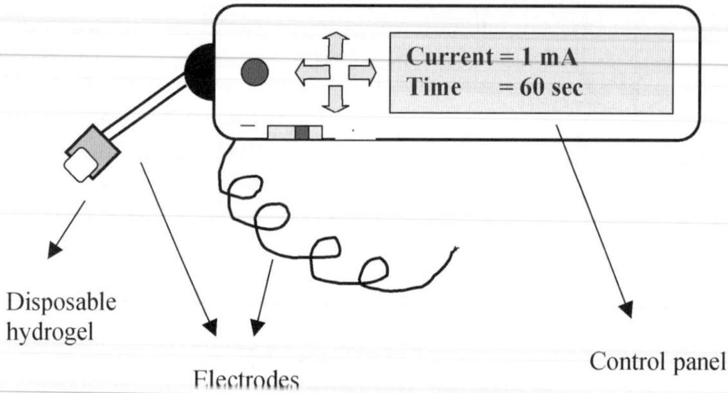

Figure 1. Schematic structure of the Mini Ion device applied for iontophoresis delivery of gentamicin. The device is composed of a cylindrical well for the insertion of a portable hydrogel, two electrodes and a control panel for time and current control.

In vitro *studies*

Preparation of hydroxyethyl methacrylate (HEMA) hydrogel discs. Hydrogels of hydroxyethyl methacrylate (HEMA) crosslinked with ethylene glycol dimethacrylate (EDGMA) were prepared at different ratios. In a typical experiment, HEMA (97% pure), EDGMA (98% pure) and deionized water (2.0, 0.04 and 6.5 ml, respectively) were mixed under a stream of dry nitrogen to produce a clear solution. The following initiator solutions were added: 2% sodium persulfate ($Na_2S_2O_8$, 0.05 ml), 2% sodium metabisulfite ($Na_2S_2O_5$, 0.05 ml) and 2% ammonium ferrous sulfate ($Fe(NH_4)_2(SO_4)_2$, 0.025 ml). The solution was rapidly mixed and divided into a Teflon tray with wells (5 mm height, 6 mm diameter) and left to polymerize over night at room temperature, under nitrogen blanket in the dark. The obtained discs were immersed in 500 ml water with stirring for 2 h, to clean the hydrogels from unreacted monomers, catalysts and low-molecular-weight by-products. The purified hydrogels were frozen by immersing in liquid nitrogen and dehydrated by lyophilization overnight to form spongy cylinders that were of similar size to the wet cylinders (5 mm height and diameter).

Preparation of AG-chitosan and dextran-chitosan hydrogel discs. Hydrogels of chitosan and oxidized arabinogalactan (AG) or dextran at different ratios were prepared. In a typical experiment, a 2% chitosan solution in 2% (v/v) acetic acid was mixed with the appropriate volume of 10% oxidized AG or dextran solutions in 0.1 M borate buffer (pH 7.5). After a short gentle stirring the solution was dispersed into 5×6 mm cylindrical wells, 0.2 ml each, and left to polymerize for 48 h at 37°C. The hydrogels obtained were frozen over night and lyophilized for 24 h.

Preparation of AG-gelatin hydrogel discs. Hydrogels of gelatin and oxidized arabinogalactan (AG) at different ratios were prepared. In a typical experiment, a 25% gelatin solution (3 ml), prepared by dissolving 0.75 g gelatin in 3 ml DDW with heating, was mixed with the appropriate volume of 5% oxidized AG, prepared by dissolving oxidized AG (0.6 g) in 12 ml borate buffer (0.1 M, pH 7.5). After a short gentle stirring the solution was dispersed into wells, 0.2 ml each, and left to polymerize as described above for chitosan–AG hydrogels.

Preparation of AG-ethylene diamine hydrogel discs. Hydrogels of ethylene diamine and oxidized arabinogalactan (AG) at different ratios were prepared. In a typical experiment, a solution of 10% w/v oxidized AG (3 ml) prepared by dissolving oxidized AG (1.2 g) in 12 ml borate buffer (0.1 M, pH 7.5), was mixed with the appropriate volume of ethylene diamine in deionized water. After a short gentle stirring the solution was dispersed into wells, 0.2 ml each, and left to polymerize, as described above.

SEM imaging of hydrogels. Pore shape and size of the different hydrogels were determined using scanning electron microscope (Philips SEM 505, Eindhoven, The Netherlands). Dry gels were sectioned horizontally and processed for SEM imaging.

Swelling of hydrogels. Dry weighted hydrogels, after an overnight lyophilization, were immersed in DDW or drug solutions (gentamicin sulfate, dexamethasone phosphate or hydrocortisone succinate, Scheme 1) for 2 h at room temperature. After 2 h the hydrogels were removed from the solution, placed between two filter papers to remove excess solution and weighed. The degree of swelling (hydration) was calculated as followed:

$$((W_w - W_d)/W_w) \times 100, \tag{1}$$

where W_w and W_d are the wet and dry hydrogel weights, respectively.

Iontophoresis onto agar gel

Agar gels prepared from 1.5% (w/v) agar (Difco Laboratories, Detroit, MI, USA) solution (3 ml) loaded into 1 cm (diameter) wells were used in this study. 2 h before treatment, the selected HEMA hydrogels were loaded with the drug by immersing them in 10% (w/v) drug solution (gentamicin sulfate, dexamethasone phosphate or hydrocortisone succinate, Scheme 1). The drug-loaded hydrogel disc was inserted into the cylindrical well of the iontophoretic device and placed onto the central agar surface. The complementary electrode was attached to the edge of the agar well by means of an alligator clip. The iontophoretic administration was performed at the current intensity and duration that was determined in the experiment. The current was performed at a negative or positive direction; depending on the charge of the drug.

After the iontophoretic treatment, the agar gels were dissolved in 1 ml water. The solutions were diluted and the drug concentration was determined using a fluorescence polarization immunoassay by TDx System Analyzer (Abbott Laboratories,

Abbott Park, IL, USA) for gentamicin sulfate analysis, or using UV absorbance at 250 nm for hydrocortisone succinate and dexamethasone phosphate. The TDx system analyzer is a precise drug analyzer (less than 3% CV) with detection limit of 0.1 μg/ml for gentamicin at 95% confidence level and average recovery of 102.6%.

Drug amounts in hydrogel. Each drug-loaded hydrogel was weighted before and after immersion in a drug solution. The absorbed drug was calculated from the solution volume absorbed using equation (2):

$$A_{\text{drug}} = (W_{\text{after}} - W_{\text{before}}) \times C_{\text{sol}}, \tag{2}$$

where A_{drug} is the amount of drug in the hydrogel in mg, W_{after} is the hydrogel weight in g after immersion, W_{before} is the hydrogel weight in g before immersion and C_{sol} is the drug concentration in the immersion solution in mg/ml.

When the drug concentration in the immersed solution changed after the hydrogels immersed in it, the drug amount in the hydrogel was calculated as follows:

$$A_{\text{drug}} = (V_{\text{ext}} \times C_{\text{before}}) - (V_{\text{ext}} - V_{\text{in}}) \times C_{\text{after}}, \tag{3}$$

where A_{drug} is the amount of drug in the hydrogel in mg, V_{ext} is the initial volume of the external immersion solution in ml, V_{in} is the volume absorbed in the hydrogel in ml, C_{before} is the drug concentration in the immersion solution before the immersion in mg/ml and C_{after} is the drug concentration in the immersion solution after the immersion in mg/ml.

All *in vitro* results are the mean of two or three measurements.

In vivo *studies*

Animals. Healthy New Zealand white male rabbits ($n = 13$), weighing 2.0–3.0 kg, were used in the study. The study confirmed to the ARVO Statement for the Use of Animals in Ophthalmic and Vision Research.

Drug administration. The animals were anesthetized by injection of ketamine and xylazine solution (i.m., 25 and 2.5 mg/kg, respectively).

Group 1 ($n = 4$) was treated with transconjunctival iontophoresis and group 2 ($n = 4$) was treated with transscleral iontophoresis of gentamicin sulfate using Mini Ion system. The hydrogel disk probe contained an average amount of 26.0 mg gentamicin sulfate. Before placing of the electrode on the ocular surface, the eye was topically anesthetized with 0.4% benoxinate eye drops (Localin®, Dr. Fischer, Israel). The gentamicin-loaded hydrogel disc was inserted into the cylindrical well of the iontophoretic device and placed onto the conjunctiva (group 1) or directly on the sclera after removing the conjunctival membrane (group 2), at the pars plana area. The complementary electrode was attached to the ear of the rabbit by means of an alligator clip. The iontophoretic treatment was performed twice for 120 s at two near places of the upper-back sclera (pars-plana area), with a current intensity of 1 mA (5.1 mA/cm^2).

Table 1.
Corneal iontophoretic parameters for the toxicity experiment (group 3)

	Rabbit No.				
	1	2	3	4	5
Current (mA)	0.5	1	0.5	1	0
Iontophoresis duration (s)	60	120	60	120	60
Time of killing after iontophoresis	5 min	5 min	8 h	8 h	8 h

Rabbits 1–5 received transcorneal iontophoresis using the Mini Ion system with gentamicin-loaded hydrogel disc probe at current intensities of 0.5–1 mA (2.55–5.1 mA/cm^2) for 60–120 s.

For the preliminary toxicity test (group 3), five rabbits underwent corneal iontophoresis using the Mini Ion system with a hydrogel disk probe containing an average amount of 26.0 mg gentamicin sulfate. The iontophoretic administration was performed on one eye with a current intensity of 0.5, or 1 mA (2.5–5.1 mA/cm^2) for 60 or 120 s. The reference animal (animal 5) was treated with mock iontophoresis (contact without current) for 60 s (Table 1).

Sample collection. Animals from groups 1–2 (after transconjunctival and transscleral iontophoresis) were killed 4 h after the iontophoresis by an intravenous injection of a lethal dose of 2% pentobarbital, in which the cornea, sclera and retina were harvested; aqueous humor and vitreous body were aspirated.

The animals in the preliminary toxicity group (group 3) were killed within 5 min from the iontophoretic treatment (rabbits 1 and 2) or 8 h after the iontophoretic treatment (rabbits 3–5) in the same way. After killing the corneas were excised and immersed in formalin for preservation until the histological test. Aqueous humor volume and cornea weight were measured following their collection from the animals. The average volume of the aqueous humor and weight of the cornea were 311 μl and 47.8 mg, respectively.

Histological test. Five cornea samples (rabbits 1–5, group 3) were treated with hematoxylin-eosin solution (Sigma Diagnostics, St. Louis, MO, USA) and observed in a light microscope (Nikon 136405, Japan).

Gentamicin assay.

Cornea and sclera. After excision, the specimen was rinsed in 5 ml of 0.9% NaCl solution, cut with a scalpel and weighed. Each sample of the specimen was transferred into the micro-centrifuge tube containing 1.0 ml of 0.01 M phosphate-buffered saline pH 7.4 (Sigma). The tubes were incubated for 18 h at 37°C with oscillation at 100 rpm followed by centrifugation at 2000 rpm for 10 min. The supernatant was collected (200 μl) and assayed for gentamicin concentration using a fluorescence polarization immunoassay by TDx System Analyzer (Abbott Laboratories).

Retina. Similar to the cornea treatment, except that the rinsing and cutting stages were not applied, due to the small amounts collected.

Aqueous humor. 100 μl aliquots of the collected samples were diluted with 100 μl of normal saline and gentamicin concentrations were determined using TDx, as above.

Vitreous body. The amount aspirated in Eppendorf tubes was micro-centrifugated at 2000 rpm for 10 min. 200 μl of the supernatant was diluted with a similar amount of saline and analyzed by TDx.

The concentration of gentamicin in the cornea, sclera and retina was calculated as follows:

$$C_{\text{sample}} = (W_{\text{sample}} + V_{\text{buffer}}) \times (C_{\text{buffer}}/W_{\text{sample}}), \tag{4}$$

where C_{sample} is gentamicin concentration in the sample in μg/g, C_{buffer} is gentamicin concentration in the buffer in μg/g, W_{sample} is weight of the sample in g and V_{buffer} is the volume of buffer in ml.

Data analysis

The Kruskal–Wallis test was used for statistical evaluation of the results obtained for the studied groups.

RESULTS

Hydrogels

Hydrogels were prepared from HEMA crosslinked with EGDMA at different water contents (53–80%) and EGDMA contents (0.5–4%) in the monomer mixture. Increasing the amount of the crosslinker (EGDMA) or decreasing the amount of water in the monomer solution produces harder and more convenient hydrogels to work with. On the other hand, the water content in the monomer mixture also has a direct effect on the water absorption properties. As the water content in the monomer mixture increases, the resulting hydrogel absorbs more aqueous solution. For example, a hydrogel prepared with 80% water can absorb about 78% (w/w) of water after 2 h of immersion (see Fig. 2).

The most suitable hydrogel for iontophoresis should be soft, solid and with maximum water absorbance capacity. Therefore, the selected optimal hydrogel in this study was prepared from HEMA, 2% (v/v) EGDMA and 75% water in the preparation solution. The solution was polymerized as described in Materials and Methods.

Of the different formulations of polysaccharides (AG or dextran) with polyamines (chitosan, gelatin or ethylene diamine) prepared, only chitosan hydrogels crosslinked with oxidized dextran (30% and 40%) or AG (10% and 20%) resulted in useful soft and stable hydrogels. AG-gelatin and AG-ethylene diamine hydrogels at various prepared ratios were brittle and easy to crumble.

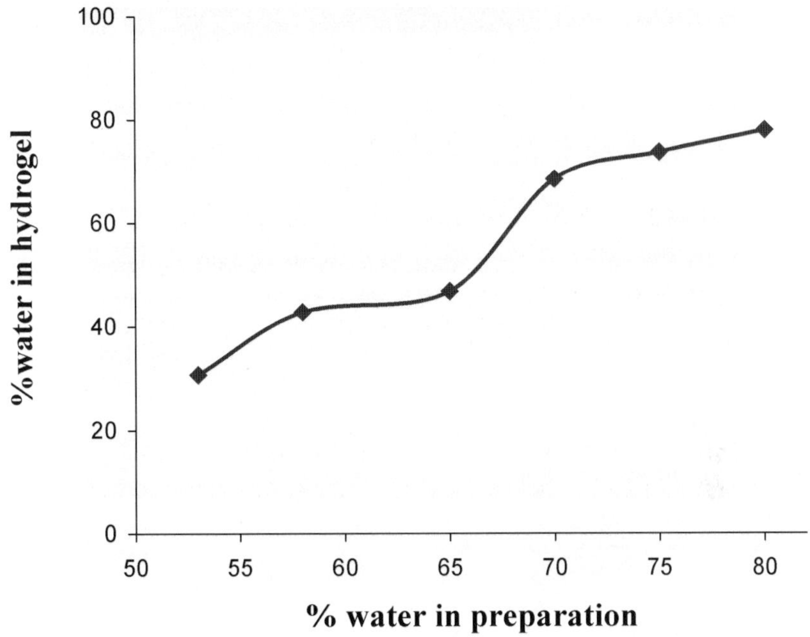

Figure 2. The influence of water content in HEMA hydrogels preparation solution on the water absorption properties of the final hydrogel. % water in hydrogel was calculated using equation (1), after 2 h of immersion.

SEM imaging of hydrogels

The pore shape and size of lyophilized HEMA and polysaccharide hydrogels as shown by SEM images are given in Fig. 3. The images from the horizontally sectioned hydrogels showed different pore shapes which allow water or solutions to absorb in the hydrogel. The pore size of 20% AG-chitosan hydrogel was about 200 μm, whereas the pores of the HEMA hydrogel crosslinked with 2% EGDMA and 75% water content in the preparation solution were only 10–20 μm in diameter.

Hydrogel absorption

The absorption capacity of seven selected hydrogel formulations (10, 20 and 40% AG-chitosan, 20, 30 and 40% dextran-chitosan hydrogels and the HEMA hydrogel) was determined after 2 h immersion in different drug solutions (Fig. 4). Polysaccharide hydrogels absorbed the drug solution to about 95% of the swollen hydrogels, independent of the drug in the solution. The swollen hydrogels were fragile, sticky and too soft to be applied on the eye surface without squeezing out some of the drug solution from the gel to the eye surface. The HEMA hydrogels absorbed the steroid solutions to about 85% of their swollen weight and only 65% for the gentamicin solution. These HEMA hydrogels were stable, non-fragile and retained the drug solution at surface contact conditions.

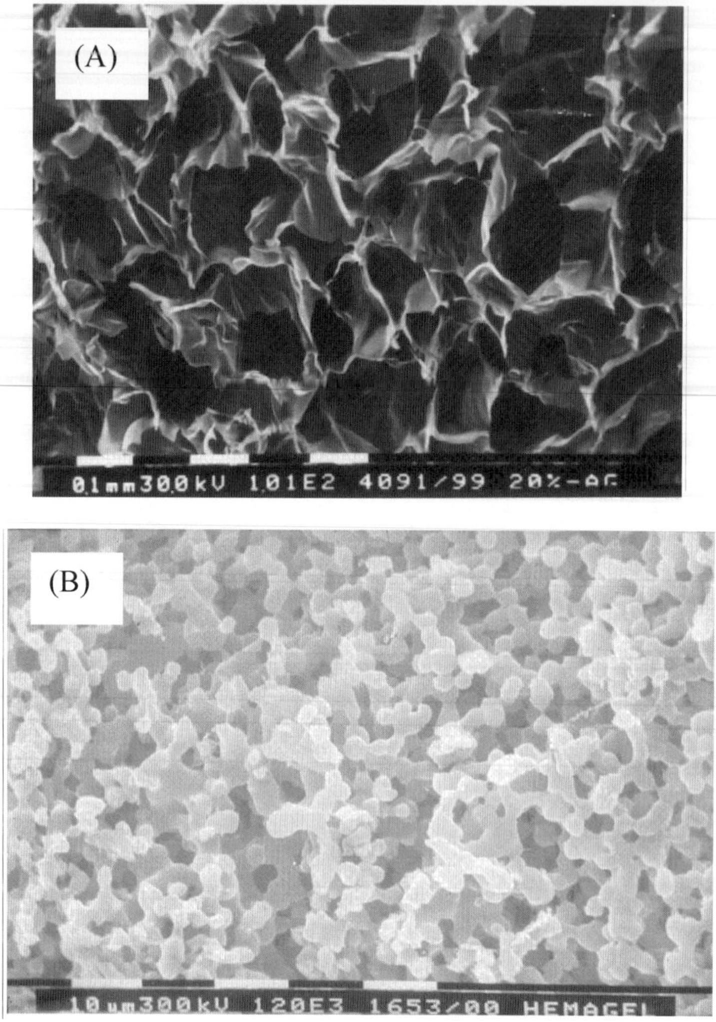

Figure 3. Representative scanning electron microscopy of 20% AG-chitosan hydrogel (A) and HEMA hydrogel crosslinked with 2% EGDMA and 75% water in preparation (B).

Drug content in HEMA hydrogel

The drug amount in the selected HEMA hydrogel (HEMA : EGDMA 98 : 2 and 75% water) was calculated after 2 h of immersion in different drug solutions at various concentrations (Table 2). The drug content in the hydrogel was dependent on the drug type and its concentration in the immersing solution. For the same hydrogel sample, steroid solutions can be absorbed in larger volumes (85% w/w in hydrogel) compared with gentamicin sulfate solution (73% w/w in hydrogel). However, the steroid amounts into the hydrogel were smaller than gentamicin due to different affinities of the hydrogel to the two types of drugs. It was found that 2 h after hydrogel immersion, the drug concentration in the immersed solution

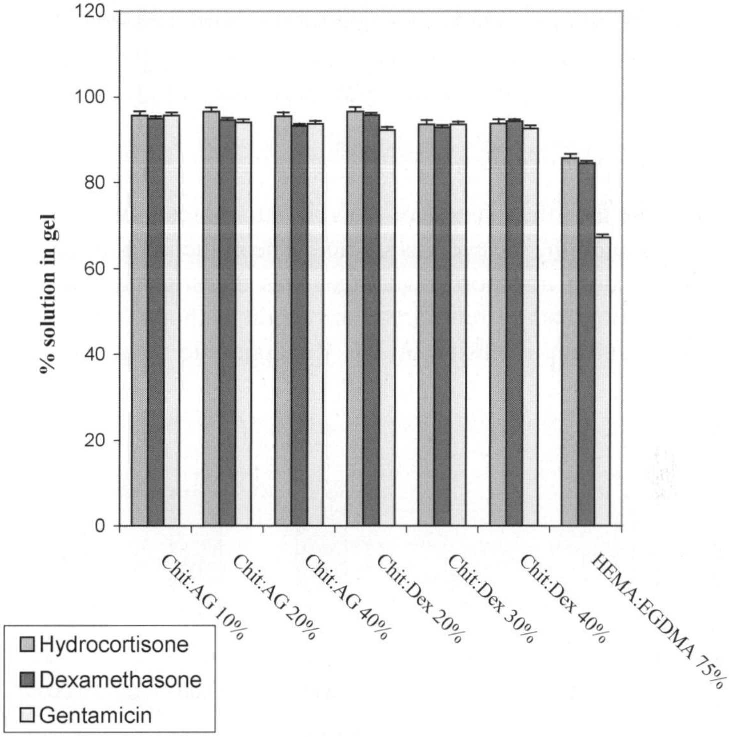

Figure 4. Different polysaccharide and HEMA hydrogel absorption of 100 mg/ml drug solutions. % solution in hydrogel was calculated using equation (1), after 2 h of hydrogels immersion in the different drug solutions.

Table 2.
Average drug amounts in HEMA hydrogels

Drug	Conc. (mg/ml)	W_d (mg)	W_w (mg)	Swelling (%)	Drug amount (mg)/hydrogel
Hydrocortisone succinate	300	37.4	264.7	85.9	68.0
Hydrocortisone succinate	200	40.4	300.8	86.6	52.1
Hydrocortisone succinate	100	37.2	250.6	85.1	21.3
Dexamethasone phosphate	100	34.2	222.1	84.6	18.8
Gentamicin sulfate	100	31.3	118.0	73.5	26.1

% swelling was determined in the drug solutions after 2 h. Swelling was calculated using equation (1).

changes, gentamicin concentration decreases by about 40% and the hydrocortisone concentration increases by 20% to 50%. This indicates that gentamicin molecules have greater affinity to the HEMA hydrogel than the hydrocortisone, which could be explained by the different polarity and ionization of the two molecules. The positively-charged gentamicin is attached by the partial negative charge of the oxygens of the hydroxyl and ester groups within the HEMA hydrogel, while the negatively-charged steroids may be repelled by this partial charge. The amount of drug absorbed in the hydrogel can be increased by increasing the drug concentration of the immersion solution; however the hydrogel dimensions may be altered.

In vitro *iontophoresis onto agar gel*

In order to evaluate the effectiveness of the gel iontophoresis system, the release of drug in agar gel resembling the eye was studied. The influence of three iontophoretic parameters on the drug transport to the agar was investigated: current intensity, iontophoretic duration and current direction (anodic/cathodic iontophoresis). The influence of the current intensity of the iontophoretic treatment on the drug

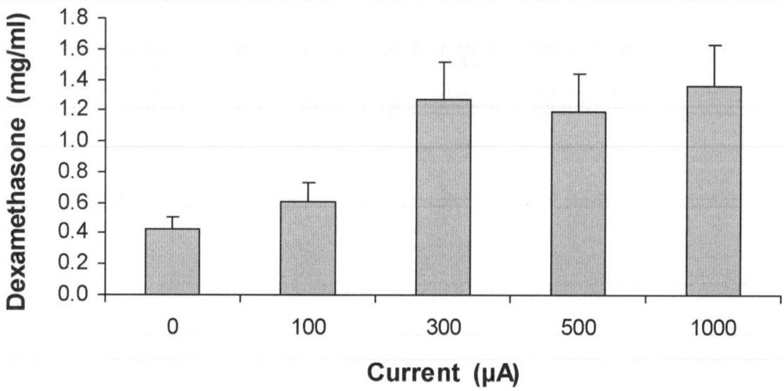

Figure 5. Dexamethasone phosphate concentration in agar after iontophoretic treatment of 120 s, applying increasing current intensity (100–1000 μA), compared with mock iontophoresis. Crosslinked HEMA hydrogel loaded with 10% (w/v) solution of dexamethasone phosphate was applied on an agar gel. The amount of drug accumulated in the agar gel was determined by UV absorbance at 250 nm.

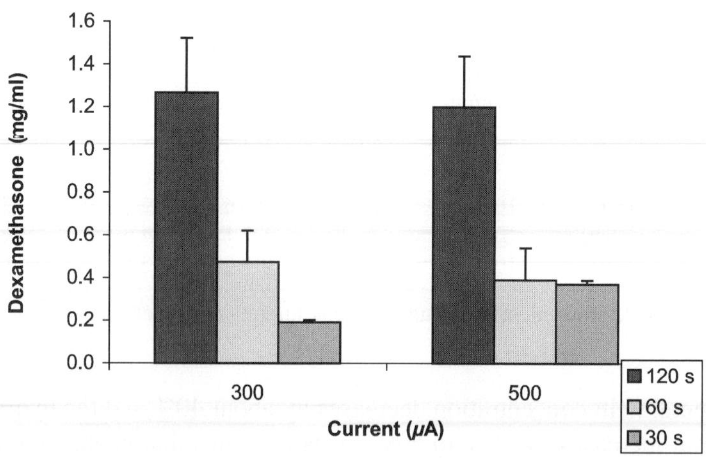

Figure 6. Dexamethasone phosphate concentration in agar after iontophoretic treatment of 30–120 s applying increasing current intensity (300–500 μA). Crosslinked HEMA hydrogel loaded with 10% (w/v) solution of dexamethasone phosphate was applied on an agar gel. The amount of drug accumulated in the agar gel was determined by UV absorbance at 250 nm.

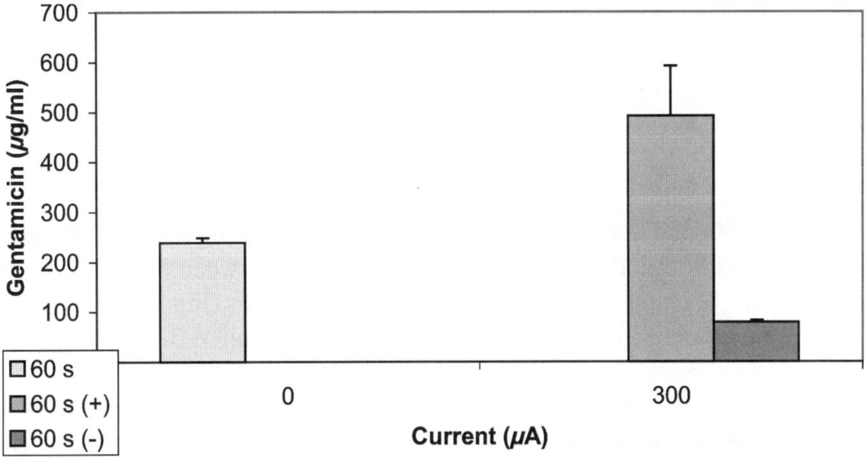

Figure 7. Gentamicin sulfate concentration in agar after iontophoretic treatment of 60 s applying a current intensity of 300 μA in both directions, compared with mock iontophoresis. Crosslinked HEMA hydrogel loaded with 10% (w/v) solution of gentamicin sulfate was applied on an agar gel. The amount of drug accumulated in the agar gel was determined by TDx.

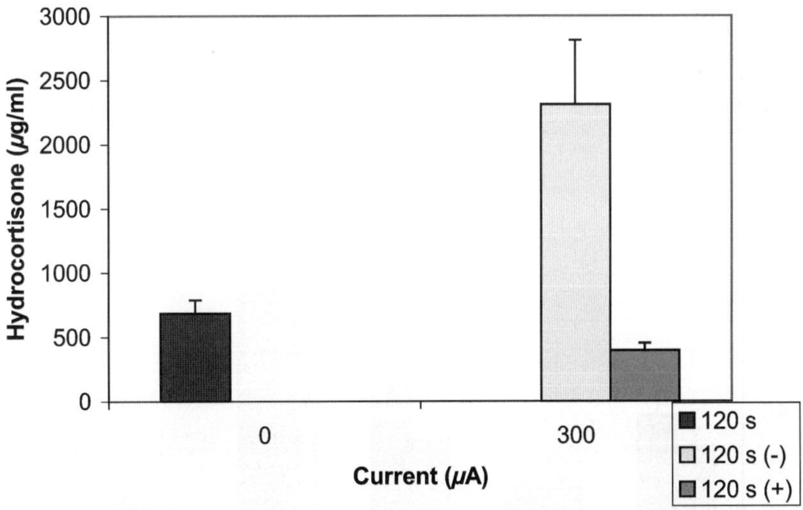

Figure 8. Hydrocortisone succinate concentration in agar after iontophoretic treatment of 120 s applying a current intensity of 300 μA in both directions, compared with mock iontophoresis. Crosslinked HEMA hydrogel loaded with 10% (w/v) solution of hydrocortisone succinate was applied on an agar gel. The amount of drug accumulated in the agar gel was determined by UV absorbance at 250 nm.

penetration was investigated using HEMA hydrogels loaded with an average of 17.1 mg dexamethasone phosphate per hydrogel disk. After 120 s of iontophoresis onto agar surface at different current intensities (0–1000 μA), it was found that there is a direct correlation between the amounts of drug found in the agar and the current applied (Fig. 5).

The duration of iontophoresis was found to be important to achieve higher agar drug concentrations. Using dexamethasone loaded hydrogels and applying a current of 300–500 μA, the duration of the treatment directly influenced the drug concentration found in the agar. The difference between 60 and 120 s was significantly higher than the difference between 30 and 60 s (Fig. 6).

In addition, it was found that the current direction applied in an iontophoretic treatment is important to be coordinated with the drug charge, as seen in Figs 7 and 8. When the current direction was the same as the drug charge (positive for gentamicin sulfate and negative for hydrocortisone succinate), the drug concentration achieved in the agar was 5–10-fold higher than with the current in the opposite direction.

In vivo *results*

Transscleral and transconjunctival iontophoresis using hydrogels loaded with gentamicin sulfate demonstrated the ability of the iontophoretic device to deliver high gentamicin concentrations to the posterior segments of the eye (Fig. 9). After 240 s

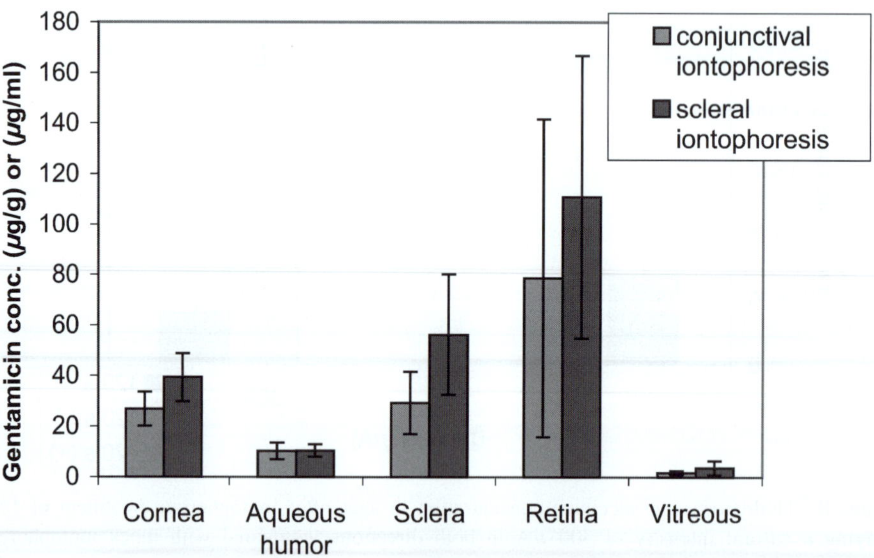

Figure 9. Gentamicin sulfate concentrations in the different segments of the eye, after a transconjunctival and transscleral iontophoresis. Crosslinked HEMA hydrogel loaded with 10% (w/v) solution of gentamicin sulfate was applied on the pars-plana area of the rabbit's sclera, at a current of 1 mA for 4 min. The amounts of drug accumulated in the different segments of the eye were determined by TDx.

of transscleral iontophoresis at 1 mA, 110.8 ± 55.9 μg and 56.3 ± 23.9 μg of gentamicin per gram of tissue were obtained in the retina and sclera, respectively. Lower concentrations were obtained in the cornea, aqueous humor and vitreous (39.3 ± 9.5 μg/g, 10.4 ± 2.5 μg/g and 3.6 ± 2.8 μg/g, respectively) due to the time of killing, as will be explained in the Discussion. Comparing the transconjunctival results with the transscleral, a non-significant difference was found (Kruskal–Wallis test), indicating that removing the conjunctival membrane was not necessary. Figure 9 presents the mean gentamicin levels in the different segments of the eye after transconjunctival (group 1) and transscleral (group 2) iontophoresis.

Preliminary toxicity experiment conducted after a single iontophoretic treatment at a current of 0.5–1 mA for 60–120 s revealed no permanent damage to the cornea (group 3). 5 min after iontophoretic treatment (rabbits 1 and 2) there are epithelial defects that are probably caused by the mechanical pressure on the cornea, and stromal edema which is probably associated with the electric current. However, 8 h after the iontophoretic treatment in a current intensity of 0.5 mA for 60 s (rabbit 3) no lesions were seen. The epithelium, stroma and endothelium layers are clearly seen in normal size similar to the histological findings of the reference rabbit (rabbit 5) which was treated with mock iontophoresis. Eight hours after an intensive iontophoretic treatment of 1 mA for 120 s, a mild stromal edema is observed (probably reversible) and small Descament's ruptures which could derive from improper sampling and storage.

DISCUSSION

For decades hydrogels have been investigated as carriers of drugs [2, 25, 26] with only a few studies reporting on iontophoresis using hydrogel probes [3–6].

This work evaluates the efficiency of a portable iontophoretic delivery system, which was developed in our lab (Fig. 1), in delivering antibiotics and steroids into the eye through drug-loaded hydrogels. Various methacrylate and polysaccharide hydrogels were evaluated as probes for this system. Their drug solution absorption and release, and physical properties were studied. A HEMA hydrogel crosslinked with 2% EGDMA was selected for further iontophoresis studies. This hydrogel has a high drug solution absorption capacity (75–85%), good mechanical strength, flexible but not sticky and predictable size and shape.

In an attempt to develop a reliable in vitro method to evaluate the usefulness of hydrogel sponges, we examined the possibility to mimic the rabbit eye surface with agar gel. In this study, iontophoresis was applied onto solid-agar gels at different conditions and the drug ejected into the agar gel after iontophoretic application was determined. The duration of iontophoresis, current intensity and current direction applied were correlated with the drug concentrations achieved in the agar gel. These results indicate the usefulness of the agar gel method for initial evaluation of the suitability of hydrogels for iontophoresis.

The effectiveness of the iontophoretic device using the selected hydrogel for delivering drugs into the eye was evaluated on rabbits. Gentamicin-loaded hydrogel was selected for transsclearal and transconjunctival iontophoresis using a current of 1 mA for 240 s. High gentamicin concentrations were found in the retina and sclera 4 h after the iontophoretic treatment (Fig. 9). The lower concentrations in the cornea, aqueous humor and vitreous could be explained by the time the animals were killed in this study. It is possible that 4 h after the iontophoretic treatment the concentrations in the cornea and aqueous humor are declining, and the concentrations in the vitreous did not get to their peak yet, as reported by Grossman (reported peak concentration after 16 h in the vitreous and after 2 h in the aqueous humor) [15]. However, all concentrations achieved at the different segments of the eye four hours after transscleral iontophoresis, are above the minimal inhibitory concentration (MIC) of gentamicin against *Pseudomonas* species (3 μg/ml) [14]. More pharmacokinetic studies should be conducted to evaluate the time course of the drug concentration after iontophoresis at the different segments of the eye.

The transscleral iontophoretic treatment indicated that this device can be used to obtain high concentrations of drugs to the posterior segments of the eye using a short transscleral iontophoresis with low current, without removing the conjunctiva. This method can be used to enable a convenient and efficient therapy of various posterior infections of the eye, i.e. posterior uveitis, scleritis or endophthalmitis, that until now have been treated with invasive intravitreous injections of antibiotics, which are associated with severe complications and pain.

Preliminary toxicity to the eye after iontophoresis indicated a reversible inflammation. The iontophoretic treatment with a current intensity of 5.1 mA/cm^2 for 2 min was found to be safe, with only mild reversible effect. This finding is supported by reported data in which transcorneal iontophoresis at a current density of 20 mA/cm^2 for 5 min was not found to be harmful to the cornea [13] and transscleral iontophoresis at a current density lower than 50 mA/cm^2 is practically safe [19]. These publications, however, were on iontophoretic devices using eye-cup solution.

To conclude, this study demonstrated the usefulness of the portable iontophoretic device (Mini-Ion) and the drug loaded hydrogel probe for enhancement of drug penetration to anterior and posterior segments of the eye using low and harmless current. The low-cost single-use disposable HEMA hydrogels were found to be suitable for the iontophoresis application and comfortable to work with.

Acknowledgements

This work was partially supported by grants from the Horowits Foundation and from the Prof. Michael Chorco and Zeev Ta-Shma fund. We would like to thank Prof. Jacob Pe'er for the histological work and to Dr. Rachamim Ezra for the SEM pictures. A. J. D. is associated with the David R. Bloom Center for Pharmacy at the Hebrew University. This work is part of the PhD thesis of E. E.-B.

REFERENCES

1. A. Kishida and Y. Ikada, in: *Polymeric Biomaterials*, S. Dumitriu (Ed.), p. 133. Marcel Dekker, New York, NY (2002).
2. S. W. Kim, Y. H. Bae and T. Okano, *Pharm. Res.* **9**, 283 (1992).
3. A. K. Banga and Y. W. Chien, *Pharm. Res.* **10**, 697 (1993).
4. J. Y. Fang, L. R. Hsu, Y. B. Huang and Y. H. Tsai, *Int. J. Pharm.* **180**, 137 (1999).
5. M. J. Alvarez-Figueroa and J. Blanco-Mendez, *Int. J. Pharm.* **215**, 57 (2001).
6. J. Y. Fang, K. C. Sung, J. J. Wang, C. C. Chu and K. T. Chen, *J. Pharm. Pharmacol.* **54**, 1329 (2002).
7. G. Erlanger, *Ophthalmologica* **128**, 232 (1954).
8. M. W. M. Bridger, M. Keene, J. M. Graham, R. Healey and M. M. Ammar, *J. Med. Eng. Technol.* **6**, 62 (1982).
9. L. E. Gibson and R. E. Cooke, *Pediatrics* **23**, 545 (1959).
10. L. P. Gangarosa, J. M. Hill, B. L. Thompson, C. Leggett and J. P. Rissing, *J. Infect. Dis.* **154**, 930 (1986).
11. L. Gangarosa and N. Park, *J. Prosthet. Dent.* **19**, 73 (1978).
12. D. M. King and S. A. Estrem, *Laryngoscope* **100**, 1112 (1990).
13. L. Hughes and D. M. Maurice, *Arch. Ophthalmol.* **102**, 1825 (1984).
14. P. H. Fishman, W. M. Jay, J. P. Rissing, J. M. Hill and R. K. Shockey, *Invest. Ophthalmol. Vis. Sci.* **25**, 343 (1984).
15. R. Grossman, D. F. Chu and D. A. Lee, *Invest. Ophthalmol. Vis. Sci.* **31**, 909 (1990).
16. J. Frucht-Pery, D. Goren, A. Solomon, C. S. Siganos, H. Mechoulam, P. Ever Hadani and M. Shapiro, *J. Ocul. Pharmacol. Ther.* **15**, 251 (1999).
17. J. Frucht-Pery, A. Solomon, R. Doron, P. Ever Hadani, O. Manor and M. Shapiro, *Graefes Arch. Clin. Exp. Ophthalmol.* **234**, 765 (1996).
18. T. T. Lam, D. P. Edward, X. A. Zhu and M. O. M. Tso, *Arch. Ophthalmol.* **107**, 1368 (1989).
19. F. F. Behar-Cohen, A. El Aouni, S. Gautier, G. David, J. Davis, P. Chapon and J. M. Parel, *Exp. Eye Res.* **74**, 51 (2002).
20. F. F. Behar-Cohen, J. M. Parel, Y. Pouliquen, B. Thillaye-Goldenberg, O. Goureau, S. Heydolph, Y. Courtois and Y. De-Kozak, *Exp. Eye Res.* **65**, 533 (1997).
21. M. Barza, *Ophthalmology* **93**, 133 (1986).
22. M. Barza, C. Peckman and J. Baum, *Invest. Ophthalmol. Vis. Sci.* **28**, 1033 (1987).
23. M. Barza, C. Peckman and J. Baum, *Arch. Ophthalmol.* **105**, 1418 (1987).
24. E. D. Millikan, in: *American Hospital Formolary Service (AHFS) Drug Information*, K. Litvak and G. K. McEvoy (Eds), p. 67. American Society of Health System Pharmacists, Bethesda, MD (2001).
25. V. Michailova, S. Titeva, R. Kotsilkova, E. Krusteva and E. Minkov, *Int. J. Pharm.* **209**, 45 (2000).
26. V. Michailova, S. Titeva, R. Kotsilkova, E. Krusteva and E. Minkov, *Int. J. Pharm.* **222**, 7 (2001).

Gels, Genes, Grafts and Giants, pp. 45–60 (2005)

Aqueous N,N-diethylnicotinamide (DENA) solution as a medium for accelerated release study of paclitaxel

NAMJIN BAEK, JAEHWI LEE and KINAM PARK *

Departments of Pharmaceutics and Biomedical Engineering, Purdue University School of Pharmacy, Room G22, 575 Stadium Mall Drive, West Lafayette, IN 47907-2051, USA

Received 24 September 2003; accepted 12 January 2004

Abstract—N,N-Diethylnicotinamide (DENA) was identified as an excellent hydrotropic agent for paclitaxel (PTX) in our previous studies. The aqueous solubility of PTX was increased by several orders of magnitude in the presence of DENA. Because of such a high hydrotropic property, DENA was used as a release medium providing a sink condition for the release of PTX from poly(lactic-co-glycolic acid) (PLGA) matrices. The release profiles of PTX from PLGA matrices into DENA, serum and phosphate-buffered saline (PBS) were compared. The stability of PTX in DENA and the degradation of PLGA molecules in DENA were examined. The degradation rate constant of PTX in 2 M DENA was similar to those in other aqueous solutions. The use of 2 M DENA as a release medium allowed differentiation of the release profiles of PTX from PLGA matrices made of different PLGA compositions. The PTX release from PLGA matrices was much faster in DENA solution than in serum or PBS, and the concentration of DENA affected the PTX release rate. The presence of DENA in the release medium increased the hydrolysis rate of PLGA polymers. The faster release of PTX from PLGA matrices in DENA solution may be due to the high PTX solubility and faster degradation of PLGA polymers in the presence of DENA. Our study suggests that the aqueous DENA solution can be used for the accelerated release study of PTX from PLGA matrices.

Key words: Hydrotropic agent; paclitaxel; *in vitro* release; PLGA; accelerated release.

INTRODUCTION

Paclitaxel (PTX) has been used as an effective agent against ovarian, breast and lung cancers [1]. Recently, PTX was also shown to be effective in preventing restenosis after it is released from the drug-eluting stents [2–5]. In one study, PTX was loaded into a poly(lactic-co-glycolic acid) (PLGA) matrix that was subsequently released from the stents into the surrounding tissue [5]. While many studies have shown that

*To whom correspondence should be addressed. Tel.: (1-765) 494-7759. Fax: (1-765) 496-1903. E-mail: kpark@purdue.edu

the release of PTX from the stents is highly beneficial in preventing restenosis, the optimum release kinetics of PTX from the stents has not been examined thoroughly. While the most realistic PTX release profile from the stents can be obtained from *in vivo* animal studies, such *in vivo* studies require the use of numerous animals that may not be realistic for testing many samples. For practical reasons, most drug release kinetics are measured from *in vitro* studies.

The *in vitro* release or dissolution testing of poorly soluble drugs poses problems due to their very low aqueous solubility. Phosphate-buffered saline (PBS) [6], PBS with surfactants [7] and PBS with albumin [8, 9] were used for the *in vitro* release of PTX from microspheres or paste. The release media were replaced with fresh media frequently to maintain the sink condition. The assay methods usually require a concentration step by lyophilization and/or extraction with organic solvents to measure the amount of paclitaxel released. Although the residual amount of PTX in microparticles can be determined [10], such a method requires a lot of samples with large volumes for each *in vitro* release experiment.

Recently, it was shown by our group that the aqueous solubility of PTX was significantly increased by using a hydrotrope, N,N-diethylnicotinamide (DENA) [11]. The aqueous solubility of PTX in DENA was much higher than that in acetonitrile or ethanol [12]. The PTX solubility in 2 M DENA is about 1 mg/ml and, thus, even a very small volume, as low as 1 ml, can be used to provide a sink condition in the *in vitro* PTX release if the total amount of PTX released is 200 μg/ml or less. In addition, a small volume of a release sample can be directly injected into an HPLC system for the determination of PTX concentration without concentration and extraction steps. Thus, use of DENA aqueous solution as a release medium is attractive. In this study, the DENA solution was used as a medium for the *in vitro* release study of PTX, and the effect of the release medium on the PTX release from model PLGA/PTX films was studied.

MATERIALS AND METHODS

Materials

Poly(D,L-lactic-co-glycolic acid) (PLGA) polymers with various co-polymer compositions were obtained from Birmingham Polymers (Birmingham, AL, USA). Calf serum bovine and thimerosal were obtained from Sigma (St. Louis, MO, USA).

HPLC assay

The concentration of PTX was determined by reversed-phase HPLC using the Agilent 1100 HPLC system (Agilent Technologies, Palo Alto, CA, USA). Each sample of 5–20 μl was injected into a Symmetry C_{18} 5 μm column (3.9 × 150 mm, Waters, Milford, MA, USA). A mixture of acetonitrile and water (45 : 55, v/v) was used as a mobile phase at a flow rate of 1.0 ml/min at 25°C. A diode array detector was set at 227 nm.

Stability of PTX in an aqueous DENA solution

PTX needs to be released for at least a few weeks from stents and, thus, PTX should be stable in the release medium for accurate determination of the amount of PTX released. In order to use an aqueous solution of DENA as a medium for PTX release, the stability of PTX in 2 M DENA (in distilled water) was studied.

A PTX solution of about 100 μg/ml concentration was prepared in 2 M DENA and dispensed into small glass vials. Sample vials were incubated at 37°C. At predetermined time intervals, duplicate samples were stored frozen until analyzed by the HPLC method.

Effect of PLGA co-polymer composition on PTX release from films

PLGA/PTX films were prepared using PLGA(50/50) (intrinsic viscosity 0.63 dl/g, weight-average molecular weight (M_w) 43×10^3, polydispersity 1.8), PLGA(85/15) (intrinsic viscosity 0.72 dl/g, M_w 105.6×10^3) and poly(D,L-lactic acid) (PLA, intrinsic viscosity 0.66 dl/g, M_w 106×10^3) to examine the effect of co-polymer compositions on the release of PTX in 2 M DENA. Solutions of 10% (w/v) PLGA/1% (w/v) PTX in methylene chloride were prepared, and 81 μl of the solution was placed on a glass slide (0.9 cm$\times$0.9 cm) to make films about 100 μm in thickness. After overnight air-drying, PLGA/PTX film samples were vacuum-dried at room temperature for 1.5 days.

PLGA/PTX films were transparent but were tightly bound to the glass surface; thus, films attached to the glass surface were used in the release study. Triplicate samples of each PLGA/PTX (10 : 1, w/w) film were immersed in 3 ml of 2 M DENA solution at 37°C. At predetermined time intervals, 0.3-ml samples were taken and stored frozen. The same volume of fresh medium was added back to the release medium. The amount of PTX in the release medium was determined by HPLC.

In vitro *PTX release in different media*

Polyethylene (PE) discs (1.12 cm diameter) were firmly attached onto the glass surface using double-sided tape. PLGA(50/50) (intrinsic viscosity 0.63 dl/g) and PTX were dissolved in acetonitrile or methylene chloride to make 10% PLGA(50/50), 10% PLGA(50/50)/1% PTX, or 10% PLGA(50/50)/3% PTX solutions. The PLGA or PLGA/PTX solution (100 μl) was placed on the PE disc. After overnight air-drying, films were vacuum-dried for 2.5 days and stored in a desiccator until use. PLGA or PLGA/PTX films could be peeled off easily from the PE disc after drying.

Different concentrations of DENA were used as release media. In the release experiments using DENA solution, the total volume of the release medium was 10 and 30 ml for film samples in 2 M and 1.5 M DENA, respectively. Volumes of all DENA solutions, including 1.2 M and 1 M DENA, were enough to maintain the total amount of the released PTX to less than 15% of the PTX solubility in the DENA solution, which ensures the sink condition. Release medium was incubated at 37°C. The samples in DENA solution were used for HPLC analysis after filtration.

For the release study in PBS, 500 ml of PBS was completely replaced with a fresh solution at every sampling point. Of those samples 10 ml was lyophilized and reconstituted in 1 ml of a acetonitrile/water mixture (25 : 75, v/v).

Film samples were also incubated in 5 ml of calf serum containing 0.02% (w/v) thimerosal as a preservative. In addition, in order to prevent the contamination in serum, glass vials or tips were autoclaved before use, and all the samples using serum were handled under aseptic conditions. Serum was replaced every 4 days. 0.3 ml of release sample was stored frozen.

After the release study, the residual PTX amounts in the films from PBS and serum were determined as follows. Film samples were washed with distilled water (DW), vacuum-dried and dissolved in 5 ml of acetonitrile. After transferring 0.5 ml of solution to a new tube, 0.5 ml of DW was added to precipitate any large MW PLGA polymers. Since the PTX solubility in 50% acetonitrile is 4.3 mg/ml, it was likely that most of the PTX remained dissolved. The mixture was sonicated for 30 s and filtered. The concentration of PTX was determined by the HPLC method. The extraction efficiency of PTX from serum was determined as shown in the next section and was used for the quantification of PTX released into serum.

To examine whether PTX crystal forms on the PLGA/PTX films upon incubation in PBS, films with different PLGA(50/50)/PTX ratios (10 : 0, 10 : 1 and 10 : 3, w/w) were incubated in two different volumes, 5 ml or 80 ml, of PBS at 37°C for 1 day. According to *in vitro* PTX release data from the PLGA(50/50)/PTX (10 : 1, w/w) films, the amount of PTX released in PBS during the first day was about 3.3 μg. Since PTX solubility in PBS is 0.45 μg/ml, 5 ml of PBS represents the non-sink environment, where the PTX concentration will exceed the PTX solubility. A larger volume (80 ml) of PBS represents the sink condition for PLGA(50/50)/PTX (10 : 1, w/w) films. After freeze-drying, microscopic images were taken for those samples using a light microscope (Labophot-2, Nikon, Japan).

Extraction of PTX from serum

For the extraction of PTX from tissue samples, organic solvent or solid phase extraction techniques were mainly used [13]. In this study, a modified method of Koshkina *et al.* [14] was used for the extraction of PTX from serum samples using ethyl acetate.

10 μl PTX working stock solution in acetonitrile was added to 0.99 ml of serum containing 0.02% (w/v) thimerosal in an Eppendorf tube to prepare solutions of 0.1, 0.2, 0.5, 1, 2, 5 and 10 μg/ml PTX in serum. Thimerosal was used as a preservative. Triplicate samples containing serum with PTX were incubated at 37°C for 12 h, in order to obtain equilibrium binding of PTX to serum proteins.

300 μl serum, containing PTX, was transferred to a new Eppendorf tube, and 0.9 ml of ethyl acetate was added. After vortex-mixing for 30 s, samples were centrifuged ($10^4 \times g$, 5 min). The ethyl acetate layer was transferred to a new glass vial. The extraction step was repeated, and the ethyl acetate supernatant layer was pooled together. Ethyl acetate from the pooled supernatant was removed by

vacuum-evaporation in about 1 h. Liquid nitrogen was used in the trap to collect evaporated ethyl acetate.

Dried serum extract was dissolved in 0.2 ml of acetonitrile/water (1 : 1, v/v) by vortexing for 30 s. After filtering through a 0.25 μm nylon filter, 20 μl of each filtrate was used for HPLC analysis. The HPLC analysis was implemented as described previously, but with a different mobile phase: 43% of acetonitrile in DW was used until 28 min to separate a PTX peak from a serum-originating peak, and then the acetonitrile concentration was gradually increased to 100% to remove other peaks originating from the serum.

After extraction and HPLC analysis, a regression line of AUC *versus* PTX concentration was prepared, and the slope of the regression line was compared with that of PTX standard solutions in acetonitrile (data not shown). PTX extracted from serum had a smaller slope than that of the PTX standard. Extraction efficiency was calculated as: ((slope of regression line of PTX extracted)/(slope of regression line of PTX standard)) $\times$ 100% = 73.4%. This value was used for the adjustment of the released PTX amount into serum.

Molecular weight determination of PLGA polymers

PLGA(50/50) or PLGA(50/50)/PTX (10 : 1, w/w) films were placed individually in glass vials containing 5 ml of 2 M DENA (in DW), 5 ml of 2 M DENA (in PBS), 15 ml of PBS, or 15 ml of calf serum containing 0.02% (w/v) thimerosal. The incubation temperature was 37°C and the medium was replaced every 5 days. At predetermined time intervals, film samples were washed with distilled water, lyophilized to remove water, and then stored frozen until analysis.

The molecular weight of PLGA polymer was determined by gel-permeation chromatography (GPC) using the Agilent 1100 HPLC system (Agilent Technologies, Palo Alto, CA, USA) with a refractive index detector. The molecular weight calibration curve was established using polystyrene standards (Polymer Laboratories, Amherst, MA, USA) covering a molecular weight range of 580 to 283.3×10^3. For GPC analysis, lyophilized polymer samples were dissolved in 0.5 ml of chloroform stabilized with 1% ethanol (Mallinckrodt Baker, Paris, KY, USA). After filtering through a 0.45 μm Teflon syringe filter (Pall Gelman Laboratory, Ann Arbor, MI, USA), 50 μl of sample was injected into a PLgel 5 μm mixed-D column (Polymer Laboratories). Chloroform was used as a mobile phase at 1 ml/min and 35°C.

RESULTS

Stability of PTX in aqueous DENA solution

Before an aqueous DENA solution was used as a release medium for PTX release, the stability of PTX in the release medium was first examined for accurate determination of the amount of the released PTX.

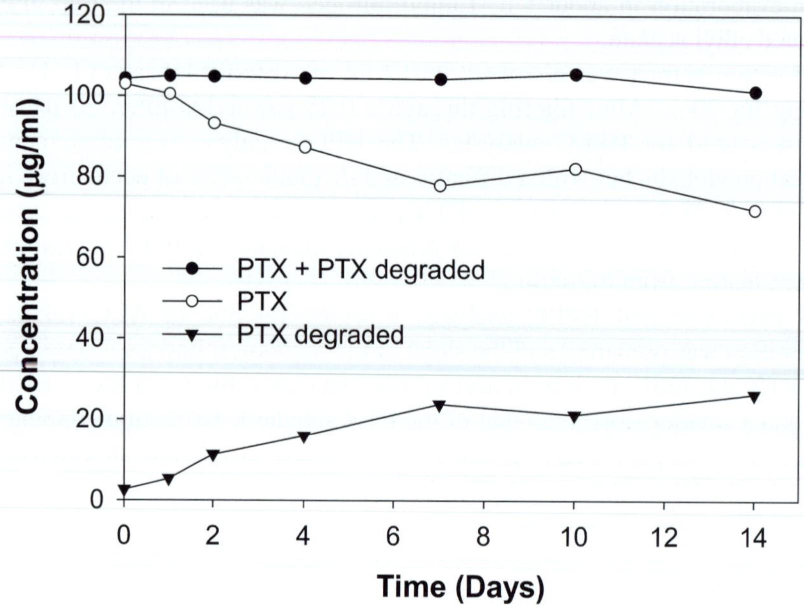

(A)

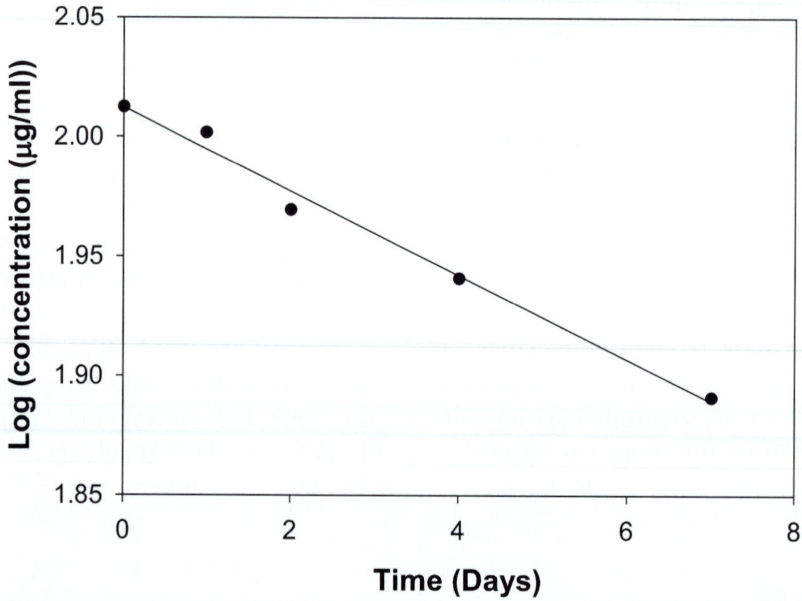

(B)

Figure 1. Stability of PTX in 2 M DENA at 37°C. The retention times in the HPLC chromatogram were 11.4 min for PTX and 21.9 min for degraded PTX. The mobile phase was a mixture of acetonitrile and water (45 : 55, v/v). Duplicate samples were used at each time point. Changes of PTX concentration in linear scale (A) and in log scale (B) as a function of time.

When PTX was incubated in 2 M DENA solution (in distilled water, pH 6.8), which has a PTX solubility of 1 mg/ml, the PTX concentration in the medium decreased gradually (Fig. 1A). A semi-log plot of the concentration *vs.* time is linear (Fig. 1B), which shows first-order degradation kinetics of PTX in 2 M DENA solution. From the slope of Figure 1B, the degradation rate constant of 1.69×10^{-3} h^{-1} was obtained. The reported degradation rate constants of PTX in phosphate buffer (pH 6) and citrate buffer (pH 6) were 1.25×10^{-2} h^{-1} and 2.97×10^{-3} h^{-1}, respectively [15]. Degradation of PTX in 2 M DENA is about the same as that in other aqueous buffer solutions.

As the PTX concentration decreased upon incubation in 2 M DENA solution, a new peak appeared with a larger retention time (21.9 min) than that of the PTX peak (11.4 min). UV spectra of those two peaks, obtained from HPLC chromatograms using a diode array detector, showed that they were similar (data not shown). Thus, it is suggested that PTX degraded in 2 M DENA to produce another product with a 21.9 min retention time. Paclitaxel converts to 7-*epi*-taxol, which is a paclitaxel isomer, in many solvents, such as DMSO, isobutyl alcohol and methanol, and 7-*epi*-taxol also has a higher retention time than PTX [16]. Considering the retention time in an HPLC chromatogram and a UV spectrum, the PTX-deg peak is expected to be 7-*epi*-taxol.

The degradation product (PTX-deg) can be counted for calculating the total amount of drug released. When the decrease in the AUC of PTX ($-AUC_{PTX}$) in 2 M DENA was compared with the increase in the AUC of PTX-deg ($AUC_{PTX\text{-}d}$), there was a linear correlation between those variables (data not shown), and the ratio of $(-AUC_{PTX})/(AUC_{PTX\text{-}d})$ was 1.158. Thus, by converting AUC of PTX-deg to AUC of PTX, it was possible to calculate the total amount of PTX in the medium, which included the amount of PTX degraded. The total concentration of PTX, which is the sum of the intact PTX concentration and the degraded PTX concentration, remained almost constant until 14 days in 2 M DENA at 37°C (Fig. 1A). Thus, the amount of PTX released into 2 M DENA could be accurately measured by combining the PTX-deg peak with the native PTX peak together.

Effect of PLGA co-polymer composition on PTX release

PLGA/PTX (10 : 1, w/w) films attached on the glass surface were used for PTX release in the 2 M DENA aqueous solution. The release profiles of PTX from PLGA/PTX films are shown in Fig. 2A. The release rate of PTX was in the order PLGA(50/50) > PLGA(85/15) > PLA. In addition, PLGA(50/50)/PTX films showed an abrupt release of PTX between 10 and 20 days, while PTX was released slowly in other films composed of PLGA(85/15) or PLA.

The release of PTX from a biodegradable PLGA matrix was governed by both diffusion and biodegradation of the polymer matrix, as was the case for mifepristone release from PLGA films [17]. The PTX release in the initial period is governed by diffusion. As shown in Fig. 2B, the amount of PTX release during the initial 4 days is linearly proportional to the square root of time, which

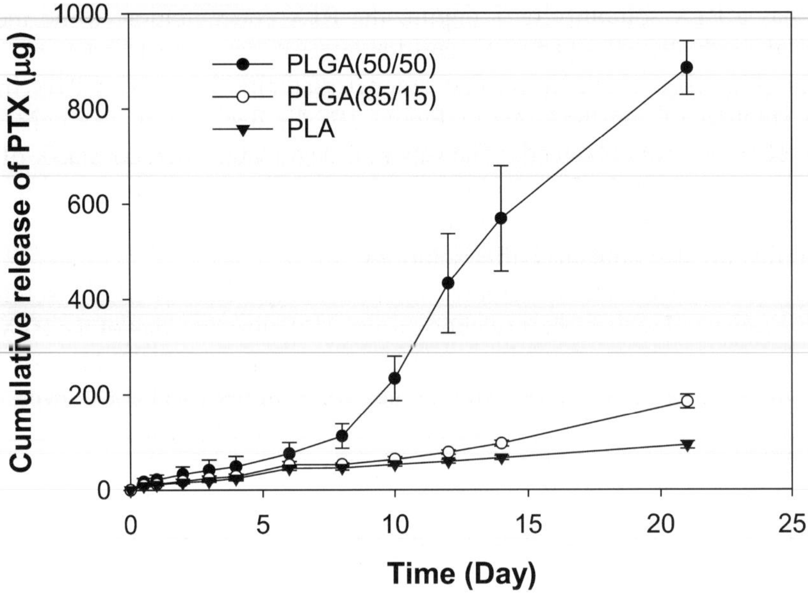

(A)

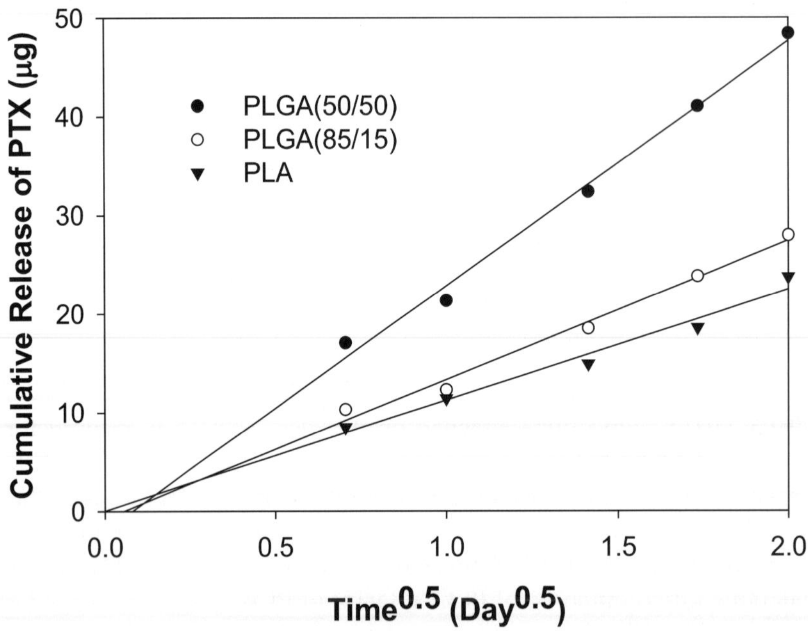

(B)

Figure 2. *In vitro* cumulative release of PTX from PLGA/PTX films of various lactide/glycolide ratios in PLGA in 2 M DENA at 37°C as a function of time (A) and $(\text{time})^{1/2}$ (B). $n = 3$.

suggests that the mechanism of PTX in this period is diffusion, according to Higuchi's model. During the initial period, the PLGA polymer matrix swells in the medium, and the polymer chains degrade slowly into smaller molecular weight chains. Although polymer degradation occurs in this period, the polymer chains are still compact enough to restrict the diffusion of drug molecules. The difference in the initial release rate between PLGA films is due to a difference in swelling of the PLGA/PTX films, since hydrophilicity of the films is in the order PLGA(50/50) > PLGA(85/15) > PLA.

After the initial period of slow release, the PLGA(50/50)/PTX film released a large amount of PTX abruptly, while the other two films still showed the slow release of PTX. This can be explained by the different susceptibility of PLGA polymers to degradation. Higher glycolide composition in PLGA leads to faster polymer degradation, because those chains are more susceptible to hydrolysis by water molecules. Thus, the degradation rate of PLGA is in the order of PLGA(50/50) > PLGA(85/15) > PLA. Both PLGA(85/15) and PLA chains degrade much slower than PLGA(50/50), while PLGA(85/15) degrades faster than PLA. Once polymer chains become very small, they cannot maintain their film structure any more, and the drug molecules are liberated into the release medium. Thus, the abrupt release of PTX from PLGA(50/50) can be explained by significant degradation of PLGA(50/50) chains. Since PLGA(85/15) and PLGA polymers are more resistant to hydrolysis than PLGA(50/50), slower PTX release was still observed until 22 days without showing an abrupt PTX release.

The increase in the surface area also played a certain role in the faster PTX release from the PLGA(50/50) film. PLGA/PTX films attached onto the glass surface were used for the release study. PLGA(50/50)/PTX films were detached from the glass surface during the release period, while PLGA(85/15) or PLA films remained attached to the glass surface. Thus, the surface area of the PLGA(50/50)/PTX film was doubled during the release period, which should result in faster release of PTX from PLGA(50/50)/PTX. Thus, for further release studies, free PLGA/PTX films were used to avoid the change in the surface area during the release study. This result suggests that aqueous solution of DENA can successfully differentiate the release of PTX from PLGA films having different co-polymer composition ratios.

Effect of release media on the PTX release

Although the 2 M DENA aqueous solution allowed differentiation of the PTX release from films with different co-polymer compositions, it was necessary to compare it with different release media that are commonly used, such as PBS, serum and lower concentrations of DENA. Figure 3 shows the release profiles of PTX from PLGA(50/50)/PTX (10 : 1, w/w) films in various media. PTX solubility was very low when the concentration of DENA was below 1 M, thus DENA concentrations between 1 M and 2 M were used in this experiment. In DENA solutions, abrupt PTX release due to polymer degradation occurred between 10 and 30 days in the order 2 M DENA in PBS (pH 7.5) > 2 M DENA in DW (pH 6.8) > 1.5 M

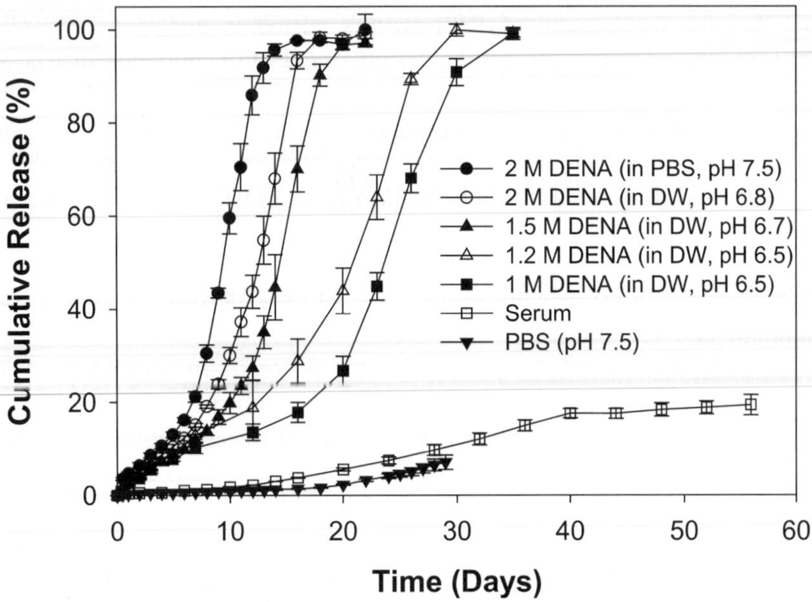

Figure 3. *In vitro* release of PTX from PLGA(50/50)/PTX (10 : 1, w/w) films in various media. $n = 3$.

DENA in DW (pH 6.7) > 1.2 M DENA in DW (pH 6.5) > 1 M DENA in DW (pH 6.5), which shows the DENA concentration-dependent profile when DENA aqueous solution was prepared in DW. The abrupt PTX release was not observed in PBS and serum until 29 and 56 days, respectively. The residual amounts of PTX were 83% and 65% for samples incubated in PBS and serum, respectively. The released amount calculated from the residual amount is much larger than that measured from the release medium. The difference in the released amount from two measurements might be due to the degradation of PTX in the serum and a very low concentration of PTX in the medium. PTX adsorption to the container surface might also have played a certain role [10].

Even though we consider the higher amounts of PTX released, which is the difference between the initial amount and the residual amount, it is evident that there is a difference in the rate of PTX release between DENA solutions and serum or PBS. In addition, a higher concentration of DENA showed a faster release of PTX. There was a possibility that PTX crystal might have formed in PLGA/PTX films when they were incubated in PBS or serum, thereby dissolution of PTX from crystals becoming the rate-limiting step in the PTX release.

Figure 4 shows the microscopic images of the films incubated in PBS. When the films were incubated in the non-sink condition, PTX crystals formed on the surface of the PLGA(50/50)/PTX films. More PTX crystals were observed in films with a higher PTX concentration (10 : 3, w/w, PLGA/PTX). However, upon incubation in 80 ml of PBS, which is a sink condition for PLGA(50/50)/PTX 10 : 1 (w/w) films, PTX crystal was not observed in PLGA(50/50)/PTX films (10 : 1 or 10 : 3, w/w).

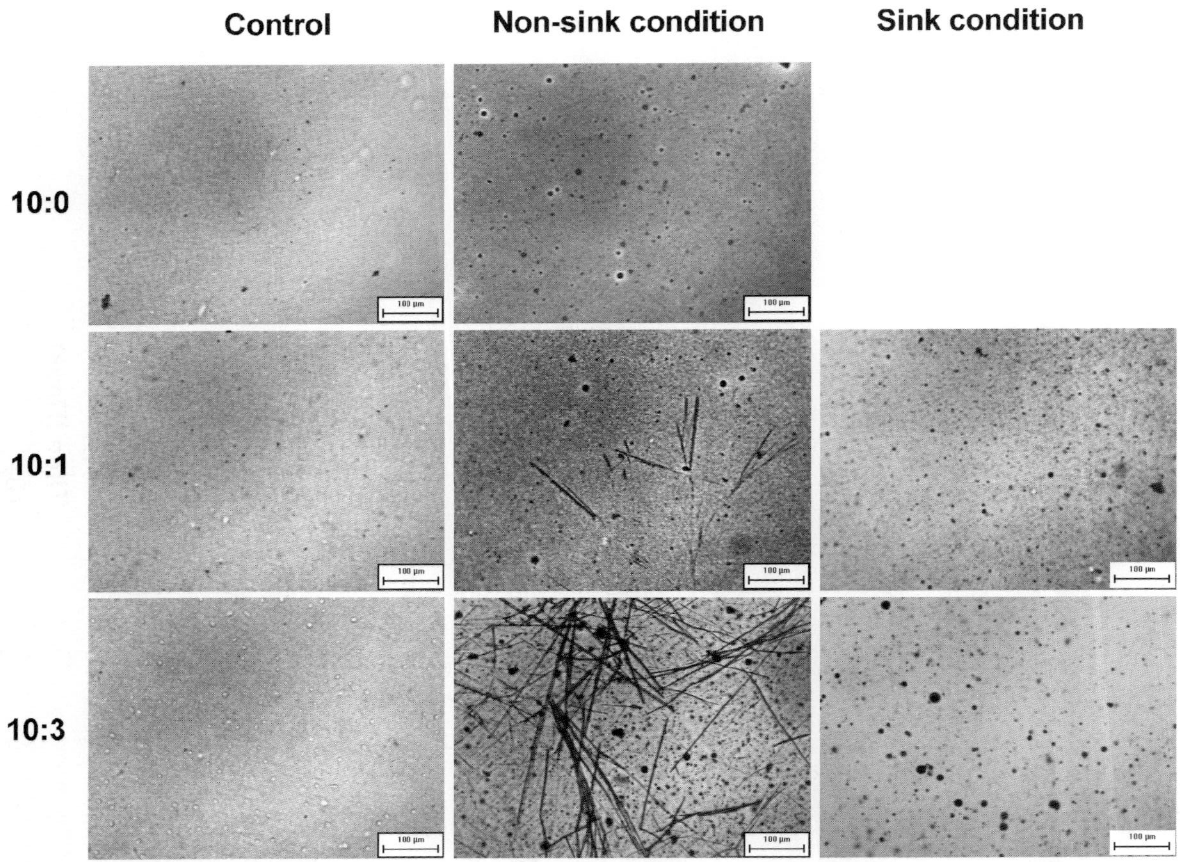

Figure 4. Microscopic images showing PTX crystals formed on the surface of PLGA/PTX films after 1-day incubation in 5 ml (non-sink condition) or 80 ml (sink condition) of PBS at 37°C. The weight ratio of PLGA to PTX in the PLGA(50/50)/PTX films was 10 : 0, 10 : 1, or 10 : 3. Control is the film before incubation in PBS.

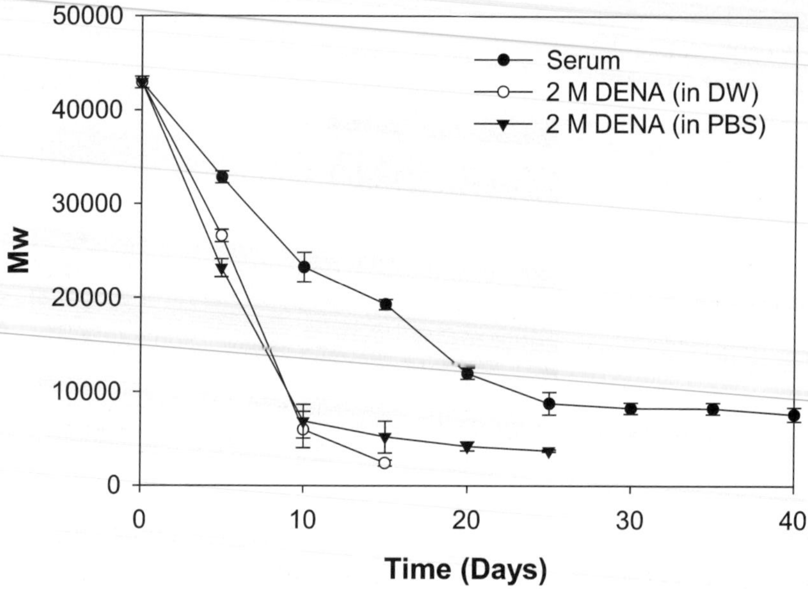

(A)

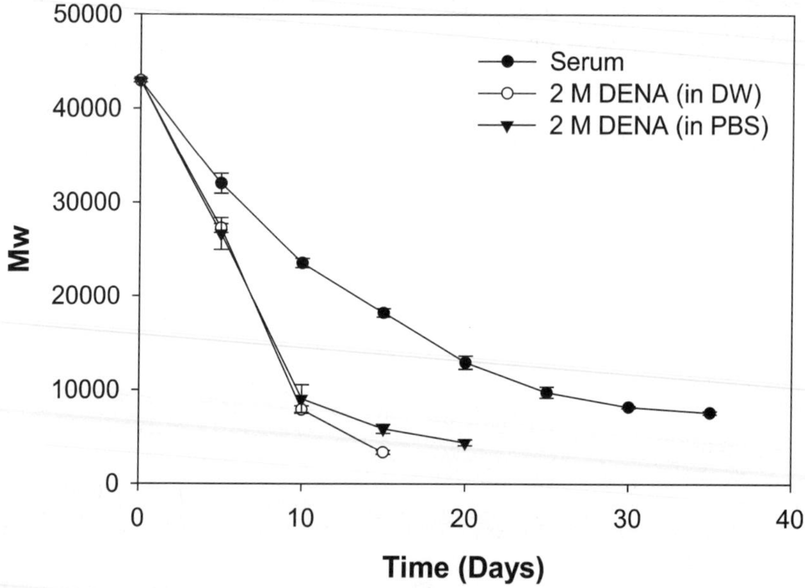

(B)

Figure 5. Changes in weight average molecular weight (M_w) of PLGA upon incubation of PLGA or PLGA/PTX films in different media. Each data point represents the mean ± SD of triplicate samples. (A), PLGA(50/50) film; (B) PLGA(50/50)/PTX (10 : 1, w/w) film.

A larger volume (80 ml) of PBS was enough to maintain the sink condition for both PLGA(50/50)/PTX films during the 1-day incubation. Thus, it is not likely that PTX crystallization and dissolution of crystal caused a slower PTX release in PBS or serum than in the DENA solutions, because a sink condition was maintained by replacing 500 ml of PBS frequently.

Effect of release media on PLGA degradation

There was also another possibility that the degradation rate of PLGA polymer was increased in DENA solutions, considering the DENA concentration-dependent PTX release. Thus, the molecular weight change of PLGA polymers upon incubation of films in different media was analyzed by GPC.

In both PLGA and PLGA/PTX films, the weight average molecular weight (M_w) of the PLGA polymer decreased much faster in 2 M DENA aqueous solutions than in serum (Fig. 5). Both 2 M DENA solutions in DW or PBS showed similar polymer hydrolysis profiles. The weight average molecular weight of PLGA from PLGA/PTX films was less than 6000 after 15 days in 2 M DENA solutions (Fig. 5), and most of the PTX was released between 10 and 16 days (Fig. 3). Thus, abrupt release of PTX from PLGA/PTX films in 2 M DENA would be related to rapid degradation of PLGA chains. In contrast, the molecular weight of PLGA in serum after 35 days was higher than 7000, which still may be high enough to hold PTX molecules in the film so as not to produce the abrupt release.

Since there was no difference in polymer molecular weight change between PLGA and PLGA/PTX films, the presence of PTX in PLGA films did not affect the degradation of PLGA polymer. This suggests that effective sink condition was maintained in both 2 M DENA and serum release media. The effective sink environment must have been helpful in removing PTX molecules from the PLGA films, thereby resulting in similar PLGA degradation profiles regardless of the presence of PTX in films.

From molecular weight analysis data, it is evident that DENA accelerated hydrolysis of PLGA chains. However, since the presence of PBS did not further facilitate the hydrolysis of PLGA in 2 M DENA solution, it is still not clear why PTX was released faster in 2 M DENA (in PBS) than in 2 M DENA (in DW).

DISCUSSION

An aqueous solution of DENA increased the PTX solubility significantly. Thus, the effect of using the DENA aqueous solution as a release medium on the PTX release from PLGA film matrices was studied. Release of PTX from film samples showed a difference in the PTX release profiles between different release media (Fig. 3). This may be explained by two factors: the solubility difference of PTX in different media and the accelerated degradation of PLGA polymers by DENA.

PTX solubility values in 2 M DENA, 1.5 M DENA, 1.2 M DENA, 1 M DENA and PBS are 980, 290, 101, 41 and 0.45 μg/ml, respectively. Effective solubility of PTX in calf serum was 171 μg/ml [18], and it is attributed to PTX binding to protein components. Since PTX solubility is higher in 2 M DENA than in lower concentrations of DENA, this might have resulted in faster dissolution of PTX molecules inside the polymer film matrix in a higher concentration of DENA than in a lower concentration of DENA. Much slower release of PTX in PBS or serum than in the DENA solutions can also be explained by the difference in PTX solubility. Although proteins in serum might have diffused into the PLGA matrix and formed PTX–protein complexes, the complex itself will be diffused out more slowly than free PTX molecules, and this would be why serum with a higher PTX solubility does not show a much faster PTX release than PBS. However, the solubility difference between media does not explain why the presence of PBS in 2 M DENA facilitated the release of PTX further than in 2 M DENA (in DW), because it is not likely that the presence of PBS in 2 M DENA increases the PTX solubility significantly.

In addition to the PTX solubility, DENA itself could have facilitated the hydrolysis of polymer chains. The PTX release experiment from films showed that the higher concentration of DENA resulted in a faster degradation-induced PTX release than the lower concentration of DENA. This result suggested that DENA facilitated the degradation of PLGA.

Some basic drugs with tertiary amine have shown the ability to accelerate the degradation of polylactide or PLGA polymers [19–22]. Poly(L-lactic acid) (PLLA) microspheres loaded with basic drugs, such as meperidine, methadone, and promethazine, showed faster drug release and faster polymer chain degradation than microspheres loaded with naltrexone, which is also a basic drug [19]. Thus, although basic drugs can accelerate the hydrolysis of PLGA chains, there was variation in the extent of acceleration between drugs. An increase in the amount of basic drug loading resulted in faster degradation of PLLA or poly(D,L-lactide) chains [19, 22]. Maulding *et al.* also observed that polymer hydrolysis occurred even during the process of thiorididazine microencapsulation using poly(D,L-lactide), but microencapsulation at 4°C or the use of the pamoate salt slowed down the polymer hydrolysis [21]. Basic drug-induced degradation of a polyester, poly(β-hydroxybutyrate), was also observed even in the films stored in the air [20]. The films, which contained thioridazine, indenorol, or clonidine, stored in the air could have been exposed to moisture in the air. Also, water absorbed from the air might have acted as a substrate for polymer hydrolysis induced by a basic drug because storage of poly(D,L-lactide) microspheres loaded with methadone did not result in polymer hydrolysis upon 1-year storage under desiccated conditions [23].

DENA itself has a tertiary amine group, and the accelerated degradation of PLGA in 2 M DENA was confirmed (Fig. 5). Since DENA can be moved into the PLGA film together with water molecules while the PLGA film matrix swells, higher local concentrations of DENA can be maintained inside the film matrix when the concentration of DENA in the release medium is higher. Thus, PLGA degradation will be increased as the concentration of DENA increases.

The observation that PLGA degrades faster in a DENA solution than in serum or PBS can be utilized for simulation of the *in vivo* release. PLGA polymers are known to degrade much faster *in vivo* than *in vitro* [24], which can be attributed to the presence of lipids or other biological compounds that increase the mobility of polymer chains [25] or the autocatalytic effect of the accumulated acidic degradation products [26]. Enzymes such as esterase can also catalyze the degradation of PLGA [27]. The presence of serum albumin alone was also shown to increase the initial rate of PLGA degradation [28]. Since implanted stents will directly contact with the vessel wall and will be exposed to blood, PTX molecules available from the stent will be removed promptly from the stent surface by binding with blood proteins such as serum albumin, which have a high affinity to PTX [29], or by partitioning easily into membranes of the cells in the blood vessel due to their hydrophobic nature. Very effective sink conditions will be maintained *in vivo*, and there would be additional factors that facilitate PLGA polymer degradation. Thus, the *in vivo* release profile of PTX from stents may not be the same as the *in vitro* release in PBS or serum. By adjusting the release medium, such as the DENA concentration or the presence of PBS, a suitable *in vitro* release profile that would be close to the *in vivo* release profile can be obtained. *In vitro–in vivo* correlation can be studied by analyzing the amount of PTX released from an *in vitro* experiment and the residual PTX from an *in vivo* experiment using PLGA/PTX-coated stents. Once *in vivo* drug release data are obtained, the concentration of DENA in the *in vitro* release medium can be varied to find a condition that produces *in vitro* release profiles matching with the *in vivo* drug release profiles.

In addition, the *in vitro* release from the PLGA matrix can be accelerated in DENA solutions, which will significantly decrease the time for the testing of sustained release formulations. An accelerated release of peptide from the PLGA depot at an elevated temperature was reported [30], but high temperatures may not be suitable for compounds that are susceptible to heat. The degradation rate of PTX in 2 M DENA was not larger than that in PBS, suggesting that the use of DENA in a release medium did not increase the degradation rate of PTX. The 2 M DENA aqueous solution was successfully used as a release medium for *in vitro* PTX release from stents of various formulations [31]. The *in vitro* release test of controlled release parenterals is usually used for quality control purposes [32] and also for differentiating different formulations [30]. Thus, it is suggested that the accelerated *in vitro* test using DENA solutions as release media can increase the efficiency of comparing different formulations composed of PLGA polymers.

Acknowledgements

This study was supported in part by National Institute of Health through grant GM 65284, Samyang Corporation, and NSF Industry/University Center for Pharmaceutical Processing Research.

REFERENCES

1. B. R. Goldspiel, *Pharmacotherapy* **17**, 110S (1997).
2. A. W. Heldman, L. Cheng, M. Jenkins, P. F. Heller, D.-W. Kim, M. J. Ware, C. Nater, R. Hruban, B. Rezai, B. S. Abella, K. E. Bunge, J. L. Kinsella, S. J. Sollott, E. G. Lakatta, J. A. Brinker, W. L. Hunter and J. P. Froehlich, *Circulation* **103**, 2289 (2001).
3. A. Farb, P. F. Heller, S. Shroff, L. Cheng, F. D. Kolodgie, A. J. Carter, D. S. Scott, J. Froehlich and R. Virmani, *Circulation* **104**, 473 (2001).
4. D. E. Drachman, E. R. Edelman, P. Seifert, A. R. Groothuis, D. A. Bornstein, K. R. Kamath, M. Palasis, D. Yang, S. H. Nott and C. Rogers, *J. Am. Coll. Cardiol.* **36**, 2325 (2000).
5. A. Finkelstein, D. Mcclean, S. Kar, K. Takizawa, K. Varghese, N. Baek, K. Park, M. C. Fishbein, R. Makkar, F. Litvack and N. L. Eigler, *Circulation* **107**, 777 (2003).
6. S. K. Dordunoo, J. K. Jackson, L. A. Arsenault, A. M. C. Oktaba, W. L. Hunter and H. M. Burt, *Cancer Chemoth. Pharm.* **36**, 279 (1995).
7. L. Mu and S. S. Feng, *J. Control. Rel.* **76**, 239 (2001).
8. R. T. Liggins and H. M. Burt, *Int. J. Pharm.* **222**, 19 (2001).
9. C. I. Winternitz, J. K. Jackson, A. M. Okataba and H. M. Burt, *Pharmaceut. Res.* **13**, 368 (1996).
10. Y. M. Wang, H. Sato, I. Adachi and I. Horikoshi, *Chem. Pharm. Bull.* **44**, 1935 (1996).
11. J. Lee, S. C. Lee, G. Acharya, C.-J. Chang and K. Park, *Pharm. Res.* **20**, 1022 (2003).
12. J. Lee, N. Baek and K. Park, in: *Proceedings of the 29th Annual Meeting of the Controlled Release Society*, Seoul, Korea, p. 487. Controlled Release Society, Minneapolis, MN (2002).
13. D. S. Sonnichsen and M. V. Relling, *Clin. Pharmacokinet.* **27**, 256 (1994).
14. N. V. Koshkina, V. Knight, B. E. Gilbert, E. Golunski, L. Roberts and J. C. Waldrep, *Cancer Chemoth. Pharm.* **47**, 451 (2001).
15. S. K. Dordunoo and H. M. Burt, *Int. J. Pharm.* **133**, 191 (1996).
16. G. J. Maceachern-Keith, L. J. W. Butterfield and M. J. I. Mattina, *Anal. Chem.* **69**, 72 (1997).
17. A. Charlier, B. Leclerc and G. Couarraze, *Int. J. Pharm.* **200**, 115 (2000).
18. M. A. Lovich, C. Creel, K. Hong, C.-W. Hwang and E. R. Edelman, *J. Pharm. Sci.* **90**, 1324 (2001).
19. Y. Cha and C. G. Pitt, *J. Control. Rel.* **8**, 259 (1989).
20. A. Kishida, S. Yoshioka, Y. Takeda and M. Uchiyama, *Chem. Pharm. Bull.* **37**, 1954 (1989).
21. H. V. Maulding, T. R. Tice, D. R. Cowsar, J. W. Fong, J. E. Pearson and J. P. Nazareno, *J. Control. Rel.* **3**, 103 (1986).
22. S. Li, S. Girod-Holland and M. Vert, *J. Control. Rel.* **40**, 41 (1996).
23. A. Delgado, C. Evora and M. Llabres, *Int. J. Pharm.* **166**, 223 (1998).
24. A. Kamijo, S. Kamei, A. Saikawa, Y. Igari and Y. Ogawa, *J. Control. Rel.* **40**, 269 (1996).
25. M. A. Tracy, K. L. Ward, L. Firouzabadian, Y. Wang, N. Dong, R. Qian and Y. Zhang, *Biomaterials* **20**, 1057 (1999).
26. L. Lu, S. J. Peter, M. D. Lyman, H.-L. Lai, S. M. Leite, J. A. Tamada, S. Uyama, J. P. Vacati, R. Langer and A. G. Mikos, *Biomaterials* **21**, 1837 (2000).
27. K. Park, W. S. W. Shalaby and H. Park, *Biodegradable Hydrogels for Drug Delivery*. Technomic, Lancaster, PA (1993).
28. E. Catiker, M. Gumusderelioglu and A. Guner, *Polym. Int.* **49**, 728 (2000).
29. K. Paal, J. Muller and L. Hegedus, *Eur. J. Biochem.* **268**, 2187 (2001).
30. M. Shameem, H. Lee and P. P. Deluca, *AAPS Pharmsci* **1** (3), article 7 (1999). Accessible at http://www.aapspharmsci.org/view.asp?art=ps010307
31. N. Baek, Controlled drug delivery from stents for anti-restenosis, PhD thesis, Department of Industrial and Physical Pharmacy, Purdue University, West Lafayette, IN (2002).
32. D. J. Burgess, A. S. Hussain, T. S. Ingallinera and M. Chen, *AAPS Pharm. Sci.* **4** (2), article 7 (2002). Accessible at http://www.aapspharmsci.org/view.asp?art=ps040207

Gels, Genes, Grafts and Giants, pp. 61–73 (2005)

Novel carboxymethylcellulose-based microporous hydrogels suitable for drug delivery

R. BARBUCCI *, G. LEONE and A. VECCHIULLO

C.R.I.S.M.A. and the Department of Chemical and Biosystem Sciences and Technologies, University of Siena, Via Aldo Moro 2, 53100 Siena, Italy

Received 24 September 2003; accepted 21 January 2004

Abstract—Several materials capable of acting as structures for controlled release were analysed for the fabrication of matrices. Among those used, hydrophilic polysaccharides appeared to be the most suitable materials. Carboxymethylcellulose (a semi-synthetic polysaccharide) was chemically cross-linked with a 60% and 90% cross-linking degree in order to obtain hydrogels and utilised as matrix for the realisation of controlled drug release systems. The morphology of the gels was changed in order to obtain a microporous structure with different porosity (14, 30 and 40 μm). The obtained porous matrices were characterised in terms of pore density, dimension and swelling behaviour. The influence of both the pore dimension and technique of loading on the release kinetics was analysed. By increasing the pore dimension the release of ibuprofen-lysin was slower. Inducing the microporous structure after the loading of the hydrogel with the drug resulted in a slower release.

Key words: Carboxymethylcellulose; microporous hydrogel; water uptake; morphology; drug release.

INTRODUCTION

These days, interest in drug-controlled release systems is increasing. Several definitions have been given. One of the most exaustive is ‘a system that is capable of releasing a carried bioactive agent in a specific location at a specific rate’. Thus, the main goal of a drug-controlled release system is to facilitate the dosage and the duration of the drug effect.

A great number of controlled release systems have been developed in the last decades. All the drug-controlled release systems can be included in one of the following categories: diffusion [1–3], solvent-activated release [4, 5] and polymeric degradation [6, 7]. In a diffusion system, the drug is either incapsulated in a polymer membrane or suspended within a polymer matrix. The driving force of this kind of

*To whom correspondence should be addressed. Tel.: (39-57) 723-4382. Fax: (39-57) 723-4383. E-mail: barbucci@unisi.it

system is the water diffusion into the membrane or matrix. The drug dissolves and then diffuses. In a membrane system the rate of release is constant and proportional to the concentration of the drug initially present [8], whereas in a matrix system the rate of release is time dependent since the release depends upon the amount of drug present at a particular time [9]. The most common example of a solvent-activated system employs a semi-permeable membrane containing a small laser-drilled hole. Within the membrane there is a high concentration of an osmotic agent, either the drug itself or salt, which causes water to enter through the membrane. The drug is forced out through the hole because of the increased pressure [10]. The last category of a drug-controlled system is that which employs the polymer degradation. Also in this case, the drug is contained in a polymer membrane or matrix. The polymer is structured in such a way that it is subjected to degradation and, consequently, the drug is released in a specific location. The release rate is proportional to the degradation rate of the polymer [11].

For all these kinds of drug-controlled release systems many families of materials were studied as vectors. Among them, polyhydroxyethylmetacrylate (PHEMA) [12–14], polyvinyl alcohol (PVA) [15], polyurethane (PU) [16, 17] and poly(ethylene oxide) (PEO) [18] play a very important role. Moreover, highly hydrophilic materials, such as polyaccharides, are utilised for their great swelling property and for the high amount and variety of functional groups present in the chains modified to obtain hydrogels [19]. Hydrogels are able to respond to the changes of the surrounding environment, such as temperature, pH or ionic strength modifications, modifying their state (swollen/shrunken) and, consequently, the amount of adsorbed water and the amount of drug dissolved and released. Among the polysaccharides, hyaluronane and alginate are the most frequently utilised, but cellulose and its derivatives (carboxymethylcellulose, hydroxypropylmethylcellulose) play an important role in the realisation of systems for the controlled release of drugs [20–23].

In this work carboxymethylcellulose (CMC)-based hydrogels (Fig. 1) with different cross-linking degrees (60% and 90%) were synthesised in order to obtain matrices for the controlled-release of the drugs. The morphology of the material was modified in order to obtain a porous structure since the presence of pores affects the release of the substance from the matrices [24–27]. Matrices with pores of three different diameters (14, 30 and 40 μm) were obtained. The obtained porous hy-

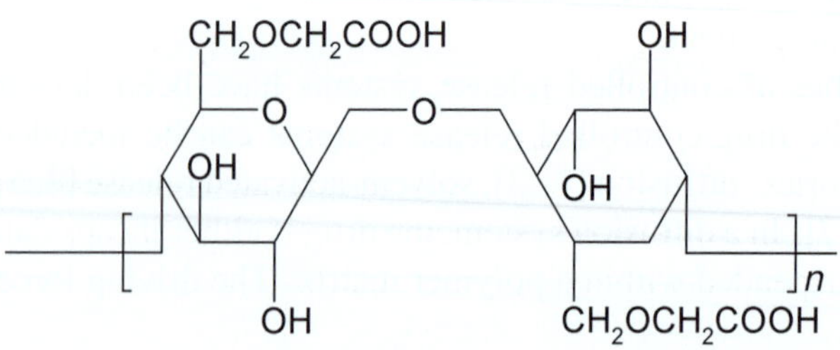

Figure 1. Disaccharidic unit of CMC (carboxymethylation degree of 0.9 ± 0.1 for monosaccharidic unit).

drogels were physico-chemically characterised in terms of swelling behaviour. The CMC-based native and porous matrices were loaded with ibuprofen-lysin, one of the most important drugs in osteoarthritis therapy. Two different techniques, which differed during the loading phase, were utilised to load the matrix with the drug. The influence of the loading technique and the presence and dimension of pores on the release kinetics was evaluated.

MATERIALS AND METHODS

Materials

The sodium salt of carboxymethylcellulose (NaCMC, viscosity 402 mPa·s in 2% (w/v) aqueous solution at 25°C and carboxymethylation degree of 0.9 ± 0.1 per monosaccharide unit, $M_w = 10^5$) was supplied by Hercules Italia (Milan, Italy). The polysaccharide was used as received after a check performed with an FT-IR spectroscopy.

All the other reagents were purchased from Fluka Chemie (Buchs, Switzerland) and used without further purification.

Hydrogel synthesis and characterisation

The polysaccharide-based 60% and 90% hydrogels were synthesised as previously described [28]. Briefly, a solution of sodium salt of polysaccharide (1.0%, w/v) was subjected to a sodium-hydrogen ionic exchange using Dowex 50WX8 resin and then added to a 5% tetrabutylammonium hydroxide solution until a pH of 8–9 was reached. The solution was then lyophilised and the tetrabutylammonium salt of the polysaccharide was obtained. The TBA salt was then dissolved in N,N′-dimethylformamide (DMF) under stirring and nitrogen flow. The solution was kept at 0°C before the activating agent 2-chloro-methyl-piridinium iodide (CMP-J) was added. The amount of CMP-J added depended on the quantity of carboxylate groups to be activated (60% and 90%). 1.3 diaminopropane was used as the cross-linking agent. The reaction was catalysed by a small amount of triethylamine as a hydrogen iodide captor. The reaction was then left at room temperature for 3–4 h. The hydrogel formed was then washed alternatively with bi-distilled water and ethanol until no more solvents or secondary products were found in the washing solution, as demonstrated by UV absorption measurements.

The ninhydrin assay was performed to check for the presence of unreacted NH_2 groups of the 1.3 diaminepropane used as the cross-linker [29]. A small amount of dried hydrogel was rinsed with 1 ml of acetate buffer (pH 5) and a small amount of ninhydrin was added. The mixture was left in boiling water for 15 min, then 15 ml of ethanol (50%) was added and protected from direct exposure to light for 1 h. The colour of the mixture became violet in the presence of free aminic groups.

To check the cross-linking degree potentiometric titration was performed utilising the procedure previously reported [30]. The obtained cross-linking degrees were

60 ± 3% and 90 ± 4% (i.e. 60% and 90% of the carboxylate groups of the polysaccharide were cross-linked).

Microporous structure formation

The microporous structure was obtained by stratifying the hydrogel onto a filter consisting of a cell-culture strainer with a defined and controlled porosity. The dimension of the cell strainer pore diameters was checked by SEM (diameters 40.7 ± 2.2 μm, 70.2 ± 10.1 μm and 99.8 ± 21.2 μm, respectively). Once obtained, the film was rinsed with several drops of chloroform in order to keep the matrix soft, but at the same time, to prevent excessive swelling which would alter the pore formation. The filter was placed on a beaker with the same diameter as the filter, which contained 1.5 g of porogen salt $NaHCO_3$ along the bottom. By slowly adding 0.1 M HCl solution (one drop at a time) with a syringe, violent effervescence was observed (Fig. 2). First, the formation of CO_2 bubbles with passage through the filter was observed; then, the matrix induced the hydrogel to assume a porous morphology [31].

Scanning electron microscopy (SEM) analysis. The morphology of native and microporous hydrogel samples was examined by Scanning Electron Microscopy (SEM, XL 20 Philips, Eindhoven, The Netherlands). 1.5 ml of distilled water was used to rinse 5 mg of each sample, once it had reached the swelling equilibrium, was frozen in liquid nitrogen and lyophilised. Samples were cut along the three axes to obtain a complete observation of the pore distribution in all the dimensions.

They were mounted on SEM stubs in order to analyse the surface and the internal structure and successively gold-sputtered with an automatic sputter-coater

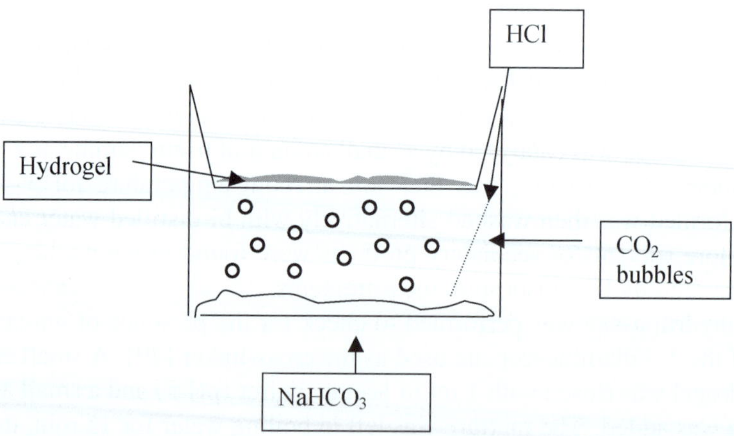

Figure 2. Disposal for the realisation of microporous structure. The filter containing the stratified hydrogel was collocated on the beaker containing the porogen salt ($NaHCO_3$). By adding HCl, CO_2 was formed and, crossing the filter and the matrix, induced the hydrogel to assume the porous morphology.

(BAL-TEC SCD 050, Balzers, Alzenau, Germany). The hydrogel morphology was then observed by SEM at 10 kV. For the statistical evaluation of the pore diameter, thickness and pore density, 15 specimens for each dimension were used.

Water uptake measurements. The water uptake (WU) was determined for each sample, as calculated with the following formula:

$$WU = ((W_s - W_d)/W_d) \times 100,$$

where W_s and W_d are the weight of the swollen and dried hydrogels respectively. In practice, known amounts of the dried samples were deposited on a propylene cell strainer with a nylon pore mesh of 100 mm and immersed several times at 25°C in 50 ml of a buffer solution (pH 2: 6.5 ml 0.2 M HCl mixed with 25 ml 0.2 M KCl; pH 7.4: PBS (phosphate buffer solution); pH 9: 50 ml 0.1 M Tris with 5.7 ml 0.1 M HCl). They were then placed on a piece of dry filter paper to remove the excess water. The filters containing the swollen hydrogels were weighed to calculate the water-uptake. Five specimens of each dimension were utilised for each pH. No time-related decrease in weight in the short time was observed [31].

Controlled release system

Loading of the matrices with ibuprofen-lysin. Two different procedures were utilised to load the matrix with the drug.

(a) The first technique consisted of immersing the already obtained microporous hydrogel in 3 mg/ml of ibuprofen-lysin salt solution for 48 h and then drying it in liquid nitrogen and lyophilising it.

(b) The second technique consisted of loading the hydrogel in 3 mg/ml of ibuprofen-lysin salt solution for 48 h. The hydrogel was then induced to assume a porous structure, utilising the technique reported above.

Determination of the amount of the drug captured by hydrogels. To determine the amount of drug captured by the hydrogels, the samples were washed and the washing solution was analysed by UV spectroscopy at a wavelength of 263 nm (maximum absorption of the drug) in order to evaluate the amount of drug adsorbed superficially. Lastly, the hydrogels were then destroyed by immersion in 0.1 M NaOH for 24 h and the obtained solution was analysed by the UV spectrometer in order to determine the amount of the drug in the matrices.

Evaluation of the release kinetics. A circuit was created for the evaluation of the release kinetics of ibuprofen-lysin from the different matrices. The hydrogel, loaded with the drug and washed to remove the superficial drug, was wrapped in a nylon envelope in order to prevent any loss of the material. The envelope containing the hydrogel was placed inside a polystyrene cell through which the washing solution flowed. The washing solution (PBS, pH 7.4) was maintained in flow condition by a peristaltic pump. At regular time intervals an aliquot of the flowing solution was analysed by the UV spectrometer at 263 nm, measuring the amount of released drug. The analysis was performed until no more drug was detached [32].

RESULTS AND DISCUSSION

Microporous hydrogel synthesis and characterisation

As previously reported [28], CMC 60% and 90% hydrogels were obtained by changing the amount of the activating agent (CMP-J). 60% and 90% refer to the amount of the carboxylate groups in the polysaccharidic chains, which were involved in the cross-linking reaction. There is in fact a stoichiometric relationship between the quantity of the carboxylate activating agent and the cross-linking degree. CMC-based hydrogels were induced to assume a porous morphology by forcing the passage of CO_2 bubbles through a filter with a known porosity containing the hydrogel. The hydrogel pore dimension can be modulated by changing the filter porosity. The pore diameter, thickness and density are given in Table 1. On the basis of these results (Table 1) and the previously reported data concerning CMC 50% hydrogels [31], the pore dimension was only affected by filter porosity. No influence of the crosslinking degree was observed. In fact, using both a 40 μm and a 70 μm filter, a very similar pore diameter was obtained for all the hydrogels (50%, 60% and 90%). An exception was observed using the 100 μm filter; in fact, 70 μm pores with a large dispersion of dimension, instead of 40 μm (as obtained for 50% and 60% hydrogels) were obtained in 90% hydrogels. The large pore dimension found for the 90% hydrogel can be related to the consistency of the 90% (Fig. 3b1) matrix, different from that of 50% and 60%. In terms of pore thickness a similar value was found for similar pore diameter. In fact in 60% hydrogel a similar value of pore thickness was found for 30 and 40 μm (2.70 and 2.30, respectively) whereas 14 μm showed a greater pore thickness (3.70 μm). In 90% hydrogel 14 and 22 μm showed a similar pore thickness (1.20 and 1.50 μm, respectively), whereas 70 μm showed a very high pore thickness value (4.80 μm).

Table 1.
Morphological characterisation of microporous 60% and 90% CMC-based matrices

Material	Filter diameter (μm)	Pore diameter (μm)	Pore thickness (μm)	Pore density (pores/mm^2)
CMC 60%	40	14 ± 4	3.70 ± 0.18	2.0 ± 0.1
CMC 60%	70	30 ± 4	2.70 ± 0.13	1.5 ± 0.1
CMC 60%	100	40 ± 9	2.30 ± 0.11	1.0 ± 0.1
CMC 90%	40	14 ± 3	1.20 ± 0.35	4.6 ± 0.4
CMC 90%	70	22 ± 4	1.50 ± 0.25	2.6 ± 0.1
CMC 90%	100	73 ± 20	4.80 ± 0.10	0.3 ± 0.1

The pore density, for hydrogels of all the crosslinking degrees, decreased when increasing the pore diameter.

The presence of pores affected the swelling behaviour of the hydrogels. In Fig. 4, the water uptake of the native and microporous 60% and 90% hydrogels as a function of pH is reported. The 60% hydrogels showed a greater increase of water uptake, increasing the pH, and higher than the corresponding 90% hydrogels. The influence of pore diameter on swelling is negligible, and at pH 7.4 all the 60% hydrogels (native and microporous) showed the same water uptake (Fig. 4a).

In Fig. 4b, the water uptake behaviour of the native and microporous 90% CMC-based hydrogels is reported. The influence of pore diameter on swelling at the three different pH values is different and depends on the dimensions of the pores. A peculiar behaviour was shown by 73 μm porous hydrogel. In fact, environmental pH did not affect the water uptake of the materials (Fig. 4b). Moreover, the dimension of pores did not affect the swelling kinetics; the time necessary to reach the swelling equilibrium was 48 h for both native and porous CMC-based hydrogels.

Evaluation of the release kinetics for the 60% CMC hydrogels

The native 60% CMC hydrogels, loaded with ibuprofen-lysin, completely released the drug in 24 h. The 60% microporous hydrogels were loaded using two different techniques. The first one consisted of immersing the already obtained microporous hydrogels in a drug solution. The latter consisted of immersing the hydrogel in an ibuprofen lysin salt solution and, subsequently, inducing the microporous structure. The loading technique drastically affects the amount of captured drug (Table 2) and its release kinetics. Loading the hydrogel with the drug after the pore formation (30 μm), the drug was completely released in 24 h. Loading the hydrogel with the drug before the pore formation, the drug had been released for seven days. This difference was entirely due to the steadier capture of the drug by the matrix in the second technique of loading. In fact, the hydrophilic drug already inside the matrix contributed to a higher swelling of the hydrogel which allowed the drug to penetrate more deeply into the matrix. With pore formation, the drug was entrapped more

Table 2.
Amount of ibuprofen-lysin salt taken up and captured by CMC hydrogels on the basis of the different techniques of loading

Material	Ibuprofen-lysin in 3 ml (mg)	Ibuprofen-lysin uptake (mg)
5 mg CMC 60%	15	14.0
5 mg CMC 30 μm[a]	15	13.5
5 mg CMC 30 μm[b]	15	11.0

[a] Pore formation and subsequent loading the matrix with the drug.
[b] Loading the matrix with the drug and subsequent pore formation.

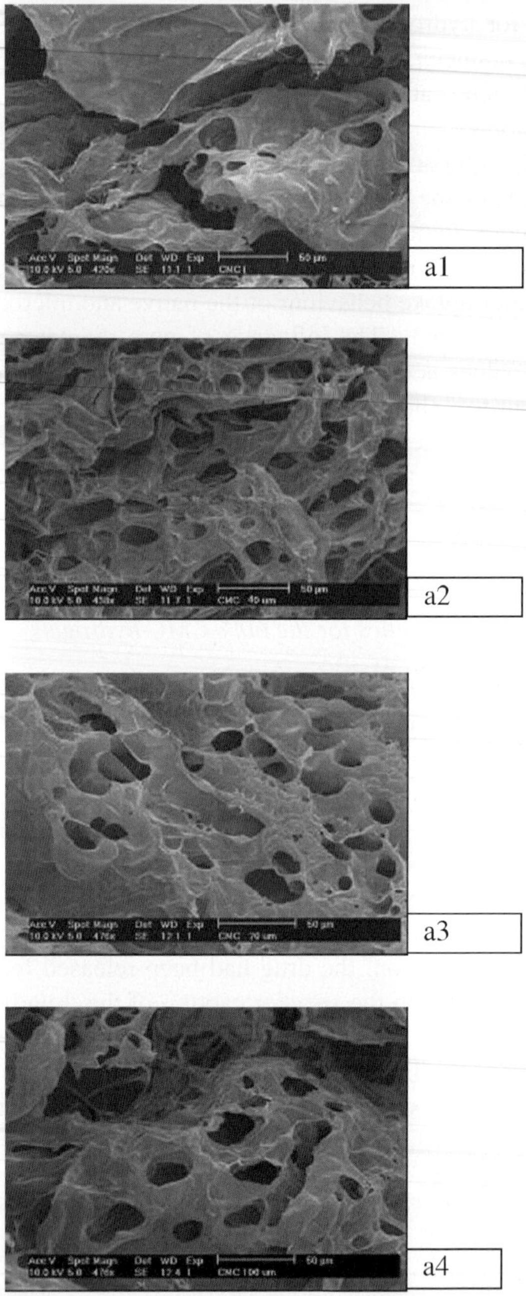

(a)

Figure 3. SEM micrographs of 60% CMC hydrogels and 90% CMC hydrogels: (a) CMC 60% hydrogels: (a1) CMC 60% hydrogel show wide fissures and holes among the laminae; (a2) CMC 60% with pores of 14 μm; (a3) CMC 60% with pores of 30 μm; (a4) CMC 60% with pores of 40 μm. (b) 90% CMC hydrogels: (b1): CMC 90% hydrogel shows a compact morphology with few fissures; (b2) CMC 90% with pores of 14 μm, (b3) CMC 90% with pores of 22 μm; (b4) CMC 90% with pores of 73 μm.

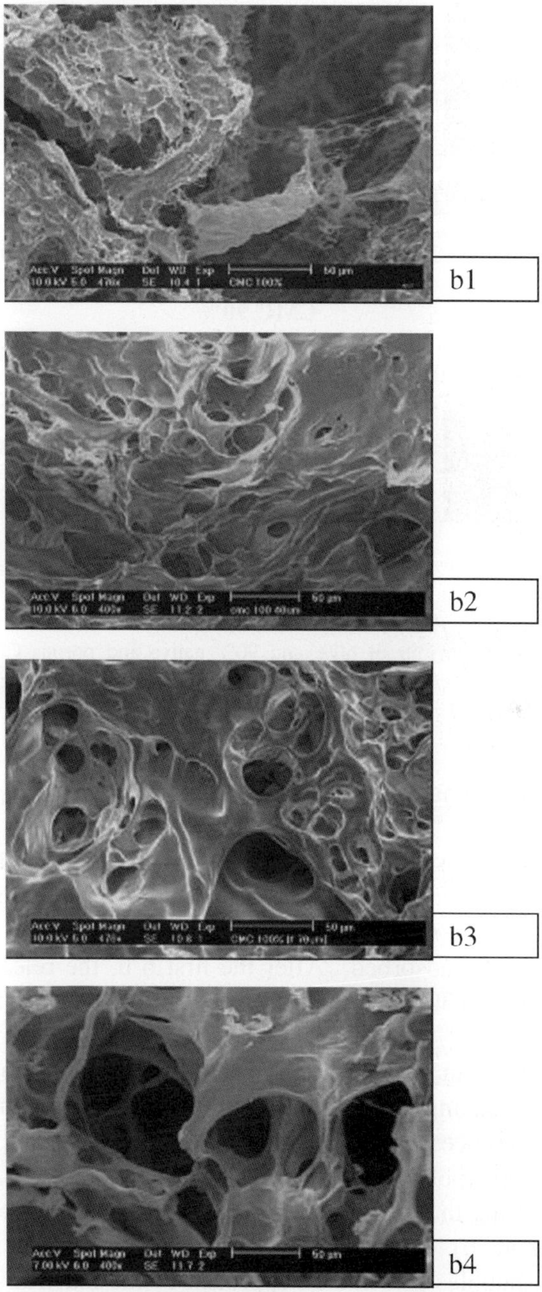

(b)

Figure 3. (Continued).

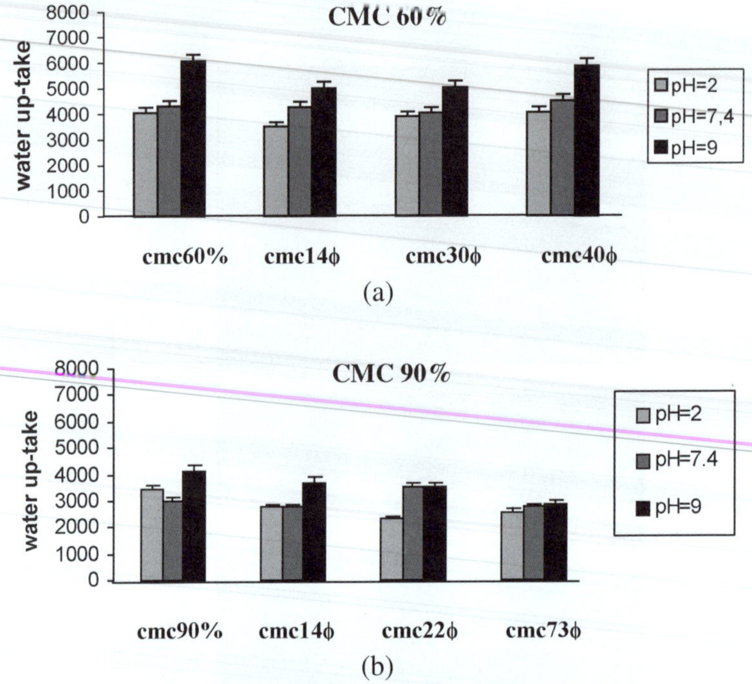

Figure 4. Water uptake behaviour of 60% and 90% native and porous CMC-based hydrogels. (a) Water uptake dependence on pH for CMC 60% native and porous hydrogels; (b) water uptake dependence on pH for CMC 90% native and porous hydrogels.

steadily for the approaching of the walls caused by the forced CO_2 bubbles crossing the hydrogel [31].

By comparing the release slopes of the native CMC hydrogel with the 30 μm microporous hydrogels, two different phases were observed (Fig. 5). During the first 6 h, for all the systems, a very rapid release was observed, which was the release of the drug superficially adsorbed. After the first 6 h, the release slope changed for CMC ibuprofen-lysin 30 μm microporous gel, indicating a diffusive mechanism (Fig. 5).

The main difference in the release rate was observed changing the loading technique. In fact, loading the hydrogel with the drug after the pore formation no significative differences were found between the native and the microporous hydrogels. As reported before, for both the matrices 24 h were necessary for the complete release. Using the second technique the release rate decreased drastically (7 days instead of 48 h). In fact, as previously reported [31], by crossing the hydrogel, the CO_2 bubbles provoke an approach of the material of the hydrogels, consequently when the pore formation is subsequent to the loading of the hydrogel with the drug, the substance is entrapped more steadily into the matrix.

The influence of the pore dimension on the release kinetics of CMC-based 14 μm, 30 μm and 40 μm was analysed by using the second loading technique (Fig. 6).

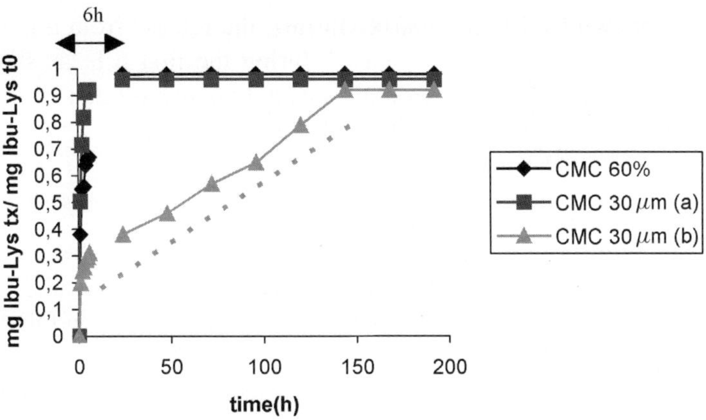

Figure 5. Dependence of the release kinetics of ibuprofen-lysin from CMC matrices using the two loading techniques. CMC release slope of ibuprofen-lysin from native 60% CMC 30 μm. (a) Release slope of ibuprofen-lysin from microporous CMC 60% loaded with the drug after the pore formation. (b) Release slope of ibuprofen-lysin from CMC 60% loaded with the drug before the pore formation.

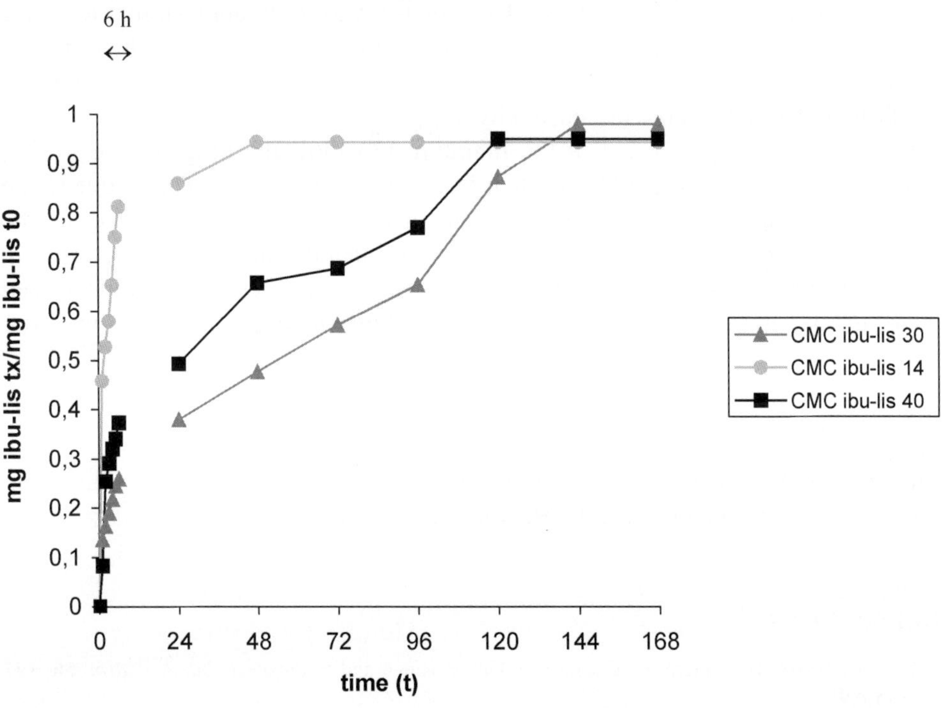

Figure 6. Dependence of the release kinetics of ibuprofen-lysin from CMC matrices on the pore dimension.

CMC-based hydrogels with pores of 14 μm released the drug in 48 h, whereas the 30 μm system needed 7 days and the 40 μm 5 days. Increasing the pore dimension (from 30 to 40 μm), the release kinetics was easier and faster, although peculiar behaviour was shown by 14 μm matrix. In fact, the release from this system was very rapid: 90% of the drug was released during the first 6 h, as shown by the release slope reported in Fig. 6. This allows us to hypothesise that all the drug was superficially distributed, as in the native CMC.

CONCLUSIONS

Carboxymethylcellulose-based 60% and 90% hydrogels were synthesised. Their morphology was modified in order to obtain a porous structure and three different pore dimensions. CMC-based native and porous hydrogels were loaded with Ibuprofen-lysin and the influence of the loading technique and pore dimension on the release kinetics was evaluated. The loading technique drastically affected the release kinetics. In fact, loading the matrix with the drug first and creating the microporous structure thereafter, it was possible to modulate the drug release for a longer period, and the time necessary for the complete release of the drug was extended from 24 h to 7 days. The pore dimension partially affected the release kinetics. Actually, the release from 14 μm matrix was very rapid, similar to native CMC (24 h), but increasing the pore dimension up to 30 and 40 μm, the time was extended. The prolongation was most likely due to the more internal distribution of the drug in the largest pore dimension hydrogels.

Finally, as widely reported in the literature [33–36], the drug release from a porous material is easier and faster than from a non-porous one. This is due to the influence of the porosity on the physico-chemical properties of the hydrogel. The pore presence generally induces faster swelling kinetics and higher water uptake. In conclusion, as shown in our samples, the presence of pores induced a decreased water uptake and had no influence on the swelling kinetics, and in turn the drug release decreased.

Acknowledgements

The authors wish to thank the COFIN 2002 of the Italian Ministry of Education, University and Research (M. I. U. R.) for financial support.

REFERENCES

1. G. Courraze, B. Leclerc, G. Conrath, F. Falson-Riegh and F. Puisieux, *Int. J. Pharm.* **56**, 197 (1989).
2. N. Marcotte, A. Polk and M. F. A. Goosen, *J. Pharm. Sci.* **79**, 407 (1990).
3. J. A. Wesselingh, *J. Control. Rel.* **24**, 47 (1993).
4. U. Conte, P. Colombo, A. Gazzaniga, M. E. Sangalli and A. La Manna, *Biomaterials* **9**, 489 (1988).
5. G. H. P. Te Wierik, A. C. Eissens, J. Bergsma, A. W. Arends-Scholte and C. F. Lerk, *J. Control. Rel.* **45**, 25 (1997).
6. F. Najafi and M. N. Sarbolouki, *Biomaterials* **24**, 1175 (2003).

7. R. S. Hasland, A. Gazzaniga, M. E. Sangani, P. Colombo and N. A. Peppas, *Pharm. Res.* **5**, 488 (1988).
8. S. Sato and S. W. Kim, *J. Pharm.* **22**, 229 (1984).
9. P. I. Lee, *J. Membr. Sci.* **7**, 255 (1980).
10. R. H. Juang and D. Storey, *J. Control. Rel.* **89**, 375 (2003).
11. B. A. Miller-Chou and J. L. Koenig, *Prog. Polym. Sci.* **28**, 1223 (2003).
12. R. M. Trigo, M. D. Blanco, J. M. Teijón and R. Sastre, *Biomaterials* **15**, 1181 (1994).
13. L. Ferreira, M. M. Vidal and M. H. Gil, *Int. J. Pharm.* **194**, 169 (2000).
14. G. H. Hsiue, J. Guu and C.-C. Cheng, *Biomaterials* **22**, 1763 (2001).
15. R. Morita, R. Honda and Y. Takahashi, *J. Control. Rel.* **90**, 109 (2003).
16. J. M. Schierholz, H. Steinhauser, A. F. E. Rump, R. Berkels and G. Pulverer, *Biomaterials* **18**, 839 (1997).
17. S. Kohjiya, Y. Ikeda, S. Takesako and S. Yamashita, *React. Polym.* **15**, 165 (1991).
18. G. Di Colo and Y. Zambito, *Eur. J. Pharm. Biopharm.* **54**, 193 (2002).
19. J.-K. F. Suh and H. W. T. Matthew, *Biomaterials* **21**, 2589 (2000).
20. S. N. Makhija and P. R. Vavia, *J. Control. Rel.* **89**, 5 (2003).
21. L. Alvarez, A. Concheiro, J. L. Gómez-Amoza, C. Souto and R. Martínez-Pacheco, *Eur. J. Pharm. Biopharm.* **55**, 291 (2003).
22. Ç. Tas, Y. Özkan, A. Savaser and T. Baykara, *Farmaco* **58**, 605 (2003).
23. S. Zuleger, R. Fassihi and B. C. Lippold, *Int. J. Pharm.* **247**, 23 (2002).
24. W. J. Lin and H.-G. Lee, *J. Control. Rel.* **89**, 179 (2003).
25. S. L. Tao and T. A. Desai, *Adv. Drug Deliv. Rev.* **55**, 315 (2003).
26. J.-H. Jang and L.-D.Shea, *J. Control. Rel.* **86**, 157 (2003).
27. R. S. Byrne and P. B. Deasy, *Int. J. Pharm.* **246**, 61 (2002).
28. R. Barbucci, R. Rappuoli, A. Borzacchiello and L. Ambrosio, *J. Biomater. Sci. Polymer Edn* **11**, 383 (2000).
29. K. S. Virender, S. B. H. Kent, J. P. Tam and R. B. Merrifield, *Anal. Biochem.* **117**, 147 (1981).
30. R. Barbucci, A. Magnani and M. Consumi, *Macromolecules* **33**, 7475 (2000).
31. R. Barbucci and G. Leone, *J. Biomed. Mater. Res. Appl. Biomater.* **68B**, 117 (2004).
32. R. Barbucci, M. Consumi, S. Lamponi and G. Leone, *Macromol. Symp.* **204**, 37 (2003).
33. M. V. Badiger, M. E. McNeill and N. B. Graham, *Biomaterials* **14**, 1059 (1993).
34. D. J. A. Netz, P. Sepulveda, V. C. Pandolfelli, A. C. C. Spadaro, J. B. Alencastre, M. V. L. B. Bentley and J. M. Marchetti, *Int. J. Pharm.* **213**, 117 (2001).
35. J. Fu, J. Fiegel, E. Krauland and J. Hanes, *Biomaterials* **23**, 4425 (2002).
36. A. Rivera, T. Farías, A. R. Ruiz-Salvador and L. C. De Ménorval, *Micropor. Mesopor. Mater.* **61**, 249 (2003).

Gels, Genes, Grafts and Giants, pp. 75–83 (2005)

IgG responses to intranasal immunization with cholera-toxin-immobilized polymeric nanospheres in mice

TATSUO KANEKO [1,2], SHOGO SHIMOMAI [1], MUTSUMI MIYAZAKI [3], MASANORI BABA [2,3] and MITSURU AKASHI [1,2,4,*]

[1] *Department of Nanostructured and Advanced Materials, Graduate School of Science and Engineering, Kagoshima University, 1-21-40 Korimoto, Kagoshima 890-0065, Japan*
[2] *CREST, JST, Kawaguchi City, Saitama, Japan*
[3] *Division of Antiviral Chemotherapy, Center for Chronic Viral Diseases, Faculty of Medicine, Kagoshima University, Kagoshima, Japan*
[4] *Department of Molecular Chemistry, Graduate School of Engineering, Osaka University, 2-1 Yamadaoka, Suita 565-0871, Japan*

Received 24 September 2003; accepted 14 January 2004

Abstract—IgG responses to antigen-nanosphere hybrids were studied in mice. Cholera toxin (CT) was covalently immobilized onto the surface of polymeric nanospheres (NS) with a nanophase-separated structure consisting of a polystyrene core and a poly(methacrylic acid) graft corona. Reaction conditions favoring the dehydroxide condensation reaction of the amino group of the CT with the carboxyl group of NS effectively immobilized CT onto their surface. When CT-immobilized nanospheres (CT-NS) were suspended in aqueous solution and administrated to mice either intranasally or intramuscularly, serum IgG titers elevated with increasing time and reached a maximum level at 8 weeks after immunization. On the other hand, intranasal administration of CT alone induced an even higher serum IgG titer than that of CT-NS at 4 weeks. However, the titer gradually decreased thereafter. Thus, polymeric NS may be an effective substrate to covalently immobilize antigen on their surface, steadily inducing a high level of IgG production in response to the intranasal administration.

Key words: Polymeric nanospheres; surface immobilization; cholera toxin; vaccine; IgG response.

INTRODUCTION

Polymeric nanospheres (NS) have a wide potential in the manufacture of biomedical wares [1, 2], optical devices [3] and other industrial materials [4] because of their extremely large surface area to volume ratios. Various substances from metal ions

*To whom correspondence should be addressed at Osaka University.
E-mail: akashi@chem.eng.osaka-u.ac.jp

to virions can be adsorbed on their surface. We have developed a method for the preparation of uniform polymeric nanospheres. The surface can be multifunctionally modified by self-organization of macromolecular chains with polymerization of monomers containing macromonomer. For instance, the copolymerization of styrene with poly(methacrylic acid) (PMAA) having a polymerizable end group in water/ethanol solution yields a surface-carboxylated NS comprised of a hydrophobic polystyrene core and hydrophilic PMAA corona chains.

We have previously shown that the mannose-specific lectin Concanavalin A (Con A) can be successfully immobilized on the surface of NS while maintaining its binding specificity. The Con A-NS hybrids efficiently captured HIV-1 particles and gp120 antigens [5, 6]. Furthermore, Con A-NS could capture HIV-1 particles irrespective of their cellular tropism (R5 or X4) and infectivity (heat-inactivated or not). Intranasal immunization with heat-inactivated HIV-1-capturing nanospheres (HIV-NS) induced HIV-1-specific IgA antibody in the murine genital tract, whereas this HIV-1-specific IgA response was not detectable with HIV-1 immunization alone [7]. The NS covered by cationic surfactants bound to HIV-1 *gag* DNA yielding a device for the DNA vaccination in mice, guinea pigs and rhesus macaques [8]. Thus, polymeric NS-antigen hybrids may similarly be effective substances for vaccination. However, multi-step methods for preparation of the antigen-NS hybrids are complicated and difficult to control, since their sensitivity changes under various experimental conditions, such as pH, temperature, ion species and ionic strength. One-step antigen immobilization on the surface of NS using covalent linkage may be an effective and facile method for preparation of vaccines without impairing antigenicity. We have previously reported on the method for the preparation of NS having a surface with a linkable amine, carboxylic acid, hydroxyl groups and other groups [9–11]. Various proteins can be linked with such NS through carboxylic acids and amines of the proteins. For instance, Con A immobilization was successfully achieved by dehydroxide condensation of its amines with the carboxylic acids on the surface of NS. In this case, as described above, the mannose-binding capacity of Con A was maintained, even if some of the amine groups were involved in amide linkages [5, 6].

Here we attempted to immobilize the cholera toxin (CT) onto the NS surface by the formation of covalent linkages, and to investigate the activity of CT-NS hybrids for their induction capacity of anti-CT IgG antibody in mice. In particular, we have attempted to perform the intranasal immunization [7], which is actually confirmed as a useful method for non-invasive vaccination.

MATERIALS AND METHODS

Nanospheres

Procedures for the synthesis of NS have been described previously [5]. The mean diameter of NS used in the present study was 310 nm (standard deviation

30 nm). CT was purchased from Vector Laboratories (Burlingame, CA, USA) and immobilized onto the NS surface by a condensation reaction using water-soluble carbodiimide (WSC) at 4°C or 20°C. The amount of immobilized CT was measured by the conventional ninhydrin method.

Animals

BALB/c mice (5 weeks old) were purchased from Charles River Japan (Yokohama, Japan). Experiments were carried out in accordance with the Guidelines for Animal Experimentation, Kagoshima University, and extra care was taken to avoid animal suffering.

Immunization

Four mice in each group were immunized four times at a 7-day interval (on days 0, 7, 14 and 21) with either CT-NS suspension in phosphate-buffered saline (PBS), CT solution in PBS, or PBS alone. An equivalent amount (1 μg) of CT was intranasally or intramuscularly administered to each mouse in the CT-NS and CT groups. Intranasal immunization was performed on anesthetized mice by placing 10 μl of inoculums into one nostril. Intramuscular immunization was carried out by injecting 10 μl of inoculums into a femoral muscle.

Sample collection

Blood was collected from the retro-orbital plexus under deep anesthesia with ether. Fecal samples were collected and extracted with PBS. One milliliter of PBS was added to 100 mg of fresh fecal pellets. After centrifugation, the supernatants were stored at −40°C until antibody assay.

Determination of anti-CT IgG antibody levels

CT-specific IgG antibody titers in serum and feces were measured by an enzyme-linked immunosorbent assay (ELISA) method. A 96-well microtiter plate (Nunc-Immuno™ Modules, Maxisorp™, Nalge-Nunc, Rochester, NY, USA) was coated with 1 μg/ml CT in CBC (35 mM Na_2CO_3 and 15 mM $NaHCO_3$) buffer (100 μl/well). After incubation for 2 h at room temperature, CT was removed. The plate was further incubated with a blocking buffer (300 μl/well) containing 3% skim milk, 0.05 M Tris, 0.138 M NaCl, 0.027 M KCl and 1% bovine serum albumin (pH 8.0) for 2 h at room temperature and washed four times. The test samples were serially diluted with the blocking buffer and applied to the plate (100 μl/well). The plate was incubated for 2 h at room temperature. After washing, alkaline phosphatase-conjugated goat anti-mouse IgG (Southern Biotechnology Associates, Birmingham, AL, USA) was added to the plate and incubated for 1 h at room temperature. The plate was washed and treated with *p*-nitrophenyl phosphate (Sigma, St. Louis, MO, USA). Specific absorbance was read at 405 nm on a microplate reader (Bio-Rad, Hercules, CA, USA).

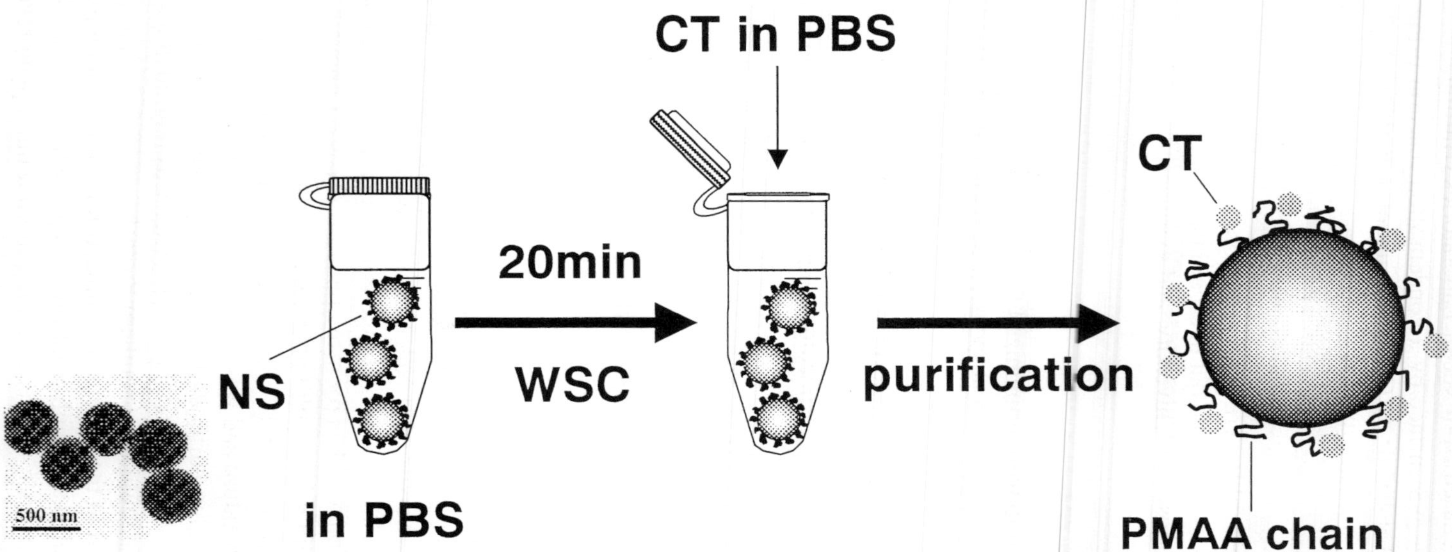

Figure 1. Schematic illustration of the procedure for the preparation of choleratoxin-immobilized NS. TEM image of NS is shown on the left.

RESULTS AND DISCUSSION

Immobilization of CT onto NS

CT is a spherical protein with a diameter of about 5 nm, and its surface has reactive groups at the lysine and arginine residues. Although there is a difference in the reactivity of these amines derived from the steric hindrance, almost all of them may be capable of reacting with carboxylic acid groups. In the present study, we used uniform NS with a PSt core and PMAA corona whose transmission electron microphotograph (TEM) taken by the previously-reported procedures was shown in Fig. 1 (left side), indicating that NS have a spherical morphology and smooth surface. Con A was successfully immobilized onto this kind of NS through the amide linkage, as shown in the previous study [5]. Carboxylic acids of the surface were activated by WSC in PBS agitating at 4°C, and then the aqueous solution of CT was added as shown in Fig. 1. The condensation reaction was made by further agitation at 4 or 20°C. Figure 2 shows the reaction-time-dependence of the weight amount of immobilized CT onto NS (1 mg). The amount of immobilized CT increased with increasing reaction time up to 50 min and reached a plateau level (14–25 μg/mg) which is considered to be high enough for vaccination. More CT was immobilized at 20°C than at 4°C. As a result, the immobilization reaction was enhanced at higher temperatures and longer reaction times, which is typical of chemical reactions dominated by the collision probability between reactive sites. However, over-regulation of reaction time and temperature may potentially induce the degradation of CT. In the next experiment, the effect of CT concentration on the amount of CT immobilized was investigated. Figure 3 shows that immobilization was enhanced by increasing the concentration of the CT solution irrespective of reaction temperature (20 or 4°C). Since the increase in polymer concentration elevated the viscosity of the polymer solution, a temperature increase should be required in order to control the reaction rate. Accordingly, elevation of the reaction temperature increased the amount of immobilized CT. Consequently, a sufficient amount of CT was immobilized on the surface of NS using 100 mg/ml CT at 20°C for 2 h, and the excellent water-dispersibility of NS was kept after CT immobilization.

Vaccination with CT-NS hybrids

When sera and feces were obtained at an interval of 7 days and examined for their anti-CT IgG antibody titers, significant anti-CT IgG antibodies were detected in the mice immunized with both CT-NS hybrids and CT alone (Fig. 4). In this experiment, the immunogens were administered either intranasally or intramuscularly. The mice immunized with CT-NS hybrids steadily produced a high level of anti-CT IgG antibody in sera, although anti-CT IgG antibody was not detected in feces to the same extent. The mice immunized with CT alone exhibited higher anti-CT IgG antibody titers than those with CT-NS hybrids. However, the serum antibody titer in these mice gradually decreased after 4 weeks and reached a level comparable to

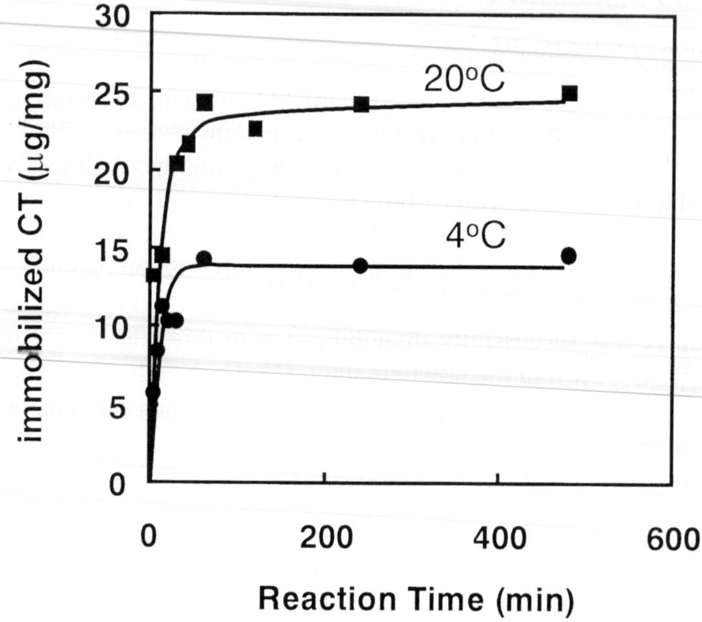

Figure 2. Reaction time dependence of the amount of CT immobilized on the NS. Concentration of CT was 100 μg/ml.

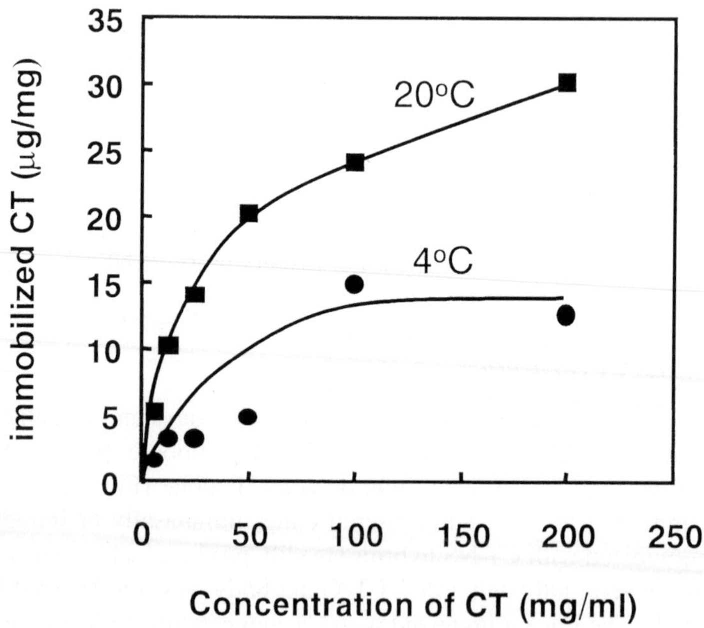

Figure 3. Effects of CT concentration in feed on amount of CT immobilized on the nanospheres.

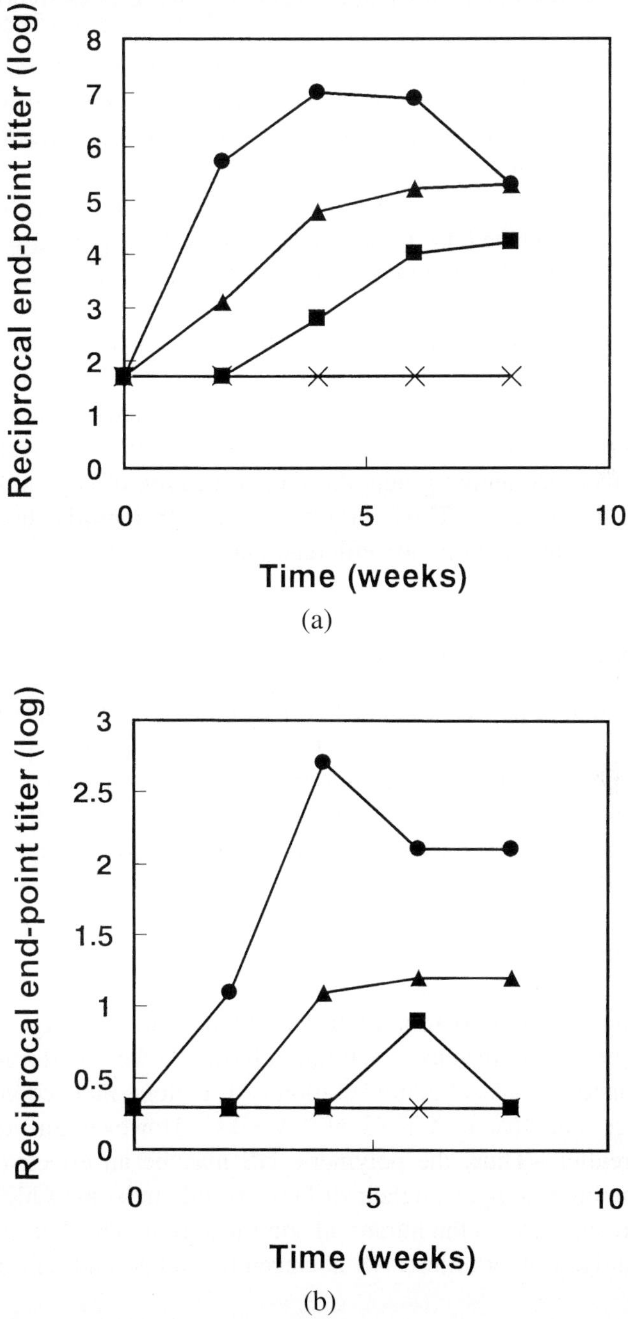

Figure 4. Anti-CT IgG responses in mice immunized with CT-NS hybrids intranasally (■) or intramuscularly (▲), CT alone intranasally (●), or PBS intranasally (×). The reciprocal titers of CT-specific IgG antibodies in (a) sera and (b) feces were measured by ELISA, as described in Materials and Methods.

that in the CT-NS-immunized mice at 8 weeks (Fig. 4). CT-NS hybrids were more efficient for a steady production of IgG in sera than CT alone. These results indicate that the immobilized CT on the surface of NS could retain sufficient antigenicity; nevertheless, the immobilization might lose some amino groups of CT. It is exaggerated that the intranasal immunization with CT-NS hybrids steadily produced a high level of anti-CT IgG antibody in sera.

In a previous study, we demonstrated that vaginal immunization with HIV capturing NS could induce HIV-1-specific IgA antibody in vaginal washes of mice, since HIV capturing NS accumulated in the folds of the mucosal membrane for a longer period than with HIV immunization alone [5, 6]. Since NS is a solid dispersing in aqueous solution rather than dissolving, it is assumed that NS are easily trapped in a narrow space such as mucosal folds. The results of the present study suggest that CT-NS may also become trapped somewhere in the vasculature or nasal mucosa, leading to the steady activation of the immune system. Since all proteins have a surface amino group, the immobilization of any antigen should be possible using this system. Thus, various antigen-NS hybrids should be further pursued for their potential as highly efficient mucosal vaccines.

CONCLUSIONS

We attempted to immobilize cholera toxin (CT) covalently onto the surface of polymeric nanospheres (NS, diameter 310 nm) with a nanophase-separated structure consisting of a polystyrene core and a poly(methacrylic acid) graft corona. Reaction conditions favoring the dehydroxide condensation reaction of the amino group of the CT surface with the carboxyl group of NS (CT concentration 100 mg/ml, temperature 20°C, reaction period 2 h) effectively immobilized a sufficient amount of CT onto the surface of NS, and the excellent water-dispersity of NS was kept after CT immobilization.

When CT-immobilized nanospheres (CT-NS) were suspended in aqueous solution and administrated intranasally or intramuscularly to mice, serum IgG titers elevated with increasing time and reached a maximum level at 8 weeks after immunization. On the other hand, intranasal administration of CT alone induced an even higher serum IgG titer than that of CT-NS at 4 weeks. However, the titer gradually decreased thereafter. Thus, the polymeric NS may be an effective substrate to covalently immobilize antigen on their surface, steadily inducing a high level of IgG production in responding to the intranasal administration. The detailed vaccination efficiency of antigen-immobilized NS is currently studied and will be reported in another paper.

Acknowledgements

This paper in part XI in the series of the study on graft copolymers having hydrophobic backbone and hydrophobic branches.

REFERENCES

1. T. Akagi, M. Kawamura, M. Ueno, K. Hiraishi, M. Adachi, T. Serizawa, M. Akashi and M. Baba, *J. Med. Virol.* **69**, 163 (2003).
2. B. Rihova, *Adv. Drug Deliv. Rev.* **54**, 653 (2002).
3. R. C. Hayward, D. A. Saville and I. A. Aksay, *Nature* **404**, 56 (2000).
4. T. Okubo, *Prog. Polym. Sci.* **18**, 481 (1993).
5. M. Akashi, T. Niikawa, T. Serizawa, T. Hayakawa and M. Baba, *Bioconjug. Chem.* **9**, 50 (1998).
6. T. Hayakawa, M. Kawamura, M. Okamoto, M. Baba, T. Niikawa, S. Takehara, T. Serizawa and M. Akashi, *J. Med. Virol.* **56**, 327 (1998).
7. M. Kawamura, T. Naitou, M. Ueno, T. Akagi, K. Hiraishi, I. Takai, M. Makino, T. Serizawa, K. Sugimura, M. Akashi and M. Baba, *J. Med. Virol.* **66**, 291 (2002).
8. D. O'hangan, M. Singh, M. Ugozzoli, C. Wild, S. Harnet, M. Chen, M. Schaefer, B. Doe, G. R. Otten and J. B. Ulmer, *J. Virol.* **75**, 9037 (2001).
9. M. Riza, S. Tokura, M. Iwasaki, E. Yashima, A. Kishida and M. Akashi, *J. Polym. Sci. A Polym. Chem.* **33**, 1219 (1995).
10. T. Serizawa and A. Mitsuru, *Chem. Lett.*, 809 (1997).
11. T. Uchida, T. Serizawa, H. Ise, T. Akaike and M. Akashi, *Biomacromolecules* **2**, 1343 (2001).

Gels, Genes, Grafts and Giants, pp. 85–92 (2005)

Water-soluble polymers with low critical solution temperature (LCST) as carriers for protein drug delivery

NICOLAI A. PLATE, IVAN L. VALUEV * and LEV I. VALUEV

Institute of Petrochemical Synthesis, Russian Academy of Sciences, 29 Leninsky prospect, 119991 Moscow, Russia

Received 29 September 2003; accepted 18 December 2003

Abstract—The copolymers of N,N-diethylacrylamide and ovomucoid from duck egg white (an inhibitor of proteolytic enzymes) with low critical solution temperature (LCST) have been synthesized. The behaviour of these copolymers at the point of phase transition was investigated. It was shown that the relation between LCST and physiological activity for these copolymers is a function of their composition. The increase of ovomucoid content leads to the increase of LCST. When the content of ovomucoid rises above 0.2 mol%, LCST disappears. At the same time the physiological activity of obtained copolymers decrease with increasing of N,N-diethylacrylamide content.

Key words: Low critical solution temperature; polymeric carriers; physiological activity; phase transition; drug targeting.

INTRODUCTION

Drug targeting is one of the most important problems in medical chemistry and pharmacology. It is not merely economic aspects that give rise to this one (it is well known that, during the conventional administration of drugs, up to 90% of the amount adminstered does not reach the damaged organ and is, in fact, wasted [1]). The main reason of the solution of this problem is the fact that many highly active drugs, for instance, enzymes, cytostatic agents, etc., are toxic to healthy tissue.

Most of the approaches to designing targeting drug-delivery systems involve the co-immobilization of drug and vector molecules capable of delivery it to a target organ. The vector molecules can be divided into two groups. The first group includes substances exhibiting an enhanced affinity to a damaged organ: hormones [2], lectins [3], proteins and enzymes [4], glycoproteins [5], etc. The second one is the compounds capable of selective concentration in a certain place

*To whom correspondence should be addressed. E-mail: ivaluev@ips.ac.ru

under the action of various external effects, such as magnetic field [6], variation in pH, temperature or concentration of specific substances [7–9].

In our previous works [10, 11] we suggested an approach to the design of a universal protein drug-targeting system based on using temperature-sensitive polymers with a low critical solution temperature (LCST). This approach involves the immobilization of protein drugs on the polymers having a LCST near physiological temperature (37°C) and followed phase transition to the non-soluble state in the organ with higher temperature. The higher temperature of damaged organ could be result of inflammatory reaction [12] or this organ could be heated by the external influence.

Hoffman was one of the first who paid attention to the possibility of using cross-linked polymers with LCST for the solution of some biomedical problems [13, 14].

It is obvious that thermoactivated transport is only possible if two main requirements are met. First, upon protein immobilization, a polymeric carrier should retain the ability to undergo phase transition at the LCST point. Second, the proteins have to retain their physiological activity after phase transition.

Therefore, the main goal of the present work is to investigate processes that take place with protein-polymeric carrier systems at the phase-transition point and to study the dependence between transport function of these systems and their physiological activity. Duck egg white ovomucoid (an inhibitor of proteolytic enzymes) was used as a model protein. Poly-N,N-diethylacrylamide was used as polymeric carrier with LCST [15–17].

MATERIALS AND METHODS

N,N-diethylacrylamide (DEAA), ammonium persulfate and N,N,N′N-tetramethyl-ethylenediamine (Serva) were used in this work. Ovomucoid from duck egg whites (OM) was isolated and separated according to Ref. [18].

Unsaturated derivative of OM (UDOM) was prepared by the reaction of OM with acryloyl chloride [19].

The numbers of double bonds in the UDOM molecules were calculated as a difference between the quantity of amino groups in native OM molecules and quantity of amino groups in UDOM molecules. The number of amino groups was examined by the titration of protein molecules with trinitrobenzenesulfoacid [20].

OM-DEAA copolymers were synthesized by the radical copolymerization of DEAA and UDOM in an aqueous solution. A redox system ammonium persulfate — N,N,N′N-tetramethylethylenediamine was used as an initiator. After the copolymerization products were separated by gel filtration on a Sephadex G-100 column (1.5 × 60 cm) equilibrated with distilled water. Spectrophotometer Hitachi-3410 was used as the detector (wavelength 220 nm).

LCST were determined by the turbidity of 1% copolymers aqueous solution measured on a Hitachi-3410 at wavelength 600 nm.

The physiological (antitryptic) activity of OM and ODOM was measured by suppressing the amidase activity of trypsin according to Kakade [21]. *p*-Nitroanilide-N-α-benzoyl-D,L-arginine was used as the substrate (temperature 41°C).

The molecular weight and molecules conformations of copolymers were determined by dynamic light scattering on a Malvern-4400. Twice distilled benzene was employed as a standard.

RESULTS AND DISCUSSION

For copolymerization with DEAA the UDOM molecules containing 4–5 double bonds were used. The biological activity of these derivatives was about 80% of native OM [18]. When UDOM with smaller content of unsaturated bonds was used, its conversion during copolymerization did not exceed 10% [22]. The greater amount of C=C bonds in UDOM molecules results in considerable reduction in its antitrypsin activity [18].

The gel formation during DEAA-UDOM copolymerization was observed. It could be explained by the presence of more than one double bond in UDOM molecules. In this case, the protein derivative may also function as a cross-linking agent. The decrease of DEAA : UDOM ratio in the initial mixture leads to the increase of non-soluble product's share (Table 1).

The soluble products of copolymerization were investigated by gel filtration (Fig. 1). As is seen, copolymerization leads to the formation of DEAA-OM copolymers (peak I), as well as to homopolymer of DEAA (peak III). The homopolymer formation is the result of high molar excess of DEAA. Peak II corresponds to unreacted OM.

The elastic light scattering data of the copolymer water solution below LCST point shows that molecular weight of copolymers is a function of OM content. Extrapolation of curves in Zimm's coordinates [23] to the zero scattering angle shows, that increasing of OM content leads to the decreasing of molecular weight (Fig. 2a). This phenomenon is the result of steric hindrances, which appears when

Table 1.
The dependence of copolymers composition and soluble: non-soluble products ratio on the composition of initial monomers mixture

The composition of initial DEAA : UDOM mixture (mol/mol)	The composition of DEAA : UDOM co-polymers (mol/mol)	Soluble products (% ±3%)	Gel (% ±3%)
2500 : 1	2500 : 1	100	—
600 : 1	510 : 1	87	13
400 : 1	280 : 1	68	32
200 : 1	130 : 1	55	45

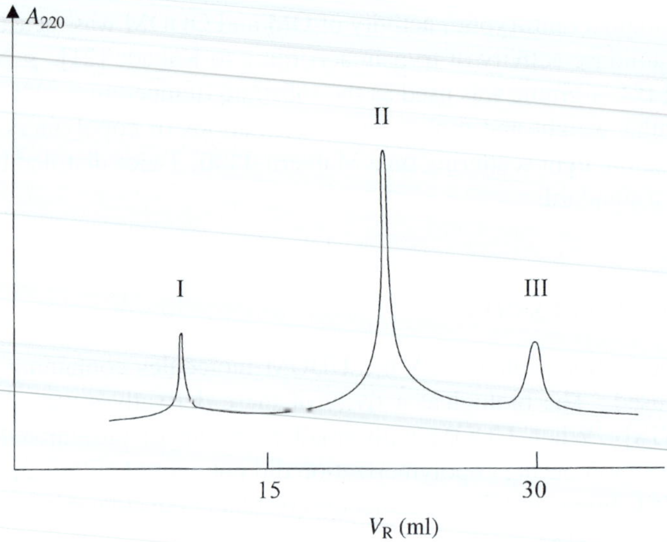

Figure 1. Chromatogram of water-soluble products of DEAA-OM copolymerization ratio DEAA/UDOM in copolymer = 280 : 1).

UDOM molecules join to the growth macroradical and complicate the formation of high molecular weight macromolecules.

As follows from the data on the composition and molecular weight of copolymers, each copolymer molecule, regardless of the composition of the initial monomer mixture, contains on average two OM units that are united by DEAA fragments with different polymerization degree. The light scattering curves in Kratki and Porod coordinates [24] demonstrate that all copolymer molecules at temperature below LCST having a statistical coil conformation (Fig. 3a).

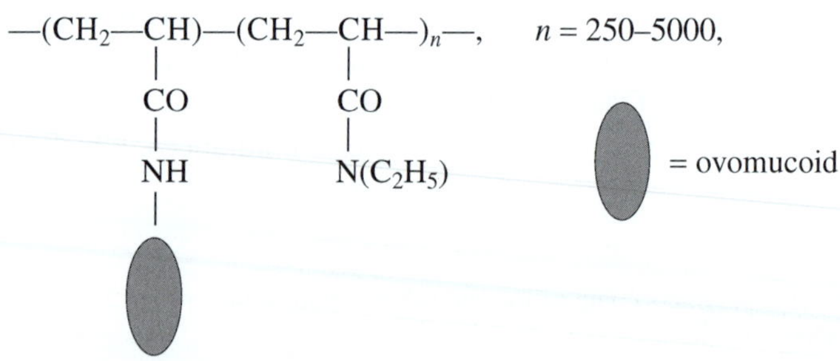

When the temperature of the copolymer solution is raised above the LCST, the measured molecular weight for all copolymers increases, i.e. at phase transition temperature we observe the association of macromolecules (Fig. 2b). The higher the degree of polymerization of polyDEAA fragments in the copolymer, the more pronounced the increase in molecular weight. The spatial structure of copolymers at temperature above LCST also depends on their composition. When the content of OM in the copolymer is above 0.2 mol% (DEAA/OM $<$ 500 : 1), copolymers

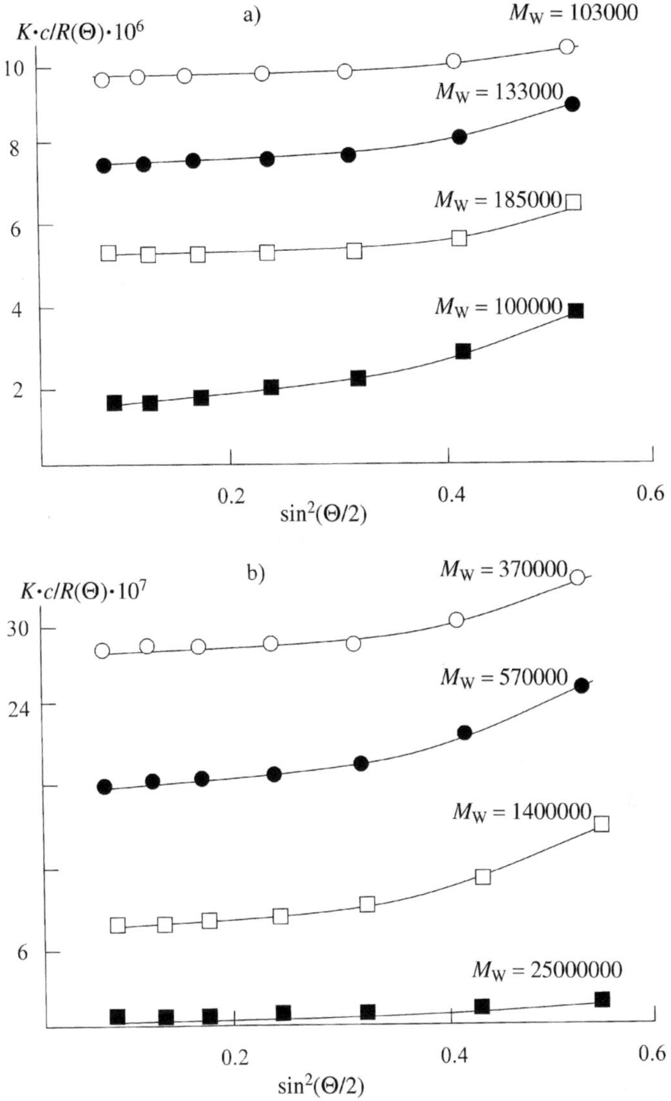

Figure 2. Light scattering curves for DEAA-OM copolymers in the Zimm's coordinates at temperature (a) below (28°C) and (b) above (48°C) LCST. The copolymer content (here and in Fig. 3): DEAA : OM = 2500 : 1 (■); 510 : 1 (□); 280 : 1 (●) and 130 : 1 (○).

still persist in solution and retain the coil structure (Fig. 3b). However, at a lower content of the protein (0.2 mol% and below), copolymers tend to form precipitates at temperature above LCST. Judging by the extremum pattern of the light scattering curves, they have a branched structure with a relatively dense packing of the segments [24].

A similar dependence could be observed in Table 2, which demonstrates the dependence of LCST on the composition of synthesized copolymers. It is seen that

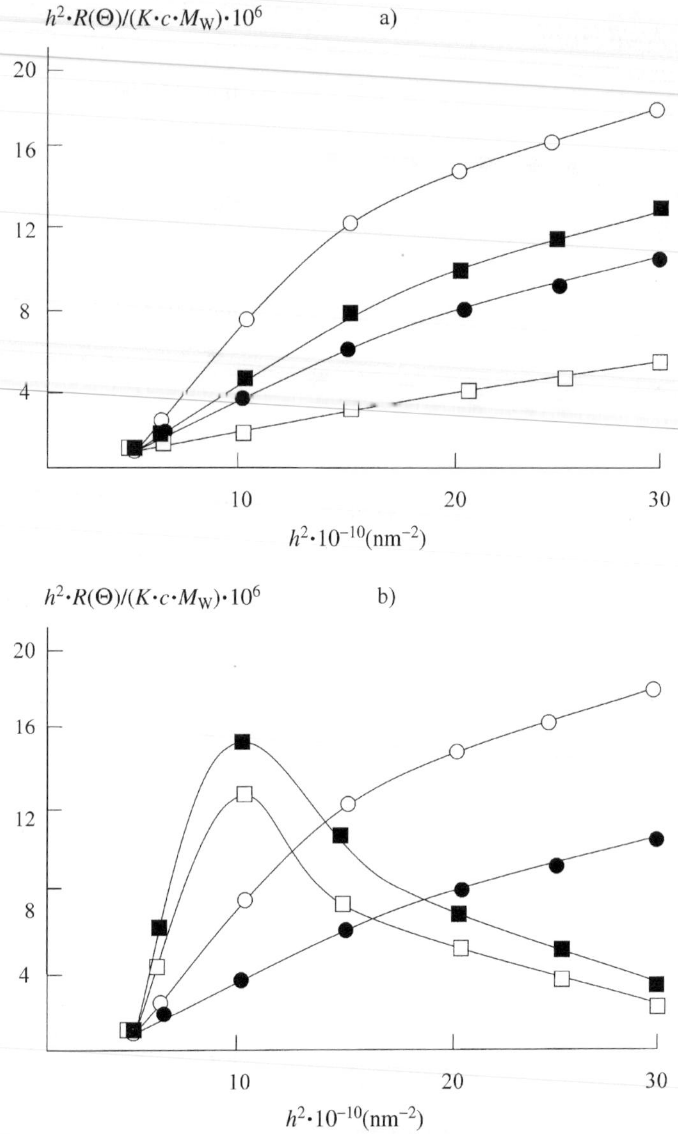

Figure 3. Light scattering curves for DEAA-OM copolymers in the Kratky and Porod's coordinates at temperature (a) below (28°C) and (b) above (48°C) LCST.

increasing the OM content in copolymer leads to increasing of LCST point. When OM content was higher than 0.2 mol%, LCST disappeared and these copolymers show complete solubility in water within a wide temperature range.

This finding is apparently due to the diphilic nature of the copolymers at temperatures above LCST. The polyDEAA fragments are hydrophobic and tend to form a new phase upon phase separation, whereas OM retains a high affinity to the water. Here the existence of two counterforces is evident. One of these

Table 2.
The dependence of LCST of OM : DEAA copolymers on their composition

The composition of DEAA : UDOM co-polymers (mol/mol)	LCST (°C ±0.5°C)
Homopolymer DEAA	31.0
2500 : 1	34.5
510 : 1	40.0
280 : 1	no LCST
130 : 1	no LCST

Table 3.
The dependence of antitrypsin activity of OM : DEAA copolymers on their composition

The composition of DEAA : UDOM co-polymers (mol/mol)	Antitrypsin activity, (% of UDOM ±4%)
2500 : 1	55
510 : 1	75
280 : 1	81
130 : 1	84

forces, which appears when the polyDEAA fragments undergo a conformational transition, favors sticking together of the copolymers macromolecules. Another force hampers their association due to the occurrence of the solubilizing effect and preventing further association of macromolecules. For systems studied the visual phase separation in the copolymer solution appears when associates are composed of more than seven macromolecules (M_w in Fig 2a and 2b).

The results of the study of antitrypsin activity of OM-DEAA copolymers are summarized in Table 3. It is seen that the increase of DEAA content in copolymer causes a decrease in the antiproteinase activity of OM. But in any case the OM polymer derivative keeps more then 50% of UDOM, which was used for copolymer synthesis.

It appears that this trend is related to the shielding of the OM active centers by the polyDEAA fragments, thus hampering formation of OM-trypsin complex. It is obvious that the longer the polyDEAA fragment, the more pronounced its effect. Hence, it is follows that short oligomeric fragments of DEAA in DEAA-OM copolymer should be preferred in order to keep the UDOM activity at a maximum level. However, as was shown above, such low-molecular-weight polyDEAA chains do not exhibit the ability to undergo phase transition at an elevated temperature, and consequently, they cannot provide the thermoactivated targeting transport of the whole systems.

Thus, the experimental data obtained in the present work demonstrate that the amount of the polymeric carrier should be minimized to retain a high physiological activity of immobilized on polymer proteins with respect to high-molecular-weight substrates. At the same time, one should keep in mind that for each thermoactivated transport system there is a limiting ratio between polymer carrier and physiological active substance. For the system studied, the increase of OM content in copolymers above 0.2 mol% leads to loss of LCST in the resulting copolymers, i.e., of their ability for thermoactivated targeting transport.

REFERENCES

1. V. P. Torchilin and A. A. Klibanov, *Zh. Vsesoyuzn. Khim. O-va im. D. I. Mendeleeva* **32**, 502 (1987).
2. S. Yoless, T. F. Morton and M. F. Sartori, *J. Polym. Sci.* **17**, 4111 (1979).
3. R. G. Anderson and M. S. Brown, *J. Cell. Biol.* **88**, 441 (1981).
4. K. Yamamoto, N. Katsuda and K. Kato, *Eur. J. Biochem.* **92**, 499 (1978).
5. J. Kopechek and P. Kopechkova-Reimanova, *J. Chromatogr. Biomed. Appl.* **376**, 221 (1986).
6. M. S. Chang, in: *Drug Carriers in Biology and Medicine*, p. 271. Academic Press, London (1981).
7. L. Brannon-Peppas and N. A. Peppas, *J. Control. Rel.* **8**, 267 (1989).
8. X. Li, O. B. Bennet, N. W. Adams and S. W. Kim, *Polym. Prepr.* **31**, 198 (1990).
9. S. Sato, S. Y. Yeong, Y. C. McRea and S. W. Kim, *J. Control. Rel.* **1**, 67 (1984).
10. L. I. Valuev, O. N. Zefirova, I. V. Obydennova and N. A. Plate, *Polym. Sci. Ser. A* **135**, 68 (1993).
11. L. I. Valuev, V. V. Chupov, T. A. Valueva, L. V. Vanchugova, V. A. Sinani, O. V. Noa and N. A. Plate, *Polym. Sci. Ser. A* **31**, 75 (1991).
12. E. I. Chazov, V. N. Smirnov and V. P. Torchilin, *Zh. Vsesoyuzn. Khim. O-va im. D. I. Mendeleeva* **32**, 485 (1987).
13. A. S. Hoffman, A. Affrasiabi and L. C. Dong, *J. Control. Rel.* **4**, 213 (1986).
14. C.-A. Cole, S. M. Schreiner, J. H. Priest, N. Monji and A. S. Hoffman, *Reversible Polymeric Gels and Related Systems, ACS Symposium Series, Volume 245.* American Chemical Society, Washington, DC (1987).
15. W. S. Cai, L. H. Gan, K. C. Tam, *Colloid Polym. Sci.* **279**, 793 (2001).
16. M. Itakura, K. Inomata, T. Nose, *Polymer* **41**, 8681 (2000).
17. J. Spevacek, D. Geshke, M. Ilavsky, *Polymer* **42**, 463 (2001).
18. L. V. Vanchugova, T. A. Valueva, L. I. Valuev, V. I. Romashkin, M. A. Rozenfel'd, V. V. Mosolov and N. A. Plate, *Biokhimiya* **22**, 1455 (1988).
19. N. A. Plate, L. I. Valuev, N. S. Egorov and M. A. Al'-Nuri, *Prikl. Biokhim. Microbiol.* **13**, 673 (1977).
20. R. Hynes, D. T. Osuga and R. E. Feeney, *Biochemistry* **6**, 541 (1967).
21. M. L. Kakade, M. Simons and Y. E. Liener, *Cereal Chem.* **46**, 518 (1969).
22. N. A. Plate, L. I. Valuev and V. V. Chupov, *Vysokomol. Soed. Ser. A* **22**, 1963 (1980).
23. B. H. Zimm, *J. Chem. Phys.* **16**, 1099 (1948).
24. O. Kratky and G. Porod, *J. Colloid Interface Sci.* **4**, 35 (1949).

Gels, Genes, Grafts and Giants, pp. 93–104 (2005)

Polymeric drugs with prolonged sustained delivery of specific anti-aggregant agents for platelets: kinetic analysis of the release mechanism

ALBERTO GALLARDO *, GEMA RODRÍGUEZ, MAR FERNÁNDEZ, MARÍA ROSA AGUILAR and JULIO SAN ROMÁN

Instituto de Ciencia y Tecnología de Polímeros, CSIC, Juan de la Cierva 3 28006 Madrid, Spain

Received 30 September 2003; accepted 11 February 2004

Abstract—The *in vitro* aqueous behaviour of a metacryloyloxyethyl [2-(acetyloxy)-4-(trifluoromethyl)]benzoate (THEMA)/N,N′-dimethylacrylamide (DMA) copolymer with a THEMA molar content of 39% (labeled THDMA39) has been investigated. This composition has been selected to achieve a system able to keep both the non-water solubility during the release and the resorbability (and the water solubility) after the completion of the drug release. This copolymer exhibited, at pH 7.4, a constant release during several months, very interesting for a long term treatments required for the application of some cardiovascular devices. A kinetic model has been developed to explain the pseudo-zero-order kinetics of the release process. This model, which considers (from the aqueous studies) a linear increase with time of the amount of water present in the polymeric matrix, has been able to fit adequately the experimental data.

Key words: Polymeric drug; Triflusal; controlled release; antithrombogenicity.

INTRODUCTION

Different polymeric systems derived from Triflusal (a well-known anti-aggregant agent for platelets [1, 2], marketed as Disgren®) have been recently developed in our laboratory. These systems are based on a methacrylic derivative of Triflusal (THEMA), the synthesis of which has been described in a previous article [3]. Preparation and some preliminary evaluations of the homopolymer poly-THEMA and copolymers of THEMA with hydrophilic components, ionic (2-acrylamido-2-metylpropanesulfonic acid, AMPS) and non-ionic (DMA), have been studied previously (Refs [3–5] and data not shown). DMA is a well-known hydrophilic

*To whom correspondence should be addressed. Tel.: (34-91) 562-2900. Fax: (34-91) 564-4853. E-mail: gallardo@ictp.csic.es

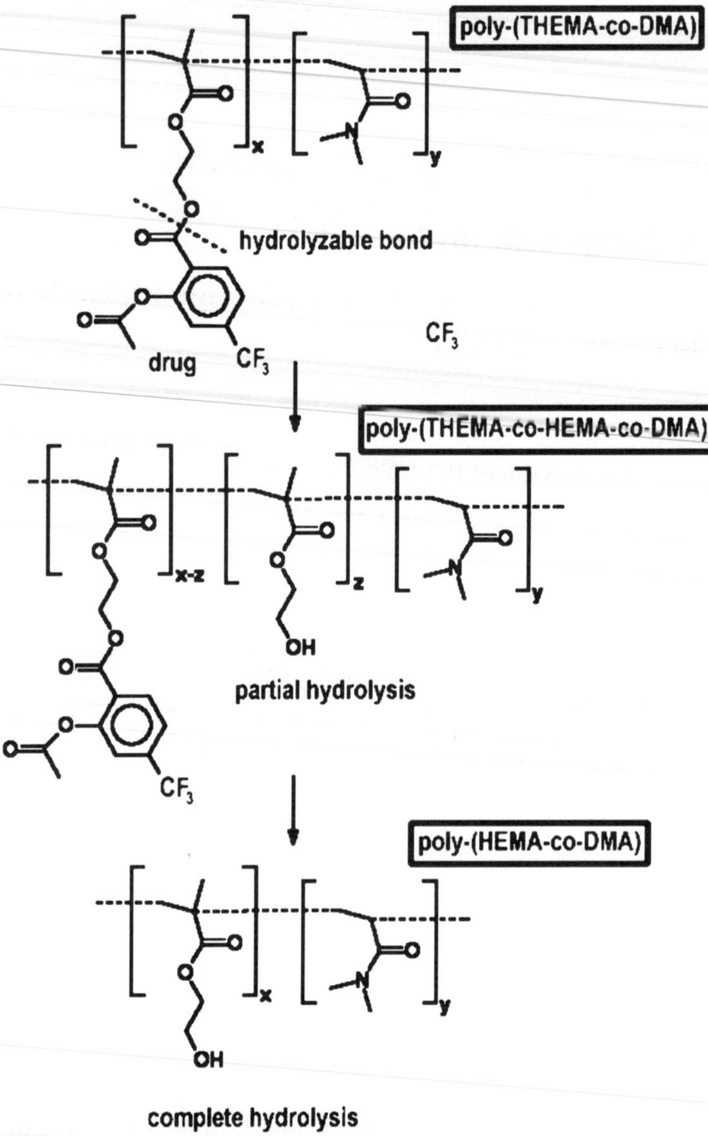

Figure 1. Schematic hydrolytical process of the THEMA-DMA copolymers.

biocompatible component. This component as well as the residue after the hydrolysis of THEMA/DMA copolymers, i.e. a HEMA/DMA copolymer (see Fig. 1), have exhibited in addition a good behaviour in contact with blood [6, 7]. Most of these polymeric drugs containing Triflusal have already shown a good potential to provide sustained drug release during several months, very attractive for long-term applications. These systems, administered parenterally or prepared as coatings, could fit adequately some current requirements associated with the use of synthetic cardiovascular devices and in particular for coronary stents.

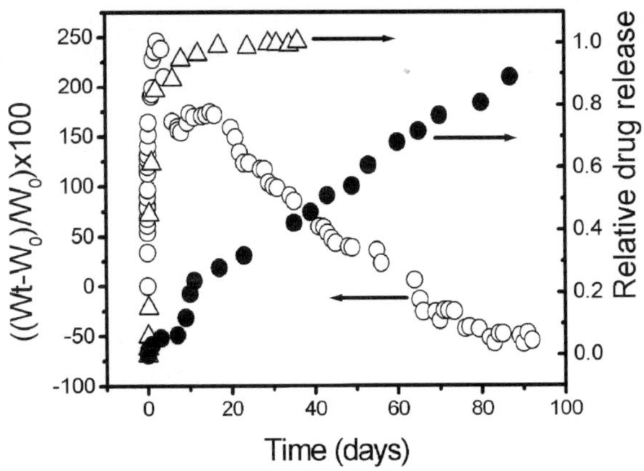

Figure 2. Gravimetrical study of the wet copolymers (○) and correlative release data (●) from the *in vitro* aqueous experiments of THDMA22 at pH 7.4. Release data at pH 10 (△) have been depicted for comparative purposes. The gravimetrical analysis has been represented in left *y*-axis as described in the Materials and methods.

Copolymerization of THEMA with DMA provides systems with a balance of hydrophilia/hydrophobia mainly controlled by the composition of the copolymeric chains. This balance exerts an important influence in the aqueous behaviour, that is, swelling, solubility and release rate of Triflusal. In another study, a highly hydrophilic poly(THEMA-*co*-DMA) with a THEMA molar content of 22%, named THDMA22, has been described (data not shown). The reported THDMA22 exhibited good *in vitro* cytocompatibility by culturing with Human Osteoblastic Cells (HOB) and a very promising *in vitro* anti-aggregating effect (data not shown). This copolymer also showed a pseudo-zero-order release of Triflusal during several weeks in physiological conditions. Besides, the complete *in vitro* study of the aqueous behaviour of the copolymer has revealed that solubilization takes place continuously during these weeks where it exhibited a constant release (see Fig. 2).

Although this composition (THDMA22) and the related behaviour (with the complete resorption of the system) are very attractive, we can argue that for some specific applications as the coating of vascular devices, a more stable copolymer would be preferable. To keep resorbability, the optimum formulation should be a copolymer that releases the drug in a sustained way, keeping the stability until the last stages of the release, where the copolymer may become hydrosoluble, rendering the entire device resorbable. Taking this approach, we have optimized the copolymer composition to obtain a copolymer with this behaviour.

MATERIALS AND METHODS

Polymer synthesis

Triflusal was a gift of Laboratorios Uriach. N,N′-dimethylacrylamide (DMA, Aldrich) was vacuum distilled. Other reagents were of extra pure grade and

Table 1.
Average composition of copolymer systems studied in this work.

	THEMA molar fraction	DMA molar fraction	HEMA molar fraction
THDMA39	0.39	0.61	—
THDMA22	0.22	0.78	—
THHD17	0.17	0.61	0.22

used as purchased. The synthesis of the methacrylic derivative of Triflusal, methacryloyloxyethyl [2-(acetyloxy)-4-(trifluoromethyl)]benzoate (THEMA) has been described elsewhere [3].

THDMA39 copolymer was prepared by the free radical polymerization of a mixture of the corresponding monomers in purified Dioxane at 50°C ([M] = 0.5 M), using benzoylperoxide as initiator ([I] $= 1.5 \times 10^{-2}$ M). Molar composition of THEMA in the feed was 0.39. Reactions were carried out in absence of oxygen by bubbling nitrogen for 30 min before the sealing of the system. The polymers were isolated by precipitation in diethyl ether and dried at reduced pressure until constant weight.

Poly(THEMA-*co*-HEMA-*co*-DMA) terpolymer (THHD17) was prepared in the same conditions as the copolymer but using the following feed molar compositions: $F_{\mathrm{THEMA}} = 0.17$, $F_{\mathrm{HEMA}} = 0.22$ and $F_{\mathrm{DMA}} = 0.61$.

^{1}H-NMR characterization

^{1}H-NMR spectroscopy was used to characterize THDMA39 and THHD17 and to calculate the molar composition of the monomers in the corresponding polymer (Table 1). The analysis was performed in a Varian Gemini 200 spectrometer working at 200 MHz. The spectra were recorded at room temperature and using 5% (w/v) deuterated chloroform solutions.

Copolymer composition analysis was carried out by comparing the integral area some characteristic peaks of both monomers, i.e. for THEMA composition we used the signals from methylenic protons (3.9–4.7 ppm) and for DMA composition we used the signal from tertiary carbon protons (2.4–3.3 ppm). The copolymer composition, f_{THEMA}, was 0.39, confirming total conversion.

In the case of the terpolymer, we used the aromatic protons to calculate THEMA composition (7.6–8.1 ppm), methylenic protons for HEMA composition (3.9–4.7 ppm) and protons signals from tertiary carbon to calculate DMA composition (2.4–3.3 ppm). The copolymer composition was $f_{\mathrm{THEMA}} = 0.17$, $f_{\mathrm{HEMA}} = 0.22$ and $f_{\mathrm{DMA}} = 0.61$, also confirming total conversion.

Release studies

Drug release from THDMA39 was studied in buffered solutions at pH 7.4 and pH 10. Thin polymer films (approx. 1 mm thick; 51,4 and 47.4 mg, respectively) were prepared by casting from ethanol (33%, w/v) and immersing in 10 ml buffer solution. The systems were introduced in a thermostatic bath (37°C) and kept under magnetic stirring; 0.3 ml of the supernatant was periodically collected for analysis and replaced by the same volume of fresh medium. The release was monitorized by means of an HPLC method. The analysis was performed by using a mixture (60 : 40) of methanol/aqueous solution of PIC A (a solution of tetrabutylammonium hydrogen sulphate in water, supplied by Waters) as the mobile phase and a flow rate of 1 ml/min. The system consisted in a Perkin Elmer LC-250 pump, a Perkin Elmer LC-95 UV/Vis detector and a Waters μBoundapak C_{18} column (3.9 × 300 mm). Measurements were done at 305 nm. All the experiments were carried out in triplicate.

Gravimetrical analysis

The absorption of water and the weight loss from THDMA39 (and the initial water uptake of THHD17) were followed gravimetrically measuring periodically the weight of thin films immersed in a buffered solution (pH 7.4 and 10) at 37°C, after carefully wiping the surface with filter paper. The films were obtained by casting from ethanol (33%, w/v). Copolymer samples weights were 48.8 mg for pH 7.4 and 44.4 mg for pH 10; terpolymer sample weights were 21.22 mg and 22 mg for pH 7.4 and 10, respectively. The following ratio, referred to the weight of the initial dry polymer, has been used to represent the gravimetrical analysis (Figs 2 and 3):

$$\frac{\text{Weight}_{\text{wet polymer at time } t} - \text{Weight}_{\text{dry polymer at time 0}}}{\text{Weight}_{\text{dry polymer at time 0}}} = \frac{W_t - W_0}{W_0}.$$

All the experiments were carried out in duplicate.

RESULTS AND DISCUSSION

According to the scheme in Fig. 1, partial hydrolysis of the THEMA units leads to the formation of the terpolymer poly(THEMA-*co*-HEMA-*co*-DMA), which overcomes a continuous drift in composition until the complete drug release where the final residue poly(HEMA-*co*-DMA) is formed.

To achieve a system stable during the release but resorbable after the completion of the hydrolysis, we need a macromolecular system where the final copolymer, but not the successive intermediate terpolymers, is water soluble. We have studied the solubilization behaviour of the HEMA-DMA copolymers (which are water soluble in all the studied compositions, from 0.2 to 0.8 molar fraction of HEMA) and we have also prepared THEMA-HEMA-DMA terpolymer (to check solubility and swelling), in order to select the appropriate compositions that fits these requirements, which we have found to be close to a poly-(THEMA-*co*-DMA) with a THEMA molar fraction of 0.39, labelled THDMA39. Copolymers with

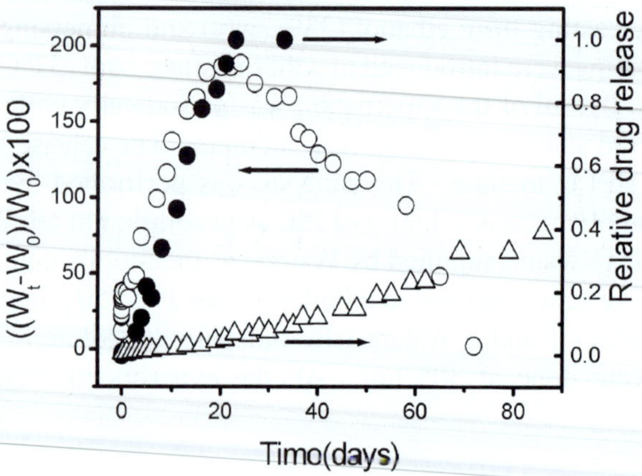

Figure 3. Gravimetrical study of the wet copolymers (○) and correlative release data (●) from the *in vitro* aqueous experiments of THDMA39 at pH 10. Release data at pH 7.4 (△) have been depicted for comparative purposes. The gravimetrical analysis has been represented in left *y*-axis as described in the Materials and methods.

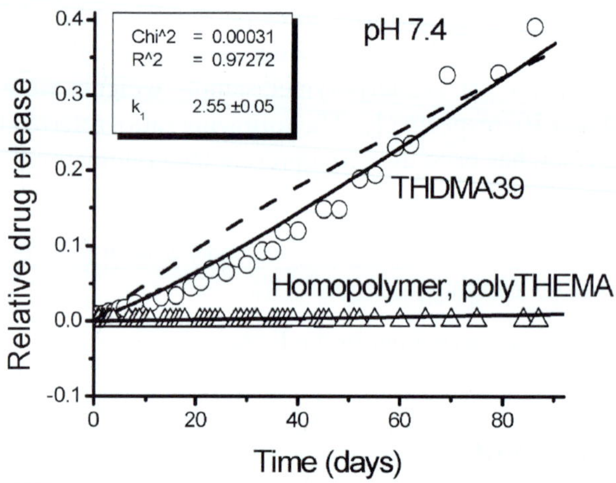

Figure 4. Release profiles of Triflusal from polyTHEMA (△) and THDMA39 copolymer (○) at pH 7.4. Solid lines correspond to the fitting obtained using equation [8], while dashed line represents the best fit to first-order kinetics.

higher DMA content (as THDMA22) become partially hydrosoluble before the complete release of the drug (see Fig. 2 for the experiment carried out at pH 7.4), while copolymers with high THEMA content are very hydrophobic, close to the homopolymer polyTHEMA, which exhibits a very slow release rate (see Figs 4 and 5, data for the homopolymer taken from Ref. [3]).

Therefore, copolymers with average composition close to THDMA39 seem to be the optimal systems in terms of hydrolysis rate and dimensional stability before final resorption. The gravimetrical study, hydrolysis and solubilization of THDMA39 at pH 10 have been depicted in Fig. 3. Solubilization starting point and 100% release

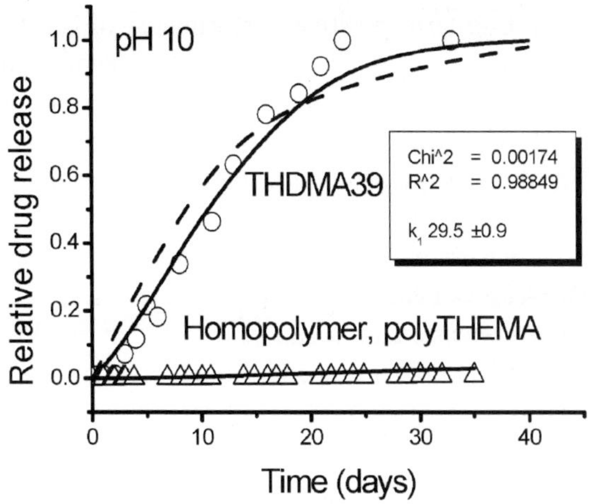

Figure 5. Release profiles of Triflusal from polyTHEMA (△) and THDMA39 copolymer (○) at pH 10. Solid lines correspond to the fitting obtained using equation [8], while dashed line represents the best fit to first-order kinetics.

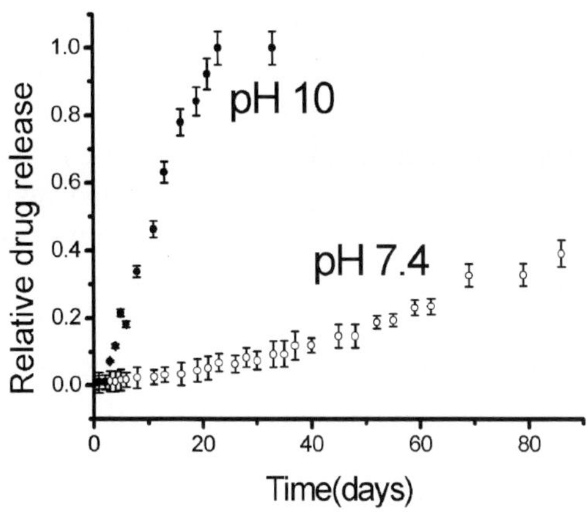

Figure 6. Release profiles of Triflusal from THDMA39 copolymer at pH 7.4 (○) and 10 (●).

take place almost at the same time (24 days). Under *in vitro* physiological conditions (pH 7.4, 37°C), the release rate is noticeable slower (see Figs 4 and 6) and we only have data of partial hydrolysis, but we should expect a similar behaviour in solubilization for the 100% of the hydrolisis. As has been discussed above, this copolymer composition may be able to keep stable on application during several months until a very high degree of hydrolysis is reached; at this time, the system is hydrophilic enough to become hydrosoluble and the support can be clearable (if the molecular weight is the appropriate [8]).

The release profile of THDMA39 at pH 7.4 (Figs 4 and 6), compared with THDMA22 at the same pH (Fig. 2), is noticeable lower. In any case, both exhibit zero- or pseudo-zero-order kinetics. As it has been noted in the Introduction, this constant release that takes place during several months in THDMA39 may be very attractive to achieve a long-term positive performance of different cardiovascular devices, although it has to be noted that this *in vitro* behaviour must be considered an approximation to its *in vivo* behaviour. The release data obtained from the hydrophobic homopolymer, which have been reported elsewhere [3], have been also depicted in Fig. 4. The incorporation of the hydrophilic component DMA significantly increases the release rate, mainly due to the higher water uptake and accessibility by the increasing hydrophilic character. The higher the hydrophilicity of the system (THDMA22 > THDMA39 ⋙ poly-THEMA), the higher the release rate. From these *in vitro* release data, it can be concluded that hydrophilic systems, as the ones studied in this work, are needed to achieve an efficient drug-delivery system.

Kinetic model

Release data from THDMA39 at pH 10 are shown in Fig. 6 together with the data obtained at pH 7.4. The hydrolysis rate increases noticeably at alkaline pH (as it should be expected) because of the ester nature of the labile bond that undergoes the hydrolysis, which is catalyzed at basic pH. It is clear from Figs 4 and 5 that the experimental points obtained at pH 7.4 and 10 do not adequately fit the first-order kinetics (dashed lines in Figs 4 and 5). We have related this 'anomalous' behaviour (very interesting for a controlled release formulation) to the increase in hydrophilicity associated to the hydrolysis of THEMA (and to the conversion of the hydrophobic THEMA in the hydrophilic HEMA). This increase in hydrophilicity leads to a increase in the content of water inside the matrix. The same deviation from the first-order kinetics happens with different copolymers of THEMA with the ionic monomer AMPS, as has been described in a previous work [5]. In this previous paper, the microstructural arrangement (with a tendency to THEMA block formation) allowed the development of a kinetic model (called Progressive Accessibility Model) that considered the THEMA units aggregated forming hydrophobic domains. However, we can not extend this study to the THDMA39 copolymer because of different reasons. On one hand, the microstructural arrangement, as theoretically obtained from the reactivity ratios ($r_{\text{THEMA}} = 1.25$, $r_{\text{DMA}} = 0.60$) (data not shown), does not exhibit a tendency towards the block formation (this can be easily seen from the reactivity ratios values, very dissimilar in the case of THEMA-AMPS ($r_{\text{THEMA}} = 4.54$, $r_{\text{AMPS}} = 0.32$) [4] and much more close to the unity in the case of THEMA-DMA). On the other hand, the THEMA-AMPS copolymers described by the previous model were THEMA-rich copolymers, while in this case THDMA39 is a DMA-rich copolymer.

In this case (THDMA39 copolymer), we have to use a different approach. We have developed a very simple kinetic model, called Swelling Gradient Model, where the increasing hydrophilicity has been associated to the obvious increase of water concentration in the matrix.

As the drug is covalently linked to the macromolecular backbone, we can use this scheme as starting point:

$$-\text{Ester} + H_2O \xrightarrow{k_1} -\text{OH} + \text{Drug}, \tag{1}$$

$$\frac{dn_{\text{drug}}}{dt} = -\frac{dn_{\text{ester}}}{dt} = k_1 n_{\text{ester}} n_{H_2O}, \tag{2}$$

where n_i is the corresponding mol number of the species i (i = drug, ester, etc.), incorporating the gradient on swelling

$$n_{H_2O} = f(\text{hydrolysis extent}), \tag{3}$$

but from data at pH 10 in Fig. 3, it seems that the dependence n_{H_2O}/hydrolysis is linear. Assuming that solubilization and release phenomena do not overlap (according to Fig. 3), we can switch to swelling data by correcting the gravimetrical data of Fig. 3 using the weight loss associated to the drug release:

$$H = \frac{W_{H_2O}}{W_{\text{polymer}}} \times 100, \tag{4}$$

and

$$W_{\text{polymer}} = (W_{\text{polymer}})_0(1 - w_{\text{THEMA}}) + (W_{\text{polymer}})_0 \frac{w_{\text{THEMA}}}{360}(130\text{DR} + (1 - \text{DR})360), \tag{5}$$

where W_{H_2O} and W_{polymer} are the weight of water and dry polymer, H the water uptake, w_{THEMA} the THEMA weight fraction in the copolymer (0.70), DR the relative drug release according to Fig. 3, and 130 and 360 the molar weights of the comonomers HEMA and THEMA, respectively.

Using this equation we obtain, at pH 10, the graph depicted in Fig. 7 (up to the 100% release time). There is a clear linear behaviour after a relatively short induction period. The triangle corresponds to the swelling obtained, at pH 7.4,

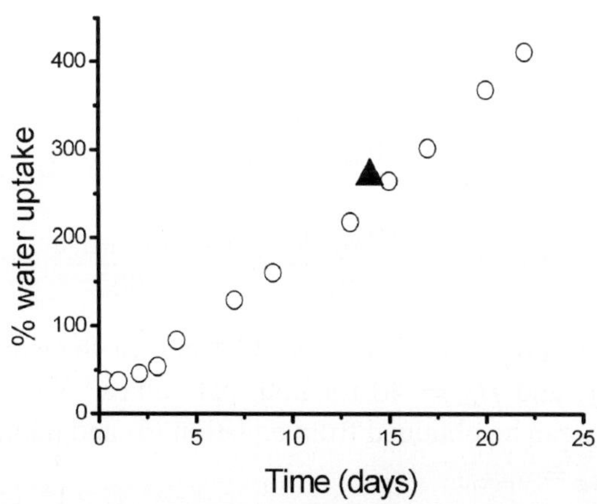

Figure 7. Initial terpolymer swelling at pH 7.4 (▲) and water uptake of THDMA39 at pH 10 (○) obtained from the gravimetrical study of Fig. 3 as discussed in the text.

by the terpolymer THHD17. This terpolymer composition corresponds to the THDMA39 residue of 56% release, which in the experiment at pH 10 (Fig. 3) is obtained at 14 days approximately. It is expected that the reactivity of THEMA and HEMA in radical polymerization is very similar and, therefore, the distribution of THEMA and HEMA has a random character. Therefore, we consider that the microstructure (i.e. distribution of comonomer units) is similar in THEMA-DMA copolymers and THEMA-HEMA-DMA terpolymers. There is a very good agreement between the swelling of the terpolymer at pH 7.4 and the corrected data of Fig. 5 that correspond to the experiment carried out at pH 10. This fact allows us to extend the linear behaviour of Fig. 7, obtained at pH 10, to pH 7.4 (also taking into account that there are no ionic groups and, therefore, no pH sensitivity associated to those). We can now use a general (pH-independent) linear form in equation (3):

$$n_{H_2O} = (n_{H_2O})_0 + k_2 n_{OH} = (n_{H_2O})_0 + k_2 n_{drug}. \tag{6}$$

On the other hand

$$n_{ester} = (n_{ester})_0 - n_{drug}, \tag{7}$$

and therefore

$$\frac{dn_{drug}}{dt} = \mathrm{k}_1\{(n_{ester})_0 - n_{drug}\}\{(n_{H_2O}) + k_2 n_{drug}\}, \tag{8}$$

and the integration leads to

$$\frac{n_{drug}}{(n_{ester})_0} = \mathrm{DR} = \frac{(1 - \mathrm{e}^{-k_1\{(n_{H_2O})_0 + k_2(n_{ester})_0\}t})}{1 + k_2 \dfrac{(n_{ester})_0 \mathrm{e}^{-k_1\{(n_{H_2O})_0 + k_2(n_{ester})_0\}t}}{(n_{H_2O})_0}}, \tag{9}$$

$(n_{ester})_0$ and $(n_{H_2O})_0$ can be obtained from the initial polymer mass ($(W_{polymer})_0$) and swelling (H_0)

$$(n_{ester})_0 = \frac{(W_{polymer})_0 w_{THEMA}}{360}, \tag{10}$$

$$(n_{H_2O})_0 = \frac{(W_{H_2O})_0}{18} = \frac{H_0 (W_{polymer})_0}{1800}, \tag{11}$$

with $w_{THEMA} = 0.70$, $(W_{polymer})_0$ 51.4 and 47.4 mg for the experiments at pH 7.4 and 10, respectively, and $H_0 = 40$ for both pH values. We obtained the values quoted in Table 2. k_2 can be obtained from equation (6) and taking into account that at 100% release $n_{drug} = n_{ester0}$.

$$k_2 = \frac{(n_{H_2O})_{100} - (n_{H_2O})_0}{(n_{ester})_0} = \frac{H_{100}(W_{polymer})_{100} - H_0(W_{polymer})_0}{1800(n_{ester})_0}, \tag{12}$$

Table 2.
Recompilation of the different parameters and experimental data used in the kinetic model

	n_{ester0} $\times 10^5$	$(n_{H_2O})_0$ $\times 10^5$	k_2	k_1 mol^{-1} day^{-1}	H_0	H_{100}	$(W_{polymer})_0$ (mg)	$(W_{polymer})_{100}$ (mg)
THDMA39 pH 7.4	9.99	114	55	2.55	40	420	51.4	28.4
THDMA39 pH 10	9.22	105	55	29.5	40	420	47.4	26.2
Homopolymer pH 7.4	55.6	22.2	2.5	2.55	2	40	200	72
Homopolymer pH 10	55.6	22.2	2.5	29.5	2	40	200	72

taking H_{100} from calculations or from Fig. 7 (420; extending the behaviour at pH 10 to pH 7.4). $W_{polymer100}$ is

$$(W_{polymer})_{100} = (W_{polymer})_0 - W_{total\ drug} = (W_{polymer})_0 \left(1 - \frac{w_{THEMA}230}{360}\right), \quad (13)$$

giving 28.4 and 26.2 mg for the experiments at pH 7.4 and 10, respectively.

Release data have been fitted to equation (9), obtaining the curves represented by solid lines in Figs 4 and 5. The slight sigmoid behaviour obtained experimentally, which is far from the first-order kinetics (dashed lines represent the fitting to first-order kinetics), is predicted by this model. The best k_1 values, obtained by a non-linear least square method, have been found to be 2.55 and 29.5 mol^{-1} day^{-1} for the experiments at pH 7.4 and 10, respectively. Fitting statistical parameters (the correlation factor r^2 and the sum of the residuals χ^2) as obtained with the software Origin 6.0® (Origin, Natick, MA, USA) have been included in the graphs. r^2 values higher than 0.97 have been obtained in both cases.

The application of this model and the obtained k_1 values to the homopolymer release data (using $H_0 = 2$ and $H_{100} = 40$, respectively, as obtained from the swelling experiments of poly-THEMA and poly-HEMA), has given rise to a good prediction of its very low release rate. According to this model, the initial release rate is governed by the initial swelling. The continuous increase in hydrophilicity and water concentration inside the matrix are responsible for the deviations from the reference first-order kinetics. The initial and the final swelling of the homopolymer are comparatively very low and the theoretically predicted release is almost negligible, in good agreement with the experimental data.

In conclusion, the composition of the copolymeric polymeric drug THEMA-DMA has been optimized in order to get a formulation that becomes water-soluble after the completion of the hydrolysis. THDMA39, a copolymer with a THEMA molar content of 39%, fits this requirement. Moreover, this system has exhibited (*in vitro* at pH 7.4) a pseudo-zero-order release profile during several months, which is very attractive for its application in long-term anti-thrombogenic treatment. The reported release data have fitted adequately a kinetic model based on the continuous water uptake of the system. *In vivo* studies are in progress.

The kinetic analysis considered in this work can be extended and applied to those polymeric systems, designed as supports for specific drugs linked covalently to the macromolecular chains, that during the release process overcome a drastic change on solubility, being very sensitive to the chemical microstructure of the polymeric chains.

Acknowledgements

Financial support from the J. Uriach & Cia group and from the grant Mat2001-1618, are gratefully acknowledged. M. F. thanks Comunidad de Madrid (Spain) for providing a post-doctoral grant.

REFERENCES

1. P. Guiteras, J. Altimiras, A. Aris, J. M. Auge, T. Bassons, J. Bonal, J. M. Casalps, C. Castellarnau, C. Crexells, M. Masotti, A. Oriol, J. M. Padró and M. Rutllant, *Eur. Heart J.* **10**, 159 (1989).
2. J. P. de la Cruz, M. A. Villalobos, P. J. García, J. M. Smith-Agreda and F. Sánchez de la Cuesta, *Eur. J. Clin. Pharmacol.* **47**, 497 (1995).
3. G. Rodriguez, A. Gallardo, J. San Román, M. Rebuelta, P. Bermejo, J. Buján, J. M. Bellón, N. Garcia Honduvilla and C. Escudero, *J. Mater. Sci. Mater. Med.* **10**, 873 (1999).
4. M. R. Aguilar, G. Rodríguez, M. M. Fernández, A. Gallardo and J. San Román, *J. Mater. Sci. Mater. Med.* **13**, 1099 (2002).
5. A. Gallardo, G. Rodriguez, M. R. Aguilar, M. Fernández and J. San Román, *Macromolecules* **36**, 8876 (2003).
6. A. A. Alencar de Queiroz, A. Gallardo, J. San Román and O. Z. Higa, *J. Biomater. Sci. Polymer Edn* **7**, 523 (1995).
7. A. A. Alencar de Queiroz, S. C. Castro and O. Z. Higa, *J. Biomater. Sci. Polymer Edn* **8**, 335 (1997).
8. L. W. Seymour, in: *Polymeric Biomaterials*, S. Dumitriu (Ed.), p. 843. Marcel Dekker, New York, NY (2002).

Gels, Genes, Grafts and Giants, pp. 105–116 (2005)

Promotion of U937 cell adhesion on polypropylene surfaces bearing phosphorylcholine functionalities

YONG-KUAN GONG [1], FACKSON MWALE [2], MICHAEL R. WERTHEIMER [3] and FRANCOISE M. WINNIK [1,*]

[1] *Department of Chemistry and Faculty of Pharmacy, University of Montreal, CP6128 Succursale Centre Ville, Montréal, Québec, Canada H3C 3J7*

[2] *Department of Engineering Physics, Ecole Polytechnique, CP 6079 Succursale Centre Ville, Montréal, Québec, Canada H3C 3A7*

[3] *Lady Davis Institute, 3755 Chemin de la Cote Ste Catherine, Montreal, Québec, Canada*

Received 20 November 2003; accepted 5 April 2004

Abstract—Phosphorylcholine (PC) groups were grafted onto ammonia plasma-treated biaxially-oriented polypropylene (BOPP) surfaces, *via* (a) reductive amination of phosphorylcholine glyceraldehyde and (b) a two-step procedure involving the chemical amplification of surface amine groups with tris(2-aminoethyl amine) and subsequent reductive amination of phosphorylcholine glyceraldehyde. The occurrence of grafting was ascertained by X-ray photoelectron spectroscopy (XPS) and attenuated total reflectance Fourier-transform infrared (ATR–FT-IR) spectroscopy. The wettability of PC-modified surfaces was assessed by dynamic contact-angle measurements using the Wilhelmy plate method. Human U937 macrophages adhered and proliferated to a significantly larger extent on PC-modified surfaces, compared to unmodified or ammonia plasma-modified BOPP.

Key words: Phosphorylcholine; reductive amination; human macrophage adhesion; plasma treatment.

INTRODUCTION

There is no doubt that total hip replacement surgery (arthroplasty) is a quality-of-life enhancing and cost effective surgical procedure. Nonetheless, survival analysis has revealed a failure rate of 1% per year for the first 10 years after surgery, and up to 5% annually thereafter. Aseptic loosening due to bone destruction around the prosthesis has emerged as the major cause of failure. Clinical observations suggest that macrophages are involved in the pathogenesis of this process, but the exact mechanism by which these inflammatory cells respond to wear particles, as

*To whom correspondence should be addressed. Tel.: (1-514) 340-5179. Fax: (1-514) 340-3245. E-mail: francoise.winnik@umontreal.ca

well as their interactions with bone cells remain poorly understood [1]. To address this issue, several laboratories have developed animal co-culture models. However, no human co-culture model exists yet. The human U937 macrophage cell line was shown to respond in a manner similar to primary cultured macrophages to metal ions [2] and polyethylene particles [3]. However, these macrophages remain in suspension and do not adhere to culture dishes currently available. This is a limitation because when particles are added to a cell suspension, they tend to sink to the bottom of the well making them unlikely to interact effectively with macrophages. In view of our interest in understanding the mechanisms involved in periprosthetic osteolysis, we have explored various chemical modifications of cell substrates, in order to have access to surfaces onto which macrophages can adhere.

The phosphorylcholine (PC) moiety, which is the major head group of the outer cell membrane phospholipids, has become the focus of increasing scrutiny as a component of biomaterials in various applications, in view of its excellent biocompatibility [4–7], low affinity towards protein adsorption [8–10] and platelet adhesion [11]. In most applications, the PC groups are linked to the structural polymer, used either alone or blended with other polymers. Since our ultimate objective was to devise a surface modification method that can be carried out on available cell-culture dishes, we selected to probe processes allowing covalent linkage of PC groups onto a polymer surface.

There have been a few reports on the preparation of PC-modified surfaces, either via graft polymerization of 2-(methacryloyloxy)ethyl phosphorylcholine [12], by photochemical reaction of PC-containing phenylazido groups [13, 14], by glow discharge in the presence of PC [15], by multi-step conversion of surface-bound hydroxyl groups to phosphorylcholine derivatives [10, 16, 17], or *via* polymerization of surface-adsorbed polymerizable phospholipids [18]. We developed recently a synthetic method to convert polyamines to PC-polymers *via* reductive amination of phosphorylcholine glyceraldehyde with amine groups linked to the polymer chain [19, 20]. The reaction occurs under mild conditions starting from readily available reagents and, when it is carried out in solution, it is nearly quantitative. We set about to apply the same methodology to prepare PC-modified surfaces by reacting phosphorylcholine glyceraldehyde with primary amine groups introduced on polymer surfaces via ammonia plasma treatment [21–24].

We describe here the preparation of surface-modified polypropylene bearing PC groups *via* a two-step procedure which involves modification of ammonia plasma-modified BOPP *via* (1) chemical amplification of the surface amine groups and (2) grafting of PC groups by reductive amination of PC-glyceraldehyde. The physico-chemical properties of the modified surfaces were characterized by X-ray photoelectron spectroscopy (XPS), attenuated-total-reflectance Fourier-transform infrared (ATR–FT-IR) spectroscopy and dynamic contact angle (DCA) measurements. The adhesion and growth of U937 cells on PC-modified surfaces were monitored by phase-contrast microscopy.

MATERIALS AND METHODS

Materials

Biaxially-oriented polypropylene (BOPP, isotactic, thickness 50 μm) was obtained from 3M. L-α-Glycerophosphorylcholine, cadmium chloride, sodium periodate, sodium cyanoborohydride, tris(2-aminoethyl)amine (96%), pentafluorobenzaldehyde and 1,6-diisocyanatohexane were purchased from Aldrich. Methanol and acetone (anhydrous grade) were dried over 4A molecular sieves prior to use. Phosphorylcholine glyceraldehyde was synthesized and characterized as described previously [19, 20]. All other reagents were commercially available and used as received.

Ammonia plasma treatment of PP surfaces

BOPP samples (1 $\times$ 1 cm^2) were placed in a cylindrical aluminum/steel chamber, 20 cm in diameter and 20 cm in height. Ammonia gas was admitted into the evacuated (0.1 mTorr) plasma chamber via electronic flow meters and controller (Vacuum General) at a flow of 10 standard cm^3 min^{-1} (sccm). A butterfly throttle valve allowed precise control of the pressure in the chamber (300 mTorr). Plasma was initiated and sustained by a radio-frequency (13.56 MHz) source capacitively coupled to a circular electrode located in the center of the chamber, the walls of the chamber being at ground potential. The BOPP specimens on the electrode were exposed to ammonia plasma for various periods of time (5 s to 240 s) with an input power of approx. 20 W. Chemical derivatization of ammonia plasma-treated surfaces, carried out for characterization, was achieved by keeping freshly-treated films in a solution of pentafluorobenzaldehyde in ethanol for a period of 2 h. The modified surfaces washed under a flow of ethanol, methanol and finally acetone, and were air-dried prior to analysis.

Phosphorylcholine grafting via reductive amination

Phosphorylcholine glyceraldehyde was reacted with ammonia plasma treated BOPP surfaces, *via* slight modifications of the method reported previously [20], as shown schematically in Fig. 1. Briefly, several ammonia plasma-modified BOPP samples were immersed in a solution of phosphorylcholine glyceraldehyde (0.250 g) in methanol (20 ml). They were kept in the solution at room temperature for 3 h under gentle stirring. The solution containing the samples was cooled to 0°C and a solution of sodium cyanoborohydride (0.350 g) in methanol (10 ml) was added dropwise. At the end of the addition, the reaction mixture was warmed to room temperature and maintained under gentle shaking for 20 h. The films were washed thrice under a flow of methanol and dried *in vacuo* at room temperature for 24 h.

Surface amines amplification

Ammonia plasma treated BOPP films were immersed in diisocyanatohexane (5 ml) and subjected to shaking for 1 h at room temperature. The films were washed

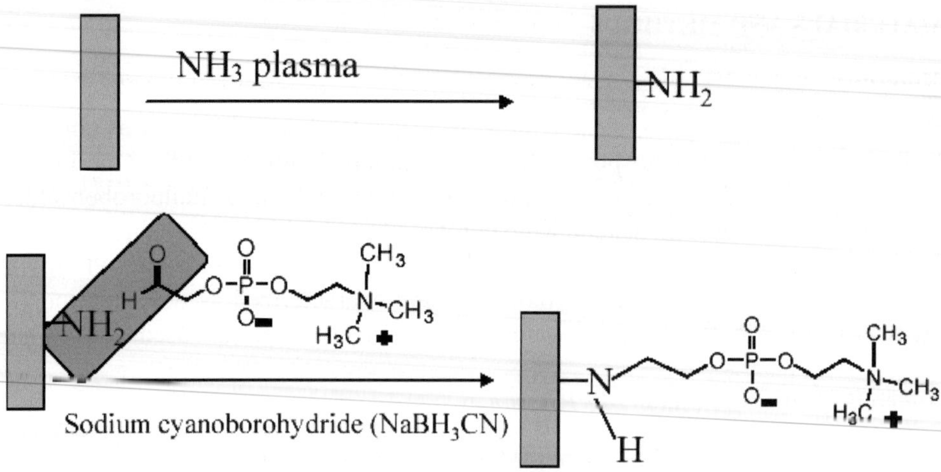

Figure 1. Schematic representation of the reductive amination of PC glyceraldehyde by surface-grafted primary amine groups.

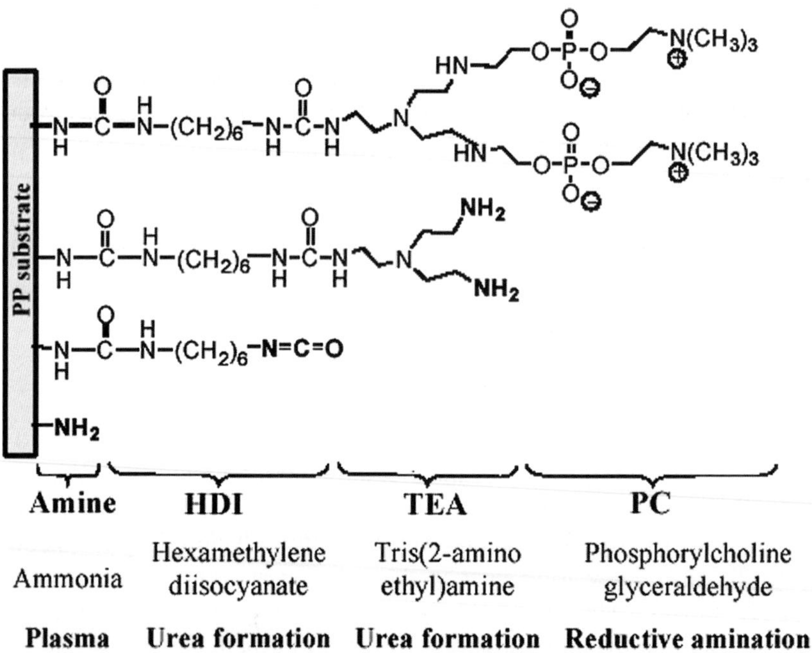

Figure 2. Schematic representation of the amine amplification/PC reductive amination sequence.

thrice first with anhydrous hexane, then with anhydrous acetone. The air-dried films were immersed in a solution of tris(2-aminoethyl)amine in acetone (8%, v/v) and shaken for 6 h at room temperature. They were washed first with acetone, then with methanol. They were air-dried and treated with phosphorylcholine glyceraldehyde as described above. The step-by-step surface modifications are illustrated in Fig. 2.

Characterization of the modified surfaces

The elemental composition of the surfaces was investigated by X-ray Photoelectron Spectroscopy (XPS) using an ESCALAB 3MkII (V. G. Scientific) system equipped with a Mg Kα radiation source operated at 12 kV and a 18 mA anode current. All spectra were collected at an electron take-off angle of 90°. The C_{1s} hydrocarbon peak, used as a reference, was set at 285.0 eV. Attenuated Total Reflectance Fourier-Transform Infrared (ATR–FT-IR) spectra were recorded on a Bio-Rad FTS-3000 Excalibur Series instrument equipped with a deuterated triglycine sulfate (DTGS) detector. The spectra were recorded at a 45° incident angle using a Ge crystal with a 45° bottom angle. The spectra were obtained from 256 consecutive scans with a resolution of 4 cm^{-1}.

Dynamic contact angle (DCA) measurements using the Wilhelmy plate method

DCA loops were obtained with a DCAT 11 tensiometer from Dataphysics Instruments. The technique, which is based on the Wilhelmy plate method, requires samples presenting identical front and back surfaces. These were obtained by gluing the backsides of two identical BOPP surfaces (approx. 2 $\times$ 2 cm^2) and cutting the resulting samples into 1 $\times$ 1 cm^2 specimens. In each case, a minimum of three specimens were analyzed to ensure reproducibility of the measurements. Deionized water (Millipore system, 18 MΩ cm) was used as probe liquid. The water surface tension was determined daily before and after the analysis of BOPP surfaces. Typical values were 72.6$\pm$0.1 dyn cm^{-1}. A dry specimen was attached to the electrobalance *via* a clip and the stage carrying a beaker of water was automatically raised and lowered (10 mm min^{-1}) to allow water to impinge upon the specimen. A dwell time of 60 s was allowed between sample immersion and withdrawal. Analysis of the resulting force *versus* distance (fdc) curves yields the advancing and receding angles, θ_{adv} and θ_{rec}, respectively. Two consecutive fdc curves were recorded for each sample. The absence of contaminants was ascertained by frequent measurements of the water surface tension, which should remain constant [25, 26].

Culture of U937 macrophages on PC-modified BOPP surfaces

PC-modified BOPP films (1 $\times$ 1 cm^2) were placed face-up on the flat bottom of a 24-well plate for cell culture. U937 cells (human macrophages, ATCC) were maintained in a RPMI supplemented with 10% FBS, 100 U ml^{-1} penicillin and 100 μg ml^{-1} streptomycin at 37°C in air containing 5% CO_2. The U937 macrophages (5$\times$10^5 cells ml^{-1} in RPMI 1640 supplemented with 10% FBS) were seeded in the 24 wells and incubated for 24 h at 37°C. At the end of the incubation time, the wells were rinsed with buffered saline to remove non-adherent cells. The morphology of the adhered cells was observed using a phase-contrast microscope. The same assay was performed on untreated and ammonia plasma-treated BOPP films, used as control substrates.

Table 1.
XPS elemental composition of various modified surfaces

Ammonia plasma exposure time (s)	BOPP-NH_2			BOPP-NH_2-PC				BOPP-DIH-PC			
	N_{1s}	O_{1s}	C_{1s}	P_{2p}	N_{1s}	O_{1s}	C_{1s}	P_{2p}	N_{1s}	O_{1s}	C_{1s}
0	0	0.6	99.4	0	0	1.11	98.9	0	0	1.11	88.9
5	9.0	6.6	84.4	0.15	0.07	4.56	95.2	0.48	0.07	4.56	94.9
10	11.4	8.0	80.6	0.30	1.75	7.98	90.0	0.52	1.75	7.98	89.7
20	14.6	8.0	77.4	0.36	1.48	7.40	90.8	0.73	0.48	7.40	90.4
40	14.4	7.9	77.7	0.54	2.13	8.86	88.5	0.53	2.13	8.86	88.5
240	17.0	6.6	76.4	0.44	1.27	7.57	90.7	0.52	1.27	7.57	90.6

RESULTS AND DISCUSSION

Surface composition of ammonia plasma-treated PP films

BOPP samples were subjected to ammonia plasma under standard conditions for various lengths of time in order to determine the optimal conditions for the grafting of primary amine groups. The overall surface elemental composition of the ammonia plasma treated BOP films was determined by XPS (Table 1). The total nitrogen surface concentration increases rapidly with increasing exposure time up to 20 s and levels off with further plasma treatment. Even though the ammonia used for the plasma was ultra-high purity grade, we note (second column of Table 1) the presence of up to 8% oxygen on the plasma-modified BOPP surface. This well-known effect is due to bond scissions in the polymer by high-energy photons and particle bombardment during plasma exposure, followed by reaction of the resulting free radical species near the surface with atmospheric oxygen and water vapor upon exposure of the specimen to the atmosphere [21, 27]. Ammonia plasma treatment produces primary amine groups on polymer surfaces [18, 21, 22, 27–29], as well as a number of other aminated functional groups. Under most circumstances, only 15–20% of the total surface nitrogen content correspond to primary amines [27]. In our hands, the surface primary amine composition, determined by XPS analysis of films treated with pentafluorobenzaldehyde, ranged from 15 to 22% of the total nitrogen surface content. In other words, we found a maximum amine content of approx. 3% NH_2/C, as reported also by Meyer-Plath and co-workers [27, 28].

Surface composition of PC-modified BOPP films

We first attempted to link PC groups to aminated surfaces obtained by ammonia plasma treatment using reductive amination of PC-glyceraldehyde (Fig. 1). The outcome of the surface modification was monitored by XPS measurements (Figs 3 and 4). The occurrence of PC grafting was confirmed by the presence of a P_{2p} signal at 133.0 eV in the XPS survey spectrum, acquired at a 60° take-off angle. The intensity of the P_{2p} signal increases with increasing ammonia plasma exposure time of virgin BOPP, but it remains weak compared to the signals due to the other

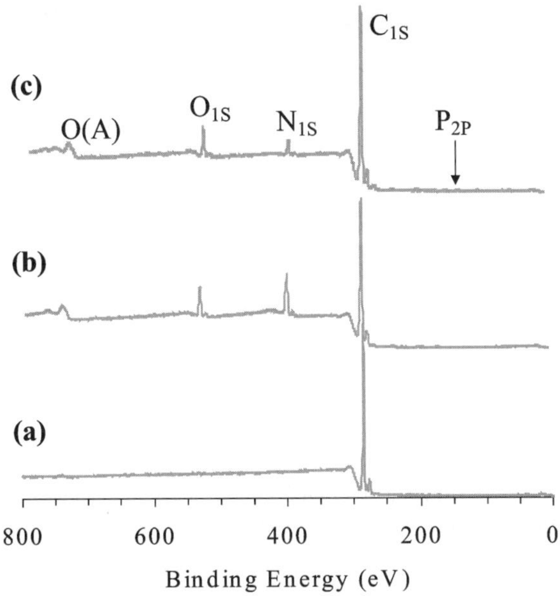

Figure 3. XPS survey spectra of (a) a BOPP sample, (b) an ammonia-plasma-treated BOPP sample and (c) a PC-grafted sample.

elements detected, O_{1s}, N_{1s} and C_{1s} (Table 1). We attribute this to the relatively low grafting yield, as anticipated given the rather low primary amine content of the starting aminated surfaces. Nonetheless, these results, corroborated by the high energy tail of the N_{1s} peak centered 400.5 eV observed in the high-resolution XPS trace of a PC-modified sample (Fig. 4) and reflecting the contribution of the phosphorylcholine ammonium nitrogen ($N^+(CH_3)_3$), confirmed that the reductive amination of PC-glyceraldehyde with surface-linked amine groups is a viable method to prepare PC-grafted samples.

We set about to increase the grafting yield by increasing the surface concentration of primary amine groups by a surface amine amplification technique depicted in Fig. 2. Thus, a diisocyanatohexane (DIH) linking unit was attached onto surface-linked primary amine and further reacted with tris(2-aminoethyl)amine. Theoretically, this reaction scheme doubles the surface amine group concentration, and consequently the surface phosphorous content of amine-amplified surfaces further treated with PC-glyceraldehyde. Indeed, the phosphorous surface concentration increased in these samples, compared to samples resulting from direct PC-modification of aminated surfaces (Table 1). We note, however, that this amplification effect vanishes in samples initially treated with ammonia plasma for more than 20 s. The phosphorous content nearly doubles after the amine amplification for samples with initially low nitrogen content. It decreases, however, for samples having an initial nitrogen content higher than approx. 15% N/C. Considering the amine amplification sequence of reactions, we propose the following mechanism to account for the existence of an optimal nitrogen content (Fig. 2). Treatment with

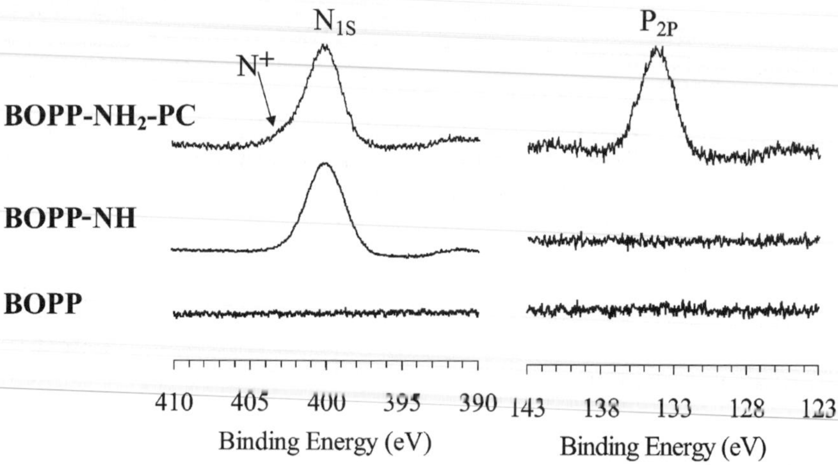

Figure 4. High-resolution XPS spectra of a BOPP sample, an ammonia plasma-treated BOPP sample and a PC-grafted sample.

diisocyanatohexane of samples of low nitrogen content results in the creation of a surface containing isocyanate groups sufficiently separated in space from each other to allow monoamination, that is, the reaction of only one amine group of tris(2-aminoethyl)amine, leaving two primary amine groups available for further grafting. In contrast, when the surface density of isocyanate groups increases beyond a certain value, the isocyanates are close enough in space to allow reaction of one tris(2-aminoethyl)amine with two adjacent −NCO groups, leaving only one primary group available for further reaction, negating the desired amplification effect (Fig. 5).

The samples were analyzed also by ATR–FT-IR spectroscopy in order to ascertain the success of the reductive amination of PC-glyceraldehyde onto the various aminated surfaces. Presented in Fig. 6 are sections of the ATR–FT-IR spectra recorded for an aminated film (bottom spectrum), a film treated with dicyanatohexane (DIH, middle spectrum) and a PC-modified sample obtained after amine amplification (top spectrum). All the spectra presented in Fig. 6 were obtained after subtraction of the ATR–FT-IR spectrum of virgin BOPP. Of particular significance is the presence, in the spectrum of the DIH-modified film, of a strong band at 2270 cm^{-1}, characteristic of the stretching vibration of the cumulated −N=C=O double bonds [30]. The lower wavenumber section of the spectrum of the PC-modified film displays evidence of the presence of phosphorylcholine groups. The two peaks at 1041 and 1076 cm^{-1} can be attributed to the symmetric O=P−O stretch, as it is known that the P=O band can often appear as a doublet [30]. Note that the C−O stretching also appears in this area at approx. 1060 cm^{-1}. A peak at 1207 cm^{-1}, ascribed to the asymmetric stretching of O=P−O, together with the band at 1450 cm^{-1}, attributed to the CH_3 bending from $N^+(CH_3)_3$, further confirm the successful grafting of PC groups.

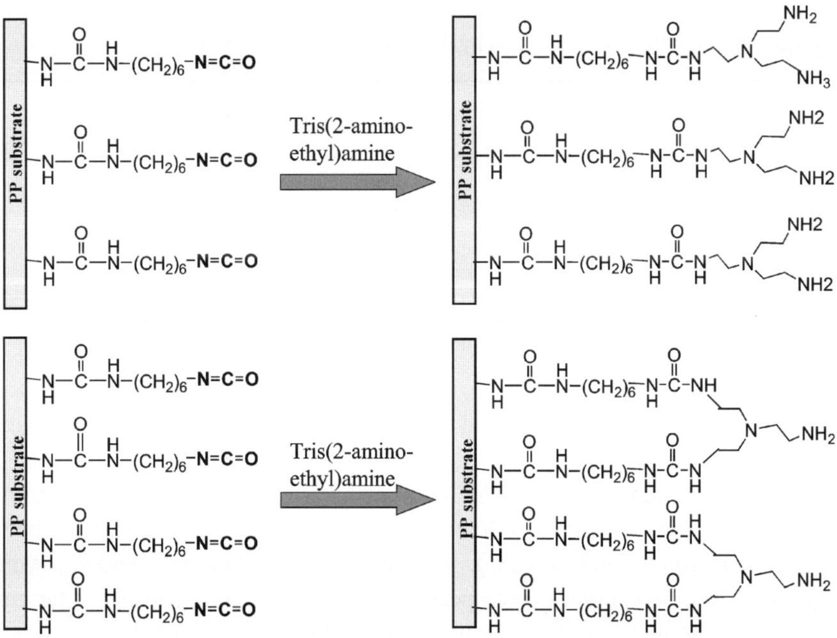

Figure 5. Schematic representation of the reaction of tris(2-aminoethyl)amine with surfaces bearing a low concentration of isocyanate groups (top) and a high concentration of isocyanate groups (bottom).

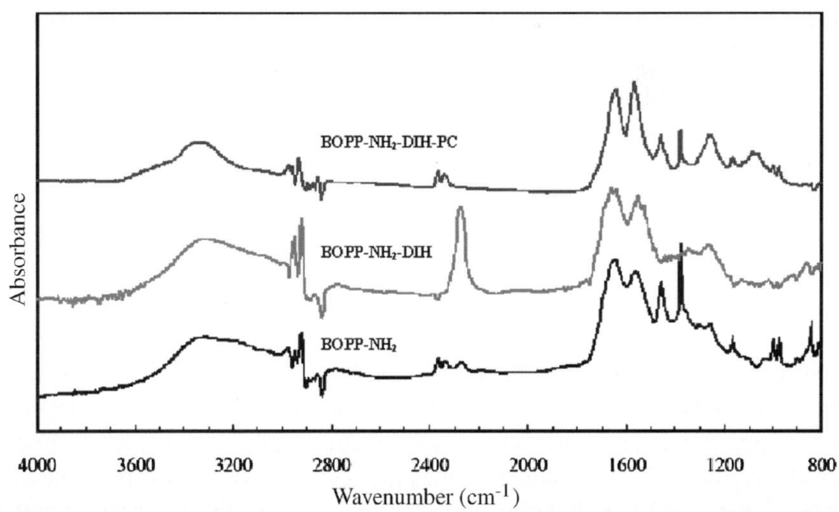

Figure 6. ATR–FT-IR spectra of an ammonia-plasma-treated BOPP sample (PP-NH_2), a DIH-modified sample (PP-NH_2-DIH) and a PC-grafted sample -NH_2-DIH-PC). The spectra were obtained after subtraction of the ATR–FT-IR spectrum of BOPP measured under identical conditions.

Dynamic contact angle (DCA) measurements

Advancing and receding contact angles were determined by DCA measurements for virgin BOPP, and for the modified films at each stage of the amine amplification

Table 2.
Water contact angles of the films measured by the Wilhelmy plate method

Sample [a]	Advancing contact angle (θ_{adv}, deg)	Receding contact angle (θ_{rec}, deg)	$\Delta\theta = \theta_{adv} - \theta_{rec}$ (deg)
BOPP	110 ± 2	82 ± 2	28
BOPP-NH_2	85 ± 1	43 ± 3	42
BOPP-NH_2-PC	88 ± 1	44 ± 2	44
BOPP-DIH-NH_2	87 ± 2	41 ± 2	46
BOPP-DIH-PC	93 ± 2	27 ± 2	66

[a] See Fig. 5 for the chemical structure of the surfaces of the various samples.

sequence, leading to the PC-modified surfaces (Table 2). We note that all samples exhibit contact-angle hysteresis $\Delta\theta$, defined as the difference between the advancing and receding contact angles. The hysteresis varies greatly among the samples, the highest value (66°) being recorded for the PC-modified surfaces, while the value of virgin BOPP was the lowest (28°). We confirmed that both θ_{adv} and θ_{rec} were the same for several consecutive cycles performed with the same sample. The contact angle hysteresis reflects, on the one hand the surface chemical heterogeneity [31], and on the other hand reorganization of surface functionalities as the sample is hydrated.

Macrophage adhesion to PC-grafted BOPP substrates: possible mechanism

Human U937 macrophages have long been used for studying periprosthetic osteolysis [2, 32, 33], although their use in co-culture studies is limited because they remain in suspension and do not adhere to cell culture dishes currently available [32, 34]. Here we explored the adhesion of these cells to PC-grafted BOPP substrates with the aim of finding a substrate to which they can adhere. The micrographs presented in Fig. 7 show that human U937 macrophages became adherent to PC-grafted surfaces after 24 h of culture. They did not adhere on unmodified BOPP surfaces and minimally attached to ammonia-plasma modified BOPP. These culture studies were performed on at least three different occasions, producing essentially the same results. It was unexpected that human U937 macrophages adhered to PC-grafted surfaces, since previous reports have shown reduced adhesion of cells to polymers by introducing phosphorylcholine moieties [35–37]. However, it is possible that the adhesion is due to the fact that phosphorylcholine mimics the outer cell membrane. Interestingly, mutating the choline binding protein, which belongs to a family of surface proteins non-covalently bound to the phosphorylcholine moiety of the cell wall of *Streptococcus pneumoniae* by a conserved choline-binding domain, led to reduced adhesion to eukaryotic cells suggesting that this might be a mechanism for cell–PC binding [38]. However, the effect of disrupting PC-binding proteins on U937 macrophages adhesion to PC-grafted surfaces is yet to be established. It should be noted as well that human monocytes/macrophages have been shown to

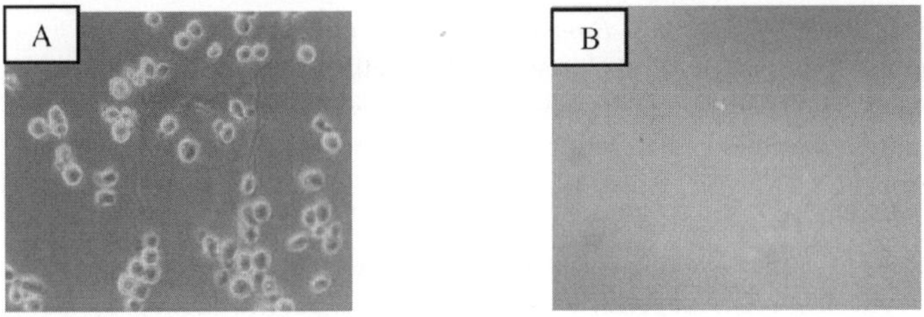

Figure 7. (A) Morphology of U937 cells cultured on a PC-grafted BOPP substrate; cells were incubated at 37°C for 24 h. (B) View of a BOPP surface placed in contact with U937 cells under identical conditions.

adhere onto surfaces coated with hexadecylphosphocholine and that such surfaces upregulate the intracellular adhesion molecule-1 and class-1 major histocompatibility complex antigen [39]. Thus, adhesion mechanisms exist whereby surfaces bearing PC functionalities interact with macrophages. Whether the process of cell adhesion with PC-grafted BOPP surface involves the same mechanisms that are observed with surfaces coated with a phospholipid remains to be determined and experiments are in progress in order to elucidate the interaction mechanisms.

CONCLUSIONS

Phosphorylcholine glyceraldehyde was grafted onto ammonia plasma treated BOPP surface by reductive amination. The surface amine groups was amplified by coupling tris(2-aminoethyl)amine with the linker of 1,6-diisocyanatohexane. The success of the surface modification is supported by evidence gathered from XPS, ATR–FT-IR and DCA measurements. We demonstrate that the PC-rich surfaces created by this procedure provide a favorable substrate for the adhesion and proliferation of human U937 macrophages.

Acknowledgements

This article is dedicated to Allan Hoffman in honor of his 70th birthday. The work was supported by NanoQuébec and by the Natural Sciences and Engineering Research Council of Canada. The authors gratefully acknowledge the expert technical support provided by Dr. Sophie Lerouge, as well as help offered by Dr. Nial A. Bullett (plasma treatment) and Ms. Florina Truica-Marasescu (ATR–FT-IR measurement).

REFERENCES

1. S. B. Goodman, *J. Long-Term Effect. Med. Implants* **6**, 91 (1996).
2. J. Y. Wang, B. H. Wicklund, R. B. Gustilo and D. T. Tsukayama, *Biomaterials* **17**, 2233 (1996).
3. J. Matthews, T. Green, M. Stone, B. Wroblewski, J. Fisher and E. Ingham, *J. Mater. Sci. Mater. Med.* **12**, 249 (2001).

4. J. A. Hayward and D. Chapman, *Biomaterials* **5**, 135 (1984).
5. N. Morimoto, Y. Iwasaki, N. Nakabayashi and K. Ishihara, *Biomaterials* **23**, 4881 (2002).
6. Y. Iwasaki, N. Nakabayashi and K. Ishihara, *J. Biomed. Mater. Res.* **57**, 74 (2001).
7. T. Hasegawa, Y. Iwasaki and K. Ishihara, *J. Biomed. Mater. Res.* **63**, 333 (2002).
8. E. J. Campbell, V. O'Byrne, P. W. Stratford, I. Quirk, T. Vick, M. C. Wiles and Y. P. Yianni, *Am. Soc. Artif. Intern. Organs J.* **40**, M853 (1994).
9. K. Ishihara, T. Hasegawa and J. Watanabe, *Artif. Organs* **26**, 1017 (2002).
10. V. A. Tegoulia and S. L. Cooper, *J. Biomed. Mater. Res.* **50**, 291 (2000).
11. K. Ishihara, Y. Iwasaki and N. Nakabayashi, *Mater. Sci. Eng.* **C6**, 253 (1998).
12. A. Korematsu, Y. Takemoto, T. Nakaya and H. Inoue, *Biomaterials* **23**, 263 (2002).
13. A. P. van der Heiden, G. M. Willems, T. Lindhout, A. P. Pijpers and L. H. Koole, *J. Biomed. Mater. Res.* **40**, 195 (1998).
14. A. P. van der Heiden, D. Goebbels, A. P. Pijpers and L. H. Koole, *J. Biomed. Mater. Res.* **37**, 282 (1997).
15. M. C. Sunny and C. P. Sharma, *Biomater. Artif. Cells Immobil. Biotechnol.* **19**, 599 (1991).
16. K. G. Mara, T. M. Winger, S. R. Hanson and E. L. Chanikof, *Macromolecules* **30**, 6483 (1997).
17. P. Chevallier, D. Mantovani and G. Laroche, in: *Advanced Materials for Biomedical Applications, Proceedings of the International Symposium on Advanced Materials for Biomedical Applications*, Montreal, Québec, August 11–14, 2002, p. 269 (2002).
18. J. R. Lu, E. F. Murphy, T. J. Su, A. L. Lewis, P. W. Stratford and S. K. Satija, *Langmuir* **17**, 3382 (2001).
19. K. Miyazawa and F. M. Winnik, *Macromolecules* **35**, 2440 (2002).
20. K. Miyazawa and F. M. Winnik, *Macromolecules* **35**, 9536 (2002).
21. J. E. Klemberg-Sapieha, O. M. Küttel, L. Martinu and M. R. Wertheimer, *J. Vac. Sci. Technol. A* **9**, 2975 (1991).
22. T. K. Markkula, J. A. Hunt, F. R. Pu and R. L. Williams, *Surf. Interface Anal.* **34**, 583 (2002).
23. J. Yang, J. Bei and S. Wang, *Polym. Adv. Technol.* **13**, 220 (2002).
24. P. Chevallier, M. Castonguay, S. Turgeon, N. Dubrulle, D. Mantovani, P. H. McBreen, J.-C. Wittmann and G. Laroche, *J. Phys. Chem. B* **105**, 12490 (2001).
25. K. G. Tingey and J. D. Andrade, *Langmuir* **7**, 2471 (1991).
26. J. M. Uilk, A. E. Mera, R. B. Fox and K. J. Wynne, *Macromolecules* **36**, 3689 (2003).
27. A. A. Meyer-Plath, K. Schröder, B. Finke and A. Ohl, *Vacuum* **71**, 391 (2003).
28. A. A. Meyer-Plath, B. Finke, K. Schröder and A. Ohl, *Surface Coating Technol.* **174–175**, 877 (2003).
29. Y. Nakayama, T. Takahagi and F. Soeda, *J. Polym. Sci. A Polym. Chem.* **26**, 559 (1988).
30. L. J. Bellamy, *The Infrared Spectra of Complex Molecules*. Chapman and Hall, London (1980).
31. R. E. Johnson and R. H. Dettre, *J. Phys. Chem.* **68**, 1744 (1964).
32. J. W. Burgess, R. S. Kiss, H. Zheng, S. Zachariah and Y. L. Marcel, *J. Biol. Chem.* **277**, 31318 (2002).
33. E. Ingham, T. R. Green, M. H. Stone, R. Kowalski, N. Watkins and J. Fisher, *Biomaterials* **21**, 1005 (2000).
34. T. Hoff, T. Spencker, A. Emmendoerffer and M. Goppelt-Struebe, *J. Leukoc. Biol.* **52**, 173 (1992).
35. K. M. DeFife, J. K. Yun, A. Azeez, S. Stack, K. Ishihara, N. Nakabayashi, E. Colton and J. M. Anderson, *J. Biomed. Mater. Res.* **29**, 43 (1995).
36. K. Ishihara, E. Ishikawa, Y. Iwasaki and N. Nakabayashi, *J. Biomater. Sci. Polymer Edn* **10**, 1047 (1999).
37. Y. Iwasaki, Y. Tojo, T. Kurosaki and N. Nakabayashi, *J. Biomed. Mater. Res.* **65A**, 164 (2003).
38. K. K. Gosink, E. R. Mann, C. Guglielmo, E. I. Tuomanen and H. R. Masure, *Infect. Immun.* **68**, 5690 (2000).
39. I. Eue, *J. Exp. Ther. Oncol.* **2**, 333 (2002).

Gels, Genes, Grafts and Giants, pp. 117–129 (2005)

Controlled release of plasmid DNA from biodegradable scaffolds fabricated using a thermally-induced phase-separation method

KI WOO CHUN, KYUNG CHUL CHO, SUN HWA KIM, JI HOON JEONG and TAE GWAN PARK *

Department of Biological Sciences, Korea Advanced Institute of Science and Technology, Daejeon 305-701, South Korea

Received 1 October 2003; accepted 8 January 2004

Abstract—Highly porous poly(D,L-lactic-co-glycolic acid) (PLGA) scaffolds were fabricated by a thermally-induced phase-separation (TIPS) method to deliver plasmid DNA in a controlled manner. A variety of TIPS parameters directly affecting pore structures and their interconnectivities of the scaffold, such as polymer concentration, solvent/non-solvent ratio, quenching methods and annealing time, were systematically examined to explore their effects on sustained release behaviors of plasmid DNA. Plasmid DNA was directly loaded into the inner pore region of the scaffold during the TIPS process. By optimizing the parameters, PLGA scaffolds releasing plasmid DNA over 21 days were successfully fabricated. DNA release profiles were mainly affected by the pore structures and their interconnectivities of the scaffolds. Plasmid DNA released from the scaffolds fully maintained its structural integrity and showed comparable transfection efficiency to native plasmid DNA. These biodegradable polymeric scaffolds capable of sustained DNA release can be potentially applied for various tissue engineering purposes requiring a combined gene delivery strategy.

Key words: Porous scaffold; plasmid DNA; sustained release; biodegradable polymer.

INTRODUCTION

A wide range of biodegradable scaffolds were fabricated to serve as templates for regeneration of various tissues. Biocompatible and biodegradable PLGA scaffolds have been prepared by using various methods, such as thermally induced phase separation [1–3], solvent casting/salt leaching, gas foaming/salt leaching [4] and expansion in supercritical fluid [5]. These polymeric scaffolds must have an

*To whom correspondence should be addressed. Tel.: (82-42) 869-2621. Fax: (82-42) 869-2610. E-mail: tgpark@kaist.ac.kr

open porous and well inter-connected three-dimensional structure for sufficient cell seeding and must provide a large inner pore surface area for cell attachment. In addition, a highly open porous structure is essentially required for facile transport of oxygen and nutrients to the seeded cells and for rapid removal of cell metabolic waste from the scaffolds. For successful tissue regeneration, various growth factors are also needed to stimulate cell proliferation, to induce angiogenesis and to guide cell differentiation. A diverse range of growth factors, such as vascular endothelial growth factor (VEGF) [6–8], platelet-derived growth factor (PDGF) [8, 9], epidermal growth factor (EGF) [10], nerve growth factors (NGF) and bone morphogenic proteins (BMPs), have been loaded within the inner pore region of PLGA scaffolds to achieve their controlled release in a local tissue defect area [11, 12]. Slowly released growth factors from the scaffolds resulted in far better tissue formation [8, 13]. PLGA microspheres encapsulating various growth factors were sometimes seeded along with cells into the porous scaffolds to attain sustained release. While the controlled release of growth factors from the PLGA scaffolds is an attractive and straightforward method, it has some serious problems that must be solved. First, it was difficult to control their release rate over an extended period required for tissue formation. Very often, high burst releases at an initial stage were reported due to the porous nature of the scaffolds [14, 15]. Second, their biological activities after the release could not be fully guaranteed due to their structural instability during harsh loading processes into the scaffolds [16, 17].

Naked plasmid DNA coding for a specific protein has been injected into a specific tissue to express the protein at the site [18, 19]. A single injection of naked plasmid DNA into cells at the target tissue results in appreciable protein expression in a detectable level, even though the transfection efficiency is low. To achieve sustainable and therapeutically effective protein expression, sustained delivery of DNA at the injection site is highly needed. Slowly released DNA from biodegradable scaffolds can provide an opportunity for continuous protein expression in the local tissue region [19, 20]. In this aspect, many porous polymeric matrices have been used to achieve sustained release of naked DNA. Poly(D,L-lactic-co-glycolic acid) (PLGA) scaffolds fabricated by gas foaming [8, 9], particulate leaching, electro-spinning [21] and fusing of microspheres [22] were used for sustained DNA delivery.

Thermally-induced phase separation (TIPS) is a useful method for fabricating polymeric foams and scaffolds. The main principle of the method is a homogeneous polymer solution attained at high temperature being phase-separated into polymer-rich and polymer-poor domains by decreasing temperature. Subsequent freeze-drying of the solvent gives microcellular structures. Uniformly porous and interconnected structure can be attained through the phase separation when the quenching temperature is located in the unstable area [23]. Previously, we have shown that various TIPS parameters, such as polymer concentration, solvent/non-solvent ratio, quenching method and annealing time, had great impacts on pore size distribution and their inter-connectivity of PLGA scaffolds. In particular, it was shown that the scaffold microstructure was highly dependent on quenching route and annealing time [24, 25].

In this study, DNA-releasing PLGA scaffolds were prepared by the TIPS process. The scaffolds prepared by the TIPS process have a versatile and controllable array of pore morphologies that can be judiciously exploited for sustained DNA release, in contrast to PLGA scaffolds prepared by other methods. Luciferase plasmid DNA was directly dissolved in a mixed solvent of dioxane/water in which PLGA was also dissolved. The mixture solution of DNA/PLGA/co-solvent was subjected to various quenching processes by changing TIPS parameters. The pore size and their morphologies of the scaffolds were examined with scanning electron microscopy (SEM). DNA release profiles from the resultant scaffolds were determined over 20 days. The structural integrity and gene transfection capacity of the released plasmid DNA were also investigated.

MATERIALS AND METHODS

Materials

Poly(D,L-lactic-co-glycolic acid) (RG756, molecular mass 100 kDa) having a lactic/glycolic molar ratio of 75 : 25 was obtained from Boehringer-Ingelheim (Ingelheim, Germany) and used for the fabrication of scaffolds. Plasmid DNA (pCMV-Luc) was purified from *Escherichia coli* DH5α using a Qiagen column (Qiagen, Boulder, CO, USA). The concentration and the purity of DNA were determined by using spectrophotometric method. Lipofectamine™ and all reagents for cell culture were purchased from Gibco-BRL (Grand Island, NY, USA). All other chemicals and reagents were of analytical grade.

Preparation of scaffolds

A series of DNA-loaded polymeric scaffolds with different processing conditions was fabricated by thermally-induced phase separation (TIPS) as described previously [1, 2].

Processing condition 1. PLGA polymer dissolved in 1,4-dioxane (5% (w/v) or 10% (w/v)) was mixed with DNA dissolved in deionized water at room temperature. The final concentration of DNA was 50 μg/ml. The solvent ratio of the co-solvent mixture (1,4-dioxane/distilled deionized water) was set to either 84 : 16 or 87 : 13 (v/v). The homogeneous mixture was poured into a metal cast for molding and quenched in liquid nitrogen or at −20°C for 24 h. Solvents were removed by freeze-drying.

Processing condition 2. To study the annealing time effect, PLGA polymer (10% (w/v)) was mixed with DNA solution at the solvent ratio of 87 : 13. The homogeneous mixture in a metal cast was maintained at −20°C for 4 h, 12 h, 24 h and 43 h, in order to induce the coarsening effect. Solvents were removed by freeze-drying.

Processing condition 3. PLGA-grafted poly(L-lysine) (PLL-g-PLGA) was used as an excipient to control release kinetics of DNA from the scaffold foam. PLL-g-PLGA (PLGA RG502H, molecular mass 14 kDa, Boehringer-Ingelheim; PLL, molecular mass 25 kDa, Sigma; 5.6% grafted) was synthesized as previously described by using DCC (dicyclohexyl carbodiimide)/NHS (*N*-hydroxyl succinimide) chemistry [26, 27]. Prior to blending with DNA solution, the desired amount of PLL-g-PLGA was added to PLGA polymer solution. The final concentration of PLL-g-PLGA was adjusted to 0.05%, 0.1%, 0.5%, 1% and 5% (w/w). 10% (w/v) PLGA polymer solution containing PLL-g-PLGA was mixed with DNA solution at the solvent ratio of 87 : 13. The polymer/DNA mixtures were kept at −20°C for 24 h and freeze-dried for solvent removal.

Scanning Electron Microscopy (SEM)

Scanning Electron Microscopy (SEM, Philips 535M) was used to observe the internal pore structure of the scaffolds. The morphology of fractured cross-section was observed after gold coating by using a sputter-coater (Hummer V, Technics, Stanford, CA, USA). Argon gas pressure was set at 5 psi and the current was maintained at 10 mA for 5 min.

DNA release measurements

The scaffold containing 100 μg of DNA was immersed in 1 ml of TE buffer (10 mM Tris-HCl, 1 mM EDTA, pH 7.4). The samples were incubated under mild shaking conditions at 37°C. The release medium was collected and replaced with fresh medium at specified time (6 h, 12 h, 1, 3, 7, 14 and 21 days). The amount of DNA was determined by PicoGreen® as fluorescent dye according to the manufacturer's recommendation (Molecular Probes, Eugene, OR, USA). The experiment was conducted in triplicate.

Agarose gel electrophoresis

The integrity of released DNA was evaluated by agarose (1.0%) gel electrophoresis. Electrophoresis was carried out with a current of 100 V for 20 min in TAE buffer solution (40 mM Tris/HCl, 1% (v/v) acetic acid, 1 mM EDTA). The DNA bands were visualized by ethidium bromide staining.

Cell culture and transfection

293 cells were purchased from Korea Cell Line Bank (Seoul, South Korea) and cultured in DMEM supplemented with 10% FBS, streptomycin at 100 μg/ml, penicillin at 100 IU/ml and 2 mM L-glutamine. Cells were maintained at 37°C in a humidified 5% CO_2 atmosphere. 2.5×10^5 cells per well were plated in a 6-well

plate (35 mm diameter, Nunc, Roskilde, Denmark) in 1.5 ml DMEM with 10% FBS 24 h before transfection. Transfection was carried out by using Lipofectamine™ according to the manufacturer's recommendation. To adjust the concentration of the samples collected at various time, the DNA fractions were pooled and concentrated using Centricon® (Amicon, Danvers, MA, USA). The concentration was confirmed by PicoGreen® assay and the same amount of DNA (1 μg) was used for each transfection. Luciferase gene expression was determined 48 h after transfection by using a commercial luciferase assay kit (Promega, Madison, WI, USA). The luciferase activity was monitored for 15 s in a Lumat LB 9501 luminometer (Berthold, Wilbach, Germany). The transfection efficiency was expressed as relative light units per mg of cell protein, of which the protein concentration was measured by using Micro-BCA assay (Pierce, Rockford, IL, USA).

RESULTS AND DISCUSSION

The TIPS process is based on a phase-separation phenomenon that a homogeneous polymer solution is separated into polymer-rich and polymer-poor phases by quenching at temperatures below a certain temperature boundary. When the temperature of a polymer solution is quenched deeply into an unstable zone below a spinodal curve, the phase separation occurs mainly by a spinodal decomposition mechanism. After a subsequent freeze drying process, a variety of porous polymer structures can be produced. The pore size and its inter-connection of the resultant scaffold are highly dependent on a slight change in various TIPS parameters including the molecular mass of the polymer, polymer concentration, co-solvent composition, quenching temperature and annealing time. In particular, the annealing time at a selected quenching temperature plays an important role in generating a macroporous structure by a coarsening effect in the later stage of the phase separation process [28].

Plasmid DNA is a hydrophilic macromolecule with a hydrodynamic molecular size as large as approx. 1 μm [29]. Once entrapped within the porous polymer scaffold, its release rate is expected to be largely influenced by a diffusion rate through aqueous fluid filled pores. In this study, PLGA with a lactic/glycolic ratio of 75 : 25 with a co-solvent mixture of dioxane/water was used for a liquid-liquid phase separation. Dioxane and water were chosen as a good and a poor solvent, respectively, for PLGA [1–3, 30]. As a preliminary study, three different formulations with two different polymer concentrations (5% and 10% (w/w)) and two different solvent/non-solvent ratios (84 : 16 and 87 : 13 (v/v)) were selected in order to pre-screen the optimized pore structures for DNA release. Plasmid DNA was directly added in the above co-solvent solution. The three different formulations were quenched at two different temperatures, liquid nitrogen and −20°C for 24 h. Figure 1 shows the internal porous structures of the completed six scaffolds after freeze-drying. It can be seen that the porosity and inter-connectivities are highly dependent on the quenching temperature rather than on the polymer

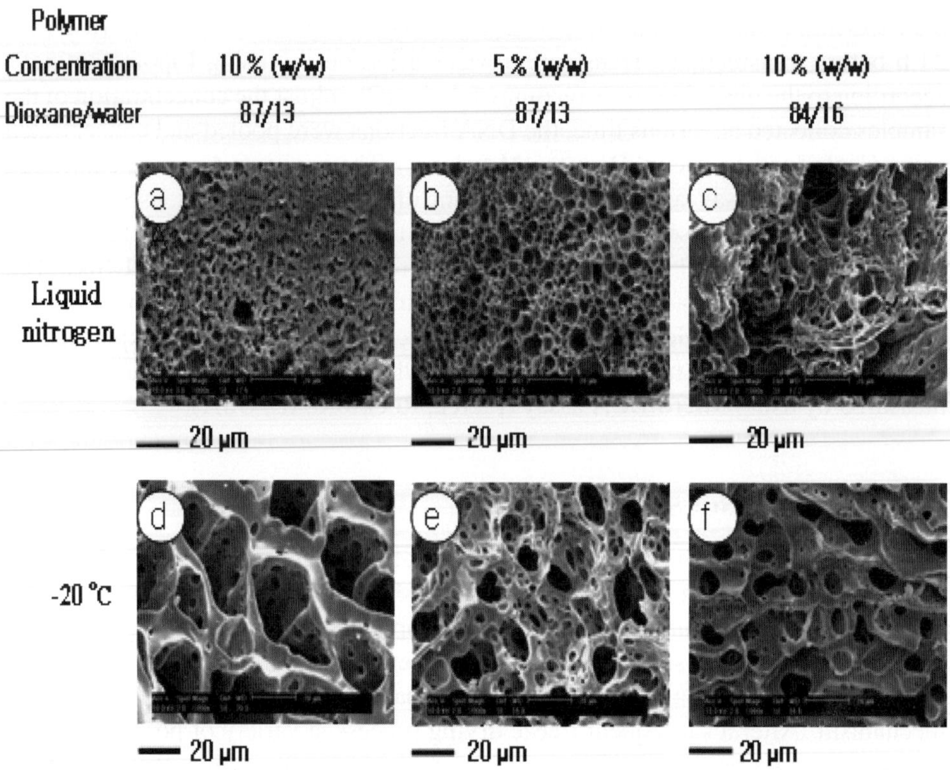

Figure 1. SEM images of cross-sectional PLGA scaffolds fabricated with various conditions of polymer concentration, solvent ratio and quenching route. The scaffolds were quenched at liquid nitrogen temperature (a, b, c) or at −20°C (d, e, f) with the following conditions; 10% (w/v) polymer concentration and 87 : 13 (dioxane : water) (a, d); 5% (w/v) polymer concentration and 87 : 13 (dioxane : water) (b, e); 10% (w/v) polymer concentration and 84 : 16 (dioxane : water) (c, f).

concentration and the co-solvent composition. Lowering the quenching temperature at −20°C produced much larger pores in their structures. The pore size of the scaffold quenched at liquid nitrogen temperature was smaller than 5 μm, but the pore size was enlarged up to 30–50 μm when quenched at −20°C for 24 h. The increased pore size was caused by the coarsening effect that phase separated polymer-poor (solvent-rich) droplets coalesce to form large droplet domain. The coarsening effect occurs more readily at higher quenching temperature because of an increased thermal energy. The obtained pore structures at lower quenching temperature are consistent with those reported previously [1, 2].

Figure 2 shows DNA release profiles from the six different porous PLGA scaffolds. The scaffolds quenched at −20°C exhibit far better DNA release profiles than those quenched at liquid nitrogen temperature. They showed more sustained release profiles with higher extents of cumulative release, although initial burst releases were observed at the level of 30–50%, depending on the scaffold. These release behaviors are well expected from the observed pore structures of

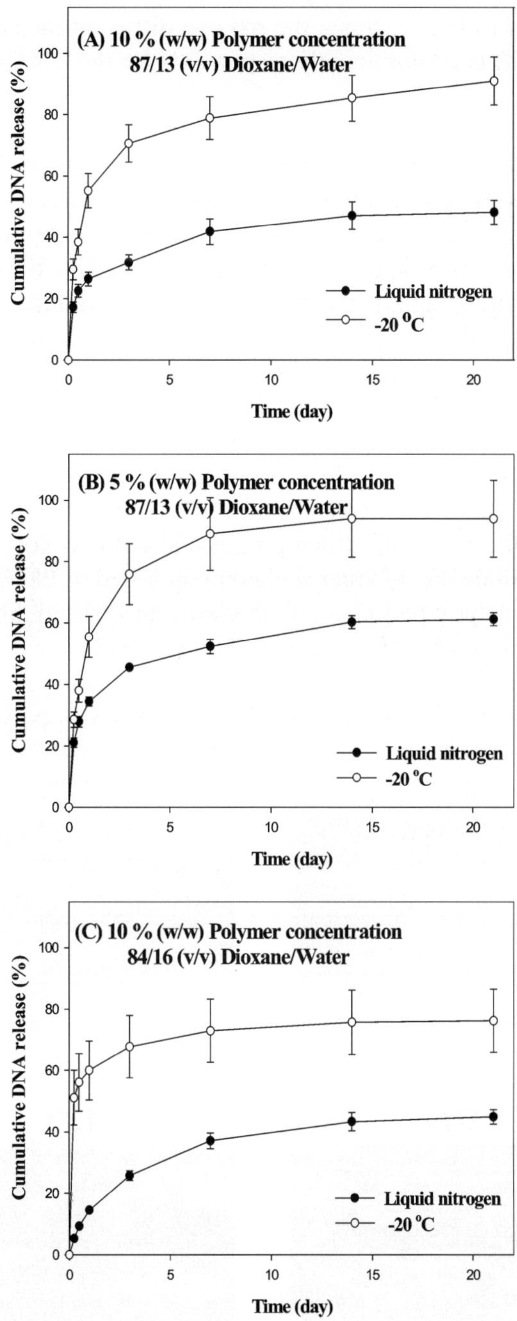

Figure 2. DNA release profiles from the scaffolds fabricated with various scaffolds prepared by changing polymer concentration, solvent ratio and quenching route; (A) 10% (w/v) polymer concentration and 87 : 13 (dioxane : water), (B) 5% (w/v) polymer concentration and 87 : 13 (dioxane : water) and (C) 10% (w/v) polymer concentration and 84 : 16 (dioxane : water). Each was quenched at liquid nitrogen temperature or −20°C.

the scaffolds shown in Fig. 1. Since the passive diffusion mechanism through the pores of the scaffolds is predominantly responsible for the DNA release, more open porous scaffolds caused more sustained DNA release. For the scaffolds (sample d in Fig. 1) prepared with the process conditions of 10% (w/w) polymer concentration, 87 : 13 dioxane/water ratio and −20°C quenching temperature, DNA was released almost 100% in a diffusion controlled manner over 15 days. The DNA release profile follows a typical Fickian diffusion pattern, strongly suggesting that the release was diffusion controlled [31, 32]. It is also noted that the cumulative release percentages at the later stage of incubation are highly variable, depending on the scaffold. This reveals that the pore inter-connectivity, in addition to the pore size, was a very critical factor in releasing large DNA molecules through the scaffolds. During the TIPS process, hydrophilic DNA molecules would be preferentially located in the polymer-poor (solvent-rich) region, which eventually results in the physical entrapment of DNA within the pores. Thus, the sustained release of plasmid DNA through the scaffolds was most likely to occur in a snake-like reptile mode by diffusion through water-filled porous and tortuous pores generated during the TIPS process. Since the polymer skeleton composed of PLGA 75 : 25 degraded very slowly over the study period, a polymer erosion-assisted diffusion mechanism may probably play a minimal role in the sustained release.

To study the effect of annealing time on the formation of the internal pore structures, the polymer solutions prepared with 10% (w/w) polymer concentration

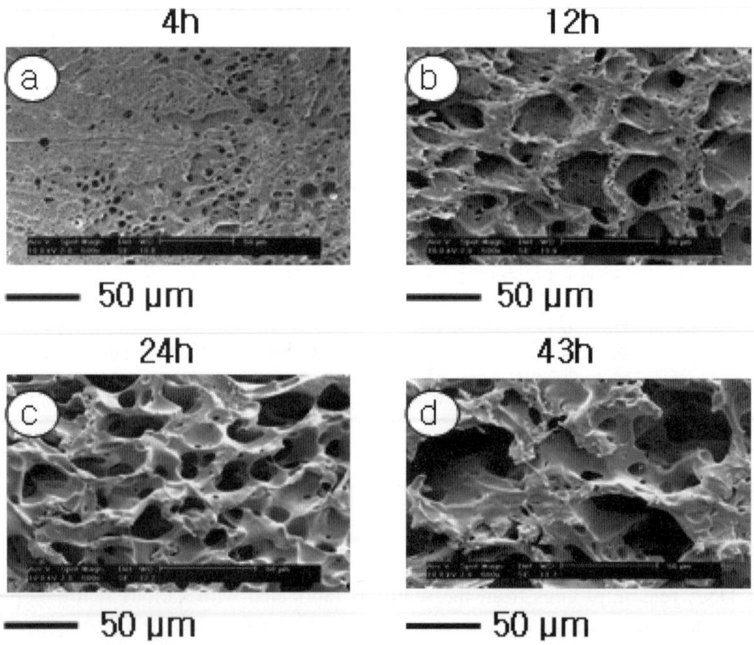

Figure 3. SEM images of cross-sectional PLGA scaffolds fabricated by annealing at −20°C for (a) 4 h, (b) 12 h, (c) 24 h and (d) 43 h.

and 87 : 13 (v/v) dioxane/water ratio were quenched at −20°C for different periods of time (4 h, 12 h, 24 h and 43 h). Previously, we reported that by increasing the annealing time, the coarsening of phase-separated domains led to dramatic enlargement of the pore sizes and the extent of inter-connection between pores was also enhanced. Figure 3 shows SEM images of cross-section for different PLGA scaffolds annealed for different periods. By increasing the annealing time to 43 h, the pores in the scaffold were enlarged to over 50 μm. Inter-connected channels between the pores were also well developed. In contrast, the pores in the scaffold annealed for 4 h have an isolated and closed structure. A diverse array of such porous morphologies was clearly developed with time by the coalescence of thermodynamically unstable polymer-poor domains. Figure 4 shows DNA release profiles from these scaffolds as a function of annealing time. It can be seen that more open porous scaffolds, prepared by annealing for a longer period of time, tend to rapidly release plasmid DNA with higher extents of initial burst and more complete release percentages can be observed at the later incubation stage. This is consistent with the aforementioned finding that open porous structure in the scaffold plays a key role in controlling the diffusion rate of plasmid DNA.

To further control the release rate of DNA from the scaffolds, PLGA-grafted poly(L-lysine) (PLL-g-PLGA) was blended at different weight ratios with PLGA to produce PLL surface coated PLGA scaffolds. By blending PLL-g-PLGA with PLGA, it was anticipated that a hydrophobic part of PLGA in the PLL-g-

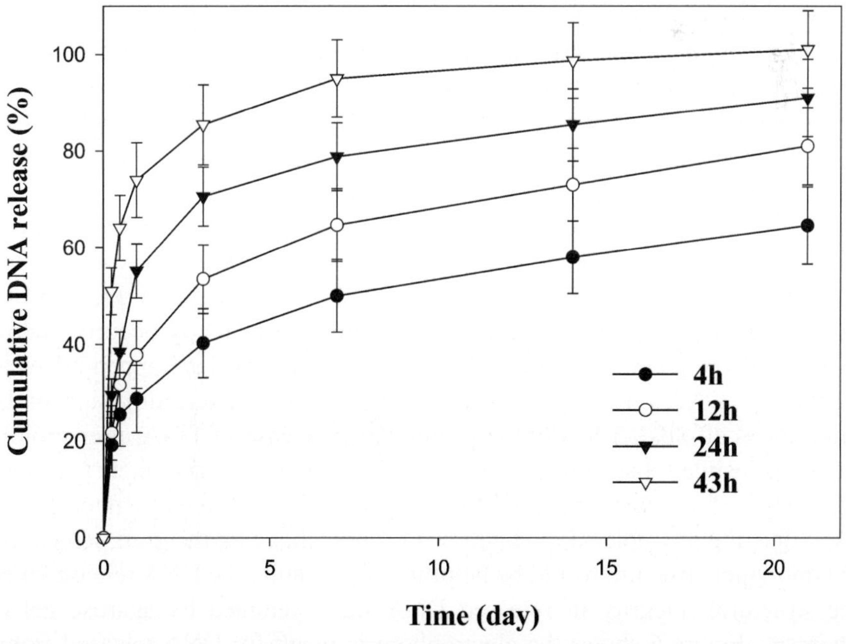

Figure 4. Cumulative DNA release from the scaffolds quenched at −20°C for 4 h, 12 h, 24 h and 43 h.

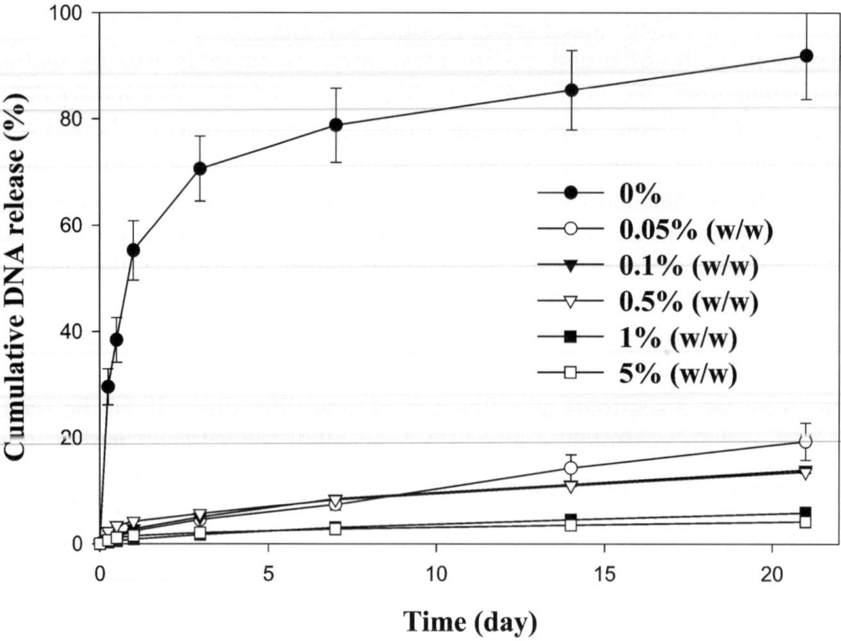

Figure 5. Cumulative DNA release from the scaffolds fabricated by blending 0.05%, 0.1%, 0.5%, 1% and 5% (w/w) PLL-g-PLGA with PLGA.

PLGA graft copolymer was anchored in the PLGA skeleton bulk phase, while a highly positively-charged PLL part was oriented on the pore surface towards the aqueous phase. By this surface coating strategy, it was postulated that physically entrapped DNA in the pores might interact with surface exposed PLL via ionic interactions, possibly resulting in more retarded diffusion through the pores. Different concentrations of PLL-g-PLGA were blended to investigate the effect of incorporation of positively-charged materials on the DNA release.

There was no significant change in the morphology of porous PLGA scaffolds when PLL-g-PLGA was blended. Figure 5 shows that even a small weight ratio of PLL-g-PLGA in the blend mixture can effectively retard the rapid release of DNA from the scaffold. This indicates that ionic interactions really occurred between PLL immobilized on the pore surface and negatively charged entrapped DNA. Ionic adsorption of DNA on the pore surface impeded the passive diffusion of DNA through the scaffolds. The very low cumulative release of DNA after prolonged incubation, despite substantial decrease of the blending ratio, was likely to be caused by effective surface clustering of PLL backbones that were strongly interacted with DNA. Adjusting the molecular weight of PLL and changing the graft percent in the graft copolymer structure would be helpful in fine-tuning the DNA release kinetics.

The structural integrity of released DNA was examined by agarose gel electrophoresis. Figure 6 shows the electrophoresis result for DNA released from the scaffold, prepared with the condition of 10% (w/v) polymer concentration, 87 : 13

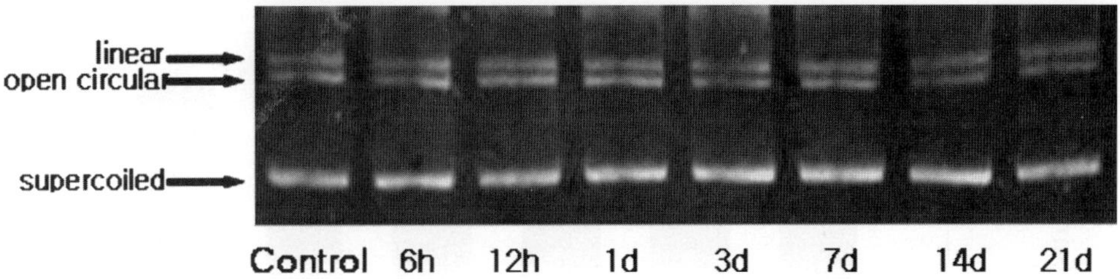

Figure 6. Agarose gel electrophoresis result of DNA fraction released from the scaffold. An equal amount of DNA was loaded into each lane.

dioxane/water ratio and quenching at −20°C for 24 h. Up to 21 days, the released DNA fraction, collected at different time intervals, maintained structurally intact forms. The released fraction had the same band densities of linear, open-circular and supercoiled DNA species as a function of incubation time. In particular, the amount of supercoiled DNA species did not change significantly for the release period. The same results were obtained in the experiments using other sets of TIPS processing parameters. Since the polymer scaffolds did not degrade significantly over the release period and they have a highly porous structure, the internal pH within the pore channels of the scaffolds was likely to be neutral, due to facile exchange of buffer medium in and out of the scaffold. This is the most possible reason for preserving the DNA structure. Often, it was reported that chemical degradation and fragmentation of entrapped DNA within fast degrading PLGA matrices took place by an acidic microenvironment created by accumulated degradation products of PLGA [11, 13, 33–35]. To investigate the structural integrity and function of released DNA, 293 cells were chosen and transfected using a commercial tranfection agent, Lipofectamine. Figure 7 shows transfection efficiencies of luciferase plasmid DNA fraction released from the scaffolds. There were no significant changes in the extent of luciferase gene expression for the released DNA fraction collected at different time intervals, supporting that the released DNA was structurally and functionally intact. While DNA was released from the scaffolds in a controlled manner, its transfection efficiency in a naked form was very low *in vitro* and as well as *in vivo*. It remains to see whether the DNA releasing scaffolds, when implanted *in vivo*, will exhibit an appreciable degree of gene transfection in the tissue. In our preliminary result, it was found that a detectable level of luciferase was expressed in a subcutaneous tissue of rats. More detailed studies to enhance the *in vivo* gene expression are now under way.

CONCLUSIONS

Open porous PLGA scaffolds capable of releasing DNA in a sustained manner were fabricated by using the thermally-induced phase-separation (TIPS) method. A diverse range of porous scaffolds with different internal morphologies could be

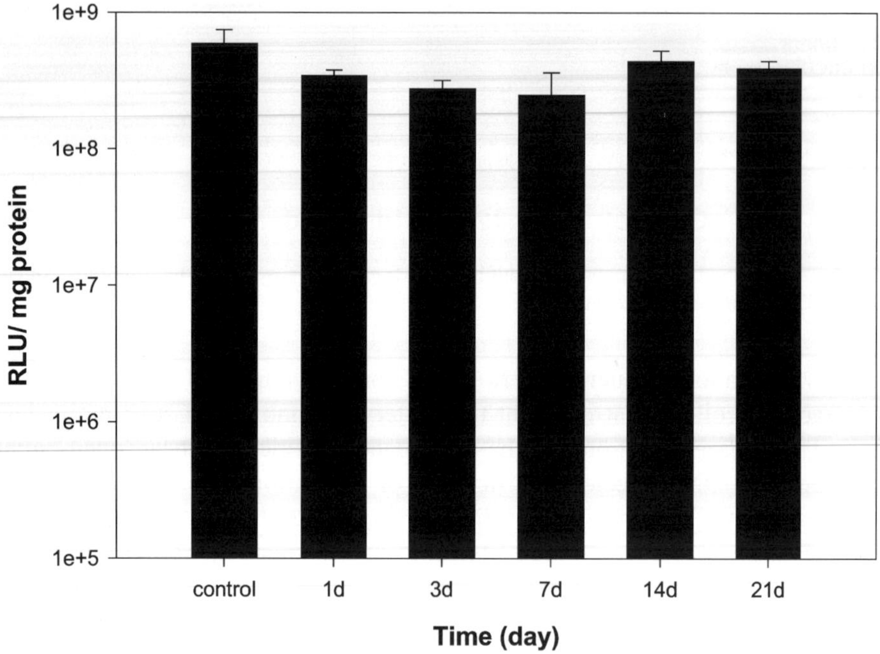

Figure 7. *In vitro* transfection efficiency of DNA fraction released from the scaffold at specified time intervals. The relative light units were normalized per milligram of protein.

produced by controlling various TIPS parameters. It was evident that the internal structures were greatly influenced by quenching route and annealing (coarsening) time. DNA was slowly released over 20 days and its profile was highly dependent on internal pore structures and their inter-connectivities. DNA release profiles could be readily controlled by varying TIPS parameters. Specifically, longer coarsening time at relatively higher quenching temperature produced larger pores with a well-developed inter-connected structure in the scaffold. The released DNA was structurally intact for 21 days and maintained its biological function as determined by measuring its gene expression degree of luciferase. DNA-releasing PLGA scaffolds can be potentially used, as a dual functional delivery vehicle of DNA and cell, for various tissue regenerations.

Acknowledgements

This work was supported by the Korea Science and Engineering Foundation, Korea (Grant # R01-2003-000-10362-0).

REFERENCES

1. Y. S. Nam and T. G. Park, *J. Biomed. Mater. Res.* **47**, 8 (1999).
2. Y. S. Nam and T. G. Park, *Biomaterials* **20**, 1783 (1999).

3. F. J. Hua, T. G. Park and D. S. Lee, *Polymer* **44**, 1911 (2003).
4. J. J. Yoon, J. H. Kim and T. G. Park, *Biomaterials* **24**, 2323 (2003).
5. D. J. Mooney, D. F. Baldwin, N. P. Suh, J. P. Vacanti and R. Langer, *Biomaterials* **17**, 1417 (1996).
6. W. L. Murphy, M. C. Peters, D. H. Kohn and D. J. Mooney, *Biomaterials* **21**, 2521 (2000).
7. M. H. Sheridan, L. D. Shea, M. C. Peters and D. J. Mooney, *J. Control. Rel.* **64**, 91 (2000).
8. T. P. Richardson, M. C. Peters, A. B. Ennett and D. J. Mooney, *Nature Biotechnol.* **19**, 1029 (2001).
9. L. D. Shea, E. Smiley, J. Bonadio and D. J. Mooney, *Nature Biotechnol.* **17**, 551 (1999).
10. J. Mayer, E. Karamuk, T. Akaike and E. Wintermantel, *J. Control. Rel.* **64**, 81 (2000).
11. D. Luo and W. M. Saltzman, *Nature Biotechnol.* **18**, 33 (2000).
12. L. G. Griffith and G. Naughton, *Science* **295**, 1009 (2002).
13. Y. Hu, C. Zhang, S. Zhang, Z. Xiong and J. Xu, *J. Biomed. Mater. Res.* **67A**, 591 (2003).
14. D. D. Hile, M. L. Amirpour, A. Akgerman and M. V. Pishko, *J. Control. Rel.* **66**, 177 (2000).
15. C. M. Agrawal, J. Best, J. D. Heckman and B. D. Boyan, *Biomaterials* **16**, 1255 (1995).
16. C. M. Agrawal and R. B. Ray, *J. Biomed. Mater. Res.* **55**, 141 (2001).
17. K. Fu, A. M. Kilbanov and R. Langer, *Nature Biotechnol.* **18**, 24 (2000).
18. J. A. Wolff, R. W. Malone, P. Williams, W. Chong, G. Acsadi and A. Jani, *Science* **247**, 1465 (1990).
19. R. J. Mumper and A. P. Rolland, *Adv. Drug Deliv. Rev.* **30**, 151 (1998).
20. Y. S. Jong, J. S. Jacob, K. Yip, G. Gardner, E. Seitelman, M. Whitney, S. Montgomery and E. Mathiowitz, *J. Control. Rel.* **47**, 123 (1997).
21. Y. K. Luu, K. Kim, B. S. Hsiao, B. Chu and M. Hadjiargyrou, *J. Control. Rel.* **89**, 341 (2003).
22. J. H. Jang and L. D. Shea, *J. Control. Rel.* **86**, 157 (2003).
23. P. van de Witte, P. J. Dijkstra, J. W. A. van den Berg and J. Feijen, *J. Membr. Sci.* **117**, 1 (1996).
24. F. J. Tsai and J. M. Torkelson, *Macromolecules* **23**, 775 (1990).
25. S. W. Song and J. M. Torkelson, *Macromolecules* **27**, 6389 (1994).
26. J. H. Jeong and T. G. Park, *J. Control. Rel.* **82**, 159 (2002).
27. J. H. Jeong, Y. R. Byun and T. G. Park, *J. Biomater. Sci. Polymer Edn* **14**, 1 (2003).
28. S. W. Song and J. M. Torkelson, *J. Membr. Sci.* **98**, 209 (1995).
29. M. S. Levy, R. D. O'Kennedy, P. Ayazi-Shamlou and P. Dunnill, *Trends Biotechnol.* **18**, 296 (2000).
30. F. J. Hua, G. E. Kim, J. D. Lee, Y. K. Son and D. S. Lee, *J. Biomed. Mater. Res. (Applied Biomaterials)* **63**, 161 (2002).
31. J. S. Park, Y. K. Oh, H. Yoon, J. M. Kim and C. K. Kim, *J. Biomed. Mater. Res.* **59**, 144 (2002).
32. J. Crank, *The Mathematics of Diffusion*. Clarendon Press, Oxford (1975).
33. M. Nof and L. D. Shea, *J. Biomed. Mater. Res.* **59**, 349 (2002).
34. E. Walter, K. Moelling, J. Pavolvich and H. P. Merkle, *J. Control. Rel.* **61**, 361 (1999).
35. C. R. Middaugh, R. K. Evans, D. L. Montgomery and D. R. Casimiro, *J. Pharm. Sci.* **87**, 130 (1998).

Gels, Genes, Grafts and Giants, pp. 131–140 (2005)

PEGylated dextran as long-circulating pharmaceutical carrier

ANATOLY N. LUKYANOV, RISHIKESH M. SAWANT, WILLIAM C. HARTNER and VLADIMIR P. TORCHILIN *

Department of Pharmaceutical Sciences, Bouve College of Health Sciences, 312 Mugar Building, Northeastern University, Boston, MA 02115, USA

Received 1 October 2003; accepted 19 January 2004

Abstract—Dextran-polyethylene glycol (PEG) conjugates were synthesized by activating dextran hydroxy groups with carbonyldiimidazole, introducing amino groups by attaching ethylenediamine, and reacting amino groups with a succinimidyl-activated derivative of PEG. Conjugates with an average of 12 and 21 PEG (5 kDa) residues per single dextran (73 kDa) molecule were prepared. These conjugates have circulation half-lives of 5.3 h and 7.0 h, respectively, compared to 4.0 h for non-PEGylated dextran. The modification of dextran with PEG inhibits the uptake of polymer by the major organ of the reticuloendothelial system, the liver. Dextran-PEG conjugates may represent a convenient platform for long-circulating pharmaceutical preparations.

Key words: PEGylation; dextran; polymeric pharmaceutical carriers; blood clearance.

INTRODUCTION

Water-soluble polymers are gaining increased popularity as carriers for various therapeutic and diagnostic agents [1, 2]. Conjugation with polymers increases the circulation lifetime of many low-molecular-mass drugs by preventing their renal filtration when the total molecular mass of the polymer/drug conjugate exceeds approx. 40 kDa and by preventing drug degradation by the action of various body enzymes [3]. In addition, polymer conjugation prevents the passive diffusion into cells typical for low-molecular-mass therapeutics, thereby increasing circulation time and decreasing the accumulation of the drug in non-targeted tissues [4]. This latter effect allows for a substantial decrease in side-effects for certain drugs [5]. The conjugation with polymers also allows for an increase of therapeutic potential of biologically active proteins and peptides [3].

*To whom correspondence should be addressed. Tel.: (1-617) 373-3206. Fax: (1-617) 373-8886. E-mail: v.torchilin@neu.edu

Drugs conjugated with water-soluble polymers are capable of spontaneous accumulation in solid tumors *via* the enhanced permeability and retention effect (EPR) [6, 7]. This effect is based on the fact that tumor vasculature, unlike the vasculature of healthy tissues, is permeable to macromolecules with a molecular mass of 50 kDa or more [8], and that the lymphatic system, responsible for the removal of macromolecules from healthy tissues, is virtually not working in the case of many tumors as the result of the disease [7]. Because of these facts, macromolecules entering tumor tissues are retained there with higher efficiency and for a prolonged time compared to normal tissues.

The use of stimuli-responsive polymers allows further refinements for the delivery of drugs to their molecular targets in the body, and control of a drug's fate at the intracellular level. Thus, pH-sensitive polymers [9, 10] or swellable dendritic polymers [11] promote the endosomal escape of attached drugs providing them with a greater likelihood for interaction with an appropriate intracellular receptor. Additionally, the use of drug-polymer conjugates with oligopeptides degradable by lysosomal enzymes allows for the adjustment of the intracellular drug release rate [12].

A number of polymers have been tested as potential drug carriers with various degrees of success [2]. Currently, PEG is one of the most popular polymers for drug delivery, in particular, for the development of protein- and peptide-based long-circulating therapeutics [1]. Using antibody fragments, it has been shown recently that an attachment of a single 60-kDa branched molecule of PEG can be sufficient for the steric protection of a typical protein molecule [13]. A disadvantage of PEGs with molecular masses exceeding 30 kDa is, however, their retarded kidney elimination. Although PEG has been proven to have a very low toxicity and has been approved for a variety of products for human use [1], this polymer is not metabolizable. While PEG molecules with a molecular mass of 30 kDa or lower are easily cleared by renal filtration, PEGs with higher molecular masses are cleared more slowly and have been found in the intact form in the urine and feces [14]. Such biologically inert polymers with high molecular mass can accumulate in kidneys, decrease the efficiency of renal filtration and cause various undesirable side effects [3].

Dextran, a natural polymer of glucose residues linked mainly with 1,6-bonds (Fig. 1) has also been used as a drug carrier for a long time. This polymer is widely available from natural sources and can be prepared in a variety of molecular masses [15]. For example, as early as 1978 it was shown that conjugation of daunomycin with dextran increased the therapeutic efficacy and decreased the toxicity of the drug [16]. However, there are certain problems associated with dextran. Thus, the degree of dextran branching (the property associated with dextran immunogenicity) depends on the source of dextran and may vary within broad limits [17]. The high-molecular-mass dextrans required for effective prolongation of drug circulation time may have a rather broad molecular size distribution and can cause anaphylactic reactions [15]. Although dextran-bound drugs demonstrated

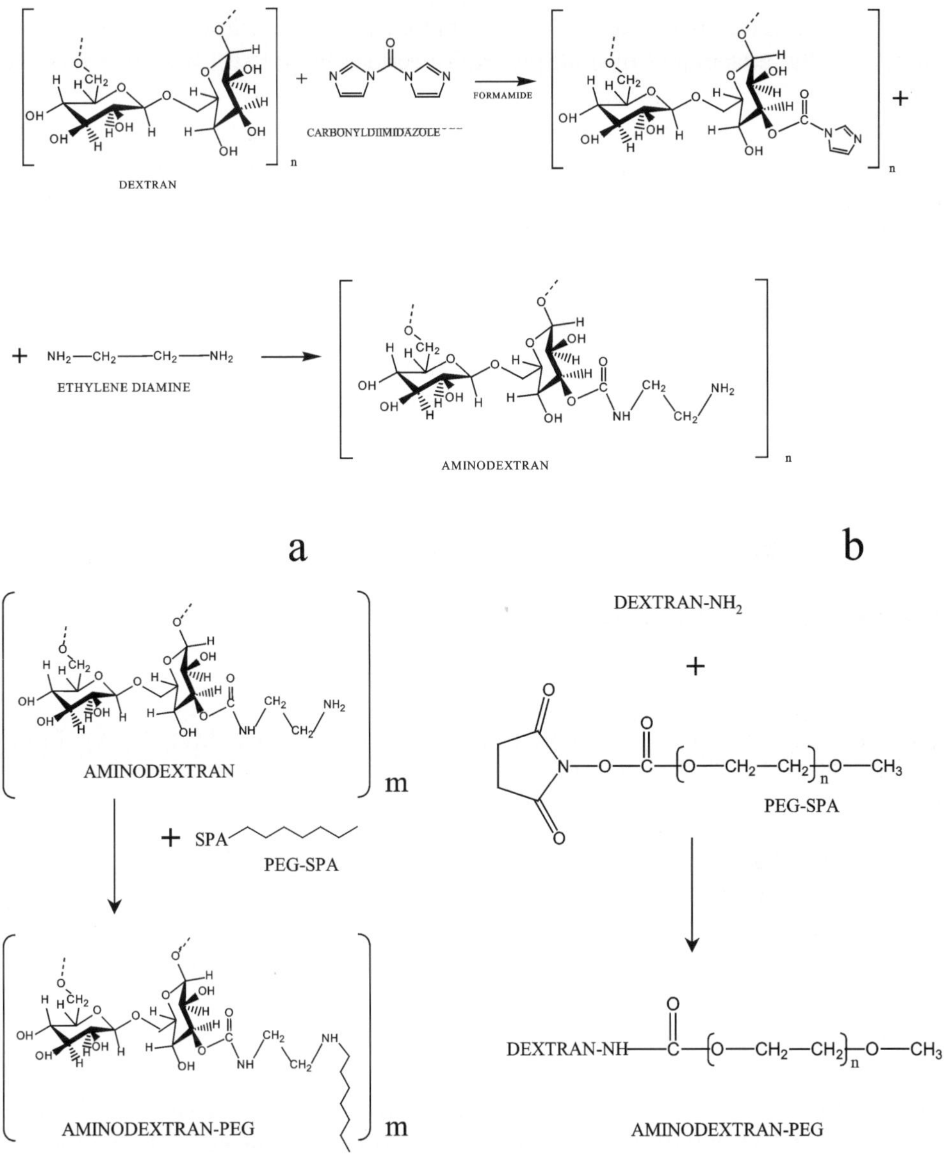

Figure 1. Synthesis of PEGylated dextran. (a) General scheme of aminodextran conversion into Dextran-PEG. (b) Chemical specifics of (a).

increased circulation times, dextrans still undergo biodegradation under the action of various dextranases found in many body tissues [18–20].

Delivery systems built of a biodegradable core polymer, such as dextran, grafted with relatively short PEG residues can serve as a good alternative to high molecular PEGs or pure dextrans. Such co-polymers that can be synthesized using mono-functional PEG derivatives with a low-to-moderate molecular mass will slowly degrade to fragments small enough to be eliminated by renal filtration. The dextran core, after the attachment of multiple PEG residues, might still bear free reactive groups suitable for the attachment or single-point modification of various drugs including proteins and peptides to obtain their long-circulating forms. Earlier, some dextran-PEG co-polymers have been described used to prepare colloidal particles with internal phase separation [21] or sterically-protective coating for polymeric microparticles [22].

In our paper, we report the results on the new synthesis and biodistribution studies of dextran (molecular mass 73 kDa) grafted with PEG with a molecular mass of 5 kDa. Such a co-polymer was prepared from dextran with introduced amino groups. Partial modification of these amino groups with PEG residues resulted in a construct with a nearly doubled circulation half-life in mice of up to 7 h. Residual dextran amino groups can be used for attachment of drugs, diagnostic agents or even stimuli-responsive polymer blocks, thus providing a convenient platform for long-circulating multifunctional polymer-drug pharmaceutical preparations.

MATERIALS AND METHODS

Dextran with average molecular mass of 73 kDa, carbonyldiimidazole (CDI), ethylenediamine (EDA), diethylene triamine pentaacetic acid anhydride (DTPA) and fluorescamine were purchased from Sigma (St. Louis, MO, USA). PEG-succinimidyl propionate (PEG-SPA) with a molecular mass of 5 kDa was from Nektar Therapeutics (San Carlos, CA, USA). All other chemicals and organic solvents were of analytical grade. Distilled and deionized water was used.

Preparation of aminodextran

Aminodextran was prepared from dextran activated with CDI by attachment of EDA (Fig. 1) using a procedure adapted from a protocol for the attachment of diamines to cross-linked agarose [23]. To study the kinetics of CDI activation, 17 μmol of dextran dissolved in 30 ml of formamide was mixed with 360 μmol of CDI dissolved in 3 ml of formamide. The reaction mixture was incubated for 15 min, 30 min, 1 h and 3 h at room temperature. The reaction was stopped by the addition of at least 100-fold molar excess (over CDI) of EDA. After the overnight incubation, resultant aminodextran was precipitated by addition of a 2-fold volume excess of acetone. The white precipitate obtained was spun down at $1500 \times g$ for 1 h, dissolved in water, dialyzed against at least a 3000-fold excess of water for 2–4 days with several changes and freeze dried. The number of EDA residues attached per dextran molecules was estimated by determining the concentration of amino groups in the samples using a fluorescamine method (see below).

In a typical preparative procedure, 68 μmol of dextran was dissolved in 125 ml of formamide. To this solution, 1.44 mmol of CDI dissolved in 10 ml of formamide was added. After a 30 min to 3 h incubation, 25 ml of EDA was added to the mixture, and it was stirred at room temperature overnight. The resultant aminodextran was purified as described above.

Introduction of DTPA residues

A 28.5 μmol sample of aminodextran was dissolved in 25 ml of formamide containing 0.8 ml of pyridine and mixed with 57 μmol of DTPA anhydride dissolved in 0.5 ml of formamide (2-fold molar excess over aminodextran). The mixture was incubated for 1–5 h at room temperature, dialyzed against water and freeze dried.

PEGylation of aminodextran

To study the kinetics of PEG attachment, a 20.5 μmol sample of aminodextran was dissolved in 50 ml of formamide containing 2 ml of pyridine. To this solution, 103 μmol of PEG-SPA dissolved in 5 ml of formamide was added and the mixture was incubated for 30 min, 2 h, 5 h and overnight at room temperature with stirring. To stop the reaction, 5 ml of water per 10 ml of the reaction mixture was added. The samples were purified by dialysis and freeze-dried as described above. The number of PEG residues attached per each aminodextran molecule was estimated by amino group loss measured by the fluorescamine method (see below).

For preparative purposes, 20 μmol of aminodextran was dissolved in 50 ml of formamide containing 1.2 ml of pyridine. To this solution, 300 μmol or 600 μmol of PEG-SPA dissolved in formamide with was added. The mixture was incubated at room temperature with stirring overnight. Organic solvents and unbound PEG were removed by dialysis against water using dialysis bags with a cutoff size of 50 kDa (Spectrum Medical Industries, Rancho Dominguez, CA, USA) and freeze-dried. For the biodistribution studies, PEGylated aminodextran samples were acetylated in formamide containing pyridine by addition of a 10-fold molar excess of acetic anhydride and purified as described above.

Amino group determination

The concentration of amino groups was measured using fluorescamine according to a procedure adopted from the previous publication [24]. The material of interest was dissolved at about 0.2 mg/ml in 10 mM Na phosphate-buffered saline, pH 7.4. A 1.5-ml aliquot of the sample obtained was mixed with 0.5 ml of 3 mg/ml fluorescamine solution in acetone. The mixture was incubated in the darkness at room temperature for 30 min. The concentration of amino groups was determined following the fluorescence at the excitation wavelength of 400 nm and the emission wavelength of 460 nm by a F-2000 spectrofluorimeter (Hitachi, Tokyo, Japan) using a sample of commercially available aminodextran (Molecular Probes, Eugene, OR, USA) with a known number of amino groups per dextran molecule as a standard.

Radiolabeling of PEGylated dextrans by transchelation

To radiolabel PEG-dextran for *in vivo* studies, a sample of PEGylated dextran prepared from aminodextran partly modified with DTPA anhydride (DTPA is a very strong chelator of ^{111}In, with $K_d = 10^{29}$ [25], so that no liberation *in vivo* could be expected) and acetylated, was dissolved in 10 mM HEPES-buffered saline, pH 7.4 (HBS) at a concentration of 10 mg/ml. To 1 ml of this solution, 10–20 μCi ^{111}In in 0.1 ml of 0.1 M Na citrate, pH 3.7 (a weak chelator for ^{111}In) was added and incubated for 1 h at room temperature. The unbound ^{111}In was removed by dialysis against HBS at 4°C overnight.

Biodistribution

Experiments were performed in female C57BL/6J mice (Charles River Laboratories, Wilmington, MA, USA). The mice were injected with 100 μl of 10 mg/ml of ^{111}In-labeled dextran or PEG-dextran *via* the tail vein. The animals were allowed free access to food and water. Mice were killed by cervical dislocation at time points between 0.5 and 17 h post injection. Blood, liver, spleen, kidney and lung samples were collected and analyzed by γ-counting for the presence of the dextran-associated ^{111}In radioactivity. Dextran circulation half-lives were determined assuming a first order elimination process from the log(% of injected dose in blood) *vs.* time plots using a computer program, Origin (Microcal Software, Northampton, MA, USA). There were 4 animals in a group for each time point.

RESULTS AND DISCUSSION

The schematics of dextran modification are outlined in Fig. 1. First, hydroxyl groups of dextran were activated with CDI. Carbonyldiimidazole forms imidazole carbamate complexes with dextran hydroxyl groups. These complexes react with amino groups forming urethane bonds. In our design, EDA was added to the activated dextran without purification of the latter. To avoid the formation of cross-linked dextrans, EDA was used in at least 100-fold molar excess over CDI. The quantification of amino groups in the sample obtained by the fluorescamine method has shown that, on average, 21 amino groups were introduced into each dextran molecule with a molecular mass of 73 kDa. This corresponds to a quantitative yield of CDI hydroxy group activation. The incubation of dextran with CDI for longer than 30 min before the addition of EDA did not result in a reaction yield increase. Incubation for 15 min, however, resulted in the introduction of substantially fewer amino groups. These data suggest the incubation time of 30 min to 5 h for the CDI activation of dextran step.

The aminodextran prepared was further modified with PEG-SPA (Fig. 1). To characterize the kinetics of this step, the number of PEG residues attached per each aminodextran molecule was quantified following the amino group loss after

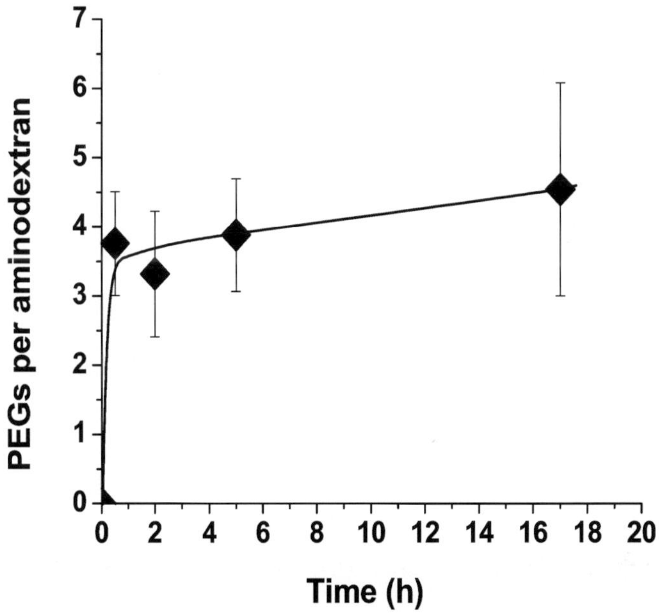

Figure 2. Attachment of PEG residues to aminodextran as a function of time ($n = 4$). See Materials and Methods for details.

Table 1.
Aminodextran PEGylation efficiency and circulation half-life

Initial PEG-SPA/aminodextran ratio (mol/mol)	PEG/aminodextran attached (mol/mol, mean ± S.D.)	Circulation half-life (h)
0 : 1 (non-modified aminodextran, Dextran(Ac))	0	4.0
15 : 1 (Dextran(PEG)$_{12}$)	12 ± 2	5.3
30 : 1 (Dextran(PEG)$_{21}$)	21 ± 2	7.0

the incubation of aminodextran in the presence of 5 molar (over aminodextran) excess of PEG-SPA for 0.5 to 17 h (Fig. 2). The results indicated that, although the substantial initial modification proceeds rather fast (within 1 h), the reaction's completion requires at least 5 h of incubation. In any case, the modification scheme suggested here is simpler, easier to control and not associated with possible dextran degradation compared to previously described protocol [21].

For the biodistribution studies, aminodextran was modified with PEG-SPA at 15 : 1 and 30 : 1 (mol/mol) ratios. The degree of PEGylation of the samples obtained is shown in Table 1. The modification at the initial PEG-SPA/aminodextran ratio of 15 : 1 (mol/mol) resulted in the introduction of 12 ± 2 PEG residues per aminodextran molecule (Table 1), indicating that the reaction yield of the PEG attachment was close to 80% (12 of possible 15 PEG residues were attached). The addition of the 30 molar excess of PEG-SPA over aminodextran resulted in the

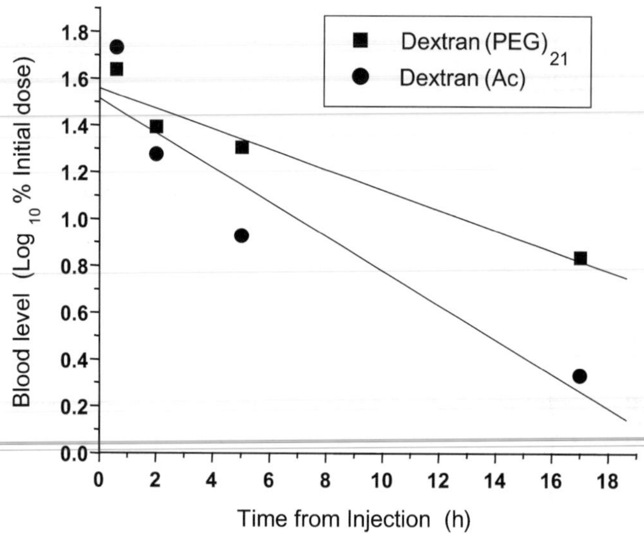

Figure 3. Blood clearance of Dextran(Ac) and Dextran$(PEG)_{21}$ in mice. See Materials and Methods for details.

PEGylation of all available amino groups (21 of possible 21), i.e. the attachment of 21 PEG residues per aminodextran molecule.

Blood clearance and biodistribution of dextran with 12 or 21 PEG (5kDa) chains attached (Dextran$(PEG)_{12}$ or Dextran$(PEG)_{21}$) in comparison with an appropriate aminodextran with all its aminogroups acetylated (Dextran(Ac)) was studied in female C57BL/6J mice. The blood concentration profiles for both samples showed a good linear fit in log(% of dose) *vs.* time coordinates ($R^2 \geqslant 0.97$), indicating that the samples followed a first-order elimination process (see the data for the Dextran$(PEG)_{21}$ in Fig. 3).

The PEGylation of dextran caused a substantial circulation lifetime increase. The circulation half-life of Dextran$(PEG)_{21}$ was 7.0 h *vs.* 4.0 h for Dextran(Ac). The circulation half-life of Dextran$(PEG)_{12}$ was in between, at 5.3 h (Table 1). The results obtained indicate that the modification of dextran with PEG residues increases the dextran circulation life and that the increase depends on the amount of PEG molecules attached. At the same time, there should be no concerns regarding the possibly decreased *in vivo* biodegradability of the PEGylated dextran, since it was shown in earlier studies that even dextran modification with large protein molecules does not prevent its degradation with the dextranase [26].

The results of the comparative biodistribution studies with Dextran(Ac) and Dextran$(PEG)_{21}$ are shown in Fig. 4. Uptake by the main organs of RES, liver, primary responsible for dextran elimination from the systemic circulation, was substantially decreased in the case of PEGylated dextran. At the same time, the uptake by spleen, lung and kidney remained on the comparable level for both Dextran(Ac) and Dextran$(PEG)_{21}$.

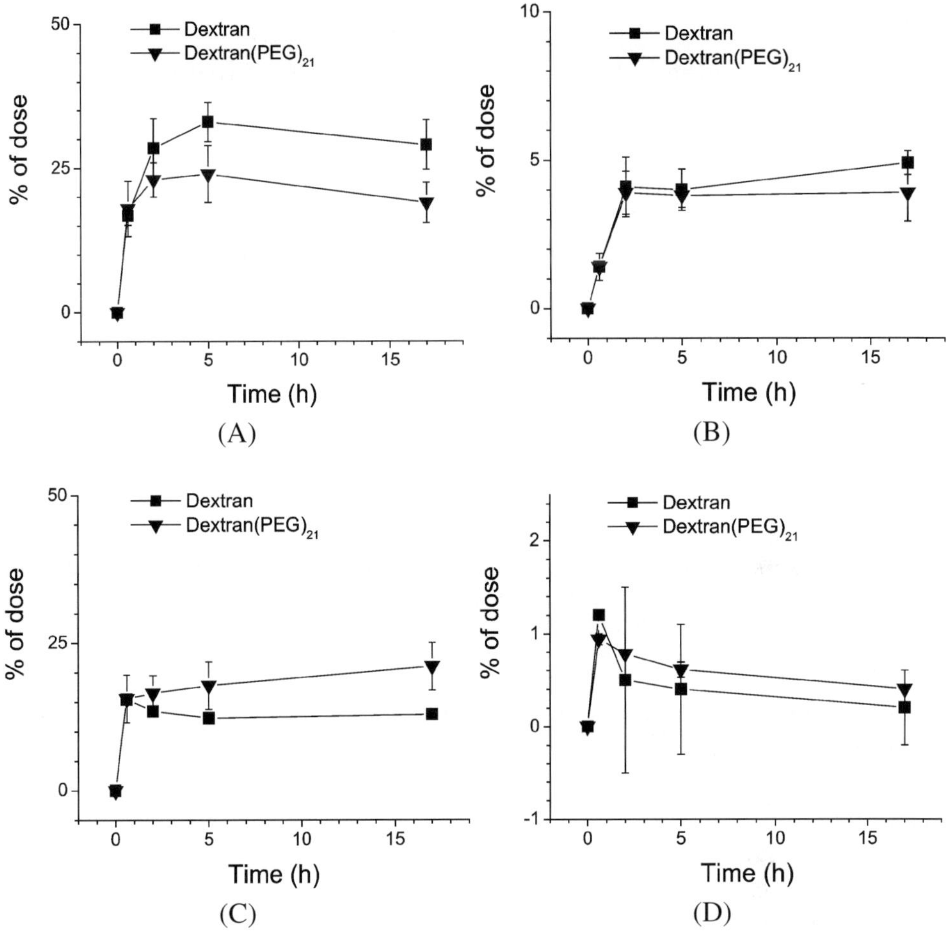

Figure 4. Biodistribution of Dextran(PEG)$_{21}$ in comparison with Dextran(Ac): (A) Liver; (B) Spleen; (C) Kidney; (D) Lung. See Materials and Methods for details.

In conclusion, the activation of dextran with CDI provides a convenient method for the preparation of aminodextran by the attachment of EDA residues. Introduced amino groups can be used for the further attachment of PEG residues to dextran using PEG-SPA with a nearly quantitative reaction yield. The modification with PEG decreases the uptake of dextran by the liver and substantially increases the dextran circulation time. The increase in the circulation time depends on the number of PEG residues introduced. The PEGylated dextrans (with residual reactive groups, such as amino groups) seem to be convenient long-circulating carriers for therapeutic and diagnostic agents.

Acknowledgements

This work was supported by the NIH Bioengineering Partnership Grant HL64395.

REFERENCES

1. J. M. Harris, N. E. Martin and M. Modi, *Clin. Pharmacokinet.* **40**, 539 (2001).
2. R. Duncan, *Nature Rev. Drug Discov.* **2**, 347 (2003).
3. V. P. Torchilin, *Immobilized Enzymes in Medicine.* Springer, Berlin (1991).
4. R. Duncan and F. Spreafico, *Clin. Pharmacokinet.* **27**, 290 (1994).
5. T. Minko, P. Kopeckova and J. Kopecek, *Int. J. Cancer* **86**, 108 (2000).
6. H. Maeda, *Adv. Drug Deliv. Rev.* **46**, 169 (2001).
7. H. Maeda, T. Sawa and T. Konno, *J. Control. Rel.* **74**, 47 (2001).
8. F. Yuan, M. Dellian, D. Fukumura, M. Leunig, D. Berk, V. Torchilin and R. K. Jain, *Cancer Res.* **55**, 3752 (1995).
9. C. A. Lackey, O. W. Press, A. S. Hoffman and P. S. Stayton, *Bioconjug. Chem.* **13**, 996 (2002).
10. P. S. Stayton, A. S. Hoffman, N. Murthy, C. A. Lackey, C. Cheung, P. Tan, L. A. Klumb, A. Chilkoti, F. S. Wilbur and O. W. Press, *J. Control. Rel.* **65**, 203 (2000).
11. O. L. Padilla De Jesus, H. R. Ihre, L. Gagne, J. M. Frechet and F. C. Szoka, *Bioconjug. Chem.* **13**, 453 (2002).
12. J. Kopecek, P. Kopeckova, T. Minko, Z. R. Lu and C. M. Peterson, *J. Control. Rel.* **74**, 147 (2001).
13. A. P. Chapman, P. Antoniw, M. Spitali, S. West, S. Stephens and D. J. King, *Nature Biotechnol.* **17**, 780 (1999).
14. T. Yamaoka, Y. Tabata and Y. Ikada, *J. Pharm. Sci.* **83**, 601 (1994).
15. R. Mehvar, *J. Contr. Rel.* **69**, 1 (2000).
16. A. Bernstein, E. Hurwitz, R. Maron, R. Arnon, M. Sela and M. Wilchek, *J. Natl. Cancer Inst.* **60**, 379 (1978).
17. G. J. Walker, in: *Biochemistry of Carbohydrates II*, D. J. Manners (Ed.), p. 75. University Park Press, Baltimore, MD (1978).
18. R. Ammon, *Enzymologia* **25**, 245 (1966).
19. M. E. Preobrazhenskaya, A. L. Minakova and E. L. Rosenfeld, *Carbohydr. Res.* **38**, 267 (1974).
20. C. Larsen, *Adv. Drug Deliv. Rev.* **3**, 103 (1989).
21. J. M. Duval, C. Delestre, M.-C. Carre, P. Hubert and E. Dellacherie, *Carbohydr. Polym.* **15**, 233 (1991).
22. A. G. Coombes, S. Tasker, M. Lindblad, J. Holmgren, K. Hoste, V. Toncheva, E. Schacht, M. C. Davies, L. Illum and S. S. Davis, *Biomaterials* **18**, 1153 (1997).
23. G. S. Bethell, J. S. Ayers, W. S. Hancock and M. T. Hearn, *J. Biol. Chem.* **254**, 2572 (1979).
24. A. Lorenzen and S. W. Kennedy, *Anal. Biochem.* **214**, 346 (1993).
25. F. Jasanada, P. Urizzi, J. P. Souchard, F. Le Gaillard, G. Favre and F. Nepveu, *Bioconjug. Chem.* **7**, 72 (1996).
26. J. J. Marshall, *Carbohydr. Res.* **49**, 389 (1976).

Gels, Genes, Grafts and Giants, pp. 141–152 (2005)

Enhanced ileal absorption of a hydrophilic macromolecule, pentosan polysulfate sodium (PPS)

LIANG DONG *, ALICIA YUM, JOE NGUYEN and PAT WONG
ALZA Corporation, 1010 Joaquin Road, P.O. Box 7210, Mountain View, CA 94039-7210, USA

Received 1 October 2003; accepted 12 January 2003

Abstract—An *in situ* gelling, bioadhesive liquid formulation was developed to enhance the bioavailability (BA) of a polysaccharide, pentosan polysulfate sodium (PPS). The formulation was tested to determine its bioavailability enhancement in a non-flush/non-ligated rat ileal model. A potent synergistic effect was found with a gelling agent Cremophor and a permeation enhancer sodium salicylate. The absolute bioavailabilities were 1.9%, 4.6%, 6.3% and 46.4%, respectively, for the PPS solution in saline, sodium salicylate/PPS, Cremophor/PPS and Cremophor/sodium salicylate/PPS. Therefore, we successfully demonstrated the approach of utilizing an *in situ* gelling/bioadhesive liquid carrier to enhancing the bioavailability of a hydrophilic macromolecule at the distal small intestine.

Key words: *In situ* gelling; bioadhesion; oral absorption; hydrophilic macromolecules; pentosan polysulfate sodium.

INTRODUCTION

Pentosan polysulfate sodium (PPS) is a polysaccharide, indicated for treatment of interstitial cystitis (IC). Its molecular mass ranges from 4000 to 6000 Da. Having dense negative charges, PPS is very soluble in water, about 50% by weight. The oral bioavailability of PPS in humans is very low, approximately 3% [1]. The low oral bioavailability of PPS can be attributed to its hydrophilicity, large molecular size and negative charges. Consequently, the low bioavailability (BA) of PPS has limited its application for other indications, including glomerulosclerosis, arteriosclerosis and vascular graft stenosis. Additionally, the low oral BA of PPS compromises its efficacy for the treatment of interstitial cystitis. Patients need to continue the chronic therapy for many months in order to achieve optimal therapeutic effect. Therefore, a novel PPS formulation is needed to significantly improve oral BA and reduce the time required for PPS to reach a clinically therapeutic level.

*To whom correspondence should be addressed. Tel.: (1-650) 564-4061. Fax: (1-650) 564-5398. E-mail: ldong2@alzus.jnj.com

Several approaches were reported to enhance the oral absorption of polysaccharides, such as unfractionated heparin and low molecular heparin (LMWH). One approach was to use mixed micelle [2] or delivery agents to form physical associates with heparin or LMWH [3–5]. The second approach was to use muco-adhesive polymers such as Carbopol and chitosan [6, 7]. The third was to use the enteric-coated tablet comprising polycarbophil-cystein conjugate [8]. All of these approaches led to absorption enhancement to certain extent. But, enhanced oral absorption of macromolecules including polysaccharides still remains to be a great challenge [9] and a subject of continuing research.

In the present study, we developed a novel formulation that is composed of a lyotropic liquid and an absorption enhancer. The lyotropic liquid converts to a muco-adhesive gel *in situ* when it absorbs water. Due to the muco-adhesion of this formulation, a high concentration gradient can be established for both PPS and the enhancer, thereby promoting PPS transport across the epithelia, and increasing its bioavailability.

MATERIALS AND METHODS

Pentosan polysulfate sodium (PPS) was provided by Baker Norton Pharmaceutical (Miami, FL, USA) and [^{3}H]PPS sodium was prepared by Amersham Pharmacia Biotech (Uppsala, Sweden). Cremophor EL and RH (ethoxylated castor oil) were obtained from BASF (Ledgewood, NJ, USA). Sodium salicylate, sodium caprate, sodium deoxycholate and mucin were all from Sigma (St. Louis, MO, USA), sodium laurate was obtained from TCI (Portland, OR, USA).

Characterization of rheological properties of the carriers

Haake RS100 RheoStress Rheometer in oscillatory mode was used to determine η (dynamic viscosity), G' (storage modulus), G'' (loss modulus), and $\tan\delta(G''/G')$ of Cremophor/water blends at various weight ratios at 37°C. A 60-mm, 2°-cone and plate were used for low viscosity samples, and a 35-mm, 4°-cone and plate were used for viscous samples. The strain of 0.001 was used for the most viscous samples (Cremophor/water, 60 : 40, 50 : 50 and 40 : 60, by weight), while the strain of 0.03 was used for the rest of the samples.

Evaluation of mucosal adhesion of the carriers

The adhesion energy of the Cremophor EL/water blends to a mucin surface was determined using a texture profile analyzer (TPA, Texture Technologies, Scarsdale, NY, USA). A 500-mg mucin tablet with a flat circular surface area of 0.096 in^2 was prepared by compressing mucin powder with 0.5 ton of force. The mucin tablet was firmly attached to the lower end of the TPA probe using a double-sided adhesive tape. Samples of Cremophor EL/water at various ratios were prepared in small vials that were affixed onto the TPA platform. The mucin tablet was moistened in simulated gastric fluid for 60 s before measurements were taken. During measurement, the TPA probe with attached mucin tablet was lowered onto the surface of each sample at a constant speed of 1 mm/s. To ensure the intimate contact between the mucin tablet and the sample, the tablet remained at the sample surface for 60 s before the probe was moved upward. The force required to detach the mucin tablet from the surface of the Cremophor El/water samples was recorded as a function of time. Adhesion energy (E) was calculated from area under curve (AUC) of the force *vs*. time curve.

$$E = \text{AUC} \times V,$$

where V is the moving speed of the probe.

Permeation through Caco-2 cells

Caco-2 cells were cultured for 18–26 days as described previously [10]. Before the permeation experiment, the cells were rinsed with a fresh Hanks balance salt solution (HBSS). All the permeation studies were conducted in HBSS at 37°C. Radio-labeled mannitol was used as a model compound to find the effect of permeation enhancers and Cremophor on mannitol transport. The amount of mannitol transported through the Caco-2 cells was quantified with a scintillation counter LS 3801 (Beckman Instruments, Fullerton, CA, USA).

In vivo *screening of the BA-enhanced formulations using flush/ligated(F/L) or non-flush/non-ligated(NF/NL) rat model*

A variety of formulations were prepared immediately before the *in vivo* rat studies. The compositions of these formulations tested in rats are listed in Table 1. The components of each formulation were first weighed into a plastic centrifuge tube, then warmed to 35–40°C in a water bath and mixed with a vortex. Two ml of saline spiked with [^{3}H]PPS was added into the plastic centrifuge tube. The blend was mixed with a vortex and/or a spatula to a homogenous formulation.

Table 1.
Composition of formulations (in wt%) for rat absorption study

Component	Formulation					
	A	B	C	D	E	F
PPS	0.14	0.14	0.14	0.14	0.14	0.14
Sodium deoxycholate		14				
Sodium caprate			14			
Sodium salicylate				14	14	
Cremophor		65.86	65.86	65.86		79.86
Saline	99.86	20	20	20	85.86	20

Male and/or female Sprague–Dawley rats (from Charles River Laboratories, Wilmington, MA, USA) weighing 200–450 g were used to evaluate these PPS formulations. All rats were food-fasted overnight and anesthetized intraperitoneally with sodium pentobarbital. Three to four rats were used to evaluate each formulation. In the F/L model (close compartment model), the ileal segment was flushed with warm saline, and both proximal and distal openings of the segment were ligated with a piece of suture.

In the NF/NL model (open compartment model), an ileal segment (6–7 cm in length) of the GI tract was isolated and cleared of surrounding omentum, following a midline abdominal incision. Lumenal content was left undisturbed and the segment was positioned for formulation injection. The PPS formulation spiked with [^{3}H]PPS was injected (with needles of varying gauges, depending on the viscosity of the formulation) directly into the lumen of the ileal segment. After dosing with formulation, the punctured site was tightly closed with a piece of suture; the closure was performed parallel to the serosal surface allowing continual flow of lumenal content.

In both models, the dose of PPS administrated to rat was 1.4 mg/kg rat body weight. The blood samples were withdrawn up to 4 h after administration of the formulations. Scintillation counting of plasma samples was performed to assess the PPS concentration in plasma.

RESULTS

It is known that non-ionic surfactants will convert to liquid crystals when they absorb water [11, 12]. The liquid crystal structure is dependent on the amount of water absorbed to the surfactant. Some liquid crystal structures, such as lamellar or cubic forms, are bioadhesive to the mucosal surface.

To characterize the rheological behavior of *in situ* gelling liquid formulation, Cremophor EL was mixed homogeneously with water at various weight ratios. The Cremophor EL/water blends were measured by the Haake RS 100 RheoStress rheometer for η (dynamic viscosity), G' (storage modulus) and G'' (loss modulus). Figure 1 shows dynamic viscosities of Cremophor EL containing various amounts of water. Once the water content is increased to 40%, the viscosity of the blend increases drastically. As the water content of a Cremophor/water blend is raised, the blend converts from reverse micelles (low viscosity) through liquid crystal phases (high viscosity) to micelles (low viscosity). The dynamic viscosity (η) of Cremophor EL/water blends was measured at a shear rate ranging from 0.0628 to 628 rad/s. As shown in Fig. 2, shear rate has an inverse effect on the dynamic viscosity of the samples containing 30% to 60% of Cremophor EL. For these compositions, η decreased as the shear rate increased, exhibiting pseudoplastic behavior of non-Newtonian fluid. Other compositions with higher water content showed dilatant properties, with dynamic viscosity increasing with the shear rate.

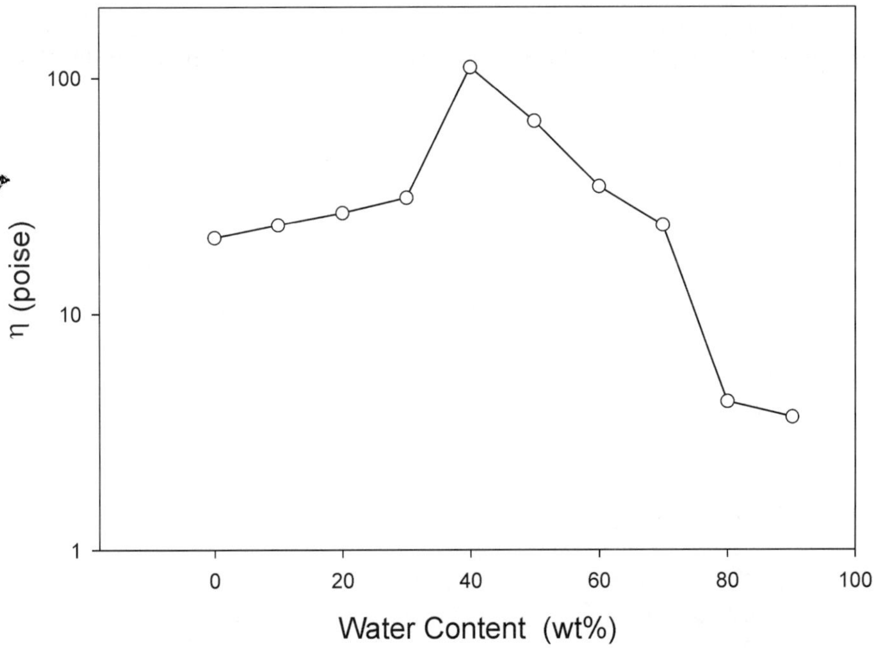

Figure 1. Dynamic viscosity of Cremophor/water blends as a function of water content at 158 rad/s and 37 °C.

Figure 3 shows the G' (storage modulus), G'' (loss modulus) and $\tan\delta(G''/G')$ as a function of the water content in the blends. The G''/G' indicates that the viscoelastic property of Cremophor EL/water blend changes from fluid type ($G''/G' > 1$) to rubber type ($G''/G' < 1$) and returns to fluid type. It appears that Cremophor EL, a low viscosity liquid, will gel when absorbing water *in situ*, with the gelling composition at a Cremophor/water weight ratio of 60 : 40.

The muco-adhesive property of Cremophor EL/water blends was also assessed using a texture profiles analyzer. Figure 4 presents the results of the measurements. The Cremophor EL/water blend at the weight ratio of 60 : 40 was most adhesive to the surface of the mucin tablet. These results showed good correlation between adhesion and viscosity. The more viscous formulation tends to be more muco-adhesive.

As indicated above, the rheological and muco-adhesive properties of the formulations comprising Cremophor as a liquid carrier for PPS are strongly dependent on the hydration. Direct administration of these formulations to the stomach *via* oral gavage would lead to extensive dilution. Therefore, these formulations would lose their absorption-enhancing potential. To screen for the most effective enhancer, a flush/ligated rat ileal model (close compartment) was used. Formulations containing sodium salicylate, sodium caprate, or sodium deoxycholate as permeation enhancers were tested. Figure 5 shows the PPS plasma concentration profiles and

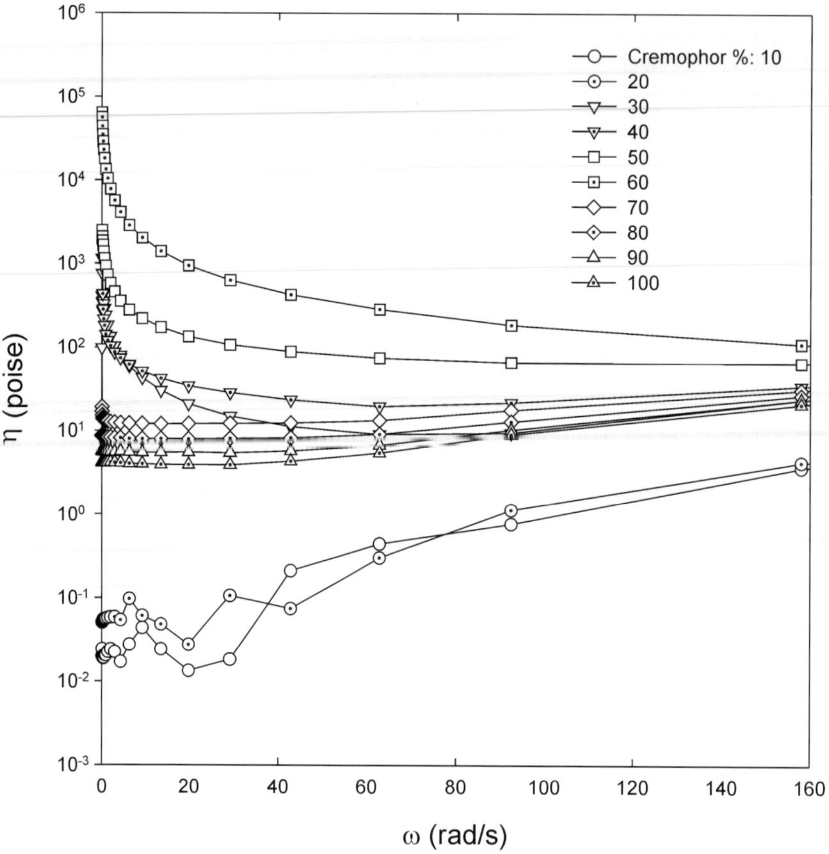

Figure 2. Shear rate effect on dynamic viscosity of Cremophor EL/water blends at 37 °C.

absolute bioavailability (BA). The BA was calculated by the following equation.

$$\mathrm{BA\%} = 100 \times \frac{\mathrm{AUC}_{\mathrm{ileal}}}{\mathrm{AUC}_{\mathrm{i.v}}} \times \frac{D_{\mathrm{i.v}}}{D_{\mathrm{ileal}}},$$

where $\mathrm{AUC}_{\mathrm{ileal}}$ and $\mathrm{AUC}_{\mathrm{i.v}}$ are the area under the drug concentration-vs-time curve for ileal dosing and intravenous dosing, respectively. AUC was estimated by trapezoidal integration to the last sampling point (t). $D_{\mathrm{i.v}}$ and D_{ileal} are, respectively, the dose for intravenous and ileal administration. The formulation containing sodium salicylate showed the highest BA of 75.3% compared to others. Other formulations containing sodium caprate or sodium deoxycholate yielded 43.6% and 27.3% of BA, respectively. All these enhancer-comprising formulations significantly enhanced the PPS absorption, with P values being, respectively, 0.004, 0.001 and 0.002 for Formulation B, C and D, compared to the control (Formulation A). This close compartment model represented a best scenario for absorption enhancement, in which high concentration of an enhancer and PPS was retained due to limited dilution effect.

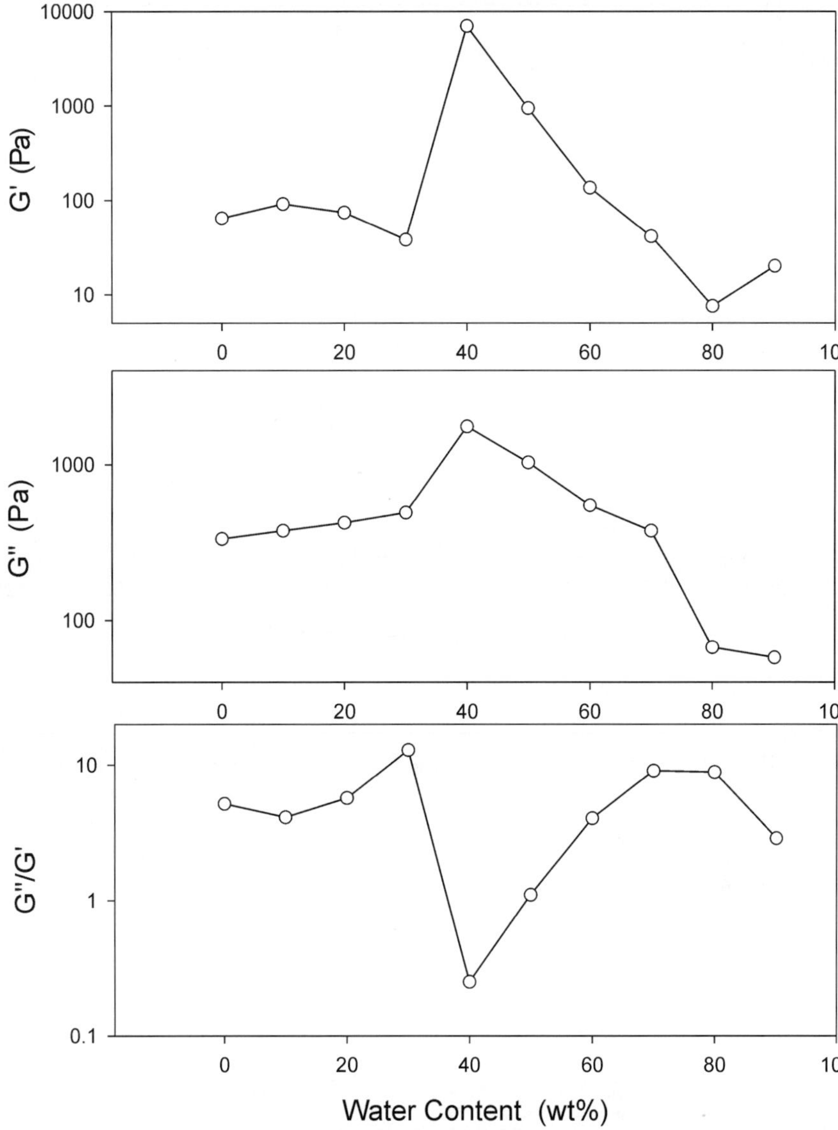

Figure 3. G', G'' and G''/G' of Cremophor/water blends as a function of water content at 158 rad/s and 37 °C.

As shown in Fig. 6, using a NF/NL rat ileal model (open compartment), low absorption enhancement (BA 4.6%) was seen for the formulation comprising sodium salicylate but not the gelling carrier. Another formulation, containing Cremophor but no sodium salicylate, led to 6.9% of BA. In sharp contrast, a potent synergistic effect (BA 46.4%, P value 0.001) was illustrated for the formulation comprising both sodium salicylate and the gelling carrier.

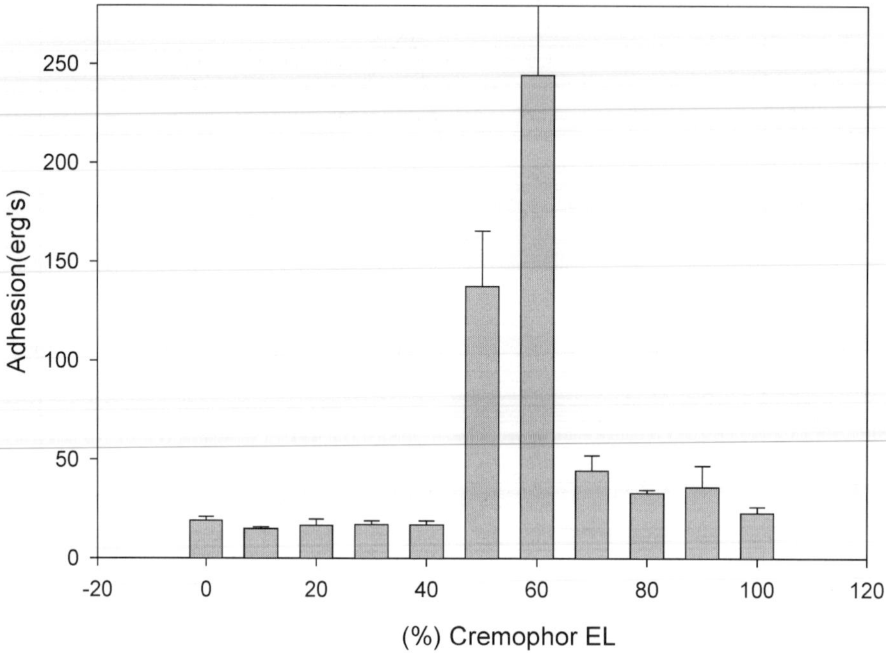

Figure 4. Mucoadhesion of Cremophor/water blends as a function of water content.

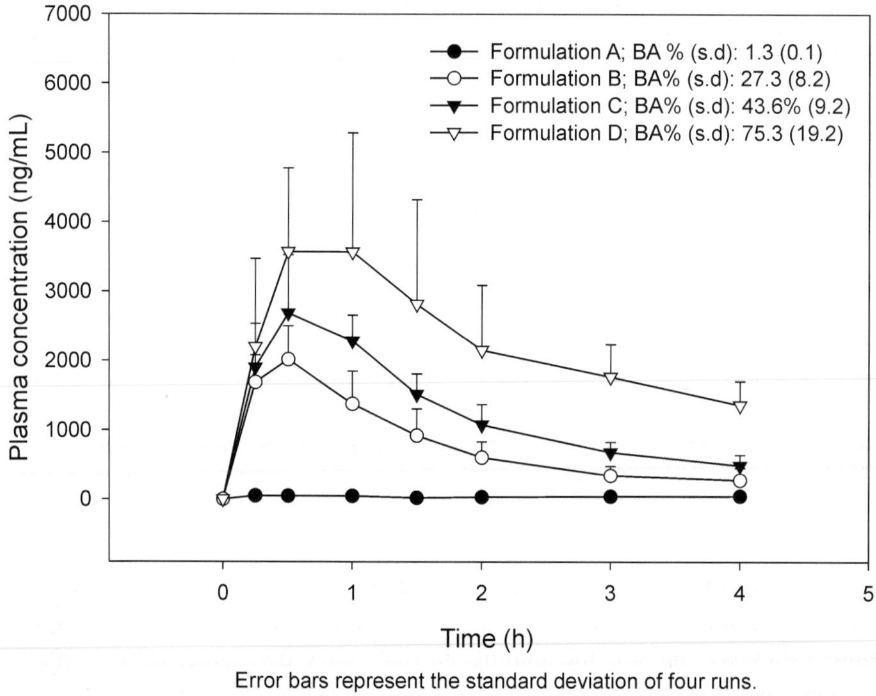

Figure 5. PPS plasma concentration profiles following administration of various formulations (F/L rat ileal model).

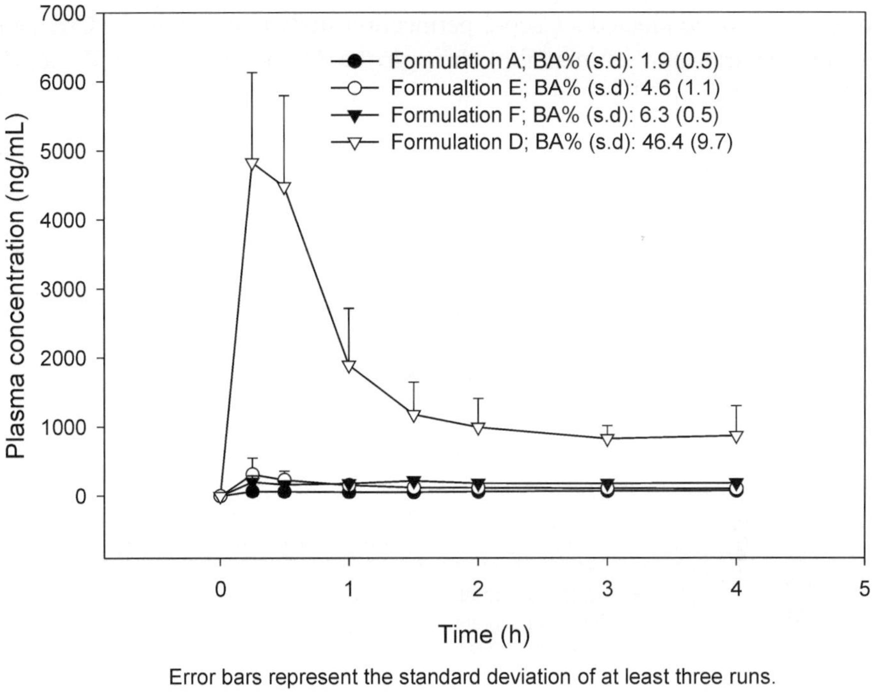

Figure 6. PPS plasma concentration profiles following administration of various formulations (NF/NL rat ileal model).

DISCUSSION

Since PPS is a hydrophilic molecule, its transport through the epithelia is mainly a passive diffusion through paracellular route. The concentration gradient of PPS is very critical, since it is the driving force for passive diffusion. Due to the low permeability of PPS through the epithelial tight junction, a permeation enhancer is needed. Numerous studies were reported to use penetration enhancers for enhancing the oral bioavailability of hydrophilic molecules. Such enhancers facilitate absorption of the hydrophilic molecules by either increasing membrane fluidity [13–15] or loosening the tight junction between cells [16, 17]. A permeation enhancer does not have significant enhancement effect on PPS permeation below its critical concentration. It is, therefore, of paramount importance to retain high concentration of a permeation enhancer as well as PPS at the mucosal surface. As we have measured, Cremophor gels *in situ* and the resultant gel can show a muco-adhesive property. It is believed that the synergistic effect can be ascribed to the *in situ* gelling property of the carrier that facilitates the PPS transport by minimizing the dilution effect and establishing high concentration gradients for PPS and sodium salicylate at the mucosal surface.

Cremphor is a non-ionic surfactant. It may enhance the PPS absorption by increasing cell membrane fluidity but not loosening the cell tight junction.

To confirm, we conducted a Caco-2 permeation study to examine the Cremophor effect on the transport of mannitol through the Caco-2 cells. As depicted in Fig. 7, the Cremophor does not enhance the mannitol transport even at a very high

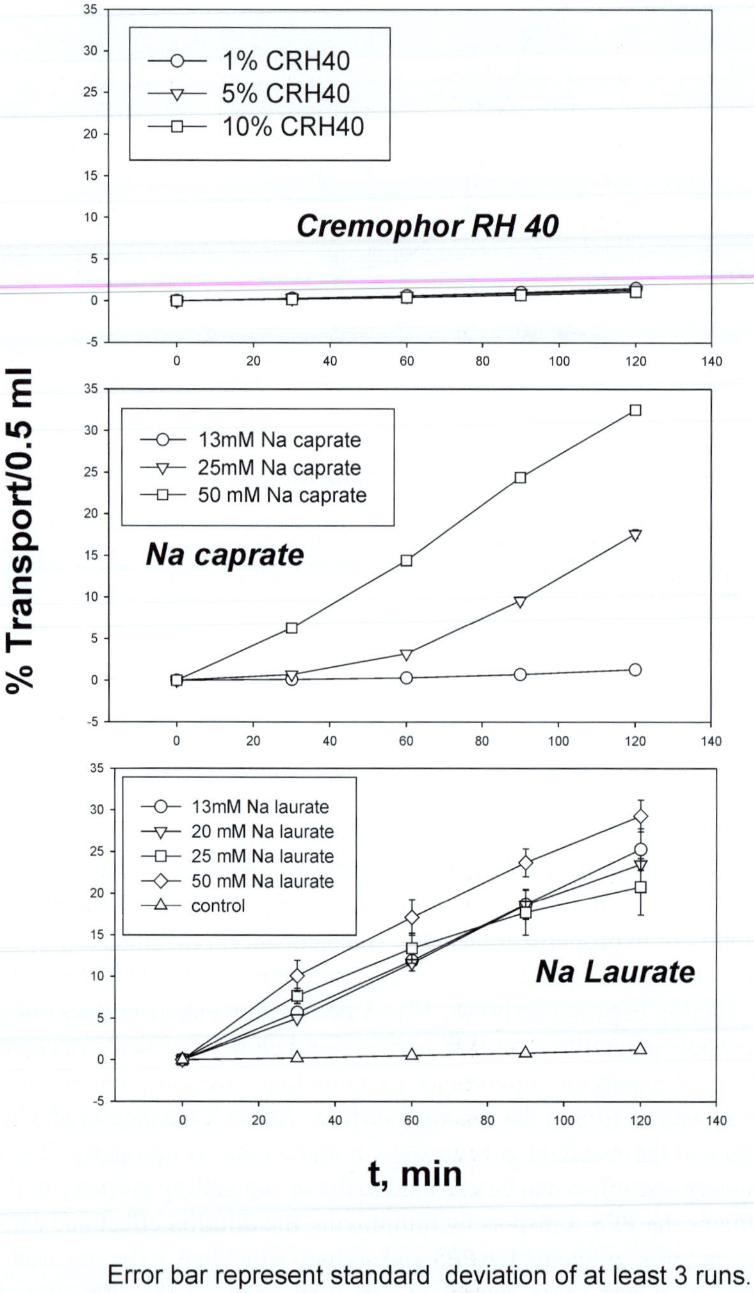

Figure 7. Enhanced transport of 0.1 mM mannitol through Caco-2 cells at 37 °C.

concentration, 10% by weight. In contrast, other two permeation enhancers, sodium caprate and sodium laurate demonstrate the enhancement at concentrations above 25 mM (0.48%, by weight) and 13 mM (0.29% by weight), respectively.

Even though our *in situ* gelling formulation has showed very encouraging results for enhancing PPS absorption, there is still a concern that the potential unmatched diffusion and absorption of PPS and sodium salicylate, due to the difference of their molecular size and physical properties, might lead to lower BA if tested in a larger animal. As reported, preferential absorption of a permeation enhancer resulted in a much-compromised absorption of hydrophilic compounds [18]. In addition, the *in situ* gelling formulation will eventually turn into a micelle after absorbing a large quantity of water. Micelle formation will expose both a hydrophilic macromolecule and an enhancer to the continuous aqueous phase, thereby causing them to be subject to extensive dilution. We are still researching to develop a more effective formulation that can further minimize the dilution effect. Because of the rheological nature of our *in situ* gelling formulation, its extensive dilution would occur in the stomach if the formulations were administered via oral gavage. To circumvent this dilution effect in the stomach, this liquid formulation should be incorporated into a delivery platform such as L-OROS™ systems for a site-specific release at the distal small intestine such as the ileum or the ascending colon to maximize the absorption of a hydrophilic macromolecule [19, 20].

CONCLUSIONS

An *in situ* gelling formulation comprising Cremophor and sodium salicylate has been demonstrated to greatly enhance PPS bioavailability in rats. In a non-flushed/non-ligated rat model (open compartment), we observed a very encouraging synergistic effect of an *in situ* gelling agent, Cremophor, and a permeation enhancer, sodium salicylate.

Acknowledgements

The authors would like to express their appreciation to J. Lee for his help with rheological measurement and to M. Broughall for her assistance with the Caco-2 cell experiment.

REFERENCES

1. *Physicians' Desk Reference*, 52th edn, p. 592. Thomson PDR, Montvale, NJ (1998).
2. K. Tanigushi, S. Muranishi and H. Sezuki, *Int. J. Pharm.* **4**, 219 (1980).
3. A. Leone-Bay, D. R. Paton, J. Freeman, C. Lercara, D. O'Toole, D. Gschneidner, E. Wang, E. Harris, C. Rosado, T. Rivera, A. DeVincent, M. Tai, F. Mercogliano, R. Agarwal, H. Leipold and R. A. Baughman, *J. Med. Chem.* **41**, 1163 (1998).
4. T. Rivera, A. Leone-Bay, D. R. Paton, H. R. Leipold and R. A. Baughman, *Pharm. Res.* **14**, 1830 (1998).

5. M. Gonze, J. D. Manord, A. Leone-Bay, R. A. Baughman, C. L. Garrard, W. C. Sternbergh and S. R. Money, *Am. J. Surg*. **176**, 176 (1998).
6. M. Thanou, J. C. Verhoef, M. T. Nihot, J. H. Verheijden and H. E. Junginger, *Pharm. Res*. **18**, 1638 (2001).
7. M. Thanou, M. T. Nihot, M. Jansen, J. C. Verhoef and H. E. Junginger, *J. Pharm. Sci*. **90**, 38 (2001).
8. C. E. Kast, D. Guggi, N. Langoth and A. Bernkop-Schnurch, *Pharm. Res*. **20**, 931 (2003).
9. M. Goldberg and I. Gomez-Orellana, *Nature Rev. Drug Discov*. **2**, 289 (2003).
10. A. Chao, J. V. Nguyen, M. Broughall, A. Griffin, J. A. Fix and P. E. Daddona, *Int. J. Pharm*. **191**, 15 (1999).
11. M. Malmsten and B. Lidman, *Macromolecules* **26**, 1282 (1993).
12. K. Zhang and A. Khan, *Macromolecules* **28**, 3807 (1995).
13. S. J. Wu and J. R. Robinson, *Pharm. Res*. **16**, 1266 (1999).
14. K. Horie, M. Hayashi and S. Awazu, *Life Sci*. **37**, 523 (1985).
15. K. Horie, M. Hayashi and S. Awazu, *J. Pharm. Sci*. **77**, 390 (1988).
16. S. Yamashita, H. Saitoh, K. Nakanishi, M. Masada, T. Nadai and T. Kimura, *J. Pharm. Pharmacol*. **37**, 512 (1985).
17. K. Sakai, T. M. Kutsuna, T. Nishino, Y. Fujihara and N. Yata, *J. Pharm. Sci*. **75**, 387 (1986).
18. K. Yakahashi, T. Murakami, R. Yumoto, T. Hattori, Y. Higashi and N. Yata, *Pharm. Res*. **11**, 1401 (1994).
19. L. Dong, K. Shafi, P. Wong and J. Wan, *Drug Deliv. Technol*. **2**, 52 (2002).
20. L. Dong, R. Zhao and P. Wong, in: *Proceedings of the 30th International Symposium on Controlled Release of Bioactive Materials*, Glasgow, p. 544 (2003).

Gels, Genes, Grafts and Giants, pp. 153–174 (2005)

Review

Gene delivery: intelligent but just at the beginning

E. PİŞKİN [1,2,*], S. DİNÇER [1,2] and M. TÜRK [1,2]

[1] *Hacettepe University, Chemical Engineering Department and Bioengineering Division, Beytepe, Ankara, Turkey*
[2] *TÜBİTAK-Center of Excellence: BİYOMÜH, Beytepe, Ankara, Turkey*

Received 13 October 2003; accepted 8 March 2004

Abstract—Gene therapy is used to treat genetic disorders, which may be achieved both *ex vivo* and *in vivo*. Gene-delivery systems usually include a carrier system which both protects the gene expression plasmid and allows its extracellular and intracellular trafficking. Viruses are used in most of the clinical trials today; however, they do have important drawbacks. Non-viral vectors based on lipids, water-soluble polycations, other non-condensing polymers and nano- or microparticles/capsules have been proposed. Cationic polymers. especially carrying novel targeting ligands. are receiving increasing attention. Intelligent polymers with temperature, pH, and light sensitivities for a controllable and effective non-viral transfection have recently been introduced but are just at the beginning. Our preliminary studies showed that block copolymers of *N*-isopropylacrylamide-acrylic acid with poly(ethylene imine) could be one example of these novel non-viral vectors.

Key words: Gene delivery; non-viral vectors; polycations; intelligent polymers; temperature sensitive copolymers; N-isopropylacrylamide; poly(ethylene imine).

INTRODUCTION

Conventionally, gene therapy is used to treat genetic disorders, such as cystic fibrosis, hemophilia and cancer, by replacement of mutated (disease coding) gene with the healthy copies in the target cells (essential, non-reproductive somatic cells) for the production of therapeutic proteins necessary for normalization of the cellular processes [1–7]. Recent interest has been expanded to the applications of gene therapy to treat disease by delivering DNA, RNA, or antisense sequences that alter gene expression within a specific cell population, thereby manipulating cellular processes and responses, which in turn stimulates immune responses or tissue regeneration, or blocks expression at the level of transcription or translation for treatment of several diseases, such as aids, cancer, and soft and hard tissue wounds.

*To whom correspondence should be addressed. E-mail: piskin@hacettepe.edu.tr

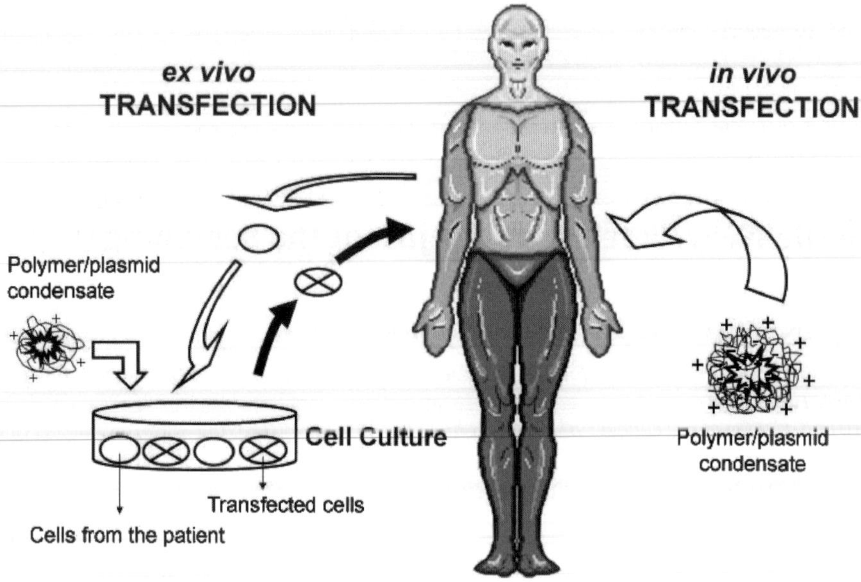

Figure 1. Gene therapy: *ex vivo* and *in vivo* transfection with polymer/plasmid condensates.

There are three components in current gene therapy which are as follows: (i) a gene that encodes a specific therapeutic protein; (ii) a gene expression plasmid that controls the functioning of a gene within a target cell [8]. Gene expression plasmid contain both the therapeutic protein encoding gene and several other genetic elements, including introns, polyadenylation sequences, and transcript stabilizers to control transcription, translation, and protein stability and secretion from the host cells; and (iii) a gene delivery system that delivers the gene expression plasmid to specific locations within the body and into the target cells [9]. Gene delivery systems are designed to control the effective distribution and access of a gene expression system to the target cell, and/or recognition by a cell-surface receptor followed by intracellular uptake and nuclear translocation [10]. Gene-delivery systems should protect the gene-expression system from premature degradation in the extracellular media and to deliver specifically to the target cells. Other elements in a gene delivery system may facilitate the intracellular trafficking of the gene expression system [11–13].

Gene therapy can be performed by manipulating gene expression either *ex vivo* or *in vivo* (Fig. 1). The *ex vivo* approach includes the following steps: (i) isolation of healthy cells (specific target cells, or even cells differentiated from stem cells) from either the patient or a donor; (ii) transfecting these cells in cell culture with the gene expression plasmid; and (iii) subsequently implanting the transfected cells back to the patient. In the *in vivo* gene therapy, the gene delivery system carrying the gene expression plasmid is directly introduced to the patient.

It is possible to obtain local transient transgene expression when naked plasmid DNA is administered to tissue, but not very effectively [14–18]. Naked plasmid

DNA is usually inable to cross biological membranes, such as the endothelium, the plasma membrane, and the nuclear membrane. Plasmid DNA is typically 10^3 to 10^4 bp in length (molecular weight is about 10^6–10^7), has a supercoiled tertiary structure in aqueous media, and has an effective hydrodynamic diameter of greater than 100 nm. The surface charge density of naked DNA, which has zeta potentials ranging from −30 to −70 mV, creates repulsion between the DNA and the negatively charged cell surface. This large size and negative surface charge density most likely limit the uptake of DNA by cells. In addition, DNA injected directly into a tissue drains rapidly into the lymphatics. Nucleases easily degrade unprotected DNA. Naked DNA is fragmented within a few minutes *in vivo* after intravenous injection. Intravenous delivery of DNA leads also to rapid clearance from the circulation owing to extensive uptake by the liver via scavenger receptors, thus preventing extravasation to any organ or tissue, which results in low levels of gene expression in all major organs.

Ex vivo gene delivery can be achieved either by applying several mechanical techniques or using a carrier (a 'vector'). Numerous mechanical techniques are being investigated for *ex vivo* gene expression usually with naked DNA, such as the gene gun, hydrostatic pressure, electroporation, continuous infusion, and sonication [19–23]. They provides a relatively high gene transfer efficiency *ex vivo*; however, these procedures are costly and may not be appropriate for all situations. Mechanical techniques are also capable of transfecting cells *in vivo*, possibly by compromising the integrity of the cell membrane, thus allowing entry of DNA into the cell. However, the details of the mechanism by which DNA is internalized for the various systems are not well understood. In addition, these techniques may create possible difficulties with patient compliance because target tissues must be surgically exposed for gene transfer.

In the *in vivo* approach, the plasmid DNA is usually delivered with a vector directly to the target site or as close to the site of pathology as possible. Here, barriers not only at the intracellular but also at the extracellular level should seriously be considered. There is almost no correlation between *ex vivo* and *in vivo* transfections. An approach of a carrier is characteristic, favoring efficient transfection *ex vivo* may be ineffective *in vivo*.

One of the most important factors for successful gene therapy is the vector (viral and non-viral) which delivers genes into the target cells. Viruses are reported to be comparatively much more efficacious in transferring genes into mammalian cells both *ex vivo* and *in vivo* [24–29]. The first clinical trials for gene therapy approaches to combat disease were carried out in 1990 using viral vectors. Over hundreds of gene therapy trials have been completed to date with the majority using viral vectors. Note that the viruses used have all been disabled of any pathogenic effects. The use of viruses is a powerful technique, because many of them have evolved a specific machinery to deliver DNA to cells. However, humans have an immune system to fight off the virus, which may reduce the effectiveness of the viral vectors. There are also several recurring issues that have led to a reconsideration of their use in human clinical trials. These include the ability of some viral vectors to integrate with the host genome and permanently alter its genetic structure, the capability of self-replication with a unique possibility of recombination and compliment activation, and an inability to bypass the host's defense mechanism. Another current drawback is the carrying capacity of the viral vectors, in other words the limitation on the amount of genomic information that can be introduced into these vectors [30].

Retroviruses have been used as very effective viral vectors [31]. A critical limitation of retroviral vectors is their inability to infect non-dividing cells. Another problem is the possibility of random integration of vector DNA into the host chromosome. This could lead to an activation of oncogenes or an inactivation of tumour suppressor genes. Although the theoretical probability of such an event is quite low, it is still of some concern. Lentiviruses also belong to the retrovirus family, but they can infect both dividing and non-dividing cells. The best-known lentivirus is the human immunodeficiency virus, which has been disabled and developed as a vector for *in vivo* gene delivery. The adeno viruses are a family of DNA viruses that can infect both dividing and non-dividing cells. There are considerable immunological problems to be overcome before adenoviral vectors can be used to deliver genes and produce sustained expression. A relative newcomer to the field, adeno-associated virus, is a simple, non-pathogenic, single-stranded DNA virus. It is believed that all of these existing viral vectors need to be reevaluated and further modified with novel targeting molecules such as oligopeptides as specific ligands to receptors to generate a safer, more specific and efficient gene delivery, and most importantly with sustained expression.

NON-VIRAL VECTORS

Several lipid-, peptide- and polymer-based non-viral vectors are currently under investigation for gene delivery [32]. We have come a long way since the first non-viral gene therapy clinical trial in 1992 by Nabel and co-workers. However, only about 20% of clinical trials reported the use of non-viral vectors today [33]. Non-viral vectors are receiving increasing attention due to compound stability and potential ease of chemical modification, low cost and consistent standard of production, higher biosafety and high flexibility [34]. However, to date, non-viral vectors are generally less efficient in delivering DNA and initiating gene expression as compared to their viral counterparts, particularly when used *in vivo*.

Cationic lipids

Felgner and co-workers pioneered gene delivery with cationic liposomes in 1987 [35]. Since then several cationic lipids systems were proposed. Lipofection is the first cationic lipid formulation composed of cationic lipids, *N*[1-(2,3-dioleyloxy)propyl]-*N*,*N*,*N*-trimethylammonium chloride and dioleoylphosphatidylethanol amine with a ratio of 1 : 1. So far, several cationic lipid systems have been investigated to deliver plasmid DNA to the lungs, brain, tumors and the skin, by local administration, or to vascular endothelial cells [36–39]. Some cationic lipid–DNA complexes have also been utilized in clinical trials for the treatment of cancer and cystic fibrosis. However, there are still serious limitations of cationic lipid systems including low transfection efficiency similar to other non-viral vectors, high toxicity on repeated use; potent anti-inflammatory activity *in vivo*, high price, etc. [40, 41]. Although the addition of polyethylene glycol (PEG), the so-called PEGylation can reduce the levels of compliment activation and binding to plasma proteins, subsequent gene expression levels of PEGylated cationic lipids are significantly reduced [42].

Polycations

The cationic polymers ('polycations'), which are frequently investigated as non-viral vectors, generally bear protonable amines, such as poly(L-lysine) (PLL), poly(L-oronithin) (PLO), both linear and branched poly(ethylene imine) (PEI), diethylaminoethyl-dextran (DMAE-dextran), poly(amidoamine) (PAMAM) dendimers, poly(dimethylaminoethyl methacrylate) (PDMAEMA) and chitosan [43–49]. These polymers vary widely in their structures, which influence their complexation with nucleic acids and their transfection efficiency. In addition to providing positive charges for DNA complexation, the primary amines also serve as functional groups which can be further chemically modified to bind specific ligands that can enhance one or more of the steps in the transfection process. Note that some polycations, such as poly(L-lysine) (PLL), are composed of linear chains, while other ones, like some PEI and PAMAM dendrimers, are highly branched polymers. The positive charges may be on the backbone (e.g. PEI) or on side groups (e.g. PLL). Moreover, both block copolymers (e.g. PEG-PLL block copolymer) and comb-type graft copolymers (e.g. PLL-*g*-dextran) have been studied [43, 50–52]. The relative number and pK_a of the protonable amines (primary, secondary or tertiary) differs between these polycationic vectors. Therefore, they have different positive charges at different pH. PLL exhibit a pK_a value around 9–10, which means that all amine groups on the side chains in PLL are protonated (positively charged) at physiological conditions (pH 7.4). Most of the others have pK_a values at acidic pH; therefore, they are partially protonated at pH 7.4, compared to lower pH values. The pK_a values of some polycations are as follows: for primary amine group PEI 5.5; for PAMAM dendimers 3.9–6.9 and for chitosan 6.5.

Due to electrostatic interactions between the positively charged amine groups and negatively charged phosphate groups of plasmid DNA, polycations spontaneously form complexes with plasmid by a spectacular compaction as schematically depicted in Fig. 2. Depending on the type and properties of the polycation and also the polymer/DNA ratio applied, these DNA condensates have different shape (e.g. toroidal, spheroidal, or even amorphous shapes) and size (from 20 nm to close to few μm) and different surface charges from negative charges (zeta potentials even less than –40 mV) to positive values (zeta potentials as high as +40 mV).

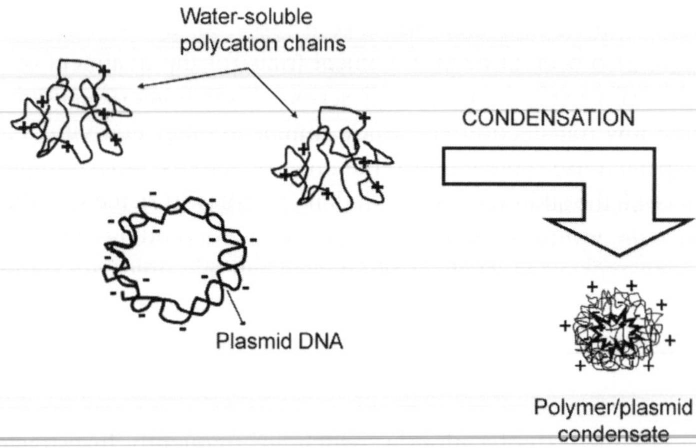

Figure 2. Condensation of plasmid DNA with polycations.

PLL is one of the most commonly used cationic polymers for gene delivery, and typically used at charge ratios ranging from 3 : 1 to 6 : 1. As increasing amounts of PLL are added to DNA, the structure changes from circular to thick, flattened to compact, and finally to toroids and rods at a charge ratio of 6 : 1. The diameter and cross section of the toroids are approximately 140 and 44 nm, respectively. The ideal molecular weight of the PLL represents a balance between two competing effects: effective condensation and cytotoxicity. The PLL with longer chains forms tighter smaller condensates that are more resistant to the effects of salt concentration and sonication. However, cytotoxicity has been found to be inversely related to particle size, likely due to the high-molecular-weight PLL.

In general, the large size condensates may contain more than one plasmid, or may be a result of clustering of smaller toroidal complexes. Electron microscopy and light scattering studies surprisingly showed that the mean size of the condensates is independent on the size (over a wide range of DNA lengths between 400 bp and 50 000 bp) and the sequence of base pairs of the DNA, and do not differ significantly when linear, supercoiled and circular DNA are condensed.

Aggregation of polycation–DNA condensates during or after complexation is one of the problems to be faced. In addition to type and properties of polycation and the polycation/DNA ratio, preparation protocol and conditions, such as pH, concentration of salt (ionic strenght) and type (usually NaCl), temperature, etc. strongly affects the aggregation behavior. The way of adding the polycation solution to the DNA solution (or *vice versa*), the DNA and salt upon mixing, diluting the complexes after their preparation, all influence the aggregation. Aggregation and therefore solubility/stability *in vivo* is a more severe problem and may strongly influence the transfection efficiency of these non-viral vectors, due to interaction of the condensates with the biological molecules in the living system, after the condensates are introduced in tissue or in blood.

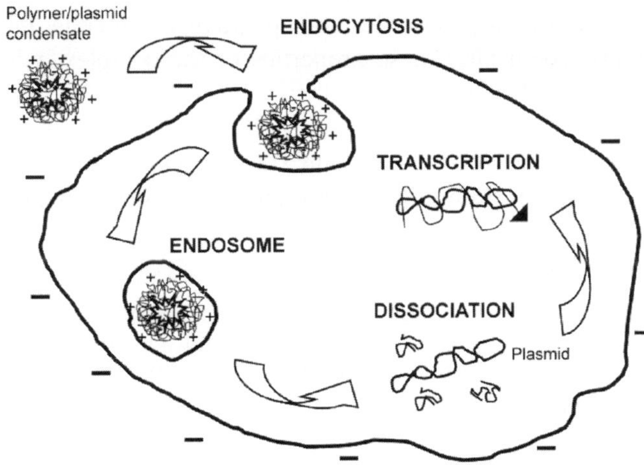

Figure 3. Cellular trafficking of polymer/plasmid DNA condensates.

There are approaches like having a hydrophilic layer around the condensate (for instance by PEGylation) which may reduce aggregation and increase solubility by reducing interactions with biomolecules. However, as mentioned above for the polycationic lipid formulations, this may decrease transfection efficiency significantly. Since there is no general strategy to prevent aggregation and to increase solubility and stability condensates, each system should be evaluated separately and optimum conditions and protocol should therefore be defined properly.

The cellular uptake process is schematically described in Fig. 3. The extra positive charge on the surfaces, which in turn allows better interaction with the negatively charged target cell membrane and therefore an enhanced uptake by endocytosis is observed. By endocytosis, the complex goes into the endosome then moves into the cytoplasm. Then, the complex translocates from the endosome to the cytoplasm by several complex mechanisms. A carrier having the capacity to buffer the pH drop in late endosomes is effective in releasing DNA into the cytoplasm, by destabilizing the endosome membrane. For instance, the polycation PEI can efficiently condense DNA and has a high intrinsic endosomal activity as compared to other polycations. PEI/DNA complexes that are taken up by the cell into acidified endosomal compartments are efficiently released into the cytoplasm *via* the 'proton sponge mechanism'. The released DNA from the complex moves into the nucleus; and the transcription process is initiated. Note that complete dissociation of the complex is required for the final transcription process for RNA polymerase reading DNA information and for expression of the target protein molecules.

In vivo gene delivery with non-viral vectors is rather difficult, and the studies are still not very satisfactory (premature) to achieve an effective transfection of the target cell population [53]. These colloidal and physical parameters are critical for efficient gene delivery *in vivo* [54–57]. The physicochemical properties of a DNA–vector complex will affect its passage through capillaries, extravasation, capture by the mononuclear phagocytes and uptake by target cells. Interaction with blood components would alter these properties of the complex, which results in altered biodistribution. It is well known that polycations exhibit serious problems when in contact with serum proteins. Especially polycations with high molecular weight (e.g. PLL, PEI and PAMAM dendimers) activate the complement system. It was suggested that the most important factor regulating complement activation and the interactions between the proteins and polycations is the number of accessible cationic charges, while type and geometry of the polycation are less importance [58–61].

As mentioned before, nucleases easily degrade unprotected DNA within a few minutes *in vivo* after intravenous injection. Condensing DNA with polycations generally improves the resistance of DNA against enzymatic breakdown. The dissociation of polycation/DNA complexes *ex vivo* as well as *in vivo* is critically important. If the affinity between the DNA on the uncomplexed polycations is too low, the complex may dissociate prematurely, e.g. in the blood stream, while a strong affinity might prevent the release of DNA intracellularly, as also mentioned above. Electrostatic forces play by far the dominant role in the dissociation of DNA to/from polycations. The interaction of polycations with blood proteins, such as albumin, can also cause the complex to disassemble exposing DNA to degrading enzymes.

The targeting of gene complexes to a desired cell population is an important subject in the field of gene therapy. Some examples to targeting ligands are as follows: lactose, mannose, galactose and derivatives, insulin-based ligands, transferin, low-density lipoprotein, monoclonal antibodies, Fab fragments of IgG, folic acid and RGD peptides. Most studies focus on the effect of targeting ligands that are covalently attached to the DNA complex and have all been demonstrated to enhance the uptake and expression of DNA by taking advantage of receptor-mediated endocytosis [62–67].

One of the drawbacks of non-viral vectors is the relatively short duration of gene expression. Unfortunately, little is understood about the fate of plasmid DNA after it reaches the nucleus. In particular, little is known about the mechanism by which gene expression is shut off. Despite these undisclosed mysteries, several studies have shown that the efficiency of expression plasmid can be significantly improved by optimizing promoters, enhancers, introns, terminator sequences and codon usage. In addition, episomal vectors have been developed to improve the expression efficiency further as well as the duration of gene expression.

Other polymers

With regard to polymers, major attention is paid to polycations which are able both to condense large genes into smaller structures and to mask the negative DNA charges, necessities for transfecting most types of cells. However, neutral polymers like polyvinyl alcohol (PVA) or polyvinylpyrrolidone (PVP) which do not condense DNA, are evaluated to protect 'naked' genes from extracellular nuclease degradation and to retain them better at the site of injection. These polymers are amphiphilic molecules, having both a hydrophilic and a hydrophobic portion. The hydrophilic portion of these polymers may interact with plasmid DNA by hydrogen bond (*via* a hydrogen-bond acceptor or donor groups), van der Waals interactions and/or by ionic interactions. PVP is a hydrogen-bond acceptor, whereas PVA is a hydrogen-bond donor. The interaction between the hydrophobic monomer units *via* van der Waals interaction can form a hydrophobic coating of vinyl backbone around the plasmid DNA and this causes a significant increase in the uptake of final condensate. The cellular uptake is due to hydrophobic interaction between the condensate and the cell wall. Commercially available surfactants poly(ethylene oxide)-*b*-poly(propylene oxide)-*b*-poly(ethylene oxide) copolymer (e.g. Pluronic P85) have also been shown to increase to enhance the transfection.

Moreover, polymeric nano or micro-particles (based on e.g. polycyanoacrylates, poly(D,L-lactic acid), poly(D,L-lactic acid-*co*-glycolic acid), gelatin, alginate, chitosan); which adsorb or encapsulate oligonucleotides or genes, are under investigation as sustained release matrices for genetic information [68–71].

STIMULI-RESPONSIVE ('INTELLIGENT') POLYMERS

Stimuli-responsive polymers also nicknamed as 'intelligent' or 'smart' polymers exhibit fast, macroscopically observable and reversible physical changes in response to external stimuli such as pH, temperature, light, ionic strength, solvents, electrical or magnetic field, etc. [72–77]. These types of polymer undergo microstructural changes stimulated by small changes in environment. These microscopic changes are apparent at the macroscopic level changes as phase, size/shape, reactivity, permeability, surface wettability, etc. The changes are reversible and the system returns to its initial state when the stimuli are removed.

These polymers may be as water-soluble chains, immobilized on solid surfaces or in solid gel form. A stimulus, for instance, increasing temperature over a certain critical value, causes a response such as polymer chain change from water-soluble coils to water-insoluble globules in aqueous media, collapse of the polymer chains on the solid surface, or shrinkage of the solid gel.

Several pH-responsive polymers have been described. They are usually composed of a hydrophobic monomer and an ionizable, more hydrophilic comonomer [72, 73]. The net charge on the polymer chain can be decreased by changing pH to neutralize the charges on the polymer chains, which in turn causes reduction in their hydrophilicity (or increase the hydrophobicity) and, therefore, the phase change. Typical examples are the copolymers of methylmethacrylate (MMA) with methacrylic acid (MAc) or dimethylaminoethyl methacrylate (DMAEMA) [78]. MMA is the hydrophobic section, while MAc is the hydrophilic part of the chains. MAc is hydrophilic at high pH when COOH groups are deprotonated but becomes more hydrophobic when COOH groups are protonated. The phase change occurs at the pK_a value of COOH groups, which is around 4.5–5.5. The copolymers of MMA

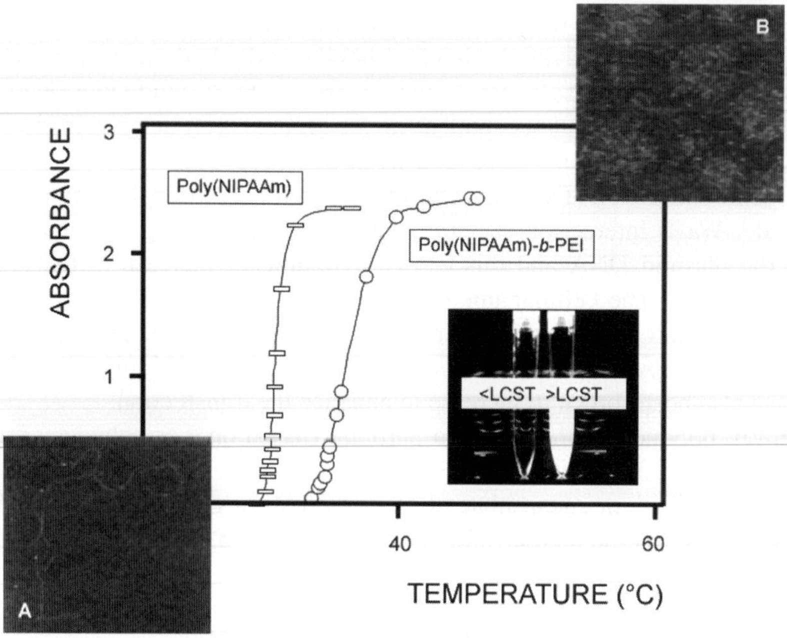

Figure 4. Phase transition of poly(NIPAAm) and its block copolymer with PEI. AFM micrographs below (A) and above (B) its LCST value.

with DMAEMA are hydrophilic at low pH when amino groups are protonated but more hydrophobic when amino groups are deprotonated. These copolymers are soluble at low pH but precipitate at slightly alkaline conditions.

Another important class of intelligent polymers are the so-called thermo-responsive ones, which are uncharged polymers and soluble in water due to the hydrogen bonding with water molecules. The efficiency of hydrogen bonding reduces with increase in temperature. The phase separation of polymer takes place when the efficiency of hydrogen bonding becomes insufficient for solubility of polymer. When the temperature is increased above a certain critical temperature, which is known as 'Lower Critical Solution Temperature' (LCST), referred to also as the 'cloud point', phase separation takes place, and the polymer chains change from water-soluble coils to water-insoluble globules (Fig. 4). A number of poly-*N*-alkylacrylamides such as poly(*N*-isopropylacrylamide) poly(NIPAAm) possesses this property, which exhibits phase transition at 32°C (its LCST) in water. Below its LCST, favorable interactions via hydrogen bonding between amide groups in polymer and water molecules lead to dissolution of polymer chains. Above the LCST, the hydrogen bonds are broken and water molecules are expelled from the polymer, resulting in the coil-to-globule transition, which was clearly demonstrated by us using AFM [80].

By copolymerization of NIPAAm with more hydrophilic (e.g. acrylamide) or more hydrophobic (e.g. n-butylacrylamide, BMA) comonomers, the LCST can

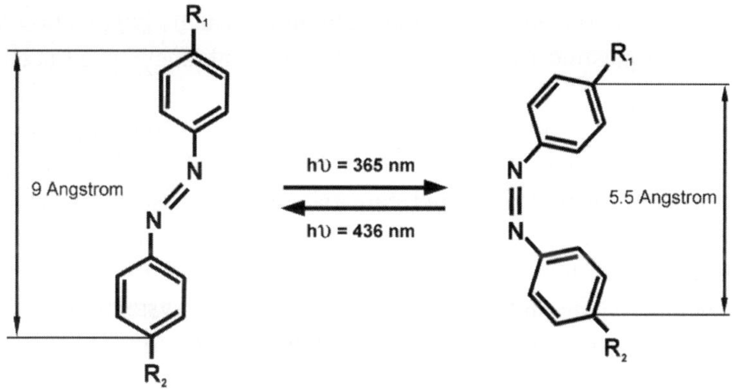

Figure 5. Light-sensitive isomers of azobenzene.

significantly be increased or reduced, respectively. Note that these copolymer chains are not sensitive to pH changes, but LCST can be reduced by increasing the ionic strength (salt concentration) in the medium. Copolymerization of NIPAAm with hydrophilic comonomers (e.g. acrylic acid (AAc) or DMAEMA) results an increase in LCST, and makes the polymer both temperature- and pH-sensitive. In addition, several molecules (including targeting bioligands, such as antibodies, oligopeptides, transferin, growth factors, folates) can be covalently bound onto the polymer backbone *via* carboxyl groups to create more intelligent carriers.

Light can also be applied as a stimulus. Various photo-responsive polymers have been extensively studied to date. The conformational changes of the polymers induced by the photopromoted or thermally promoted isomerization enable it to tailor the physical and chemical properties, including viscosity, refractive index, conductivity, pH, solubility, wettability, mechanical properties, polymer morphology and so on. This concept has stimulated many chemists to create a wide variety of photoresponsive polymers, especially those with azobenzene moieties either in the side chains or in the main chains, with the motivation of producing the light-controllable functional materials. Azobenzene is a well-known photoresponsive chromophore, and its photoinduced and thermal geometric isomerizations have been extensively studied over four decades. Azobenzene and its derivatives take both *trans* and *cis* structures with respect to the azo linkage and normally exist in the more stable trans form. The transition from *trans* to *cis* isomer is observed when UV light is applied (365 nm), while the cis confuguration returns back to trans form upon photoirradiation with visible light around 436 nm (Fig. 5).

Gene delivery with intelligent polymers

As described by Yokoyama in his recent review using intelligent polymers as novel vectors for gene delivery, two benefits can be obtained [74]. The first benefit can be on high gene expression efficiency. In an ideal non-viral vector system, the polymer/plasmid complex needs to fulfill the two functions simultaneously:

(i) protection from extracellular and intracellular enzymatic degradation and allow trafficking of the plasmid (both through the cell membrane and intracellular) and (ii) safe dissociation of the gene expression plasmid in the cell for free access to RNA polymerase. However, it is almost imposible to fulfill the two opposing phenomena: tight complex formation, which allows a favorable cell uptake and protection of DNA degradation, and ease of complex dissociation for favorable transcription by RNA polymerase. To reach an optimal 'intermediate' tightness with these conventional vectors is difficult and cannot afford the maximum efficiency for both functions. The aim with the intelligent vector systems is to fulfill both functions simultaneously, in which a tight complex can be formed to ensure evading DNA degradation and high cellular uptake, but at the later step, by introducing a stimulus which can maximize by complex dissociation for high transcription. The second benefit of the stimuli-responsive vectors is selective gene expression. Site-, timing- and duration-period-specific gene expression can be achieved with the intelligent synthetic vectors by releasing DNA from the DNA–carrier complex by inducing stimuli (local changes in temperature, pH, applying light, etc.) with a controllable manner.

Recently several stimuli-responsive non-viral vectors have been reported. Hinrich and coworkers synthesized a series of thermosensitive copolymers of NIPAAm and DMAEMA with various monomer ratios and molecular weights by free radical polymerization in organic or aqueous media by using AIBN and APS, as initiators, respectively [80]. The number average molecular weights of the copolymers prepared in organic solvents were in the range of $(9–22) \times 10^3$ and higher molecular weights were achieved in benzene. Molecular weights achieved in water were much higher ($(75–138) \times 10^3$). All copolymers, even with a low DMAEMA content of 15 mol%, were able to bind to DNA at 25°C. The copolymer/plasmid ratio at which complexes with a size of around 200 nm were formed increased with increasing NIPAAm content of the copolymer and was independent of molecular weight of the copolymer. The complexes using high-molecular-weight copolymers or lower ratios of NIPAAm with plasmid DNA were relatively stable at 37°C, when compared to the others. The zeta potential of the complexes was also independent of molecular weight of the copolymer and increased with increasing copolymer/plasmid ratio until a plateau value was reached. The plateau values decreased from around 26 mV to around 13 mV when the NIPAAm content of the copolymer was increased from 0 to 85 mol%. The cytotoxicity of the copolymers decreased with increase in concentration of NIPAAm. The transfection efficiency was a function of the copolymer/plasmid ratio and showed a bell-shaped curve. The copolymer/plasmid ratio at which the transfection efficiency was maximal increased with increasing NIPAAm content, while the maximum transfection efficiency strongly decreased with increasing NIPAAm content of the copolymer. The formation of stable copolymer/plasmid condensates with a size of around 200 nm is a prerequisite for efficient transfection. It was concluded that besides particle size, the zeta potential plays an important role in transfection. With decreasing zeta potential both transfection efficiency and cytotoxicity strongly decrease.

Kurisawa and co-workers synthesized another thermosensitive NIPAAm-based terpolymer, using DMAEMA and butylmethacrylate (BMA) as comonomers, and evaluated its transfection efficiency at different incubation temperatures [73]. In this terpolymer, BMA is the hydrophobic component, and thus the solubility of terpolymer/DNA complexes is probably regulated by both ionic and hydrophobic interactions. A terpolymer containing 8 mol% of DMAEMA and 11 mol% of BMA had a phase transition temperature of 21°C, which was found to be the same after complex formation with DNA. The terpolymer/DNA complexes showed partial dissociation at 20°C, but no dissociation at 37°C, suggesting that the formation/dissociation of the complexes was also modulated by temperature. The transfection efficiencies of terpolymer/DNA complexes incubated at lower temperatures were much higher than for those incubated at higher temperatures even for longer times. The increase in transfection by lowering temperature is due to the formation/dissociation control of thermosensitive polymer. Terpolymer/DNA complex can easily be dissociated for transcription below the LCST, while above the LCST, these complexes are tightly formed by additional hydrophobic interaction. The enhancement ratio (activity with the cooling procedure/ activity without the cooling procedure) reached 8.6 times in the 3 h incubation period case. It was concluded that such an enhancement could be useful in obtaining site- and timing-specific expression of foreign genes in biological and medical applications.

Nakasaki and colleques synthesized a novel water-soluble polyazobenzene dendimer modified with L-lysine at the periphery and investigated interaction of this polycation with plasmid DNA that is photoregulated by radiation [81]. Light scattering and gel-filtration chromatography studies showed that particle size is controllable by UV and visible light irradiation. The affinity of this cationic dendimer toward DNA was photo-controllable due to changes of zeta potential alteration on the dendimer's surface. In the transfection studies, interestingly, UV light irradiation after the polymer/plasmid condensate was uptaken in the cells caused about 50% increase in the transfection efficiency compare with the negative control, which was explained by the fact that UV radiation promoted dissociation of the complex in the cytoplasm.

OUR EXPERIENCE

We have also synthesized several water-soluble intelligent polymers based on NIPAAm for diverse biomedical applications: Conjugates of poly(*N*-isopropyl-acrylamide-co-acrylic acid) with peptides (or oligopeptides) as novel prodrugs [82–84]; poly(*N*-isopropylacrylamide-co-*p*-vinylphenyl boronic acid) for boron neutron capture therapy (data not shown); poly(*N*-isopropylacrylamide-co-*N*-[3-(dimethylamino)propyl] methacrylamide) for albumin determination in aqueous media [85]; poly(*N*-isopropylacrylamide-*co*-maleic anhydride) derivetized with poly(ethylene oxide) and poly(ethylene imine) for thermal stabilization of enzymes and as novel non-viral vectors [86–88]; poly(*N*-isopropylacrylamide-co-2-methacryloamido histidine) for identification and separation of immunoglobulins and also for gene delivery [89, 90]. Recently, we have investigated poly(*N*-isopropylacrylamide-*block*-polyethyleneimine) copolymers as intelligent polycationic DNA carriers for gene delivery and reported their synthesis, characterization, and use in *ex vivo* transfection of both cell lines and primary cells in detail elsewhere (Refs [90–92] and data not shown). Here, we briefly discuss some important aspects of these studies.

Polymer synthesis/characterization

The carboxylic acid-ended poly(NIPAAm) was synthesized by a free radical polymerization of *N*-Isopropylacrylamide (NIPAAm) in ethanol, by using selected amounts of the chain transfer agent (mercaptoacetic acid, MAA) and initiator (2,2′-azoisobutyronitrile, AIBN) under nitrogen-atmosphere at 50°C for 24 h with a shaking rate of 400 cpm (Fig. 6A). The carboxyl-ended poly(NIPAAm) was copolymerized with PEI by the reaction of the carboxylic acid groups activated with a water-soluble carbodiimide (1-ethyl-3-(3-dimethylamino)propyl carbodiimide EDAC) (Fig. 6B). A typical recipe and reaction conditions were as follows: 1.5 g carboxylic acid-ended poly(NIPAAm) and 2.2 g branched polyethyleneimine

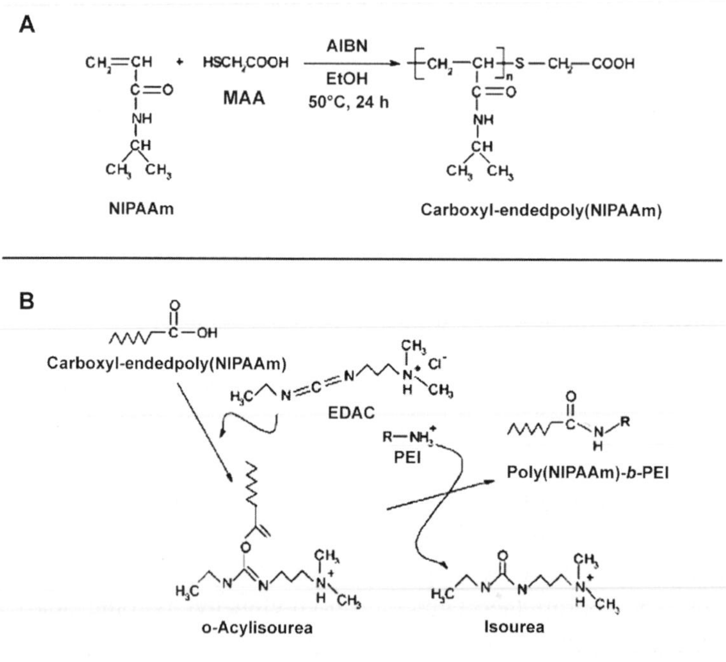

Figure 6. Synthesis of (A) carboxylic acid-ended poly(NIPAAm) and (B) poly(NIPAAm)-*b*-PEI copolymer.

(PEI, MW 2000) and 0.21 g EDAC; solvent (water) 50 ml; temperature 25°C; pH 5.0 (adjusted with 1 M HCl); duration 5 h).

FT-IR and ^{1}H-NMR data confirmed the homo- and copolymer formation [90, 91]. Satisfactorily high copolymer yields, in the range of 78–95%, were achieved in the homopolymerization of NIPAAm, which was similar reported in the related literature. However, in the case of block copolymerization of poly(NIPAAm) and PEI, polymer yields were relatively lower, and were in the range of 42–56%.

A series of carboxylic acid-ended poly(NIPAAm) was synthesized changing the concentrations of both the initiator (AIBN) and the chain-transfer agent (MAA) within the polymerization medium. The viscosity-average molecular weights of these polymers were in the range of $(13\text{–}40)\times10^3$. The initial concentration of AIBN did not change the molecular weight of the homopolymer, while the chain-transfer agent, MAA was found as the main component controlling the chain length. It was possible to increase the molecular weight significantly by decreasing the concentration of the chain-transfer agent.

Sharp phase transitions were achieved especially in the case of carboxylic acid-ended poly(NIPAAm) as given in the previous section (see Fig. 4). There was no significant difference in the LCST values obtained at two different pH values (pH 4.0 and 7.4), which indicated that the carboxylic acid-end groups were not enough to add a pH-sensitivity to the temperature-sensitive poly(NIPAAm) chains. The increase in the molecular weight slightly increased the temperature sensitivity of the chains which may be due to dominating hydrophobic interactions between the poly(NIPAAm) chains at higher temperatures. Copolymerization of poly(NIPAAm) chains with more hydrophilic PEI chains caused significant changes in the LCST behavior of the virgin poly(NIPAAm) chains. These results indicated that copolymer solubility was higher in the aqueous medium when poly(NIPAAm) with lower molecular weight was used. This is expected because the hydrophilicity of the copolymer chain increases with addition of more hydrophilic PEI chains. In addition, the solubility of copolymer was higher due to the protonation of basic amino groups of PEI at low pH. Therefore, higher LCST observed at acidic pH can also be explained by the protonation effect on the copolymer. The relative size of the blocks (poly(NIPAAm) and PEI blocks) on the copolymer chains is important in the resultant hydrophobicity of the copolymer. The change in the LCST behavior of the copolymers was much less when poly(NIPAAm) with higher molecular weights were used, in which most probably poly(NIPAAm) chains are dominating the LCST behavior.

A series of poly(NIPAAm)-*b*-PEI block copolymers, namely poly(NIPAAm)/PEI2B, poly(NIPAAm)/PEI25B, and poly(NIPAAm)/PEI25L were synthesized and use *in vitro* transfection studies, as briefly described below [90]. Here, B stands for branch and L for linear PEI, and the number at the end gives the molecular weight (dimensionless). Note that copolymerization of poly(NIPAAm) with more hydrophilic PEI chains caused increases in the LCST of the poly(NIPAAm) chains, from 31°C to around body temperature (36–39°C) [90].

Polymer/plasmid DNA condensates and in vitro transfection

Human cervix epithelioid carcinoma cell line (HeLa, obtained from Şap Institute, Ankara, Turkey) and two primary cells, namely human umbilical vein endothelial cell (PHUVEC, obtained from TÜBİTAK-Research Institute, GMBAE, Gebze, Turkey) and rabbit aortic smooth muscle cells (PRASMC, isolated in our labs from Newzealand rabbits), were used. The model plasmid used for transfections was PEGFP-N2 (Clontech, Palo Alto, CA, USA), which carries a green fluorescent protein (GFP) expressing gene. To follow transfections, fluorescein (Aldrich, Milwaukee, WI, USA) and CYBR-green I (Sigma, St. Louis, MO, USA) were used for labelling of the copolymers and the plasmid DNA, respectively.

PEGFP-N2 plasmid DNA was condensed by incubation with the polymer solutions containing different amounts of polymer (Ref. [92] and data not shown). Both PEI homopolymers PEI2, PEI25B and PEI25L and their copolymers with poly(NIPAAm) were included in these studies. Sizes and zeta potentials of the homo and copolymers, plasmid and copolymer–plasmid DNA complexes were measured at 25°C and 37°C using a Zetamaster HSA3000 (Malvern Instruments, Orsay, France).

Both HeLa cell line and primary cells were used in cell-toxicity tests, as desribed in detail elsewhere (Ref. [92] and data not shown). For transfection experiments, the cell line and primary cells were incubated with polymer/plasmid conjugates in cell culture, and transfections were followed by fluorescence microscopy, 4 h after transfection for copolymer and complex uptake, and 12–120 h after transfection GFP expression. Transfection and gene expression efficiencies were calculated as the percentage of the cells transfected or GFP expression was observed.

Zeta potentials of plasmid DNA, homo- and copolymers, at 37°C are given in Table 1. As seen here, the plasmid DNA was negatively charged while all

Table 1.
Zeta potential and average particle size of plasmid DNA and homo- and copolymers and their plasmid condensates at 37°C

Sample	Zeta potential (mV)	Average particle size (nm)
Plasmid DNA	−21	
PEI2B	+4 (+1)	992 ± 140
PEI25L	+38 (+13)	543 ± 40
PEI25B	+46 (+19)	345 ± 40
Poly(NIPA)/PEI2B	+3 (+1)	714 ± 120 (743 ± 100)
Poly(NIPA)/PEI25L	+30 (+10)	305 ± 80 (350 ± 80)
Poly(NIPA)/PEI25B	+28 (+14)	190 ± 40 (253 ± 40)

Zeta potentials of the polymer/plasmid condensates (with a polymer/plasmid ratio : 6) are given in parantheses. Average values are given here and the standard deviations were in the range of ±0.3–2 mV. Average particles sizes (and standard deviations) of the polymer/plasmid condensates (with a polymer/plasmid ratio of 6) measured at 37°C. Note that the values in parentheses are the particle sizes measured at 25°C.

Table 2.
Toxicity, transfection and gene expression efficiencies obtained in HeLa cell cultures

Sample	Toxicity (No. of viable cells in a well)		Transfection ratio (%)	Gene expression ratio (%)
	Condensates	Polymers		
PEI2B	245×10^3	240×10^3	10 ± 5	5 ± 1
PEI25L	225×10^3	210×10^3	40 ± 3	20 ± 3
PEI25B	190×10^3	130×10^3	60 ± 5	15 ± 5
Poly(NIPAAm)/PEI2B	250×10^3	248×10^3	20 ± 3	7 ± 2
Poly(NIPAAm)/PEI25L	245×10^3	230×10^3	60 ± 6	30 ± 5
Poly(NIPAAm)/PEI25B	220×10^3	200×10^3	70 ± 5	20 ± 5

In control wells there were neither polymers nor condensates. The cell count was 250×10^3.

other homo and copolymers were positively charged. Higher zeta potentials were observed when PEIs with higher molecular weights and branched are used, as expected. Copolymerization of PEIs with NIPA caused observable decreases in the zeta potential values. The zeta potentials of these complexes were between −3.1 and +21.3 mV. Higher values were observed for the complexes prepared with branched and higher-molecular-weight PEIs.

Particle sizes of the polymer/plasmid condensates were in the range of 190–992 nm (Table 1). The smallest complexes were obtained with the copolymer prepared with branched PEI with a molecular weight of 25×10^3. Copolymers were able to sequeeze plasmid DNA more at the body temperature due to phase transition of the copolymer.

Cytotoxicity of the homo and copolymers used in this study are given in Table 2. Cytotoxicity was observed with PEIs especially with the branched higher molecular weights. Copolymerization reduced the cytotoxicities. Primary cells (especially endothelial cells) were more sensitive to the polymers comparing the cell line (Table 3).

The naked plasmid DNA molecules were not able to enter the cells as expected, while transfection efficiencies in the range of 5–70% were observed with polymer/plasmid complexes that we have used (Table 2). Figure 7A gives a representative optical micrograph showing uptake of condensates within the HeLa cells. Transfections with linear PEI homopolymer (molecular weight 2000) were around 10–20%. Much higher transfections were reached with the complexes prepared with the copolymers with higher molecular weights (25×10^3), especially in the case where we used the branched PEI.

From this part of study, we can conclude that complexes prepared with copolymers (carrying PEI25L or PEI25B) are highly effective for transfection of cells *in vitro* with high efficiencies. A positive charge around 10–14 mV and complex size in the range of 200–300 nm seems optimal to reach high transfection efficiencies. The complexes prepared with linear PEI25L less effective comparing to the

Table 3.
Comparison of toxicity and gene expression efficiencies in cultures with HeLa cell line and primary cells

Sample	Toxicity (No. of viable cells in a well)		Gene expression ratio (%)	
	Condensates	Polymers	with Poly(NIPAAm)/PEI25L/palsmid condensates	with Poly(NIPAAm)/PEI25L copolymer
HeLa cell line	245×10^3	240×10^3	30 ± 5	20 ± 5
Primary smooth muscle cells	225×10^3	210×10^3	30 ± 5	15 ± 5
Primary endothelial cells	190×10^3	130×10^3	20 ± 5	8 ± 2

In control wells there were neither polymers nor condensates. The cell count was 250×10^3.

ones prepared with the branched PEI25B. However, the later one exhibited higher cytotoxicity; therefore, the linear one seems the best.

In the gene expression studies no fluorescent label was used. Green fluorescent protein (GFP) expression in HeLa cells transfected *in vitro* cell cultures was followed both by light and fluorescent microscopies. Representative micrographs shown in Fig. 7B–D clearly demonstrate gene expression.

The GFP expression efficiencies were in the range of 5–35% with the polymer–plasmid complexes that we have used (Table 2). It means that the complexes can enter into the cells but the plasmid cannot find the host DNA and integrated with it for GFP expression. Rather low transfection efficiencies were observed with the copolymer prepared with low-molecular-weight PEI (poly(NIPA)/PEI2B) in parallel to low transfection efficiencies. The most successful gene expressions were achieved with the poly(NIPA)/PEI25L, about 35% expression (the maximum) was observed at a polymer/plasmid ratio of 6. Note that this is almost half of the transfection efficiency (around 60%) observed with the same complex and with the same polymer/plasmid ratio. Surprisingly, however, gene expressions reached with the high-molecular-weight branched copolymer (PEI25B) was around 20%, comparing to about 70% transfections (which was the maximum) observed with the complexes prepared with this copolymer. Gene expressions were lower with the primary cells, especially with endothelial cells (Table 3).

As a final conclusion, we can say that complexes prepared with copolymers (carrying PEI25L or PEI25B) are highly effective for transfection of cells *in vitro*. A positive charge around 10–14 mV and complex size in the range of 200–300 nm seems optimal to reach high transfection efficiencies. Even though the lower gene expression (compared to corresponding transfection efficiencies), the complexes

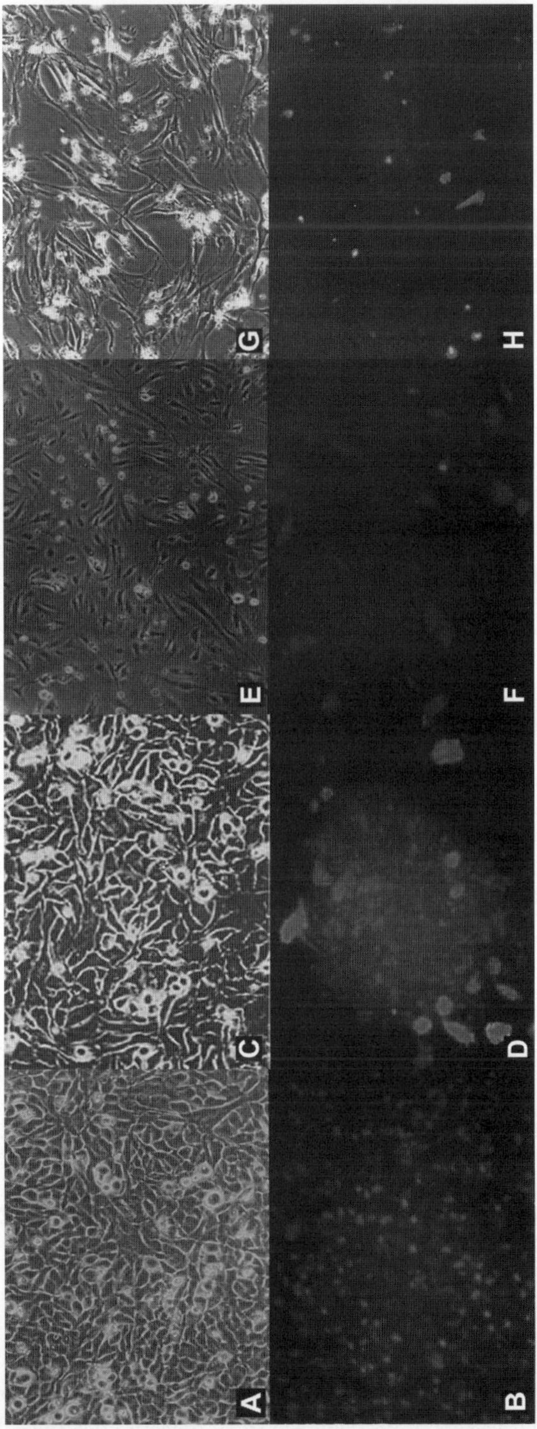

Figure 7. Representative micrographs showing transfection and gene expression. (A) Light micrroscopy image, transfected with poly(NIPA)PEI25L/plasmid DNA in HeLa cell line; (B) fluorescence microscopy image, transfected with poly(NIPA)PEI25L/plasmid DNA in HeLa cell line; (C) light micrroscopy image, transfected with poly(NIPA)PEI25L/plasmid DNA in HeLa cell line; (D) fluorescence microscopy image, gene expression, transfected with poly(NIPA)PEI25L/plasmid DNA in HeLa cell line; (E) light micrroscopy image, transfected with poly(NIPA)PEI25L/plasmid DNA in primary endothelial cells; (F) fluorescence microscopy image, gene expression, transfected with poly(NIPA)PEI25L/plasmid DNA in primary endothelial cells; (G) light microscopy image, transfected with poly(NIPA)PEI25L/plasmid DNA in primary smooth muscle cells; (H) fluorescence microscopy image, gene expression, transfected with poly(NIPA)PEI25L/plasmid DNA in primary smooth muscle cells.

prepared with a poly(NIPA)/PEI25L copolymer with a polymer/plasmid ratio of 6 seems safe (low cytoxicities) and therefore quite satisfactory potential polycationic non-viral vector system to the alternative existing ones.

Acknowledgements

We greatly acknowledge Prof. Allan Hoffman's contributions to our studies related to gene delivery, which were initiated after his excellent lecture delivered in our department few years ago. E. P. is supported by Turkish Academy of Sciences as a full member.

REFERENCES

1. M. Verma and N. Somia, *Nature* **389**, 239 (1997).
2. W. F. Anderson, *Nature* **392**, 25 (1998).
3. S. Han, R. I. Mahato, Y. K. Sung and S. W. Kim, *Mol. Ther.* **2**, 302 (2000).
4. O. Boussif, M. A. Zanta and J. P. Behr, *Gene Ther*. **3**, 1074 (1996).
5. B. Abdallah, L. Sachs and B. A. Demeneix. *Biol. Cell.* **85**, 1 (1995).
6. D. J. Porteous and J. R. Dorin, *Hum. Mol. Genet.* **2**, 211 (1993).
7. A. G. Schätzlein, *Anticancer Drugs* **12**, 275 (2001).
8. J. Vacik, B. Dean, W. Zimmer and D. Dean, *Gene Ther*. **6**, 1006 (1999).
9. D. V. Schaffer and D. A. Lauffenburger, *J. Biol. Chem.* **273**, 28004 (1998).
10. D. T. Curiel, M. L. Birnstiel, M. Cotton, E. Wagner, K. Zatloukal, C. Lank, B. Oberhauser and W. G. M. Schmidt, US Patent No. 5,547,932 (1996).
11. D. Lechardeur, K. L. Sohn, M. Haardt, P. B. Joshi, M. Monck, R. W. Graham, B. Beatty, J. Squire, H. O'Brodovich and G. L. Lukacs, *Gene Ther.* **6**, 482 (1999).
12. B. E. Houk, G. Hochhaus and J. A. Hughes, *AAPS PharmSci.* **1**, 3 (1999). Accessible at http://www.aapspharmsci.org/view.asp?art=ps010309
13. R. I. Mahato, *J. Drug Target.* **7**, 241 (1999).
14. M. E. Barry, D. Pinto-Gonzales, F. M. Orson, G. L. Mckenzie, G. R. Petry and M. A. Barry, *Hum. Gene Ther.* **10**, 2461 (1999).
15. G. Zhang, D. Vargo, V. Budker, N. Armstrong, S. Knechtle and J. A. Wolff, *Hum. Gene Ther*. **8**, 1763 (1997).
16. J. P. Yang and L. Huang, *Gene Ther*. **3**, 542 (1996).
17. J. A. Wolff, R. W. Malone, P. Williams, W. Chong, G. Acsadi, A. Jani and P. L. Felgner, *Science* **247**, 1465 (1990).
18. Y. K. Song, F. Liu and D. Liu, *Gene Ther*. **5**, 1531–1537 (1998).
19. M. A. Hickman, R. W. Malone, K. Lehmann-Bruinsma, T. R. Sih, D. Knoell, F. C. Szoka, R. Walzem, D. M. Carlson and J. S. Powell, *Hum. Gene Ther.* **5**, 1477 (1994).
20. J. M. Wells, L. H. Li, A. Sen, G. P. Jahreis and S. W. Hui, *Gene Ther*. **7**, 541 (2000).
21. A. L. Rakhmilevich, J. G. Timmins, K. Janssen, E. L. Pohlmann, M. J. Sheehy and N. S. Yang, *J. Immunother.* **22**, 135 (1999).
22. M. A. Stackhouse and D. J. Buchsbaum, *Gene Ther.* **7**, 1085 (2000).
23. S. Gottschalk, R. J. Cristiano, L. C. Smith and S. L. Woo, *Gene Ther.* **1**, 185 (1994).
24. E. Smith, *Annu. Rev. Microbiol.* **49**, 807 (1995).
25. E. Marshall, *Science* **286**, 2244 (2000).
26. E. Raz, D. A. Carson, S. E. Parker, T. B. Parr, A. M. Abaı, G. Aıchınger, S. H. Gromkowski, M. Sıngh, D. Lew, M. A. Yankauckas, A. M. Baırd and G. H. Rhodes, *Proc. Natl. Acad. Sci. USA* **91**, 9519 (1994).

27. K. Mechtler and E. Wagner, *New J. Chem.* **21**, 105 (1997).
28. M. Nakanishi, *J.Control. Rel.* **54**, 61 (1998).
29. J. D. Boyer, M. A. Chattergoon, K. E. Ugen, A. Shah, M. Bennett, A. Cohen, S. Nyland, K. E. Lacy, M. L. Bagarazzi, T. J. Higgins, Y. Baine, R. B. Ciccarelli, R. S. Ginsberg, R. R. MacGregor and D. B. Weiner, *Clin. Immunol.* **90**, 100 (1999).
30. N. Muzyczka, *Curr. Top. Microbiol. Immunol.* **158**, 97 (1992).
31. H. Varmus, *Science* **240**, 1427 (1988).
32. S. A. Rosenberg, R. M. Blaese, M. K. Brenner, A. B. Deisseroth, F. D. Ledley, M. T. Lotze, J. M. Wilson, G. J. Nabel, K. Cornetta, J. S. Economou, S. M. Freeman. S. R. Riddell, M. Brenner, E. Oldfield, B. Gansbacher, C. Dunbar, R. E. Walker, F. G. Schuening, J. A. Roth, R. G. Crystal, M. J. Welsh, K. Culver, H. E. Heslop, J. Simons, R. W. Wilmott, R. C. Boucher, H. F. Siegler, J. A. Barranger, S. Karlsson, D. Kohn, J. E. Galpin, C. Raffel, C. Hesdorffer, J. Ilan, P. Cassileth, J. O'Shaughnessy, L. E. Kun, T. K. Das, F. Wong-Staal, R. E. Sobol, R. Haubrich, M. Sznol, J. Rubin, E. J. Sorcher, J. Rosenblatt, R. Walker, K. Brigham, N. Vogelzang, E. Hersh and S. L. Eck, *Hum Gene Ther.* **11**, 919 (2000).
33. G. J. Nabel, E. G. Nabel, Z. Y. Yang, B. A. Fox, G. E. Plautz, X. Gao, L. Huang, S. Shu, D. Gordon and A. E. Chang, *Proc. Natl. Acad. Sci. USA* **90**, 11307 (1993).
34. M. R. Dyer and P. L. Herrling, *Mol. Ther.* **1**, 213 (2000).
35. P. L. Felgner, T. R. Gadek, M. Holm, R. Roman, H. W. Chan, M. Wenz, J. P. Northrop, G. M. Ringold and M. Danielsen, *Proc. Natl. Acad. Sci. USA* **84**, 7413 (1987).
36. G. McLachalan, H. Davidson, D. Davidson, P. Dickinson, J. Dorin and D. Porteous, *Biochemistry* **31**, 19 (1994).
37. X. A. Gao and L. Huang, *Biochem. Biophys. Res. Commun.* **179**, 280 (1991).
38. T. Ono, Y. Fujino, T. Tsuchıya and M. Tsuda, *Neurosci. Lett.* **117**, 259 (1990).
39. H. Farhood, N. Serbina and L. Huang, *Biochim. Biophys. Acta* **1235**, 289 (1995).
40. M. C. Filion and N. C. Phillips, *Int. J. Pharm.* **162**,159 (1998).
41. K. Crook, G. McLachlan, B. J. Stevenson and D. L. Porteous, *Gene Ther.* **3**, 834 (1996).
42. K. Hong, W. Zheng, A. Baker and D. Papahadjopoulos, *FEBS Lett.* **400**, 233 (1997).
43. Y. H. Choi, F. Liu, J. S. Choi, S. W. Kim and J. S. Park, *Hum. Gene Ther.* DNA **10**, 2657 (1999).
44. A. U. Bielinska, J. F. Kukowskalatallo and J. R. Baker, *Biochim. Biophys. Acta* **1353**, 180 (1997).
45. O. Boussif, M. A. Zanta and J. P. Behr, *Gene Ther.* **3**, 1074 (1996).
46. P. Erbacher, S. Zou, T. Bettinger, A. M. Steffan and J. S. Remy, *Pharm. Res.* **15**, 1332 (1998).
47. S. Ferrari, E. Moro, A. Pettenazzo, J. P. Behr, F. Zacchello and M. Scarpa, *Gene Ther.* **4**, 1100 (1997).
48. L. C. Smith, J. Duguid, M. S. Wadhwa, M. J. Logan, C. H. Tung, V. Edwards and J. T. Sparrow, *Adv. Drug Deliv. Rev.* **30**, 115 (1998).
49. W. E. Hennink, J. Y. Cherng, W. Jiskoot, H. Talsma and D. J. A. Crommelin, *STP Pharma Sci.* **11**, 11 (2001).
50. A. Maruyama, T. Ishihara, J. S. Kim, S. W. Kim and T. Akaike, *Bioconj. Chem.* **8**, 735 (1997).
51. D. W. Lim, Y. I. Yeom and T. G. Park, *Bioconj. Chem.* **11**, 688 (2000).
52. P. Erbacher, T. Bettinger and V. P. Belguise, *J. Gene Med.* **1**, 210 (1999).
53. C. W. Pouton, P. Lucas, B. J. Thomas, A. N. Uduehi, D. A. Milroy and S. H. Moss, *J. Control. Rel.* **53**, 289 (1998).
54. R. Kircheis, *J. Gene Med.* **1**, 111 (1999).
55. M. Nishikawa, S. Takamura, F. Yamashita, D. K. Meijer, M. Hashida and D. J. Swart, *J. Drug Target.* **8**, 29 (2000).
56. T. Takai and H. Ohmori, *Biochim. Biophys. Acta* **1048**, 105 (1990).
57. M. Ogris, S. Brunner, S. Schuller, R. Kircheis and E. Wagner, *Gene Ther.* **6**, 595 (1999).
58. D. Goula, J. S. Remy and P. Erbacher, *Gene Ther.* **5**, 712 (1998).
59. S. Ferrari, A. Pettenazzo, N. Garbati, F. Zacchello, J. P. Behr and M. Scarpa, *Biochim. Biophys. Acta* **1447**, 219 (1999).

60. D. Fischer, T. Bieber, Y. Li, H. P. Elsasser and T. Kissel, *Pharm. Res.* **16**, 1273 (1999).
61. R. I. Mahato, S. Takemura, K. Akamatsu, M. Nishikawa, Y. Takakura and M. Hashida, *Biochem. Pharmacol.* **53**, 887 (1997).
62. E. Wagner, M. Cotton, R. Foisner and M. Birnstiel, *Proc. Natl. Acad. Sci. USA* **88**, 4255 (1991).
63. W. J. Guo and R. J. Lee, *AAPS PharmSci.* **1**, 19 (1999). Accessible at http://www.aapspharmsci.org/view.asp?art=ps010419
64. S. Li, Y. D. Tan, E. Viroonchatapan, B. R. Pitt and L. Huang, *Am. J. Physiol. Lung Cell. Mol. Physiol.* **278**, L504 (2000).
65. M. Cotten, F. Langle-Rouault, H. Kirlappos, E. Wagner, K. Mechtler, M. Zenke, H. Beug and M. L. Birnstiel, *Proc. Natl. Acad. Sci. USA* **87**, 4033 (1990).
66. M. A. Zanta, P. Belguise-Valladier and J. P. Behr, *Proc. Natl. Acad. Sci. USA* **96**, 91 (1999).
67. J. A. Reddy, D. Dean, M. D. Kennedy and P. S. Low, *J. Pharm. Sci.* **88**, 1112 (1999).
68. C. Plank, K. Mechtler, F. C. Szoka Jr. and E. Wagner, *J. Drug Target.* **7**, 313 (1999).
69. P. Harvie, F. M. P. Wong and M. B. Bally, *J. Pharm. Sci.* **89**, 652 (2000).
70. D. Wang, D. R. Robinson, G. S. Kwon and J. Samuel, *J. Control. Rel.* **57**, 9 (1999).
71. Y. Capan, B. H. Woo, S. Gebrekidan, S. Ahmed and P. P. DeLuca, *Pharm. Res.* **16**, 509 (1999).
72. J. M. Benns, J. S. Choi, R. I. Mahato, J. S. Park and S. W. Kim, *Bioconjug. Chem.* **11**, 637 (2000).
73. M. Kurisawa, M. Yokoyama and T. Okano, *J. Control. Rel.* **69**, 127 (2000).
74. M. Yokoyama, *J. Artif. Organs* **4**, 138 (2001).
75. A. Izumi, R. Nomura and T. Masuda, *Macromolecules* **34**, 4342 (2001).
76. J. E. Chung, M. Yokoyama, M. Yamato, T. Aoyagi, Y. Sakurai and T. Okano, *J. Control. Rel.* **62**, 115 (1999).
77. A. Bochot, E. Fattal, A. Gulik, G. Couarraze and P. Couvreur, *Pharm. Res.* **15**, 1364 (1998).
78. J. S. Park, Y. B. Lim, Y. M. Kwon, B. Jeong, Y. H. Choi and S. W. Kim, *Polym. Chem.* **37**, 2305 (1999).
79. H. M. Zareie, V. Bulmuş, A. P. Gunning, A. S. Hoffman, E. Pişkin and V. J. Morris, *Polymer* **41**, 6723 (2000).
80. W. L. J. Hinrichs, N. M. E. Schuurmans-Nieuwenbroek, P. van de Wetering and W. E. Hennink, *J. Control. Rel.* **60**, 249–259 (1999).
81. T. Nagasaki, K. Atarashi, K. Makino, A. Noguchi and S. Tamagaki, *Mol. Cryst. Liq. Cryst.* **345**, 227 (2000).
82. V. Bulmuş, S. Patır, S. A. Tuncel and E. Pişkin, *J. Control. Rel.* **76**, 265 (2001).
83. C. Koçum, V. Bulmuş and E. Pişkin, *J. Bioact. Compat. Polym.* **17**, 239 (2002).
84. V. Bulmuş, S. Patır, S. A. Tuncel and E. Pişkin, *J. Appl. Polym. Sci.* **88**, 2012 (2003).
85. A. Tuncel, D. Demirgöz, S. Patır and E. Pişkin, *J. Appl. Polym. Sci.* **84**, 2060 (2002).
86. S. Dinçer, V. Köseli, H. Kesim, Z. M. O. Rzaev and E. Pişkin, *Eur. Polym. J.* **38**, 2143 (2003).
87. H. Kesim, Z. M. O. Rzaev, S. Dinçer and E. Pişkin, *Polymer* **44**, 2897 (2003).
88. V. Köseli, Z. M. O. Rzaev and E. Pişkin, *J. Polym. Sci. Part A Polym. Chem.* **41**, 1580 (2003).
89. E. Kalaycıoğlu, S. Patır and E.Pişkin, *Langmuir* (2004) (in press).
90. S. Dinçer, A. Tuncel and E. Pişkin, *Macromol. Chem. Phys.* **203**, 1460 (2002).
91. M. H. Zareie, S. Dinçer and E. Pişkin, *J. Coll. Interface Sci.* **251**, 424–428 (2002).
92. M. Türk, S. Dinçer, I. G. Yuluğ and E. Pişkin, *J. Control. Rel.* **96**, 325 (2004).

Gels, Genes, Grafts and Giants, pp. 175–183 (2005)

Anti-inflammatory drug delivery from hyaluronic acid hydrogels

SEI K. HAHN [1,*], SANDRA JELACIC [2], RONALD V. MAIER [2], PATRICK S. STAYTON [1] and ALLAN S. HOFFMAN [1]

[1] *Department of Bioengineering, University of Washington, Seattle, WA 98195, USA*
[2] *Department of Surgery, University of Washington, Seattle, WA 98195, USA*

Received 30 October 2003; accepted 2 March 2004

Abstract—Two different types of hyaluronic acid (HA) hydrogels were synthesized by crosslinking HA with divinyl sulfone (DVS) and poly(ethylene glycol)-divinyl sulfone (VS-PEG-VS). Vitamin E succinate (VES), an anti-inflammatory drug, and bovine serum albumin (BSA), a model of anti-inflammatory protein drugs, were loaded into the gels and their release kinetics were measured *in vitro*. VES and BSA released with a burst from both HA hydrogels during the first few hours, and release continued gradually for several days. The rate of release from HA-VS-PEG-VS-HA hydrogels was faster than that from HA-DVS-HA hydrogels, presumably due to the lower crosslink density in the former. The anti-inflammatory action of released VES was tested by incubating peripheral blood mononuclear cells (PBMC) on HA hydrogels with and without VES in the gel. The number of cells adhering on HA hydrogels was very low compared to that on tissue culture polystyrene (TCPS), which might be one of the important advantages of using HA hydrogels for implant coatings or tissue engineering applications. ELISA test results showed that the tumor necrosis factor-α (TNF-α) concentration was very low in the supernatant of the wells containing the HA hydrogel with VES in contact with the activated macrophages compared to that without VES. This is probably the effect of the released VES reducing the production of anti-inflammatory cytokine, TNF-α. HA hydrogels containing anti-inflammatory drugs may have potential for use in tissue engineering and also as biocompatible coatings of implants.

Key words: Hyaluronic acid; hydrogel; drug delivery; anti-inflammatory; TNF-α.

INTRODUCTION

Hyaluronic acid (HA) is a natural linear polysaccharide composed of alternating disaccharide units of D-glucuronic acid and N-acetyl-D-glucosamine with $\beta(1 \rightarrow 4)$

*To whom correspondence should be addressed. Current address: Chugai Pharmaceutical Co., Fuji Gotemba Research Labs, 1-135 Komakado, Gotemba, Shizuoka 412-8513, Japan. Fax: (81-550) 87-5397. E-mail: sekanhn@chugai-pharm.co.jp

interglycosidic linkages [1]. This polyanionic biopolymer is biocompatible and biodegradable and can be used in a wide molecular mass range, from 10^3 to 10^7 Da. HA is the only non-sulfated glycosaminoglycan (GAG) that is abundant in synovial fluid and extracellular matrix (ECM) [1]. HA acts to control tissue hydration and is present in hydrated networks with collagen fibers in the ECM [2]. It also constitutes the backbone of cartilage proteoglycan [3]. HA plays pivotal roles in wound healing and in promoting cell motility and differentiation during development [4]. HA has unique viscoelastic properties, which contribute to the lubrication functions of synovial joint fluids and vitreous humor in the eye [5, 6]. Because of its various functions, physical properties and biocompatibility, HA and modified HA have been extensively investigated, and are used for arthritis treatment [5], ophthalmic surgery [6], drug delivery [7] and tissue engineering [8, 9]. A number of strategies for the modification of the carboxyl and hydroxyl groups of HA have been developed, including esterification of HA [10], chemical modification of HA with amine compounds after derivatization with carbodiimide [11], and crosslinking of HA using divinyl sulfone [12], glycidyl ether [13], or dialdehyde after modification of HA with adipic dihydrazide [14]. However, chemical modification of HA has generally been carried out in highly alkaline or acidic solutions and at elevated temperatures. Due to these extreme conditions, whenever protein drugs or living cells are to be incorporated into the HA gels, they must be loaded after preparation of the polymer network [14].

Vitamin E is a biological anti-oxidant, which prevents peroxides from accumulating and protects cells from the damaging effects of free radicals [15]. Vitamin E and its derivatives have also been described as anti-inflammatory agents that inhibit the cytokine response [16]. A short-term, high-dose enteral supplement of vitamin E, vitamin E acetate and vitamin E succinate (VES) can modulate the monocyte and macrophage response to endotoxin [16]. Recently, the effects of vitamin E on monocytic cell adhesion have been studied extensively [17–19]. Aside from its ability to protect low-density lipoprotein (LDL) from oxidation, vitamin E has been shown to inhibit monocyte adhesion by inhibiting the surface expression of endothelial cell surface adhesion molecules [20]. VES can be synthesized by esterification of vitamin E with succinic anhydride and has been shown to be more efficiently taken up in the body compared with conventional vitamin E, which is very hydrophobic [21].

In this work, two different HA hydrogels were synthesized using crosslinkers of divinyl sulfone (DVS) and poly(ethylene glycol) divinyl sulfone (VS-PEG-VS). We studied the release kinetics of VES and BSA from these gels. We also studied the *in vitro* action of VES as an anti-inflammatory drug in the presence of lipopolysaccharide (LPS)-activated peripheral blood mononuclear cells (PBMC). The number of cells adhering on the gel surface in each well was determined by measuring the optical density (OD) of lactate dehydrogenase (LDH) and the concentration of tumor necrosis factor α (TNF-α) secreted from the macrophages in each well was measured with an ELISA kit.

MATERIALS AND METHODS

HA hydrogel synthesis

Two crosslinking reagents were used for the preparation of HA hydrogels, in order to compare the effect of the chain length of the crosslinkers on the release profiles of VES and BSA: DVS (Aldrich, St. Louis, MO, USA) and VS-PEG-VS, MW = 2000 (Shearwater Polymer, now Nektar Pharmaceuticals, Huntsville, AL, USA) (Fig. 1). One mmol of HA (MW $= 1.7 \times 10^6$, Genzyme, Cambridge, MA, USA) was dissolved in 10 ml of 0.2 M NaOH solution (pH 13). After stirring for 60 min, 0.6 mmol of one or the other crosslinking agent was added into the HA solution [12]. In both cases, strong hydrogels were formed within 30 min. In both cases, the hydrogel was allowed to react completely for 1 h before the gel was neutralized with 0.2 M HCl, washed by immersion in a large volume of phosphate-buffered solution (PBS, pH 7.4) for 30 min each time for 5 times, and finally dehydrated with ethanol. The HA hydrogels were lyophilized while water swollen and then visualized with a scanning electron microscope (SEM, ElectroScan Model 2020, Portland, OR, USA).

In vitro *release kinetics*

VES or BSA were loaded into the HA hydrogels by allowing the dehydrated gels to swell in VES (0.4 mg/ml) or BSA (2 mg/ml) PBS solutions for a day. VES was predissolved in ethanol (0.5 wt%) and diluted with PBS. For the *in vitro* release tests, the concentration of VES corresponding to 100% release was adjusted to be 160 μg/ml by controlling the volume of the release medium, and the concentration of BSA corresponding to 100% release was adjusted to be 400 μg/ml in a similar way. After the drug loading, the HA hydrogels were lyophilized for 2 days. *In vitro*

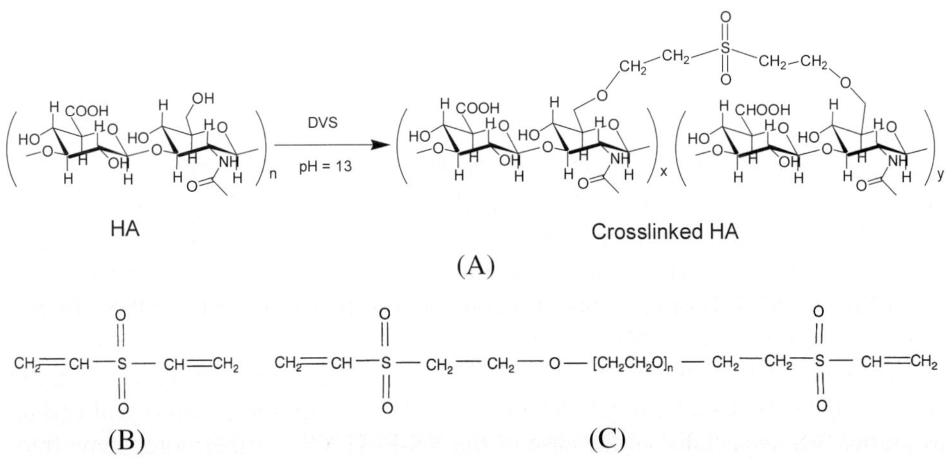

Figure 1. Schematic crosslinking reaction of divinyl sulfone (DVS) with HA in the synthesis of HA hydrogels (A) and chemical structures of the two different crosslinking reagents. (B) DVS and (C) poly(ethylene glycol) divinyl sulfone (VS-PEG-VS).

release was carried out in PBS at 37°C in a shaking water bath. VES release was quantified by measuring the OD at 286 nm and BSA release was quantified by using the BCA total protein assay (Pierce, Rockford, IL, USA) [22].

Macrophage experiment

In order to investigate the anti-inflammatory effect of released VES, hydrogels loaded with or without VES were deposited in the bottom of wells of tissue culture polystyrene plates (TCPS, Corning, New York, NY, USA). PBMC were isolated from the blood of healthy volunteers by density centrifugation, the cells were seeded at a density of 2×10^6/well and allowed to adhere to the different gel surfaces for 4 h in RPMI-1640 medium (Biowhitaker, Walkersville, MD, USA). Medium was then aspirated and changed to RPMI-1640 plus 10% adult bovine serum (HyClone, Logan, UT, USA). For macrophage activation, the PBMC were treated with 10 ng/ml LPS for 24 h. The supernatants were harvested the following day and analyzed to measure the concentration of TNF-α with an ELISA kit (R&D Systems, Minneapolis, MN, USA).

Meanwhile, cell numbers adhering on TCPS and HA hydrogels with and without VES were determined by measuring the OD at 405 nm of LDH with a cytotoxicity detection kit (Roche Molecular Biochemicals, Grenzach-Wyhlen, Germany). The LDH assay was based on the measurement of cytoplasmic enzyme activity released by damaged cells in the supernatants after activation with and without LPS. The amount of enzyme activity detected in the supernatant represents the cell numbers adhering on TCPS and HA hydrogels.

RESULTS AND DISCUSSION

HA hydrogels, HA-DVS-HA and HA-VS-PEG-VS-HA, were successfully synthesized. As soon as the crosslinking reagent was added to the HA solution, the solution became viscous and gels started to form in 5 min. The crosslinking reaction, based on the Michael addition of the HA —OH groups to the VS vinyl groups, appeared to be almost completed within 15 to 30 min. The HA-VS-PEG-VS-HA hydrogel formed relatively slowly after 30 min, while the HA-DVS-HA hydrogel was already strong after only 15 min, perhaps due to the higher reactivity of the small DVS crosslinker compared to the macromolecular VS-PEG-VS crosslinker. The HA-VS-PEG-VS-HA hydrogels exhibited higher swelling ratios (wet wt/dry wt in PBS at pH 7.4) of *ca.* 280–300 compared to those of the HA-DVS-HA hydrogels, which were *ca.* 200–240. The somewhat higher swelling of the former is probably related to its lower crosslink density. In addition, the crosslinkers will not react completely. That is, some fraction of the VS crosslinking groups will end up as grafted VS, especially, in the case of the VS-PEG-VS. Furthermore, some fraction of the crosslinking agents may not react at all and may be washed out of the gels during the final wash steps. It is very difficult to accurately quantitate the final composition and structure of these HA gels.

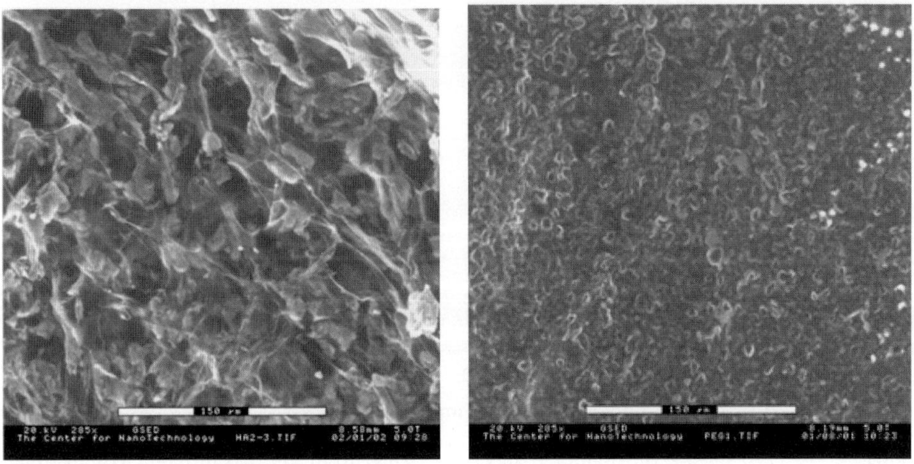

(A) (B)

Figure 2. Scanning electron micrographs of the two different HA hydrogels that have been lyophilized while swollen. (A) HA-DVS-HA and (B) HA-VS- PEG-VS-HA. Bar = 150 μm.

Figure 2 shows SEM photos of the different morphologies of the two HA hydrogels (black bar = 150 μm). Both appeared porous, as expected, since they were lyophilized in the swollen condition. It can be seen that the HA-DVS-HA gel has pore sizes in the range of 70–150 μm, which is almost the same as published by Luo *et al.* [14], who crosslinked with dialdehyde after modification of HA with adipic dihydrazide. In the case of HA-VS-PEG-VS-HA, however, the pores were relatively small, with a size of *ca.* 15 μm. According to the literature [23], a mixed solution of HA and PEG phase-separates into two phases after two days, with the top phase containing PEG and the bottom phase containing HA. Furthermore, thermal analysis of PEG-grafted HA by Moriyama *et al.* [23] suggests the presence of crystallized PEG domains that may have formed, depending on the PEG content. Considering such phase separation between PEG and HA, the smaller pores of HA-VS-PEG-VS-HA hydrogel might be ascribed to the formation during lyophilization of phase-separated domains of grafted PEG-VS. Pores may also form due to extraction of unreacted VS-PEG-VS crosslinking agent, but this is not considered to be an important contributor to the pore formation.

Figure 3 shows the *in vitro* release kinetics of VES. In both gels there was a rapid release of about 50–60% of the VES in the first few hours, while most of the remainder gradually released during the next several days. The rate of release from HA-VS-PEG-VS-HA hydrogels was faster than that from HA-DVS-HA hydrogels. This result might be ascribed to the postulated lower crosslink density of the HA-VS-PEG-VS-HA hydrogels. It should be pointed out that in the release experiment the medium was not changed during the release experiment, simulating the likely *in vivo* condition of a stagnant boundary layer. When the HA hydrogel with remaining VES was inserted into fresh PBS, almost all the remaining VES was released from the hydrogels. It is likely that *in vivo* a drug such as VES would

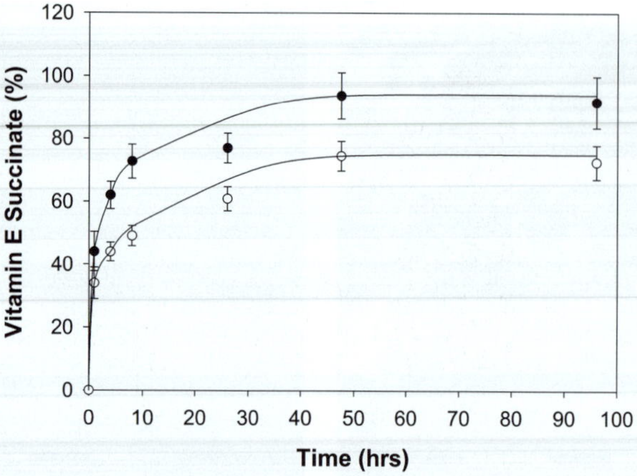

Figure 3. *In vitro* release of vitamin E succinate (VES) from the two different HA hydrogels: (open circles) HA-DVS-HA and (black circles) HA-VS-PEG-VS-HA. The concentration of VES corresponding to 100% release is 160 μg/ml.

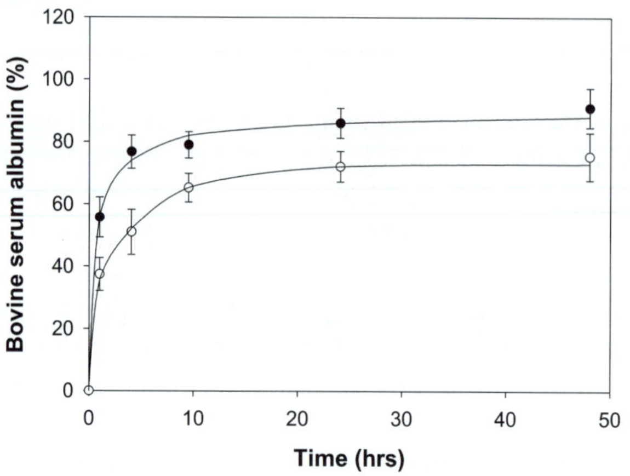

Figure 4. *In vitro* release of bovine serum albumin (BSA) from the two different HA hydrogels: (open circles) HA-DVS-HA and (black circles) HA-VS-PEG-VS-HA. The concentration of BSA corresponding to 100% release is 400 μg/ml.

be completely released from the hydrogel as long as the released VES was able to gradually diffuse away from the gel surface.

Figure 4 shows the *in vitro* release of BSA from two different HA hydrogels. The BSA is used in our studies as a model of anti-inflammatory proteins, cytokines or growth factors. The release patterns were similar to those of VES in Fig. 3, and our conclusions about these release kinetics are also similar. From the combined results shown in Figs 3 and 4 for release of VES and BSA, we conclude that HA hydrogels may be useful as efficient carriers, not only for the delivery of small size drugs like VES but also for the delivery of anti-inflammatory proteins or growth factors.

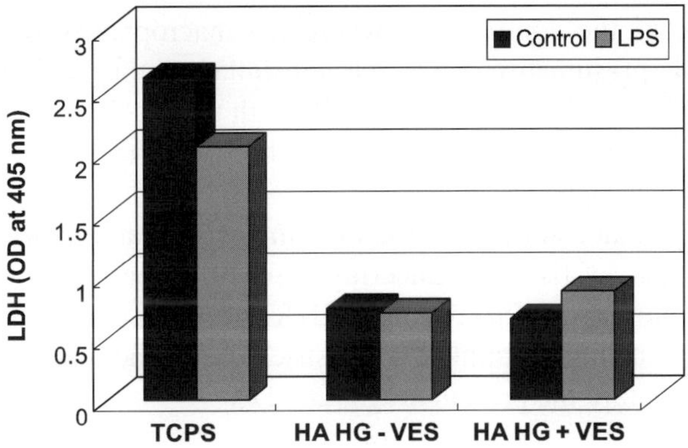

Figure 5. The optical densities (OD) of lactate dehydrogenase (LDH) from normal and LPS-activated macrophages, using a cytotoxicity detection kit. The LDH values represent the cell numbers adhering on surfaces of tissue culture polystyrene (TCPS) and the HA hydrogel crosslinked with DVS.

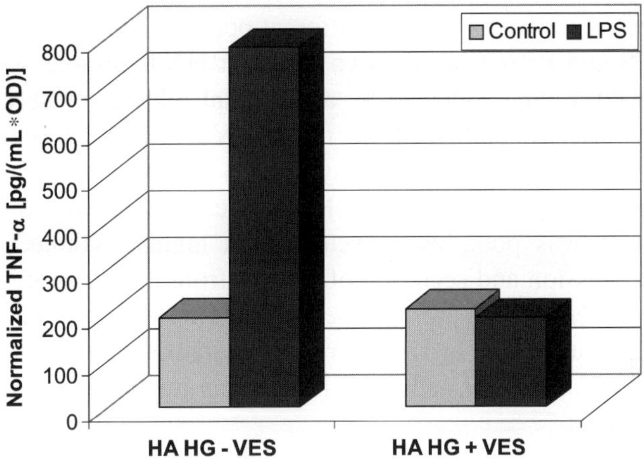

Figure 6. Effect of releasedVES on tumor necrosis factor α (TNF-α) production by lipopolysaccharide (LPS)-activated macrophages. TNF-α concentration was measured by an ELISA test kit and normalized with the cell numbers adhering on the surfaces of the DVS-crosslinked HA hydrogels with and without VES (see Fig. 5).

The effect of VES release on TNF-α production by LPS-activated macrophages was investigated using monocytes seeded onto the different HA hydrogels in wells (Figs 5 and 6). The OD of LDH represents the cell numbers in each well that were adhering on TCPS or HA hydrogels. It is well known in the literature that HA can act as a non-fouling surface [24, 25]. We also noted that cell adherence on our HA gel surfaces was poor, regardless of whether VES was in the gels or not (Fig. 5). This demonstrates one of the important advantages of HA hydrogels for tissue engineering applications and also as prosthetic implant coatings.

The TNF-α ELISA test results were normalized to the cell contents in each well. They showed that the TNF-α concentration in the supernatant of the wells containing the HA-VES hydrogel in contact with the activated macrophages was almost the same as the control well where the macrophages were not activated with LPS. This is presumably due to the anti-inflammatory action of VES on the activated macrophages. In contrast, the wells with the macrophages in contact with the HA hydrogels without VES showed a very high TNF-α concentration after LPS activation of the macrophages (Fig. 6).

Our results show that HA hydrogels containing anti-inflammatory drugs should be promising materials for tissue-engineering applications and implant coatings. We are currently developing *in situ* crosslinked HA hydrogels containing both chemical and protein drugs, using crosslinker chemistries that do not chemically modify the small drugs or denature the protein drugs [26, 27].

CONCLUSIONS

We have synthesized two types of HA hydrogels with different crosslink chemistries, different crosslink densities and different pore structures. The *in vitro* release kinetics of VES and BSA from the two different HA hydrogels were all similar; they each showed a burst followed by a continual release for several days. The rates of release of VES or BSA from the HA-VS-PEG-VS-HA hydrogels were faster than the rates from HA-DVS-HA hydrogels, presumably due to the postulated lower crosslink density of the former. Monocyte adherence on the different HA hydrogel surfaces was poor, as expected for a highly hydrated, 'non-fouling' surface. The expression and secretion of TNF-α from LPS-activated macrophages was significantly reduced by the released VES. This anti-inflammatory effect of the released VES suggests that HA hydrogels containing anti-inflammatory drugs may be promising for use as coatings of biomedical implants or tissue-engineering scaffold components.

Acknowledgements

We thank the NIH (grant # R24-HL64387) and NSF (grant # EEC-9529161) for their financial support, and Genzyme Corp. (Cambridge, MA, USA) for the contribution of HA. In addition, we are grateful to Chad Brown and Jason Hwang for valuable discussions and experimental assistance.

REFERENCES

1. T. C. Laurent, *The Chemistry, Biology and Medical Applications of Hyaluronan and its Derivatives*. Portland Press, London (1998).
2. J. R. E. Fraser, T. C. Laurent and U. B. G. Laurent, *J. Intern. Med.* **242**, 27 (1997).
3. K. Fukuda, H. Dan, M. Takayama, F. Kumano, M. Saitoh and S. Tanaka, *J. Pharmacol. Exp. Ther.* **277**, 1672 (1996).
4. K. L. Goa and P. Benfield, *Drugs* **47**, 536 (1994).
5. E. A. Balazs and J. L. Denlinger, *J. Rheumatol.* **20**, 3 (1993).

6. E. A. Balazs, in: *Healon (Sodium Hyaluronate). A Guide to its Use in Ophthalmic Surgery*, D. Miller and R. Stegmann (Eds), p. 5. John Wiley, New York, NY (1983).
7. K. P. Vercruysse and G. D. Prestwich, *Crit. Rev. Ther. Carr. Sys.* **15**, 513 (1998).
8. R. N. Feinberg and D. C. Beebe, *Science* **220**, 1177 (1983).
9. D. C. West, I. N. Hampson, F. Arnold and S. Kumar, *Science* **228**, 1324 (1985).
10. D. Campoccia, P. Doherty, M. Radice, P. Brun, G. Abatangelo and D. F. Williams, *Biomaterials* **19**, 2101 (1998).
11. J. W. Kuo, D. A. Swann and G. D. Prestwich, *Bioconj. Chem.* **2**, 232 (1991).
12. E. A. Balazs and A. Leshchiner, US Patent No. 4,582,865 (1986).
13. E. A. Balazs and A. Leshchiner, US Patent No. 4,713,448 (1987).
14. Y. Luo, K. Kirker and G. Prestwich, *J. Control. Rel.* **69**, 169 (2000).
15. G. Scott, *Antioxidants in Science, Technology, Medicine and Nutrition.* Albion Publishing, Chichester (1997).
16. E. M. Bulger, W. S. Helton, C. M. Clinton, R. P. Roque, I. Garcia and R. V. Maier, *Arch. Surg.* **132**, 1337 (1997).
17. S. Devaraj, D. Li and I. Jialal, *J. Clin. Invest.* **98**, 756 (1996).
18. A. Martin, T. Foxall, J. B. Blumberg and M. Medyani, *Arterioscler. Thromb. Vasc. Biol.* **17**, 429 (1997).
19. D. Wu, T. Koga, K. R. Martin and M. Meydani, *Atherosclerosis* **147**, 297 (1999).
20. I. Jialal, C. J. Fuller, B. A. Huet, *Arterioscler. Thromb. Vasc. Biol.* **15**, 190 (1995).
21. M. W. Fariss, F. A. Nicholls-Grzemski, M. A. Tirmenstein and J. Zhang, *Free Radic. Biol. Med.* **31**, 530 (2001).
22. P. K. Smith, R. I. Krohn, G. T. Hermanson, A. K. Mallia, F. H. Gartner, M. D. Provenzano, E. K. Fujimoto, N. M. Goeke, B. J. Olson and D. C. Klenk, *Anal. Biochem.* **150**, 76 (1985).
23. K. Moriyama, T. Ooya and N. Yui, *J. Control. Rel.* **59**, 77 (1999).
24. M. Hu, E. E. Sabelman, C. Tsai, J. Tan and V. R. Hentz, *Tiss. Eng.* **6**, 585 (2000).
25. V. B. Lokeshwar, N. Iida and L. Y. W. Bourguignon, *J. Biol. Chem.* **271**, 23853 (1996).
26. S. K. Hahn, T. Shimobouji and T. Nakamura, Japanese Patent No. 2002-338167 (2002).
27. S. K. Hahn and T. Shimobouji, Japanese Patent No. 2003-385054 (2003).

Gels, Genes, Grafts and Giants, pp. 185–199 (2005)

Design and characterisation of new nanoparticulate polymer blends for drug delivery

N. CSABA, L. GONZÁLEZ, A. SÁNCHEZ and M. J. ALONSO *

Department of Pharmacy and Pharmaceutical Technology, School of Pharmacy, University of Santiago de Compostela, 15782 Santiago de Compostela, Spain

Received 4 November 2003; accepted 4 March 2004

Abstract—The aim of the present work was the design of novel nanoparticle compositions based on poly(lactic acid/glycolic acid) (PLGA) : poloxamer and PLGA : poloxamine blend matrices. For this purpose, we have applied a modified solvent diffusion technique that allows the preparation of the nanoparticles without the use of high energy sources. Nanoparticles have been prepared with different PLGA : poloxamer and PLGA : poloxamine ratios using PEO-derivatives with different molecular weights (M_w) and hydrophilia–lipophilia balance (HLB) values. Our results show that the physico-chemical characteristics of the nanoparticles, such as size and zeta potential, are influenced by the type of PEO-derivative associated to the PLGA matrix. The ^{1}H-NMR analysis of the different nanoparticle compositions showed that the extent of incorporation of the PEO-derivative depends strongly on its HLB and also on the nanoparticles preparation conditions. The capacity of these nanoparticles as drug delivery devices was evaluated using bovine insulin as a model drug. The insulin-encapsulation efficiency was shown to be dependent on the composition of the nanoparticles, those containing hydrophilic PEO-derivatives being the most effective in entrapping the drug molecules. The formation of the blend system displayed positive effects on the release characteristics of the nanoparticles. Nanoparticles exhibited a reduced initial burst and a nearly linear, constant release rate over a time period of two weeks.

Key words: PLGA; polyoxyethylene derivatives; nanoparticle; polymer blend; protein delivery.

INTRODUCTION

Biodegradable particulate carrier systems have gained considerable interest in the recent years because of their potential use for the encapsulation, protection and controlled release of biologically active agents. The most commonly used materials for the preparation of these vehicles are poly(lactic acid) (PLA) and its copolymers with glycolic acid (PLGA), polyesters that are well known for their safety and

*To whom correspondence should be addressed. E-mail: ffmjalon@usc.es

biocompatibility. Micro- and nanoparticulate carriers made of PLA or PLGA have already been used for the delivery of many therapeutic molecules and there is a large number of studies directed to the formulation and release characteristics of these systems [1–4].

A particularly important feature of PLA and PLGA particles is that their degradation characteristics give the ability to control the release time and rate of the encapsulated molecules. This degradation is governed by hydrolytic processes and leads to the formation of acidic oligo- and monomers that cause a significant pH decrease inside the particles [5, 6]. In the course of time, this acidic microclimate and the presence of an increased number of carboxylic end groups can affect the stability and biological activity of the encapsulated molecules [7]. This acidification-induced inactivation or degradation presents an obstacle in the use of these delivery systems that still has to be overcome.

It has already been proposed by a number of authors that the co-encapsulation of protective excipients into PLGA particles could possibly prevent the unwanted interactions between the drug and the polymer as well as neutralize the acidity generated in the course of polymer degradation [7–10]. Among the alternatives, the use of block copolymers containing polyoxyethylene and polyoxypropylene units (i.e. poloxamers and poloxamines) (Fig. 1) could be a possible choice for pro-

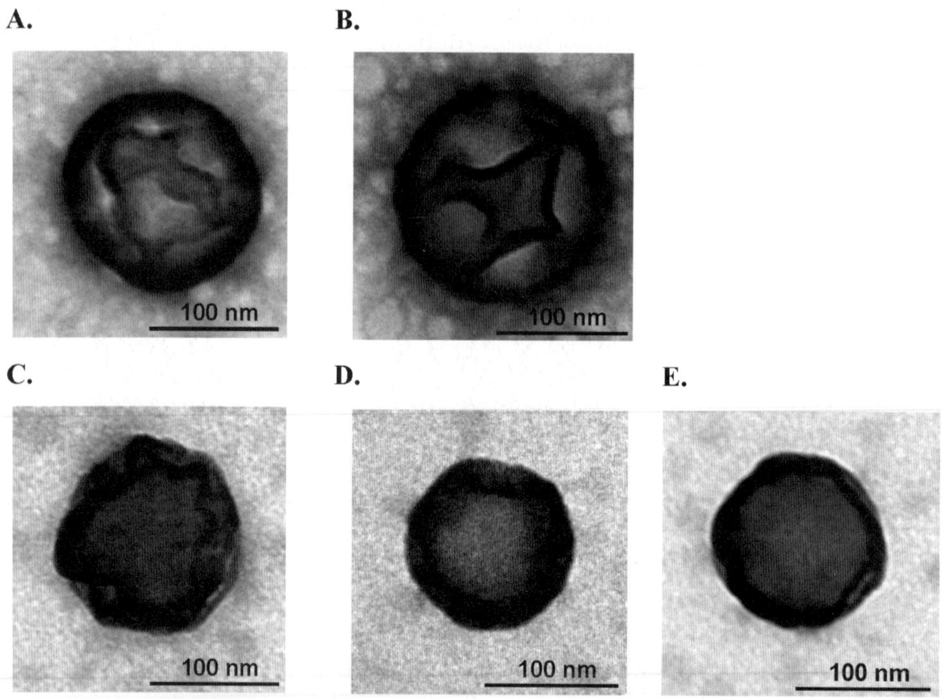

Figure 1. TEM micrographs of different blend nanoparticles with polymer ratios 50 : 50. (A) PLGA : Pluronic® F68; (B) PLGA : Pluronic® L121; (C) PLGA : Tetronic® 908; (D) PLGA : Tetronic® 904; (E) PLGA : Tetronic® 901.

viding the encapsulated macromolecules with a more stable environment. These polyoxyethylene-polyoxypropylene block copolymers have already been widely used for particle engineering as surface active agents and coating materials in the preparation of surface modified, long circulating vehicles [4, 11–13]. Moreover, poloxamer-type copolymers have extensively been investigated for other pharmaceutical applications, mainly in form of gels and micelles. The results of these investigations also include substantial information on the biocompatibility and safe use of these polyoxyethylene derivatives *in vivo* [14–17].

Previous studies carried out in our laboratory have shown that the incorporation of poloxamer-type polyoxyethylene-polyoxypropylene block copolymers into poly (lactic acid/glycolic acid) microparticles and nanoparticles is a useful approach for improving the stability of proteins encapsulated into them. In these previously reported nano- and microparticles the poloxamer was co-encapsulated in small amounts with proteins, such as BSA and interferon alpha [18, 19]. As an alternative approach, we have also developed a composite microparticulate system that is based on a blend structure of PLGA and poloxamer [20]. This blend system exhibited improved encapsulation efficiency while providing a controlled, pulsatile release profile for the protein in its active form. The formation of similar polyester–poloxamer blends in the form of films and microparticles has also been described as an approach for improving the release behaviour of these systems [21, 22]. Surprisingly, despite the recognized value of these blend matrices, this strategy has not been applied yet to the design of nanostructures. Based on this previous information, the main goal of the present work was the design of novel nanoparticle compositions based on blend matrices of PLGA : poloxamer and PLGA : poloxamine. For the preparation of the particles, we have optimised a new, modified solvent diffusion technique. The resulting blend nanoparticles have been characterised with respect to their physico-chemical characteristics, morphology, composition and also to their suitability for encapsulation and controlled release of model compounds such as bovine insulin.

MATERIALS AND METHODS

Materials

The polymer poly (D,L-lactic acid/glycolic acid) 50 : 50 (PLGA) was purchased from Boehringer-Ingelheim (Ingelheim, Germany) under the commercial name of Resomer® RG 503. The poloxamer Pluronic® F-68 with HLB = 29 and poloxamer Pluronic® L121 with HLB = 1 were obtained from Sigma Aldrich (Madrid, Spain) and BASF (Ludwigshafen, Germany). The poloxamines Tetronic® 901, 904 and 908 (HLB = 2.5, 14.5 and 30.5) were kindly donated by Bascom Belgium (Brussels, Belgium). Bovine insulin was purchased from Sigma Aldrich. All other solvents and chemicals used were of the highest grade commercially available.

Preparation of blank PLGA blend nanospheres

Nanoparticles with different PLGA : poloxamer and PLGA : poloxamine ratios were prepared by a modified emulsion–solvent diffusion technique. First, 50 mg of PLGA and the different quantities of Pluronic® or Tetronic® polymer (0, 25, 50 and 75 mg, respectively) were dissolved in 2 ml of methylene chloride and this organic solution was mixed for 30 s with a small aqueous phase by vortex (2400 min^{-1}, Heidolph, Germany). Then, the obtained emulsion was poured under moderate magnetic stirring onto a larger polar phase (25 ml ethanol), leading to immediate polymer precipitation in the form of nanoparticles. The formulations were diluted with 25 ml MilliQ water and the stirring was maintained for 10 min more. After solvent evaporation under vacuum at 30°C (Rotavapor, Büchi R-114, Flawil, Switzerland) nanoparticles were collected and concentrated in an aqueous medium. For further analysis, nanoparticles were isolated by centrifugation (1 h, 8000×g, 15°C, Avanti 30, Beckman, Barcelona, Spain) and optionally freeze dried (primary drying step for 48 h at −34°C, secondary drying step until the temperature gradually rose to 25°C, Labconco, Kansas City, MO, USA).

Physicochemical characterisation of PLGA blend nanoparticles

Nanoparticle size, polydispersity and zeta potential were determined by photon correlation spectroscopy and laser-Doppler anemometry, respectively (Zetasizer 3000 HS, Malvern Instruments, Malvern, UK). Mean values were obtained from 4 different batches, each measured 3 times. In the case of the determination of zeta potential, measurements were performed in 1 mM KCl.

The morphology of the nanoparticles was examined by transmission electron microscopy (CM 12 Philips, Eindhoven, The Netherlands) using samples stained with a 2% phosphotungstic acid solution.

The composition of the nanoparticle matrix was analysed by ^{1}H-NMR (Bruker AMX-300, Bruker, Barcelona, Spain) studies using samples of isolated and freeze-dried formulations dissolved in deuterated chloroform. Quantitative analysis was performed by the integration of the characteristic peaks in the spectra of the dissolved particles, using the software MestRe-C version 2.3. The amount of the poloxamer or poloxamine associated to the nanoparticles, calculated from the ^{1}H-NMR spectra was represented as a percentage with respect to the total mol quantity of PLGA and PEO derivative in the nanoparticles.

The thermal behaviour of the blend matrices was examined by differential thermal calorimetry (DSC) between −30°C and 80°C at a rate of 5°C/min (DSC-50, Shimadzu, Tokyo, Japan). Sample analysis was performed after nanoparticle isolation and freeze-drying.

Encapsulation of bovine insulin, determination of the encapsulation efficiency

Bovine insulin was introduced into the internal aqueous phase (200 μl) of the formulations before the emulsification step and encapsulated with 1% theoretical loading. In order to provide conditions for the complete dissolution of insulin, in this case the internal aqueous phase of the formulation also comprised 0.01 M HCl. Theoretical loadings were calculated with respect to the total weight of the PLGA in the formulation. Physico-chemical characterisation of the insulin loaded formulations was performed as described previously for the blank nanoparticles.

Encapsulation efficiency was calculated from the amount of insulin that remained in the supernatant samples recollected by the centrifugation (1 h, 8000×*g*, 15°C, Avanti 30, Beckman) of the nanoparticle suspension. The amount of this insulin was determined by the standard micro BCA protein assay (Pierce, Rockford, IL, USA). For these evaluations we used calibration curves prepared from the supernatant of the corresponding blank formulations.

In vitro *release study*

Nanoparticles were prepared with 1% theoretical loading of insulin, according to the solvent diffusion technique described above. After the solvent evaporation step, nanoparticles were recollected and concentrated in phosphate-buffered saline (PBS, pH 7.4). The final concentration of this suspension was adjusted to 2 mg particles/ml release medium. This nanoparticle suspension was divided in volumes of 4 ml into individual tubes, each corresponding to a time interval of the release experiment ($t = 3$ h and 1, 3, 7, 14 days). These release samples were incubated at 37°C with horizontal shaking (120 min^{-1}, Promax 1020, Heidolph, Schwabach, Germany). Samples were recollected at the predetermined time intervals by centrifugation for 1 h at 8000×*g* and 15°C (Avanti 30, Beckman). The amount of insulin released at each time point was determined from the isolated supernatants by the micro BCA protein assay as described above. For the calculation of the cumulative release profiles we used calibration curves prepared from the supernatant of the corresponding blank formulation.

RESULTS AND DISCUSSION

In the present work, we have prepared biodegradable nanoparticles using a solvent diffusion technique that was conveniently adapted for the encapsulation of hydrosoluble macromolecules while minimizing the use of high energy sources. Indeed, the incorporation of an aqueous phase (in which the macromolecule can be dissolved) into the polymer (PLGA) solution was achieved by means of vortex mixing, and the formation of nanostructures was a consequence of the controlled diffusion of the polymer dissolving phase (methylene chloride) into the external polar phase (ethanol).

Besides its mildness and simplicity, the technique developed in the present work allowed us to incorporate several types of PEO-PPO-PEO block copolymers with different characteristics (Table 1) into the nanoparticles structure, forming an intimate blend with PLGA. Based on previous studies performed by our group,

Table 1.
Composition, molecular weight, HLB value and cloud point of the poloxamers and poloxamines used in this study

Commercial name	Composition	M_w ($\times 10^{-3}$)	HLB	Cloud point (°C)
Pluronic® F68	2×75 EO + 30 PO	8.5	29	>100
Pluronic® L121	2×5 EO + 68 PO	4.4	1	14
Tetronic® 901	4×3 EO + 4×17 PO	4.7	2.5	20
Tetronic® 904	4×15 EO + 4×17 PO	6.7	14.5	74
Tetronic® 908	4×119 EO + 4×17 PO	25	30	>100

we hypothesized that these modifications would display positive effects on the stability and controlled release of protein drugs encapsulated into them [13]. Hence, the influence of the incorporation of PEO derivatives on the physico-chemical characteristics of the nanoparticles, as well as on their encapsulation capacity and controlled release properties was investigated.

Physico-chemical characteristics of the nanoparticles

By adjusting the preparation conditions, the above described method leads to the formation of particles of different composition in the desired nanometric size range, between 150 and 350 nm (Tables 2–6). The size distribution of the nanoparticles was, in general, homogeneous, having polydispersity values generally lower than 0.20. An exception was represented by the PLGA : Tetronic® 901 blend series, which displayed the largest particle size and broader size distribution (polydispersity index between 0.162 and 0.308). This could be explained by the fact that the cloud point of this particular poloxamine is close to the ambient temperature (20°C) and, consequently, this could affect particle formation.

The yield of the preparation process, based on the PLGA component, was always higher than 75%. Morphological analysis by transmission electron microscopy (TEM) showed that irrespective to their composition, nanoparticles had spherical shape (Fig. 1). The TEM images also show some irregularities in the nanoparticles structure, which could be related to the presence of two differentiated polymers (PLGA and a PEO derivative) in the nanostructures.

Tables 2–6 show the influence of the quantity and type of the polyether component on the particle size and zeta potential of the nanoparticles. As compared to the pure PLGA nanoparticles, an initial decrease in particle size due to the incorporation of the PEO derivative could be observed in all series. This effect was previously described for PLGA microspheres containing surfactants such as polyoxyethylenes or polysorbates [23] and was attributed to the surface active properties of the polyethers present in the formulations. After this initial decrease, the size of the particles containing more hydrophilic block copolymers was either maintained (Pluronic® F68, Tetronic® 904) or further decreased (Tetronic® 908). On the contrary, in the case of the most hydrophobic block copolymers (Pluronic® L121,

Table 2.
Mean particle size, polydispersity index, zeta potential and relative percentage of Pluronic® F68 (average ± SD, $n = 4$) in the nanoparticles prepared with different theoretical polymer ratios

PLGA : Pluronic® F68	Size (nm)	P.I.	ζ-potential (mV)	Pluronic® F68	
				Mol%	Weight%
50 : 0	191.5 ± 7.1	0.046	−60.1 ± 7.4	—	—
50 : 25	162.8 ± 4.4	0.079	−50.2 ± 0.8	7.9 ± 0.7	2.0 ± 0.2
50 : 50	163.2 ± 5.1	0.135	−43.1 ± 6.4	10.8 ± 0.5	2.8 ± 0.1
50 : 75	159.8 ± 6.5	0.163	−38.5 ± 0.6	10.6 ± 0.6	2.7 ± 0.2

Table 3.
Mean particle size, polydispersity index, zeta potential and relative percentage of Pluronic® L121 (average ± SD, $n = 4$) in the nanoparticles prepared with different theoretical polymer ratios

PLGA : Pluronic® L121	Size (nm)	P.I.	ζ-potential (mV)	Pluronic® L121	
				Mol%	Weight%
50 : 0	191.5 ± 7.1	0.046	−60.1 ± 7.4	—	—
50 : 25	164.5 ± 6.3	0.156	−27.3 ± 7.1	45.4 ± 3.5	9.5 ± 1.2
50 : 50	185.5 ± 6.0	0.195	−30.0 ± 8.0	58.3 ± 1.5	15.0 ± 0.8
50 : 75	257.3 ± 0.0	0.179	−24.5 ± 5.5	63.0 ± 1.5	17.6 ± 0.9

Table 4.
Mean particle size, polydispersity index, zeta potential and relative percentage of Tetronic® 908 (average ± SD, $n = 4$) in the nanoparticles prepared with different theoretical polymer ratios

PLGA : Tetronic® 908	Size (nm)	P.I.	ζ-potential (mV)	Tetronic® 908	
				Mol%	Weight%
50 : 0	191.5 ± 7.1	0.046	−60.1 ± 7.4	—	—
50 : 25	189.2 ± 4.6	0.202	−30.9 ± 3.9	3.9 ± 0.5	2.8 ± 0.3
50 : 50	174.0 ± 5.4	0.271	−26.9 ± 1.2	5.5 ± 0.1	4.0 ± 0.1
50 : 75	171.2 ± 3.2	0.235	−24.1 ± 1.0	6.4 ± 0.5	4.6 ± 0.4

Table 5.
Mean particle size, polydispersity index, zeta potential and relative percentage of Tetronic® 904 (average ± SD, $n = 4$) in the nanoparticles prepared with different theoretical polymer ratios

PLGA : Tetronic® 904	Size (nm)	P.I.	ζ-potential (mV)	Tetronic® 904	
				Mol%	Weight%
50 : 0	191.5 ± 7.1	0.046	−60.1 ± 7.4	—	—
50 : 25	160.2 ± 5.6	0.188	−40.0 ± 4.6	14.7 ± 1.5	3.2 ± 0.4
50 : 50	168.7 ± 9.4	0.179	−38.4 ± 3.3	18.8 ± 2.0	4.2 ± 0.5
50 : 75	168.8 ± 2.5	0.160	−39.6 ± 2.0	19.5 ± 0.6	4.4 ± 0.1

Table 6.
Mean particle size, polydispersity index, zeta potential and relative percentage of Tetronic® 901 (average ± SD, $n = 4$) in the nanoparticles prepared with different theoretical polymer ratios

PLGA : Tetronic® 901	Size (nm)	P.I.	ζ-potential (mV)	Tetronic® 901	
				Mol%	Weight%
50 : 0	191.5 ± 7.1	0.046	−60.1 ± 7.4	—	—
50 : 25	205.3 ± 54.5	0.162	−25.4 ± 5.0	23.3 ± 0.2	3.9 ± 0.0
50 : 50	277.4 ± 102.9	0.308	−28.9 ± 5.4	26.3 ± 6.1	4.6 ± 1.3
50 : 75	333.7 ± 82.1	0.275	−38.2 ± 8.3	28.8 ± 1.1	5.1 ± 0.3

Tetronic® 901), the nanoparticle size gradually increased as more polyether was incorporated into the formulations.

With respect to the surface characteristics of the formulations, all nanoparticles had negative surface charge. However, it can also be seen that all PLGA : poloxamer and PLGA : poloxamine nanospheres have less negative zeta potentials than the PLGA ones. This suggests that a part of the polyoxyethylene component can be situated on-, or close to the surface of the nanoparticles. As has been described for other nanoparticles with PEG, poloxamer or poloxamine coatings, these polyoxyethylene derivatives can displace the shear plane of the diffuse layer to a larger distance from the particle surface and, thus, reduce the zeta potential [24–26].

Analysis of the particle composition

The composition of the nanoparticle series was determined by ^{1}H-NMR studies, using samples of isolated and freeze-dried formulations. The spectrum of the control particles prepared from pure PLGA contains the three characteristic peaks corresponding to the $-CH-$ (5.2 ppm) and $-CH_3$ (1.6 ppm) groups of the lactic acid and to the $-CH_2-$ (4.8 ppm) group of the glycolic acid component of the polymer. The presence of new peaks was noted between 1.1 and 1.2 ppm, and between 3.3 and 3.7 ppm in the spectra of the blend nanoparticles (Fig. 2). These peaks correspond to the $-CH-$ and $-CH_2-$ groups of the PEO-PPO blocks and indicate the successful incorporation of poloxamer or poloxamine into the nanoparticles. Moreover, the intensity of these peaks increased with respect to those of PLGA when the amount of the PEO–PPO block copolymer incorporated into the nanoparticle matrix was increased. This suggests that the composition of the nanoparticles can be modified by adjusting the initial polymer ratios.

The quantitative analysis of the spectra of the formulations containing poloxamers showed that Pluronic® F-68 can be incorporated up to 11 mol% with respect to the total quantity of PLGA and poloxamer in the particle. This percentage was achieved at the theoretical PLGA : poloxamer ratio of 50 : 50 and remained unmodified for the 50 : 75 ratio. A very different situation was observed in the case of Pluronic®

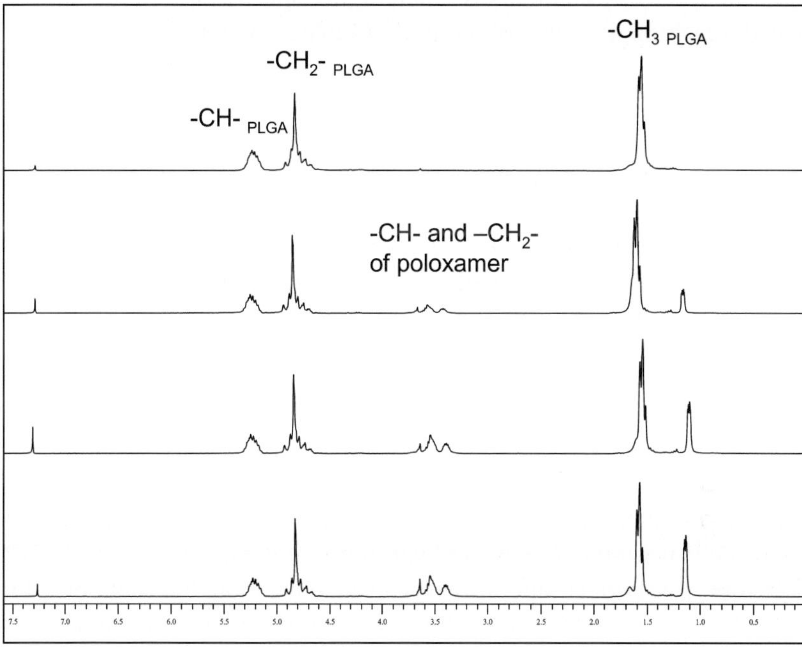

(A)

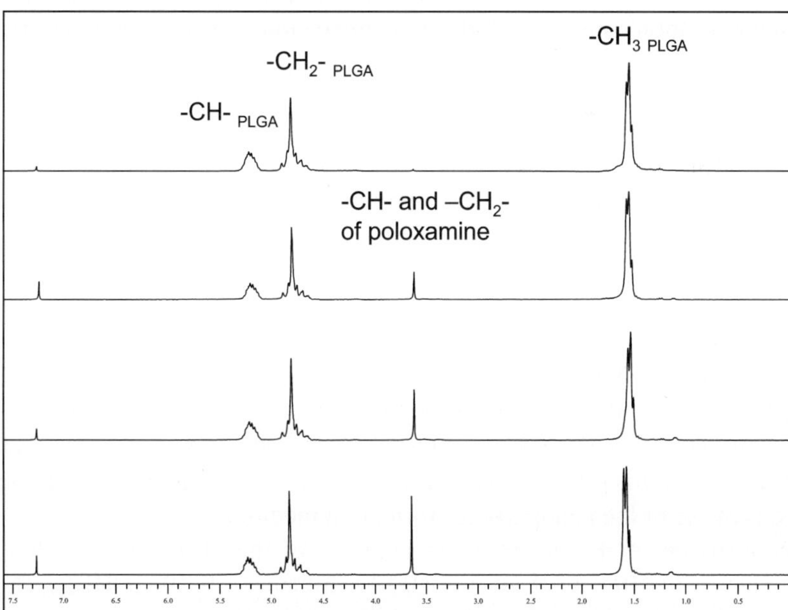

(B)

Figure 2. Typical ^{1}H-NMR spectra of PLGA : poloxamer and PLGA : poloxamine blend formulations. (A) PLGA : Pluronic® L121; (B) PLGA : Tetronic® 908.

L121, for which the association was much higher and was not limited at the ratios investigated, reaching the value of 63 mol% (Tables 2 and 3).

These differences in composition for the two types of nanoparticles could be explained by the different thermodynamic compatibility of these polyoxyethylene derivatives with the PLGA component. As described previously, the organization of these blend structures relies basically on chain entanglement and the formation of hydrogen bonds between the carboxylic end group of the polyester and the hydroxyl terminal of the polyether [27]. Characteristics such as the PEO : PPO ratio, the chain lengths and the hydrophobicity of the poloxamer/poloxamine is expected to determine its miscibility with PLGA and, thus, it will also influence the structure and composition of the particle matrix. In addition, the different nanoparticle sizes, as indicated above, could be considered as a consequence of the formation of different blend structures.

These results are in accordance with those of a previous quantitative FT-IR study performed by Yeh *et al.* [21], in which they analysed the composition of blend PLGA microparticles containing different types of poloxamers. In this previous work it was shown that the hydrophilic poloxamer Pluronic® F127 was less effectively incorporated into the particles than the more hydrophobic Pluronic® L121. This tendency underlines the importance of the HLB as a parameter for the prediction of the degree of miscibility of the two polymers to be blended.

Apart from the influence of the thermodynamic compatibility, the differences in the incorporation amounts of the polyethers into the nanoparticle structures may also be attributed to the preparation conditions. In our case, the different solubility of the polyoxyethylene derivatives in the external medium of the particle formation should logically affect their incorporation within the matrix. The low solubility of the poloxamer Pluronic® L121 prevents its partition towards the external ethanol/water phase and facilitates its incorporation at higher amounts. However, the association of the poloxamer Pluronic® F68 into the nanoparticle matrix is probably limited by its hydrophilic character [27].

Tendencies similar to those observed for the poloxamer-containing nanomatrices could be identified following the analysis of the poloxamine-containing nanostructures (Tables 4–6). The hydrophobic character of the poloxamines Tetronic® 904 and 901 allows their efficient association to the polyester matrix while in the formulations comprising of the very hydrophilic Tetronic® 908 this efficiency was limited by the tendency of this poloxamine to diffuse from the organic phase to the aqueous continuous phase of the nanoparticles formation medium.

The comparison of the incorporated amounts of the different PEO derivatives (Fig. 3) shows that this parameter is clearly influenced by their HLB. Thus, the minimum value was observed for Tetronic® 908 (HLB = 30) and the maximum incorporation was achieved for the lipophilic Pluronic® L121 (HLB = 1). The tendency of the poloxamines to associate to the polyester matrix could also be affected by the molecular weight. In fact, as seen in Fig. 3, their association went from less than 10% for Tetronic® 908 (M_w 25 × 10^3) to nearly 30% for Tetronic®

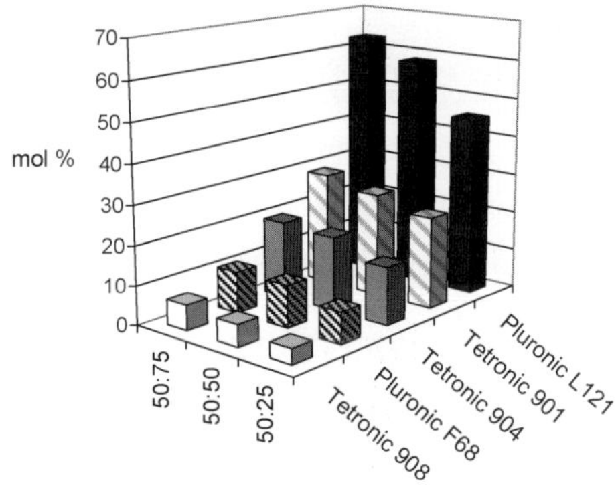

Figure 3. Incorporated amounts determined by ^{1}H-NMR for the different poloxamers and poloxamines into the nanoparticle series with different initial polymer ratios.

901 (M_w 4700). However, since this parameter is not independent of the HLB, it is not possible to draw a conclusion about its possible role in the incorporation of the PEO derivatives into the PLGA matrix. On the other hand, the comparison of the behaviour of Tetronic® 901 and Pluronic® L121, both of them hydrophobic and with a similar M_w, led us to suggest that not only the hydrophobicity but also their different chemical composition may affect their entanglement within the PLGA network.

Finally, further evidence of the incorporation of the PEO derivatives in the nanomatrices was found in the DSC thermograms (Fig. 4). The processing of the co-dissolved PLGA : PEO derivatives to form nanoparticles led to a shift in the glass transition temperature (T_g) from 51.34°C, corresponding to the PLGA nanoparticles, to 48.61°C/50.27°C observed for the blend PLGA : poloxamer/poloxamine nanoparticles, respectively. Additionally, a significant change in the shape of the endotherms could be noted. Instead of the marked T_g characteristics of PLGA, the nanoparticles containing poloxamer or poloxamine displayed a broad endotherm. The reduction in the glass transition temperature could be attributed to the PLGA : poloxamer/poloxamine miscibility [27]. In fact, the more hydrophobic Pluronic® L121 can be more readily intermixed with the PLGA polymer chains than Tetronic® 908, thus leading to a higher modification in the thermal event. This would also agree with the more important association of Pluronic® L121 to the nanomatrix as compared to Tetronic® 908.

Insulin encapsulation into the nanoparticles

In order to investigate the potential of our blend nanoparticles for the delivery of macromolecules, we encapsulated bovine insulin as a model compound. For these studies we selected the formulations with theoretical polymer ratio 50 : 50 from the following compositions : PLGA : Pluronic® F68 and L121, PLGA : Tetronic® 908 and 904. The results clearly show the influence of the different PEO derivatives on the success of the encapsulation of insulin (Table 7). In the case of the

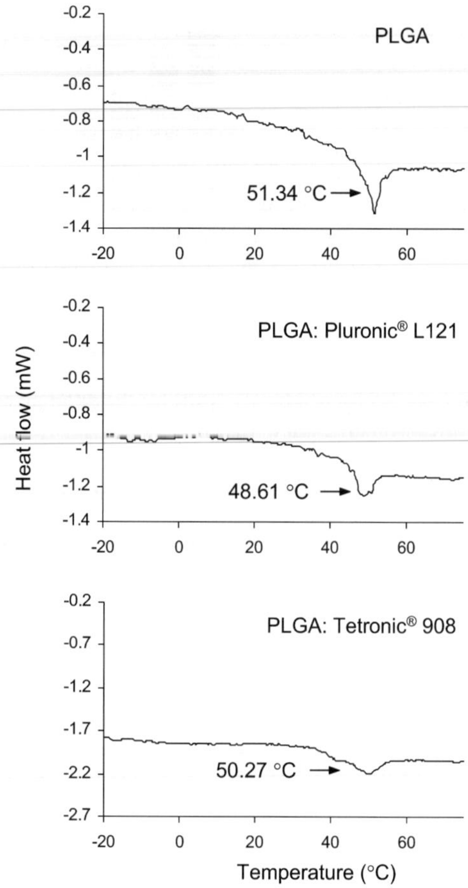

Figure 4. DSC flow curves of PLGA, PLGA : Pluronic® L121 and PLGA : Tetronic® 908 nanoparticle formulations with polymer ratio 50 : 50 (T_g values of the different formulations marked by arrows).

Table 7.
Mean particle size, zeta potential and encapsulation efficiency of insulin-loaded nanoparticles with polymer ratio 50 : 50 ($n = 3$)

Polymer type	Size (nm)	ζ-potential (mV)	Encapsulation efficiency (%)
Pluronic® F68	310.3 ± 1.9	-29.9 ± 5.5	39.4 ± 12.2
Pluronic® L121	—[a]	—	—
Tetronic® 908	269.2 ± 0.4	-14.4 ± 6.0	44.2 ± 0.9
Tetronic® 904	183.6 ± 3.0	-30.8 ± 6.5	19.3 ± 0.4

[a] No determination possible due to precipitation.

more hydrophobic blend system comprising the poloxamer Pluronic® L121, the insulin-loaded nanoparticles precipitated rapidly during the solvent evaporation step after particle formation. The characteristics of the insulin-loaded particles improved when the more hydrophilic PEO derivatives were present in the blend. Simultaneously, a remarkable increase could be observed in the encapsulation efficiencies. More specifically, the poloxamer Pluronic® F68 and the poloxamine Tetronic® 908 (both having HLB values around 30) led to the most effective insulin association reaching encapsulation efficiencies of around 40%. These values were also accompanied by a significant increase in particle size (between 100 and 150 nm) and by a reduction of about 10 mV in the zeta potential. This could be considered as a further confirmation of the successful incorporation of the peptide. Additionally, the decrease of the negative zeta potential also suggests that insulin is partially located on the surface of the nanoparticles. On the other hand, it was interesting to note that PLGA nanoparticles containing insulin could not be formed due to the uncontrolled precipitation of the polymer caused by the presence of insulin. Consequently, the presence of PEO derivatives in the formulation process was critical for the association of insulin.

Previous studies directed to the definition of the drug incorporation characteristics of polymeric blend microparticles also led to the conclusion that the presence of the polyether component has beneficial effects on the encapsulation of proteins such as ovalbumine and tetanus toxin [20, 21, 28]. This positive effect has been attributed to an interaction between the protein and the poloxamer [21]. In the present study this effect could not be observed given the unfeasibility to form PLGA nanoparticles containing insulin without using PEO derivatives. Consequently, it appears that in the experimental conditions defined in the present work, the presence of insulin interferes with the particle formation giving an uncontrolled precipitate. However, this interference could be overcome by facilitating the accommodation of insulin due to the presence of the PEO derivatives.

In vitro *release of insulin from the nanospheres*

Polymeric micro- and nanoparticles typically exhibit a biphasic release profile consisting of an initial rapid release, followed by a sustained release phase [29, 30]. This initial, so-called burst effect is attributed to the release of the surface-associated molecules which are easily detached upon dilution of the particles. This initial burst is more pronounced in the case of nanometric vehicles where the large surface area facilitates the loss of the associated drug molecules. Following this initial burst, the release is controlled by the polymer degradation and the diffusion of the polymer degradation products. A different behaviour has been reported for polymeric blend films and microparticles, for which an improvement of the controlled release properties was observed [21, 27, 28]. More specifically, these previous studies showed a reduced burst, followed by a continuous release of the associated protein. This different release pattern, as compared to the typical one observed for simple polyester matrices, was attributed to the matrix hydration process, followed by a retarded release of the PEO derivative due to its entanglement with the polyester chains.

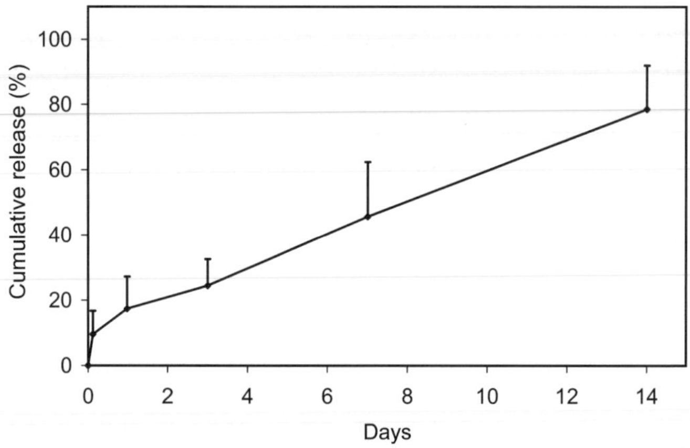

Figure 5. Cumulative release profile of insulin from PLGA : Tetronic® 908 50 : 50 nanoparticles ($n = 3$).

Taking the previous information into account, our approach was to evaluate the possibility of controlling the release of the model peptide insulin from the newly developed nanomatrices. With this purpose in mind, we selected the formulation PLGA : Tetronic® 908 for investigation of its release behaviour, based on its favourable physico-chemical properties and insulin encapsulation efficiency. Figure 5 illustrates the *in vitro* release profile obtained for this formulation, representing the percent of release with respect to the total amount of insulin encapsulated. It can be seen that these nanoparticles exhibit a controlled release of insulin with a negligible burst effect and a very constant, closely linear release rate over a time period of two weeks.

This release behaviour, which agrees with those previously reported for PEO-containing microspheres [21, 28] suggests that the release of the encapsulated macromolecule is governed by the erosion of the nanomatrice, this process being not simply controlled by the polyester degradation but also by the swelling and release of the associated PEO. Therefore, these results led us to confirm that the modification of the particle matrix by a PEO derivative is a useful approach for the adjustment of the release profile of polyester-based nanoparticles.

CONCLUSIONS

We have developed a new nanoparticulate drug carrier consisting of a blend nanomatrix of PLGA and a PEO derivative using an optimised solvent diffusion method. The amount of PEO derivative incorporated into the nanomatrix can be modulated by selecting the derivative with the adequate HLB. These new nanoparticles could associate insulin and release it in a controlled manner, avoiding the burst effect typically observed for protein containing PLGA micro- and nanospheres.

Acknowledgements

This work was supported by the Xunta de Galicia, Spain (PGIDT00BIO20301PR). The first author would like to thank the Spanish Agency of International Collaboration (AECI) for her Ph.D. grant.

REFERENCES

1. P. Johansen, Y. Men, H. P. Merkle and B. Gander, *Eur. J. Pharm. Biopharm.* **50**, 129 (2000).
2. D. Quintanar-Guerrero, E. Alleman, H. Fessi and E. Doelker, *Drug Dev. Industr. Pharm.* **24**, 1113 (1998).
3. M. J. Alonso, in: *Microparticulate Systems for the Delivery of Proteins and Vaccines*, S. Cohen and H. Bernstein (Eds), p. 203. Marcel Dekker, New York, NY (1996).
4. K. S. Soppimath, T. M. Aminabhavi, A. R. Kulkarni and W. E. Rudzinski, *J. Control. Rel.* **70**, 1 (2001).
5. A. Belbella, C. Vauthier, H. Fessi, J. Devissaguet and F. Puisieux, *Int. J. Pharm.* **129**, 95 (1996).
6. K. Fu, D. W. Pack, A. M. Klibanov and R. Langer, *Pharm. Res.* **17**, 100 (2000).
7. G. Zhu, S. R. Mallery and S. P. Schwendeman, *Nature Biotech.* **18**, 52 (2000).
8. A. Sanchez, M. Tobio, L. Gonzalez, A. Fabra and M. J. Alonso, *Eur. J. Pharm. Sci.* **18**, 221 (2003).
9. A. Sanchez, B. Villamayor, Y. Guo, J. McIver and M. J. Alonso, *Int. J. Pharm.* **185**, 255 (1999).
10. M. Tobio, S. P. Schwendeman, Y. Guo, J. McIver, R. Langer and M. J. Alonso, *Vaccine* **18**, 618 (2000).
11. S. M. Moghimi and A. C. Hunter, *Trends Biotechnol.* **18**, 412 (2003).
12. H. M. Redhead, S. S. Davis and L. Illum, *J. Control. Rel.* **70**, 353 (2001).
13. S. M. Moghimi, A. E. Hawley, N. M. Christy, T. Gray, L. Illum and S. S. Davis, *FEBS Lett.* **344**, 25 (1994).
14. T. P. Johnston and S. C. Miller, *J. Parenter. Sci. Technol.* **39**, 83 (1985).
15. V. A. Kabanov and V. Alakhov, *Crit. Rev. Ther. Drug Carrier Syst.* **19**, 1 (2002).
16. V. A. Kabanov, E. V. Bartakova and D. W. Miller, *Adv. Drug Deliv. Rev.* **55**, 151 (2003).
17. V. A. Kabanov, E. V. Batrakova and V. Alakhov, *J. Control. Rel.* **82**, 189 (2002).
18. D. Blanco and M. J. Alonso, *Eur. J. Pharm. Biopharm.* **45**, 285 (1998).
19. D. Blanco and M. J. Alonso, *Eur. J. Pharm. Biopharm.* **43**, 287 (1997).
20. M. Tobio, J. Nolley, Y. Guo, J. McIver and M. J. Alonso, *Pharm. Res.* **16**, 682 (1999).
21. M. Yeh, S. S. Davis and A. G. A. Coombes, *Pharm. Res.* **13**, 1693 (1996).
22. T. G. Park, S. Cohen and R. Langer, *Pharm. Res.* **9**, 37 (1992).
23. G. De Rosa, R. Iommelli, M. I. La Rotonda, A. Miro and F. Quaglia, *J. Control. Rel.* **69**, 283 (2000).
24. S. Stolnik, S. E. Dunn, M. C. Garnett, M. C. Davies, A. G. A. Coombes, D. C. Taylor, M. P. Irving, S. C. Purkiss, T. F. Tadros, S. S. Davis and L. Illum, *Pharm. Res.* **11**, 1800 (1994).
25. A. E. Hawley, L. Illum and S. S. Davis, *Pharm. Res.* **14**, 657 (1997).
26. R. Gref, A. J. Domb, P. Quellec, T. Blunk, R. H. Muller, J. M. Verbavatz and R. Langer, *Adv. Drug Deliv. Rev.* **16**, 215 (1995).
27. T. G. Park, S. Cohen and R. Langer, *Macromolecules* **25**, 116 (1992).
28. M. Yeh, P. Jenkins, S. S. Davis and A. G. A. Coombes, *J. Control. Rel.* **37**, 1 (1995).
29. A. Sanchez, J. L. Vila-Jato and M. J. Alonso, *Int. J. Pharm.* **99**, 263 (1993).
30. T. Niwa, H. Takeuchi, T. Hino, N. Kunou and Y. Kawashima, *J. Control. Rel.* **25**, 89 (1993).

Gels, Genes, Grafts and Giants, pp. 201–213 (2005)

Semi-interpenetrating network of poly(ethylene glycol) and poly(D,L-lactide) for the controlled delivery of protein drugs

CHAD D. BROWN, PATRICK S. STAYTON and ALLAN S. HOFFMAN *

University of Washington, Department of Bioengineering, Box 352255, Seattle, WA 98195, USA

Received 10 November 2003; accepted 24 May 2004

Abstract—We have prepared a semi-interpenetrating network (IPN) of poly(ethylene glycol) dimethacrylate (PEGDMA) with entrapped poly(D,L-lactide) (PLA) using photochemical techniques. These IPNs were developed for the controlled delivery of protein drugs such as growth factors. The PEG component draws water into the network, forming a hydrogel within the PLA matrix, controlling and facilitating release of the protein drug, while the PLA component both strengthens the PEG hydrogel and enhances the degradation and elimination of the network after the protein drug is released. The rate and extent of swelling and the resultant protein release kinetics could be controlled by varying the PEG/PLA ratio and total PLA content. These IPNs were prepared using a biocompatible benzyl benzoate/benzyl alcohol solvent system that yields a uniform, fine dispersion of the protein throughout the PEG/PLA IPN matrix. IPNs composed of high molecular mass PLA and lower PEG/PLA ratios exhibited lower equilibrium swelling ratios. The release of bovine serum albumin (BSA), a model protein, from these IPNs was characterized by a large initial burst, regardless of the PEG/PLA ratio, due to the entrapment of residual solvent within the network. Microparticles of the PEG/PLA IPNs were also prepared using a modified Prolease® strategy. Residual solvent removal was significantly enhanced using this process. The microparticles also exhibited a significant reduction in the initial burst release of protein. Mixtures of different compositions of PEG/PLA microparticles should be useful for the delivery of a variety of protein drugs with different release kinetics from any tissue-engineering matrix.

Key words: Interpenetrating networks; protein drugs; microparticles; tissue engineering.

INTRODUCTION

The development of devices for the timely delivery of active and stable growth factors is a major area of current research in tissue engineering [1–7]. Growth factors

*To whom correspondence should be addressed. Tel.: (1-206) 543-9423. Fax: (1-206) 543-6124. E-mail: hoffman@u.washington.edu

can direct many complex cellular functions, such as inflammation, wound healing, angiogenesis, immune regulation and cellular proliferation and differentiation [3]. Thus, the ability to deliver protein drugs such as growth factors in a controlled and stable manner from tissue-engineering scaffolds can have a dramatic impact on the ultimate wound healing response and viability of engineered tissues. However, precise control over the release kinetics and timing of delivery is needed in order to mimic conditions seen during embryonic development and wound repair. Conventional methods for the delivery of therapeutics are inadequate for the delivery of growth factors in tissue engineering due to the short plasma half-life and the large size of proteins. Further, due to slow tissue penetration of growth factors and rapid diffusion from the site of injection, high concentrations of protein are needed when delivered systemically, leading to high cost and potential systemic toxicity [2, 4]. Growth factor delivery from tissue engineering scaffolds would allow for reduced systemic administration of growth factors, while locally providing the biological stimuli necessary for cellular processes such as angiogenesis and innervation [8]. Further, the ability to deliver multiple growth factors at various rates from the same tissue engineering construct would allow the orchestration of complex tissue development, as well as control over wound healing.

Many traditional drug-delivery and tissue-engineering matrices, including poly(lactide) (PLA), poly(glycolide) (PGA), and their co-polymers poly(lactide-co-glycolide) (PLGA), require processing conditions such as organic solvents or elevated temperatures and pressures that can lead to the denaturation of proteins. Further, the more hydrophobic nature of these polymers can lead to surface aggregation of the protein, which can cause protein denaturation and non-ideal release profiles (high initial burst). In addition, these polymers typically follow a bulk degradation process leading to acidic microenvironments inside the matrix, which can be detrimental to the bioactivity of the growth factor [9, 10]. Finally, some formulation strategies involve organic/aqueous interfaces that lead to unfolding of the protein at the interface and leaching of the protein, thus decreasing the efficacy of these expensive biomolecules. On the other hand, hydrogel scaffolds are processed under aqueous conditions making them conducive to entrapping proteins in a stable manner [11]. However, these scaffolds swell extensively, generate large pores, and release entrapped proteins quickly. Thus, the development of formulations with extended and controlled release of biologically-active growth factors is challenging. By combining traditional synthetic, biodegradable and hydrophobic polymers with hydrogels we hypothesize that matrices with desirable properties of each type of polymer can be obtained. Thus, the mechanical strength and degradability provided by the synthetic, hydrophobic and degradable component can be combined with the stable and non-denaturing environment provided by the hydrogel component. Further, the swelling of the hydrogel component should lead to increased transport of acidic by-products out of the matrix, allowing for more uniform degradation in a protein-friendly environment.

To this end, we have prepared semi-interpenetrating polymer networks (IPNs) comprised of poly(ethylene glycol)dimethacrylate (PEGDMA) and poly(D,L-lactide) (PLA). The IPNs were synthesized using photochemical techniques in a mixture of benzyl benzoate and benzyl alcohol (BB/BA), which is a co-solvent for the PEGDMA and the PLA and, it is presumed, for the protein as well. This solvent system has been utilized with protein therapeutics as a carrier for injectable protein delivery depots and has been found to be biocompatible and to have no detrimental effects on the protein [12–14]. The use of this solvent system for preparing our IPNs allows for uniform protein dispersion throughout the matrix, which may lead to more uniform release kinetics (elimination of initial burst). The hydrophilic/hydrophobic balance of the IPN and, thus, the swelling and release rate can be manipulated by varying the polymer molecular weights and the PEG/PLA ratio. The effects of these parameters on IPN swelling and release of a model protein, bovine serum albumin (BSA), have been investigated here, with the goal of developing a tunable protein delivery system having compositional variables that permit control of the rate of protein release. Further, microparticles of these PEG/PLA IPNs were prepared using a modified Prolease® strategy [15–18]. There has recently been much interest in encapsulating cells in photo-polymerizable hydrogel scaffolds such as hyaluronic acid, alginate and polyethylene glycol [19–22]. We envision a micro-particulate formulation that could be suspended in these scaffolds and entrapped along with cells during the initial cell encapsulation step. To this end, microparticles comprised of PLA/PEG IPNs with different polymer molecular weights and polymer ratios, and containing 'finely dispersed' protein, could be utilized to release multiple growth factors at different rates from the same tissue engineering scaffold. Together, this would lead to a versatile method for the delivery of combinations of growth factors at designed rates from a variety of tissue engineering scaffolds. Thus, in this study, cross-linked PEG/PLA IPN microparticles have been investigated for their ability to deliver a model protein drug, bovine serum albumin.

MATERIALS AND METHODS

Materials

Poly(D,L-lactide) with inherent viscosities of 0.36 dl/g (38 kDa), 0.49 dl/g (52 kDa) and 0.67 dl/g (78 kDa) were purchased from Birmingham Polymers (Birmingham, AL, USA). Poly(ethylene glycol) diacrylate (3400 Da) was donated by Shearwater (Huntsville, AL, USA). Poly(ethylene glycol) dimethacrylate (550 Da), anhydrous benzyl alcohol (99.8%), benzyl benzoate (>99%), 2′,4′-dimethoxy-2-phenylacetophenone (98%, Igracure 651), hexane (HPLC grade), tetrahydrafuran (HPLC grade) and NaN_3 were all obtained from Aldrich (Milwaukee, WI, USA). Physiological phosphate-buffered saline (10 mM PBS, 138 mM NaCl, 2.7 mM KCl, pH 7.4) and bovine serum albumin (BSA, fraction V) were purchased from Sigma

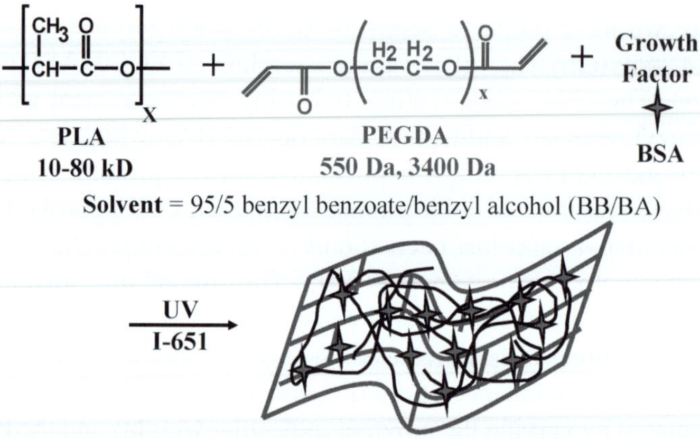

Figure 1. Schematic representation of the photo-cross-linking of PEGDMA/PLA IPNs loaded with BSA in BB/BA.

(St. Louis, MO, USA). NaOH (1 M) was obtained from VWR Scientific Products (Plainfield, NJ, USA). The Bio-Rad DC protein assay was purchased from Bio-Rad Life Sciences (Hercules, CA, USA).

Synthesis of PEG/PLA IPNs in the form of gel discs

IPNs of various mass ratios and molecular masses of PEG and PLA were synthesized by dissolving PLA and PEGDMA in a 95:5 mixture of benzyl benzoate/benzyl alcohol (BB/BA) to give a total polymer concentration of 300 mg/ml (Fig. 1). Various PEG/PLA mass ratios were prepared by adjusting the respective amounts of each polymer in the solution while maintaining a constant total polymer concentration. BSA was added to each solution at a concentration of 5% by mass (based on total polymer mass). Photo cross-linked IPNs were formed by adding 350 μl of the polymer solution to a Teflon mold (14 mm in diameter). To each mold 40 μl Igracure 651 (100 mg/ml) was added followed by exposure to UV at 365 nm for 1 h (Blakray B100AP). The resulting hydrogel discs were allowed to react overnight before being removed from the Teflon molds and used in the release study. Blank IPNs containing no protein were prepared in the same manner without the protein addition step. Due to the extremely low volatility of BB/BA, the solvent cannot be efficiently removed from the hydrogel prior to delivery. Thus, the solvent was released along with the entrapped protein from our delivery device. However, BB/BA has been utilized as a biocompatible carrier for injectable depot formulations for the delivery of proteins and the inclusion of the solvent in our delivery device is not of concern [12–14].

Swelling and release profiles of BSA from PEG/PLA IPNs

The swelling and BSA release profiles of PEG/PLA IPNs of various mass ratios and molecular weights of PEG and PLA were examined using the procedure detailed below. Approximately 3 ml of release buffer (10 mM PBS, 138 mM NaCl, 2.7 mM KCl, pH 7.4) was added to a 20-ml glass scintillation vial. PEG/PLA IPNs were then added to each vial and the vials were incubated at 37°C in a shaking water bath. After various time periods the release buffer was removed and replaced with fresh buffer. The release medium was centrifuged to remove any particulates and stored at 4°C until analyzed for protein concentration. After completion of the release study, the swollen PEG/PLA IPN was weighed (wet mass), lyophilized and reweighed (dry mass). The equilibrium swelling ratio (Q) was calculated using the following relationship:

$$Q = (\text{wet mass})/(\text{dry mass}).$$

Protein loading was verified by degrading the remaining PEG/PLA IPN in 1 M NaOH and measuring the resulting protein concentration. The amount of BSA released at each time point was determined using the Bio-Rad DC protein assay. Triplicate samples were measured and expressed as average ± 1.0 standard deviation. Results were presented as a percentage of the amount of loaded protein released at each time point compared to blank IPNs containing no protein.

Degradation of PEG/PLA IPNs

PEG/PLA IPNs consisting of various molecular weights of PLA were prepared in order to analyze the degradation profile of the IPNs. Briefly, IPNs were photo crosslinked in the same manner as described for the release study without protein loading. The cross-linked IPN was placed into physiological PBS (pH 7.4) and incubated at 37°C with constant shaking. At time points of 1, 5, 10 and 20 days, a PEG/PLA IPN was removed from the release media and lyophilized. Next, the PLA portion of the dry PEG/PLA IPN was extracted into tetrahydrafuran. The extracted sample was analyzed by gel-permeation chromatography (GPC) against polystyrene standards in order to track the degradation of the PLA component.

Synthesis of PEG/PLA IPNs in the form of microparticles

Cross-linked microparticles of PEGDMA and PLA were prepared using a modified Prolease® strategy [15–18]. PLA (38 kDa) and PEGDMA (550 Da) were dissolved at various mass ratios of PEG/PLA at a total polymer concentration of 300 mg/ml in benzyl benzoate. To this solution, BSA (5%, w/w) and Igracure 651 photoinitiator (4%, w/w) were added. This solution was atomized through an ultrasonic nozzle (model #8700-120, Sonotek, Milton, NY, USA) at a volumetric flow rate of 0.8 ml/min and a power of approx. 6.0 W. The atomized stream was exposed to UV radiation at 365 nm before being immediately frozen in a non-solvent bath containing hexane (approx. 400 ml) maintained at −80°C. The hexane bath was placed at −80°C for 1 day to ensure freezing of the particles. After 1 day, the particle suspension was stirred to break up any aggregates and the mixture was placed at −20°C where the benzyl benzoate solvent began to slowly thaw and then was extracted into the hexane non-solvent. After 5 days the particles were placed at 4°C and stirred continuously to extract any remaining benzyl benzoate for 5 days. On the fifth day, 1 g Igracure 651 was dissolved in the hexane non-solvent and the particle suspension was exposed to UV radiation at 365 nm for 30 min with stirring. Approximately 100 ml of absolute ethanol was then added to extract any unreacted PEGDMA cross-linker or initiator fragments from the microparticles. After 1 day, the particles were filtered in a dry ice box and lyophilized. After lyophilization, a free-flowing powder was produced that did not dissolve completely in aqueous or organic solvent, suggesting successful cross-linking of the microparticles.

Release of protein from PEG/PLA microparticles

Release of protein from the PEG/PLA IPN microparticles was investigated by suspending 10 mg of particles in 1 ml of physiological PBS (pH 7.4) containing 0.05% NaN_3 as an antibacterial. Ultrasonication with a Branson Sonifier 450 ultrasonic probe (10 pulses, 10% duty) was utilized to uniformly disperse the microparticles at the onset of the release study. The microparticles were then incubated at 37°C in microcentrifuge tubes mounted on a rotating shaker. At various time-points, the release media was centrifuged to pellet the microparticles and the medium was removed and stored at −20°C until further analysis. Fresh PBS was added at each time-point to resuspend the microparticles and continue the release study. At the end of the study the particles were degraded by the addition of 1 M NaOH to verify the total amount of loaded protein. The concentration of BSA in the release media was determined using the Bio-Rad DC colorimetric protein assay. Triplicate samples were measured and expressed as an average ±1.0 standard deviation. Results were presented as a percentage of the amount of loaded protein released at each time point compared to blank microparticles containing no protein.

Microscopic imaging of PEG/PLA microparticles

Microparticles of PEG/PLA IPNs were imaged using bright field light microscopy. Microparticles were resuspended in cold physiological PBS (pH 7.4) containing 0.05% NaN_3 and ultrasonicated using a Branson Sonifier 450 ultrasonic probe (10 pulses, 10% duty). In order to observe the cross-linked network of the PEG/PLA IPNs, microparticles were resuspended in benzyl benzoate. This allowed for dissolution of the PLA component and imaging of the remaining cross-linked PEG network. The particles were imaged using an inverted Nikon TE200 microscope with a Nicon 10× objective. The images were captured using a Hamamatsu digital camera (model# C4742-98) and Metamorph 4.6r3 software.

RESULTS AND DISCUSSION

Swelling and BSA release profiles of PEG/PLA IPN gel discs

Semi-interpenetrating networks of PEG and PLA were successfully synthesized using photochemical techniques in a unique BB/BA solvent system. This process yielded hydrogel discs that incorporated the entire volume of their reaction solvent. As a result, the hydrogel discs were easily handled and removed from their casting molds allowing for the generation of hydrogel discs of reproducible size and shape. The IPNs were synthesized at PEG/PLA mass ratios of 100 : 0, 75 : 25 and 50 : 50; however, cross-linking was unsuccessful at a 25 : 75 ratio, due to insufficient density of vinyl groups in the solution.

In order to evaluate the effect of PEG molecular mass on the swelling and protein release profiles of the PEG/PLA IPNs, hydrogel discs consisting of 15 kDa PLA and 550 Da PEGDMA or 3400 Da PEGDA were allowed to swell in release media for 10 days and the equilibrium swelling was determined. As either the PEG molecular mass or the PEG content increased, the swelling of the IPN increased, as expected (Table 1). Use of the higher molecular mass PEG (3400 Da) would be expected to create larger pore sizes in the swollen hydrogel, resulting in a greater uptake of water. The increased swelling of the IPNs containing 3400 Da PEGDA led to rapid release of BSA, with a substantial initial burst and complete release after 8 days, regardless of the PEG/PLA mass ratio (Fig. 2A). The similar release rates seen with IPNs comprised of 3400 Da PEGDA in spite of differences in the PEG/PLA ratio suggests that the hydrated pore size of these IPNs is too large to allow for tuning of protein release rates by variation of the PEG/PLA composition. On the other hand, IPNs formed with 550 Da PEGDMA demonstrated better control over the rate of protein release and allowed for variation in release rates with changes in the PEG/PLA ratio (Fig. 2B). IPNs with lower 550 Da PEG/PLA mass ratios exhibited a decreased initial burst and extended release times. The reduction in the burst of protein released from the 550 Da PEG/PLA IPNs is probably due to the smaller pore size in this gel. This results from its higher cross-link density in comparison to the 3400 Da PEG gel, and leads to a decrease in swelling volume and an increase in diffusional resistance for the protein as it is released from the hydrogel disc. The rate of protein release slows dramatically at later time points, suggesting that the remaining protein is entrapped in the more hydrophobic PLA domains. As a result, the overall release of protein from the hydrogel is influenced by the gradual degradation of this more hydrophobic PLA component. Due to the ability of PEG/PLA IPNs formulated with 550 Da PEGDMA to provide control over

Table 1.
Swelling ratio of PEG/PLA (15 kDa) IPNs containing 5% (w/w) BSA placed in release buffer (10 mM PBS, 138 mM NaCl, 2.7 mM KCl, pH 7.4) at 37°C

PEGDMA/PLA mass ratio	Swelling ratio (Q, swollen mass/dry mass)	
	550 Da PEGDMA	3400 Da PEGDA
100 : 0	4.4 ± 0.5	5.7 ± 1.0
75 : 25	3.1 ± 0.5	4.5 ± 0.5
50 : 50	2.6 ± 0.4	4.0 ± 0.2

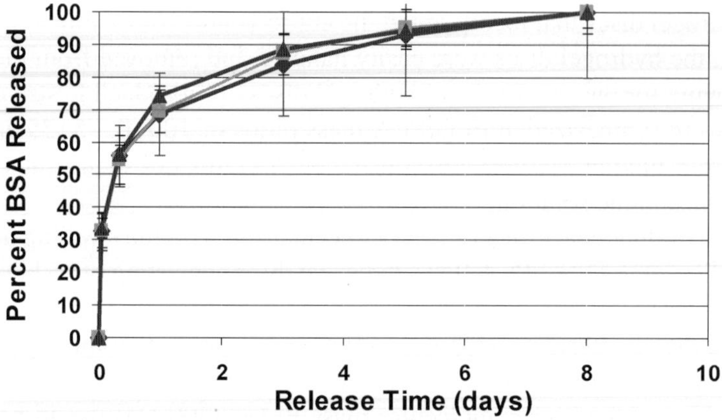

(A)

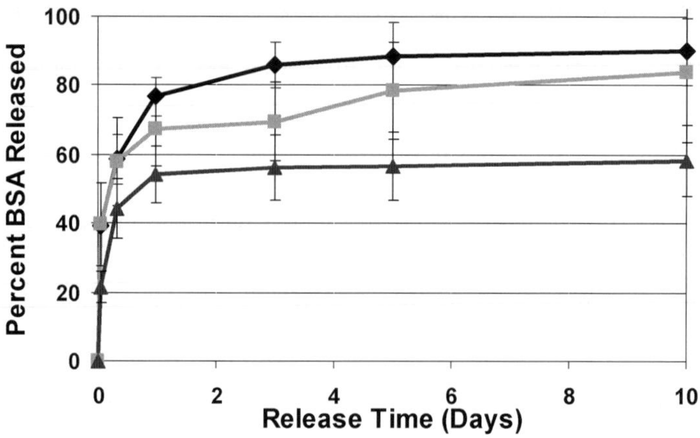

(B)

Figure 2. Release profiles of BSA from (A) 3400 Da PEG/15 kDa PLA IPNs and (B) 550 Da PEG/15 kDa PLA IPNs containing 5% (w/w) BSA placed in release buffer (10 mM PBS, 138 mM NaCl, 2.7 mM KCl, pH 7.4) at 37°C. PEG/PLA mass ratios: 100 : 0 (◆), 75 : 25 (■) and 50 : 50 (▲).

the protein release rate with changes in the PEG/PLA ratio, only IPNs composed of 550 Da PEGDMA were investigated further.

Next, the effects of PLA molecular mass and polymer concentration on protein release rates from 50 : 50 PEG/PLA IPNs were investigated in an attempt to further control the protein release rates and decrease the initial burst. IPNs consisting of 550 Da PEGDMA and various molecular weights of PLA were prepared. Additionally, 50 : 50 PEG/PLA IPNs with a higher vinyl group density were fabricated by increasing the total polymer concentration from 300 mg/ml to 600 mg/ml. As before, the IPNs contained 5% (w/w) BSA and were examined for their protein release rates by allowing the IPN to swell in physiological PBS (pH 7.4). The PLA molecular mass and total polymer concentration had a dramatic

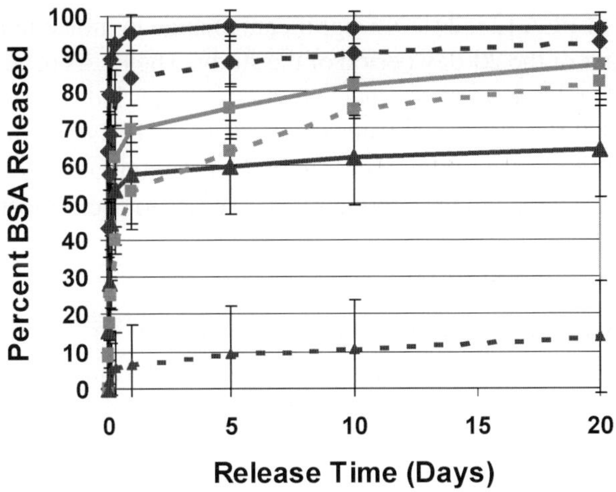

Figure 3. PLA molecular mass- and polymer concentration-dependent release of BSA from 550 Da PEG/PLA IPNs (50 : 50 mass ratio) containing 5% (w/w) BSA placed in release buffer (10 mM PBS, 138 mM NaCl, 2.7 mM KCl, pH 7.4) at 37°C. PLA molecular mass: 38 kDa (◆), 52 kDa (■) and 78 kDa (▲). Total polymer concentrations: 300 mg/ml (solid) and 600 mg/ml (dashed).

effect on the release of BSA from the 550 Da PEG/PLA (50 : 50 mass ratio) IPNs (Fig. 3). As the molecular mass of the more hydrophobic PLA component was increased, the release rate of BSA decreased. Likewise, at a given PLA molecular mass the release rate of BSA decreased significantly with an increase in total polymer concentration, suggesting a successful increase in cross-link density. For the highest molecular mass PLA (78 kDa) and highest total polymer concentration (600 mg/ml), a significant decrease in both the initial burst and release rate was observed, as approx. 15% of the loaded protein was released over the 20-day period. However, most formulations still exhibited a 50% or greater initial burst followed by slower release rates over the remainder of the study. This is most likely due to the hydrogel discs being pre-swollen in BB and BA at the onset of the study. This leads to rapid penetration of water into the IPN and extraction of the dispersed protein along with some of the reaction solvent upon placement in physiological PBS. It was concluded that a strategy for removing residual BB and BA would, therefore, be important for reducing the rapid initial release of protein and providing control over the protein release rate.

Degradation of PLA in the IPNs

The degradation of the PLA in the 550 Da PEG/PLA (50 : 50 mass ratio) IPNs was analyzed by extracting the PLA component into tetrahydrofuran after various time periods in physiological PBS (pH 7.4). The extracted PLA was analyzed by gel-permeation chromatography against polystyrene standards in order to determine the percent decrease in PLA molecular mass. With all formulations, there was an initial 10–20% decrease in PLA molecular mass, which may be due to UV scission of the polyester bonds of PLA [23, 24]. No significant change in molecular mass was seen over the remainder of the 20-day period of the study. The PLAs used in these studies are hydrophobic and known to degrade slowly over a period of months to years. It was speculated that the increased water uptake of these IPNs due to the hydrophilic PEG component might lead to an increase in degradation rate. However, this does not appear to be the case and the protein release seems to occur mostly by diffusion.

Fabrication of cross-linked PEG/PLA microparticles

Microparticle IPNs comprised of 40 : 60, 50 : 50 and 60 : 40 mass ratios of 550 Da PEGDMA and 38 kDa PLA were successfully prepared using a modified Prolease® technique [15–18]. Formulations containing less than 40% PEG or more than 60% PEG were unsuccessful in producing microparticles. In these cases, a cloudy immiscible liquid was obtained during the 4°C extraction. For formulations containing less than 40% PEG, the PLA concentration in the frozen microparticles was probably high enough to impede the efficient extraction of the BB/BA solvent, whereas for formulations containing more than 60% PEG the concentration of PLA was probably insufficient to maintain the structural integrity of the microparticles. Successful formulations resulted in a free-flowing powder after lyophilization that did not fully dissolve in aqueous or organic solvents, suggesting that the microparticles had been cross-linked. The swollen microparticles ranged in size from 50 to 100 μm and were irregular in shape (Fig. 4A). The irregularity may be due to difficulty in atomizing the high viscosity polymer solutions or to extraction of unreacted PEG from the microparticles. This may lead to pore formation, which in return may enhance the initial burst of protein release as well as increase the overall protein release rate. It should be possible to obtain more uniform particles either by using lower molecular mass polymers or by decreasing the polymer concentration, to provide a less viscous solution for atomization. This strategy should also aid in the production of microparticles with more diverse PEG/PLA ratios, since a decrease in solution viscosity should aid in extraction of the benzyl benzoate from the microparticles during fabrication. When the PLA component of the microparticles was extracted by dissolution in benzyl benzoate, a PEG network remained, demonstrating successful cross-linking of the IPN microparticles (Fig. 4B).

Release of BSA from cross-linked PEG/PLA microparticles

Microparticles comprised of a 40 : 60, 50 : 50, or 60 : 40 mass ratio of 550 Da PEGDMA and 38 kDa PLA and containing 5% (w/w) BSA were placed into physiological PBS (pH 7.4) and analyzed for their ability to control the release of BSA (Fig. 5). These microparticle formulations provided good control over the rate of release of BSA with a dramatic decrease in the extent of the initial burst of protein release (approx. 15–30%) seen previously with the matrix IPNs. This reduction in initial burst is presumably due to the more efficient extraction of the

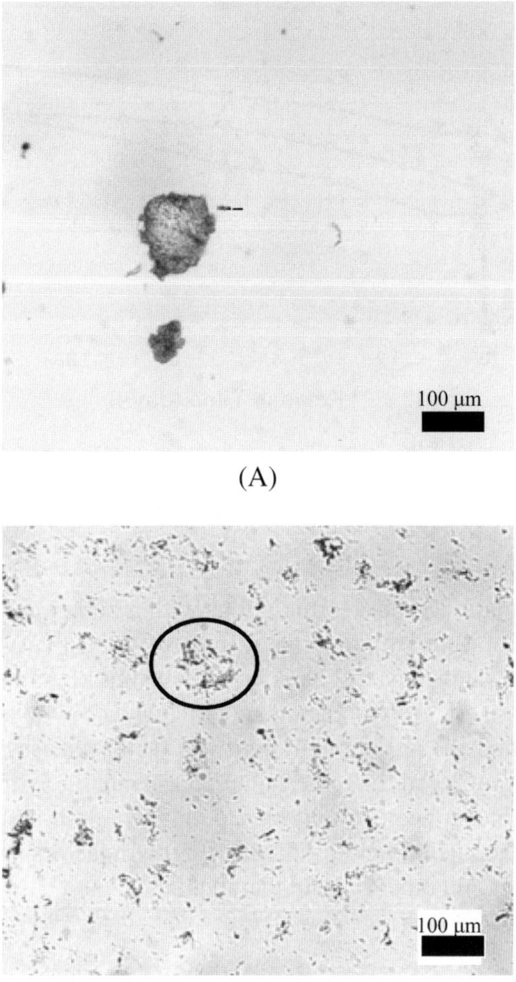

Figure 4. IPN microparticles comprised of a 50 : 50 mass ratio of PEGDMA 550 Da and 38 kDa PLA loaded with BSA (5%, w/w) and resuspended in (A) PBS (pH 7.4) and (B) after extraction of PLA component into benzyl benzoate (B). Images were taken using light transmittance microscopy at 10× magnification. Scale bar equals 100 μm.

benzyl benzoate component from the release system into the hexane bath during microparticle fabrication. This moderate burst was followed by a slow controlled release of BSA over the 10-day period of the study in which 30–40% of the BSA was released. A higher release of BSA was observed as the PEGDMA/PLA mass ratio increased. This is as expected due to the increase in swelling and diffusion anticipated with the more hydrophilic formulation. Interestingly, it appears that the rate of release of BSA is independent of the PEGDMA/PLA mass ratio with variations in total BSA release being attributed solely to differences in the extent of the initial burst. This could be the case if unreacted PEG is acting as a pore

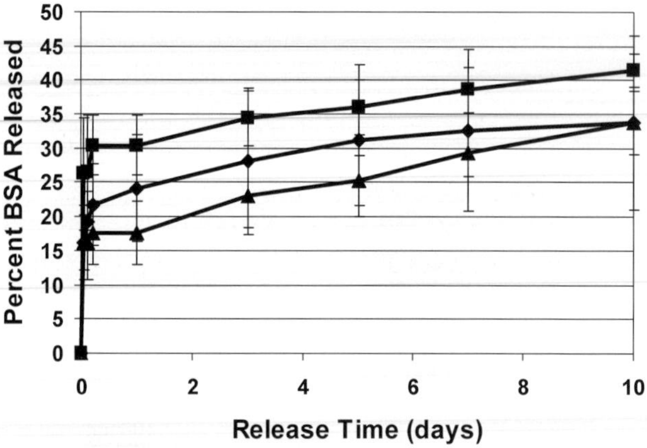

Figure 5. Release profile of BSA from IPN microparticles comprised of 550 Da PEGDMA and 38 kDa PLA. Particles loaded with 5% (w/w) BSA were placed in release buffer (10 mM PBS, 138 mM NaCl, 2.7 mM KCl (pH 7.4) and 0.05% NaN_3) at 37°C. PEG/PLA mass ratios: 60 : 40 (■), 50 : 50 (◆) and 40 : 60 (▲).

former in the formulation. Thus, formulations with a higher content of PEG would have more surface area for initial protein release. PEG/PLA IPN microparticles with various release rates may still be achievable using PLAs of various molecular masses. The release of growth factor should increase with a decrease in PLA molecular mass. Thus, the initial burst of growth factor release could be modulated using the PEG/PLA ratio and the underlying release rate finely tuned using the PLA molecular mass. This strategy should provide a versatile protein delivery strategy for many applications, including the delivery of growth factors in tissue engineering, vaccine development and other protein therapeutics.

CONCLUSIONS

Semi-interpenetrating networks (IPN) of poly(ethylene glycol) dimethacrylate (PEGDMA) and poly(D,L-lactide) (PLA) containing a model protein drug (BSA) were successfully prepared using photochemical techniques. This IPN was prepared using a unique, biocompatible benzyl benzoate/benzyl alcohol solvent system. The rate and extent of swelling and the resultant protein release kinetics of the IPNs could be manipulated by varying the PEG/PLA ratio and total PLA content. In addition, microparticles of the PEG/PLA IPNs prepared using a modified Prolease® strategy demonstrated significant improvements in the initial burst of protein released in comparison to hydrogel discs due to more efficient removal of residual solvent during fabrication. We believe that mixtures of these PEG/PLA IPN microparticles of various compositions should be useful for the delivery of a variety of protein drugs having variable and controlled release kinetics, for applications in tissue engineering, vaccine development and other protein therapeutics.

Acknowledgements

C. B. acknowledges the Whitaker Foundation for financial support through a Whitaker Foundation Graduate Fellowship. Further funding was provided by the NIH (R24-HL64387, BEAT). We also thank Nektar, Inc. for donating the 3400 Da poly(ethylene glycol) diacrylate used in these studies. Finally, the authors would like to express their sincere gratitude to Jeff Anderl for his helpful discussions and expertise in developing the Prolease® strategy used to prepare the PEG/PLA IPN microparticles.

REFERENCES

1. J. L. Cleland, A. Daugherty and R. Mrsny, *Curr. Opin. Biotechnol.* **12**, 212 (2001).
2. Y. Tabata, *Pharm. Sci. Technol. Today* **3**, 80 (2000).
3. M. J. Whitaker, R. A. Quirk, S. M. Howdle and K. M. Shakesheff, *J. Pharm. Pharmacol.* **53**, 1427 (2001).
4. J. E. Babensee, L. B. McIntire and A. G. Mikos, *Pharm. Res.* **17**, 497 (2000).
5. S. P. Baldwin and W. M. Saltzman, *Adv. Drug Deliv. Rev.* **33**, 71 (1998).
6. W. L. Murphy and D. J. Mooney, *J. Periodont. Res.* **34**, 413 (1999).
7. T. P. Richardson, W. L. Murphy and D. J. Mooney, *Crit. Rev. Eukaryot. Gene Expr.* **11**, 47 (2001).
8. C. E. Krewson and M. W. Saltzman, *Brain Res.* **727**, 169 (1996).
9. R. Baker, *Controlled Release of Biologically Active Agents*. Wiley, New York, NY (1987).
10. G. Zhu, S. R. Mallery and S. P. Schwendeman, *Nature Biotechnol.* **18**, 52 (2000).
11. I. Molina, S. Li, M. B. Martinez and M. Vert, *Biomaterials* **22**, 363 (2001).
12. J. Street, M. Bao, L. deGuzman, S. Bunting and F. V. Peale, Jr., *Proc. Natl. Acad. Sci. USA* **99**, 9656 (2002).
13. F. W. Okumu, *et al.*, in: *27th International Meeting of the Controlled Release Society*. Paris (2000).
14. F. W. Okumu, *et al.*, in: *Millennial World Congress of Pharmaceutical Sciences*. San Francisco, CA (2000).
15. W. R. Gombotz, M. S. Healy and L. R. Brown, US Patent No. 5,019,400 (1991).
16. H. J. Lee, G. Riley, O. Johnson, J. L. Cleland, N. Kim, M. Charnis, L. Bailey, E. Duenas, A. Shahzamani, M. Marian, A. J. S. Jones and S. D. Putney, *J. Pharmacol. Exp. Ther.* **281**, 1431 (1997).
17. O. L. Johnson, J. L. Cleland, H. J. Lee, M. Charnis, E. Duenas, W. Jaworowicz, D. Shepard, A. Shahzamani, A. J. S. Jones and S. D. Putney, *Nature Med.* **2**, 795 (1996).
18. M. A. Tracy, *Biotechnol. Prog.* **14**, 108 (1998).
19. K. A. Smeds, A. Pfister-Serres, D. L. Hatchell and M. W. Grinstaff, *Pure Appl. Chem.* **A36**, 981 (1999).
20. K. A. Smeds and M. W. Grinstaff, *J. Biomed. Mater. Res.* **54**, 115 (2001).
21. S. Bryant and K. S. Anseth, *J. Biomed. Mater. Res.* **59**, 63 (2002).
22. D. L. Elbert and J. A. Hubbell, *Biomacromolecules* **2**, 430 (2001).
23. A. Copinet, C. Bertrand, S. Govindin, V. Coma and Y. Couturier, *Chemosphere* **55**, 763 (2004).
24. C. Fischbach, J. Tessmar, A. Lucke, E. Schnell, G. Schmeer, T. Blunk and A. Göpferich, *Surf. Sci.* **491**, 333 (2001).

Gels, Genes, Grafts and Giants, pp. 215–226 (2005)

Preparation of cationic comb-type copolymer having guanidino moieties and its interaction with DNAs

SUNG WON CHOI [1], YUICHI SATO [2], TOSHIHIRO AKAIKE [1] and ATSUSHI MARUYAMA [1,2,*]

[1] *Department of Biomolecular Engineering, Graduate School of Bioscience and Biotechnology, Tokyo Institute of Technology, 4259 Nagatsuta, Midori, Yokohama 226-8501, Japan*

[2] *Precursory Research for Embryonic Science and Technology, Japan Science and Technology Agency, 4-1-8, Honcho, Kawaguchi, Saitama Pref. 332-0012, Japan*

Received 14 November 2003; accepted 25 February 2004

Abstract—In order to evaluatet effects of different cationic moieties, such as primary amino and guanidino groups, on interactions between DNAs and cationic comb-type copolymers, comb-type copolymers having guanidino groups were prepared. The copolymers (GPLL-*g*-Dex) were obtained by guanidination of poly(L-lysine)-graft-dextran copolymers (PLL-*g*-Dex) using 1-guanyl-3,5-dimethylpyrazole nitrate under weak basic conditions. The resulting copolymers were characterized using NMR spectroscopy and size-exclusion chromatography-multiangle light scattering (SEC-MALS). The primary amino groups of the PLL backbones were thoroughly replaced with guanidino ones without any detectable side reactions, including fragmentation of PLL or Dex chains. The interactions of GPLL-*g*-Dex and PLL-*g*-Dex with DNAs were assessed by UV-melting curve measurements. These copolymers diversly affected the melting behavior of double-stranded DNA (dsDNA). GPLL-*g*-Dex has a lower ability to increase the T_m of dsDNA than PLL-*g*-Dex and also exhibits a higher affinity for dsDNA. The results suggest that the stability of dsDNA may be affected not only by ionic interaction between the copolymers and DNAs, but also by other modes of interactions, such as hydrogen-bondinginteractions.

Key words: Cationic comb-type copolymer; guanidination; polyelectrolyte complex; nucleic-acid transitions; DNA hybrids.

INTRODUCTION

Some of nucleic-acid-binding proteins, such as helicases, recombinases and nucleic acid chaperones, are involved in regulating structural transitions of nucleic acid hybrids *in vivo*. Retroviral nucleocapsid (NC) proteins transfer DNA strands from

*To whom correspondence should be addressed. Tel.: (81-45) 924-5791. Fax: (81-45) 924-5815. E-mail: amaruyam@bio.titech.ac.jp

one hybrid to a more stable hybrid and play several roles in viral infection [1, 2]. They have an activity to stimulate breakage and reassociation of base pairs, thus acting as nucleic-acid chaperones that support proper hybridization of nucleic acids [3–5]. *Escherichia coli* RecA protein is an essential protein in homologous recombination process and promotes homologous pairing by a strand-exchange mechanism [6–9]. The RecA protein has multivalent binding sites with DNAs and forms an intermediate complex with both double-stranded DNA (dsDNA) and single-stranded DNA (ssDNA).

Sequence-specificity and well-characterized assembling properties of nucleic acids allow us to utilize them for molecular construction. Various types of DNA-based molecular devices, such as DNA molecular machines [10, 11], nanoassembly [12, 13] and molecular computing [14, 15] have been proposed. Hence, an artificial agent that is capable of manipulating structural transition of nucleic-acid would be a convenient tool to elaborate not only biological systems but also these nanotechnology devices.

We have reported unique properties of cationic comb-type copolymers (CCCs), composed of a cationic poly(L-lysine) (PLL) backbone and water-soluble side chains of dextran in their interaction with DNAs [16, 17]. The comb-type copolymers with a higher degree of grafting accelerate DNA hybridization [18, 19] and increase stability of DNA duplexes and triplexes [16, 17]. Furthermore, CCCs are found to stimulate the DNA strand-exchange reaction between dsDNA and its homologous ssDNA at a much higher rate (50 000 times at 37°C) than spermine and N,N,N-trimethylhexadecylammonium bromide (cetyltrimethylammonium bromide (CTAB)) [20]. The copolymers were demonstrated to exhibit nucleic acid chaperoning activity and to be useful for nucleic-acid analysis [21]. While CCCs have shown a prospect of regulating transition of DNA assemblies, the structural effect of the cationic moieties has not been clearly understood.

The nucleic-acid-binding proteins commonly possess one or more clusters of basic amino acids, i.e. lysine and arginine. However, compositions of basic amino acids in these clusters are irregular and differ drastically among the proteins. For example, histone proteins contain appreciable amounts of lysine and arginine residues, while protamines contain little or no lysine residues [22]. These facts suggest that lysine and arginine have different inherent interactions with nucleic acids. Lysine has a primary amino group as a basic functional moiety, whereas arginine has a guanidino one. Both the primary amino and guanidino groups bear positive charges at physiologic pH. It was reported that while lysine- or arginine-rich peptides interact with DNAs predominantly through ionic interactions at physiologic pH, stronger hydrogen bonding interaction is involved in the interaction between arginine-rich peptides and DNAs or RNAs [23–29]. Hence, there a particular interest to assess different roles between these cationic groups in the comb-type copolymer/DNA interaction. To this end, we have prepared the comb-type copolymers having guanidino groups. The quanidination method to convert the primary amino groups of lysine moieties into guanidino ones without changing the comb-type structures of the copolymers was employed. These copolymers would allow us to exploit the effect of different cationic moieties on nucleic-acid transitions and assemblings.

MATERIALS AND METHODS

Materials

Poly(L-lysine) (PLL, $M_n = 27\,600$) was purchased from Nacalai Tesque (Kyoto, Japan). Dextran T-10 (Dex, $M_n = 8700$) was obtained from Pharmacia Biotech (Uppsala, Sweden). The guanidination reagent, 1-guanyl-3,5-dimethylpyrazole nitrate (GDMP), was purchased from Sigma-Aldrich (St. Louis, MO, USA). Oligonucleotides (ODNs) of a HPLC-purified grade were obtained from FASMAC (Kanagawa, Japan) and their purity was checked by reverse-phase HPLC. Nucleotide sequences of the ODNs used for UV-melting curve measurements are as follows; ODN1: 5′-TCC TCG CCC TTG CTC ACC AT-3′, ODN2: 5′-ATG GTG AGC AAG GGC GAG GA-3′. DNA duplexes were obtained by mixing ODN1 and ODN2 in equimolar amounts and annealing at 95°C for 5 min, followed by slow cooling to room temperature over 16 h. Other solvents and chemicals of reagent grade were obtained from Wako Pure Chemical (Osaka, Japan) and were used without further purification.

Preparation of the poly(L-lysine)-graft-dextran (PLL-g-Dex) comb-type copolymer

Preparation of the PLL-*g*-Dex comb-type copolymer has been described in detail previously [17, 18]. Briefly, the copolymer was prepared by a reductive amination reaction of PLL · HBr with Dex in borate buffer and purified by ultrafilteration using a Sartorius ultrafilter device (MWCO 20 000, Sartorius, Göttingen, Germany). The resulting copolymer was finally lyophilized followed by calculation of its molecular weight and composition by SEC-MALS and ^{1}H-NMR ($M_n = 95\,000$ as free salt, Dex content = 82.9 wt%, grafting $\text{ratio}_{\text{Dex}} = 7.4\%$).

Guanidination of the PLL-g-Dex comb-type copolymer

Scheme 1 shows the guanidination of PLL-*g*-Dex. PLL-*g*-Dex (10 mg, 13.3 μmol for the primary amino groups) and GDMP (13.4 mg, 66.6 μmol) were each dissolved in 400 μl D_2O. The solutions were adjusted to pH 9.5 with 5 M NaOH in D_2O (final concentration of 13 mM for the primary amino groups and 67 mM for GDMP). The solutions were mixed and then placed into a NMR sample tube. To evaluate the time-course changes in the guanidination of PLL-*g*-Dex, ^{1}H-NMR spectra were recorded while the reaction was carried out at 37°C in a NMR spectrometer. The resulting copolymer was isolated by dialysis against Milli-Q water using a Spectra/Por 7 membrane of MWCO 25 000 (Spectrum, Rancho Dominguez, CA, USA) to remove unreacted GDMP. Then the guanidinated PLL-*g*-Dex (GPLL-*g*-Dex) was obtained by freeze-drying (84.3% yield).

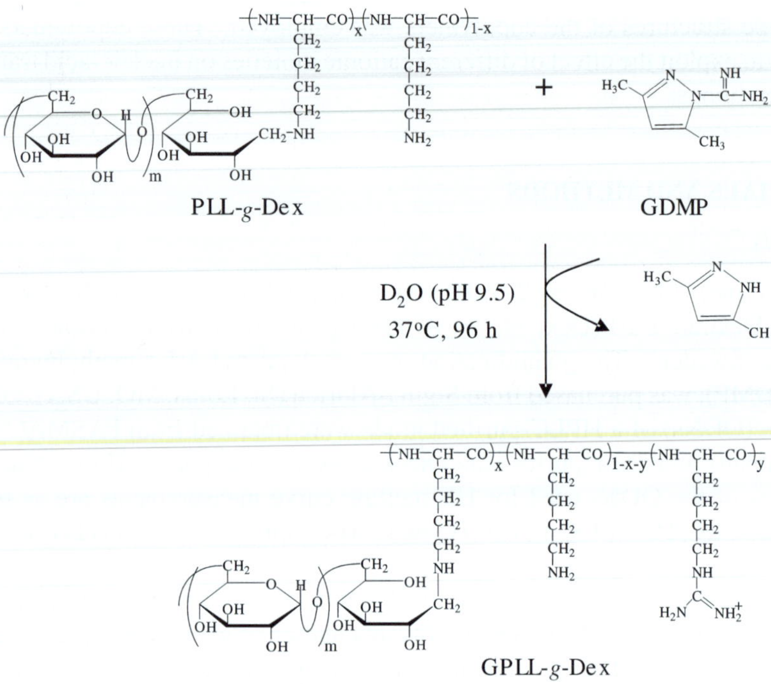

Scheme 1. Guanidination of PLL-*g*-Dex comb-type copolymer.

Size-exclusion chromatography-multiangle light scattering (SEC-MALS)

The resulting copolymers were analyzed using a size-exclusion chromatography (SEC) system (Jasco model 800, Tokyo, Japan) on Shodex OHpak SB-804 and 806M columns connected with a multiangle light scattering (MALS) detector (Dawn-EOS, Wyatt Technology, Santa Barbara, CA, USA) and a differential refractive index (RI) detector (Jasco, model 830-RI). An aqueous solution of 0.5 M acetic acid and 0.2 M Na_2SO_4 was used as mobile phase at a flow rate of 0.8 ml/min at 25°C. An aliquot (100 μl) of graft copolymer solution (10 mg/ml) was injected into the columns. The number-average molecular weight (M_n) and weight-average molecular weight (M_w) were calculated from RI and MALS signals using Astra for Windows Version 4.90 (Wyatt Technology). The dn/dc value was determined by assuming 100% recovery of samples from the columns.

NMR spectroscopy

The resulting copolymer was dissolved in D_2O (99.95 at% D, Merck, Darmstadt, Germany). ^{1}H-NMR spectra (270 MHz) were obtained with a Jeol JNM-EX270 spectrometer (Jeol, Tokyo, Japan) at a probe temperature of 25°C or 37°C. The chemical shifts are expressed as parts per million using sodium 3-trimethylsilyl-propionated-2,2,3,3-d_4 (TSP) ($\delta = 0.0$ ppm in D_2O) as an external reference. The conversion of the guanidination was calculated from the peak areas corresponding to the ε methylene groups of PLL. After purification of GPLL-*g*-Dex, the ^{13}C-NMR spectrum (125 MHz) was measured with a Jeol JNM-GX500 spectrometer (Jeol) at 25°C, using TSP as an external reference.

UV melting curve measurements

Stock solutions were prepared by dissolving nucleotides in Buffer I (10 mM PBS (pH 7.2) containing 150 mM NaCl and 0.5 mM EDTA). Concentrations of the stock solutions were calculated using molar extinction coefficients of 1.65×10^5 M^{-1} cm^{-1} for ODN1 and 2.11×10^5 M^{-1} cm^{-1} for ODN2 at 260 nm. Solutions for UV melting curve measurements were prepared by diluting the stock nucleotide solutions (final concentration of 27.8 μmol) in Buffer I. The mixture of the diluted nucleotide solution was heated at 90°C for 5 min, gradually cooled and allowed to stand for 16 h at room temperature. The stock solutions of the copolymers were also diluted in Buffer I at various concentrations. The diluted solutions of the nucleotide and the copolymers were mixed with a micro pipette at various N/P ratios ($=$ [amino groups]$_{\text{copolymer}}$/[phosphate groups]$_{\text{DNA}}$).

UV spectra and UV melting curves were recorded with a Shimadzu UV-1600 PC spectrometer equipped with a TMSPC-8 temperature controller (Shimadzu, Kyoto, Japan). The UV melting curves were recorded at 1 K/min as a heating rate. The differential absorbance ($\Delta A = A_{260} - A_{340}$) was calculated to correct the baseline shift. The first derivative ($d(\Delta A)/dT$) was calculated from the melting curve data. Peak temperatures in the derivative curves were designated as melting temperatures (T_m).

RESULTS AND DISCUSSION

Guanidination of PLL-g-Dex

In this study, among reagents, GDMP was used for the guanidination of the primary amino groups of PLL-*g*-Dex. The reagent permits the guanidination to occur under more gentle conditions (pH 9.5) and in less time than the *O*-methylisourea (about pH 10.5–11.0) [30–34] employed until now. The guanidination was carried out in a NMR sample tube at 37°C for 96 h. The time-course changes in ^{1}H-NMR spectra of the copolymer are shown in Fig. 1. Here, attention was given to the 2.7–3.2 ppm region, where the peak of ε-methylene protons was expected to be shifted by the guanidination. While the ε-methylene peak at 2.8 ppm decreased as the reaction progressed, the peak at 3.2 ppm, corresponding to ε'-methylene protons of guanidinated lysine moieties, increased gradually. Finally, the ε-methylene peak at 2.8 ppm disappeared. This result indicated that the primary amino groups were thoroughly replaced by the guanidino ones with 96 h reaction. Figure 2 shows ^{13}C-NMR spectra of the resulting copolymer after purification. The appearance of the new peak (g') at 156.6 ppm clearly indicated that the guanidino groups were

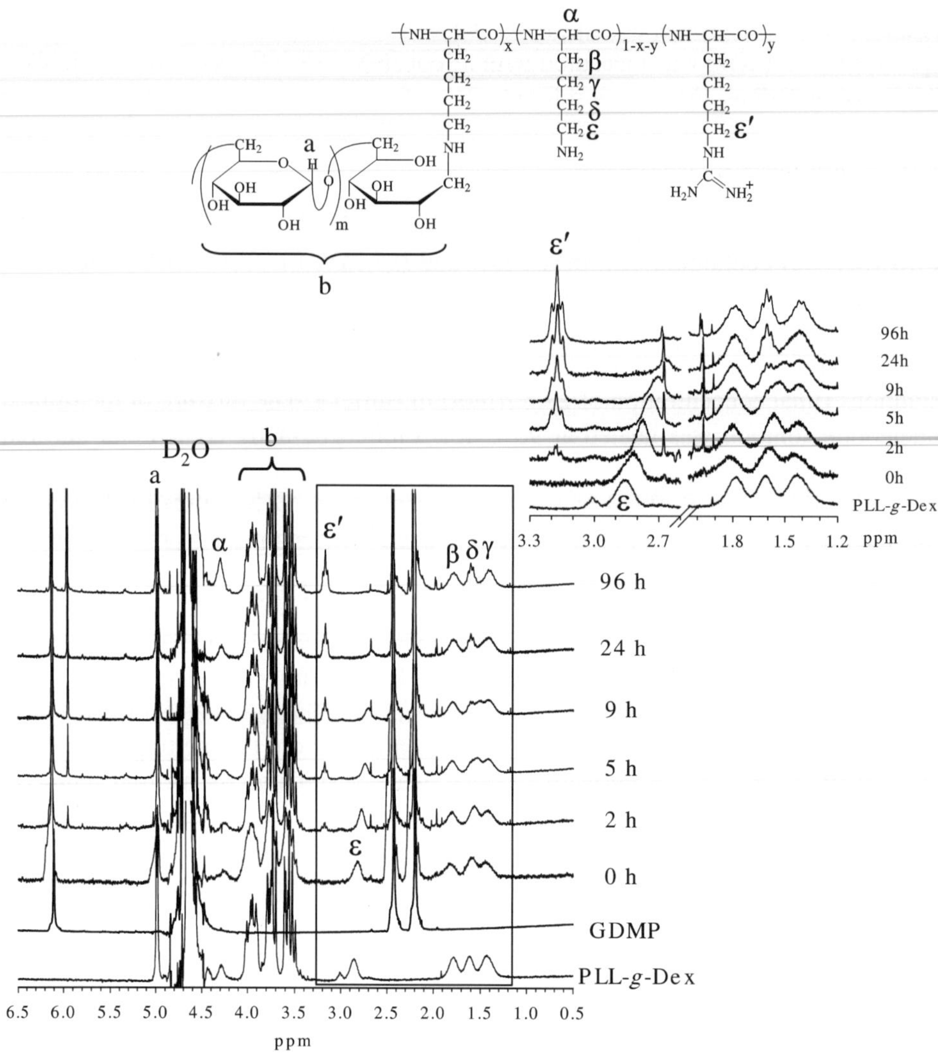

Figure 1. Time-course changes in 270 MHz ^{1}H-NMR spectra of PLL-*g*-Dex during guanidination at 37°C in D_2O (pH 9.5).

incorporated in the copolymer. Furthermore, ^{1}H- and ^{13}C-NMR spectra implicated negligible side reaction during the guanidination reaction.

Figure 3 shows the results of SEC-MALS measurement of GPLL-*g*-Dex and PLL-*g*-Dex, respectively. There was almost no change in the molecular weight and its distribution as determined by MALS (Fig. 3b). Theoretically the guanidination of PLL-*g*-Dex results in a 5% increase in molecular weight, because the PLL content of the copolymer is only 17.1 wt%. Therefore, it is reasonable that the guanidination of the copolymer was hardly detected by the molecular weight change. On the other hand, Fig. 3b clearly reveals that there was no significant fragmentation in PLL or

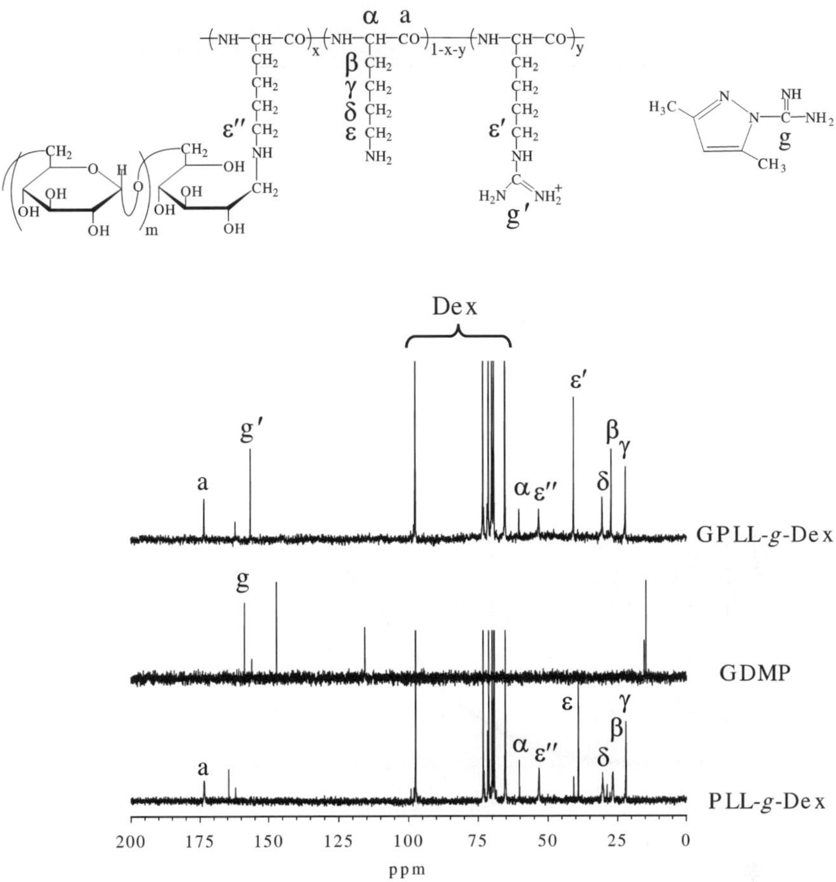

Figure 2. 125 MHz ^{13}C-NMR spectra of PLL-*g*-Dex, GDMP and GPLL-*g*-Dex in D_2O.

Dex chains during the guanidination. The elution of GPLL-*g*-Dex was, however, slightly retarded by the guanidination (Fig. 3a). The retarded elution might be due to change in the copolymer conformation or interactions between the copolymer and the column packings.

The time-course changes of the guanidination of PLL-*g*-Dex copolymer and PLL homopolymer ($M_n = 27\,600$) at 37°C are shown in Fig. 4. The conversion values for both PLL and PLL-*g*-Dex reached more than 96% after 45 h reaction. However, the reaction rate of PLL was higher than that of PLL-*g*-Dex. Since GDMP is a low-molecular-weight reagent, it is unlikely that the reaction should be affected by the steric hindrance of the grafted Dex chains. We, therefore, suggest that the behavior might result from the change in microenvironmental factors, such as dielectric constants around the PLL backbone, owing to the grafted Dex chains. On the base of the above results, it was also possible to obtain copolymers with a desired degree of guanidination (data not shown).

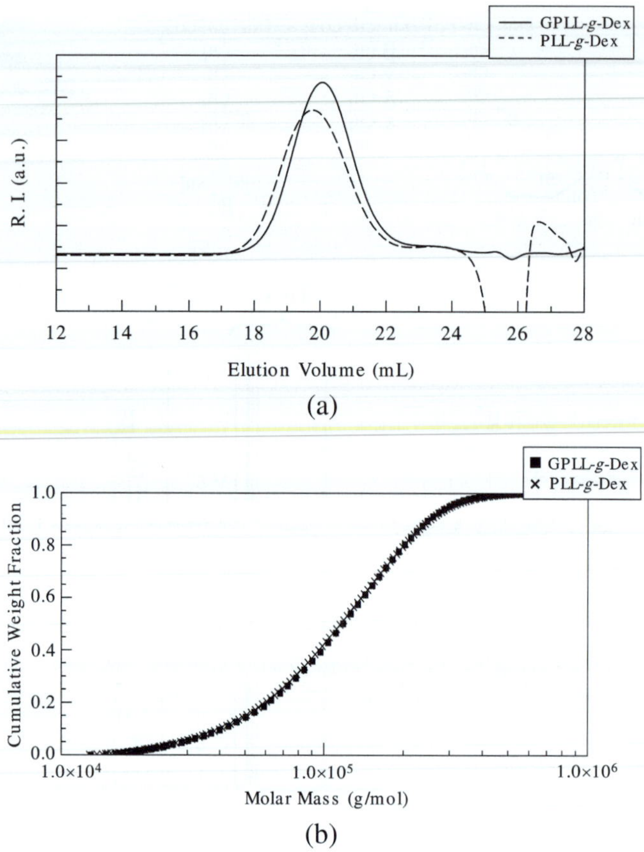

Figure 3. (a) Size-exclusion chromatogram and (b) molecular weight and its distribution of PLL-*g*-Dex and GPLL-*g*-Dex.

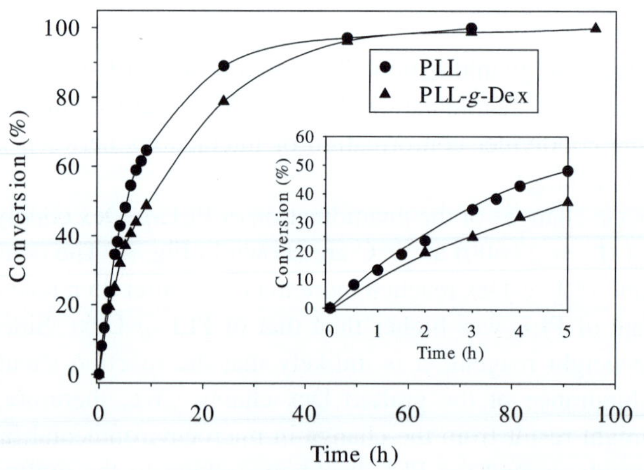

Figure 4. Time-course of the guanidination of PLL (●) and PLL-*g*-Dex (▲). Inset: time-course of the initial stage.

Effect of the guanidination on DNA/copolymer interaction

Melting analysis of DNA duplexes was employed to elucidate the effect of the guanidination on DNA/copolymer interaction. Figure 5 shows the UV melting curves and T_m values of 20-bp dsDNA in the absence or presence of the comb-type copolymers at various N/P ratios ranging from 0 (DNA only) to 10. DNA alone undergoes double helix to single-strand transition at 70.3°C at physiological ionic strength (10 mM PBS/150 mM NaCl). In the presence of either PLL-*g*-Dex or GPLL-*g*-Dex, an increase in the T_m of dsDNA was observed. The increase in the T_m value reached to a plateau at an N/P ratio of 2, and remained constant up to an N/P ratio of 10 (Fig. 5b). While PLL-*g*-Dex increased the T_m by 16°C, GPLL-*g*-Dex only caused a 7°C increase. It has been considered that the comb-type copolymer stabilizes the DNA duplex and triplex mainly by suppressing the electrostatic repulsive forces among DNA strands [17, 35]. From the results shown in Fig. 5, it was supposed that the ionic affinity of GPLL-*g*-Dex towards dsDNA is weaker than that of PLL-*g*-Dex. To assess the affinity of the copolymers for DNAs, we carried out competitive T_m measurements (Fig. 6). The PLL-*g*-Dex and GPLL-*g*-Dex solutions were put together into the DNA solutions at given N/P ratios; the N/P ratio of GPLL-*g*-Dex was kept at 2, while those of PLL-*g*-Dex were adjusted to 2, 4 and 8. It should be noted that T_m in the presence of GPLL-*g*-Dex hardly increased, regardless of the presence of an excess concentration of PLL-*g*-Dex (up to 4-times higher than that of GPLL-*g*-Dex, Fig. 6b). As shown in Fig. 5, the T_m values in the presence of either PLL-*g*-Dex or GPLL-*g*-Dex are very distinctive, especially higher in the presence of PLL-*g*-Dex than of GPLL-*g*-Dex under their complete complex formation with the dsDNA. Therefore, if the affinity of PLL-

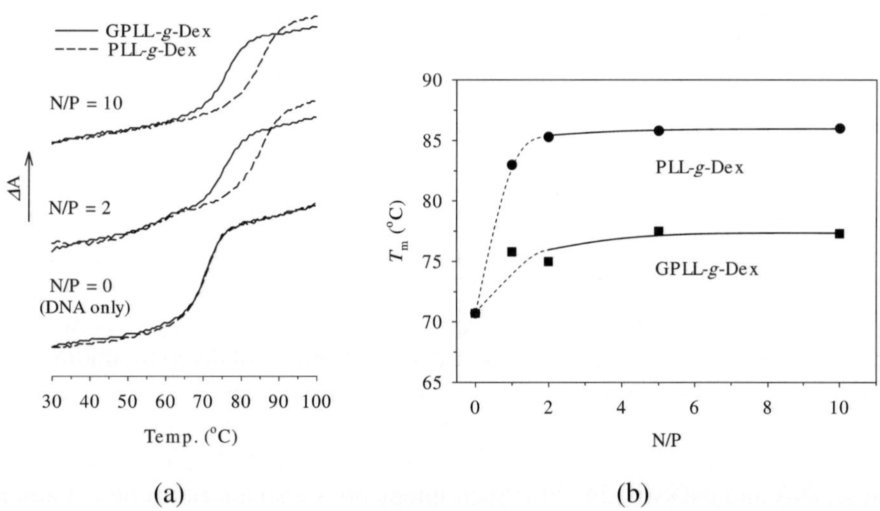

Figure 5. UV–T_m profiles (a) and T_m values (b) of 20-bp dsDNA in the presence of PLL-*g*-Dex or GPLL-*g*-Dex at various N/P ratios ranging from 0 to 10 in Buffer I (10 mM PBS (pH 7.2) containing 150 mM NaCl and 0.5 mM EDTA).

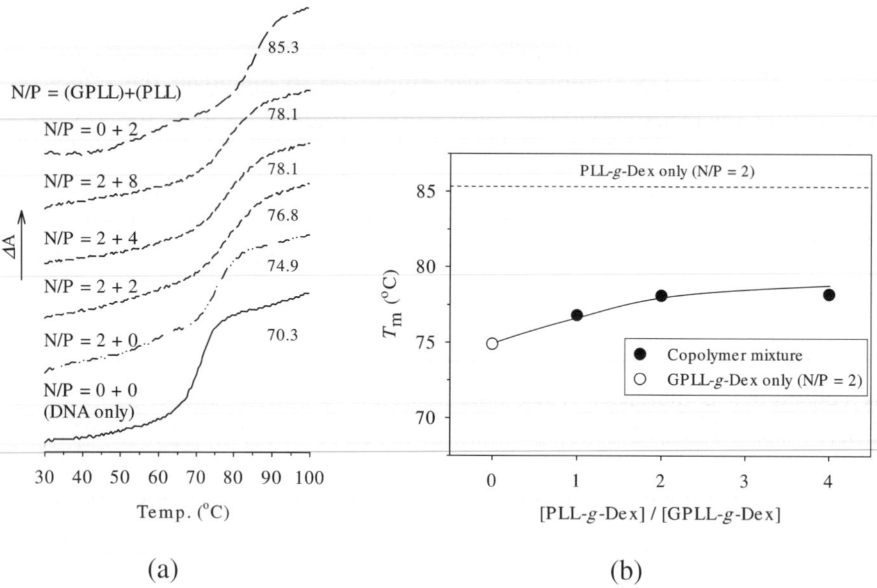

Figure 6. UV–T_m profiles (a) and T_m values (b) of 20 bp dsDNA in the presence of PLL-*g*-Dex and GPLL-*g*-Dex at given N/P ratios; the N/P ratio of GPLL-*g*-Dex was kept at 2, while those of PLL-*g*-Dex were adjusted to 2, 4 and 8 in Buffer I. For comparison, the T_m value (85.3°C) in the presence of PLL-*g*-Dex at a N/P ratio of 2 is indicated with a dashed line. The measurements were performed according to the same procedure described in Fig. 5, except for the case of PLL-*g*-Dex and GPLL-*g*-Dex at given P/D ratios.

g-Dex were higher than that of GPLL-*g*-Dex, the T_m should increase, even in the ternary mixtures. The results in Fig. 6 indicate a higher affinity of GPLL-*g*-Dex for dsDNA than PLL-*g*-Dex. The above results imply that the stability of dsDNA may be affected not only by the affinity for dsDNA but also by models of interactions. It has been reported that oligoarginines have a higher affinity for DNA than oligolysines [25, 36, 37]. Both oligoarginines and oligolysines interact with DNA predominantly through ionic interaction. In addition to carrying a positive charge, arginine also possesses a guanidino group capable of forming bifurcated hydrogen bonds resulting in its higher binding affinity for DNAs than lysine [23]. This is demonstrated by the fact that the interaction of inosine hexamers with oligoarginines is higher than that with oligolysines [37]. The interactions between arginines and phosphates were also reported, e.g. in the binding interaction between the guanidino moiety of arginine and the 5′-diphosphate group of deoxythymidine-3′,5′-diphosphate [23, 26]. These observations are consistent with the studies reporting that arginine, within the dipeptide Arg–Glu, can form hydrogen bonds with the nucleic-acid phosphates, as well as the cytosines within ssDNA and guanosines in both ssDNA and dsDNA [24, 29]. Such interactions also appear to play dominant roles in the binding of a number of proteins to RNA [27] and DNA [28]. On the basis of these studies, it could be concluded that arginines bind DNAs with higher affinity than lysines do. The stronger affinity of GPLL-*g*-Dex over PLL-*g*-

Dex might result from hydrogen bonding interaction between guanidino groups and DNAs. The binding selectivity of the copolymers to ssDNA compared to dsDNA is considered to be another important factor affecting melting behavior of DNAs. For example, a stronger interaction of the copolymers with ssDNA than dsDNA would result in a decrease of the T_m. The effect of the copolymers in the ssDNA/dsDNA equilibrium should be considered. Further study on GPLL-*g*-Dex/DNA interaction is now on going in our laboratory.

CONCLUSIONS

In this study, we have prepared a guanidinated PLL-*g*-Dex comb-type copolymer with a simple reaction method in basic conditions, where the primary amino groups of PLL were replaced with guanidino ones without any fragmentation of PLL or Dex. The time-course of the guanidination of PLL-*g*-Dex was evaluated by ^{1}H-NMR measurement throughout the reaction. The dextran graft chains of the copolymer seemingly affected the kinetics of the guanidination and did not influence the final degree of the guanidination, leading to thorough conversion of primary amino groups into guanidino ones. UV melting curve measurements showed that the guanidination of the copolymer resulted in weaker stabilization of dsDNA. However, the higher affinity of the guanidinated copolymer for DNAs than PLL-*g*-Dex was indicated bx the competitive T_m measurements of the ternary mixtures of these copolymers and DNAs. Modification of cationic groups of the comb-type copolymers would enable us to regulate CCC/DNA interactions and their double-strand/single-strand selectivity.

Acknowledgements

This work was supported in part by Grants-in-aid (Nos. 11167225 and 12480260) for scientific research from Ministry of Education, Culture, Sports, Science, and Technology of Japan.

REFERENCES

1. Z. Tsuchihashi and P. O. Brown, *J. Virol.* **68**, 5863 (1994).
2. M. Lapadat-Tapolsky, C. Pernelle, C. Borie and J.-L. Darlix, *Nucleic Acids Res.* **23**, 2434 (1995).
3. W. Wu, L. E. Henderson, T. D. Copeland, R. J. Gorelick, W. J. Bosche, A. Rein and J. G. Levin, *J. Virol.* **70**, 7132 (1996).
4. X. Ji, G. J. Klarmann and B. D. Preston, *Biochemistry* **35**, 132 (1996).
5. V. Tanchou, C. Gabus, V. Rogemond and J.-L. Darlix, *J. Mol. Biol.* **252**, 563 (1995).
6. J. W. Roberts, C. W. Roberts, N. L. Craig and E. M. Phizicky, *Cold Spring Harbor Symp. Quant. Biol.* **43**, 917 (1979).
7. T. Shibata, C. DasGupta, R. P. Cunningham and C. M. Radding, *Proc. Natl. Acad. Sci. USA* **76**, 1638 (1979).
8. K. Mcentee, G. M. Weinstock and I. R. Lehman, *Proc. Natl. Acad. Sci. USA* **76**, 2615 (1979).

9. E. Cassuto, S. C. West, J. Mursalim, S. Conlon and P. Howard-Flanders, *Proc. Natl. Acad. Sci. USA* **77**, 3962 (1980).
10. B. Yurke, A. J. Turberfield, A. P. Mills, Jr., F. C. Simmel and J. L. Neumann, *Nature* **406**, 605 (2000).
11. H. Yan, X. Zhang, Z. Shen and N. C. Seeman, *Nature* **415**, 62 (2002).
12. B. Liu, N. B. Leontis and N. C. Seeman, *Nanobiology* **3**, 177 (1994).
13. E. Winfree, F. Liu, L. A. Wenzler and N. C. Seeman, *Nature* **394**, 539 (1998).
14. L. M. Adleman, *Science* **266**, 1021 (1994).
15. K. Sakamoto, H. Gouzu, K. Komiya, D. Kiga, S. Yokoyama, T. Yokomori and M. Hagiya, *Science* **288**, 1223 (2000).
16. A. Maruyama, M. Katoh, T. Ishihara and T. Akaike, *Bioconj. Chem.* **8**, 3 (1997).
17. A. Maruyama, H. Watanabe, A. Ferdous, M. Katoh, T. Ishihara and T. Akaike, *Bioconj. Chem.* **9**, 292 (1998).
18. H. Torigoe, A. Ferdous, H. Watanabe, T. Akaike and A. Maruyama, *J. Biol. Chem.* **274**, 6161 (1999).
19. A. Ferdous, T. Akaike and A. Maruyama, *Bioconj. Chem.* **11**, 520 (2000).
20. W. J. Kim, T. Ishihara, T. Akaike and A. Maruyama, *Chem. Eur. J.* **7**, 176 (2001).
21. W. J. Kim, Y. Sato, T. Akaike and A. Maruyama, *Nature Mater.* **2**, 815 (2003).
22. H. Busch, in: *Histones and Other Nuclear Proteins*, p. 28. Academic Press, New York, NY (1965).
23. F. A. Cotton, V. W. Day, E. E. Hazen, Jr. and S. Larsen, *J. Am. Chem. Soc.* **95**, 4834 (1973).
24. G. Lancelot, R. Mayer and C. Helene, *Biochim. Biophys. Acta* **564**, 181 (1979).
25. K. M. Weeks, C. Ampe, S. C. Schultz, T. A. Steitz and D. M. Crothers, *Science* **249**, 1281 (1990).
26. A. Arnone, C. J. Bier, F. A. Cotton, V. W. Day, E. E. Hazen, Jr., D. C. Richardson, J. S. Richardson and A. Yonath, *J. Biol. Chem.* **246**, 2302 (1971).
27. D. J. Kenan, C. C. Query and J. D. Keene, *Trends Biochem. Sci.* **16**, 214 (1991).
28. E. M. Bradbury, *Methods Cell Biol.* **16**, 179 (1977).
29. C. Helene, *FEBS Lett.* **74**, 10 (1977).
30. W. L. Hughes, Jr., H. A. Saroff and A. L. Carney, *J. Am. Chem. Soc.* **71**, 2476 (1949).
31. C. H. Chervenka and P. E. Wilcox, *J. Biol. Chem.* **222**, 635 (1959).
32. W. A. Klee and F. M. Richards, *J. Biol. Chem.* **229**, 489 (1957).
33. G. S. Shields, R. L. Hill and E. L. Smith, *J. Biol. Chem.* **234**, 1747 (1959).
34. A. F. S. A. Habeeb, *Can. J. Biochem. Physiol.* **38**, 493 (1960).
35. A. Maruyama, Y.-I. Ohnishi, H. Watanabe, H. Torigoe, A. Ferdous and T. Akaike, *Colloids Surf. B* **16**, 273 (1999).
36. D. P. Mascotti and T. M. Lohman, *Biochemistry* **36**, 7272 (1997).
37. D. Pörschke, *Eur. J. Biochem.* **86**, 291 (1978).

Gels, Genes, Grafts and Giants, pp. 227–244 (2005)

Effect of a dextran derivative associated with TGF-β1 or FGF-2 on dermal fibroblast behaviour in dermal equivalents

LAETITIA FRANK [1,*], CORINNE LEBRETON-DECOSTER [2], GASTON GODEAU [3], BERNARD COULOMB [2] and JACQUELINE JOZEFONVICZ [1]

[1] *Laboratoire de Recherches sur les Macromolécules, CNRS FRE 2314, Université Paris 13, 93430 Villetaneuse, France*
[2] *Laboratoire de Dermatologie, INSERM U532, Institut de Recherche sur la Peau, Hôpital Saint-Louis, 1 avenue Claude Vellefaux, 75475 Paris cedex 10, France*
[3] *Laboratoire de Physiopathologie des Tissus non Minéralisés, Faculté d'Odontologie, Université René Descartes, 92120 Montrouge, France*

Received 16 November 2003; accepted 30 April 2004

Abstract—Dextran derivatives that mimic the action of heparin have been shown to protect heparin-binding growth factors, such as Transforming Growth Factor-β1 (TGF-β1) and Fibroblast Growth Factor-2 (FGF-2). The aim of this study was to investigate the effect of LS21 DMCBSu, a dextran derivative which contains methylcarboxylate, benzylamide and sulfate groups, both by itself and when combined with TGF-β1 and FGF-2, on the behaviour of fibroblasts. Two systems were assessed: a monolayer culture and three-dimensional collagenous matrices (dermal equivalent). Polymeric biomaterial LS21 DMCBSu and LS21 DMCBSu associated with either TGF-β1 or FGF-2, were added to the monolayer culture on day 3. After 7 days of culture the number of cells was determined. Two treatments were carried out on the dermal equivalents: 9 days of treatment from day 0 to day 9 of culture and 9 days of treatment from day 21 to day 30 of culture for the premature and the mature dermal equivalents respectively. In the monolayer culture, the bioactive polymer produced a slight increase in fibroblast growth (10% with 10 μg/ml of LS21 DMCBSu) and promoted the stimulating effect of the growth factors on cell growth. In the premature dermal equivalents growth was stimulated by 20% when 10 μg/ml LS21 DMCBSu was added. The dextran derivative mixed with TGF-β1 slightly inhibited the growth effect of the growth factor in the dermal equivalents. The functionalized dextran with FGF-2 enhanced the stimulating effect of the growth factor in the premature dermal equivalent. A significant increase in cell growth was observed with the fibroblasts treated with the FGF-2 LS21 DMCBSu mixture and FGF-2 (51% and 40%, respectively). However, none of the described treatments affected the cell growth in the mature dermal equivalent. Furthermore,

*To whom correspondence should be addressed. Tel.: (33-1) 4940-3672. Fax: (33-1) 4940-2037. E-mail: laetitia.frank@free.fr

the dextran derivative had no effect on dermal contraction under these experimental conditions (3D culture).

Key words: Derivatized dextran; TGF-β1; FGF-2; human dermal fibroblasts; dermal equivalents.

INTRODUCTION

Cutaneous wound healing is a complex biological process which involves inflammatory, proliferative and remodeling phases. After injury, haemostasis is achieved by clot formation. Then inflammatory cells, such as neutrophils, monocytes and macrophages, invade the injury site. Inflammatory cells secrete cytokines which initiate the proliferative phase of repair. Mesenchymal cells proliferate and synthesize a new connective tissue matrix, replacing the lost dermis [1]. The healing of the cutaneous wound is partially regulated by the growth factors secreted by the cells involved in tissue repair [2]. Transforming Growth Factor-β1 (TGF-β1) and Fibroblast Growth Factor-2 (FGF-2) are heparin- and heparin-binding growth factors that exhibit mitogenic and angiogenic activity [3, 4]. TGF-β1 is a 25-kDa homodimeric peptide [5, 6]. This multifunctional peptide is able to either stimulate or inhibit the proliferation of many cell lines; it is also implicated in the differentiation of the proliferating cells [2]. TGF-β1, which is present in a wide variety of tissues, plays a key role in wound healing. TGF-β1 released by an inflammatory cell has been shown to stimulate the synthesis of collagen by fibroblasts [2, 7], to affect the proliferation of fibroblasts [7] and to induce the secretion of other growth factors such as interleukin 1, fibroblast growth factors (FGFs) and platelet-derived growth factor (PDGF) [8]. Similar to TGF-β1, FGFs constitute a family of homologous proteins which are potent mitogens for most mesodermal-derived cells and for numerous cells of ectodermal embryogenic origin [9]. FGF-2, identified as a 18-kDa protein, is involved in the development and growth of new blood vessels (angiogenesis) [4, 10] and wound healing [11, 12]. Previous investigations have demonstrated that a heparin-like family of dextran derivatives was able to protect TGF-β1 and FGF-2 from enzymatic degradation *in vitro* [11, 13–15]. Soluble functionalized polysaccharides substituted with methylcarboxylate (MC), benzylamide (B) and sulfate (Su) groups, named DMCBSu, exhibit various types of biological activity which are dependent on the relative proportions of each of the chemical groups [16]. The ability of polymeric biomaterial to act as stabilizer, protector and promoter of heparin-binding growth factors was shown by experiments performed on a monolayer cell culture. Hence, it was important to evaluate if the capacity of dextran derivatives to interact with the growth factors was different or comparable in three-dimensional collagenous matrix cultures. This dermal equivalent collagen culture contains human dermal fibroblasts to mimic an *in vitro* cutaneous wound [17–21].

In the present work, we have investigated the capacity of a polymeric biomaterial, named LS21 DMCBSu, to modulate the proliferation of human fibroblasts in the presence of two growth factors involved in wound healing. Three different models have been employed: a monolayer culture, a dermal equivalent treated from day 0 to 9 (premature dermal equivalent) and a dermal equivalent treated from day 21 to 30 (mature dermal equivalent). In the latter two models, which utilize a three-dimensional culture, dermal contraction was also studied. TGF-β1 and FGF-2 were chosen for their known effect on fibroblasts and for their capacity to form a complex with heparin or with dextran derivatives. The stimulating effect of the bioactive polymer was assessed by investigating its capacity to act on fibroblast growth under the chosen experimental conditions. Then these results were compared to those obtained when the dextran derivative was associated either with TGF-β1 or FGF-2.

MATERIALS AND METHODS

Materials

Dextran T40 ($\overline{M_n} = 28\,800$, $\overline{M_w} = 40\,400$, batch 228 608) was purchased from Amersham Pharmacia Biotech (Saclay, France). All of the analytical grade chemicals and solvents were purchased from Aldrich (St. Quentin Fallavier, France), Acros (Noisy le Grand, France), Interchim (Montluçon, France) and Sigma (St. Quentin Fallavier, France).

The derivatized dextran, named LS21 DMCBSu, was synthetised by Biodex (St. Amand, France) by performing a statistical substitution of chemical groups within the glucose units and along the polymer chain, as has been previously described [22]. This sample was prepared using good manufacturing practice. The synthesis of LS21 DMCBSu involved the carboxymethylation of native dextran with monochloroacetic acid at 60°C for 1 h. The carboxylic groups of the carboxylated dextran were coupled with benzylamine using as coupling agent 1-cyclohexyl-3-(2-morpholino-ethyl)carbodiimide in water, at room temperature. Sulfate groups were fixed by sulfatation of hydroxyl groups with pyridine-sulfur trioxide complex in dimethyl formamide at room temperature. The solution was purified with an ultrafiltration device (cut-off 5000 Da; Filtron, St. Quentin en Yvelyne, France) and finally lyophilized. Acidimetric titration and elemental analysis of the content of nitrogen and of sulfur were used to determine the degree of substitution (ds) of methylcarboxylate, benzylamide and sulfate groups on the LS21 DMCBSu chain ($ds_{CM} = 0.55$, $ds_B = 0.11$, $ds_{Su} = 0.33$).

Fibroblast culture

Primary cultures of the fibroblasts were obtained from 6 foreskins taken from healthy donors aged between 2 and 4 years old. The fibroblasts were grown in Dulbecco's Modified Eagle's Medium (DMEM) with a high glucose concentration (4500 mg/l) (Gibco-BRL, Life Technology, Cergy-Pontoise, France) supplemented with 10% Fetal Calf Serum (FCS), penicillin (100 U/ml), streptomycin (100 μg/ml) and fungizone (2 μg/ml) (all Gibco-BRL) and pooled. Human fibroblasts, named

HFP-A (Human Fibroblasts Pool A), were routinely cultured at 37°C in a 5% CO_2 incubator with a humidified atmosphere. All the experiments were carried out between passages 3 and 8.

Monolayer culture: cell proliferation tests

Cells (10^4/well) were seeded onto a 24-well plate in DMEM supplemented with 10% FCS. After 48 h, the culture medium was renewed with DMEM + 0.1% FCS. 24 h later, human dermal fibroblasts were incubated in DMEM with 2% FCS to which a dextran derivative with or without growth factors was added concurrently. FGF-2 and TGF-β1 were purchased from Invitrogen Life Technology (Cergy-Pontoise, France). Control cells were incubated in DMEM with 2% FCS. After 96 h, cells were trypsinized and counted with a Coulter counter model ZM (Coultronics, Evry, France). The procedure was repeated three times. ED_{50} was measured as the concentration of growth factor required to yield half the maximum number of cells. Dose–response curves for dextran (range 0–100 μg/ml) and growth factors (range 0–5 ng/ml for TGF-β1 and 0–20 ng/ml for FGF-2) were used to determine the levels for the two combination studies. A summary of all the experiments carried out is presented in Table 1.

Human dermal equivalent collagen matrices

Human dermal equivalents were prepared employing to the method described by Bell *et al.* [23]. HFP-A was treated with trypsin–ethylendiamine tetraacetic acid (EDTA) (Gibco-BRL). Fibroblasts were suspended at a concentration of 3×10^5 cells/ml in DMEM supplemented with 10% FCS. Human dermal equivalents were manufactured by mixing 2.75 ml of a 1.76-fold concentrate of the DMEM medium combined with 10% FCS, 1.5 ml of type-I collagen solution extracted from rat tail in 0.1 M acetic acid, 0.25 ml 0.1 M NaOH and 0.5 ml of fibroblast suspension, leading to a final density of 1.5×10^5 cells/gel. Petri dishes (5 cm diameter, Merck Eurolab, Strasbourg, France) containing this mixture were placed in an incubator with a 5% CO_2 atmosphere at 37°C. After a few minutes at 37°C, the well-formed gel was detached from the wall of the Petri dish by gentle shaking. All the experiments were repeated three times (Table 1).

Human dermal equivalent tests

Premature human dermal equivalent collagen matrices. Derivative dextran and/or growth factor were added concurrently to the collagen gel containing the human dermal fibroblasts at various times: day 0, 3, 6 and 8. Control cells were renewed with DMEM 10% FCS on days 3, 6, 8. On day 9, the number of cells was determined. The latter was achieved by extracting cells from the collagen by proteolysis with 2 mg/ml of collagenase (Sigma type 1, No. 69891) in phosphate-buffered saline (PBS, Gibco-BRL) and by gentle shaking until the

Table 1.
Summary of the various experiments carried out

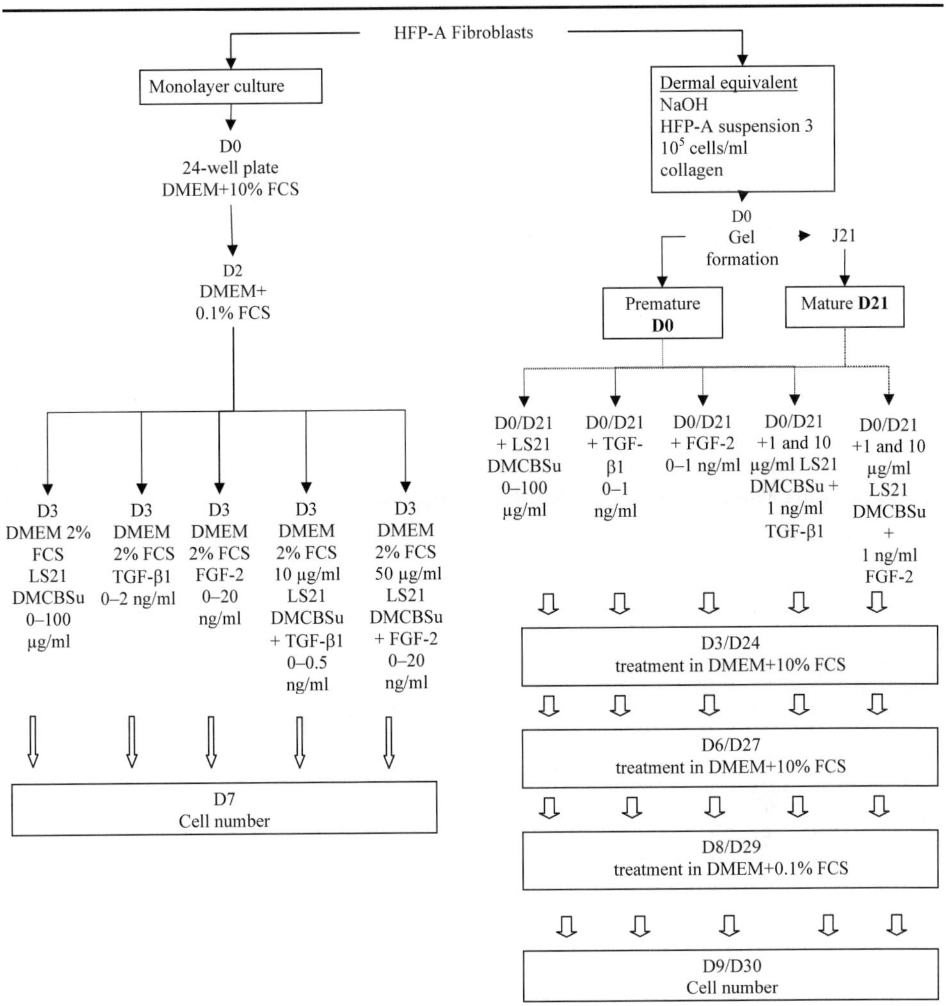

collagen dissolved (15 min maximum). Then, the solution was centrifuged at $800 \times g$ for 15 min. Fibroblasts were suspended in 1 ml of PBS and the number of cells was determined using a model ZM Coulter counter. The diameters of the collagen matrices were observed and measured daily employing optical microscopy (Axiovert 100, Zeiss, Jena, Germany) (Table 1).

Mature human dermal equivalent collagen matrices. The dermal equivalent medium was changed on days 6 and 14. LS21 DMCBSu and/or the growth factor were added concurrently on days 21, 24, 27 and 29. Control cells were renewed with

DMEM 10% FCS on days 21, 24, 27 and 29. The dermal equivalent was dissolved on day 30 and the number of cells was determined as previously described (Table 1).

Cellular viability. HFP-A was extracted by proteolysis as previously described. Fibroblasts were suspended in 1 ml of culture medium. 50 μl of the cellular suspension was mixed with 50 μl 0.4% Trypan blue (Gibco-BRL). The number of living cells was measured using a Malassez cell (Polylabo, Strasbourg, France). A fibroblast suspension of 10^4 cells/ml was prepared and seeded onto a 24-well plate in DMEM supplemented with 10% FCS.

Statistical analysis

A statistically analysis of the results was accomplished by using XLstat-pro (version 7). Comparison of data was carried out by analysis of variance. When the overall F-test showed a significant difference among groups, individual group means were compared using the Student's t-test. All statistically significant differences are reported at 95% confidence intervals ($P < 0.05$).

RESULTS

Monolayer culture

Effect of TGF-β1, FGF-2 and LS21 DMCBSu on HFP-A proliferation in a monolayer culture. The effect of TGF-β1 and FGF-2 on the proliferation of HFP-A was tested in a monolayer culture (Fig. 1). An effect on HFP-A cell growth of increasing concentrations of the growth factor (TGF-β1 or FGF-2) added simultaneously to the culture medium was observed. The greatest effect on the proliferation was induced when incubating the cells with 0.5 ng/ml TGF-β1 or 2 ng/ml FGF-2. The ED_{50} values were 0.15 or 0.5 ng/ml for TGF-β1 and FGF-2, respectively. Greater biological activity was observed for TGF-β1. The ED_{50} value of 0.5 ng/ml FGF-2 was similar to that measured by Khouw *et al.* [24].

Concentrations of LS21 DMCBSu between 1 and 50 μg/ml only induced a very slight increase in proliferation of HFP-A (Fig. 2A). Cells were treated with fixed concentration of TGF-β1 (2 ng/ml) or FGF-2 (2 ng/ml) at variable concentrations of LS21 DMCBSu. The maximum effect of this bioactive polymer on cell proliferation was observed when cells were incubated with a concentration of 10 μg/ml or 50 μg/ml of LS21 DMCBSu, respectively, in the presence of TGF-β1 or FGF-2 (result not shown). Hence, LS21 DMCBSu concentrations of 10 and 50 μg/ml were chosen to carry out the following experiments on the proliferation of fibroblasts in a monolayer culture.

Effect of TGF-β1 and FGF-2 combined with LS21 DMCBSu on HFP-A proliferation in a monolayer culture. The ability of LS21 DMCBSu to modulate HFP-A

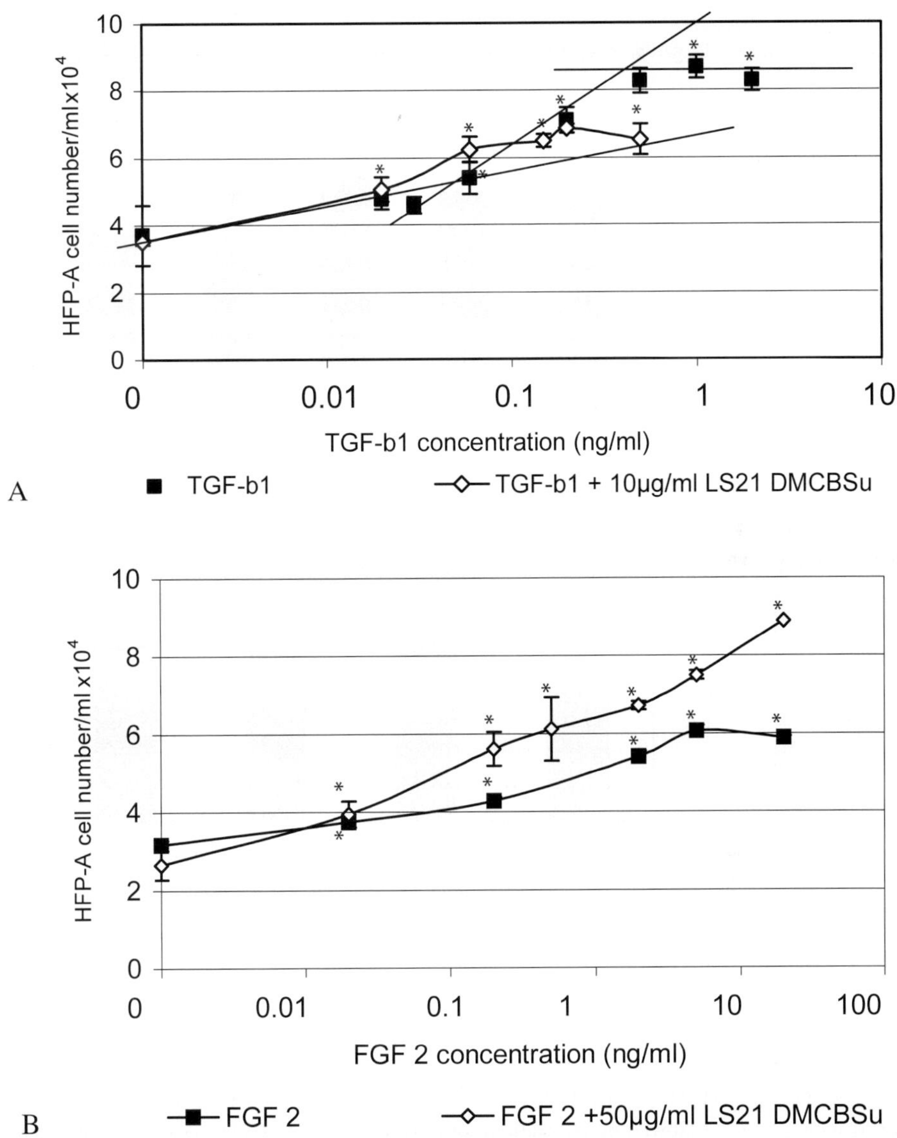

Figure 1. Dose-dependent effect of TGF-β1 (A) and FGF-2 (B) alone and with LS21 DMCBSu on the HFP-A fibroblast proliferation in a monolayer culture. Cells (10^4/well) were seeded onto a 24-well plate in DMEM supplemented with 10% FCS. After 48 h, the growth of cells was arrested by placing them in DMEM + 0.1% FCS for 24 h. Proliferation of cells was initiated by incubation in DMEM with 2% FCS followed by the addition of the growth factor. After 96 h, cells were treated with trypsin and counted with a Coulter counter. The same lot of serum and medium was used throughout all the experiments. (A) Effect of TGF-β1 alone and with 10 μg/ml of LS21 DMCBSu. Control represents the growth of fibroblasts not treated with the LS21 DMCBSu and TGF-β1 mixture. (B) Effect of FGF-2 with 50 μg/ml LS21 DMCBSu. Control represents the growth of fibroblast not treated with the LS21 DMCBsu and FGF-2 mixture. Each point represents 3 measurements: $^*P < 0.05$, LS21 DMCBSu *vs.* control.

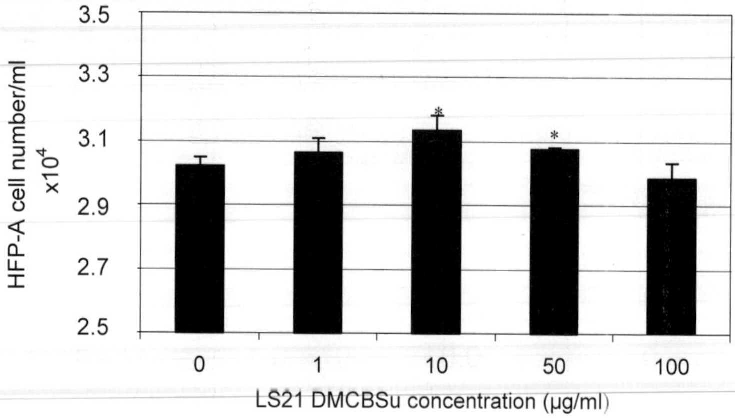

A : Monolayer culture

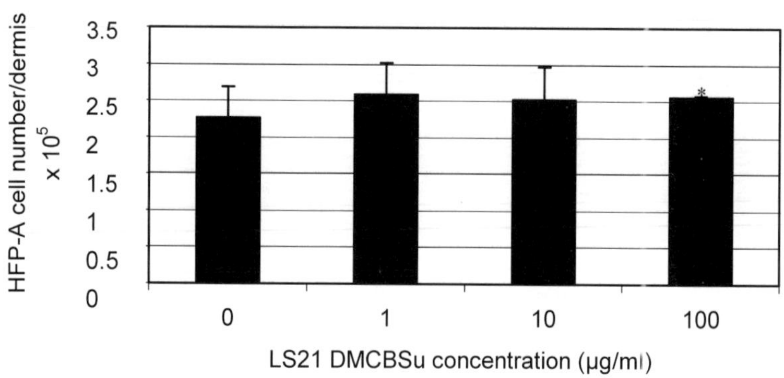

B : dermal equivalent

Figure 2. Effect of LS21 DMCBSu on cell proliferation in the monolayer culture and the premature human dermal equivalent collagen matrices. (A) Cells (10^4/well) were seeded onto a 24-well plate in DMEM supplemented with 10% FCS. After 48 h, the growth of human dermal fibroblasts was stopped by placing them in DMEM + 0.1% FCS for 24 h. Cell proliferation was initiated by incubation in DMEM with 2% FCS followed by the addition of dextran derivative LS21 DMCBSu named LS21. After 96 h, cells were treated with trypsin and then were counted with a Coulter counter. The same lot of serum and medium was used throughout all the experiments. Control represents the growth of fibroblast not treated with LS21 DMCBSu. Each point represents 3 measurements. $^*P < 0.05$, LS21 DMCBSu *vs.* control. (B) Derivative dextran was added to the premature human dermal equivalent collagen matrices at different times: day 0, 3, 6 and 8. On day 9, the number of cells was determined by extracting fibroblasts from the collagen by liaising the collagen with 2 mg/ml collagenase (Sigma type 1, 69891) in PBS. The HFP-A fibroblasts were suspended in 1 ml of PBS and the number of cells was determined using a Coulter counter. The same lot of serum and medium was used throughout all the experiments. Control represents the growth of fibroblast untreated with LS21 DMCBSu. Each point represents 3 measurements: $^*P < 0.05$ LS21 DMCBSu *vs.* control.

cell growth in the presence of increasing growth factor concentrations ranging from 0 to 0.5 ng/ml and 0 to 20 ng/ml for TGF-β1 and FGF-2, respectively, was investigated (Fig. 1). When TGF-β1 was added to LS21 DMCBSu (Fig. 1A), the lowest ED_{50} was obtained (0.06 ng/ml and 0.15 ng/ml with or without LS21 DMCBSu, respectively). Cells treated with a mixture of FGF-2 and LS21 DMCBSu were stimulated with concentrations from 0 to 20 ng/ml (Fig. 1B). No difference was detected between FGF-2 by itself or when combined with the polymer. Hence, it was shown that the interaction between LS21 DMCBSu and TGF-β1 follows a different dose-dependent relationship than that of these polymers mixed with FGF-2. It also appeared that the modulating effect of LS21 DMCBSu corresponds to different interactions between the polymer and either TGF-β1 or FGF-2.

Premature dermal equivalent collagen matrices

Fibroblast behaviour in premature dermal equivalent collagen matrices in the presence of LS21 DMCBSu. The effect of LS21 DMCBSu on the proliferation of HFP-A in premature dermal equivalent was tested, with no treatment as control (Fig. 2B). Similar to the monolayer culture, a very slight increase in the number of fibroblasts was observed when 1 and 10 μg/ml LS21 DMCBSu were incorporated into the dermal equivalents (an increase of 20%). Hence, LS21 DMCBSu concentrations of 1 and 10 μg/ml were chosen to carry out the experiments on the proliferation of fibroblasts in premature and mature dermal equivalent collagen matrices.

Fibroblast behaviour in premature dermal equivalent collagen matrices in the presence of TGF-β1 with or without addition of LS21 DMCBSu. Initially, the HFP-A cells were incorporated within the collagen lattices at the time of freezing, at the rate of 1.5×10^5 cells/gel in the presence of various concentrations of TGF-β1. This process enables the determination of the optimum concentration of TGF-β1 for use in the investigation of the effect of LS21 DMCBSu. The number of fibroblasts per dermal equivalent increased very slowly in the presence of TGF-β1 concentrations ranging from 0 to 1 ng/ml (26%) and a plateau was reached for growth factor concentrations above 0.5 ng/ml (Fig. 3A). Moreover, the contraction of the collagen governed by the fibroblasts was stimulated by TGF-β1 (Fig. 3B). Interestingly, a plateau was reached at the same concentration (0.5 ng/ml). The diameter of the collagen gels appeared to be inversely proportional to the increasing concentration of TGF-β1. This data indicated that a concentration of 1 ng/ml of TGF-β1 was appropriate to investigate the effect of mixture of LS21 DMCBSu and TGF-β1 on cell growth and dermal contraction in premature dermal equivalent collagen matrices.

When TGF-β1 was added with a concentration of 1 μg/ml of LS21 DMCBSu, a slight inhibition of HFP-A growth was observed (20%, Fig. 4A). The ability of the dextran derivative to mimic heparin, in terms of complex formation with TGF-β1, may explain the limited action of the growth factor when mixed with the polymer

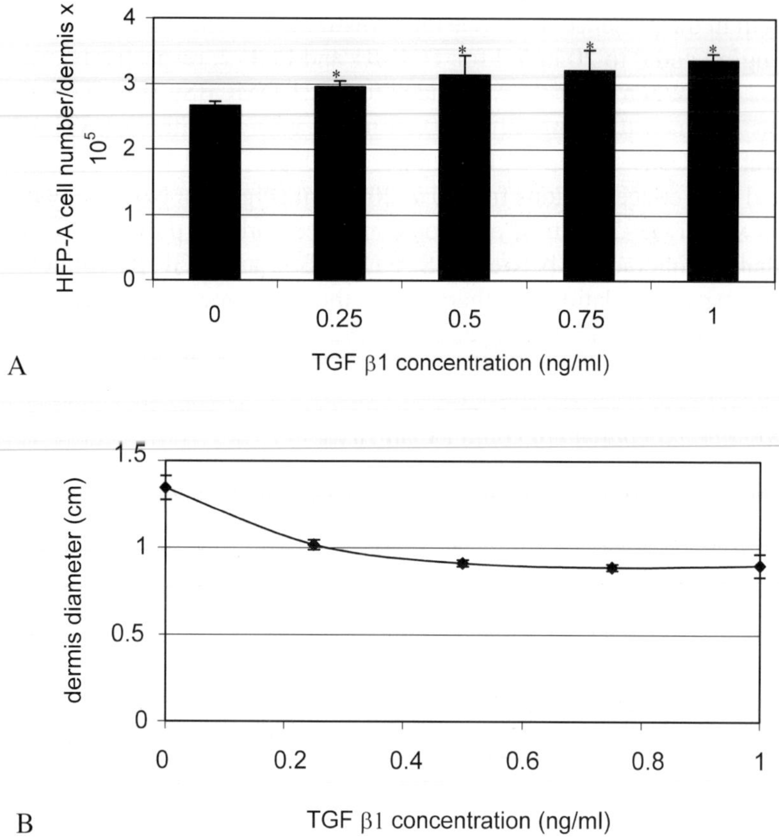

Figure 3. Effect of TGF-β1 on cell proliferation and dermal contraction in premature human dermal equivalent collagen matrices. TGF-β1 was added to the collagen gel including cells, at various times: day 0, 3, 6 and 8. On day 9, the number of cells was determined by extracting fibroblasts from the collagen by lysing the collagen with 2 mg/ml collagenase in PBS. The same lot of serum and medium was used throughout all the experiments. (A) Determination of the number of HFP-A cells. Control represents the growth of fibroblast not treated with TGF-β1. (B) Measurement of the diameter after 9 days of incubation. Each point represents 3 measurements. Control represents the diameters of premature human dermal equivalent collagen matrices untreated with TGF-β1. $^*P < 0.05$ TGF-β1 *vs.* control.

produced. Moreover, the contraction produced by TGF-β1 was not affected by the presence of the polymer (Fig. 4B). The polymer did not have any apparent effect on the dermal contraction, indicating that the association of the growth factor and the polymer appeared to only slightly inhibit the growth of the fibroblasts without affecting the size of collagen lattice.

In order to test the cytotoxic effect of LS21 DMCBSu with or without added TGF-β1, the fibroblasts were stained with Trypan blue (0.4%). More than 95% of cells were shown to be alive. In addition, HFP-A, after treatment, exhibited normal proliferative activity, similar to that obtained without treatment.

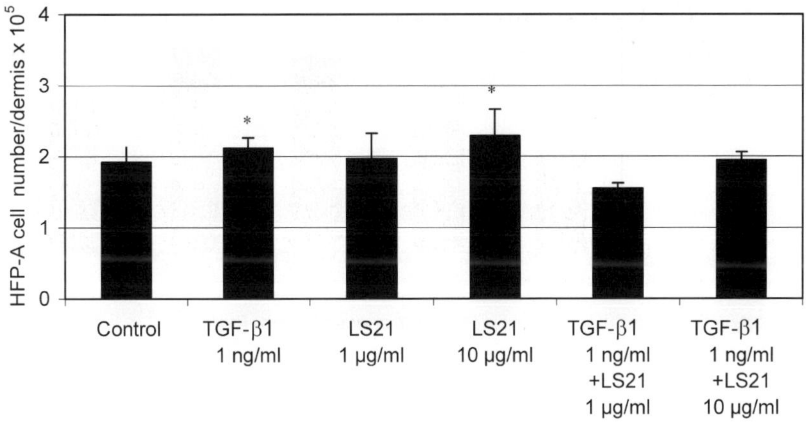

A

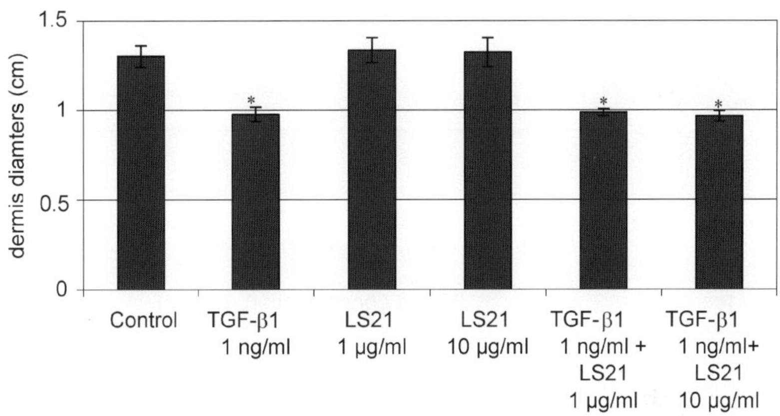

B

Figure 4. Effect of TGF-β1 and LS21 DMCBSu both individually and when combined on HFP-A proliferation and dermal contraction in premature human dermal equivalent collagen matrices. TGF-β1, LS21 DMCBSu or a mixture of both compounds containing 1 ng/ml of TGF-β1 was added concurrently to the premature human dermal equivalent collagen matrices at various times: day 0, 3, 6 and 8. On day 9, the number of cells was determined by extracting fibroblasts from the collagen by lysing the collagen with 2 mg/ml collagenase in PBS. The same lot of serum and medium was used throughout all the experiments. (A) Determination of the number of HFP-A cells. Control represents the growth of fibroblast not treated with LS21 DMCBSu, TGF-β1 or a mixture of both compounds. (B) Measurement of the diameter after 9 days of incubation. Each point represents 3 measurements. Control represents the diameters of premature human dermal equivalent collagen matrices not treated with LS21 DMCBSu, TGF-β1 or a mixture of both compounds. $^*P < 0.05$ TGF-β1, LS21 DMCBSu and/or mixture of both compounds *vs.* control.

Fibroblast behaviour in premature dermal equivalent collagen matrices in the presence of FGF-2 with or without added LS21 DMCBSu. HFP-A cells were cultured in premature dermal equivalents in the presence of FGF-2. In order to evaluate the effect of FGF-2, the number of HFP-A cells per dermis and dermis diameter were studied to define an optimum mixture (Fig. 5A and 5B). The number of fibroblasts per dermal equivalent increased when FGF-2 (0–1 ng/ml) was added

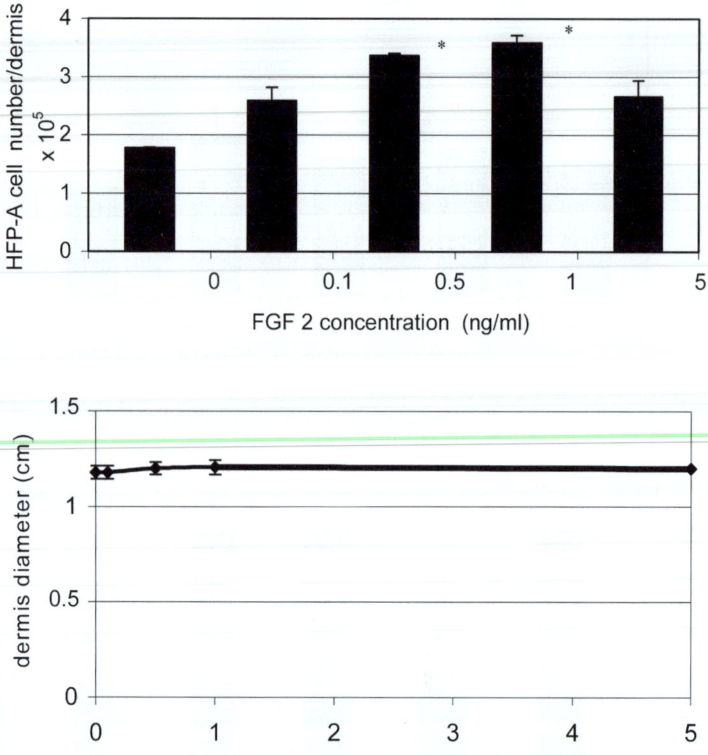

A

B

Figure 5. Effect of FGF-2 on cell proliferation and dermal contraction in premature human dermal equivalent collagen matrices. FGF-2 was added to the collagen gel including fibroblasts at various times: day 0, 3, 6 and 8. On day 9, the number of cells was determined by extracting cells from the collagen by lysing the collagen with 2 mg/ml collagenase in PBS. (A) Evaluation of the number of HFP-A cells. The same lot of serum and medium was used throughout all the experiments. Control represents the growth of fibroblast untreated with FGF-2. (B) Measurement of the diameter after 9 days of incubation. Control represents the diameters of premature human dermal equivalent collagen matrices untreated with FGF-2. Each point represents 3 measurements. $^{*}P < 0.05$ FGF-2 *vs.* control.

to the culture medium. A two-fold increase in the number of fibroblasts was measured with 1 ng/ml of FGF-2. Dermal contraction was not affected by presence of FGF-2 (Fig. 5B). For a concentration of FGF-2 higher than 1 ng/ml, a decrease of the cell growth was observed. Hence, a 1 ng/ml solution of FGF-2 was employed to investigate the effect of LS21 DMCBSu mixed with FGF-2 on both cell proliferation and dermal contraction. It is interesting to note that this optimum concentration was similar to that of TGF-β1, as shown in Fig. 3A. The effect of FGF-2 with or without LS21 DMCBSu on cell proliferation and dermal contraction is shown in Fig. 6. The presence of LS21 DMCBSu slightly (10%) increased the capacity of FGF-2 to stimulate the proliferation of fibroblasts. No modification of diameters was detected after contraction whether FGF-2 was utilized by itself or mixed with LS21 DMCBSu (Fig. 6B).

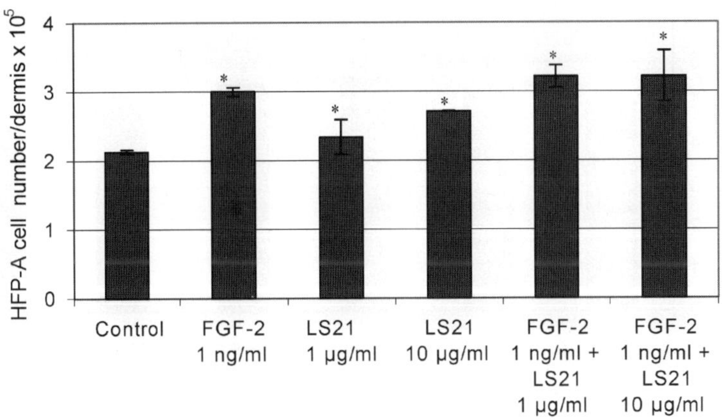

A

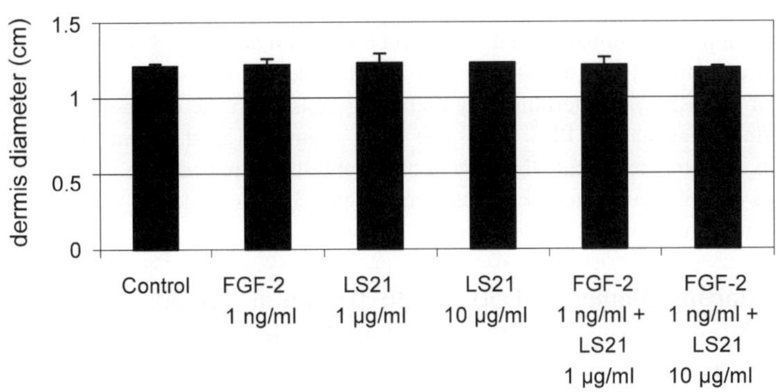

B

Figure 6. Effect of FGF-2 and LS21 DMCBSu both individually and when combined on HFP-A proliferation and dermal contraction in premature human dermal equivalent collagen matrices. FGF-2, LS21 DMCBSu or a mixture of both substances containing 1 ng/ml of FGF-2 was added concurrently to the collagen gel including cells at various times: day 0, 3, 6 and 8. On day 9, the number of cells was determined by extracting fibroblasts from the collagen by lysing the collagen with 2 mg/ml collagenase in PBS. The same lot of serum and medium was used throughout all the experiments. (A) Evaluation of the number of HFP-A cells. Control represents the growth of fibroblast not treated with LS21 DMCBSu, FGF-2 or a mixture of both compounds. (B) Measurement of the diameter after 9 days incubation. Control represents the diameters of premature human dermal equivalent collagen matrices not treated with LS21 DMCBSu, FGF-2 or a mixture of both compounds. Each point represents 3 measurements: $^*P < 0.05$ FGF-2, LS21 DMCBSu and/or mixture of both compounds *vs.* control.

Mature dermal equivalent collagen matrices

Fibroblast behaviour in mature dermal equivalent collagen matrices in the presence of TGF-β1 with or without addition of LS21 DMCBSu. The effect of TGF-β1, LS21 DMCBSu and a mixture of both of these substances on the mature dermal equivalents was investigated (Fig. 7). The very complex geometry of the mature dermal equivalent rendered measurement of its diameters impossible. Indeed, the maximum contraction of this type of lattice resulted in a decrease

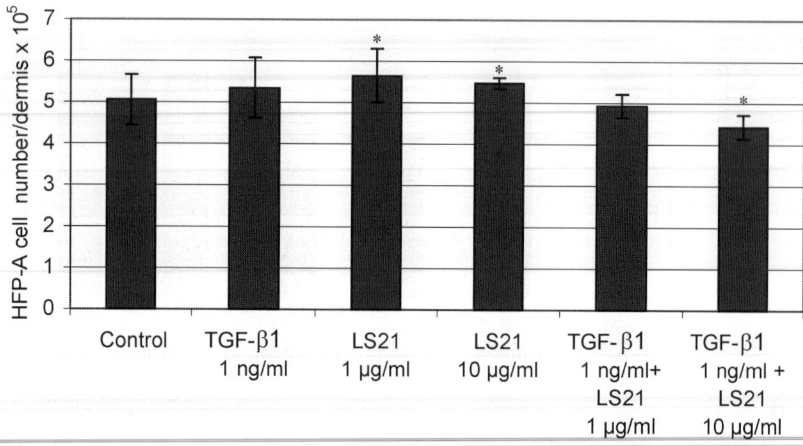

Figure 7. Effect of TGF-β1 and LS21 DMCBSu both individually and when combined on HFP-A proliferation in mature human dermal equivalent collagen matrices. LS21 DMCBSu and/or TGF-β1 were added concurrently on day 21, 27 and 29. The dermal equivalent was dissolved on day 30. The number of cells was determined by extracting cells from the collagen by lysing the collagen with 2 mg/ml collagenase in PBS. The same lot of serum and medium was used throughout all the experiments. Control represents the growth of fibroblast not treated with LS21 DMCBSu, TGF-β1 or a mixture of both compounds. Each point represents 3 measurements. $^*P < 0.05$ TGF-β1, LS21 DMCBSu and/or mixture of both compounds *vs.* control.

from the starting diameter of 50 mm to 10 mm at day 30. A slight effect on cell proliferation was observed with either 1 μg/ml or 10 μg/ml of LS21 DMCBSu. A similar result was observed for TGF-β1. A slight decrease in the number of cells was detected when the growth factor was added to the dextran derivative.

Fibroblast behaviour in mature dermal equivalent collagen matrices in the presence of FGF-2 with or without addition of LS21 DMCBSu. FGF-2, LS21 DMCBSu and a mixture both compounds were tested on mature dermal equivalent collagen matrices (Fig. 8). It is interesting to note that the growth factor and the polymer had only a slight effect on fibroblast proliferation and no effect on the contraction of the matrix (results not shown).

DISCUSSION

Heparin-binding growth factors, such as TGF-β1 and FGF-2, play a key role in wound healing. Previously, it has been reported that some dextran derivatives could stimulate tissue repair for *in vivo* models [25, 26]. It has been established that these polymeric biomaterials could enhance the biological activity of several heparin-binding growth factors on various cell lines, such as vascular cells [27, 28], osteoblasts [29] and tumoral mammary epithelial cell lines [30]. Depending on the proportions of the chemical functions substituted on the macromolecular chains of original dextran, the final dextran derivative may specifically act as ligands to cell

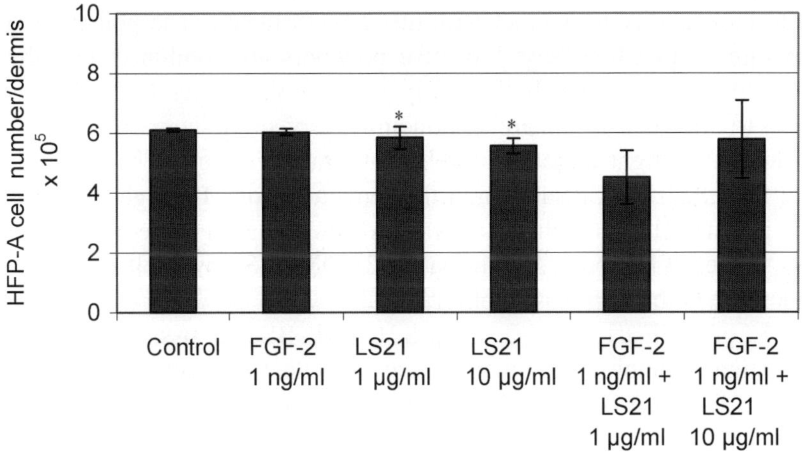

Figure 8. Effect of FGF-2 and LS21 DMCBSu both individually and when combined on HFP-A proliferation in mature human dermal equivalent collagen matrices. LS21 DMCBSu and/or FGF-2 were added concurrently on day 21, 27 and 29. The dermal equivalent was dissolved on day 30 and the number of cells was determined by extracting fibroblasts from the collagen by lysing the collagen with 2 mg/ml collagenase in PBS. The same lot of serum and medium was used throughout all the experiments. Control represents the growth of fibroblast not treated with LS21 DMCBSu, FGF-2 or a mixture of both compounds. Each point represents 3 measurements. $^*P < 0.05$ FGF-2, LS21 DMCBSu and/or mixture of both compounds *vs.* control.

membrane receptors [16]. *In vitro*, these polymers act as protectors and promoters of the released growth factors [31]. They also act on proteins such as collagen, osteocalcin and plasmin, which are involved in tissue remodelling [29, 32, 33]. Derivative dextrans may be used to replace the function of heparin by binding to the two growth factors, TGF-β1 and FGF-2, increasing stability and increasing the subsequent biological response.

In the present study, we report that a dextran derivative is able to modulate the effect of both TGF-β1 and FGF-2 on human dermal fibroblast proliferation. It has been established that several dextran derivatives with varying substitution ratios of active chemical groups were able to act as promoting agents for FGF-2 [11, 14]. It has been demonstrated that dextran derivatives which contain a relatively high substitution ratio of sulfate groups were the most efficient in facilitating the proliferation of fibroblasts. The polymer LS21 DMCBSu (prepared using good manufacturing practice) has been selected for its effect on the growth of cells in the presence of the growth factors.

In the monolayer culture, the inciting effect on the proliferation of fibroblasts was significant when the dextran derivative was combined with either TGF-β1 or FGF-2. The cell proliferation depended on the growth factor concentration. A concentration of 10 or 50 μg/ml of LS21 DMCBSu added to the optimum dose of TGF-β1 or FGF-2, respectively, led to a similar degree of stimulation of the cell proliferation. Recently, Logeart-Avramoglou *et al.* [15] have shown that the potentiating effect of dextran derivatives on the TGF-β1 biological response is related to their ability to interact with this growth factor. The potentiating effect is more significant when these bioactive polymers are combined with the lower growth factor concentration. In this study, we have shown that LS21 DMCBSu has the ability to bind TGF-β1, thereby inducing a very slight increase in the fibroblast proliferation. A slight increase of cell proliferation seems to be induced by the capacity for LS21 DMCBSu to bind the growth factor. Through this association with the polymer, the growth factor was protected against proteolytic degradation and was stable. It has been shown that LS21 DMCBSu by itself produced a very slight increase in the fibroblast proliferation.

The role of TGF-β1, FGF-2 and LS21 DMCBSu in tissue remodelling has been investigated employing human dermal equivalent collagen matrices containing fibroblasts. The use of three-dimensional models in fibroblast culture provides a cellular environment which is comparable to the *in vivo* matrices. According to our previous work [34], which demonstrated that fibroblasts were viable and proliferated over a period of 21 days of culture, the experiments were carried out on two types of collagen matrices. These matrices were premature dermal equivalents cultured over 9 days (treatment once every 3 days during the 9-day period) and mature dermal equivalents cultured over 30 days (treatment began at day 21 and lasted for 9 days). On day 9, the fibroblasts progressively reorganized the lattice, whilst on day 30 the collagen matrix was already reorganized.

These data were different from that obtained for the monolayer culture and were dependent on the growth factors used. Dermal fibroblasts in the premature dermal equivalent were significantly affected by both TGF-β1 and FGF-2 without the addition of the polymer. The dose–response variation for both growth factors reached maximum value at approximately 1 ng/ml, as had been previously reported with FGF-1 [17]. Therefore, the growth factor induction of mitosis was accompanied by a decrease in the diameter of the matrices treated with TGF-β1. This effect correlates with the data given particularly for TGF-β1. TGF-β1 has been reported to promote collagen matrix contraction by fibroblasts [35]. The promotion of this contraction by TGF-β1 is not strictly dependent on mitosis. Moreover, when FGF-2 was added to the culture medium, a difference in the diameter in the matrices was not observed. Previous data reported that high concentrations of FGF-2, which does not stimulate fibroblast division, did not influence the dermal contraction [17].

The optimum conditions have been selected to study the effect of growth factors mixed with the dextran derivative. The concentration of TGF-β1 and of FGF-2 was fixed at 1 ng/ml containing doses of LS21 DMCBSu of 1 or 10 μg/ml, respectively, which have a slight effect on the fibroblast in the premature dermal equivalent. When LS21 DMCBSu was added to 1 ng/ml of TGF-β1 or FGF-2 the fibroblast proliferation in the premature dermal equivalents was significantly modified. Surprisingly, fibroblast growth was inhibited when TGF-β1 was added to LS21 DMCBSu, while it was almost stimulated when FGF-2 was added to the bioactive polymer. The inhibition of the cell growth in the presence of TGF-β1 was accompanied by a decrease in the mean diameter. Although the mechanism of this inhibition is unknown, the data may be explained in terms of formation of a complex between TGF-β1 and LS21 DMCBSu. The complex will have the capacity of masking the active site able to interact with the cell receptors. As a consequence, the ability of TGF-β1 to stimulate the fibroblast growth will be affected. The presence of the dextran derivative at the higher dosage of 10 μg/ml induced a mitotic effect, probably due to the availability of a quantity of the polymer which was not involved in the complex reaction. Interestingly, the slight stimulation effect on the cell growth by FGF-2 in the presence of LS21DMCBSu was not accompanied by a variation in the mean diameter of the collagen matrices. This led to the conclusion that the FGF-2/LS21 DMCBSu complex acted on the metabolism of the fibroblasts in a similar manner to FGF-2 by itself. It has been demonstrated that LS21 DMCBSu produced biological effects depending on whether the dextran derivative was combined with TGF-β1 or FGF-2. In spite of the differences found in amino-acid composition, both TGF-β1 and FGF-2 contain sequences able to interact with LS21 DMCBSu. However, TGF-β1 and FGF-2 differ in their ability to form the dextran derivative complex. These observations have been confirmed by studying the effect of LS21 DMCBSu on fibroblast behaviour in matrice dermal equivalent collagen matrices containing 1 ng/ml TGF-β1 or FGF-2. It was noted that LS21 DMCBSu did not induce any toxicity over the 30-day culture period.

The major conclusion to be drawn from the presented data is that the bioactive polysaccharide was able to slightly stimulate the growth of dermal fibroblast in the presence of FGF-2. In contrast, the presence of 1 μg/ml LS21 DMCBSu in the culture medium of premature dermal equivalent collagen matrices inhibited the stimulating effect of TGF-β1. This dextran derivative exhibited a slight effect on dermal fibroblast proliferation and no effect on contraction of the collagen matrix. Therefore, it is expected that these bioactive polymers can promote tissue remodelling by accelerating the process of wound healing. Hence, data in relation with the effect on metalloproteinases and their inhibitors are currently under investigation.

Acknowledgements

This work was supported by the Centre National de la Recherche Scientifique and by a grant awarded to Paris 13 University by THERAPOL S A. We thank Mr. Dubois for the helpful critical reading of manuscript.

REFERENCES

1. H. P. Ehrlich, *Adv. Wound Care* **11**, 326 (1998).
2. M. McGrath, *Clin. Plast. Surg.* **17**, 421 (1990).
3. T. McCaffrey, D. Falcone and B. Du, *J. Cell Physiol.* **152**, 430 (1992).
4. M. Nugent and R. Iozzo, *Int. J. Biochem. Cell Biol.* **32**, 115 (2000).
5. M. Sporn, A. Roberts, L. Wakefield and A. Assoian, *Science* **233**, 532 (1986).
6. D. Cromack, G. Pierce and T. Mustoe, *Progr. Clin. Biol. Res.* **365**, 359 (1991).

7. E. Amento and L. Beck (Eds), in: *Clinical Applications of TGF-β*, p. 115. Wiley, Chichester (1991).
8. S. Wahl, H. Wong and N. McCartney-Francis, *J. Cell Biochem.* **40**, 193 (1989).
9. T. Mellin, R. Mennie, D. Cashen, J. Ronan, J. Capparella, M. James, J. Disalvo, J. Frank, D. Linemeyer, G. Gimenez-Gallego and K. Thomas, *Growth Factors* **7**, 1 (1992).
10. K. Takehara, *J. Dermatol. Sci.* **24** (Suppl. 1), 70 (2000).
11. M. Tardieu, C. Gamby, T. Avramoglou, J. Josefonvicz and D. Barritault, *J. Cell Physiol.* **150**, 194 (1992).
12. W. Burgess and T. Maciag, *Annu. Rev. Biochem.* **58**, 575 (1989).
13. A. Meddahi, J. Benoit, N. Ayoub, A. Sézeur and D. Barritault, *J. Biomed. Mater. Res.* **31**, 293 (1996).
14. P. Bittoun, R. Bagheri-Yarmand, F. Chaubet, M. Crépin, J. Jozefonvicz and S. Fermandjian, *Biochem. Pharmacol.* **57**, 1399 (1999).
15. D. Logeart-Avramoglou, R. Huynh, F. Chaubet, L. Sedel and A. Meunier, *Biochem. Pharmacol.* **63**, 129 (2002).
16. D. Logeart-Avramoglou and J. Jozefonvicz, *J. Biomed. Mater. Res.* **48**, 578 (1999).
17. L. Dubertret, F. Brunner-Ferber, J. Misiti, K. Thomas and M. L. Dubertret, *J. Invest. Dermatol.* **79**, 793 (1991).
18. E. Gentilhomme, Y. Neveux, J. Lebeau, A. Desmoulière, J. Bergier, D. Schmitt and M. Haftek, *Cell Biol. Toxicol.* 15, 229 (1999).
19. P. G. Genever, E. J. Wood and W. J. Cunliffe, *Exp. Dermatol.* **2**, 266 (1993).
20. M. Cordeiro, S. Bhattacharya, G. Schultz and P. Khaw, *Invest. Ophtalmol. Vis. Sci.* **41**, 756 (2000).
21. V. Moulin, G. Castilloux, F. A. Auger, D. Garrel, M. O'Connor-McCourt and L. Germain, *Exp. Cell Res.* **238**, 283 (1998).
22. R. Huynh, F. Chaubet and J. Jozefonvicz, *Carbohydr. Res.* **332**, 75 (2001).
23. E. Bell, B. Ivarsson, C. Merrill, *Proc. Natl. Acad. Sci. USA* **76**, 1274 (1979).
24. I. Khouw, P. Van Wachem, J. Plantinga, Z. Vujaskovic, M. Wissink, L. Leij and M. Van Luyn, *Biomaterials* **20**, 1815 (1999).
25. A. Meddahi, F. Brée, D. Papy-Garcia, J. Gautron, D. Barritault and J. P. Caruelle, *J. Biomed. Mater. Res.* **62**, 525 (2002).
26. A. Meddahi, C. Alexakis, D. Papy, J. P. Caruelle and D. Barritault, *J. Biomed. Mater. Res.* **60**, 497 (2002).
27. D. Letourneur, J. Champion, F. Slaoui and J. Jozefonvicz, *In Vitro Cell Dev. Biol.* **29**, 67 (1993).
28. D. Letourneur, D. Logeart, T. Avramoglou and J. Jozefonvicz, *J. Biomater. Sci. Polymer Edn* **4**, 431 (1993).
29. S. Berrada, J. Amedee, T. Avramoglou, J. Jozefonvicz and M. F. Harmand, *J. Biomater. Sci. Polymer Edn* **6**, 211 (1994).
30. R. Bagheri-Yamand, Y. Kourbali, J. F. Morère, J. Jozefonvicz and M. Crépin, *Cell Growth Diff.* **9**, 497 (1998).
31. T. Kajio, K. Kawahara and R. Kato, *FEBS Lett.* **306**, 243 (1992).
32. Y. Benazzoug, D. Logeart, J. Labat-Robert, L. Robert, J. Jozefonvicz and P. Kern, *Biochem. Pharmacol.* **49**, 847 (1995).
33. D. Ledoux, D. Papy-Garcia, Q. Escartin, M. A. Sagot, Y. Cao, D. Barritault, J. Courtois, W. Hornebeck and J. P. Caruelle, *J. Biol. Chem.* **275**, 29383 (2000).
34. L. Besseau, B. Coulomb, C. Lebreton-Decoster and M. M. Giraud-Guille, *Biomaterials* **23**, 27 (2002).
35. R. Montesano and L. Orci, *Proc. Natl. Acad. Sci. USA* **85**, 4894 (1988).

Gels, Genes, Grafts and Giants, pp. 245–260 (2005)

Novel biodegradable cholesterol-modified polyrotaxane hydrogels for cartilage regeneration

WANPEN TACHABOONYAKIAT [1], TOMOYUKI FURUBAYASHI [1], MASAKAZU KATOH [2], TOORU OOYA [3] and NOBUHIKO YUI [3,*]

[1] *Innovation Plaza Ishikawa, Japan Science and Technology Agency, 2-13 Asahidai, Tatsunokuchi, Ishikawa 923-1211, Japan*
[2] *Japan Tissue Engineering Co., Ltd., 6-209-1 Miyakitadori, Gamagori, Aichi 443-0022, Japan*
[3] *School of Materials Science, Japan Advanced Institute of Science and Technology, 1-1 Asahidai, Tatsunokuchi, Ishikawa 923-1292, Japan*

Received 14 October 2003; accepted 5 April 2004

Abstract—Cholesterol was introduced to a hydrolyzable polyrotaxane (PRx), not only to improve cell proliferation and glycosaminoglycan (GAG) production, but also to control the degradation rate of the hydrogels. The cholesterol was introduced to hydrolyzable PRx species by threading many α-cyclodextrins (α-CDs) on a poly(ethylene glycol) (PEG) chain having hydrolyzable ester linkages at the terminals; the PRx species were then cross-linked with other PEGs to prepare cholesterol-modified PRx hydrogels. The degree of cholesterol substitution was varied in the range of 1–25%. These hydrogels were examined to clarify the effect of cholesterol groups on mechanical properties, erosion time and chondrocyte proliferation. Highly porous biodegradable cholesterol-modified PRx hydrogels were fabricated using a combination of potassium hydrogen carbonate (as an effervescent salt) and citric acid. This fabrication process enabled the homogeneous expansion of pores within the polymer matrices, leading to well-interconnected macroporous hydrogels with a mean pore size of around 200–400 μm, ideal for high-density chondrocyte seeding. Time to complete degradation of the hydrogels was shortened by increasing the degree of substitution due to the aggregation of α-CDs through hydrophobic interaction of cholesterol groups. The presence of approx. 10% cholesterol improved the chondrocyte proliferation and GAG production. The modification of cholesterols to PRx is a good approach for creating new biodegradable hydrogels in terms of chondrocyte culture and controlling degradation time of the hydrogels.

Key words: Macroporous hydrogel; cholesterol; hydrolyzable polyrotaxane; chondrocyte.

*To whom correspondence should be addressed. E-mail: yui@jaist.ac.jp

INTRODUCTION

Tissue engineering of cartilage, in which chondrocytes are cultured on biocompatible matrices to prepare transplantable tissue, is a promising strategy to restore articular cartilage [1–4]. The concept of tissue engineering is based on cell–scaffold systems, in which biodegradable polymers are used as templates to promote tissue formation [5]. Highly open porous polymer matrices are required for high-density cell seeding, so that efficient nutrient and oxygen can be supplied to the cells cultured on the three-dimensional matrices. We have studied polyrotaxane (PRx) hydrogels prepared by linking the poly(ethylene glycol)bisamine (PEG-BA) with α-cyclodextrins (α-CDs) threaded onto a PEG chain having terminal ester linkages (hydrolyzable PRx) [6]. The erosion time of the interlocked hydrogels is controlled by inclusion complexation of α-CDs with the terminal ester linkages: both the maintenance of high water content and prolonged erosion time is achieved in these hydrogels [7]. This unusual finding is due to the structural feature of the mechanical locks between the α-CDs and the PEG chain [8]. The PRx hydrogels are completely eroded, which produces the degradation products of the PEG, the α-CDs and an amino-acid derivative. In those degradation products, the amino acid derivative does not change the pH of the medium, so the influence of the acids as a degradation product can be negligible. This property is a significant advantage compared to poly(lactic-co-glycolic acid) (PLGA). Recently, we have demonstrated that interlocked hydrogels with controlled degradation properties have good potential to serve as biodegradable scaffolds for chondrocyte cultures [9]. However, optimum conditions for cell culture in these scaffolds are unknown.

An important characteristic of our designed PRx is the easy introduction of functional groups and/or biological ligands on to the hydroxyl groups of the α-CDs threaded on the PEG chain, which can control cellular response in the cell attachment and the proliferation process. Of special interest regarding this inclusion, supramolecular PRx, is the ability of the α-CDs on the PEG chain to slide and rotate. We suggest this motion assists the ligands in finding suitable contact sites on the cell membrane [10]. It is well established that hydrophobic moieties affect cell adhesion and proliferation, and cell adhesion is better on hydrophobic than on hydrophilic surfaces [11, 12]. Cell attachment and internalization are correlated well to hydrophobicity [13]. Cholesterol is mainly found as a component in cell membranes [14]. One investigation supports the ability to insert cholesterol into phospholipid membranes [15].

In this study, cholesterol groups were attached to the PRx. The cholesterol-modified PRx hydrogels were prepared by linking the PEG-BA with the α-CDs threaded onto the PEG chain. Macroporous structures were formed by a gas foaming technique using a combination of potassium hydrogen carbonate as an effervescent salt and citric acid. The effect of the cholesterol modification on erosion time was evaluated in 0.1 M NaOH and phosphate buffer. The effect of the degree of cholesterol substitution for PRx on chondrocyte proliferation and production of glycosaminoglycans (GAGs) was investigated.

MATERIALS AND METHODS

Materials

Poly(ethylene glycol) (PEG) with a number-average molecular weight of 3300 was a gift from Sanyo Chemical (Kyoto, Japan). Amino-terminated poly(ethylene glycol) (poly(propylene glycol)-block-poly(ethylene glycol)-block-poly(ethylene glycol)bis(2-aminopropyl ether), PEG-BA, M_n 2000) was a gift from Suntechno Chemical (Tokyo, Japan). PEG-BA was dehydrated before use by an azeotropic distillation in toluene, precipitated in diethyl ether, filtered and dried at room temperature *in vacuo*. The following chemicals were reagent grade and used without further purification: α-cyclodextrin (α-CD) (Bio-Research Corporation of Yokohama, Yokohama, Japan); benzyloxycarbonyl (Z)-L-phenylalanine (Z-L-Phe) (Kokusan Chemical, Tokyo, Japan); succinic anhydride, ethylenediamine, N,N'-dicyclohexylcarbodiimide (DCC), sodium chloride, sodium hydroxide and citric acid anhydrous (Nacalai Tesque, Kyoto, Japan); N,N'-carbonyldiimidazole (CDI) (Aldrich, Milwaukee, WI, USA); N-hydroxysuccinimide (Peptide Institute, Osaka, Japan); cholesterol chloroformate (Tokyo Kasei Kogyo, Tokyo, Japan); triethylamine (Sigma, St. Louis, MO, USA); Potassium hydrogen carbonate (Kanto Chemical, Tokyo, Japan); Dimethylsulfoxide containing 0.03% trimethylsilane ([D_6]DMSO, Arcos Organics, Geel, Belgium). Dimethylsulfoxide (DMSO) and dimethylformamide (DMF) were purchased from Nacalai Tesque and purified before use.

Synthesis of cholesterol-modified PRx

The hydrolyzable PRx was synthesized by our established method [7, 8]. PRx (1 g, 24 μmol) with an α-CDs threading number of 38 was dissolved in 40 ml DMSO/DMF (1 : 1). Cholesterol chloroformate (54.7–1370 μmol) was added to the solution and stirred at 60°C for 3 h under N_2 atmosphere. The reaction mixture was poured into an excess amount of ether to obtain the crude product. The resulting white precipitant was washed with ether several times, centrifuged and dried *in vacuo*. The hydrolyzable PRx and cholesterol-modified PRx were characterized by ^{1}H-NMR (Varian 300 MHz FT-NMR, Gemini300, Palo Alto, CA, USA). ^{1}H-NMR (DMSO-d_6, ppm) of PRx: 5.80–5.20 (d, OH-2, OH-3 of α-CD), 4.79 (s, H-1 of α-CD), 4.48 (b, OH-6 of α-CD), 3.80–3.25 (m, H-3, H-5, H-6 of α-CD), 3.51 (s, CH_2 of PEG) ^{1}H-NMR (DMSO-d_6, ppm) of cholesteroyl derivative: $\delta = 5.80$–5.20 (b, 12H, OH-2, OH-3 of α-CD/1H, H-6 of Chol), 4.80 (s, 1H, H-1 of α-CD), 4.65–4.20 (b, 6H, OH-6 of α-CD/2H, H-4, H-4$'$ of Chol), 3.80–3.60 (m, 24H, H-3, H-5, H-6, H-6$'$ of α-CD), 3.51 (m, 4H, H-1, H-1$'$, H-2, H-2$'$ of PEG), 3.40–3.20 (m, 12H, H-2, H-4 of α-CD) and 1.30–1.17 (m, 41H, H of Chol). Degree of substitution was calculated from the integral between 41H of Chol and H-1 of α-CD. By changing the feed ratio of cholesterol chloroformate to hydroxyl groups of α-CDs, the degree of substitution could be varied in the range of 1–25%. The schematic structure of the cholesterol-modified PRx is shown in Fig. 1.

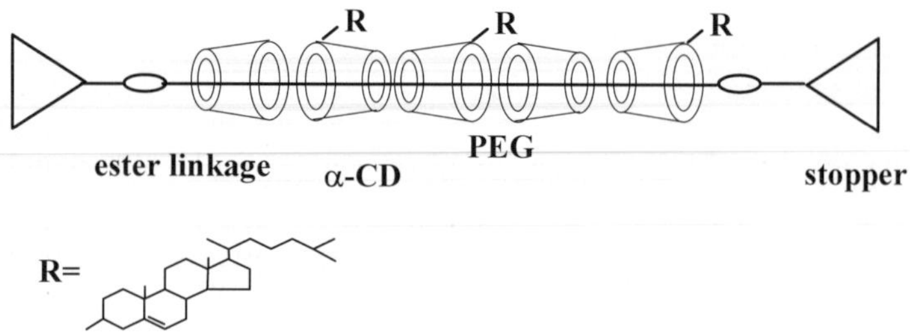

Figure 1. Schematic structure of the cholesterol-modified PRx.

Preparation and characterization of macroporous hydrogels using cholesterol-modified hydrolyzable PRx species

Briefly, hydroxyl groups of α-CDs in the choresterol-modified PRx species were activated with 1,1-carbonyldiimidazole (CDI) in DMSO [7]. PEG-BA (MW 2000) was added to the solution so as the feed ratio of PEG-BA and α-CD (PEG-BA/α-CD) was 1.0. Two fabrication processes were carried out to prepare macroporous hydrogels. (1) Salt-leaching process: 600 wt% NaCl, passed through a sieve with holes 300–500 μm in diameter, was added to the mixture and mixed homogeneously. NaCl (1.2 g) containing the solution was put into a mold, compressed and incubated at 40°C for 24 h. The top and bottom surfaces of the pellet were then sliced and the pellet was salt-leached in a large amount of distilled water. The distilled water was changed several times during the salt-leaching process, which was carried out for 48 h. (2) Gas-foaming process: 600 wt% $KHCO_3$, passed through a sieve with holes 300–500 μm in diameter, was added to the mixture and mixed homogeneously. Similarly, 1.2 g of $KHCO_3$ containing the solution was put into a mold, compressed and incubated at 40°C for 24 h. The top and bottom surfaces of the pellet were then sliced and the pellet was then put in a large amount of 3 M citric acid solution for 12 h, then washed with distilled water that was changed several times. The washing process was carried out for 36 h. The hydrogels had a 19.65-mm^2 cross-section and a height of about 4 mm; Young's modulus was determined in a compression mode using a thermomechanical analysis (TMA/SS150 analyzer, SII Seiko Instruments, Japan), operated on a SSC/5200H computer at 37°C. The surface and cross-section structures of the dehydrated hydrogels were observed using scanning electron microscopy (JSM-5510, Jeol, Japan) with magnification of 100× under a 25 kV electron beam. The dehydrated hydrogels were cut with a razor blade and sputter coated with platinum (Pt) in an electric field of 30 mA for 30 s under vacuum, using a JFC-1600 Auto Fine Coater (Jeol) before the observation.

In vitro degradation behavior of cholesterol-modified PRx hydrogels

The swollen cholesterol-modified PRx hydrogels were immersed in 0.1 M NaOH aq. to record accelerated degradation times, in order to predict long-term degradation characteristics. The erosion behavior was also observed by immersing the hydrogels in 0.1 M phosphate buffer (pH 7.4) containing 0.02 wt% sodium azide as a bacteriostatic agent at 37°C and shaking them at 130 rpm with a universal shaker SHK-U4 (Iwaki Asahi Techno Glass, Japan). The erosion process was evaluated by determining the residual weight of the hydrogels. Both experiments were carried out in triplicate. The gel weight (%) was calculated as follows.

$$\text{Gel weight (\%)} = (W_t / W_0) \times 100,$$

where W_t is the wet weight of the hydrogels at time t and W_0 is the wet weight of the hydrogels before degradation.

Chondrocyte culture and determination of glycosaminoglycan production

Chondrocytes were isolated from Japanese white rabbit cartilage (allogenic model). The macroporous hydrogels, divided into 4 parts, were washed and stirred in distilled water four times, then sterilized by immersion in 70% ethanol for 30 min. After that, the macroporous hydrogels were placed in 24-well plates. A 10-μl cell suspension containing 1×10^7 cells/ml was seeded on the hydrogels and left for 30 min. Chondrocytes were cultured at 37°C, 5% CO_2 and 95% humidity for 14 and 28 days. The culture medium (DMEM, 10% FBS, 50 μg/ml ascorbic acid) was changed every 3 or 4 days. The living cells were counted with a colorimetric microtiter (MTT assay, Dojindo, Japan). This assay uses microtiter plates and measures the ability of viable cells to reduce a tetrazolium salt (MTT) to an insoluble formazan salt. Glycosaminoglycan (GAG) production was determined by rapid assay for GAG using a rapid assay kit for GAG (Hokudo, Japan), which specifically stained acidic-GAG.

Histochemical analysis

Histological specimens were fixed in 10% formaldehyde neutral buffered solution and processed for paraffin embedding. The blocks were sectioned using an automated microtome, mounted on glass slides. The slides were stained with Alcian blue to visualize cell and tissue components, e.g., extracellular matrix, by light microscopy (TE300 inverted microscope, Nikon, Japan).

RESULTS AND DISCUSSION

Preparation of macroporous hydrogels using cholesterol-modified hydrolyzable PRx

Cholesterol-modified PRx hydrogels were prepared by crosslinking the hydroxyl groups of α-CDs on the cholesterol-modified PRx with PEG-BA (M_n 2000). As described in our previous study [7], the physiological condition of the hydrogels prepared with a PEG-BA/α-CD ratio above 2 was not eroded, since on the average two molecules of PEG-BA were introduced onto one α-CD molecule, forming an insoluble network, even after hydrolyzing the terminal ester linkage. Therefore, the PEG-BA/α-CD ratio for preparing cholesterol-modified PRx hydrogels was adjusted to be 1. Macroporous structures in hydrogels are necessary, since in most cases chondrocytes will change their phenotype to fibroblast-like cells when cultured on monolayer substrates [16, 17]. In general, pore size of the scaffolds for chondrocyte cultivation is also an important factor to allow nutrient transport which is required for cell proliferation. Chondrocytes (size 15–20 μm) seeded in high concentration can become necrotic due to their covering of the pores if the pore size is not large enough. Therefore, the pore size of the scaffolds should be large enough for long-term cultivation. Thus, macroporogens with a sieving size of 300–500 μm were considered in this experiment. The salt-leaching technique [18–20] has been extensively utilized for cell seeding and transplantation. The scaffolds prepared by this method always exhibited a dense layer, which hampered *in vitro* cell seeding into the scaffolds and tissue in-growth after *in vivo* implantation. The gas-foaming fabrication process produces highly porous scaffolds. The foaming process uses potassium hydrogen carbonate as the effervescent salt along with citric acid to improve the porous interconnectivity of the scaffolds.

Surface and cross-sectional morphologies of the resultant dehydrated hydrogels using cholesterol-modified PRx are shown in Fig. 2. The surface of gas-foamed dehydrated hydrogel has a highly porous structure and a less dense skin layer than that observed in hydrogels prepared using the salt-leaching method. Pores on the surface and in the internal region of the dehydrated hydrogels were uniformly macroporous, with sizes ranging from 200 to 400 μm. The pores also are well interconnected throughout the dehydrated hydrogel matrix. It is interesting to consider the effect of citric acid concentration in the bubbling process: the citric acid concentrations were varied from 1, 2, 3 M to saturated (approx. 3.5 M). While the weight ratio of potassium hydrogen carbonate to cholesterol-modified PRx was set at 600%; higher concentrations of citric acid caused a gradual increase in porosity, as observed by SEM measurements (data not shown). This result suggests that the porosity of the dehydrated hydrogels can indeed be modulated by controlling the extent of acid–base reaction between citric acid and potassium hydrogen carbonate. More vigorous gas evolution was visually observed at higher citric acid concentrations.

Figure 3 shows Young's modulus of non-porous hydrogels (Fig. 3a, control), the cholesterol-modified PRx hydrogels prepared by salt-leaching (Fig. 3b) and those prepared by gas foaming (Fig. 3c). Young's modulus decreased with increasing degree of cholesterol substitution in both the non-porous and salt-leaching prepared hydrogels. One possible reason is intramolecular hydrophobic association in the higher degree of cholesterol substitution hydrogels due to clustering in the crosslinking regions, which resulted in low stiffness regions of low crosslinking. On the other hand, the hydrogels prepared by gas foaming resulted in a much smaller decrease in Young's modulus, even at increased degrees of substitution. This result

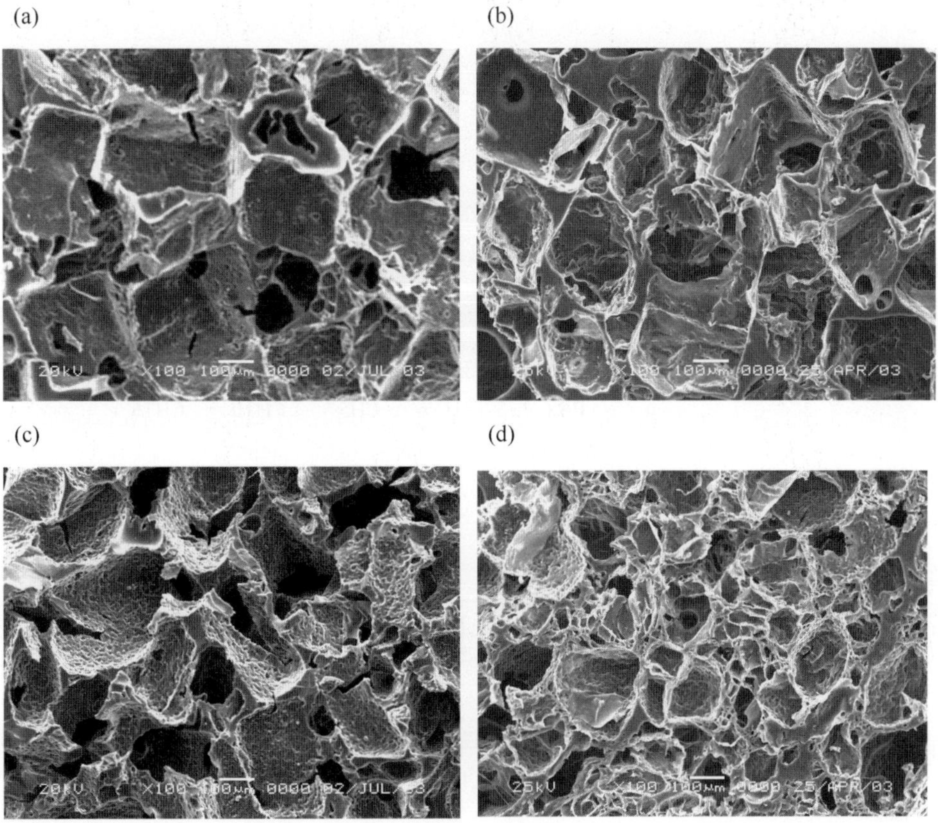

Figure 2. Scanning electron micrographs of the macroporous cholesterol-modified PRx dehydrated hydrogels. Prepared by salt-leaching (NaCl): (a) top surface, (b) cross section. Prepared by gas foaming ($KHCO_3$): (c) top surface, (d) cross section.

indicates that the elastic modulus of the macroporous hydrogels was not affected by the degree of substitution when the pores become interconnected.

In vitro degradation behavior of cholesterol-modified PRx hydrogels

Previously, we reported *in vitro* degradation of the hydrolyzable PRx species [21] and non-porous PRx hydrogels [7]. Dispersing α-CDs along with the PEG chain in the PRx species prolonged the degradation time. In this study, cholesterol groups were introduced to the hydroxyl groups of α-CDs, and then cross-linked with PEG-BA *via* urethane linkages. If the cholesterol molecules associate intra- or inter-molecularly in the hydrogels, we expect that the terminal ester linkages will be exposed to the aqueous medium, due to the moving of the α-CDs. Degradation behavior was monitored under two conditions; an accelerated condition of 0.1 M NaOH solution and a physiological condition of a phosphate buffer at pH 7.4. The accelerated degradation time was measured after placing the hydrogels into a 0.1 M NaOH solution until they were completely dissolved, as shown in Fig. 4. As

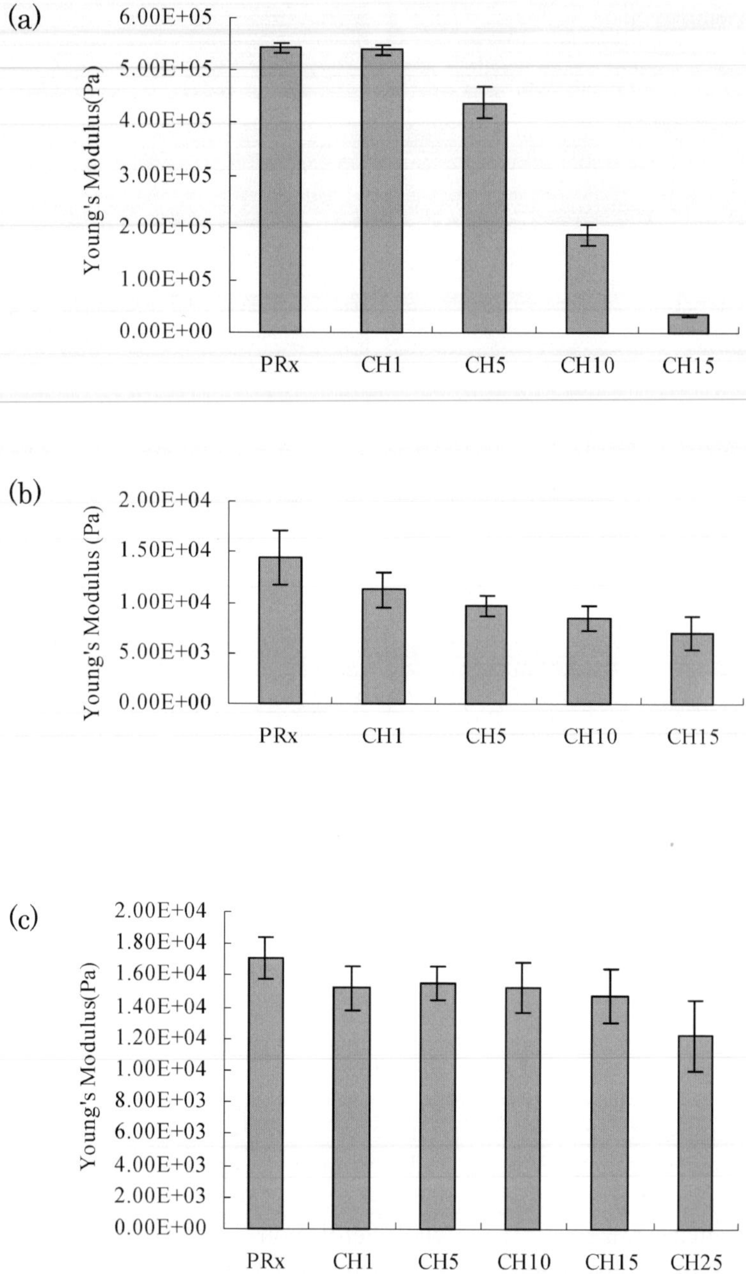

Figure 3. Elastic modulus. (a) Non-porous cholesterol-modified PRx hydrogels, macroporous cholesterol-modified PRx hydrogels prepared by (b) salt-leaching (NaCl) and (c) gas foaming ($KHCO_3$). Error bars represent standard deviation, $n = 3$. CH and the number after CH refer to cholesterol-modified PRx and its degree of substitution.

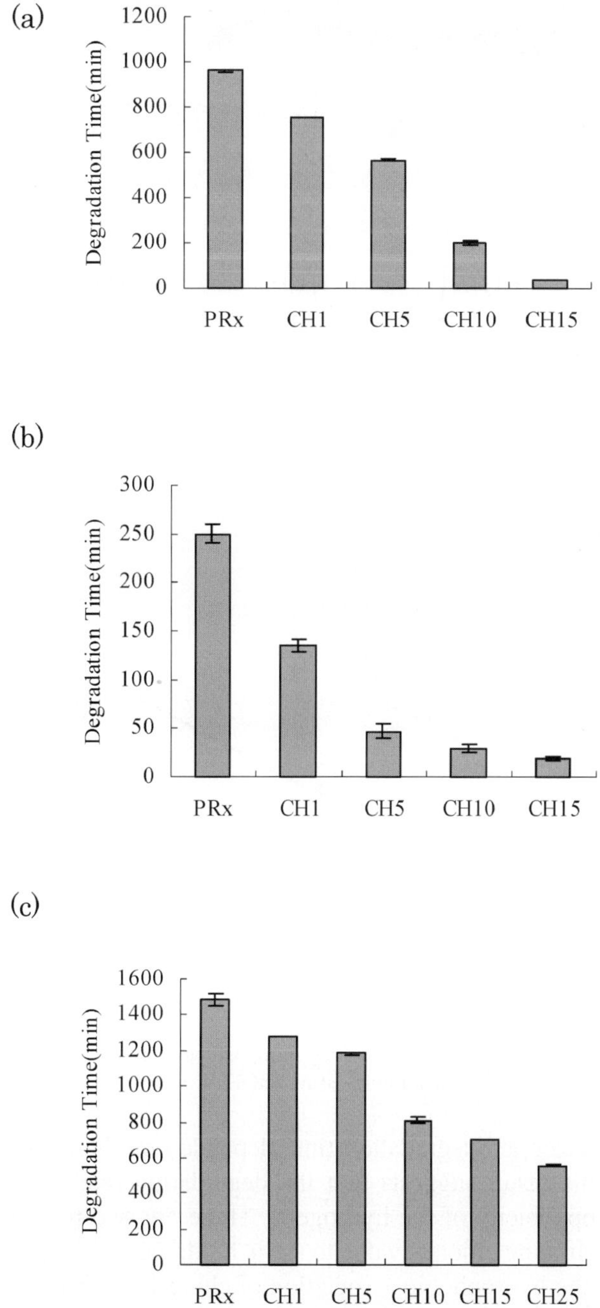

Figure 4. Time to complete degradation. (a) Non-porous cholesterol-modified PRx hydrogels, macroporous cholesterol-modified PRx hydrogels prepared by (b) salt-leaching (NaCl) and (c) gas foaming ($KHCO_3$) observed in 0.1 M NaOH. Error bars represent standard deviation, $n = 3$. CH and the number after CH refer to cholesterol-modified PRx and its degree of substitution.

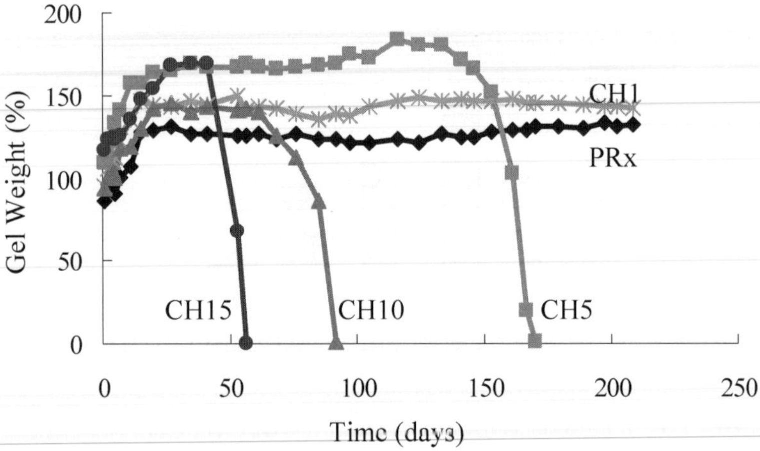

Figure 5. Weight change of cholesterol-modified PRx hydrogels incubated in 0.1 M phosphate buffer, pH 7.4.

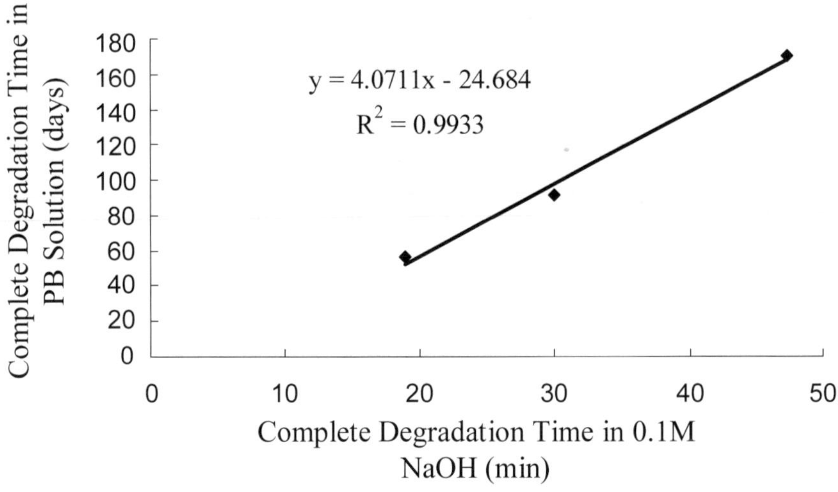

Figure 6. Calibration curve for estimated long-term complete degradation time in PB solution, derived from time to complete degradation in 0.1 M NaOH.

expected, the accelerated degradation time depended on the degree of cholesterol substitution. This result suggests that the degradation rate becomes faster with increasing hydrophobicity of the hydrogels. There are two possible reasons: one is lowered cross-linking efficiency due to steric hindrance of the cholesterol groups, and the other is the hydrophobic association of cholesterol groups, resulting in the movement of α-CDs to assemble in another region, thus exposing the terminal ester linkages to the aqueous media. The difference in degradation rate of the dense hydrogels (Fig. 4a), salt-leached porous hydrogels (Fig. 4b) and gas-foamed porous hydrogels (Fig. 4c) can be described as follows. The dense hydrogels and salt-leached porous hydrogels exhibited similar cross-linking efficiency. Only

Table 1.
Predicted degradation time of the hydrogels in PB solution

Hydrogels	Observed complete degradation time in 0.1 M NaOH (min)	Predicted degradation time in PB solution (days)
Dense hydrogels		
PRx	960.7	3900
CH1	754	3000
CH5	564.7	2300
CH10	200	800
CH15	37.7	130
CH25	—	—
Salt-leaching porous hydrogels (NaCl)		
PRx	250.3	1000
CH1	135	500
CH5	47.3	170[a]
CH10	30	92[a]
CH15	19	56[a]
CH25	—	—
Bubbling porous hydrogels ($KHCO_3$)		
PRx	1482.3	6000
CH1	1275.7	5200
CH5	1182	4800
CH10	808.7	3300
CH15	703.7	2800
CH25	558	2200

[a] Complete degradation time in PB solution, determined by measuring weight of the hydrogels (Fig. 5), i.e., those data were used to make the calibration curve (Fig. 6).

the hydrogels at higher degrees of cholesterol substitution (CH25) could not be formed. This may be due to lower cross-linking efficiency accompanied by the aggregation of the cholesterol-modified PRx species *via* hydrophobic interactions. The degradation rate of the cholesterol-modified PRx should be related to the opportunity for water to contact its terminal ester linkages. Compared to the salt-leached porous hydrogels, the dense hydrogels have less opportunity to contact water than the porous hydrogels, resulting in a longer time to complete degradation. We consider that the difference in degradation time between the salt-leached porous hydrogels and the gas foamed porous hydrogels is due to the distribution of the salts during cross-linking. $KHCO_3$ exhibited good dispersion compared to NaCl, as observed in these experiments. This may support the dissociation of α-CDs at high hydrophobic substitution and increase efficiency in forming a regular network structure. The good dispersion may also decrease opportunities for interaction between water and the terminal ester linkages. Thus, the degradation rate of macroporous cholesterol-modified PRx hydrogels prepared using $KHCO_3$ is slower than that of hydrogels prepared by NaCl (salt-leaching) and dense hydrogels.

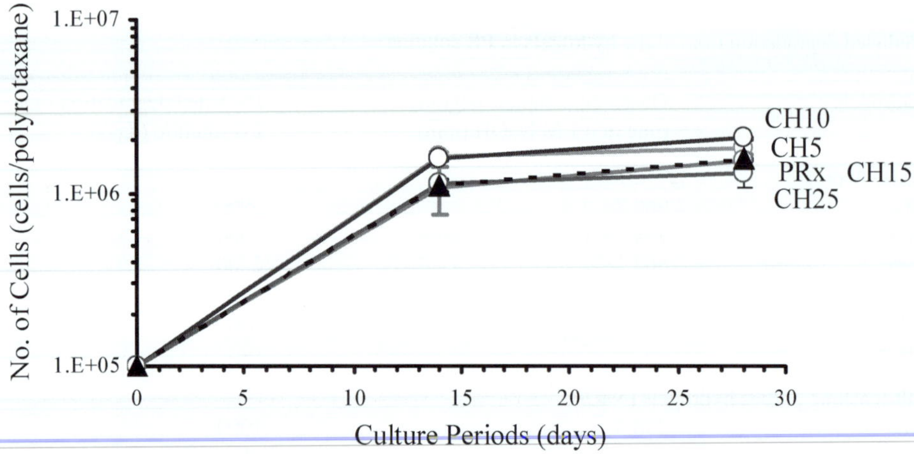

Figure 7. Proliferation of rabbit chondrocytes in cholesterol-modified PRx hydrogels. CH and the number after CH refer to cholesterol-modified PRx and its degree of substitution.

Figure 5 shows the degradation behavior of macroporous cholesterol-modified PRx hydrogels prepared using the salt-leaching process and studied in a 0.1 M phosphate buffer (PB) solution, pH 7.4. The hydrogel weight was observed to increase in the initial stage until reaching a maximum, then to decrease until the hydrogel was completely eroded. This phenomenon is due to water absorption at the initial stage, followed by the hydrolysis of the terminal ester linkages. From Figs 4 and 5 an accelerated degradation condition performed in 0.1 M NaOH solution can be utilized to predict the degradation time of the hydrogels in the PB solution at pH 7.4. The data are summarized as a calibration curve in Fig. 6. From this calibration curve, we can estimate the complete degradation time in PB solution over 50 days. The degradation time of dense hydrogels, salt-leached porous hydrogels and gas-foamed porous hydrogels can be predicted as summarized in Table 1.

Chondrocyte culture and determination of GAG production

Since the hydrogels prepared from mixed $KHCO_3$ salt (gas foaming) gave better interconnectivity than those prepared from mixed NaCl salt (salt-leaching), the gas foamed hydrogels were utilized as substrates for chondrocyte cultivation, to clarify the effect of cholesterol groups on chondrocyte proliferation. The results of cell proliferation and GAG production of rabbit chondrocytes on the cholesterol-modified PRx hydrogels are shown in Figs 7 and 8, respectively. GAG production is known as a specific phenotype of cartilage-like tissue formation. The cell number of chondrocytes was increased on all the hydrogels during 28 days, indicating that the cholesterol groups did not exhibit cytotoxicity. Both cell proliferation and GAG production from the cultured chondrocytes on the cholesterol-modified PRx hydrogels (CH5 and CH10) were higher than the control (PRx only). These results suggest that an appropriate amount of cholesterol contributes to attractive interactions with the chondrocyte membrane and the resulting cartilage-like tissue

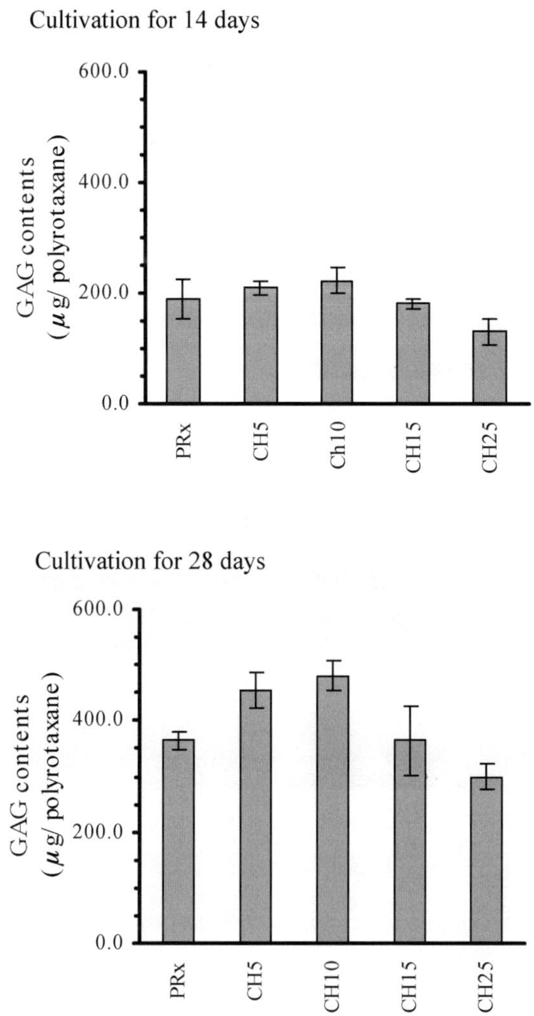

Figure 8. GAG production from rabbit chondrocytes in cholesterol-modified PRx hydrogels under cultivation for 14 and 28 days. Error bars represent standard deviation, $n = 3$. CH and the number after CH refer to cholesterol-modified PRx and its degree of substitution.

formation. Histochemical analysis indicated similar results corresponding to cell proliferation and GAG production, suggesting that the chondrocytes had populated the cartilage on the hydrogels and deposited an extracellular matrix containing glycosaminoglycans, detected by Alcian blue staining as shown in Fig. 9.

CONCLUSIONS

In this study, cholesterol was introduced to a hydrolyzable polyrotaxane (PRx), not only to improve cell proliferation and GAG production, but also to control the degradation rate of the hydrogels. The cholesterol-modified PRx hydrogels were

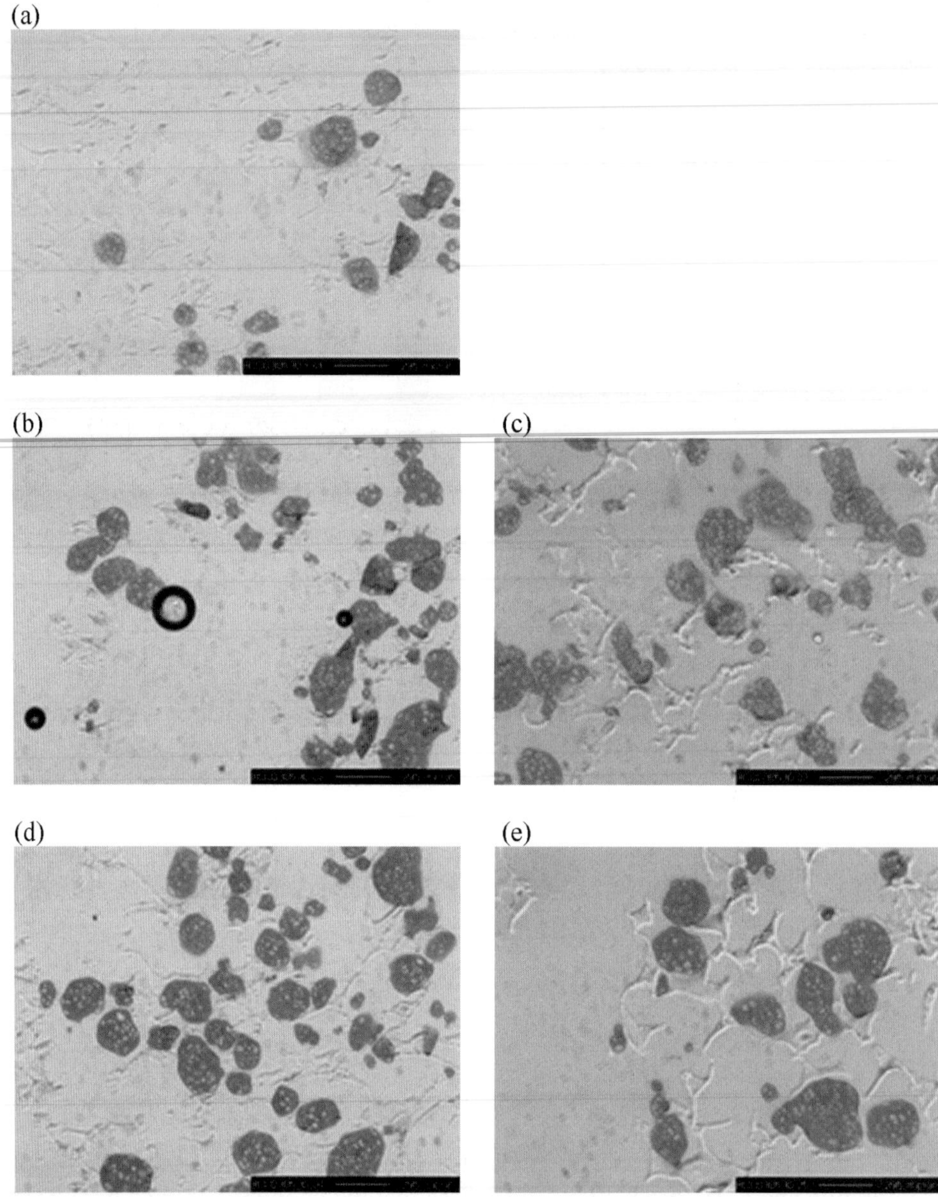

Figure 9. Alcian blue stained histochemical sections of chondrocytes after a 28-day cultivation in cholesterol-modified PRx hydrogels. (a) PRx (control), (b) CH5, (c) CH10, (d) CH15 and (e) CH25. CH and the number after CH refer to cholesterol-modified PRx and its degree of substitution.

prepared by cross-linking PEG-BA with the hydroxyl groups of α-CDs threaded onto a PEG chain. An appropriate macroporous structure was fabricated by a gas-foaming process. The degradation rate of cholesterol-modified PRx hydrogels was controlled by inclusion-based hydrolytic susceptibility of the terminal ester linkages. The degradation time decreased with increasing degree of cholesterol substitution. This is possibly due to the intra- or inter-molecular hydrophobic interaction attributed to the aggregation of α-CDs, exposing the terminal ester linkages to hydrolysis. The presence of moderate amounts of cholesterol improved cell proliferation and GAG production. Thus, the cholesterol-modified PRx hydrogels may prove to be advantageous for cartilage regeneration in terms of enhanced cartilage-like tissue formation and degradation of the hydrogels.

Acknowledgements

This study was supported by a Grant-in-Aid for Scientific Research (B) (No. 14380397), from the Ministry of Education, Science, Sports and Culture, Japan, and by the 21st century COE program "Scientific Knowledge Creation Based on Knowledge Science", Japan Advanced Institute of Science and Technology. The authors also would like to thank Dr. Takahiro Ichi for his helpful discussions.

REFERENCES

1. D. A. Grande, C. Halberstadt, G. Naughton, R. Schwartz and R. Manji, *J. Biomed. Mater. Res.* **34**, 211 (1997).
2. P. Brun, G. Abatangelo, M. Radice, V. Zacchi and D. Guidolin, *J. Biomed. Mater. Res.* **46**, 337 (1999).
3. L. Freed, J. Marquis, A. Nohria, J. Emmanuel, A. Mikos and R. Langer, *J. Biomed. Mater. Res.* **27**, 11 (1993).
4. G. Bentley and R. Greer, *Nature* **230**, 383 (1971).
5. S. J. Peter, M. J. Miller, A. W. Yasko, M. J. Yaszemski and A. G. Mikos, *J. Biomed. Mater. Res.* **43**, 422 (1998).
6. J. Watanabe, T. Ooya and N. Yui, *Chem. Lett.*, 1031 (1998).
7. T. Ichi, J. Watanabe, T. Ooya and N. Yui, *Biomacromolecules* **2**, 204 (2001).
8. T. Ichi, T. Ooya and N. Yui, *Macromol. Biosci.* **3**, 373 (2003).
9. W. K. Lee, T. Ichi, T. Ooya, T. Yamamoto, M. Katoh and N. Yui, *J. Biomed. Mater. Res.* **67A**, 1087 (2003).
10. T. Ooya, M. Eguchi and N. Yui, *J. Am. Chem. Soc.* **125**, 13016 (2003).
11. W. G. Brodbeck, M. S. Shive, E. Cotton, Y. Nakayama, T. Matsuda and J. M. Anderson, *J. Biomed. Mater. Res.* **55**, 661 (2001).
12. A. G. Karakecili and M. D. Gumu, *J. Biomater. Sci. Polym. Edn.* **13**, 185 (2002).
13. T. LeDoan, F. Etore, J. P. Tenu, Y. Letourneux and S. Agrawal, *Bioorg. Med. Chem.* **7**, 2263 (1999).
14. C. Rodriguez, N. Naito and H. Kunieda, *Colloids Surfaces A: Physicochem. Eng. Aspects* **181**, 237 (2001).
15. R. Auzely-Valty, B. Perly, O. Tache, T. Zemb, P. Jehan, P. Guenot, J. P. Dalbiez and F. Djedaini-Pilard, *Carbohydr. Res.* **318**, 82 (1999).
16. H. M. van Beuningen, R. Stoop, P. Buma, N. Takahashi, P. M. van der Kraan and W. B. van den Berg, *Osteoarthritis Cartil.* **10**, 977 (2002).
17. J. Farjanel, G. Schijrmann and P. Bruckner, *Osteoarthritis Cartil.* **9**, S55 (2001).
18. Q. Hou, D. W. Grijpma and J. Feijen, *Biomaterials* **24**, 1937 (2003).
19. A. P. Pêgo, A. A. Poot, D. W. Grijpma and J. Feijen, *J. Control. Rel.* **87**, 69 (2003).

20. Y. Deng, K. Zhao, X. Zhang, P. Hu and G. Q. Chen, *Biomaterials* **23**, 4049 (2002).
21. J. Watanabe, T. Ooya and N. Yui, *J. Artif. Organs* **3**, 136 (2000).

Gels, Genes, Grafts and Giants, pp. 261–287 (2005)

Biodegradable poly(terephthalate-*co*-phosphate)s: synthesis, characterization and drug-release properties

HAI-QUAN MAO [1,2], IRINA SHIPANOVA-KADIYALA [1], ZHONG ZHAO [3], WENBIN DANG [3], ANGELA BROWN [1] and KAM W. LEONG [1,*]

[1] *Department of Biomedical Engineering, Johns Hopkins University School of Medicine, 726 Ross Building, 720 Rutland Avenue, Baltimore, MD 21205, USA*
[2] *Department of Materials Science and Engineering, Johns Hopkins University, Baltimore, 102 Maryland Hall, 3400 N. Charles Street, MD 21218, USA*
[3] *Guilford Pharmaceuticals Inc., 6611 Tributary Street, Baltimore, MD 21224, USA*

Received 19 November 2003; accepted 23 April 2004

Abstract—To develop biodegradable polymers with favorable physicochemical and biological properties, we have synthesized a series of poly(terephthalate-*co*-phosphate)s using a two-step polycondensation. The diol 1,4-bis(2-hydroxyethyl) terephthalate was first reacted with ethylphosphorodichloridate (EOP), and then chain-extended with terephthaloyl chloride (TC). Incorporation of phosphate into the poly(ethylene terephthalate) backbone rendered the co-polymers soluble in chloroform and biodegradable, lowered the T_g, decreased the crystallinity and increased the hydrophilicity. With an EOP/TC molar feed ratio of 80 : 20, the polymer exhibited good film-forming property, yielding at 86.6 ± 1.6% elongation with an elastic modulus of 13.76 ± 2.66 MPa. This polymer showed a favorable toxicity profile *in vitro* and good tissue biocompatibility in the muscular tissue of mice. *In vitro* the polymer lost 21% of mass in 21 days, but only 20% for up to 4 months *in vivo*. It showed no deterioration of properties after sterilization by γ-irradiation at 2.5 Mrad on solid CO_2. Release of FITC-BSA from the microspheres was diffusion-controlled and exceeded 80% completion in two days. Release of the hydrophobic cyclosporine-A from microspheres was however much more sustained and near zero-ordered, discharging 60% in 70 days. A limited structure–property relationship has been established for this co-polymer series. The co-polymers became more hydrolytically labile as the phosphate component (EOP) was increased and the side chains were switched from the ethoxy to the methoxy structure. Converting the methoxy group to a sodium salt further increased the degradation rate significantly. The chain rigidity as reflected in the T_g values of the co-polymers decreased according to the following diol structure in the backbone: ethylene glycol > 2-methylpropylene diol > 2,2-dimethylpropylene diol. The wide range of physicochemical properties obtainable from this co-polymer series should help the design of degradable biomaterials for specific biomedical applications.

Key words: Poly(terephthalate-*co*-phosphate); biodegradability; polycondensation; chain extension.

*To whom correspondence should be addressed. Tel.: (1-410) 614-3741. Fax: (1-410) 955-0549. E-mail: kleong@bme.jhu.edu

INTRODUCTION

Advances in drug and gene delivery and tissue engineering continue to fuel the development of new biodegradable polymers. Polyesters [1], poly(amino acid)s [2, 3], polyamides [4], polyurethanes [5], poly(orthoester)s [6], poly(anhydride)s [7], poly(carbonate)s [8, 9], poly(iminocarbonate)s [10] and poly(phosphazene)s [11, 12] have been the prominent players. In drug delivery and tissue engineering scaffolding applications, poly(lactide-*co*-glycolide) co-polymers still dominate the field [13–16]. However, with an increasingly broadening utility for biodegradable drug-carriers and scaffolds, no one single material can be expected to satisfy all requirements of different applications. For instance, a coating material for a drug-eluting stent might require the polymer to be film-forming and elastomeric; or a fibrous tissue-engineering scaffold would demand a polymer with fiber-forming and strong mechanical properties. In this study, we evaluated the potential of a new series of poly(ester–phosphoester) co-polymers.

Polyphosphoesters represent a class of biodegradable polymers having a phosphoester linkage in the backbone, which would include poly(phosphate)s, poly(phosphonate)s and poly(phosphite)s [17–19]. Much of the research into the basic science of these polymers has been pioneered by Penczek and his colleagues in the 1970s [20–22]. The interesting characteristics of these polymers include: (1) versatility and tunable properties, the physicochemical and degradation properties can be adjusted by changing the backbone and the side-chain structures; (2) favorable physicochemical properties, the flexible phosphoester bond in the backbone provide the plasticizing effect to lower the glass transition temperature and confer good solvent solubility and processibility of the polymers; (3) biocompatibility, the ultimate hydrolytic breakdown products of a polyphosphoester are phosphates, alcohols and diols. A biocompatible polyphosphoester can be prepared from innocuous alcohol and diol building blocks.

We have previously tested the concept of improving the solubility and degradation behavior of poly(D,L-lactide) (PDLA) by forming a co-polymer between PDLA and polyphosphate [18, 23, 24]. The poly(lactide-*co*-ethylphosphate) shows a faster and more linear degradation profile than PDLA, and is soluble in non-halogenated solvents such as ethyl acetate [23]. As a drug carrier for paclitaxel (PACLIMER®), it showed promise in preclinical studies by improving the pharmacokinetics and toxicity of paclitaxel when the drug-loaded microspheres were injected intraperitoneally into dogs. The efficacy of this delivery system was also demonstrated in the local-regional therapy of lung cancer tumor nodules in mice [25]. Using similar approach and with an interest of tissue engineering scaffolding applications, we aimed to develop a co-polymer between polyphosphate and a polyester with strong mechanical properties. For that we targeted the extensively studied poly(ethylene terephthalate) (PET).

Poly(ethylene terephthalate) (PET) is one of the most commonly used biomedical polyesters, such as in vascular graft applications [26, 27]. Its appeal partly lies in its superb mechanical properties, which are derived from the liq-

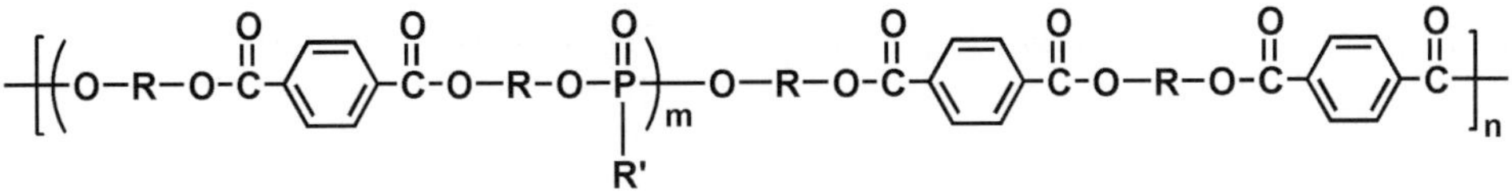

R= $-CH_2CH_2-$, $-CH_2CH_2CH_2CH_2-$, $-CH_2CH(CH_3)CH_2-$, $-CH_2C(CH_3)_2CH_2-$;
R'=$-OCH_2CH_3$, $-OCH_3$, $-O^-Na^+$.

Figure 1. Structures of designed poly(terephthalate-*co*-phosphate)s, P(BHET-EOP/TC).

uid crystalline characteristics of the ethylene terephthalate structure. Hypothesizing that a biodegradable polymer with good mechanical properties can be built on this structure, we introduced phosphate units into PET. The general structure of poly(terephthalate-*co*-phosphate)s is shown in Fig. 1. In this paper, we primarily focus on poly(terephthalate-*co*-phosphate)s synthesized from BHET (Fig. 1, R $= -CH_2CH_2-$). The synthesis, characterization and structure–properties relationships have been investigated in this study. The *in vitro* cellular toxicity and *in vivo* tissue compatibility have been studied in details. Illustrative applications of poly(terephthalate-*co*-phosphate) for drug delivery in the form of microspheres and film are also included.

MATERIALS AND METHODS

Terephthaloyl chloride, ethyl phosphorodichloridate, cobalt(II) acetate tetrahydrate, calcium acetate hydrate and all other reagents (if not specified) and solvents were purchased from Sigma-Aldrich (St. Louis, MO, USA). Terephthaloyl chloride was recrystallized from hexane before use, and ethyl phosphorodichloridate was redistilled. Dimethyl terephthalate and 4-dimethylaminopyridine (DMAP) were obtained from TCI America (Portland, OR, USA) and used without further purification. Cyclosporine A was obtained from Sigma.

Synthesis of 1,4-bis(2-hydroxyethyl) terephthalate (BHET)

Dimethyl terephthalate (277 g, 1.4 mol) and ethylene glycol (445 g, 7.2 mol) were weighed into a 1-l round bottom flask connected to a vacuum line. A catalytic amount of cobalt(II) acetate tetrahydrate (180 mg, 0.5 mol) and calcium acetate hydrate (90 mg, 0.4 mol) was added to prevent transesterification. The reaction mixture was heated at 160°C in an oil bath under mild vacuum and terminated after 18 h. While still molten, the mixture was precipitated into cold water. The precipitate was collected, dried under vacuum, redissolved into warm methanol and filtered to remove the sludge (oligomers). A precipitate formed when the solution was cooled to −20°C. After recrystallization from ethyl acetate and methanol, BHET was obtained as a white powder, mp. 108–110°C; Lit. 109–110°C [28], yield 68%.

Synthesis of poly(terephthalate-co-phosphate) P(BHET-EOP, 80 : 20)

Under argon stream, 10 g 1,4-bis(hydroxyethyl) terephthalate (BHET), 9.61 g 4-dimethylaminopyridine (DMAP) and 70 ml methylene chloride were transferred to a 250-ml flask equipped with an addition funnel. After BHET and DMAP were dissolved, the solution was cooled down to −40°C. A solution of 5.13 g ethyl phosphorodichloridate (EOP) in 20 ml methylene chloride was added dropwise through the funnel. Following the addition of EOP, the mixture was stirred at room temperature for 4 h and then a solution of 1.60 g terephthaloyl chloride (TC) in 20 ml methylene chloride was added drop-wise to the flask. The temperature was gradually brought up to 50°C, and allowed to react overnight. The solvent was removed by evaporation, and the residue was re-dissolved in 200 ml chloroform, washed with NaCl saturated solution three times and dried over Na_2SO_4. A white precipitate was obtained upon quenching the solution into cold methanol. The purification process was repeated twice, and then dried under vacuum at 60°C. The poly(terephthalate-*co*-phosphate) P(BHET-EOP, 80 : 20) was obtained as a tough, off-white solid, with a yield of 82%. Elemental analysis: Found % (Theoretical value, %): C: 49.61 (51.82); H: 4.95 (4.81); P: 6.24 (7.03). $M_w = 7920$, $M_n = 6120$ (GPC, polystyrene as standards).

A series of other P(BHET-EOP/TC) co-polymers with different molar ratios of phosphates in the backbone were prepared according to the procedure described above. These poly(terephthalate-*co*-phosphate)s were defined by the different feed ratios of EOP to TC, $x/(100-x)$ (see Table 1).

Table 1.
Composition of poly(terephthalate-*co*-phosphate)s calculated from ^{1}H-NMR spectra

Poly(terephthalate-*co*-phosphate)	EOP/TC ratio (Charged)	BHET-EOP/TC ratio* (Found)
P(BHET-EOP/TC, 75 : 25)	75 : 25	75.6 : 24.4
P(BHET-EOP/TC, 80 : 20)	80 : 20	81 : 19
P(BHET-EOP/TC, 82.5 : 17.5)	82.5 : 17.5	82.5 : 17.5
P(BHET-EOP/TC, 85 : 15)	85 : 15	84 : 16
P(BHET-EOP/TC, 87.5 : 12.5)	87.5 : 12.5	87 : 13
P(BHET-EOP/TC, 90 : 10)	90 : 10	89 : 11
P(BHET-EOP/TC, 95 : 5)	95 : 5	96 : 4

*For the convenience of calculation and peak assignment, the structure was rearranged as (also see Fig. 3 for details):

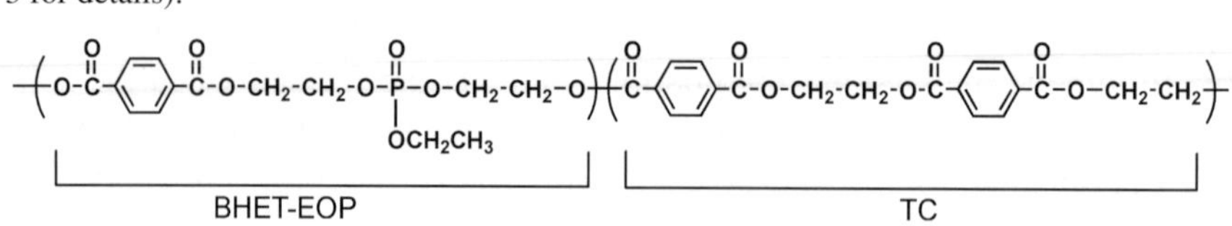

Synthesis of P(BHET-EOP/MOP/TC, 60 : 20 : 20)

To a cold (−40°C) solution of 20 g (78.66 mmol) BHET and 19.22 g (157.36 mmol) DMAP in 150 ml methylene chloride, was added drop-wise a solution of 7.69 g (47.20 mmol) ethyl phosphorodichloridate (EOP) and 2.34 g (15.73 mmol) methyl phosphorodichloridate (MOP) in 25 ml methylene chloride. The temperature of the mixture was brought up to refluxing gradually and maintained overnight. Another solution of 3.20 g terephthaloyl chloride (TC) in 20 ml methylene chloride was added through a funnel, and the mixture was refluxed for another 18 h. The mixture was washed three times with 200 ml of 0.1 M HCl saturated with NaCl, once with saturated NaCl solution and dried over anhydrous Na_2SO_4. The polymer was quenched out with ether and hexane, and obtained as white powder. Yield 79%, $M_w = 9630$, $M_n = 4580$ (GPC, polystyrene as standards).

Synthesis of P(BHET-EOP/MOP/TC, 40 : 40 : 20) was performed according to the same procedure. Yield 72%, $M_w = 8020$, $M_n = 4580$ (GPC, polystyrene as standards).

Synthesis of poly(terephthalate-co-phosphate) with negative charge

To a solution of 5 g P(BHET-EOP/MOP/TC, 60 : 20 : 20) in 50 ml chloroform, was added a solution of 0.858 g NaI in 25 ml acetone. The solution was refluxed under dried nitrogen or argon atmosphere overnight. The product was gradually precipitated. The pellet was washed and extracted extensively with acetone in a Soxhlet extractor. White powder was obtained after vacuum drying for 24 h.

Characterization

Chemical analysis. Gel-permeation chromatography (GPC) was performed on a Hewlett Packard 1090M liquid chromatograph. Chloroform was used as the mobile phase to elute the sample at a concentration of 2 mg/ml through a series of three PL gel polystyrene columns (10^2 Å, 10^3 Å and 10^5 Å pore size) (Hewlett Packard). Analysis was performed at 1 ml/min, 40°C and 270 nm (UV detection). The system was calibrated with monodispersed polystyrene standards (Polymer Laboratories). Fourier transfer infrared spectroscopy (FT-IR) was performed on a Perkin-Elmer 1600 Series machine. Samples were prepared by film casting on KBr pellets. ^{1}H-NMR and ^{31}P-NMR spectra were recorded on a Bruker AMX300 in $CDCl_3$ or DMSO-d_6, using H_3PO_4 as an external standard. Elemental analysis was performed by Galbraith Laboratories (Knoxville, TN, USA). The intrinsic viscosity in chloroform was measured in Ubbelohde viscometer (ASTM size OC) at 40°C.

Thermal properties. Differential scanning calorimetry was performed on a Seiko DSC220 at a heating rate of 10°C/min in sealed pans. Analysis was performed on a Seiko SDM 5500 Rheostation. The glass transition temperatures (T_g) were measured from the inflection points.

Contact angle analysis. Polymer thin film was deposited on coverslips (diameter of 10 mm, Corning) by spin-coating using 5% polymer solution in chloroform. Water-in-air contact angle was measured using a goniometer.

Tensile test. Polymer films were fabricated by casting polymers solution (15–20% in chloroform) in a Teflon mold. The film was air dried first, followed by extensive vacuum drying at 60°C for three days. The films were then cut into dumb-bell-shaped specimens and the tensile properties were analyzed on an Instron 5544 material testing system according to ASTM (American Society for Testing and Materials) method D638M.

Cytotoxicity of poly(terephthalate-co-phosphate)

Cytotoxicity of co-polymer by direct contact. The cytotoxicity of P(BHET-EOP, 80 : 20) was assessed by culturing HEK 293 (human embryonic kidney) cells on the P(BHET-EOP, 80 : 20) coated glass cover slips. Cell morphology and growth rate were determined after 72 h. As a control, HEK cells were also cultured directly on TCPS.

Cytotoxicity of co-polymer microspheres on GT3TKB tumor cells. P(BHET-EOP/TC, 80 : 20) microspheres were added to 96-well tissue-culture plates at different concentration, and then human gastric carcinoma cells (GT3TKB, RIKEN cell bank, Tsukuba, Japan) at 104 cells/well density were seeded. The cells were then incubated for 48 h at 37°C. Cell proliferation rate was analyzed by MTT assay. The results were plotted as % relative growth *vs* concentration of polymer microspheres in the tissue culture well.

Cytotoxicity of co-polymer degradation products. About 100–150 mg of P(BHET-EOP, 80 : 20) were degraded in 20 ml 1 M NaOH at 37°C for 1–2 days to ensure complete degradation. The solution was then neutralized with 1 M HCl. Poly(L-lactide) (PLLA, $M_w = 14\,000$, Sigma) and poly(L-lysine) (PLL, $M_w = 88\,000$, Sigma) were treated similarly and used as controls. About 200 μl of various concentrations of the degraded polymer products were placed in 96-well tissue culture plates, which were seeded with human gastric carcinoma cells (GT3TKB) at a density of 1×10^4 cells/well one day prior to the experiment. The degraded polymer products were incubated with the GT3TKB cells for 48 h. The results of the assay were plotted as % relative growth *vs* concentration of degraded polymer in the culture well.

Ames test and ISO agarose overlay tests. These two tests were performed at NamSA Laboratories (Northwood, OH, USA) according to the International Organization for Standardization ISO 10993-5 standard.

Degradation of poly(terephthalate-co-phosphate)s in PBS

P(BHET-EOP/TC, 80 : 20) and P(BHET-EOP/TC, 85 : 15) films were made by solution casting (15–20% in chloroform) and dried under vacuum for 3 days. Discs with 1 mm in thickness and 6 mm in diameter were cut from these sheets. Three discs of each co-polymer were placed in 4 ml of phosphate-buffered saline (PBS, 0.1 M; pH 7.4) at 37°C. The discs were taken out of the PBS at different points in time, washed with distilled water, and freeze-dried overnight. The samples were analyzed for change in molecular weight and mass loss over time.

In vivo *tissue compatibility and degradation*

P(BHET-EOP/TC, 80 : 20) discs, 8 mm in diameter, approx. 1 mm in thickness and 60–80 mg in weight, were prepared by solvent casting as described above. Poly(lactide-*co*-glycolide) (PLGA, RG755, Birmingham Polymers, Birmingham, AL, USA) was used as a control. PLGA discs, 8 mm in diameter, approx. 1.5 mm in thickness and 100 mg in weight were prepared by compression molding at room temperature using a manual Carver press. All discs were individually weighed, packaged in aluminum foil pouch and sterilized by γ-irradiation at 2.5 Mrad on dry ice. The initial molecular weights of the polymers were measured by GPC. Specific pathogen free (SPF) Sprague–Dawley rats (200–250 g) were randomly grouped and weighed individually before surgery. The sterilized discs were implanted into the muscles at the right hind limb of the rats under general anesthesia. The fascia over the implantation site was closed with a couple of stitches using 4-O Nylon suture to mark the location. The subcutaneous layer and skin were closed with absorbable suture (4-O Dexon). The rats were monitored for a few h post-operation and daily. At predetermined time points (days 3, 7, 14 and 1 month, 2, 4 and 6 months), three rats from each group were killed by CO_2 inhalation. Blood was collected by cardiac puncture. Tissue samples from the implantation site were harvested, fixed in buffered formalin (10%), sectioned and stained with H&E for histological examination. The tissue reactions were evaluated using the criteria including inflammation and damage to the surrounding tissue. The extent of inflammation response at the implantation site was assessed by an independent pathologist. At each time point, polymer implants were retrieved from the rat muscle, rinsed with cold DI water several times to clean any attached tissue. Samples were lyophilized to constant weight and weight loss of the polymer disc comparing to its original weight was calculated.

Preparation of microspheres encapsulating fluorescein isothiocyanate-labeled bovine serum albumin (FITC-BSA)

FITC-BSA solution (10 mg/ml, 100 μl) was added to a solution of 100 mg P(BHET-EOP, 80 : 20) in 1 ml methylene chloride, and emulsified with a vortexer for 15 s on ice. The resulting emulsion was immediately poured into 5 ml 1% polyvinyl alcohol (PVA)–5% NaCl solution. The vortexing was continued for 1 min. The resulting emulsion was poured into 20 ml of an aqueous solution of 0.3% PVA and 5% NaCl, which was being stirred vigorously. Isopropanol solution (2%, 25 ml) was added, and the mixture was kept stirring for 1 h to ensure complete extraction. The microspheres were collected by centrifugation at 3000 × g, washed three times with water and lyophilized. Empty microspheres were prepared in the same way except that water was used as the inner aqueous phase.

Encapsulation efficiency was determined by comparing the quantity of FITC-BSA entrapped with the amount added to the emulsion. The amount entrapped was determined by measuring the quantity of FITC-BSA (assayed at 492 nm on a Shimadzu UV 160 spectrophotometer, Shimadzu North America, Columbia, MD, USA) remaining in the supernatant following the collection of the microspheres and subtracting it from the initial FITC-BSA amount.

To determine the loading level, 5 mg microspheres were digested in 1 ml 0.5 M NaOH for 6–8 h and assayed for FITC-BSA content at 492 nm in a UV spectrophotometer. The loading level was determined by assaying for FITC-BSA after hydrolyzing the microspheres with 0.5 M NaOH for 8 h. The loading level was calculated from the amount of drug incorporated and the initial weight of microspheres.

In vitro *release of FITC-BSA from microspheres*

5 mg P(BHET-EOP, 80 : 20) microspheres containing FITC-BSA was suspended into 1 ml PBS (pH 7.4) and placed into a 37°C shaker. At various time points, the suspension was spun at 3000 × g for 10 min and 500 μl of the supernatant was replaced by fresh PBS. The release of FITC-BSA from the microspheres was followed by measuring the fluorescence intensity at 519 nm on a Shimadzu UV 160 spectrophotometer.

Preparation of microspheres containing cyclosporine A

In order to determine the loading level and release kinetics of cyclosporine A, tritium-labeled cyclosporine A (Amersham-Pharmacia) was included in the microspheres. A solution of 200 mg P(BHET) in 1 ml methylene chloride was added to 20 mg cyclosporine A with 10 μCi radioactivity. The mixture was vortexed for 1 min, then added to 5 ml 1% PVA–5% NaCl solution while vortexing for 3 min. This mixture was added to 8 ml 0.3% PVA–5% NaCl solution while vortexing for 3 min. The microspheres were washed 5 times by centrifugation and resuspension in deionized water. The suspension was stirred in a magnetic stirrer overnight. The microspheres were suspended in 2% mannitol and lyophilized for at least 48 h.

Approximately 30 mg of polymer microspheres was weighed and placed in 10 ml 0.1 M PBS (pH 7.4) and placed in a shaking incubator at 37°C. At various time points, microspheres were centrifuged and the supernatant was replaced by fresh PBS. 200 μl of the supernatant was added to 2 ml of scintillation fluid and counted in a scintillation counter under a tritium specific program measuring counts per min. At the end of the study, the microspheres were dissolved in DMSO. 200 μl of this solution was added to 2 ml of scintillation fluid and counted similarly. The standard curve constructed with known amount of tritiated cyclosporine A was used to calculate the concentration of the released cyclosporine A.

Preparation of P(BHET-EOP/TC, 80 : 20) film containing paclitaxel

Solvent casting method was used to prepare the co-polymer film with or without 10% nominal paclitaxel loading. The film was air-dried followed by vacuum dried at 60°C for two days to remove any residue solvent that might influence the drug release rate (<0.05 to 5 ppm, depending upon the thickness of the film). The film was then cut into small discs with 6 mm in diameter and incubated in 1.2 ml of PBS (pH 7.4) layered with 0.3 ml of octanol. Discs with paclitaxel loaded were used as the negative control, and paclitaxel floated to the octanol layer was used as the positive control. At various time points, the octanol layer was collected and diluted 10 times with methanol, and analyzed on a HPLC system (Agilent Technologies) equipped with Zorbax Rx C_{18} column (4.6 mm × 15 cm) using a mixture mobile phase (water/methanol/acetonitrile, 37.5 : 50 : 12.5 (v/v/v)) at a flow rate of 1.5 ml/min. The release rate was calculated from the amount of paclitaxel released and the total amount of drug loaded.

RESULTS AND DISCUSSION

Synthesis and structural characterization of poly(terephthalate-co-phosphate)s

Our initial attempt was to synthesize a polyphosphoester (P(BHET-EOP)) by polycondensing 1,4-bis(hydroxyethyl) terephthalate (BHET) with ethyl phosphorodichloridate (EOP) at stoichiometric ratios (Fig. 2). The resulting P(BHET-EOP) was a very viscous semi-solid at room temperature. The T_g of P(BHET-EOP) was 19°C, with a water-in-air contact angle of 8°. To raise the T_g and improve mechanical properties of the polymer, the scheme shown in Fig. 3 was developed to reduce the phosphoester content in the polymer. This two step polymerization scheme led to the synthesis of a series of poly(terephthalate-*co*-phosphate)s with different contents of phosphoester component (EOP). In this scheme, BHET was first reacted with EOP to yield a hydroxyl-terminated prepolymer, which was further polymerized with terephthaloyl chloride (TC). The BHET/EOP/TC ratio was controlled to be $100 : x : (100 - x)$. The ratio in the notation of the P(BHET-EOP/TC) series (Fig. 1) refers to the molar ratio of EOP to TC ($x/(100 - x)$).

The addition sequence and the ratio of reactive chlorides affected the molecular weight and solubility of the polymers. Adding the terephthaloyl chloride first or adding the two reactive dichlorides at the same time yielded a polymer that was only slightly soluble in DMF and DMSO, but not in chloroform or methylene chloride.

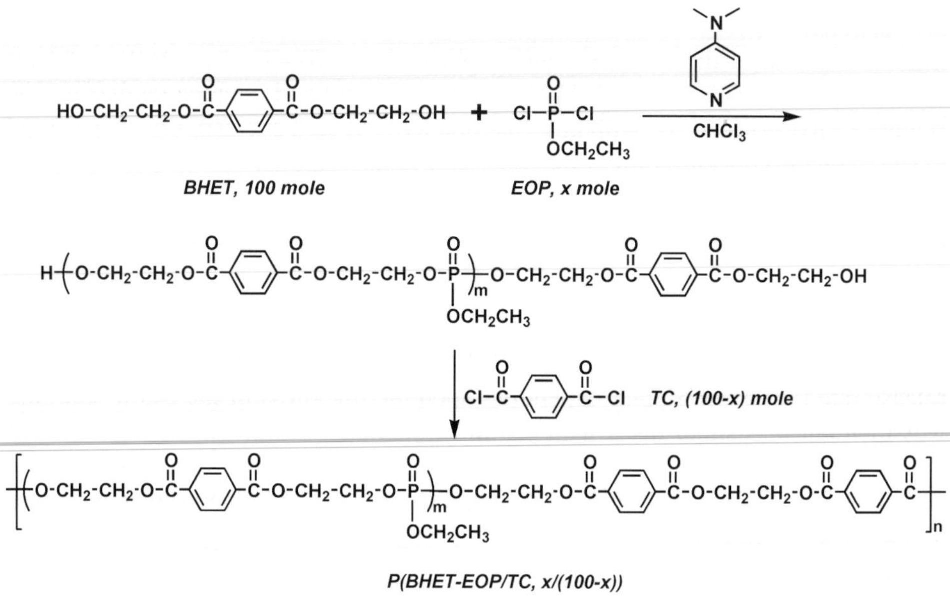

Figure 2. Synthetic scheme of P(BHET-EOP).

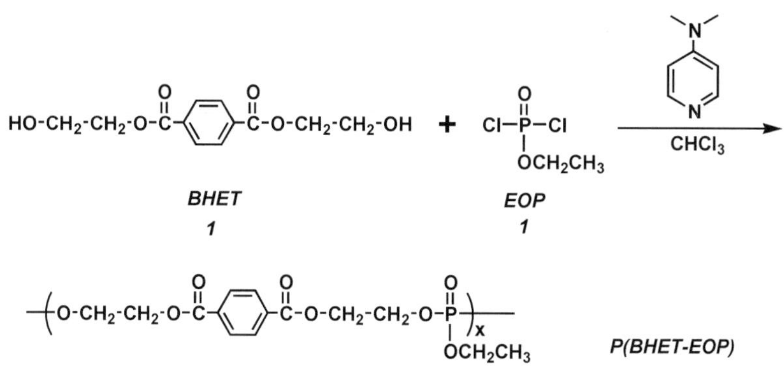

Figure 3. Synthetic scheme of poly(terephthalate-*co*-phosphate)s, P(BHET-EOP/TC).

The polymers had high crystallinity with a T_m of 184°C. We hypothesize that the higher reactivity of the TC monomer compared with EOP towards the hydroxyl group might have produced longer oligo-terephthalate blocks in the backbone to account for this high T_m. The two-step polycondensation shown in Fig. 3 prevented the formation of long segments of oligo-terephthalate, rendering the co-polymers with good solubility in common organic solvents (chloroform, methylene chloride and NMP) and relatively higher molecular weights. The solubility of the co-polymers decreased as the phosphate content decreased. When the EOP to TC feeding ratio was 1, the resulting co-polymer P(BHET-EOP/TC, 50 : 50) became insoluble in chloroform. Table 2 summarizes the solubility property *vs* EOP/TC feeding ratio.

Table 2.
Solubility of poly(terephthalate-*co*-phosphate)s

Co-polymer	$CHCl_3$	CH_2Cl_2	NMP	DMF	DMSO
P(BHET-EOP)	+++	+++	++	++	++
P(BHET-EOP/TC, 95 : 5)	+++	+++	++	++	++
P(BHET-EOP/TC, 90 : 10)	+++	+++	++	++	++
P(BHET-EOP/TC, 85 : 15)	++	++	++	++	++
P(BHET-EOP/TC, 80 : 20)	++	++	++	++	++
P(BHET-EOP/TC, 50 : 50)	—	—	+/Δ	+/Δ	+/Δ

NMP, N-methylpyrrolidone; DMF, dimethylformamide; DMSO, dimethylsulfoxide; +++, readily soluble; ++, soluble; +/Δ, soluble on heating.

Table 3.
Intrinsic viscosity and molecular weights of poly(terephthalate-*co*-phosphate)s determined by GPC analysis using polystyrene as standards

Poly(terephthalate-*co*-phosphate)	Batch size	M_w	M_n	$[\eta]$ (dl/g)
P(BHET-EOP/TC, 80 : 20)	S	7920	6120	0.180
	M	15 520	6322	N.D.
	L	62 585	32 940	N.D.
P(BHET-EOP/TC, 85 : 15)	S	5180	2270	0.146
P(BHET-EOP/TC, 90 : 10)	S	4200	1590	0.148
	M	14 050	4790	N.D.
P(BHET-EOP/TC, 95 : 5)	S	1480	600	0.089

Batch size was defined according to the amount of BHET used as starting material. S: 10–20 g of BHET, M: 50 g of BHET and L: 100 g of BHET. M_w and M_n were calculated by GPC using polystyrene as standard. Intrinsic viscosity was measured at 40°C in chloroform using an Ubbelohde viscometer.

The structures of these co-polymers were confirmed by ^{1}H-NMR, ^{31}P-NMR and FT-IR spectra. A set of spectra for P(BHET-EOP : TC, 80/20) is shown in Fig. 4. The ^{1}H-NMR chemical shift assignments were listed in Fig. 4a. Using ratio of integrations of peaks δ 4.5 ppm (or δ 4.4 ppm) and δ 4.7 ppm, we could calculate the phosphate to terephthalate contents. The compositions of poly(terephthalate-*co*-phosphate)s as calculated by this method were fairly consistent with the feeding ratios of the monomers (Table 1). The ^{31}P-NMR spectrum of P(BHET-EOP/TC) showed one single peak at δ 2.49 ppm. This indicated that there were very few terminal phosphate groups. The FT-IR spectrum of P(BHET-EOP/TC, 80 : 20) revealed the characteristic absorption peaks for carbonyl groups (1721 cm^{-1}, stretching), P=O groups (1274 cm^{-1}, stretching) and P—O—C groups (1035 cm^{-1}, P—O stretching) (data not shown). All other co-polymers showed similar spectrum characteristics.

GPC analysis of the co-polymer showed a polydispersed distribution typical of polycondensation products (Fig. 4c). The average molecular weight of poly(terephthalate-*co*-phosphate)s increased with the TC content, which was also confirmed by the intrinsic viscosity data (Table 3). This confirms that the reaction of the extended diol with terephthaloyl chloride is more energetic than that with ethyl

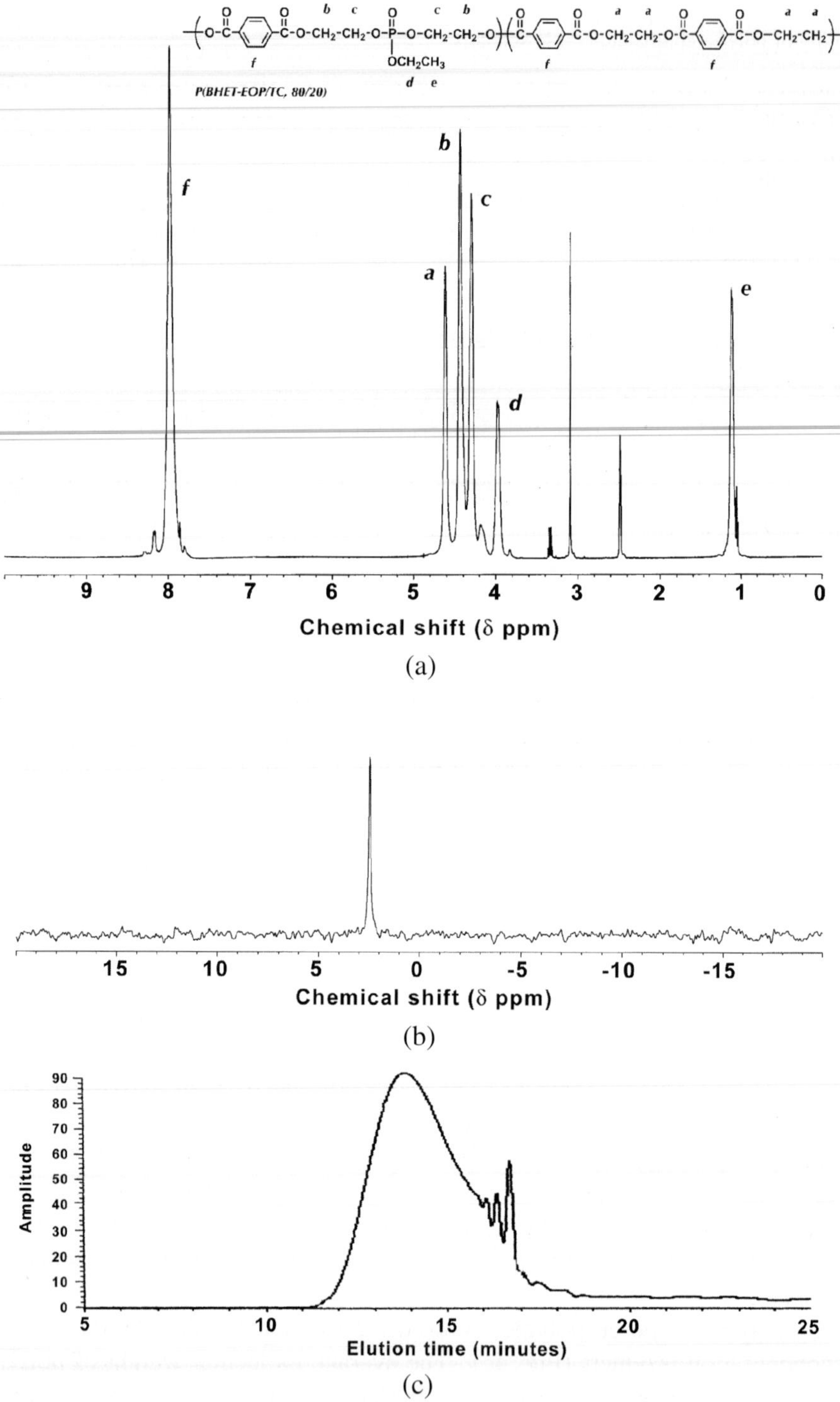

Figure 4. (a) ^{1}H-NMR spectrum for P(BHET-EOP/TC, 80 : 20). (b) ^{31}P-NMR spectrum for P(BHET-EOP/TC, 80 : 20). (c) GPC chromatogram for P(BHET-EOP/TC, 80 : 20).

phosphorodichloridate. Interestingly, we found that a 2–5% excess of EOP yielded a higher molecular weight polymer. This might be due to the high susceptibility of EOP degradation to moisture. In addition, increasing the batch size significantly increases the molecular weight of the resulting polymer (Table 3). For example, in the synthesis of P(BHET-EOP/TC, 80 : 20), increasing the amount of BHET from 10 g to 100 g increased the weight average molecular weight by 7-fold to 62 585. This is most likely because at larger batch size, it is much easier to maintain the stoichiometry of BHET : (EOP + TC), which is crucial to obtaining a high-molecular-weight polymer through a polycondensation scheme.

A unique feature of this series of co-polymers, as compared to other commonly used biodegradable polymers such as poly(lactide-co-glycolide)s, is the availability of functional side groups. This is an advantage of the pentavalency of the phosphorus atom in the backbone, allowing the chemical linkage of drug molecules or other agents to the polymer [29]. For instance, drugs with hydroxyl groups may be coupled to the phosphorus via an ester bond, which is hydrolyzable. Cell-adhesion molecules can be conjugated to the surface of the poly(terephthalate-*co*-phosphate) scaffolds, improving the cell adhesion for tissue-engineering applications [30]. These poly(phosphoester)s are natural candidates for a completely biodegradable pendant delivery system. In this study, we investigated the feasibility of introducing phosphate ion to the polymer chains.

Phosphate ion incorporating polymer was prepared by synthesizing co-polymers with methoxy side chains and then cleaving the methoxy side chain with NaI [30]. Replacing all the EOP with methyl phosphorodichloridate (MOP) yielded a co-polymer P(BHET-MOP/TC, 80 : 20) with relatively low molecular weight, maybe due to the increased instability of MOP compared with EOP (reacting to moisture). This change also dramatically increased the hydrophilicity of the polymer (data not shown). We here reported two co-polymers prepared with MOP equivalent to 40% (mol) and 20% (mol) of BHET used in the polymerization, respectively (Fig. 5). The molecular weights of these two co-polymers were slightly lower than P(BHET-EOP/TC, 80 : 20). Furthermore, converting the methoxy groups into sodium salt renders the polymer less soluble in chloroform and methylene chloride [30].

Physicochemical properties of poly(terephthalate-co-phosphate)s

Dacron (PET) is a highly crystallized polymer (T_m = 265°C, T_g = 69°C) [31]. Introduction of ethyl phosphate bond into the backbone significantly changes the thermal properties of the polymer. In this P(BHET-EOP/TC) co-polymer series, when the EOP/TC ratio was higher than 50 : 50, the PPEs were amorphous. No melting point (T_m) was observed in differential scanning calorimetric (DSC) analysis. The co-polymers began to soften at around 150°C. When the EOP/TC charging ratio was reduced to 50 : 50, DSC graph showed a melting peak at 201°C, indicating the presence of crystalline phase.

As expected, the glass transition temperature (T_g) increased as the EOP/TC feeding ratio decreased. Nevertheless, the T_g increment was minor until the EOP/TC

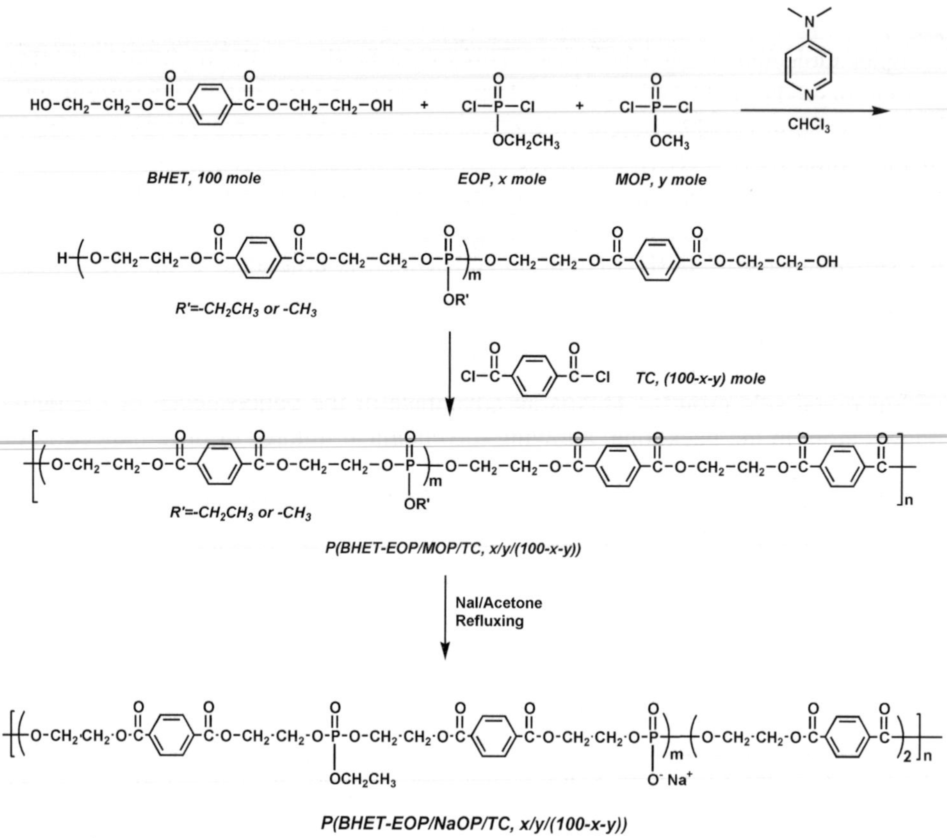

Figure 5. Synthetic scheme of P(BHET-EOP/MOP/TC) and P(BHET-EOP/NaOP/TC).

ratio decreased to below 75 : 25. There was a dramatic increase of T_g from P(BHET-EOP/TC, 80 : 20) ($T_g = 25°C$) to P(BHET-EOP/TC, 50 : 50) ($T_g = 64°C$). This significant increase is hypothesized to be related to the crystalline phase in P(BHET-EOP/TC, 50 : 50). Figure 6a and 6b shows the DSC curves for these two co-polymers. The T_g values of a series of other P(BHET-EOP/TC) co-polymers of differing EOP/TC feed ratios were determined, and the results are shown in Fig. 6c and in Table 4.

Introducing sterically hindered diol to the polymer backbone, by substituting ethylene glycol with 2-methylpropylene diol (MPD) or 2,2-dimethylpropylene diol (DMPD) residues, reduced its glass transition temperature (Fig. 6c). The T_g values of the co-polymers with different aliphatic diol residues decreased in the order of diol residue sequence: EG > DMPD > MPD. This result indicated that the co-polymer chain rigidity reduced in the same order.

The relative hydrophobicity of the P(BHET-EOP/TC) series was assessed by measuring the water-in-air contact angle. The contact angle decreased dramatically from 65° for P(BHET-EOP/TC, 80 : 20) to 8° for P(BHET-EOP). While the charging molar ratio of EOP/TC was maintained at 80 : 20, substituting the ethoxy side

Table 4.
Glass transition temperatures and water-in-air contact angles of selected poly(terephthalate-*co*-phosphate)s

Poly(terephthalate-*co*-phosphate)	T_g (°C)	Water-in-air contact angle (deg)
P(BHET-EOP/TC, 50 : 50)	64	N.D.
P(BHET-EOP/TC, 80 : 20)	25	65
P(BHET-EOP/TC, 85 : 15)	28	54
P(BHET-EOP/TC, 90 : 10)	21	N.D.
P(BHET-EOP/TC, 95 : 5)	21	10
P(BHET-EOP/TC, 100 : 0)	19	8
P(BHET-EOP/MOP/TC, 60 : 60 : 20)	32	14
P(BHET-EOP/MOP/TC, 40 : 40 : 20)	39	9
P(BHET-NaOP/TC, 80 : 20)	58 (T_m = 180)	N.D.
P(BHET-EOP/NaOP/TC, 60 : 20 : 20)	52	N.D
P(BHET-EOP/NaOP/TC, 40 : 40 : 20)	63	N.D.

chain with methoxy side chain (MOP) also markedly increased the hydrophilicity (Table 4). At the ratio of EOP/MOP/TC of 40 : 40 : 20, the contact angle decreased to a value close to that of P(BHET-EOP).

Furthermore, introducing negative charges to phosphate sidechains by converting the methoxy groups into sodium salt rendered the polymer less soluble in chloroform and methylene chloride. The negatively charged P(BHET-ENaOP/TC)s became hydrophilic by gross observation. As expected, the methoxy group substituted P(BHET-EOP/TC) and the negatively charged poly(terephthalate-*co*-phosphate)s had much higher T_g values than their parent polymers (Table 4). P(BHET-NaOP/TC, 80 : 20) has a clear sharp melting peak at 180°C as determined by DSC.

Poly(terephthalate-*co*-phosphate)s in general had good film-forming properties, particularly at EOP/TC feeding ratios of 80 : 20 and 85 : 15. These two co-polymers formed elastomeric films with different thickness by solvent casting. Poly(terephthalate-*co*-phosphate)s showed typical tensile properties for amorphous or low crystalline thermoplastics. At an EOP/TC ratio higher than 90 : 10, the film became weak and difficult to handle. P(BHET-EOP/TC, 80 : 20) film yielded at 86.6 ± 1.6% elongation (Fig. 7), with a modulus of elasticity of 13.76±2.66 MPa. Substituting EOP with MOP partially (40 : 40 or 60 : 20, see Table 5) did not affect the film forming property significantly, except that the film became more brittle and less ductile. P(BHET-EMOP/TC)s had nearly 100% higher moduli of elasticity than P(BHET-EOP/TC, 80 : 20). In addition, fibers could be drawn from co-polymer melts at 160–180°C from all three co-polymers listed in Table 5 (with a 20% molar ratio of TC).

Due to the above described favorable properties of P(BHET-EOP/TC, 80 : 20) and P(BHET-EOP/TC, 85 : 15), we focused on this polymer in the evaluation of cytotoxicity, degradation, tissue compatibility and drug-release kinetics.

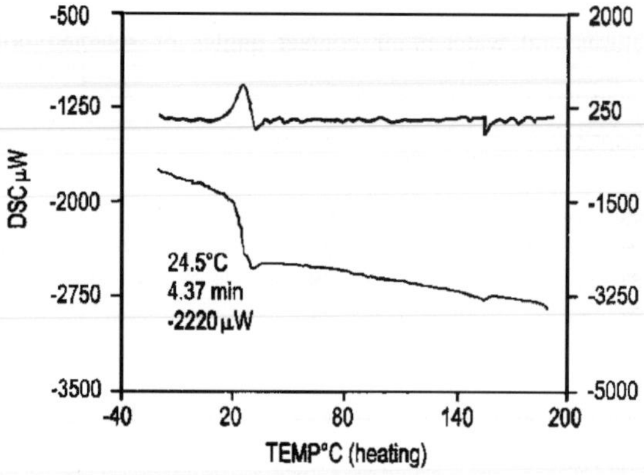

(a)

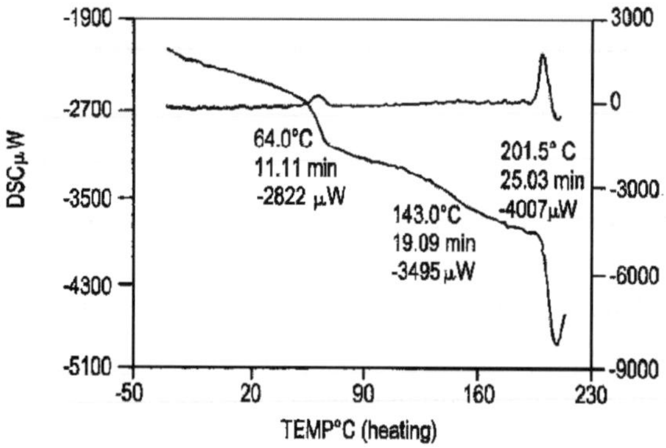

(b)

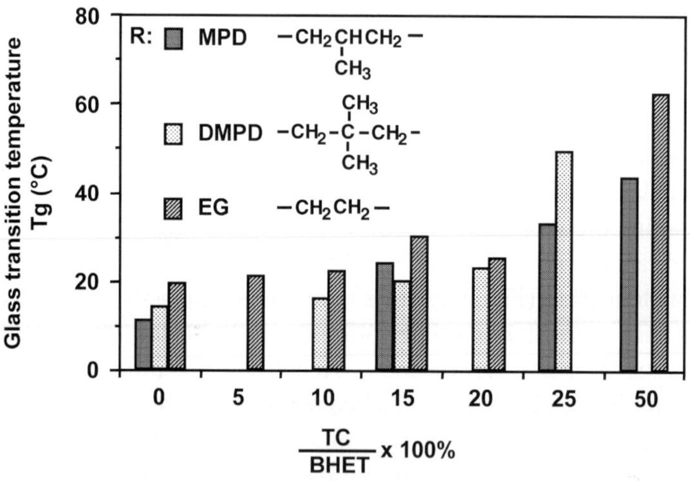

(c)

Figure 6. (a) DSC graph for P(BHET-EOP/TC, 80 : 20). (b) DSC graph for P(BHET-EOP/TC, 50 : 50). (c) Glass transition temperature as a function of terephthalate (or phosphate) content in poly(terephthalate-*co*-phosphate) and the structure of the diol residue in the co-polymer backbone.

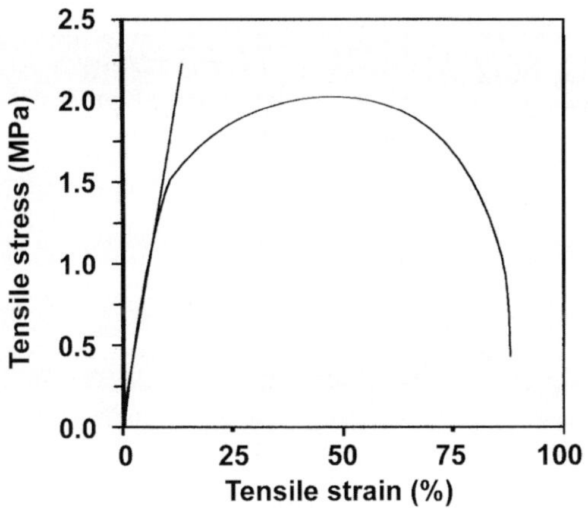

Figure 7. Tensile stress–strain curve for P(BHET-EOP/TC, 80 : 20) thin film prepared by the solvent casting method.

Table 5.
Tensile properties of selected poly(terephthalate-*co*-phosphate)s

Poly(terephthalate-*co*-phosphate)	Tensile stress at yield (MPa)	Elongation at break (%)	Modulus of elasticity (MPa)
P(BHET-EOP/TC, 80 : 20)	1.47 ± 0.31	86.6 ± 1.6	13.76 ± 2.66
P(BHET-EOP/MOP/TC, 60 : 20 : 20)	2.55 ± 0.47	74.6 ± 8.7	26.77 ± 2.19
P(BHET-EOP/MOP/TC, 40 : 40 : 20)	2.79 ± 1.01	48.9 ± 1.6	27.59 ± 9.36

Cytotoxicity of poly(terephthalate-co-phosphate)s

The cytotoxicity of P(BHET-EOP/TC, 80 : 20), as a representative, was assessed by three cell culture assays: (1) cytotoxicity of the polymer itself by directly culturing cells on the co-polymer thin film; (2) cytotoxicity of the polymer microspheres to cells because of potential applications in drug delivery; (3) cytotoxicity of the degradable products of the co-polymer.

The first assay was performed by culturing human embryonic kidney cells (HEK293 cell line, ATCC) on co-polymer-coated coverslips. After 3 days of culture, HEK293 cells exhibited a normal morphology that was comparable to the TCPS control group; they also proliferated at a rate comparable to the control (Fig. 8a).

The second assay on cytotoxicity of co-polymer microspheres was most relevant to particulate-mediated drug delivery, where the polymer microspheres might or might not be taken up by the cells. When cultured with polymer microspheres (<10 μm), the proliferation assay on human gastric carcinoma cells (GT3TKB) showed no inhibition of cell growth for concentration of P(BHET-EOP/TC, 80 : 20) microspheres up to at least 1.0 mg/ml (Fig. 8b).

The degradation products of P(BHET-EOP/TC, 80 : 20) and poly(L-lactide) did not show any inhibitory effect on cellular proliferation at doses lower than 125 μg/ml

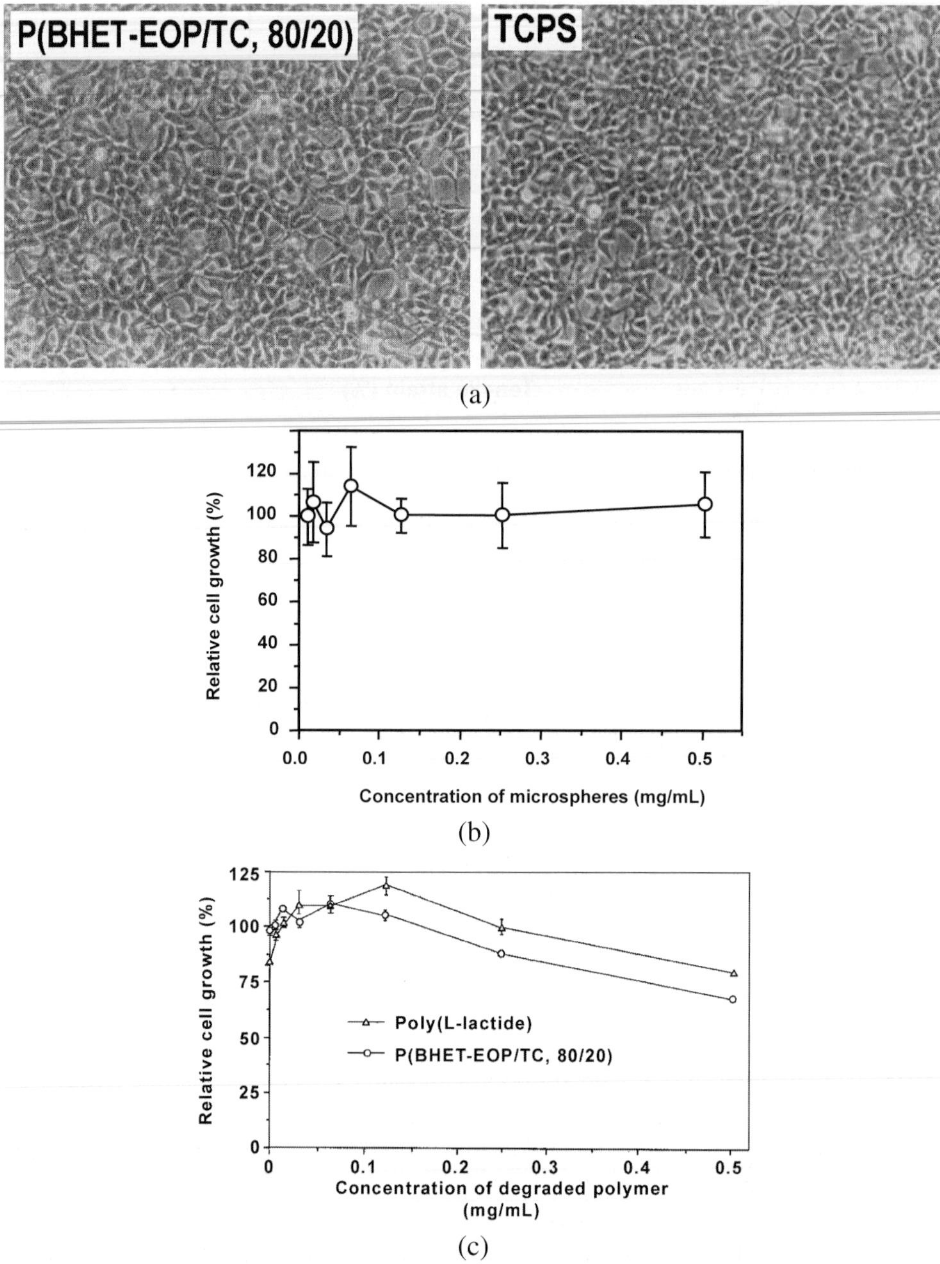

Figure 8. (a) Phase-contrast images of HEK293 cells cultured on P(BHET-EOP/TC, 80 : 20) coated cover slip in comparison with that cultured on tissue-culture polystyrene surface. (b) Effect of P(BHET-EOP/TC, 80 : 20) microspheres (2–10 μm) on the proliferation of human GT3TKB gastric tumor cells in reference to cells cultured on tissue culture plate. (c) Effect of the degradation products of P(BHET-EOP/TC, 80 : 20) and PLLA on the proliferation of human GT3TKB gastric tumor cells in reference to cells cultured on tissue culture plate.

(Fig. 8c). At higher doses (0.25–0.5 mg/ml) both P(BHET-EOP/TC, 80 : 20) and PDLA did show some inhibition. In separate Ames test and ISO agarose overlay test, neither mutagenicity nor cytotoxicity was observed for P(BHET-EOP/TC, 80 : 20), P(BHET-EOP/TC, 85 : 15) and the BHET monomer (data not shown). These results indicated that there no toxic leachable or degradation products produced from polymers and BHET, or the leached products could not diffuse through the agarose layer.

Degradation and stability of poly(terephthalate-co-phosphate)s

Figure 9a shows the cumulative mass loss of P(BHET-EOP/TC, 80 : 20) and P(BHET-EOP/TC, 85 : 15) discs at 37°C in 0.1 M PBS. P(BHET-EOP/TC, 80 : 20) degraded at a slower rate than P(BHET-EOP/TC, 85 : 15), reflecting the influence of the phosphate in accelerating the degradation. The former showed a gradual increase of mass loss over time, losing 21% total mass after 18 days of incubation. The latter had a 22% mass loss after 24 h of incubation, followed by a gradual increase to about 43% on day 18. SEM analysis of the cross-sections of the samples on day 18 indicated that P(BHET-EOP/TC, 80 : 20) swelled much less than P(BHET-EOP/TC, 85 : 15). P(BHET-EOP/TC, 85 : 15) discs developed a thick swollen degradation layer, compared with P(BHET-EOP/TC, 80 : 20) discs (Fig. 9b). GPC analysis of the recovered samples on day 18 showed that the molecular weights of P(BHET-EOP/TC, 80 : 20) and P(BHET-EOP/TC, 85 : 15) discs had declined by 21% and 43% of their original values, respectively. Figure 9c shows a GPC chromatograph of P(BHET-EOP/TC, 80 : 20) sample after incubation for 2 weeks at 37°C in 0.1 M PBS. The GPC trace of the degraded sample resembled that of the original sample, but shifted towards the lower molecular weight direction. This result suggested that the degradation of P(BHET-EOP/TC, 80 : 20) occurred as a relatively uniform random chain scission. Given the fact that the phosphoester bond in the backbone is more susceptible to hydrolysis than the terephthalate ester bond, this finding is consistent with the previous conclusion that this co-polymer is largely amorphous and the distribution of the phosphate groups in the backbone is relatively uniform and random throughout the polymer chain.

This degradation study highlighted the hydrolytic sensitivity of the co-polymer to the phosphate content in the backbone. Increasing the TC content in the polymer would significantly retard the degradation rate. It is worth noting that molecular weight is also an important parameter affecting the degradation rate of the co-polymer. Apparently, higher molecular weight PPEs with the same composition degraded at a slower rate than that with a lower M_w. For example, P(BHET-EOP/TC, 80 : 20) with a M_w of 14 300 showed only a 5% weight loss in 14 days of incubation under the same condition (data not shown).

As discussed above, in the P(BHET-EOP/TC, 80 : 20) series, substituting ethoxy with methoxy group yielded co-polymers with increased hydrophilicity, which in turn accelerated the *in vitro* degradation as demonstrated in Fig. 10. Furthermore,

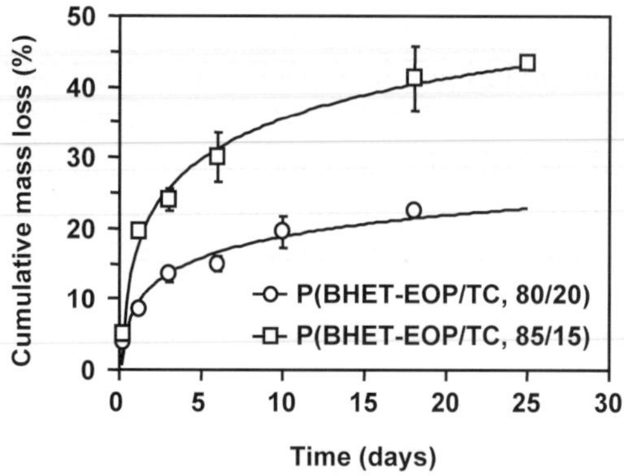

(a)

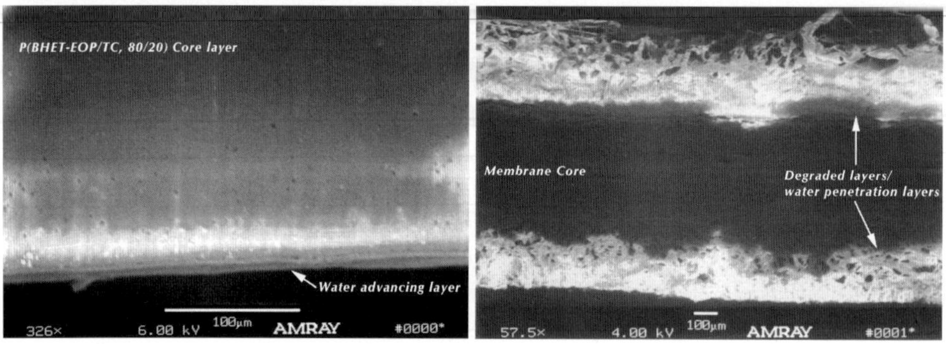

(b)

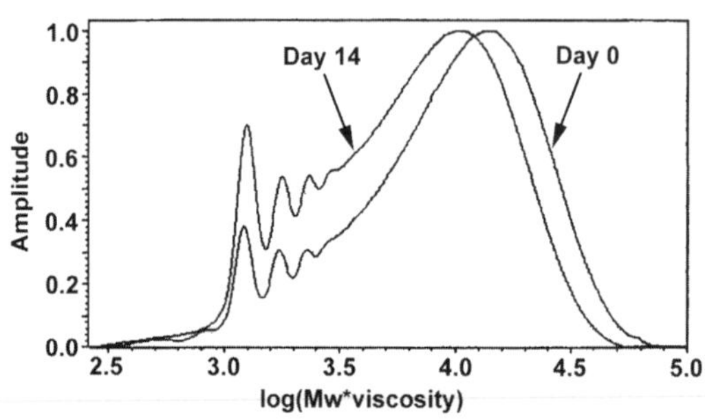

(c)

Figure 9. (a) *In vitro* degradation of P(BHET-EOP/TC) discs in 0.1 M PBS at 37°C. (b) SEM images of P(BHET-EOP/TC, 80 : 20) (left panel) disc and P(BHET-EOP/TC, 85 : 15) disc (right panel) after incubation in 0.1 M PBS at 37°C for 18 days. (c) GPC analysis of P(BHET-EOP/TC, 80 : 20) disc after incubation in 0.1 M PBS at 37°C for 14 days, in comparison with the original sample.

converting the methoxy group to a sodium salt dramatically increased the degradation rate.

These data demonstrated the feasibility of fine-tuning the degradation rate of the co-polymers by adjusting the EOP/TC ratio, changing the side-chain structure (ethyl group *vs* methyl group) and controlling the molecular weight of the co-polymers.

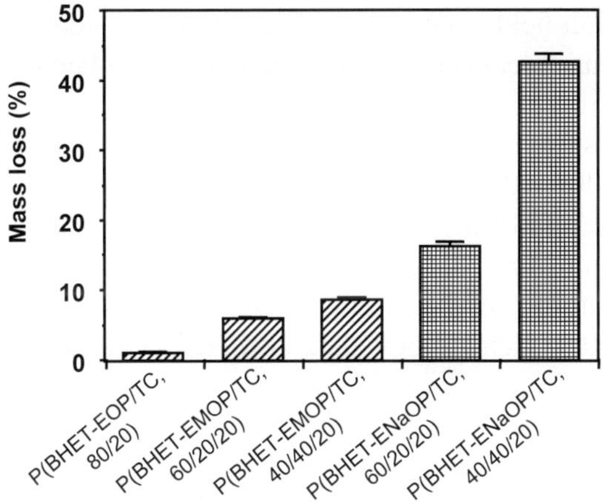

Figure 10. Mass loss of P(BHET-EOP/TC, 80 : 20) discs after incubation in 0.1 M PBS at 37°C. Molecular weights of the sample tested are: P(BHET-EOP/TC, 80 : 20), M_w = 12 500; P(BHET-EOP/MOP/TC, 60 : 20 : 20), M_w = 9350; P(BHET-EOP/MOP/TC, 40 : 40 : 20), M_w = 7830. No M_w data were available for P(BHET-EOP/NaOP/TC, 60 : 20 : 20) and P(BHET-EOP/NaOP/TC, 40 : 40 : 20).

The co-polymers became more hydrolytically labile as the phosphate component (EOP) was increased and the side chains were switched from the ethoxy to the methoxy structure.

Storage stability and sterilization stability

Designed to be biodegradable, poly(terephthalate-*co*-phosphate)s are naturally susceptible to hydrolysis, even in air (to moisture in the air). Nevertheless, compared with other degradable polymers, for example, aliphatic polyanhydrides, poly(terephthalate-co-phosphate)s were relatively stable when stored under dry condition, even without vacuum or inert gas packaging. When P(BHET-EOP/TC, 80 : 20) was stored in a desiccator at room temperature, the molecular weight drop was insignificant after one month (data not shown).

Sterilizability of a biomedical polymeric material often determines its clinical utility. When P(BHET-EOP/TC, 80 : 20) was γ-irradiated at 2.5 Mrad in an aluminum foil pouch on solid CO_2, neither the chemical structure nor the molecular weight were altered, as monitored by GPC, intrinsic viscosity measurement and FT-IR (data not shown).

In vivo *degradation and biocompatibility*

The biodegradation and biocompatibility of poly(terephthalate-*co*-phosphate)s were evaluated in a SPF Sprague–Dawley rat model by implanting polymer samples in the right hind limb muscle. P(BHET-EOP/TC, 80 : 20) was investigated as a representative compared with PLGA (75 : 25, RG755). Tissue reaction surrounding the implantation site was characterized by the average monocyte and macrophage count per observation field under 400×. Four areas per side were counted and the arithmetic mean was calculated and plotted against implantation time in Fig. 11. The inflammation response in the surrounding tissue was rated according to the following scoring system: Score 0 for 'no irritation'; Score 1–200 for 'slight irritation'; Score 201–400 for 'mild irritation'; Score 401–600 for 'moderate irritation' and score of 601 and more for severe irritation. As illustrated in Fig. 11, inflammation reactions to P(BHET-EOP/TC, 80 : 20) implants were similar to that of PLGA wafers, and were characterized as slightly irritating from day 3 to the 6-month time point. Hematology data, differential leukocyte counts, cellular

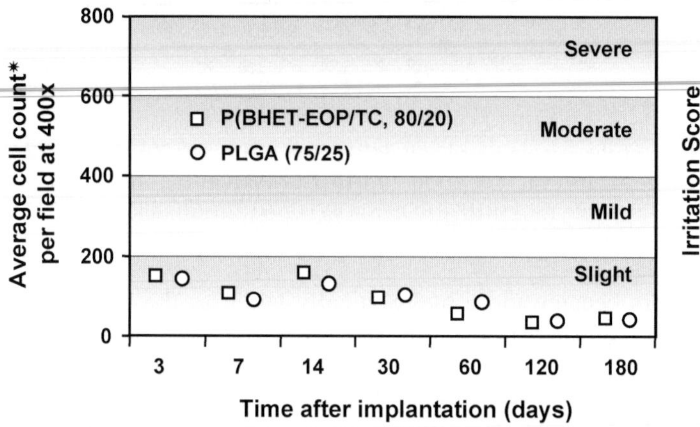

Figure 11. Tissue reaction scores (average monocyte and macrophage counts per field, magnification 400×) for poly(phosphoester) discs implanted between the muscle layers of the hind limb of male Sprague–Dawley rats were plotted as a function of implantation time. Four areas per side were counted and the arithmetic mean was shown in the figure. Tissue reaction was graded according to the following system: no irritation (0); slight irritation (1–200); mild irritation (201–400); moderate irritation (401–600); severe irritation (>601).

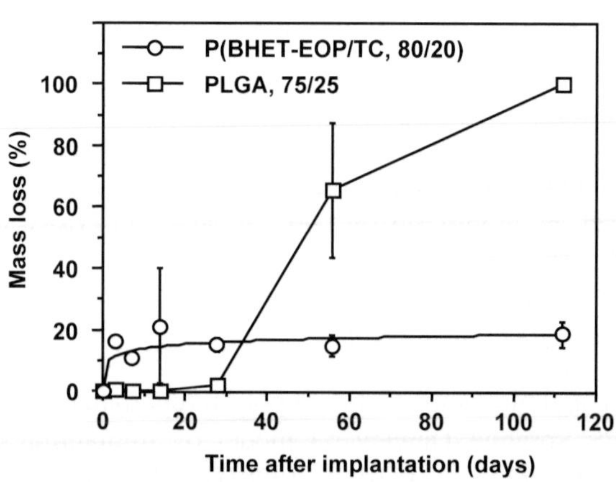

Figure 12. Mass loss of P(BHET-EOP/TC, 80 : 20) and PLGA (75 : 25) discs following implantation in the hind limb of male Sprague–Dawley rats.

morphology findings and clinical chemistry values were overall unremarkable and comparable between the two groups at each interval. No histomorphological change in all major organs at the 6-month time point was observed (data not shown).

The analysis of polymer discs retrieved at different time points showed a typical biphasic degradation behavior of PLGA (Fig. 12). P(BHET-EOP/TC, 80 : 20), on the other hand, showed approx. 20% mass loss in 2 weeks, and remained unchanged for up to 4 months. This degradation behavior is speculated to be a reflection of the poor solubility of the di-terephthalate units in the physiological fluid. With higher phosphate content in the backbone, as in P(BHET-EOP/TC, 85 : 15), which not only increases the hydrophilicity but also reduces the di-terephthalate segments in the breakdown products, the *in vitro* degradation study showed a mass loss of over 40% in 25 days. As discussed above for the *in vitro* degradation rate of the P(BHET-EOP/TC)s, it is expected that their *in vivo* degradation rates would be accelerated by increasing the phosphate content in the backbone.

FITC-labeled BSA-containing microspheres: preparation and release kinetics

P(BHET-EOP/TC, 80 : 20) microspheres were prepared *via* a double-emulsion solvent-extraction method using FITC-BSA as a model protein drug. The preparation procedure was adapted from Cleek *et al.* [32] and Uchida *et al.* [33] with some modifications. The preparation parameters were optimized to increase the encapsulation efficiency, improve the microsphere morphology and minimize the burst release of protein drug. Three different batches of microspheres with different loading levels (1.5, 14.1 and 22.8 wt%) were prepared. Under the optimal conditions as described in Materials and Methods, the size of the microspheres prepared with P(BHET-EOP/TC, 80 : 20) ranged between 2 and 20 μm with a smooth surface morphology (Fig. 13a). At lower loading level (1–5%), the protein (FITC-

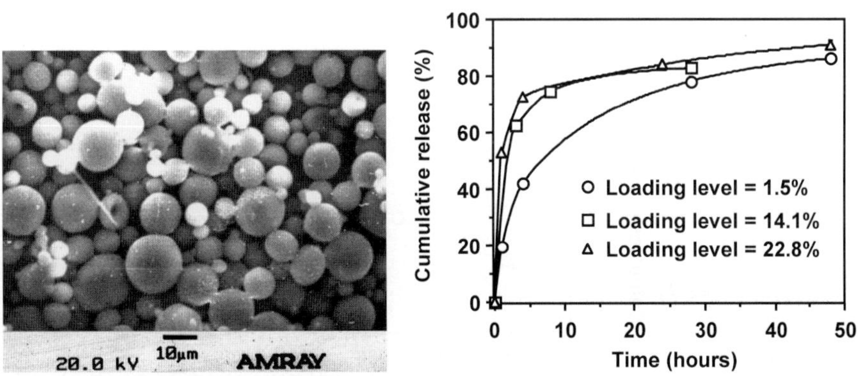

(a) (b)

Figure 13. (a) SEM image of P(BHET-EOP/TC, 80 : 20) microspheres with 1.5% loading of FITC-BSA. (b) Effect of loading level on the release kinetics of FITC-BSA from P(BHET-EOP/TC, 80 : 20) microspheres. M_w of P(BHET-EOP/TC, 80 : 20) used in this study was 8200 (polystyrene as standards).

BSA) encapsulation efficiency was about 90%. At higher loading level (10–20%), the encapsulation efficiency decreased slightly to 70–80%.

Figure 13b shows the release profiles of FITC-BSA from the three batches of P(BHET-EOP/TC, 80 : 20) microspheres with different loading levels. The *in vitro* release of FITC-BSA was relatively fast for all three batches of microspheres. Microspheres with loading levels of 14.1% or 22.8% released about 80% of the encapsulated protein within the first day, with a burst of 60–70% in the first 4 h. This was followed by a much lower release rate during the next 24 h. There remained about 5% of FITC-BSA in the microspheres, even after 10 days of release at 37°C in PBS (data not shown in the figure). Microspheres with a loading level of 1.5% showed a more gradual release kinetics, with minimal burst.

Cyclosporine-A-containing microspheres: preparation and release kinetics

The release kinetics of cyclosporine A are shown in Fig. 14a. The release was sustained with no burst effect. About 60% of the encapsulated cyclosporine A was

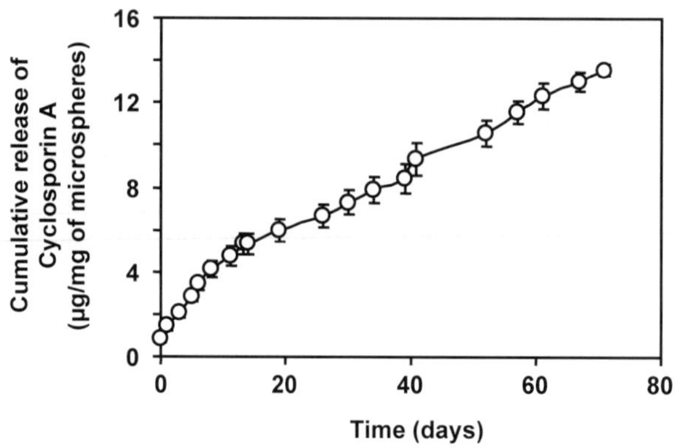

(a)

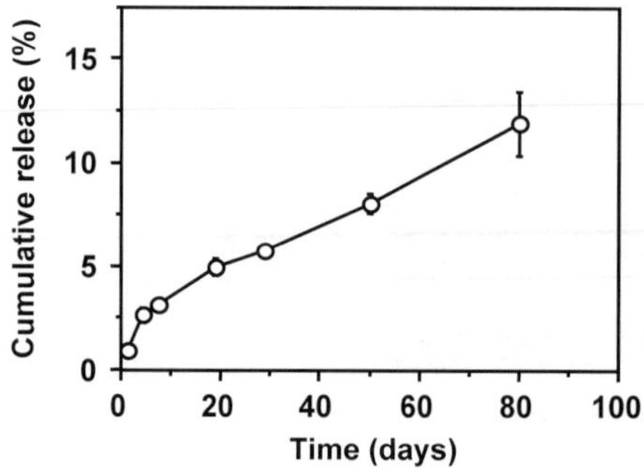

(b)

Figure 14. (a) Cumulative release of cyclosporine A (loading level of 2.3%) from P(BHET-EOP/TC, 80 : 20) microspheres in 0.1 M PBS at 37°C. (b) Cumulative release of paclitaxel (loading level of 10%) from P(BHET-EOP/TC, 80 : 20) microspheres in 0.1 M PBS at 37°C.

released by day 70. This led to a projected release time of approx. 120 days in PBS at an average release rate of 0.2 μg/day per mg microspheres.

Cyclosporine-A-encapsulated P(BHET-EOP/TC, 80 : 20) microspheres were prepared for a potential application of treating chronic uveitis, which is a blinding disease that may require long-term treatment [34]. Cyclosporine A is a 1.2-kDa cyclic peptide that is poorly insoluble in water, facilitating the nearly linear release profile observed in the study. The drug has been shown in several human trials to be effective in treating chronic uveitis, especially that associated with Bechet's disease [34]. However, side effects from systemic administration, including nephrotoxic effects and hypertension, limit its use [35]. The systemic toxic effects of cyclosporine A could be minimized by delivering the drug locally. Direct intravitreal injection has been shown to prevent experimental autoimmune uveitis, but the intravitreal half-life of 6 h [36, 37] would require repeated injections of cyclosporine A. A non-biodegradable sustained release device for implantation to the vitreous has been developed [36, 38]. There are several limitations of this system primarily relating to the size (2.5 mm). The implantation procedure carries a significant risk of retinal detachment and requires replacement after the drug is depleted. The replacement surgery is further complicated by a high rate of tissue encapsulation of the device.

A microsphere formulation described here would offer an attractive alternative. Microspheres prepared by this method were small in size ($<$10 μm) with an aim to reduce the blockage of vision when injected in the vitreous. Using a single emulsion procedure, a drug loading level of 2.3% was obtained. The release rate of 0.2 μg/day per mg microspheres obtained in this study is comparable to that achieved by a non-biodegradable system implanted in the vitreous, which maintains a vitreous cyclosporine A level between 0.3 and 0.5 μg/ml [36, 38]. In addition, we have established that the presence of up to 5 mg of P(BHET-EOP/TC, 80 : 20) microspheres in the vitreous was biocompatible and did not affect retinal function (data not shown). These results demonstrated that this poly(terephthalate-*co*-phosphate) microsphere system would be a promising approach to delivery cyclosporine A intravitreously.

Paclitaxel-containing film: preparation and release kinetics

Paclitaxel (10 wt%) was embedded in P(BHET-EOP/TC, 80 : 20) film (approx. 0.7 mm) during solvent casting. The film was flexible with a smooth surface. DSC and X-ray analyses did not show any crystallinity of the drug in the film. This film could deliver paclitaxel for several months (Fig. 14b) at a constant rate of 0.15% per day. This corresponded to an average of 15 μg paclitaxel per day for a 100-mg film (approx. 0.7 mm). A total of 12% of paclitaxel was released for 80 days. During the same period, less than 30% of mass loss of the co-polymer was observed.

This paclitaxel release profile remarkably resembled the release profile of cyclosporine A from P(BHET-EOP/TC, 80 : 20) microspheres, despite the vastly different surface area and different drug loading levels. This indicates that the release mechanism of these two drugs is drug-dissolution controlled, suggesting that many drug release configurations would work for sustained release of paclitaxel.

CONCLUSIONS

In summary, a series of new biodegradable poly(terephthalate-*co*-phosphate)s with chemical structures analogous to PET, one of the most popular polymeric biomaterials in medical use today, was developed. Incorporation of phosphate into the poly(ethylene terephthalate) backbone rendered the co-polymers soluble in chloroform and biodegradable, lowered the T_g, decreased the crystallinity and increased the hydrophilicity. The co-polymers became more hydrolytically labile as the phosphate component (EOP) was increased and the side chains were switched from the ethoxy to the methoxy structure. Detailed physico-chemical and biological characterization was performed on a co-polymer synthesized with an EOP/TC molar feed ratio of 80 : 20. This polymer showed good film-forming property, stability against γ-irradiation, a favorable toxicity profile *in vitro* and good tissue biocompatibility in the muscular tissue of mice. With interesting physico-chemical and biological properties, this class of biodegradable and versatile poly(ester-phosphoester) co-polymers should find utility in a diverse range of biomedical applications, particularly drug delivery and tissue engineering.

Acknowledgements

This research was partially supported by Guilford Pharmaceuticals and by NIH/R21EB003203. The authors would like to thank the technical assistance of Bart S. Hendriks, David P. Kao (Johns Hopkins University) and Dr. Bing Wang (Guilford Pharmaceuticals Inc.).

REFERENCES

1. T. W. Atkins and S. J. Peacock, *J. Biomater. Sci. Polymer Edn* **7**, 1065 (1996).
2. J. Kohn and R. Langer, *J. Am. Chem. Soc.* **109**, 817 (1987).
3. Y. Ogawa, *J. Biomater. Sci. Polymer Edn* **8**, 391 (1997).
4. I. Gachard, S. Bechaouch, B. Coutin and H. Sekiguchi, *Polym. Bull.* **38**, 427 (1997).
5. B. I. Dahiyat, E. Hostin, E. M. Posadas and K. W. Leong, *J. Biomater. Sci. Polymer Edn* **4**, 529 (1993).
6. J. Heller, R. V. Sparer and G. M. Zentner, in: *Biodegradable Polymers as Drug Delivery Systems*, M. Chasin and R. Langer (Eds), p. 121. Marcel Dekker, New York, NY (1990).
7. K. W. Leong, A. Domb, E. Ron and R. Langer, in: *Encyclopedia of Polymer Science Engineering*, J. I. Kroschwitz (Ed.). Wiley, New York, NY (1989).
8. A. Sen Gupta and S. T. Lopina, *Mater. Sci. Forum* **426-4**, 3261 (2003).
9. T. Kojima, M. Nakano, K. Juni, S. Inoue and Y. Yoshida, *Chem. Pharm. Bull. (Tokyo)* **33**, 5119 (1985).
10. S. Pulapura, C. Li and J. Kohn, *Biomaterials* **11**, 666 (1990).

11. H. R. Allcock, in: *Bioerodible Polymers as Drug Delivery Systems*, M. Chasin and R. Langer (Eds), p. 163. Marcel Dekker, New York, NY (1990).
12. H. R. Allcock, S. R. Pucher and A. G. Scopelianos, *Biomaterials* **15**, 563 (1994).
13. C. G. Pitt, F. I. Chasalow and Y. M. Hibionada, *J. Appl. Polym. Sci.* **26**, 3779 (1981).
14. S. M. Li, *J. Biomed. Mater. Res.* **48**, 342 (1999).
15. S. S. Davis, L. Illum and S. Stolnik, *Curr. Opin. Colloid Interf. Sci.* **1**, 660 (1996).
16. U. Edlund and A. C. Albertsson, *Adv. Polym. Sci.* **67**, 229 (2002).
17. M. Richards, B. Dahiyat, D. M. Arm, P. Brown and K. W. Leong, *J. Biomed. Mater. Res.* **25**, 1151 (1991).
18. Z. Zhao, J. Wang, H. Q. Mao and K. W. Leong, *Adv. Drug Deliv. Rev.* **55**, 483 (2003).
19. K. W. Leong, H. Q. Mao and R. X. Zhuo, *Chin. J. Polym. Sci.* **13**, 289 (1995).
20. S. Penczek, J. Pretula and K. Kaluzynski, *Polish J. Chem.* **75**, 1171 (2001).
21. S. Penczek, G. Lapienis, K. Kaluzynski and A. Nyk, *Polish J. Chem.* **68**, 2129 (1994).
22. S. Penczek, P. Klosinski, A. Narebska and R. Wodzki, *Makromol. Chem. Macromol. Symp.* **48-49**, 1 (1991).
23. H.-Q. Mao, W. Dang, I. Shipanova-Kadiyala, Z. Zhao, D. Cline and K. W. Leong, in: *Proceedings of the American Institute of Chemical Engineers Topical Conference: Biomaterials Carriers, Drug Delivery Scaffolds, Tissue Engineering*, p. 193 (1997).
24. M. V. Chaubal, G. Su, E. Spicer, W. B. Dang, K. E. Branham, J. P. English and Z. Zhao, *J. Biomater. Sci. Polymer Edn* **14**, 45 (2003).
25. E. Harper, W. Dang, R. G. Lapidus and R. I. Garver, Jr., *Clin. Cancer Res.* **5**, 4242 (1999).
26. A. Tunstall, R. C. Eberhart and M. D. Prager, *J. Biomed. Mater. Res.* **29**, 1193 (1995).
27. M. S. Aronoff, *J. Biomater. Appl.* **9**, 205 (1995).
28. W. S. Lyoo, S. G. Lee, W. S. Ha, J. Lee and J. H. Kim, *Polym. Bull.* **42**, 9 (1999).
29. J. C. Brosse, D. Derouet, L. Fontaine and S. Chairatanathavorn, *Makromol. Chem.* **190**, 2339 (1989).
30. A. C. A. Wan, H.-Q. Mao, S. Wang, S. H. Phua, G. P. Lee, J. Pan, S. Lu, J. Wang and K. W. Leong, *J. Biomed. Mater. Res. Appl. Biomater.* **70B**, 91 (2004).
31. J. Brandrup and E. H. Immergut (Eds), *Polymer Handbook*. Wiley-Interscience, New York, NY (1975).
32. R. L. Cleek, A. A. Rege, L. A. Denner, S. G. Eskin and A. G. Mikos, *J. Biomed. Mater. Res.* **35**, 525 (1997).
33. T. Uchida, K. Yoshida, A. Ninomiya and S. Goto, *Chem. Pharm. Bull. (Tokyo)* **43**, 1569 (1995).
34. Y. Sullu, I. Oge, D. Erkan, N. Ariturk and F. Mohajeri, *Acta Ophthalmol. Scand.* **76**, 96 (1998).
35. Y. Kato, J. Numaga, S. Kato, T. Kaburaki, H. Kawashima and Y. Fujino, *Clin. Exp. Ophthalmol.* **29**, 335 (2001).
36. G. J. Jaffe, C. S. Yang, X. C. Wang, S. W. Cousins, R. P. Gallemore and P. Ashton, *Ophthalmology* **105**, 46 (1998).
37. P. A. Pearson, G. J. Jaffe, D. F. Martin, G. J. Cordahi, H. Grossniklaus, E. T. Schmeisser and P. Ashton, *Arch. Ophthalmol.* **114**, 311 (1996).
38. B. C. Gilger, E. Malok, T. Stewart, D. Horohov, P. Ashton, T. Smith, G. J. Jaffe and J. B. Allen, *Vet. Immunol. Immunopathol.* **76**, 239 (2000).

Gels, Genes, Grafts and Giants, pp. 289–300 (2005)

Synthesis and characterization of stimuli-sensitive hydrogels having a different length of ethylene glycol chains carrying phosphate groups: loading and release of lysozyme

KATSUHIKO NAKAMAE [1,*], TAKASHI NISHINO [2], KOICHI KATO [2], TAKASHI MIYATA [2] and ALLAN S. HOFFMAN [3]

[1] *Japan Synchrotron Radiation Research Institute (JASRI), Mikazuki, Sayo, Hyogo 679-5198, Japan*
[2] *Department of Chemical Science and Engineering, Kobe University, Rokko, Nada, Kobe 657-8501, Japan*
[3] *Department of Bioengineering, University of Washington, Seattle, WA 98195, USA*

Received 29 September 2003; accepted 12 April 2004

Abstract—In order to prepare a polymer matrix capable of loading protein at high density, anionic hydrogels were synthesized by copolymerizing a monomer carrying a pendant phosphate group, methacryloyl-polyoxyethyl phosphate, with N-isopropylacrylamide and N,N′-methylene-bis-acrylamide, and the stumuli-sensitivity of hydrogels was characterized. The number of repeating ethylene glycol units in the phosphate carrying monomer was 1, 2, 5 or 8. Lysozyme bearing a positive net charge was immobilized in the hydrogel through formation of polyelectrolyte complex. It was shown that the amount of complexed lysozyme reached to 1.7 g/g dry gel, when high content of a phosphate-carrying monomer with 5 ethylene glycol units was incorporated into a hydrogel. It was further found that lysozyme complexed with phosphate-carrying network could be released by immersion of the lysozyme/hydrogel composite in a phosphate buffer solution of pH 7.4 owing to the pH-sensitivity of the hydorgel but no lysozyme was released at pH 1.4. The initial rate of lysozyme release was varied depending on the length of the ethylene glycol chains possessed by a network polymer and the content of the phosphate-carrying monomer unit. Lysozyme released from the phosphate-carrying hydrogel was proved to retain enzymatic activity at a level similar to the activity of lysozyme, which had been kept in buffer solution.

Key words: Phosphate-carrying polymer; stimuli-sensitive hydrogel; polyelectrolyte complex; protein immobilization; drug-delivery system.

*To whom correspondence should be addressed. Tel.: (81-791) 58-0833. Fax: (81-791) 58-0830. E-mail: nakamae@spring8.or.jp

INTRODUCTION

Stimuli-sensitive hydrogels have received much attention for their potential in a broad range of applications, including biosensors, clinical diagnostics, soft actuators, energy transducing devices and controlled drug-delivery systems [1]. They are designed to exhibit reversible phase transition in response to a small change in environmental factors, such as temperature, pH, ionic strength, solvent composition, introduction of specific molecules, electric field and magnetic fields [2].

In a previous study [3] we synthesized stimuli-sensitive hydrogels containing a phosphate-carrying polymer by a free radical coplymerization of (2-methacryloyloxy) ethyl phosphate (Phosmer M^R) with N-isopropyl acrylamide (NIPAAm) in the presence of a crosslinking agent (N,N′-methylen-bis-acrylamide, MBA). It was shown that the hydrogels swelled and collapsed reversibly in response to small changes in the external pH, salt concentration and solvent composition [4].

It was further reported that the incorporation of phosphate groups in the gel network allowed us to load basic proteins through formation of a polyelectrolyte complex (PEC) [5]. Lysozyme, a model protein bearing a net positive charge under physiological conditions, was loaded in the phosphate-carrying network. The hydrogels released lysozyme with an increase in the ionic strength of the medium.

Application of such pH-sensitive hydrogels may be in the intestine targeting of biomacromolecules such as proteins and peptides. When administered to the gastro-intestinal tract *via* oral administration, bare biomacromolecules would be exposed to the highly acidic, gastric environment to reach the target site in the intestine, which may lead to appreciable reduction in their biological activities. The simplest design for protecting proteins and peptides from the gastric environment may be an enteric coating using a variety of polymers, which disintegrate or dissolve under a neutral, intestinal condition [6]. In some cases, these coatings have been used in combination with inhibitors toward proteolytic enzymes [7] and other approach is to chemically modify peptides, such as insulin and calcitonin, with monosaccharides [8] and fatty acids [9]. These modifications are expected to reduce the accessibility of the peptides to digestive enzymes and, in the case of fatty acid derivatization, to enhance absorption by making drugs more lipophilic. Using this approach, however, one cannot avoid the alteration of biological activity during the modification process.

Recently, polymeric hydrogels capable of reversible swelling depending on charges in environmental pH have been considered as a novel matrix for drug-delivery systems [10–12]. It is possible that such polyelectrolyte hydrogels may be designed to incorporate the oppositely-charged proteins into the gels. Dong and Hoffman [13] prepared pH-sensitive hydrogels made of NIPAAm, acrylic acid and vinyl functionalized polydimethylsiloxane, and examined their use in enteric drug delivery. Katono *et al.* [14] synthesized interpenetrating polymer networks composed of poly(acylamide-butyl methacrylate) and polyacrylic acid. These studies employed polyelectrolytes carrying carboxylic acids as an anionic component of hydrogels.

In this study, phosphate-carrying hydrogels with loaded lysozyme have been subjected to an *in vitro* release test at different pH values. We also studied the effect of the number of repeating ethylene glycol units on the loading and release of lysozyme. The external solution pH values were chosen to be 1.4 and 7.4, according to the consideration that these are the pH values of the stomach and the intestine, respectively [6].

MATERIALS AND METHODS

Preparation of hydrogels

The preparation of phosphate-carrying hydrogels and the subsequent loading of lysozyme into the hydrogel were carried out as described in a previous report [5]. Phosphate-carrying monomers, supplied by Unichemical (Nara, Japan), were purified by the extraction of divinyl derivatives with n-hexane. The chemical structure of the monomers is shown in Fig. 1. The monomer was co-polymerized with NIPAAm and MBA in N,N-dimethyl sulfoxide (DMSO) using 1,1′-azoisobutyronitrile (AIBN) as the free radical initiator. The nominal concentration of monomer was fixed at 20 wt% of DMSO solution. The hydrogels formed were washed and equilibrated with distilled water. The swelling ratios of hydrogels were determined from the weight differences of the swollen hydrogels and dried gels by the following equation:

$$\text{Degree of swelling} = (W_\text{s} - W_\text{d})/W_\text{d},$$

where W_s and W_d are the weight of the hydrogel in the swollen and the dry states, respectively.

The compressive modulus of the hydrogels was measured by using an uniaxial compression apparatus. The physical and chemical cross-linking density can be estimated by compressive modulus and compressive stress [3].

Lysozyme from egg white was purchased from Sigma (St. Louis, MO, USA) and was used without further purification.

Lysozyme loading and release

The phosphate-carrying hydrogel was immersed in a 10 mg/ml aqueous lysozyme solution at pH 3.5 and complexation reaction was allowed to proceed at 30°C after 3 days. Then the hydrogel was washed extensively with distilled water for 1 day at 30°C to remove any loosely-bound lysozyme. The amount of lysozyme loaded in hydrogel was determined by gravimetry after vacuum drying. The lysozyme-loaded hydrogel was subjected to a release test in the different pH solutions and an electrolyte solution. The hydrogel (*ca.* $1.0 \times 1.0 \times 0.2$ cm) was immersed in 8 ml of different pH solutions or different KCl solutions at 37°C. The concentration of lysozyme released into the external solution was determined by measuring

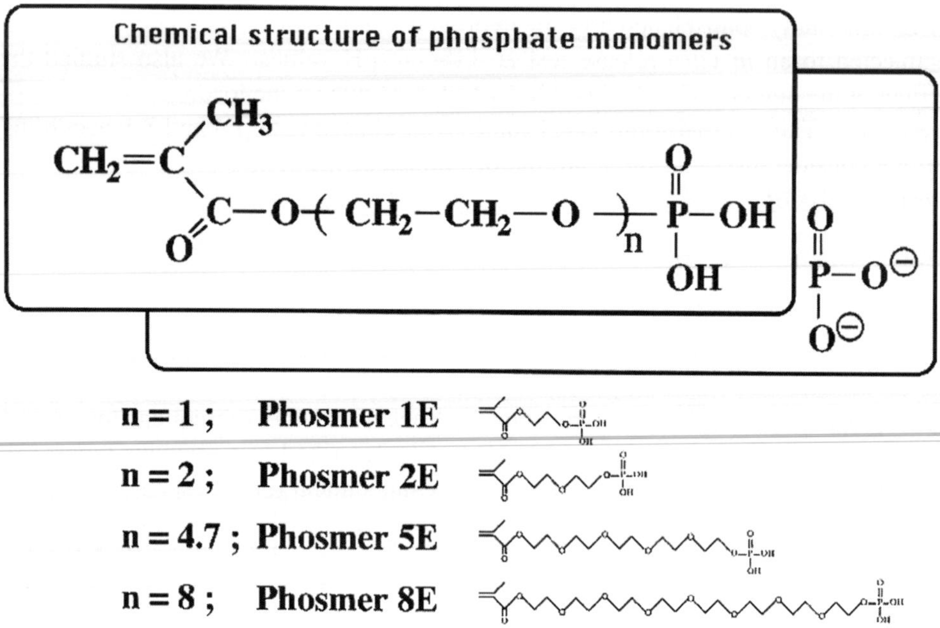

Figure 1. Chemical structure of the monomers used for hydrogel preparation.

absorbance at 280 nm on the supernatant with a spectrophometer (U-2000, Hitachi, Tokyo, Japan). The external buffer solution was not changed during the release test.

Titration of polymer having phosphate group

The water-soluble homopolymer of phosphate-carrying monomer (Phosmer 1E) was synthesized by free radical polymerization. The homopolymer was recovered by precipitation in *tert*-butylalcohol and then freeze-dried. Hydrogen ion titration of the homopolymer and lysozyme was carried out using 0.1 M aqueous NaOH as titrant.

Lysozyme assay

The enzymatic activity of released lysozyme was determined using *Micrococcus lysodeikticus* as substrate [15]. The sample solution (200 μl) containing released lysozyme was added to 2.0 ml of phosphate buffer solution (pH 7.4) containing 0.6 mg of *M. lysodeikticus* suspension. A decrease in suspension turbidity, attributed to the degradation to the substrate, was recorded with a spectrophotometer as a function of digestion time. The enzymatic activity was calculated from the initial slope of the course of turbidity *vs.* time. The activity of lysozyme which had been kept dissolved in solution at 30°C for 3 days and then at 37°C for various periods of time was determined for comparison.

RESULTS AND DISCUSSION

In Fig. 2, the degree of hydrogel swelling is plotted against the mol fraction of Phosmer in the hydrogels. The equilibrium swelling of the phosphate-carrying hydrogels exhibited unusual dependence on the fraction of phosphate groups. The degree of the swelling increases with the phosphate groups in the hydrogel up to a mole fraction of 20 mol% because of the hydrophilicity of phosphate groups. However, above this point, the degree of swelling decreases with an increase in the content of phosphate groups, forming a stiffer physical network structure [3]. Phosphate monomer is a very viscous liquid and only little soluble in water because the phosphate groups associate by the strong hydrogen bonds. Phosphate groups in copolymer chain act as physical cross-linking agent, which lowers the swelling ratio over the 20 mol%. When a swollen hydrogel was immersed in a lysozyme solution, the transparent hydrogel gradually turned opaque from the surface to the core. During this loading, the swelling ratio of hydrogel decreases to near 1 or 2. However, the hydrogel still has enough flexibility and a lot of water in the gel. This result suggests that lysozyme was diffusing into the hydrogel network, and a lysozyme produces the polymer complex with phosphate ions which did not associate each others in the gels. The swelling ratio of hydrogels decreased considerably in aqueous lysozyme solution by the complex formation of lysozyme to phosphate group when the phosphate group fraction was in the range of 20 to 50 mol%, shown in Fig. 2. Some phosphate groups in the range 20–50 mol% can not complex with lysozyme because of self-association of phosphate groups in the gels. The decrease of swelling ratio is probably due to the ionic cross-linking of the phosphate groups in the hydrogel by the multiple cationic groups in lysozyme

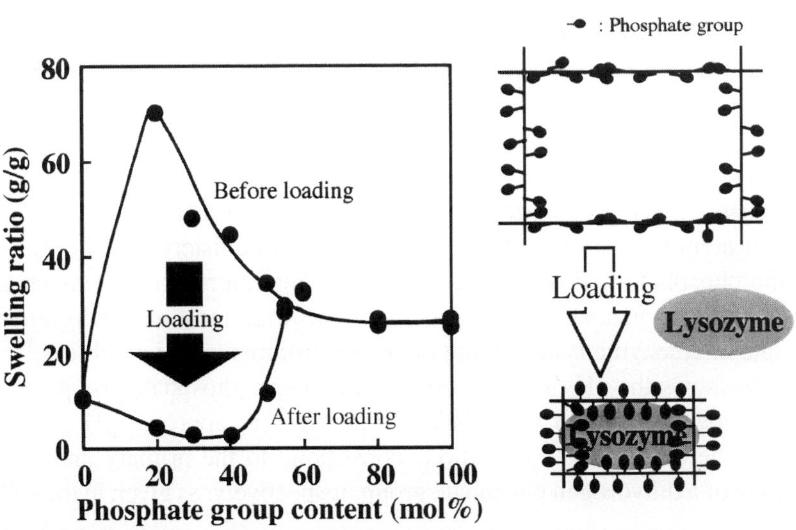

Figure 2. Swelling behavior of hydrogel before lysozyme loading and after loading, and schematic representation.

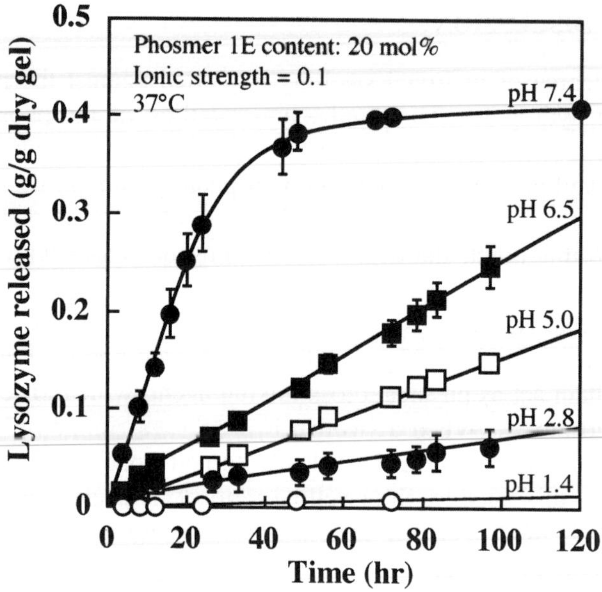

Figure 3. The amount of lysozyme released from the poly(Phosmer E1) at different pH values.

(lysozyme contains approx. 15 positively charged groups in an aqueous solution of pH 3.5 with an ionic strength of 0.15), schematically shown in Fig. 2 [16]. This phenomenon is the so-called complex formation.

Figure 3 shows the cumulative amount of lysozyme released into a buffer solution at different pH values. It can be seen that lysozyme is released into the external solution at pH 7.4. No "burst effect" was observed and overall amount of lysozyme released reached 95% of the theoretical value after 50 h. The release profile was found to synchronize with the swelling of the hydrogels associated with the osmotic expansion of the ionic network. In contrast, a negligible amount of lysozyme was detected in the external buffer solution of pH 1.4 and the gel volume remained unchanged all through the test. These experimental results clearly demonstrate that the phosphate-carrying hydrogel is practically effective as a matrix for intestine-targeting protein delivery by which protein drugs administered orally are retained within the phosphate hydrogel under the acidic condition of the stomach and are released rapidly upon contacting with the more neutral, intestinal environment. The release rate of lysozyme is depending on the environmental pH condition.

Figure 4a shows the pH titration curve of polymeric phosphate group, exhibiting step-wise dependence with two distinct transitions at approx. pH 5 and 9. It can be considered that these transitions correspond to the primary and secondary dissociation of a dihydrogen phosphate group, respectively, as given in the following scheme:

$$\mathrm{R-OPO(OH)_2} \overset{K_1}{\longleftrightarrow} \mathrm{R-OPO(OH)O^-} \overset{K_2}{\longleftrightarrow} \mathrm{R-OPO_3^{2-} + 2H^+}$$

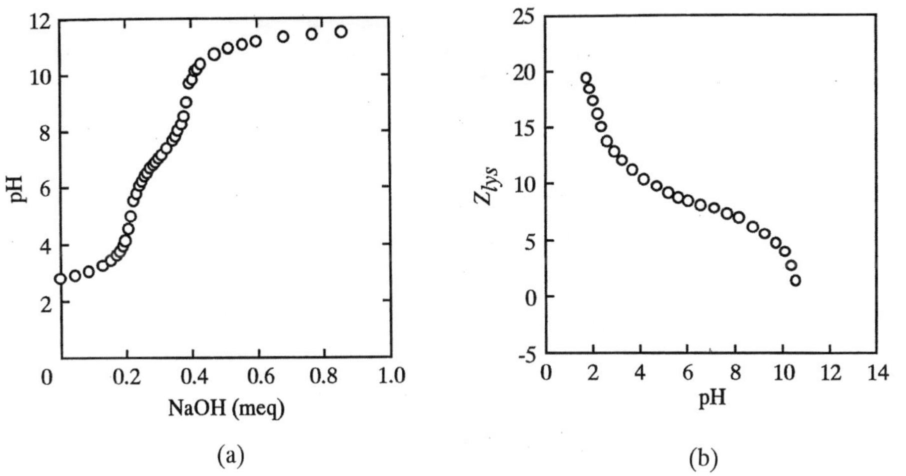

Figure 4. pH titration curves of poly(Phosmer E1) (a) and lysozyme (b).

From the titration curve, apparent dissociation constants of the polymeric dihydrogen phosphate, pK_1 and pK_2, were estimated to be 2.8 and 6.4. This result implies that the most of the phosphate groups in polymer chains possess divalent anions at pH 7.4, whereas protonation of the phosphates predominates at pH 1.4. Assuming that the activity coefficient of hydrogen ions is unity, it can be roughly calculated that 90% of POH dissociates at pH 7.4, while 4% dissociates at pH 1.4.

Figure 4b shows the pH titration curve of lysozyme. The valency of a lysozyme molecule, Z_{lys}, was converted from the amount of titrant consumed. As is seen, Z_{lys} exhibits almost zero net charge at pH 11, which is in accordance with the isoelectric point of lysozyme [16]. With an increase the acidity of medium, lysozyme gains its positive charge. The valency of lysozyme at pH 2 ($Z_{lys} = 19$) was almost identical to the value expected from the primary structure of this protein [17]. A lysozyme molecule contains 18 residues of basic amino acids and a terminal amine. This result is in good agreement with the conclusion by Tanford and Roxby [18] that lysozyme carries no inaccessible α- and γ-amines. The pH-dependent dissociations shown in Fig. 4a and 4b greatly implicated in the formation of a polyion complex as discussed below.

Figure 5 shows the release of the lysozyme from the Phosmer E1 hydrogel in solutions of different ionic strength by KCl [5]. It can be seen that the loaded lysozyme is released with increasing with the ionic strength. However, essentially none is released when the hydrogel is immersed in distilled water. This result indicates that lysozyme is fixed in the phosphate hydrogel with ionic complex formation and release dose not occur by the squeezing effect of hydrogel but the dissociation of ionic polymer complex.

Figure 6 shows the swelling ratio of hydrogel in distilled water before and after lysozyme loaded at 30°C as a function of the ethylene glycol length. The swelling

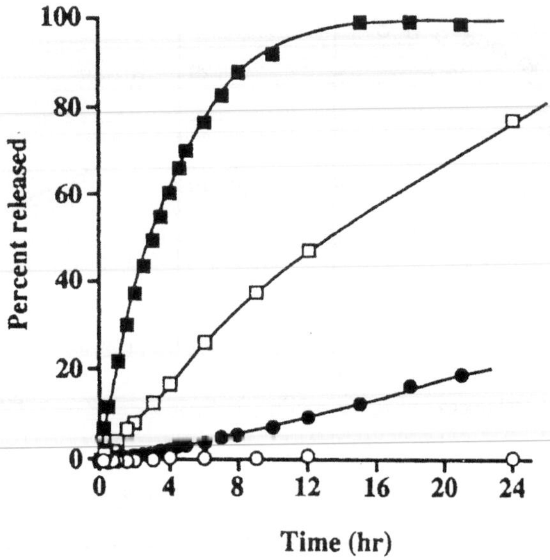

Figure 5. The amount of lysozyme release from the poly(Phosmer E1) at different ionic strength. KCl concentrations (○): 0; (●): 0.05; (□): 0.1; (■): 0.5 M.

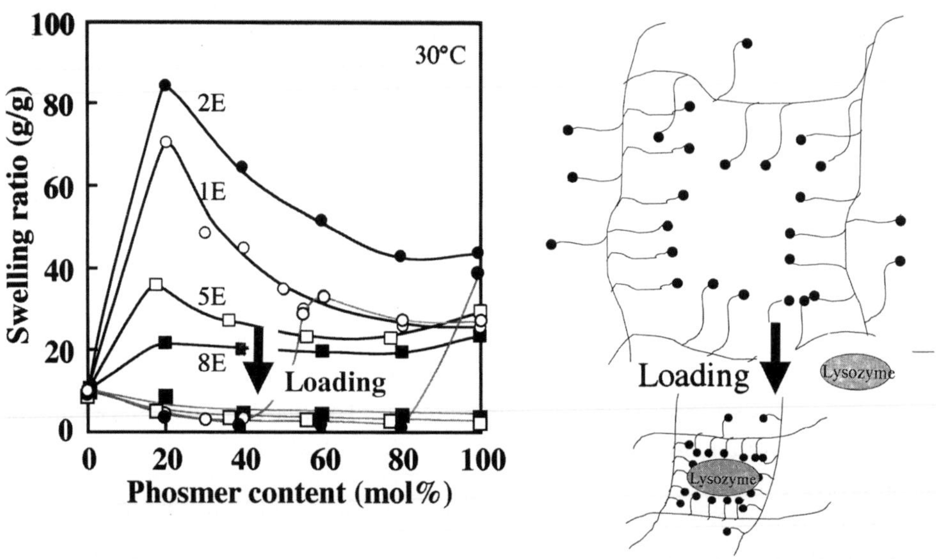

Figure 6. Swelling behavior of hydrogel having a different length of ethylene glycol chains carrying phosphate group before lysozyme loading and after loading.

ratio of hydrogel increased up to 20% of phosphate groups and then decrease. The reason of the decrease of swelling ratio over the 20 mol% is the formation of the self-association of phosphate groups (physical cross-linking) in the polymer chain, already shown in Fig. 2. The value of swelling ratio depends on the physical

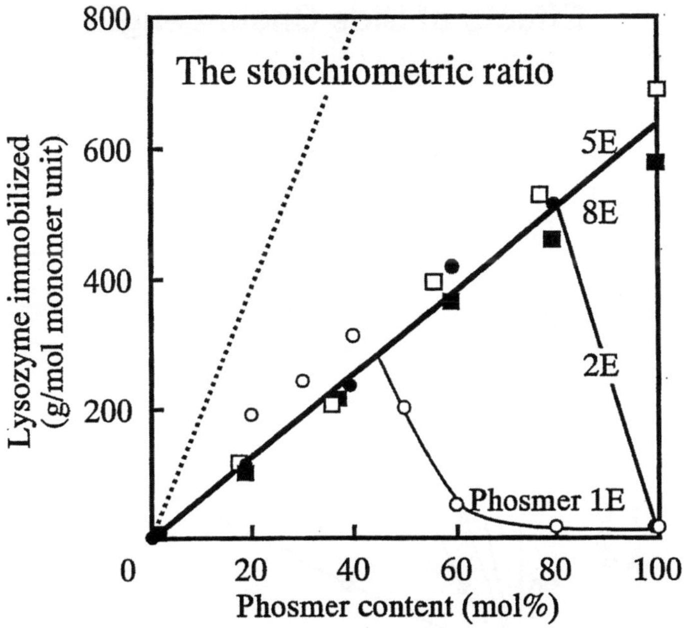

Figure 7. The amount of lysozyme loaded in the hydrogel having a different length of ethylene glycol chain carrying a phosphate group. The dotted line represents the amount of lysozyme calculated under the assumption the positive charge of lysozyme compensate the negative charge of a hydrogel.

and chemical cross-linking density, and also on the ethylene glycol length. The swelling ratio decreases after the loading of lysozyme by physical cross-linking (ionic complex formation) of lysozyme to phosphate group in the same way as for Fig. 2. From these results the amount of the lysozyme loaded in these hydrogels can be calculated.

Figure 7 shows the amount of lysozyme loaded in the hydrogels *vs.* phosphate group content in various co-polymers. The dotted line represents the amount of lysozyme calculated under the assumption that the positive charge of lysozyme compensates the negative charge of hydrogels, which can be calculated from the amount of phosphate groups. The solid line shows the amount of lysozyme loaded in the hydrogels, which have a different ethylene glycol length. Each polymer has a same amount of lysozyme loaded at the same phosphate group content in copolymers up to the maximum loading of each co-polymer. It does not depend on the length of the ethylene glycol because the amount of lysozyme loaded depends on the ionic complex formation to phosphate groups. However, the maximum value of lysozyme loaded does depend on the ethylene glycol length. As the ethylene glycol length increases the amount of lysozyme loaded increases up to 1.7 g/g dry gel (Phosmer E5). This value, however, is one-third of the stoichiometric ratio shown by the dotted line. This means the phosphate groups in the hydrogel network have a relatively high steric hindrance for the complex formation. Probably some

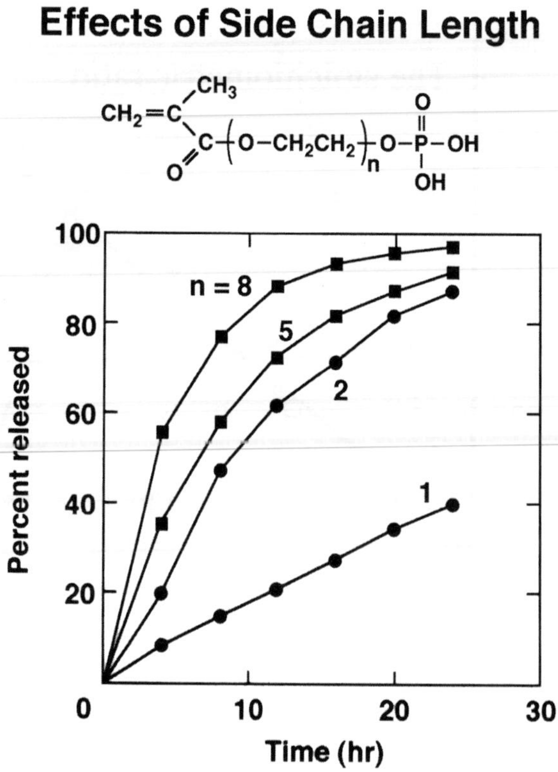

Figure 8. The amount of lysozyme released from the hydrogel having a different length of ethylene glycol chain carrying a phosphate group. The content of phosphate groups in the co-polymer hydrogel is fixed to 20 mol%.

of phosphate groups were not involved in ionic binding with lysozyme because of the hydrogel network structure reported [20]. With an increasing spacer length of the phosphate group introduced in the polymer side chain, the ratio of cationic and anionic ions complex approaches unity because of side-chain flexibility.

Figure 8 shows the amount of lysozyme released from a hydrogel with different ethylene glycol length. The initial rate of lysozyme release was varied depending on the length of the ethylene glycol chains of a network polymer and the content of the phosphate-carrying monomer unit. With increasing length of ethylene glycol chains in monomer, the rate of lysozyme release increases because the phosphate groups in the side chains may become more mobile, up to 20 mol% phosphate groups in the co-polymer hydrogels.

Table 1 shows the enzymatic activity of immobilized and free lysozyme. There was no change in the amount of lysozyme released from the hydrogel and the enzymatic activity was retained at a level similar to the activity of lysozyme which had been kept in buffer solution.

Table 1.
Enzymatic activity of immobilized and free lysozyme[a]

Phosphate-carrying monomer	Remaining activity[b] (units/mg lysozyme)
Phosmer 1E	943 ± 159
Phosmer 2E	1113 ± 103
Phosmer 5E	723 ± 244
Phosmer 8E	917 ± 68
Lysozyme free in solution	832 ± 48

[a] Initial activity = 899 ± 82 units/mg lysozyme, phosmer content = 20 mol%.
[b] Values are mean ± SD ($n = 3$).

CONCLUSIONS

A hydrogel carrying dihydrogen phosphate groups provided an anionic network in which cationic proteins were immobilized efficiently through polyion complexation. The phosphate hydrogels had a higher binding capacity for cationic proteins than hydrogels with carboxylate groups. The loaded lysozyme was released from the phosphate-carrying hydrogel. The rate of release of lysozyme depended on the pH, ionic strength of a medium and the ethylene glycol length of phosphate-carrying groups. These results suggest that the release mechanism is an ion exchange between phosphate groups with the cationic charge of lysozyme and small electrolytes. The release rate depends on the mobility of the phosphate groups in the side chain of hydrogel. It was further found that the loaded lysozyme retained its enzymatic activity during incubation at 37°C for 24 h. These results suggest that the system studied here will be useful for the controlled release of protein drugs in response to a change pH and ionic strength of external environments.

REFERENCES

1. A. S. Hoffman, *Artif. Organs* **16**, 43 (1992).
2. H. Park and K. Park, *CS Symp. Ser.* **627**, 2 (1996).
3. K. Nakamae, T. Miyata and A. S. Hoffman, *Macromol. Chem.* **193**, 983 (1992).
4. T. Miyata, K. Nakamae, A. S. Hoffman and Y. Kanzaki, *Macromol. Chem. Phys.* **195**, 1111 (1994).
5. K. Nakamae, T. Nizuka, T. Miyata, M. Furukawa, T. Nishino, K. Kato and T. Inoue, *J. Biomater. Sci. Polymer Edn.* **9**, 43 (1997).
6. A. Agyilirah and G. S. Banker, in: *Polymers for Controlled Drug Delivery*, P. J. Tarcha (Ed.), p. 39. CRC Press, Boca Raton, FL (1991).
7. A. Yamamoto, T. Taniguchi, K. Rikyuu, T. Tsuji, T. Fujita, M. Murakami and S. Muranishi, *Pharm. Res.* **11**, 1496 (1994).
8. M. Haga, K. Saito, T. Shimaya, Y. Maezawa, Y. Kato and S. W. Kim, *Chem. Pharm. Bull.* **38**, 1983 (1990).
9. H. Asada, T. Douen, M. Waki, S. Adachi, T. Fujita, A. Yamamoto and S. Muranishi, *J. Pharm. Sci.* **84**, 682 (1995).

10. Y. H. Kim, Y. H. Bae and S. W. Kim, *J. Control. Rel.* **28**, 143 (1994).
11. A. Serres, M. Baudys and S. W. Kim, *Pharm. Res.* **13**, 196 (1996).
12. L. C. Dong, Q. Yan and A. S. Hoffman, *J. Control. Rel.* **19**, 171 (1992).
13. L. C. Dong and A. S. Hoffman, *J. Control. Rel.* **15**, 141 (1991).
14. H. Katono, A. Maruyama, K. Sanui, T. Ogata, T. Okano and Y. Sakurai, *J. Control. Rel.* **16**, 215 (1991).
15. D. Shugar, *Biochim. Biophys. Acta* **8**, 302 (1952).
16. C. Tanfor and M. L. Wagner, *J. Am. Chem. Soc.* **76**, 3331 (1954).
17. R. E. Canfield, *J. Biol. Chem.* **238**, 2698 (1963).
18. C. Tanford and R. Roxby, *Biochemistry* **11**, 2192 (1972).
19. G. S. Kwon, Y. H. Bae, H. Cremers, J. Feijen and S. W. Kim, *J. Control. Rel.* **22**, 83 (1992).
20. J. M. Park, B. B. Muhoberac, P. L. Dubin and J. Xia, *Macromolecules* **25**, 290 (1992).

Gels, Genes, Grafts and Giants, pp. 301–317 (2005)

Poly(propylacrylic acid)-mediated serum stabilization of cationic lipoplexes

CHARLES Y. CHEUNG, PATRICK S. STAYTON* and ALLAN S. HOFFMAN *

Department of Bioengineering, Box 352255, University of Washington, Seattle, WA 98195, USA

Received 8 December 2003; accepted 4 May 2004

Abstract—The serum instability associated with cationic lipoplexes represents one of the major obstacles for the *in vivo* delivery of nonviral gene therapy vectors. Recently, we have shown that poly(propylacrylic acid) (PPAA), a pH-sensitive polyanionic polymer, can significantly improve the *in vitro* serum stability of DOTAP lipoplexes and enhance transfection (Cheung *et al.*, *Bioconjug. Chem.* **12**, 906 (2001)). We investigated this serum-stabilizing effect provided by PPAA using methods to identify the specific serum proteins that interact with DOTAP/DNA and DOTAP/DNA/PPAA lipoplexes and determined their modes of interaction with these lipoplexes. Studies showed that only low-density lipoprotein (LDL) caused significant decondensation of DNA from lipoplexes lacking PPAA, but that fully condensed DNA was retained within lipoplexes incorporating PPAA. Another major factor in the loss of transfection activity was due to the reduced cellular uptake of DOTAP lipoplexes upon exposure to serum, with bovine serum albumin (BSA) and high-density lipoprotein (HDL) acting as major contributors to this reduction in vector internalization. In contrast, lipoplexes containing PPAA maintained high levels of uptake into cells in the presence of these proteins. Transfection results generally concurred with the mechanistic studies, suggesting that maintaining effective cellular delivery of intact lipoplexes in the presence of serum proteins is important to retain high transfection efficiencies. These results indicate that the addition of PPAA as a ternary component in DOTAP lipoplexes can overcome some of the serum-related deficiencies encountered with these lipoplexes to provide efficient transfection.

Key words: Poly(propylacrylic acid); nonviral gene therapy; serum stability; serum–protein interactions.

INTRODUCTION

Therapeutic applications for gene therapy require effective and non-toxic delivery systems for the successful treatment of disease. High transfection efficiencies have been attained *in vivo* using attenuated viruses for DNA delivery, but significant

*To whom correspondence should be addressed. Tel.: (1-206) 543-9423. Fax: (1-206) 543-6124. E-mail: hoffman@u.washington.edu or stayton@u.washington.edu

concerns over the safe application and potential immunogenic response may limit the utility of these delivery systems [1–4]. Cationic lipids represent a promising alternative to viruses for DNA delivery. Although safer than viruses, the low transfection efficiency associated with most cationic lipid formulations in the presence of serum has limited their appeal for *in vivo* applications [5–7]. Also, the potential for opsonization of cationic lipid/DNA complexes (i.e., cationic lipoplexes) by plasma proteins and rapid clearance by the reticuloendothelial system (RES) upon intravenous administration may reduce particle circulation times [8–10]. Indeed, complement activation can occur with nonviral complexes formed at high cation/anion (+/−) charge ratios [8, 9].

Exposure of lipoplexes to serum has been shown to cause aggregation of the complexes [11], as well as impart a negative surface charge to these particles [12]. These changes in particle characteristics may have an adverse effect on transfection due to decreased association of particles with cell membranes and subsequently reduced cellular entry. For example, serum-protein interactions with cationic lipoplexes reduced the ability of these lipoplexes to agglutinate red blood cells (RBCs), most likely by preventing lipoplex association with the RBCs [13].

Attempts have been made to identify the individual serum components that may interact with lipoplexes and causing reduction in DNA delivery and transfection. Separation of serum into positive and negative components by ion-exchange chromatography showed that inhibition in transfection was mainly due to the negative components within serum [5]. Zelphati *et al.* showed that the nuclear localization of fluorescently-labeled oligodeoxynucleotides decreased significantly in the presence of either 2.5% FBS or BSA, whereas low-density lipoprotein (LDL), high-density lipoprotein (HDL), α_2-macroglobulin (A2M), fibrinogen, heparin, IgG and β_2-glycoprotein had either minimal or no effect on nuclear uptake [14]. Gel electrophoresis has also been used to visualize protein binding onto nonviral complexes. Separation and analysis of plasma proteins bound to DNA/transferrin–PEI complexes by SDS–PAGE led to speculation that the major proteins bound to these particles may include IgM, fibronectin, fibrinogen and complement C3 [10]. SDS–PAGE applied to LPD particles showed higher levels of protein binding to complexes containing DOPE compared to those containing cholesterol, with especially high levels of binding to albumin [12].

Investigators have attempted to overcome the serum-instability of nonviral vectors by altering formulation compositions as well as changing formulation parameters. So-called ‘stealth’ liposomes have been constructed to reduce the *in vivo* aggregation of lipoplexes as well as reduce protein binding by conjugating hydrophilic polymers, such as poly(ethylene glycol) (PEG) and N-(2-hydroxypropyl)methacrylamide (HPMA), to lipids [10, 15]. These changes would effectively reduce blood clearance and increase circulation times for these complexes, allowing for the potential for increased delivery of lipoplexes to target tissues and organs. However, conjugation to hydrophilic polymers can reduce the cellular internalization of certain complexes [16]. In addition, altering the helper lipid composition of lipoplexes from DOPE to cholesterol helped to maintain complex integrity in the presence of serum [7, 11]. Preparation of lipoplexes at higher charge ratios can overcome the serum-inhibition associated with transfection with DC-Chol [5], DOTAP/DOPE [5] and transferrin–DOTAP lipoplexes [17], but as stated previously, this may induce greater clearance of these lipoplexes from the RES, as well as increase the cytotoxicity of these formulations. Finally, physical additives to lipoplex formulations such as transferrin, which aids in the endocytosis of lipoplexes [17], were capable of providing serum-stabilization for DOTAP lipoplexes and allow for more effective transfection.

Recently, synthetic, pH-sensitive polymers, such as poly(propylacrylic acid) (PPAA), have been utilized to enhance the cytoplasmic delivery of macromolecular therapeutics [18–20]. PPAA is a highly anionic polymer containing titratable carboxyl moieties under acidic conditions. The unique structure of PPAA allows the initially hydrophilic polymer molecules to undergo conformation changes and become more hydrophobic upon cellular entry and trafficking within acidic endosomes, which may lead to membrane destabilization and release of macromolecules from endosomal compartments into the cell cytoplasm. PPAA has been shown to enhance transfection efficiencies of DOTAP lipoplex formulations prepared at low charge ratios in NIH3T3 murine fibroblast cells. Interestingly, efficient *in vitro* transfection in the presence of serum was achieved with DOTAP lipoplexes containing PPAA, with high levels of transfection in the presence of up to 50% serum [21]. The mechanism for the serum stability conferred by PPAA to DOTAP has not yet been elucidated and was, therefore, the focus of this investigation.

MATERIALS AND METHODS

Materials

PPAA (molecular mass 61 kDa, Fig. 1) was synthesized according to the protocols published previously [18, 21]. Dioleyltrimethylammonium propane (DOTAP) was purchased from Roche Molecular Biochemicals (Indianapolis, IN, USA). pCMVβ plasmid DNA (Clontech, Palo Alto, CA, USA) was amplified in TOP10 *E. coli* (Invitrogen, Carlsbad, CA, USA) and purified by anion-exchange chromatography (Qiagen, Valencia, CA, USA). TO-PRO-1 was purchased from Molecular Probes

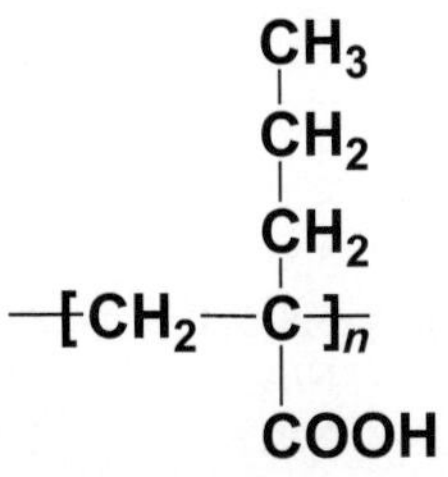

Figure 1. Chemical structure of poly(propylacrylic acid).

(Eugene, OR, USA). Serum proteins (BSA, A2M, IgG and transferrin) were purchased from Sigma (St. Louis, MO, USA), whereas lipoproteins (HDL and LDL) were purchased from Calbiochem (San Diego, CA, USA). [^{35}S]dCTP used for nick translation was obtained from NEN (Boston, MA, USA). Dulbecco's Modified Eagle Medium (DMEM) and fetal bovine serum (FBS) were purchased from Gibco (Carlsbad, CA, USA).

Fluorescence techniques for probing DNA condensation

DNA binding studies using TO-PRO-1 were performed to investigate the stability of plasmid DNA within lipoplex structures upon exposure of the lipoplexes to PPAA and/or serum proteins. Ternary physical mixtures of lipoplexes containing DOTAP, pCMVβ plasmid DNA and PPAA were prepared for condensation studies. DOTAP was first mixed with the pCMVβ plasmid DNA in HEPES-buffered saline (HBS, 20 mM HEPES, 150 mM NaCl, pH 7.4) and incubated for 30 min, subsequently followed by the addition of PPAA and incubation for an additional 30 min. DOTAP/DNA control lipoplexes were prepared in a similar manner, with HBS added to the formulation instead of PPAA. DOTAP/DNA control lipoplexes were prepared at different charge ratios (+/−) by varying the amount of DOTAP added to a fixed quantity of DNA (1.6 μg/ml). DOTAP/DNA/PPAA lipoplexes were prepared by varying the amount of PPAA added to fixed quantities of DOTAP (10 μg/ml) and DNA (1.6 μg/ml). 1 μl TO-PRO-1 was added to each sample and dye binding occurred for 10 min, with subsequent addition of the appropriate media condition. The media conditions tested in these condensation studies consisted of either serum-free DMEM, DMEM/10% FBS, or DMEM supplemented with individual serum proteins, which are listed in Table 1. 350 μl media was added to 150 μl of lipoplex formulation, to maintain the same ratio of particle solution to media used in transfections. The proteins chosen for testing represent the major proteins present within human serum [14, 22], utilizing either human or bovine sources for these proteins. Lipoplexes were exposed to protein levels that were estimated to be the average levels present in 10% serum [14]. Lipoplex particles

Table 1.
Serum proteins and concentrations utilized in experiments

Protein	Molecular mass (kDa)	p*I*	Physiological concentration (mg/ml)	Concentration in media (10% serum) (mg/ml)
BSA	66.3	4.7	35–55	2.4
α_2-Macroglobulin	725	5.4	1.5–4.2	0.1
Transferrin	76.5	5.5	2–4	0.3
IgG	160	5.8–7.3	8–18	1
LDL	N/A	N/A	1.9–6	0.5
HDL	N/A	N/A	2.9–7.7	0.5

Values taken from Ref. [22].

were incubated with media for up to 4 h. Fluorescence measurements were taken using a Perkin-Elmer LS-5B Luminescence Spectrometer (Shelton, CT, USA) at $\lambda_{ex} = 509$ nm and $\lambda_{ex} = 533$ nm with 5 nm excitation and emission slit widths. Fluorescence measurements were taken for all samples at time points ranging from 0–4 h. Controls consisted of TO-PRO-1 added to either DOTAP alone (negative) or DNA alone (positive) in the presence of the appropriate media condition tested (serum-free, 10% FBS, or supplemented with the appropriate serum protein). The fluorescence data was standardized by subtracting the background fluorescence from the negative controls, and then normalized to the fluorescence of the positive control.

Cellular uptake of lipoplex particles

The cellular uptake of DOTAP lipoplexes containing radiolabeled plasmid DNA in the presence of various media conditions was examined. 1 μg pCMVβ plasmid DNA was labeled with [^{35}S]dCTP by nick translation using the Nick Translation System (Invitrogen). Purification of labeled plasmid DNA from unincorporated nucleotides and enzymes was performed by centrifugation with Microspin S-400 HR columns (Amersham Pharmacia, Piscataway, NJ, USA). Uptake studies were performed by incubating NIH3T3 cells with either DOTAP/DNA or DOTAP/DNA/PPAA lipoplexes incorporating ^{35}S-labeled plasmid DNA for 4 h at 37°C under different media conditions. Following the incubation period, the cells were washed twice with PBS, then trypsinized from the substrate surface. The trypsinized cells were centrifuged to recover the cells, then lysed with 0.3 M NaOH/0.1% Triton X-100 for 15 min. The cell lysates were then neutralized with 0.9 M HCl, re-centrifuged to pellet any remaining insoluble components and the level of radioactivity of the supernatant was determined by liquid scintillation counting. Concomitantly, the protein content within cell lysates was determined using the BCA assay (Pierce, Rockford, IL, USA). The level of radioactivity recovered within the cell lysates was normalized to both the total amount of radioactivity applied to the cells and the total protein content.

In vitro *transfection*

NIH3T3 murine fibroblast cells were seeded into 12-well tissue culture plates the day prior to transfection experiments at 7.5×10^4 cells/well. Transfection mixtures were prepared at theoretical charge ratios of 1.3 (+/−) for DOTAP/DNA/PPAA lipoplexes and 1.9 (+/−) for DOTAP/DNA lipoplexes, which represented the particles that provided the optimal transfection efficiency levels for this system [21]. All formulations were tested in triplicate. For cell transfections, 0.7 ml serum-free DMEM, DMEM 10% FBS or DMEM containing individual serum proteins and 0.3 ml of solutions containing ternary mixtures of DOTAP/DNA/PPAA or binary control mixtures of DOTAP/DNA were added to each well. The cells were incubated with transfection mixtures for 4 h at 37°C, after which the media was removed and replaced with 2 ml DMEM/10% FBS. The cells were then incubated for an additional 48 h, then lysed to recover the β-galactosidase produced. The β-galactosidase activity from the cell extracts was quantified using *o*-nitrophenyl-β-D-galactopyranoside (ONPG) (Promega, Madison, WI, USA) as the enzymatic substrate. Concurrently, the total protein produced by the cells was determined by measuring the protein content in the cell extracts by the BCA method. Gene expression levels were, therefore, normalized to total protein content.

RESULTS

DNA condensation within DOTAP lipoplexes

The ability of DOTAP to condense plasmid DNA has been tested with TO-PRO-1, a cyanine dye with a cationic side chain that binds to double-stranded DNA with high affinity. Low intrinsic levels of fluorescence with are observed when the dye is not bound to nucleic acids. However, binding of TO-PRO-1 to DNA results in a significant enhancement in fluorescence. TO-PRO-1 was used to investigate the level of condensation for DNA within DOTAP lipoplexes upon exposure to PPAA under a variety of media conditions. The extent of DNA condensation for both DOTAP/DNA and DOTAP/DNA/PPAA lipoplexes was initially tested in serum-free conditions. Decondensation of DNA from lipoplexes would be represented by an increase in TO-PRO fluorescence as a greater fraction of DNA would become accessible for binding by the fluorophore. The fraction of DNA remaining in a condensed state and retained within lipoplex particles was examined for lipoplexes prepared at charge ratios ranging from 0–2.5 (+/−) over a 4-h time-course, which represented the incubation period used in transfection experiments (Fig. 2). The results in Fig. 2 show that both types of lipoplexes were capable of condensing DNA in the absence of serum. Maximum condensation of DOTAP/DNA/PPAA lipoplexes occurred at a lower charge ratio (>1.0) than for DOTAP/DNA formulations (>2.0), which was probably due to the higher quantity of DOTAP present within the former lipoplexes. No significant decondensation of DNA from lipoplexes was evident over the 4-h incubation period with either type of formulation tested. Condensation experiments with lipoplex formulations were also performed in the presence of media containing 10% FBS (Fig. 3). Significant levels of decondensation of DNA from most DOTAP/DNA lipoplexes occurred throughout the duration of the 4-h incubation with serum-containing media, indicating that lipoplex interactions with serum components was promoting the partial release of DNA from these particles. Only DOTAP/DNA lipoplexes prepared at high charge ratios showed minimal levels of decondensation in the presence of serum. In contrast, no significant amount of decondensation was observed with DOTAP/DNA/PPAA formulations prepared at any charge ratio over the time period tested. These results indicated that PPAA was stabilizing the DOTAP lipoplexes, allowing the DNA to be retained within the lipoplex structure enhance the delivery of intact DNA to cells.

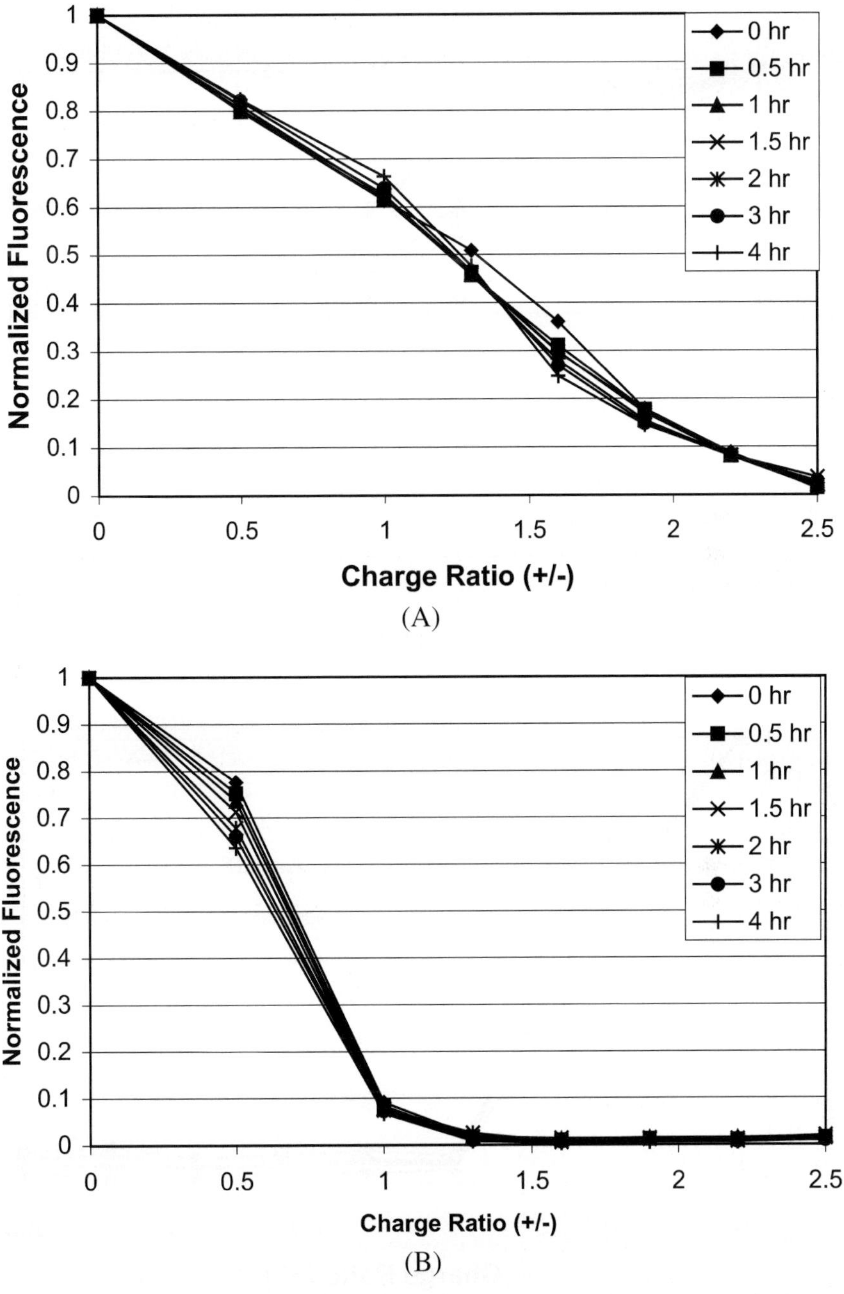

Figure 2. Condensation of (A) DOTAP/DNA and (B) DOTAP/DNA/PPAA lipoplexes in the presence of TO-PRO-1 in serum-free DMEM. The extent of DNA condensation was determined by measuring TO-PRO-1 fluorescence in the presence of the different lipoplex formulations. DOTAP lipoplexes were prepared at theoretical particle charge ratios ranging from 0 to 2.5 (+/−). Fluorescence intensity measurements were taken for a 0–4 h time-course. Background fluorescence from DOTAP-negative controls were subtracted from the fluorescence data, which was then normalized to the fluorescence obtained from DNA-positive controls in serum-free conditions.

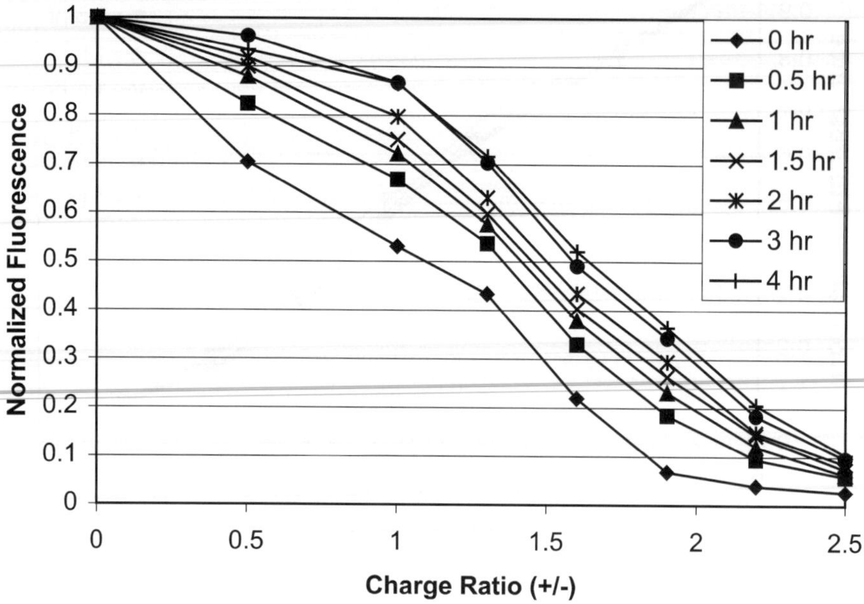

(A)

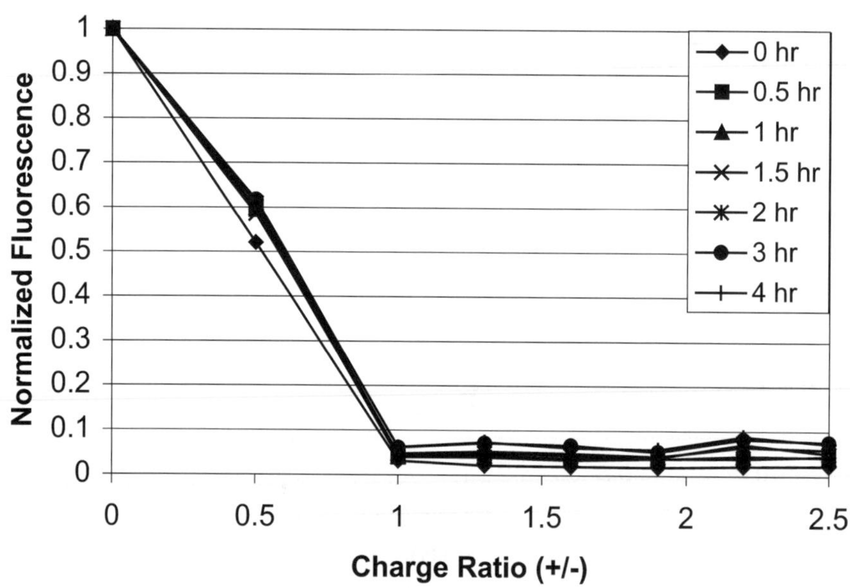

(B)

Figure 3. Decondensation of (A) DOTAP/DNA and (B) DOTAP/DNA/PPAA lipoplexes in the presence of TO-PRO-1 in DMEM/10% FBS. Conditions tested and normalization of the data were the same as in Fig. 2, except that lipoplexes were exposed to DMEM supplemented with 10% FBS.

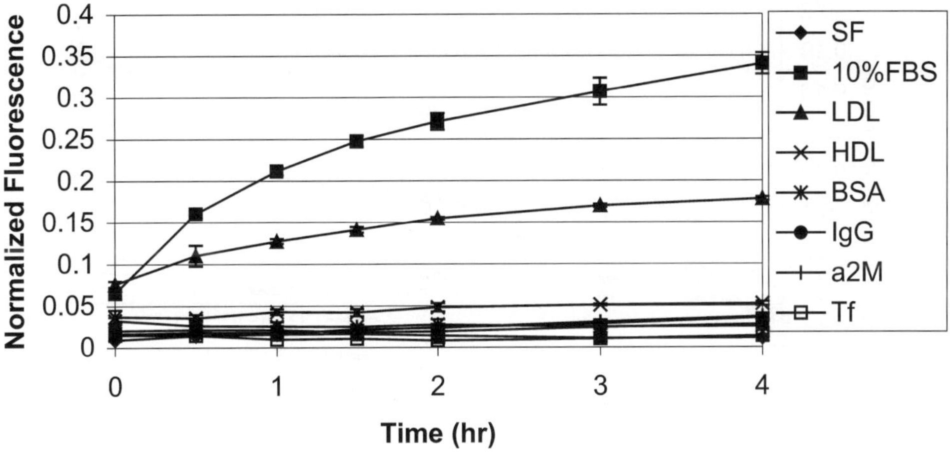

Figure 4. Decondensation of DOTAP/DNA 1.9 (+/−) lipoplexes with TO-PRO-1 in DMEM supplemented with individual serum proteins. Serum proteins tested included LDL (▲), HDL (×), BSA (∗), IgG (●), α_2-macroglobulin (|) and transferrin (□). The extent of DNA condensation for the DOTAP/DNA 1.9 lipoplexes was measured over a 4-h time-course. Decondensation of DOTAP/DNA 1.9 lipoplexes was also tested in serum-free DMEM (SF, ◆) and DMEM supplemented with 10% FBS (10% FBS, ■) to show the minimum and maximum levels of decondensation expected at each time point. Background fluorescence from DOTAP-negative controls were subtracted from the fluorescence data, which was then normalized to the fluorescence obtained from DNA-positive controls at the varying media conditions. Data points represented samples tested in triplicate.

The decondensation of DNA from DOTAP/DNA lipoplexes in 10% FBS can be attributed to the interaction with specific factor(s) within serum, resulting in lipoplex instability. We attempted to identify the individual components within serum that may be interacting with these lipoplexes. The proteins tested represent those that were most abundantly found in human serum. Table 1 indicates the proteins tested and concentrations used in experiments. The levels of proteins tested were adjusted to the average levels expected in 10% serum. Initial studies with individual serum proteins examined their effect on the condensed state of DNA only in DOTAP/DNA control lipoplexes, as no decondensation was evident with DOTAP/DNA/PPAA lipoplexes in 10% FBS. DOTAP/DNA 1.9 (+/−) lipoplexes, which provided the highest level of serum-free transfection [21], were utilized in preliminary tests. Figure 4 shows that a high degree of DNA decondensation from DOTAP/DNA lipoplexes was observed only in the presence of LDL. The level of decondensation attained with LDL (approx. 18%) represented approximately half the decondensation observed in the presence of 10% FBS (approx. 34%). Background levels of decondensation were observed for all other proteins tested.

Further condensation experiments focused on lipoplex stability in the presence of LDL. Lipoplex stability was maintained with DOTAP lipoplexes containing PPAA when exposed to 10% serum, so the ability of PPAA to stabilize lipoplexes upon exposure to LDL was examined (Fig. 3). Figure 5 shows condensation results for both DOTAP/DNA 1.9 (+/−) and DOTAP/DNA/PPAA 1.3 (+/−) lipoplexes, ratios

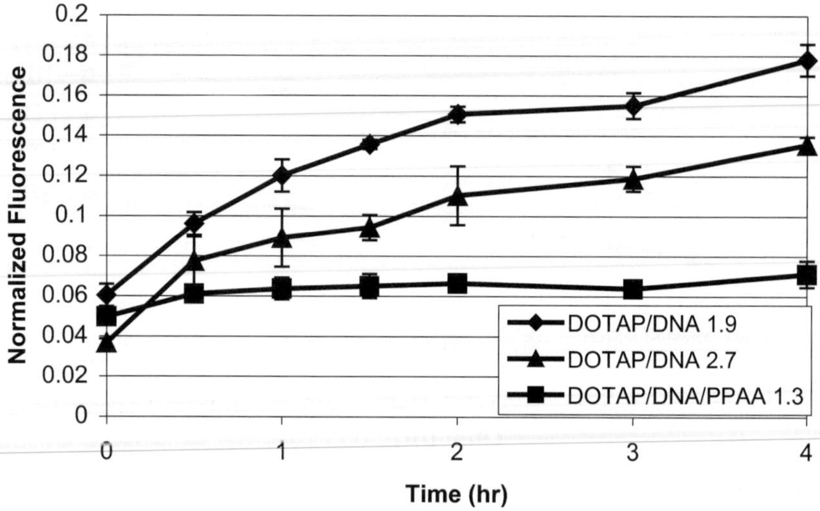

Figure 5. Decondensation of DOTAP/DNA 1.9 (+/−) (◆), DOTAP/DNA 2.7 (▲) and DOTAP/DNA/PPAA 1.3 (■) lipoplexes with TO-PRO-1 in DMEM supplemented with LDL. DOTAP/DNA 2.7 control lipoplexes contained the same quantities of DOTAP (10 μg) and DNA (1.6 μg) as found within DOTAP/DNA/PPAA 1.3 lipoplexes, allowing for more appropriate comparison with DOTAP/DNA/PPAA 1.3 lipoplexes and the effects of PPAA on decondensation. Decondensation of DNA from these lipoplexes was measured over a 4-h time-course. Fluorescence measurements were normalized in the same manner as in Fig. 4. Data points were tested in triplicate.

that provided optimal transfection efficiencies from previous studies [21], in the presence of DMEM containing LDL. Little decondensation of DNA was observed with DOTAP/DNA/PPAA 1.3 lipoplexes, whereas significant decondensation of DNA from DOTAP/DNA 1.9 lipoplexes was evident in the presence of LDL. Decondensation of DNA increased over time with DOTAP/DNA 1.9 lipoplexes, while the level of DNA condensation remained within background levels with DOTAP/DNA/PPAA 1.3 lipoplexes in the presence of LDL. An additional DOTAP control formulation, DOTAP/DNA 2.7, which contained the same quantity of DOTAP present in DOTAP/DNA/PPAA 1.3 lipoplexes, was tested to determine if the retention of condensed DNA was due to the increased amount of DOTAP present in lipoplexes containing PPAA. DOTAP/DNA 2.7 lipoplexes showed a significant level of decondensation as well, although values were slightly lower than levels attained with DOTAP/DNA 1.9 lipoplexes.

Quantitation of radiolabeled lipoplex uptake

The cellular internalization of lipoplexes was quantitated utilizing ^{35}S-radiolabeled plasmid DNA incorporated into the lipoplexes. Lipoplex uptake was determined for both DOTAP/DNA 1.9 and DOTAP/DNA/PPAA 1.3 lipoplexes in the presence of various media conditions. As in condensation experiments, media containing individual serum proteins were tested at protein concentrations present in 10% serum. Figure 6 illustrates varying effects for the different proteins tested. An

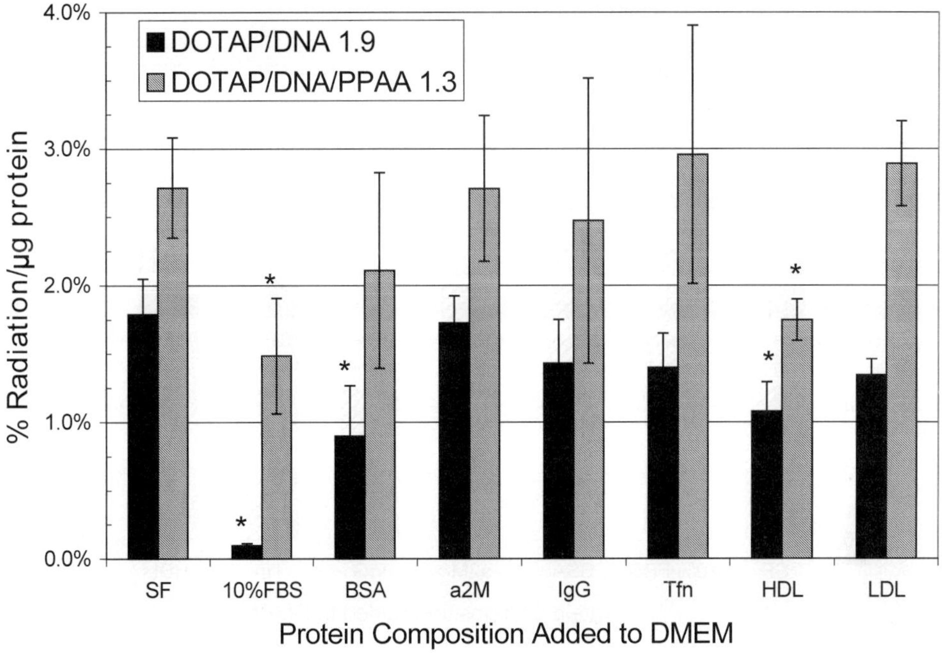

Figure 6. Cellular uptake of DOTAP/DNA 1.9 and DOTAP/DNA/PPAA 1.3 lipoplexes containing ^{35}S-labeled pCMVβ plasmid DNA into NIH3T3 cells in serum-free DMEM, DMEM/10% FBS, or DMEM supplemented with individual serum proteins (see Table 1). Cells were exposed to radiolabeled lipoplexes for 4 h prior to quantitation of uptake. Uptake values were normalized both to the total radioactivity applied to cells and to total protein content per sample. Samples were tested in triplicate. An asterisk represents statistically significant reduction in uptake (P-value <0.05) of the lipoplex formulation at the specified media condition when compared to uptake for the same formulation in serum-free conditions.

approx. 95% reduction in particle uptake was observed with DOTAP/DNA 1.9 lipoplexes incubated in 10% FBS when compared to serum-free conditions, whereas only an approx. 45% reduction is observed for DOTAP/DNA/PPAA 1.3 lipoplexes. This indicates that serum proteins are interfering with DOTAP lipoplex entry into cells, but that the addition of PPAA into DOTAP formulations significantly enhances uptake. Uptake studies in the presence of individual serum proteins indicated that nearly all proteins tested caused a reduction in DOTAP/DNA lipoplex uptake, with statistically significant decreases due to BSA and HDL ($P < 0.03$, t-test). DOTAP/DNA/PPAA lipoplexes also show a slight reduction in uptake, due mainly to HDL, but the loss was not as severe as with DOTAP/DNA lipoplexes ($P < 0.02$, t-test).

Transfection of DOTAP lipoplexes exposed to serum proteins

Transfection experiments were performed in the presence of individual serum proteins to determine if the observations from DNA condensation and uptake experiments translated to comparable functional effects. Transfections were performed

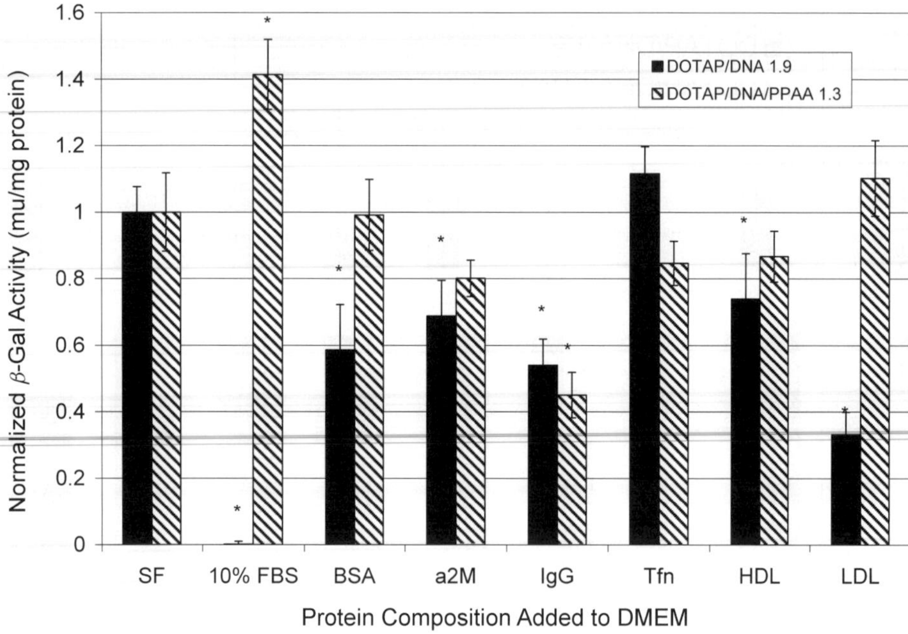

Figure 7. β-Galactosidase gene expression levels following transfection of NIH3T3 cells with DOTAP/DNA 1.9 and DOTAP/DNA/PPAA 1.3 lipoplexes in serum-free DMEM, DMEM/10% FBS, or DMEM supplemented with individual serum proteins. Transfections with DOTAP lipoplexes proceeded for 4 h, and gene expression levels were assessed 48 h post-transfection. The β-gal activity for each lipoplex formulation at the specified media condition was normalized to the level of gene expression obtained for the same formulation in serum-free conditions. All formulations were tested in triplicate at each condition. An asterisk represents statistically significant variation in gene expression (P-value <0.05) for the lipoplex formulation at the specified media condition when compared to the level of gene expression for the same formulation in serum-free conditions.

with either DOTAP/DNA 1.9 or DOTAP/DNA/PPAA 1.3 lipoplexes. These experiments were performed with serum-free DMEM supplemented with individual serum proteins at concentrations that would mimic conditions expected using 10% serum. Results in Fig. 7 were normalized to the levels of transfection observed for the same lipoplexes in serum-free conditions. The presence of all proteins, except transferrin, led to a statistically significant reduction in transfection efficiency of DOTAP/DNA 1.9 lipoplexes ($P < 0.05$, t-test) compared to serum-free levels, with the greatest reductions observed with LDL, IgG and BSA (>40% reduction). In contrast, only IgG reduced gene expression levels with DOTAP/DNA/PPAA 1.3 lipoplexes in a statistically significant manner ($P < 0.01$, t-test).

DISCUSSION

The *in vivo* efficacy of cationic liposomes to deliver nucleic acid therapeutics requires the retention of activity in the presence of serum proteins. Many studies have looked at the ability of different nonviral vector formulations to transfect cells in the presence of serum [23–25], but few studies have attempted to investigate the mechanism behind the reduced efficacy observed with most nonviral formulations. Previous studies have shown that DOTAP lipoplexes were ineffective at transfection in the presence of even low levels of serum [7, 26]. However, the introduction of a ternary component to DOTAP lipoplexes, such as pH-sensitive anionic polymers or transferrin, have provided these lipoplexes with the ability to transfect cells effectively in the presence of high concentrations of serum [17, 21]. The work presented in this study showed that individual serum proteins have varying effects on DOTAP lipoplex stability and efficacy for plasmid DNA delivery.

Lipoplex instability may be a major cause of serum inhibition of transfection. Serum protein binding may influence the structural integrity of cationic lipoplexes. Retaining DNA in a condensed state within the lipoplex is imperative for the intracellular delivery of functional plasmid DNA. Decondensation of DNA may explain the inability of some lipoplexes to transfect cells in the presence of serum, since uncoupling of the DNA from cationic particles may make the DNA more susceptible to enzymatic cleavage and degradation in circulation with systemically-delivered vectors [12]. Vector decondensation, by individual components such as cell surface proteoglycans (heparan sulfate, chondroitin sulfate and dextran sulfate) [27, 28], or in the presence of complete serum [11], has been shown previously, but the specific soluble components within serum that may cause decondensation were not identified. Condensation experiments in this study showed that of all the serum proteins tested, only LDL causes significant release of the DNA from DOTAP/DNA lipoplexes. However, the DNA was retained within the complex upon addition of PPAA to DOTAP lipoplexes in the presence of LDL. Transfection results in the presence of LDL corroborate the condensation results, indicating the importance of lipoplex stability in the presence of serum proteins.

Other investigators have shown that serum lipoproteins significantly decreased transfection efficiencies for DC-Chol lipoplexes in all cell lines tested, and that the use of delipidated serum, which only contained 5–15% of the serum lipoproteins present in normal serum, abolished the inhibitory effects observed in transfection [29]. LDL consists of a hydrophobic core of triglycerides and cholesterol esters, surrounded by cholesterol, phospholipids and apolipoproteins [30]. Serum lipoproteins have been shown to destabilize liposomal membranes, resulting in release of liposomal content [31]. Potential instability of cationic lipoplexes exposed to LDL may be due to phospholipid interaction with the cationic lipoplexes, similar to interactions with cell membranes. Zelphati *et al.* postulated that anionic phospholipids in vesicular membranes can catalyze the release of ODNs from cationic complexes after cellular internalization [32]. This may lead to destabilization of the lipid bilayer structure by proposed mechanisms such as lipid flip–flop [28], resulting in DNA release from these complexes. The increased cationic charge of the DOTAP/DNA 1.9 (+/−) lipoplexes compared to the DOTAP/DNA/PPAA 1.3 (+/−) lipoplexes may also contribute to increased interaction between the anionic phospholipids within

LDL, leading to further destabilization. HDL did not cause significant decondensation of DNA from DOTAP/DNA lipoplexes. The unique decondensation observed with LDL, yet not with HDL, may be due to greater electrostatic interaction between the components of the DOTAP lipoplexes and LDL. The higher lipid content in LDL (78% for LDL *vs.* 50% for HDL) may allow for increased interactions between the cationic DOTAP and anionic phospholipids in LDL, while greater attraction may exist between DNA and apolipoprotein B-100, the predominant protein found in LDL, which is more cationic than the apolipoproteins found in HDL (mainly apo A-I and A-II) [33]. These competing attractions may lead to greater destabilization of the DNA from lipoplexes. Although significant destabilizing effects were observed, LDL itself only accounted for approximately half of the decondensation observed with DOTAP/DNA lipoplexes in the presence of 10% FBS. As only minimal levels of decondensation were observed with the other serum proteins tested, the residual decondensation observed in the presence of FBS is most likely due to decondensation caused by other proteins or factors present in whole serum that were not tested.

Protein interactions with lipoplex particles may also have a substantial effect on the cellular uptake of lipoplexes. Cellular uptake studies of DOTAP lipoplexes have shown that exposure to serum reduced cellular binding of the particles compared to serum-free conditions [21, 26]. Overall, higher levels of uptake were maintained with DOTAP/DNA/PPAA lipoplexes compared to DOTAP/DNA lipoplexes for each media condition tested. Uptake of DOTAP/DNA/PPAA lipoplexes in all media conditions was equivalent to the uptake in serum free conditions with the exception of HDL, which may have accounted for the reduced uptake observed in 10% FBS. Lower levels of DOTAP/DNA lipoplex uptake were observed with most proteins tested, with statistically significant reductions in the presence of BSA and HDL. This reduction in uptake with BSA and HDL cannot completely account for the extreme reduction in lipoplex internalization observed in 10% FBS, indicating that other untested factors in serum may also reduce DOTAP/DNA internalization. Therefore, the addition of PPAA to DOTAP lipoplexes may reduce protein interactions with these lipoplexes, thereby allowing for more effective cellular entry of these lipoplexes.

Transfection results were in general agreement with the trends observed from DNA condensation and lipoplex uptake studies with most of the serum proteins tested. Lower transfection efficiencies with DOTAP/DNA 1.9 lipoplexes may be attributed to the reduced condensation (with LDL) or cellular uptake (with BSA, HDL) of these lipoplexes. Interestingly, no reduction in uptake of DOTAP/DNA lipoplexes was observed in the presence of LDL, even though significant decondensation of DNA was observed with these lipoplexes. A possible explanation may be that the DNA is only partially released from these lipoplexes, which still allows the DNA to enter into cells, albeit in a partially decondensed state which is ineffective at transfecting cells. High levels of transfection were maintained with DOTAP/DNA/PPAA 1.3 lipoplexes in the presence of LDL, BSA and HDL, which correlated well with condensation and uptake results for these complexes. In addition, the presence of transferrin did not alter the condensation, uptake or transfection of either DOTAP/DNA or DOTAP/DNA/PPAA lipoplexes. However, no correlation between condensation and/or uptake and the lower levels of transfection could be determined with either lipoplex formulation in the presence of A2M and IgG. This discrepancy in results could not be accounted for by differences in the physical characteristics of A2M and IgG and the other proteins tested, like the molecular weight or pI of the protein. This type of inconsistency between particle uptake and transfection efficiency has been observed in other studies, where exposure of amphiplex vectors to BSA increased vector uptake but inhibited gene expression [27]. One possible explanation for this inconsistency may involve the need for the proper and timely decoupling of the DNA from cationic carrier after cytoplasmic delivery, which may be important for nuclear entry and transcription of the DNA. Microinjection of cationic lipid/DNA complexes into cell nuclei exhibited low transfection efficiencies, indicating that DNA uncoupling from lipoplexes is required prior to entry into the nucleus for efficient gene expression [34]. Potential association of A2M and IgG to DOTAP lipoplexes may inhibit transfection by interfering with the latter stages of lipoplex trafficking and processing, such as the intracellular decoupling of DNA from the lipoplexes, rather than the early steps in trafficking that were tested in this study. These results show that PPAA has the ability to assist in the early steps in transfection, promoting greater cellular delivery of DOTAP lipoplexes, as well as maintaining greater lipoplex stabilization in the presence of serum proteins. However, transfection efficiency cannot solely be predicted by the occurrences within these early stages of transfection. Later stages within the transfection process, such as complex decondensation and nuclear localization, are also important determinants influencing transfection efficiency in the presence of serum proteins.

CONCLUSIONS

These studies showed that serum proteins reduce transfection efficiencies with DOTAP lipoplexes by interfering with the intracellular delivery of intact lipoplexes. Protein association with DOTAP lipoplexes can result in the loss of lipoplex structural stability, as well as reduce the cellular internalization of lipoplexes. DOTAP/DNA lipoplexes appeared more susceptible to association by various serum proteins and resulted in a loss in transfection efficiency, whereas DOTAP/DNA/PPAA lipoplexes maintained high levels of transfection by minimizing the inhibitory effects of the serum proteins tested. Therefore, the addition of PPAA enhanced the internalization and stabilization of DOTAP lipoplexes. However, these two parameters alone were not completely capable of predicting transfection efficacy in the presence of individual serum proteins. Protein association with lipoplexes may also strongly influence other latter steps within the lipoplex trafficking pathway such as complex uncoating and/or nuclear entry, steps in which the benefits of PPAA incorporation remain undetermined. Further investigations into these mechanisms will provide a more complete picture of the effects of protein association with lipoplexes on transfection.

Acknowledgements

The authors gratefully acknowledge the NIH (Grant No. GM-53771), University of Washington Office of Technology Transfer, and the Washington Research Foundation. C. Y. C. also gratefully acknowledges fellowship support from Center for Nanotechnology (NSF IGERT Grant No. DGE-9987620). We would also like to thank the University of Washington Engineered Biomaterials Research Center (UWEB) (NSF Grant No. EEC-9529161) for use of their cell culture facilities.

REFERENCES

1. H. M. Temin, *Hum. Gene Ther.* **1**, 111 (1990).
2. K. C. Gunter, A. S. Khan and P. D. Noguchi, *Hum. Gene Ther.* **4**, 643 (1993).
3. W. K. O'Neal, H. Zhou, N. Morral, C. Langston, R. J. Parks, F. L. Graham, S. Kochanek and A. L. Beaudet, *Mol. Med.* **6**, 179 (2000).
4. Y. Yang, Q. Li, H. C. Ertl and J. M. Wilson, *J. Virol.* **69**, 2004 (1995).
5. J. P. Yang and L. Huang, *Gene Ther.* **4**, 950 (1997).
6. S. Audouy, G. Molema, L. de Leij and D. Hoekstra, *J. Gene Med.* **2**, 465 (2000).
7. K. Crook, B. J. Stevenson, M. Dubouchet and D. J. Porteous, *Gene Ther.* **5**, 137 (1998).
8. C. Plank, K. Mechtler, F. C. Szoka, Jr. and E. Wagner, *Hum. Gene Ther.* **7**, 1437 (1996).
9. L. G. Barron, K. B. Meyer and F. C. Szoka, Jr., *Hum. Gene Ther.* **9**, 315 (1998).
10. M. Ogris, S. Brunner, S. Schuller, R. Kircheis and E. Wagner, *Gene Ther.* **6**, 595 (1999).
11. S. Li, W. C. Tseng, D. B. Stolz, S. P. Wu, S. C. Watkins and L. Huang, *Gene Ther.* **6**, 585 (1999).
12. S. Li, M. A. Rizzo, S. Bhattacharya and L. Huang, *Gene Ther.* **5**, 930 (1998).
13. H. Eliyahu, N. Servel, A. J. Domb and Y. Barenholz, *Gene Ther.* **9**, 850 (2002).
14. O. Zelphati, L. S. Uyechi, L. G. Barron and F. C. Szoka, Jr., *Biochim. Biophys. Acta* **1390**, 119 (1998).
15. D. Oupicky, M. Ogris, K. A. Howard, P. R. Dash, K. Ulbrich and L. W. Seymour, *Mol. Ther.* **5**, 463 (2002).
16. C. Monfardini and F. M. Veronese, *Bioconjug. Chem.* **9**, 418 (1998).
17. C. Tros de Ilarduya and N. Duzgunes, *Biochim. Biophys. Acta* **1463**, 333 (2000).
18. N. Murthy, J. R. Robichaud, D. A. Tirrell, P. S. Stayton and A. S. Hoffman, *J. Control. Rel.* **61**, 137 (1999).
19. C. A. Lackey, N. Murthy, O. W. Press, D. A. Tirrell, A. S. Hoffman and P. S. Stayton, *Bioconjug. Chem.* **10**, 401 (1999).
20. C. A. Lackey, O. W. Press, A. S. Hoffman and P. S. Stayton, *Bioconjug. Chem.* **13**, 996 (2002).
21. C. Y. Cheung, N. Murthy, P. S. Stayton and A. S. Hoffman, *Bioconjug. Chem.* **12**, 906 (2001).
22. F. W. Putnam, *The Plasma Proteins: Structure, Function, and Genetic Control*. Academic Press, New York, NY (1975).
23. E. Dodds, M. G. Dunckley, K. Naujoks, U. Michaelis and G. Dickson, *Gene Ther.* **5**, 542 (1998).
24. L. Vitiello, A. Chonn, J. D. Wasserman, C. Duff and R. G. Worton, *Gene Ther.* **3**, 396 (1996).
25. L. Vitiello, K. Bockhold, P. B. Joshi and R. G. Worton, *Gene Ther.* **5**, 1306 (1998).
26. P. C. Ross and S. W. Hui, *Gene Ther.* **6**, 651 (1999).
27. Z. Hyvonen, M. Ruponen, S. Ronkko, P. Suhonen and A. Urtti, *Eur. J. Pharm. Sci.* **15**, 449 (2002).

28. Y. Xu and F. C. Szoka, Jr., *Biochemistry* **35**, 5616 (1996).
29. K. K. Son, D. Tkach and J. Rosenblatt, *Biochim. Biophys. Acta* **1511**, 201 (2001).
30. P. C. Rensen, R. L. de Vrueh, J. Kuiper, M. K. Bijsterbosch, E. A. Biessen and T. J. van Berkel, *Adv. Drug Deliv. Rev.* **47**, 251 (2001).
31. F. Bonte and R. L. Juliano, *Chem. Phys. Lipids* **40**, 359 (1986).
32. O. Zelphati and F. C. Szoka, Jr., *Pharm. Res.* **13**, 1367 (1996).
33. A. S. Kuznetsov and B. V. Missyul, *Ukr. Biokhim. Zh.* **64**, 3 (1992).
34. J. Zabner, A. J. Fasbender, T. Moninger, K. A. Poellinger and M. J. Welsh, *J. Biol. Chem.* **270**, 18997 (1995).

Surfaces, modification and characterization

Gels, Genes, Grafts and Giants, pp. 321–344 (2005)

Platelet adhesion to radiofrequency glow-discharge-deposited fluorocarbon polymers preadsorbed with selectively depleted plasmas show the primary role of fibrinogen

W.-B. TSAI [1], Q. SHI [1], J. M. GRUNKEMEIER [1], C. McFARLAND [3] and T. A. HORBETT [1,2,*]

[1] *Department of Bioengineering, University of Washington, Seattle, WA 98195, USA*
[2] *Department of Chemical Engineering, Box 351750, University of Washington, Seattle, WA 98195, USA*
[3] *CSIRO Molecular Science, No 2 Richardson Place, Riverside Corporate Park, Delhi Road, North Ryde, NSW 2113, Australia*

Received 19 August 2003; accepted 28 January 2004

Abstract—Fluorocarbon radio-frequency glow-discharge (RFGD) treatment has previously been shown to cause decreased platelet adhesion despite the presence of adsorbed fibrinogen on the surfaces. In this study platelet adhesion to fluorocarbon RFGD-treated surfaces preadsorbed with human plasma was further examined. A series of plasma deposited fluorocarbon thin films were made by varying the C_3F_6/CH_4 ratio in the monomer feed. The surfaces were preadsorbed with plasma, serum, or plasma selectively depleted of fibronectin, vitronectin, or Von Willebrand factor, and platelet adhesion was measured. We also measured fibrinogen adsorption to the surfaces from plasma, monoclonal antibody binding to adsorbed fibrinogen and SDS elutability of the adsorbed fibrinogen. The antibodies used bind to the three putative platelet binding sites on fibrinogen, namely, M1 antibody binds to the dodecapeptide at the C-terminus of the gamma chain, γ(402–411), R1 antibody binds to a sequence in the Aα chain (87–100) which includes RGDF at Aα(95–98) and R2 antibody binds a sequence in the Aα chain (566–580) which includes RGDS at Aα(572–575). Fibrinogen was found to play a decisive role in mediating platelet adhesion to the fluorocarbon surfaces contacting plasma. Few platelets adhered to the fluorocarbon surfaces preadsorbed with serum, while preadsorption with plasma selectively-depleted of either fibronectin, vitronectin, or von Willebrand factor did not decrease platelet adhesion significantly. Replenishment of exogenous fibrinogen to serum restored platelet adhesion, while replenishment of the other proteins had no effect. Platelet adhesion to the fluorocarbon surfaces was lower than to PET or the methane glow-discharge-treated PET. However, there was no apparent correlation between platelet adhesion and the amount of fibrinogen adsorption or monoclonal antibody binding to surface-bound fibrinogen.

*To whom correspondence should be addressed. E-mail: horbett@uweb.engr.washington.edu

Key words: Platelet adhesion; fibrinogen adsorption; glow discharge; fluorocarbon; monoclonal antibody binding; biomaterials; von Willebrand factor; fibronectin; vitronectin.

INTRODUCTION

After contact with blood, synthetic biomaterials are coated with an unwanted layer of plasma proteins, which later elicit adverse reactions such as platelet adhesion and activation, and consequent blood clotting. Platelet adhesion to artificial surfaces that have been exposed to plasma is thought to be mainly mediated by fibrinogen (FGN), as a critical role for FGN has been shown for several materials, including glass [1], polystyrene [2, 3], Biomer [4] and radio-frequency glow discharge/allylamine substrates [5]. Platelet adhesion was much lower if these surfaces were preadsorbed with afibrinogenemic plasma instead of normal plasma. Furthermore, when afibrinogenemic plasma was reconstituted with exogenous FGN prior to adsorption, platelet adhesion was restored in a dose-dependent manner [2, 3, 5]. Recently, we have shown that although fibronectin (FN), vitronectin (VN) and von Willebrand factor (vWF) mediated platelet adhesion when they were adsorbed as purified proteins to four kinds of modified polystyrene surfaces that vary widely in surface chemistry, selective removal of these proteins from plasma did not reduce platelet adhesion to surfaces preadsorbed with these plasmas. In contrast, FGN played a decisive role in mediating platelet adhesion to plasma-preadsorbed polystyrene surfaces, because few platelets adhered to these surfaces when they were preadsorbed with serum. Furthermore, we recently found that the ability of the adsorbed FGN to mediate platelet adhesion was determined by the conformation and/or orientation of the adsorbed FGN as detected using monoclonal antibody binding to the platelet binding sites in FGN [6].

It has been suggested that tight binding of FGN to a surface may directly decrease the ability of the adsorbed FGN to interact with platelets. It is known that proteins bind extremely tightly to certain types of fluorocarbon coatings made by glow-discharge treatment, namely those that are especially rich in CF_3 groups [7]. Kiaei *et al.* showed that these fluorocarbon surfaces exhibited lower platelet adhesion *in vitro* and *in vivo* [8]. In another study, Kiaei *et al.* showed that the surfaces deposited with such fluorocarbons exhibited low platelet adhesion, even when preadsorbed with purified FGN [9]. This property has been found to be controllable by varying the conditions in a plasma reactor and that the tightness of binding of adsorbed FGN increases with the increasing surface CF_3 content. In this study, it was also found that platelet adhesion appeared to correlate with the SDS elutability of adsorbed FGN, but not with the amount of adsorbed FGN or monoclonal antibody to the three platelet-binding domains of FGN. Nevertheless, that conclusion was only applicable to surfaces preadsorbed with FGN or FGN/IgG solution as done in that study [9]. It is not known if this conclusion applies to the CF_3-rich fluorocarbon surfaces if they are preadsorbed with plasma.

The first purpose of this study was to show that FGN is the key plasma protein in mediating platelet adhesion to fluorocarbon radio-frequency glow discharge (RFGD)-treated surfaces. A series of RFGD-treated poly(ethylene terephthalate) films was synthesized by varying the C_3F_6/CH_4 ratio in the monomer feed. By comparing platelet adhesion to the surfaces preadsorbed with normal human plasma and serum or plasma selectively depleted in FGN, FN and VN or plasma inherently deficient in vWF, the roles of these proteins in mediating platelet adhesion were evaluated. The second aspect of this study was to further examine the role of binding strength and platelet epitope expression in fibrinogen adsorbed to CF_3-rich fluorocarbon surfaces from plasma in regards to platelet adhesiveness.

MATERIALS AND METHODS

Materials

Rectangular poly(ethylene terephthalate) (PET) sheets (untreated Lux Thermanox, 22 × 60 mm, catalog No. 5412) were purchased from Nunc (Naperville, IL, USA). Argon (Ar) was obtained from Praxair (Danbury, CT, USA, catalog No. UN1006); Hexafluoropropene (HFP, C_3F_6) was obtained from Lancaster Synthesis (Windham, NH, USA, catalog No. 10422). Methane (M, CH_4) was received from Matheson (Newark, CA, USA). Methylene chloride, methanol and acetone came from Fisher Chemical (Fair Lawn, NJ, USA, HPLC grade).

Biochemicals

Human FGN was purchased from Sigma (St. Louis, MO, USA, catalog No. F4883). Human FN was purchased from Life Technologies (Grand Island, NY, USA, catalog No. 33016-015). Human VN was purified from pooled human normal plasma using immuno-affinity chromatography. Purified human vWF was provided by Dr. Gerald Roth (School of Medicine, University of Washington, Seattle, WA, USA).

Three types of mouse anti-human monoclonal IgG antibodies raised against the putative platelet binding sites of FGN were provided by several investigators. M1 was provided by Dr. Gary R. Matsueda (Bristol Meyers Squibb, New York, NY, USA). R1 and R2 were provided by Dr. Z. Ruggeri (Scripps Clinic, La Jolla, CA, USA). M1 binds to the C-terminus of the γ chain, which contains a platelet binding region (402–411). R1 binds to $A\alpha$(87–100), near the N-terminus of the $A\alpha$ chain, which includes RGDF at the $A\alpha$(95–98) platelet binding region. R2 binds to $A\alpha$(566–580) near the C-terminus of the $A\alpha$ chain, which includes RGDS at the $A\alpha$(572–575) platelet binding region. Goat anti-mouse IgG (H + L) conjugated to horseradish peroxidase was obtained from Bio-Rad (Richmond, CA, USA, catalog #172-1011). Antibodies were either stored frozen at −70°C or 4°C.

Preparation of human normal plasma, serum and plasmas selectively-depleted of FN, VN or vWF

ACD anticoagulated human blood (blood/ACD ratio 9 : 1 (v/v)) was centrifuged at 187 × *g* for 20 min at room temperature to generate platelet rich plasma (PRP). Human plasma was then prepared by centrifuging PRP at 30 000 × *g* at 4°C for 30 min. The plasma from several donors was pooled (pooled normal human plasma), and stored at −70°C until use.

To prepare serum, blood was drawn into a 50 ml centrifuge tube without adding ACD and the tube was set on ice for 2 h before centrifuging at 1500 × *g* at room temperature for 20 min. The supernatant was carefully removed and centrifuged again at 30 000 × *g* at 4°C for 30 min. The resulting supernatant (serum) was then collected and stored at −70°C until use.

Fn-depleted human plasma was prepared by using immobilized gelatin affinity chromatography as reported previously [2]. VN-depleted and FN/VN-depleted human plasmas were prepared by using immunoaffinity chromatography as described previously [10]. Human plasma severely deficient in vWF was purchased from George King Biomedical (Overland Park, KS, USA) and was from a donor with severe vWF-deficient disease.

Buffers

Sterile anticoagulant (acid-citrate dextrose (ACD), NIH formula A containing citric acid dihydrate 7.3 g, sodium citrate 22 g and dextrose monohydrate 24.5 g in 1 liter deionized water) was supplied by the Department of Pharmacy Drug Services of the University of Washington Medical Center. Other buffers were made freshly from reagent-grade chemicals. Citrate phosphate-buffered saline (CPBS, pH 7.4) contained 10 mM Na_2HPO_4, 10 mM citric acid and 120 mM NaCl. Phosphate-buffered saline (PBS, pH 7.4) contained 137 mM NaCl, 2.7 mM KCl, 10 mM Na_2HPO_4 and 1.8 mM KH_2PO_4. BSA (2%) in PBS was used as the buffer for antibody binding studies. Platelet suspension buffer (pH 7.4) contained 137 mM NaCl, 2.7 mM KCl, 5.5 mM glucose, 0.35% BSA, 3 mM NaH_2PO_4, 3.5 mM HEPES and 0.1 U/ml apyrase. The buffer for protein elution was 3% SDS in PBS (pH 7.4).

RFGD polymerization

PET sheets were cut into 10 × 10 mm (for platelet adhesion experiments) or 6.5 × 6.5 mm (for fibrinogen adsorption and ELISA) squares. A hole was pierced in a corner of every square sample with a needle, and the perforated samples were strung together with steel wires. All samples were then soaked in a RBS (Pierce, Rockford, IL, USA, catalog No. 27950) detergent solution and sonicated for 15 min and rinsed with water. Then the samples were sonicated for 15 min in a series of three solvents: methylene chloride, acetone and methanol. The cleaned samples were allowed to air dry before use. The samples were then hung on a glass frame and put into a plasma reactor for deposition.

Plasma deposition was conducted in an RFGD system illustrated in Fig. 1. A tubular glass reactor (10 cm i.d.) was capacitively coupled to a 13.56 MHz radio frequency (RF) generator (Model HF-650, E.N.I. Power Systems, Rochester, NY, USA) *via* a home-made matching network and two external, symmetrical, brass capacitor rings (1 inch width). Capacitors were typically separated by eight inches. The vacuum system was pumped down by a Stokes rotary vane pump (Model V-009-2, Stokes-Penwalt, Philadelphia, PA, USA). Flow of the monomer gas was controlled by a mass flow controller (Model 80-4, Vacuum General, San Diego, CA, USA). System pressure was maintained by a motorized throttle valve (model MDV-015, Vacuum General) connected to a capacitance manometer (model CML, Vacuum General) in a feedback loop. Samples were placed in the after-glow region. The reactor chamber was evacuated to less than 10 mTorr and argon was introduced to 200 mTorr for 2 min at 50 W to clean the surface prior to plasma deposition (plasma etching). To prepare hexafluoropropene plasma treated PET (HFP/PET), C_3F_6 gas was introduced at a flow rate of 4 standard cm^3/min (sccm). The plasma was generated at 20 W power for 5 min while the system pressure was maintained at 200 mTorr. After turning off the power to end the plasma reaction, the chamber was quenched by continuing to flow the gas for one minute, followed by re-evacuation. The system was brought to atmospheric pressure with

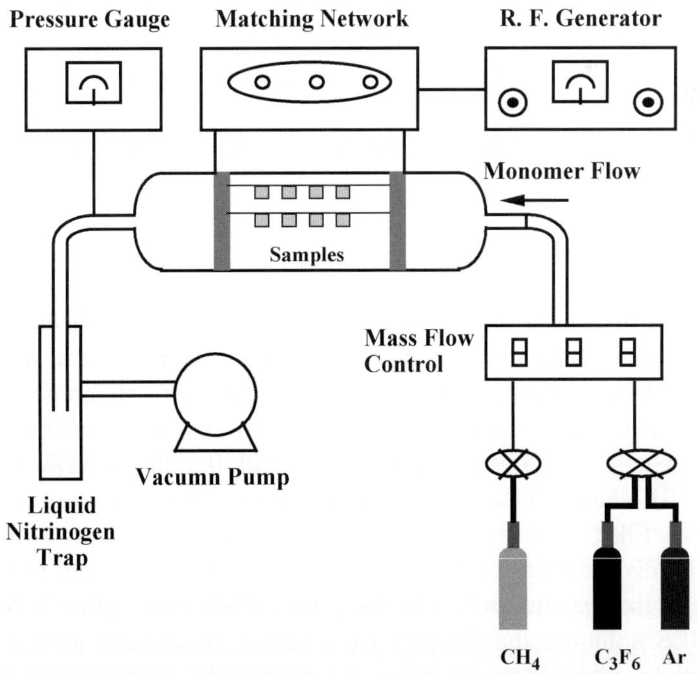

Figure 1. Schematic diagram of the RFGD reactor for CH_4/C_3F_6 deposition.

air. The conditions for the other batches of samples are as follows: M/PET, 4 sccm, 150 mTorr/20 W/5 min; M-HFP(2 : 1)/PET, 2.67 sccm/1.33 sccm, 150 mTorr/20 W/5 min; M-HFP(1 : 2)/PET, 1.33 sccm/2.67 sccm, 150 mTorr/10 W/5 min.

Surface characterization

The treated substrates were characterized with an SSX100 ESCA spectrometer (Surface Science Laboratories, Mountain View, CA, USA) at the National ESCA and Surface Analysis Center for Biomedical Problems (NESAC/BIO) at the University of Washington. Survey scans of 0 to 1000 eV binding energy were run to determine the elemental compositions of the surfaces. High-resolution scans of the C_{1s} region were also recorded. The take-off angle, the angle between the detector and axis normal to the surface, was 55°.

Surface hydrophobicity was determined by water contact angle measurement using a Rame-Hart goniometer (Rame-Hart, Mountain Lakes, NJ, USA).

Determination of FGN adsorption

Iodination of human FGN was performed using the iodine monochloride (ICl) technique [11] as modified by Horbett [12]. The details were described previously [2]. The iodinated FGN was stored at −70°C and used within two weeks.

Enough ^{125}I-labeled FGN was added to normal plasma to give a specific activity of 2000 cpm/μg FGN. The specific activity was calculated by dividing the cpm per ml of plasma by the total amount of unlabeled and labeled FGN per ml plasma. ^{125}I-labeled FGN comprised 5% or less (weight fraction) of the total amount of FGN in the plasma. The plasma was then diluted to the desired concentration with CPBS prior to plasma adsorption.

The untreated or plasma-treated PET samples (6.5 × 6.5 mm) were placed vertically in microtiter wells, and then incubated with ^{125}I-labled FGN-containing normal human plasma at various plasma dilutions in quadruplicate at 37°C for 2 h (350 μl/well). After 2 h protein adsorption, the samples were rinsed 5 times with PBS and then equilibrated with 2% BSA in PBS at room temperature for one hour. After rinsing, the samples were transferred to gamma counter tubes (Becton Dickinson, Oxnard, CA, USA, Falcon #2025) with 0.5 ml 3% SDS solution in it. The bound ^{125}I-labeled FGN was measured using a gamma counter (model 1185R, TM Analytic, Elk Grove, IL, USA). The total amount of the adsorbed FGN was calculated by dividing the retained radioactivity (after background subtraction) of the sample by the specific activity of the plasma FGN pool. After 24 h incubation with 3% SDS solution, the samples were rinsed, transferred to new tubes and the radioactivity retained on the films was counted to determine the retention of adsorbed FGN.

Enzyme-linked immunosorbent assay (ELISA)

The ELISA procedure for measuring monoclonal antibody binding to surface-bound fibrinogen was as follows. The untreated or RFGD plasma-treated PET samples (6.5×6.5 mm) were placed vertically in microtiter wells and incubated with 350 μl of 100% or 1% normal human plasma. After rinsing with PBS, the samples were incubated with 2% BSA solution (350 μl/well) at room temperature for 1 h. The samples were rinsed with PBS again. The first antibody, M1, R1, or R2, was then added and incubated with the plasma-preadsorbed surfaces at 37°C for 60 min (350 μl/well). The dilutions for the antibodies were $1:1 \times 10^6$ for M1, $1:2 \times 10^4$ for R1 and $1:2 \times 10^4$ for R2. The samples were rinsed with PBS five times and then incubated with 350 μl of HRP-conjugated goat-anti-mouse antibody (1 : 3000) at 37°C for 60 min. After rinsing, the samples were then transferred to 10×75 mm tubes (VWR, cat. no. 60825-538). TMB substrate solution (250 μl) was then added to each tube and allowed to react for 10 min at room temperature on a shaker which rotated at 200 rpm (Gyrotory Shaker, model G2, New Brunswick Scientific, Edison, NJ, USA). The enzymatic reaction was stopped by addition of 100 μl of 1 M sulfuric acid. 100 μl of each TMB solution (350 μl in total) was transferred to a well in a 96-well microtiter plate. The optical density (OD) of the TMB solution was read at 450 nm with the VMAX™ kinetic microplate reader. The OD developed in TMB solution without any HRP-conjugated antibody was used as the blank. The OD for each sample was corrected by subtraction of the blank and multiplied by a factor of 3.5 to obtain the total optical density for the whole TMB solution.

Platelet preparation and adhesion

Washed platelets were prepared by passing PRP through a Sepharose 2B column that had been pre-equilibrated with platelet suspension buffer which contained 0.35% BSA as a stabilizer and blocking agent. The fractions containing platelets were collected from the column and pooled. Bulk phase platelet concentration was determined using a Coulter counter (Coulter Electronics, Hialeah, FL, USA). The platelet concentration was adjusted to 1×10^8 platelets/ml with platelet suspension buffer and kept in a 37°C water bath until use. Fifteen minutes prior to platelet adhesion, platelet suspensions were supplemented with both Ca^{2+} and Mg^{2+} to a final concentration of 1 mM.

The untreated or RFGD plasma-treated PET samples (10×10 mm) were placed horizontally at the bottom of wells of 24-well plates and incubated with 500 μl of plasma or serum at 37°C for 2 h. After rinsing with PBS, the samples were transferred to other wells which had been previously filled with 2% BSA, and incubated at room temperature for 1 h. The BSA solution was aspirated from the plates and rinsed with PBS, and platelet suspension was then added (500 μl/well) and allowed to settle for 1 h at 37°C. After rinsing 5 times with PBS, the samples were transferred to new plates. Platelet adhesion was determined by measuring lactate dehydrogenase (LDH) activity after lysing the platelets with Triton X-100, using a modification of Tamada's method [13], as described previously [2, 14]. The number of adherent platelets was determined from a calibration curve generated by plotting OD *versus* platelet concentration for a series of suspensions of known platelet concentration.

RESULTS

Surface analysis

A series of surfaces varying in fluorine content were prepared by mixing methane and HFP in various ratios for the glow discharge treatment of PET films. The surface elemental compositions of the untreated and the RFGD-treated PET films determined by ESCA are shown in Table 1. The fluorine content increased with increasing HFP ratio in the feed gas, from 0% for the untreated or methane treated PET, to 56.6% for films treated with HFP only. The oxygen content of the PET films was 27.0% and that of methane-treated films was 10.9%, whereas the oxygen content of the other treated surfaces was below 3%. Our results were similar to the study by Kiaei *et al.* [9], in which PET films were treated with ethylene and tetrafluoroethylene in various ratios.

The effect of varying the proportion of HFP in the feed mixture on the high resolution ESCA C_{1s} spectra of RFGD-treated and untreated PET films is shown in Fig. 2. The C_{1s} spectrum of untreated PET was composed of three peaks centered at 285, 286.6 and 289.0 eV, representing the $\underline{C}{-}C$, $\underline{C}{-}O$ and $O{-}\underline{C}{=}O$ functional groups, respectively, which is consistent with the chemical structure of PET. Methane treatment of PET increased $\underline{C}{-}C$ species (285 eV), and the intensities of two other C_{1s} peaks representing $\underline{C}{-}O$ (286.6 eV) and $O{-}\underline{C}{=}O$ (289.0 eV) groups were decreased. Increasing the HFP proportion in the feed increased the intensity of fluorocarbon species ($\underline{C}F_3$ (294 eV), $\underline{C}F_2$ (292 eV), $\underline{C}F{-}CF_n$ (289.5 eV) and $\underline{C}{-}CF_n$ (287.5 eV)) peaks, and decreased the $\underline{C}{-}C$ peak (Fig. 2). The percentages of carbon associated with $\underline{C}F_3$ and $\underline{C}F_2$ groups in the total surface carbon content increased with increasing HFP feeding (Table 2).

The water contact angles for the RFGD-treated and the untreated PET films are shown in Fig. 3. The values of the advancing contact angles are in the order of

Table 1.
Surface elemental compositions determined by ESCA

	C	F	O	N
PET	73.1	—	27.0	—
M/PET	88.2	1.0	10.9	—
M-HFP(2 : 1)/PET	61.0	36.2	2.7	—
M-HFP(1 : 2)/PET	47.6	51.3	1.1	—
HFP/PET	41.9	56.6	1.1	0.4

HFP/PET > M-HFP(1 : 2)/PET > M/PET > M-HFP(2 : 1)/PET > PET, the same order of the receding contact angles.

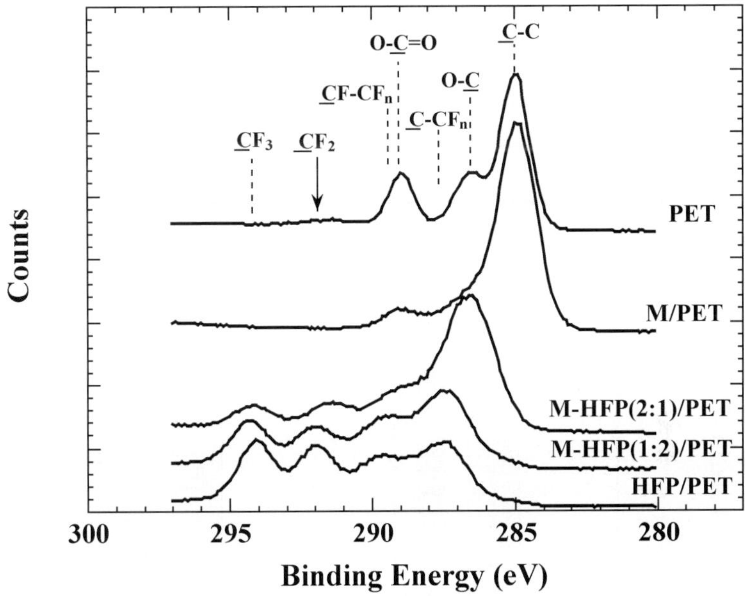

Figure 2. High-resolution ESCA C_{1s} spectra for RFGD-treated and untreated PET.

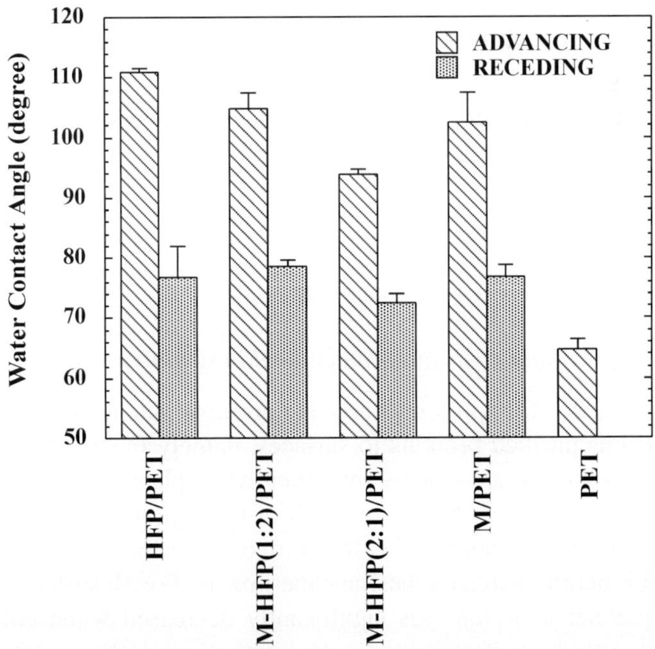

Figure 3. Water contact angles of RFGD-treated and untreated PET.

Table 2.
Carbon concentrations of $\underline{C}F_3$ and $\underline{C}F_2$ for HFP glow-discharged PET

	$\underline{C}F_3$	$\underline{C}F_2$
M/PET	0	0
M-HFP(1 : 2)/PET	7.4	9.6
M-HFP(1 : 2)/PET	17.4	17.4
HFP/PET	23.8	20.6

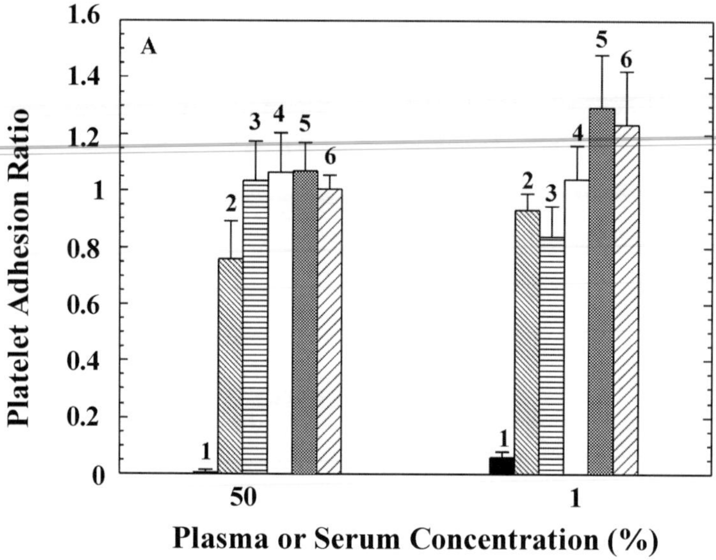

Figure 4. Platelet adhesion to the surfaces preadsorbed with serum or selectively-depleted plasma: (A) M/PET; (B) M-HFP(2 : 1)/PET; (C) M-HFP(1 : 2)/PET; (D) HFP/PET; (E) PET. The ordinate represents the ratio of platelet adhesion after preadsorption with a depleted plasma to platelet adhesion after preadsorption with normal plasma at the same plasma concentration. For each plasma concentration, the bars represent, from left to right: (1) serum, (2) FN-depleted plasma, (3) VN-depleted plasma, (4) VN, FN-depleted plasma, (5) vWF-depleted plasma and (6) vWF-depleted plasma replenished with vWF to 25 μg/ml (for 50% plasma) or 0.5 μg/ml (for 1% plasma). Error bars indicate SEM, $n = 4$.

Determination of the major plasma protein mediating platelet adhesion

Blood plasma contains four proteins known to promote platelet adhesion when they are preadsorbed as purified proteins to surfaces, namely FGN, FN, VN and vWF, but this does not mean all are equally important when plasma rather than a purified protein is used. To examine the role of each of these proteins, the plasma-treated and control surfaces were incubated with serum or a plasma selectively-deficient in FN, VN or vWF before platelet adhesion experiments. For all surfaces preadsorbed with serum, platelet adhesion was significantly decreased when compared with platelet adhesion to the surfaces preadsorbed with normal plasma ($P < 0.001$ for a two-sided t-test; Fig. 4A–E, bar 1). In contrast, platelet adhesion to the surfaces

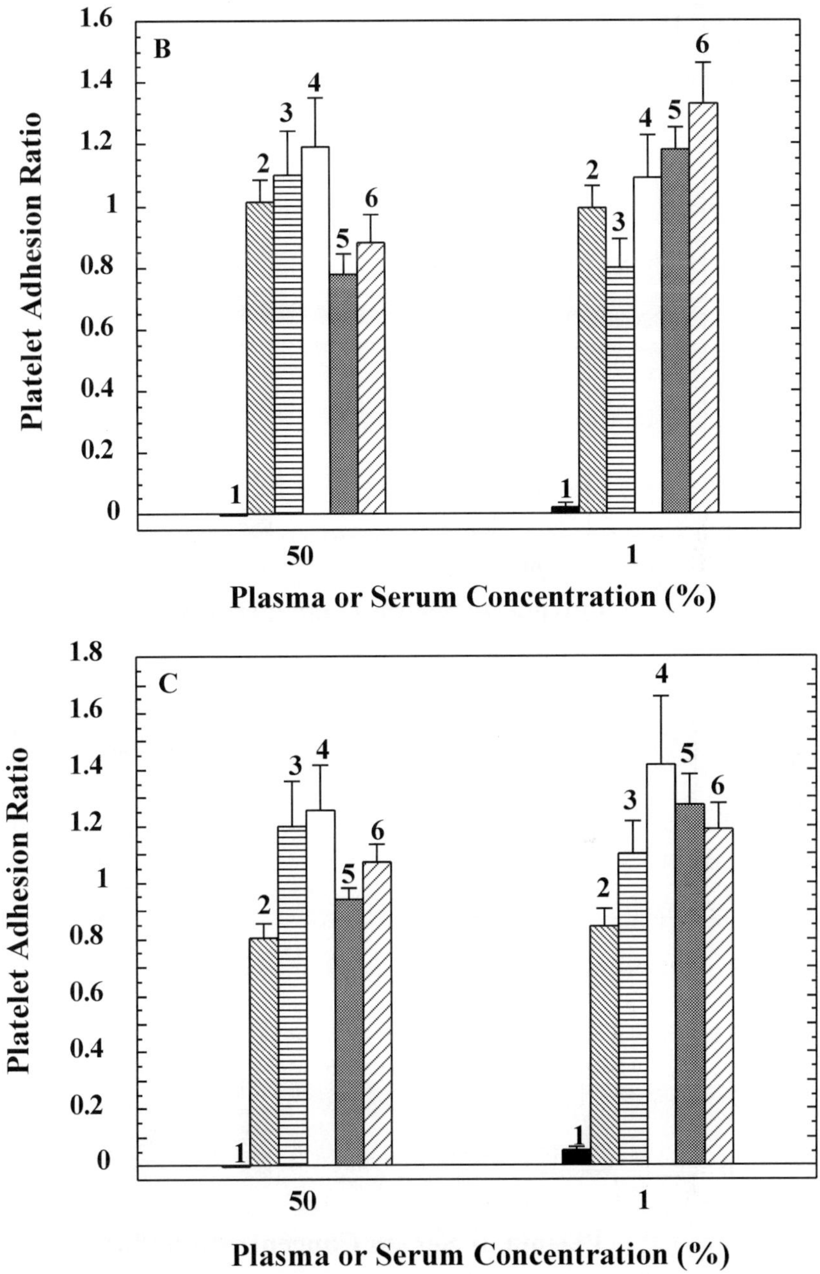

Figure 4. (Continued).

preadsorbed with plasma selectively deficient in FN, VN, or vWF did not show any significant decrease when compared to platelet adhesion to the surfaces preadsorbed with normal plasma ($P > 0.1$ for a two-sided t-test; Fig. 4A–E, bars 2–6).

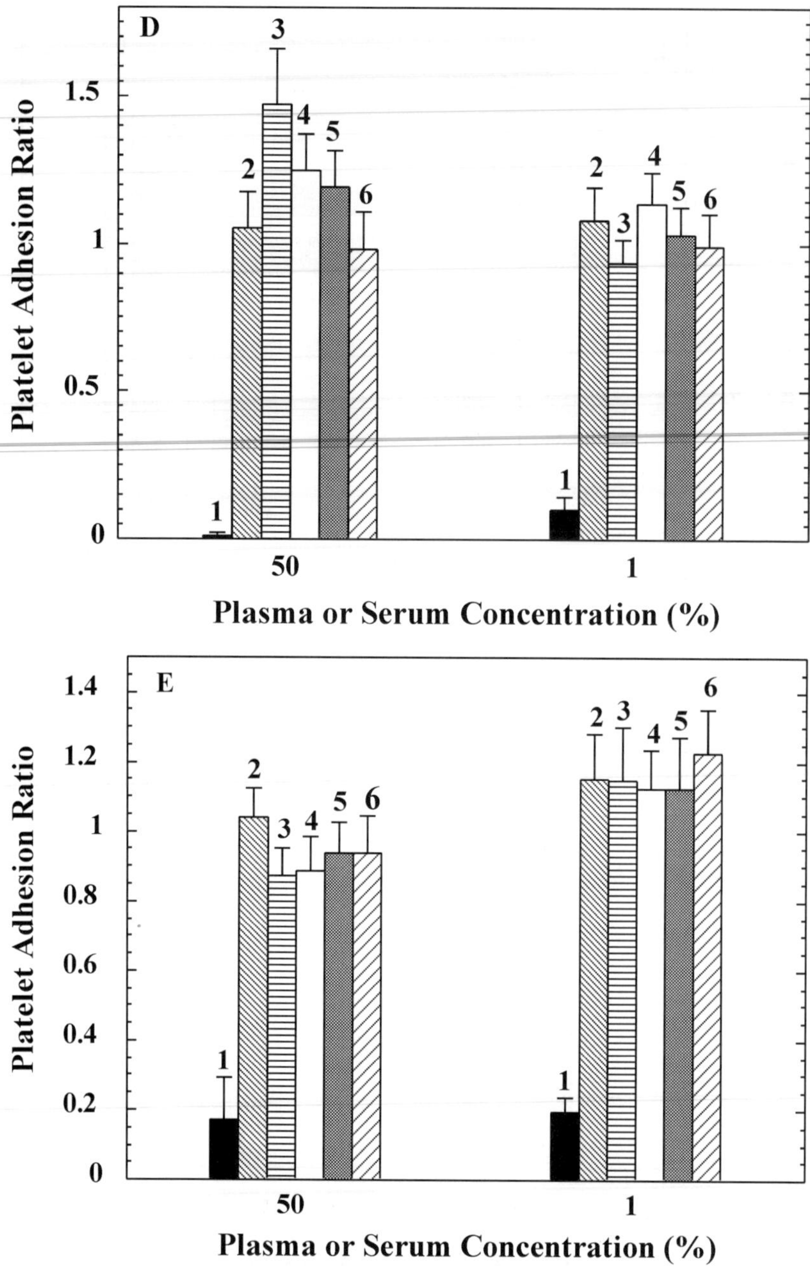

Figure 4. (Continued).

We have previously shown that besides FGN, the concentrations of FN and vWF were also reduced in serum when compared to those in normal plasma [3]. Therefore, the role of FGN, FN and vWF in mediating platelet adhesion was further investigated on three of the five surfaces studied by adding exogenous FGN, FN, or

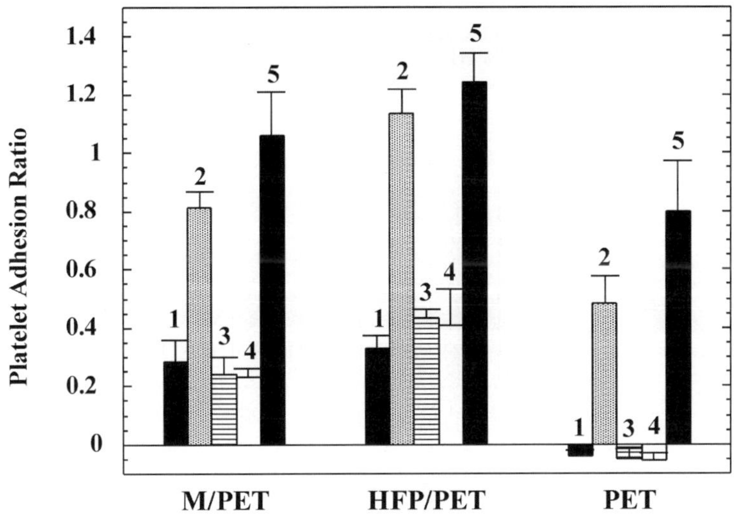

Figure 5. Platelet adhesion to the surfaces preadsorbed with serum replenished with fibrinogen, fibronectin, or vWF. For each material, the bars represent, from left to right, preadsorption with: (1) 1% serum, (2) 1% serum + FGN, (3) 1% serum + FN, (4) 1% serum + vWF, (5) 1% serum + FGN, FN and vWF. The ordinate represents the ratio of platelet adhesion for each condition to the same surface preadsorbed with 1% normal plasma. Error bars indicate SEM, $n = 4$.

vWF to serum in order to restore their normal plasma concentrations prior to plasma adsorption. Replenishment of FGN alone restored platelet adhesion to the normal level (bar 2 in Fig. 5, $P < 0.001$ for a two-sided t-test), while replenishment of either FN or vWF alone did not increase platelet adhesion (bars 3 and 4 in Fig. 5, $P > 0.1$ for a two-sided t-test). However, adding all the three proteins elicited more platelet adhesion than adding FGN alone (compare bars 2 and 5 in Fig. 5).

Platelet adhesion to surfaces preadsorbed with normal plasma

Platelet adhesion to the surfaces preadsorbed with 100% or 1% normal plasma is shown in Fig. 6. Platelet adhesion to each surface was normalized by dividing by the platelet adhesion to 1% plasma preadsorbed M/PET. The values of platelet adhesion in Fig. 6 represent the average values from three separate experiments, each of which was done in quadruplicates. The average platelet adhesion to M/PET preadsorbed with 1% plasma was $(2.84 \pm 1.57) \times 10^6$ platelets/cm^2 (mean ± one standard deviation, $n = 4$ for every experiment).

In general, the differences in platelet adhesion among the various surfaces was relatively small. Thus, the difference between platelet adhesion to different surfaces preadsorbed with the same concentration of normal plasma was not significant ($P > 0.1$ for a two-sided t-test). Similarly, the difference between platelet adhesion to each surface preadsorbed with 100% or 1% plasma was insignificant ($P > 0.1$ for a two-sided t-test).

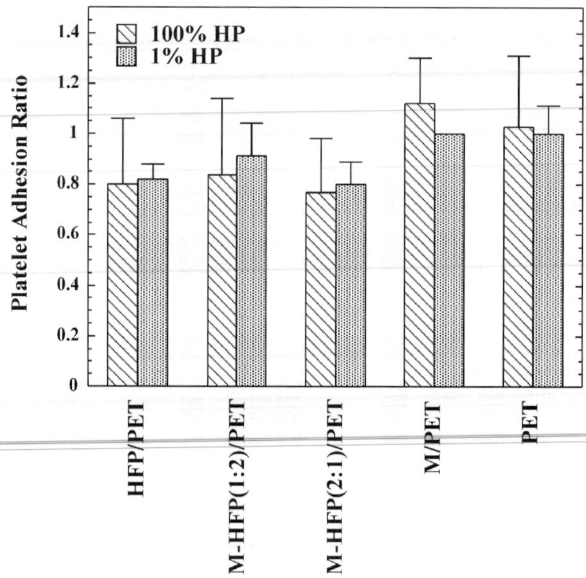

Figure 6. Platelet adhesion to the surfaces preadsorbed with normal plasma. Substrates were contacted with 100% or 1% human normal plasma for 2 h prior to platelet contact. Error bars are ±1 SEM obtained from three experiments which were each done in quadruplicate. The average platelet adhesion to M/PET preadsorbed with 1% plasma was 2.84×10^6 platelets/cm^2. The ordinate represents the ratio of platelet adhesion for each condition to the platelet adhesion to M/PET preadsorbed with 1% normal plasma. Error bars indicate SEM, $n = 4$.

Fibrinogen adsorption to surfaces

Fibrinogen adsorption and retention after SDS elution to the untreated and the fluorocarbon RFGD-treated PET samples are shown in Fig. 7. When FGN was adsorbed from 100% plasma, the highest adsorption appeared on M-HFP(2 : 1)/PET with a value of 572 ± 38 ng/cm^2 (mean ± SEM, $n = 4$), and the lowest FGN adsorption appeared on M/PET and PET, around 325 ng/cm^2 (Fig. 7A). When FGN was adsorbed from 1% plasma, PET exhibited the largest amount of FGN (183 ± 13 ng/cm^2, $n = 4$; Fig. 7A), but the difference in FGN adsorption from 1% plasma among the surfaces was small. The retention of adsorbed FGN after SDS elution was higher on the fluorocarbon-containing surfaces (>80%) than M/PET and PET (<65%) (Fig. 7B). In summary, FGN adsorption from 100% plasma was higher than from 1% plasma, FGN retention was higher when FGN was adsorbed from more diluted plasma and to the fluorocarbon-treated surfaces.

Monoclonal antibody binding to surface-bound fibrinogen

M1 antibody binding to the surfaces preadsorbed with 100% or 1% normal plasma is shown in Fig. 8A. M1 antibody binding to surface-bound FGN was higher when FGN was adsorbed from plasma to the fluorocarbon-treated surfaces in comparison to the methane-treated surfaces. The M1 antibody binding to adsorbed FGN was

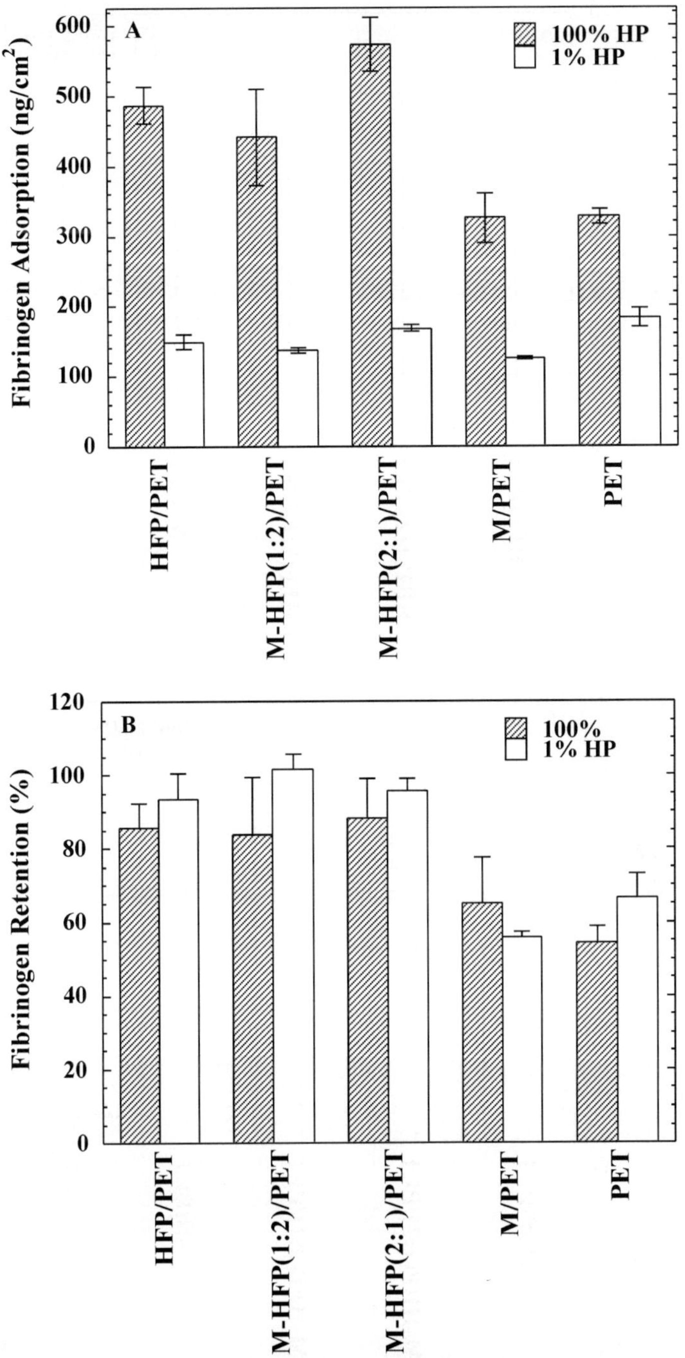

Figure 7. Fibrinogen adsorption (A) and retention after SDS elution (B) on surfaces. Error bars indicate SEM, $n = 4$.

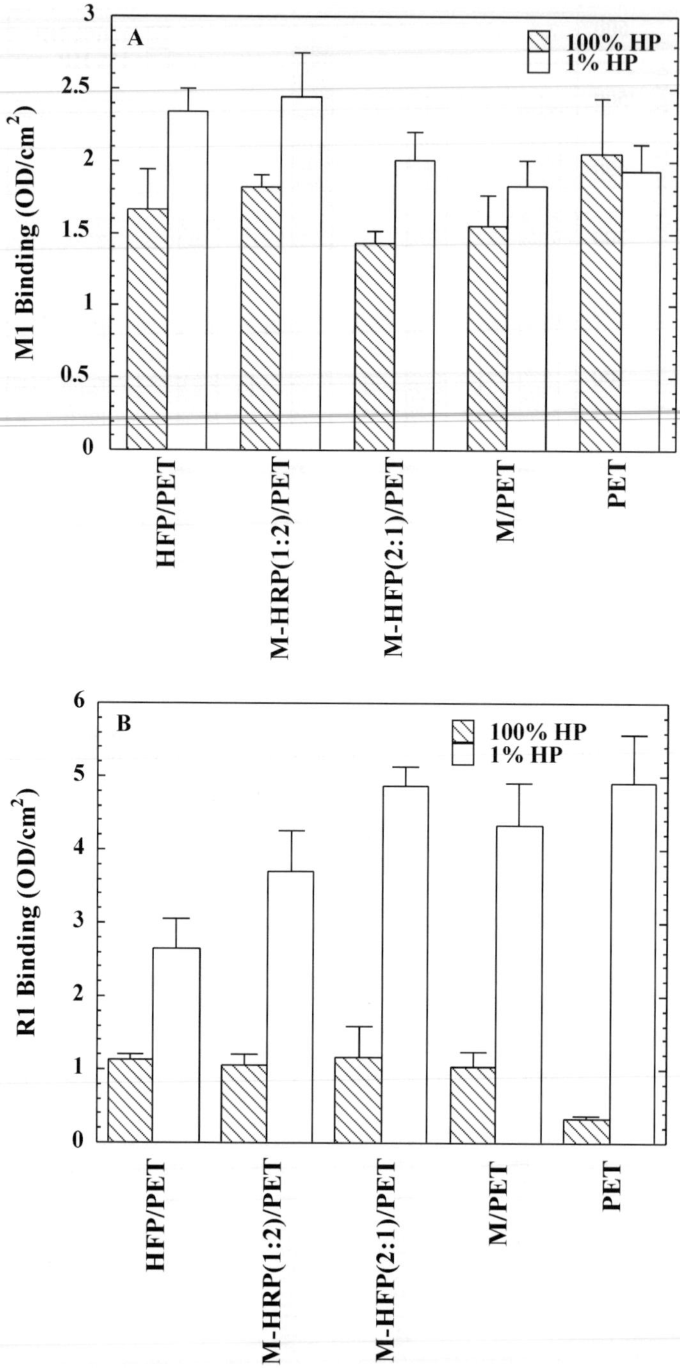

Figure 8. Monoclonal antibody binding to surface-bound fibrinogen: (A) M1; (B) R1; (C) R2. Error bars indicate SD, $n = 4$.

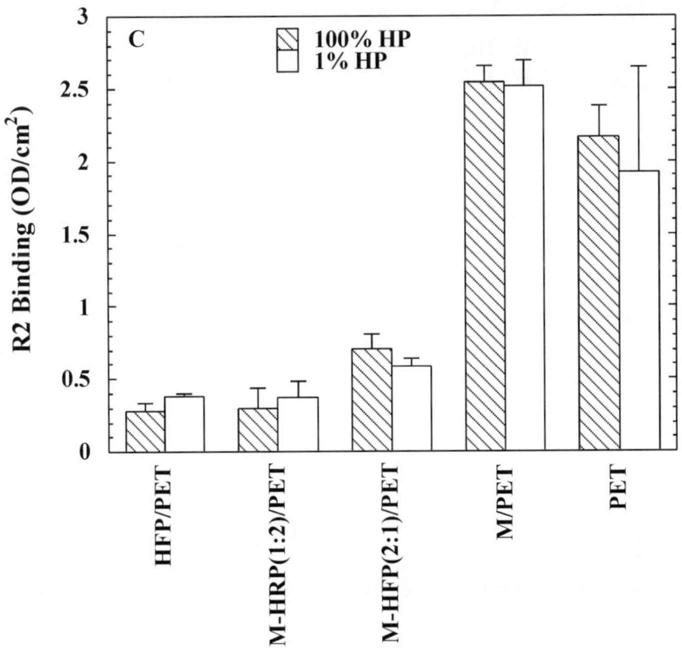

Figure 8. (Continued).

generally greater when these surfaces were preadsorbed with 1% plasma than when adsorbed from 100% plasma, except on the PET surfaces (Fig. 8A).

Figure 8B shows the R1 antibody binding to the surfaces preadsorbed with 100% or 1% normal plasma. Compared to the surfaces preadsorbed with 1% plasma, R1 antibody binding to the surfaces preadsorbed with 100% plasma was much lower. R1 antibody binding to the four RFGD-treated surfaces preadsorbed with 100% plasma was very close, but was higher than that to the untreated PET.

R2 antibody binding to the surface-bound FGN adsorbed with 100% or 1% normal plasma is shown in Fig. 8C. R2 antibody binding to adsorbed FGN was much lower on the three types of fluorocarbon-treated surfaces than methane-treated or untreated PET. On the other hand, the concentration of plasma preadsorbed did not make any significant difference in R2 antibody binding to these surfaces.

DISCUSSION

Surface analysis

Among the polymers made with the RFGD HFP/M treatment, those made with more HFP in the feed had greater fluorine and $\underline{C}F_3$ content on their surfaces (Fig. 9 and Tables 1 and 2). The atomic percentage of fluorine in the HPF/M RFGD polymer series was similar to the atomic percentage of fluorine reported by Kiaei *et al.* for a C_2F_4/C_2H_4 RFGD polymer series [9] and by Grunkemeier *et al.* for a C_3F_6/C_2H_4

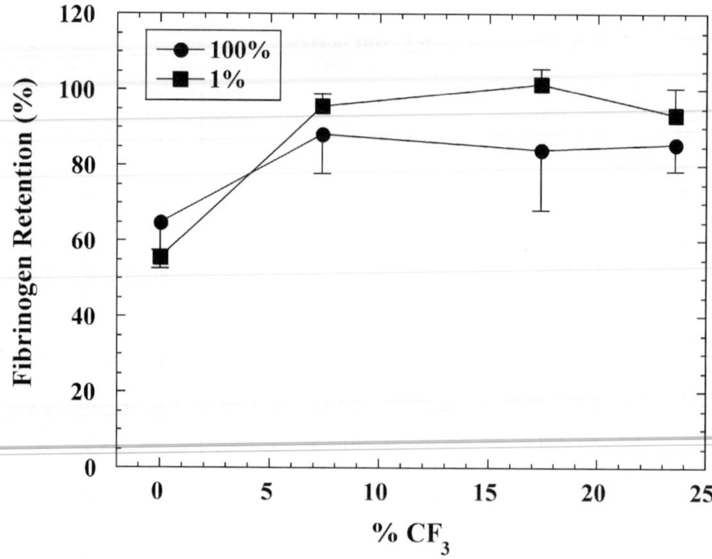

Figure 9. Correlation between fibrinogen retention and the contents of the surface CF_3 groups. Error bars indicate SEM, $n = 4$.

RFGD polymer series [15], suggesting that these polymer series were similar in surface chemistry.

Fibrinogen mediates platelet adhesion to plasma preadsorbed fluorocarbon RFGD-treated surfaces

In previous studies, platelet adhesion to several types of artificial surfaces preadsorbed with plasma deficient in FGN was greatly reduced [1–4]. Here, similar results were found for PET and methane/fluorocarbon-RFGD treated PET. When these surfaces were preadsorbed with serum, platelet adhesion was very low compared to adhesion to the same surfaces preadsorbed with normal plasma (bar 1 in Fig. 4A–E). Furthermore, platelet adhesion to these surfaces after preadsorption with plasma selectively depleted of FN, VN, or VWF was not significantly decreased (bars 2–6 in Fig. 4A–E). When serum with added FGN was used for preadsorption, platelet adhesion was restored to the level of platelet adhesion to the surfaces preadsorbed with normal plasma (comparing bar 1 and bar 2 in Fig. 5). The replenishment of FN or vWF alone did not increase platelet adhesion (bars 3 and 4 in Fig. 5). Thus, it appears that FGN is the dominant plasma protein that mediates platelet adhesion to these substrates. However, platelet adhesion on the surfaces treated with plasma replenished with the three adhesion proteins FGN, FN and vWF was higher than that on the surfaces replenished with FGN only (comparing bars 2 and 5 in Fig. 5), suggesting that FN and vWF might still have auxiliary effects on platelet adhesion to synthetic biomaterials when FGN is present.

The influence of fluorocarbon RFGD treatments of PET on FGN adsorption, elutability and monoclonal antibody binding

The fact that fluorocarbon RFGD treatment significantly influences protein adsorption and the state of adsorbed proteins has been reported in several studies [9, 15–18]. Previously, it was shown that RFGD fluorocarbon treatment decreased SDS elutability of surface-bound FGN when adsorbed from either pure FGN solution [16] or a FGN/γ-globulin mixture [9]. In this study we showed that the same conclusion also applied when FGN was adsorbed from plasma. The apparently increased strength of FGN interactions with the RFGD fluorocarbon treated surfaces was shown to be correlated with surface energy [9]. The low surface energy (in air) may promote strong hydrophobic interactions through formation of multiple protein–surface contact points, leading to the observed strong binding. Previous work by Kiaei *et al.* with a series of tetrafluoroethylene glow-discharge-treated polytetrafluoroethylene showed that the SDS elutability of the surface-bound FGN decreased with an increase in the amount of surface CF_3 species (from 5 to 10.5%) [9]. In this study, however, the tightness of FGN bound to the surfaces was not changed by an increase in surface CF_3 content (from 7.4 to 23.8%). It is possible that when the content of CF_3 groups is beyond a certain value, the influence of the amount of CF_3 groups on FGN binding is insignificant.

Antibody binding to adsorbed proteins is thought to be another way to examine the conformation and/or orientation of adsorbed proteins [6, 19]. Three different types of monoclonal antibodies that bind to the three regions in FGN thought to be involved in platelet adhesion were used to measure the reactivity of adsorbed FGN. M1 binds to the C-terminus of the γ-chain, which contains a platelet-binding region (402–411). R1 binds to Aα(87–100), near the N-terminus of the Aα chain, which includes the platelet-binding region RGDF at Aα(95–98). R2 binds to Aα(566–580) near the C-terminus of the Aα chain, which includes platelet binding region RGDS at Aα(572–575). Monoclonal antibody binding data were normalized by dividing the OD values by the total amount of adsorbed FGN, as displayed in Table 3. Normalized antibody binding can be used as an index of the recognizability of the adsorbed FGN by the monoclonal antibodies, although this is true only when antibody concentration is in great excess over the molar dis-association constant for antibody binding to the antigen or if the dis-association constant (K_D) is known to be constant for all the conditions compared. However, the antibody concentrations we used here were below the K_D values we determined for these antibodies in a previous study [19]. In addition, we also reported in that study that K_D can vary by up to three fold, depending on the adsorbing surfaces or postadsorptive conditions [19]. Thus, the antibody binding data we obtained here could be affected by both the amount of the available binding sites as well as the affinity of the antibody for those sites.

Fibrinogen adsorbed from plasma to the RFGD fluorocarbon treated surfaces studied here generally exhibited decreased M1, R1 and R2 antibody binding, compared to FGN adsorbed to surfaces treated with RFGD methane, except for

Table 3.
Normalized monoclonal antibodies

	M1 (OD/μg)		R1 (OD/μg)		R2 (OD/μg)	
	100% plasma	1% plasma	100% plasma	1% plasma	100% plasma	1% plasma
PET	6.3	10.5	1.0	26.8	6.6	10.5
M/PET	4.8	14.6	3.2	34.6	7.8	20.1
M-HFP(2 : 1)/PET	2.5	11.9	2.0	28.9	1.2	3.5
M-HFP(1 : 2)/PET	4.1	17.9	2.4	27.0	0.7	2.7
HFP/PET	3.4	15.7	2.3	17.8	0.6	2.6

The values in the table are the ratio of measured antibody binding (in units of optical density (OD) from the ELISA measurement) to the amount of adsorbed fibrinogen for each surface.

the M1 antibody binding to the surfaces preadsorbed with 1% plasma. Besides surface chemical properties, plasma concentration also influences the conformation of adsorbed FGN. The normalized binding of the three monoclonal antibodies to surface-bound FGN was relatively high when FGN was adsorbed from 1% plasma compared to FGN adsorbed from 100% plasma, especially for R1 antibody binding (Table 3). These findings, combined with the facts that the low elutability of the FGN adsorbed to the fluorocarbon RFGD-treated PET or adsorbed from low plasma concentration suggest that the natural conformation of FGN is disturbed by the increasing contacts between the adsorbed proteins and the surfaces.

Platelet adhesion

Several studies have suggested that tight binding of FGN to a surface may directly reduce the ability of FGN to interact with the platelet receptors [9, 20] and this was also seen in this study (Fig. 10E). This phenomenon has been explained by noting that platelet interaction with adsorbed FGN may require partial removal of the parts of the FGN molecule from the substrate [20], because of the observation that FGN was moved after being bound to the GPIIb/IIIa receptor [21]. Following this principle, tetrafluoroethylene (TFE) discharge treatment has been shown to improve the thromboresistance of Dacron grafts as measured using an *ex vivo* shunt in baboons, and decreased platelet adhesion *in vitro* [7–9]. In this study we also found that platelet adhesion to fluorocarbon-treated surfaces was lower than that to the methane-treated or untreated PET (Fig. 6), and platelet adhesion was correlated negatively to the retention of surface-bound FGN from SDS elution (Fig. 10E). Thus, the tightness of FGN adsorption on surfaces could be an important factor in affecting platelet interactions with adsorbed FGN.

In a previous study, we showed that platelet adhesion to other surfaces preadsorbed with human plasma was mainly determined by the exposure of platelet binding sites in adsorbed FGN, especially the dodecapeptide in the γ-chain (indicated by M1 antibody binding), not by the amount of adsorbed FGN [6]. In this study, a similar comparison was made by plotting platelet adhesion against FGN adsorption,

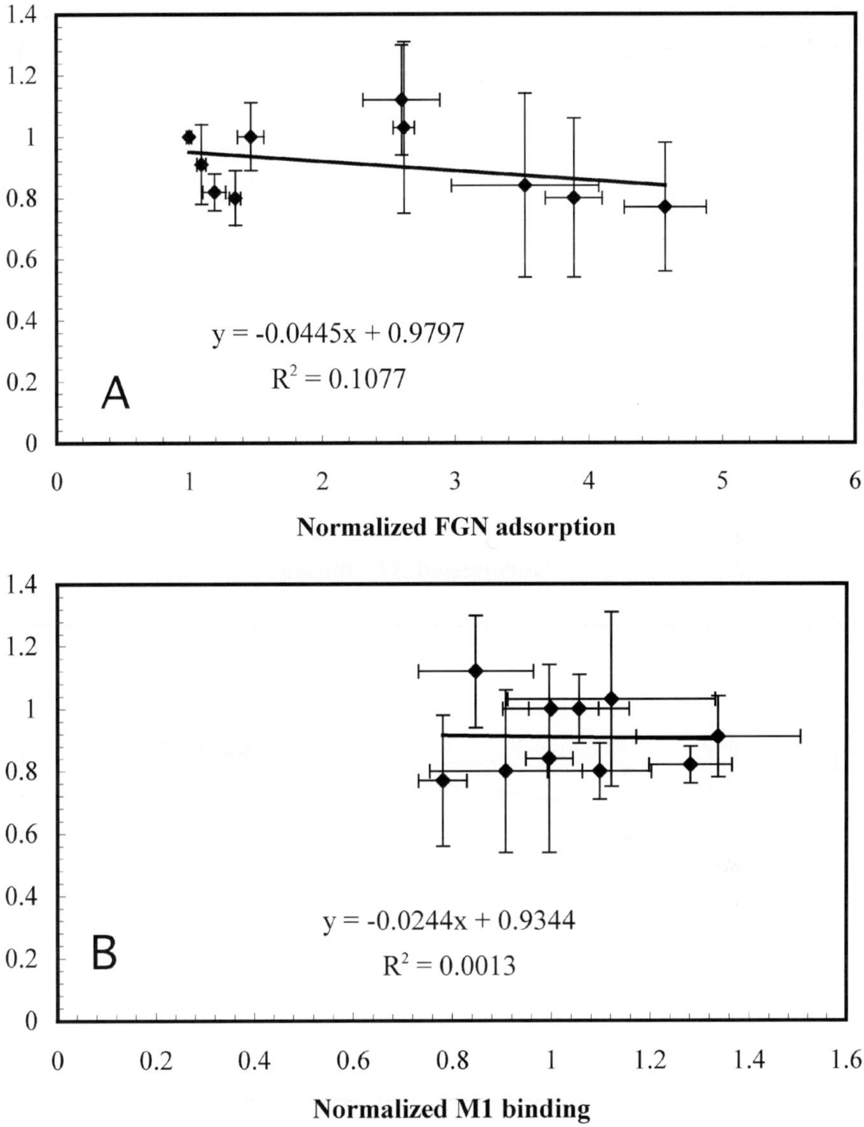

Figure 10. Correlation between platelet adhesion and fibrinogen adsorption: (A) the amount of fibrinogen adsorption; (B) M1 antibody binding; (C) R1 antibody binding; (D) R2 antibody binding; (E) fibrinogen retention after SDS elution. All values represent the ratio to those on M/PET preadsorbed with 1% plasma, except fibrinogen retention.

M1, R1 and R2 monoclonal antibody binding, and FGN retention from SDS elution (Fig. 10A–E). However, no apparent correlation appeared between platelet adhesion to the amount of FGN adsorption (Fig. 10A), M1 antibody binding (Fig. 10B) and R1 antibody binding (Fig. 10C), while platelet adhesion seemed to increase with R2 antibody binding (Fig. 10D) or decrease with FGN retention (Fig. 10E). The lack of

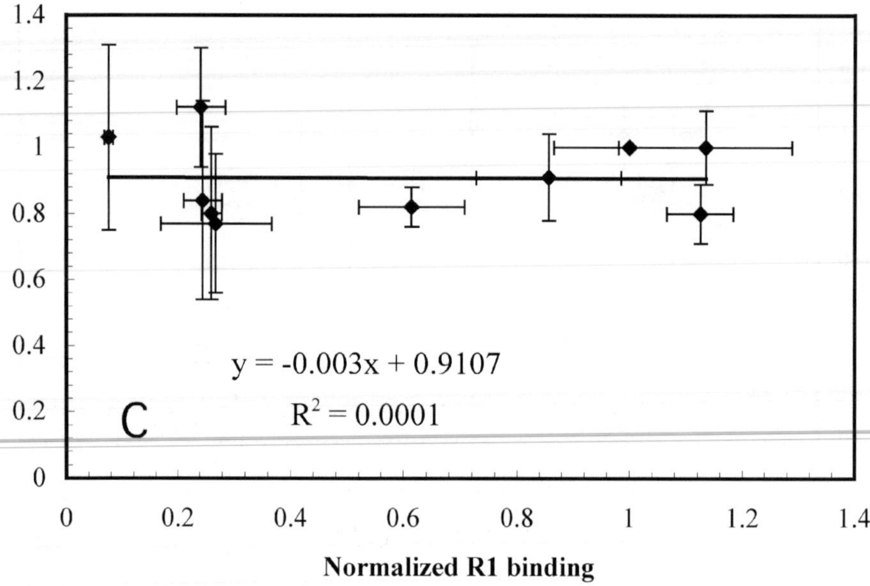

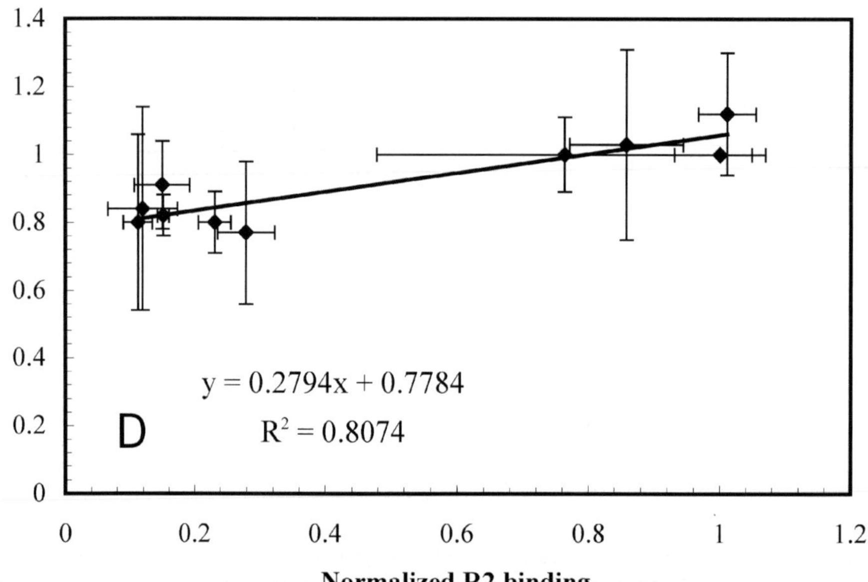

Figure 10. (Continued).

an apparent correlation between platelet adhesion and M1 antibody binding might be due to the fact that the differences in M1 antibody binding and platelet adhesion among the different surfaces was not large and the standard deviations were substantial, so any correlation between FGN adsorption and platelet adhesion would be difficult to discern.

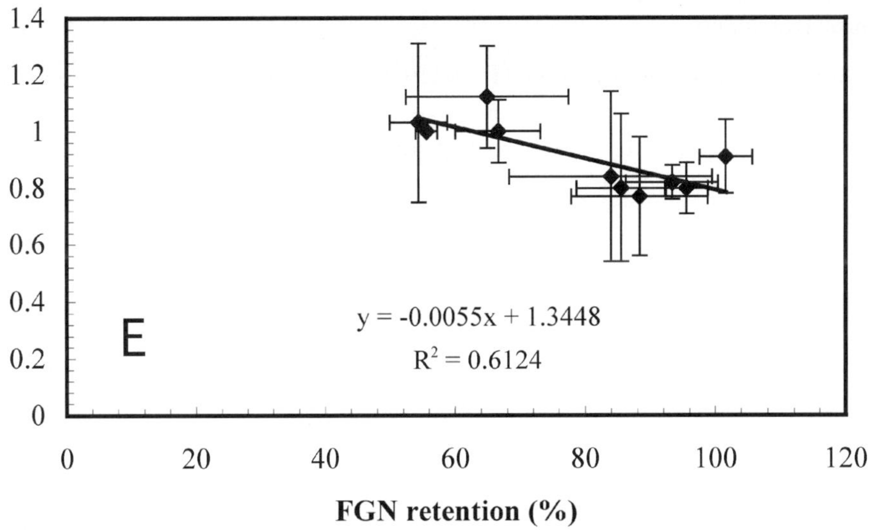

Figure 10. (Continued).

CONCLUSIONS

Fibrinogen was shown to be the major plasma protein that mediated platelet adhesion to plasma-preadsorbed PET, RFGD methane-treated PET and RFGD fluorocarbon-treated PETs. Platelet adhesion to the surfaces preadsorbed with serum was very low compared to the same surfaces after preadsorption with normal plasma. On the other hand, depletion of FN, VN, or vWF from the plasma used to preadsorb the surfaces did not significantly decrease subsequent platelet adhesion. Platelet adhesion was restored when surfaces were preadsorbed with serum that had been replenished with FGN, but the replenishment of FN or vWF had no effect on platelet adhesion.

Fluorocarbon RFGD treatment increased the hydrophobicity of PET, which resulted in increasing FGN adsorption and retention of adsorbed FGN after SDS elution. The conformation and/or orientation of the adsorbed FGN appeared to be significantly affected by fluorocarbon RFGD treatment, as indicated by the SDS elution and monoclonal antibody binding assays. The conformational difference in adsorbed FGN probably influenced its ability to mediate platelet adhesion. Platelet adhesion was lower on the fluorocarbon surfaces than on the methane treated PET or the untreated PET, although FGN adsorption on the fluorocarbon surfaces was higher than that on the methane-treated or the untreated PET. Our findings confirmed that fluorocarbon RFGD treatment affected the ability of FGN to mediate platelet adhesion, whether adsorbed from FGN/IgG solutions as in the earlier studies [8, 9] or from plasma as done in this study.

Acknowledgements

The financial assistance of grant HL19419 from the National Heart, Lung, and Blood Institute is gratefully acknowledged. The ESCA studies were done at NESAC/Bio, supported by RR01296.The authors are grateful to Dr. G. Roth (UW, School of Medicine, Seattle) for the gift of purified human vWF.

REFERENCES

1. M. B. Zucker and L. Vroman, *Proc. Soc. Exp. Biol. Med.* **131**, 318 (1969).
2. W. B. Tsai, J. M. Grunkemeier and T. A. Horbett, *J. Biomed. Mater. Res.* **44**, 130 (1999).
3. W. B. Tsai, J. M. Grunkemeier, C. D. McFarland and T. A. Horbett, *J. Biomed. Mater. Res.* **60**, 348 (2002).
4. J. A. Chinn, T. A. Horbett and B. D. Ratner, *Thromb. Haemost.* **65**, 608 (1991).
5. J. A. Chinn, B. D. Ratner and T. A. Horbett, *Biomaterials* **13**, 322 (1992).
6. W.-B. Tsai, J. Grunkemeier and T. A. Horbett, *J. Biomed. Mater. Res.* **67A**, 1255 (2003).
7. A. M. Garfinkle, A. S. Hoffman, B. D. Ratner, L. O. Reynolds and S. R. Hanson, *Trans. Am. Soc. Artif. Intern. Organs* **30**, 432 (1984).
8. D. Kiaei, A. S. Hoffman and S. R. Hanson, *J. Biomed. Mater. Res.* **26**, 357 (1992).
9. D. Kiaei, A. S. Hoffman, T. A. Horbett and K. R. Lew, *J. Biomed. Mater. Res.* **29**, 729 (1995).
10. R. A. Underwood and F. A. Bennett, *J. Cell Sci.* **93**, 641 (1989).
11. R. W. Helmkamp, R. L. Goodland, W. F. Bale, I. L. Spar and L. E. Mutschler, *Cancer Res.* **20**, 1495 (1960).
12. T. A. Horbett, *J. Biomed. Mater. Res.* **15**, 673 (1981).
13. Y. Tamada, E. A. Kulik and Y. Ikada, *Biomaterials* **16**, 259 (1995).
14. J. M. Grunkemeier, W. B. Tsai and T. A. Horbett, *J. Biomed. Mater. Res.* **41**, 657 (1998).
15. J. M. Grunkemeier, W. B. Tsai, M. R. Alexander, D. G. Castner and T. A. Horbett, *J. Biomed. Mater. Res.* **51**, 669 (2000).
16. J. L. Bohnert, B. C. Fowler, T. A. Horbett and A. S. Hoffman, *J. Biomater. Sci. Polymer Edn* **1**, 279 (1990).
17. D. Kiaei, A. S. Hoffman and T. A. Horbett, *J. Biomater. Sci. Polymer Edn* **4**, 35 (1992).
18. A. Safranj, D. Kiaei and A. S. Hoffman, *Biotechnol. Prog.* **7**, 173 (1991).
19. J. Grunkemeier, C. Wan and T. Horbett , *J. Biomater. Sci. Polymer Edn* **8**, 189 (1996).
20. J. A. Chinn, S. E. Posso, T. A. Horbett and B. D. Ratner, *J. Biomed. Mater. Res.* **25**, 535 (1991).
21. J. C. Loftus and R. M. Albrecht, *J. Cell Biol.* **99**, 822 (1984).

Gels, Genes, Grafts and Giants, pp. 345–360 (2005)

pH-dependent elution profiles of selected proteins in HPLC having a stationary phase modified with pH-sensitive sulfonamide polymers

SEONG IL KANG [1] and YOU HAN BAE [2,*]

[1] *Center for Biomaterials and Biotechnology, Department of Materials Science and Engineering, Kwangju Institute of Science and Technology, 1 Oryong-dong, Puk-gu, Gwangju 500-712, South Korea*

[2] *Department of Pharmaceutics and Pharmaceutical Chemistry, University of Utah, 421 Wakara Way, Suite 315, Salt Lake City, UT 84108, USA*

Received 23 September 2003; accepted 4 February 2004

Abstract—An approach to controlling protein interactions with silica (glass) beads grafted with pH-sensitive copolymers of a sulfonamide and *N,N*-dimethylacrylamide is proposed. The bead surface was modified with poly(*N,N*-dimethyl acrylamide-co-methacryloyl sulfadimethoxine), which exhibits hydrophobic/hydrophilic switching around pH 7.0. Phase transition in the aqueous solution, the wettability change of the modified surface and surface tension of the polymer-grafted glass surface was characterized by changes in turbidity and the dynamic contact angle as a function of pH. The discontinuous phase transition in the copolymer solution shifted to a higher pH region as the composition of sulfadimethoxine monomer (SDM) was increased. However, the transition pH in continuous wettability changes on the modified surface was not affected by polymer composition due to differences in the dynamic motion. The ionized SDM of the copolymers enhanced wettability. Therefore, the advancing contact angle of the surface decreased by 45° above pH 7.0, whereas wettability at low pH values was low due to the hydrophobic surface. Elution behavior of the selected four model proteins with different p*I* (bovine serum albumin (p*I* 5.0), insulin (p*I* 5.3), fibrinogen (p*I* 6.1) and myoglobin (p*I* 7.1)) was investigated by using pH-sensitive copolymer-modified glass beads at four pH values (5.0, 6.0, 7.0 and 8.0). Retention time of the proteins in pH-sensitive aqueous chromatography was easily regulated by eluent pH that controls the proportion of electrostatic and hydrophobic interactions between the stationary phase and the protein. At pH 5.0, hydrophobic interaction on the deionized polymer-grafted surface is an important force in separating the proteins. However, identification of the proteins at pH 7.0 was effectively facilitated due to the electrostatic interaction between the negatively charged polymer support and positive charges on protein.

Key words: High-performance liquid chromatography; p*I*; pH-sensitive sulfonamide copolymers; elution profiles.

---

*To whom correspondence should be addressed. Tel.: (1-801) 585-1518. Fax: (1-801) 585-3614. E-mail: you.bae@m.cc.utah.edu

INTRODUCTION

In conventional bioseparation, based on hydrophobic interaction and ion-exchange mechanism, biomolecules adsorb on the surface differently, exhibiting different elution times. This technology allows for the purification of proteins and other biochemicals and the manufacturing of pharmaceutical and food products [1–3]. The use of reverse-phase chromatography (RPC) has been especially popular in pharmaceutics and biochemistry because analytes are partitioned easily by altering the polarity of the mobile phase [1, 4]. However, the bioactivity of proteins and peptides in RPC is influenced by organic solvents in the mobile phase. Furthermore, the use of organic solvents is environmentally controversial and expensive to prepare [5–7]. Recently, a newly developed type of chromatography, using aqueous mobile phases, has been introduced for the separation of bioactive compounds. Okano and co-workers reported chromatography systems composed of thermosensitive polymer-modified surfaces. Typically, oligomers of *N*-isopropylacrylamide (NiPAAm) or its copolymers with *N*-hydroxyethylacrylamide, butyl methacrylate, or glycidyl methacrylate were covalently attached to silica beads [5, 8]. Their hydrated forms on the bead surfaces can switch to the dehydrated state by heating [9]. However, these systems have limitations, such as the denaturation of heat-sensitive proteins from the external column temperature despite the easy control of the stationary properties [10, 11].

We have chosen to explore a new strategy for the stable separation of bioactive compounds, i.e. pH-sensitive aqueous chromatography operated by a multimode that combines the hydrophobic interactions and electrostatic interactions around the physiological conditions. Previously, we reported that anionic polymers and gels polymerized with comonomers and hydrophobic sulfonamide monomers with a pK_a in the range 6 to 8, showing a pH-regulated sharp transition of solubility and swelling [12–14]. It is expected that an aqueous chromatography system using our pH-sensitive polymers is a desirable method to separate proteins and peptides stably because properties of the packing material in the column are modulated effectively at neutral pH.

In this study, stationary phases were modified with hydrophobic anionic sulfonamide monomer and hydrophilic *N,N*-dimethylacrylamide (DMAAm) in order to investigate their interaction with four different proteins. By means of ester–amine coupling reactions, pH-sensitive copolymers were grafted to (aminopropyl)silica glasses or beads. Both in a polymer solution and on polymer-grafted glass, sensitivity around neutral pH was estimated from the magnitude of the change of turbidity and wettability, respectively. To choose the most effective pH for bioseparation, the chromatographic elution behaviors of four different proteins on the column packed with modified beads were demonstrated by comparing the retention times at four pH levels of the aqueous mobile phase.

MATERIALS AND METHODS

Materials

N,N-Dimethylacrylamide (DMAAm; Aldrich, Milwaukee, WI, USA) was purified by vacuum distillation at 20°C/12 mmHg. Methacryloyl chloride (Aldrich) and *N,N*-dimethylformamide (DMF, Aldrich) were distilled at 30°C/10 mmHg and 70°C/8 mmHg, respectively. 2,2′-Azobisisobutyronitrile (AIBN, Acros, Geel, Belgium) was recrystallized from methanol (Acros) twice. 3-Mecaptopropionic acid (MPA, Aldrich) was distilled under reduced pressure (5 mmHg) and 95°C. *N,N*′-dicyclohexylcarbodiimide (DCC) and *N*-hydroxysuccinimide (NHS) were purchased from Acros. The aminopropyl-functionalized silica beads (purchased from Aldrich) contained 1.5 mmol of amino functional group per gram. Sulfo-succinimidyl-4-O-(4, 4′-dimethoxytrityl)butyrate (s-SDTB) was purchased from Pierce (Rockford, IL, USA). (3-Aminopropyl)-triethoxysilane (APTES; Aldrich) was distilled under reduced pressure (85°C/8 mmHg). Four proteins, bovine serum albumin (BSA, p*I* 5.0), insulin (INS, p*I* 5.3), fibrinogen (FB, p*I* 6.1) and myoglobin (MYG, p*I* 7.1), were obtained from Sigma (St. Louis, MO, USA). The molecular weights of selected model-proteins (FB, BSA, MYG and INS) are 40×10^3, 69×10^3, 17.4×10^3 and 5.7×10^3, respectively. Phosphate-buffered saline (PBS) solution (0.15 M ionic strength) was prepared from 12 mM of monopotassium phosphate (Sigma) and 8 mM dipotassium phosphate (Sigma), 150 mM sodium chloride (Aldrich). The buffer pH was adjusted, as desired, with 0.1 M NaOH (Aldrich) or 0.1 M HCl (Aldrich) solutions.

Synthesis of carboxyl-terminated polymers

The protocol used to prepare sulfadimethoxine monomer (SDM) was based on the synthetic method described in our previous report [14]. Then the carboxyl group-terminated poly(DMAAm-co-SDM) (CDSD) was prepared by radical polymerization of DMAAm and SDM in the presence of telogen (MPA; 1.0 mmol) and initiator (AIBN; 0.5 mmol) in DMF after the solution was degassed by repeated freeze-thaw cycles and the flask was sealed under reduced pressure. The objective degree of polymerization was 300. The reaction was allowed to proceed at 65°C for 10 h and the polymers were precipitated in a 10-fold amount of non-solvent, ethyl acetate. The precipitate was filtrated, dried *in vacuo* and redissolved in aqueous NaOH solution (pH 9.0). For further purification, the solution was dialyzed against deionized water for 2 weeks using dialysis tubing (Spectra, molecular weight cut-off 2000) to remove unreacted monomer, AIBN and MPA. Purified copolymers were finally obtained by freeze-drying. Similarly, carboxyl group-terminated poly(DMAAm) (CDMAAm) was prepared by radical polymerization and purified by dialysis.

Polymers were characterized by FT-IR (Perkin-Elmer 2000) to confirm the structure. UV/Vis-spectrophotometer (Varian CARY 1E) was used for the measurement of pH sensitivity. The optical transmittance of polymer solution (1 g/200 ml) was measured at 500 nm and at various pH values. SDM composition in the copolymer was calculated by the previous method using the absorbance–content relationship [14]. The molecular weight of polymers was determined by gel-permeation chromatography (GPC, Waters, Milford, MA, USA).

Amination of glass surfaces

Glass (2 cm×4 cm×1 mm in thickness) was cleaned with 3% hydrofluoric acid for a few seconds and 10% hydrogen peroxide for 30 min, and washed with deionized water. The glass was then dried completely at 40°C for 12 h under vacuum in a flask. The glass in flask was mixed with toluene (250 ml) and APTES (5 ml) under a nitrogen gas stream, followed by the silanization reaction under reflux at 100°C for 7 days. Aminated glass was then washed with toluene, methanol and distilled water, respectively. It was finally dried at 40°C for 24 h *in vacuo* and stored in a desiccator under vacuum before use.

The degree of introduction of polymer on an aminated surface was calculated from the subtraction of the amino groups consumed from the total amino groups. A spectrophotometric method using s-SDTB was used to determine the amount of amino groups on the grafted surface [15, 16]. This method was for semi-quantitative analysis of amino groups on the surface with a detection limit of 0.1 nmol/cm^2. This reagent was prepared by dissolution of s-SDTB in 10 ml of dry DMF. Dry polymer support (4 mg) was suspended in a 4-ml reaction mixture that consisted of reagent, catalyst DMAP (10 mg) and triethyl amine (200 μl). The suspension in a vial was vigorously stirred at room temperature for 30 min. The optimum duration of reaction time was determined by pre-experiments with 10-min increments of the reaction time (10–90 min). The reaction mixture was washed with DMF (20 ml), methanol (20 ml) and finally deionized water (40 ml). s-SDTB reacted quantitatively with an amino group on the solid surface by the formation of amide bonds. After drying under vacuum, a weighed polymer support (2 mg) was treated with 70% perchloric acid solution in a flask (10 ml) to release the dimethoxytrityl cation from the support. The consumption of surface amino groups by s-SDTB can be estimated from the absorbance of the dimethoxytrityl cation at 498 nm using a UV/Vis-spectrophotometer. The concentration of the amino groups on the polymer support (Q) is then calculated from the following formula.

$$Q \text{ (mol/g solid support)} = 14.3 \times \Delta_{498} \times V/W, \quad (1)$$

where V and W are the volume of the perchloric acid solution and weight of support, respectively.

Modification of aminated surface with polymers

The coupling reaction of activated carboxyl-group-terminated copolymers to aminated glass or aminopropyl silica is shown in Fig. 1. First of all, the terminal carboxyl functional group (1 mol) of CDSD was activated with NHS (2 mol) and DCC

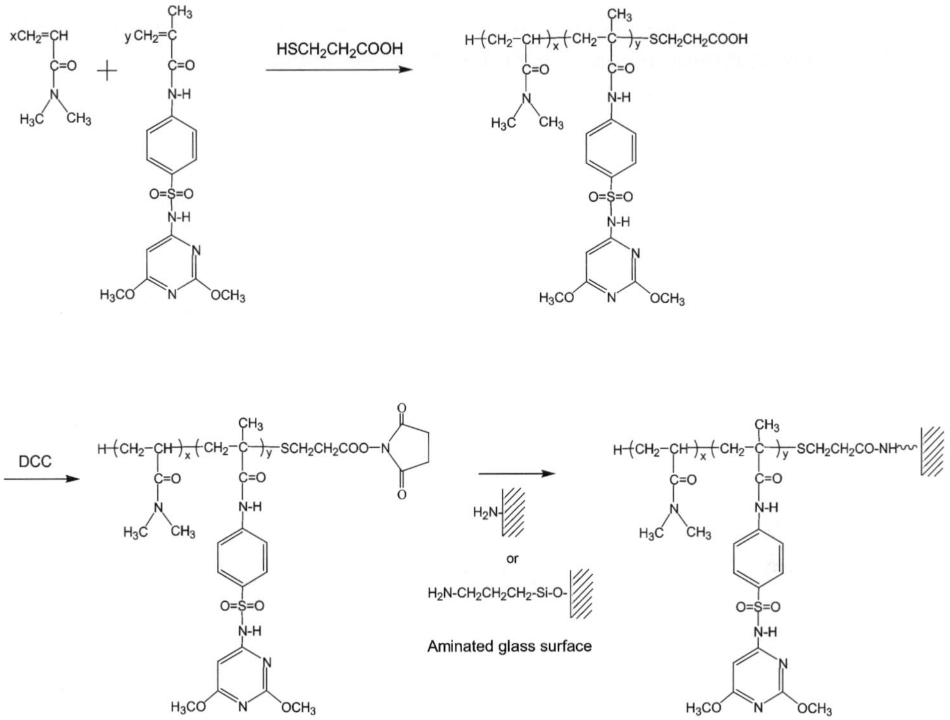

Figure 1. Schematic representation for the structure of synthetic copolymer and the modification of glass surface with copolymer. The activated carboxyl group of polymer was chemically coupled with amine group of glass or glass bead.

(2 mol) in DMF at 25 °C for 24 h. After reaction, *N,N′*-dicyclohexylurea (DCU) was removed by filtration and the activated CDSD extracted by the use of cyclohexane. Then aminated glass surface was reacted with activated carboxyl-group-terminated copolymer. Aminated glass surface or glass bead was immersed in solution containing activated copolymers in DMF. The reaction proceeded at room temperature for 24 h. Modified glass or silica was washed with deionized water and ethanol before drying *in vacuo*. This coupling process was repeated three times. Finally, the s-SDTB method was used to evaluate the degree of substitution in the glass beads.

Surface characterization of polymer-modified glass

In order to minimize environmental effect and to obtain reliable data, preparation and contact angle measurements for all samples were carried out under the same conditions using the dynamic contact angle device (DCA 322, Cahn Instrument, Madison, WI, USA). Polymer-grafted slide glasses cleaned by deionized water were dried for 1 day at 25 °C. Advancing and receding contact angles were calculated by changes of total force during the immersion. Five consecutive strokes were conducted in each measurement. The data of the last three strokes were used to obtain average contact angles with the standard calculation procedure. The contact angle results before drying and after rehydration were not noticeably different. The effect of sample preparation on contact angle was not apparent.

Modified glass surfaces were also investigated by atomic force microscopy (AFM, Park Scientific, Sunnyvale, CA, USA) with tapping mode AFM in a dry state. Contact mode images were made using a contact silicon nitride ultralevers with a spring constant of 0.06 N/m. Surface roughness was analyzed by PSI Proscan software (Park Scientific) and an Autoprobe CP system (Park Scientific). Images presented were obtained after several repetitions of the samples, which revealed similar features with regard to the depth of holes and the root-mean-square (rms) roughness.

The base glass surface was very smooth because the rms roughness is within 0.5 nm.

Surface tension of copolymers was obtained by the two-liquid method. The values were calculated by using the following equation [17]. γ_S is the sum of surface tension components with contribution form dispersion and polar parts.

$$\gamma_S = \gamma_S^d + \gamma_S^p. \tag{2}$$

Using two liquids with known γ_L^d and γ_L^p for contact angle measurements, one can easily calculate γ_S^d and γ_S^p by solving the following two equations.

$$\gamma_{LV1}(1 + \cos\theta_1) = 2\left(\gamma_S^d \gamma_{LV1}^d\right)^{1/2} - 2\left(\gamma_S^p \gamma_{LV1}^p\right)^{1/2}, \tag{3}$$

$$\gamma_{LV2}(1 + \cos\theta_2) = 2\left(\gamma_S^d \gamma_{LV2}^d\right)^{1/2} - 2\left(\gamma_S^p \gamma_{LV2}^p\right)^{1/2}. \tag{4}$$

The values of γ_L^d and γ_L^p of reference liquids have been taken from Kaelble [18].

Parameters for well-defined liquids are given in previous reports [19, 20]. Water and glycerol were used in this study because only water, glycerol, formamide and CH_2I_2 are considered to have sufficiently reliable and well-tested surface tension values, among the contact liquids.

Elution test using modified silica beads

The copolymer modified silica beads were packed into a stainless steel column for liquid chromatography. The column (80 mm×7 mm i.d.) was connected to an HPLC system including pump (Waters, USA), data processor (Perkin-Elmer Systems, Wellesley, MA, USA) and UV wavelength absorbance detector (Young Lin Sciences, South Korea) set at 254 nm. The elution behavior of four proteins was monitored at a flow rate of 1 ml/min at 20°C. To simplify the study of retention mechanism, PBS was used as the eluent. The phosphate buffer was adjusted to pH 5.0–8.0 with NaCl or NaOH solutions. Although the weak buffer made from phosphate salts is appropriate in the pH range 6.0–8.0, the same buffer salts used for pH 5.0 were made with the same phosphate salts to minimize the salt effect on protein–surface interactions.

RESULTS AND DISCUSSION

Synthesis of the carboxy-terminated polymers

SDM was obtained after the substitution reaction and a detailed characterization of it is reported elsewhere [14]. The purity of the product calculated by NMR peaks was over 99% after purification.

Table 1 lists the information on the synthetic polymers, including the homopolymer (CDMAAm) and copolymers. In the FT-IR spectra of the CDSD copolymers shown in Fig. 2, the peaks of C=O and aromatic C=C are observed at 1690 and 1593 cm^{-1} and the typical two peaks of the sulfonyl group appear at 1134 and 1353 cm^{-1}. The existence of these peaks is evidence of an SDM structure in the copolymers. The composition and relative peak intensity of SDM increased with the increase in the initial feed mol% of SDM. The mol% of SDM in CDSD10, CDSD20 and CDSD30 was 9.9, 18.7 and 28.1, respectively. The pK_a values for

Table 1.
Characterization of carboxyl group-terminated polymers

Code	SDM fraction (mol%)		Molecular weight ($\times 10^{-3}$)	pK_a	Transition point (0.5%, w/v)
	Feed ratio	Observed[a]			
CDMAAm	0	0	27.26	—	—
CDSD10	10	9.9	24.20	6.51	pH 5.55
CDSD20	20	18.7	23.50	6.68	pH 6.23
CDSD30	30	28.1	21.40	6.75	pH 6.71

[a] SDM fraction (mol%) was determined by using a standard curve of concentration of SDM *versus* UV absorbance at 270 nm (r^2 of standard curve = 0.99) [14].

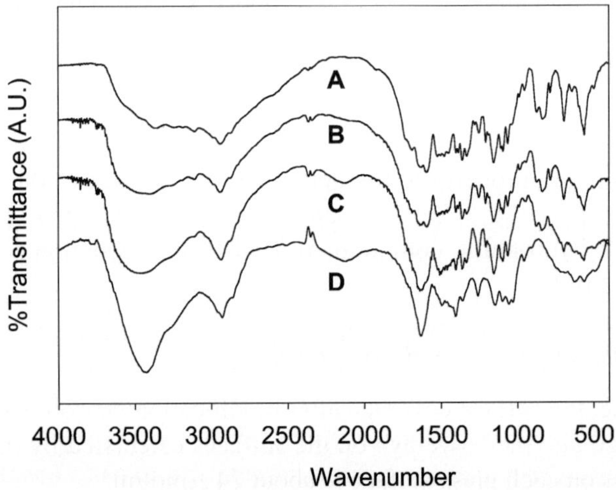

Figure 2. IR spectra of synthesized carboxyl group-terminated homopolymer and copolymers (A, CDSD30; B, CDSD20; C, CDSD10; D, CDMAAm). Two sulfonyl group peaks at 1134 and 1353 cm^{-1} were intensified with increasing SDM content.

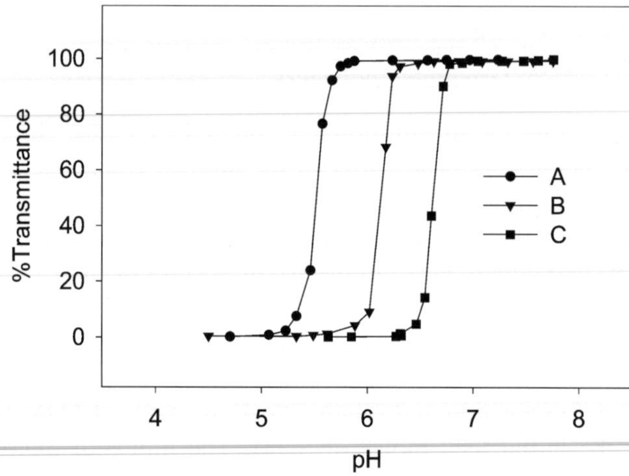

Figure 3. pH-sensitive behavior of synthesized copolymers as a function of pH at 25°C (A, CDSD10; B, CDSD20; C, CDSD30).

the copolymers are slightly higher than the pK_a of intrinsic SDM (pH 6.17). The number-average molecular weight of the polymers estimated from GPC ranges from 21×10^3 to 27×10^3 g/mol.

The phase transition of the pH-sensitive copolymers was characterized by turbidity changes at 500 nm before they were attached to silica glass or beads. As the pH of the solution was decreased, the turbidity of the copolymer solution increased rapidly to near its pK_a point (see Fig. 3) [12, 21]. Sharpness of the pH-sensitive transition was independent of the SDM composition. Since the sulfonyl group in SDM attracts adjacent electrons, the hydrogen of the amide nitrogen is attached less tightly. Therefore, the ionized SDM in the copolymers has a negative charge above its pK_a [22]. In contrast, no pH sensitivity was observed in the solution dissolving CDMAAm polymers.

As the DMAAm composition in the copolymer is increased, the pH-induced transition (0.5% (w/v) solution) shifts to a lower pH. The incorporation of DMAAm imparts a hydrophilic property to the copolymer and increases the local dielectric constant of the polymer. Several studies have reported that polyelectrolytes containing more hydrophilic comonomers lead to the reduction of the acidity of weak acid [12, 23].

DCA test of glass surface grafted with polymers

For the first time, the surface coverage after modification was observed as shown in Table 2. Average polymer coverage on the surfaces calculated by the consumption of amino groups on each glass surface is about 24 μmol/m^2.

Surface properties can reflect both hydrophobic and ionic features of polymers [24]. Therefore, wettability changes at 25°C were estimated by using a dynamic contact angle. The phase transition of the pH-sensitive polymer attached

Table 2.
The average coverage of copolymers on aminopropyl silanized glass surfaces

Surface	Degree of substitution	
	μmol/m^2	%
A	23.9	85
B	23.7	84
C	24.4	87
D	23.2	82

A, CDMAAm-grafted glass surface; B, CDSD10-grafted glass surface; C, CDSD20-grafted glass surface; D, CDSD30-grafted glass surface.

to the glass surface was different from that in the polymer solution. This difference is attributed to the restriction of movement of the linked chain on surface. Therefore, a continuous pH-dependent transition pattern was observed in the case of the polymer-attached glass. Figure 4 shows the changes in the dynamic contact angle on the modified surface as a function of the environment pH. Three kinds of surfaces containing the CDSD series showed hydrophilic/hydrophobic switching around their pK_a, except for the surface grafted with CDMAAm. It was observed that the wettability of surfaces B, C and D started to change dramatically near pH 6.0. The modified surface showed a more effective response to the solution pH in the higher SDM composition. In Fig. 4a, surface D shows an advancing contact angle (θ) of 89° at pH 4.9 and 45° at pH 8.0, while surface A shows an almost constant contact angle (θ) of around 60°. The receding contact angle of surface D changed sharply after wetting, from 33° at pH 4.9 to 0° at pH 8.0, as shown in Fig. 4b. The ionizable polymer grafts in the osmotic region, the layer thickness, assuming that layer thickness is independent of the grafting density, obey the scaling law. At pH levels higher than pK_a, most of the polymer chains are stretched because of the increased osmotic pressure from counterions. The hydrophobicity/hydrophilicity of the copolymer-grafted surfaces is determined by the ionization state of SDM. Thus, the modified surfaces turn hydrophilic and disrupt hydrophobic aggregation when copolymers on the glass surface were exposed to a high pH solution. On the other hand, the intra/intermolecular interactions among the chains were stronger than the solvent–polymer interactions at lower pH. Therefore, the formation of polymer aggregates was accelerated and the flexibility of the chains was reduced when the polymer chains were in the process of deionizing. The transition of the contact angle for all CDSD-grafted glass occurred at one pH unit below the pK_a of the polymers and did not shift to a higher pH when the chemical composition of SDM was increased. The effect of grafting density on the transitions was not considered in this experiment because the grafting density was almost the same in all polymer surfaces.

The surface energy parameters of the polymers were also estimated from interfacial tensions in at least two different solvents. Figure 5 presents the surface free

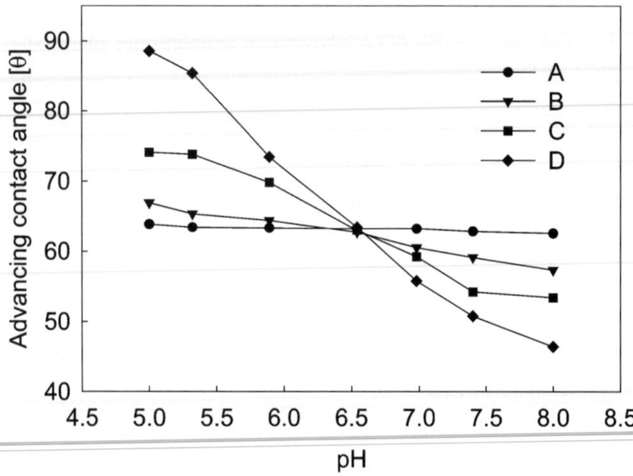

(a)

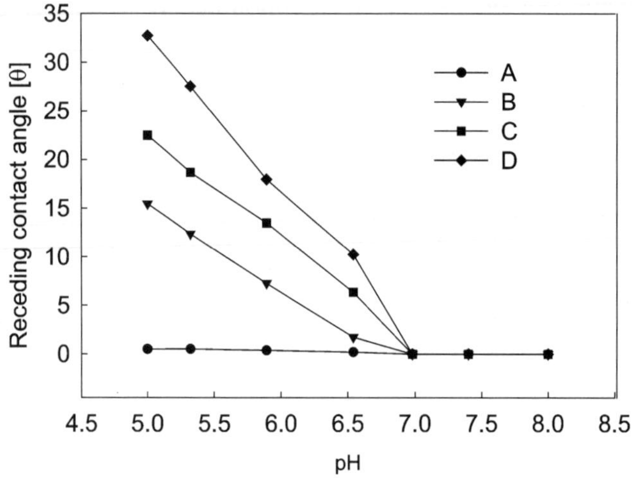

(b)

Figure 4. Changes of (a) dynamic advancing contact angle and (b) receding contact angle of modified glass as a function of pH (A: CDMAAm-grafted glass surface; B: CDSD10-grafted glass surface; C: CDSD20-grafted glass surface; D: CDSD30-grafted glass surface). Depending on solution pH, surfaces grafted by sulfonamide copolymers were switched from hydrophobic/hydrophilic to hydrophilic/hydrophobic.

energy calculated by two-liquid geometric methods to determine the effects of polymer introduction on surface tension. It becomes obvious that the surface modified with a polymer of higher SDM content leads to an essential increase in the surface free energy. The experimental evidence is related to the exposed concentration of sulfonamide groups on the surface layers. It is remarkable that the surface modification almost exclusively influences the dispersive portions, which is to be expected from the increase of SDM functional groups. The dispersive components were sig-

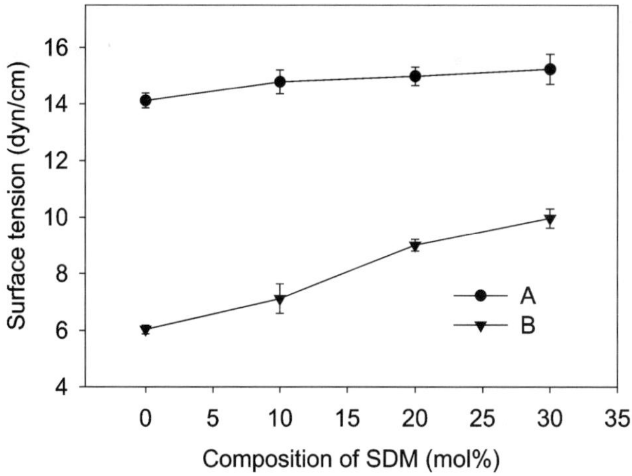

Figure 5. Surface tension depending on the chemical composition (mol%) of SDM in polymer (A, polar component; B, dispersive component) related with the interaction of deionized polymers with proteins.

nificantly influenced by the effect of the benzene structure in SDM. However, the polar component had an almost constant value around 14 dyne/cm. It indicates that hydrophobic force becomes the most critical factor when deionized polymers are interacted with protein.

From atomic force microscopic photographs of the dried glass surface grafted with the copolymers in a dry state, it can be seen that the surface topography did not undergo significant changes as a result of grafting reactions and all surfaces were flat with a roughness less than 5 nm. However, the CDMAAm-grafted surface seemed homogeneous, but more rms roughness (4.3 nm) than those of glasses grafted with CDSD10 (3.3 nm), CDSD20 (2.8 nm) and CDSD30 (2.7 nm). The extent of roughness be probably affected by the composition and configuration, as well as the molecular chain length and density of the copolymer chains [24].

pH-dependent retention time

Several factors must be taken into consideration for the use of the mobile phase. The buffer provides the proper pH environment for the protein of interest. When no particular pH is required, the pH where the protein exhibits optimum stability is best. There is a contribution from the increase of charge repulsion at the protein surface. To be used as the mobile phase, buffer solutions with concentrations ranging from 20 to 100 mM may require the addition of other components such as salt (150 mM for saline) for proteins sensitive to ionic strength [10].

The effects of the mobile phase on the elution behavior of four proteins with different p*I* were examined at four selected pH modes and 25°C. Figure 6 shows the retention times of four proteins in aqueous mobile phases. Two types of glass beads, modified with CDMAAm and CDSD30, were used for comparison between

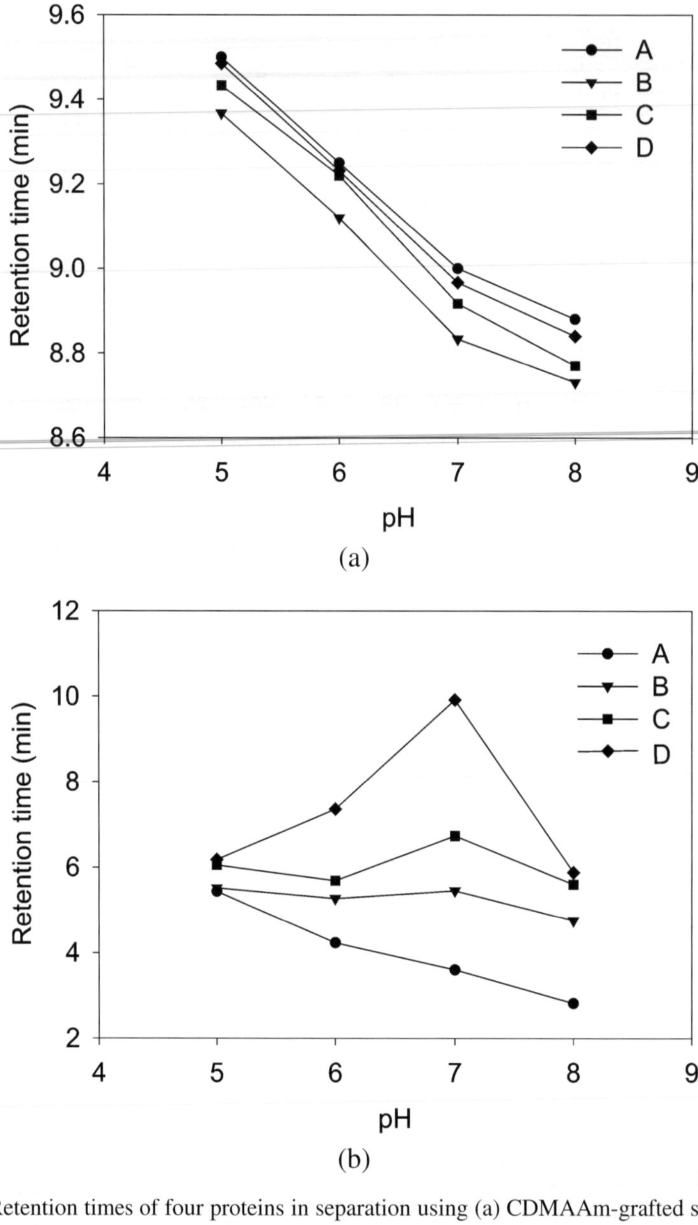

(a)

(b)

Figure 6. Retention times of four proteins in separation using (a) CDMAAm-grafted silica supports and (b) CDSD30-grafted silica supports as a function of pH (A, BSA; B, INS; C, FB; D, MYG).

the nonionic and the ionic polymers. Retention time is defined as the time between the injection of a sample and the appearance of a solute peak at the detector of a chromatographic column. Deuterium oxide was used as the mobile phase to confirm the void volume change of the modified column. The retention time of deuterium oxide was constant in all the columns regardless of pH, indicating that a thin layer grafted with the polymer is well modified on the silica surface and the void volume was not changed by the pH of the mobile phase. The important interactions in pH-sensitive chromatography are electrostatic and hydrophobic interactions rather than irreversible, covalent and disulfide bondings [8]. Therefore, the retention of protein is influenced by the presence of hydrophoibic or charged groups on interfaces between proteins and polymers.

Figure 6a shows different pH-dependent retention times for four model compounds at four pH values (5.0, 6.0, 7.0 and 8.0). A protein is minimally soluble near its pI and the solubility of proteins is gradually increased as the solution pH moves away from the pI value. Hydrophobic proteins easily interact with the hydrophobic stationary phase as the solution pH was increased from its pI to a higher pH. In chromatography using nonionic CDMAAm, the retention time of the protein mainly depends on the hydrophobic interaction force between the stationary phase and analytes, which increases partitioning from the mobile phase to the stationary phase. However, the resolution among the peaks is not good because CDMAAm was not so hydrophobic. The order of the retention time is in accordance with that of protein hydrophobicity (INS $<$ FB $<$ MYG $<$ BSA) [1]. It can be concluded that the four proteins are mainly separated by the differences in hydrophobic interactions with the stationary phase.

In contrast, the retention time of solutes in the CDSD30 column varied according to the solution pH, as presented in Fig. 6b. The charges on the protein influence retention behavior in chromatography. Most bioactive compounds can have electrostatic interactions with CDSD30 because proteins have many ionizable groups on the side chains of their amino acids, as well as their amino- and carboxyl-termini. These include basic groups on the side chains of lysine, arginine and histidine, and acidic groups on the side chains of glutamate, aspartate, cysteine and tyrosine. Therefore, solution pH, the pK of the side chain and the side chain's environment affect the ionized state of each side chain. The relationship between pH and charge for individual amino acids can be described in detail by the Henderson–Hasselbalch equation [14]. Figure 7a shows the charge of the protein depending on the solution pH. In proteins, the pI is defined as the pH at which a protein has no net charge. In general terms, as the pH of a solution increases, deprotonation of the acidic and basic groups in proteins occurs. Therefore, carboxyl groups are converted to carboxylate anions ($R{-}COOH$ to $R{-}COO^-$) and ammonium groups are converted to amino groups ($R{-}NH_3^+$ to $R{-}NH_2$). When pH $>$ pI, a protein has a net negative charge; when pH $<$ pI, the protein has a net positive charge. The positive charges of a protein are enhanced as the solution pH is decreased from the pK_b of amino acids in the protein. Negative charges of a stationary surface that can interact with basic groups on proteins are higher than the pK_a of CDSD30. Therefore, the electrostatic attractions between negatively charged polymers and cationic amino acids in proteins are enhanced within the range from the pK_a of CDSD polymers to the pK_b of amino acids in the protein, as shown in Fig. 7b. Above the pK_b of amino acids in the protein, the electrostatic interaction between the protein surface and the anionic polymer is decreased due to a strong repulsion force among opposite charges. This

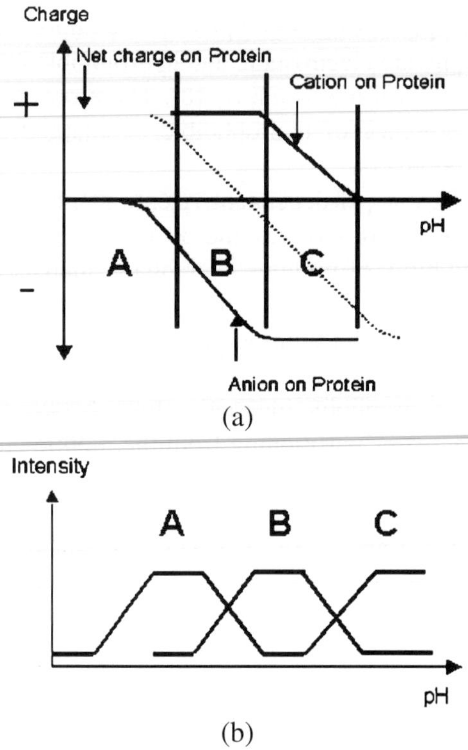

Figure 7. (a) Protein charges and (b) interaction changes between column packing material modified with CDSD30 and protein as a function of pH (A, hydrophobic interaction; B, electrostatic attraction; C, electrostatic repulsion).

repulsion decreases the partitioning of the proteins to the stationary phase. Below the pK_a of the polymer, analytes are separated by nonionic hydrophobic interactions due to the exposure of deionized hydrophobic groups on the polymer-grafted surfaces.

In chromatography using CDSD30, the retention times of the four proteins were increased in the order BSA (pI 5.0) < INS (pI 5.3) < FB (pI 6.1) < MYG (pI 7.1). This order is in agreement with that of the pI values of the analytes. MYG has the highest pI among the four proteins tested [5]. Electrostatic interactions between the anionic charges of the polymer and positive charges of MYG are maximized in the range from the pK_a of the polymer to the pK_b of the amino acids of MYG. The retention time of MYG had the highest value at pH 7.0.

To measure how well the column separates the two proteins, selectivity between two samples is defined as the ratio of the capacity factors of both samples. Protein selectivity in the CDMAAm-grafted column is independent of pH changes of the mobile phase. For example, the selectivity between BSA and MYG is almost constant (1.00–1.02) at all pH values. On the other hand, selectivity in the CDSD30-grafted column is strongly affected by the pH environment of the mobile phase. The selectivity factor between MYG and BSA decreases in the order pH 5.0

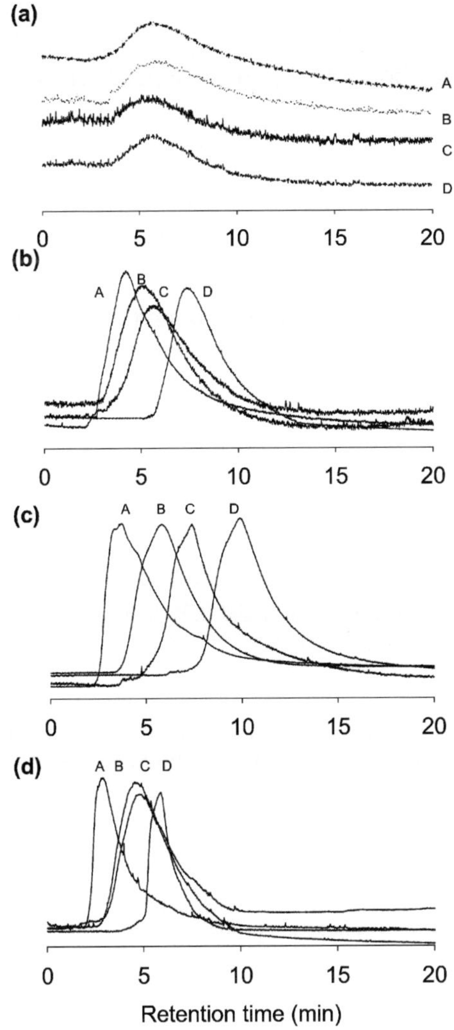

Figure 8. Chromatograms of four proteins using CDSD30-grafted silica supports as stationary phase at four PBS of (a) pH 5.0, (b) pH 6.0, (c) pH 7.0 and (d) pH 8.0 (A, BSA; B, INS; C, FB; D, MYG).

(1.06), pH 6.0 (1.97), pH 7.0 (3.43) and pH 8.0 (2.68). If the selectivity of two components is equal to one, there is no way to separate them by improving the column efficiency [1]. Therefore, the CDSD30-grafted column with a high pH value is more favorable than the others in terms of selectivity.

Figure 8 shows the chromatograms of the four proteins in the column modified with CDSD30. In the case of CDMAAm-grafted silica column, the peaks of the proteins could not be resolved at any pH level (data not shown). The chromatogram using the CDSD30-grafted column at pH 7.0 shows relatively good selectivity among the proteins, while the chromatograms at pH 5.0 and 6.0 showed overlapping unresolved peaks and poor resolution with observable peak tailing.

CONCLUSIONS

Stationary-phase, surface-bonded, pH-sensitive polymers were prepared to separate different proteins in aqueous mobile phases. Chromatograms using non-ionic CD-MAAm showed insufficient resolution due to weak interactions with the proteins. However, protein separation was achieved easily by the use of pH-dependent attraction/repulsion with CDSD. Retention behavior of the proteins was controlled by eluent pH related to interaction between the pH-responsive stationary phase and the protein. Therefore, this separation method using the aqueous mobile phase can be useful for biological and bio-related compounds due to its stable conditions and lack of a need for an organic solvent if sulfonamide is used.

Acknowledgements

This work was financially supported by BK21 program (South Korea).

REFERENCES

1. W. G. Burton, K. D. Nugent, T. K. Slattery and B. R. Summers, *J. Chromatogr.* **443**, 363 (1988).
2. H. Ichijo, R. Kishi and O. Hirasa, *Biotechnol. Bioeng.* **2**, 315 (1994).
3. W.-C. Lee, *J. Chromatogr. B* **699**, 29 (1997).
4. A. A. Gorbunov, L. Y. Solovyova and A. M. Skvortsov, *Polymer* **39**, 697 (1998).
5. J. Kobayashi, A. Kikuchi and T. Okano, *Anal. Chem.* **73**, 2027 (2001).
6. H. Kanazawa, K. Yamamoto, Y. Matsushima, N. Takai, A. Kikuchi, Y. Sakurai and T. Okano, *Anal. Chem.* **68**, 100 (1996).
7. H. Kanazawa, Y. Matsushima and T. Okano, *Trends Anal. Chem.* **17**, 435 (1998).
8. H. Kanazawa, Y. Kashiwase, K. Yamamoto, Y. Matsushima, A. Kikuchi, Y. Sakurai and T. Okano, *Anal. Chem.* **69**, 823 (1997).
9. C. Boutris, E. G. Chatzi and C. Kiparissides, *Polymer* **38**, 2567 (1997).
10. M. Kaufmann, *J. Chromatogr. B* **699**, 347 (1997).
11. G. Chen and A. S. Hoffman, *Polym. Gels Networks* **196**, 1251 (1995).
12. S. Y. Park and Y. H. Bae, *Macromol. Rapid Commun.* **20**, 269 (1999).
13. S. I. Kang and Y. H. Bae, *Macromolecules* **34**, 8173 (2001).
14. S. I. Kang and Y. H. Bae, *J. Control. Rel.* **80**, 145 (2002).
15. S. S. Lateef, S. Boateng, T. J. Hartman, C. A. Crot, B. Russell and L. Hanley, *Biomaterials* **23**, 3159 (2002).
16. H. Kanazawa, T. Sunamoto, E. Ayano, Y. Matsushima, A. Kikuchi and T. Okano, *Anal. Sci.* **18**, 45 (2002).
17. J. D. Andrade, S. M. Ma, R. N. King and D. E. Gregonis, *J. Colloid Interface Sci.* **72**, 488 (1979).
18. D. H. Kaelble, *J. Adhesion* **2**, 66 (1970).
19. C. J. van Oss, R. F. Giese and R. J. Good, *Langmuir* **6**, 1711 (1990).
20. D. E. Sullivan, *J. Chem. Phys.* **74**, 2604 (1981).
21. A. Cammarata and R. C. Allen, *J. Pharm. Sci.* **56**, 640 (1967).
22. V. de Zea Bermudez, C. Poinsignon and M. B. Armand, *J. Mater. Chem.* **7**, 1677 (1997).
23. R. A. Siegel and B. A. Firestone, *Macromolecules* **21**, 3254 (1988).
24. T. Yakushiji and K. Sakai, *Langmuir* **14**, 4657 (1998).

Gels, Genes, Grafts and Giants, pp. 361–395 (2005)

Review

Molecular composition and mechanical properties of biopolymer interfaces studied by sum frequency generation vibrational spectroscopy and atomic force microscopy

TELLY S. KOFFAS [1], ELLA AMITAY-SADOVSKY [1,2], JOONYEONG KIM [1] and GABOR A. SOMORJAI [1,*]

[1] *Department of Chemistry, University of California, Berkeley and Materials Sciences Division, Lawrence Berkeley National Laboratory, Berkeley, CA 94720, USA*
[2] *Department of Mechanical Engineering, University of California, Berkeley, CA 94720, USA*

Received 30 September 2003; accepted 9 December 2003

Abstract—Sum frequency generation (SFG) vibrational spectroscopy and atomic force microscopy (AFM) have been used to study the surface structure and surface mechanical behavior of biologically-relevant polymer systems. These techniques have emerged as powerful surface analytical tools to deduce structure/property relationships *in situ*, at both air/solid and air/liquid interfaces. SFG and AFM studies have been performed to understand how the surface properties of polymers are linked to polymer bulk compositions, changes in the ambient environment, or the degree of mechanical strain. Specifically, this review discusses (1) the macroscopic- and molecular-level tracking of small end groups attached to polyurethane blends, engineered to reduce blood clotting; (2) the role of ambient humidity on the surface mechanics of soft contact lenses possessing different water content in the bulk; (3) the affect of cyclic stretch on the molecular surface structure of polyurethane films, designed to mimic the mechanical deformation caused by heartbeat; and (4) the molecular ordering of functional groups at the polystyrene–protein interface. The correlation of spectroscopic and mechanical data by SFG and AFM is a powerful methodology to study and design materials with tailored surface properties.

Key words: Sum frequency generation (SFG) vibrational spectroscopy; atomic force microscopy; contact lens; protein adsorption; surface mechanics; polyurethane; molecular ordering.

INTRODUCTION

Biological reactions occur at interfaces. The fact that surfaces have been exploited and optimized by nature should come as no surprise. Despite the inherent

*To whom correspondence should be addressed. Tel.: (1-510) 642-4053. Fax: (1-510) 543-9668. E-mail: somorjai@socrates.berkeley.edu

complexity of biological systems and their interactions with synthetic biomaterials, the basic concepts of surface science still apply [1, 2]. Surfaces offer a readily accessible and low-energy pathway for chemical conversion that is unavailable in aqueous solution [3]. High surface area geometries can be used to drive complex or energetically unfavorable biological reactions toward higher turnover rates [4]. Surfaces also offer opportunities for precise molecular self-assembly and epitaxial growth [5]. A more complete understanding of the molecular surface structure and mechanical properties of biopolymers and how they are affected by protein adsorption, mechanical deformation, or changes in the local environment is needed because these factors define what it means for a material to be biocompatible [6–9].

A major hurdle in developing a surface science model for biomedical systems has been the limitation imposed by traditional electron-based, surface analytical tools [10, 11]. Atomic and molecular-level studies of the surface monolayers of solids often require the use of very low background pressures or even ultrahigh vacuum to allow electrons and other probes to reach or leave the surface unimpeded. Biology, however, involves solid-liquid or solid-air interfaces under more routine pressures. The development of optical and scanning probe techniques which could be conducted outside of a vacuum has allowed researchers to study more interesting systems under biologically-relevant conditions [12–18]. Sum frequency generation (SFG) vibrational spectroscopy and atomic force microscopy (AFM) have emerged as powerful *in situ* surface analytical tools to deduce structure/property relationships in a host of model polymer systems [19–25]. SFG provides a chemical description of the surface — what functional groups are present, their orientation and the general state of molecular order. In turn, polymer surface chemistry underlies phenomena such as adsorption, wettability, friction, lubrication, adhesion and, ultimately, biocompatibility [1, 26–30]. By determining the morphology and mechanical properties associated with certain surface chemical compositions and environmental variables, AFM can complement the spectroscopic data collected by SFG and paint a clearer portrait of what is occurring at the surface.

This review summarizes recent studies in our group concerning the chemical and mechanical interrogation of polymer, copolymer, polymer blend and hydrogel surfaces at air, water, or partially humidified interfaces [24, 31–35]. The systems discussed include (a) a comparison of the surface chemistry and structure of two polyurethane blends, Biospan-SP/Phenoxy (BSP/PHE) and Biospan-F/Phenoxy (BF/PHE), capped with different surface modifying end groups; (b) differences in the surface mechanical properties of neutral poly(HEMA) (HEMA, hydroxyethyl methacrylate) and ionic poly(HEMA + MA) (MA, methacrylic acid) soft contact lenses; (c) changes in the surface structure of stretched polyurethane block copolymers; and (d) the structure of adsorbed protein and the effects of protein adsorption on the orientation and ordering of functional groups present on polystyrene surfaces. The relevant molecular structures for these polymers are displayed in Fig. 1.

MATERIALS AND METHODS

Several surface sensitive techniques have already been used for studies of biological interfaces, each having certain advantages and drawbacks. Contact angle measurement [36] is a standard method to monitor macroscopic changes in surface energy, but yields no molecular information. Fourier transform infrared (FT-IR) and Raman spectroscopies possess good wavenumber resolution but lack the inherent surface sensitivity of SFG [37–39]. X-ray photoelectron spectroscopy (XPS) and secondary ion mass spectroscopy (SIMS) are surface sensitive but require a vacuum, even when used with a cold stage [40–42]. Described below are the two major analytical methods used in our laboratory to collect molecular and macroscopic information at interfaces under natural ambient conditions.

Sum frequency generation

SFG is a non-linear spectroscopic method that provides a vibrational spectrum of functional groups at a surface. In an SFG setup (see Fig. 2), a pulsed visible laser beam (frequency ω_{vis}) is overlapped at the surface with a tunable pulsed IR laser beam (frequency ω_{ir}) and the light emitted, by the second-order nonlinear optical process, at the sum frequency, $\omega_{\text{sum}} = \omega_{\text{vis}} + \omega_{\text{ir}}$, is detected by a high-gain photodetector (such as a photomultiplier). The intensity of the light at ω_{sum} is proportional to the intensities of the visible and IR lasers respectively and the square of the second-order susceptibility $\chi^{(2)}$ of the surface. $\chi^{(2)}$ vanishes for a material with inversion symmetry; thus, bulk materials do not usually generate sum frequency output. $\chi^{(2)}$ is non-zero for a surface (or an interface) because surface molecules lack inversion symmetry. A plot of SFG intensity *versus* the frequency of the IR laser yields a vibrational spectrum of surface molecules. SFG provides surface information *in situ* because it can be applied to any interface accessible by light. Different polarization combinations of visible, IR and sum frequency light access different components of $\chi^{(2)}$, which arise from the surface molecular orientations. Therefore, information concerning the orientation of surface molecules can be obtained by SFG.

The SFG experiment is performed by overlapping a visible and tunable IR beam on a polymer surface at incident angles of 40° and 50°, respectively. The visible beam, 532 nm, is a small portion of the frequency doubled 1064 nm output from a Continuum YAG-PY61 laser (generating approx. 20 ps pulses at 20 Hz). The IR beam, tunable from 1100 to 4000 cm^{-1}, is generating from a combined OPG/OPA (optical parametric generator/amplifier) and DFM (difference frequency mixing) system composed of KTP (potassium titanyl phosphate), KTA (potassium titanyl arsenate) and $AgGaSe_2$ crystals pumped by the fundamental output of the YAG. The SFG output signal is collected by a photon counting system and a gated integrator.

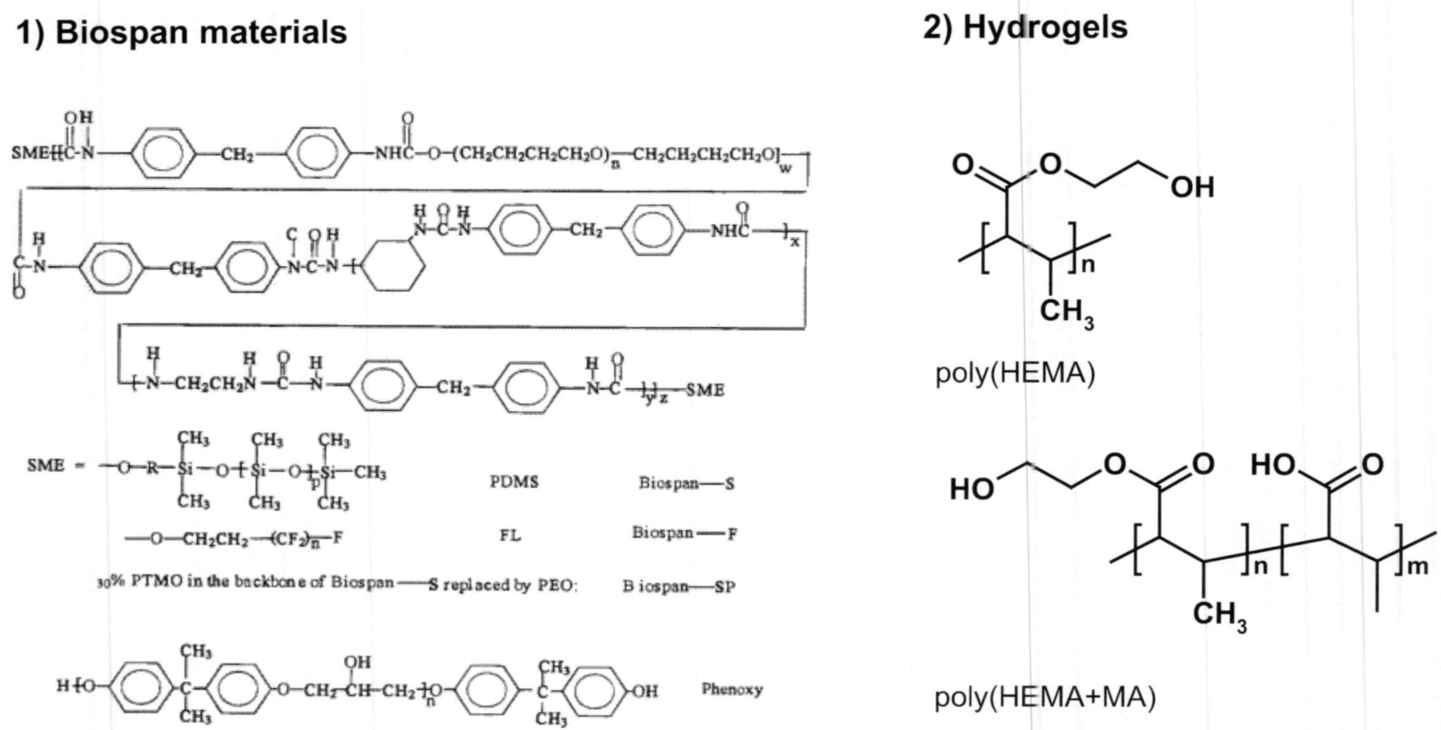

Figure 1. Molecular structures of polymers referenced in the review: (1) Biospan (BS) and phenoxy (PHE) materials with surface modifying end groups, (2) hydrogel materials poly(2-hydroxyethyl methacrylate) (poly(HEMA)) and poly(HEMA) copolymerized with a small amount ($n \ll m$) of methacrylic acid (poly(HEMA + MA)), (3) hard and soft polyurethane segments and (4) partially deuterated polystyrene (PS-d_3) and perdeuterated polystyrene (PS-d_8).

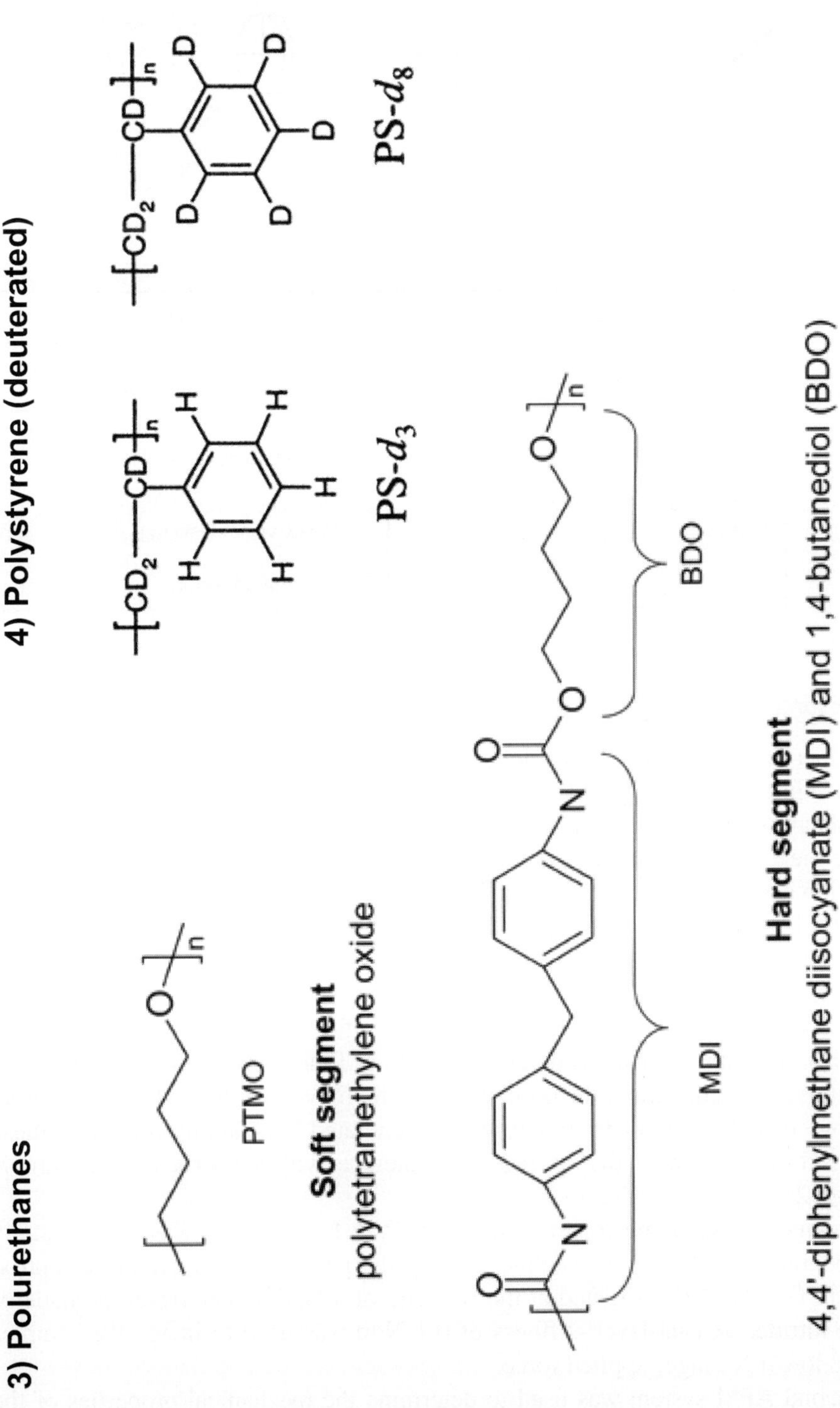

Figure 1. (Continued).

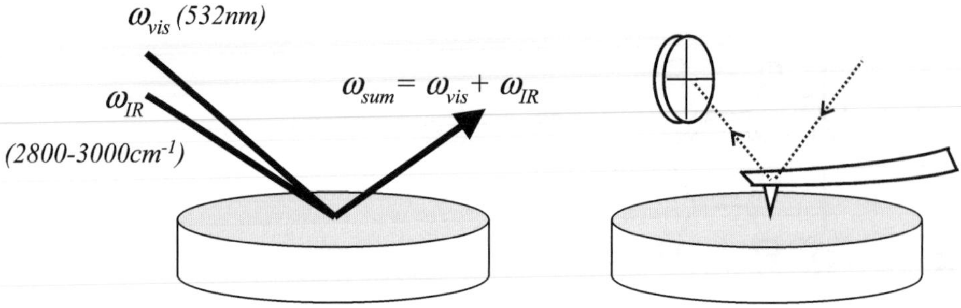

SFG	AFM
Surface sensitivity: monolayer	Surface sensitivity: subnanometer to several hundred nanometers
Surface monolayer composition Interface ordering of molecular groups	Imaging of surface lateral morphology (submicron to several micron) Surface mechanical behavior (stiffness, viscoelasticity) Friction and adhesion

Figure 2. Schematic of SFG surface vibrational spectroscopy and AFM experiments.

Atomic force microscopy

AFM operates by monitoring the deflection of a cantilever due to the attractive and repulsive intermolecular forces (e.g. van der Waals forces) acting between the AFM tip and sample. In the contact regime, the tip is gently lowered into the sample via the piezoelectric material to which the cantilever arm is attached. Cantilever deflection can be calibrated to a force by adjusting the slope of the curve measured against a hard reference material to the nominal spring constant of the cantilever. As the tip is rastered across a surface, the horizontal and vertical deflections of the cantilever are monitored using a photodiode. In this setup, a laser beam is bounced off the top-side of the cantilever and reflected into a photodiode which measures the beam intensity separately in each of four areas. By comparing beam intensities in each quadrant, the normal and lateral movement of the tip is measured. A topographical image or frictional map is generated by spatially recording these motions during a scan, as the tip traces and interacts with the surface of the sample (see Fig. 2).

A commercial scanning force microscope (Park Instruments) was used to study the topography of all polymer samples using a sharp tip (radius of curvature approx. 20 nm). The AFM tip, situated at the free end of a flexible cantilever, is made of silicon nitride. A cantilever stiffness of 0.1 N/m was used to image the samples under a low (nN range) applied force.

A second AFM system was used to determine the mechanical properties of the polymer samples. This homebuilt, walking-style AFM scanning head, controlled by

RHK electronics (Troy, MI, USA), has been previously described and is completely enclosed within a 30-l glass bell jar [43]. Relative humidity (RH) can be varied in this system by balancing the evaporation of water from a reservoir within the jar with a steady flow of nitrogen through the chamber. Decreasing the flow rate of nitrogen increases the experimental humidity, which is measured by a digital hydrometer placed inside the chamber.

RESULTS AND DISCUSSION

Surface-modified polyurethane blends

Polyurethane-based polymers are currently used in a number of blood-contacting implants such as heart-valve leaflets, stents and catheter tubing [44]. The advantages of polyurethanes include their favorable mechanical properties and ease of molding [45–47]. The bulk mechanical properties of polyurethanes can be modified to have a glass transition temperature slightly above room temperature — meaning that the implant will soften when it is placed in the body. However, most polyether polyurethanes are susceptible to surface oxidation and environmental stress cracking. Many past attempts to remedy these problems by replacing the bulk ether soft segment with siloxane have failed because of the degraded mechanical properties and *in vitro* instability.

The Polymer Technology Group has developed a method of altering the surface properties of the polymer by covalently attaching an oligomeric surface-modifying small monofunctional end (SME) group ($\leqslant$3 wt%), which is surface active, to the polyurethane backbone [48]. The SME groups occupy the termini of linear polyurethane chains and are, therefore, more flexible than the backbone groups. When implanted, the SME groups of these materials migrate to the polymer surface, depending on the hydrophobicity of the chemical environment at the interface. Methyl groups, for instance, prefer to interact with a hydrophobic medium like air, whereas hydroxyl groups are stabilized in aqueous environments. A small amount of SME groups will not significantly alter bulk mechanical properties; they will, however, completely dictate surface properties. This method more effectively controls the surface density of the modifier than other, more conventional, surface treatment methods (e.g. plasma and glow discharge), since the groups are bonded to specific molecules within the polymer. The use of these SME groups also allows more precise control of the surface composition than relying on block polyurethane blends. In addition, only small changes in polymer development are required to make many different types of SME polyurethanes.

Biocompatiblity testing of SME polyurethanes has shown a significant decrease in both surface oxidation and environmental stress cracking compared to unmodified control polyurethanes [49]. Further application of SME technology would include designing end groups that not only prevent oxidation and cracking, but also direct favorable serum protein/cell adhesion responses. For example, polyurethanes that have polyethyleneoxide (PEO) end-groups are of significant interest since PEO has been shown to resist protein adsorption [50–52].

The mechanism by which protein adsorption is prevented is still not clear, but one possible explanation is based on steric repulsion theory [53, 54]. This theory suggests that when a protein approaches a surface covered by water-soluble PEO, the surface grafted chains are compressed. Compression results in decreased conformational entropy for the linked PEO chains, which is energetically unfavorable. The system tends to move back to the higher entropy state by removing adsorbed protein. The chain length of PEO and spacing of PEO segments at the surface are considered to be important variables in preventing protein adsorption, however the specific details of spatial orientation and conformation are unclear and require systematic investigation.

A series of polyurethanes containing different monofunctional SME groups (<3.0 wt%) were prepared by coupling end groups to the backbone of a polyurethane. The surface activity is controlled by the appropriate selection of SME groups. The SME groups studied in our laboratory are polydimethylsiloxane (PDMS), polyethyleneoxide (PEO) and sulfonate SO_3^- (SO). The names and molecular structures of these samples are shown in Fig. 1. The SFG surface spectra of these modified polyurethanes in air are shown in Fig. 3. The SFG spectra of polytetramethyleneoxide (PTMO), the parent backbone of the BioSpan and PDMS, the end group of Biospan-S (BS), are also shown in Fig. 3a and 3b for comparison. SFG spectra for these modified polyurethanes show pronounced spectral differences, despite nearly identical bulk compositions. The strong peak at about 2915 cm^{-1} in BioSpan-S (BS, Fig. 3d) is the symmetric C−H stretch of Si−CH_3 in PDMS (Fig. 3b), indicating that the PDMS end groups cover most of the BS surface. In contrast, the SFG spectrum of BioSpan-P (BP) does not show the strong PEO peak expected close to 2865 cm^{-1}. This suggests that the hydrophilic PEO end groups of BP do not exist at the air interface.

The similarity of the SFG spectra of BP and PTMO (BioSpan backbone) indicates that the BP surface is covered by PTMO backbone. For BP, the hydrophilic end-groups (PEO) stay in the bulk and the surface is covered by PTMO backbone. Interestingly, ATR-FTIR spectra for these compounds are nearly indistinguishable, indicating that SFG is sufficiently surface-sensitive to detect end groups that are not probed by other conventional techniques [55, 56].

Solvent contact angle measurements were performed to analyze the SME-modified polyurethane surfaces. The results are in good agreement with the SFG studies and show that the surface tensions of BioSpan derivatives change as the SME content changes [56]. XPS results are also compatible with the SFG data [55]. Further, these studies reveal that grafted end groups incorporated into the polymer will not necessarily remain at the surface.

Binary polymer blends with different compositions. Generally, polymer blends provide a way to improve upon the properties of a pure polymer *via* mixing with

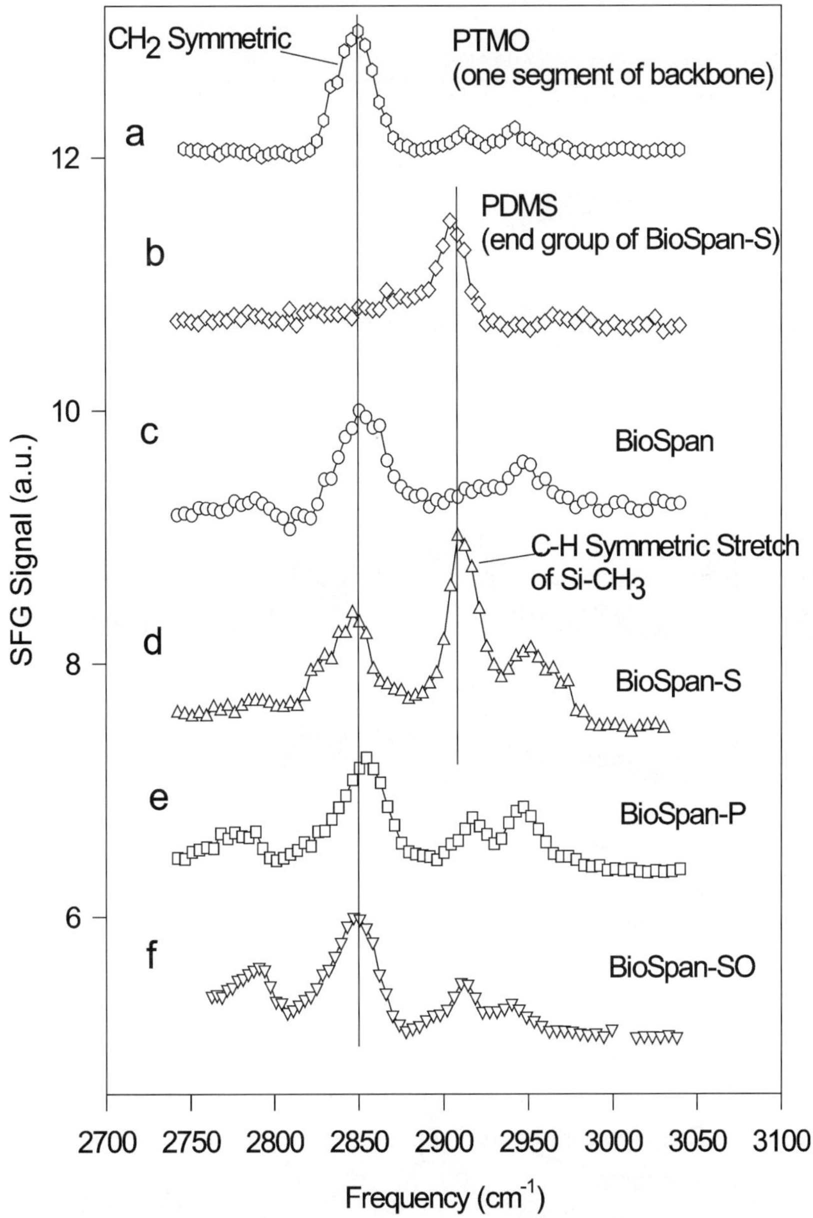

Figure 3. SFG spectra of polyurethane derivatives with different SME groups at the air interface.

one or more miscible polymers [57, 58]. One of the components is typically more surface active and will tend to segregate to the surface. We prepared binary polymer blends containing a phenoxy base polymer (PHE) as the major component and one of BS, BSP and BF as the minor component (see Fig. 1). The bulk concentration of the minor component is less than 4 wt% in all cases. The surface properties of three

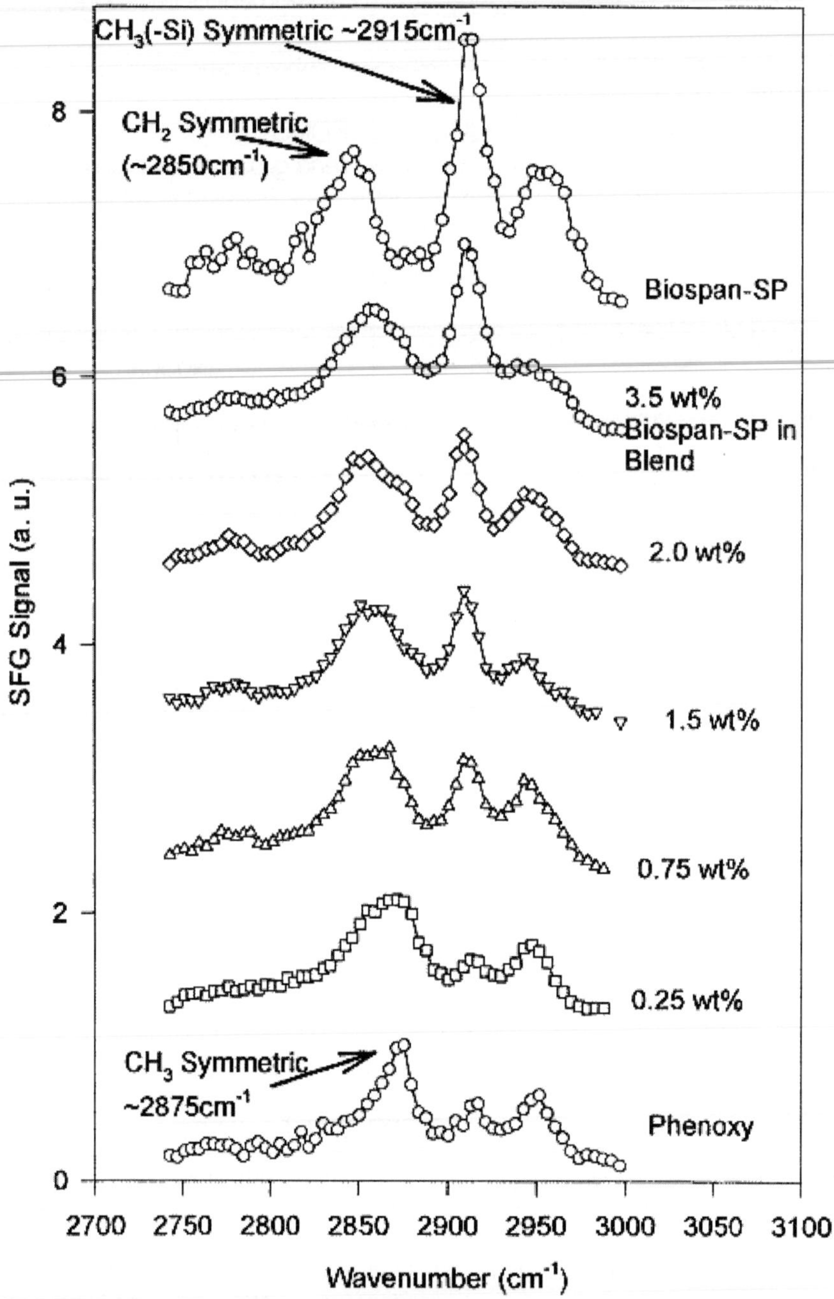

Figure 4. SFG spectra of BS/PHE at the air interface.

polymer blends, BS/PHE, BSP/PHE and BF/PHE, were studied by SFG, AFM, XPS and contact angle goniometry. Comparison of the results for these three polymer blends shows that the surfaces of polymer blends in air are sensitive to both the bulk concentrations and surface energies of the minor component used in the blend. From the SFG results (Fig. 4), it was found that the surface of BSP/PHE is fully covered by BSP when its bulk concentration is about 3.5 wt%. On the other hand, the air/polymer interface becomes saturated by BS and BF for BS/PHE and BF/PHE binaries when bulk concentrations are 1.7 wt% (BS) and 1.0 wt% (BF), respectively. The surfaces, however, are covered by PHE when the bulk concentrations of BSP, BS and BF are less than 0.25, 0.17 and 0.125 wt%, respectively (Fig. 4, BS/PHE as an example). The surface compositions change sharply from one limit to the other in the intermediate range, which means that polymer blend surface can be modified in a step wise fashion by changing the bulk concentration. The contact angle measurements and XPS studies give compatible results. The polymer with a lower surface tension (BF = 16 dyne/cm, BS = 22 dyne/cm, BSP = 26 dyne/cm) saturates the surface at a lower bulk concentration (BS (1.7 wt%)/PHE, BSP (3.5 wt%)/PHE, BF (1.0 wt%)/PHE).

AFM images of the blend surfaces for different bulk BS concentrations are shown in Fig. 5. For BS concentrations lower than 0.17 wt%, (Fig. 5-1c), the surface appears featureless and resembles that of pure PHE (Fig. 5-1a). Fig. 5-1d is an AFM image obtained at a concentration higher than 1.7 wt% and appears similar to the surface of pure BS (Fig. 5-1b). The AFM results are in good agreement with SFG data. AFM images of blend surfaces for BS bulk concentrations in the range of 0.17 and 1.7 wt% exhibit a domain structure (Fig. 5-2). Consequently, we attribute these domains to the preferential segregation of the BS component to the surface. In doing so, the PHE component is displaced at the surface and forms the observed ridges in the images. From the shape of these patterns one can conclude that the segregation is isotropic in the plane of the surface, since the polygons show no preferential orientation. By using a combination of different surface-sensitive techniques, more complete pictures of polymer blend surface structure are obtained. These findings confirm that the surface is not necessarily representative of the bulk in terms of either composition or structure.

Reconstruction of polymer surfaces. Another issue in studying polymer surfaces is related to the rearrangement of particular functional groups when exposed to different environments. For example, it is sensible to study the surfaces of biopolymers in a hydrated state because implanted devices are usually in contact with blood or living tissues. SFG data reveal that the BioSpan-S (BS, Fig. 1) surface undergoes a significant reordering when exposed to water [59]. In air, the SFG spectrum shows that the hydrophobic PDMS end-groups cover most of the surface (Fig. 3d, consistent with the XPS results) [60]. When immersed in water, the PDMS component gradually retreats from the surface and the initially 'buried', more hydrophilic part of the polymer, appears at the surface. Specifically,

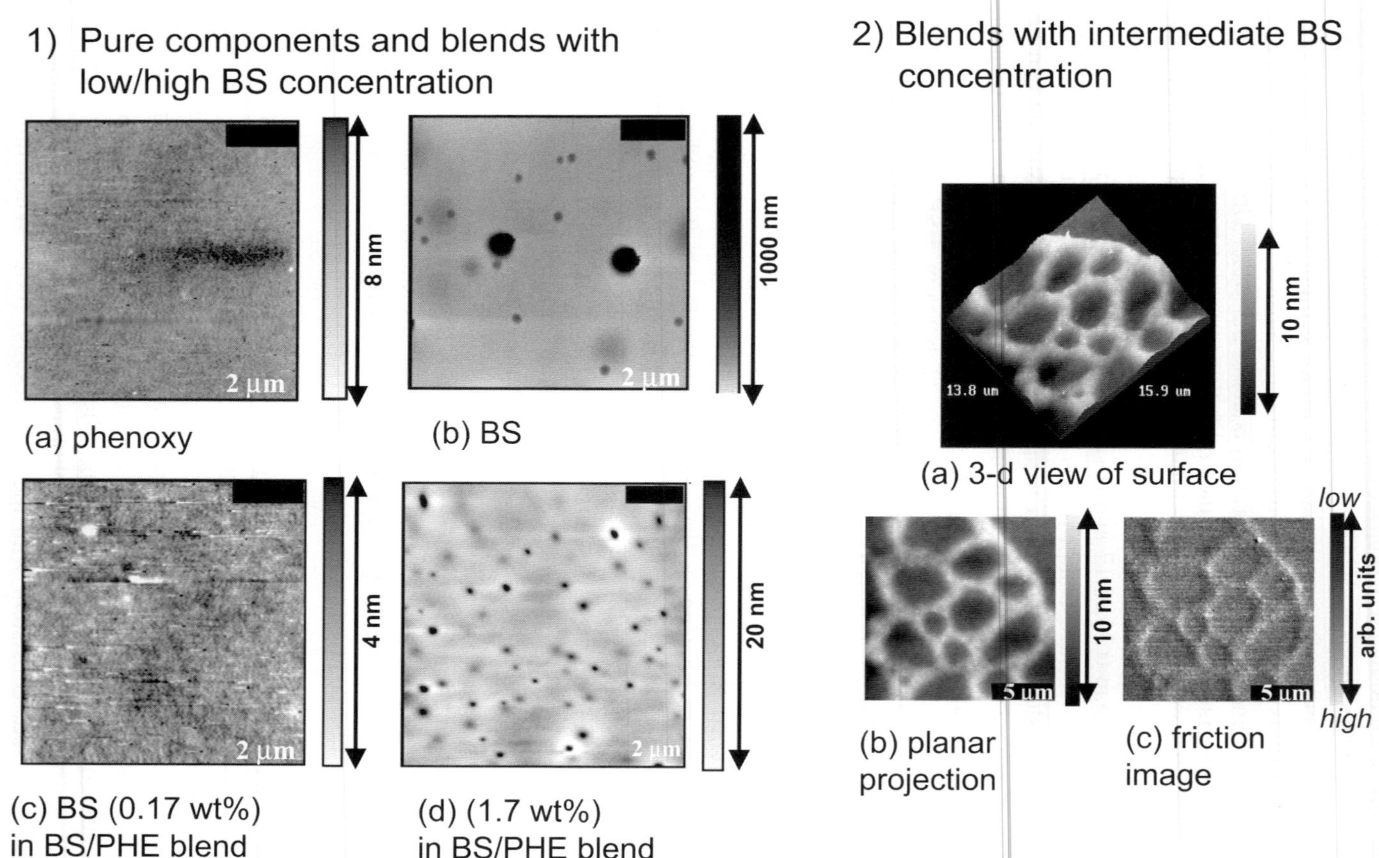

Figure 5. AFM images for BS/PHE blends. 5-1: AFM images of pure PHE, pure BS and the blends at low and high BS concentration (left). 5-2: AFM images of BS/PHE blend at intermediate BS concentration (right).

the CH_3 group in the PDMS end group has an average tilting angle of $\theta = 35 \pm 5°$ with respect to the surface normal in air. This angle changes to 60° when BS is in contact with water for 51 h, i.e. CH_3 groups incline more towards the surface plane. This results in the attenuation of peak intensity at 2918 cm^{-1}, with respect to 2851 cm^{-1} (Fig. 3d). Figure 6 shows the ratios of these two peaks as a function of hydration time (open circles). The surface structural transformation in response to the environmental change from air to water takes about 25 h at 300 K. The contact angle measurement also reveals time-dependent variations in the BS surface in water and the results are consistent with SFG data (closed circles in Fig. 6).

Polymer blends respond to an aqueous environment by inducing the more hydrophilic component of the blend to segregate to the surface [31, 61]. BSP dominates the surface of 3.5 wt% BSP/PHE in both air and water, but its hydrophilic and hydrophobic components reorient after exposure to water. The BF (1.0 wt%)/PHE blend surface is completely covered by BF in air, but some of the PHE component diffuses to the surface after contacting water. At 5.0 wt% BF, the surface is fully covered by BF in both air and water.

SFG, AFM and contact angle goniometry have been used in concert to study the BS/BP blend surface. As expected, the surface structure of the polymer blend differs from the bulk. The BS surface content becomes noticeable at BS bulk concentrations as low as 0.17 wt%. It increases with the BS concentration and reaches full coverage at 1.7 wt%. There is a strong tendency for BS to segregate to the surface. When BS and BP coexist on the surface, AFM images reveal a pattern

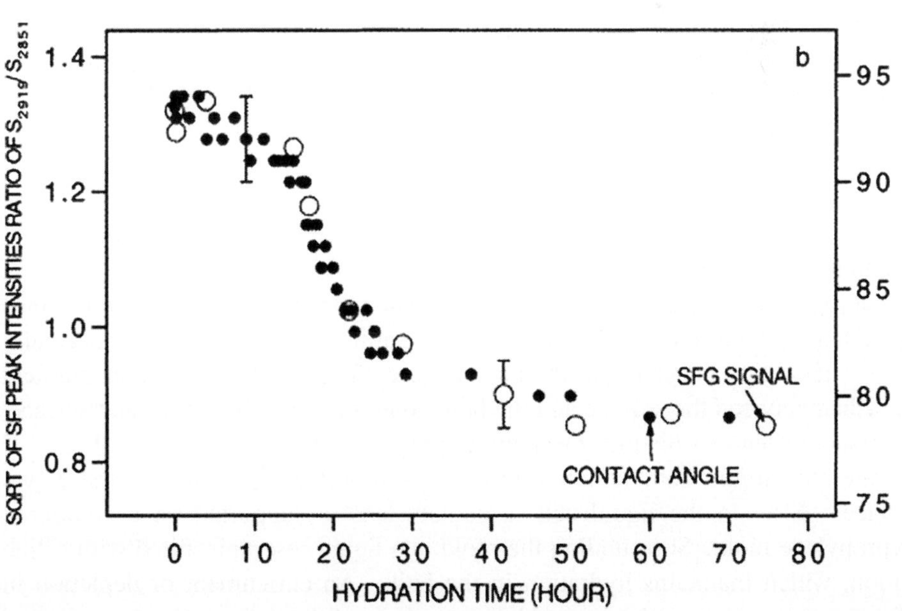

Figure 6. Surface reconstruction of BS under water measured by SFG and contact angle goniometry.

of domains. It is clear that by using SFG and AFM in combination, deeper insight into *in situ* phenomena occurring at biopolymer interfaces can be obtained.

Surface mechanical properties of hydrophilic contact lenses under variable humidity

The mechanical and chemical properties of the near-surface region of hydrogels determine, in large measure, the comfort of their use as contact lenses [25, 62, 63]. Consequently, the lens surface chemistry is often modified to tune the interfacial water content, lubricate against shearing friction and render the contact lens more biocompatible [64, 65]. The chemical composition and water content near the hydrogel surface, however, are altered on exposure to external environmental variables (e.g. temperature and humidity) [62, 66]. As the interfacial water content re-equilibrates with changes in the environment, the surface mechanical properties of the contact lens will be affected. Changes in the surface mechanical response of the lens reflect the degree of interfacial hydration.

Water not only influences surface stiffness and tribology, it is also involved in the biodeposition of protein and lipid from tear fluid, mediates oxygen transport to the cornea and promotes the dimensional stability of the lens [67–70]. Thus, losses in interfacial water content compromise mechanical and physiological performance and lead to decreases in visual acuity — all issues of major concern to contact lens wearers. While many synthetic strategies are aimed at improving lens comfort, the state of bulk and interfacial hydration and its impact on surface mechanics is not well understood [71–73].

Synthetic hydrogels for biomedical use were developed during the 1960s, with derivatives of hydroxyethyl methacrylate (HEMA) remaining the most popular contact lens material [74, 75]. Cross-linked poly(HEMA) networks are soft and flexible materials, possessing excellent mechanical and optical properties. Figure 1 displays the chemical structures of poly(HEMA) and its more hydrophilic copolymer derivative, poly(HEMA) copolymerized with methacrylic acid (poly(HEMA + MA)). The ionic nature of methacrylic acid increases the equilibrium water content in the bulk from 38 to 55%, by weight.

In hydrogels, water (de)hydration is regulated at two regions: the lens/air interface and the lens/eye interface. Bulk rehydration of saline provided by the eye normally offsets evaporation occurring at the air interface. The steady-state water gradient established between the surface and the bulk creates a partially dehydrated surface, which should lead to distinct mechanical properties for both regions. Figure 7 illustrates the liquid cell, which was used to simulate the action of these two interfaces [63]. In the liquid cell, a contact lens is supported on a contoured polypropylene mold. Surrounding the mold is a liquid reservoir, filled with saline solution, which maintains hydration in the bulk. An enrichment or depletion in surface water content can be gauged by an analysis of the deformation experienced by contact lenses under the tip of an Atomic Force Microscope (AFM). This

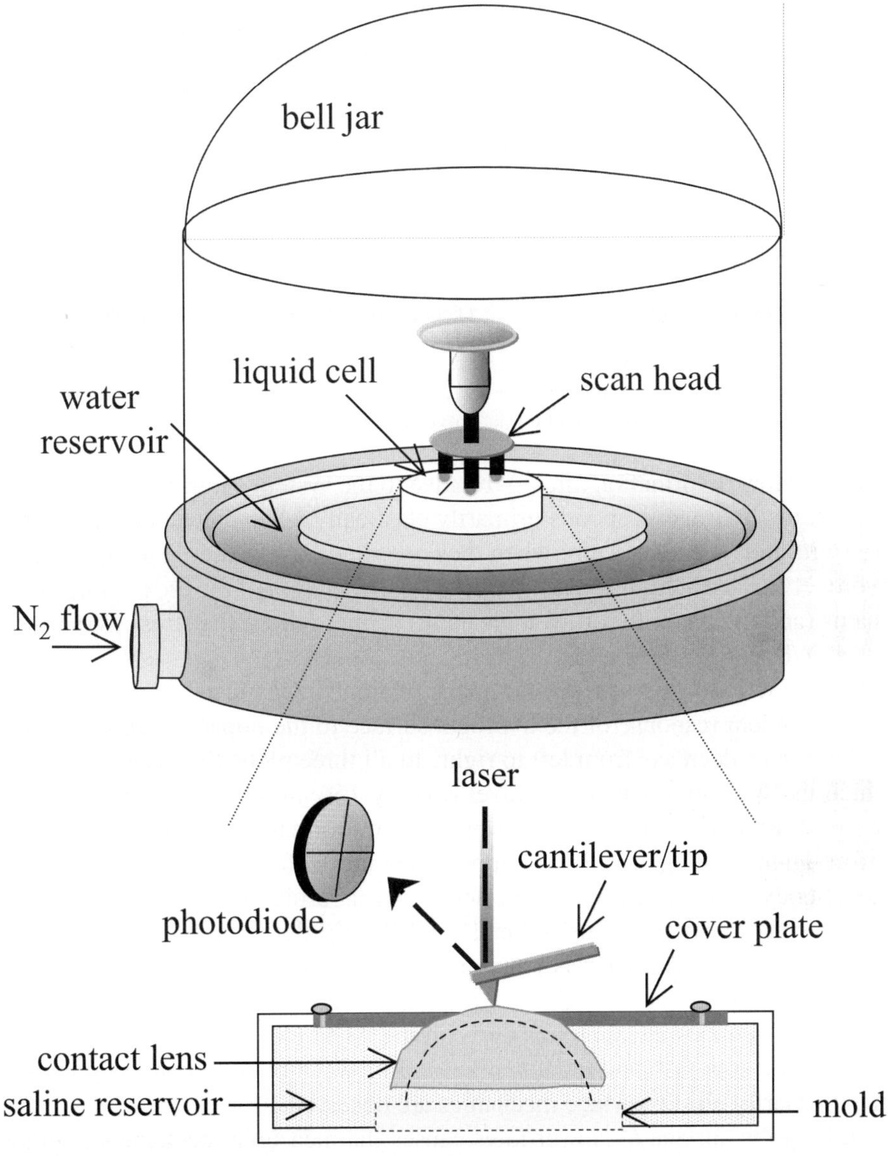

Figure 7. AFM experimental setup for hydrogel experiments. A bell-jar encloses the scan head and Teflon liquid cell. The liquid cell cross-section (in detail) shows a contact lens supported on a polypropylene mold and bulk-hydrated from the surrounding saline solution.

deformation is a combination of the viscoelastic strain of the polymer chains and the viscous compression of the surrounding water [66, 73].

Characterization of viscoelasticity in AFM load vs. displacement measurements. The presence of water in the hydrophilic polymer network complicates the physical properties of a contact lens. It responds as a solid to fast rates of deformation or as a

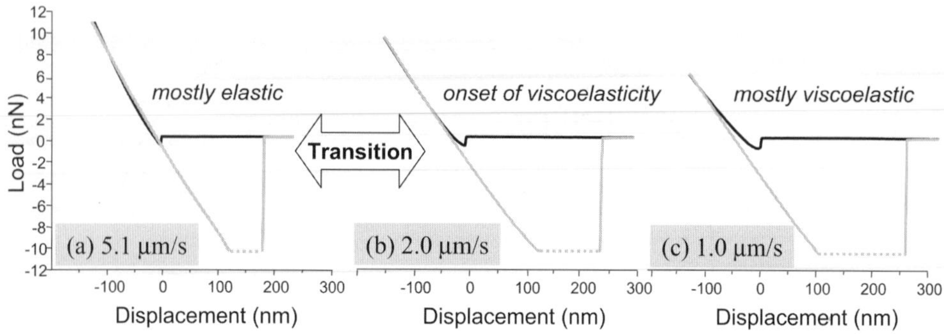

Figure 8. AFM load *vs.* displacement plots taken at (a) 5.1, (b) 2.0 and (c) 1.0 μm/s for bulk-hydrated poly(HEMA + MA) ($k = 100$ mN/m, $T = 25\,^{\circ}$C, RH $= 72\%$). The force is truncated below -11 nN. Hysteresis between the approach and retract curves indicates the onset of viscoelastic mechanics.

liquid to slow deformation rates — relative to the system's intrinsic relaxation time. That is, the hydrogel responds primarily elastically when mechanical perturbations are swift and viscoelastically when they are slow. When the AFM tip impacts the hydrogel surface at high speeds, the cooperative motion of the cross-linked polymer chains (and the water that surrounds them) cannot relieve the stress as quickly as it is applied.

The AFM load (F) *vs.* displacement (z) plots, shown in Fig. 8, highlight the rate-dependent response of the hydrogel surface to the impact of an AFM tip. The probing rates decrease from left to right. In all three plots, the piezoelectric legs to which the AFM tip is attached travel roughly 150 nm after contacting the surface at $z = 0$ nm. This distance represents the sum of cantilever bending and sample deformation. In Fig. 8a, the tip approaches the surface at a velocity higher than the viscous relaxation time of the lens. As a result, the deformation is mostly elastic. The cantilever mostly bends and the tip deforms the surface by only a few nanometers. The approach curve in Fig. 8a, from the point of snapping into contact ($z = 0$ nm), is linear (i.e. $\partial F/\partial z$ remains constant). Additionally, the retract curve fully retraces the approach curve in the AFM plot. The absence of hysteresis between the approach and retract curves in Fig. 8a implies that, at a certain probing rate, primarily elastic surface mechanics are observed.

Probing the surface at slower rates causes the measured mechanics to gradually shift from being primarily elastic to viscoelastic. In Fig. 8b, the hysteresis between approach and retract curves near the contact region indicates viscoelastic deformation. In Fig. 8c, the slowest rate, the hysteresis is more pronounced; in addition, there is a large degree of curvature in the approach curve. When the experiment is repeated over the same area, identical data is acquired, indicating that the lens surface is recovered for each indentation cycle.

Surface viscoelastic behavior of poly(HEMA) and poly(HEMA + MA). The data points in Fig. 9 mark the probing rates which transition between the elastic and viscoelastic measured response for a given humidity. Faster probing causes the lens

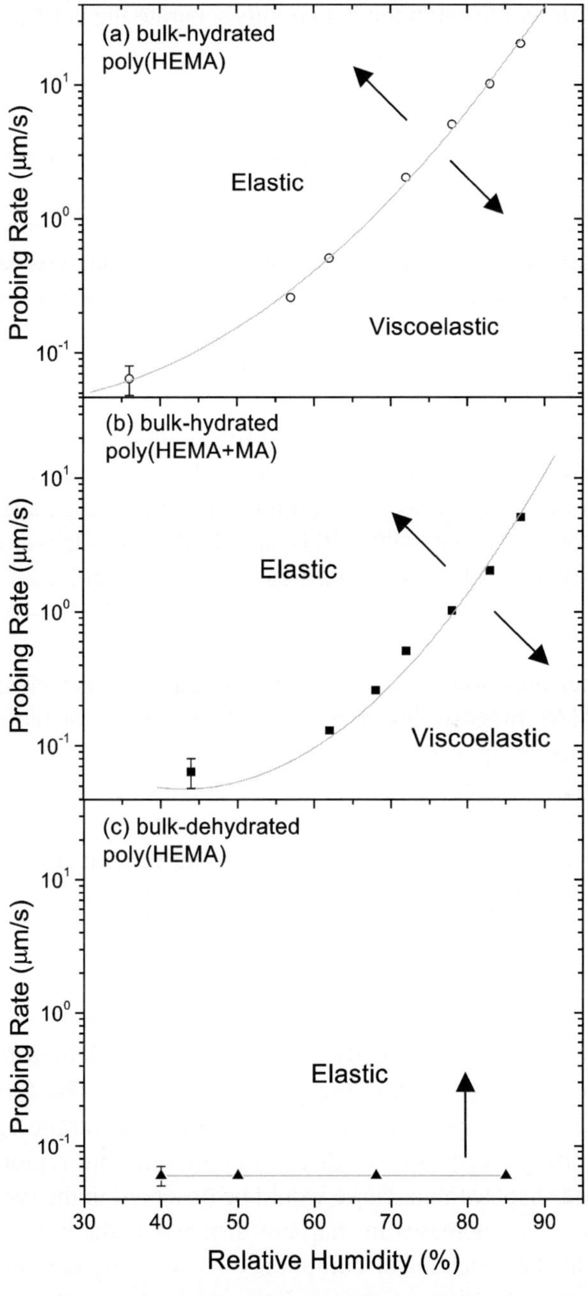

Figure 9. Humidity and probing rate dependence on the transition from viscoelastic to elastic behavior for bulk-dehydrated (a) poly(HEMA) and (b) poly(HEMA + MA), contrasted with (c) bulk-dehydrated poly(HEMA). The curves divide the elastic from the viscoelastic regimes under a load of 2.5 nN.

to respond elastically, while slower probing causes it to respond viscoelastically. Thus, the curves drawn through the data points separate the elastic and viscoelastic regimes. The onset of viscoelasticity is mapped out under a 2.5-nN load for both the bulk-hydrated and bulk-dehydrated contact lenses. The data indicate that the transition between elastic and viscoelastic mechanics depends not only on the probing rate, but also on the humidity and bulk water content of the system.

The contact lens surface is dehydrated, relative to the bulk, below about 35% RH for poly(HEMA) and 50% RH for poly(HEMA + MA). Even with the slowest probing rate available in the AFM experimental setup, no viscoelastic behavior can be measured in bulk-hydrated lenses at these low humidites. Above 60% RH, there is sufficient water present in the near-surface region of both lenses to mediate viscoelastic mechanics, as shown in previously reported results [66]. Further, the bulk-dehydrated lens in Fig. 9c shows no signs of viscoelastic behavior — even at high relative humidity. Atmospheric moisture, alone, condensing through the surface of the bulk-dehydrated lens is not enough to plasticize the surface. Instead, the condensed water is likely sequestered into the bulk. Given adequate time in a well-humidified environment (>90% RH), an initially bulk-dehydrated lens should absorb enough water to behave viscoelastically at the surface.

The onset of surface viscoelastic behavior in poly(HEMA + MA) is delayed by 15% RH, relative to poly(HEMA), despite the fact that poly(HEMA + MA) possesses a higher bulk water content. The difference in viscoelastic behavior for poly(HEMA + MA) suggests lower relative water content at the surface for any given humidity above approx. 40% RH. This observation can also be made by comparing the relative areas separated by the curves. Comparing Fig. 9a to 9b, the viscoelastic regime for poly(HEMA) encompasses more area, relative to the same regime in poly(HEMA + MA). These results agree with clinical studies, which show that lenses with high bulk water content dehydrate more quickly and to a greater extent than those with lower bulk water content [76].

Surface adhesion effects. The location of water in a hydrogel is of concern when addressing mechanical properties. The AFM data demonstrates that water does not exist as a thin layer atop the polymer surface. The forces required to snap out of water are significantly less than 100 nN and argue against capillary adhesion. In addition, successive or discontinuous snap-out regions in the retract portion of the load-displacement curve would be observed in the case of two phase-separated regions. The increases in snap-out distance as the humidity is increased reflect changes in the mechanical compliance of the polymer-water system, not the presence of a water film at the surface. Indeed, no changes in the snap-out distance are observed for a hydrophilic glass reference as the humidity is changed. The relative surface tensions of poly(HEMA) (dry: 37 dyne/cm, bulk-hydrated: 69.0 dyne/cm) and water (72.8 dyne/cm) support this conclusion, as do previously published surface-specific sum-frequency generation (SFG) vibrational spectra. SFG spectra indicate the presence of only polymer at the hydrogel-air interface [77].

The work of adhesion, calculated from the area constituting the snap-in portion of the AFM approach curve, is presented in Fig. 10. Although the radius of curvature of the AFM tip is known (approx. 50 nm), the work of adhesion has not been normalized by the maximum area of contact between the sample and AFM tip. This is because the area of contact is likely to be significantly less than the maximum possible (2500π nm^2) for data taken at the fast probing rates. The hydrogel surface is not given adequate time to absorb and relieve the stress at these fast rates. Consequently, the work of adhesion is presented as an absolute value in J.

The surface of poly(HEMA) is most compliant between 80–85% RH for each rate. This is where the product of tip sinking and adhesion forces is maximized (approx. 6.5×10^{-17} J). Slower probing rates tend to higher values for the work of adhesion. This is expected, as the surface is given adequate time to reconstruct around the AFM tip. The work of adhesion, however, does not approach an asymptotic value as the humidity is increased. Instead, there is a drop-off at 87% RH. Because the tip penetrates so quickly, this may be a non-equilibrium process, in which case the work is overestimated.

The work of adhesion values for poly(HEMA + MA) follow a similar trend, although they are typically lower than the adhesion values of poly(HEMA). A maximum of about 3.5×10^{-17} J is expended for the slowest rate. All of the work values below 80% RH are quite similar for any given rate, however and suggest a less compliant surface. Smaller snap-out distances for poly(HEMA), which give a direct indication of the adhesive forces acting between tip and sample, also bear this out. The fastest probing rate (20.4 μm/s) appears to have gone through its maximum (0.5×10^{-17} J), but the data for any of the remaining rates have not. Thus, relative humidities exceeding 87% are required to achieve a maximum work value for the slower rates. The required increase in humidity for poly(HEMA + MA) also suggests that its surface is water-deprived and less compliant relative to poly(HEMA).

Correlation of surface mechanical to chemical properties. The mechanical data confirms earlier investigations that indicate a surface water deficit exists for poly(HEMA) contact lenses, relative to the water content in the bulk. In addition, as the bulk water content is increased by making the lens material more hydrophilic, the interfacial water content actually declines. In order to rationalize this behavior, the control of bulk and interfacial water content by the hydrogel chemistry must be considered. Further experimentation using other techniques (e.g. NMR) must be conducted to follow water transport in a hydrogel and characterize the relative amounts of free and hydrogen-bonded water as a function of the polymer formulation and humidity.

Although manufacturers can alter the copolymer formulation of their contact lenses to improve water binding in the bulk, the surface composition and chemistry

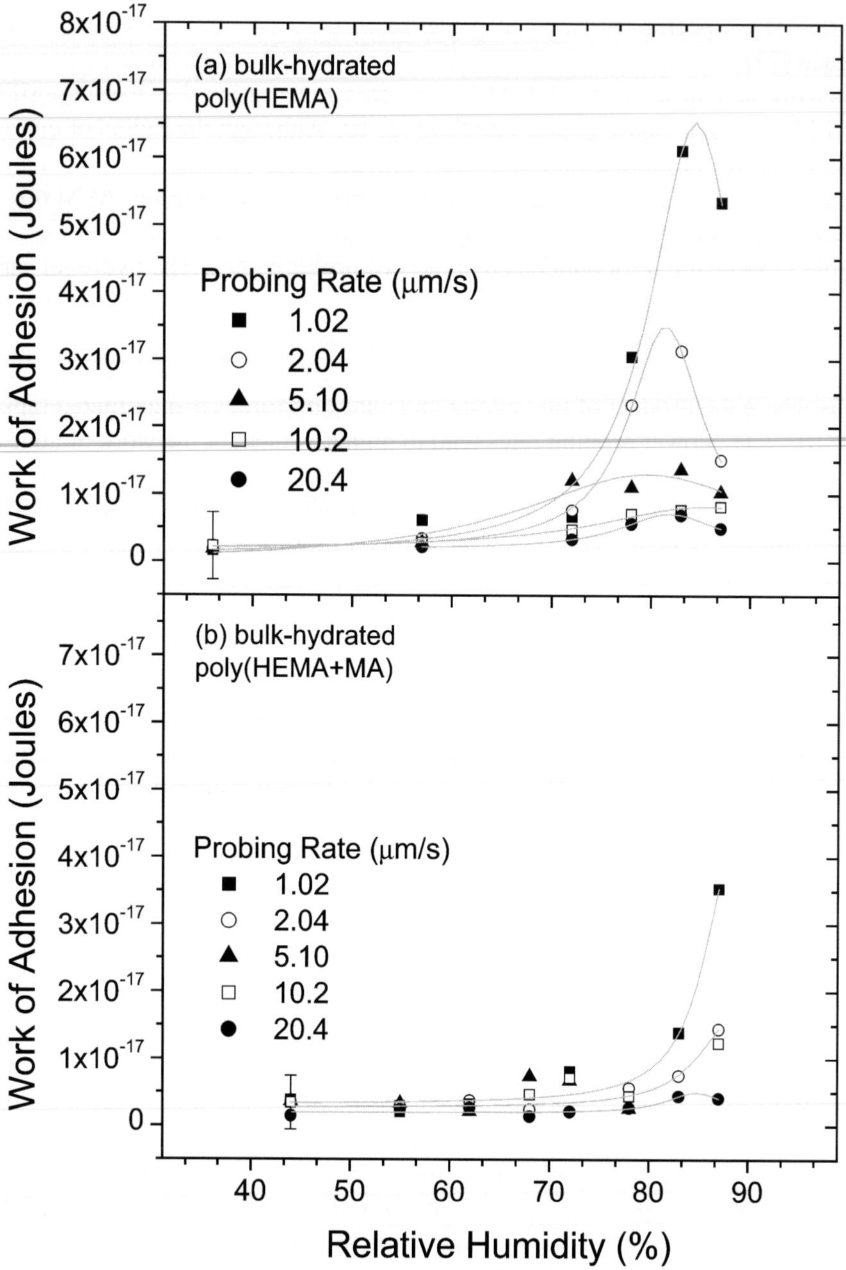

Figure 10. Absolute values of the work of adhesion as a function of humidity for bulk-hydrated (a) poly(HEMA) and (b) poly(HEMA + MA) under a load of 2.5 nN.

of the hydrogel dictate the state of interfacial hydration. AFM and SFG studies have shown that a surface-enrichment in the hydrophilic component of a copolymer system or in the fraction of hydrophilic pendant groups occurs at a wetted polymer interface [21, 24, 78]. For instance, as the humidity is increased, poly(HEMA) chains at the surface reorient to allow their hydrophilic hydroxyethyl groups to replace pendant methyl groups [77]. Long-range migration of hydrophilic polymer segments to a water interface, however, is not likely in cross-linked hydrogel networks. As a result, monomers may be locked in place, unable to minimize the surface energy of the hydrogel. SFG has not been able to detect a peak corresponding to the carboxyl group of the methacrylic acid component of poly(HEMA + MA) in high humidity. This suggests that either carboxyl groups are not present at the hydrogel surface or that they are not sufficiently ordered to generate SFG. A greater understanding of chain mobility and morphology in cross-linked hydrogels is needed to determine how the chemistry of copolymers controls interfacial water content.

Ongoing work in finding novel methods to retain surface wettability and biocompatibility for high water content lenses (e.g. preferential segregation or grafting of the hydrophilic component to the surface) is needed [79]. Future studies should address the role of an adsorbed protein layer to both surfaces of a contact lens, as this may affect water uptake into the contact lens. If water transport is inhibited by a protein layer, this may lead to a decreased evaporative flux and an enrichment in the near-surface water content.

Molecular surface structure of stretched polyurethanes

The control and stability of the molecular surface structure of polymers under external force are of critical importance, such as in applications involving biomedical implants or thin polymer films used in transistors [80, 81]. The distribution and orientation of various groups at the surface of a polymer control its physical, chemical and biological interactions with the environment. Generally, most elastomers possess a rubbery, flexible phase that absorbs strain energy and is altered under deformation and a crystalline (or rigid) phase that can be stressed without undergoing significant deformation. In general, polymer deformation mechanisms depend on the size of the flexible and rigid phases [82, 83]. To examine deformation at small scales, the polymer should consist of small-size phases on the order of several nanometers.

Segmented PUR is one of the most commonly used biomaterials, due to its excellent mechanical properties and good biocompatibility. PUR block copolymers are characterized by a two-phase microstructure consisting of two immiscible hard and soft segments [84]. In the first PUR composition, the molecular weights of the soft and hard segments are 710 and 1000 g/mol (about 2–10 nm in length), respectively (42 wt% hard segments, Polymer Technology Group). The hard segment consists of alternating 4,4′-diphenylmethane diisocyanate (MDI) and 1,4-butanediol (BDO) (chain extender) with a glass transition temperature of approx. 50°C. The soft segment is polytetramethylene oxide (PTMO) with a glass transition temperature of approx. −40°C. Polymer structures are displayed in Fig. 1. In the second PUR composition, the molecular weight of the hard segment is doubled to 1450 g/mol (69 wt% hard segments), while that of the soft segment is maintained equal to 1000 g/mol. Hereafter, the two PUR compositions containing 710 and 1450 g/mol hard segments will be referred to as short and long hard segment PUR, respectively. Further information regarding the synthesis and bulk properties of this PUR can be found elsewhere [84, 85]. Two approaches are used to examine the surface composition of PUR in this study. The first method involves reversibly stretching the polymer, while the second involves changing the length of only the rigid segment in the copolymer.

PUR films cast from solution were cut into dog-bone shaped specimens using a specimen preparation procedure described elsewhere [33, 86]. The PUR films were stretched to an elongation of 10%, which is within the bulk elastic regime [87], and then relaxed to their original lengths using a custom-made stretching device [82]. SFG vibrational spectra of the PUR films in the CH stretch region were obtained at each stretch cycle. Three complete stretch cycles of duration equal to 30 min were applied before allowing the samples to relax for 12 h. Spectra were collected using the $s_{\text{sum-output}}s_{\text{vis-input}}p_{\text{IR-input}}$ (ssp) polarization combination.

Surface characterization of films with different hard segment compositions. Figure 11 shows typical AFM images and SFG spectra of the as-cast PUR films with short and long hard segments. Prior to elongation, the copolymer with 69% hard segments possesses lower surface roughness compared to PUR films with 42% hard segments. Nanoindentation measurements reveal that the copolymer with 69% hard segments has a higher surface stiffness and lower adhesion than the copolymer with 42% hard segments. However, it was found that the surfaces of both PUR compositions exhibit higher adhesion and lower stiffness compared to the bulk. SFG spectra exhibit similarities and contain the same vibrational CH modes [88]. The relatively high intensity of the $O-CH_2$(s) peak, which is characteristic only of the soft segments, is attributed to the CH_2 symmetric stretch near the ether group. Spectra of both PUR compositions, in conjunction with the nanomechanical and topographical measurements obtained with AFM, suggest that lower surface energy soft segments dominate the surface composition of the as-cast PUR films [89]. While the CH_2(s) peak is the strongest feature in the spectrum of the PUR film with short hard segments, the CH_2(a) peak is the strongest feature in the spectrum of PUR with long hard segments. Since the infrared beam was p-polarized in all experiments, a large portion of the dipole vibrational resonance of the molecules contributing to the SFG signal must occur in a direction nearly normal to the surface in order to be detected. The molecular average orientation of the contributing methylene groups was estimated by using a procedure described elsewhere [90].

The decrease of the CH_2(s)/CH_2(a) mode strength ratio of PUR with long hard segments relative to that with short hard segments (Fig. 11) implies a high CH_2 orientation angle θ and a tilted molecular orientation. The difference between the

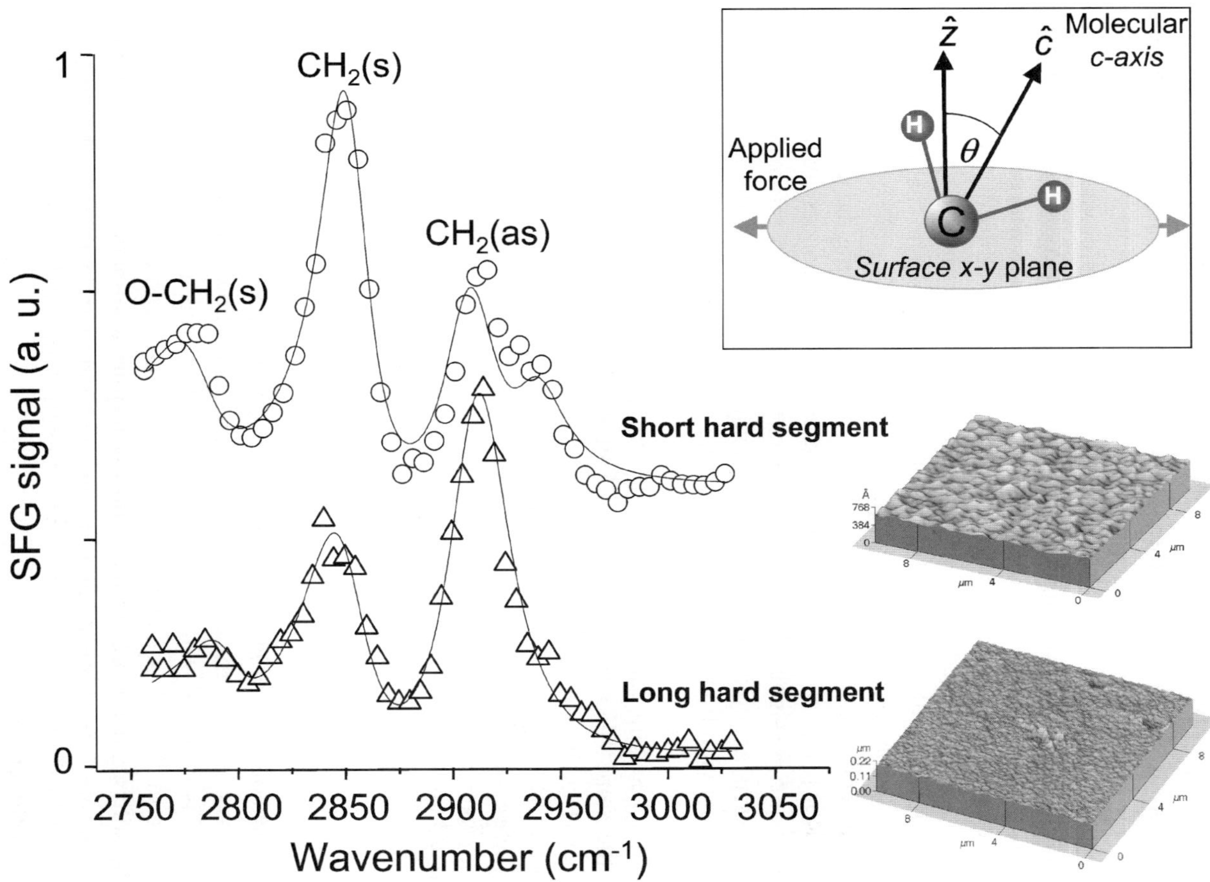

Figure 11. AFM images and SFG spectra of as-cast PUR films with short (upper image, circles) and long (lower image, triangles) hard segments, respectively.

two PUR compositions is attributed to the coverage of the hard segments by soft segments, which is expected to decrease with increasing hard-segment length.

Surface characterization of stretched films. Figure 12 shows SFG results from cyclic stretching of PUR films with short hard segments. To independently examine the behavior of the polymer surface during loading (stretching) and unloading (relaxation), SFG spectra of stretched and relaxed PUR films are contrasted in Fig. 12a and 12b, respectively. Significantly different behavior was observed after relaxation (Fig. 12b). Although after the first stretch/relaxation cycle the SFG spectrum does not change, the CH_2(a) mode strength increases with each subsequent stretch cycle. After three consecutive stretch cycles followed by a 12 h relaxation, the mode strength of the CH_2(a) stretch intensifies further and the spectrum of the relaxed PUR film with short hard segments becomes similar to that of the as-cast PUR film with long hard segments (Fig. 11).

Stretching enhances polymer chain ordering, thus increasing the number of trans sequences [91]. Due to symmetry considerations, trans sequences are not expected to produce an SFG signal, as opposed to gauche sequences. Therefore, the intensity of the peaks is anticipated to vary with elongation. Studying the ratio between the methylene vibrational peaks can eliminate overall SFG signal intensity changes between scans. The average reorientation of the contributing methylene groups of the stretched film can be quantified by comparing ratios of the mode strength A_q in the same SFG spectrum [90]. Figure 13 shows a comparison between the effect of cyclic stretching and relaxation on the CH_2(s)/CH_2(a) mode strength ratios of PUR films. For both compositions, the CH_2(s)/CH_2(a) ratio increases slightly with stretching. However, the two PUR compositions exhibit different relaxation behavior. The CH_2(s)/CH_2(a) ratio of PUR with short hard segments decreases continuously after each stretch cycle. An additional pronounced decrease occurs after relaxation for 12 h, resulting in a mode strength ratio close to that of PUR with long hard segments. In contrast to PUR with short hard segments, the CH_2(s)/CH_2(a) ratio of PUR with long hard segments does not change after relaxation.

The effects of segment length (i.e. composition) and stretching on the molecular-level polymer surface structure can be explained as follows. Before deformation, the surface is enriched with soft segments. Considering the low glass transition temperature of the soft segments relative to the hard segments, it is inferred that only the soft segments are stretched during testing at ambient conditions. Upon elongation, the soft segments align almost parallel to the surface. This leads to an increase in the CH_2(s)/CH_2(a) mode strength ratio, indicating a more favored upward methylene orientation.

For PUR with short hard segments, the average angle θ of the contributing CH_2 groups decreases after relaxation, while it remains constant in the case of PUR with long hard segments. After only three stretch cycles, the surface composition of PUR with short hard segments appears to be similar to that of PUR with long hard

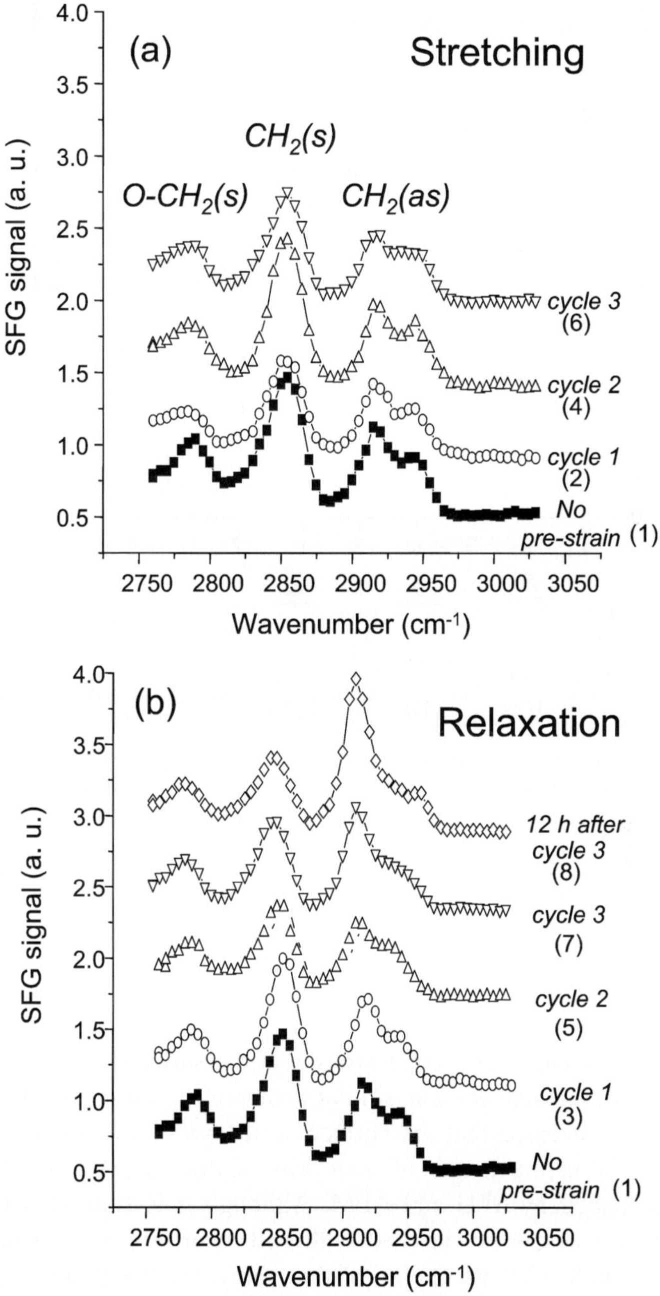

Figure 12. Typical SFG spectra of elastically (macroscopically) stretched and relaxed PUR films with short hard segments: (a) stretching to 10% elongation and (b) relaxation to the initial specimen length *versus* stretching and relaxation cycles, respectively. Numbers in parentheses correspond to the loading/unloading steps shown in Fig. 13.

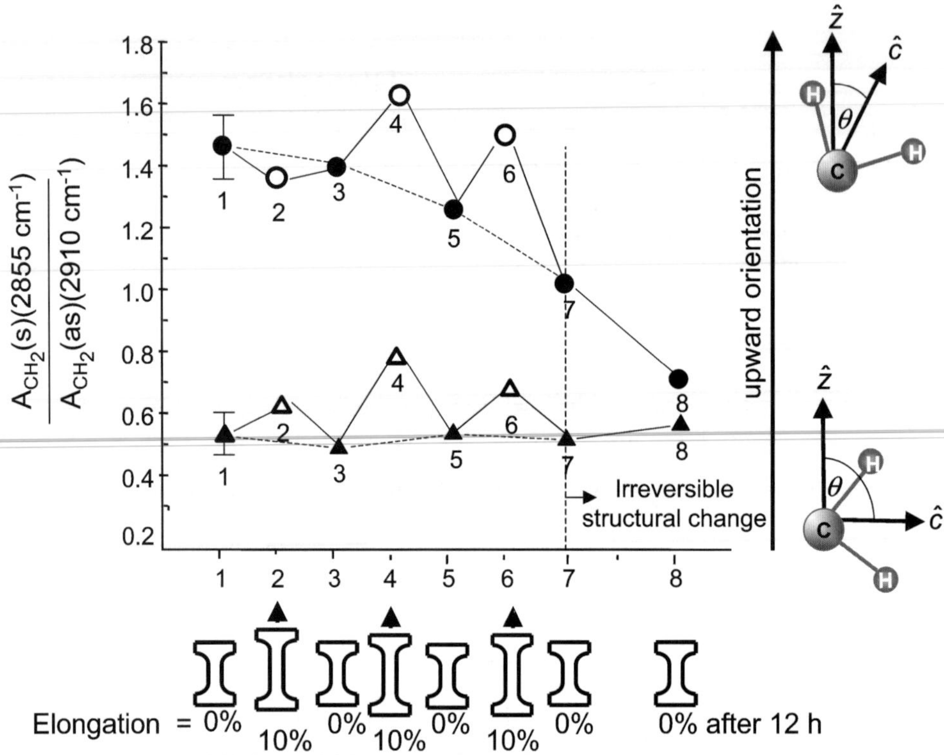

Figure 13. CH_2(s) (2855 cm^{-1})/CH_2(a) (2915 cm^{-1}) mode strength ratios of three stretching and relaxation cycles of PUR films with short and long hard segments. Numbers on the horizontal axis denote the loading/unloading steps in each stretch cycle, illustrated below the graph. Open (closed) symbols correspond to stretching (relaxation).

segments. This implies that after relaxation the surface of PUR with short hard segments cannot recover its original molecular arrangement. Therefore, stretching the soft segments beyond their elastic limit permanently changes their original arrangement and reduces their concentration at the surface. This yields a surface with a higher hard segment content, similar to that of PUR with long hard segments.

This study has revealed that the surface structures of two PUR compositions, which differ only in the length of their hard segments (710 *vs.* 1450 g/mol), can be distinguished by SFG and AFM. Although it is suggested that both as-cast polymer surfaces were enriched with soft segments, the results indicate a greater upward molecular orientation of the methylene groups in PUR with short hard segments. For both PUR compositions, the CH_2(s)/CH_2(a) mode strength ratios increased with stretching, indicating an increase in the upward orientation of the methylene groups that contribute to the SFG signal. The CH_2(s)/CH_2(a) ratio decreased with subsequent relaxation. Based on these results, two methods for producing similar surface compositions of PUR block copolymers can be proposed: (a) macroscopic elastic cyclic stretching and relaxation of the polymer or (b) increasing the molecular weight of the hard segment in the copolymer chain.

Characterization of polystyrene surfaces with adsorbed proteins

In this study, SFG is used to probe the structure of adsorbed protein and the effects of protein adsorption on the orientation and ordering of functional groups present on polystyrene surfaces [34]. Polystyrene has been a popular model surface for controlled protein adsorption studies and several groups have characterized its bare surface structure with SFG [92–94]. Lysozyme, fibrinogen and bovine serum albumin (BSA) have been chosen as model proteins, since their adsorption behavior, layer formations and degrees of denaturation on polystyrene surfaces have been well studied. Lysozyme is regarded as a hard protein because it resists denaturation and exhibits concentration-dependent layer formation. Soft proteins like BSA, however, undergo denaturation relatively easily upon adsorption [95–99]. Fibrinogen is a larger molecule with high surface affinity and usually displaces pre-adsorbed proteins [100, 101].

SFG analysis of proteins adsorbed on hydrophobic polystyrene surfaces. The relative amounts of protein adsorbed on hydrophobic polystyrene surfaces were determined by fluorescence measurements [101–103]. As the labeled protein molecules contain a nearly identical number of conjugated fluorophores and their quantum yields are nearly constant regardless of protein conformation, the measured fluorescence intensities are proportional to the amount of adsorbed protein for a given area [101, 102]. Our results are presented in Fig. 14. For lysozyme, the relative amounts adsorbed on polystyrene surfaces increase in proportion to the bulk protein concentration. Similarly, more fibrinogen and BSA molecules are present on polystyrene surfaces when adsorption is carried out at higher bulk protein concentrations. This result is in good agreement with previous reports [95, 102, 104–106].

Figure 15 shows SFG spectra in the CH stretch region for polystyrene surfaces with and without adsorbed lysozyme, fibrinogen and BSA. Partially deuterated polystyrene (PS-d_3, see Fig. 1 for structure) has been used to separate SFG signals from the adsorbed proteins from those generated by the underlying polystyrene surfaces. Protein adsorption has been conducted using solutions with different bulk protein concentrations (1.0, 0.2, 0.04 and 0.008 mg/ml). The polarization combination used was ssp, with respect to SFG signal, visible input and infrared input components, respectively. The solid lines represent Lorentzian least-square fits to the raw spectral data to obtain oscillator strengths (A_q), peak positions and widths.

In Fig. 15, SFG spectra contain several features in the 2800–3000 cm^{-1} range, which are responsible for the CH_2 symmetric (CH_2(s)), CH_3 symmetric (CH_3(s)), CH_3 Fermi resonance (CH_3(F)) and CH_3 asymmetric (CH_3(a)) stretches [107–110]. The SFG intensities of these features generally increase in proportion to the surface lysozyme concentration. SFG spectra for fibrinogen and BSA on hydrophobic

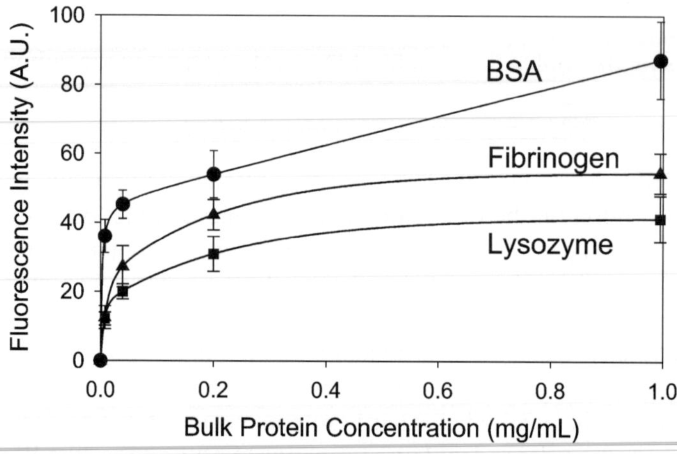

Figure 14. Relative amounts of protein adsorbed on hydrophobic polystyrene surfaces as a function of bulk concentration for lysozyme (■), fibrinogen (▲) and BSA (●). These results are obtained by measuring the fluorescence intensity of tetramethylrhodamine isothiocyanate (TRITC)-conjugated lysozyme, Alexa Fluor 594-conjugated fibrinogen and Texas Red-conjugated BSA by fluorescence microscopy.

surfaces are somewhat similar compared to those for lysozyme, except for the fact that fibrinogen molecules adsorbed at lower bulk concentrations (0.04 and 0.008 mg/ml) do not exhibit observable SFG signal intensity (Fig. 15b).

The SFG signal intensity, I_{sfg}, for the CH stretch modes in the 2800–3000 cm^{-1} range is governed by the surface number density and orientation average for a given vibrational mode. Thus, we propose that SFG data and relative surface concentrations obtained from fluorescence microscopy provide structural information regarding amino acid residues containing methyl groups in adsorbed proteins [34]. At present, however, we cannot identify the amino acids which are responsible for the SFG signal intensity, since all proteins contain a number of amino-acid residues with methyl groups [111–113].

In general, hydrophobic interactions between proteins and polystyrene surfaces are more likely to cause a significant breakdown in protein secondary structure. In fact, neutron reflectivity and calorimetry studies have independently shown that lysozyme loses a considerable amount of secondary structure on hydrophobic surfaces, despite its structural stability on hydrophilic surfaces [97, 114]. Fibrinogen and BSA are also known to lose a larger amount of α-helical structure on hydrophobic surfaces [114, 115]. Therefore, we assume that the orientation of SFG-active methyl groups in these proteins is more randomly distributed and less concentration-dependent. Therefore, as more protein adsorbs on a polystyrene surface, the increase in the surface number density of the methyl groups is responsible for the simultaneous enhancement in SFG signal intensity for all CH_x ($x = 2$ and 3) stretch modes in Fig. 15.

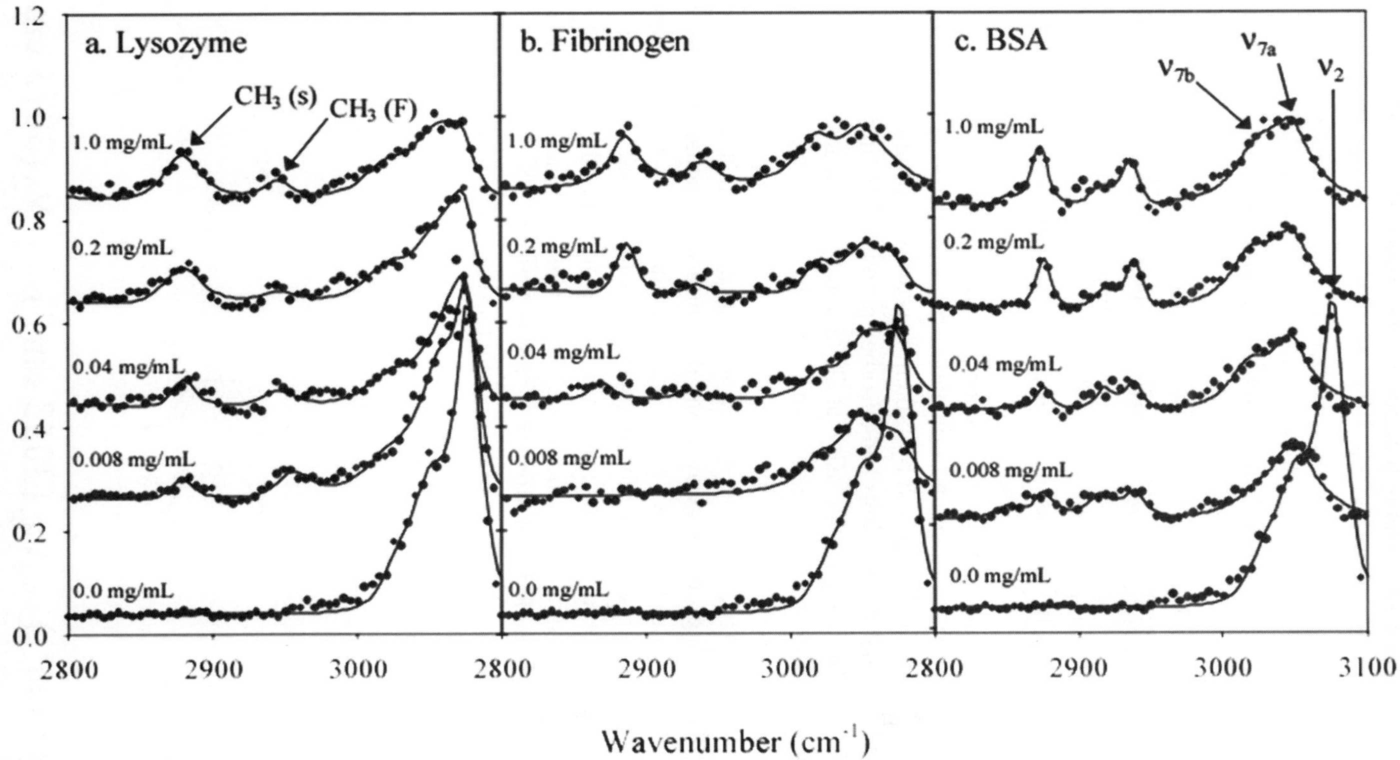

Figure 15. SFG spectra of (a) lysozyme, (b) fibrinogen and (c) BSA adsorbed on hydrophobic polystyrene (PS-d_3) surfaces from solutions with indicated bulk protein concentrations.

Molecular interactions at the protein/polystyrene interface. In addition, SFG spectra in Fig. 15 contain several features in the 3000–3100 cm^{-1} range that are responsible for the CH stretch modes of the phenyl ring in the underlying polystyrene. This is confirmed by experiments using *per*-deuterated polystyrene (PS-d_8, see Fig. 1 for structure) that does not contain CH bonds. Without adsorbed proteins, a large peak at 3073 cm^{-1} and a shoulder around 3053 cm^{-1} are observed. They are responsible for the ν_2 and ν_{7a} modes, respectively (see Fig. 16) [92, 93]. It was found that the SFG intensity of the ν_2 mode slowly attenuates, as more lysozyme molecules are present. Fibrinogen adsorption yields similar effects on the intensity of the ν_2 mode. However, the SFG intensity of the ν_2 mode is more rapidly suppressed and a new mode around 3025 cm^{-1} appears, as more fibrinogen molecules are present. In both cases, the intensity of the ν_2 mode is less affected, compared to the case of BSA adsorption, where even a small amount of BSA deposition completely suppresses the intensity of the ν_2 mode (Fig. 15c). The results indicate that the structure and orientation of the underlying phenyl ring in the polystyrene substrate is affected by protein deposition, presumably due to the interaction between the phenyl rings and adsorbed protein molecules.

In Fig. 15, the SFG signal intensity at 3073 cm^{-1} attenuates when protein molecules are present on the surface. This peak is responsible for the totally symmetric stretch mode of the phenyl ring (ν_2, see Fig. 16). Previous SFG studies have characterized free polystyrene surfaces at the air interface and have shown that phenyl rings order away from the surface, although the tilt angle is in question [92–94]. Our SFG results indicate that the ν_2 mode is partially blocked after protein deposition. Considering the reduction of SFG signal intensity, BSA binds to polystyrene surfaces in such a way as to affect the ν_2 mode more significantly than do fibrinogen and lysozyme. Curve fitting shows that the magnitude of the ν_{7a} mode is not affected by protein adsorption. Interestingly, the ν_{7b} mode around 3025 cm^{-1} appears with an increasing number of fibrinogen and BSA molecules on the surface.

These results show interesting trends. Although the relative amounts of proteins which actually adsorb on the polystyrene surfaces cannot be compared from the

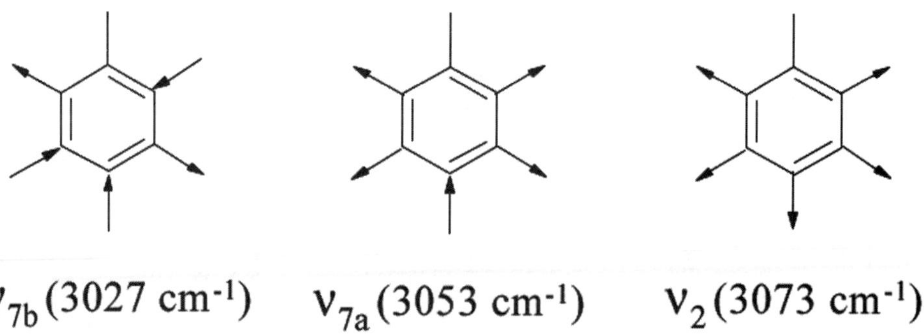

Figure 16. Normal modes of vibration for phenyl rings in polystyrene.

same bulk concentration, the greatest reduction in SFG signal intensity for the ν_2 mode was achieved by BSA adsorption, followed by fibrinogen and lysozyme. The results suggest that each protein molecule interacts with the phenyl rings in a different way, giving rise to different changes in the orientation and ordering of the underlying phenyl ring. The adsorption behavior of fibrinogen, in terms of spreading on hydrophilic surfaces and interacting with phenyl rings, is somewhat similar to that of BSA, but quite different from that observed for lysozyme.

Our SFG data show that the orientation of the methyl groups in the adsorbed proteins is intimately affected by the degree of denaturation and the difference in the layer formation, which is governed by the surface concentration of adsorbed protein, hydrophobicity of the solid substrate and the nature of the protein used. Amino acids in protein adsorbed on hydrophobic surfaces appear more randomly oriented regardless of surface concentration. In addition, SFG results reveal that the orientation and ordering of the phenyl rings in hydrophobic polystyrene surfaces are affected by protein deposition.

CONCLUSIONS

SFG and AFM, together with complementary surface diagnostic tools such as XPS and contact angle goniometry, were used to study the surface structure and surface mechanical properties of model biopolymer materials. These techniques confirm that polymer surfaces restructure in response to bulk polymer composition, surface energy differences, mechanical strain and chemical environment. Aqueous environments, for instance, stabilize different functional groups in Biospan materials; consequently, if one expects to acquire meaningful data, surfaces must be studied *in situ*, under the conditions in which biopolymers may be employed.

The ocular comfort of contact lenses is tied into their surface mechanical and chemical properties. In this example, AFM was used to appraise the role of humidity and, consequently, interfacial water content in controlling surface viscoelastic and compliance effects. In concert with mechanical studies of the surface, SFG may also help assess synthetic schemes which attempt to improve water retention by altering the chemistry of the hydrogel surface.

Correlation of the surface molecular composition with the nanoscale elastic behavior and topography of stretched polyurethane films was also made by AFM and SFG. Here, the length of the hard segment was found to determine the surface properties of the copolymer. In the final case, SFG was used to study the adsorption of proteins on hydrophobic polystyrene surfaces and to follow molecular-level changes that occur in the polymer substrate. The spectra also revealed that the orientation of methyl groups in adsorbed protein depends on the surface concentration and nature of the protein used. To connect molecular orientation effects to the strength of the protein–polymer interaction, binding forces between each protein and polymer substrate are currently being measured by AFM.

These examples illustrate the advantages of connecting data gathered by SFG surface spectroscopy and AFM probe microscopy to interpret interfacial phenomena occurring in biopolymer systems. Efforts must be made, however, to overcome certain experimental limitations in SFG. The time resolution of SFG must be decreased to seconds to be able to follow more dynamic phenomena and the frequency range for a given scan must also be expanded to access greater molecular structure information. To strengthen correlation arguments, the spatial resolution of SFG experiments (several hundred square microns) must also be improved to match the <50 nm spatial resolution of AFM experiments. Despite these hurdles, SFG and AFM remain among the most powerful technologies to study biomedical surfaces in realistic environments.

Acknowledgements

This work was supported by the Director, Office of Science, Office of Basic Energy Sciences, Division of Materials Sciences and Engineering, of the US Department of Energy under Contract No. DE-AC03-76SF00098. The authors acknowledge support from The Polymer Technology Group, Ocular Sciences, Inc. and Roche Diagnostics, Inc.

REFERENCES

1. D. G. Castner and B. D. Ratner, *Surf. Sci.* **500**, 28 (2002).
2. B. Kasemo, *Surf. Sci.* **500**, 656 (2002).
3. A. Dabrowski, *Adv. Colloid Interface* **135**, 93 (2001).
4. O. G. Berg, *Interfacial Enzyme Kinetics*. Wiley, New York, NY (2002).
5. H. Otsuka, Y. Nagasaki and K. Kataoka, *Curr. Opin. Colloid. Interf.* **6**, 3 (2001).
6. B. D. Ratner and D. G. Castner, *Surface Modification of Polymeric Biomaterials.* Plenum, New York, NY (1996).
7. A. F. Recum, *Handbook of Biomaterials Evaluation: Scientific, Technical and Clinical Testing of Implant Materials.* Macmillan, New York, NY (1986).
8. J. B. Park and R. S. Lakes, *Biomaterials: An Introduction.* Plenum, New York, NY (1992).
9. M. Tirrell, E. Kokkoli and M. Biesalski, *Surf. Sci.* **500**, 61 (2002).
10. D. Woodruff and T. Delchar, *Modern Techniques of Surface Science.* Cambridge University Press, Cambridge (1986).
11. G. A. Somorjai, *Introduction to Surface Chemistry and Catalysis.* Wiley, New York, NY (1994).
12. Y. R. Shen, *The Principle of Nonlinear Optics.* Wiley, New York, NY (1984).
13. Y. R. Shen, *Nature* **337**, 519 (1989).
14. K. B. Eisenthal, *Chem. Rev.* **96**, 1343 (1996).
15. G. Binnig, C. F. Quate and C. Gerber, *Phys. Rev. Lett.* **56**, 930 (1986).
16. S. H. Cohen and M. L. Lightbody, *Atomic Force Microscopy/Scanning Tunneling Microscopy 3.* Kluwer, New York, NY (1999).
17. K. B. Rider, K. S. Hwang, M. Salmeron and G. A. Somorjai, *Phys. Rev. Lett.* **86**, 4330 (2001).
18. S. M. Lindsay, L. A. Nagahara, T. Thundat, U. Knipping and R. L. Rill, *J. Biomol. Struct. Dyn.* **7**, 279 (1989).
19. T. Miyamae and H. Nozoye, *Surf. Sci.* **532**, 1045 (2003).
20. S. Ye, H. Noda, S. Morita, K. Uosaki and M. Osawa, *Langmuir* **19**, 2238 (2003).

21. C. Chen, J. Wang, S. E. Woodcock and Z. Chen, *Langmuir* **18**, 1302 (2002).
22. E. Amitay-Sadovsky, K. Komvopoulos, Y. Tian and G. A. Somorjai, *Appl. Phys. Lett.* **80**, 1829 (2002).
23. M. Salmeron, *Tribol. Lett.* **10**, 69 (2001).
24. D. Zhang, D. H. Gracias, R. Ward, M. Gauckler, Y. Tian, Y. R. Shen and G. A. Somorjai, *J. Phys. Chem. B* **102**, 6225 (1998).
25. S. H. Kim, C. Marmo and G. A. Somorjai, *Biomaterials* **22**, 3285 (2001).
26. Z. W. Wicks, *Organic Coatings: Science and Technology*. Wiley, New York, NY (1992).
27. J. W. Goodwin and R. Buscall, *Colloidal Polymer Particles*. Academic Press, London (1995).
28. H. Kawaguchi, *Prog. Polym. Sci.* **25**, 1171 (2000).
29. J. Israelachvili, *J. Vac. Sci. Technol. A* **10**, 2961 (1992).
30. T. Aoike, H. Uehara, T. Yamanobe and T. Komoto, *Langmuir* **17**, 2153 (2001).
31. Z. Chen, R. Ward, Y. Tian, A. S. Eppler, Y. R. Shen and G. A. Somorjai, *J. Phys. Chem. B* **103**, 2935 (1999).
32. T. S. Koffas, A. Opdahl, C. Marmo and G. A. Somorjai, *Langmuir* **19**, 3453 (2003).
33. E. Amitay-Sadovsky, B. Ward, G. A. Somorjai and K. Komvopoulos, *J. Appl. Phys.* **91**, 375 (2002).
34. J. Kim and G. A. Somorjai, *J. Am. Chem. Soc.* **125**, 3150 (2003).
35. T. S. Koffas, J. Kim, C. C. Lawrence and G. A. Somorjai, *Langmuir* **19**, 3563 (2003).
36. F. M. Fowkes, *Contact Angle, Wettability and Adhesion. Am. Chem. Soc. Adv. Chem. Ser. 43*. American Chemical Society, Washington, DC (1964).
37. K. C. Cole, J. Guevremont, A. Ajji and M. M. Dumoulin, *Appl. Spectrosc.* **48**, 1513 (1994).
38. N. Yui, Y. Suzuki, H. Mori and M. Terano, *Polym. J.* **27**, 614 (1995).
39. B. J. Kip and M. C. P. van Eijk, *J. Polym. Sci. B* **29**, 99 (1991).
40. D. G. Castner and B. D. Ratner, *Surf. Interface Anal.* **15**, 479 (1990).
41. D. Briggs, *Br. Polym. J.* **21**, 3 (1989).
42. A. Brown and J. C. Vickerman, *Surf. Interface Anal.* **8**, 75 (1986).
43. D. H. Gracias, D. Zhang, H. Lianos, W. Ibach, Y. R. Shen and G. A. Somorjai, *Chem. Phys.* **245**, 277 (1999).
44. M.-C. Belanger, Y. Marois, R. Roy, Y. Mehri, E. Wagner, Z. Zhang, M. W. King, M. Yang, C. Hahn and R. Guidoin, *Artif. Organs* **24**, 879 (2000).
45. R. Yoda, *J. Biomater. Sci. Polym. Edn.* **9**, 561 (1998).
46. M. D. Lelah and S. L. Cooper, *Polyurethanes in Medicine*. CRC Press, Boca Raton, FL (1986).
47. E. Rechavia, F. Litvack, M. C. Fishbien, M. Nakamura and N. Eigler, *Catheter. Cardiol. Diagn.* **45**, 202 (1998).
48. R. S. Ward and K. A. White, US Patent No. 5,589,563.
49. A. B. Mathur, T. O. Collier, W. J. Kao, M. Wiggins, M. A. Schubert, A. Hiltner and J. M. Anderson, *J. Biomed. Mater. Res.* **36**, 246 (1997).
50. S. I. Jeon, J. H. Lee, J. D. Andrade and P. G. de Gennes, *J. Colloid Interface Sci.* **142**, 149 (1991).
51. E. Osterberg, K. Bergstrom, K. Holmberg, T. P. Schman, J. A. Riggs, N. L. Burns, J. M. Vanalstine and J. M. Harris, *J. Biomed. Mater. Res.* **29**, 741 (1995).
52. K. L. Prime and G. M. Whitesides, *J. Am. Chem. Soc.* **115**, 10714 (1993).
53. R. C. Chatelier and A. P. Minton, *Biophys. J.* **71**, 2367 (1996).
54. T. B. McPherson, S. J. Lee and K. Park, in: *Proteins at Interfaces II: Fundamentals and Applications,* T. A. Horbett and J. L. Brash (Eds), p. 395. American Chemical Society, Washington, DC (1995).
55. Z. Chen, D. H. Gracias and G. A. Somorjai, *Appl. Phys. B* **68**, 549 (1999).
56. Z. Chen, R. Ward, Y. Tian, S. Baldelli, A. Opdahl, Y. R. Shen and G. A. Somorjai, *J. Am. Chem. Soc.* **122**, 10615 (2000).

57. C. D. Han, *Advances in Chemistry Series, Vol. 206*. American Chemical Society, Washington, DC (1984).
58. B. R. Liang, L. H. Pan and X. J. He, *J. Appl. Polym. Sci.* **66**, 217 (1997).
59. D. Zhang, R. S. Ward, Y. R. Shen and G. A. Somorjai, *J. Phys. Chem. B* **101**, 9060 (1997).
60. J. M. Wen, G. A. Somorjai, F. Lim and R. Ward, *Macromolecules* **30**, 7206 (1997).
61. C. Y. Chen, M. A. Even, J. Wang and Z. Chen, *Macromolecules* **35**, 9130 (2002).
62. S. H. Kim, A. Opdahl, C. Marmo and G. A. Somorjai, *Biomaterials* **23**, 1657 (2002).
63. G. L. Grobe, P. L. Valint and D. M. Ammon, *J. Biomed. Mater. Res.* **32**, 45 (1996).
64. C. R. Arciola, M. C. Maltarello, E. Cenni and A. Pizzoferrato, *Biomaterials* **16**, 685 (1995).
65. M. Wahlgren and T. Arnebrant, *Trend Biotechnol.* **9**, 201 (1991).
66. A. Opdahl, S. H. Kim, T. S. Koffas, C. Marmo and G. A. Somorjai, *J. Biomed. Mater. Res.* **67A**, 350 (2003).
67. G. M. Bruinsma, H. C. van der Mei and H. J. Busscher, *Biomaterials* **22**, 3217 (2001).
68. J. A. Bonanno, T. Stickel, T. Nguyen, T. Biehl, D. Carter, W. J. Benjamin and P. S. Soni, *Invest. Ophthalmol. Vis. Sci.* **43**, 371 (2002).
69. M. G. Harris and A. Mock, *Am. J. Optom. Phys. Opt.* **50**, 747 (1973).
70. P. C. Nicolson and J. Vogt, *Biomaterials* **22**, 3273 (2001).
71. R. Barbieri, M. Quaglia, M. Delfini and E. Brosio, *Polymer* **39**, 1059 (1998).
72. P. McConville and J. M. Pope, *Polymer* **41**, 9081 (2000).
73. M. S. Jhon, S. M. Ma, S. Hattori, D. E. Gregonis and J. D. Andrade, in: *Hydrogels for Medical and Related Applications*, J. D. Andrade (Ed.), p. 37. American Chemical Society, Washington, DC (1976).
74. O. Wichterle and D. Lim, *Nature* **185**, 117 (1960).
75. B. D. Ratner and A. S. Hoffman, in: *Hydrogels for Medical and Related Applications*, J. D. Andrade (Ed.), p. 1. American Chemical Society, Washington, DC (1976).
76. J. A. Kohler and G. W. Flanagan, *Int. Contact Lens Clin.* **12**, 152 (1985).
77. Q. Chen, D. Zhang, G. A. Somorjai and C. R. Bertozzi, *J. Am. Chem. Soc.* **121**, 446 (1999).
78. A. Opdahl, T. S. Koffas, E. Amitay-Sadovsky, J. Kim and G. A. Somorjai, *J. Phys: Cond. Matt.* (2004) (in press).
79. D. L. Elbert and J. A. Hubbell, *Annu. Rev. Mater. Sci.* **26**, 365 (1996).
80. T. P. Russell, *Science* **297**, 964 (2002).
81. T. Wu, K. Efimenko and J. Genzer, *Macromolecules* **34**, 684 (2001).
82. A. Opdahl and G. A. Somorjai, *J. Polym. Sci. B Polym. Phys.* **39**, 2263 (2001).
83. J. Rottler and M. O. Robbins, *Phys. Rev. Lett.* **89**, 195501 (2002).
84. B. Chu, T. Gao, Y. J. Li, J. Wang, C. R. Desper and C. A. Byrne, *Macromolecules* **25**, 5724 (1992).
85. E. Amitay-Sadovsky, K. Komvopoulos, R. Ward and G. A. Somorjai, *Appl. Phys. Lett.* **83**, 3066 (2003).
86. ASTM, *1997 Annual Book of American Society for Testing and Materials Standards, Volume 8.01*. ASTM, West Conshohocken, PA (1997).
87. X. Wei, S. Hong, X. Zhuang, T. Goto and Y. R. Shen, *Phys. Rev. E* **64**, 5160 (2000).
88. Z. Chen, R. Ward, Y. Tian, F. Malizia, D. H. Gracias, Y. R. Shen and G. A. Somorjai, *J. Biomed. Mater. Res.* **62**, 254 (2002).
89. E. Amitay-Sadovsky, K. Komvopoulos, R. Ward and G. A. Somorjai, *J. Phys. Chem. B* **107**, 6377 (2003).
90. A. Opdahl, R. A. Phillips and G. A. Somorjai, *J. Phys. Chem. B* **106**, 5212 (2002).
91. P. de Gennes, *Scaling Concepts in Polymer Physics*. Cornell University Press, Ithaca, NY (1979).
92. K. S. Gautam, A. D. Schwab and A. Dhinojwala, *Phys. Rev. Lett.* **85**, 3854 (2000).
93. D. Zhang, S. M. Dougal and M. S. Yeganeh, *Langmuir* **16**, 4528 (2000).

94. K. A. Briggman, J. C. Stephenson, W. E. Wallace and L. J. Richter, *J. Phys. Chem. B* **105**, 2785 (2001).
95. W. Norde and J. P. Favier, *Colloids Surf.* **64**, 87 (1992).
96. W. Norde and C. E. Giacomelli, *Macromol. Symp.* **145**, 125 (1999).
97. J. R. Lu, T. J. Su, P. N. Thirtle, R. K. Thomas, A. R. Rennie and R. Cubitt, *J. Colloid Interface Sci.* **206**, 212 (1998).
98. T. J. Su, J. R. Lu, R. K. Thomas, Z. F. Cui and J. Penfold, *J. Colloid Interface Sci.* **203**, 419 (1998).
99. T. J. Su, J. R. Lu, R. K. Thomas, Z. F. Cui and J. Penfold, *Langmuir* **14**, 438 (1998).
100. P. Schaaf, P. Déjardin and A. Schmitt, *Langmuir* **3**, 1131 (1987).
101. C. F. Wertz and M. M. Santore, *Langmuir* **18**, 706 (2002).
102. D. Horsley, J. Herron, V. Hlady and J. D. Andrade, in: *Proteins at Interfaces: Physicochemical and Biochemical Studies*, J. L. Brash and T. A. Horbett (Eds), p. 290. American Chemical Society, Washington, DC (1986).
103. C. M. Cuppett, L. J. Doneski and M. J. Wirth, *Langmuir* **16**, 7279 (2000).
104. J. Chen, D. E. Dong and J. D. Andrade, *J. Colloid Interface Sci.* **89**, 577 (1982).
105. G. B. Sigal, M. Mrksich and G. M. Whitesides, *J. Am. Chem. Soc.* **120**, 3464 (1998).
106. W. Norde and J. Lyklema, *J. Colloid Interface Sci.* **66**, 257 (1978).
107. D. E. Gragson, B. M. McCarty and G. L. Richmond, *J. Am. Chem. Soc.* **119**, 6144 (1997).
108. D. E. Gragson and G. L. Richmond, *J. Phys. Chem. B* **102**, 3847 (1998).
109. M. Himmelhaus, F. Eisert, M. Buck and M. Grunze, *J. Phys. Chem. B* **104**, 576 (2000).
110. J. Wang, S. M. Buck and Z. Chen, *J. Phys. Chem. B* **106**, 11666 (2002).
111. A. Haeberli (Ed.), *Human Protein Data*. Wiley-VCH, Weinheim (1998).
112. E. F. Osserman, R. E. Canfield and S. Beychok (Eds), *Lysozyme*. Academic Press, New York, NY (1974).
113. T. J. Peters, *All about Albumin Biochemistry, Genetics and Medical Applications*. Academic Press, San Diego, CA (1996).
114. G. Yan, J.-T. Li, S.-C. Huang and K. D. Caldwell, in: *Proteins at Interfaces II: Fundamentals and Applications*, T. A. Horbett and J. L. Brash (Eds), p. 256. American Chemical Society, Washington, DC (1995).
115. D. R. Lu and K. Park, *J. Colloid Interface Sci.* **144**, 271 (1991).

Gels, Genes, Grafts and Giants, pp. 397–410 (2005)

Surface characteristics of block-type copolymer composed of semi-fluorinated and phospholipid segments synthesized by living radical polymerization

YUUKI INOUE, JUNJI WATANABE, MADOKA TAKAI
and KAZUHIKO ISHIHARA *

Department of Materials Engineering, School of Engineering, The University of Tokyo, 7-3-1, Hongo, Bunkyo-ku, Tokyo 113-8656, Japan

Received 1 October 2003; accepted 5 March 2004

Abstract—A series of random and block copolymers composed of hydrophilic and hydrophobic monomer units have been synthesized by the free and living radical polymerization methods, respectively. The hydrophilic monomer unit, 2-methacryloyloxyethyl phosphorylcholine (MPC), was selected because the MPC polymers are well-known for their excellent bio- and blood compatibilities. The semi-fluorinated monomer, 2,2,2-trifluoroethyl methacrylate (TFEMA), was used as the hydrophobic monomer unit. Several analyses of the copolymer surface showed that the TFEMA unit was concentrated at the outermost surface on the random copolymer surface and characteristics of MPC unit was dominant on the block copolymer with a low-MPC-unit composition in the dry state. A reorientation of the MPC unit occurred dynamically in the wet state because of the strong hydrophilicity of the MPC units. In the case of the block copolymer with a low-MPC-unit composition, the surface was covered with the MPC units in the wet state. As a result, the amount of the adsorbed bovine plasma fibrinogen and bovine serum γ-globulin on the block copolymer surface was reduced dramatically.

Key words: 2-Methacryloyloxyethyl phosphorylcholine; semi-fluorinated polymer; biocompatibility; protein adsorption; living radical polymerization; block copolymer; surface tension.

INTRODUCTION

Several surface modifications to conventional polymeric materials have been made using newly designed polymers because the functional groups at the surface play an important role in the interactions between the materials and biological molecules or cells. Much attention has been paid to the control of the microdomain

*To whom correspondence should be addressed. Tel.: (81-3) 5841-7124. Fax: (81-3) 5841-8647. E-mail: ishihara@bmw.t.u-tokyo.ac.jp

structures [1] or the assignment of functionality [2] at the surface to improve the biocompatibility of the conventional polymeric materials. In particular, the study of surface modification with block copolymers indicates that the morphology, size, and chemical structure of the monomer units of the microdomain structure are important to obtain biocompatibility of the surface [3, 4].

To create biocompatible surfaces on the conventional polymeric materials by surface modification with other polymers, we prepared polymers using a bioinspired concept. If a biomembrane-like structure could be prepared on the surface, the surface would become biocompatible. Thus, we designed bioinspired polymers, which have 2-methacryloyloxyethyl phosphorylcholine (MPC) units as a component [5–9]. The MPC unit has a phosphorylcholine (PC) group that exists mainly at the outer surface of the biomembrane and the MPC polymers could provide the improved biocompatibilities. The protein-resistant property of the MPC polymers could be explained thermodynamically and electrostatically, that is, the MPC polymer has little bound water causing protein adsorption and its surface potential is close to zero [10, 11].

The (semi-)fluorinated polymers, such as 2,2,2-trifluoroethyl methacrylate (TFEMA), were widely studied because they possess several unique properties, such as low surface tension, water- and oil repellency, and chemical and thermal stability [12]. However, it is also known that the solubility of fluorinated polymers in several solvents is often low and their biocompatibility is not very high. Thus, the semi-fluorinated polymer with the amphiphilic and biocompatible MPC unit could improve the solubility and biocompatibility of the fluorinated polymers. The unique functionalities could also be assigned to the surface by the copolymers composed of hydrophilic/hydrophobic monomer units with extremely large differences in the surface tensions.

It should be essential to modify the biomaterial surface with well-defined polymers in order to control the microscopic interactions between the polymer surface and highly structured biomolecules. Recent progress in controlled living radical polymerization (LRP) makes it possible to obtain polymers with narrow molecular weight distribution, active terminal groups and well-defined structures. Therefore, LRP could be an indispensable way to synthesize well-defined polymers for controlling the structure of the biomaterial surface. Several mechanisms in LRP have been proposed such as nitroxide-mediated polymerization (NMP) [13], photo-induced LRP with an iniferter (initiator-transfer agent-terminator) [14], atom transfer radical polymerization (ATRP) with a metal complex [15] and reversible addition-fragmentation chain transfer radical polymerization (RAFT) with thiocarbonylthio compounds [16]. Photo-induced LRP is a relatively facile method to synthesize the block copolymer and can be applied to a large range of the monomer units.

In this study, we synthesized block copolymers composed of MPC and TFEMA by LRP, focusing on the large differences in their surface tensions. In particular, to investigate the influence of the monomer unit sequence on the surface properties, the block copolymer surface and the corresponding random copolymer surface were investigated.

MATERIALS AND METHODS

Materials

2-Methacryloyloxyethyl phosphorylcholine (MPC) was synthesized using a method described in detail elsewhere [17]. 2,2,2-Trifluoroethyl methacrylate (TFEMA) was obtained from Tosoh/F-TECH (Tokyo, Japan) and was used without further purification. Other reagents and solvents were commercially available in extra-pure grade and were used without further purification.

Synthesis of photoiniferter

Benzyl N,N-diethyldithiocarbamate (BDC) as a photoiniferter was prepared according to the method previously reported by Otsu *et al.* [18]. The pure synthesis of BDC was confirmed by ^{1}H-NMR (α-300, Jeol, Tokyo, Japan) and FT-IR (FT-IR-615, Jasco, Tokyo, Japan).

Synthesis of PTFEMA macroiniferters

The poly(TFEMA) macroiniferter (PTFEMA-BDC), which means the TFEMA homopolymer introduced the iniferter site at the polymer termination, was synthesized by living radical polymerization. TFEMA was placed in a flask and BDC was added to this flask. The mixture was then diluted with 1,4-dioxane to adjust the concentration of TFEMA to 0.50 mol/l. The concentrations of the BDC were adjusted to 17, 5.0 and 2.5 mmol/l. Argon was bubbled into the solution to eliminate oxygen.

The polymerization was carried out at room temperature under irradiation by UV light (UVL-400HA, Riko, Chiba, Japan) ($\lambda = 360 \pm 50$ nm). After polymerization for 12 h, the content was poured into a large amount of hexane to precipitate the polymer. The precipitate was filtered off and dried *in vacuo*. The chemical structure of PTFEMA-BDC was confirmed by ^{1}H-NMR and FT-IR. The molecular weight of PTFEMA-BDC was determined by both ^{1}H-NMR and gel-permeation chromatography (GPC). Tetrahydrofuran (THF) was used as an eluent for the GPC measurement at room temperature (flow rate 1.0 ml/min; column: KF-804L, Shodex, Tokyo, Japan). Polystyrene standard samples were used to determine the apparent molecular weight. The synthetic result for PTFEMA-BDC is shown in Table 1.

Synthesis of block copolymers

A procedure similar to the synthesis of the macroiniferter was used for block polymerization with PTFEMA100-BDC as the macroiniferter. The PTFEMA100-BDC was synthesized when the concentration of the BDC was 5.0 mmol/l. The methanol/1,4-dioxane mixture (7 : 3 by volume) was used as a solvent and the concentration of MPC was adjusted to 0.2 mol/l. The polymerization was carried out with various concentrations of PTFEMA100-BDC. After the reaction, the polymers formed were purified by reprecipitation. In the case of PMFb100-6 and PMFb100-20, a water/acetone mixture (1 : 1 by volume) was poured into a

Table 1.
Synthetic results of poly(TFEMA)(PTFEMA-BDC)

Abbreviation	[Monomer]/[I]	M_n ($\times 10^3$)		M_w/M_n [a]	Yield (%)
		NMR	GPC [a]		
PTFEMA30-BDC	30	6.0	8.0	1.4	33
PTFEMA100-BDC	100	8.0	12.0	1.4	32
PTFEMA200-BDC	200	16.0	14.0	1.8	32

[Monomer] = 0.5 mol/l, I = photoiniferter (BDC), solvent = dioxane, polymerization time = 12 h.
[a] PSt standard, THF.

Table 2.
Synthetic results of poly(MPC-*block*-TFEMA)

Abbreviation	[Monomer]/[S]	MPC unit mol fraction		Yield (%)
		In feed	In copolymer [a]	
PMFb100-6	6	0.11	0.25	20
PMFb100-20	20	0.30	0.38	41
PMFb100-50	50	0.53	0.46	41
PMFb100-200	200	0.81	0.64	44

[Monomer] = 0.2 mol/l, polymerization time = 9 h, S = chain transfer agent (PTFEMA100-BDC).
[a] Determined by phosphorus analysis.

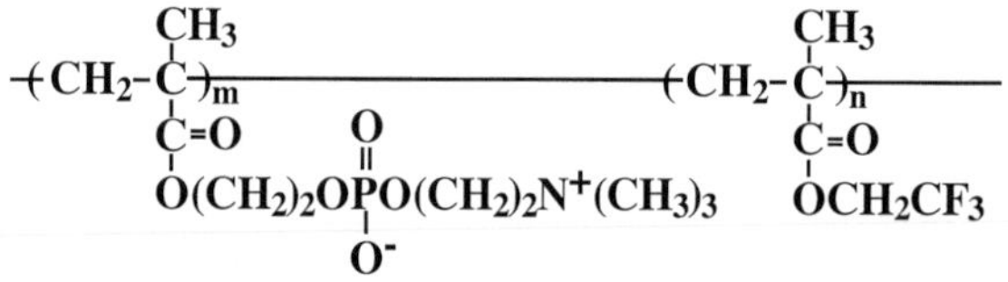

Figure 1. Chemical structure of poly(MPC-*co*-TFEMA).

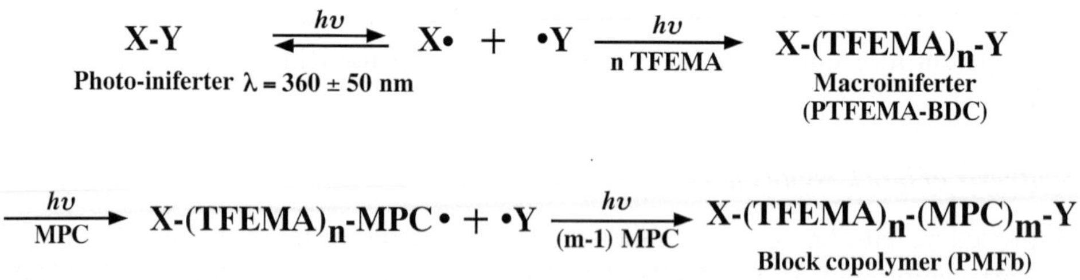

Scheme 1.

solvent for precipitation, and in the case of PMFb100-50 and PMFb100-200 a methanol/acetone mixture (1 : 20 by volume) was used. After these mixtures were centrifuged, the solvent was removed, and the precipitated polymers were then dried *in vacuo*. The chemical structure of poly(MPC-*block*-TFEMA) (PMFb, Fig. 1) was confirmed by ^{1}H-NMR and FT-IR. The compositions of each component in the PMFb were determined from the phosphorous analysis. The synthetic route and results for PMFb are shown in Scheme 1 and Table 2, respectively.

Synthesis of random copolymers

Poly(MPC-*random*-TFEMA) (PMTF) was synthesized by free radical polymerization. MPC and TFEMA were placed in a glass tube and 2,2′-azobisisobutyronitorile (AIBN) was added. The mixture was then diluted with ethanol to adjust the total volume to 30 ml, and the concentrations of the total monomer and AIBN were 0.5 mol/l and 5 mmol/l, respectively. Argon was bubbled into the solution to eliminate oxygen and then the glass tube was sealed.

The polymerization was carried out at 60°C for 24 h. After cooling, the reaction mixture was poured into a large amount of a diethyl ether/chloroform mixture (8 : 2 by volume) to remove any unreacted monomers and to precipitate the polymer. The precipitate was filtered off and dried *in vacuo*. Poly(MPC) (PMPC) and poly(TFEMA) (PTFEMA) were also synthesized using ethanol and 1,4-dioxane as the polymerization solvents, while chloroform and hexane were used as the solvents for the precipitation, respectively. The chemical structures of PMTF, PMPC and PTFEMA were confirmed by ^{1}H-NMR and FT-IR. The mole fraction of each component in the PMTF was determined from the ^{1}H-NMR. The apparent molecular weight was determined by the GPC measurement. The mixture of methanol/water (7 : 3) containing 10 mmol/l of lithium bromide was used as an eluent for the GPC measurement at room temperature for PMPC and PMTF (flow rate; 0.5 ml/min; column: SB-804 HQ, Shodex). A poly(ethylene glycol) (PEG) was used as a standard sample. The synthetic results for PMTF are shown in Table 3.

Polymer coating on substrates

Prior to the coating, $40 \times 10 \times 1.1$ mm^3 glass plates were cleaned with hexane, acetone and methanol or ethanol, and $40 \times 10 \times 0.95$ mm^3 quartz plates with fuming nitric acid, water and acetone. These substrates were immersed in the polymer solutions (PMFb, 1.0 wt% in methanol; PMTF and PMPC, 0.5 wt% in ethanol; PTFEMA100-BDC and PTFEMA, 2.0 wt% in chloroform) twice (the interval was 30 min), dried under a solvent atmosphere overnight and then vacuum-dried overnight. These samples were used for the analysis in the dry state. For the analysis in the wet state, every sample was stored in distilled water or phosphate-buffered solution (PBS, pH 7.4) for 1 day to equilibrate the surface. The samples were frozen with liquid nitrogen and then the sample was freeze-dried when it was needed.

Table 3.
Synthetic results of poly(MPC-*random*-TFEMA)

Abbreviation	MPC unit mol fraction		Molecular weight[b]		Yield (%)
	In feed	In copolymer[a]	M_n ($\times 10^3$)	M_w/M_n	
PTFEMA	0	0	33[c]	1.9[c]	55
PMTF20	0.20	0.23	6.0	2.2	66
PMTF40	0.40	0.43	9.0	3.1	73
PMTF60	0.60	0.64	12	4.4	64
PMTF80	0.80	0.76	12	4.3	98
PMPC	1.0	1.0	5.0	3.3	89

[Monomer] = 0.5 mol/l, [AIBN] = 5 mmol/l, polymerization time = 24 h.
[a] Determined by ^{1}H-NMR measurement.
[b] PEG standard, MeOH/H_2O = 7:3, [LiBr] = 10 mM.
[c] PSt standard, THF.

X-ray photoelectron spectroscopic measurement

An X-ray photoelectron spectroscope (XPS, AXIS-HSi, Shimadzu/Kratos, Kyoto, Japan) was used to determine the monomer unit composition at the polymer surface. The take-off angle of photoelectrons was 15 degrees. The measurement was performed in both the dry state and the wet state after freeze-drying.

Static contact angle measurement

The sessile drop technique (CA-W, Kyowa Interface Science, Saitama, Japan) was used to measure the static contact angle (SCA) of liquids with known surface tension on the polymer surfaces. The liquids used in this study were water and formamide. The polar and non-polar components of water and diiodomethane are 43.7 dyn/cm and 29.1 dyn/cm, and 4.0 dyn/cm and 46.8 dyn/cm, respectively [19]. The geometric mean method was used to deduce the surface tension of the sample surfaces. The surface tension values are the sum of polar and non-polar components calculated from the static contact angles by water and diiodomethane.

ζ-potential measurement

The glass plates and the polymer surfaces were subjected to ζ-potential measurement using an ELS-800 electrophoretic light-scattering spectrophotometer (Otsuka Electronics, Osaka, Japan) equipped with a plate sample cell. The measurement was carried at 25°C in water containing 10 mmol/l sodium chloride.

Evaluation of protein adsorption on the polymer surface

The micro BCA method was used to determine the amount of adsorbed proteins on the surface of the samples [20]. The proteins used in this study were bovine plasma fibrinogen (Fib, Sigma-Aldrich, St. Louis, MO, USA, F 8630) and bovine serum γ-globulin (IgG, Sigma-Aldrich, G 5009) because the Fib and IgG are the representative plasma proteins. Single protein adsorption was investigated to clarify the biological interactions between the surface and each protein. The concentrations of the Fib and IgG in PBS were 0.03 and 0.16 g/dl, respectively. These protein solutions were placed in contact with the surface of samples at 37°C for 2 h. To rinse these samples, they were immersed in PBS repeatedly 50 times. To detach the adsorbed proteins on the surface completely, the samples were immersed in a 1 wt% PBS solution of sodium dodecylsulfate (SDS) and applied sonication for 20 min using an ultrasonic water bath. The protein analysis kit (micro BCA protein assay reagent kit, #23235, Pierce, Rockford, IL, USA) was used to determine the concentration of the proteins in the SDS solution, and the amount of proteins adsorbed on the surface was calculated.

All the measurement was repeated at least 3 times, and the mean value was shown as data point. The t-test method was used to estimate statistical significance for all analysis.

RESULTS AND DISCUSSION

Characterization of block and random copolymers

Figure 2 shows the time–M_n relation in the polymerization of TFEMA when the concentration of the BDC was 5 mmol/l. The M_n of the polymer increased with the polymerization time. Therefore, it was clear that this polymerization was performed by a living radical polymerization mechanism [14]. Table 1 shows the synthetic results for PTFEMA when the concentration of the photoiniferter (BDC) was varied. It was revealed that the molecular weight of the polymers could be controlled by the concentration of the BDC.

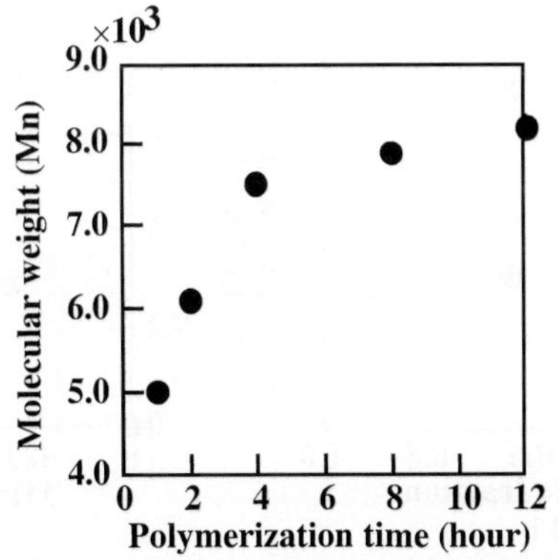

Figure 2. Time–M_n relation for photopolymerization of TFEMA with BDC. ([TFEMA] = 0.5 mol/l, [BDC] = 5 mmol/l).

The block copolymers were synthesized with PTFEMA100-BDC as the macro-iniferter. Table 2 shows the synthetic results for the block copolymers. The MPC unit composition of the PMFb increased with that of the feed. All the PMFbs were insoluble in water.

Random copolymers composed of MPC and TFEMA (PMTF) could be obtained by conventional radical polymerization. Table 3 shows the synthetic results for PMTF. The MPC unit composition of the PMTF was almost the same as that in the feed. The PMTFs with a high-MPC-unit composition were soluble in water.

Chemical analysis of the polymer surface

The samples prepared in this study were all transparent. The surface chemical structure of the samples was analyzed by XPS. Figure 3 shows the typical XPS charts of the PMTF and PMFb surfaces in the dry state when the takeoff angle of photoelectrons was 15 degrees. For all samples, the XPS signals were observed at 285.0 eV (CH_3, $-CH_2-$), 286.0 eV ($-C-O-C-$), 288.5 eV ($C=O$) and

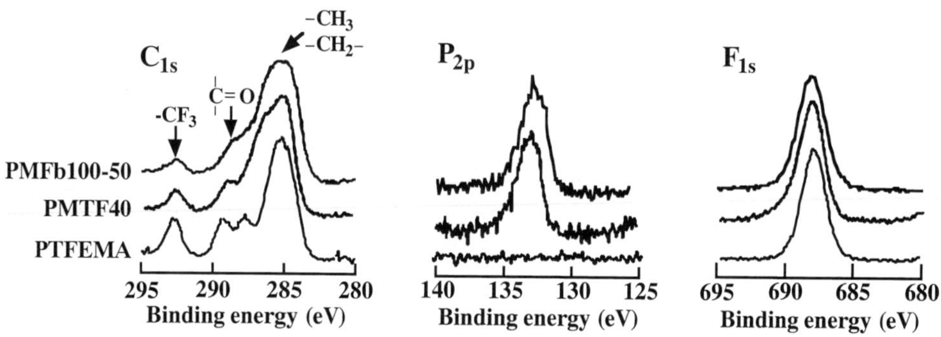

Figure 3. Representative XPS charts of polymer surfaces.

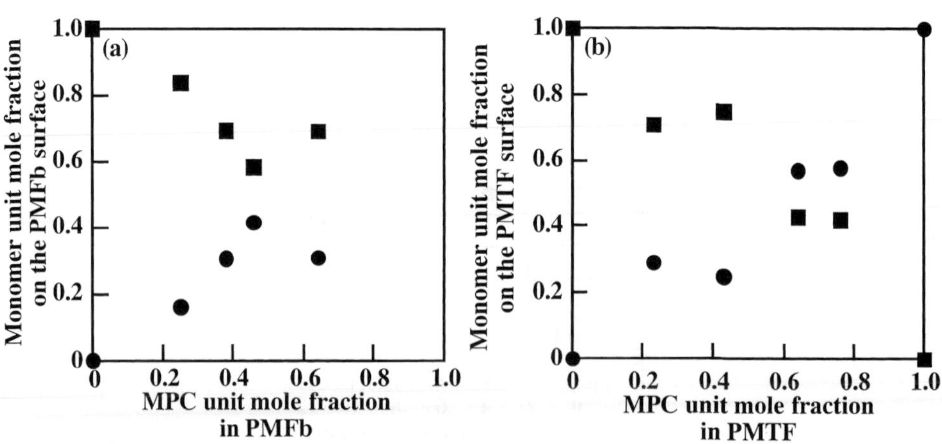

Figure 4. Monomer unit composition on the surface in the dry state. (●) MPC unit; (■) TFEMA unit. (a) PMFb; (b) PMTF.

292.0 eV (CF_3) in the C_{1s} region [21]. Moreover, the peaks observed at 133.4 eV and 689.0 eV were attributed to the phosphorus of the phosphate group and fluorine of the CF_3 group, respectively. The peaks attributed to the silicon of the glass plate were not observed for all samples. Figure 4 shows the composition of each monomer unit on the outer surface of PMFb and PMTF. The TFEMA unit was concentrated at the surface of PMFb100-200, PMTF40 and PMTF80, that is, the TFEMA unit composition at the surface was higher than that in the bulk. Figure 5 represents the relationship between the monomer unit composition of the bulk and that on the surface in the wet state. When surfaces with the same monomer unit composition were compared, the MPC unit composition on the wet surface was larger than that on the dry surface regardless of the copolymer. This result indicated that the reorientation of the MPC unit occurred on the sample surface when they were soaked in water.

Wettability of the polymer surface was analyzed by the static contact angle by water and diiodomethane. Figure 6 shows the static contact angles on the sample surfaces. There were sufficient statistical significances between the static contact angles by water and diiodomethane for the PMFb surfaces with a low-MPC-unit composition ($P < 0.05$); however, there was no significance for the PMFb surfaces with a high-MPC-unit composition ($P > 0.05$). Therefore, the static contact angles by two liquids on the PMFb surface with a low-MPC-unit composition decreased with an increase in the MPC unit composition and took the constant values on the surface with a high-MPC-unit composition. On the other hand, in the case of the PMTF, the static contact angle for water was maximum on the PMTF60 surface, and that of diiodomethane decreased with the increase in the MPC unit composition of PMFb. The surface tension of the polymer samples and its polar and non-polar components were calculated from the static contact angles by water and diiodomethane [22]. Figure 7 shows the surface tensions of the sample surfaces,

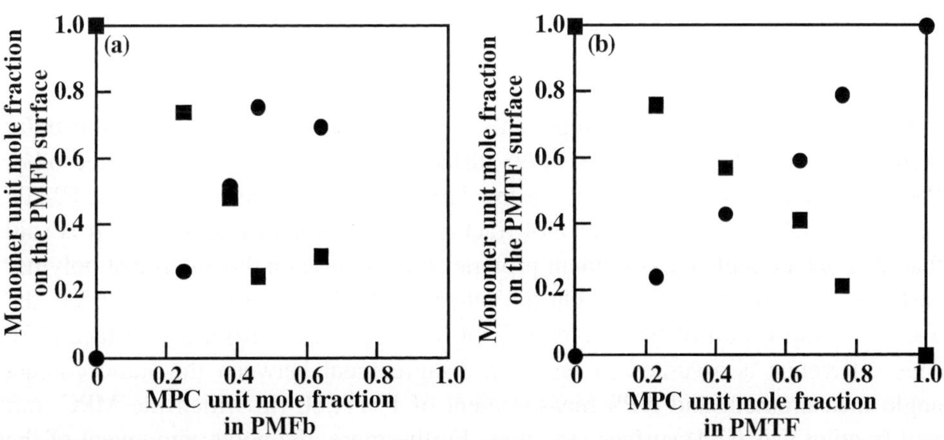

Figure 5. Monomer unit composition on the surface in the wet state. (●) MPC unit; (■) TFEMA unit. (a) PMFb; (b) PMTF.

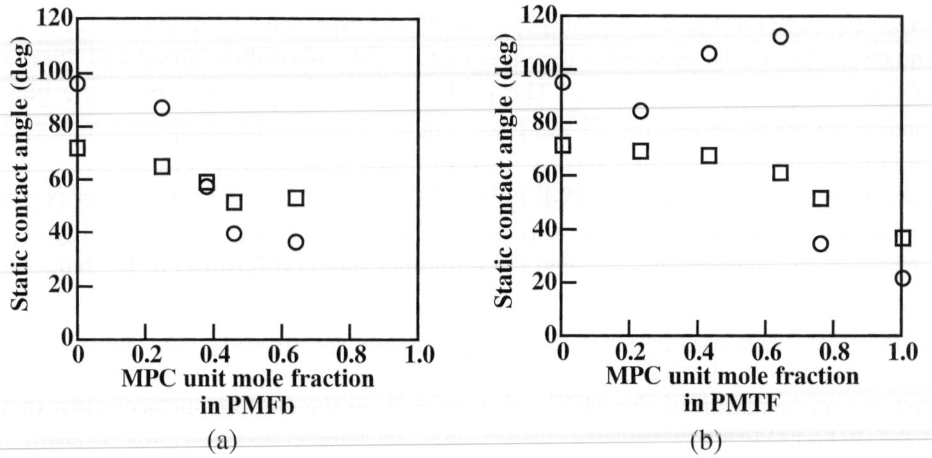

Figure 6. Static wettability of the surface. (○) Water; (□) diiodomethane. (a) PMFb; (b) PMTF.

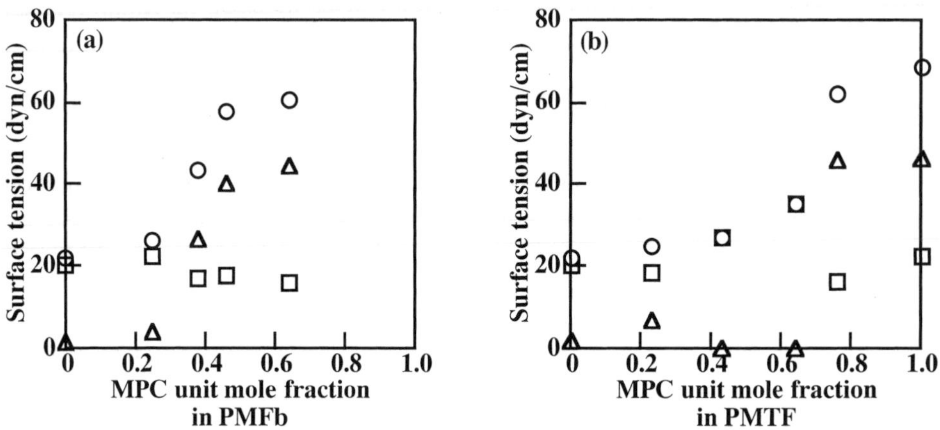

Figure 7. Surface tension. (○) Surface tension; (□) non-polar component; (△) polar component. (a) PMFb; (b) PMTF.

their polar, and non-polar components. As shown in Fig. 7b, the non-polar component of PMTF40 and PMTF60 surfaces took higher values than the other PMTF surface. It was supposed that the TFEMA unit was enriched on the PMTF surfaces of which the MPC unit mol fraction of the bulk was about 0.5. It is said that the contact angle measurement provides information on the outermost polymer surfaces of a few Ångstroms. On the other hand, XPS measurement reflects the average composition of the polymer for a few nanometers from the surface [23]. This difference is assumed to result in disagreement between the static contact angle measurement and XPS measurement of PMTF60 (in which the MPC unit mol fraction was 0.64) surface structure. Furthermore, the polar component of the PMFb surface began to increase with a low-MPC-unit composition, while that of the PMTF surface began to increase with a high-MPC-unit composition. This result

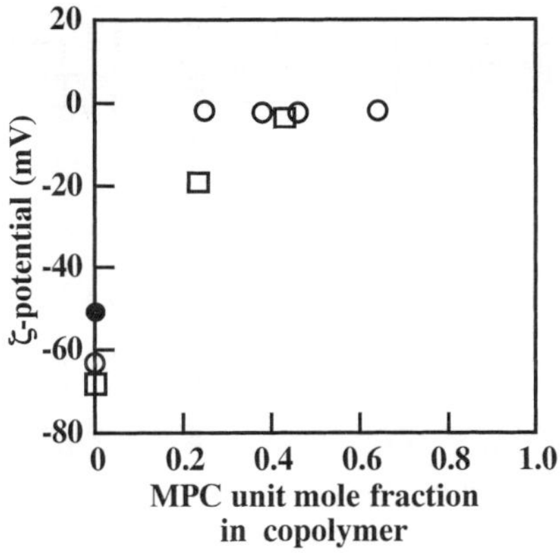

Figure 8. Surface potential. (●) Glass plate; (○) PMFb; (□) PMTF.

supposed that the characteristics of MPC unit were dominant on the PMFb surface with a low-MPC-unit composition compared to the PMTF surface in the dry state.

To evaluate the surface properties in the wet state, ζ-potential measurement was carried out using PMFb and PMTF with a lower MPC unit composition. It is known that the ζ-potential of the surface covered with the MPC unit is about zero because the ionic group in the MPC unit forms an inner salt and the electrostatic effects are diminished [9, 10]. This property is the same as that of the natural phospholipid, phosphatidylcholine. Figure 8 shows the relationship between the MPC unit composition of the polymers and the ζ-potential of the surface. The ζ-potential of the glass plate and the PTFEMA surfaces was both strongly negative. As the MPC unit composition of the PMTF increased, the ζ-potential of the surfaces became gradually close to zero. However, the ζ-potential of the PMFb surfaces became zero even on the surface with a low-MPC-unit composition. This result suggested that all PMFb surfaces were covered with the MPC unit.

Protein adsorption on the surface

The amount of adsorbed proteins on the surfaces was determined by a micro BCA method after detachment of the adsorbed proteins. The model proteins used in this study were typical plasma proteins, that is, Fib and IgG. As shown in Figs 9 and 10, the amount of adsorbed proteins on the surfaces was significantly reduced by the introduction of an MPC unit for each protein compared with that on a PTFEMA, quartz or glass surface ($P < 0.05$). However, the amount of adsorbed proteins made a difference between the PMFb and PMTF surfaces. The theoretical amounts of the monomolecular Fib and IgG adsorption are 0.27 $\mu g/cm^2$ and 0.18 $\mu g/cm^2$ for the side-on-type adsorption, and 1.70 $\mu g/cm^2$ and 1.85 $\mu g/cm^2$ for the end-on-type adsorption, respectively [24, 25]. The amount of the adsorbed IgG was at most the same as the theoretical value both on PMFb and PMTF surfaces. On the other hand, the amount of the adsorbed Fib was suppressed on the PMFb surface with a low-

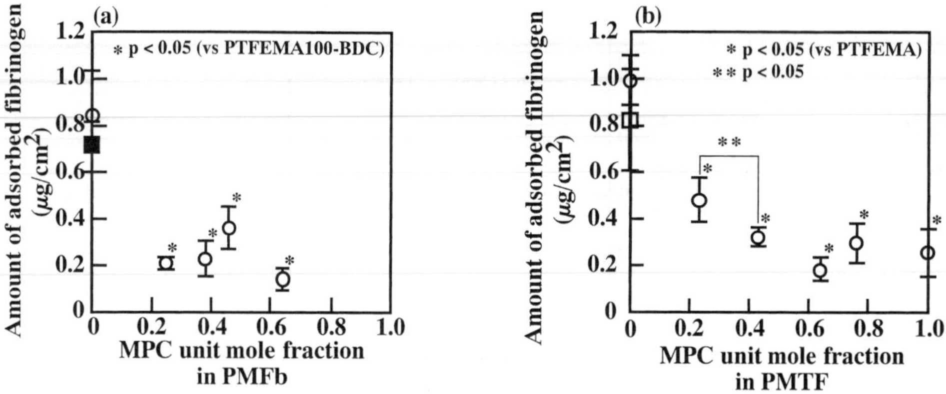

Figure 9. Amount of adsorbed fibrinogen. (■) Quartz plate; (□) glass plate. (a) PMFb; (b) PMTF.

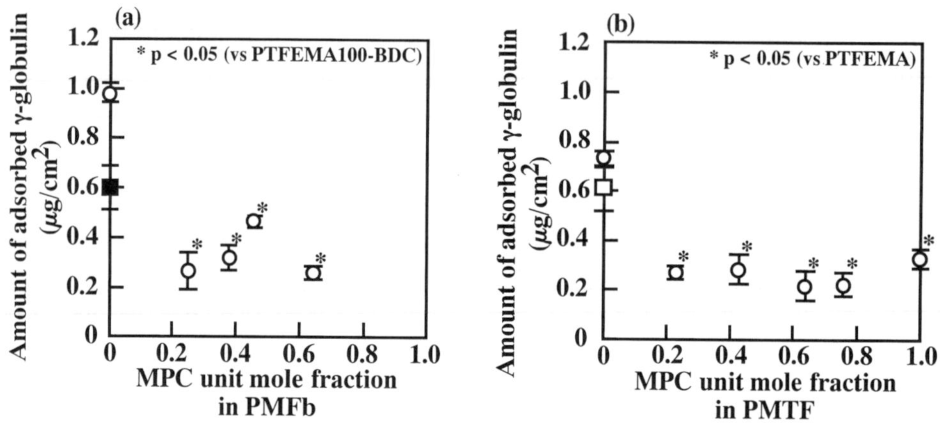

Figure 10. Amount of adsorbed γ-globulin. (■) Quartz plate; (□) glass plate. (a) PMFb; (b) PMTF.

MPC-unit composition, which was different from that on the PMTF surface with almost the same MPC unit composition ($P < 0.05$).

We have considered that the protein adsorption was closely related to the state of the water molecules on the material surface [9, 26]. Displacement between the bound water around the proteins and that around the surface could induce protein adsorption. Therefore, the surface with little bound water could suppress protein adsorption. The PTFEMA surface has a hydrophobic character and strongly negative charge measured by static contact angle and ζ-potential measurement, respectively. Therefore, the hydrophobic and electrostatic interactions were supposed to induce much protein adsorption on PTFEMA surface. As mentioned above, the ζ-potential of the PMTF surface with a low-MPC-unit composition was relatively negative compared with the other surfaces in the wet state. It was suggested that the PMTF surface with a low-MPC-unit composition was partially covered with TFEMA unit in the wet state. Therefore, the amount of adsorbed Fib on this surface was larger than the other surfaces with MPC unit. Among a number of proteins in blood, Fib dominated clot formation and subjected to adsorb on the surface. Therefore, Fib adsorbed more on the surface with a low-MPC-unit than other PMTF surface, while IgG did not.

PMFb could suppress the adsorption of adhesive and clot-inducing Fib; therefore, it may provide material improved biocompatibility.

CONCLUSIONS

Block copolymers, poly(MPC-*block*-TFEMA) (PMFb) composed of hydrophilic phospholipid and hydrophobic fluorinated segments, could be synthesized by living radical polymerization. According to the surface analysis in the dry state, the characteristics of MPC unit were indicated to be dominant on the PMFb surface. In the wet state, the reorientation of the MPC unit occurred. The results suggested that the surface was covered with MPC units, even in the case of the PMFb surface with a low-MPC-unit composition. Therefore, the amount of the adsorbed proteins was lowered significantly on the PMFb compared with that on the corresponding random copolymer with the same MPC unit composition. We concluded that the PMFbs are promising materials to modify the surface of conventional polymers to obtain improved biocompatibility compared to PMTF polymers.

Acknowledgements

The present research was supported in part by a grant for the 21st Century COE Program 'Human-Friendly Materials based on Chemistry' from the Ministry of Education, Culture, Sports, Science, and Technology of Japan.

REFERENCES

1. T. Okano, S. Nishiyama, I. Shinohara, T. Akaike and Y. Sakurai, *Polym. J.* **10**, 223 (1977).
2. S. Nagaoka and A. Nakao, *Biomaterials* **11**, 119 (1990).
3. T. Okano, S. Nishiyama, I. Shinohara, T. Akaike, Y. Sakurai, K. Kataoka and T. Tsuruta, *J. Biomed. Mater. Res.* **15**, 393 (1981).
4. T. Okano, K. Kataoka, Y. Sakurai, M. Shimada, T Akaike and I. Shinohara, *Artificial Organs* **5** (Suppl.), 468 (1981).
5. K. Ishihara, R. Aragaki, T. Ueda, A. Watanabe and N. Nakabayashi, *J. Biomed. Mater. Res.* **24**, 1069 (1990).
6. K. Ishihara, N. P. Ziats, B. P. Tierney, N. Nakabayashi and J. M. Anderson, *J. Biomed. Mater. Res.* **25**, 1397 (1991).
7. K. Ishihara, H. Oshida, T. Ueda, Y. Endo, A. Watanabe and N. Nakabayashi, *J. Biomed. Mater. Res.* **26**, 1543 (1992).
8. K. Ishihara, T. Tsuji, T. Kurosaki and N. Nakabayashi, *J. Biomed. Mater. Res.* **28**, 225 (1994).
9. K. Ishihara, H. Nomura, T. Mihara, K. Kurita, Y. Iwasaki and N. Nakabayashi, *J. Biomed. Mater. Res.* **39**, 323 (1998).
10. K. Ishihara and Y. Iwasaki, *J. Biomater. Appl.* **13**, 111 (1998).
11. K. Ishihara, *Sci. Technol. Adv. Mater.* **1**, 131 (2000).

12. T. Imae, *Curr. Opin. Colloid Interface Sci.* **8**, 307 (2003).
13. M. K. Georges, R. P. N. Veregin, P. M. Kazmaier and G. K. Hamer, *Macromolecules* **26**, 2987 (1993).
14. T. Otsu, M. Yoshida and T. Tazaki, *Makromol. Chem. Rapid Commun.* **3**, 133 (1982).
15. J. S. Wang and K. Matyjaszewski, *Macromolecules* **28**, 7901 (1995).
16. G. Moad, J. Chiefari, Y. K. Chong, J. Krstina, R. T. A. Mayadunne, A. Postma, E. Rizzardo and S. H. Thang, *Polym. Int.* **49**, 993 (2000).
17. K. Ishihara, T. Ueda and N. Nakabayashi, *Polym. J.* **22**, 355 (1990).
18. T. Otsu and A. Kuriyama, *Polym. Bull.* **11**, 135 (1984).
19. D. K. Owens and R. C. Wendt, *J. Appl. Polym. Sci.* **13**, 1741 (1969).
20. K. P. Smith, I. P. Krohn, T. G. Hermonson, K. A. Mallia, H. F. Gartner, D. M. Provenzano, K. E. Fujimoto, M. N. Goeke, J. B. Olsen and C. D. Klenk, *Anal. Biochem.* **150**, 76 (1985).
21. C. M. Kassis, J. K. Steehler, D. E. Betts, Z. Guan, T. J. Romack, J. M. DeSimone and R. W. Linton, *Macromolecules* **29**, 3247 (1996).
22. F. M. Fowkes, *Ind. Eng. Chem.* **56**, 40 (1964).
23. P. M. Sean and A. W. Dench, *Surf. Interface Anal.* **1**, 2 (1979).
24. A. Baskin and D. J. Lyman, *J. Biomed. Mater. Res.* **14**, 393 (1980).
25. M. Tanaka, A. Mochizuki, T. Motomura, K. Shimura, M. Onishi and Y. Okahata, *Colloids Surf. A* **193**, 145 (2001).
26. D. R. Lu, S. J. Lee and K. Park, *J. Biomater. Sci. Polymer Edn* **3**, 127 (1991).

Gels, Genes, Grafts and Giants, pp. 411–419 (2005)

Grafting copolymerization of 2-methacryloyloxyethyl phosphorylcholine (MPC) onto pre-irradiated cellulose films

JIE CHEN [1,*], YOUNG CHANG NHO [2] and ALLAN S. HOFFMAN [3]

[1] *Department of Chemical Engineering and Technology, Shanghai University (Jiadiang Campus), Jiading, Shanghai 201800, China*

[2] *Radiation Application Division, Korea Atomic Energy Research Institute, P. O. Box 105, Yusong, Taejon, South Korea*

[3] *Bioengineering Department, Box 352255, University of Washington, Seattle, WA 98159, USA*

Received 3 October 2003; accepted 7 January 2004

Abstract—Grafting of 2-methacryloyloxyethyl phosphorylcholine (MPC) and poly(ethylene glycol mathacrylate) (PEGMA) onto cellulose films was performed using the pre-irradiation grafting method. The effects of monomer concentration, solvent system and co-solvent composition, and reaction time on the degree of grafting were determined. The grafted samples were confirmed by FT-IR–ATR spectra. The blood compatibilities of the grafted cellulose were evaluated by platelet-rich plasma contact studies and viewed by scanning electron microscopy; non-grafted cellulose film sample was used as references. As a result, MPC and MPC/PEGMA were grafted on the surface of cellulose films. It was found that fewer platelets adhered to the MPC-grafted surfaces and that they showed less shape variation than the ungrafted references.

Key words: Radiation grafting; cellulose film; blood compatibility; MPC; PEGMA.

INTRODUCTION

The 2-methacryloyloxyethyl phosphorylcholine (MPC, Scheme 1) polymer having phospholipid polar group shows excellent blood compatibility.

Some experiments were reported that the suppression of platelet adhesion and activation was observed when the MPC polymer came in contact with human whole blood, even without anticoagulant. This is due to reduction of protein adsorption on the MPC polymer surface from human plasma [1–3].

MPC can be polymerized in water or grafted on a polymer substrate surface in a heterogeneous system. It is reported that the introduction of poly(MPC)

*To whom correspondence should be addressed. E-mail: chenjjd@online.sh.cn

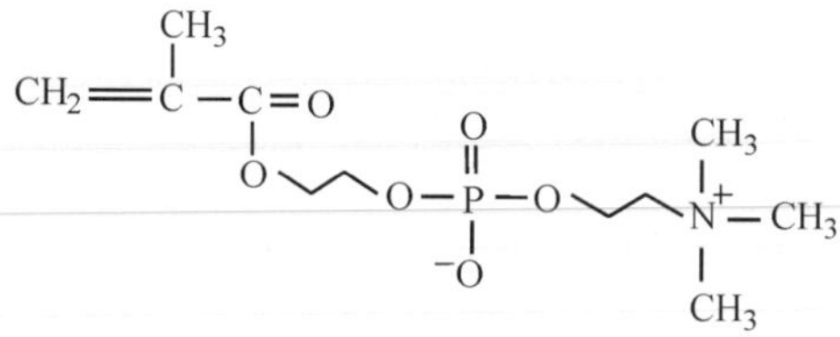

Scheme 1. Structure of MPC.

chains on the substrate surface was effective in preventing platelet adhesion and activation [4–6]. Some works have been done on the introducing of poly(MPC) onto cellulose [7, 8], chitosan [9] membrane and other substrates by various methods, such as blending, coating, ozonizing, etc. [10–12]. Co-polymers of the hydrophilic monomer (MPC) with hydrophobic monomers have been reported, such as poly(MPC-co-BMA) and others [13, 14].

Radiation grafting is an effective method for the modification of polymeric surface, which has been widely applied for the preparation of biomaterials [15, 16]. To date, there is almost no report on the grafting of MPC onto a polymer substrate using the radiation technique. This paper reports the grafting reaction of MPC onto cellulose films by means of a pre-irradiation method. The effects of solvent system, monomer concentration, as well as reaction time on the degree of grafting were investigated.

MATERIALS AND METHODS

Materials

Commercial cellulose acetate film (10 μm thickness) was used as a substrate for the grafting reaction.

2-Methacryloyloxyethyl phosphorylcholine (MPC, a gift from Biocompatibles, UK) and poly(ethylene glycol methacrylate) (PEGMA) with 4–5 polyethylene oxide repeat units (purchased from Nippon Oil & Fats, Japan) were used without further treatment. Other chemicals were reagent grades and used without further treatment.

Grafting procedure

The gamma-ray irradiation from a ^{60}Co source was carried out at an exposure rate of 4.56 kGy/h in the presence of air to a total dose of 30 kGy. The irradiated cellulose films were stored in a refrigerator and kept at -130°C until the grafting reaction was performed. The grafting reaction was conducted in a Pyrex ampoule with a vacuum cock. Solvent was added first, followed by the monomer. Nitrogen was bubbled for 10 min to remove the oxygen in the solution. Irradiated sample films (2×5 cm) were immersed in the 30-ml monomer solution, and then the system was purged again by bubbling nitrogen for 10 min. Ferrous sulfate (1×10^{-3} M) was added to inhibit homopolymerization. The grafting reaction was carried out by placing the ampoule in a water bath at a relevant temperature. After the grafting reaction, the sample films were taken out from the monomer solution in the ampoule and washed with a solvent mixture of methanol and distilled water (1 : 1, v/v) for 24 h; the solvent solution was changed frequently to remove the remaining monomer and homopolymer.

The degree of grafting was determined by the following equation:

$$\text{Degree of grafting (\%)} = \frac{W_g - W_o}{W_o} \times 100,$$

where W_g and W_o are the weights of the sample before and after grafting, respectively.

FT-IR analysis

Besides by gravimetry, the grafted sample films were also verified by a Fourier transform infrared (FT-IR) spectrophotometer (Nicolet 5SXC, USA) with a nominal 45 degree attenuated total reflectance (ATR) to examine the functional group of carbonyl (>C=O) on the grafted films.

Platelet adhesion

Human blood from a healthy volunteer was collected with a polypropylene syringe containing a 3.8% sodium citrate solution. Platelet-rich plasma (PRP, 2.43×10^5 cells/μl) was obtained by centrifuging human blood at 2300 rpm for 4 min at 4°C. The ungrafted and grafted cellulose film samples were hydrated by placing them in a 24-well polystyrene plate filled with phosphate-buffered saline (PBS, pH 7.4) for 10 min, after which the PBS was changed. This cycle was repeated 5 times. Each hydrated film was transferred into a PRP that was pre-warmed to 37°C for 30 min. After incubation at 37°C for 60 min, the samples were washed carefully with PBS to remove weakly adhered platelets. The platelets adherent on the films were fixed with 2.5% glutaraldehyde in PBS for 10 min at room temperature, and then dehydrated in an ethanol-grade series (50, 60, 70, 80, 90 and 100%, 10 min at each concentration), after each film was washed with PBS and allowed to dry on a clean hood at room temperature. The platelets attached to the cellulose films were examined by a scanning electron microscope (SEM, JSM-840A, Jeol, Japan) with a tilt angle of 45 degrees after gold deposition in a vacuum.

RESULTS AND DISCUSSION

Effect of solvent system

The distilled water and the co-solvent system of MeOH/H_2O were used as the diluent for MPC and PEGMA. The effect of H_2O composition on the degree of

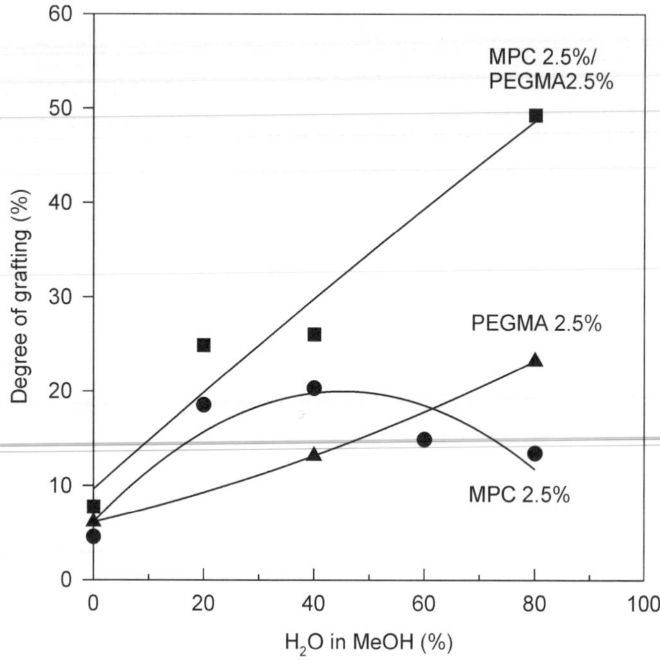

Figure 1. Effect of H_2O composition in MeOH on the degree of grafting of various monomer systems. Grafting reaction was performed at 70°C for 5 h. Monomer concentrations: MPC, 2.5%; PEGMA, 2.5%; MPC + PEGMA, 2.5% + 2.5%.

grafting of various monomers was studied as shown in Fig. 1. It appeared that the grafting yield of MPC/PEGMA and PEGMA increased with the fraction of H_2O in the co-solvent system. In the case of MPC grafting, a higher grafting yield was obtained when the concentration of H_2O in the co-solvent is about 40%.

Effect of monomer concentration

It was found that the homopolymer was formed in large amounts, especially when the grafting of MPC was performed in aqueous solution. Therefore, $FeSO_4$ (1×10^{-3} M) was used to inhibit the homopolymerization in aqueous solution system. Considering the solubility of PEGMA in water, the MeOH/H_2O (1 : 1) co-solvent system was selected for the grafting reaction of PEGMA/MPC (1 : 1). The effect of monomer concentration on the degree of grafting was determined, as shown in Fig. 2. The grafting yield of MPC/PEGMA (1 : 1) increased very fast with the monomer concentration. In the case of grafting reaction of MPC in the same co-solvent system, the grafting yield increased slowly with the monomer concentration at first and then went to level off. Comparing Fig. 1 and Fig. 2, it was found that the solution mixture of MPC and PEGMA gave the higher grafting yield than that of the total yield of the monomers when they were grafted individually.

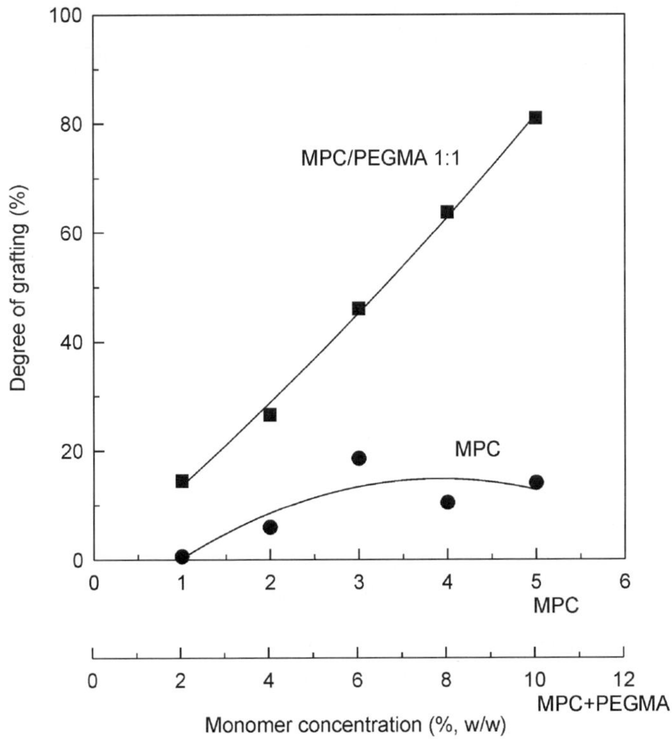

Figure 2. Effect of monomer concentration on the degree of grafting. Grafting reaction was performed in (a) aqueous solution of MPC with 1.0×10^{-3} M $FeSO_4$ and (b) MPC + PEGMA in $MeOH/H_2O$ (1 : 1) at 70°C for 5 h.

Effect of reaction time

The effect of reaction time on the degree of grafting of MPC was studied in the $MeOH/H_2O$ (1 : 1) co-solvent system with the monomer concentration of 1%, as shown in Fig. 3. We found that the degree of grafting increased with the reaction time, indicating that extending the reaction time would be useful in obtaining a higher grafting yield when the other reaction conditions are fixed.

FT-IR analysis of the samples

Figures 4 and 5 are the FT-IR–ATR spectra of MPC-grafted cellulose films, which were grafted in $MeOH/H_2O$ co-solvent system and aqueous solution system, respectively. The strength of carbonyl group (>C=O, belong MPC) peak around the wavenumber of 1720 cm^{-1} increased with grafting yield. Compared with the similar grafting yield of the sample grafted in $MeOH/H_2O$ (Fig. 4), the carbonyl peak at 1720 cm^{-1} was relatively stronger when the grafting reaction was carried out in aqueous solution (Fig. 5). This might be due to different surface

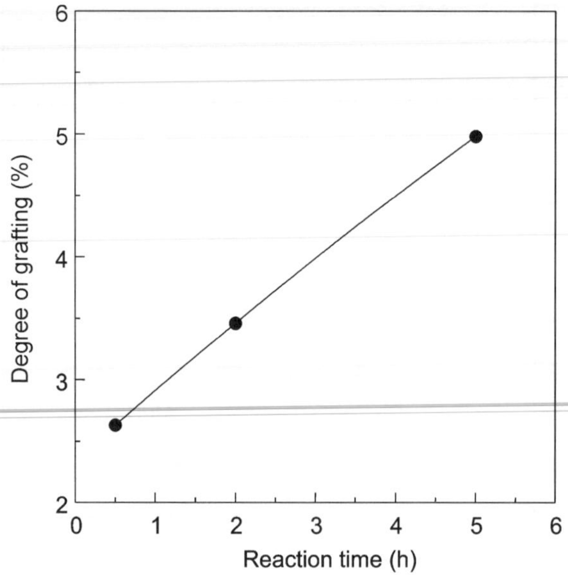

Figure 3. Effect of reaction time on the degree of grafting. Grafting reaction was performed in MeOH/H_2O (1 : 1) solution system with 1% MPC at 70°C.

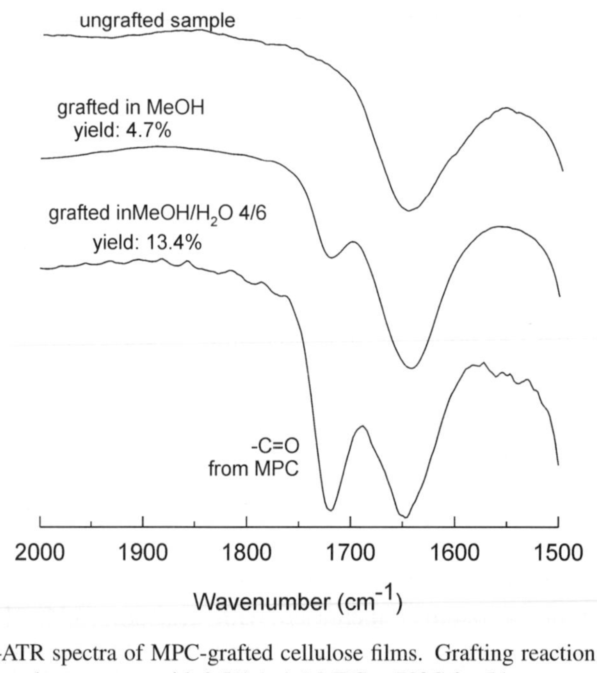

Figure 4. FT-IR–ATR spectra of MPC-grafted cellulose films. Grafting reaction was performed in the MeOH/H_2O co-solvent system with 2.5% (w/w) MPC at 70°C for 5 h.

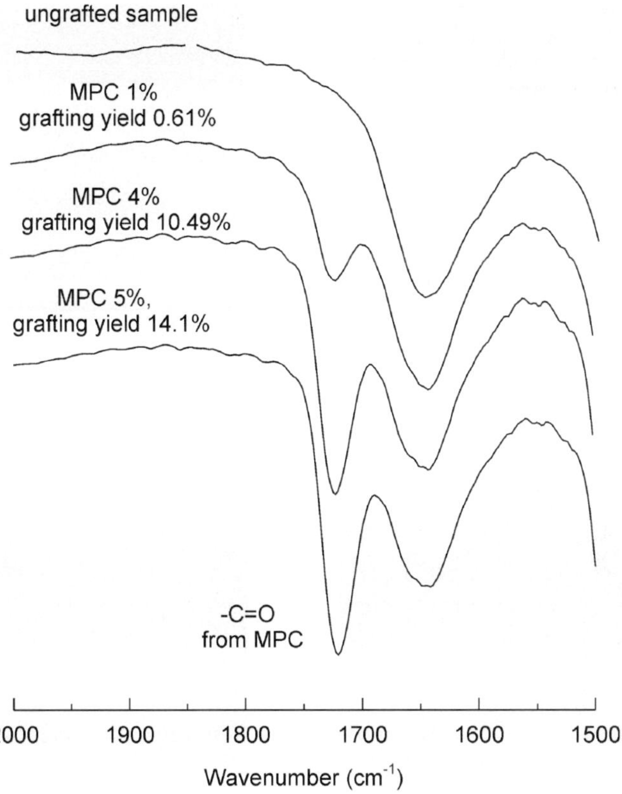

Figure 5. FT-IR–ATR spectra of MPC-grafted cellulose films. Grafting reaction was performed in aqueous solution of MPC (1%) in the presence of 1×10^{-3} M $FeSO_4$ at 70°C for 5 h.

properties in the different grafting solutions, similar to our previous experimental results [17].

Platelet adhesion studies

When materials are in contact with blood, proteins are first absorbed instantaneously onto surfaces and deformed, and then platelets are absorbed, activated and aggregated, so that platelets play a major role in the thrombus formation [18]. Therefore, the modified surfaces were assessed as biomaterials mainly by the degree and nature of platelet adhesion resulting from incubation in fresh human PRP. Platelet adhesion results viewed by SEM for these cellulose samples are shown in Fig. 6. It shows that the adhesion of platelet on the MPC- and MPC/PEGMA-grafted sample was clearly less than on the ungrafted reference. There was little platelet adhesion (yield 13.45%) on the MPC-grafted cellulose films carried out in aqueous solution (b) and in MeOH/H_2O (c) (10.49%). Also, there were few platelets on the cellulose film, which was grafted with MPC/PEGMA (yield 4.58%).

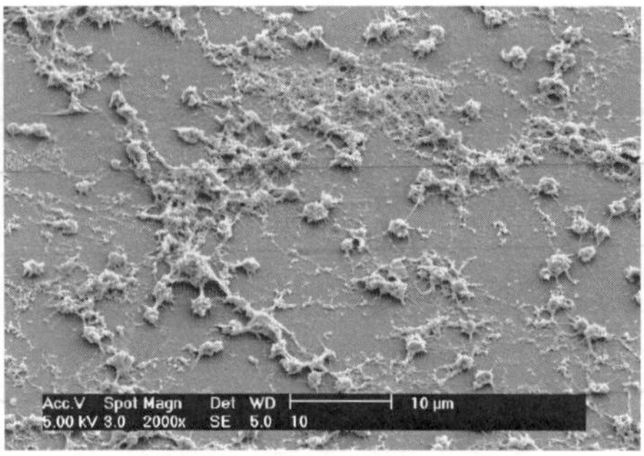

(a) Ungrafted cellulose film

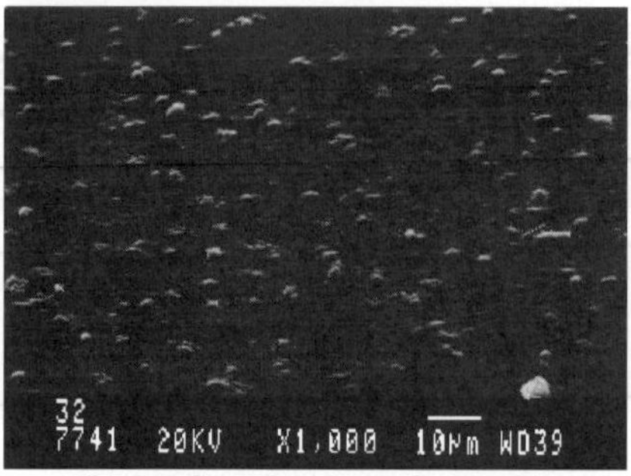

(b) MPC grafted in H_2O (13.5%)

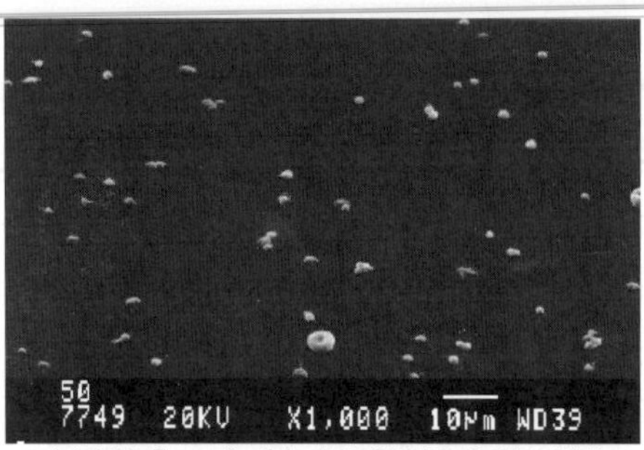

(c) MPC grafted in MeOH/H_2O (10.5%)

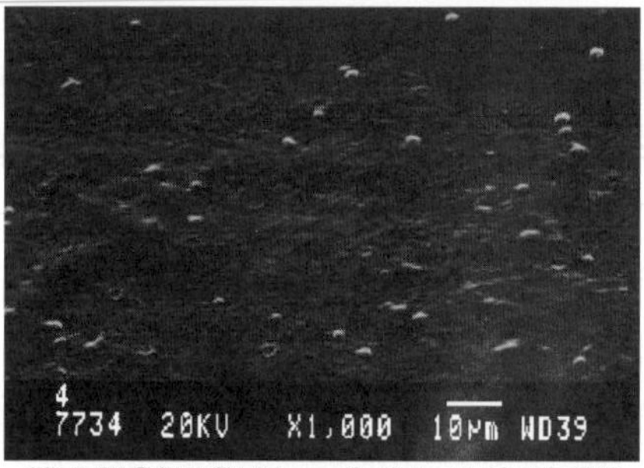

(d) MPC/PEGMA grafted in H_2O (4.6%)

Figure 6. Scanning electron microscope pictures of cellulose film, MPC- and MPC/PEGMA-grafted cellulose after interaction with PRP.

CONCLUSSIONS

1. It is possible to graft MPC and MPC/PEGMA onto cellulose film in aqueous solution and MeOH/H_2O co-solvent system.
2. The grafting yield of MPC and MPC/PEGMA is adjustable by the changing of reaction conditions such as monomer concentration, reaction time, etc.
3. Platelet adhesion on the surface of cellulose was decreased obviously after the grafting of MPC and MPC/PEGMA.

Acknowledgements

The authors acknowledge the Biocompatibles Ltd. UK for the gift of MPC. The authors also thank the IAEA Research Fund (IAEA Research Contract No. 10350/Japan) and the National Natural Science Foundation of China (to J. C.) for financial support.

REFERENCES

1. T. Yoneyama, K. Sugihara, K. Ishihara, Y. Iwasaki and N. Nakabayashi, *Biomaterials* **23**, 1455 (2002).
2. A. Korematsu, Y. Takemoto, T. Nakaya and H. Inoue, *Biomaterials* **23**, 263 (2002).
3. S. H. Ye, J. Watanabe, Y. Iwasaki and K. Ishihara, *Biomaterials* **24**, 4143 (2003).
4. K. Ishihara, K. Fukomoto, J. Aoki and N. Nakabahashi, *Biomaterials* **13**, 145 (1992).
5. K. Fukomoto, K. Ishihara, R. Takayama, J. Aoki and N. Nakabahashi, *Biomaterials* **13**, 235 (1992).
6. K. Ishihara, K. Fukomoto, M. Miyazaki and N. Nakabahashi, *Artif. Organs* **18**, 559 (1994).
7. M. K. Lee, H. S. Park, E. Y. Kim and S. M. Park, *J. Kor. Soc. Dyers Finishers* **7**, 63 (1995).
8. S. H. Ye, J. Watanabe, Y. Iwasaki and K. Ishihara, *J. Membr. Sci.* **210**, 411 (2002).
9. K. Ishihara, T. Tsuji, Y. Sakai and N. Nakabayashi, *J. Polym. Sci. Part A Polym. Chem.* **32**, 859 (1994).
10. J. Xu, Y. Yuan, B. Shan, J. Shen and S. Lin, *Colloids Surfaces B: Biointerfaces* **30**, 215 (2003).
11. K. Ishihara, K. Fukumoto, Y. Iwasaki and N. Nakabayashi, *Biomaterials* **20**, 1545 (1999).
12. A. L. Lewis, Z. L. Cumming, H. H. Goreish, L. C. Kirkwood, L. A. Tolhurst and P. W. Stratford, *Biomaterials* **22**, 99 (2001).
13. K. Ishihara, A. Fujiike, Y. Iwasaki, K. Kurita and N. Nakabayashi, *J. Polym. Sci. Part A Polym. Chem.* **34**, 199 (1996).
14. A. Yamasaki, Y. Imamura, K. Kurita, Y. Iwasaki, N. Nakabayashi and K. Ishihara, *Colloids Surfaces B Biointerfaces* **28**, 53 (2003).
15. Y. Ikada, M. Suzuki, M. Taniguchi and H. Iwata, *Radiat. Phys. Chem.* **18**, 1270 (1981).
16. K. Saito, S. Tsuneda, M. Kim, N. Kubata, K. Sugita and T. Sugo, *Radiat. Phys. Chem.* **54**, 517 (1999).
17. J. Chen, Y. C. Nho, O. H. Kwon and A. S. Hoffman, *Radiat. Phys. Chem.* **55**, 87 (1999).
18. Y. J. Li, T. Nakaya and Z. Zhang, *J. Biomater. Appl.* **12**, 167 (1997).

Cells on surfaces

Gels, Genes, Grafts and Giants, pp. 423–438 (2005)

Cell-adhesive and blood-coagulant properties of ultrathin poly(methyl methacrylate) stereocomplex films

TAKESHI SERIZAWA, KEIKO YAMASHITA and MITSURU AKASHI *

Department of Nanostructured and Advanced Materials, Graduate School of Science and Engineering, Kagoshima University, 1-21-40 Korimoto, Kagoshima 890-0065, Japan

Received 24 September 2003; accepted 7 January 2004

Abstract—Isotactic (it) and syndiotactic (st) poly(methyl methacrylate)s (PMMAs) were assembled on a poly(ethylene terephthalate) cell disk (as a solid substrate), which is normally used for cell culture, to produce ultrathin stereocomplex films. Static contact angles of the films using air bubbles in aqueous phase revealed the stepwise assembly of both polymers. More L929 fibroblast cells adhered and were flattened on the ultrathin stereocomplex films, compared to homogeneous PMMA films and to spin-coated films comprised of a mixed solution of it- and st-PMMAs with a molar ratio of 1 : 2, which is the stoichiometry of a stereocomplex. The difference in the number of adherent cells was greatest after the first 3 h of incubation. Pre-adsorption of proteins, which are related to cell adhesion, onto the stereocomplex film potentially modulated the cell-adhesive properties. Fresh whole human blood did not coagulate on the complex film for 20 min, although blood coagulated after 15 min on the homogeneous films and on the 1 : 2 mixed film. Platelets did not adhere onto the complex films after 15 min of incubation, which was consistent with the results of blood coagulation. These observations indicate that the stereocomplex formation of PMMAs on the surface of films strongly affects the bioactivity of these films.

Key words: Poly(methyl methacrylate); stereocomplex; ultrathin film; cell adhesion; blood coagulation.

INTRODUCTION

Protein- and cell-adhesive properties of the surfaces of polymeric materials are important when the materials initially make contact with bio-systems, and determine potential applications of the materials in biomedical fields. It is therefore, essential to analyze the properties of these materials using appropriate techniques. Either

*To whom correspondence should be addressed. Present address: Department of Molecular Chemistry, Graduate School of Engineering, Osaka University, Yamada-oka, Suita 565-0871, Japan. E-mail: akashi@chem.eng.osaka-u.ac.jp

bioactive or bioinert materials are required depending on the application. It is however, difficult to systematically evaluate the properties of these polymeric materials, as these properties are governed by physicochemical surface characteristics, as well as by the chemical compositions of the materials. It is therefore, attractive to compare the properties of polymeric materials with the same chemical composition but assembled by different methods.

Poly(methyl methacrylate) (PMMA), as well as derivative polymers of methacrylates, are well-known synthetic polymers frequently used for biomedical applications, due to their stable chemical structures in bio-related circumstances and their inertness against bio-systems. The PMMA stereocomplex [1, 2] comprises a double-stranded helical assembly between isotactic (it) and syndiotactic (st) PMMAs with a respective molar stoichiometry of 1 : 2 based on van der Waals interactions. These stereocomplexes have been applied to hollow fiber membranes used for artificial dialysis through a process that involves blending of the complexes with cellulose membranes [3]. Although the bioactivities of the stereocomplex surface should be significant for determining the potential for biomedical applications, it is difficult to regulate the surface structure by simply blending both stereoregular PMMAs using conventional solution- or spin-casting methods, because of self-association (or crystallization) of the polymers [1].

We have previously prepared ultrathin films composed of double-stranded PMMA stereocomplex [4], as well as stereocomplexes of other methacrylate polymers [5–7] by the simple alternate immersion of a solid substrate into solutions of stereoregular polymers. Assemblies were demonstrated to occur in a layer-by-layer (LbL) method, which is similar to the method used for the preparation of polyelectrolyte multilayers based on electrostatic interactions [8]. Accordingly, it has been demonstrated that not only electrostatic interactions but also weaker van der Waals ones could be utilized for LbL assembly. With immersion of a solid substrate into it-PMMA solution the first monolayer is deposited and with subsequent immersion into st-PMMA solution the double-stranded stereocomplex is completed [4]. The repetition of this cycle successfully fabricates ultrathin stereocomplex films. Since these films have a surface which is the product of regulated assembly, it is possible to understand the effect of assembly structures on the bioactivity of PMMA. Presently however, little is known about the cell-compatible properties of stereoregular PMMAs.

In our previous study, the physical adsorption of bovine serum albumin (BSA) onto the film surface of the PMMA stereocomplex and various stereoregular PMMAs were analyzed quantitatively using a quartz crystal microbalance (QCM) technique [9]. BSA adsorbed onto the complex in an end-on manner without denaturing based on a smaller apparent adsorption constant. On the other hand, BSA adsorbed onto each PMMA in a side-on manner with denaturing based on a greater adsorption constant. Furthermore, other proteins such as human serum albumin, fibrinogen and lysozyme also adsorbed onto the complex with smaller adsorption constants, compared to the adsorption onto homogeneous PMMA films (data not shown). These observations strongly indicated that the bioactivity of stereoregular PMMA films was strongly affected by the surface assembly structure. As such the assembly method might potentially change not only the protein adsorptive but also the cell-adhesive properties of the stereocomplex film.

In the present study, the adhesion/proliferation of L929 fibroblast cells and the coagulation of human blood on ultrathin stereocomplex PMMA films and various stereoregular PMMA films were analyzed. We will demonstrate, for the first time, significant cell-related bioactivities of PMMAs dependent on stereoregularity and the assembly structure.

MATERIALS AND METHODS

Assembly

it- and st-PMMAs were synthesized, following methods described in previous studies [10, 11]. The tacticities of the polymers *(mm : mr : rr)* were measured using ^{1}H-NMR signals from the α-methyl protons. A cell disk 13.5 mm in diameter (normally utilized as a cell-culture disk made from poly(ethylene terephthalate)), on which the static contact angle of water droplets was 48.8 ± 0.2 degrees, was used as a solid substrate. The assembly process was essentially as described in a previous study [4]. The substrate was immersed into an acetonitrile/water (10 : 1.5, v/v) solution of it-PMMA at 1.7 mg/ml for 15 min at ambient temperature, rinsed with the same solvent and dried. Subsequently, the substrate was immersed into a solution of st-PMMA, and the same procedure was repeated. This alternate cycle was repeated for the given number of assembly steps. The obtained films were immediately used for the following characterization and bioactivity experiments. For static contact angle measurements and cell adhesion tests, a combination of it- (M_n 19 700, M_w/M_n 1.15, $mm : mr : rr = 97 : 1 : 2$) and st-PMMAs ($M_n$ 24 200, M_w/M_n 1.28, $mm : mr : rr = 1 : 9 : 90$) was used ($M_n$ mean number molecular weight; M_w mean weight molecular weight). For blood coagulation tests, a combination of it- (M_n 21 900, M_w/M_n 1.16, $mm : mr : rr = 99 : 1 : 0$) and st-PMMAs ($M_n$ 24 200, M_w/M_n 1.28, $mm : mr : rr = 1 : 9 : 90$) was used, with acetonitrile/water (10 : 1, v/v) as a solvent. Due to the solubility of the PMMAs used for blood coagulation tests, the solvents for the assemblies were changed from acetonitrile/water (10 : 1.5, v/v) to acetonitrile/water (10 : 1, v/v). Atactic (at) PMMA (M_n 22 200, M_w/M_n 1.07) was purchased from Polysciences (USA). The static contact angles were measured with a Kyowa Interface Science CA-X (Japan) using air bubbles in water at ambient temperature, and the measurements were repeated three times.

Cell culture

The cell disks coated with PMMA films were placed in 24-well tissue culture plates. A suspension of mouse fibroblast cells (L929, donated by Japanese Cancer

Research Resources Bank, Foundation for Promotion of Cancer Research, Tokyo, Japan) at a concentration of 1.0×10^5 cells/ml was added into each well, and then incubated for 3 h (initial adhesion test) or for a given time (proliferation test) at 37°C under a 5% CO_2, atmosphere in 10% Eagle's minimum essential medium (Eagle's MEM) in the absence of serum. After the films with adherent cells were rinsed with PBS, the cells were peeled off from the films using a trypsin solution. Subsequently, the number of living cells was counted using a colorimetric assay (Cell Counting Kit; Dojindo Laboratories, Kumamoto, Japan) that can determine living cells on the basis of quantification of NADH using 2-(4-iodophenyl)-3-(4-nitrophenyl)-5-(2,4-disulfophenyl)-2H-tetrazolium, monosodium salt and l-methoxy-5-methylphenazinium methylsulfate. The adherent cells were then treated with formalin and hematoxylin–eosin staining, and then photographed. For pre-adsorption of proteins for the initial adhesion test, fibronectin and human serum albumin (HSA) were used (Sigma, USA; the proliferation test was not performed by the pre-adsorption). The proteins were pre-adsorbed onto films from PBS solutions at respective concentrations of 0.1 and 1 μM for 1 h at 37°C. The concentrations of aqueous fibronectin and HSA solutions were estimated by assuming the molecular weights to be 460 000 and 69 000, respectively. All experiments were repeated for three times. For the analysis of cell morphology, 150–200 cells in a given area were counted for each film.

Blood pro-coagulation

The anti- versus pro-coagulant activity was analyzed essentially as reported in previous studies [12, 13]. The cell disk was trimmed down to 10.5 mm diameter, and was used as the substrate. Fresh whole human blood was supplied by Mr. H. Sakaguchi (22 years old, Kagoshima University) in Venoject II (4.5 ml, Terumo) containing 3.13% sodium citrate (0.5 ml). The film-coated substrate was then immersed in whole human blood (1 ml) in a polypropylene tube (13.5 mm diameter, Maruemu, Japan) at 37°C and coagulation was initiated by the addition of 0.2 M aqueous $CaCl_2$ solution (100 μl). After a prescribed incubation time, the substrate was removed, gently rinsed in PBS and then photographed. For the platelet adsorption test, platelet-rich plasma (PRP) was obtained as the supernatant after centrifugation of whole blood at 1000 rpm for 10 min. The PRP was mounted on the films for a given time, and platelets adsorbed were dehydrated using graded alcohol/saline solutions (25, 50, 70, 90, 95, 99.5% ethanol, and *t*-butanol). *t*-Butanol solutions containing the films were freeze-dried. Then, gold was sputtered onto the films at a thickness of approx. 20 nm, and the platelets were observed by a Hitachi S-4100H scanning electron microscope (SEM) at an accelerated voltage of 5 kV. All experiments were repeated for two times, to ensure that the resulting data were reproducible.

RESULTS AND DISCUSSION

Assembly

Stereocomplex formation between it- and st-PMMAs can be identified by suitable physicochemical techniques [1, 2]. In previous papers [4–6], complex formation on ultrathin films prepared by stepwise deposition of each stereoregular polymer of methacrylates on a gold-coated QCM substrate was potentially analyzed by static contact angle measurements in air using water droplets, in conjunction with quantitative QCM detection and infrared reflection absorption spectra. The contact angle was sensitive to the surface assembly structure of stereoregular PMMAs [4]. The contact angle at the deposition step of it-PMMA was the same as that of cast it-PMMA; however, the angle at the st-PMMA step was slightly smaller (more hydrophilic) than that of cast st-PMMA. These observations suggested that the stereocomplex had been formed at the st-PMMA step after physical adsorption of it-PMMA. These analyses of static contact angles were thus used to analyze complex formation on the present substrate. However, since bioactivities are analyzed in aqueous media, static contact angles were measured in water using air bubbles.

Figure 1 shows the changes in the static contact angles during steps 11–16 of the assembly process for it- and st-PMMAs. The angle changed in a zigzag fashion, indicating a stepwise and regular assembly of the two polymer components. The

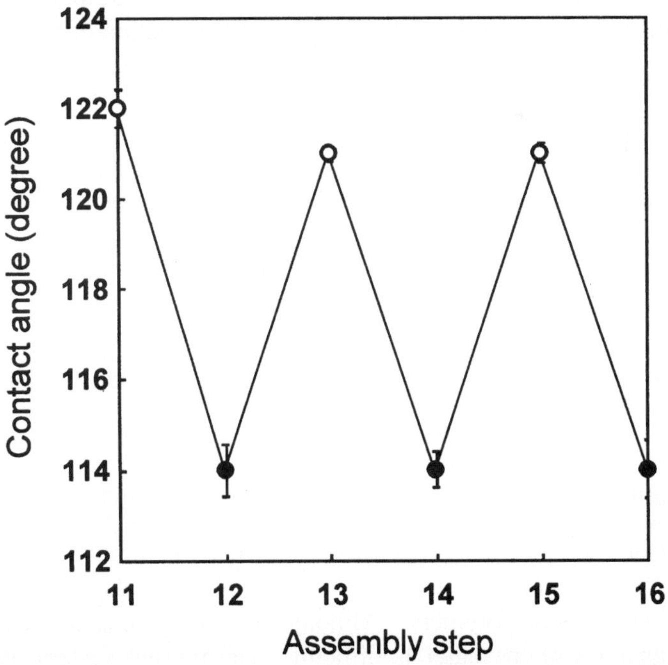

Figure 1. Static contact angles of a stereocomplex film for it- and st-PMMAs using air bubbles in aqueous phase during assembly steps 11–16. Open and closed symbols indicate it- and st-PMMA steps, respectively.

mean contact angle on odd it-PMMA steps was 121.3 ± 0.6 degrees and was slightly smaller than that of a cast it-PMMA film (125.1 ± 0.6 degrees). This observation was not consistent with the angles analyzed using water droplets in air, as described above [4]. Since a different substrate (the cell disk) was used in the present study, the conformation of the adsorbed it-PMMA may be slightly different. Although we do not demonstrate bioactivities of the odd-step surface of the it-PMMA monolayer adsorbed on the stereocomplex surface in the present study, those activities were significantly similar to those of the stereocomplex surfaces. On the other hand, the mean angle on even st-PMMA steps was 113.8±0.5 degrees, which was much smaller than that of a cast st-PMMA film (130.0±0.2 degrees), indicating that the surface of the assembly was much more hydrophobic. It is known that the analysis medium alters the static contact angles of films of stereoregular PMMAs. In aqueous media, lateral ester groups of st-PMMA, of which the main chain is in a *trans*-zigzag form, should be pointed out toward the water to obtain a mostly hydrophilic surface [14]. Accordingly, when the stereocomplex was formed between it- and st-PMMAs at the st-PMMA step, the surface of the complex might be more hydrophobic than the surface of the cast st-PMMA film, due to the formation of double strands in the stereocomplexes, in which the methyl ester groups of st-PMMA are pointed out toward the water [2]. All angles measured were unchanged in aqueous phase even after 48 h, although rearrangement of polymers at the near surface might occur to minimize interfacial free energy in the aqueous phase. This suggests the formation of a stable assembly between it- and st-PMMAs. Furthermore, the zigzag change in the contact angles was not observed when immersion into a single polymer solution was repeated. From these observations, we concluded that the cell disk was successfully coated with the stereocomplex and that the assembly structure was sufficiently stable for the analysis of bioactivity. In addition, the above contact angle analysis suggests that the PMMAs could be assembled on the cell disk on the basis of the similar assembly manner to the assembly on the gold-coated QCM substrate [4]. Although it is unfortunately difficult to determine the film thickness of the stereocomplex assembly on the cell disk, the thickness was expected to be approx. 25 nm for the 16-step assembly, assuming that the polymers were assembled similarly to that on the surface of QCM [4].

Cell-compatibility test

Cell adhesion and proliferation properties of polymeric materials should be affected not only by the chemical structure of the materials, but also by the assembling structures of constituent polymers. Although it is easy to analyze the effect of the former, analysis of the latter is difficult. The present system uses PMMAs with different stereoregularity and different assembling structures, although the chemical components were the same in all cases. Accordingly, the effect of the stereoregularity and assembling structures of the PMMAs on surfaces could be

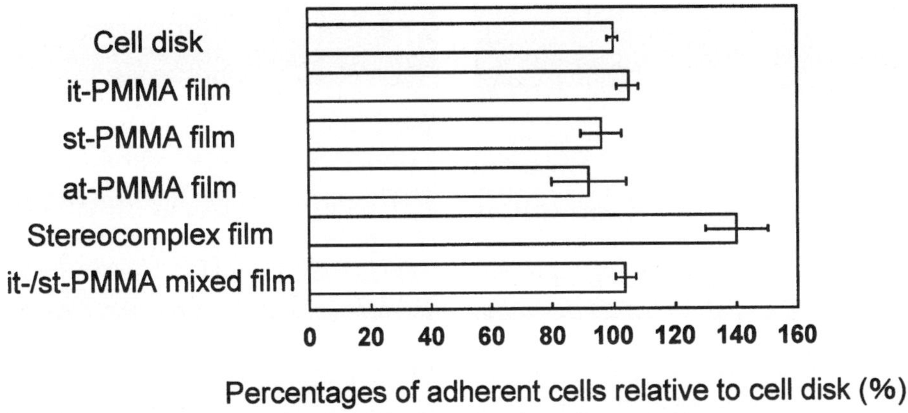

Figure 2. Percentages of L929 fibroblast cells adhering onto a cell disk, it-, st-, at-PMMA spin-coated films, a 16-step stereocomplex film and an it-/st-PMMA (1 : 2, w/w) mixed film.

analyzed by cell-adhesion and proliferation tests using the stereocomplex films prepared and spin-coated films of each PMMA.

Figure 2 shows the percentages of cells that adhered to each film compared to the number of cells that adhered to the cell disk substrate as a control after 3 h of incubation ($(3.0 \pm 0.1) \times 10^4$ cells adhered on the cell disk). The cells adhered well to each spin-coated homogeneous film and to the cell disk. The number of cells that adhered to homogeneous films comprised of it-, st- and at-PMMAs was almost the same. This indicates that the number of adherent cells is independent of the stereoregularity of PMMA. On the other hand, cell-adhesive properties were significantly enhanced by stereocomplex formation between it- and st-PMMAs. Greater numbers of cells clearly adhered to the 16-step stereocomplex film, compared to the numbers that adhered to the cell disk, the homogeneous films of PMMAs and the spin-coated film of a mixture of it- and st-PMMAs (1 : 2, w/w). In addition, the numbers of cells adhered onto the 12- and 14-step complex films were similar to the number adhering to the 16-step film.

In order to analyze the cell adhesive properties in detail, the cells adhering after 3 h were observed directly by photography, as shown in Fig. 3. The shape of the adherent cells was classified into the difference in the aspect ratio, which was estimated by the ratio of long axis divided by short axis for each shape of adherent cells. Accordingly, the aspect ratio 1 indicates cells to be round shape, and the ratio approx. 6 indicates cells to be fully flattened. The resulting data are statistically shown as a histogram in Fig. 4. In this analysis, the tacticity of the film clearly affected the shape of the adherent cells. Although the cells were not flattened on the st-PMMA film, approx. 35% and 20% of the cells were flattened (aspect ratio > 4) on the it- and at-PMMA films, respectively. Interestingly, approx. 60% of the cells were flattened (aspect ratio > 4) on the stereocomplex film, and this percentage was maximal among the films analyzed. It is necessary to discuss why we obtained the above difference in cell-adhesive properties between the films. The hydrophobicity

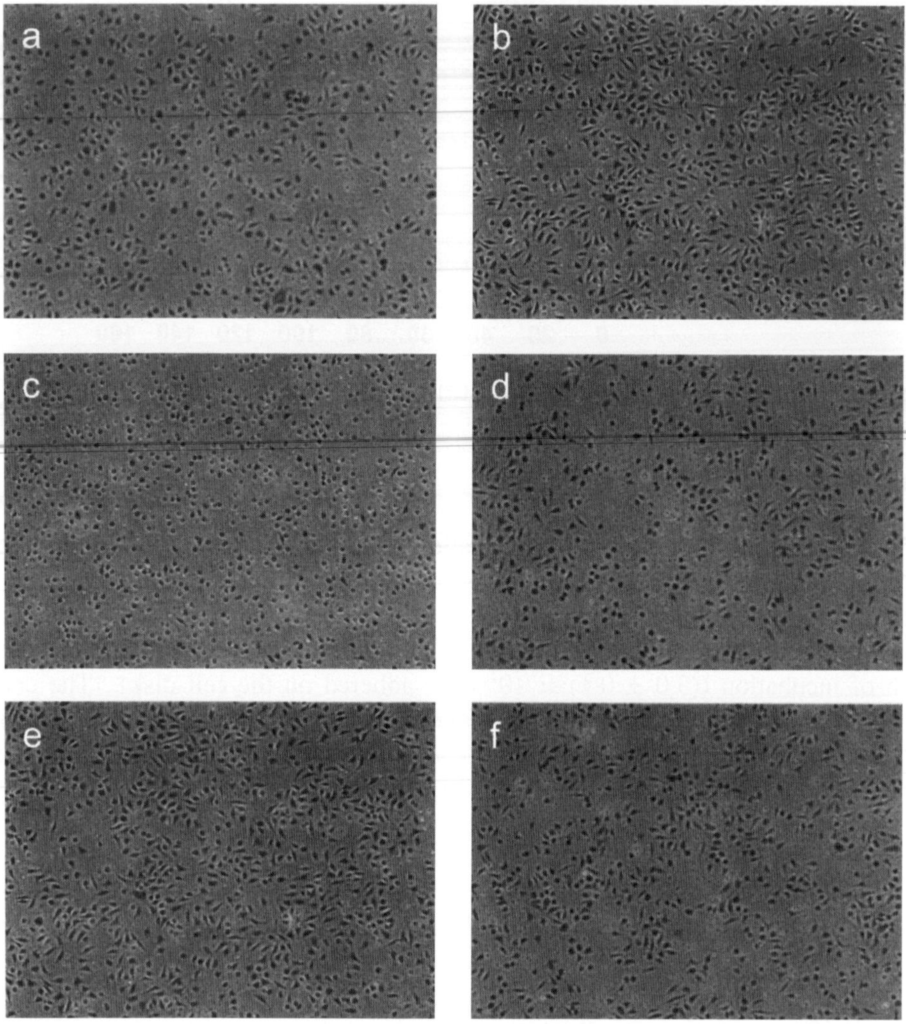

Figure 3. Photographs of the cells adhered onto (a) a cell disk, (b) it-, (c) st-, (d) at-PMMA spin-coated films, (e) a 16-step stereocomplex film and (f) an it-/st-PMMA (1 : 2, w/w) mixed film after 3 h incubation.

of the films is one well-known factor that affects cell adhesion. However, the present data could not be explained by differences in the static contact angles, as described above. Accordingly, it was strongly suggested that the formation of double-strands between it- and st-PMMAs on the surface accelerated cell-adhesive properties. In particular, the two-dimensional disposition of the polymer main and side chains, differences in hydration of the film surface, and/or macromolecular mobility on the film surface, which are dependent both on assembly structure and polymer stereoregularity, possibly affected the cell-adhesive properties. Our previous study demonstrated that BSA adsorbed physically onto the complex films without any denaturing, and the apparent adsorption constant was smaller than that adsorbed on

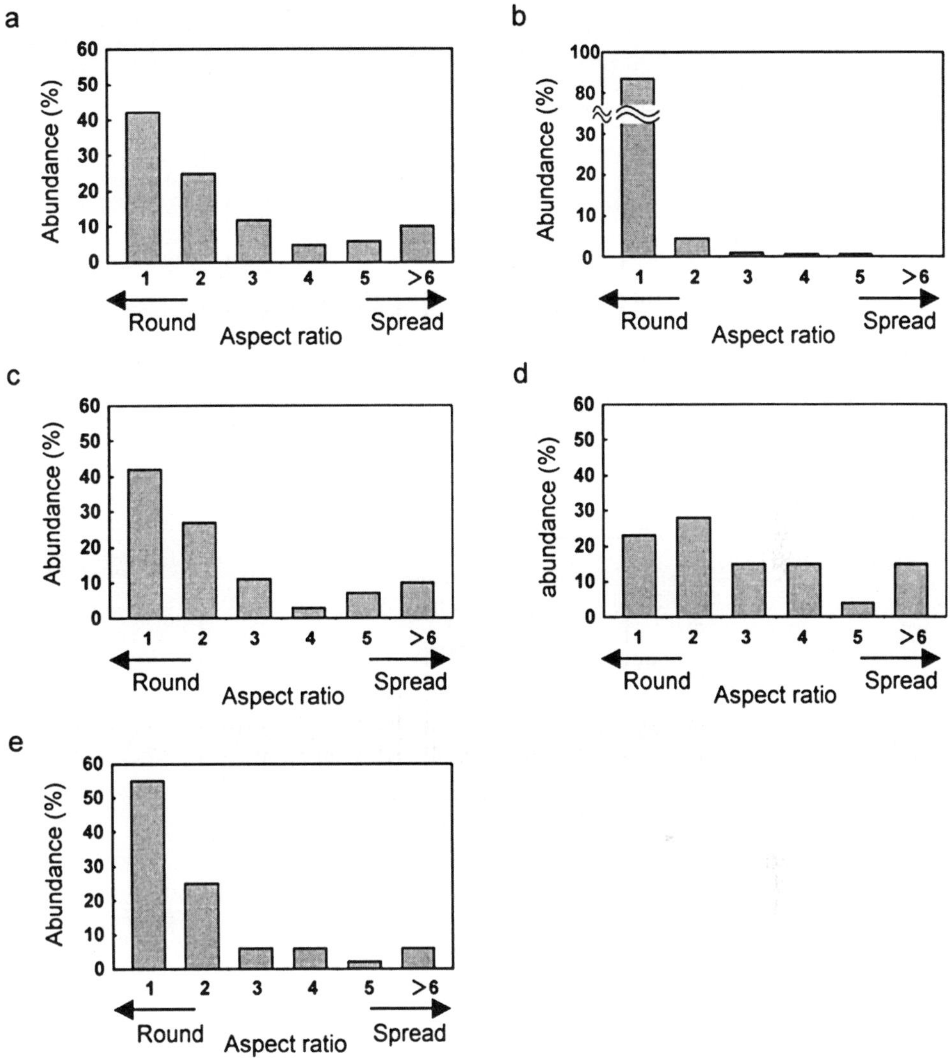

Figure 4. Histogram of shapes of cells adhered on (a) it-, (b) st-, (c) at-PMMA spin-coated films, (d) a 16-step stereocomplex film and (e) an it-/st-PMMA (1 : 2, w/w) mixed film after 3 h incubation. Aspect ratios were estimated by the ratio of long axis divided by short axis for each shape of adherent cells. Diamond-shaped cells were included as a most expanded population (>6). Figures 5 and 6 were analyzed in a similar fashion.

each homogeneous film [9]. In the present case, the extra-cellular matrices of the cells seemed to adsorb similarly, thus resulting in flattening of the cells.

The initial cell adhesion after pre-adsorption of a cell-adhesive protein, fibronectin, and a non-cell-adhesive protein, HSA, was analyzed, and the resulting data are summarized in Figs 5 and 6. In the former case, slightly greater numbers of cells adhered onto the stereocomplex film and a greater number of cells were flattened in shape on the complex film. In the latter case, fewer cells were flattened in

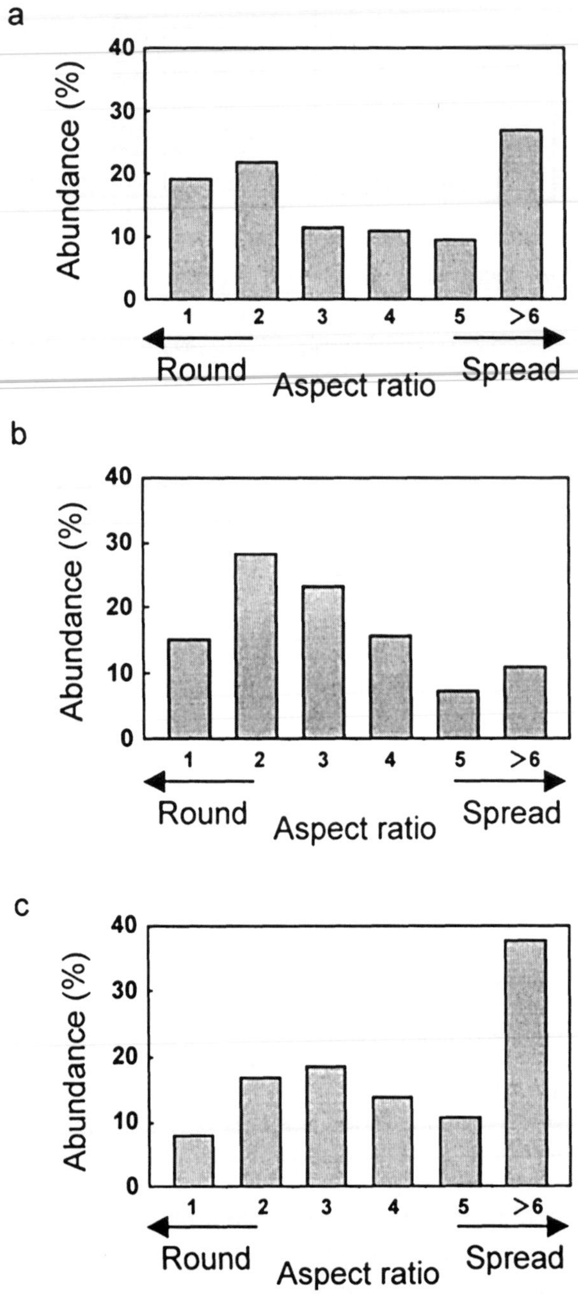

Figure 5. Histogram of shapes of cells adhered on (a) it-, (b) st- and (c) a 16-step stereocomplex film, onto which fibronectin had been pre-adsorbed, after 3 h incubation.

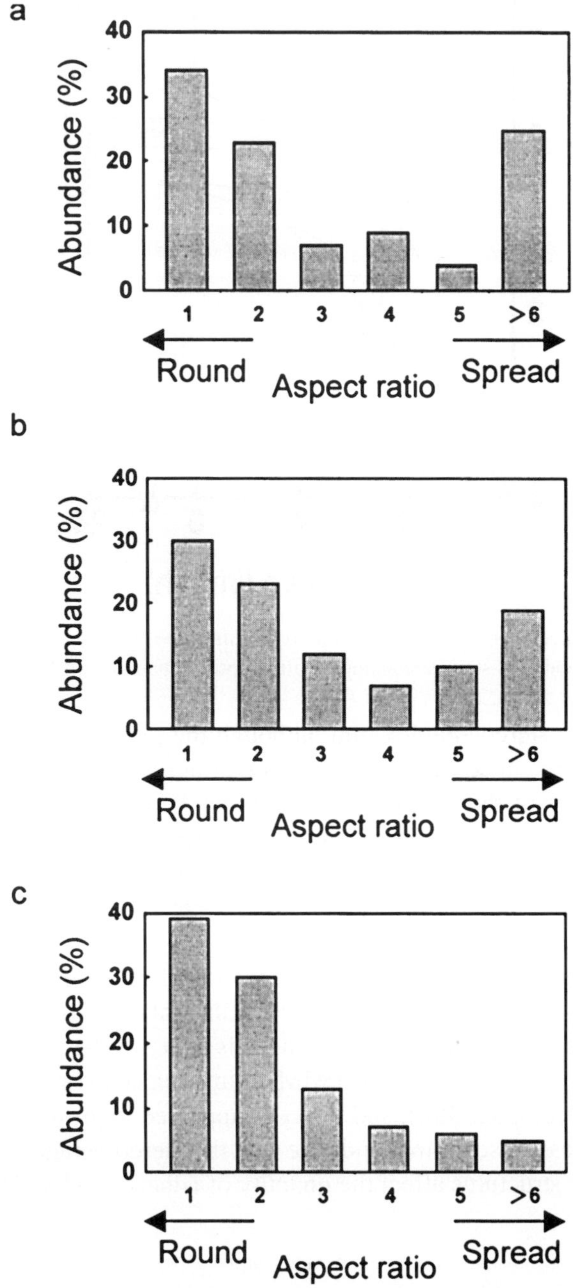

Figure 6. Histogram of shapes of cells adhered on (a) it-, (b) st- and (c) a 16-step stereocomplex film, onto which HSA had been pre-adsorbed, after 3 h incubation.

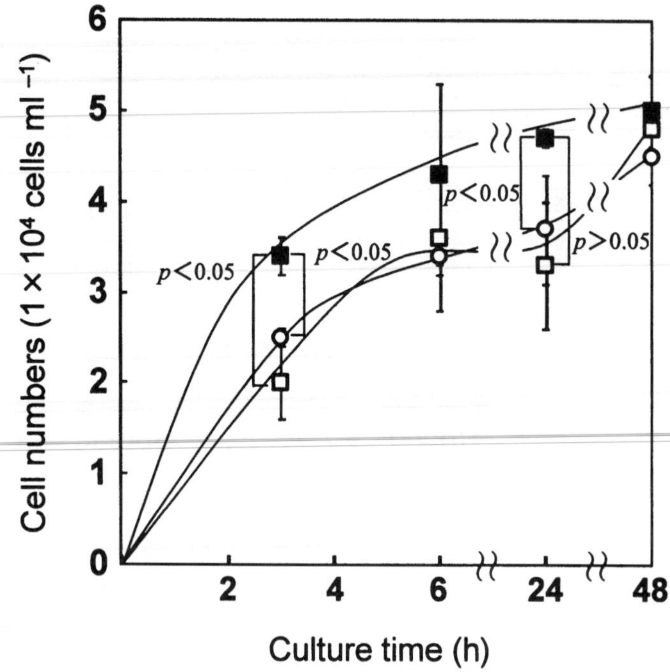

Figure 7. Growth of cells on an it-PMMA spin-coated film (open circles), an st-PMMA spin-coated film (open squares) and a 16-step stereocomplex film (closed squares).

appearance on the stereocomplex film, although the total number of adherent cells was almost the same in all films. These observations strongly suggest that the cell-adhesive properties of the stereocomplex film can be modulated by pre-adsorption of adequate proteins and that the essential characteristics of the proteins tend to be effectively expressed on the complex film. These properties are possibly due to the non-denaturing physical adsorption of proteins onto the stereocomplex films (Ref. [9] and data not shown).

Cell proliferation was analyzed by the incubation of cells for 48 h on each film, as shown in Fig. 7. Differences in the number of adherent cells were apparent until 3 h. After 6 h, however, the numbers of adherent cells were similar for all films (although the number of adherent cells on the it-PMMA film seemed to be slightly smaller than that on the stereocomplex film), and the cells appeared to grow reasonably well on all the films. These observations indicate that the stereoregularity of the PMMAs and the assembly structures affect the quantity of adherent cells at the initial stages of cell interaction.

Coagulation of human blood

In the previous section, we demonstrated that model fibroblast cells demonstrated preferential adherence after 3 h of incubation to the surfaces of the stereocomplex. Adhesion of fibroblast cells can be easily examined due to the absence of serum. It is

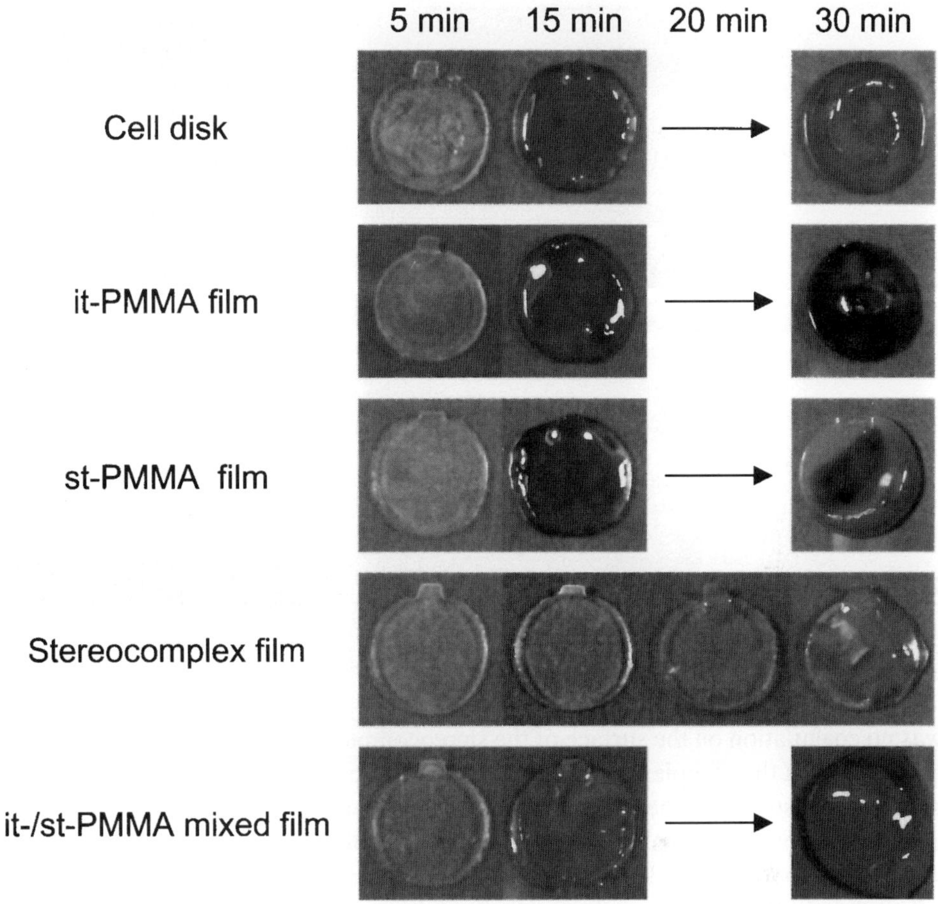

Figure 8. Coagulation of whole human blood on a cell disk, it- and st-PMMA spin-coated films, a 16-step stereocomplex film and an it-/st-PMMA (1 : 2, w/w) mixed film after given incubation times.

also significant to determine whether other bioactivities of the films are regulated by nanostructures. As another assay of stereocomplex films, pro-coagulation of human whole blood on the films was examined. This assay is also based on a simple system: the adsorption of fibrinogen and subsequent cross-linking of activated platelets with fibrin derived from fibrinogen. In addition, since the blood coagulation assay is very rapid compared to cell adhesion assays, initial bioactivity can be analyzed.

Figure 8 shows photographs taken after the cell disks coated with various PMMA films were immersed into whole blood for 30 min, of which the incubation time had been determined because blood coagulation was clearly different from each other in the present static system. Blood coagulation on PMMA films was clearly dependent on the assembly structure. Blood coagulated after 15 min of incubation on the surface of the cell disk, homogeneous films of it- or st-PMMA, and on the surface of a spin-coated film of a mixed solution of it- and st-PMMAs with a molar ratio of 1 : 2, which is the stoichiometry of a stereocomplex [1, 2]. However, there

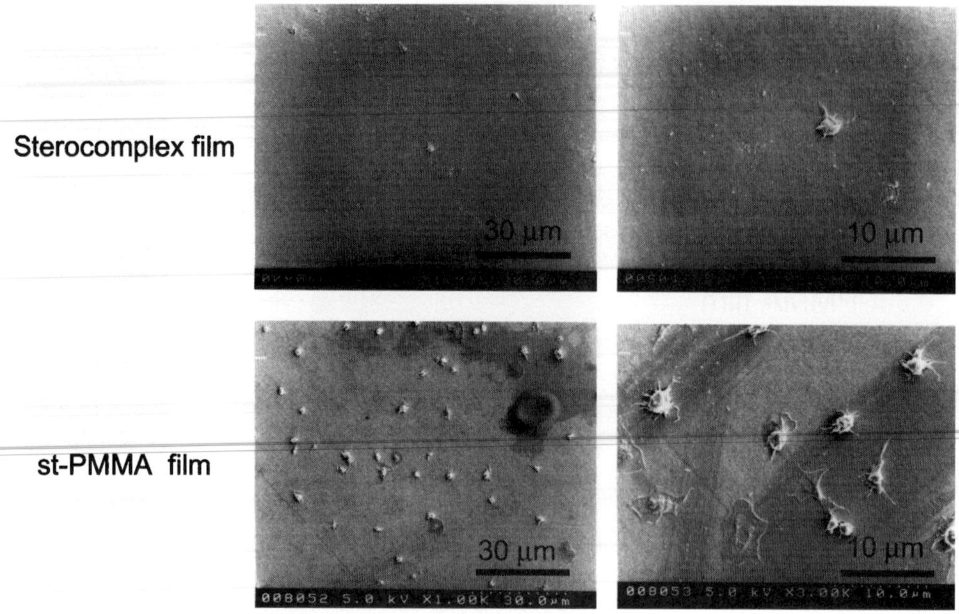

Figure 9. Photographs of platelets adhering from platelet rich plasma (PRP) onto a 16-step stereocomplex film and onto a st-PMMA spin-coated film.

was no coagulation on the surface of the stereocomplex film (16 steps) after 20 min, indicating that the complex surface had the greatest coagulation-resistant activity of the films investigated. These observations strongly indicated that blood coagulation was affected by the assembly structure of stereoregular PMMAs.

In order to further understand blood coagulation on the films, platelets were adhered to films of the complex and of a spin-coated st-PMMA from PRP after 15 min of incubation; anti- and pro-coagulation were observed, respectively. These observations were made using SEM, as shown in Fig. 9. Almost no platelets adhered to the complex surface, while many platelets adhered and deformed on the homogeneous st-PMMA film, indicating that activated platelets adhered to the homogeneous st-PMMA film to induce blood coagulation. Since fibrinogen, which is directly involved in the process of blood coagulation, was adsorbed without denaturing on the complex surface (data not shown), blood coagulation was possibly suppressed. It was found that not only *in vitro* cell-adhesive properties but also blood coagulation activity varied on PMMA stereocomplex films prepared by a stepwise adsorption process.

CONCLUSIONS

Alternate immersion of a cell disk into it- and st-PMMAs has been used to fabricate ultrathin stereocomplex films. L929 fibroblast cells preferentially adhered and were flattened on the stereocomplex films. The homogeneous films showed less flattening. Cell proliferation demonstrated that the cells recognized the surface structures of the films after 3 h, at which time we observed maximum differences in the number of adherent cells dependent on the assembly structure. Coagulation of human whole blood was suppressed on the complex film, compared to the higher level of coagulation on the homogeneous films. The number of platelets that adhered onto the complex film was also smaller than the number that adhered onto homogeneous films. These observations indicate that the bioactivities of polymeric materials are strongly affected by the nanostructures of constituent polymers on the film surface. Not only the stereoregularities of polymers but also the control of assembly structures are important in producing biomedical materials. The present system was suitable for understanding the effects of bioactivities on delicate assembly structures due to simple and regulated chemical structures of stereoregular PMMAs. Although it was difficult to propose the satisfactory mechanism of novel bioactivities of stereocomplex films, a hydration structure or macromolecular dynamics before or after interactions with biomolecules might be essential for the present differences. We have already fabricated a stereocomplex (racemic crystal) film of enantiomeric poly(lactides) [15], and its hydrolytic rate was much faster than that of homocrystals [16]. Further research into the analysis of protein adsorption, as well as adhesion tests for other cells using stereocomplex films, is now in progress.

Acknowledgements

This work was financially supported in part by Grant-in-Aid for Scientific Research in the Priority Area of 'Molecular Synchronization for Design of New Materials System' (No. 404/13022258) and by Grant-in-Aid for Scientific Researches (No. 851/14780643) from the Ministry of Education, Culture, Sports, Science and Technology, Japan.

REFERENCES

1. J. Spěváček and B. Schneider, *Adv. Colloid Interface Sci.* **27**, 81 (1987).
2. E. Schomaker and G. Challa, *Macromolecules* **22**, 3337 (1989).
3. B. Bikson, J. K. Nelson and N. Muruganandam, *J. Membr. Sci.* **94**, 313 (1994).
4. T. Serizawa, K.-I. Hamada, T. Kitayama, N. Fujimoto, K. Hatada and M. Akashi, *J. Am. Chem. Soc.* **122**, 1891 (2000).
5. T. Serizawa, K.-I. Hamada, T. Kitayama, N. Fujimoto, K. Hatada and M. Akashi, *Langmuir* **16**, 7112 (2000).
6. K.-I. Hamada, T. Serizawa, T. Kitayama, N. Fujimoto, K. Hatada and M. Akashi, *Langmuir* **17**, 5513 (2001).
7. T. Serizawa, K.-I. Hamada, T. Kitayama and M. Akashi, *Angew. Chem. Int. Ed.* **42**, 1118 (2003).
8. G. Decher and J. B. Schlenoff (Eds), *Multilayer Thin Films*. Wiley-VCH, Weinhein, (2003).
9. K.-I. Hamada, K. Yamashita, T. Serizawa, T. Kitayama and M. Akashi, *J. Polym. Sci. Part A: Polym. Chem.* **41**, 1807 (2003).
10. K. Hatada, K. Ute, Y. Okamoto and T. Kitayama, *Polym. J.* **18**, 1037 (1986).
11. T. Kitayama, T. Shinozaki, T. Sakamoto, M. Yamamoto and K. Hatada, *Makromol. Chem.* **15**, 167 (1989).

12. T. Serizawa, M. Yamaguchi, T. Matsuyama and M. Akashi, *Biomacromolecules* **1**, 306 (2000).
13. T. Serizawa, M. Yamaguchi and M. Akashi, *Biomacromolecules* **3**, 724 (2002).
14. O. N. Tretinnikov, *Langmuir* **13**, 2988 (1997).
15. T. Serizawa, H. Yamashita, T. Fujiwara, Y. Kimura and M. Akashi, *Macromolecules* **34**, 1996 (2001).
16. T. Serizawa, Y. Arikawa, K.-I. Hamada, H. Yamashita, T. Fujiwara, Y. Kimura and M. Akashi, *Macromolecules* **36**, 1762 (2003).

Gels, Genes, Grafts and Giants, pp. 439–456 (2005)

Macrophage behavior on surface-modified polyurethanes

JACQUELINE A. JONES [1], MAHROKH DADSETAN [2], TERRY O. COLLIER [2], MICHAEL EBERT [3], KEN S. STOKES [3], ROBERT S. WARD [4], P. ANNE HILTNER [2] and JAMES M. ANDERSON [1,2,5,*]

[1] *Department of Biomedical Engineering, Case Western Reserve University, Cleveland, OH 44106, USA*
[2] *Department of Macromolecular Science, Case Western Reserve University, Cleveland, OH 44106-7202, USA*
[3] *Medtronic, Inc., 7000 Central Avenue NE, Minneapolis, MN 55432, USA*
[4] *Polymer Technology Group, 2810 7th Street, Berkeley, CA 94710, USA*
[5] *The Institute of Pathology, Case Western Reserve University, Room 306, 2085 Adelbert Road, Cleveland, OH 44106-2622, USA*

Received 1 October 2003; revised 16 January 2004; accepted 21 January 2004

Abstract—Adherent macrophages and foreign body giant cells (FBGCs) are known to release degradative molecules that can be detrimental to the long-term biostability of polyurethanes. The modification of polyurethanes using surface modifying endgroups (SMEs) and/or the incorporation of silicone into the polyurethane soft segments may alter macrophage adhesion, fusion and apoptosis resulting in improved long-term biostability. An *in vitro* study of macrophage adhesion, fusion and apoptosis was performed on polyurethanes modified with fluorocarbon SMEs, polyethylene oxide (PEO) SMEs, or poly(dimethylsiloxane) (PDMS) co-soft segment and SMEs. The fluorocarbon SME and PEO SME modifications were shown to have no effect on macrophage adhesion and activity, while silicone modification had varied effects. Macrophages were capable of adapting to the surface and adhering in a similar manner to the silicone-modified and unmodified polyurethanes. In the absence of IL-4, macrophage fusion was comparable on the modified and unmodified polyurethanes, while macrophage apoptosis was promoted on the silicone modified surfaces. In contrast, when exposed to IL-4, a cytokine known to induce FBGC formation, silicone modification resulted in more macrophage fusion to form foreign body giant cells. In conclusion, fluorocarbon SME and PEO SME modification does not affect macrophage adhesion, fusion and apoptosis, while silicone modification is capable of mediating macrophage fusion and apoptosis. Silicone modification may be utilized to direct the fate of adherent macrophages towards FBGC formation or cell death through apoptosis.

Key words: Macrophage adhesion; FBGC formation; macrophage apoptosis; surface-modifying endgroups; polyurethanes.

*To whom correspondence should be addressed at The Institute of Pathology. Tel.: (1-216) 844-1012. E-mail: jma6@po.cwru.edu

INTRODUCTION

When a biomaterial is implanted into the body, the inflammatory and wound healing responses are initiated causing monocytes to infiltrate the injured site, adhere to the implant surface and differentiate into macrophages [1–3]. These macrophages, adherent to the biomaterial, can subsequently fuse to form multi-nucleated foreign body giant cells (FBGCs) [3, 4]. Upon activation, the adherent macrophages and FBGCs can release degrading molecules such as surface-damaging oxygen radicals (i.e. superoxide anions) that will cause oxidative degradation of the polyurethane [5–7]. Over time, it has been shown that the oxidation of the polyether segments can lead to environmental stress cracking and permanent damage of the polyurethane material [1, 8, 9].

Modulation of macrophage adhesion, formation into FBGCs and apoptosis activity at the implant/tissue interface provides a reasonable attempt to increase the *in vivo* lifetime of biomaterials. Induction of apoptosis may be the most efficient mechanism for minimizing adherent cell activities, because it results in the removal of cells without inducing an inflammatory response in the surrounding tissue [10]. Apoptosis is a specialized programmed cell death, which requires active gene expression and is mediated by signal transduction unlike normal cell death, i.e. necrosis. It is easily characterized by distinct chemical and morphological features such as the translocation of phosphatidylserine to the outer leaflet of the cell membrane, cell shrinkage, condensed chromatin and the presence of apoptotic bodies.

Polyether and polycarbonate urethanes have been used for years in a number of biomedical applications due to their favorable biocompatibility and mechanical properties [11, 12]. However, it has been shown that polyurethane materials are prone to degradation via surface oxidation, environmental stress cracking (ESC) and metal ion oxidation (MIO) [6, 13, 14]. Polyurethanes have the potential to degrade via oxidative and hydrolytic chain cleavage. Specifically, oxidative chain cleavage can be reduced by a reduction in reactive oxygen radicals through the minimization of adherent macrophages.

Polymer Technology Group, Inc. (PTG) has developed a series of polyurethanes chemically designed to be more oxidative and hydrolytically biostable. In one system, surface modifying endgroups (SMEs) of materials that are known to be more biostable or less biointeractive are covalently bonded to the ends of the polyurethane chains. SME segments are capable of migrating to the surface of the material in order to produce a more energetically favorable surface without diffusing from the surface of the biomaterial. This chain rearrangement can affect cellular–biomaterial interface interactions, while having no effect on the bulk properties of the polyurethane.

Surfaces rich in polyethylene oxide (PEO) have been shown to reduce protein adsorption and cellular adhesion [15–18]. It is believed that the inhibition of these cellular–biomaterial interactions is due to the high association of PEO with water molecules that provide a barrier to protein adsorption. Research by Jenny *et al.*

indicates that long-term macrophage adhesion and IL-4-induced FBGC formation can be significantly reduced by PEO SME-modified surfaces [15].

Due to their extremely hydrophobic nature and ability to inhibit the adsorption of blood clotting proteins, fluorocarbon-based materials have been used in various biomedical applications, such as vascular grafts and other anti-thrombogenic applications [19]. It has been shown that, although the initial adsorption of proteins commonly involved in macrophage adhesion was high on PTFE surfaces, the strength of adhesion was low and proteins were easily removed with a SDS wash [20]. Jahangir *et al.* showed that a system of PEUs containing fluorine surface-modifying macromolecules (SMMs) created a more hydrophobic surface chemistry, decreased fibrinogen protein adhesion and altered platelet adhesion [21]. These fluorine SMMs materials were also shown to inhibit enzyme-induced biodegradation [22]. These material characteristics suggest that the addition of PEO and fluorocarbon SMEs to polyurethanes may alter macrophage adhesion, fusion and apoptosis and ultimately improve the biostability of polyurethanes.

A second approach that has been investigated to produce a more oxidative and hydrolytically biostable polyurethane involves the addition of poly(dimethylsiloxane) (PDMS) to the soft segment and SME unit [23–26]. PDMS materials have excellent flexibility, good blood biocompatibility, low toxicity and an anti-adhesive nature. It has been shown that the modification of polyurethanes with PDMS resulted in a silicone-rich surface due to the high mobility and low surface energy of the PDMS segment [27]. A previous study by our group has shown an enhanced *in vivo* biostability of PDMS end-capped polyurethanes in comparison to the uncapped polyurethanes [14]. In addition, Martin *et al.* reported that the incorporation of PDMS into polyurethane soft segment increased the *in vivo* biostability of polyurethane [26]. These findings suggest that the PDMS incorporation into both the soft segment and SME unit may alter the cellular-biomaterial interactions and may be beneficial to polyurethane biostability.

In this study, we investigated the effect of the addition of fluorocarbon and PEO SME to polyurethanes on monocyte/macrophage adhesion, fusion and apoptosis. Also, we have examined the simultaneous effect of PDMS in the soft segments and SME unit on macrophage adhesion and activity.

MATERIALS AND METHODS

Biomaterials

Polyurethane films, listed in Table 1, were synthesized by Polymer Technology Group (Berkeley, CA, USA) and extruded by Medtronic (Minneapolis, MN, USA) [28]. Elasthane 80A is a polyether urethane with an aromatic hard segment of 4,4′-methylene bisphenyl diisocyanate (MDI) chain extended with 1,4-butanediol (BD) and a polytetramethylene oxide (PTMO) soft segment. Elasthane 80A with surface-modifying endgroup (SME) modification has either polyethylene oxide

Table 1.
Chemical description of polyurethane materials

Material	Description	Soft segment	Hard segment	SME content	SME molecular mass (Da)
Elasthane 80A	E80A	PTMO	MDI/BD	—	—
Bionate 80A	B80A	PHECD	MDI/BD	—	—
Elasthane 80A with Fluorocarbon SME	E80A-F	PTMO	MDI/BD	6 wt% Fluorocarbon	500
Elasthane 80A with PEO SME	E80A-P	PTMO	MDI/BD	6 wt% PEO	2000
PurSil 20 80A	PSil80A	PTMO, 20 wt% PDMS	MDI/BD	0.2 wt% PDMS	2000
CarboSil 20 90A	CSil90A	PHECD, 20 wt% PDMS	MDI/BD	0.2 wt% PDMS	2000

Abbreviations: MDI, 4,4′-methylene bisphenyl diisocyanate; BD, 1,4-butanediol; PTMO, polytetramethylene oxide; PHECD, poly(1,6-hexyl-1,2-ethyl carbonate) diol; PDMS, polydimethylsiloxane. SME molecular mass as reported by Medtronic.

(PEO) or polytetrafluoroethylene (PTFE) SMEs covalently bonded to the ends of the polyether urethane chain. The SME average molecular masses for PEO and PTFE are 2000 and 500 Da, respectively. PurSil 20 80A has the same composition as Elasthane 80A; however, it contains a 20% PDMS co-soft segment and 0.2 wt% PDMS SMEs. Bionate 80A is a polycarbonate urethane with a MDI hard segment chain extended with BD and a polycarbonate soft segment of poly(1,6-hexyl-1,2-ethyl carbonate) diol (PHECD). CarboSil 20 90 A is similar to Bionate 80A; however, it has a PHECD and 20 wt% PDMS co-soft segment, 0.2 wt% PDMS SMEs and a different hard to soft segment ratio. All polyurethane films were extruded as previously described by Patel *et al.* [29].

Polydimethyl siloxane (PDMS) films were prepared from SILASTIC® BioMedical Grade ETR Elastomer Q7-4765 (Dow Corning, Midland, MI, USA) according to manufacturer's instruction. PDMS film samples were extracted overnight with two changes of hexane to remove silicone oil residues after vulcanization. Prior to cultures, air bubbles were removed from the PDMS surfaces using the method previously described by Kalman *et al.* [30].

The polyurethane films were cut into 15-mm diameter disks, wiped with 100% ethanol and a kimwipe, sonicated in 100% ethanol for 10 min, rinsed in distilled water twice, air dried and then gas sterilized with ethylene oxide. Silicone tubing (Cole-Parmer, Vernon Hills, IL, USA) was sectioned into rings, 8 mm in length, sonicated and autoclaved for 1 h. Under sterile conditions, the material samples were inserted into 24-well tissue culture polystyrene (TCPS) plates (Fisher Scientific, Pittsburgh, PA, USA) and secured with the silicone rings. The resultant surface area for each culture well was 0.71 cm^2.

Surface characterization

Contact angle analysis. Advancing contact angle measurements were used to determine the hydrophobic/hydrophilic nature of the materials using a goniometer (Edmund Scientific, Barrington, NJ, USA) at 22°C, room temperature. Contact angle measurements were performed in order to understand the surface effect of the modified chemistries and the surface effect of an aqueous environment on the material. Two systems were utilized in the contact angle analysis to study these effects, respectively; (A) in air with a deionized, distilled water bubble probe and (B) under aqueous conditions in a deionized, distilled water bath with a hydrophobic methylene iodide probe. It is important to note that more hydrophobic surfaces produce a greater contact angle degree in the first system, while more hydrophilic surfaces produce a greater contact angle degree in the second system.

In vitro *cell culture.* Human peripheral monocytes and serum were obtained from the whole, venous blood of healthy, unmedicated donors as described by McNally *et al.* [31]. Isolated monocytes were cultured at a concentration of 5×10^5 cells/well in 0.5 ml culture media containing 20% autologous serum and RPMI 1650 cell media (Gibco-BRL Gaithersburg, MD, USA). Cultures were incubated at 37°C with a 5% CO_2 environment. After the initial 2 h, non-adherent cells were removed by rinsing with warmed Dulbecco's phosphate-buffered saline (PBS) with Ca^{2+} and Mg^{2+} (Gibco-BRL) and fresh culture media was added. At days 3 and 7, the remaining cultures were re-fed with a culture media of RPMI and 20% autologous, heat-inactivated serum. Serum was heat inactivated at 56°C for 1 h. Initially at day 3, IL-4 was added to specific cultures at a concentration of 15 ng/ml to induce FBGC formation [32]. Cultures were terminated at days 0 (2 h), 3, 7 and 10. Each sample was rinsed with warm PBS and treated according to the procedure for adhesion, fusion and apoptosis analysis.

Analysis

Adhesion and fusion analysis. For optical microscopy (OM), samples were rinsed with PBS, fixed in 100% methanol for 5 min and air-dried. Adherent cells were stained with May–Grunwald for 1 min, rinsed with PBS twice, stained with Giemsa for 5 min, rinsed twice with distilled water and air dried overnight. Total number of adherent cells and foreign body giant cells (FBGC) were counted from 5–20× objective fields (463×463 μm) on each sample. These cell densities were averaged to determine the adherent macrophage density and percentage of nuclei within FBGC at all time points.

Apoptosis analysis. At the appropriate timepoints, each sample was rinsed twice with PBS and then stained using an Annexin V fluorescein isothiocyanine (FITC) conjugate reagent kit according to the instructions provided by the manufacturer (R&D Systems, Minneapolis, MN, USA) [33]. Samples were then removed from the 24-well plate, mounted to glass slides with gel mount (Biomeda, Foster City, CA, USA) and imaged using CSLM. A minimum of 200 cells was counted per sample to obtain the total number of adherent cells/mm^2. Positively stained cells were considered to be apoptotic. The data is presented as a percentage of apoptotic macrophages of the total number of adherent macrophages. Separate samples were stained with May–Grunwald and Giemsa as previously described and examined for the classical morphological features of apoptosis within 5–20× objective fields (463 × 463 μm) using OM. Results are shown as the total number of apoptotic and viable cells/mm^2.

Statistical analysis. For the macrophage adhesion, fusion and apoptosis studies, measurements were taken from 5–8 fields on duplicate samples, yielding 10–16 datapoints per experiment and a total of 30–48 datapoints. The data was collected from three experiments and analyzed using an one-way factorial ANOVA post-hoc test (Bonferoni–Dunn) with a 95% confidence level (5% alpha value).

For contact angle analysis, three fields on triplicate samples were measured for advancing and receding contact angles. The advancing contact angle measurements were obtained from an incremental addition of 5–2 μl droplets. The 5 measurements were averaged together for the 3 fields to produce the mean advancing water contact angle of the single experiment. The presented data is the mean of 3 experiments in which each experiment is the result of the mean of 15 datapoints. The receding contact angles were obtained in a similar manner. Statistical analysis was performed using an unpaired t-test with a 95% confidence level.

All statistical analysis was performed using the StatView 5.0.1.0 software (SAS Institute, Cary, NC, USA). All data is presented as the mean ± standard error of the mean (SEM) ($n = 3$).

RESULTS

Surface characterization

Contact angle measurements performed in both systems, (A) air surrounding with a water probe and (B) water surrounding with a methylene iodide probe, are shown in Table 2. System A was used as a standard contact angle analysis system. Under dry conditions, the fluorocarbon SME modification (109 ± 0) caused the surface chemistry of the unmodified Elasthane 80A (61 ± 2) to be more hydrophobic, while the PEO SME modification (62 ± 3) had no effect on the surface chemistry. After hydrating the sample for 1 h, the fluorocarbon SME-modified surfaces became more hydrophilic (96 ± 5), but still statistically different ($P < 0.0001$) from the Elasthane 80A (70 ± 4). This increase in hydrophilicity may be due to chain rearrangement and the migration of the hydrophobic fluorocarbon SMEs away from the surface and into the bulk of the material. Hydration of the PEO SME-modified polyurethanes for 1 h resulted in a more hydrophilic surface chemistry (56 ± 1).

Table 2.
Water contact angle measurements on polyurethane materials

Materials and conditions	Water probe in air (System A)		Methylene iodide probe in water (System B)	
	Advancing contact angle (°)	Receding contact angle (°)	Advancing contact angle (°)	Receding contact angle (°)
Elasthane 80A				
Unhydrated	61 ± 2	50 ± 1	68 ± 5	57 ± 7
Hydrated for 1 h	70 ± 4^{a}	60 ± 3^{a}	74 ± 3	55 ± 4
Hydrated for 24 h	—	—	77 ± 5	54 ± 6
PurSil 20 80A				
Unhydrated	$90 \pm 5^{a,b}$	$74 \pm 2^{a,b}$	—	—
Hydrated for 1 h	$88 \pm 5^{a,b}$	$73 \pm 9^{a,b}$	—	—
Elasthane 80A–Fluorocarbon				
Unhydrated	109 ± 0^{b}	95 ± 2^{b}	77 ± 2	56 ± 4
Hydrated for 1 h	$96 \pm 5^{a,b}$	$77 \pm 11^{a,b}$	71 ± 4	53 ± 4
Hydrated for 24 h	—	—	73 ± 3	51 ± 5
Elasthane 80A–PEO				
Unhydrated	62 ± 3	42 ± 1^{b}	96 ± 4^{b}	89 ± 4^{b}
Hydrated for 1 h	56 ± 1^{a}	$38 \pm 9^{a,b}$	98 ± 4^{b}	98 ± 6^{b}
Hydrated for 24 h	—	—	103 ± 3^{b}	96 ± 6^{b}
Bionate 80A				
Unhydrated	62 ± 1	48 ± 0	66 ± 5	46 ± 4
Hydrated for 1 h	69 ± 4^{a}	59 ± 3^{a}	75 ± 2	55 ± 6
Hydrated for 24 h	—	—	80 ± 2	55 ± 7
CarboSil 20 90A				
Unhydrated	$86 \pm 8^{a,c}$	$71 \pm 4^{a,c}$	—	—
Hydrated for 1 h	$91 \pm 6^{a,c}$	$73 \pm 13^{a,c}$	—	—

The data represents the mean±SEM of three experiments. Surfaces were hydrated in PBS for 0, 1 and/or 24 h. For system A, contact angle measurements were performed in air with a deionized, distilled water probe to investigate the effect of the modifications on the hydrophobicity/hydrophilicity of the surface. More hydrophobic surfaces have a greater contact angle degree, while more hydrophilic surfaces have a lower contact angle degree. For system B, contact angle measurements were performed submerged in deionized, distilled water with a hydrophobic methylene iodide probe to determine the hydrophobicity/hydrophilicity of the surfaces in culture media. More hydrophilic surfaces have a greater contact angle degree, while more hydrophobic surfaces have a lower contact angle degree.

[a] Data referenced from past studies in our laboratory by Patel *et al.* [27].

[b] Statistically significant from Elasthane 80A ($P < 0.05$).

[c] Statistically significant from Bionate 80A ($P < 0.05$).

Chain rearrangement of the hydrophilic PEO SMEs towards the water–surface interface may have caused the increase in hydrophilicity. The increase in the contact angle of Elasthane 80A after 1 h of hydration (61 ± 2 *vs.* 70 ± 4) may have resulted from an interference of the water molecules with the hydrogen bonds within the polyether urethane resulting in chain reorientation. Silicone modification provided a more hydrophobic surface chemistry for PurSil 20 80A (90 ± 5) and CarboSil 20 90A (86 ± 8) when unhydrated. Hydrating the silicone-modified materials for 1 h did not affect the hydrophobicity of either silicone modified polyurethane.

System B, the methylene iodide probe in water, was utilized to investigate the effect of an aqueous environment on the materials as seen in culture. After hydrating the samples for 1 h, Elasthane 80A-F (71 ± 4), Elasthane 80A (74 ± 3) and Bionate 80A (75 ± 2) had similar hydrophilicity as seen by the contact angles, while Elasthane 80A-P (98 ± 4) was more hydrophilic. This indicates that in culture the fluorocarbon SME modification no longer increases the hydrophobicity of Elasthane 80A, while the PEO SME modification resulted in a more hydrophilic surface. The greatest change in contact angle was seen between the unhydrated and hydrated for 1-h samples, while no difference was found between the hydrated for 1 and 24 h samples for any of the materials. This suggests that the majority of the chain rearrangements occurred during the first hour of hydration.

Macrophage adhesion

As in the normal progression of adherent monocytes and macrophages, cell morphology progressed from the expected rounded monocyte morphology into the spread cytoplasm morphology of macrophages between days 0 and 10. Figure 1 shows monocyte/macrophage adhesion on the silicone-modified, SME-modified and unmodified polyurethanes. Adherent cell densities decreased from day 0 to day 10 as expected [15, 34]. The total number of adherent cells was greatest at day 0. A large decrease in the adherent cell density was seen between days 0 and 3. After day 3, the adherent cell densities decreased linearly.

Initially at day 0, the fluorocarbon and PEO SME-modified Elasthane 80A appeared to support slightly more macrophage adhesion than Elasthane 80A; however, one-way factorial ANOVA statistical analysis revealed that the materials were not significantly different on a 5% confidence level. PurSil 20 80A and CarboSil 20 90A also did not promote or inhibit macrophage adhesion in comparison to Elasthane 80A and Bionate 80A, respectively. After day 3, all materials supported similar numbers of adherent macrophages. Macrophage adhesion was similar on PDMS, silicone-modified polyurethane and the unmodified polyurethanes, indicating that silicone did not have an effect on macrophage adhesion.

Macrophage fusion

The capacity for macrophages to fuse into FBGC was examined on each biomaterial surface (Fig. 2). In the presence of IL-4, macrophages aggregated on the surface of all materials by days 7 and 10. At day 7, little to no fusion was seen on all polyurethane materials (<5%). PDMS promoted 13% of the macrophages to fuse into FBGCs. Percent fusion tended to be greater on PurSil 20 80A and CarboSil 20 90A in comparison to Elasthane 80A and Bionate 80A, respectively. The trend for FBGC formation at day 10 is as follows: E80A-F = E80A-P $\leqslant$ E80A $\leqslant$ B80A <

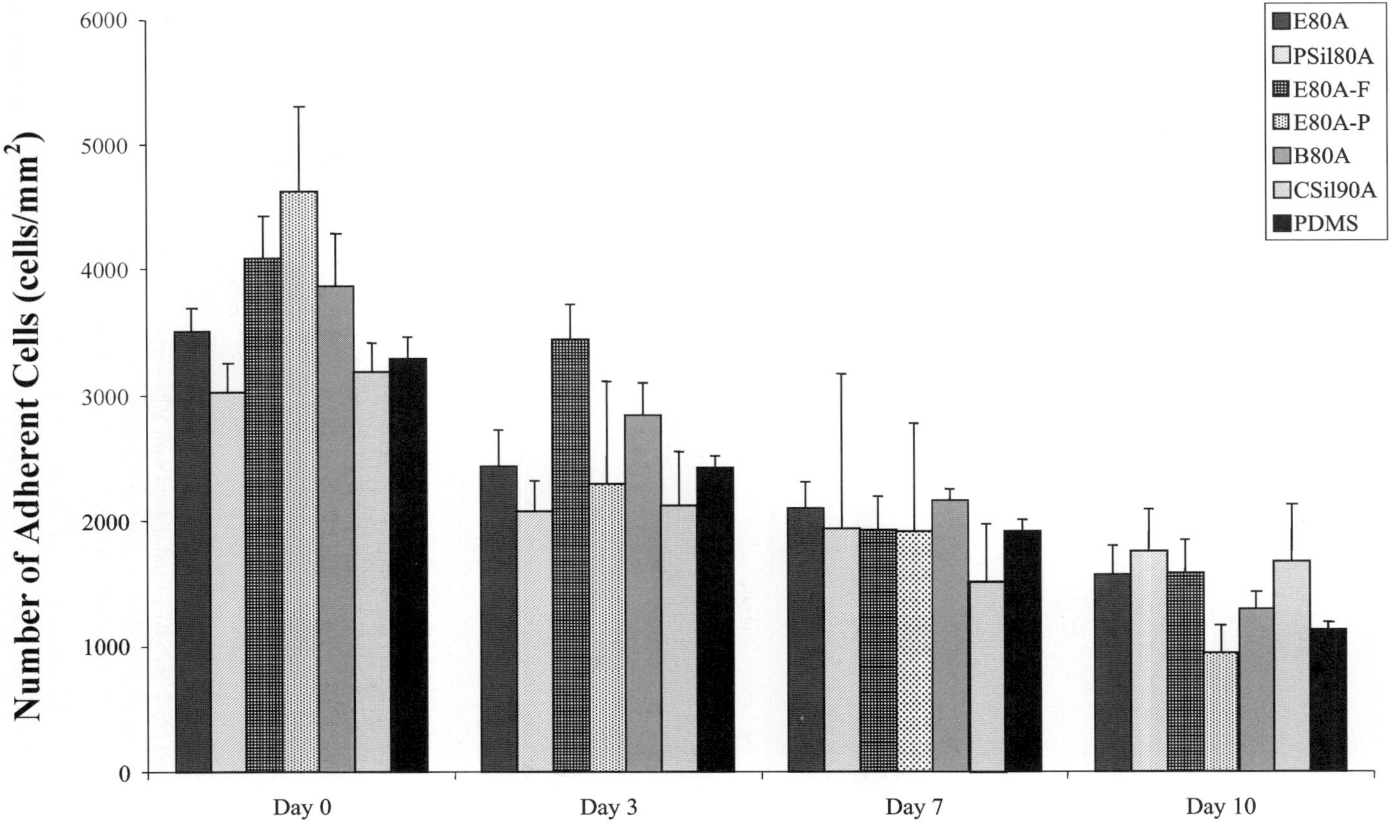

Figure 1. Monocyte/macrophage adhesion on modified and unmodified polyurethane materials. Samples were fixed with methanol at days 0 (2 h), 3, 7 and 10 and stained with May–Grunwald and Giemsa for OM analysis. Densities were obtained from 5–20× objective fields per sample. The data represents the mean±SEM of three experiments.

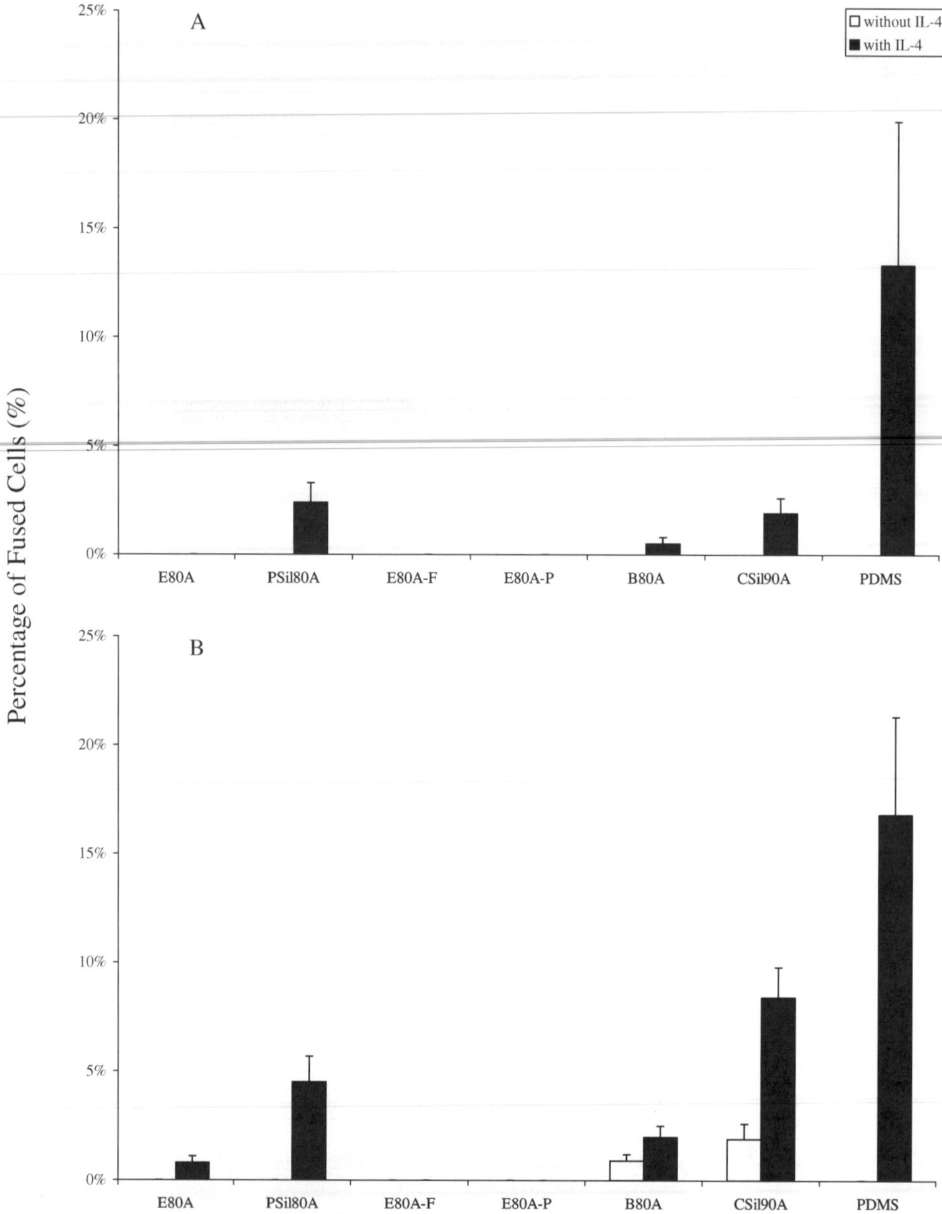

Figure 2. FBGC formation on modified and unmodified polyurethane materials. Samples were fixed with methanol at day (A) 7 and (B) 10 and stained with May–Grunwald and Giemsa for OM analysis. The data represents the mean±SEM of three experiments. Surfaces with missing data points (i.e. E80A, E80A-F, E80A-P at day 7) had less than 1% of fused cells on the surface.

PSil80A $<$ CSil90A $\ll$ PDMS. These significant differences in FBGC formation on PDMS and the silicone-modified and unmodified polyurethanes at day 10 can be seen in the photomicrographs in Fig. 3.

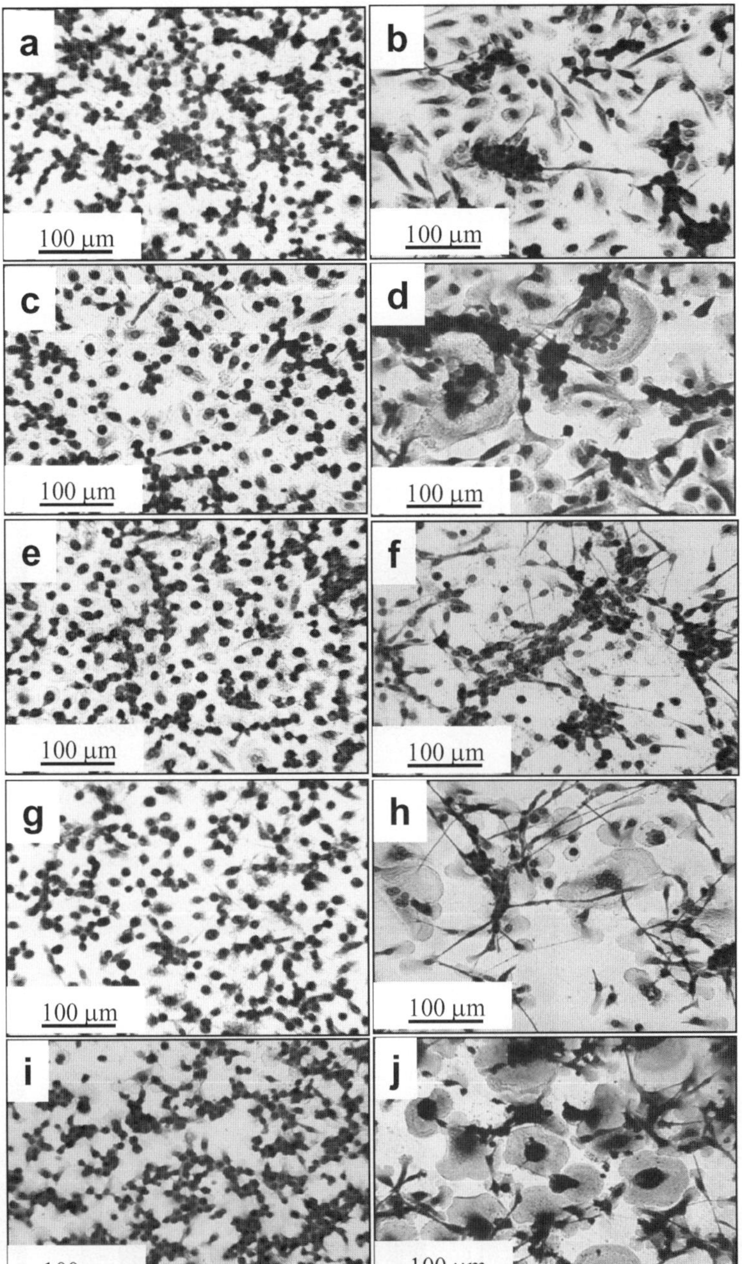

Figure 3. Optical micrographs of adherent macrophages and IL-4-induced FBGC on PDMS and the silicone-modified and unmodified polyurethanes at day 10. Adherent cells on the surfaces of (a, b) Elasthane 80A, (c, d) PurSil 20 80A, (e, f) Bionate 80A, (g, h) CarboSil 20 90A, (i, j) PDMS were stained with May–Grunwald and Giemsa for OM at day 10. IL-4 (15 ng/ml) was added to the culture media of some samples (b, d, f, h, j) at days 3 and 7 in order to promote the formation of IL-4-induced FBGC. All images were viewed with a 20× objective and the scale bars represent 100 μm.

Macrophage apoptosis

Figure 4 shows the percentage of apoptotic macrophages, while Fig. 5 shows the surface densities of apoptotic and viable adherent macrophages for the polyurethane

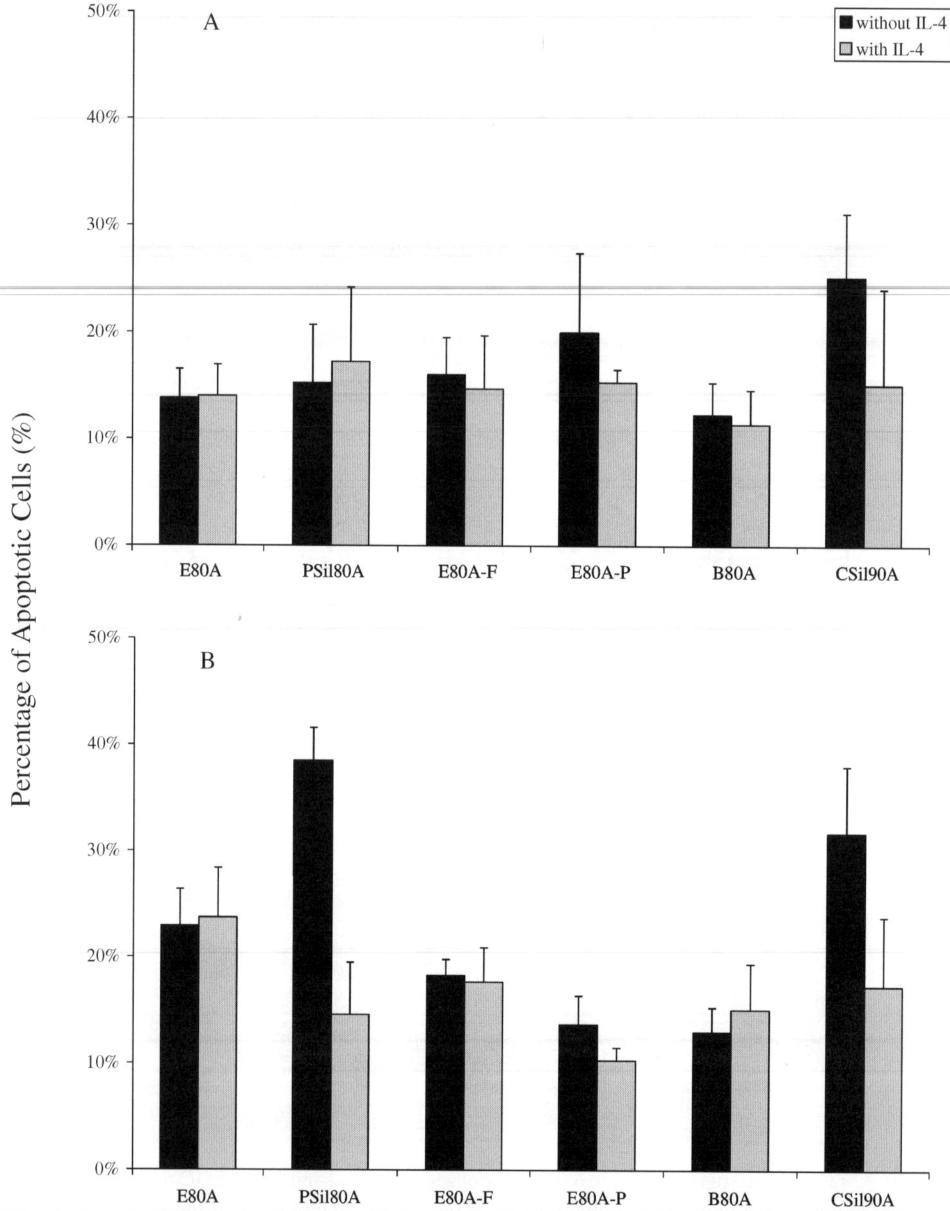

Figure 4. Annexin V apoptosis analysis of cells adherent to unmodified and modified polyurethane materials. Samples were retrieved at days (A) 7 and (B) 10 and stained with Annexin V-FITC for CSLM. The percentage of apoptotic cells was determined from 200 adherent cells. The data represents the mean±SEM of three experiments.

materials. E80A, E80A-F, E80A-P and B80A materials promoted similar minimal amounts of macrophage apoptosis at days 7 and 10 in the presence and absence of IL-4. At day 7, PurSil 20 80A promoted a similar percentage of apoptotic macrophages as Elasthane 80A, while CarboSil 20 90A promoted more apoptosis than Bionate 80A. By day 10, PurSil 20 80A and CarboSil 20 90A promoted more apoptotic cells than Elasthane 80A and Bionate 80A, respectively, in the absence of IL-4. Also, the percentage of apoptotic cells decreased on PurSil 20 80A and

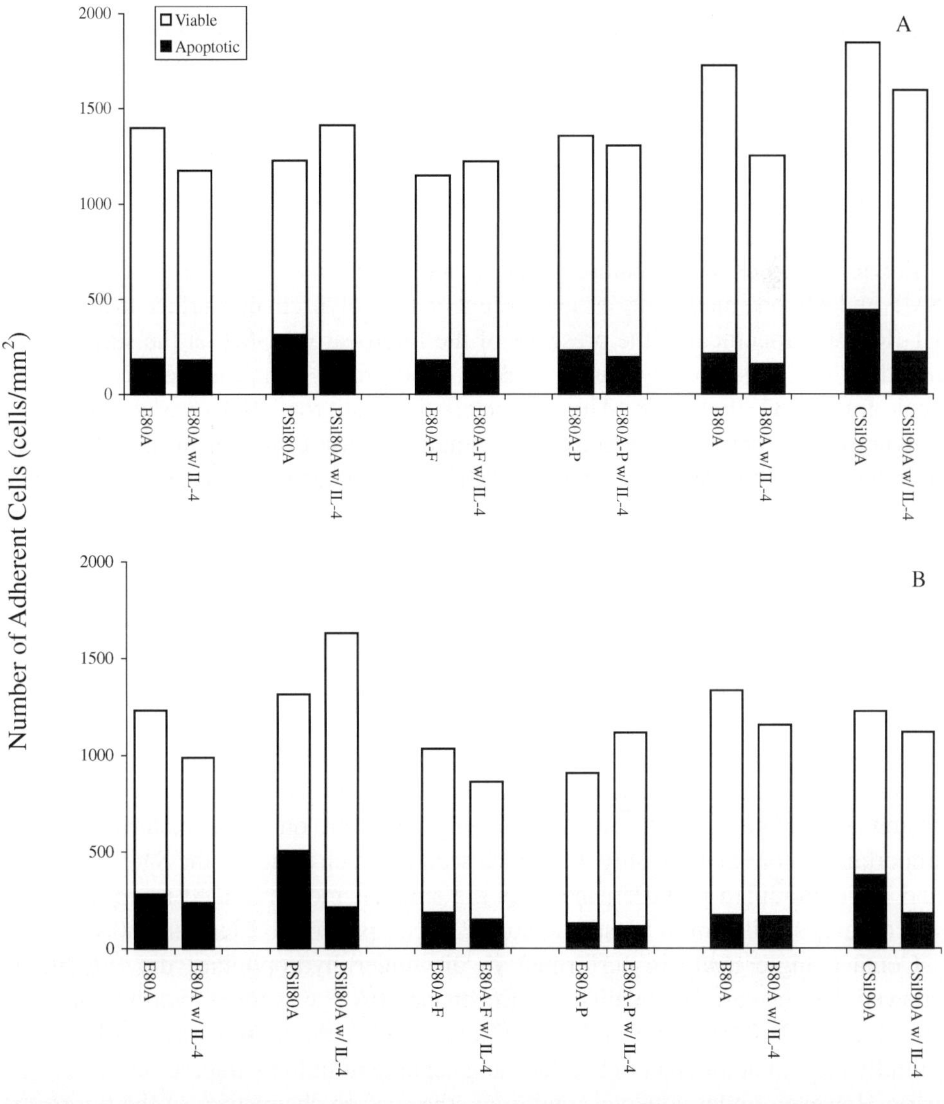

Figure 5. Viable and apoptotic adherent cell densities on the unmodified and modified polyurethane materials. Samples were retrieved at days (A) 7 and (B) 10 and stained with Annexin V-FITC for CSLM. Viable and apoptotic cell densities were obtained from 200 adherent cells. The data represents the mean±SEM of three experiments.

CarboSil 20 90A at day 10, due to the addition of IL-4. This decrease in apoptotic cells on PurSil 20 80A and CarboSil 20 90A is correlated to higher level of fusion on silicone-modified polyurethanes.

DISCUSSION

Adherent macrophages can be detrimental to a polyurethane surface by releasing degradative enzyme and oxygen radicals that result in chain scission and the eventual failure of the device. Modulation of cell-biomaterial interactions such as adhesion, fusion and apoptosis could be a possible mechanism of reducing these degradative effects. Incorporation of more biostable and less biointeractive materials into the polyurethane polymer chain may provide a means to modulate the cell-biomaterial interactions without compromising the mechanical properties of the polyurethanes.

Previous studies in our laboratory have shown that the fluorocarbon SME, PEO SME and silicone modifications are present on the polyurethane surface using ATR–FT-IR [29]. Specifically, the presence of the fluorocarbon SME at the surface was noted by the presence of a small peak at 1150 cm^{-1}, which is the characteristic peak for the C—F bond. The characteristic peak for PEO overlaps with the polyurethane ether peak; therefore, the small increase observed in the ether peak is indicative of the presence of PEO SME segments at the surface. Three peaks (800 cm^{-1}, 1253 cm^{-1}, 1018 cm^{-1}) indicate the presence of the PDMS groups at the polyurethane surface.

Although past studies have demonstrated that fluorocarbon surfaces inhibit the adsorption of proteins required for macrophage adhesion and PEO-rich surfaces inhibit protein adsorption and cellular adhesion, Jenney and others showed that macrophages are capable of adhering to a wide variety of surface chemistries in a similar manner [15–17, 19]. Therefore, it was not surprising to find no significant difference in macrophage adhesion between Elasthane 80A-F, Elasthane 80A-P and Elasthane 80A by days 7 and 10. However, it was unique to find that macrophage fusion and apoptosis was not significantly different on these materials. It is important to note that the only difference seen between either of the SME-modified and the unmodified polyurethanes was the smaller, more rounded morphology of macrophages adherent to Elasthane 80A-P in comparison to Elasthane 80A.

Contact angle analysis performed in air under hydrophobic, dry conditions showed that the surface chemistry of Elasthane 80A-F was more hydrophobic than Elasthane 80A. This confirmed Patel's findings that the fluorocarbon SMEs were initially present at the material surface and capable of influencing the surface chemistry. However, under aqueous conditions, the surface chemistries of the fluorocarbon SME-modified and unmodified polyurethanes were comparable. This suggests that in culture media fluorocarbon SMEs rearrange away from the surface in order to avoid the energetically unfavorable interface with the aqueous environment. As a result, this allows the polyurethane hard and soft segments to dominate the surface producing a surface similar to the unmodified polyurethane, which explains our similar macrophage adhesion, fusion and apoptosis findings.

Unlike with the fluorocarbon SMEs, contact angle analysis in both air and aqueous conditions indicate an increase in hydrophilicity on Elasthane 80A-P in comparison to Elasthane 80A. In aqueous conditions, the PEO SMEs migrate to the polyurethane surface to produce a more energetically favorable interface. The confirmed presence of the PEO SMEs at the surface both in air and aqueous environments further suggest that the PEO SMEs would affect macrophage adhesion and activity on the polyurethane surface; however, no significant differences were seen between the PEO SME-modified and unmodified polyurethanes. Past studies demonstrated that not only was the presence of PEO segments important to macrophage adhesion but the molecular mass of the PEO segments [15]. They found that PEO segments with molecular masses between 200 and 4600 Da were capable of shifting the surface chemistry while not altering macrophage adhesion or FBGC formation. Macrophage adhesion was reduced only when the PEO segments had a molecular mass of 4.6 kDa or 18.5 kDa, while FBGC formation was reduced with the longer chains of 18.5 kDa. In our study, the PEO SMEs have a molecular mass of 2 kDa, which is sufficient to alter surface chemistry; however, not sufficient enough to reduce macrophage adhesion or FBGC formation. Possibly an increase in the molecular mass of the PEO SME segments to at least 18.5 kDa would reduce the cell-biomaterial interactions.

Research by Patel *et al.* in our laboratory confirmed the presence of silicone on the surface of polyurethanes with ATR–FT-IR, while contact angle measurements indicated an increase in hydrophobicity of the silicone-modified *versus* unmodified polyurethanes [29]. Water contact angles on silicone modified samples did not change following 1 h hydration, indicating surface reorientation in water is minimal to very slow. These results are in agreement with Ward's findings that silicone modification of polyurethanes promotes a silicone-rich surface due to high mobility and low surface energy of PDMS.

Macrophages have been shown to be adaptable to various surface chemistries, allowing for similar adhesion to a wide range of surface chemistries [34]. Previous studies in our laboratory on alkyl-silane surfaces have revealed that monocyte/macrophage adhesion over a 10-day culture was unaffected by surface chemistry, while the density of IL-4 induced FBGCs was correlated to carbon content. Surfaces with shorter alkyl chains (DM and C1) appeared to promote FBGC formation. As expected, macrophage adhesion was not affected by silicone modification after day 3. Also, FBGC formation was promoted by the silicone modified polyurethanes similarly to dimethyl silane-modified glass (DM) [35]. These results are also consistent with the findings that hydrophobic surfaces promote FBGC formation [10].

Finally, the percentage of apoptotic cells on the silicone modified polyurethanes (PurSil 20 80A and CarboSil 20 90A) tended to be greater than on the unmodified polyurethanes (Elasthane 80A and Bionate 80A, respectively). Also, a greater number of apoptotic cells were seen on silicone modified polyurethane after 10 days. This confirms previous results suggesting that macrophage apoptosis is surface dependent [10]. Apoptosis results in this study contradict Brodbeck's findings that hydrophobic surfaces inhibit apoptosis; however, the increase in apoptotic macrophages may instead be due to differences in surface compositions.

An investigation into the effect of PDMS on macrophage adhesion, fusion and apoptosis as compared to the silicone-modified and unmodified polyurethane was performed to further understand the effect of PDMS modification (apoptosis data not shown for PDMS). PDMS results indicated that macrophage adhesion was similar on PDMS, PurSil 20 80A and Elasthane 80A. Macrophage fusion was significantly greater on PDMS in comparison to PurSil 20 80A and Elasthane 80A, which explains the promotion of fusion seen on the silicone modified polyurethanes. In the presence of IL-4, macrophage fusion decreased on PDMS and the silicone-modified polyurethanes. The decrease in apoptosis correlates to an increase in fusion seen in the presence of IL-4. This correlation supports Brodbeck's hypothesis that macrophages fuse into FBGC in order to escape apoptosis.

CONCLUSIONS

Modulation of macrophage adhesion, fusion and apoptosis is an effective mechanism to ultimately control the biostability of polyurethanes over long periods of time. It was hypothesized that the incorporation of more biostable and/or less biointeractive chemistries (i.e. fluorocarbon, PEO and PDMS) into the polymer chain may modify the surface chemistry of the material and ultimately have an effect on macrophage adhesion, fusion and apoptosis. In this study, it was found that none of the modifications altered macrophage adhesion and only the silicone modification promoted fusion in comparison to the unmodified polyurethane. The fluorocarbon SME-modified polyurethanes did not have an effect on macrophage adhesion, fusion, or apoptosis, due to the rearrangement of the hydrophobic SME segments away from the hydrophilic surroundings into the bulk of the polyurethane. In contrast, the PEO SME segments were still present at the polyurethane surface in the culture media; however, the PEO molecular mass was insufficient to induce a cellular response as confirmed by past studies in this laboratory. Finally, macrophages were adaptive enough to still adhere to the silicone-modified surfaces; however, the silicone modification was effective at modifying macrophage fusion and apoptosis. In the absence of IL-4, macrophage apoptosis was promoted with little FBGC formation. When induced with IL-4, FBGC formation increased significantly and macrophage apoptosis was inhibited. Silicone modification provides a possible means for modulation the ultimate fate of adherent macrophages. If unstimulated, adherent macrophages will undergo apoptosis on the silicone modified surfaces; however if stimulated by the proper cytokines, macrophages will be more inclined to fuse into FBGCs. Future *in vivo* studies will eventually confirm the ultimate fate of the macrophage. In conclusion, polyurethanes incorporated with surface-modifying endgroups may still affect macrophage adhesion and activity with a few alterations in the SME segments. Silicone modification does affect macrophage adhesion and activity and may be an effective method of controlling the fate of macrophages *in vivo*.

Acknowledgements

J. A. J. and M. D. provided equivalent contribution to this manuscript. The authors would like to thank Erica Colton for her aid with the monocyte isolations, Elizabeth Christenson for her contribution with the FT-IR surface analysis and polyurethane materials, Jasmine Patel for her input with the contact angle analysis and William Brodbeck for his instruction with the apoptosis analysis. NIH-NIBIB Grants EB-00272 and EB-00279 as well as a NIH Minority Supplement Award to J. A. J. funded this project.

REFERENCES

1. J. M. Anderson, *Cardiovasc. Pathol.* **2**, 33 (1993).
2. J. M. Anderson, K. DeFife, A. McNally, T. Collier and C. Jenney, *J. Mater. Sci. Mater. Med.* **10**, 579 (1999).
3. J. M. Anderson, *Annu. Rev. Mater. Res.* **31**, 81 (2001).
4. J. M. Anderson, *Hematology* **7**, 40 (2000).
5. W. J. Kao, A. Hiltner, J. M. Anderson and G. A. Lodoen, *J. Biomed. Mater. Res.* **28**, 819 (1994).
6. K. S. Stokes, A. Coury and P. Urbanski, *J. Biomater. Appl.* **1**, 411 (1987).
7. Q. Zhao, N. Topham, J. M. Anderson, A. Hiltner, G. Lodoen and C. R. Payet, *J. Biomed. Mater. Res.* **25**, 177 (1991).
8. M. J. Wiggins, B. Wilkoff, J. M. Anderson and A. Hiltner, *J. Biomed. Mater. Res.* **58**, 302 (2001).
9. M. A. Schubert, M. J. Wiggins, J. M. Anderson and A. Hiltner, *J. Biomed. Mater. Res.* **34**, 493 (1997).
10. W. Brodbeck, M. Shive, E. Colton, Y. Nakayama, T. Matsuda and J. M. Anderson, *J. Biomed. Mater. Res.* **55**, 661 (2001).
11. J. M. Brown, *J. Intraven. Nursing* **18**, 120 (1995).
12. S. H. Hsu, H. J. Tseng and M. S. Wu, *Artif. Organs* **24**, 119 (2000).
13. K. S. Stokes, R. McVenes and J. M. Anderson, *J. Biomater. Appl.* **9**, 331 (1995).
14. A. B. Mathur, T. O. Collier, W. J. Kao, M. J. Wiggins, M. A. Schubert, A. Hiltner and J. M. Anderson, *J. Biomed. Mater. Res.* **36**, 246 (1997).
15. C. Jenney and J. M. Anderson, *J. Biomed. Mater. Res.* **44**, 206 (1999).
16. L. G. Cima, *J. Cell Biochem.* **56**, 155 (1994).
17. W. R. Gombotz, W. Guanghui, T. A. Horbett and A. S. Hoffman, *J. Biomed. Mater. Res.* **25**, 1547 (1991).
18. J. H. Lee, B. J. Jeong and H. B. Lee, *J. Biomed. Mater. Res.* **34**, 105 (1997).
19. G. A. Boffa, N. Lucien, A. Faure and M. C. Boffa, *J. Biomed. Mater. Res.* **11**, 317 (1977).
20. C. R. Jenney and J. M. Anderson, *J. Biomed. Mater. Res.* **49**, 435 (2000).
21. A. Jahangir, W. McClung, R. Cornelius, C. McCloskey, J. Brash and J. Santerre, *J. Biomed. Mater. Res.* **60**, 135 (2002).
22. Y. Tang, J. Santerre, R. Labow and D. Taylor, *J. Biomed. Mater. Res.* **35**, 371 (1997).

23. R. Ward, *Med. Plastics Biomater.* **2**, 34 (1995).
24. R. Hergenrother, X. Yu and S. Cooper, *Biomaterials* **15**, 635 (1994).
25. F. Lim, C. Yang and S. Cooper, *Biomaterials* **15**, 408 (1994).
26. D. Martin, W. Poole, P. Gunatillake, S. McCarthy, G. Meijs and K. Schindhelm, *Biomaterials* **21**, 1021 (2000).
27. R. Ward, K. White, R. Gill and F. Lim, *ASAIO J.* **47**, 17 (1996).
28. R. Ward and K. White, US Patent No. 5,589,563 (1996).
29. J. Patel, M. Ebert, K. Stokes, R. Ward and J. M. Anderson, *J. Biomater. Sci. Polymer Edn* **14**, 279 (2003).
30. P. G. Kalman, C. A. Ward, N. B. McKeown, D. McCullough and A. D. Romaschin, *J. Biomed. Mater. Res.* **25**, 199 (1991).
31. A. K. McNally and J. M. Anderson, *Proc. Natl. Acad. Sci. USA* **91**, 10119 (1994).
32. A. McNally and J. M. Anderson, *Am. J. Pathol.* **147**, 1487 (1995).
33. M. Shive, W. Brodbeck, E. Colton and J. M. Anderson, *J. Biomed. Mater. Res.* **60**, 148 (2002).
34. K. DeFife, M. S. Shive, K. M. Hagen, D. L. Clapper and J. M. Anderson, *J. Biomed. Mater. Res.* **44**, 298 (1999).
35. C. Jenney, K. M. DeFife, E. Colton and J. M. Anderson, *J. Biomed. Mater. Res.* **41**, 171 (1998).

Gels, Genes, Grafts and Giants, pp. 457–471 (2005)

Oral administration of biochemically active microcapsules to treat uremia: new insights into an old approach

JILL A. O'LOUGHLIN *, JAN M. BRUDER and MICHAEL J. LYSAGHT

Center for Biomedical Engineering, Brown University, Box G-B395, Providence, RI 02912-G, USA

Received 3 December 2003; accepted 15 April 2004

Abstract—This paper begins with an extensive review of previous research on the degradation of non-protein nitrogen compounds for improved therapy of renal failure. During the 1970s, Malchesky established that naturally occurring strains of microorganisms were highly effective for the *in vitro* degradation of urea and other compounds found in urine, and that these bacteria could be conditioned with selected media to enhance growth and degradation efficiency. A few years later, Setala introduced the concept of oral delivery of lyophilized bacteria, harvested from soil, to uremic patients, for degradation of non-protein nitrogen compounds. In the 1990s, Chang proposed delivery of encapsulated genetically modified bacteria for removal of uremic waste products *in vitro* and *in vivo*. Recently, our group has pursued the idea of orally delivering formulated combinations of enzymes or modified bacteria. A new study is also described, which characterizes the capacity of a single alginate microcapsule containing a mixture of genetically modified cells and enzyme to degrade urea, uric acid and creatinine. The combination capsules were found to be effective *in vitro* and *in vivo* in a rodent model of chemically-induced renal failure. Reduction of urea concentration *in vivo* required co-administration of a cation exchange resin to adsorb ammonia. Increased investigative effort is warranted for these approaches which offer significant potential as an adjunct to conventional forms of dialysis.

Key words: Uremic toxins; encapsulation; solute removal; bacteria; enzymes; ESRD; degradation.

INTRODUCTION

The usual therapeutic objective of encapsulation is often to provide controlled and effective treatment of degenerative and disabling disorders, by using microcapsules to release a bioactive agent from a living cell within the capsule. Some examples include potential treatments for diabetes [1], liver failure [2] and pain [3]. In

*To whom correspondence should be addressed. Present address: Karolinska Institute, Division of Matrix Biology, Department of Medical Biochemistry and Biophysics, Karolinska Institutet, SE-171 77 Stockholm, Sweden. E-mail: Jill.Oloughlin@mbb.ki.se or Michael_Lysaght@brown.edu

contrast, the therapy format described in this paper involves oral delivery of a combination of encapsulated cells and enzymes as bioactive agents for metabolic degradation of harmful substances, in particular wastes which accumulate during renal failure.

The primary excretory function of the healthy kidney is removal of fluid and metabolic waste products. When the kidneys are damaged to an extent beyond which they can no longer maintain body fluid homeostasis, renal disease is described as having progressed to acute or chronic renal failure. Our work is focused on the chronic renal failure and the End Stage Renal Disease (ESRD) patient population. There are currently approx. 325 000 such patients in the USA and only about 14 000 receive a kidney transplant each year [4]. Worldwide, there are about 1.3 million chronic renal failure patients and approx. 30 000 annual kidney transplants. The remaining patients must undergo maintenance dialysis. 85% of chronic renal failure patients receive hemodialysis treatment, which is life sustaining and adequate, but far from an optimal therapy. The remaining 15% of the patient population is on continuous ambulatory peritoneal dialysis (CAPD), in which the patient manually exchanges the dialysate 4–5 times each day. Both hemodialysis and peritoneal dialysis are burdensome to perform, costly and incomplete in their rehabilitative effect. It is widely agreed that affordable and efficient alternative treatment options are highly desirable.

Nature has developed elaborate biochemical pathways to recycle urea and other nitrogenous metabolic waste products, including uric acid and creatinine. The most common of these is the nitrogen cycle. Nitrogen is an essential component of amino acids and proteins and could not be recycled in nature without certain bacteria which convert organic nitrogen to inorganic form. For example, bacterial species such as *Klebsiella aerogenes*, found in soil, contain the enzyme urease, which breaks down urea from animal waste into ammonia and carbon dioxide. The ammonia is subsequently converted into nitrates by nitrifying bacteria. The nitrates are taken up and metabolized by plants as amino acids and then proteins, which may serve as a food source for animals and eventually are broken down to urea repeating the cycle.

Further motivation to look at pathways in an *in vivo* model comes from hibernating bears, who have a unique capacity to recycle urea formed during their long hibernations [5, 6]. These animals appear in good clinical condition and are capable of going three to five months without eating or drinking and without excreting any waste products. Studies by Nelson [7, 8] reported that urea and uric acid blood concentration levels remain unchanged during hibernation. Urea continues to be produced from the breakdown of stored protein, but urea concentrations in blood and plasma do not increase because the animals have an as yet undefined pathway which recycles urea to more metabolically complex amino acids.

Our proposed therapy format is illustrated in Fig. 1. It consists of oral administration of microencapsulated enzymes and/or bacteria which take up and degrade urea, uric acid and creatinine (rather than remove them as in dialysis) while passing through the intestine. The targeted solutes are generated throughout the body,

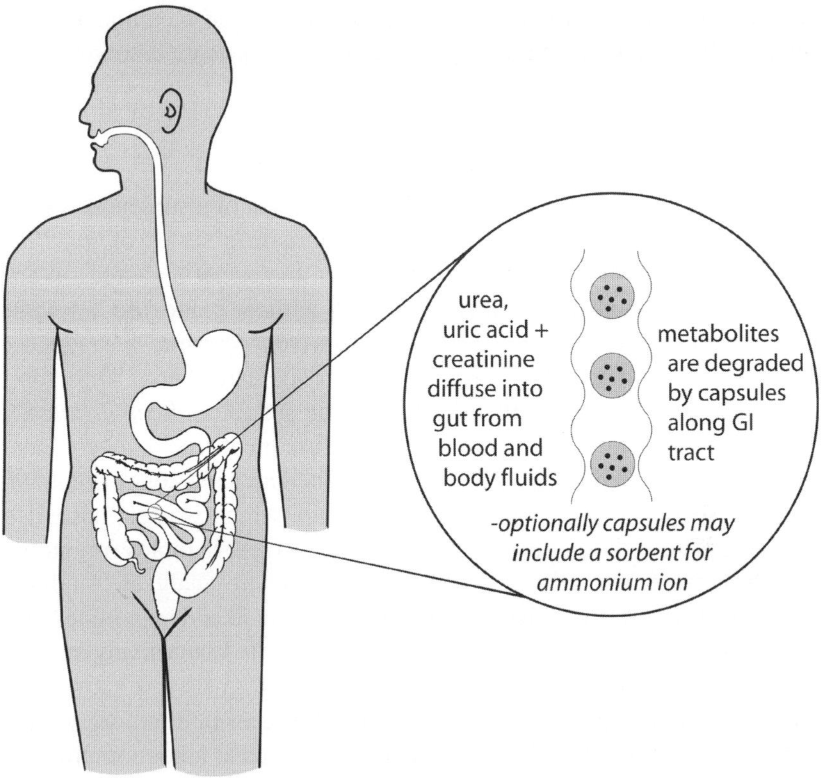

Figure 1. Proposed therapy format. The targeted metabolites urea, uric acid and creatinine diffuse into the gut and are degraded by the capsules along the gastrointestinal tract. Most degradation products are excreted as waste. One exception is ammonia, produced by the degradation of urea: it could potentially diffuse from the intestine to the liver and be converted back into urea. For this reason capsules may contain a sorbent for ammonia.

especially the liver, and diffuse into the capsules where they are degraded upon conversion of: (1) urea into ammonia and carbon dioxide (2) uric acid into allantoin and hydrogen peroxide and (3) creatinine and water into creatine. As described in Fig. 1, the degradation products are excreted as waste, with the possible exception of ammonia.

We are designing an adjunctive therapy which might be useful in conjunction with hemodialysis to decrease the frequency and duration of treatments, or to provide improved outcomes for existing therapeutic regimens. Alternatively, this treatment might allow a postponement of the initiation of dialysis in patients with early stage renal failure. This paper begins with an extensive review of previous research in the area of degradation of non protein nitrogen compounds, by the research groups of Malchesky, Setala and Chang. This body of work has not been reviewed previously, and the investigators largely worked independently of one another. This paper also includes a summary of our own previously published work, followed by a report with new data on the development and characterization of a combination capsule containing both cells and enzymes for degradation of waste metabolites.

REVIEW OF PUBLISHED RESULTS

In the mid-1970s, Malchesky quantified the degradation of nitrogenous metabolites [9–11] by selectively culturing wild type bacteria in a series of *in vitro* batch reactor trials (Fig. 2). To start the reactor system, the bacteria source, an activated sludge solution obtained from a local sewage treatment plant (in Cleveland, OH, USA) and containing approx. 5×10^5 mixed microorganisms, was added to 900 ml of normal concentrated urine from males on standard diets. Three dominant wild-type organisms were identified in the reactor mix (*Pseudomonas alcaligenes, Diphtheroid bacillus* and *Alpha streptococci*). Subsequent reactors were inoculated with bacterial mass taken from earlier reactors. The batch reactors were maintained for less than five days, with the exception of one, which was supported for two months by periodic addition of urine and bacteria species. Samples from the reactor and fresh urine were routinely analyzed on a clinical auto-analyzer. Analysis of 70 reactors showed significant consumption of urea, uric acid and creatinine: decreases over five days were on average of 74%, 64% and 28%, respectively, as shown in Fig. 3, while ammonia and sludge mass were generated.

Malchesky demonstrated that naturally occurring strains of microorganisms could be conditioned to be highly effective at degrading urea and other compounds found in urine. Wild-type cells cultivated in tissue culture supplemented with urine naturally adapted to degrade non-protein nitrogen compounds. He concluded that a mixture of several bacteria species would likely be required for the degradation of the most prominent nitrogenous wastes normally excreted in the urine.

In the late-1970s Setala advanced the concept of 'microbial anti-azotemic agents' by isolating new soil microbial enzyme systems, harvested from bacteria in soils rich in organic nitrogen, to convert nitrogenous wastes including urea, creatinine and other non-protein nitrogen compounds to essential nutrients [12, 13]. Bacterial strains were not identified or well characterized. The bacteria were lyophilized, packed in gelatin capsules and administered orally to a small number of uremic patients in short term, 10–14-day clinical trials (Fig. 2). The concentration of non-protein nitrogen compounds in the test subject's plasma was measured and decreases in blood urea and creatinine, were reported but not quantified. This research is important because it introduced the concept of orally delivering lyophilized bacteria to uremic patients, and using the gut as a 'bioreactor' for degradation of NPN compounds.

A third researcher, Chang, made major advances in this technology in the mid-1990s [14–21]. Using alginate and polylysine, Prakash and Chang encapsulated *Escherichia coli* which had been genetically modified to express urease and orally delivered capsules to 5/6-nephrectomized rats (male Wistar, 300–325 g) (Fig. 2). The duration of his study was 60 days with a 21-day treatment interval, bracketed

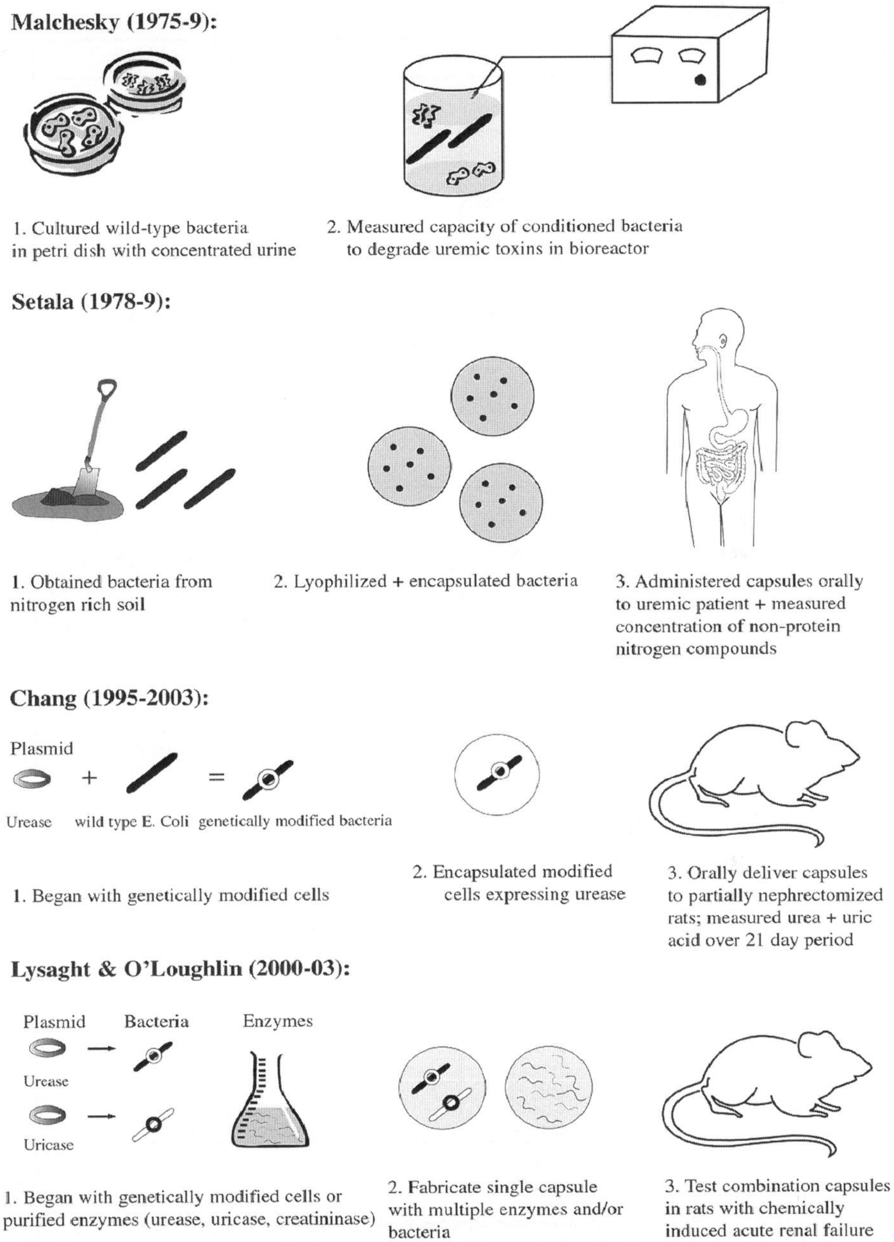

Figure 2. Previous research design. Wild-type bacteria were conditioned in a bioreactor and Malchesky observed decreases in urine components. Setala delivered encapsulated microbial anti-azotemic agents and found decreases in blood urea and creatinine; however, quantitative results were not reported. On a daily basis, Prakash and Chang delivered encapsulated urease cells to partially nephrectomized rats and dramatic decreases in urea and uric acid over 21 days were observed. Lastly, our group delivered a cocktail of encapsulated modified cells and enzymes to rats with chemically induced acute renal failure and found decreased hyperazotemia. See text for references to all four investigators' work.

A. Malchesky:

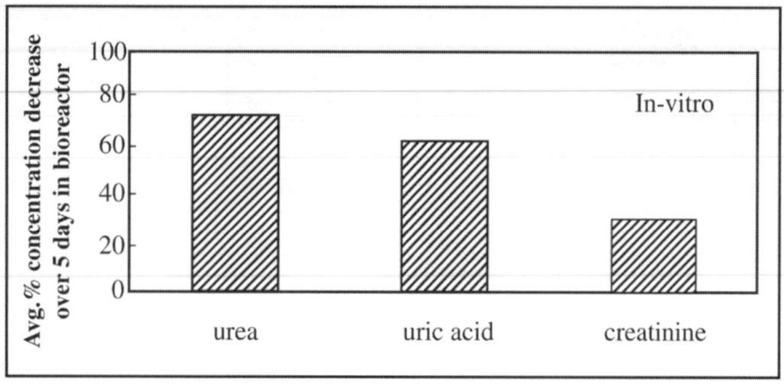

B. Chang:

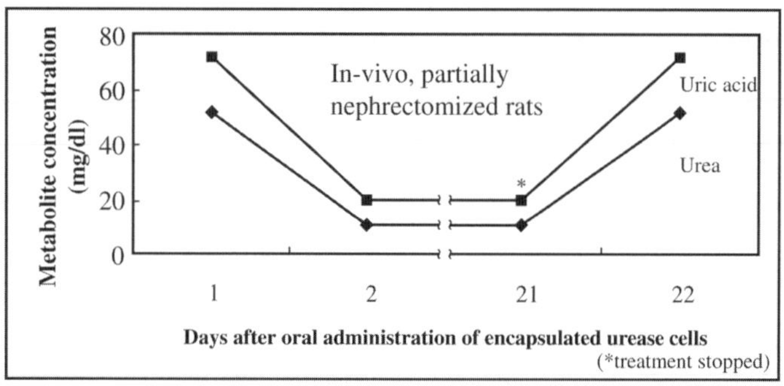

C. Lysaght & O'Loughlin

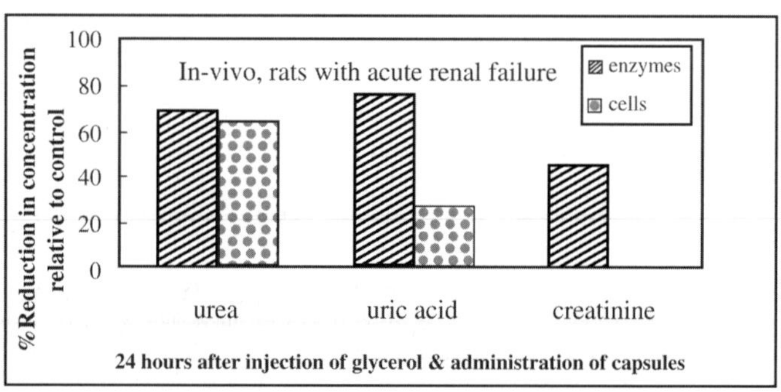

Figure 3. Previous research results. Malchesky showed significant percent decrease in urea, uric acid, and creatinine concentration *in vitro* over 5 days in a bioreactor charged with selected wild type bacteria. Prakash and Chang demonstrated lessening of elevated urea and uric acid concentration in a 21-day *in vivo* model with partially nephrectomized rats upon daily delivery of encapsulated cells genetically modified to contain urease, and return to normal after discontinuation of therapy. In an *in vivo* acute renal failure model in rats, we have reported significant lowering of elevated urea, uric acid and creatinine levels upon administration of capsules containing a combination of genetically modified cells or enzymes. Setala did not report quantitative results. See text for references.

by pre- and post-control periods. Rats were divided into three cohorts: unlesioned controls, lesioned controls receiving daily oral administration of empty capsules and lesioned study subjects receiving daily administration of capsules containing approx. 11 mg/kg body weight of encapsulated log-phase *E. coli* cells expressing urease. Significant and impressive reductions in circulating urea were reported in the animals receiving the genetically modified cells but not in the controls. As shown in Fig. 3, by the second day following administration of capsules, elevated urea levels were lowered from 52 mg/dl to 11 mg/dl by the encapsulated bacteria. When oral delivery was terminated, urea concentration returned to elevated levels. In contrast, in the lesioned and control cohorts' urea concentration stayed above 50 mg/dl 26–32 days after nephrectomy. The control group of unlesioned rats had a urea concentration of 10 mg/dl throughout. Prakash and Chang presented analysis suggesting that these procedures would be readily scalable to humans with reasonable and practical doses of cells and capsules.

Further related work by Chang includes the use of encapsulated enzymes for the degradation of nitrogenous metabolites. For example, a microcapsule containing the enzyme urease and zirconium phosphate was studied for reduction of urea concentration and adsorption of ammonia, and suggested that an oral sorbent might make it possible to decrease the dialysis treatment time from 3.5 to 1.8 h [22]. Prakash and Chang demonstrated the effectiveness of orally administered, genetically modified bacteria for uric acid degradation [23]. In both *in vitro* and *in vivo* studies uric acid levels were lowered upon administration of *E. coli* DH5 cells expressing urease. This finding is surprising, since the urease enzyme is specific for urea and no effect on uric acid is expected. No clear biochemical explanation was proposed for this finding. In most of the *in vivo* experiments, Prakash and Chang did not use an ion-exchange resin to absorb ammonia produced by urea breakdown; instead they suggested that bacteria utilized the ammonia as a nitrogen source. Overall, this group enhanced the development of bacterial degradation from an *in vitro* concept to a practical therapy format, successfully evaluated *in vivo*.

In earlier publications [24, 25] we have reported development of an alginate microcapsule containing either two strains of genetically modified bacteria, one expressing urease and the other uricase, or three isolated enzymes (urease, uricase and creatininase) capable of degrading, respectively, urea, uric acid and creatinine. *In vitro*, capsules incorporating cells or enzymes effectively degraded approx. 100% of the uric acid, 97% of the urea and 70% of the creatinine within 24 h in a test solution simulating the concentration of these solutes in uremic plasma. Enzyme degradation of urea followed Michaelis–Menten kinetics, and the Lineweaver–Burk plots for both encapsulated enzymes and unencapsulated controls were superimposable, indicating that mass transfer through the capsules was not rate limiting in the degradation process.

Additional *in vitro* trials confirmed cell specificity and demonstrated intracellular degradation. In contrast to Prakash and Chang's observations, in our hands cells containing urease had no effect on uric acid. Urea degradation was found to occur intracellularly. After 200 passages, both wild-type and genetically-modified cells cultured in media supplemented with urea and uric acid acquired no additional capacity to degrade either metabolite: unmodified and genetically modified *E. coli* DH5 cells could not be conditioned. These results suggest that, in the context of novel therapy development, genetic manipulation of cells is a more controllable and reliable technique than selection or conditioning.

In vivo, orally administered capsules were observed to remain in the gastrointestinal (GI) tract of male Wistar rats from about 2 to 7 h, and had essentially disappeared at 24 h [24]. Oral administration of capsules containing either cells or enzymes following induction of acute renal failure [26] reduced the severity of azotemia when a cation-exchange resin was co-administered. In addition, uric acid and creatinine concentrations were lowered from elevated levels whether or not ion-exchange resin was present.

THE PRESENT STUDY

Our previous studies have focused on the effects of either encapsulated cells or enzymes. In this report we describe development and characterization of a microcapsule containing a combination of both cells and enzymes. This new approach is important for several reasons. First, it is a convenient method for targeting our three metabolites of choice; since a genetically-modified cell expressing creatininase is not readily available, the isolated enzyme may be utilized. Secondly, it represents a model study for treatments combining removal of non-protein nitrogen compounds with other solutes, such as β_2 microglobulin, for which bacterial degradation is not likely to be practical. Finally, it serves to confirm that the combination of enzymes and cells in a single capsule, an approach which had not been previously evaluated, is practical and feasible. This approach permits greater design flexibility when targeting uremic metabolites and allows the most effective reagent to be utilized for each targeted uremic molecule.

Methods

As reported previously [24, 25], genetically-modified *E. coli* DH 5 cells expressing urease were obtained from S. Mulrooney (Michigan State University, East Lansing, MI, USA) [27] and *E. coli* JM109 cells expressing uricase from Y. Koyama (Kikkoman, Tokyo, Japan) [28]. Luria–Bertani (LB) growth medium was used for cell cultivation with a composition of 10 g/l sodium chloride (Sigma, St. Louis, MO, USA), 10 g/l tryptone and 5 g/l yeast (Difco, Franklin Lakes, NJ, USA). Cell concentration was obtained by measurement of optical density at 600 nm according to the following formula: 1 optical density unit (O.D.) $= 1 \times 10^6$ cells. Enzymes were sourced as follows: urease from jack beans, uricase from *Arthrobacter globiformis* and creatininase from *Flavobacterium* (all from Sigma).

A test solution consisting of metabolites at levels comparable to those found in the blood of kidney failure patients was prepared: urea (100 mg/dl) (Fisher Scientific,

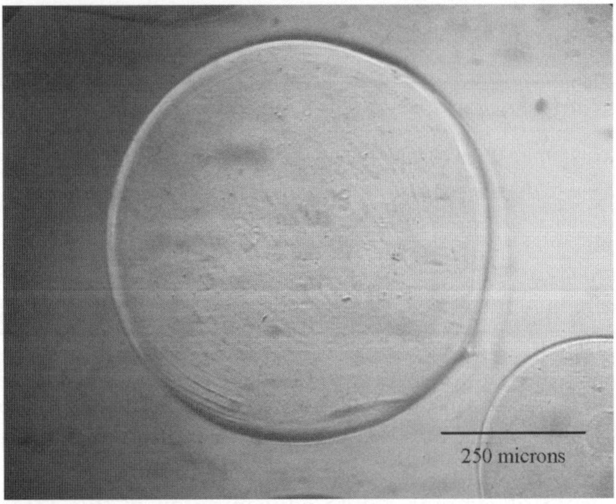

Figure 4. Microcapsule prepared from alginate, an anionic polysaccharide which forms hydrogels when placed in contact with and cross-linked by calcium chloride. The photo shows a light micrograph of an alginate bead, approx. 600 μm diameter.

Fairlawn, NJ, USA), uric acid (20 mg/dl) (Sigma) and creatinine (10 mg/dl) (Fluka, Buchs, Switzerland). Throughout, the reduction of the metabolites by the enzymes was measured by spectrophotometric assay (using kits 535, 686 and 555 from Sigma). All experiments were conducted at 37°C, solutions were maintained in an orbital shaker, and samples were stored in a −80°C freezer between collection and measurement, except for the final time point.

For kinetic experiments, 80 ml of each cell type were grown solely in LB medium (no repeated exposure to uremic components in the media as in Chang's protocol) along with ampicillin (1 mg/dl) overnight in a 37°C orbital shaker. Just prior to capsule fabrication, cell density was measured and the cells were centrifuged to form a pellet. Cells and enzyme were suspended in 5 ml of 1.8% alginate solution (low viscosity alginic acid, Sigma), loaded into a syringe and then extruded through a 300 μm diameter vibrational nozzle (frequency 1500 Hz), and formed into solid capsules upon contact with a 1.4% solution of calcium chloride (Sigma) (Fig. 4). This protocol, described in more detail elsewhere [29], resulted in capsules with approx. 600 μm diameter. Each capsule contained known concentrations of the two bacteria types, as well as the enzyme creatininase.

For *in vitro* studies alginate capsules (5 ml, approx. 50 000 capsules), containing 35 million urease cells, 20 million uricase cells and 40 units (0.14 mg) of creatininase, were added to 100 ml of the test solution. 1-ml samples from the solution were taken at intermediate points every 4 h from 0 to 12 h and finally at 24 h, and monitored for changes in metabolite concentration by assay. For dose–response studies, the study was repeated but the quantities of cells were reduced to 40 and 10% of the amounts used in the initial experiments and the enzyme concentration was reduced by 10- and 100-fold.

Table 1.
Composition of capsules delivered *in vivo*

Quantity per 2-mail dose (approx. 20 000 capsules)	Formulation A	Formulation B
No. of cells containing urease gene	65×10^6	65×10^6
No. of cells containing uricase gene	45×10^6	45×10^6
Urease (units)	1600	—
Uricase (units)	20	—
Creatininase (units)	60	60
Ion-exchange resin AG 50W-X8 (g)	3	—

In vivo, three cohorts of rats received identical intramuscular glycerol injection at a dose of 10 ml/kg body weight of 50% solution to produce acute renal failure. The first group of four rats received only the lesion but no capsules, thus serving as a measure of average solute concentration in this form of acute renal failure. The second group of two rats received an identical lesion, followed by administration of 2 ml of capsules (Table 1, Formulation A) containing a mixture of cells and enzymes, co-administered with cation-exchange resin AG 50W-X8 (Bio-Rad, Hercules, CA, USA). The third group of two rats received the lesion, followed by administration of 2 ml of capsules (Table 1, Formulation B) containing a mixture of cells and creatininase.

Except for the contents of the capsules, the overall procedure was identical to that previously reported by our group [24, 25]. Each cohort of rats was deprived of water overnight prior to glycerol injection. Capsules were administered immediately after glycerol injection. For oral delivery of the capsules, the rats were anesthetized using isoflurane gas and microcapsules suspended in maple syrup and fed through an orogastric tube. Rats were subsequently allowed free access to food and water. Rats were again sedated for blood collection. Small samples of 0.3 ml were collected from the tail at 0, 1, 3, 5 and 24 h for urea analysis. Larger samples of 0.7 ml were collected only at 0 and 5 h for determination of, uric acid and creatinine concentration. All samples were collected in heparinized tubes, centrifuged and the plasma analyzed by standard spectrophotometric methods. At the completion of the study, rats were killed in a CO_2 chamber. All animal experiments were conducted according to written protocols which had been approved in advance by Brown University's Institutional Animal Care and Use Committee (IACUC).

Results

Figure 5 demonstrates the rate of *in vitro* degradation of urea, uric acid and creatinine by encapsulated cells and enzyme exposed to the test solution for 24 h. The concentration of the test solution is plotted *versus* time at varying capsule dosages. The graph demonstrates effectiveness of metabolite degradation by a system in which cells and enzymes were co-encapsulated. 5 ml of these capsules completely cleared >99% of the urea, 100% of the uric acid and 58% of the

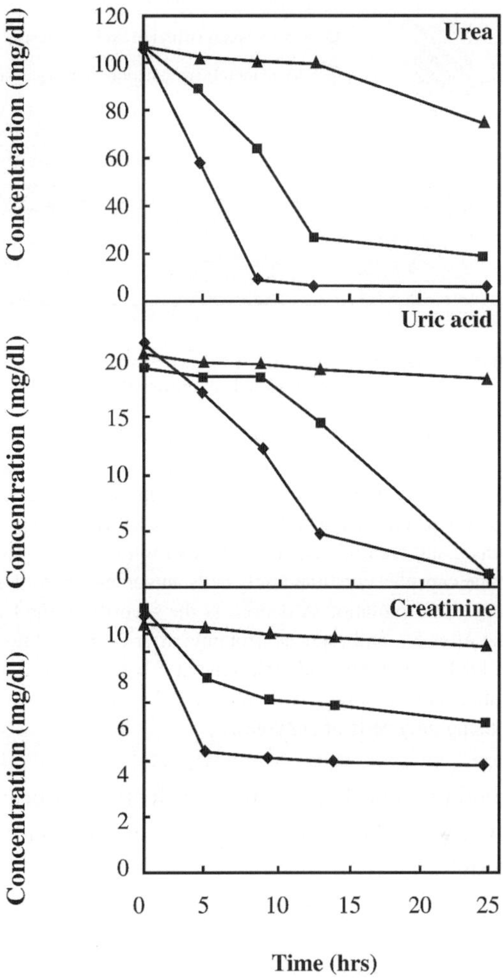

Figure 5. Rate of *in vitro* degradation of urea, uric acid and creatinine when a single 100 ml solution consisting of urea (100 mg/dl), uric acid (20 mg/dl) and creatinine (10 mg/dl) is exposed to capsules. The standard dose (◆) consists of 5 ml of capsules (approx. 50 000) containing an initial concentration of about 55 million cells (35 million *E. coli* DH5 and 20 million *E. coli* JM109 cells) and 40 units of creatininase (0.14 mg) added to 100 ml of test solution. (■) and (▲) represent 40% and 10% fewer cells and enzyme encapsulated in the same volume of alginate. Based upon single determinations.

creatinine from 100 ml of a challenge solution formulated to the concentration of these solutes in a presenting hemodialysis patient.

In *in vivo* trials with Wistar rats, orally administered capsules decreased the severity of azotemia, hyperuricemia and elevated creatinine following chemical induction of acute renal failure. Figure 6 contains a bar graph plotting the ratio of solute concentration at different time intervals in animals receiving active capsules to unlesioned controls for two different capsule formulations. Both capsule

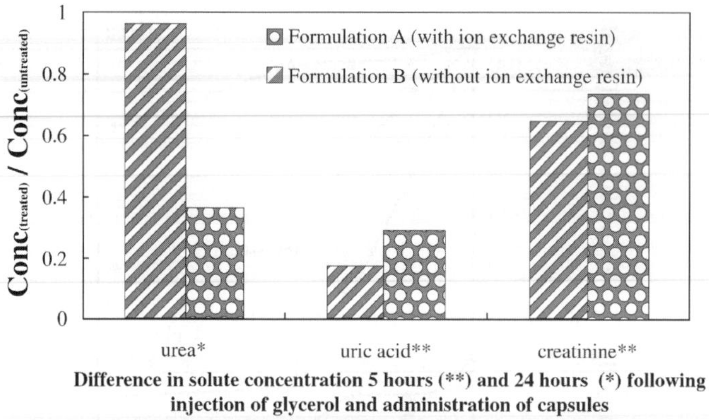

Figure 6. *In vivo* delivery of combination capsule. Circulating urea, uric acid and creatinine concentration was measured at 24 h post lesion and capsule delivery for urea and 5 h after for uric acid and creatinine. The results are reported as the ratio of concentration in treated rats to that in control rats. Two trials were performed: one with Formulation A, in which the capsules contained cells and the enzymes urease, uricase, and creatininase as well as an ion-exchange resin and the second with Formulation B, in which the capsules contained only cells and creatininase. Detailed formulations are given in Table 1. Delivery of Formulation A decreases the severity of the hyperuricemic condition, lowers elevated creatinine after 5 h and reduces azotemia after 24 h. Results after administration of Formulation B were similar for uric acid and creatinine at 5 h and showed no effect on azotemia at 24 h. The graph represents data from 8 rats: 4 controls and 2 each in the two study groups. See Refs [24, 25] for control data using only cells or enzymes.

formulations, described in Table 1, were highly effective for uric acid and creatinine degradation. However, only Formulation A, containing ion-exchange resin, was effective in lowering urea.

Discussion

This is the first report of a combination of modified cells and enzymes, co-encapsulated in a single capsule and shown capable of efficiently degrading the most abundant uremic toxins (urea, uric acid and creatinine) *in vitro* and lessening the elevation of the concentration of metabolites *in vivo*. This therapy format takes advantage of the intestinal tract as an effective route for the degradation of uremic waste. No apparent interference or blockage among the cells and enzymes was identified.

Figure 5 demonstrates a clear dose–response effect, with degradation rates decreasing with the quantity of cells and enzymes provided. As expected, the degradation rates greatly decreased with dilution of cell and enzyme quantity. Figure 6 confirms that *in vivo* metabolite concentrations are lowered from the elevated levels following induction of acute renal failure. In addition, a cation-exchange resin is required for effective urea removal, while having no impact upon uric acid or creatinine removal. This is likely because free (unadsorbed) ammonia could diffuse from the intestine to the liver where it would be converted back to urea by a complex enzymatic pathway. However, the sorbent does not affect uric acid or creatinine degradation, since ammonia is not a by-product of the enzymatic degradation of these compounds. Ion-exchange resins are not the only sorbent for urea, other possibilities include: oxystarch and zirconium phosphate.

In this therapy format, cells and enzymes each have their own advantages and disadvantages. Enzymes purchased from a commercial source are less complicated in terms of biocompatibility, immune reactions and overall safety concerns. They require much less effort in storage, packaging, and transportation and are likely to need less time to obtain FDA approval. However, bacteria can grow and expand during their passage through the gut and they might be able to metabolize some of the breakdown products from the enzymatic reaction. Bacteria are readily available, since there are easy to grow quickly in large amounts and less expensive than enzymes. Furthermore, bacteria have the advantage of metabolizing uremic compounds intracellularly, so that the metabolizing enzymes are protected from the external environment.

In the combination capsule described, cells and enzymes displayed no synergistic or antagonistic effects. The results, in terms of metabolite degradation were similar to our previous findings. Creatininase has operated effectively in both the presence of enzymes (see our previous study [24]) and cells (the present study); in both cases the normal rise in creatinine following induction of renal failure was reduced by about 30–45%. This capsule design could be useful for delivery of agents for which genetically modified cells are not available or difficult to create, such as β_2-microglobulin. A mixture of cells and enzymes in a delivery system will allow for greater flexibility in future therapy development.

Follow-up studies

Several approaches could be pursued to enhance the design of the capsules. First, bacteria or enzymes targeted toward other molecules elevated in uremia, such as β_2-microglobulin, could be identified and added to the capsule. Secondly, rather than simply adsorb urea, bacteria capable of recycling ammonia into amino acid precursors could be designed perhaps helping to counteract the malnutrition which accompanies renal failure.

Further *in vivo* characterization and testing include maintenance experiments which would demonstrate clinical benefit, rather than simple reduction of metabolite concentration. One such experiment would be based on the 5/6 nephrectomy model in which progression of chronic renal failure *via* the 'final common pathway' could be compared in rats receiving capsules with lesioned controls. Metabolic cage studies could be performed to allow long-term observation of the effect of capsule administration on nitrogen balance in chronically uremic animals. Eventually, oral administration of capsules needs to be evaluated in large animals of chronic renal failure.

CONCLUDING PERSPECTIVES

Research based on the degradation of non-protein nitrogen compounds prominent in uremia has been approached in many ways, including the use of wild-type bacteria *in vitro* by Malchesky, lyophilized bacteria delivered to patients in gelatin capsules by Setala and genetically modified bacteria administered to uremic rats by Chang and ourselves. Taken together the studies convincingly demonstrate the ability of bacteria (either wild-type and conditioned, or genetically modified) and enzymes to degrade non-protein nitrogen compounds *in vitro* or when orally delivered in capsule form *in vivo*. In newly reported experiments, a combination capsule containing both free enzymes and bacteria expressing urease and uricase effectively degraded target metabolites *in vitro* and *in vivo* over 24 h, with co-administration of an ion-exchange resin being necessary for *in vivo* removal of urea. Continued development of oral delivery of encapsulated cells and enzymes as an adjunctive therapy in the treatment of ESRD is highly warranted.

Acknowledgements

J. A. O'L. was the recipient of the P. Galletti Fellowship sponsored by Sorin Biomedica. Special thanks to Scott Mulrooney, PhD and Yasuji Koyama, PhD for the generous donation of engineered cells, to Roxanne Burrill and Narongrit Our for animal care assistance and to David Humes, MD for advice on the animal model.

REFERENCES

1. A. Sambanis, *Diabetes Technol. Ther.* **5**, 665 (2003).
2. Z. C. Liu and T. M. S. Chang, *Artif. Cells Blood Substit. Immobil. Biotechnol.* **30**, 99 (2002).
3. H. Uludag, P. De Vos and P. Tresco, *Adv. Drug Deliv. Rev.* **42**, 29 (2000).
4. United Network for Organ Sharing, accessible at http://www.unos.org (2003).
5. R. Nelson, H. Wahner, J. Jones, R. Ellefson and P. Zollman, *Am. J. Physiol.* **224**, 491 (1973).
6. R. Nelson, J. Jones, H. Wahner, D. McGill and C. Code, *Mayo Clin. Proc.* **50**, 141 (1975).
7. R. Nelson, *Kidney Int.* **13**, S177 (1978).
8. R. Nelson, *Fed. Proc.* **39**, 2955 (1980).
9. P. S. Malchesky and Y. Nose, *Cleveland Clin. Q.* **42**, 267 (1975).
10. P. S. Malchesky, *Trans. Am. Soc. Artif. Intern. Organs* **23**, 726 (1977).
11. P. S. Malchesky and Y. Nose, *Artif. Organs* **3**, 8 (1979).
12. K. Setala, *Clin. Nephrol.* **11**, 156 (1979).
13. K. Setala, *Kidney Int.* (Suppl.) **13**, S194 (1978).
14. T. M. S. Chang, in: *Biotechnology Annual Review*, Volume I, M. R. El-Gewely (Ed.), p. 267. Elsevier, Elkridge, MD (1995).
15. S. Prakash and T. M. S. Chang, *Biotechnol. Bioeng.* **46**, 621 (1995).
16. S. Prakash and T. M. S. Chang, *Nature Med.* **2**, 883 (1996).
17. S. Prakash and T. M. S. Chang, *Artif. Cells Blood Substit. Immobil. Biotechnol.* **26**, 35 (1998).
18. T. M. S. Chang, *Artif. Organs* **22**, 958 (1998).
19. T. M. S. Chang and S. Prakash, in: *Handbook of Cell Encapsulation*, W. M. Kuhtreiber, R. P. Lanza and W. L. Chick (Eds), p. 379. Birkhauser-Springer, Boca Raton, FL (1999).
20. T. M. S. Chang, *Ann. N.Y. Acad. Sci.* **875**, 71 (1999).

21. T. M. S. Chang, *Blood Purif.* **18**, 91 (2000).
22. E. Wolfe and T. M. S. Chang, *Int. J. Artif. Organs* **10**, 269 (1987).
23. S. Prakash and T. M. S. Chang, *Int. J. Artif. Organs* **23**, 429 (2000).
24. J. A. O'Loughlin, J. M. Bruder and M. J. Lysaght, *ASAIO Trans.* **50**, 253 (2004).
25. J. A. O'Loughlin, J. M. Bruder and M. J. Lysaght, *Tissue Eng.* **10** (2004) (in press).
26. D. Wilson, G. Thiel, M. Arce and D. Oken, *Nephron* **4**, 337 (1967).
27. S. Mulrooney, H. Pankratz and R. Hausinger, *J. Gen. Microbiol.* **135**, 1769 (1989).
28. Y. Koyama, T. Ichikawa and E. Nakano, *J. Biochem.* **120**, 969 (1996).
29. J. A. O'Loughlin and M. J. Lysaght, in: *Proceedings of the 28th Meeting of the Controlled Release Society*, p. 694, Abstract 6042 (2001).
30. M. J. Lysaght and J. A. O'Loughlin, *ASAIO Trans.* **46**, 515 (2000).

Gels, Genes, Grafts and Giants, pp. 473–489 (2005)

Formulation of chitosan-DNA nanoparticles with poly(propyl acrylic acid) enhances gene expression

TINA KIANG [1], CORINNE BRIGHT [1], CHARLES Y. CHEUNG [2], PATRICK S. STAYTON [2], ALLAN S. HOFFMAN [2] and KAM W. LEONG [1,*]

[1] *Department of Biomedical Engineering, Johns Hopkins University, Baltimore, MD 21205, USA*
[2] *Department of Bioengineering, University of Washington, Seattle, WA 98195, USA*

Received 19 November 2003; accepted 5 April 2004

Abstract—Poly(propyl acrylic acid) (PPAA) is a polymer specifically designed to disrupt lipid bilayer membranes within a sharply defined pH range. The pH sensitivity can be used to enhance the release of endocytosed drugs into the cytoplasmic compartment of the cell. By incorporating this polymer in a polymeric gene carrier, chitosan, the release of plasmid DNA from the endosomal compartment was enhanced. *In vitro* transfection studies confirmed that the incorporation of PPAA into the chitosan-DNA nanoparticles enhanced gene expression in both HEK293 and HeLa cells compared to chitosan nanoparticles alone. The dose and time at which PPAA was incorporated during the complex formation affected the release of DNA and transfection efficiency. The optimal dose of PPAA incorporated into the chitosan nanoparticles was determined to be 10 μg, corresponding to a PPAA/DNA weight ratio of 1 : 1. At this dose, the ternary complexes are approx. 400 nm in size with a net negative surface charge of -17.4 mV. Intracellular trafficking studies confirmed the association of PPAA, DNA and chitosan at 24 h post-transfection and the subsequent release of DNA and PPAA from the chitosan at 48 h. The diffuse appearance of the majority of the DNA and the PPAA at later time points suggests that the PPAA triggered membrane disruption resulting in the release of DNA from the endosomal compartment. Finally, the lack of colocalization between PPAA and Lysotracker indicated that the PPAA-loaded nanoparticles were not trafficked through a lysosomal pathway. This study suggests the promising strategy of including PPAA in the formulation of polymer–DNA complexes for non-viral gene delivery.

Key words: Poly(propyl acrylic acid); chitosan; DNA; gene therapy.

INTRODUCTION

Non-viral gene therapy has shown clinical potential for the safe and biocompatible delivery of genes, both locally and systemically [1, 2]. However, both lipid and polymer-based gene carriers suffer from low transfection efficiency *in vivo*. The

*To whom correspondence should be addressed. E-mail: kleong@bme.jhu.edu

low efficiency is the result of a series of barriers that the complexes encounter upon internalization, one of the most important of which is release of DNA from the endosome to the cytoplasm [3]. One approach to trigger release of the DNA from the endosome is to incorporate a membrane disruptive peptide, drug, or polymer in the gene delivery system. Poly(propyl acrylic acid) (PPAA), a highly pH-sensitive polymer, exhibits maximal membrane disruption ability at below pH 6.0 in red blood cells, but is non-lytic at pH 7.4 [4]. The pendant propyl group is thought to confer the unique pH-dependency and hydrophobic membrane destabilizing properties [5]. Within the acidic endosomal environment, protonation of the carboxylic groups leads to a more hydrophobic polymer that can disrupt the endosomal membrane to release the vesicle contents into the cytosol. PPAA has recently been demonstrated to selectively disrupt late endosomes, not lysosomes [6]. The effectiveness of this polymer in enhancing cytoplasmic delivery has been established both for drugs and genes. PPAA releases the small molecule calcein (622 Da) [6] and the large biotinylated antibody/streptavidin/biotinylated PPAA complex (approx. 265 kDa) [7] from vesicles to effectively bypass the lysosomal trafficking and degradation of the drugs. PPAA has also enhanced the gene expression in several lipid-based gene-delivery systems [5, 6, 8, 9].

Chitosan is a biodegradable polymer that has shown promise as a non-viral gene carrier *via* mucosal routes [10, 11]. Chitosan nanoparticles are taken up primarily through non-specific endocytosis, at which point they are localized to endosomes. Chitosan nanoparticles formulated with 83% deacetylation are found to stably bind DNA, even after several days in cell culture (unpublished data). Unlike polyethylenimine (PEI), which buffers the low pH environment of the endosome (proton sponge effect) [12], chitosan does not have a high buffering capacity [13] and therefore may benefit from the addition of a membrane disruptive agent. The inclusion of PPAA in the formulation of the chitosan-DNA nanoparticles may provide a means to facilitate endosomal disruption. We also hypothesized that the PPAA might reduce the stable association between chitosan and DNA to facilitate DNA release.

In this study, the size, zeta potential and stability of PPAA/chitosan/DNA ternary complexes were investigated, as well as the ability of PPAA to enhance the transfection efficiency of chitosan-DNA nanoparticles. Intracellular trafficking studies were employed to provide insight into the mechanism by which the PPAA enhances transfection.

MATERIALS AND METHODS

PPAA synthesis

Poly(propylacrylic acid) (PPAA) was synthesized by chain transfer free radical polymerization according to the protocols published previously [4, 5]. Polymerizaton of propylacrylic acid was performed using 2,2′-azobibisobutyronitrile (AIBN) as the initiator and cystamine as a chain-transfer agent. Polymerization reactions were performed at 60°C for 24 h. The resulting polymer was dissolved in methanol and precipitated three times using diethyl ether. The average molecular mass of PPAA was 38 kDa, as determined by GPC using PEG standards.

Plasmids

The plasmid pEGFP-C1 encoding green fluorescent protein (GFP), a 4.7-kb pcDNA driven by a cytomegalovirus immediate early promoter (Clontech, Palo Alto, CA, USA), was selected for the determination of percent cells transfected. The plasmid pVR1255 encoding firefly luciferase, a 6.4-kb pcDNA driven by a cytomegalovirus promoter (kindly provided by Dr. Carl Wheeler, Vical, San Diego, CA, USA), was selected for the analysis of the level of gene expression. The plasmids were expanded in an *Escherichia coli* DH5α strain and isolated using a Qiagen Giga-Prep kit according to the suppliers' protocol (Qiagen, Missisauga, Ontario, Canada). The quantity and quality of the purified plasmid was assessed by UV spectroscopy at 260 and 280 nm and by electrophoresis in 1% agarose gel. The purified plasmids were resuspended in sterile deionized water and stored in aliquots at a concentration of 4–5 mg/ml. The pGeneGrip Rhodamine/GFP Vector (Gene Therapy Systems, San Diego, CA, USA) was selected for *in vitro* trafficking experiments.

Preparation of chitosan-DNA nanoparticles

Chitosan-DNA nanoparticles (binary complexes) were prepared as previously described [13]. Chitosan (390 kDa), kindly donated by Vanson (Redmond, WA, USA), was used from a 0.02% (w/v) stock solution in 5 mM sodium acetate buffer (pH 5.5). The chitosan solution and a DNA solution of 100 μg/ml in 50 mM of sodium sulfate were heated to 50–55°C separately. Equal volumes of both solutions were quickly mixed under high vortexing for 30 s. Various doses of PPAA (0.01, 0.1, 1, 10, 50 μg, a molar range of 0.087 to 435 nmol) were added to the mixture for evaluation. PPAA was incorporated into the ternary PPAA-chitosan-DNA complexes by two different methods: prior to nanoparticle formation (pre-addition), where PPAA solution was added to the DNA-sodium sulfate solution, or after nanoparticle formation (post-addition), where PPAA solution was added during the final 5 s of particle formation while still under vortexing. The amounts of chitosan and DNA used remained constant for all formulations.

Size and zeta potential measurements

Particles were characterized by size and zeta potential using the Zetasizer 3000 (Malvern Instruments, Southborough, MA, USA). Size measurements were performed at 25°C at 90° scattering angle and recorded for a time determined by the zetasizer program in automatic mode. Number average measurements were obtained. Zeta potential measurements were performed using the aqueous flow cell in the automatic mode at 25°C.

DNA retention and release

DNA retention in nanoparticles was assessed by gel electrophoresis in 0.8% agarose (Invitrogen, Carlsbad, CA, USA) in Tris-acetate EDTA (TAE) buffer with ethidium bromide. Gels were run at 100 mV for 1 h and DNA retention was visualized under UV light. All nanoparticle formulations were tested in concentrated form to assess DNA retention immediately after particle formation. For DNA release studies, nanoparticle solutions were incubated with an equal volume of either phosphate-buffered saline (PBS) or cell culture medium (minimum essential medium with 10% fetal bovine serum) for 3, 24 or 72 h and then analyzed by gel electrophoresis under the conditions described.

Fluorescence labeling

A 0.1% chitosan (w/v) solution in 25 mM sodium acetate was incubated with rhodamine-B isothiocyanate (RITC) (Sigma, St. Louis, MO, USA) at a ratio of 1 fluorescent molecule per 10 chitosan amine groups for 2 h at room temperature. The solution was dialyzed against deionized water at 4°C for 3 days, with at least 3 changes to isolate the labeled chitosan from the unreacted fluorophore. The acetic acid concentration was adjusted back to 25 mM after dialysis was completed. The final chitosan concentration was determined by the ninhydrin assay based on the standard curve for unreacted chitosan. Amine-terminated PPAA (4.5 mg) was labeled with 4× molar excess of Oregon Green succinimidyl ester ($\lambda_{ex} = 495$ nm, $\lambda_{em} = 521$ nm) in 100 mM sodium bicarbonate buffer (pH 8.5) for 2 h. The reaction was quenched by the addition of 1.5 M hydroxylamine. The Oregon Green-labeled PPAA was purified by size-exclusion chromatography using a PD-10 desalting column with 50 mM sodium bicarbonate (pH 8.5) as the elution buffer. The first 4 ml collected from the column were pooled and lyophilized, then re-dissolved at a 1 mg/ml concentration. The labeling efficiency was determined by quantifying the amine content of the PPAA before and after the labeling reaction using the TNBSA reagent. Approximately 31% of the amines within PPAA were labeled as determined by TNBSA with glycine calibration standards. Oregon-Green-labeled PPAA was used at a concentration of 1 mg/ml in 50 mM sodium bicarbonate buffer (pH 8.5).

In vitro gene expression

The cells used for *in vitro* gene expression studies were human embryonic kidney 293 cells (HEK293) and human cervical adenocarcinoma cells (HeLa) from American Type Culture Collection (Manassas, VA, USA). 293 cells were cultured in minimum essential medium with 10% heat inactivated horse serum and HeLa cells in minimum essential media with 10% fetal bovine serum, both at 37°C and 5% CO_2. pcDNA encoding for luciferase was used for these studies. Replicate wells ($n = 8$) of 60–70% confluent cells, in 24-well plates, were incubated with the nanoparticles containing 1 μg of pcDNA for 4 h in Opti-MEM (Gibco) at 37°C, after which the Opti-MEM was replaced with full growth medium. The cells were assayed for luciferase expression after 72 h using standard protocols (Promega). Luciferase levels were normalized against total protein content determined by BCA Protein Assay Kit (Pierce, Rockford, IL, USA).

In vitro trafficking

Labtech II chambered coverglass (Nalge Nunc International, Rochester, NY, USA) were seeded with HeLa cells two days prior to nanoparticle transfection. Nanoparticles were synthesized using rhodamine-labeled chitosan, Oregon-Green-labeled PPAA and unlabeled plasmid DNA or the combination of unlabeled chitosan, Oregon-Green-labeled PPAA and rhodamine-labeled plasmid DNA (pGeneGrip rhodamine/green fluorescence protein (GFP), Gene Therapy Systems, San Diego, CA, USA). Pre-addition PPAA dose of 0.1 μg and post-addition PPAA dose of 1 μg were used for these studies. Intracellular particle translocation and GFP expression were assessed 1 and 2 days after transfection. Lysotracker Red DND-99 (Molecular Probes, Eugene, OR, USA), used to observe acidic vesicles, and Hoechst nuclear stain were incubated with the cells during the final 15 min prior to observation by laser confocal microscopy (Carl Zeiss, Chester, VA, USA).

RESULTS

PPAA-chitosan-DNA nanoparticle characterization

PPAA was incorporated in the nanoparticles either pre- or post-nanoparticle formation, hereafter referred to as pre- and post-addition. During pre-addition nanoparticle formation, PPAA was incorporated within the nanoparticles *via* charge interaction with the chitosan, while in post-addition nanoparticle formation PPAA coated the surface of the formed nanoparticles, utilizing the remaining charged amines on the nanoparticle surface.

Average particle size of the ternary nanoparticles increased compared to the binary nanoparticles at high doses of PPAA. Pre-addition of 10 or 50 μg of PPAA dramatically increased the particle size, resulting in aggregation and precipitation of the particles, while the addition of 1, 0.1 or 0.01 μg of PPAA had little effect on particle size (Fig. 1) compared to chitosan-DNA nanoparticles alone. Post-addition of 10 or 50 μg of PPAA to nanoparticle formulations also increased average particle size, although not as dramatically as the pre-addition. Similar to pre-addition, post-addition of 1, 0.1 or 0.01 μg of PPAA resulted in little change in the average particle size.

The zeta potentials of the nanoparticles after the pre- or post-addition of small amounts of PPAA (1, 0.1 or 0.01 μg) were not significantly different than that of chitosan-DNA nanoparticles alone (15 mV; Fig. 2a). The minimal impact

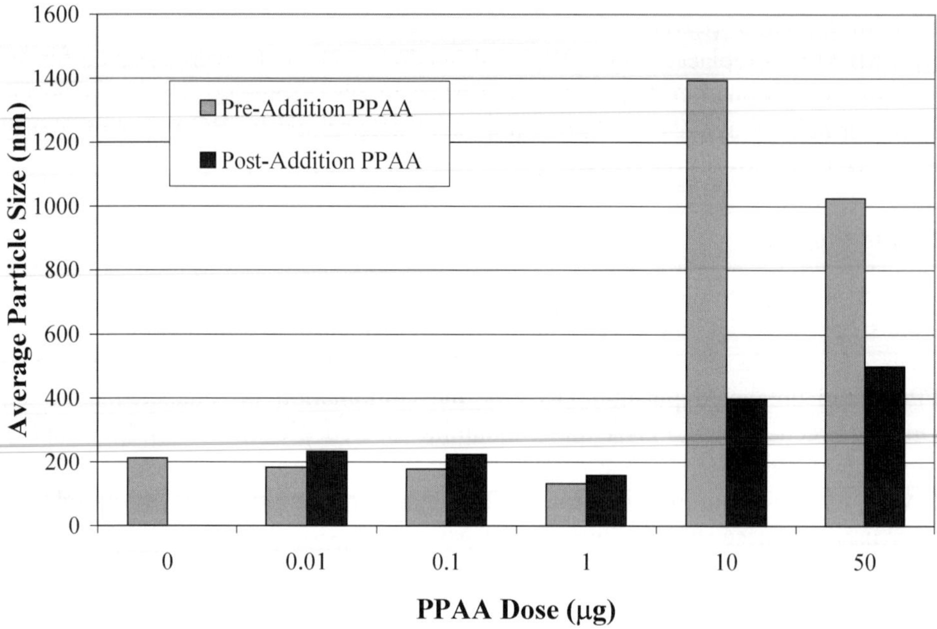

Figure 1. Number average particle size of ternary (PPAA-chitosan-DNA) nanoparticles after the pre- or post-addition of PPAA based on dose of PPAA added to the formulation, where a dose of 0 μg PPAA is the chitosan-DNA binary complex.

on zeta potential is in agreement with the theoretical net charge ratio, which is approximately that of the binary complexes. In contrast, the addition of 10 or 50 μg of PPAA shifted the zeta potential from positive to negative values, coinciding with the theoretical calculation (Fig. 2b). Furthermore, for the same PPAA doses at 10 and 50 μg, the net negative zeta potentials were significantly greater with post-addition than pre-addition (Fig. 2a), consistent with the expectation that more PPAA would be localized at the nanoparticle surface in the former case.

The DNA complexation efficiency of the ternary nanoparticles was tested *via* gel electrophoresis immediately after particle formation ($t = 0$ h). Complete DNA retention was observed with all nanoparticle formulations after the pre- and post-addition of PPAA, with the exception of complexes formed after the pre-addition of 50 μg of PPAA (Fig. 3). The DNA release from the nanoparticle formulations was assessed after incubation with either PBS or complete cell culture medium at 3, 24, 72 h and 10 days (Fig. 4). At all time points, DNA was retained in all nanoparticle formulations incubated in PBS with the exception of the complexes formed after the pre-addition of 10 or 50 μg of PPAA (Fig. 4a). The DNA released from the PBS incubated complexes remained non-degraded for the duration of the study. Incubation of nanoparticle formulations in complete culture medium resulted in both an increase in DNA release and degradation of the released DNA. At 3 h, DNA release was evident in all pre-addition PPAA complexes, with increasing DNA release correlating with higher PPAA doses (Fig. 4b). DNA was released

a

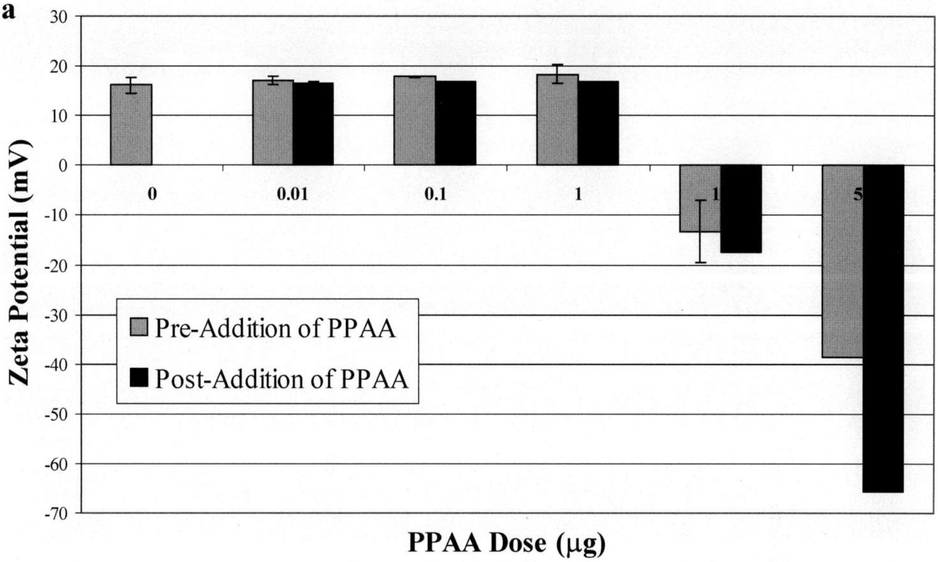

b

PPAA dose (μg)	Net chitosan : PPAA : DNA charge ratio	Total net +/− ratio
0	3.3 : 0 : 1	3.30 : 1
0.01	3.3 : 0.0029 : 1	3.29 : 1
0.10	3.3 : 0.029 : 1	3.21 : 1
1.0	3.3 : 0.29 : 1	2.53 : 1
10	3.3 : 2.9 : 1	0.83 : 1
50	3.3 : 14.5 : 1	0.21 : 1

Figure 2. Zeta potential and charge ratios of PPAA-chitosan-DNA complexes based on dose of PPAA added. (a) Zeta potential values of chitosan nanoparticle after the pre- or post-addition of PPAA. (b) Net charge ratios of ternary complexes with PPAA addition.

from post-addition PPAA complexes at doses of 1, 0.1 or 0.01 μg of PPAA but not at the higher doses of 10 and 50 μg (Fig. 4b). At 24 h, the trend of DNA release from all pre-addition complexes was consistent with observations at 3 h and DNA was released from all post-addition PPAA complexes (Fig. 4c). DNA degradation became detectable at this time, as evidenced by the loss of integrity of the DNA bands. At 72 h, the DNA released from all pre- and post-addition formulations was degraded (Fig. 4d), although a substantial portion of DNA still remained complexed with chitosan, with the exception of the pre-addition PPAA dose of 50 μg, which already showed significant release. At day 10, continued DNA release was observed, following the trend that the cumulative amount released increased with the PPAA doses and pre-addition induced higher release than the post-addition (Fig. 4e). In contrast, chitosan nanoparticles alone were not observed

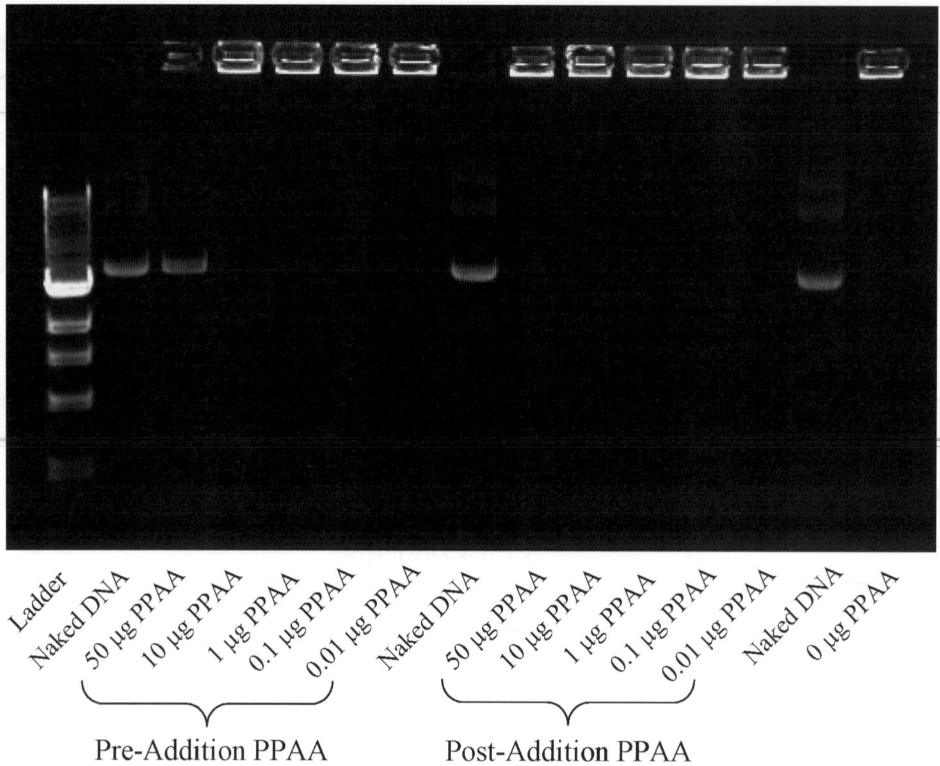

Figure 3. DNA retention in ternary complexes with the pre- or post-addition of PPAA.

to release DNA at any time point. At day 10, the DNA released into the culture medium was completely degraded.

In vitro gene expression

All PPAA-chitosan-DNA nanoparticle formulations were tested for gene transfection efficiency in HEK293 and HeLa cells. Replicate wells of cells ($n = 8$) were incubated with the nanoparticle formulations for 4 h and assayed for luciferase gene expression 3 days post-transfection. The pre-addition of 0.1 μg PPAA to the nanoparticles enhanced luciferase gene expression in HEK293 cells approx. 1.75 times over the unmodified chitosan-DNA nanoparticles (Fig. 5). This was the only dose in the case of pre-addition that had some enhancing effect on transfection efficiency. The pre-addition of higher doses of PPAA (10, 50 μg) resulted in a significant negative effect on transfection efficiency, which may be due to particle precipitation mentioned earlier. In contrast, the post-addition of PPAA to the nanoparticles enhanced the luciferase expression in HEK293 cells at all PPAA doses compared to unmodified chitosan-DNA nanoparticles, with a broad peak at a dose of 10 μg of PPAA (Fig. 5). Expression levels after the post-addition of 10 μg PPAA were similar to that obtained with the pre-addition of 0.1 μg PPAA. Typically, unmodified

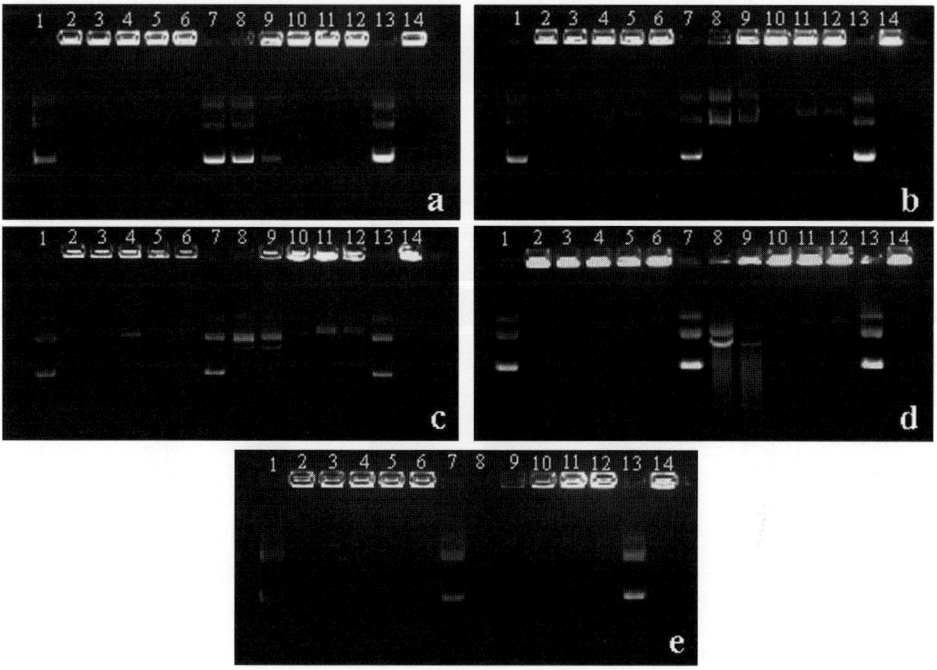

Figure 4. DNA release from ternary complexes after incubation in serum containing cell culture medium. For all gels, lanes 1, 7 and 13 are naked DNA controls, lanes 2–6 are post-addition PPAA nanoparticles in doses of 50, 10, 1, 0.1 and 0.01 μg, respectively, lanes 8–12 are pre-addition PPAA nanoparticles in doses of 50, 10, 1, 0.1 and 0.01 μg, respectively, lane 14 is chitosan-DNA nanoparticles with no PPAA added. (a) DNA release from formulations incubated in PBS for 3 h. Gels for all time points of PBS incubation appear similar to the gel at 3 h, with no additional DNA release. DNA release from nanoparticle formulations after incubation with MEM + 10% FBS is shown for (b) 3 h, (c) 24 h, (d) 72 h and (e) 10 days.

chitosan nanoparticles yield only modest luciferase gene expression in HeLa cells; therefore, Hela cells were selected to assess the effectiveness of PPAA to enhance transfection. The pre-addition of PPAA did not result in a significant enhancement in luciferase expression at any PPAA dose (Fig. 6). However, the post-addition of PPAA at doses of 10 and 50 μg increased gene expression by 11- and 13-fold, respectively, compared to unmodified chitosan nanoparticles.

Confocal microscopy

The trafficking of rhodamine-labeled (Rh-) chitosan nanoparticles containing Oregon-Green-labeled (Gr-) PPAA in HeLa cells was observed through confocal microscopy at 4, 24 and 48 h. At 4 h post-transfection with pre-or post-addition PPAA nanoparticles, Gr-PPAA was co-localized with Rh-chitosan nanoparticles in punctate perinuclear vesicles (data not shown). The intracellular distribution was consistent with that observed with unmodified Rh-chitosan nanoparticles. At 24 h, the fluorescence distribution of PPAA and chitosan was not significantly

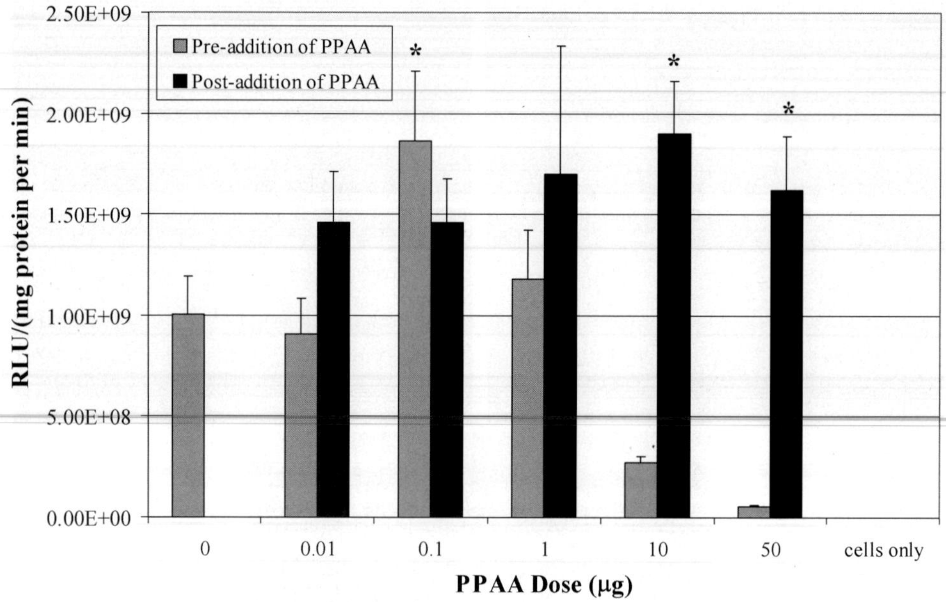

Figure 5. Luciferase gene expression in HEK293 cells after the delivery of PPAA-chitosan-DNA ternary complexes, 3 days post-transfection ($n = 8$). The error bars represent S.E.M. $^*P < 0.05$ as determined by paired Student's t-test compared to PPAA dose of 0 μg.

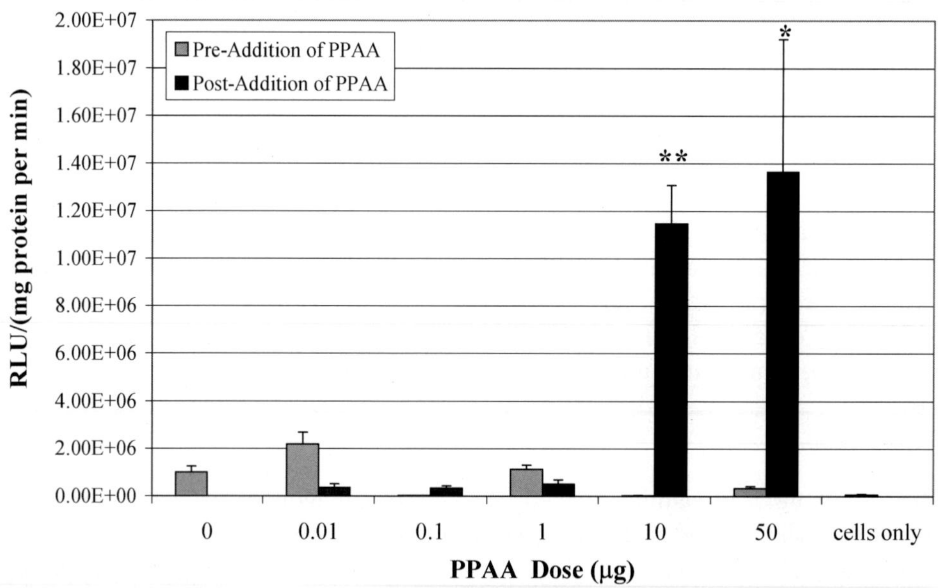

Figure 6. Luciferase gene expression in HeLa cells after the delivery of PPAA-chitosan-DNA ternary complexes, 3 days post-transfection ($n = 8$). The error bars represent S.E.M. $^*P < 0.05$ and $^{**}P < 0.01$ as determined by paired Student's t-test compared to PPAA dose of 0 μg.

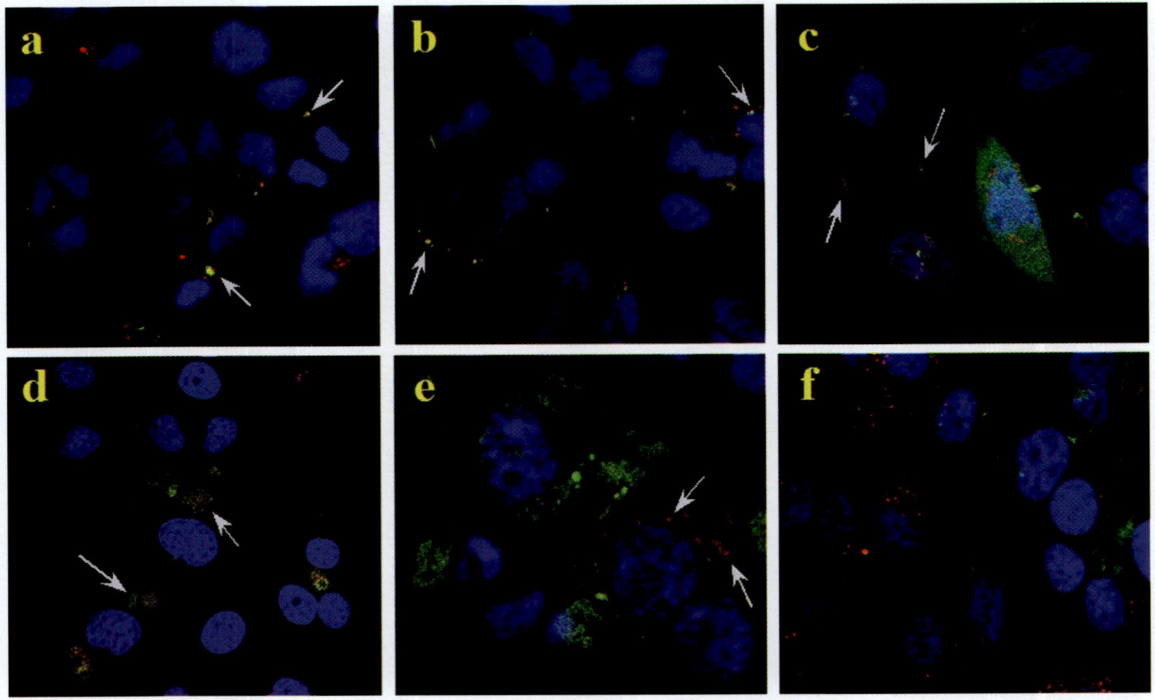

Figure 7. *In vitro* fluorescence confocal images of PPAA-chitosan-DNA ternary nanoparticles in HeLa cells. (a) Nanoparticles were formulated with Rh-chitosan, Gr-PPAA and unlabeled GFP plasmid. Arrows show punctate co-localization of Rh-chitosan with Gr-PPAA 24 h after incubation with cells. (b) Nanoparticles were formulated with unlabeled chitosan, Gr-PPAA and Rh-GFP plasmid. Arrows show punctate co-localization of Gr-PPAA with Rh-GFP plasmid 24 h after incubation. (c) GFP-expressing cells 48 h after nanoparticle incubation. Rh-DNA can be seen in vesicles of expressing cells. Arrows show the beginning of diffuse co-localization of Gr-PPAA with Rh-DNA, suggesting release from nanoparticles. (d) Arrows show diffuse co-localized Gr-PPAA and Rh-DNA signals 48 h after nanoparticle incubation. Free and diffuse Gr-PPAA signal is also apparent here. (e) Large amount of free Gr-PPAA is found in cells 48 h after transfection, not co-localized with Rh-chitosan. Arrows show Rh-chitosan still sequestered to intracellular vesicles. (f) HeLa cells incubated with Lysotracker Red are punctate in appearance and not co-localized with Gr-PPAA at 4 h after nanoparticle incubation.

altered, indicating that the pre- and post-addition PPAA remained complexed with the chitosan nanoparticles (Fig. 7a). The distribution of Gr-PPAA and Rh-DNA at 24 h is comparable to that of the Gr-PPAA and Rh-chitosan distribution (Fig. 7b). Hoechst staining confirmed the triple colocalization of the Rh-chitosan, Gr-PPAA and plasmid DNA at 24 h post-transfection. Thus, the ternary chitosan-DNA-PPAA complexes were intact and successfully internalized in a similar route to that of the binary Rh-chitosan-DNA complexes. At 48 h, the majority of Gr-PPAA and Rh-DNA took on a diffuse cytoplasmic distribution, though Gr-PPAA and Rh-DNA were still observed colocalized in punctate vesicles (Fig. 7c). The Gr-PPAA and Rh-chitosan no longer appeared co-localized. Frequently, the Gr-PPAA was observed in large cytoplasmic vesicles approximately the diameter of the nucleus as opposed to filling the entire cytoplasm (Fig. 7d and 7e). Gr-PPAA and Rh-DNA were frequently found to be co-localized, yet had become diffuse in appearance (Fig. 7d). Rh-chitosan, however, remained punctate at 48 h (Fig. 7e). These observations indicated that the PPAA and DNA were partially released from the chitosan nanoparticles, while the chitosan remained trapped in the vesicles. Incubation of the Gr-PPAA-loaded nanoparticles with Lysotracker Red DND-99 at 4 h demonstrated that both free PPAA and PPAA-loaded nanoparticles did not co-localize with acidic vesicles (Fig. 7f). The observation at 24 and 48 h was similar (data not shown). These observations are consistent with the lack of colocalization observed between the Rh-chitosan and Lysotracker Green (data not shown). The observations made between pre- and post-addition of PPAA to the chitosan nanoparticles were not qualitatively different.

DISCUSSION

One of the major barriers to non-viral gene delivery is the cytoplasmic release of DNA from the endosomes. Polycationic lipid [14–16] and polymer [17, 18] based gene complexes are typically internalized into cells through endocytosis. However, the majority of these complexes remain entrapped in the endosomal compartment or are trafficked through lysosomes [19]. In order to facilitate the efficient release of DNA from the endosome, researchers have incorporated fusogenic peptides derived from viruses such as hemagluttinin HA-2 [20] or drugs like chloroquine [17, 21] into non-viral gene carriers to disrupt the endosomal lipid bilayer. Recently, poly(propyl acrylic acid) (PPAA) has been observed to disrupt late endosomes and enhance the transfection efficiency of several lipid-based carriers including DOTAP [5] and lipid/protamine-DNA complexes [9]. In an *in vivo* murine excisional wound-healing model, PPAA has also been shown to accelerate wound healing after inclusion in a DOTAP-based formulation to deliver the antisense for the angiogenesis inhibitor thrombospondin-2 [8].

PPAA is readily incorporated into chitosan-DNA nanoparticles through the electrostatic interaction between the anionic PPAA and the cationic chitosan. The characteristics of the resulting ternary nanoparticles are dependent on the dose and time at which PPAA is added during the formulation process. The pre-addition of PPAA, that is, the addition of PPAA to the DNA solution prior to complex coacervation with chitosan, yields nanoparticles with a homogenous mixture of chitosan, DNA and PPAA. The large increase in particle size in high pre-addition PPAA dose formulations (10, 50 μg) is likely the result of a disruption in the optimal charge ratio (N/P $= 3.3:1$) required for stable chitosan nanoparticle formation. At high doses, the negative charges in ionized PPAA compete with the negatively charged phosphate groups in the DNA for the protonated amines in chitosan, resulting in unstable aggregates. In contrast, the post-addition of PPAA, that is the addition of PPAA to the chitosan-DNA nanoparticles after complex coacervation, yields heterogeneous nanoparticles with the majority of PPAA confined to the particle surface. We hypothesize that the post-addition of PPAA at the doses tested does not result in particle aggregation because the particles were initially formed at the optimal charge ratio prior to the addition of PPAA. Charge neutralization has already taken place and the rise in pH due to the mixing of the chitosan and DNA solutions has stabilized the nanoparticles. As a result, the PPAA interacts with the positive zeta potential on the nanoparticle surface to form an outer coating. The increase in particle size at high PPAA doses may be a result of the additional PPAA coating on the particle surface. The pre- or post-addition of 0.01, 0.1, or 1 μg of PPAA had no effect on particle size because the additional negative charges did not significantly contribute to the net charge ratio (Fig. 2b).

The zeta potential, or surface charge, of the ternary complexes is consistent with the theoretical charge ratios of chitosan to DNA/PPAA. Up to a dose of 1 μg of PPAA, the size and charge ratios of the ternary mixtures are not significantly different from that of unmodified chitosan nanoparticles. The greater negative surface charge on the post-addition *versus* pre-addition PPAA nanoparticles at doses of 10 and 50 μg is consistent with the hypothesis that PPAA coats the surface of post-addition PPAA complexes, while PPAA is incorporated into pre-addition PPAA complexes.

Gel electrophoresis confirms that pre- and post-addition PPAA do not affect the binding of chitosan and DNA, with the exception of the pre-addition PPAA dose of 50 μg. At this dose, the negatively charged PPAA may have displaced some of the DNA from complexation to chitosan, resulting in a less compact nanoparticle, or even free DNA in the solution. At lower pre-addition PPAA doses, the charge imbalance is not sufficient to prevent the electrostatic interaction between DNA and chitosan.

In the presence of complete cell culture medium, unmodified chitosan nanoparticles do not release DNA, even at 72 h [22]. Therefore, we hypothesize that the incorporation of PPAA into the nanoparticle may act to destabilize the particles slightly by reducing the strong electrostatic interaction between the chitosan and DNA. In this manner, we hope that the addition of PPAA would facilitate plasmid DNA release from the chitosan nanoparticles in the endosome. As hypothesized, both the pre- and post-addition of PPAA to the chitosan nanoparticles facilitated the release of plasmid DNA from the nanoparticles even in the presence of serum containing medium. The pre-addition PPAA complexes released DNA as early as 3 h after incubation in media, consistent with the speculation that the homogenous incorporation of PPAA reduces the electrostatic interaction between DNA and chitosan. This is also consistent with the observation that increasing the dose of PPAA incorporated by pre-addition into the nanoparticles increases the amount of DNA released (relative to the decreasing strength of the chitosan-DNA complex). The post-addition PPAA complexes released the DNA at the later time point of 72 h, in agreement with the hypothesis that PPAA, as a coating would have less of an influence in disrupting the tight charge interaction between the chitosan and DNA.

While our experiments demonstrate that the incorporation of PPAA has a destabilizing effect on the chitosan nanoparticles incubated in serum compared with PBS, previous work has demonstrated that PPAA stabilizes DOTAP/DNA lipoplexes in the presence of serum [5]. Two different mechanisms are in effect to enhance the gene transfection. In the presence of serum, DOTAP/DNA complexes cannot efficiently transfect cells without the post-addition of PPAA because the serum components would destabilize the lipid-DNA complexes. The hydrophilicity of PPAA at neutral pH is hypothesized to aid in inhibiting serum protein adsorption in a similar fashion to PEG, though the exact mechanism is unknown [5]. In the case with chitosan, the destabilizing effect is beneficial since the chitosan-DNA nanoparticles are too stable in cell culture and in many tissues (data not shown).

Although previous data suggest that the addition of an endosomolytic agent to chitosan-DNA complexes does not significantly affect gene transfection [13, 17], the effect of addition of PPAA to the chitosan-DNA nanoparticles in our *in vitro* gene expression studies shows the contrary. The incorporation of PPAA enhances the transfection efficiency of chitosan in both HEK293 and HeLa cells to varying degrees. In HEK293 cells, nanoparticles formed with post-addition of PPAA increased gene expression for all formulations compared to unmodified nanoparticles. This result is unexpected from the standpoint that negative surface charges are thought to reduce the interaction of particles with the negatively charged cell surface, therefore reducing total cellular uptake. Furthermore, post-addition PPAA complexes do not release DNA as rapidly as the pre-addition complexes. One explanation for the effectiveness of the post-addition PPAA complexes is that the PPAA is readily available, in sufficient concentration, to interact with the endosomal membrane, triggering membrane disruption and DNA release. After pre-addition complex transfection, gene expression was only increased in the presence of 0.1 μg PPAA, while doses of 1 and 0.01 μg PPAA did not significantly affect gene expression. Plausible factors contributing to this effect of PPAA are insufficient PPAA availability and unstable nanoparticle formulation. At the low PPAA dose, there may not be sufficient PPAA available to interact with the endosome membrane to trigger destabilization. At the high PPAA dose, there was incomplete DNA retention (observed by gel electrophoresis) and particle instability (observed by aggregation during particle sizing). The unstable and large ternary aggregates would be poorly internalized by the cells. They would also offer poor protection against enzymatic degradation of the DNA before internalization.

In HeLa cells, only the post-addition of high doses of PPAA to chitosan nanoparticles results in an increase in gene expression. Again, PPAA on the particle surface can interact with endosomal membranes to trigger endosomal release. High doses of PPAA may be necessary, because for unknown reasons these chitosan-DNA nanoparticles are highly inefficient in transfecting HeLa cells [13]. The minimal effect of pre-addition of PPAA to the complexes can be explained by reasons similar to that described for the transfection of 293 cells. High doses of pre-addition PPAA create unstable ternary nanoparticles and low doses of PPAA may not provide a strong enough membrane disruption effect.

The overall lower gene expression exhibited by HeLa compared to HEK293 cells is likely due to the difference in potency of chitosan nanoparticles for gene transfer [13]. The effectiveness of PPAA may be directly related to its proposed function as an endosome disruptor, which may have some cell-type dependency as well. An optimal balance between the dose of PPAA required to disrupt the endosomal membrane and the dose of PPAA needed to reduce the interaction between the chitosan and DNA may be necessary.

Overall, 10 μg post-addition PPAA dose was the most effective in both cell lines. At this dose, the ratio of PPAA to DNA is 1 : 1 (w/w), which is considerably lower than the optimal dose of PPAA required in the lipid-protamine-DNA (LPD) system of 3 : 1 [9]. Post-addition of 61 kDa [5] and 44 kDa [6] PPAA to DOTAP/DNA lipoplexes was shown to enhance gene expression in NIH3T3 fibroblasts 20-fold and 25-fold, respectively, compared to lipoplexes without PPAA [5, 6]; pre-addition of PPAA was not tested. The addition of PPAA to the chitosan/DNA complexes reduces the theoretical charge ratio of the chitosan/DNA complexes from 3.3 : 1 to 0.83 : 1 and the particles had a net negative surface charge. In contrast, at the optimal PPAA dose for DOTAP/DNA complexes, the surface charge remains positive. The addition of 61 kDa and 44 kDa PPAA to the DOTAP/DNA complexes reduces the theoretical positive charge ratio from 1.9 : 1 to 1.3 : 1 and 1 : 1, respectively. The average size of the ternary chitosan/DNA complexes at this dose is approx. 400 nm, which is smaller than the size of the LPD complexes at 800 nm.

Fluorescence confocal microscopy experiments verified that PPAA was incorporated in the chitosan-DNA nanoparticles and co-internalized into punctate perinuclear vesicles at 24 h post-transfection and was further confirmed by double labeling experiments. The initial uptake and trafficking of PPAA loaded nanoparticles is similar to that observed with unmodified Rh-labeled nanoparticles (data not shown). The trafficking experiments were conducted using the pre-addition of 0.1 μg PPAA and post-addition of 1 μg PPAA. Despite differences in nanoparticle size, zeta potential and PPAA distribution, no qualitative differences were observed in the co-localization of PPAA with DNA or chitosan between the two formulations.

At later time points, the Gr-PPAA and Rh-DNA signals are increasingly diffuse and the PPAA is less frequently co-localized with the vesicular-localized Rh-chitosan. Qualitative observations suggest that more Rh-DNA was released from the nanoparticles than was observed with FITC-chitosan nanoparticles co-localized with Rh-DNA (data not shown). The trafficking study confirms that the PPAA is rapidly released from the chitosan nanoparticles and facilitates the release of DNA from endosomes within the same time frame as that observed after ternary complexes were incubated with serum-containing medium in gel-electrophoresis studies. These observations correlate with previous observations in Jurkat T-lymphoma cells of diffuse cytoplasmic fluorescence of biotinylated PPAA/Gr-labeled streptavidin/biotinylated antibody complexes [7]. The lack of colocalization between Gr-PPAA and Lysotracker Red confirms that the PPAA is not trafficked through lysosomes. This lack of colocalization of the PPAA with lysosomes is consistent with recent evidence suggesting that PPAA does not disrupt early endosomal or lysosomal compartments but does disrupt the late endosomal compartments [6]. GFP expression on day 2 confirms higher percentage of transfected cells after transfection with the PPAA-loaded chitosan nanoparticles compared to chitosan nanoparticles alone.

CONCLUSIONS

The incorporation of PPAA into chitosan-DNA nanoparticles enhances the transfection efficiency in both HEK293 and HeLa cells facilitating vector unpacking from the complexes and enhancing endosomal release. Gel electrophoresis and confocal microscopy experiments confirm the enhanced kinetics of plasmid DNA release from the nanoparticles and the trafficking of the plasmid DNA into the cytosol. The addition of PPAA provides a promising strategy for the enhancing gene transfer with polymer-DNA complexes.

REFERENCES

1. M. D. Brown, A. G. Schatzlein and I. F. Uchegbu, *Int. J. Pharm.* **229**, 1 (2001).
2. F. D. Ledley, *Hum. Gene Ther.* **6**, 1129 (1995).
3. R. Wattiaux, N. Laurent, S. Wattiaux-De Coninck and M. Jadot, *Adv. Drug Deliv. Rev.* **41**, 201 (2000).
4. N. Murthy, J. R. Robichaud, D. A. Tirrell, P. S. Stayton and A. S. Hoffman, *J. Control. Rel.* **61**, 137 (1999).
5. C. Y. Cheung, N. Murthy, P. S. Stayton and A. S. Hoffman, *Bioconjug. Chem.* **12**, 906 (2001).
6. R. A. Jones, C. Y. Cheung, F. E. Black, J. K. Zia, P. S. Stayton, A. S. Hoffman and M. R. Wilson, *Biochem. J.* **372**, 65 (2003).
7. C. A. Lackey, O. W. Press, A. S. Hoffman and P. S. Stayton, *Bioconjug. Chem.* **13**, 996 (2002).
8. T. R. Kyriakides, C. Y. Cheung, N. Murthy, P. Bornstein, P. S. Stayton and A. S. Hoffman, *J. Control. Rel.* **78**, 295 (2002).
9. R. W. Paul, B. Dutzar, P. S. Stayton, A. S. Hoffman, P. Anklesaria and P. Harvie, *Proc. ASGT: 6th Annual Meeting*, Washington, DC (2003).
10. M. Kumar, A. K. Behera, R. F. Lockey, J. Zhang, G. Bhullar, C. P. De La Cruz, L. C. Chen, K. W. Leong, S. K. Huang and S. S. Mohapatra, *Hum. Gene Ther.* **13**, 1415 (2002).

11. K. Roy, H. Q. Mao, S. K. Huang and K. W. Leong, *Nature Med.* **5**, 387 (1999).
12. O. Boussif, F. Lezoualc'h, M. A. Zanta, M. D. Mergny, D. Scherman, B. Demeneix and J. P. Behr, *Proc. Natl. Acad. Sci. USA* **92**, 7297 (1995).
13. H. Q. Mao, K. Roy, V. L. Troung-Le, K. A. Janes, K. Y. Lin, Y. Wang, J. T. August and K. W. Leong, *J. Control. Rel.* **70**, 399 (2001).
14. J. Zabner, A. J. Fasbender, T. Moninger, K. A. Poellinger and M. J. Welsh, *J. Biol. Chem.* **270**, 18997 (1995).
15. F. Labat-Moleur, A. M. Steffan, C. Brisson, H. Perron, O. Feugeas, P. Furstenberger, F. Oberling, E. Brambilla and J. P. Behr, *Gene Ther.* **3**, 1010 (1996).
16. I. Wrobel and D. Collins, *Biochim. Biophys. Acta* **1235**, 296 (1995).
17. P. Erbacher, S. Zou, T. Bettinger, A. M. Steffan and J. S. Remy, *Pharm. Res.* **15**, 1332 (1998).
18. T. Bieber, W. Meissner, S. Kostin, A. Niemann and H. P. Elsasser, *J. Control. Rel.* **82**, 441 (2002).
19. A. Coonrod, F. Q. Li and M. Horwitz, *Gene Ther.* **4**, 1313 (1997).
20. E. Wagner, C. Plank, K. Zatloukal, M. Cotten and M. L. Birnstiel, *Proc. Natl. Acad. Sci. USA* **89**, 7934 (1992).
21. V. L. Truong-Le, S. M. Walsh, E. Schweibert, H. Q. Mao, W. B. Guggino, J. T. August and K. W. Leong, *Arch. Biochem. Biophys.* **361**, 47 (1999).
22. T. Kiang, J. Wen, H. Lim and K. W. Leong, *Biomaterials* (2004) (in press).

Gels, Genes, Grafts and Giants, pp. 491–505 (2005)

Monocyte activation on polyelectrolyte multilayers

JASON J. HWANG [1], SANDRA JELACIC [2], NEWTON T. SAMUEL [3], RONALD V. MAIER [2], CHARLES T. CAMPBELL [4], DAVID G. CASTNER [1,3], ALLAN S. HOFFMAN [1] and PATRICK S. STAYTON [1,*]

[1] *Department of Bioengineering, University of Washington, Box 351721, Seattle, WA 98195-1721, USA*
[2] *Harborview Medical Center, University of Washington, Box 359796, Seattle, WA 98104, USA*
[3] *Department of Chemical Engineering, University of Washington, Box 351750, Seattle, WA 98195-1750, USA*
[4] *Department of Chemistry, University of Washington, Box 351700, Seattle, WA 98195-1700, USA*

Received 8 December 2003; accepted 23 August 2004

Abstract—The adherence and activation of primary human monocytes was investigated on a polyelectrolyte multilayer film containing hyaluronic acid (HA) and poly-L-lysine (PLL). The sequential layer-by-layer deposition of the multilayer film was characterized by surface plasmon resonance. Eight alternating bilayers displayed an effective thickness of 16.15 nm with a total polymer coverage of 2.10 μg/cm^2. For cell studies, HA-PLL multilayers were constructed on tissue culture polystyrene (TCPS) substrates and characterized by time of flight second ion mass spectrometry (ToF-SIMS) analysis. Principal component analysis of the ToF-SIMS spectra resolved no significant difference in surface chemistry between PLL-terminated and HA-terminated multilayer surfaces. Monocyte adhesion on PLL- and HA-terminated surfaces was measured by the lactate dehydrogenase assay and showed a significant decrease in cell adhesion after 24 h incubation. Cell viability measured by Live/Dead fluorescent staining showed significant cell death in the adherent cell population over these 24 h. Tumor necrosis factor-α (TNF-α) production, a measure of monocyte activation, was quantified by ELISA and normalized to the number of adherent monocytes. The activation of monocytes on PLL-terminated and HA-terminated surfaces was nearly identical, and both surfaces had TNF-α levels that were 8-fold higher than TCPS. These results demonstrate that sufficient PLL had diffused into the surface layer to direct monocyte adherence and to induce cytokine activation and cell death on the HA-terminated multilayer films. The diffusion of the second multilayer component to the coating surface should, thus, be taken into account in the design of polyelectrolyte-based biomaterial coating strategies.

Key words: Polyelectrolyte; hyaluronic acid; poly-L-lysine; monocyte; biomaterial; time of flight second ion mass spectroscopy; surface plasmon resonance.

---

*To whom correspondence should be addressed. Tel.: (1-206) 685-8148. Fax: (1-206) 685-8256. E-mail: stayton@u.washington.edu

INTRODUCTION

Biomaterials elicit a foreign body response mediated by monocyte-derived cells when implanted in the body. Upon implantation, monocytes move from the vasculature to the tissue surrounding the biomaterial where they mature to the macrophage phenotype, and can further fuse to form multinucleated foreign body giant cells [1, 2]. These cells mediate the initial and chronic inflammatory responses [3, 4]. Encapsulation of an implanted device can adversely affect performance, for example, by limiting transport in drug-release systems [5] and implanted sensors [6]. Underlying the inflammatory response is the activation of pro-inflammatory signaling pathways that result in the production of cytokines and chemokines such as interleukin-1β and Tumor Necrosis Factor-α (TNF-α) [7].

In this study, we have used a monocyte activation assay to evaluate the monoctye response to a polyelectrolyte complex (PEC) coating constructed by sequential, layer-by-layer deposition of oppositely charged biopolymers [8, 9]. There has been considerable interest in these coatings in the biotechnology and biomaterials fields [10–16]. This technique may be particularly useful to construct surface coatings with natural or synthetic polymers that display favorable biological surface properties. One such natural polymer is hyaluronic acid (HA), or hyaluronan, which is a non-sulfated glycosaminoglycan made up of repeating disaccharide units of β-1,4-glucuronic acid linked to β-1,3-N-acetyl glucosamine. HA has been widely studied as a biomaterial coating due to its stealth-like and biological effector properties [17, 18]. Picart *et al.* have previously characterized the build-up of HA–poly-L-lysine (PLL) multilayers [19, 20]. Here, monocyte adhesion and activation was studied on multilayers that were terminated with an HA layer. We have found that the monocyte response is sensitive to polymer diffusion in the multilayers that leads to significant levels of the cationic PLL at the surface. Our results suggest that the choice of the cationic polymer and the multilayer surface composition and architecture will be critical for controlling monocyte responses to these coatings.

MATERIALS AND METHODS

Materials

Phosphate-buffered saline (PBS, Sigma, St. Louis, MO, USA) solution (137 mM NaCl, 2.7 mM KCl, 10 mM phosphate buffer, pH 7.4) was adjusted to 150 mM NaCl with pure sodium chloride (Sigma) then to pH 8 using 1 M NaOH (VWR, West Chester, PA, USA) and sterile filtered. HA (1700 kDa, Lifecore Biomedical, Chaska, MN, USA) and PLL (87.4 kDa, Sigma) were both dissolved in the adjusted PBS buffer at concentrations of 1 mg/ml. These polymer solutions were sterile filtered using a 0.22-μm syringe filter and were used to prepare multilayers in all of the following experiments. All solutions and samples were handled using standard tissue culture technique to maintain sterility throughout sample preparation and testing.

Endotoxin quantification

Lipopolysaccharide (LPS) levels for HA and PLL in buffer solution were quantified by Charles River Endosafe (Charleston, SC, USA) using the LAL chromogenic kinetic assay.

Preparation of gold surface for surface plasmon resonance (SPR) analysis

Clean glass microscope slides (Fisher Scientific, Hampton, NH, USA) were first coated by electron beam evaporation with approx. 15 Å of Ti as an adhesive layer followed by approx. 475 Å of gold at pressures below 1×10^{-6} Torr. To create a negatively charged surface to aid in multilayer deposition, the gold substrates were immersed in solutions of 1 mM 11-mercaptoundecanoic acid (11-MUA, Aldrich, St. Louis, MO, USA) in absolute ethanol overnight before being rinsed with ethanol then deionized (DI) water to produce a self-assembled monolayer of a −COOH-terminated alkylthiolate, which deprotonates when placed in the buffer used here [21].

SPR analysis of PLL and HA multilayer formation

The home-built SPR spectrometer used in this study has been described and characterized elsewhere [22]. A plexiglass flow cell along with an 80 μl flow channel was used to direct polyelectrolyte solutions onto the gold coated glass slide, which acts as the SPR sensor surface. Adsorption on this surface was observed by monitoring the shifts of the SPR minimum wavelength. During SPR measurements, the 11-MUA-modified gold surface was first equilibrated with degassed buffer solution. Then, PLL solution (80 μl, 0.5 ml/min) was injected into the flow cell and allowed to adsorb onto the gold surface for 5 min. Buffer solution (100 μl, 0.5 ml/min) was injected to wash unbound PLL and allowed to equilibrate for 2 min. This process was alternated with injections of HA solution for a total of 8 bilayers.

The increase in SPR wavelength observed during each such injection was converted into effective adlayer thickness and coverage using previously described methods (equations (10b), (11) and (13) in Ref. [22]), taking into account the change in SPR sensitivity with both refractive index (RI) and film thickness. The SPR wavelength response was first converted to an effective RI response, using bulk sucrose solutions to calibrate the SPR sensitivity (change in SPR minimum wavelength with refractive index). The effective thickness of each layer was then calculated following Ref. [22] using the RI values for PLL (1.57), HA (1.60) and buffer (1.331). These RI values were calculated using: (1) the RI increments ($\mathrm{d}n/\mathrm{d}C$) in aqueous solutions: $\mathrm{d}n/\mathrm{d}C_{\text{protein}} = 0.18$ ml/g [23], $\mathrm{d}n/\mathrm{d}C_{\text{salt}} = 0.16$ ml/g [23] and $\mathrm{d}n/\mathrm{d}C_{\text{HA}} = 0.15$ ml/g (as measured by Lifecore Biomedical); (2) the partial specific volumes ($\bar{v}_2^o$): $\bar{v}_{2\ \text{protein}}^o = 0.77$ ml/g [24] and $\bar{v}_{2\ \text{HA}}^o = 0.55$ ml/g [25] in aqueous solution. Coverage, or adsorbed mass per area, was calculated by multiplying the effective thickness of each layer by the effective density of the adsorbate (ρ_{eff}), which was derived from their specific volumes ($\rho_{\text{eff}} = 1/\bar{v}_2^o$).

Preparation of samples for time of flight second ion mass spectrometry (ToF-SIMS)

Tissue-culture polystyrene (TCPS) multiwell plates (BD Biosciences, Bedford, MA) were broken up into small flat pieces (approx. 1 cm^2). TCPS samples were cleaned with water and dilute detergent followed by extensive rinsing in sterile DI water. The samples were then dried under nitrogen flow and kept in covered Petri dishes. Clean TCPS samples were first placed in a PLL solution for 5 min, followed by fresh DI water. This process was alternated with HA solution for a total deposition of 7 or 8 bilayers. After the multilayer deposition steps, the samples were dipped in either PLL or HA solution two additional times. The samples were dried under nitrogen and stored for ToF-SIMS analysis. Three surfaces were prepared for this study: (A) plain TCPS, (B) TCPS coated with 7 bilayers terminated with a PLL layer (TCPS-$(PLL\text{-}HA)_7$-PLL_3) and (C) 8 bilayers terminated with an HA layer (TCPS-$(PLL\text{-}HA)_8$-HA_2). To prepare the sample for the reference HA spectrum, a 1 mg/ml HA solution was passed through a desalting column and then air-dried as a thick film on a silicon wafer.

ToF-SIMS analysis

ToF-SIMS data was acquired using a Model 7200 Physical Electronics Reflectron time of flight secondary ion mass spectrometer (PHI, Eden Prairie, MN, USA). The total primary ion flux from an 8 keV Cs^+ primary ion source was maintained below 10^{12} ions/cm^2 to ensure static conditions [26]. A pulsed low-energy electron flood gun was used for charge neutralization. High resolution mass spectra were acquired over an area of 200×200 μm^2. Duplicate samples were measured for each group and at least three spectra were recorded on each sample. The mass resolution ($m/\Delta m$) at the $C_2H_3^+$ ($m/z = 27$) and C_2H^- ($m/z = 25$) peaks were typically above 3000 in the positive and negative ion spectra. Positive and negative ion ToF-SIMS spectra were calibrated to the CH_3^+, $C_2H_3^+$, $C_3H_5^+$, $C_7H_7^+$ peaks and CH^-, CN^-, CNO^- peaks, respectively. Principal component analysis (PCA), a multivariate algorithm that separates samples based on peak intensities in the ToF-SIMS spectrum, was used to analyze the differences between the PLL- and HA-terminated surface chemistries. Details for implementing PCA have been reported elsewhere [27].

Preparation of multilayer films to evaluate monocyte response

Three surfaces were prepared for this evaluation: TCPS, PLL-terminated multilayers and HA-terminated multilayers. PEC multilayer films for analysis of cytokine production from adherent monocytes were prepared in sterile 24-well TCPS multiwell plates that have a growth area of 2.0 cm^2 per well. Films were prepared by alternating addition and aspiration of PLL and HA solutions (500 μl for 5 min each) with washes of DI water (1 ml) after each polymer deposition step. Sequential depositions from PLL and HA solutions were made for a total of seven and a half bilayers for the PLL-terminated multilayer and a total of 8 bilayers for the HA-terminated multilayer film. After the 15th and 16th deposition step for PLL- and HA-terminated multilayer films, respectively, two additional deposition and wash steps with the desired polymer solution were conducted. Samples were kept overnight in PBS at 4°C prior to use.

Isolation of primary human monocyte cells

The methods for isolating primary human monocytes have been reported elsewhere [28]. Briefly, whole blood was collected from healthy male donors into polypropylene syringes containing 3.6% sodium citrate. Peripheral blood mononuclear cells (PBMCs) were separated out in the buffy coat by Ficoll-Paque (Amersham Pharmacia Biotech, Piscataway, NJ, USA) density gradient centrifugation. Monocytes were isolated from PBMCs using a magnetic bead negative selection kit (Miltenyi Biotec, Auburn, CA, USA). The monocyte population was further purified by spinning the monocyte suspension through a fetal bovine serum gradient twice followed by three washes in Dulbecco's PBS (Biowhitaker, Walkersville, MD, USA). The purified monocytes were resuspended in RPMI 1640 media (Biowhitaker) with 100 μg/ml of gentamycin at a concentration of 1×10^6 cells/ml prior to plating onto multilayers. Cell viability was confirmed by Trypan blue exclusion on a sample population where >90% cells were viable.

Evaluation of monocyte activation

1.5×10^5 cells in 1 ml serum-free RPMI 1640 media (Biowhitaker) were plated on TCPS and 1×10^6 cells in 1 ml serum-free RPMI 1640 media were plated on both PLL-terminated and HA-terminated multilayers for 2 h. Following this adhesion step, media and non-adherent cells were aspirated, and RMPI 1640 media with 10% heat-inactivated adult bovine serum (ABS) was gently added to each well. After a 24-h incubation period, media in each well were spun down and supernatants were collected. TNF-α concentrations were quantified using a Quantikine human TNF-α enzyme-linked immunosorbent assay (ELISA) kit (R&D systems, Minneapolis, MN, USA).

Cell lysates were collected by adding and vigorously pipetting lysis buffer (20 mM Tris, 137 mM NaCl, 2 mM EDTA, 10% glycerol, 1% Triton X-100, pH 8) in each well. Cell numbers were quantified by lactate dehydrogenase (LDH) activity (Roche Molecular Biochemicals, Indianapolis, IN, USA). Methods for quantifying cell count using LDH activity have been described elsewhere [29]. Briefly, viable cell numbers for each sample were determined by measuring the LDH activity in the cell lysates compared to a standard curve. A standard curve was made for each blood donor. In each well, the TNF-α concentration was normalized to the measured cell count.

Microscopy imaging of cells

Phase-contrast and fluorescence microscopy images were taken with a 10× objective on a Nikon TE200 inverted microscope. A Live/Dead fluorescent stain (Molecular Probes, Portland, OR, USA) was used to evaluate the viability of the adherent cells. Representative images of the adherent monocyte populations on TCPS, PLL- and HA-terminated surfaces made in 24-well TCPS multiwell plates were captured at 2, 8 and 24 h time points. The numbers of live and dead cells in each of these images were counted by hand. Non-fluorescent cells were counted as viable.

Statistical analyses

Statistical analyses were conducted using a two-tailed Student's t-test for significance, $\alpha = 0.05$. Three replicate samples per surface for three blood donors were used in each evaluation unless specified.

RESULTS

Endotoxin quantification

LPS levels were measured to be <0.5 EU/mg for PLL in the buffer solution and <0.05 EU/mg for HA in the buffer solution by the LAL kinetic chromogenic assay.

SPR characterization of multilayer buildup

The stepwise PLL-HA PEC multilayer growth onto the $-COO^-$-terminated surface was followed by SPR (Fig. 1). The growth curve shows a sequential, non-linear growth of the film for a total of 8 bilayers. The kinetics of binding changed with the number of layers. For the initial layers, there was a rapid increase in SPR wavelength followed by a plateau, which indicated that all of the polymer deposition occurred in the initial binding step. The wash steps removed a small fraction of the polymer, as shown by a decrease in the SPR wavelength. The deposition steps in bilayers 5–8 showed a rapid binding step followed by a non-linear increase in deposition over the 5 min deposition step. The wash steps no longer removed polymer from the surface, but appeared to increase multilayer thickness, as shown by an increase in the SPR wavelength during the wash step. For deposition of HA in the upper bilayers, the rapid initial binding step was smaller in value compared to the lower bilayers. The effective multilayer thickness and coverage calculated from the SPR data are shown in Fig. 2. The effective thickness is defined as the thickness that the mass of adsorbed polymer would have if adsorbed as a pure, uniform film with no intermixed water. As the film thickness increased, the amount of polymer adsorbed in each bilayer increased markedly. After deposition of 8 bilayers, the effective thickness was 16.15 ± 3.41 nm and the total polymer coverage was 2.10 ± 0.44 $\mu g/cm^2$.

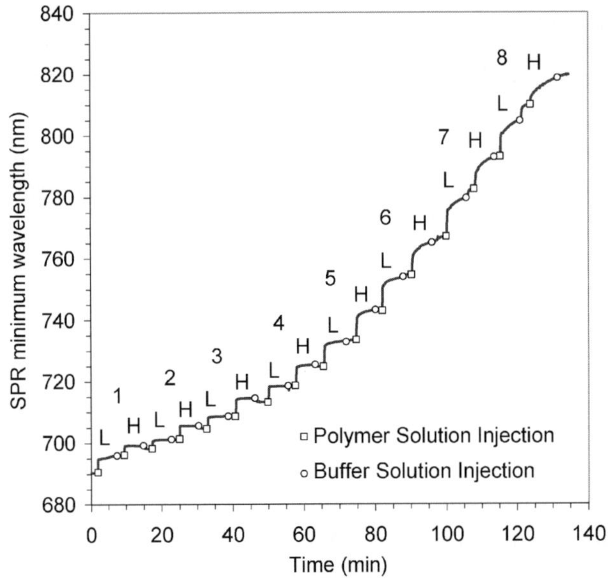

Figure 1. Representative SPR analysis of the deposition of eight PLL (L) and HA (H) bilayers on gold modified with 11-MUA. Each adsorption step following polymer injection (□) was 5 min. The surfaces were allowed to equilibrate for 2 min following buffer injection (○) before the next deposition step. Polymer solution concentrations of PLL and HA were both 1 mg/ml in 150 mM NaCl, PBS buffer at pH 8.

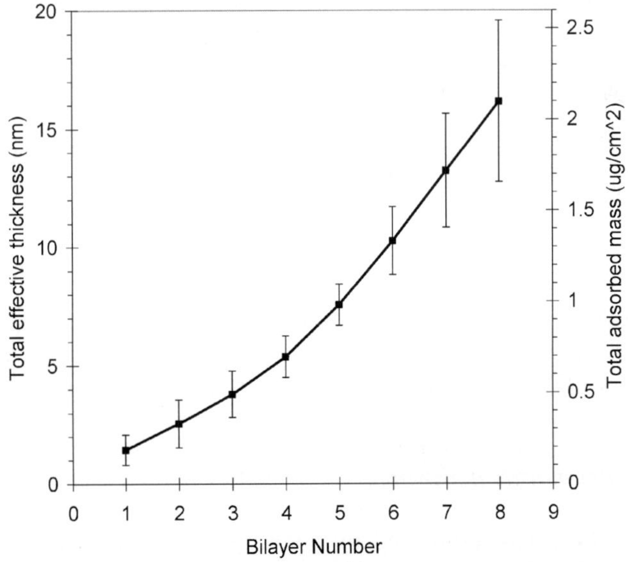

Figure 2. Effective thickness and coverage calculations from SPR analysis ($n = 3$). The effective thickness and coverage (mass per unit area), which do not take into consideration the presence of water, are calculated from shifts in the SPR minimum wavelength in response to polymer deposition. The following indices of refraction increments were used in calculations: $\mathrm{d}n/\mathrm{d}C_{\mathrm{PLL}} = 0.18$ ml/g, $\mathrm{d}n/\mathrm{d}C_{\mathrm{HA}} = 0.15$ ml/g and partial specific volumes: $\bar{v}^{o}_{2\ \mathrm{PLL}} = 0.77$ ml/g, $\bar{v}^{o}_{2\ \mathrm{HA}} = 0.55$ ml/g.

ToF-SIMS surface analysis

The positive ion spectrum of the outermost 10–20 Å of the TCPS surface (data not shown) contained peaks indicative of polystyrene ($C_6H_5^+$, $C_7H_7^+$), which was clearly altered after multilayer deposition steps. The positive ion spectrum of the PLL-terminated multilayer surface (Fig. 3A) contained peaks indicative of PLL: $C_5H_{10}N^+$, $C_3H_6N^+$ and CH_4N^+. The negative ion spectrum of the same surface (Fig. 3B) was similar to the fragmentation pattern of a reference HA spectrum (Fig. 3D, inset) and contained peaks indicative of HA: CNO^-, CHO_2^-, $C_2H_3O_2^-$, $C_3H_3O_2^-$ and $C_3H_3O_3^-$. The positive and negative ion spectra for the HA-terminated multilayer surface (Fig. 3C and 3D) were similar to the spectra for the PLL-terminated film based on PCA analysis. The individual points in the PCA scores plot, which represent entire PLL ML and HA ML spectrum, were randomly scattered, indicating that PLL- and HA-terminated surfaces were similar in chemical composition (Fig. 3E). Dissimilar chemical compositions would have resulted in distinct groupings of spectra in the PCA scores plot.

Cell adhesion and viability

Equivalent levels of adherent cells ($P = 0.95$) were observed for the PLL-terminated multilayer and the HA-terminated multilayer surfaces after a 24-h incubation period (Fig. 4). Evaluation of the cell binding time dependence at 2, 8 and 24 h showed that approx. 5.5×10^5 of the original 1×10^6 cells seeded on both PLL-terminated and HA-terminated surfaces were bound at the 2 h time point (Fig. 5). However, a significant percentage of these adherent cells detached or died between the 2 and 8 h time points, as measured by LDH activity. There were no significant differences between adherent cell numbers on PLL-terminated and HA-terminated multilayer surfaces at each of the time points ($P > 0.15$).

The adherent cells were subsequently tested for viability with the Live/Dead fluorescent stain at the 2, 8 and 24 h time points (Fig. 6). Approximately 96% of the monocytes adherent on TCPS appeared to be viable throughout the 24 h time-course in these representative images (Table 1). A low percentage of viable cells was observed for both PLL-terminated and HA-terminated multilayer surfaces.

Measurement of monocyte activation

Monocyte activation was monitored by quantitating normalized levels of TNF-α production (Fig. 7). The average TNF-α levels for the PLL-terminated and HA-terminated multilayer surfaces at the 24 h time point were 6114 pg and 6651 pg, respectively, compared to 740 pg for TCPS ($P < 0.023$). There was no significant difference between the TNF-α levels from monocytes on PLL-terminated and HA-terminated multilayer surfaces after the 24 h incubation period ($P = 0.94$).

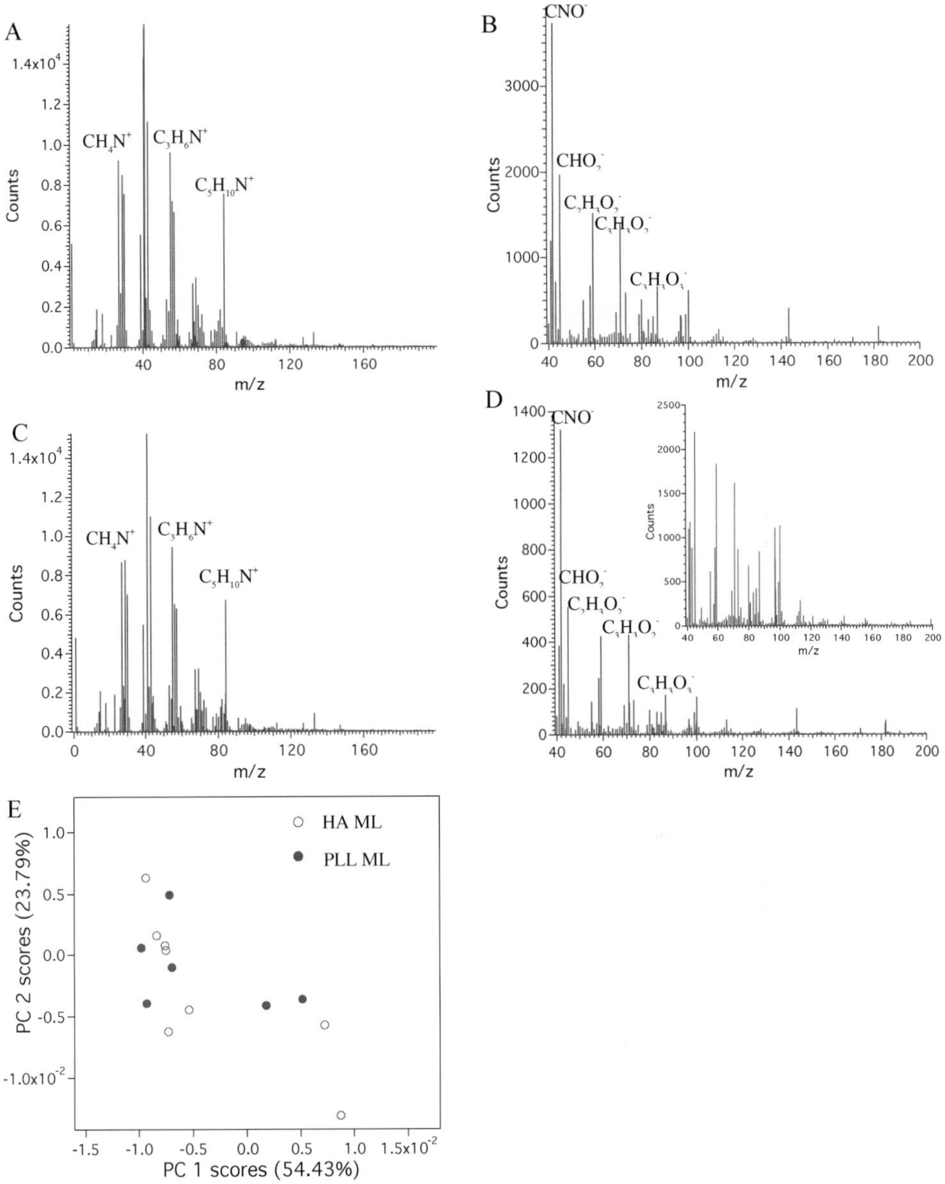

Figure 3. ToF-SIMS analysis of PLL-terminated and HA-terminated multilayer surface chemistry. Seven bilayers deposited on TCPS with a terminating layer of PLL (PLL ML) and eight bilayers terminated with HA (HA ML) were evaluated by ToF-SIMS analysis. The positive ion spectra for both PLL-terminated (A) and HA-terminated (C) multilayer surfaces show peaks indicative of PLL surface presence. The negative ion spectra for PLL-terminated (B) and HA-terminated (D) multilayer surfaces show peaks indicative of HA surface presence with a fragmentation pattern similar to the HA reference spectrum (inset in 3D). The PCA scores plot (E) quantitatively evaluates the peak intensities of the replicate positive spectra measured for each sample. A random scatter from at least three spectra taken from two samples per group indicated there were no significant differences in surface chemistry between the two groups.

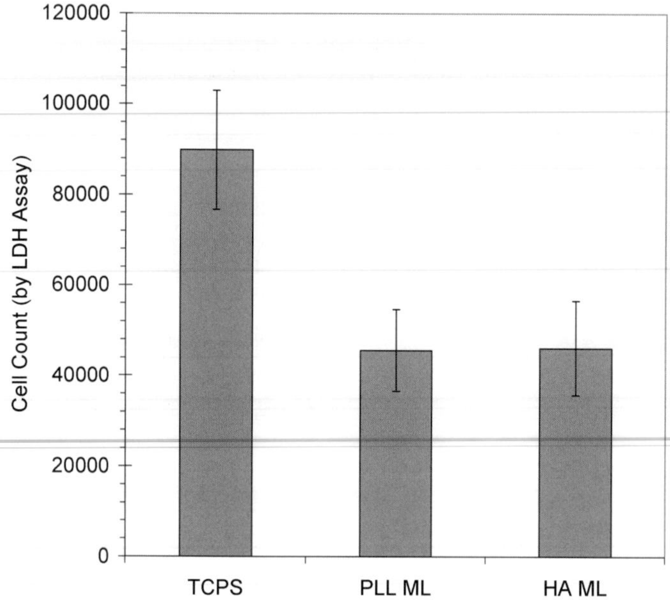

Figure 4. Monocyte cell adhesion after 24 h incubation as measured by LDH activity. 1.5×10^5 cells were plated on the TCPS and 1×10^6 cells were plated on PLL-terminated (PLL ML) and HA-terminated (HA ML) multilayer surfaces and incubated in RPMI media with 10% heat inactivated ABS. The viable cell count for triplicate samples in each group for three donors was measured to quantify cell numbers on each surface.

Table 1.
Quantification of monocyte viability from Live/Dead fluorescence images

Surface	Time point (h)	Live cell count	Total cell count	Viable cells (%)
TCPS	2	412	430	95.8
	8	113	117	96.6
	24	173	180	96.1
PLL ML	2	304	1425	21.3
	8	82	760	10.8
	24	29	946	4.5
HA ML	2	207	630	32.9
	8	41	723	5.7
	24	14	363	3.9

One representative image was captured for each surface at each time point. Non-fluorescent cells were counted as alive. PLL ML, PLL-terminated multilayer surface; HA ML, HA-terminated multilayer surface.

DISCUSSION

In this study we have evaluated the monocyte inflammatory response to a PEC multilayer coating containing hyaluronic acid (HA). These multilayers represent an interesting biomaterial surface coating strategy, particularly for the construction of

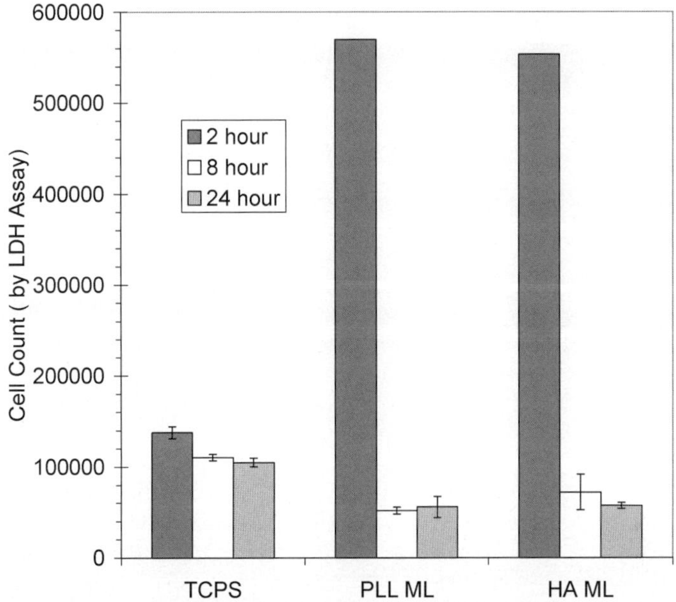

Figure 5. Time-course evaluation of monocyte adherence as measured by LDH activity. The number of viable cells was determined from the original 1.5×10^5 cells on TCPS, 1×10^6 cells on PLL-terminated (PLL ML) and 1×10^6 cells on HA-terminated (HA ML) multilayer surfaces at 2, 8 and 24 h time points. Data from only one donor was utilized for each time point. 2 replicate samples for the 2 h time point for PLL ML and HA ML surfaces were taken and, therefore, no standard error bars are shown for that time point.

stealth backgrounds onto which bioactive signals could be incorporated. The construction of PLL/HA alternating multilayer coatings has been previously described by Picart *et al.* [19], where the detailed growth mechanism and surface architecture were characterized. Here, we have used SPR to measure multilayer deposition and calculate theoretical effective thickness and coverage values. We also used ToF-SIMS surface analysis techniques to look at the upper layer composition, to relate the surface properties to their effects on monocyte adhesion and activation.

SPR analysis showed that the film thickness was indeed increasing with each deposition step, indicating that PEC multilayers were being formed at the surface as previously reported [19]. The calculated mass per area of the multilayer film was found to be within the same order of magnitude as previously reported [19]. The details of hydration, substrate charge density and polymer molecular weights may account for the observed differences. The ToF-SIMS analysis showed a significant change in the surface chemistry of the TCPS substrate with PEC multilayer formation. The analysis, however, indicated that significant levels of PLL chain segments were found in the top surface of the HA-terminated multilayers, despite multiple HA additions in the final deposition steps. This provides evidence for PLL diffusion throughout PLL/HA multilayer films during the multilayer formation process, as described previously for PEL layers [20]. Quantitative analysis by

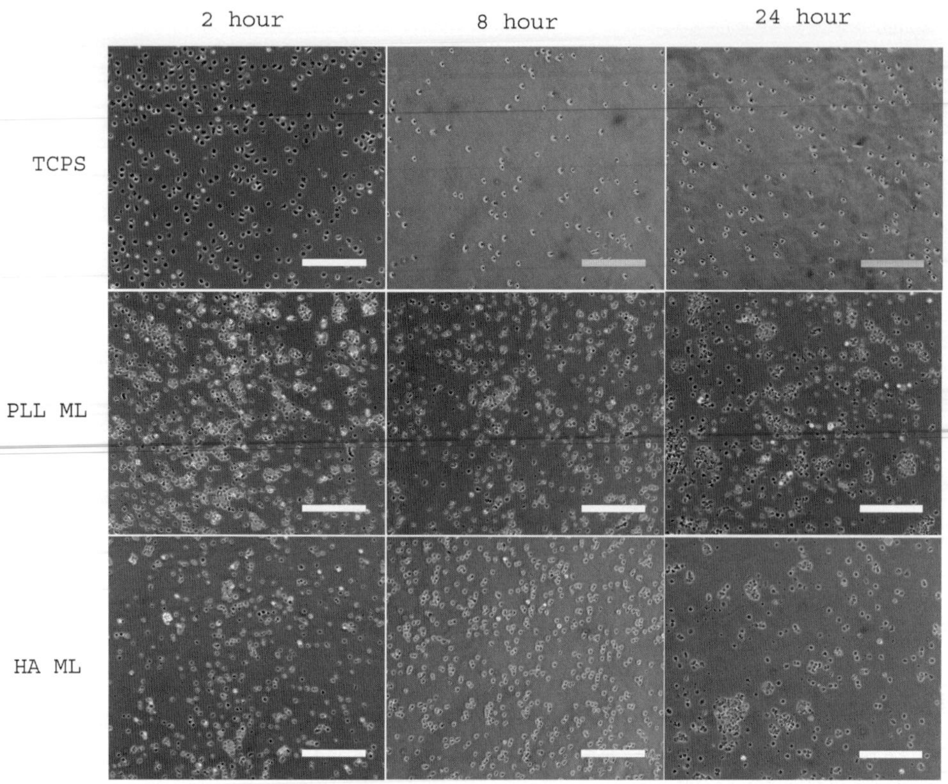

Figure 6. Live/Dead fluorescence microscopy images of adherent monocytes at three time points. A Live/Dead stain overlaid on a phase-contrast image using a 10× objective was used to evaluate the viability of adherent monocytes on TCPS, PLL-terminated (PLL ML) and HA-terminated (HA ML) multilayer surfaces made in 24-well TCPS plates. Red indicates dead cells and green indicates live cells. The quantification of viable cell percentages for each image is found in Table 1. Calibration bar = 200 μm.

PCA of the PLL- and HA-terminated film determined that the two surfaces had indistinguishable surface chemistries.

The extent of surface reorganization observed in the ToF-SIMS studies was likely affected by surface restructuring when water was removed under high vacuum, but strong evidence for significant PLL diffusion to the surface was also found in the macrophage response to the multilayer coatings. There was a high level of cell detachment or death between 2–8 h after initial seeding from both multilayer coatings. Microscopic analysis of the representative adherent cell populations demonstrated that there was significant cell death on both PLL- and HA-terminated surfaces. Both the PLL- and HA-terminated coatings displayed similarly high degrees of TNF-α up-regulation compared to the TCPS control. These results strongly suggest that the inflammatory monocyte response is due to the significant presence of PLL in the surface layer of the HA-terminated multilayer coating.

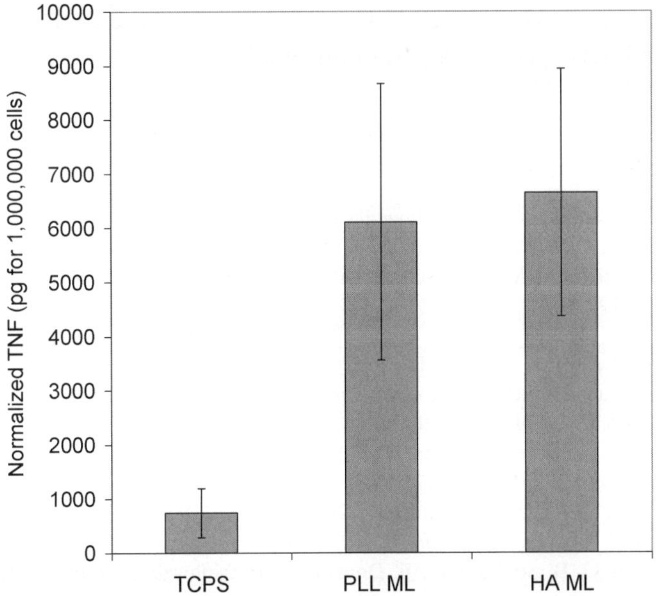

Figure 7. Monocyte activation on multilayer surfaces. Normalized TNF-α production was quantitated by ELISA and normalized to cell number after a 24 h incubation period. 1.5×10^5 cells were initially plated on TCPS and 1×10^6 cells were initially plated on PLL-terminated (PLL ML) and HA-terminated (HA ML) surfaces. Data from triplicate samples per group for three donors are presented.

Soluble PLL has been shown to elicit a strong TNF-α response from monocytes [30] and polycations in general have well-known cytotoxic potential [30–32].

Morra *et al.* have recently shown that fibroblast adherence to a single HA coating layer that was electrostatically bound onto a cationic surface was comparable to that on the bare cationic surface, while covalent grafting of HA to the polycation strongly inhibited fibroblast adherence [33]. HA surface density and presentation relative to polycationic layers and surfaces are thus critical to controlling biological responses. The appropriate choice of chain lengths and charge densities could lead to lower polycation surface diffusion, and different polycations with more favorable biological properties, such as chitosan, could also be used to create multilayer films with HA [31, 34]. The details of the polyelectrolyte multilayer architecture, and especially the surface composition as affected by inter-layer diffusion, must be carefully controlled when using layer-by-layer strategies for biomaterial coatings.

Acknowledgements

We thank the University of Washington Engineered Biomaterials (UWEB) Center and its Microscopy Facilities (EEC-9872882 and EEC-9529161), the NIH BEAT program (R24-HL64387), National ESCA and Surface Analysis Center for Biomedical Problems (NESAC/BIO, NIH EB 002027) for surface analysis and Dr. Gibum Kim for his assistance with SPR calibration. The Whitaker Foundation provided fellowship support for J. J. H. C. T. C. would like to thank DOE-BES, Chemical Sciences Division for partial support of this work.

REFERENCES

1. J. M. Anderson and K. M. Miller, *Biomaterials* **5**, 5 (1984).
2. J. M. Anderson, *Annu. Rev. Mater. Res.* **31**, 81 (2001).
3. Q. Zhao, N. Topham, J. M. Anderson, A. Hiltner, G. Lodoen and C. R. Payet, *J. Biomed. Mater. Res.* **25**, 177 (1991).
4. R. S. Labow, E. Meek and J. P. Santerre, *J. Biomed. Mater. Res.* **54**, 189 (2001).
5. B. D. Ratner, *J. Control. Rel.* **78**, 211 (2002).
6. N. Wisniewski and M. Reichert, *Colloid. Surf. B Biointerf.* **18**, 197 (2000).
7. C. A. Dinarello, *Chest* **118**, 503 (2000).
8. G. Decher, J. D. Hong and J. Schmitt, *Thin Solid Films* **210**, 831 (1992).
9. G. Decher, *Science* **277**, 1232 (1997).
10. G. Ladam, P. Schaaf, F. J. G. Cuisinier, G. Decher and J. C. Voegel, *Langmuir* **17**, 878 (2001).
11. F. Caruso and C. Schuler, *Langmuir* **16**, 9595 (2000).
12. A. A. Antipov, G. B. Sukhorukov, E. Donath and H. Mohwald, *J. Phys. Chem. B* **105**, 2281 (2001).
13. O. P. Tiourina, A. A. Antipov, G. B. Sukhorukov, N. L. Larionova, Y. Lvov and H. Mohwald, *Macromol. Biosci.* **1**, 209 (2001).
14. D. Vautier, V. Karsten, C. Egles, J. Chluba, P. Schaaf, J. C. Voegel and J. Ogier, *J. Biomater. Sci. Polymer Edn* **13**, 713 (2002).
15. L. Richert, P. Lavalle, D. Vautier, B. Senger, J. F. Stoltz, P. Schaaf, J. C. Voegel and C. Picart, *Biomacromolecules* **3**, 1170 (2002).
16. D. L. Elbert, C. B. Herbert and J. A. Hubbell, *Langmuir* **15**, 5355 (1999).
17. T. Matsuda, H. Miwa, M. J. Moghaddam and F. Iida, *ASAIO J.* **39**, M327 (1993).
18. M. Morra and C. Cassineli, *J. Biomater. Sci. Polymer Edn* **10**, 1107 (1999).
19. C. Picart, P. Lavalle, P. Hubert, F. J. G. Cuisinier, G. Decher, P. Schaaf and J. C. Voegel, *Langmuir* **17**, 7414 (2001).
20. C. Picart, J. Mutterer, L. Richert, Y. Luo, G. D. Prestwich, P. Schaaf, J. C. Voegel and P. Lavalle, *Proc. Natl. Acad. Sci. USA* **99**, 12531 (2002).
21. B. L. Frey, C. E. Jordan, S. Kornguth and R. M. Corn, *Anal. Chem.* **67**, 4452 (1995).
22. L. S. Jung, C. T. Campbell, T. M. Chinowsky, M. N. Mar and S. S. Yee, *Langmuir* **14**, 5636 (1998).
23. R. Barer and S. Joseph, *Q. J. Microsc. Sci.* **95**, 339 (1954).
24. T. E. Leslie and T. H. Lilley, *Biopolymers* **24**, 695 (1985).
25. S. Gomez-Alejandre, E. S. de la Blanca, C. A. de Usera, M. F. Rey-Stolle and I. Hernandez-Fuentes, *Int. J. Biol. Macromol.* **27**, 287 (2000).
26. J. B. Lhoest, M. S. Wagner, C. D. Tidwell and D. G. Castner, *J. Biomed. Mater. Res.* **57**, 432 (2001).
27. M. S. Wagner and D. G. Castner, *Langmuir* **17**, 4649 (2001).
28. M. R. Rosengart, S. Arbabi, G. J. Bauer, I. Garcia, S. Jelacic and R. V. Maier, *Shock* **17**, 109 (2002).
29. M. C. Shen and T. A. Horbett, *J. Biomed. Mater. Res.* **57**, 336 (2001).
30. B. L. Strand, L. Ryan, P. I. Veld, B. Kulseng, A. M. Rokstad, G. Skjak-Braek and T. Espevik, *Cell Transplant.* **10**, 263 (2001).
31. A. Prokop, D. Hunkeler, S. DiMari, M. A. Haralson and T. G. Wang, *Adv. Polym. Sci.* **136**, 1 (1998).

32. S. Choksakulnimitr, S. Masuda, H. Tokuda, Y. Takakura and M. Hashida, *J. Control. Rel.* **34**, 233 (1995).
33. M. Morra, C. Cassinelli, A. Pavesio and D. Renier, *J. Colloid Interf. Sci.* **259**, 236 (2003).
34. B. Thierry, F. M. Winnik, Y. Merhi and M. Tabrizian, *J. Am. Chem. Soc.* **125**, 7494 (2003).

Smart polymers, surfaces and hydrogels

Gels, Genes, Grafts and Giants, pp. 509–518 (2005)

The effect of pH on the LCST of poly(N-isopropylacrylamide) and poly(N-isopropylacrylamide-co-acrylic acid)

YONG PEI, JEI CHEN *, LIMING YANG, LILI SHI, QIONG TAO, BAOJUN HUI and JIAN LI

Department of Chemical Engineering and Technology, School of Environmental and Chemical Engineering, Shanghai University (Jiading Campus), Shanghai 201800, China

Received 26 September 2003; accepted 23 December 2003

Abstract—Poly(N-isopropylacrylamide) (PNIPAAm) and poly(N-isopropylacrylamide-co-acrylic acid) (P(NIPAAm-co-AAc)) hydrogels are synthesized by irradiating the aqueous solutions of NIPAAm and NIPAAm/AAc with ^{60}Co γ-ray. The effects of pH on the swelling ratio and on the lower critical solution temperature (LCST) are studied by determining the dependence of swelling ratio on temperature in different pH butter solutions. Differential scanning calorimetry (DSC) is applied in determination of the LCST of the hydrogels. Fourier transform infrared (FT-IR) spectrometry is used in the comparison of hydrogels swelled in various pH conditions. As a result, PNIPAAm was found to be a pH-sensitive hydrogel and the LCST of the PNIPAAm and P(NIPAAm-co-AAc) hydrogels are influenced by pH.

Key words: Temperature sensitivity; pH sensitivity; LCST; PNIPAAm; P(NIPAAm-co-AAc).

INTRODUCTION

Crosslinked poly(N-isopropylacrylamide) (PNIPAAm) is a temperature-sensitive hydrogel, which has a lower critical solution temperature (LCST) at about 32°C in pure water. PNIPAAm chains hydrate to form extended structures in water when it is below LCST. When heated up above the LCST, they become compacted by dehydration. Much attention has focused on the LCST since the first observation of gel collapse by Tanaka [1–3]. Using ideal parameters, the LCST is calculated as discontinuous volume transition of non-ionized PNIPAAm for the very first time, but it does not match with the experimental results very closely. Afterwards, several explanations about the LCST are reported, such as 'aqueous polymer solutions

*To whom correspondence should be addressed. E-mail: chenjjd@online.sh.cn

(e.g. NIPAAm gels) representing systems in which specific (hydrogen-bonding) interactions are the driving forces for LCST behavior [4]' or an 'interaction occurring in polymer and solvent (hydrophobic force) may be the only driving force [5]'. With the enhancement of the research, more and more people have come to agree that 'hydrogen bond' and 'hydrophobic force' are of the same importance in the transition process of the LCST [6, 7].

Because of the potential applications in biomedical materials, based upon the reversible phase transition behavior, many strategies have been reported to modify the properties of PNIPAAm. For example, Beltran *et al.* [8] studied the influence of pH, ionic strength and temperature on the swelling equilibrium of PNIPAAm hydrogels containing weakly ionizable groups such as sodium acrylate (SA) and 2-(dimethylamino)ethyl methacrylate (DMA). Hoffman *et al.* [9–14] conjugated PNIPAAm with proteins or other biomaterials to research the applications of controlled release, affinity separation and so on.

In this paper, the pH-sensitivity of PNIPAAm and P(NIPAAm-co-AAc) and their LCSTs are studied. Differential scanning calorimetry (DSC) is used to verify the LCST of the hydrogel in different pH solutions. Fourier transform infrared (FT-IR) spectrometry is used in the comparison of the differences among hydrogels swelled in various pH conditions.

MATERIALS AND METHODS

Materials

N-Isopropylacrylamide (NIPAAm), obtained as a gift from Professor Kaetsu, Kinki University, is made by Wako Pure Chemical. Prior to the reaction, NIPAAm is purified by recrystalliztion form the solvent mixture of n-hexane/benzene. Acrylic acid, purchased from Shanghai Reagent, is used without any purification. Other chemicals are all of regent grade and are used without any purification.

Synthesis of hydrogels

Hydrogels of PNIPAAm and P(NIPAAm-co-AAc) are prepared by radiation polymerization by ^{60}Co γ-ray (300 kCi, Shanghai Hexing Radiation Factory) with the dose of 25 kGy in moving condition. The aqueous solutions containing 10% monomer of NIPAAm or NIPAAm/AAc with different feed ratios, varying from 100:0 to 39.25:60.75 in molar ratio, are prepared respectively. The crosslinked hydrogels are obtained by exposing the solution in the irradiation field in moving state [15, 16].

The hydrogels are cut into small pieces with a size of $10 \times 5 \times 2$ mm approximately and are soaked in deionized water for 24 h with stirring and water changing periodically to remove non-crosslinked compounds.

Determination of swelling ratio

The gels synthesized are characterized by their equilibrium degree of swelling and rates of weight change. Swelling measurements are done using a gravimetric method. The swelling characteristics of the hydrogels are investigated by immersing the gels in different pH buffer solutions at desired temperature. Because of the effects of ionic strength on the swelling behavior [17], the ionic strength of the swelling medium is adjusted to 0.4 M by calculated NaCl. Samples are removed after 1 h intervals, blotted free of surface water with two pieces of filter paper, weighed with a balance (accuracy ±0.0001 g) and returned to swelling medium. This procedure is repeated until the samples attain constant weight. Each sample is determined three times and the average value is listed. The swelling ratio of the gel is defined as follows:

$$\text{Swelling ratio (times)} = \frac{W_s - W_d}{W_d},$$

where W_s and W_d are the weights of the swelled and dried samples, respectively.

DSC

Differential scanning calorimetry (DSC 200PC, Phox, Germany) is used to determine the LCST directly. The swelled hydrogel is put into the sample crucible and equilibrated at 2°C for 5 min. After removing the superabundant water, the hydrogels heated up from 2°C to 60°C at the rate of 2°C/min in N_2 atmosphere.

FT-IR

At swelling equilibrium in pH buffer solutions, the samples are cut into small pieces with a size of approx. 5 × 5 × 1 mm and then dried in a vacuum dryer at 50°C. The samples are determined by Fourier transform infrared (FT-IR) spectrometry (Avatar 370, Nicolet, USA) in the attenuated total reflectance (ATR) mode.

RESULTS AND DISCUSSION

The effect of pH on PNIPAAm

The effects of pH on the swelling ratio of PNIPAAm hydrogel are determined in various pH buffer solutions, as shown in Fig. 1. The swelling ratio of PNIPAAm decreases with the increase of pH within the range of 1–7 and then increases in the pH range 7–12, at 10°C. In comparison, the swelling ratio of PNIPAAm decreases with pH in the range 1–4, and then it becomes nearly zero at pH 4–7 at 26°C. It might be considered that, at pH 4–7 (with an ionic strength of 0.4 M), 26°C exceeds the LCST of the PNIPAAm hydrogel. So it can be supposed that PNIPAAm is a pH-sensitive hydrogel and its LCST affected by pH.

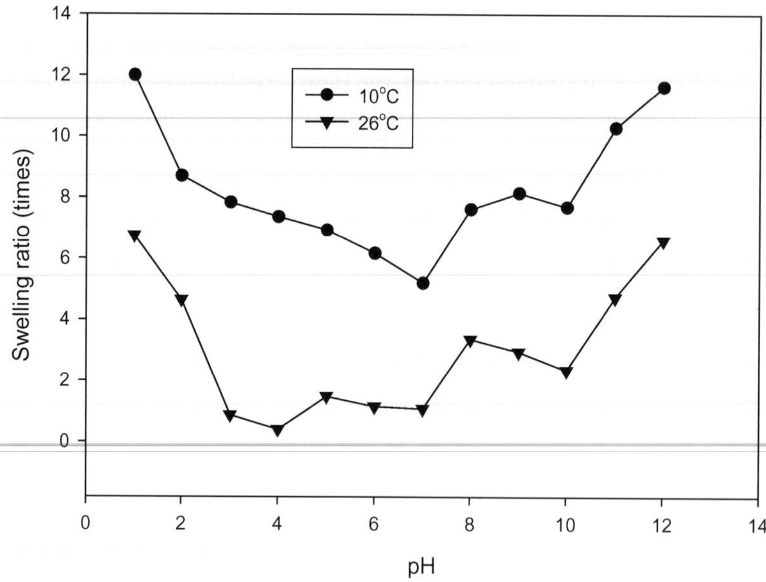

Figure 1. Effect of pH on swelling ratio of PNIPAAm at 10 and 26 °C.

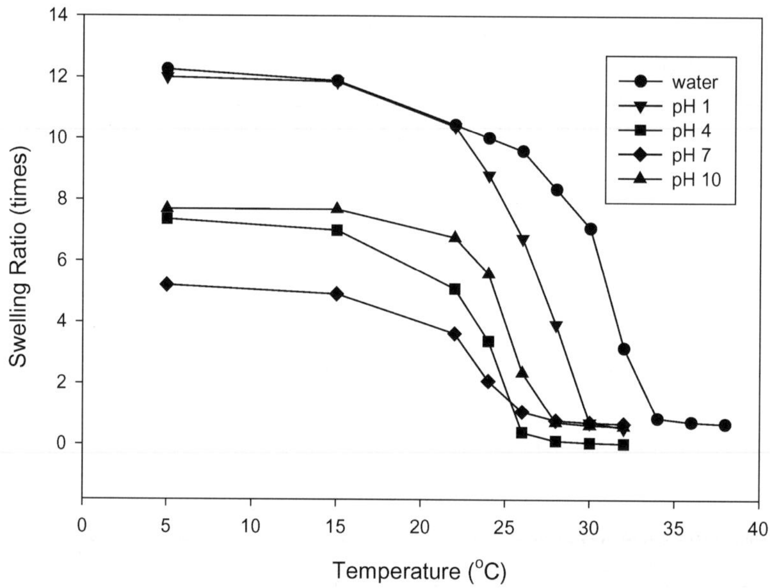

Figure 2. Effect of temperature on swelling ratio of PNIPAAm at various pH values.

The dependence relationship between swelling ratio and temperature in different pH buffer solutions is determined, as shown in Fig. 2. Similar to the swelling behavior in deionized water, the PNIPAAm hydrogel in all pH buffer solutions appears a sudden drop of swelling ratio in a narrow temperature scope. This narrow scope is different in various pH buffer solutions, which also indicates that the LCST

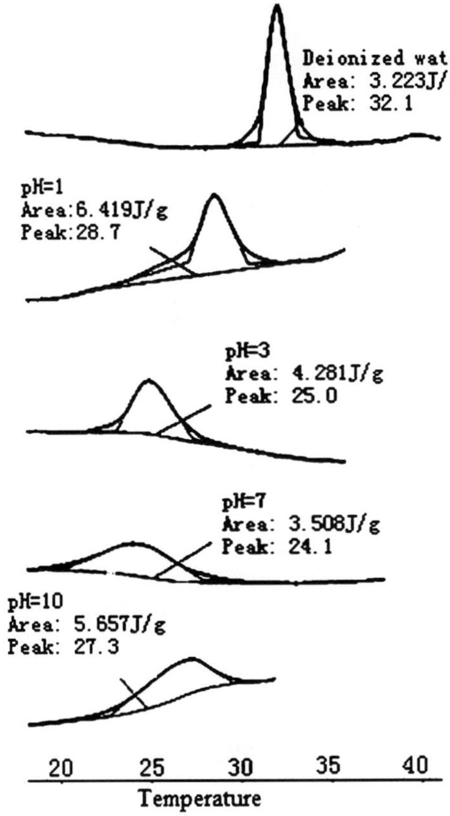

Figure 3. DSC tests of PNIPAAm in solutions varying in pH.

of the PNIPAAm is affected by pH. Relatively, the LCST decreases with the increase of pH in the range 1–7; however, it increases instead when pH $>$ 7.

Usually, the phase transition temperature (or LCST) of PNIPAAm hydrogel can be directly determined by differential scanning calorimetry (DSC). The swollen PNIPAAm samples in different pH buffer solutions were determined by DSC in an attempt to study the LCST, as shown in Fig. 3. The DSC spectra show that the phase transition peak moves to lower temperature with the increase of pH in the range 1–7. The peaks are at 28.7, 25.0 and 24.1 °C, corresponding to pH 1, 3 and 7, respectively. At pH 10, the phase transition peak moves to 27.3°C. This also indicates that the LCST of PNIPAAm is affected by pH.

Until now, it may be reasonable to illustrate the results as in Fig. 4. In weak acid/base solutions or pure water, there are hydrogen bonds in the PNIPAAm hydrogel network due to the existence of isopropyl substituted amide group. So the hydrogel cannot swell further. But in strong acid/base solutions, the hydrogen bonds are destroyed and the network is partly ionized. Then, the hydrogel network expends because of the interaction between water and the hydrogel network.

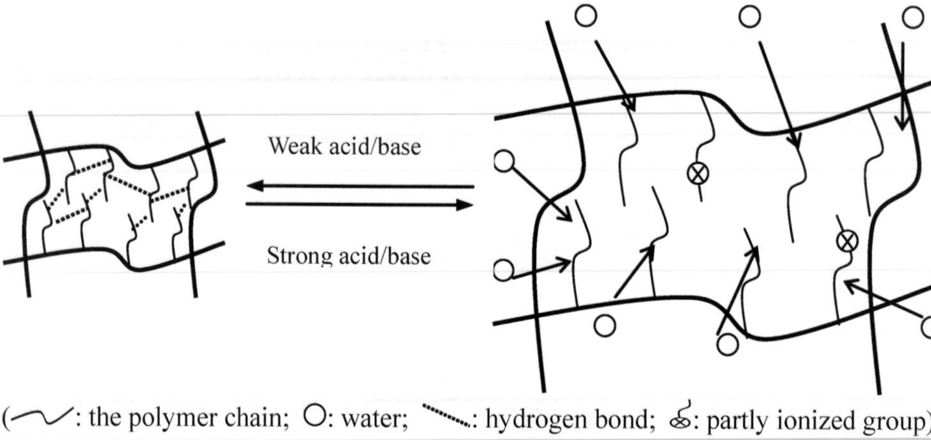

Figure 4. Scheme of the swelling change of the PNIPAAm hydrogel at various pH values.

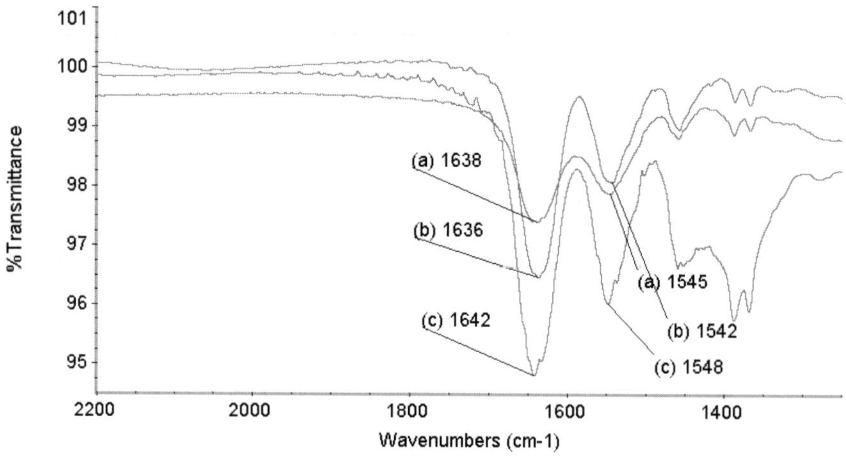

Figure 5. FT-IR spectra of PNIPAAm: (a) in pH 12 buffer solution; (b) in pH 1 buffer solution; (c) in pH 7 buffer solution.

Treated with different pH buffer solutions, the PNIPAAm hydrogels are determined by FT-IR spectrometry in the ATR mode. The comparisons of FT-IR spectra of swollen PNIPAAm hydrogels in pH 1, 7 and 12 buffer solutions are shown in Fig. 5. In pH 1, 7 and 12 buffer solutions, the amide-I peaks are at 1636 cm^{-1}, 1642 cm^{-1} and 1637 cm^{-1}, respectively; meanwhile, the amide-II peaks are at 1542 cm^{-1}, 1548 cm^{-1} and 1545 cm^{-1}. This red shift of the amide group peaks can partly verify the ionization of the amide group, as reported by Qu *et al.* [18].

The swelling reversibility of PNIPAAm hydrogel is determined by measuring the swelling ratio of the sample in the pH 1 and 4 buffer solutions at 26°C. The swelling ratio is tested with the time. The dried samples are weighted and first immersed into the pH 1 buffer solution, and then are weighed every 10 min after removing the surplus water. Swelling equilibrated, the samples are washed with deionized water

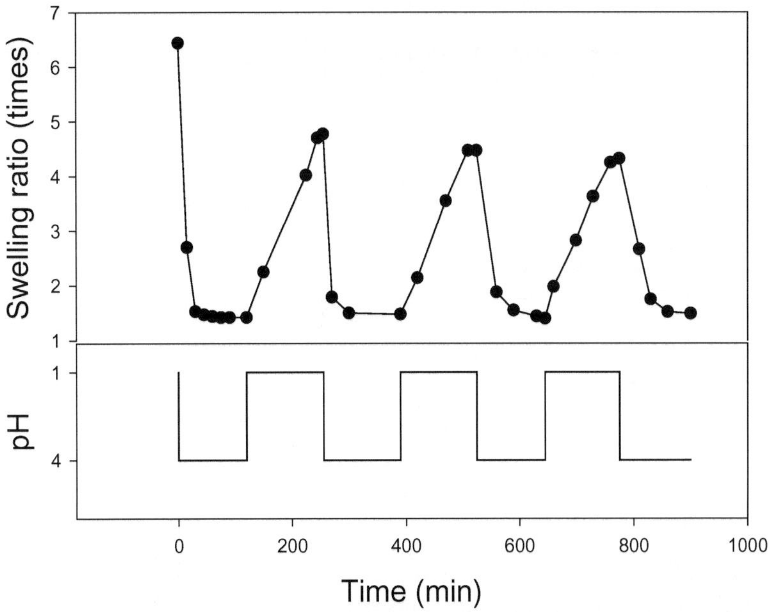

Figure 6. Swelling ratio of PNIPAAm as a function of time under repeated abrupt change of pH between 1 and 4 at 26°C.

and transferred into pH 4 for the same treatment. The whole process is repeated for several times. As shown in Fig. 6, The PNIPAAm hydrogels swell in pH 1 and deswell in pH 4 buffer solutions. The results demonstrate that the samples change their absorbability when environmental pH is altered and the change is reversible. This may be because 26°C does not exceed the LCST of PNIPAAm at pH 1, but does exceed the LCST at pH 4, which can verify that the LCST of PNIPAAm is affected by pH.

The effect of pH on P(NIPAAm-co-AAc)

The effect of pH on swelling ratio of P(NIPAAm-co-AAc) with various feed ratios, measured at about 15°C, is shown in Fig. 7. They exhibit the same swelling tendency compared with the pure PNIPAAm hydrogels. Meanwhile, when AAc, containing a negative group ($-COO^-$), is added the swelling ratio increases greatly.

It is supposed that swelling ratio of this kind of ionic gel is determined by some major factors [19, 20], such as the hydrophilic/hydrophobic balance in the polymer network, the electrostatic repulsion between the carboxylic acid polymer side chain and ions present in the buffer solutions, and ionic osmotic pressure in the network, etc. In lower pH solutions, NIPAAm unit of structure exhibits slight ionization, as shown in Fig. 4, which makes the chain extend. Meanwhile, it is the strong water absorbability of the AAc unit that makes the swelling ratio greater than that of pure PNIPAAm; with the increase of pH below the LCST of PNIPAAm, the dissociation degree of the AAc component increases gradually, accompanied

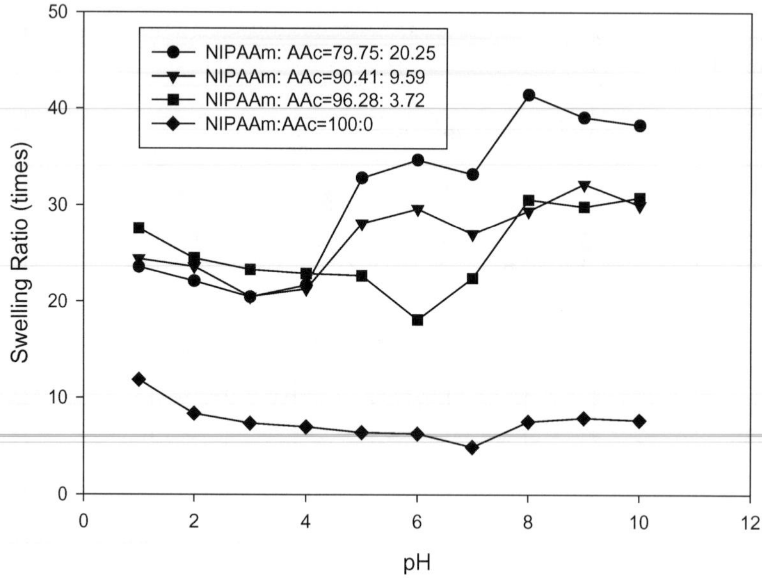

Figure 7. Effect of pH on the swelling ratio of P(NIPAAm-co-AAc) with various feed ratios at about 15°C.

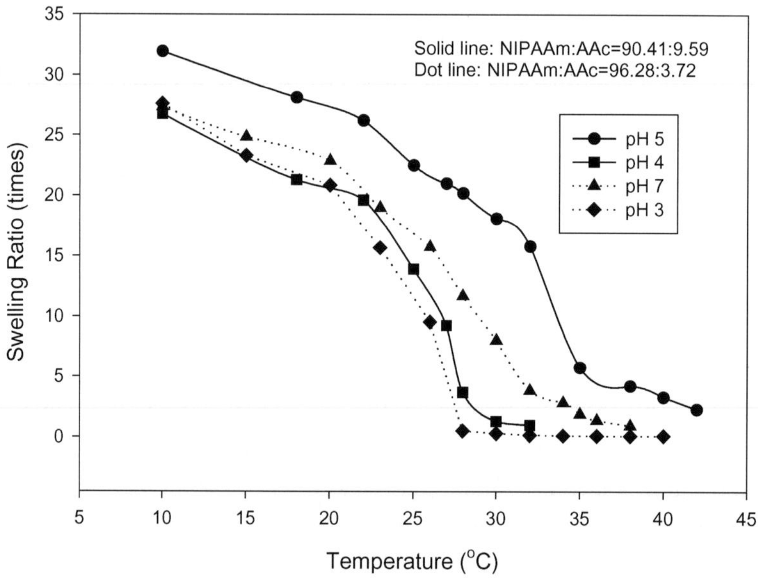

Figure 8. Effect of temperature on the swelling ratio of P(NIPAAm-co-AAc) at various feed ratios in different pH solutions.

by the disappearance of ionization. The driving force may due to the dynamic equilibrium between the disappearance of ionization, which leads to the shrinkage of the polymer network, and electrostatic repulsion of the carboxylic group, which

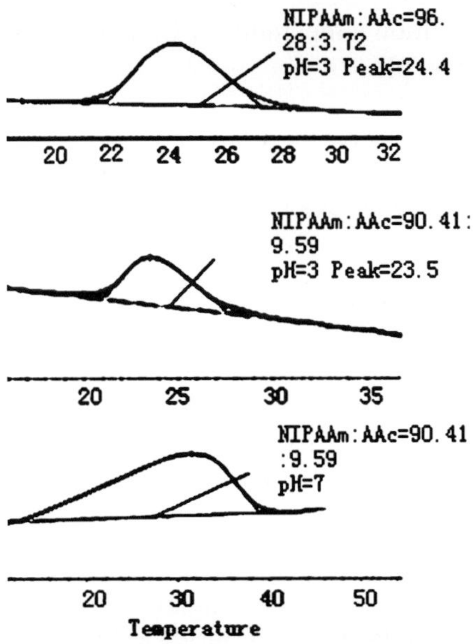

Figure 9. DSC tests of two kinds of P(NIPAAm-co-AAc) in various pH solutions.

makes the polymer chain diffuse; above the LCST of PNIPAAm, electrostatic repulsion becomes the major force.

The pH and feed ratio may be two major effective factors on the LCST of P(NIPAAm-co-AAc). At lower pH, the hydrophilic/hydrophobic balance becomes the major acting force since the dissociation of carboxylic acid is restrained by the presence of a counterion (H^+). Then, under such conditions, the LCST of P(NIPAAm-co-AAc) appears the same as that of PNIPAAm. But at the same pH, the LCST of the hydrogel with various feed ratios appears different. As shown in Fig. 8, in pH 3 buffer solutions, the LCST of the hydrogel, NIPAAm/AAc = 96.28 : 3.72, is 24.4°C. Meanwhile, that of NIPAAm/AAc = 90.41 : 9.59 is at 23.5°C. With the increase of pH, electrostatic repulsion is directly proportional to the dissociation degree. LCST exists until the AAc unit can convey sufficient solubility to offset the aggregation caused by the hydrophobic effect of the temperature-sensitive unit [14]. So pH and feed ratio can be regarded as the major factors of the existence of LCST. The curves in Fig. 9 can verify the explanation above. The LCSTs of P(NIPAAm-co-AAc) with different feed ratios are directly measured by DSC, as shown in Fig. 9.

CONCLUSIONS

(1) PNIPAAm and P(NIPAAm-co-AAc) hydrogels are obtained by irradiating the aqueous solution of NIPAAm and NIPAAm/AAc with γ-rays from ^{60}Co.

(2) PNIPAAm is proved to be a pH-sensitive hydrogel.

(3) The lower critical solution temperature (LCST) of PNIPAAm and P(NIPAAm-co-AAc) is affected by pH.

Acknowledgements

The authors thank the Shanghai Science and Technology Committee for financial support.

REFERENCES

1. T. Tanaka, *Phys. Rev. Lett.* **40**, 820 (1978).
2. Y. Hirokawa and T. Tanaka, *J. Chem. Phys.* **81**, 6379 (1984).
3. S. Hirotsu, Y. Hirokawa and T. Tanaka, *J. Chem. Phys.* **87**, 1392 (1987).
4. S. Beltran, H. H. Hooper, H. W. Blanch and J. M. Prausnitz, *J. Chem. Phys.* **92**, 2061 (1990).
5. S. Fujishige, K. Kubota and I. Ando, *J. Chem. Phys.* **94**, 5154 (1990).
6. F. M. Winnik, *Macromolecules* **23**, 233 (1990).
7. X. Z. Zhang and R. X. Zhuo, *Chem. J. Chin. Univ.* **21**, 1309 (2000) (in Chinese).
8. S. Beltran, J. P. Baker, H. H. Hooper, H. W. Blanch and J. M. Prausnitz, *Macromolecules* **24**, 549 (1991).
9. P. S. Stayton, T. Shimoboji, C. Long, A. Chilkoti, G. H. Chen, J. M. Harris and A. S. Hoffman, *Nature* **378**, 472 (1995).
10. Z. L. Ding, C. J. Long, Y. Hayashi, E. V. Bulmus, A. S. Hoffman and P. S. Stayton, *Bioconjug. Chem.* **10**, 395 (1999).
11. J. P. Chen, H. J. Yang and A. S. Hoffman, *Biomaterials* **11**, 625 (1990).
12. J. P. Chen and A. S. Hoffman, *Biomaterials* **11**, 631 (1990).
13. S. Anastase-Ravion, Z. Ding, A. Pelle, A. S. Hoffman and D. Letourneur, *J. Chromatogr. B* **761**, 247 (2001).
14. G. H. Chen and A. S. Hoffman, *Nature* **373**, 49 (1995).
15. A. Panda, S. B. Manohar and S. Sabharwal, *Radiat. Phys. Chem.* **58**,101 (2000).
16. J. M. Rosiak and P. Ulański, *Radiat. Phys. Chem.* **55**, 139 (1999).
17. H. G. Schild and D. A. Tirrell, *J. Phys. Chem.* **94**, 4352 (1990).
18. X. Qu, A. Wirsen and A. C. Albertson, *Polymer* **41**, 4589 (2000).
19. M. K. Yoo, Y. K. Sung, Y. M. Lee and C. S. Cho, *Polymer* **41**, 5713 (2000).
20. C. S. Brazel and N. A. Peppas, *J. Control. Rel.* **39**, 57 (1996).

Gels, Genes, Grafts and Giants, pp. 519–528 (2005)

A hydrogel system for stimulus-responsive, oxygen-sensitive *in situ* gelation

ANDREAS GOESSL [1,2,*], NICOLA TIRELLI [1,3]
and JEFFREY A. HUBBELL [1,4]

[1] *Institute of Biomedical Engineering, Department of Materials, Swiss Federal Institute of Technology (ETH), Zurich, Switzerland*
[2] *Baxter Bioscience, Industriestr. 131, A-1220 Vienna, Austria*
[3] *School of Pharmacy and Molecular Materials Centre, University of Manchester, Manchester, UK*
[4] *Institute for Biological and Chemical Engineering, Swiss Federal Institute of Technology Lausanne (EPFL), Ecublens, Switzerland*

Received 1 October 2003; accepted 6 February 2004

Abstract—A hydrogel system that is designed to gel due to oxidation by molecular oxygen is described. This is achieved by the use of a branched poly(ethylene glycol) modified with thiol end groups. A stable precursor molecule, starPEG thioacetate, was synthesized by radical addition of thioacetic acid to an intermediate starPEG allyl ether. Rapid deprotection of the thiol can be achieved using a base, e.g. sodium hydroxide, quantitatively liberating a thiol group and a non-toxic acetate ion. This step can be carried out under anaerobic conditions, yielding a solution with known thiol content that can be stored. The reaction with oxygen is accelerated by the use of a catalyst based on Fenton chemistry, which makes the material useful for biomedical applications where *in situ* polymerization of an injectable material is beneficial. This gelation takes place under near physiological conditions without the need for a cross-linking agent.

Key words: Hydrogel; stimulus-responsive; thiol; disulfide; *in situ* polymerization.

INTRODUCTION

There are a number of biomedical applications that can benefit from the use of *in situ* solidifying biomaterials, ranging from tissue bulking or drug delivery to the formation of solid, load-bearing hard tissue substitutes [1–3]. The liquid precursors of these materials can be directly injected into the site of application and they conform to the shape of the application site without the need to prefabricate a device to the often complex shapes of the defects or sites of applications. Several systems

*To whom correspondence should be addressed at Baxter. E-mail: andreas_goessl@baxter.com

for the formation of a covalent hydrogel network *in situ* have been described, but either the toxicity of the cross-linking process, the multi-component nature of the system, or the variability in the physico-chemical properties of the precursors (e.g. with natural biopolymers) can limit their usefulness [4, 5].

In this study we describe a novel one-component hydrogel system that is distinguished by the ability to gel in response to a chemical stimulus inherent to many biological systems. The sol–gel transition (gelation) of this system occurs under mild conditions and only in the presence of molecular oxygen supplied, for example, by oxygenated tissues. This oxidizes multifunctional poly(ethylene glycol) (PEG) thiols to the corresponding intermolecular disulfides and leads to the formation of a biocompatible, chemically cross-linked gel.

The use of multifunctional thiols, however, poses significant challenges. Uncontrolled dimerization upon synthesis and storage is likely, and the formation of both inter- and intra-molecular disulfide bonds during the gelation is possible, limiting the number of 'useful' disulfide bonds that can contribute to gelation and gel stability. The issue of uncontrolled dimerization can be overcome by the use of a protected thiol derivative. Thiols should be quantitatively liberated just prior to application. The by-products of deprotection have to be non-toxic and must not interfere with gelation. These requirements pointed to the use of thioacetate as a stable, protected thiol group (with acetate as the leaving group) [6, 7].

If the oxygen-free precursor solution is injected into the site of application, oxygen diffusion from the surrounding tissue was expected to lead to rapid gelation *in situ*. Thiol oxidation by oxygen was initially found to be too slow to be clinically useful. Therefore, we have developed a 'Fenton'-type catalyst system which accelerates oxygen dependent gelation to only a few minutes [8, 9].

We expect that materials created by this chemistry will degrade over the time course of several days by thiolysis, i.e. the reductive cleavage of disulfide bonds by physiological thiol carrying molecules. The degradation products, non-toxic multi-armed PEGs, are expected to be cleared *via* the renal system, comparable to PEGs of similar molecular weight [10].

The transient presence of these gels together with the possibility to tailor the properties of the resulting gels by varying the molecular weight and architecture of the starting starPEG molecules make these materials interesting for drug- or gene-delivery applications. Another possible application is the formation of a thin, anti-adhesive and protein repellant hydrogel on a tissue surface. This can be formed for example on the luminal surface of an arterial blood vessel, protecting it against thrombosis after dilation of the vessel lumen in an angioplasty procedure. Furthermore, this thin hydrogel barrier can serve in the prevention of surgical adhesions following interperitoneal surgery [11].

MATERIALS AND METHODS

All reagents and solvents were used as received, unless otherwise noted. A linearly branched poly(ethylene glycol) with 4 arms and a total molecular weight of 10^4 g/mol was purchased from Shearwater (now Nektar, Huntsville, AL, USA). This material will be referred to as starPEG. Prior to derivatization it was dried by refluxing its toluene solution in a Soxhlet apparatus containing a 4 Å molecular sieve. Sodium hydride, allyl bromide and 2,2′-azo-*bis*-isobutyrylnitrile (AIBN) were purchased from Aldrich (Milwaukee, WI, USA). AIBN was used after recrystallization from ethanol. Thioacetic acid and ferrous lactate were purchased from Fluka.

^{1}H-NMR analysis was carried out on a 300 MHz Bruker DPX 300 instrument. ATR–FT-IR analysis was performed on a Perkin Elmer Spectrum One system.

StarPEG thioacetate synthesis

The synthesis protocol was modified from a literature procedure [7] and will be described briefly below (Fig. 1). To the dried starPEG solution in toluene (5 g or

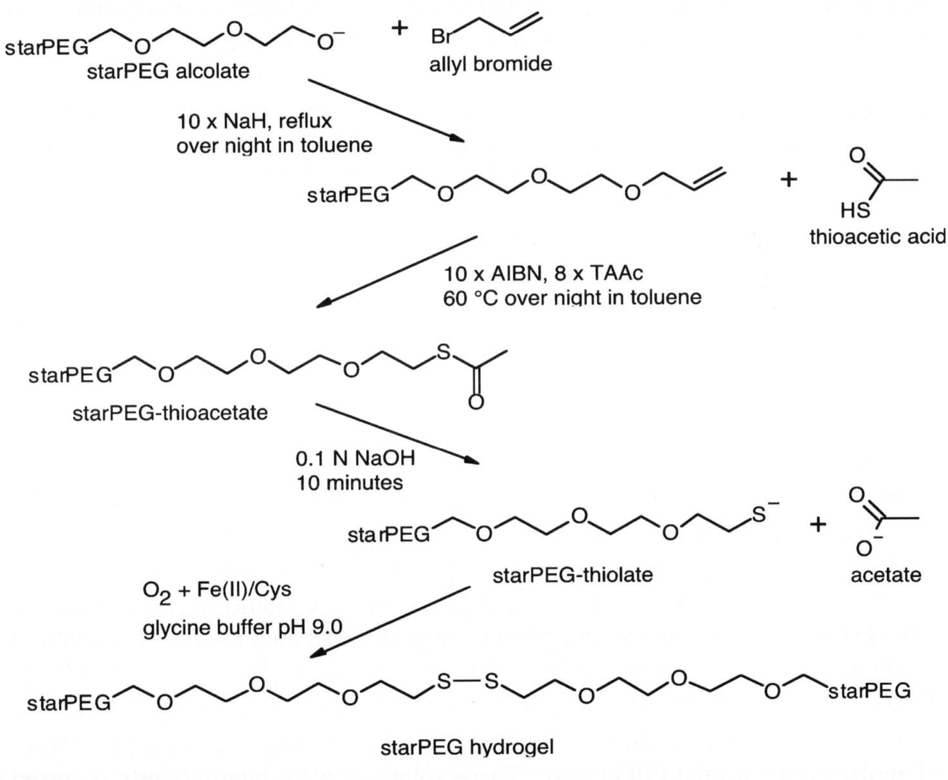

Figure 1. Reaction scheme for the synthesis and gelation of a starPEG thioacetate. Step I: substitution of the PEG alcolate with allyl bromide. Step II: radical addition of the thioacetic acid radial to the allyl double bond. Step III: hydrolysis of the starPEG thioacetate to yield the thiol. Step IV: disulfide bond formation between starPEG-SH monomers.

2 mmol end-groups in 200 ml) a 10× molar excess (over −OH end-groups) of sodium hydride and an 8× excess of allyl bromide was added with due care under nitrogen atmosphere, and the solution was refluxed over night. After removal of the residual sodium hydride by water extraction, the starPEG allyl ether was recovered by repeated precipitation in ice-cold diethyl ether.

Conversion rate was calculated based on the ratio of the allyl ^{1}H-NMR peaks at 5.2 and 5.58 ppm to the PEG methylene peak.

The reaction conditions for the radical addition of thioacetic acid (quantities of AIBN and thioacetic acid, kinetics of the additions) were optimized by performing parallel experiments on a FirstMate parallel synthesis equipment (Argonaut Technologies, Foster City, CA, USA), using the thioacetic ester IR absorption at 1691 cm^{-1} as an indication of the conversion. In a typical synthesis, the starPEG allyl ether was dissolved in toluene and a 10× molar excess of 2,2′-azo-*bis*-isobutyrylnitrile (AIBN) was added. The solution was degassed by Ar purging in a Schenk tube, which was then closed with a septum and heated to 60°C in an oil bath. Over the course of 24 h thioacetic acid was added to a total of an 8-fold excess in 5 equal aliquots.

The starPEG thioacetate was recovered and purified by repeated precipitations in ether. For details on the characterization of the polymer, see Napoli *et al.* [7].

Thiol deprotection kinetics

Equal aliquots of the starPEG thioacetate were dissolved in NaOH ranging in concentration from 0.01 M to 1 M. Small samples were taken at time points between 1 and 60 min and diluted in 0.1 M phosphate buffer pH 7.5. The thiol content of these dilutions was measured using Ellman's reagent (5,5′-dithio bis(2-nitrobenzoic acid).

Gelation kinetics

The 'Fenton'-catalyst was prepared by dissolving 0.1 mol/l of ferrous lactate and citric acid in water and then slowly adjusting the pH of the solution to 9.0 using NaOH. This solution was aliquoted, stored at −20°C and freshly thawed for every application to prevent oxidation, which happens rapidly in air. The second component of the catalyst system was a 0.1 M cysteine solution in glycine buffer (pH 9.0). The gelation experiments investigating the bulk gelation time as a function of the pH and the catalyst concentration were performed on a 8% (w/v) solution of quantitatively hydrolyzed tetra-armed PEG thioacetate in glycine buffers (0.05 M glycine and 0.1 M NaCl, pH 9.0 and pH 10). Gelation was initiated by the addition of equimolar amounts of Fe(II)-citrate/lactate complex and of cysteine in 0.001–0.1 molar ratio with the PEG thiols. These solutions were intermittently vortexed (5 s on/10 s off) to ensure oxygen saturation and complete mixing without undue mechanical disruption of the forming gel. The samples were classified as gelled when no more vortex formation could be observed.

RESULTS

Synthesis

For the reaction of allyl bromide with the PEG alcolate, the conversion rate was 100% based on the ratio of the allyl ^{1}H-NMR peaks at 5.2 and 5.58 ppm to the PEG methylene peak. A additional indication for the quantitative conversion was the disappearance of the PEG $-$OH ^{1}H-NMR peak at 4.56 ppm in d_6-DMSO.

For the radical addition of thioacetic acid to the allyl PEG, the conversion rate was 94% based on the ratio of the thioacetate methyl and methylene ^{1}H-NMR peaks at 2.86 and 2.32 ppm, respectively, to the PEG methylene peak at 3.5 ppm. Almost quantitative conversion was also indicated by the disappearance of the ^{1}H-NMR allyl peaks in d_6-DMSO.

The synthesis yielded a white powder (total yield 95%) with a slight sulfurous odor which could not be removed by extensive purification. The material is stable for an indefinite time span under ambient conditions (no free thiols detectable).

Deprotection

The reaction kinetics were found to be pseudo-first order with the base concentration (Fig. 2). If 1 M NaOH is used, the deprotection is completed within 1 min. The minimum obtainable reaction time using 0.1 M NaOH for the hydrolysis is 10 min.

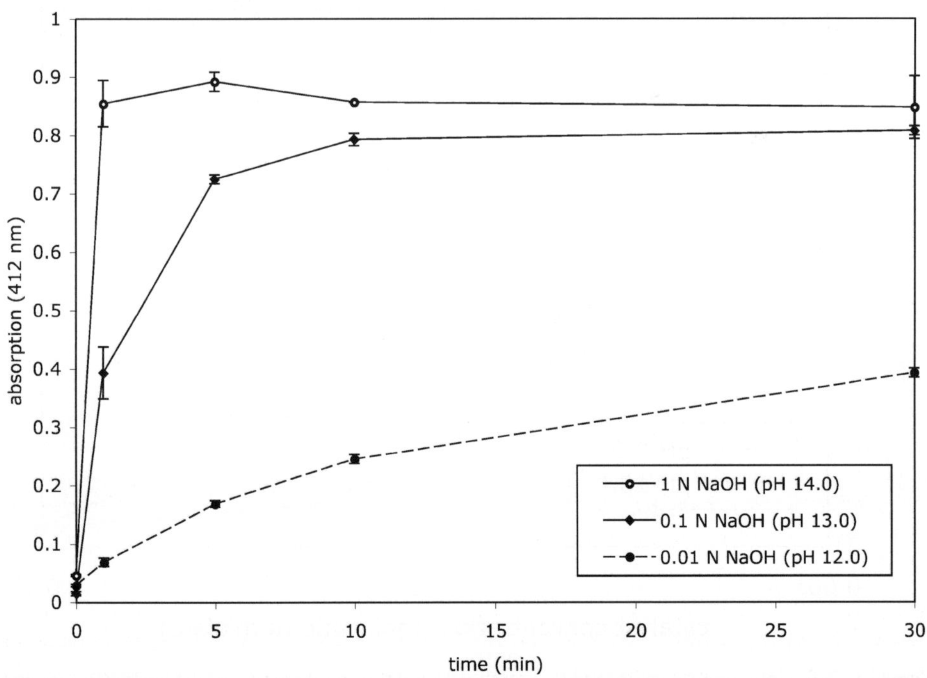

Figure 2. Hydrolysis kinetics of the starPEG thioacetate at varying base concentrations. The thiol concentration was measured by adding Ellman's reagent and measuring the absorption at 412 nm.

For actual gelation experiments, where the polymer was dissolved in a buffer, which lowers the effective pH of the system, hydrolysis times of 20 min were typically used.

Gel properties

The gels that formed were clear and highly elastic. Due to the narrow MW dispersity of the precursors and to the quantitative character of the reaction (no thiols detectable in the gels), we assume them to be characterized by very regular mesh size. No detailed mechanical investigation is provided in this report, but elastic moduli are in the range 8–20 kPa (depending on the solid content).

Gelation kinetics

Without addition of the catalyst, gelation by air oxidation was observed after more than 1 h. With the catalyst, the gelation time decreases with increasing (Fe : thiol) ratio, giving a minimum gelation time of about 90 s if the concentration of the iron catalyst is 10% of the thiol(ate) concentration.

The gelation kinetics we observed differed at pH 9.0 and 10.0. While gelation time varies linearly with log(Fe : thiol) at pH 10.0, at pH 9.0 a different reaction mechanism is suggested by the shape of the curve (Fig. 3).

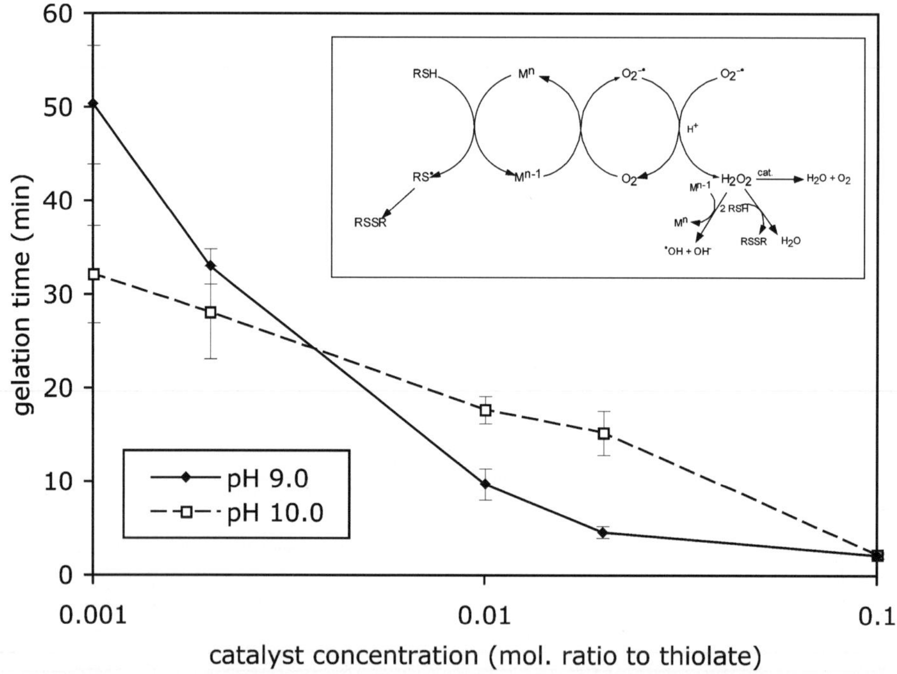

Figure 3. Bulk gelation time in dependence on the pH and the concentration of the Fe(II)/Cys catalyst present. Inset: Catalytic thiol oxidation in a cyclic 'Fenton-type' reaction catalyzed by iron (modified from Ref. [9]).

DISCUSSION

We found that the challenges usually associated with the synthesis, storage and handling of multifunctional thiols are efficiently addressed by the use of the proposed protected starPEG thioacetate as starting material. The thioacetate group is unreactive to itself and stable under normal storage conditions (room temperature, no protecting atmosphere).

For both steps of the synthesis, the molar ratios, temperatures and reaction times required for complete conversion were all higher than for the modification of linear low-MW PEGs [7]. As a starting material, we have used a tetra-armed PEG, which possesses less accessible termini than the corresponding linear polymers. We assume that the more intense reaction conditions were necessary to account for that fact.

During hydrolysis of the protected thiol polymers, a compromise had to be found between the desired speed of the reaction and the osmolarity of the salt produced during the neutralization of the base using HCl. Isotonic solute concentration (0.15 M NaCl) is a limiting concentration for solutes in the gel, if cell and tissue damage due to hypertonic solute concentration should be avoided. Therefore, we used 0.1 M NaOH which was added to a 0.05 M glycine buffer and neutralized with an equimolar amount of HCl, resulting in a physiological buffer with 0.1 M NaCl and 50 mM glycine. The corresponding typical hydrolysis time was 20 min.

Gelation mechanism and kinetics

In the gelation of materials such as the one presented here, the formation of both intra- and inter-molecular disulfide bonds can occur. The intra-molecular reaction is likely thermo-dynamically favored (entropically due to a higher degree of freedom of the molecule). However, from our gelation experiments we conclude that high concentration and rapid reaction can drive the system towards the formation of inter-molecular bonds. We hypothesize that this occurs thanks to the freezing of the partial coil overlap typically found in a semi-dilute or concentrated polymer solution.

Disulfide bonding is generally assumed to proceed through the deprotonated form of thiols [12]. An effective reaction and, thus, a quick gelation, can be obtained only if a sufficiently high concentration of thiolates is provided (pH $\approx$ pK_a). Assuming the pK_a of PEG thiols to be 9.5–10.0 (using β-mercaptoethanol, pK_a = 9.88 [13], as a model), we used a glycine solution as the only physiologically acceptable buffer at similar pH. However, even with oxygen saturation an 8% (w/v) solution of the tetra-armed PEG thiol showed gelation only after about 1 h, too long to be clinically useful. The desired simplicity of application of these hydrogels and the toxicity of many oxidizing agents precluded the use of oxidizing agents other than oxygen [5]. The required acceleration was achieved with a biocompatible Fe(II)-based catalyst in a Fenton-type reaction. This system relies on a bidentate chelate complex of Fe(II) with cysteine, which can easily be oxidized by O_2 to Fe(III). This in turn oxidizes the PEG thiolates to S radical, and the radicals eventually recombine to form a disulfide bond (Fig. 3, inset). Since Fe(III) is reduced to Fe(II) by oxidation of a thiolate a catalytic 'Fenton' cycle is formed. A previous complexation of the iron ion was necessary to prevent precipitation of ferrous hydroxide at pH 9.

Several other oxygen-containing reactive species, including $^{\cdot}O_2^-$ and $^{\cdot}OH$, are also formed and can oxidize the thiolates. The actual number and activity of the respective oxidizing species is important, because it will influence the gel formation kinetics and possibly determine the toxicity of the system due to the cytotoxicity of the hydroxyl radicals.

The hypothesis that the Fenton cycle is indeed driven by molecular oxygen was verified by the fact that the precursor solutions containing the Fe(II)/Cys catalyst did not gel at all if maintained under argon atmosphere. When carrying out experiments in presence or absence of *tert*-butanol (10%, v/v), which is an efficient hydroxyl radicals scavenger [14], a 8% (w/v) solution gelled in 3.75 and 3.0 min, respectively. This fairly small difference let us conclude that hydroxyl radicals are not a prominent species in the oxidation reaction and not a major concern with respect to cytotoxicity. Additionally, their small amount will most likely be neutralized inside the gel due to their high reactivity.

While we did not further investigate the reaction mechanisms in detail, we hypothesize the difference in reaction kinetics at pH 9.0 and 10.0 to be attributed to the following reasons. The overall gelation speed is a balance between the formation of disulfide bonds by the catalytic Fenton reaction and the thiolysis or simple reduction of thiol bonds. The disulfide bond cleavage is significantly faster at a pH above the pK_a of the PEG thiol, as can be seen from the fast gel dissolution at pH 10 (data not shown). Therefore, disulfide formation and gelation is opposed by disulfide reduction and gel dissolution (reverse gelation) to a different degree at pH values below and above the thiol pK_a.

Another possible contribution to the difference in gelation kinetics is hypothesized to result from the difference in the concentration of catalytically active Fe^{3+} molecules at different pH values. The exact nature and composition of the iron containing complexes will be dependent on the pH of the system, hence the reaction kinetics will also be effected by the pH [15].

A high degree of cross-linking and the homogeneity of the internodal distances between cross-links resulted in a low propensity for crack propagation and good mechanical stability of the gels. This could allow easy suturing with the standard non-resorbable polypropylene threads and a high mechanical stability of the suturing points.

Applications

The gel formation *via* disulfide bonds offers other potential applications as well. It is well known that thiols and disulfides bind strongly to some metals such as gold, silver and platinum. If a gel is formed on the surface of such a metal (or on a surface coated with such a metal), we hypothesize that a chemically bonded, tightly adherent hydrogel layer can be formed. This hydrogel layer could impart non-fouling properties to these metal surfaces for applications such as endovascular stents, or implantable biosensors, which oftentimes have a limited useful lifetime due to fouling of the surface [16–18]. If gelation occurs in the presence of nanoparticles of these materials, we envision the formation of a regular lattice of these nanoparticles in a hydrogel with potentially interesting, novel properties [19]. Silver nanoparticles, which would release silver ions over a prolonged period of time, could impart antibiotic properties to these hydrogels for applications, e.g. in the urogenital tract, where device-related infection poses a serious limitation. Covalent incorporation of molecules into the gel matrix — e.g. for drug or gene delivery applications — can easily be achieved as well. Any molecule that carries a free thiol group will be incorporated into the matrix *via* formation of a mixed disulfide. This feature can be used to impart specific cell adhesive properties to the otherwise non-fouling PEG hydrogel, e.g. by incorporating a peptide containing a cell surface receptor ligand sequence [20].

CONCLUSIONS

In conclusion, we describe an advanced hydrogel system that exhibits sol–gel transformation by the reaction with molecular oxygen. We have synthesized the hydrogel precursor, a multi-armed polyethylene glycol, bearing 4 stable thioacetate groups per molecule. Rapid deprotection of the thioacetate termini can be achieved using sodium hydroxide, quantitatively liberating a thiol group and a non-toxic acetate ion. The oxidation of these thiol groups to disulfides leads to the formation of a 3-dimensional hydrogel network with an equilibrium water content of 94%. To achieve rapid polymerization of this hydrogel in an aerobic environment, we have developed a catalyst system that accelerates gelation by the reaction with molecular oxygen to a clinically useful time scale of several minutes. We discuss the use of such a hydrogel system for biomedical applications such as the protection of an injured blood vessel wall after an angioplasty procedure.

REFERENCES

1. A. S. Hoffman, *Ann. N.Y. Acad. Sci.* **944**, 62 (2001).
2. R. T. Kershen, R. R. Dmochowski and R. A. Appell, *Urol. Clin. North Am.* **29**, 559 (2002).
3. P. F. Heini and U. Berlemann, *Eur. Spine J.* **10** (Suppl. 2), S205 (2001).
4. K. Lee and D. Mooney, *Chem. Rev.* **101**, 1869 (2001).
5. N. Hisano, N. Morikawa, H. Iwata and Y. Ikada, *J. Biomed. Mater. Res.* **40**, 115 (1998).
6. C. Woghiren, B. Sharma and S. Stein, *Bioconj. Chem.* **4**, 314 (1993).
7. A. Napoli, N. Tirelli, G. Kilcher and J. A. Hubbell, *Macromolecules* **34**, 8913 (2001).
8. P. Wardman and L. P. Candeias, *Radiat. Res.* **145**, 523 (1996).
9. K. D. Held, F. C. Sylvester, K. L. Hopcia and J. E. Biaglow, *Radiat. Res.* **145**, 542 (1996).
10. T. Yamaoka, Y. Tabata and Y. Ikada, *J. Pharm. Sci.* **83**, 601 (1994).

11. J. L. Hill-West, S. M. Chowdhury, A. S. Sawhney, C. P. Pathak, R. C. Dunn and J. A. Hubbell, *Obstet. Gynecol.* **83**, 59 (1994).
12. G. Capozzi and G. Modena, in: *The Chemistry of Thiol Group, Part 2,* S. Patai (Ed.), p. 785. Wiley, New York, NY (1974).
13. J. A. Dean, *Lange's Handbook of Chemistry*, 14th edn. McGraw-Hill, New York, NY (1992).
14. A. Jorns, M. Tiedge, S. Lenzen and R. Munday, *Free Radic. Biol. Med.* **26**, 1300 (1999).
15. E. Y. Yang, A. Campbell and S. C. Bnody, *Redox Rep.* **5**, 371 (2000).
16. Y. Nakayama, K. Nakamata, Y. Hirano, K. Goto and T. Matsuda, *Langmuir* **14**, 3909 (1998).
17. A. Laschewsky, O. Oari, C. Mangeney and L. Jullien, *Macromol. Symp.* **164**, 323 (2001).
18. F. Zhang, E. T. Kang, K. G. Neoh and W. Huang, *J. Biomater. Sci. Polymer Edn* **12**, 515 (2001).
19. A. K. Boal, F. Ilhan, J. E. DeRouchey, T. Thurn-Albrecht, T. P. Russell and V. M. Rotello, *Nature* **404**, 746 (2000).
20. B. K. Mann and J. L. West, *J. Biomed. Mater. Res.* **60**, 86 (2002).

Gels, Genes, Grafts and Giants, pp. 529–542 (2005)

Preparation of reversibly glucose-responsive hydrogels by covalent immobilization of lectin in polymer networks having pendant glucose

TAKASHI MIYATA [1,2,3,*], ATSUSHI JIKIHARA [1], KATSUHIKO NAKAMAE [1] and ALLAN S. HOFFMAN [4]

[1] *Department of Chemical Science and Engineering, Faculty of Engineering, Kobe University, 1-1 Rokkodai, Nada, Kobe 657-8501, Japan*
[2] *PRESTO, Japan Science and Technology Agency, 4-1-8 Honcho, Kawaguchi, Saitama, Japan*
[3] *Unit of Chemistry, Faculty of Engineering, Kansai University, Suita, Osaka 564-8680, Japan*
[4] *Department of Bioengineering, University of Washington, Seattle, WA 98195, USA*

Received 2 October 2003; accepted 25 February 2004

Abstract—Glucose-responsive hydrogels were prepared by copolymerization of a monomer having a pendant glucose with modified lectin (concanavalin A (ConA)) having vinyl groups. Swelling behavior of ConA-copolymerized glucosyloxyethyl methacrylate (GEMA) hydrogels was discussed from the viewpoint of their synthetic condition and structures. The swelling ratio of the ConA-copolymerized GEMA hydrogels was strongly dependent on the glucose concentration in a buffer solution. As free glucose resulted in the dissociation of the complex between ConA and pendant glucose in the networks and the cross-linking density in the hydrogel decreased, the hydrogels swelled gradually in the presence of free glucose. Leak of ConA from the ConA-entrapment hydrogel and ConA-copolymerized hydrogel was examined in order to discuss the reversible changes of their glucose-responsive behavior. During swelling in the presence of free glucose, ConA leaked out of the ConA-entrapment GEMA hydrogel but did not out of the ConA-copolymerized GEMA hydrogel. As a result, the ConA-copolymerized GEMA hydrogel showed the reversible swelling changes in response to a stepwise change in the glucose concentration. This study revealed that covalent immobilization of ConA in the GEMA networks is very important for development of reversibly glucose-responsive hydrogels.

Key words: Hydrogel; glucose; lectin; concanavalin A; glucose-responsive hydrogel; complex; pendant glucose.

*To whom correspondence should be addressed at Kansai University.
E-mail: tmiyata@ipcku.kansai-u.ac.jp

INTRODUCTION

Stimuli-responsive hydrogels undergo volume changes in response to renvironmental changes such as pH [1–3], temperature [4–13] and electric field [14–16]. Such stimuli-responsive hydrogels have been extensively investigated as smart materials in the biochemical and biomedical fields, since they can sense environmental changes and induce structural changes by themselves [17]. For example, as pH and temperature are the most widely utilized triggering signals for modulated systems, pH- and temperature-responsive hydrogels have many potential applications in self-regulated drug delivery, sensor, cell cultures, etc. [18, 19]. In addition, biomolecule-responsive hydrogels, which recognize specific biomolecules and undergo volume changes in response to them, are promising materials as the next generation of biomaterials and drug-delivery devices [20, 21]. A few biomolecule-responsive hydrogels, such as glucose-responsive hydrogels [22–33] and antigen-responsive hydrogels [34, 35], have been reported since such biomolecule-responsive hydrogels became increasingly important in the biomedical field.

Several investigations have been undertaken for the development of glucose-simulated, feed-back insulin-delivery systems. For example, in the treatment of diabetes caused by the inability of the pancreas, glucose-responsive hydrogels that undergo swelling changes in response to glucose can provide the tools for constructing self-regulated insulin-delivery systems, in which a necessary amount of insulin can be administered in response to the blood glucose concentration. Combining glucose oxidase with pH-responsive hydrogels to sense glucose and regulate insulin release is the method that many researchers have used to develop glucose-responsive insulin-delivery systems [22–26]. Kataoka *et al.* [28, 29] have developed totally synthetic, glucose-responsive hydrogels using temperature-responsive behavior of poly(N-isopropylacrylamide) and the complex formation between phenylboronic acid and glucose. Some researchers have focused on glucose-binding properties of a lectin (concanavalin A (ConA)) possessing four binding sites to develop glucose-responsive insulin release [30–33, 36–39].

Previously we studied the complex formation between lectin (ConA) and a polymer having pendant glucose groups (poly(glucosyloxyethyl methacrylate) (poly(GEMA))) in order to design a glucose-responsive polymer systems [36]. Poly(GEMA) formed a complex with ConA and then the aqueous solution became turbid due to the multiple associations between them. Furthermore, the addition of free glucose made the aqueous solution transparent again because of dissociation of the complex. The previous studies revealed that the complex between poly(GEMA) and ConA is a kind of glucose-responsive device. Based on the glucose-responsive complex between poly(GEMA) and ConA, we have tried to prepare novel glucose-responsive hydrogels by using the complex as a glucose-responsive cross-linking point [33]. The ConA-entrapment GEMA hydrogels, in which ConA was entrapped within GEMA networks, were prepared by the copolymerization of GEMA and methylene bisacrylamide in the presence of ConA. The swelling ratio of the ConA-entrapment GEMA hydrogels increased with an increase in free glucose concentration because their cross-linking density decreased due to the dissociation of the complex between poly(GEMA) and ConA. However, as ConA is not covalently bound to the GEMA networks, it may leak out of the hydrogel during the glucose-responsive swelling. The leak of ConA prevents the hydrogel from reversibly glucose-responsive swelling–shrinking. If insulin is continuously released from the hydrogel into blood and the concentration of blood sugar becomes too low, the body falls into a dangerous state. Therefore, in the application of a glucose-responsive hydrogel as a switching device for insulin therapy, reversible swelling–shrinking changes in response to glucose concentration are required. Since glucose-responsive hydrogels prepared by our previous method have a problem in reversible response, another method to prepare reversibly glucose-responsive hydrogels must be proposed.

This paper describes the preparation of a reversibly glucose-responsive hydrogel (ConA-copolymerized GEMA hydrogel) by copolymerizing a monomer having a pendant glucose with chemically-modified ConA having vinyl groups. In addition, glucose-responsive behavior of the ConA-copolymerized GEMA hydrogel was compared with that of the ConA-entrapment GEMA hydrogel from the viewpoint of hydrogel structures.

MATERIALS AND METHODS

Materials

A monomer having a pendant glucose (glucosyloxyethyl methacrylate (GEMA)) [40], as shown in Scheme 1, was supplied as an aqueous solution of 53.5 wt% GEMA by Nippon Fine Chemical.

Lectin concanavalin A (ConA) and N-ethyl-N'-(3-dimethylaminopropyl)-carbodiimide-hydrochloride (EDC) were purchased from Wako Pure Chemical. Acrylic acid (AAc) was purified by distillation under reduced pressure in a nitrogen atmosphere. 2,2′-Azobis(2-amidinopropane) dihydrochloride (AIBA) as an initiator in the polymerization and other analytical-grade reagents were used without further purification. Deionized water was distilled before use.

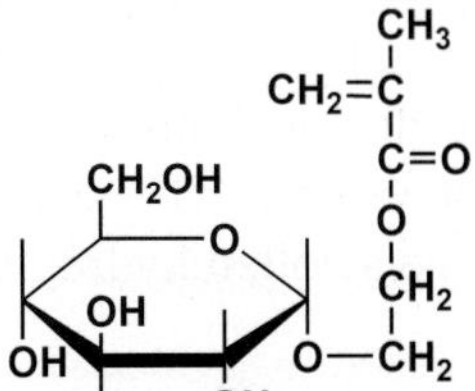

Scheme 1. Structure of GEMA.

Preparation of ConA-copolymerized GEMA hydrogel

Vinyl groups were introduced into ConA by covalent coupling of carboxyl groups in AAc with amino groups in ConA using the carbodiimide technique [41] as follows: EDC (0.0532 g) was added to 1 ml of an aqueous solution containing various amounts of ConA and 0.01 g of AAc, and then the reaction was carried out at 25°C for 1 h. An aqueous GEMA solution were mixed with the resulting aqueous AAc-modified ConA solution to afford a 20 wt% solution. After AIBA (5 wt% relative to the monomers) was added to the mixture, they were degassed sufficiently and transferred into a mold. Two glass plates with poly(vinylidene chloride) were separated by a 1-mm spacer to form the mold. The polymerization was carried out by irradiation of ultraviolet light (UV) of wavelength 280 nm at room temperature for 30 min. After gelation of the solution, the hydrogels were immersed in a large amount of 0.1 M tris HCl buffer solution (pH 7.5) containing 1 mM $MnCl_2$ and 1 mM $CaCl_2$ to remove residual chemicals and unreacted monomer at 5°C for 24 h. At the recognition site of ConA, there are Mn^{2+} and Ca^{2+} that play important roles in its complex formation with saccharide [42]. Lack of Mn^{2+} and Ca^{2+} lowers recognition function of ConA. In order to depress lowering its recognition function, therefore, Mn^{2+} and Ca^{2+} are generally added in an aqueous ConA solution when complex formation between ConA and saccharide is investigated [43]. In this study, therefore, a buffer solution containing 1 mM $MnCl_2$ and 1 mM $CaCl_2$ was used to remove residual chemicals and unreacted monomer from the resultant hydrogels without lowering recognition function of ConA. Total amount of ConA immobilized in the hydrogel was determined from the difference between the initial amount of ConA and the amount of ConA removed from the hydrogel in the buffer solution, using ultraviolet spectrophotometer (Shimadzu) at 280 nm. In this study, to investigate the effect of the covalent binding of ConA to the networks on the swelling behavior of the glucose-responsive hydrogels, we prepared ConA-entrapment GEMA hydrogel, in which ConA was entrapped within the GEMA networks, by the entrapping method reported previously [33].

Swelling ratio

The ConA-copolymerized GEMA hydrogels were kept immersed in 0.1 M Tris-HCl buffer solution (pH 7.5) containing 1 mM $MnCl_2$ and 1 mM $CaCl_2$, until equilibrium was reached at 5°C. After that, the hydrogels were transferred and kept immersed in the buffer solution containing a desired amount of glucose at 5°C. The surface water of the hydrogel was removed by lightly blotting with a laboratory tissue and the hydrogel was weighed. The disc weights were individually monitored in this way until they reached constant values. The swelling ratio of the hydrogel was determined from the weight of the swollen hydrogel (W) and dried gel (W_0) using equation (1):

$$\text{Swelling ratio} = \frac{W - W_0}{W_0}. \tag{1}$$

Measurement of compressive modulus

Compression tests were performed in order to determine compressive modulus of the hydrogels using the apparatus shown in Fig. 1 [12]. The swollen hydrogel is compressed by the crosshead (a) of the apparatus. Then the relationship between the compressive stress and strain of the hydrogel is output in recorder (b). The compressive modulus can be determined by equation (2) from the compressive stress and strain of the hydrogels. From the compressive modulus, the cross-linking density of the hydrogel can be calculated by equation (3) [44–48]:

$$\tau = G(\alpha - \alpha^{-2}), \tag{2}$$

$$G = RT\nu_e V_0^{2/3} V_2^{1/3}, \tag{3}$$

where τ is the compressive stress (Pa), G is the compressive modulus (Pa), R is gas constant, T is absolute temperature (K), α is the ratio of the thickness of the hydrogel before and after the compression, ν_e is the efficient cross-linking density (mol/l), V_0 is the volume fraction of the polymer during network formation and V_2 is the volume fraction of polymer in the hydrogel, which is obtained by the swelling measurement using equation (4);

$$\frac{1}{V_2} = 1 + (\text{Swelling ratio}) \times \frac{\rho_g}{\rho_s}, \tag{4}$$

where ρ_g and ρ_s are the density of the dried gel and the solvent, respectively.

Measurement of the amount of ConA leaking from the hydrogel

ConA-copolymerized GEMA hydrogel and ConA-entrapment GEMA hydrogel were immersed in 3 ml 0.1 M Tris-HCl buffer solution (pH 7.5) containing 1 mM $MnCl_2$ and 1 mM $CaCl_2$. The concentration of ConA leaking into the buffer solution was measured by high-performance liquid chromatography (HPLC) (Hitachi; Gelpack GL-W550) using UV spectroscopy at 280 nm.

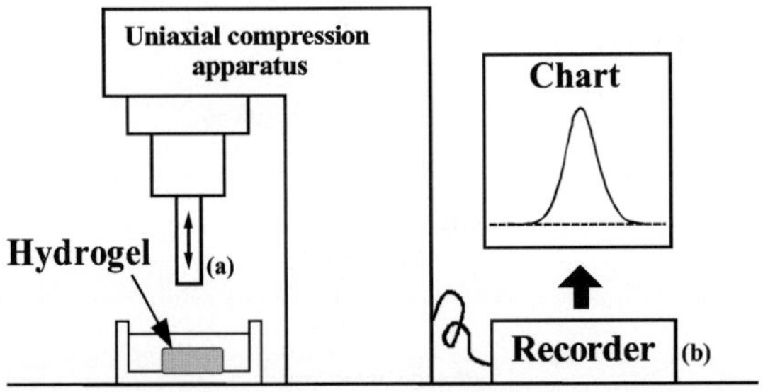

Figure 1. Uniaxial compression apparatus for the measurement of stress–strain data on hydrogels.

RESULTS AND DISCUSSION

Synthesis of ConA-copolymerized GEMA hydrogels

The first step to prepare ConA-copolymerized GEMA hydrogels is to introduce vinyl groups into ConA in order to synthesize a monomer having a pendant ConA. ConA has been covalently immobilized on a substrate by various methods (cyanuric chrolide method, epoxy method, carbodiimide method and so on) for the development of a column of affinity chromatography [41]. In the present study, the carbodiimide method was selected for binding of AAc to ConA. As shown in Scheme 2, the carboxyl group of AAc was bound to the amino group of ConA by condensation reaction with *N*-ethyl-*N*′-(3-dimethyl aminopropyl)-carbodiimide-hydrochloride (EDC). In this method, generally, urea derivative and *N*-acryl derivative are produced through intermediates. Vinyl groups were introduced into ConA by the condensation reaction of the urea derivative and ConA. Since ConA modified by AAc can act as a cross-linking agent, the hydrogel was prepared without other cross-linking agent. Figure 2 shows the relationship between initial concentration of AAc-modified ConA in polymerization solution and the concentration of ConA immobilized in the resultant ConA-copolymerized GEMA hydrogels. The amount of ConA immobilized in the hydrogel increased gradually with an increase in the initial concentration of AAc-modified ConA. Consequently, about seventy percent of AAc-modified ConA in the feed was copolymerized with GEMA in preparing ConA-copolymerized GEMA hydrogels.

The effects of the amount of ConA in the ConA-copolymerized GEMA hydrogels on their swelling ratio and cross-linking density are shown in Fig. 3. With increasing concentration of ConA immobilized in the ConA-copolymerized GEMA hydrogel, its swelling ratio decreased and cross-linking density increased. Generally, swelling behavior of a hydrogel is strongly dependent upon the cross-linking density [45]. Therefore, Fig. 3 demonstrates that decreasing swelling ratio of the ConA-copolymerized GEMA hydrogels is due to increasing cross-linking density with an increase in ConA concentration. In our previous studies, ConA formed com-

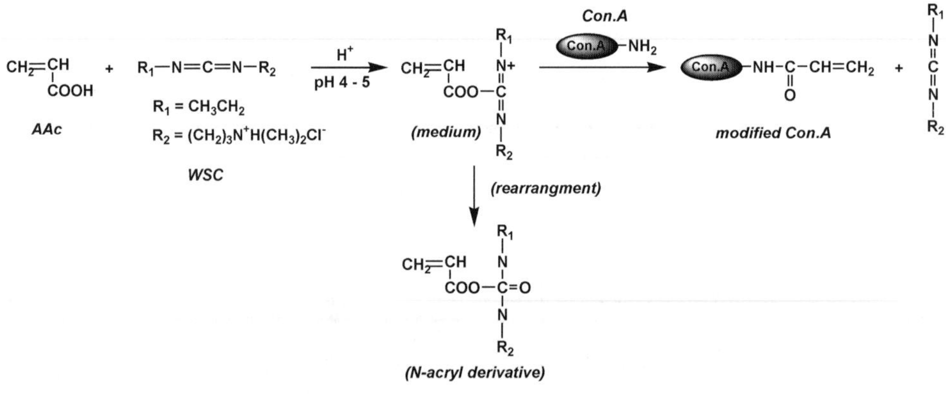

Scheme 2. Synthesis of modified ConA having vinyl groups.

plexes with pendant glucose in poly(GEMA) in water and then the aqueous solution became turbid due to the multiple associations between them [33, 36]. Based on the previous studies, we can conclude that the complex formation between ConA and the pendant glucose in ConA-copolymerized GEMA hydrogel causes an increase in its cross-linking density. This means that a complex between ConA copolymerized with GEMA and the pendant glucose of GEMA networks plays an important role as a cross-linking point.

Compared with the complex formation between ConA and poly(GEMA) in water, the cross-linking structure of the ConA-copolymerized GEMA hydrogel can be represented by the tentative model shown in Fig. 4. ConA is a tetramer having four saccharide-recognition sites and can form a complex with four saccharide

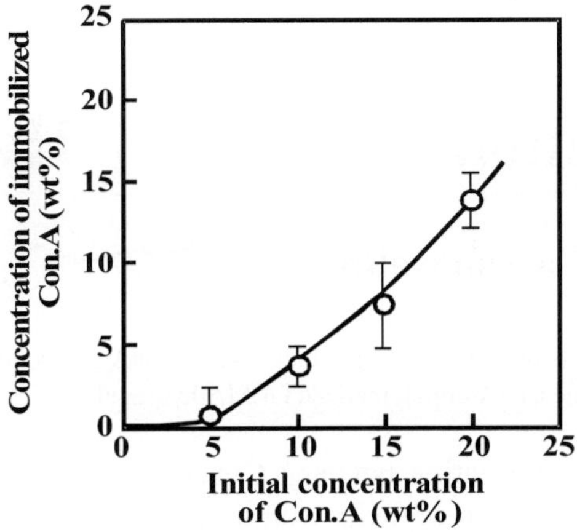

Figure 2. Relationship between initial concentration of ConA relative to monomer in polymerization solution and the concentration of ConA immobilized in ConA-copolymerized GEMA hydrogel.

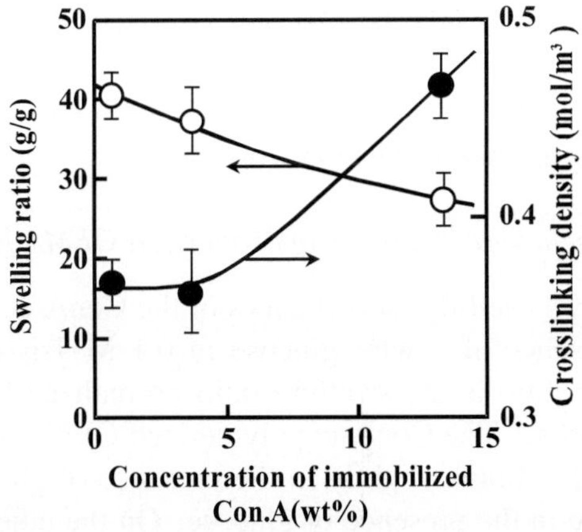

Figure 3. Effects of the concentration of ConA immobilized in the hydrogels on their swelling ratios and cross-linking density. The hydrogels were prepared by the copolymerization of 19 wt% GEMA and ConA modified by 1 wt% AAc with 5 wt% AIBA.

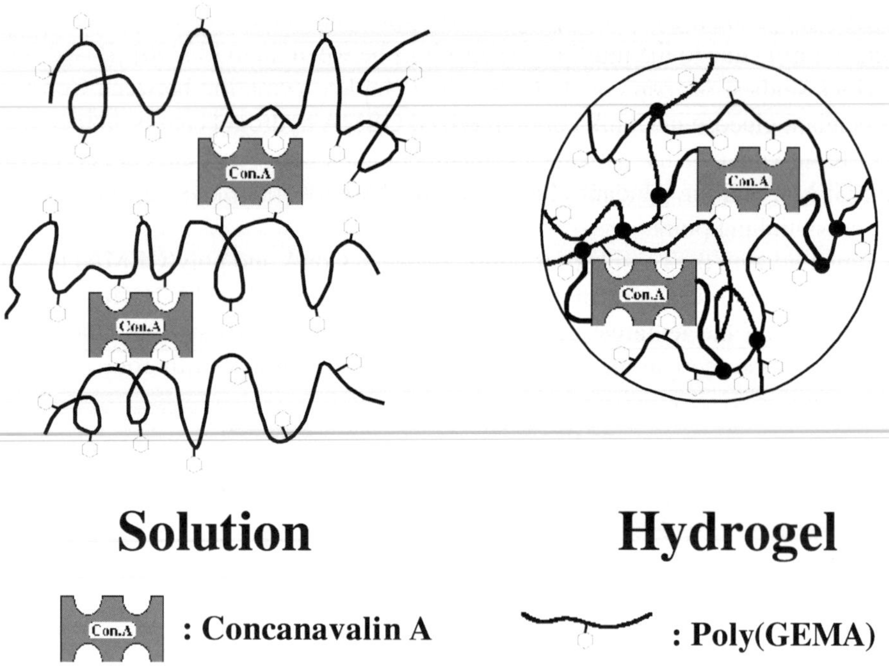

Figure 4. Tentative models of complex formation between ConA and pendant glucose of GEMA in an aqueous solution and the ConA-copolymerized GEMA hydrogel.

chains. The complex formation between ConA and saccharides often results in aggregation in water. In an aqueous solution, therefore, the complex between poly(GEMA) and ConA is formed and then the aqueous solution becomes turbid due to the multiple associations between them. An increase in the cross-linking density with the ConA concentration is due to the fact that the complex between ConA and pendant glucose of the GEMA networks acts as a cross-linking point in the ConA-copolymerized GEMA hydrogel. As a result, the increase in the cross-linking density with the ConA concentration results in a decrease in the swelling ratio of the ConA-copolymerized GEMA hydrogel.

Glucose-responsive behavior of ConA-copolymerized GEMA hydrogels

Figure 5 shows relative swelling ratio changes of the ConA-copolymerized GEMA hydrogels in the presence of 1 wt% glucose in 0.1 M Tris-HCl buffer solution. The relative swelling ratio is the swelling ratio normalized by the swelling ratio in the absence of glucose. The ConA-copolymerized GEMA hydrogel with a very small amount of ConA shrank slightly in the presence of glucose. This is due to high osmotic pressure in the presence of glucose. On the other hand, the swelling ratio of ConA-copolymerized GEMA hydrogels with a ConA concentration of 3.9 and 13.3 wt% increased gradually in the presence of glucose, in contrast to the hydrogel with 0.4 wt% ConA. The higher the ConA concentration in the ConA-

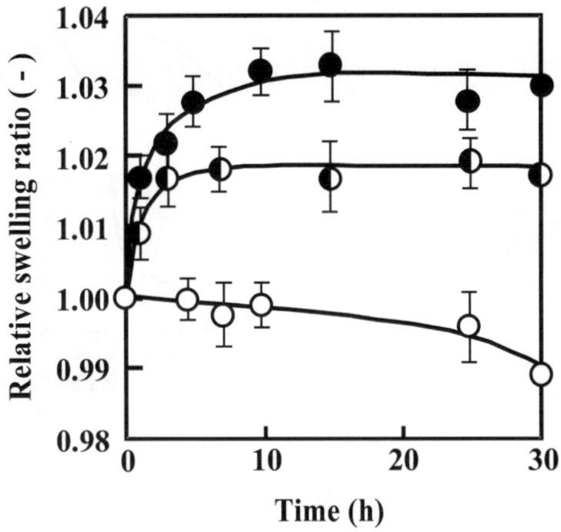

Figure 5. Relative swelling ratio changes of ConA-copolymerized GEMA hydrogels as a function of time, when the hydrogels are immersed in 0.1 M Tris-HCl buffer solution (pH 7.5) containing 1 wt% glucose. Relative swelling ratio is the swelling ratio normalized by the swelling ratio in the absence of glucose. The hydrogels were prepared by the copolymerization of 19 wt% GEMA and ConA modified by 1 wt% AAc with 5 wt% AIBA. Concentration of ConA coupled with the hydrogel networks: (○) 0.4 wt%; (◐) 3.9 wt%; (●) 13.3 wt%.

copolymerized GEMA hydrogel became, the more glucose-responsive swelling the hydrogel exhibited. These results mean that the ConA-copolymerized GEMA hydrogels have a glucose-responsive function. As described in the previous section, the complex between ConA and pendant glucose in the hydrogel networks plays an important role as cross-linking points. Therefore, the glucose-responsive swelling behavior of the ConA-copolymerized GEMA hydrogel is schematically illustrated in Fig. 6. In the previous study [36], we investigated the association and dissociation of the complex between ConA and pendant glucose of a linear poly(GEMA) in water. The previous study revealed that the addition of free glucose resulted in the dissociation of the complex. The results implies that the free glucose can induce the dissociation of the complex between ConA and pendant glucose of GEMA networks when the ConA-copolymerized GEMA hydrogels are immersed in an aqueous glucose solution. The formation of the complex between ConA and pendant glucose is in the equilibrium state. When the ConA-copolymerized GEMA hydrogels are immersed in an aqueous glucose solution, a large amount of free glucose approaches the complex between ConA and pendant glucose of GEMA networks. Then a large amount of free glucose can induce the dissociation of the complex, based on the exchange of pendant glucose with free glucose. Therefore, the cross-linking density in the hydrogel is reduced by the presence of free glucose as the glucose results in the dissociation of the complex acting as a cross-linking point. We can conclude that the ConA-copolymerized GEMA hydrogels undergo swelling changes in response to free glucose because of decreasing cross-linking density. Consequently, the copolymerization of modified ConA having vinyl groups

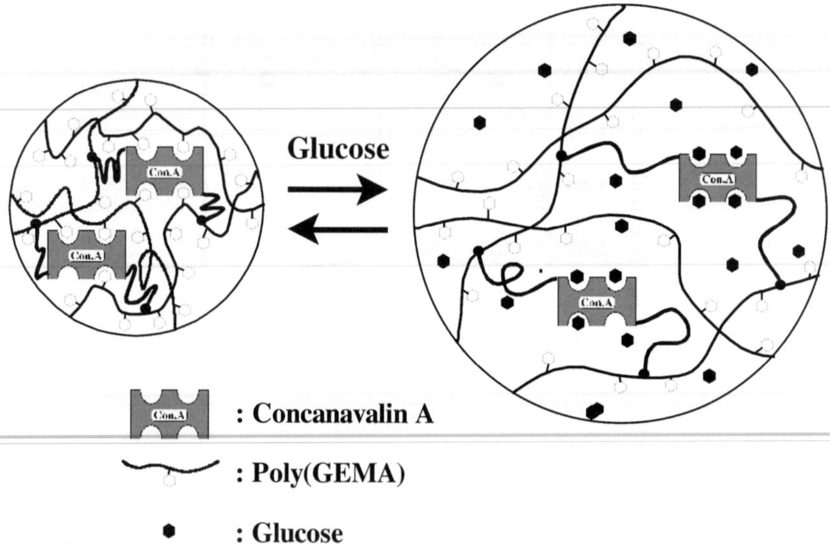

Figure 6. Schematic representation of glucose-responsive swelling of the ConA-copolymerized GEMA hydrogel.

with a monomer having a pendant glucose enables us to prepare glucose-responsive hydrogels.

Reversibly glucose-responsive swelling changes

In applications of a glucose-responsive hydrogel as a smart device for insulin release in insulin therapy, if insulin is continuously released from the hydrogel into blood and the value of blood sugar concentration becomes too low, the body falls into dangerous state. Therefore, most potential applications of the glucose-responsive hydrogel in insulin therapy require it to possess reversible swelling properties according to the concentration of blood glucose. In other words, the hydrogel should be able to not only swell in the presence of free glucose, but also shrink again in its absence. This section focuses on the effect of structure of the glucose-responsive hydrogels on reversible changes in swelling ratio in response to the glucose concentration. In this study, we prepared two types of glucose-responsive hydrogels having different structures by entrapping method and copolymerization method.

The ConA-entrapment GEMA hydrogels reported in the previous paper [33] were prepared using the entrapping method. In ConA-entrapment GEMA hydrogels ConA may leak out of them during swelling in an aqueous glucose solution, because ConA is not covalently immobilized in the hydrogel networks. The leak of ConA may prevent the hydrogel from reversible swelling changes in response to changes in the glucose concentration. In this study, therefore, ConA was covalently immobilized in the hydrogel networks for the development of a reversibly glucose-responsive hydrogel. Figure 7 shows the amounts of ConA leaking out of the ConA-

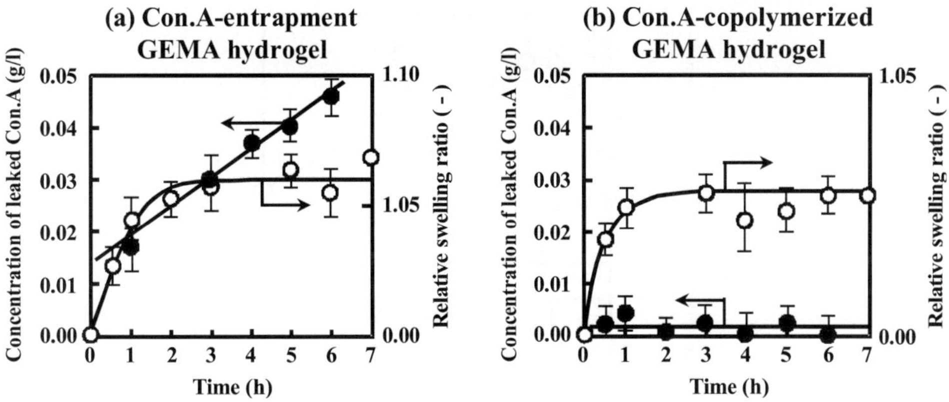

Figure 7. Changes in the relative swelling ratio and the concentration of ConA leaked out of the ConA-entrapment GEMA hydrogel (a) and the ConA-copolymerized GEMA hydrogel (b).

entrapment and ConA-copolymerized hydrogels in a buffer solution containing free glucose, together with their swelling ratio changes as a function of time. When the ConA-entrapment GEMA hydrogel swelled in the presence of free glucose, the amount of ConA leaking out of the hydrogel increased gradually (Fig. 7a). In the ConA-entrapment hydrogel, ConA is entrapped within the hydrogel network but is not immobilized covalently. Therefore, when the ConA-entrapment GEMA hydrogel swells in the presence of free glucose, ConA easily leaks out of the swollen network. This means that the ConA-entrapment GEMA hydrogel cannot reversibly respond to glucose due to a decrease in the amount of ConA within the hydrogel. On the contrary, no ConA leaked out of the ConA-copolymerized GEMA hydrogel in spite of the swelling in the presence of free glucose (Fig. 7b). This is due to covalent immobilization of ConA in hydrogel networks. Consequently, the ConA-copolymerized GEMA hydrogel has the possibility of reversible swelling changes in response to the glucose concentration. Next, we investigated the reversible swelling changes of the ConA-copolymerized GEMA hydrogel in response to stepwise changes in the glucose concentration.

Figure 8 shows reversible swelling change of the ConA-copolymerized GEMA hydrogel in response to stepwise changes in the glucose concentration between 0 and 1.0 wt%. The ConA-copolymerized GEMA hydrogel swelled gradually in the presence of free glucose, but shrank in its absence. Since ConA in the ConA-copolymerized GEMA hydrogel is bound to the polymer networks, no ConA leaks out of the hydrogel during its glucose-responsive swelling. Therefore, ConA bound to the networks can associate with pendant glucose of the GEMA networks in the absence of free glucose, as shown in Fig. 6. The ConA-copolymerized GEMA hydrogel can shrink again in buffer solution without free glucose due to the complex formation between ConA and pendant glucose of the hydrogel network. As a result, Fig. 8 demonstrates that the ConA-copolymerized GEMA hydrogel undergoes reversible swelling–shrinking changes in response to the glucose concentration.

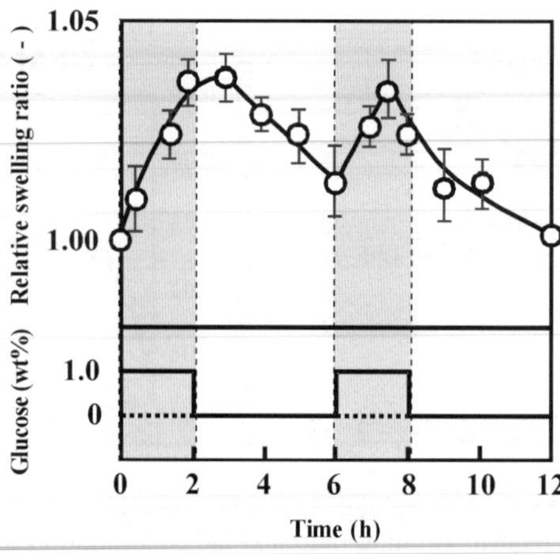

Figure 8. Reversible swelling change of ConA-copolymerized GEMA hydrogel in response to stepwise changes in the glucose concentration between 0 and 1.0 wt% in Tris-HCl buffer solution (pH 7.5).

However, the kinetics of swelling–shrinking response of the reversibly glucose-responsive hydrogels are also very important in their application for insulin delivery systems. Generally, the kinetics of the swelling–shrinking response are strongly dependent upon the hydrogel structures. In future studies, therefore, the hydrogel structures must be designed to prepare the reversibly glucose-responsive hydrogels that can control insulin release rapidly.

This paper provides a fundamental concept for developing reversibly glucose-responsive hydrogels. However, the reversibly glucose-responsive behavior of the hydrogel must be improved for its application to a device for insulin release in response to blood glucose. Designing and controlling the structures of the ConA-copolymerized GEMA hydrogel are of great importance for improving its glucose responsive behavior. We are currently improving the glucose-responsive behavior of the ConA-copolymerized GEMA hydrogel by introducing IPN structures.

CONCLUSIONS

Modified ConA having vinyl groups was synthesized by coupling ConA with AAc. The ConA-copolymerized GEMA hydrogels were prepared by the copolymerization of AAc-modified ConA with GEMA. The measurements of their cross-linking density revealed that the complex between ConA and pendant glucose in GEMA networks plays an important role as a cross-linking point. The ConA-copolymerized GEMA hydrogel swelled gradually in the presence of free glucose. This is due to the fact that free glucose results in a dissociation of the complex between ConA and pendant glucose and leads to a decrease in the cross-linking density. Although ConA leaked out of the ConA-entrapment GEMA hydrogel prepared by entrapping method, no ConA leaked out of the ConA-copolymerized GEMA hydrogel. As a result, ConA-copolymerized GEMA hydrogel shows reversible swelling–shrinking changes in response to the glucose concentration.

Acknowledgements

The authors are grateful for the financial support from Grants-in-Aid for Scientific Research on Priority Areas (A), 'Dynamic Control of Strongly Correlated Soft Materials' from the Ministry of Education, Science, Sports, Culture, and Technology, and for Scientific Research (B) from the Japan Society for Promotion of Science (JSPS).

REFERENCES

1. T. Tanaka, *Phys. Rev. Lett.* **40**, 820 (1978).
2. T. Tanaka, D. J. Fillmore, S.-T. Sun, I. Nishio, G. Swislow and A. Shah, *Phys. Rev. Lett.* **45**, 1636 (1980).
3. R. A. Siegl and B. A. Firestone, *Macromolecules* **21**, 3254 (1988).
4. Y. Hirokawa and T. Tanaka, *J. Chem. Phys.* **81**, 6379 (1984).
5. A. S. Hoffman, A. Afrassiabi and L. C. Dong, *J. Control. Rel.* **4**, 213 (1986).
6. A. S. Hoffman, *J. Control. Rel.* **6**, 297 (1987).
7. Y. H. Bae, T. Okano, R. Hsu and S. W. Kim, *Makromol. Chem. Rapid Commun.* **8**, 481 (1987).
8. Y. H. Bae, T. Okano and S. W. Kim, *Makromol. Chem. Rapid Commun.* **9**, 185 (1988).
9. G. Chen and A. S. Hoffman, *Nature* **373**, 49 (1995).
10. R. Yoshida, K. Uchida, T. Kaneko, K. Sakai, A. Kikuchi, Y. Sakurai and T. Okano, *Nature* **374**, 240 (1995).
11. K. Nakamae, T. Miyata and A. S. Hoffman, *Makromol. Chem.* **193**, 983 (1992).
12. T. Miyata, K. Nakamae, A. S. Hoffman and Y. Kanzaki, *Macromol. Chem. Phys.* **195**, 1111 (1994).
13. T. Aoki, M. Kawashima, H. Katono, K. Sanui, N. Ogata, T. Okano and Y. Sakurai, *Macromolecules* **27**, 947 (1994).
14. T. Tanaka, I. Nishio, S.-T. Sun and S. Ueno-Nishio, *Science* **218**, 467 (1982).
15. I. C. Kwon, Y. H. Bae and S. W. Kim, *Nature* **354**, 291 (1991).
16. Y. Osada, H. Okuzaki and H. Hori, *Nature* **355**, 242 (1992).
17. T. Miyata, in: *Supramolecular Design for Biological Applications*, N. Yui (Ed.), Chapter 9, p. 191. CRC Press, Boca Raton, FL (2002).
18. H. G. Schild, *Prog. Polym. Sci.* **17**, 163 (1992).
19. T. Okano, *Adv. Polym. Sci.* **110**, 179 (1993).
20. T. Miyata and T. Uragami, in: *Polymeric Biomaterials*, S. Dumitriu (Ed.), Chapter 36, p. 959. Marcel Dekker, New York, NY (2002).
21. T. Miyata, T. Uragami and K. Nakamae, *Adv. Drug Delivery Rev.* **54**, 79 (2002).
22. K. Ishihara, M. Kobayashi, N. Ishimaru and I. Shinohara, *Polym. J.* **16**, 625 (1984).
23. K. Ishihara and K. Matsui, *J. Polym. Sci. Polym. Lett. Edn.* **24**, 413 (1986).
24. G. Albin, T. A. Horbett and B. D. Ratner, *J. Control. Rel.* **2**, 153 (1985).
25. S. Cartier, T. A. Horbett and B. D Ratner, *J. Membr. Sci.* **106**, 17 (1995).
26. C. M. Hassan, F. J. Doyle, III and N. A. Peppas, *Macromolecules* **30**, 6166 (1997).
27. T. Aoki, Y. Nagao, K. Sanui, N. Ogata, A. Kikuchi, Y. Sakurai, K. Kataoka and T. Okano, *Polym. J.* **28**, 371 (1996).
28. K. Kataoka, H. Miyazaki, M. Bunya, T. Okano and Y. Sakurai, *J. Am. Chem. Soc.* **120**, 12698 (1998).

29. A. Matsumoto, S. Ikeda, A. Harada and K. Kataoka, *Biomacromolecules* **4**, 1410 (2003).
30. S. J. Lee and K. Park, *J. Mol. Recogn.* **9**, 549 (1996).
31. A. A. Obaidat and K. Par, *Pharm. Res.* **13**, 989 (1996).
32. A. A. Obidat and K. Park, *Biomaterials* **18**, 801 (1997).
33. T. Miyata, A. Jikihara, K. Nakamae and A. S. Hoffman, *Macromol. Chem. Phys.* **197**, 1135 (1996).
34. T. Miyata, N. Asami and T. Uragami, *Nature* **399**, 766 (1999).
35. T. Miyata, N. Asami and T. Uragami, *Macromolecules* **32**, 2082 (1999).
36. K. Nakamae, T. Miyata, A. Jikihara and A. S. Hoffman, *J. Biomater. Sci. Polymer Edn* **6**, 79 (1994).
37. M. Brownlee and A. Cerami, *Science* **206**, 1190 (1979).
38. L. A. Seminoff, G. B. Olsen and S. W. Kim, *Int. J. Pharm.* **54**, 241 (1989).
39. K. Makino, E. J. Mack, T. Okano and S. W. Kim, *J. Control. Rel.* **12**, 235 (1990).
40. S. Kitazawa, M. Okumura, K. Kinomura and T. Sakakibara, *Chem. Lett.*, 1733 (1990).
41. A. Tengblad, *Biochim. Biophys. Acta* **578**, 281 (1979).
42. G. N. Reeke, Jr., J. W. Becker and G. M. Edelman, *J. Biol. Chem.* **250**, 1525 (1975).
43. Y. Oda, K. Kasai and S. Ishii, *J. Biochem. (Tokyo)* **69**, 285 (1981).
44. P. J. Flory, *Principle of Polymer Chemistry*. Cornell University Press, Ithaca, NY (1953).
45. T. Miyata, in: *Supramolecular Design for Biological Applications*, N. Yui (Ed.), Chapter 6, p. 95. CRC Press, Boca Raton, FL (2002).
46. T. P. Davis, M. B. Huglin and D. C. F. Yip, *Polymer* **29**, 701 (1988).
47. A. V. Tobolsky, D. W. Carlson and N. Indictor, *J. Polym. Sci.* **54**, 175 (1961).
48. K. Dusek and W. Prins, *Adv. Polym. Sci.* **6**, 1 (1969).

Gels, Genes, Grafts and Giants, pp. 543–556 (2005)

Poly(N-isopropylacrylamide)-based semi-interpenetrating polymer networks for tissue engineering applications. Effects of linear poly(acrylic acid) chains on rheology

RANEE A. STILE [1], EUGENE CHUNG [2], WESLEY R. BURGHARDT [3] and KEVIN E. HEALY [2,4,*]

[1] *Department of Biomedical Engineering, Robert R. McCormick School of Engineering and Applied Sciences, Northwestern University, Evanston, IL 60208, USA*
[2] *Department of Bioengineering, University of California at Berkeley, 370 Hearst Memorial Mining Building, No. 1760, Berkeley, CA 94720-1762, USA*
[3] *Department of Chemical Engineering, Robert R. McCormick School of Engineering and Applied Sciences, Northwestern University, Evanston, IL 60208, USA*
[4] *Department of Materials Science and Engineering, University of California at Berkeley, 370 Hearst Memorial Mining Building, No. 1760, Berkeley, CA 94720-1760, USA*

Received 9 October 2004; accepted 30 January 2004

Abstract—Semi-interpenetrating polymer networks (semi-IPNs), comprised of poly(N-isopropylacrylamide-*co*-acrylic acid) (p(NIPAAm-*co*-AAc)) hydrogels and linear p(AAc) chains, were synthesized, and the effects of the p(AAc) chains on semi-IPN rheology were examined. Oscillatory shear rheometry studies were performed and the rheological data were analyzed as a function of temperature, frequency, and p(AAc) chain amount (weight average molecular weight (M_w) 4.5×10^5 g/mol). At 22°C, the semi-IPNs, as well as control p(NIPAAm-*co*-AAc) hydrogels, demonstrated rheological data that were representative of soft, loosely cross-linked solids. Furthermore, only the highest p(AAc) chain amount tested affected the rigidity of the p(NIPAAm-*co*-AAc)-based semi-IPNs, as compared to the p(NIPAAm-*co*-AAc) hydrogels. At 37°C, the complex shear moduli (G^*) demonstrated by the p(NIPAAm-*co*-AAc)-based semi-IPNs were significantly greater than G^* exhibited by the p(NIPAAm-*co*-AAc) hydrogels, and the semi-IPN G^* values significantly increased with increasing p(AAc) chain amount. These results can be used to develop p(NIPAAm)-based semi-IPNs with tailored mechanical properties that may function as scaffolds in tissue engineering initiatives.

Key words: Semi-IPN; N-isopropylacrylamide; rheology; oscillatory shear rheometry; acrylic acid.

*To whom correspondence should be addressed at the Department of Materials Science and Engineering. Tel.: (1-510) 643-3559. Fax: (1-510) 642-5792. E-mail: kehealy@socrates.berkeley.edu

INTRODUCTION

Poly(N-isopropylacrylamide) (p(NIPAAm)) chains and cross-linked p(NIPAAm) hydrogels exhibit unique phase properties in aqueous media when heated above the lower critical solution temperature (LCST) [1–5], which is approx. 32°C [4]. At temperatures below the LCST, p(NIPAAm) chains are soluble in water and cross-linked p(NIPAAm) hydrogels swell [6–8], while at the LCST the chains precipitate out of solution and the hydrogels demonstrate a volume-phase transition, during which they collapse considerably, expel a large amount of pore water and become stiff and opaque [6, 7]. This behavior is reversible [8] and can be modified by polymerizing the NIPAAm monomer with more hydrophobic or more hydrophilic comonomers [3, 5]. p(NIPAAm)-based systems (e.g. copolymer chains, cross-linked hydrogels, interpenetrating polymer networks and grafted surfaces) have been developed for use in a number of diverse applications, including tissue engineering, cell encapsulation, cell culture, solute delivery and molecular separation [6, 8–22].

Previously, we developed NIPAAm and acrylic acid (AAc) copolymer hydrogels for use in tissue engineering applications [15, 23]. These p(NIPAAm-*co*-AAc) hydrogels were loosely cross-linked, which allowed the matrices to be injected through a small-diameter aperture at 22°C. When heated to body temperature (i.e. 37°C), the hydrogels demonstrated a significant increase (e.g. 10×) in rigidity (i.e. complex shear modulus, G^*), without exhibiting a change in the volume or water content of the matrix [15]. In addition, the injectable p(NIPAAm-*co*-AAc) hydrogels supported bovine articular chondrocyte viability and promoted articular cartilage-like tissue formation *in vitro* [15]. Furthermore, peptide-modified p(NIPAAm-*co*-AAc) hydrogels supported rat calvarial osteoblast spreading and proliferation *in vitro* [23]. While these results were encouraging, the peptide functionalization process adversely affected the phase behavior of the p(NIPAAm-*co*-AAc) hydrogels (i.e. the matrices collapsed considerably when heated to 37°C), limiting their clinical use in tissue-engineering initiatives.

To circumvent this functionalization issue, we developed injectable semi-interpenetrating polymer networks (semi-IPNs), comprised of p(NIPAAm-*co*-AAc) hydrogels with linear peptide-modified p(AAc) chains physically entangled within the network [24]. These semi-IPNs were synthesized by first grafting peptides to the $-COO^-$ groups in the linear p(AAc) chains, and then simultaneously polymerizing and cross-linking NIPAAm and AAc in the presence of the peptide-functionalized p(AAc) chains. This approach offers important clinical and practical advantages. First, since the p(AAc) chains are modified prior to the semi-IPN synthesis, the p(NIPAAm-*co*-AAc) hydrogel is unaffected by the functionalization process, decreasing the potential that the matrix will collapse once injected into the body. Additionally, in practice, it is harder to modify three-dimensional (3D) polymer matrices than linear polymer chains (e.g. degree of substitution, removal of reactants, etc.). Furthermore, if different types of peptides or macromolecules are required in a single hydrogel, it is possible that different functionalization schemes would be needed, thus requiring that the reactions be performed in series. It is also possible that subsequent reactions could adversely affect the peptides or macromolecules that were previously grafted. In the case of a semi-IPN, the peptides or macromolecules may be attached to the linear p(AAc) chains in parallel prior to the semi-IPN synthesis, creating user-defined admixtures of functionalized p(AAc) chains, and one functionalization scheme would not affect any other. Using this methodology, we showed that p(NIPAAm-*co*-AAc)-based semi-IPNs modified with peptides that bind to hyaluronic acid (HA) adhered to HA-grafted substrates in an aqueous environment [24].

Recently, we analyzed the effects of linear non-functionalized p(AAc) chains on the injectability and phase behavior (i.e. transparency, volume change, LCST and phase transition) of p(NIPAAm-*co*-AAc)-based semi-IPNs [25] and we found that the p(AAc) chains did not affect the LCST or volume change of the semi-IPNs, as compared to control hydrogels. As a continuation of this work, we performed the current study, in which we used rheological methods to examine the effects of linear non-functionalized p(AAc) chains on the rigidity of p(NIPAAm-*co*-AAc)-based semi-IPNs. These rheological experiments are significant from a tissue-engineering perspective. If the p(NIPAAm-*co*-AAc)-based matrices are to be used as injectable scaffolds in a clinical setting, they must be injectable at temperatures below body temperature, which requires that the materials demonstrate low rigidity. Once being injected into the body, the matrices must provide adequate mechanical integrity to support tissue formation, which necessitates that the materials exhibit increased rigidity. While it is widely known that cross-linked p(NIPAAm) networks demonstrate a change in rigidity at the LCST, relatively few reports have been published detailing the use of rheological techniques to measure p(NIPAAm) hydrogel rigidity [15, 26–29]. More commonly, p(NIPAAm)-based rheological studies have focused on p(NIPAAm) microgels (i.e. cross-linked p(NIPAAm) hydrogel particles dispersed in a liquid medium) [30–33] and aqueous p(NIPAAm) solutions [34–38].

One rheological technique that has been used to examine the rigidity of p(NIPAAm)-based hydrogels is oscillatory shear rheometry [15, 28], in which an oscillatory shear stress or shear strain is imposed on a material at frequency ω and the resulting oscillatory shear strain or shear stress, respectively, is determined [39, 40]. The functions that are generally examined in oscillatory shear rheometry studies are the shear storage modulus (G', a measure of the energy stored and recovered per cycle), the shear loss modulus (G'', a measure of the energy dissipated or lost as heat per cycle), the complex shear modulus (G^*, defined mathematically as $G^* = G' + iG''$, $|G^*| = (G'^2 + G''^2)^{1/2}$) and the phase angle ($\delta$, the phase difference between the input variable and the output variable; $\tan\delta = (G''/G')$) [39–43]. Importantly, these functions can be used to characterize materials as viscoelastic solids or liquids. For example, viscoelastic solids exhibit G' values that are both independent of ω and considerably larger than G'' at low ω [42, 44, 45]. In contrast, viscoelastic liquids demonstrate G' and G'' values that approach 0 with decreasing ω, scaling as $G' \approx \omega^2$ and $G'' \approx \omega$; hence $G'' \gg G'$ as ω approaches 0 [40, 42]. This means that $\delta \rightarrow 90°$ as $\omega \rightarrow 0$ for liquids, which is in sharp contrast to the corresponding result, $\delta \rightarrow 0°$ as $\omega \rightarrow 0$, for ideal solids.

In this work, oscillatory shear rheometry studies were performed on p(NIPAAm-*co*-AAc) hydrogels and p(NIPAAm-*co*-AAc)-based semi-IPNs. In the p(NIPAAm-*co*-AAc) hydrogel studies, the molar ratio of NIPAAm : AAc in the feed was varied, and in the p(NIPAAm-*co*-AAc)-based semi-IPN studies, the amount of the p(AAc) chains (weight average molecular weight (M_w) 4.5×10^5 g/mol) was varied. The rheological data were examined as a function of both temperature (i.e. 22°C and 37°C) and ω (i.e. 0.001 Hz to 10 Hz).

MATERIALS AND METHODS

Materials

NIPAAm, AAc, *N,N′*-methylenebisacrylamide (BIS) Chemzymes™ Ultrapure grade, ammonium peroxydisulfate (AP) Chemzymes™ Ultrapure grade, *N,N,N′,N′*-tetramethylethylenediamine (TEMED) Chemzymes™ Ultrapure grade and linear p(AAc) chains (M_w 4.5×10^5 g/mol, acid form) were obtained from Polysciences (Warrington, PA, USA). Dulbecco's phosphate-buffered saline (PBS; 1.51 mM KH_2PO_4, 155 mM NaCl and 2.7 mM Na_2HPO_4; without $CaCl_2$, without $MgCl_2$; pH 7.2 ± 0.1; ionic strength 0.165 M) was obtained from Gibco-BRL (Grand Island, NY, USA). All materials were used as received.

p(NIPAAm-co-AAc) hydrogel synthesis

The p(NIPAAm-*co*-AAc) hydrogels were synthesized according to methods published previously [15, 23] as shown in Scheme 1. NIPAAm, AAc, 0.005 g (0.0325 mmol) of BIS and 50 ml of PBS were bubbled with dry nitrogen gas in a two-neck flask for 15 min to remove dissolved oxygen. Two NIPAAm : AAc molar ratios were used: 96.25 : 3.75 (2.4395 g (21.59 mmol) of NIPAAm and 0.0605 g (0.84 mmol) of AAc) and 96.5 : 3.5 (2.443 g (21.62 mmol) of NIPAAm and 0.057 g (0.79 mmol) of AAc). Following the nitrogen gas purge, 0.020 g (0.0876 mmol) of AP and 200 μl (1.3 mmol) of TEMED were added as the initiator and accelerator, respectively. The mixture was stirred vigorously for 15 s and allowed to polymerize at 22°C for 19 h under regular fluorescent lighting in a 250-ml glass beaker covered with a glass plate. Following the polymerization, the p(NIPAAm-*co*-AAc) hydrogels were washed three times, 15–20 min each, in excess ultrapure water (UPW; 18 MΩ-cm; pH between 5.5 and 7) to remove unreacted compounds.

p(NIPAAm-co-AAc)-based semi-IPN synthesis

The p(NIPAAm-*co*-AAc)-based semi-IPNs were synthesized as reported previously (Scheme 2) [25]. All samples contained 2.4395 g (21.59 mmol) of NIPAAm and

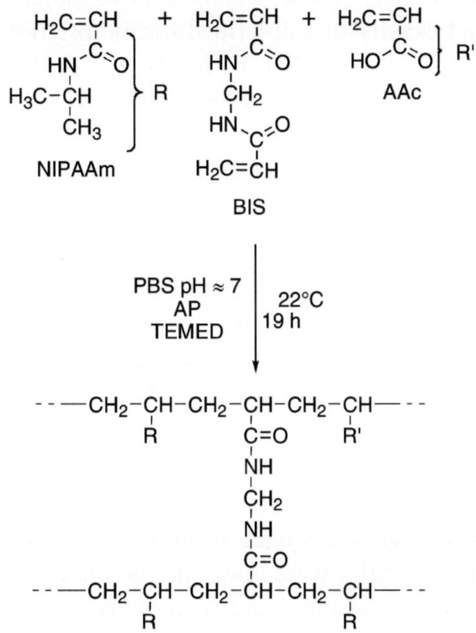

Scheme 1. Synthesis of the p(NIPAm-*co*-AAc) hydrogel.

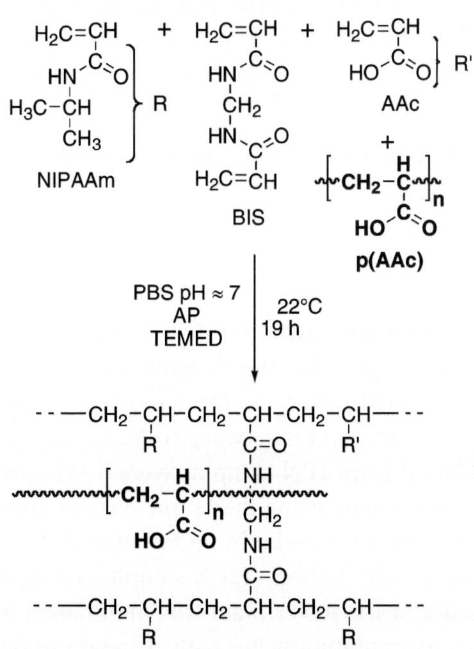

Scheme 2. Synthesis of the p(NIPAm-*co*-AAc)-based semi-IPN.

0.0605 g (0.84 mmol) of AAc (i.e. a NIPAAm : AAc molar ratio of 96.25 : 3.75). The synthesis followed exactly that described for the p(NIPAAm-*co*-AAc) hydrogels, with the addition of linear p(AAc) chains (M_w 4.5 × 10^5 g/mol) in amounts of 0.001 g (2.22 nmol), 0.0055 g (12.2 nmol) and 0.013 g (28.9 nmol).

Rheology studies

The rheology studies were performed based on methods used previously [15].

p(NIPAAm-co-AAc) hydrogels. Small samples (approx. 0.5 ml) of the p(NIPAAm-*co*-AAc) hydrogels were collected from the top of the matrices using a 3-ml syringe with a 2 mm diameter aperture. The 'top' is defined as the side of the matrix that did not polymerize in contact with the glass beaker. The rheology of the hydrogels was then examined on a Bohlin VOR rheometer (Bohlin Rheologi, Cranbury, NJ, USA) using a parallel plate (30 mm diameter) configuration in oscillatory mode. Sample thickness ranged from 0.782 to 1.128 mm, depending on the sample volume. The samples were injected onto the lower plate using the 3 ml syringe, the upper fixture was lowered, and a humidity chamber was placed around the sample to prevent dehydration during data collection. Each sample was tested at 22°C and then at 37°C. The temperature of the lower plate was maintained with a recirculating water bath. Linear viscoelastic data were collected using a 3.61 g cm torsion bar over a ω range of 0.001–10 Hz. Depending on the sample thickness, the shear strain ranged from 18.1 to 28.6%. Preliminary experiments demonstrated that the rheological properties were independent of the applied strain in this range.

p(NIPAAm-co-AAc)-based semi-IPNs. 3-ml samples of the p(NIPAAm-*co*-AAc)-based semi-IPNs were collected from the top of the matrices using a 10-ml syringe with a 2 mm diameter aperture. The rheology of the semi-IPNs was then examined on a Paar Physica MCR 300 rheometer (Paar Physica, Ashland, VA, USA) using a parallel plate (50 mm diameter) configuration in oscillatory mode. Two different rheometers were used due to Dr. Healy's move to Berkeley. Data collected using the same p(NIPAAm-*co*-AAc)-based semi-IPNs on both instruments were nearly identical and fell almost exactly on the same lines, indicating that the results were independent of the instrument they were collected on, as they should be. The p(NIPAAm-*co*-AAc)-based semi-IPN samples were 1.00 mm thick. The samples were injected onto the lower plate using the 10-ml syringe, the upper plate was lowered, excess sample was trimmed and a humidity chamber was placed around the sample to prevent sample dehydration. Each sample was tested at 22°C and then at 37°C. The temperature of the lower plate was maintained with a Peltier heating element connected to a recirculating water bath. Linear viscoelastic data were collected over a frequency range of 0.001–10 Hz. The shear strain ranged from 18 to 25%.

Statistical analysis

Analysis of variance (ANOVA) tests and Newman-Keuls post-hoc analyses were performed using StatSoft Statistica 5.0 (StatSoft, Tulsa, OK, USA) and SAS JMP 5.0 (SAS, Cary, NC, USA). Statistical significance was determined at a value of $P < 0.05$.

RESULTS AND DISCUSSION

At 22°C, the p(NIPAAm-*co*-AAc) hydrogels and p(NIPAAm-*co*-AAc)-based semi-IPNs demonstrated G^* and δ curves that were representative of extremely soft, loosely cross-linked solids (Figs 1 and 2, solid lines), consistent with p(NIPAAm-*co*-AAc) hydrogel data published previously [15]. With decreasing ω, G^* approached a constant value in the range of 3–10 Pa (Figs 1a and 2a, solid lines), and δ decreased (Figs 1b and 2b, solid lines), indicating that G' was considerably greater than G'' at low ω (i.e. at least 2× greater; Fig. 3, solid lines). When heated from 22°C to 37°C, the hydrogels and semi-IPNs became more rigid and solid-like, as signified by the significant increase in G^* ($P < 0.05$ *versus* the 22°C data; Figs 1a and 2a, dashed lines) and the decrease in δ (i.e. δ closer to 0°; Figs 1b and 2b, dashed lines). At 37°C, G' was roughly an order of magnitude greater than G'' (Fig. 3, dashed lines). This change in rigidity is commonly demonstrated by p(NIPAAm)-based hydrogels when the temperature is increased above the LCST [6, 7]. From a tissue-engineering perspective, the rigidity of the p(NIPAAm-*co*-AAc) hydrogels and p(NIPAAm-*co*-AAc)-based semi-IPNs at 37°C was comparable to that of type-I collagen gels (Table 1).

Interestingly, the shapes of the G^* and δ curves at 37°C were more representative of those obtained from amorphous glassy and crystalline linear polymers, rather than those obtained from cross-linked viscoelastic solids [42]. G^* was not constant over a wide range of ω but demonstrated a weak power law dependence on ω (Figs 1a and 2a, dashed lines), such that δ was relatively constant (Figs 1b and 2b,

Table 1.
G' values for p(NIPAAm)-based matrices and collagen gels at 37°C

Formulation	G'	Reference
p(NIPAAm-*co*-AAc) hydrogels[a]	68–128 Pa at 0.1 Hz	This paper and Ref. [15]
p(NIPAAm-*co*-AAc)-based semi-IPNs[b]	99–169 Pa at 0.1 Hz	This paper
Type-I collagen gel (1.6 mg/ml)	78 Pa at 0.3 Hz	[54]
Type-I collagen gel (2.3 mg/ml)	30 Pa at 0.3 Hz	[55]
Type-I collagen gel (3 mg/ml)	20 Pa at 0.02 Hz	[56]

[a] The p(NIPAAm-*co*-AAc) hydrogels contained NIPAAm : AAc ratios of 96 : 4, 96.25 : 3.75 and 96.5 : 3.5.

[b] The p(NIPAAm-*co*-AAc)-based semi-IPNs contained a NIPAAm : AAc ratio of 96.25 : 3.75 and p(AAc) amounts of 0.001 g, 0.0055 g and 0.013 g.

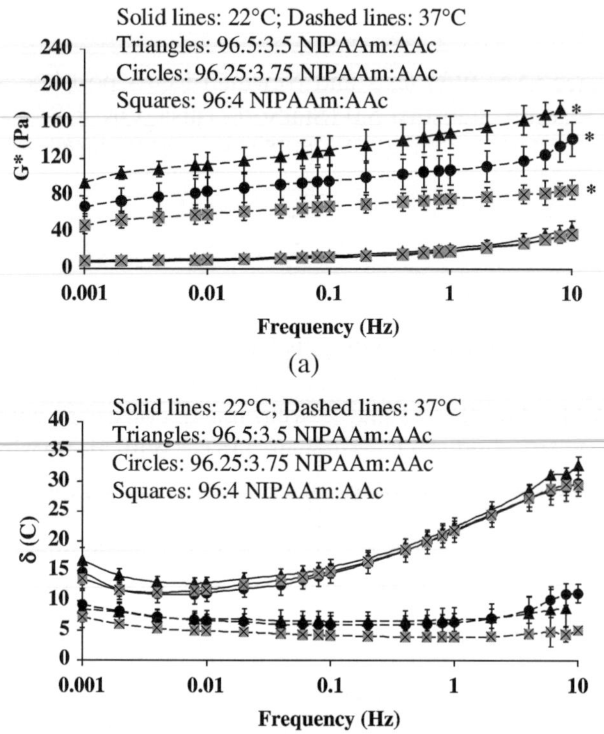

(b)

Figure 1. (a) G^* and (b) δ as a function of frequency and temperature for p(NIPAAm-*co*-AAc) hydrogels with a NIPAAm : AAc molar ratio of 96.5 : 3.5 (22°C, —▲—; 37°C, - -▲- -) and 96.25 : 3.75 (22°C, —●—; 37°C, - -●- -). Previously published rheology data [15] are also included for p(NIPAAm-*co*-AAc) hydrogels with a NIPAAm : AAc molar ratio of 96 : 4 (22°C, —⊠—; 37°C, - -⊠- -). The 22°C G^* data sets overlap each other. Frequency is on a log scale. Each data set is the average of at least 3 experiments. The lines were added for illustration purposes only and do not indicate modeling of the data. If error bars cannot be seen, they are smaller than the size of the symbol. * indicates that the data sets are significantly different from all other 37°C G^* data sets at $P < 0.05$.

dashed lines). These results were likely influenced by the complex interactions that occur during the phase transition when p(NIPAAm)-based materials are heated above the LCST [5, 7, 13, 46–48].

The p(NIPAAm-*co*-AAc) hydrogel G^* data were not affected by the NIPAAm : AAc molar ratio at 22°C ($P > 0.05$; Fig. 1a, solid lines); however, the G^* data were affected by the NIPAAm : AAc molar ratio at 37°C (Fig. 1a, dashed lines). As the NIPAAm : AAc molar ratio increased from 96 : 4 to 96.25 : 3.75 to 96.5 : 3.5, G^* significantly increased ($P < 0.05$), consistent with the addition of NIPAAm monomer to the p(NIPAAm-*co*-AAc) hydrogel. The more NIPAAm (and consequently, the less AAc) present in the hydrogel, the more hydrophobic aggregates present at 37°C (i.e. above the LCST) and the more rigid the matrix.

Only the highest p(AAc) chain amount tested (i.e. 0.013 g) significantly affected the rigidity of the p(NIPAAm-*co*-AAc)-based semi-IPNs at 22°C ($P < 0.05$;

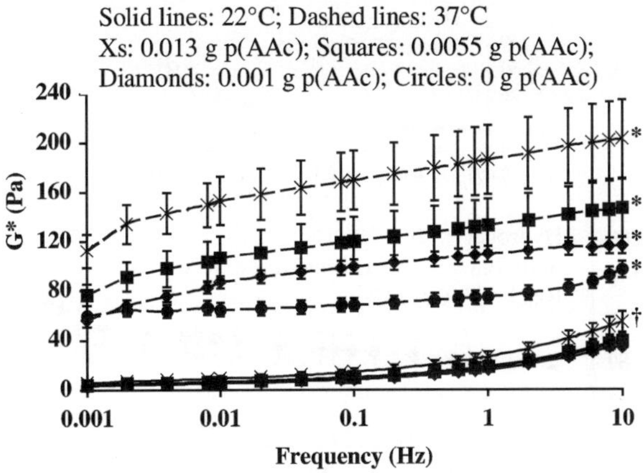

(a)

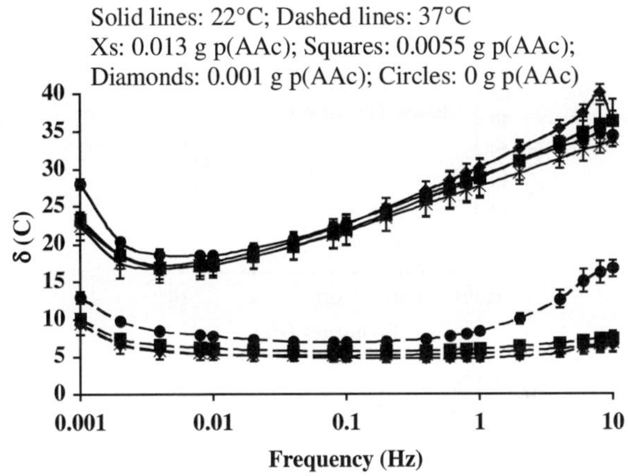

(b)

Figure 2. (a) G^* and (b) δ as a function of frequency and temperature for p(NIPAAm-*co*-AAc) hydrogels (22°C, —●—; 37°C, - -●- -), and p(NIPAAm-*co*-AAc)-based semi-IPNs containing 0.001 g (22°C, —◆—; 37°C, - -◆- -), 0.0055 g (22°C, —■—; 37°C, - -■- -) and 0.013 g (22°C, —×—; 37°C, - -×- -) of p(AAc) chains (M_w 4.5 × 10^5 g/mol). The 22°C G^* data sets overlap each other. Frequency is on a log scale. All hydrogels and semi-IPNs contained a NIPAAm : AAc molar ratio of 96.25 : 3.75. Each data set is the average of at least 3 experiments. The lines were added for illustration purposes only and do not indicate modeling of the data. If error bars cannot be seen, they are smaller than the size of the symbol. † indicates that the data set is significantly different from all other 22°C G^* data sets at $P < 0.05$. * indicates that the data sets are significantly different from all other 37°C G^* data sets at $P < 0.05$.

Fig. 2a, solid lines). At lower p(AAc) chain amounts (i.e. 0.001 g and 0.0055 g), the p(NIPAAm-*co*-AAc)-based semi-IPN G^* data were not significantly different from the p(NIPAAm-*co*-AAc) hydrogel G^* data ($P > 0.05$; Fig. 2a, solid lines). We hypothesized that the p(AAc) chains would increase the rigidity of the semi-IPNs at 22°C, due to $-COO^-$ repulsion (e.g. repulsion between the p(AAc) chains and the AAc groups in the p(NIPAAm-*co*-AAc) hydrogel and/or between adjacent

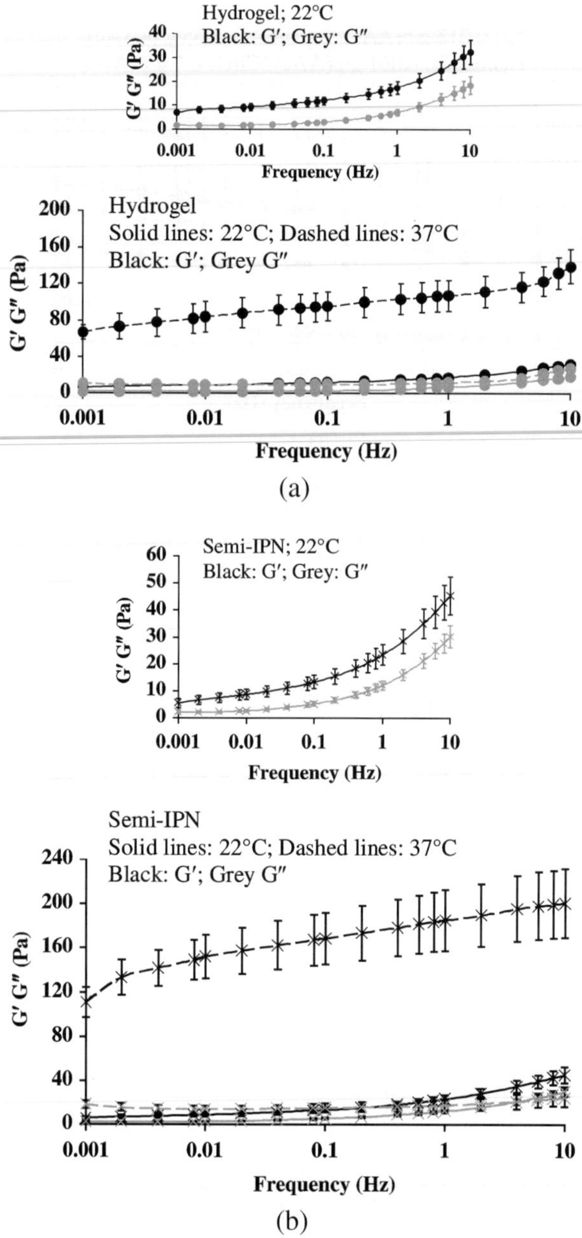

Figure 3. (a) G'(22°C, —●—; 37°C, - -●- -) and G'' (22°C, —●—; 37°C, - -●- -) for p(NIPAAm-*co*-AAc) hydrogels and (b) G'(22°C, —×—; 37°C, - -×- -) and G'' (22°C, —X—; 37°C, - -X- -) for p(NIPAAm-*co*-AAc)-based semi-IPNs containing 0.013 g of p(AAc) chains (M_w 4.5 × 10^5 g/mol). For clarity, panels (a) and (b) include insets with a different scale to show the 22°C data. The hydrogels and semi-IPNs contained a NIPAAm : AAc molar ratio of 96.25 : 3.75. Frequency is on a log scale. Each data set is the average of at least 3 experiments. The lines were added for illustration purposes only and do not indicate modeling of the data. If error bars cannot be seen, they are smaller than the size of the symbol.

p(AAc) chains) and processes analogous to second-phase strengthening [49, 50]. As discussed in our previous report [25], two-polymer systems, such as semi-IPNs, generally consist of two immiscible phases [51]. Thermodynamically, the immiscibility is driven by the low entropy of mixing that ensues when two polymers are blended. The resulting phase separation leads to the formation of phases that vary in amount, size, shape, interfacial sharpness and degree of continuity. These phases constitute the microscopic and macroscopic morphology of the material, and the morphology then directly influences the material properties of the system. In our previous work [25], we observed phase separation in the p(NIPAAm-*co*-AAc)-based semi-IPNs at 22°C, and the opacity of the matrices increased with increasing p(AAc) chain amount (i.e. semi-IPNs containing 0.0055 g and 0.013 g of p(AAc) chains were opaque at 22°C). Changing the p(AAc) chain amount most likely altered the p(AAc)-rich phase within the semi-IPNs (e.g. in amount, size, continuity, etc.), leading to changes in the microscopic and macroscopic morphology of the system (i.e. the opacity changed). The morphology changes then altered the material properties of the semi-IPNs, as the injectability of the matrices at 22°C decreased with increasing p(AAc) chain amount (i.e. the semi-IPNs containing 0.013 g of p(AAc) chains were not injectable at 22°C) [25]. This decrease in injectability may have been due to the increase in rigidity demonstrated by the p(NIPAAm-*co*-AAc)-based semi-IPNs containing higher amounts of p(AAc) chains (i.e. 0.013 g). The data in both the previous and present studies suggest that the effects of the p(AAc) chains on the p(NIPAAm-*co*-AAc)-based semi-IPNs at 22°C, whether related to morphology (e.g. opacity) or material properties (e.g. injectability and rigidity), are not discernable until a critical amount of p(AAc) chains is present within the matrices.

In contrast to the 22°C data, the 37°C G^* data were affected by the presence of the p(AAc) chains at all amounts tested (Fig. 2a, dashed lines). The p(NIPAAm-*co*-AAc)-based semi-IPNs were significantly more rigid than the p(NIPAAm-*co*-AAc) hydrogels at 37°C ($P < 0.05$), independent of p(AAc) chain amount. The 37°C data suggest that the increase in G^* resulted from interactions between the p(AAc)-rich phase and phase-separated segments of the p(NIPAAm-*co*-AAc) hydrogel. Given that the p(NIPAAm) phase transition in aqueous media has not been completely elucidated [5, 13], the ternary system that comprises the p(NIPAAm-*co*-AAc)-based semi-IPNs (i.e. p(NIPAAm-*co*-AAc) hydrogels, PBS solvent and p(AAc) chains) undoubtedly increased the complexity of the environment within the matrices at temperatures above the LCST, making it difficult to interpret the nature of the interactions at 37°C.

As well as being affected by the presence of the p(AAc) chains, G^* was also a function of p(AAc) chain amount at 37°C, since the p(NIPAAm-*co*-AAc)-based semi-IPN G^* values significantly increased with increasing p(AAc) chain amount ($P < 0.05$; Fig. 2a, dashed lines). In addition to the aforementioned interactions between the p(AAc)-rich phase and phase-separated components of the p(NIPAAm-*co*-AAc) hydrogel at 37°C, these data indicate that as the p(AAc) chain amount

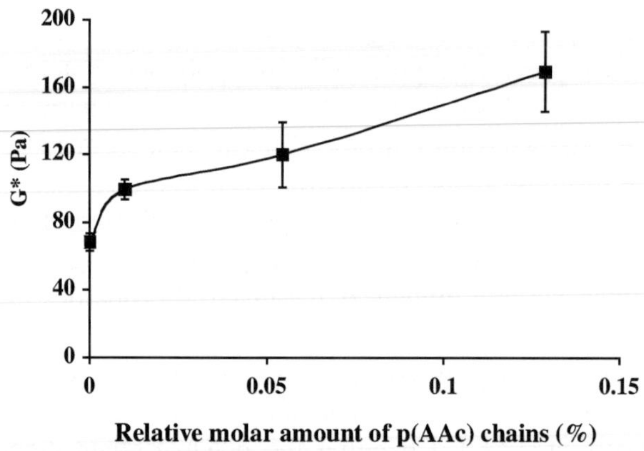

Figure 4. G^* as a function of the relative molar amount of p(AAc) chains (M_w 4.5 × 10^5 g/mol) in the p(NIPAAm-*co*-AAc)-based semi-IPNs at 37°C and 0.1 Hz. The relative molar amount of p(AAc) chains was calculated as follows: (mol p(AAc) chains/(mol NIPAAm + mol AAc in the hydrogel))×100%. Each data point is the average of at least 3 experiments. The line was added for illustration purposes only and does not indicate modeling of the data. The semi-IPNs contained a NIPAAm : AAc molar ratio of 96.25 : 3.75.

increased, alterations in the p(AAc)-rich phase led to changes in the material properties of the semi-IPNs (i.e. G^* increased); however, the mechanism by which the p(AAc)-rich phase varies with p(AAc) chain amount at 37°C is unknown. Figure 4 shows a plot of G^* as a function of the relative molar amount of p(AAc) chains in the p(NIPAAm-*co*-AAc)-based semi-IPNs at 37°C and 0.1 Hz. The relative amount of p(AAc) chains was calculated as follows: (mol p(AAc) chains/(mol NIPAAm + mol AAc in the hydrogel))×100%. The effect of the p(AAc)-rich phase on the rheology of the p(NIPAAm-*co*-AAc)-based semi-IPNs is comparable to that observed in other two-phase systems. For example, changing the content of styrene-butadiene block copolymers changes the properties of the glassy and elastomeric phases, altering the mechanical properties of the system [52]; additionally, adjusting the epoxy content in acrylic-epoxy IPNs affects the phase continuity within the system, leading to changes in the swelling behavior of the matrix [53].

CONCLUSIONS

Poly(NIPAAm-*co*-AAc) hydrogels and p(NIPAAm-*co*-AAc)-based semi-IPNs were synthesized, and the effects of numerous reaction conditions on the rheology of the matrices were analyzed. In the p(NIPAAm-*co*-AAc) hydrogel studies, the NIPAAm : AAc molar ratio in the feed was varied, and in the p(NIPAAm-*co*-AAc)-based semi-IPN studies, the amount of the linear p(AAc) chains (M_w 4.5 × 10^5 g/mol) was modified. Oscillatory shear rheometry studies were performed on the hydrogels and semi-IPNs as a function of temperature (i.e. 22°C and 37°C) and ω (i.e. 0.001–10 Hz). At 22°C, the rigidity of the p(NIPAAm-*co*-AAc) hydrogels was not affected by the NIPAAm : AAc molar ratio, and only the highest p(AAc) chain amount tested significantly affected the rigidity of the p(NIPAAm-*co*-AAc)-based semi-IPNs. At 37°C, the p(NIPAAm-*co*-AAc) hydrogel rigidity significantly increased with increasing NIPAAm : AAc molar ratio, the p(NIPAAm-*co*-AAc)-based semi-IPNs were significantly more rigid than the p(NIPAAm-*co*-AAc) hydrogels, and the rigidity of the p(NIPAAm-*co*-AAc)-based semi-IPNs significantly increased with increasing p(AAc) chain amount.

REFERENCES

1. E. F. Casassa, in: *Fractionation of Synthetic Polymers: Principles and Practices*, L. H. Tung (Ed.), p. 3. Marcel Dekker, New York, NY (1977).
2. K. Kamide, *Thermodynamics of Polymer Solutions: Phase Equilibria and Critical Phenomena.* Elsevier, New York, NY (1990).
3. L. D. Taylor and L. D. Cerankowski, *J. Polym. Sci. Polym. Chem. Edn.* **13**, 2551 (1975).
4. M. Heskins and J. E. Guillet, *J. Macromol. Sci. Chem. Edn.* **A2**, 1441 (1968).
5. H. G. Schild, *Prog. Polym. Sci.* **17**, 163 (1992).
6. A. S. Hoffman, A. Afrassiabi and L. C. Dong, *J. Control. Rel.* **4**, 213 (1986).
7. Y. Hirokawa and T. Tanaka, *J. Chem. Phys.* **81**, 6379 (1984).
8. A. S. Hoffman, *MRS Bull.* **16**, 42 (1991).
9. B. Vernon, A. Gutowska, S. W. Kim and Y. H. Bae, *Macromol. Symp.* **109**, 155 (1996).
10. J. H. Priest, S. L. Murray, R. J. Nelson and A. S. Hoffman, *ACS Symp. Ser.* **350**, 255 (1987).
11. A. S. Hoffman, G. H. Chen, S. Y. Kaang, Z. L. Ding, K. Randeri and B. Kabra, in: *Advanced Biomaterials in Biomedical Engineering and Drug Delivery Systems*, N. Ogata, S. W. Kim, J. Feijen and T. Okano (Eds), p. 62. Springer, Tokyo (1996).
12. T. Aoyagi, M. Ebara, K. Sakai, Y. Sakurai and T. Okano, *J. Biomater. Sci. Polymer Edn* **11**, 101 (2000).
13. B. Vernon, S. W. Kim and Y. H. Bae, *J. Biomed. Mater. Res.* **51**, 69 (2000).
14. R. Yoshida, K. Sakai, T. Okano and Y. Sakurai, *J. Biomater. Sci. Polymer Edn* **6**, 585 (1994).
15. R. A. Stile, W. R. Burghardt and K. E. Healy, *Macromolecules* **32**, 7370 (1999).
16. L. C. Dong and A. S. Hoffman, *J. Control. Rel.* **4**, 223 (1986).
17. Y. H. Bae, T. Okano and S. W. Kim, *Pharm. Res.* **8**, 531 (1991).
18. H. Feil, Y. H. Bae, J. Feijen and S. W. Kim, *J. Membrane Sci.* **64**, 283 (1991).
19. G. Chen and A. S. Hoffman, *Nature* **373**, 49 (1995).
20. S. K. Vakkalanka, C. S. Brazel and N. A. Peppas, *J. Biomater. Sci. Polymer Edn* **8**, 119 (1996).
21. E. Kokufuta, B. Wang, R. Yoshida, A. R. Khokhlov and M. Hirata, *Macromolecules* **31**, 6878 (1998).
22. M. Ebara, M. Yamato, M. Hirose, T. Aoyagi, A. Kikuchi, K. Sakai and T. Okano, *Biomacromolecules* **4**, 344 (2003).
23. R. A. Stile and K. E. Healy, *Biomacromolecules* **2**, 185 (2001).
24. R. A. Stile, K. R. Shull and K. E. Healy, *Langmuir* **19**, 1853 (2003).
25. R. A. Stile and K. E. Healy, *Biomacromolecules* **3**, 591 (2002).
26. M. Shibayama, M. Morimoto and S. Nomura, *Macromolecules* **27**, 5060 (1994).
27. A. Grosse-Sommer, M. In and R. K. Prud'homme, *J. Polym. Sci. Polym. Chem.* **34**, 1447 (1996).
28. T. L. Lowe, M. Benhaddou and H. Tenhu, *Macromol. Chem. Phys.* **200**, 51 (1999).
29. R. Schinner, T. Wolff and D. Kuckling, *Ber. Bunsen-Ges. Phys. Chem. Chem. Phys.* **102,** 1710 (1998).
30. T. L. Lowe, M. Benhaddou and H. Tenhu, *J. Polym. Sci. Polym. Phys.* **36**, 2141 (1998).
31. H. Senff, W. Richtering, C. Norhausen, A. Weiss and M. Ballauff, *Langmuir* **15**, 102 (1999).
32. H. Senff and W. Richtering, *Coll. Polym. Sci.* **278**, 830 (2000).

33. I. Deike and M. Ballauff, *J. Rheol.* **45,** 709 (2001).
34. H. Yoshioka, M. Mikami and Y. Mori, *J. Macromol. Sci. Pure Appl. Chem.* **A31**, 113 (1994).
35. I. V. Berlinova, I. V. Dimitrov, N. G. Vladimirov, V. Samichkov and Y. Ivanov, *Polymer* **42**, 5963 (2001).
36. A. Durand and D. Hourdet, *Macromol. Chem. Phys.* **201**, 858 (2000).
37. G. Bokias, D. Hourdet and I. Iliopoulos, *Macromolecules* **33**, 2929 (2000).
38. S. H. Hsu and T. L. Yu, *Macromol. Rapid Commun.* **21**, 476 (2000).
39. H. A. Barnes, J. F. Hutton and K. Walters, *An Introduction to Rheology*. Elsevier, New York, NY (1989).
40. G. M. Kavanagh and S. B. Ross-Murphy, *Prog. Polym. Sci.* **23**, 533 (1998).
41. F. Rodriguez, *Principles of Polymer Systems*, 3rd edn. Hemisphere, New York, NY (1995).
42. J. D. Ferry, *Viscoelastic Properties of Polymers*, 3rd edn. Wiley, New York, NY (1980).
43. R. W. Whorlow, *Rheological Techniques*, 2nd edn. Ellis Horwood, New York, NY (1992).
44. W. Burchard and S. B. Ross-Murphy, *Physical Networks: Polymers and Gels*. Elsevier Applied Science, London (1990).
45. K. Almdal, J. Dyre, S. Hvidt and O. Kramer, *Polym. Gels Networks* **1**, 5 (1993).
46. S. Hirotsu, *J. Chem. Phys.* **88**, 427 (1988).
47. S. Hirotsu, Y. Hirokawa and T. Tanaka, *J. Chem. Phys.* **87**, 1392 (1987).
48. M. Shibayama and T. Tanaka, *Adv. Polym. Sci.* **109**, 1 (1993).
49. L. H. Van Vlack, *Elements of Materials Science and Engineering*. Addison-Wesley, Reading, MA (1989).
50. L. H. Sperling, *Introduction to Physical Polymer Science*. Wiley, New York, NY (1992).
51. L. H. Sperling, *Interpenetrating Polymer Networks and Related Materials*. Plenum, New York, NY (1981).
52. H. R. Allcock and F. W. Lampe, *Contemporary Polymer Chemistry*, 2nd edn. Prentice-Hall, Englewood Cliffs, NJ (1990).
53. S. Ma and X. Tang, in: *Interpenetrating Polymer Networks*, D. Klempner, L. H. Sperling and L. A. Utracki (Eds), p. 405. American Chemical Society, Washington, DC (1994).
54. J. Tan and M. Saltzman, *J. Biomed. Mater. Res.* **46**, 465 (1999).
55. D. Velegol and F. Lanni, *Biophys. J.* **81**, 1786 (2001).
56. V. H. Barocas, A. G. Moon and R. T. Tranquillo, *J. Biomech. Eng.* **117**, 161 (1995).

Gels, Genes, Grafts and Giants, pp. 557–565 (2005)

Poly(*N*-isopropylacrylamide)-grafted gelatin as a thermoresponsive cell-adhesive, mold-releasable material for shape-engineered tissues

TAKEHISA MATSUDA *

Division of Biomedical Engineering, Graduate School of Medical Sciences, Kyushu University, Maidashi 3-1-1, Higashi-ku, Fukuoka 812-8582, Japan

Received 9 October 2003; accepted 25 February 2004

Abstract—Poly(*N*-isopropylacrylamide)-grafted gelatin (PNIPAM-gelatin) can be used as a thermoresponsive cell-adhesive matrix and mold-releasable material for shape-engineered tissues. An example of such a tissue with predetermined shape is realized by fabrication of a tubular endothelial cell construct. A solution containing a mixture of PNIPAM-gelatin and PNIPAM was coated on the luminal surface of a glass capillary tube. After air-drying, endothelial cells were seeded and cultured for four days at 37°C. The infusion of the culture medium into the tube at 20°C resulted in the spontaneous detachment and removal of a tubular vascular tissue composed of endothelial cells and supramolecularly organized extracellular matrices produced by endothelial cells on the basal side of the tissue. The potential use of a thermoresponsive cell-adhesive and mold-releasable material in complex-shaped tissue-engineered devices is discussed.

Key words: Thermoresponsiveness; poly(*N*-isopropylacrylamide); gelatin; tissue engineering; mold-releasable material.

INTRODUCTION

The unique feature of inverse temperature phenomena of solution-to-gel or solution-to-precipitation phase transition of an aqueous solution of poly(*N*-isoporopylacrylamide), PNIPAM, has taken considerable attention in biomedical applications. For example, the PNIPAM-conjugate protein, developed by Hoffman in the mid-1980s [1], caused a temperature-induced on-off switching of biochemical activity for specific immunoassays and protein isolations. Okano *et al.* have extensively studied and explored PNIPAM-based drug delivery systems with temperature-induced on-off switching of drug releasing [2] and further developed for 'cell sheet

*E-mail: matsuda@med.kyushu-u.ac.jp

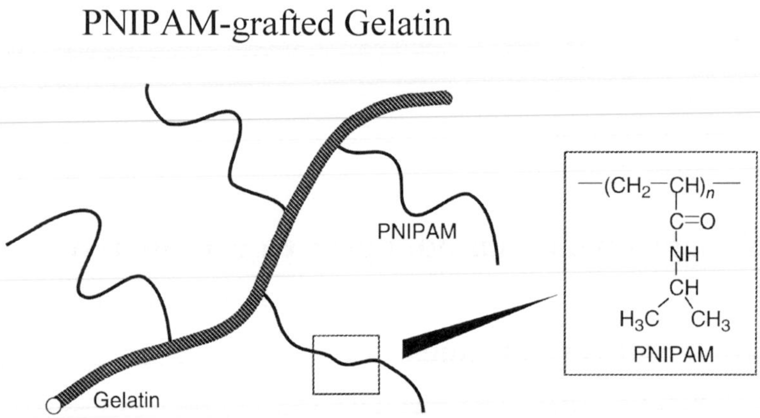

Figure 1. Schematics of PNIPAM-gelatin (molecular weight of gelatin 9.5×10^4 g/mol; average molecular weight of PNIPAM graft chain 5.0×10^4 g/mol; number of graft chains, 25.9 per gelatin molecule).

engineering' [3–5] that is realized on an electron-beam-induced PNIPAM-grafted surface. The methodology they developed eventually opened up a new field in tissue engineering that enables the formation and harvesting of intact cell sheet and overlayered cell constructs. Functional tissues thus prepared include beating cardiac tissue composed of multiple-layered cardiac cells and cornea tissues.

Previously, we have prepared a PNIPAM copolymer partially derivatized with a cell-adhesive active core peptidyl sequence (RGD; Arg-Gly-Asp) common to adhesive proteins such as fibronectin, enabling two-dimensional (2D) and 3D tissue formation [6, 7]. Recently, we have prepared a PNIPAM-grafted gelatin (PNIPAM-gelatin) molecule *via* quasi-living radical polymerization initiated by iniferter derivatized on gelatin (Fig. 1) [8, 9]. The molecular parameters for determining the lower critical solution temperature (LCST), cell adhesiveness and ease of detachment from substrates such as glass or plastic dishes are the number of graft chains per gelatin molecule (graft density) and average graft chain length (or molecular weight). The higher the graft density and the higher the molecular weight of graft chain, the lower the LCST, the lesser the cell adhesiveness and the easier the detachment [8]. Ease of detachment of a formed cell sheet from the substrate strongly depends on the type of substrate, which needs fine-tuned graft architecture (graft density and graft chain length) specific to the type of the substrate used. Instead of individual molecular design of PNIPAM-gelatin specific to each substrate to meet high cell adhesiveness and ease of detachment or high detachablity, the authors utilized the mixture of a PNIPAM-gelatin and PNIPAM. The mixed ratio of PNIPAM to a PNIPAM-gelatin was able to easily altered the target properties specific to a substrate as listed above. Since an aqueous solution of PNIPAM-gelatin and/or a mixture with PNIPAM can be cast in molds of any shape, the potential application of PNIPAM-gelatin as a temperature-induced cell-adhesive and mold-releasable material is explored in this study. An example is given as a tubular vascular tissue composed of endothelial cells (ECs) and extracellular matrices produced by endothelial cells during monolayered endothelialization. Spontaneous removal of a tubular vascular tissue upon lowering the temperature is demonstrated. Potential usefulness of solution-coatable and cell-adhesive thermoresponsive matrix in tissue engineering is discussed.

MATERIALS AND METHODS

Materials

The details of the preparation and characterization methods for PNIPAM and PNIPAM-gelatin were described in our previous paper [6]. Briefly, PNIPAM (molecular weight 4.8×10^4 g/mol) was prepared by redox polymerization of a NIPAM monomer in water, and PNIPAM-gelatin was prepared by the UV-light-induced quasi-living radical polymerization of NIPAM on dithiocarbamyl-group-derivatized gelatin (molecular weight 9.5×10^4 g/mol; extracted from bovine bone). Dithiocarbamation was carried out in an aqueous solution by the condensation reaction between the amino groups of lysine residues of gelatin and *N*,*N*-dithiocarbamyl benzoic acid, a water-soluble condensation agent. The number of graft chains in the PNIPAM-grafted gelatin was estimated to be 25.9 per gelatin molecule, and the molecular weight of graft chains was estimated to be 5×10^4 g/mol. The low critical solution temperatures (LCSTs) were 32.0°C for PNIPAM and 34.3°C for PNIPAM-gelatin. A mixture of PNIPAM (97.1 wt%) and PNIPAM-gelatin (2.9 wt%) was used for coating (total concentration 2.5% in water).

Tubular graft

A glass capillary tube (inner diameter 0.15 mm, length 3.0 cm, Iwaki, Tokyo, Japan) was coated with the aqueous solution described above. After air-drying, Dulbecco's modified Eagle's medium (DMEM; Gibco, Grand Island, NY, USA) supplemented with 15% fetal bovine serum (FBS, Gibco) and antibiotics (penicillin, streptomycin and amphotericin B). The average thickness of the coating was approx. 54 μm, which was calculated from the difference in weight before and after coating. The formulation of the medium has been described in detail in our previous paper [9]. The culture medium containing bovine aortic endothelial cells (ECs; 1.5×10^5 cells/ml; harvested from bovine thoracic aorta by the collagenase digestion method) was gently infused in a tube using a syringe at 37°C, and the tube was immersed in the culture medium in a dish, which was maintained in a fully humidified atmosphere of 5% CO_2/95% air. After culture for four days without rotation, the dishes were left to stand at room temperature (20°C), and a tubular construct was detached from the capillary tube upon gentle infusion of DMEM (20°C) using a syringe.

RESULTS

Figure 2 shows the sequential procedure for *in vitro* preparation of an endothelial cell-lined tubular tissue. (1) The aqueous mixed solution of PNIPAM and PNIPAM-gelatin was used to coat the luminal surface of a glass capillary tube (inner diameter; 0.15 mm, length 3.0 cm and coating thickness approx. 54 μm), (2) ECs were seeded on the coated luminal surface at 37 °C, and the capillary tube was immersed in the medium, and cultured under static condition without rotation. (3) Initially, cell adhered on the bottom of the capillary tube but eventually proliferated with incubation time to form a monolayer in the entire luminal surface of the tube at four days-culture at 37 °C. (4) A tubular tissue was detached and removed from the capillary tube by gentle infusion of the culture medium at 20 °C.

Phase-contrast microscopy observation shows that a confluently monolayered EC tissue was formed on the luminal surface of the capillary tube (Fig. 3A). The detached tubular tissue, fragile but maintained its tubular shape (Fig. 3B), and the

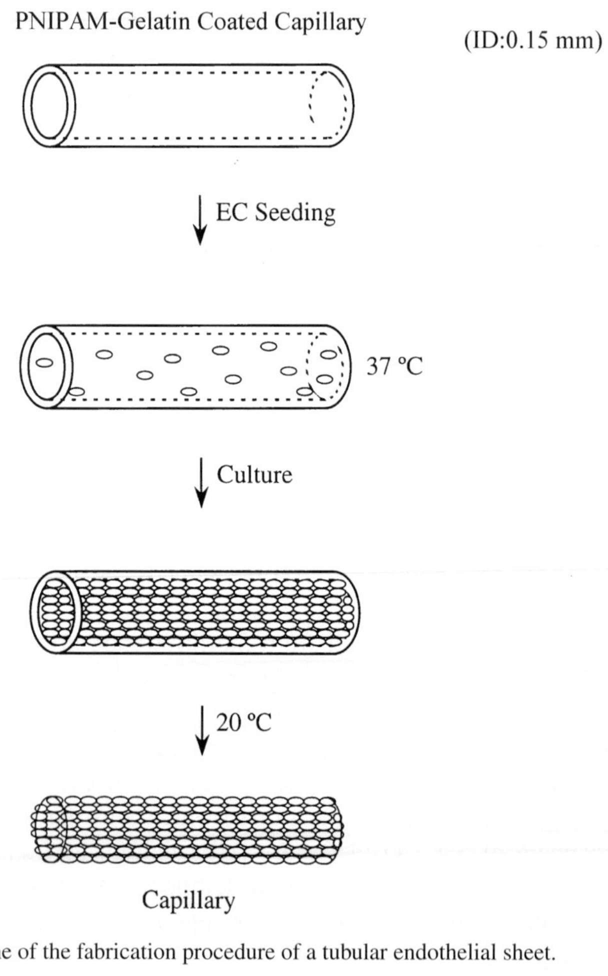

Figure 2. Scheme of the fabrication procedure of a tubular endothelial sheet.

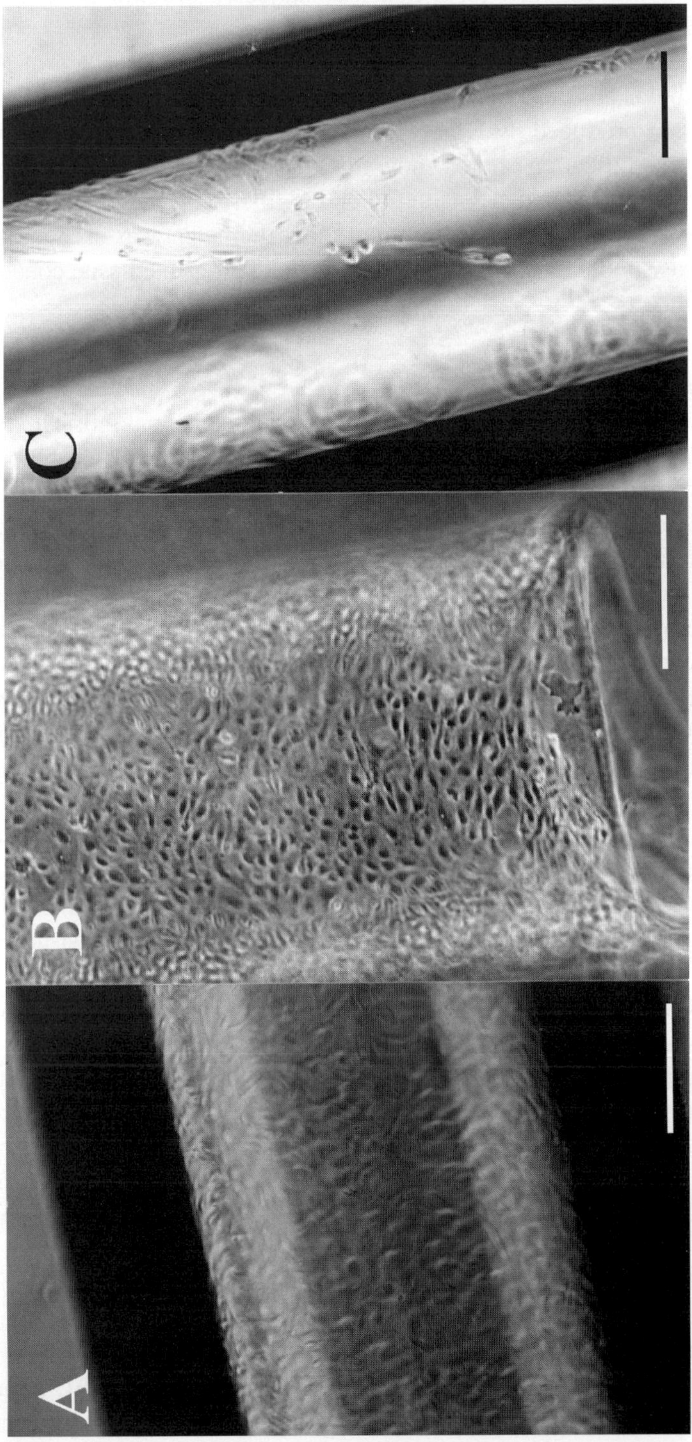

Figure 3. Phase-contrast micrographs of confluent endothelial cells sheet formed on luminal surface of glass capillary tube (A), detached and removed from the tube upon lowering temperature (B) and cell-free luminal surface after removal of tubular cell sheet (C).

luminal surface of the capillary tube, that was subjected to removal of the tissue, was cell-free (Fig. 3C).

The scanning electron microscopy observation of the basal surface of the EC-monolayered tubular construct thus obtained showed that fibrous extracellular matrix substances are produced by ECs (Fig. 4A). Tight cell-to-cell contact was observed. Higher-magnification photos show that these fibrous meshes are composed of fibers and tiny globules that are scattered on the fibers (Fig. 4B and 4C). When a larger-sized glass tube (for example, 3 mm in inner diameter) was used, a detached tubular tissue was spontaneously collapsed to form a string-like tissue construct.

DISCUSSION

Diseased or lost tissues may be regenerated by a tissue equivalent fabricated *ex vivo* using cells, scaffolds and extracellular matrices. Native tissues are composed of cells and an extraxellular matrix, both of which are specific to target tissues. In addition, tissues or organs have their own specific shapes: for example, vascular graft is tubular and skin tissues are mostly planar. More-complex-shaped and higher-ordered structural tissues must therefore require either a predetermined synthetic or biologically derived scaffold or a shape-determined mold. In the latter case, a mold-releasable material is required to remove a formed tissue from the mold without mechanical and biological damage. Such a mold-releasable material should be (1) solution-coatable and film-formable, (2) cell-adhesive and (3) easily detachable from the mold without any damage of tissue or loss of functional cellular activity. To meet these requirements, the bioconjugation of PNIPAM, which exhibits reversible thermoresponsiveness of phase transition between solution and precipitation, and cell-adhesive gelatin may be suited for this purpose.

Our previous study [8] showed that high cell adhesion at 37°C and complete cell detachment at 20°C can be achieved using a mixture of PNIPAM-gelatin and PNIPAM at an appropriate mixed ratio, which is specific to the type of substrates: The coating of PNIPAM-gelatin exhibited high cell adhesion but reduced cell detachment from glass surface probably due to ionic interaction between gelatin molecule and glass surface, whereas the coating of PNIPAM induces little cell adhesion but easily desorbed from the glass substrate at 20°C. This is probably due to hydrated non-ionic character of PNIPAM molecule. The coating mixture containing a small fraction (2.9 wt%) of PNIPAM-gelatin and a major fraction of PNIPAM (97.1 wt%) eventually induced high cell adhesion and complete detachment of cell sheet from glass substrate. The advantage of using a mold-releasable material, which is solution-coatable and exhibits temperature-dependent cell adhesiveness and tissue detachment, enables us to fabricate functional and complex-shaped tissue as discussed.

The tubular vascular tissue formed in this study is composed of an EC monolayer that is structurally enforced by cell-to-cell contact and cell-to-ECM interaction in which extracellular matrix secreted from ECS is supramolecularly organized to

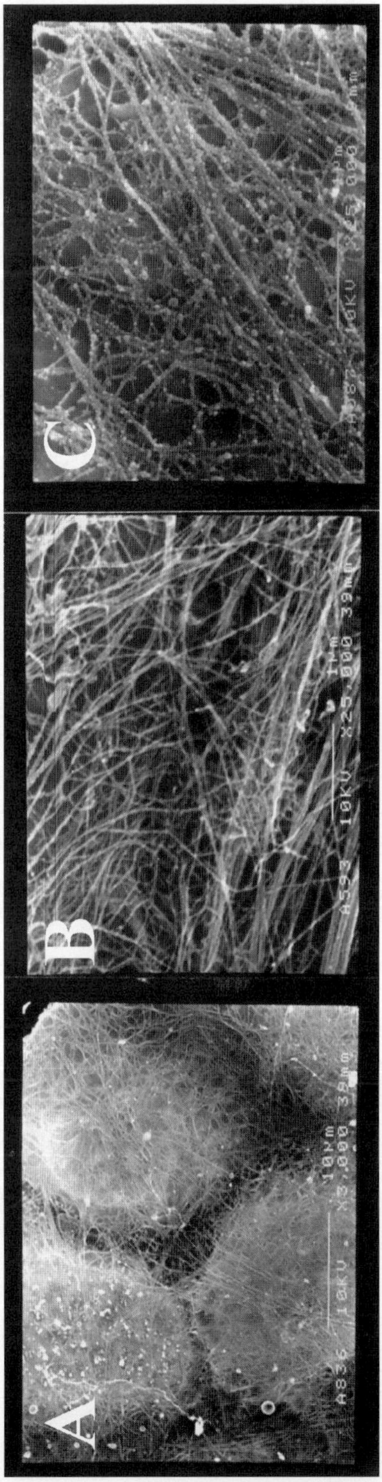

Figure 4. Scanning electron micrographs of the basal surface of a confluently monolayered endothelial cell sheet.

form fibrous meshes. Thus, a tubular endothelium, which resembles probably a part of the 'native' intimal layer without use of any supportive scaffolds was obtained. Since there is no need for enzymatic degradation of cell-extracellular matrix bond using trypsin to remove a tissue from the substrate, the physico-chemically driven detachment of tissues from the mold does not cause any cellular and ECM network damage, which was already demonstrated by Okano's group [3]. Since a mold of any substrate, any shape and any size can be coated with PNIPAM-gelatin and/or mixture with PNIPAM, fabrication of predetermined complex-shaped tissue is feasible in principle. Although an immunochemical study to characterize the nature and composition of ECMs formed on the basal side of a tubular tissue has not been done, *ex vivo* prepared intimal tissue may provide a good biological biomechanical model including the composition of ECMs and the mechanical properties of intimal layer such as tensile strength and pressure-dependent inflation characteristics. As for dissolved PNIPAM and PNIPAM-gelatin at 20°C, their fate was not clear. Probably some of them were dissolved away from both tissue and substrate surfaces into the medium, but a minor portion of them may be incorporated into a newly generated ECM network.

When multilayers of smooth muscle cells (SMCs) are formed on the luminal surface of a capillary mold and a subsequent EC monolayer is overlayered, a more-complex-shaped tubular vascular tissue, in which EC-containing intimal and SMC-containing medial layers are hierarchically structured, is formed. A study on such engineered vascular tissue using a mixture of PNIPAM-gelatin and PNIPAM and a custom-designed tissue roll-up apparatus operated by computer-assisted manufacturing (CAM) is now ongoing. In addition, the tubular intimal tissue thus formed may serve in vascular biomechanics focusing on the imitated, intimal tissue biomechanics, which is also under study.

Acknowledgements

This article is dedicated to Professor Allan Hoffman (in celebration of his 70th birthday) who has always been a pioneer in biomedical research, a very good teacher and an honest mentor to young scientists. Technical assistance by Ms. K. Shoda and Ms. Y. Miura is appreciated. The authors thank the Promotion of Fundamental Studies in Health Science of the Organization for Pharmaceutical Safety and Research (OSPR) for the financial support of this work under Grant No. 97-15, and part by a Grant-in-Aid for Scientific Research (A2-12358017, B2-12470277) from the Ministry of Education, Culture, Sports, Science, and Technology of Japan.

REFERENCES

1. L. C. Dong and A. S. Hoffman, *ACS Symp. Ser.* **350**, 216 (1987).
2. T. Okano, N. Yamada, H. Sakai and Y. Sakurai, *J. Biomed. Mater. Res.* **27**, 1243 (1993).
3. A. Kushida, M. Yamato, C. Konno, A. Kikuchi, Y. Sakurai and T. Okano, *J. Biomed. Mater. Res.* **45**, 355 (1999).

4. T. Shimizu, M. Yamato, A. Kikuchi and T. Okano, *Biomaterials* **24**, 2309 (2003).
5. M. Yamato, T. Shimizu, M. Harimoto, M. Hirose, A. Kushida, O. Kwon, A. Kikuchi and T. Okano, *Tissue Eng.* **7**, 473 (2001).
6. T. Matsuda and M. J. Moghaddam, *Mater. Sci. Eng.* **C1**, 37 (1993).
7. M. J. Moghaddam and T. Matsuda, *ASAIO Trans.* **37**, 437 (1991).
8. N. Morikawa and T. Matsuda, *J. Biomater. Sci. Polymer Edn* **13**, 167 (2002).
9. K. Takamizawa, K. Shoda and T. Matsuda, *J. Biomater. Sci. Polymer Edn* **13**, 81 (2002).

Gels, Genes, Grafts and Giants, pp. 567–582 (2005)

Characteristics of chemically cross-linked myosin gels

AKIRA KAKUGO [1], SHIN SUGIMOTO [1], KAZUHIRO SHIKINAKA [1], JIAN PING GONG [1,2] and YOSHIHITO OSADA [1,*]

[1] *Graduate School of Science, Hokkaido University, Sapporo 060-0810, Japan*
[2] *Presto, JST, Hokkaido University, Sapporo 060-0810, Japan*

Received 10 October 2003; accepted 21 June 2004

Abstract—Myosin gels, 10 mm × 10 mm × 1 mm in size, were obtained by chemical cross-linking of scallop myosin using 1-ethyl-3-(3-dimethylaminoprolyl) carbodiimide hydrochloride (EDC), glutaraldehyde (GA), or transglutaminase (TG). All myosin gels showed high Mg-ATPase activity, although it sensitively depended on the species of cross-linker used. Among cross-linkers used, myosin gel cross-linked by TG showed the highest sensitivity, almost as high as that of native myosin. The motility assay of native actin filament on the myosin gel showed that all these myosin gels can give motion to actin filament. Among them the one cross-linked by TG had the highest average velocity and the lowest threshold concentration of ATP for movement of the actin filament and the values are nearly the same as that of native myosin. In order to give the actin filament motion with preferential direction, we attempted to make myosin gel with oriented structure by applying a shear stress. Myosin gel with oriented filament array 1 cm long and 50 μm in diameter was obtained. We found that actin filaments prefer to move along the axis of the oriented myosin gel with an increased velocity.

Key words: Actin; myosin; nano-machine; ATPase activity; motility assay; gel machine; motion.

INTRODUCTION

Not only metals, ceramics, synthetic polymers and gels, but also polypeptides have been used as materials to construct machines, actuators, or sensors [1–10]. Polypeptides from living organisms are attractive materials because of their high functionality, such as enzyme activity, ability of self-organization, or selective recognition. Some groups have realized devices by depositing or arranging polypeptides on substrates. For example, thin films composed of bateriorhodopsin [11] have been successfully used as photoelectric devices, and thin films of glucose oxidase used as glucose-response biosensors [12]. Furthermore, various kinds of motor proteins, actin myosin [13–15] or microtubule kinesin [16], as well as other proteins

*To whom correspondence should be addressed. Tel./Fax: (81-11) 706-4994.
E-mail: osada@sci.hokudai.ac.jp

that generate rotational motion like F1-ATPase [17], also have been used as micromotors [18–20]. These devices using polypeptides would find applications in the medical or pharmaceutical field. However, most of these devices have been exploited only in one aspect of biological function, partly because polypeptides were immobilized or arranged two-dimensionally on the surface of substrate. By using the chemical cross-linking technique, polypeptides can have a highly-ordered structure as three-dimensional gels.

In a previous paper, we reported on the adenosine triphosphate (ATP)-fueled soft gel machine constructed from chemically cross-linked actin gel and myosin gel [21]. We have found that chemically cross-linked giant actin gel filaments, several tens of times the length of native actin filaments, move along a chemically cross-linked myosin fibrous gel with a velocity as high as that of native actins, by coupling to ATP hydrolysis. This result indicates that muscle proteins can be tailored into the desired shape and size without sacrificing their bioactivities. When myosin gel is stretched in the course of gelation to give oriented fibrils and actin gel is allowed to move on it, the actin gel preferentially moved in a direction parallel to the fibrils. These results arises several important questions: why the chemical cross-linking reaction does not reduce the ATPase activity? Cross-linking reaction of myosin might lead to reduction of physiological activity, attributing to denaturation of the active site of hydrolyzed ATP. The cross-linking reaction also might result in reduced steric flexibility which is important to interact with actin [22–24]. In other words, the effects of chemical cross-linking on the physiological activity of myosin gel and of the orientation of myosin molecules on the direction of actins movement have not been clarified yet.

In this paper we studied the effects of cross-linking on enzyme activity of myosin and of morphological features of myosin gel on the actins movement. For this study, myosin gel was chemically cross-linked using 3 different types of reagents: 1-ethyl-3-(dimethylaminopropyl) carbodiimide hydrochloride (EDC) [25–27], which has no carbon atom between reacting moieties; glutaraldehyde (GA) [28], which has carbon atoms; and transglutaminase (TG) [29, 30], which is a single polypeptide enzyme capable of covalently cross-linking between protein-bound glutamine residues and amino groups, resulting in protein–protein cross-linking. The effect of cross-linking reaction of on ATPase activity of myosin was assayed by actin-activated Mg-ATPase activity. The motility assay on chemically cross-linked myosin gel was examined and compared with that of native one. The effect of orientation of myosin fibrous gel on motion of actin was also examined changing the extent of orientation of myosin gel fibers.

MATERIALS AND METHODS

Materials

Myosin was obtained from scallops by the method of Barany *et al.* [31]. Actin was obtained from scallops by the method of Spudich *et al.* [32].

Preparation of myosin gel (Type 0)

Myosin (10 mg/ml) dissolved in 50 mM MOPS (pH 7.0) containing 600 mM KCl (high salt buffer) was chemically cross-linked by EDC, GA and TG with a prescribed molar ratio of [cross-linker·mol]/[myosin·mol] and unit par per weight ratio of [TG·unit]/[myosin·mg] at 13°C for 30 min. Cross-linking using TG was done in the presence of 5 mM $CaCl_2$. Aliquots of the reaction solutions were removed and terminated by adding 300 mM 2-mercaptethanol/50 mM KSCN and boiling for minutes, respectively. A part of the aliquot was served for SDS–PAGE analysis to determine the extent of the cross-linking. The obtained myosin gels were rinsed and swelled in high salt buffer for several hours at 4°C to remove soluble myosin and non-reacted cross-linkers.

The degree of swelling (q) was defined as the ratio of the weight of swollen gel to that of dry gel.

Preparation of oriented myosin gel with oriented fibers (Type 1)

Myosin (10 mg/ml) dissolved in high salt buffer was spread on a glass slide plate and then dipped into 50 mM MOPS (pH 7.0) containing 10 mM KCl (low salt buffer) for 10 min at 4°C, keeping the angle of slide glass at 60° against the water surface. Using this method, globular myosin transfers to myosin filament to form a cloudy film on the buffer surface. Part of the film edge was slowly picked up using tweezers, stretched with a velocity of 3–4 mm/s, and dipped in TG solution of 1 [unit]/[myosin mg] for 30 min at 13°C to give chemical cross-links.

Preparation of oriented myosin gel with oriented fibers (Type 2)

Another type of oriented fiber was obtained by applying shear stress to the myosin solution in the course of the cross-linking reaction (Fig. 1). Myosin (10 mg/ml) in high salt buffer was injected through a syringe into a cylindrical container with low salt buffer and 125 mM of EDC at 13°C for 30 min. Shear stress is controlled by changing the rotational velocity raging from 0–400 rpm. Shear stress (Pa) is calculated from the formula: $F = \eta R\omega/d$, where F is shear stress (Pa), η is the viscosity of solution (Ns/m^2), R is the radius of rotator (m), ω is the angle velocity (m/s) and d is the distance between the container and the rotator.

Gel electrophoresis

Polyacrylamide gel electrophoresis (PAGE) in the presence of 0.1% SDS was carried out by the method of Porzio and Person [33] in 50 mM Tris/150 mM glycine. The gel was stained with 0.2% Coomassie brilliant blue R-250 in 50% methanol and 10% acetic acid, and it was destained with 7% acetic acid and 5% methanol.

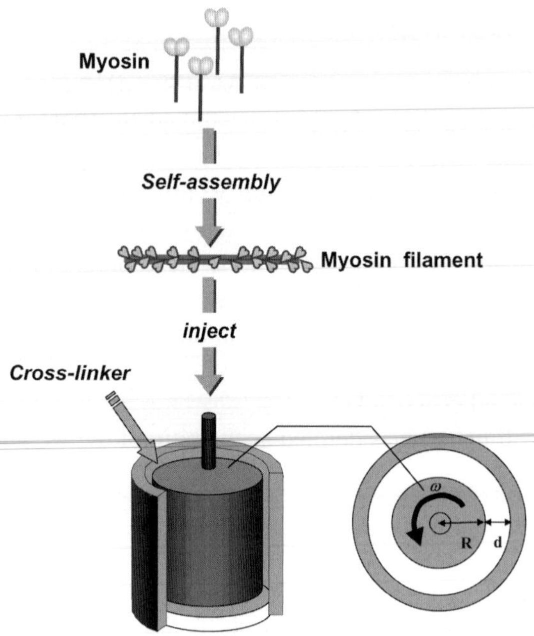

Figure 1. Schematic illustration of the formation of oriented myosin gel.

Mechanical properties and SEM observation

Young's modulus, E, of the myosin gel was measured using a tensile tester (TOM-500, Shinko Communication Industry). The molecular orientation of each filament was confirmed by the strong IR dichroism of carbonyl absorption at 1600 cm^{-1}. Morphological features were observed with scanning electron microscopy (SEM, Hitachi S-2250N) at 25 kV acceleration voltage. A specimen of myosin gel, chemically cross-linked in the presence of 2500 times molar ratio of EDC against myosin molecules, was quickly frozen by using liquid N_2 and then coated with platinum using Hitachi E-1010 ion sputter.

Mg-ATPase activity

The actin activated Mg-ATPase activity was measured by determining the amount of inorganic phosphate liberated from ATP in buffer for 8 min at 15°C (50 mM KCl, 20 mM Tris (hydroxymethyl) aminomethane (pH 7.4), containing 2 mM $MgCl_2$, 0.3 mM $CaCl_2$ and 1 mM ATP) in the presence of myosin gels (0.66 mg/ml) and 0.5 wt% of native actin, or actin gel in the absence of monomeric actin and myosin. The reaction was terminated by adding 10% trichloroacetic acid (TCA). The liberated inorganic phosphate was determined by the method of Youngburg and Youngburg [34].

Motility assay [13–15]

Myosin gel was placed on a cover glass, which was pre-coated with 0.03% nitrocellulose. The cover glass was then placed on a slide glass equipped with spacers 1.1–1.4 mm high placed at both sides. The flow-cell made in this way was filled with 50 mM MOPS buffer (pH 7.0) and placed on ice. The flow-cell was washed with 600 mM KCl and 50 mM MOPS (pH 7.0) and was then filled with buffer (50 mM KCl, 25 mM DTT, 2 mM $MgCl_2$, 0.3 mM $CaCl_2$ and 30 mM imidazole (pH 7.6)) containing 0.3 wt% methylcellulose. 2 μl of actin (10 μM) labeled with rhodamine–phalloidin (Molecular Probes) in the assay buffer was introduced into the flow-cell. The cell was placed on the stage of a fluorescence microscope (Olympus BX 50) and the fluorescent-labeled actins were observed using a $\times$60 objective. The movement of F-actin was initiated by filling the cell with assay buffer containing a prescribed amount of ATP at 20°C. The fluorescent image was obtained with an inverted CCD-camera (Olympus Color Camera HCC-3900) and recorded on a video recorder (Digital Videocassette Recorder DSR-30). The velocity of native F-actins or actin gels on myosin gel was evaluated by the displacement of their center position in 3 s. The effect of the fiber orientation of the myosin gel on the direction of actin movement was evaluated by the degree of anisotropy (D.A.), which is defined as follows:

$$\text{D.A.} = \frac{\bar{x}}{\bar{y}},$$

where

$$\bar{x} = \sqrt{\sum_{i=1}^{N} \frac{x_i^2}{N}}, \quad \bar{y} = \sqrt{\sum_{i=1}^{N} \frac{y_i^2}{N}}.$$

Here x_i^2 and y_i^2 are the respective components of displacement parallel and vertical to the myosin fiber axis in 3 s, respectively, and N is the number of samples.

RESULTS AND DISCUSSION

ATPase activity of myosin gels

Chemical cross-linking of scallop myosin using 1-ethyl-3-(3dimethylaminoprolyl) carbodiimide hydrochloride (EDC), glutaraldehyde (GA), or transglutaminase (TG), was done according to the type-0 method (see Materials and Methods). The degree of gelation was monitored by SDS–PAGE and estimated from the amount of residual myosin monomer. As shown in Fig. 2, the amount of residual myosin monomer decreases with increase of the amount of cross-linkers and all myosin monomer transfers to the gel in the presence of [EDC]/[myosin] = 2500 mol/mol, [GA]/[myosin] = 250 mol/mol and [TG]/[myosin] = 1.5 unit/mg (hereafter, myosin gel prepared under this condition is denoted as 'M-gel (0)'). The obtained

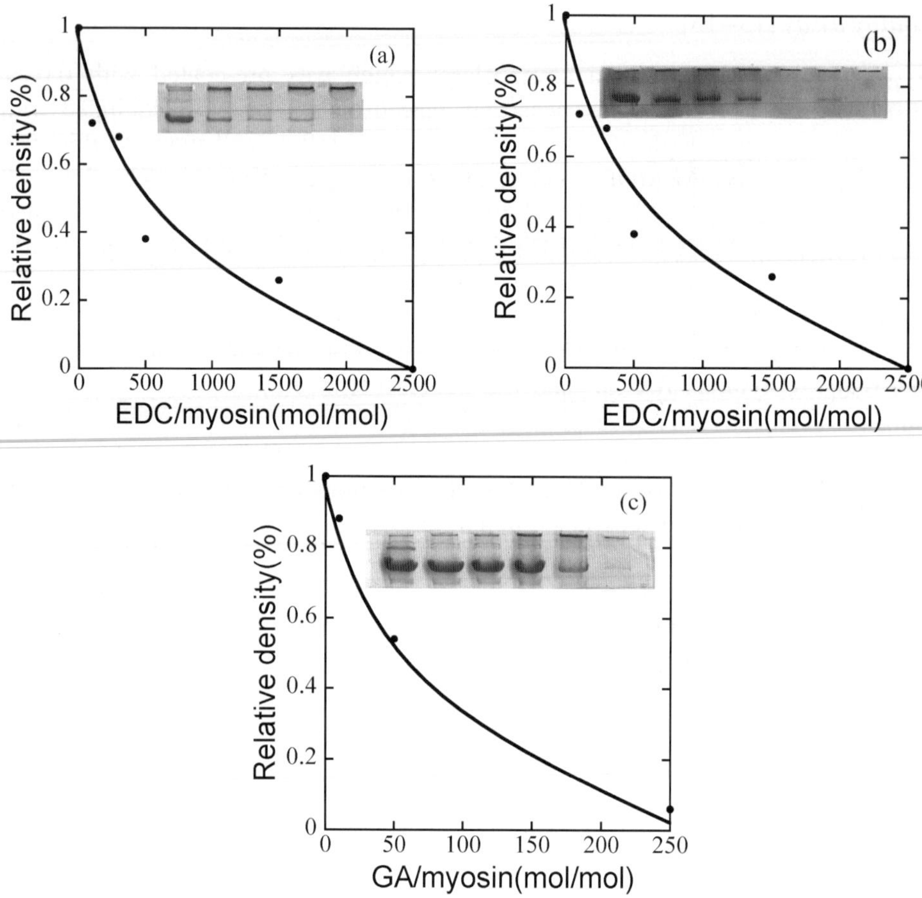

Figure 2. Analysis of SDS–PAGE density. Myosin was chemically cross-linked by using TG (a), EDC (b) and GA (c).

M-gels, 10 mm × 10 mm × 1 mm in size, are highly swollen and their volumes change with changing ionic strength in buffer. The gel is transparent in high ionic condition (600 mM KCl) and swelled as high as 200%, while it became semi-transparent in low ionic condition (60 mM KCl) and swelled about 100%. SEM of M-gels shows a porous surface texture (Fig. 3). The Young's modulus of the M-gel changed with change in the degree of swelling and showed about 100 Pa in high ionic condition and 200 Pa in low ionic condition. The swelling nature and Young's modulus might be associated with the change of myosin from the molecular state in high ionic condition to filament state in low ionic condition through electrostatic interaction. In fact, a fibrous network structure was observed by using SEM for M-gel in low ionic condition (Fig. 3a).

We attempted to prepare M-gel with oriented structure according to methods type 1 and type 2. The chemically cross-linked myosin gel with oriented filament array, 1 cm long and 50 μm in diameter, was obtained by reacting the scallop myosin

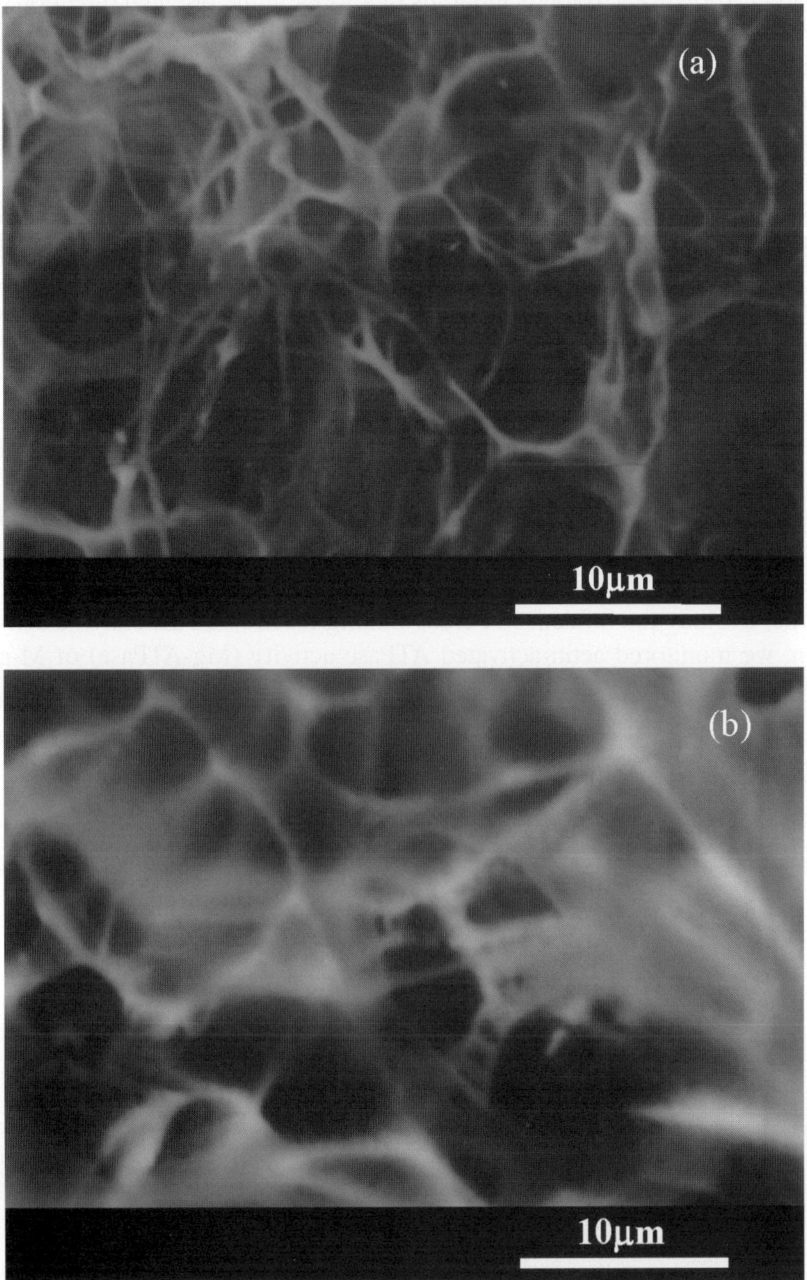

Figure 3. SEM images of the myosin gel chemically cross-linked by EDC. Myosin gel was swelled in low ionic (0.06 M KCl) (a) and high ionic (0.6 M KCl) condition (b).

using with transglutaminase (TG) under stretching. The oriented M-gel (1) is semi-transparent, showing a Young's modulus of 190 Pa in the oriented direction, which is more than two times larger than that of the M-gel prepared without the stretching. Figure 4a and 4b shows SEM and atomic force microscopy (AFM) images of an oriented M-gel (1), respectively. The surface roughness profile obtained at the cross-section along the solid white line in Fig. 4b shows a maximum at approx. 400 nm (Fig. 4c). Distinct bundles of regularly oriented filaments, *ca.* 1.54 μm in diameter, are observed, indicating that the rod-like myosin molecules are self-organized with orientation to form a hierarchical structure. The orientation of filaments was confirmed on a molecular level by the strong IR dichroism of carbonyl absorption at 1600 cm^{-1}, which could not be observed in the absence of stretching.

Another oriented filament M-gel (2) could be obtained by applying an external shear stress using a syringe. Myosin (10 mg/ml) solution was quickly injected to a large amount of buffer solution (low salt condition) containing cross-linking agent through a needle of 1.8 mm diameter. The strong IR dichroism of the M-gel (2) was increased from 1.1 to 2.8 when shear stress increases from 0 Pa to 0.066 Pa. However, when the shear stress increases to 0.126 Pa, the dichroism somehow decreased to about 1.7 Pa (Fig. 5). Thus, we found that the optimum shear stress for the orientation of M-gel seems to exist around 0.066 Pa.

In order to study the influence of cross-linking reaction on the enzyme activity of myosin, we monitored actin-activated ATPase activity (Mg-ATPase) of M-gels in

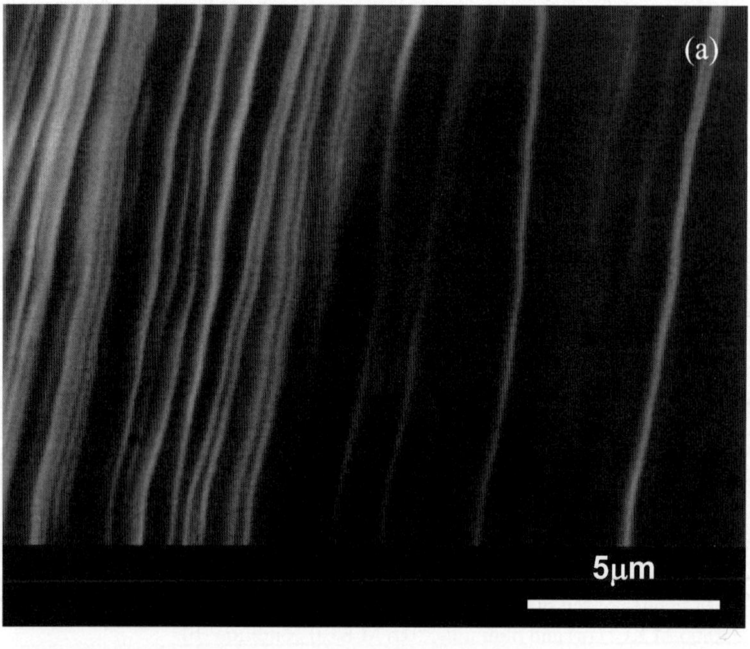

Figure 4. Morphology of oriented myosin gel. SEM image of oriented myosin gel (a), AFM image (b) and cross-section along the white line in panel b (c).

the presence of 0.5% (w/w) native actin at low ionic condition buffer (50 mM KCl). All M-gels (0) showed high Mg-ATPase activity although it sensitively depended on the species of cross-linkers used (Fig. 6). Among the cross-linkers used, M-gel (0) cross-linked by TG (hereafter TG M-gel (0)) showed the highest sensitivity, almost as high as that of native myosin. The high activity of M-gel (0) cross-linked by TG might be associated with the increased molecular size compared to the other two,

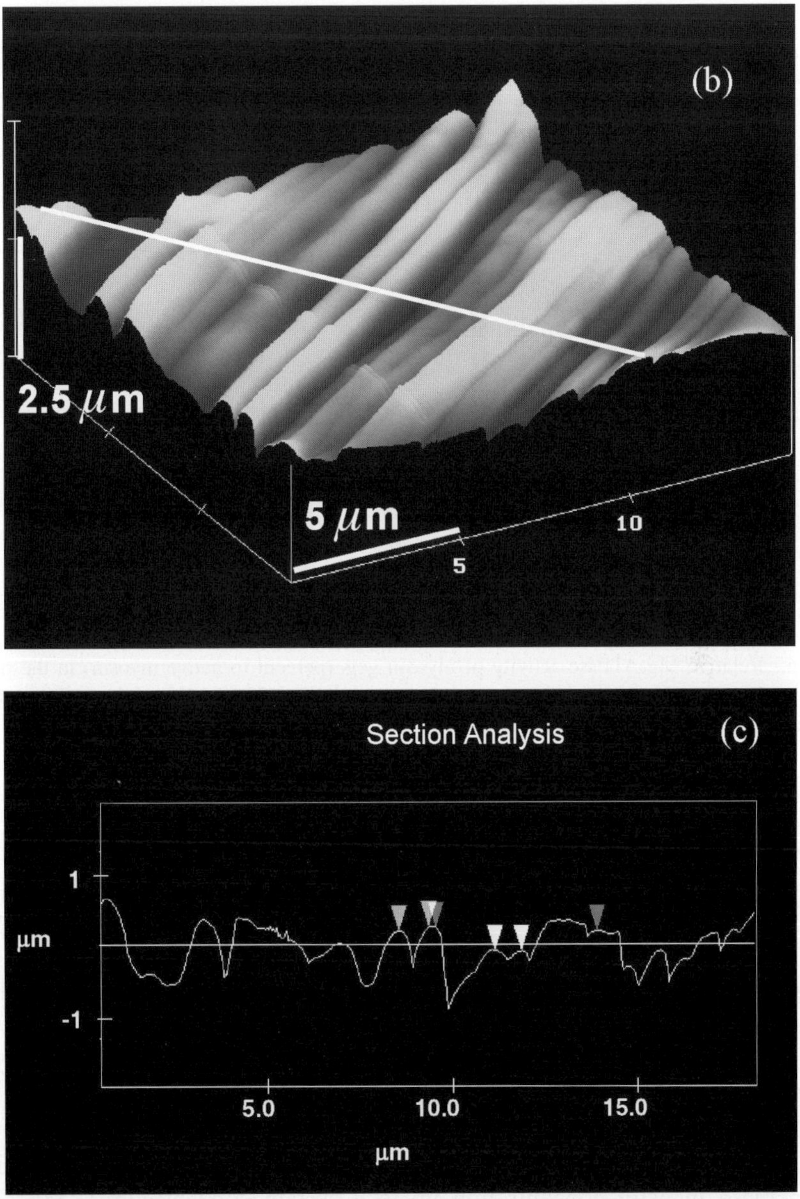

Figure 4. (Continued).

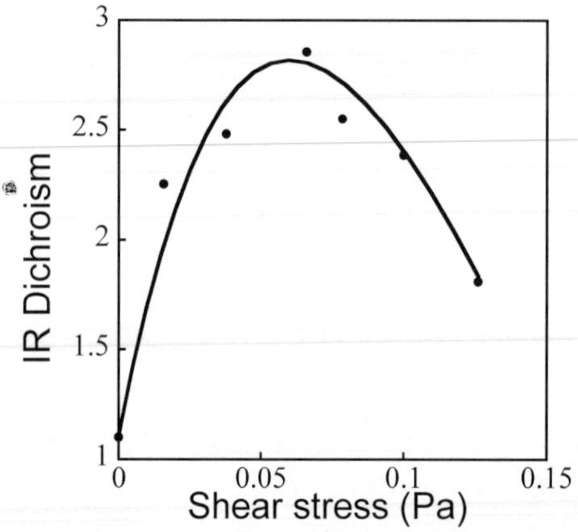

Figure 5. Strong IR dichroism at 1600 cm^{-1} of the myosin gel as a function of applying shear stress.

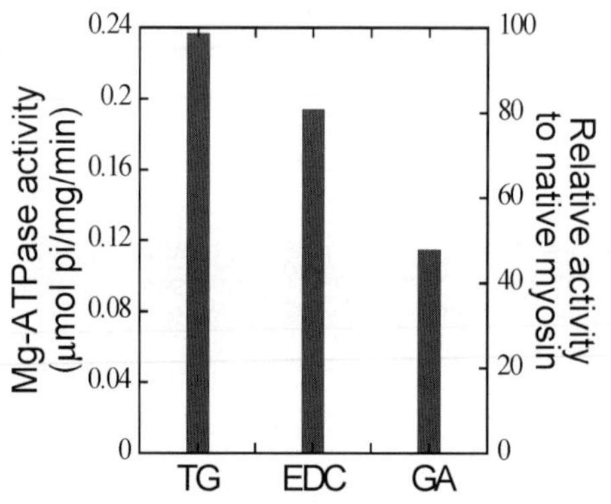

Figure 6. Relative Mg-ATPase activity of myosin gels (percent to native myosin) in the presence of 0.5% (w/w) of native actin. Assay was carried out in buffer (50 mM KCl, 20 mM Tris (pH 7.4)), containing 2 mM $MgCl_2$, 0.3 mM $CaCl_2$ and 1 mM ATP at 15°C.

i.e. TG does not make cross-linking reaction between inside residues of the myosin head which having active site, but between some other places of myosin rods which is not directly related to the ATPase activity. Usually, a substantial decrease in activity is expected when the enzyme is immobilized by chemical cross-linking, due to denaturation of the active site, and considerable decrease in effective surface volume and mobility both are important for the enzymatic reaction. The effect of shear stress on Mg-ATPase activity of M-gels (2) was also investigated (Fig. 7). We found that the activity of myosin gel, somehow decreases decreased linearly with increase in shear stress. Thus, the oriented structure on M-gel leads to deactivation of the myosin gel.

Mg-ATPase activity is enhanced in the presence of actin [35], and it increases with increase in actin concentration up to a certain concentration. Mg-ATPase activity is also sensitive to ionic strength. At low ionic strength (50 mM KCl),

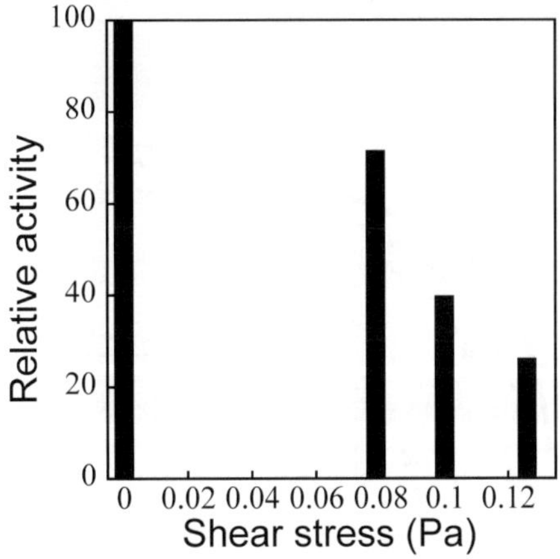

Figure 7. Effect of shear stress on Mg-ATPase activity of myosin gels. Assay was carried out in buffer (50 mM KCl, 20 mM Tris (pH 7.4)), containing 2 mM $MgCl_2$, 0.3 mM $CaCl_2$ and 1 mM ATP at 15°C.

Mg-ATPase gives a maximum value (0.24 μmol P_i/mg per min) and then decreases with increasing ionic strength up to 600 mM KCl (0.05 μmol P_i/mg per min).

It is considered that, at low ionic strength, long-range interactions between actins and myosins play an important role that disappears at high ionic strength. In other word, the affinitive interaction between actins and myosins is higher at low ionic condition and for the high Mg-ATPase activity. We examined the activity under high ionic condition of myosin gels on the Mg-ATPase activity. Different from native myosin which aggregates or dissociates by changing the ionic condition [36], the chemically cross-linked myosins maintain the gel state, regardless of the ionic condition and, therefore, it might be interesting to investigate how Mg-ATPase activity of M-gels is changed by change in the ionic condition. We used M-gel, chemically cross-linked with EDC in 125 mM KCl, and measured its Mg-ATPase in 600 mM KCl. We found that Mg-ATPase activity of M-gel is as high as 0.20 μmol P_i/mg per min, despite the high ionic condition (activity of native myosin under the same ionic condition is 0.05 μmol P_i/mg per min). This activity value is almost the same as that of M-gel in low ionic condition and about 80% of native myosin in low ionic condition. These results indicate that myosin is able to interact with actin to hydrolyze ATP, even in high ionic condition, due to chemical cross-linking.

Motility of myosin gels

The motility assay of native actin filament labeled with rhodamine–phalloidin on the M-gel (0) was done in the presence of various amounts of ATP (low ionic condition: 50mM KCl). We found that every all M-gels prepared by three cross-linkers: EDC, GA, and TG can give motion to actin filament. The average velocity of actin filament, calculated from the mean displacement in 3 s over 30–50 samples, strongly depended on ATP concentration; in all M-gels, the velocity sharply increased with

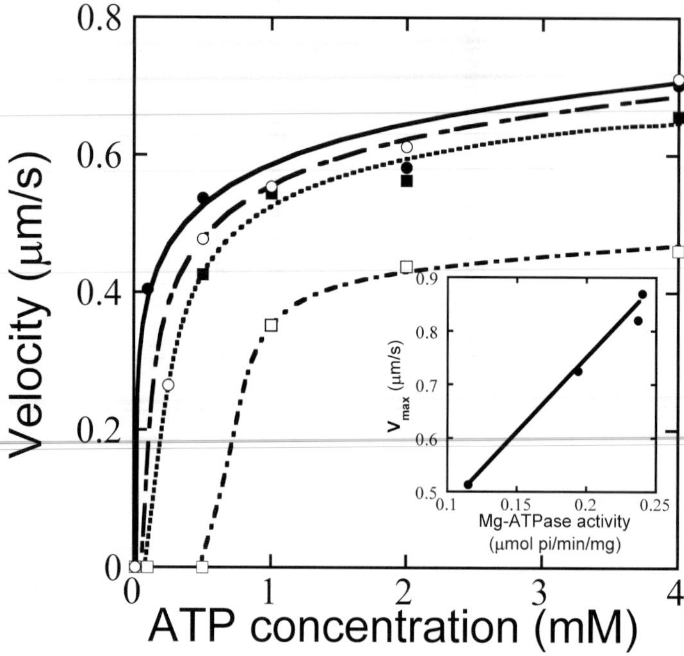

Figure 8. Velocity of native actin filament on myosin gels. Closed circles, native myosin; open circle, TG-M-gel; closed squares, EDC-M-gel; open squares, GA-M-gel. Plots represent mean velocity of 30 or more actin filaments. Inset: Maximum velocity of native actin filament on myosin gels as a function Mg-ATPase activity of myosin gels.

an increase in low ATP concentration range, and then saturated. These the profiles are essentially the same as that of native myosin (Fig. 8). Among them, however, the one cross-linked by TG had the highest average velocity and the lowest threshold concentration of ATP to give the movement to actin filament and the values are nearly the same as that of native myosin. A reciprocal plot of velocity (V) against ATP concentration (Lineweaver–Burk plot) showed a linear relationship for all M-gels, indicating that essentially the same ATP coupling process takes place. Here, if the V_{max} calculated from the Lineweaver–Burk plot is plotted as a function of Mg-ATPase activity (A) of the M-gels (0) shown in Fig. 3, a linear relationship expressed as $V_{max} = 2.70A + C$ is obtained (Fig. 8, inset), i.e., the velocity of actin filaments is proportional to the ATPase activity of M-gels (0).

Thus, we have confirmed that high motility of actin filament can be obtained on the M-gels. The motion of these actin filaments was in random direction because no specific orientation was given to these M-gels.

Next we attempted to give the actin filament the motion with preferential direction controlled by the morphological feature of M-gel with oriented fibril array. Figure 9a and 9b shows the displacement of native actin filaments in 3 s on the random and oriented M-gels (1), respectively. We observed a preferential motion

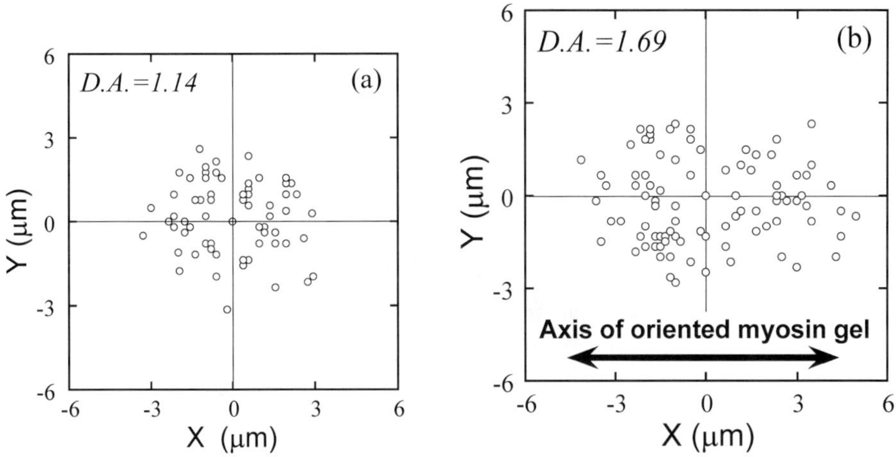

Figure 9. Sequential images of the direction of actin movement on non-oriented myosin gel (a) and on oriented myosin gel (b). Each plot indicates the position of each action filament after 3 s.

of F-actin along the axis of oriented M-gel (1), as shown in Fig. 9b. The effect of orientation on the direction of actin movement was characterized by the degree of anisotropy (D.A.), which is defined as the ratio of the square-root average velocity in the fiber direction to that perpendicular to the fiber direction. D.A. moved on the non-oriented M-gel (0) was 1.1 (average over 66 samples) and that on the oriented M-gel (1) was 1.7 (average over 91 samples). The mean velocity on the non-oriented M-gel (0) was 0.69 μm/s with a standard deviation of 0.24 μm/s, while that on the oriented M-gel (1) was 0.83 μm/s with a standard deviation of 0.30 μm/s. Thus, we confirmed that actin filaments prefer to move along the axis of the oriented M-gel with an increased velocity.

In order to know the dependence of the preferential motion of actin filament on the degree of orientation of M-gel, the motility assay of actin filament was done in the presence of 4 mM ATP using the M-gel (2) prepared by changing the shear stress from 0 Pa to 0.126 Pa. Figure 10 shows the D.A. of actin movement on that of oriented M-gel (2). It is seen that D.A. increases as shear stress increases up to 0.066 Pa, after which it decreases. This D.A. profile is well correlated with the change of IR dichroism against shear stress shown in Fig. 7 and indicates that the actin movement is strongly affected by the molecular orientation of M-gel.

The mean velocity of actin filament on the oriented M-gel (2) is shown in Fig. 11. The velocity of actin filament increases with increase of shear stress up to 0.08 Pa showing a maximum at 0.08 Pa, which is 1.4-times faster than that on M-gel (0) without shear stress. The velocity decreased with further increase in the shear stress and actin filament did not move at all on the myosin gel prepared by applying the shear stress of 0.126 Pa.

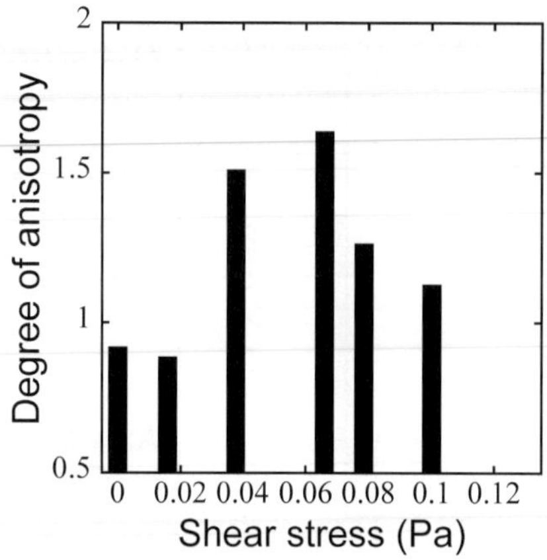

Figure 10. Effect of the fiber orientation of the myosin gels on the direction of actin movement as characterized by degree of anisotropy (D.A.). M-gel (10 mg/ml) was chemically cross-linked by EDC.

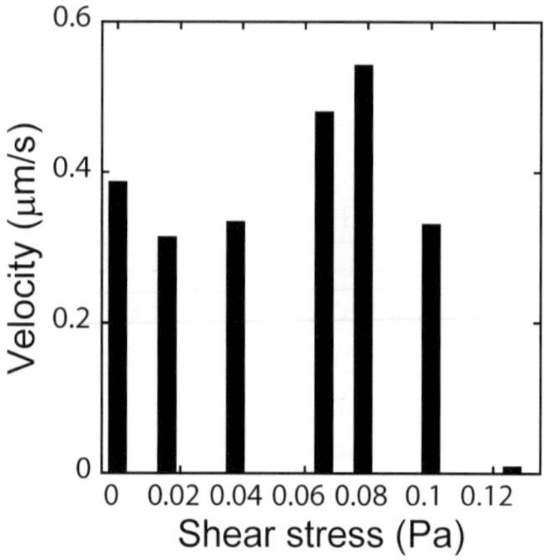

Figure 11. Effect of the fiber orientation of the myosin gels on the velocity of actin filament. The velocity of the actin filament was evaluated by the displacement of their center position in 3 s (average over 40 samples).

CONCLUSIONS

As shown in this paper, myosin gel having a three-dimensional structure can be obtained by using three cross-linkers, TG, EDC and GA, without sacrificing Mg-ATPase activity. Actin filaments can exhibit sliding motion on all of these M-gels and their maximum velocity is proportional to the Mg-ATPase activity of M-gels. These gels have an ability to move actin filaments as almost the same velocity as native myosins deposited on 2-dimensional substrate. Furthermore, 3-dimensional gels having an oriented myosin fiber structure can be induced easily by applying an external shear stress during the cross-linking reaction. Actin filaments on oriented M-gels show preferential motion, depending on degree of molecular orientation of M-gels, with higher velocity than that on non-oriented M-gels.

These results encourage us to fabricate soft machines using proteins as basic building blocks; that is, we can build up new machines with desired architecture by decomposing the muscle protein lattice into its component molecules, self-assembling them, and then connecting each protein filament by chemical cross-linking.

Acknowledgements

This research is financially supported by PRESTO, JST and the Ministry of Education, Science, Sports, and Culture, Japan (Grand-in-Aid of Creative Scientific Research).

REFERENCES

1. Y. Osada, H. Okuzaki and H. Hori, *Nature* **355**, 242 (1992).
2. T. Mitsumata, K. Ikeda, J. P. Gong and Y. Osada, *Appl. Phys. Lett.* **73**, 2366 (1998).
3. Y. Osada and A. Matsuda, *Nature* **376**, 219 (1995).
4. Y. Osada and S. B. Ross-Murphy, *Sci. Am.* **268**, 82 (1993).
5. E. W. H. Jager, E. Smela and O. Inganas, *Science* **290**, 1540 (2000).
6. D. J. Beebe, J. S. Moore, J. M. Bauer, Q. Yu, R. H. Liu, C. Devadoss and B.-H. Jo, *Nature* **404**, 588 (2000).
7. S. R. Quake and A. Scherer, *Science* **290**, 1536 (2000).
8. Y. Kaizawa and K. Kataoka, *Adv. Drug Deliv. Rev.* **54**, 203 (2002).
9. S. V. Vinogradov, T. K. Bronich and A. V. Kabanov, *Bioconjug. Chem.* **9**, 80 (1998).
10. T. Miyata, N. Asami and T. Uragami, *Nature* **399**, 766 (1999).
11. T. Koyama, N. Yamaguchi and T. Miyasaka, *Science* **265**, 762 (1994).
12. B. A. Gregg and A. Heller, *Anal Chem.* **62**, 258 (1990).
13. T. Q. P. Uyeda, H. M. Warrich, S. J. Korn and J. A. Spudich, *Nature* **352**, 307 (1991).
14. H. E. Huxley, *J. Biol. Chem.* **265**, 8347 (1990).
15. S. J. Korn and J. A. Spudich, *Proc. Natl. Acad. Sci. USA* **83**, 6272 (1986).
16. R. D. Vale, T. S. Reese and M. P. Sheetz, *Cell* **42**, 39 (1985).
17. H. Noji, R. Yasuda, M. Yoshida and K. Kinosita, Jr., *Nature* **386**, 299 (1977).
18. M. Muratsugu, S. Kurosawa and N. Kamo, *J. Colloid Interf. Sci.* **147**, 378 (1991).
19. T. Miyasaka, K. Koyama and I. Itoh, *Science* **255**, 342 (1992).
20. H. Suzuki, K. Oiwa, A. Yamada, H. Sakakibara, H. Nakayama and S. Mashiko, *Jpn. J. Appl. Phys.* **43**, 3937 (1995).
21. A. Kakugo, S. Sugimoto, J. P. Gong and Y. Osada, *Adv. Mater.* **14**, 1124 (2002).
22. H. E. Huxley, *Science* **164**, 1356 (1969).
23. A. F. Huxley, *Reflections on Muscle*. Liverpool University Press, Liverpool (1980).
24. G. H. Pollack, *Cells, Gels and the Engines of Life*. Ebner and Sons, Seattle, WA (2001).
25. D. Mornet, R. Bertrand, P. Pantel, E. Audemard and R. Kassab, *Nature* **292**, 301 (1981).
26. D. Mornet, R. Bertrand, P. Pantel, E. Audemard and R. Kassab, *Biochemistry* **20**, 2110 (1981).
27. K. Tawada and M. Kimura, *J. Muscle Res. Cell Motil.* **7**, 339 (1986).
28. D. H. Hu, S. Kimura and K. Maruyama. *J. Biochem. (Tokyo)* **104**, 509 (1988).
29. I. Cohen, L. Young-Bandala, T. A. Blankenberg, G. E. Siefring and J. Bruner-Lorand, *Arch. Biochem, Biophys.* **192**, 100 (1979).
30. H. Yi-Ping, S. Katsuya, M. Masao and T. Katsuhisa, *J. Biochem.* **112**, 229 (1992).
31. M. Barany and K. Barany, *Biochem. Z.* **345**, 37 (1966).

32. J. A. Spudich and S. Watt, *J. Biol. Chem.* **246**, 4866 (1971).
33. M. A. Porzio and A. M. Pearson, *Biochim. Biophys. Acta* **490**, 27 (1977).
34. G. E. Youngburg and M. N. Youngburg, *J. Lab. Clin. Med.* **16**, 158 (1930).
35. K. Nishita, *Bull. Jpn. Soc. Sci. Fish.* **43**, 335 (1977).
36. J. H. Sinard and T. D. Pollard, *J. Cell Biol.* **109**, 1529 (1989).

Gels, Genes, Grafts and Giants, pp. 583–596 (2005)

Tissue responses to thermally-responsive hydrogel nanoparticles

HONG WENG [1], JUN ZHOU [2], LIPING TANG [1,*] and ZHIBING HU [2,3]

[1] *Biomedical Engineering Program, University of Texas at Arlington, PO Box 19138, Arlington, TX 76019, USA*

[2] *Department of Physics, University of North Texas, PO Box 311427, Denton, TX 76203, USA*

[3] *Department of Materials Science, University of North Texas, PO Box 311427, Denton, TX 76203, USA*

Received 20 October 2003; accepted 5 March 2004

Abstract—Thermally-responsive hydrogel nanoparticles of poly(N-isopropylacrylamide) (PNIPAM) and hydroxypropyl cellulose (HPC) have been synthesized. The particle size has been correlated to surfactant concentration and polymer concentration using dynamic laser light scattering techniques. The tissue compatibility of these hydrogel nanoparticles has been evaluated by comparing with poly-L-lactic acid and polystyrene nanoparticles using a mouse implantation model. Our results suggested that both PNIPAM and HPC nanoparticles triggered lesser inflammatory and fibrotic responses among all nanoparticles tested. It is likely that these hydrogel nanoparticles may be suitable for tissue augmentation or drug-delivery devices.

Key words: Poly(N-isopropylacrylamide); hydroxypropyl cellulose; poly-L-lactic acid; polystyrene biocompatibility; mice; inflammation; fibrosis.

INTRODUCTION

Hydrogels are useful in biomedical and pharmaceutical applications due to their high water content and rubbery nature, which is similar to natural tissue [1, 2]. In response to environmental stimuli, hydrogels can change their volume reversibly by several orders of magnitude [3, 4], resulting in diverse applications from sensors to biomedical devices [5–9]. Specially, since Hoffman's pioneering work [10–12] thermally-responsive hydrogels, including poly-N-isopropylacrylamide (PNIPAM) [13, 14] and hydroxypropyl cellulose (HPC) hydrogels [15–17], have been studied intensively for controlled drug delivery. Both PNIPAM and HPC hydrogels undergo the volume phase transition in response to a temperature change. The phase

*To whom correspondence should be addressed. E-mail: lipingt@comcast.net

transition temperatures (T_c) of PNIPAM and HPC are 34 [4] and 41°C [15], respectively. Below T_c, the hydrogels are hydrophilic and swell in water, while above T_c they become hydrophobic and collapse into a small volume. Using the thermal-responsive properties of these hydrogels, pulsatile drug delivery may be intelligently triggered by a minor temperature change. An on–off drug-release profile in response to a stepwise temperature change has been obtained using PNIPAM and its derivatives [13, 14].

Despite this unique triggered release property, very little is known of the tissue compatibility of these hydrogel nanoparticles. The lack of such knowledge hinders the development of clinical application of thermally responsive hydrogels [18]. For example, it is known that N-isopropyl acrylamide (NIPA) monormers may be carcinogenic or teratogenic. But the toxicity of PNIPAM nanoparticles are not known to our knowledge. Although HPC polymer has been approved by the United State Food and Drug Administration (FDA) for use in food and drug formulations [15], the biocompatibility of HPC nanoparticles has yet to be determined. To find the answer, PNIPAM and HPC nanoparticles were synthesized and implanted in the subcutaneous cavity. For comparison, hydrophobic polymeric materials, specifically poly-L-lactic acid (PLA) and polystyrene (PS) nanoparticles, were also included in this study. PLA nanoparticles have been manufactured for different therapeutic applications, including sustained drug, oligonucleotide, vaccine, proteins and gene therapy [19–21]. We also include PS nanoparticles in our *in vivo* studies, since PS nanoparticles are often used as test subjects for studying nanoparticle distribution in tissue [22–24]. After implantation for different periods of time, the implants and surrounding tissues were recovered for histological analyses. In general, we found that HPC and PNIPAM nanoparticles triggered minimal inflammatory cell accumulation and fibrotic capsule formation surrounding the implants. The results suggest that both hydrogel nanoparticles may serve as controlled delivery devices and tissue prostheses with improved tissue compatibility.

MATERIALS AND METHODS

Materials

Dry hydroxypropyl cellulose (HPC) powder (average $M_w = 1 \times 10^6$), dodecyltrimethyl-ammonium bromide (DTAB), divinyl sulfone (DVS), styrene, sodium hydroxide (NaOH) pellets and sodium chloride (NaCl) were purchased from Aldrich (Milwaukee, WI, USA). The substitution level of the HPC polymer for this study was MS $= 3.9$, where MS is the average number of molecules of propylene oxide combined per anhydroglucose unit. High-molecular-weight poly(L-lactic acid) (PLA, 1.37×10^6), was purchased from Birmingham Polymers (Birmingham, AL, USA). N-Isopropylacrylamide (NIPA) was purchased from Polyscience. Sodium acrylate (NaAc), N,N′-Methylene-bis-acrylamide (BIS) and potassium persulfate (KPS) were purchased from Bio-Rad (Hercules, CA, USA). All materials were used without further purfication. Water for all reactions and solution preparation was distilled and purified to a resistance of 18.2 MΩ using a Millipore system and filtered through a filter to remove particulate matter. Hematoxylin–Eosin (H&E) stain, potassium phosphate monobasic, hydrogen peroxide, sodium phosphate dibasic, goat serum and 3,3′-diaminobenzidine were purchased from Sigma (St. Louis, MO, USA). Rat anti-mouse CD 11b was obtained from Serotec (Oxford, UK). Goat anti-Rat IgG with HRP conjugation was bought from Jackson Immuno Research Labs (West Grove, PA, USA).

Sample preparation

HPC nanoparticle synthesis. HPC hydrogel nanoparticles were synthesized using precipitation polymerization [25, 26]. The desired amounts of HPC powders were dispersed in aqueous NaOH solution (pH 12) by gentle stirring for a period of 4–6 days until HPC powders were thoroughly hydrolized. The solution was then mixed with 1.43 g DTAB as a surfactant and stirred for 60 min. The amounts of DVS of 50 wt% HPC were added to the HPC solution. After mixing completely, the HPC solution was heated up to about 65°C. The crosslinked HPC nanoparticles were formed after 2.5 h at about 65°C. The resultant nanoparticles were dialysized for 10 days to remove the surfactant and NaOH. The HPC nanoparticle size was measured using dynamic light scattering. The recipes for the HPC nanoparticles are listed in Table 1.

PNIPAM nanoparticle synthesis. PNIPAM-co-NaAc nanoparticles were synthesized by precipitation polymerization [27]. Amounts of NIPA, NaAc and BIS were added to 240 g deionized water in a 500-ml reactor. To control particle size, different amounts of SDS as a surfactant were used. The solution was stirred and deoxygenated by bubbling it with nitrogen for 40 min and then heated to 70°C under positive nitrogen pressure. 20 g of 0.8 wt% potassium persulfate (KPS) water solution was added to the reactor to initiate polymerization. The reaction was kept at 68–70°C under nitrogen atmosphere for 4 h to ensure that all monomer was reacted. The PNIPAM-co-NaAc nanoparticle size was measured by light scattering method. The recipes for the (PNIPAM-co-NaAc) nanoparticles are listed in Table 2.

Table 1.
Synthesis parameters for HPC nanoparticles

	HPC powder (g)	DTAB (g)	DVS (g)	Water (pH 12) (g)	R_h (nm)[a]	
					23°C	37°C
Batch 1	0.4	1.43	0.2	200	158	153
Batch 2	0.6	1.43	0.3	200	303	
Batch 3	1.2	1.43	0.6	200	336	300

[a] Hydrodynamic radii of HPC nanoparticles in water, where $C = 1.5 \times 10^{-5}$ g/g and the scattering angle is 90°.

Table 2.
Synthesis parameters for PNIPAM-co-NaAc nanoparticles

	NIPA (g)	NaAc (g)	BIS (g)	SDS (g)	KPS (g)	Water (g)	R_h (nm)[a]	
							23°C	37°C
Batch 1	3.9	0.15	0.07	0.30	0.16	240	88	40
Batch 2	3.9	0.15	0.07	0.16	0.16	240	171	
Batch 3	3.9	0.15	0.07	0.01	0.16	240	241	122

[a] Hydrodynamic radii of PNIPAM nanoparticles in water, where $C = 1.5 \times 10^{-5}$ g/g and the scattering angle is 90°.

PLA nanoparticle synthesis. PLA nanoparticles were prepared according to a modified precipitation method [28]. 0.1 g of PLA was dissolved in 5 ml of CCl_4. The organic solution was added dropwise to 100 ml of an aqueous phase. The mixture was stirred magnetically at room temperature for 30 min and then evaporated with a rotary evaprator to remove organic solvent. To further clean PLA nanoparticles, deionized water was added to the PLA solution and then the sulution was ultracentrifigued. This procedure was repeated several times to remove residual solvent. The PLA nanoparticles were finally dispersed by sonication bath.

PS nanoparticle synthesis. PS nanoparticles were synthesized by precipitation polymerization [29, 30]. 3.8 g styrene, 0.06 g BIS as a crosslink and 0.05 g SDS as a surfanctant were added to 240 g deionized water in a 500-ml reactor. The solution was stirred and deoxygenated by bubbling it with nitrogen for 40 min and then heated to 70°C under positive nitrogen pressure. 20 g of 0.8 wt% KPS water solution was added to the reactor to initiate polymerization. The reaction was kept at 68–70°C under nitrogen atmosphere for 4 h to ensure that all monomer was reacted.

Material purification and characterization. Prior to *in vitro* and *in vivo* analyses, all nanospheres were exhaustively dialyzed with pyrogen free deionized water change twice daily for 10 days. Then the nanoparticles were further purified and concentrated using a ultracentrifuge with the speed of 40 000 rpm for 4 h. A laser light scattering (LLS) spectrometer (ALV/DLS/SLS-5000) equipped with an ALV-5000 digital time correlator was used with a Helium-Neon laser (Uniphase 1145P, output power 22 mW and wavelength 632.8 nm) as the light source. The incident light was vertically polarized with respect to the scattering plane and the light intensity was regulated with a beam attenuator (Newport M-925B). The scattered light was conducted through a thin (approx. 100 μm in diameter) optic fiber leading to an active quenched avalanche photo diode, the detector. The Laplace inversion or cumulant analysis of the intensity-intensity time correlation function obtained using dynamic light scattering was performed to obtain the translational diffusion coefficient distribution $G(D)$ and the hydrodynamic radius distribution $f(R_h)$. The diffusion constant (D) is related to hydrodynamic radius by the Stokes–Einstein equation, $R_h = k_B T/(6\pi \eta D)$, where k_B, η and T are the Boltzmann constant, the solvent viscosity and the absolute temperature, respectively [31]. The dynamic light scattering measurements were carried out at an angle of 90°.

Implantation of materials. As an *in vivo* model for evaluating biomaterial-mediated tissue reactions, various nanoparticles were implanted subcutaneously in Swiss Webster mice (male, about 20 g body weight) from Taconic Farms (Germantown, NY, USA). Briefly, test nanoparticles (3 mg dry weight/300 μl per mouse) were administrated into subcutaneous space on the back. After implantation for 4 and 14 days, the implants and surrounding tissues were recovered, frozen sectioned, and then histological analyses. In general, we found that the differences of implant-mediated tissue responses are very consistent among 4-day and 14-day implants. Furthermore, tissue reactions (inflammatory cell infiltration and accumulation surrounding implants) to 4-day implants are stronger than 14-day implants. Thus, 4 days was chosen as the implantation time for assessing the extent of tissue reactions to different nanoparticles.

Histochemical analyses. After sectioning with a Leica Cryostate (CM 1850), some slides were subjected to H&E staining for assessing gross tissue reactions to nanoparticle implants, including cell accumulation and fibrotic capsule formation. Some slides of 4-day implants were stained with inflammatory (CD11b positive) cells using immunohistochemistry technique. Briefly, the slides were dipped for 5 min in 0.6% hydrogen peroxide in phosphate-buffered saline (PBS) to quench endogenous peroxidase activity, washed twice for 5 min with PBS and incubated in 10% goat serum. The slides were then incubated overnight with 20 μg/ml of primary antibody at 4°C in a humidified chamber. After thorough washing with PBS, the slides were incubated for 1 h in goat anti-Rat IgG(H + L) HRP-conjugated antibody at room temperature, followed by 3,3′-diaminobenzidine substrate system, and washed. The slides were counterstained with hematoxylin and mounted with mounting medium. Omission of the primary or secondary antibody was used as a negative control. To assess the tissue reactivity of these particles, all slides were then examined and quantified under microscope. All such experiments were repeated at least three times to ensure the reproducibility of our results.

RESULTS AND DISCUSSION

Physical properties of hydrogel nanoparticles

HPC nanoparticles. HPC nanoparticles with different sizes were obtained by varying polymer concentrations during particle formation process. Figure 1 shows hydrodynamic radius distributions ($f(R_h)$) of HPC nanoparticles ($C = 1.5 \times 10^{-5}$ g/ml) in de-ionized water at 23°C. As the HPC polymer concentration increases from 0.2 to 0.6 wt%, while the DTAB concentration and the reaction temperature were fixed at 0.713 g and 65°C, respectively, the average radius $\langle R_h \rangle$ of the nanoparticles becomes larger and its distribution becomes broader. This result may be explained in terms of interaction between the surfactant DTAB with the HPC. As the HPC concentration increases, the average number of absorbed surfactant aggregates on each HPC polymer chain should decrease, reducing the inter-aggregate electrostatic repulsion force [25]. This causes HPC linear chains to become more aggregated at higher HPC concentration. When the temperature is increased to 37°C (<LCST of HPC), particle sizes of HPC shrink slightly, as shown in Fig. 1b. The temperature dependent particle sizes of the HPC nanoparticles with room temperature sizes at 158 and 340 nm, respectively, are shown in Fig. 1c. The HPC particles clearly exhibit a shrinkage at transition temperature about 38°C. This

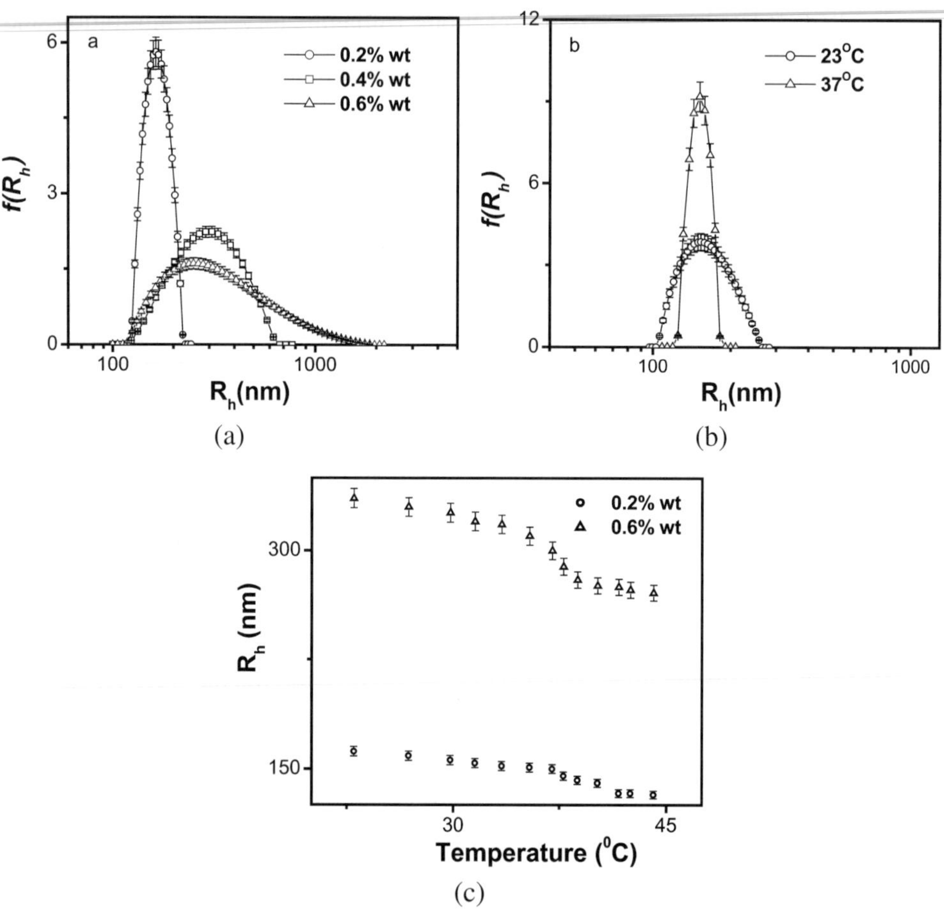

Figure 1. (a) Hydrodynamic radius distributions ($f(R_h)$) of HPC particles ($C = 8.94 \times 10^{-5}$ g/ml) in deionized water at 23°C. Their composition is shown in Table 1. (b) Hydrodynamic radius distributions ($f(R_h)$) of HPC nanoparticles of Batch 1 in water at 23°C (circles) and 37°C (squares), where $C = 1.37 \times 10^{-5}$ g/g and the scattering angle is 90°. (c) Temperature-dependent hydrodynamic radius of HPC nanoparticles with two different particle sizes.

temperature is lower than the value reported [15] because our crosslinker (DVS) concentration is much higher. The increase of the hydrophobic component (such as DVS) in a thermally responsive gel can result in the decrease of the volume phase transition tempearture [4].

PNIPAM nanoparticles. The PNIPAM-co-NaAc nanoparticles with different sizes were obtained by varying surfactant concentrations during the particle formation process. A small amount of NaAc was incorporated into PNIPAM particles to prevent possible aggregation of particles at higher colloidal concentration. As shown in Fig. 2a, the average radius (R_h) of the PNIPA-co-NaAc nanoparticles decreases with the increase of SDS from 0.01 g to 0.30 g. With the increase of the

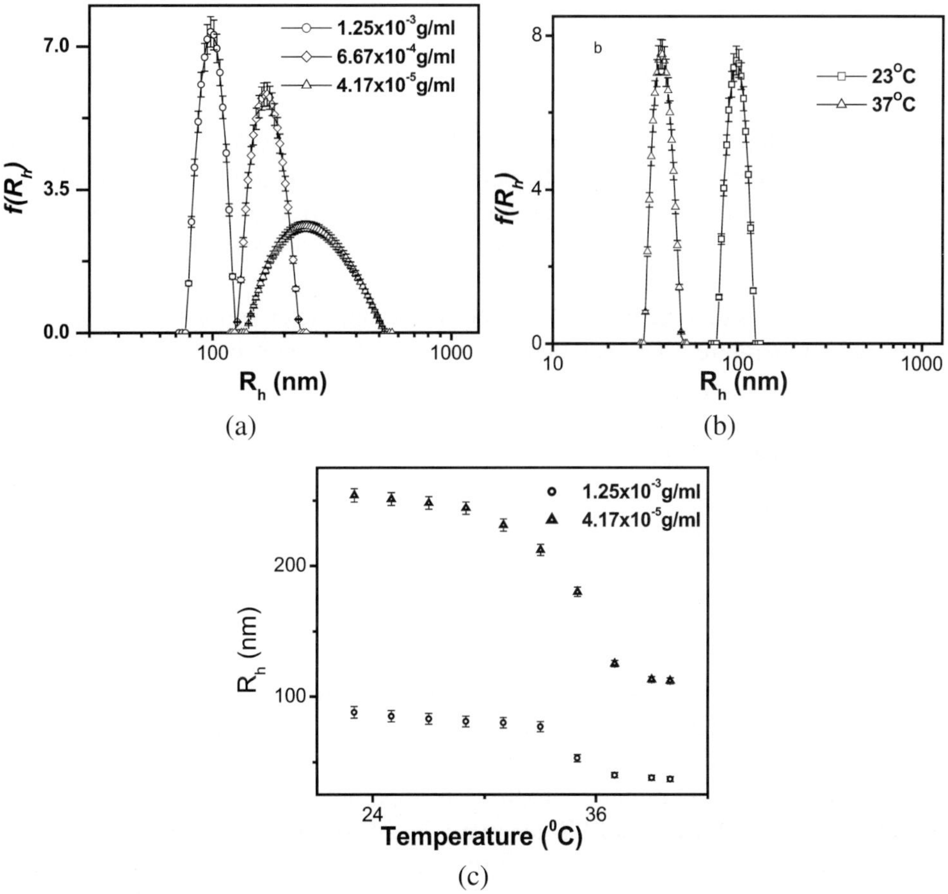

Figure 2. (a) Hydrodynamic radius distributions ($f(R_h)$) of PNIPAM nanoparticles ($C = 8.94 \times 10^{-5}$ g/ml) in deionized water at 23°C. Their composition is shown in Table 2. (b) Hydrodynamic radius distributions ($f(R_h)$) of PNIPAM nanoparticle spheres of Batch 1 in water at 23°C (circles) and 37°C (squares), respectively, where $C = 1.37 \times 10^{-5}$ g/g, and the scattering angle is 90°. (c) Temperature-dependent hydrodynamic radius of PNIPAM-co-NaAc nanoparticles with two different particle sizes.

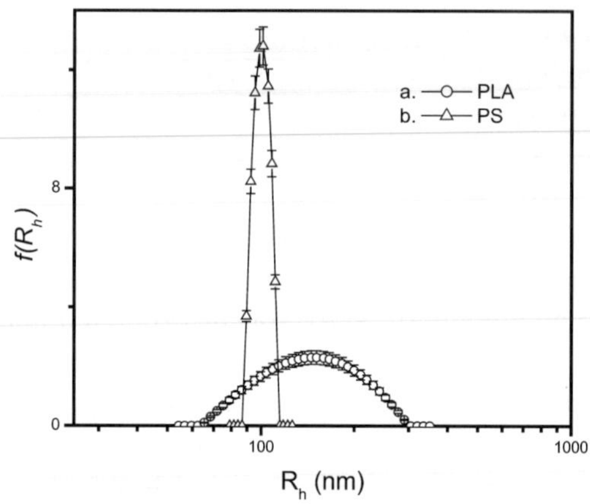

Figure 3. Hydrodynamic radius distributions ($f(R_{\mathrm{h}})$) of PLA and PS nanoparticles ($C = 8.94 \times 10^{-5}$ g/ml) in deionized water at 23°C and the scattering angle is 90°.

surfactant amount, the radius distribution ($f(R_{\mathrm{h}})$) of PNIPAM-co-NaAc particles also becomes narrower. This may be because of the formation of more micelles in higher surfactant concentration. The higher micelles are charged, the stronger the electrostatic repulsion. It helps to stabilize the colloidal dispersion and make the particle size distribution narrower.

Figure 2b shows that the PNIPAM-co-NaAc nanoparticles shrink sharply with increase of temperature. The detailed temperature-dependent particle size changes for PNIPAM-co-NaAc particles with two different sizes are shown in Fig. 2c. It is apparent that the size difference between these two particles does not change the temperature responsive property: both exhibit a volume phase-transition temperature around 34°C. Figure 3 compares hydrodynamic radius distributions of PLA and PS nanoparticles. The average sizes of PLA and PS particles are 143 and 100 nm, respectively.

Tissue reactions to nanoparticles

To determine tissue responses to different materials, a mouse subcutaneous implantation model was used [32–34]. In fact, subcutaneous cavity has been used as one of the main implantation for controlled drug delivery [35–38]. To assess the tissue compatibility of hydrogel hydrophilic nanoparticles, two commonly used hydrophobic materials, PLA and PS nanoparticles, were included in the study as the control materials for the following reasons. First, biodegradable PLA nanoparticles have been synthesized as controlled delivery devices for a variety of drugs, oligonucleotides, vaccines and proteins [19–21]. Second, non-degradable monodispersed PS nanoparticles have been used to study nanoparticle distribution in animals [22–24]. Finally, although both PLA and PS nanoparticles have been employed in several in vivo studies, the tissue compatibility to these nanoparticles have not been systematically evaluated and compared. After implantation for 4 days, the implants and surrounding tissues were recovered for histological analyses and quantification.

We found that, shortly after implantation, nanoparticle implants trigger the recruitment and accumulation of inflammatory cells. The implant-mediated acute inflammatory reactions are often followed with the formation of fibrotic capsule. Generally speaking, stronger inflammatory responses (CD11b-positive cell accumulation) lead to more severe fibrotic reactions (fibrotic capsule formation). We have observed that the extent of tissue responses to nanoparticle implants is in the following order: PLA > PS > HPC > PNIPAM. Specifically, both PLA and PS nanoparticles prompt significant cell immigration and capsule formation surrounding the implants (Fig. 4a and 4b). Foreign body reactions are likely caused by the accumulation and activation of inflammatory cells, because most of cells resided fibrotic capsule have proinflammatory cell marker, CD11b (Fig. 5a and 5b). Subsequently, the dense cellular fibrotic tissue would be replaced with thick collagen matrix which prevents the release of nanoparticle-loaded drugs to the surrounding tissue and circulation (data not shown). On the other hand, we have also observed that there are minimal cell accumulation and fibrotic reactions to both PNIPAM and HPC nanoparticles (Fig. 4c and 4d). Since weak inflammatory reactions are often associated with thin fibrotic tissue formation, it is likely that both hydrogel nanoparticles prompt minimal inflammatory reactions. Indeed, fewer CD11b-positive cells were found surrounding hydrogel nanoparticle implants (Fig. 5c and 5d) than PLA and PS implants.

The extent of implant-mediated tissue responses has also been quantified for better comparing the tissue responses among different nanoparticle. Specifically, we found that the cell densities of fibrotic tissues surrounding PLA and PS implants were 2–3-times higher than those nearby PNIPAM and HPC nanoparticles (Fig. 6a). There is no significant difference among the cell density in the tissue capsule adjacent to PNIPAM and HPC implants (Fig. 6a). As expected, we also find that the thicknesses of fibrotic capsule neighboring PLA and PS implants were approx. 2-times higher than those surrounding PNIPAM and HPC nanoparticles (Fig. 6b) and the capsules nearby PNIPAM and HPC implants have similar thickness (Fig. 6b). Finally, CD11b-positive cells surrounding implants were also quantified for assessing the extent of nanoparticle implant-induced tissue reactions. As expected, by comparing with PNIPAM implants, PLA and PS nanoparticles trigger the accumulation of CD11b-positive inflammatory cells more than 4-fold and approx. 2-fold, respectively (Fig. 6c). Rather surprisingly, although both HPC and NIPA nanoparticles prompt similar extent of tissue reactions (such as similar cell densities and capsule thicknesses), HPC implants triggered 70% more CD11b positive cell immigration than did PNIPAM nanoparticles (Fig. 6c).

It is noted that both HPC and PNIPAM nanoparticles could be biodegradable by incorporating biodegradable crosslinkers into the polymer network [17] and is currently under investigation. A major advantage of a biodegradable system is the

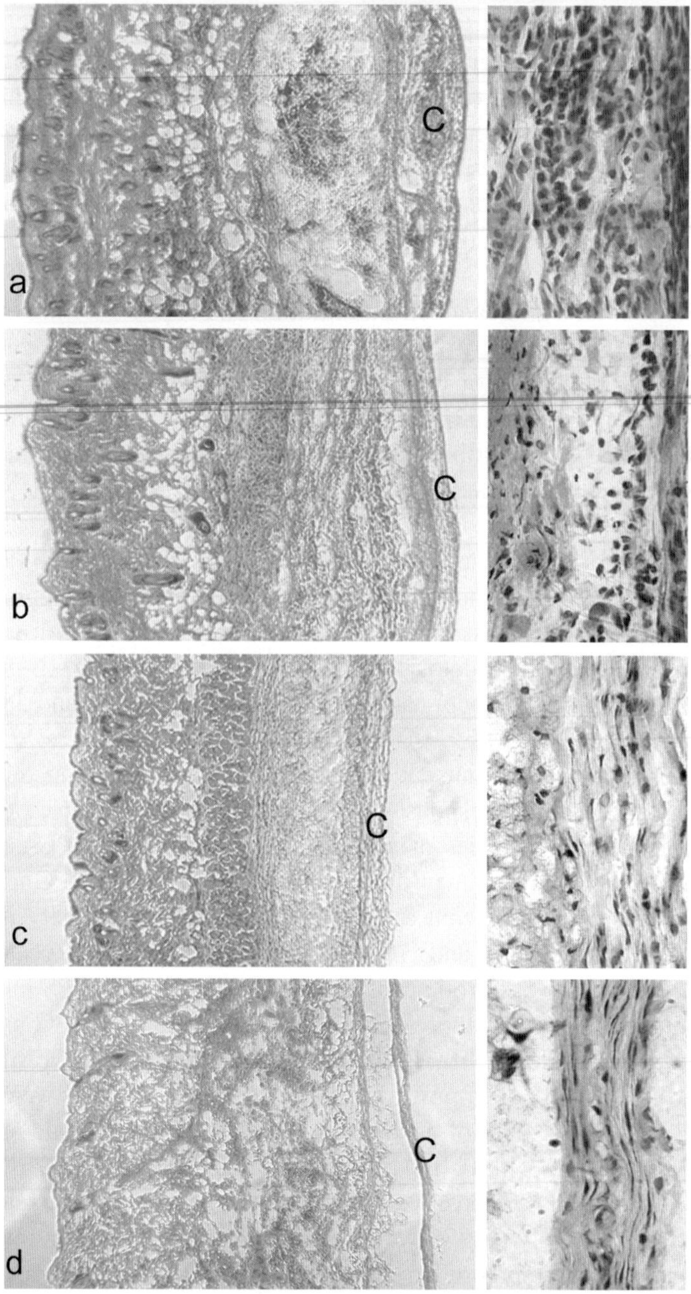

Figure 4. Histopathological analysis of injection sites (H&E stain). On the left side representative skin sections (magnification 10×) from PLA (a), PS (b), HPC (Batch 1) (c) and PNIPAM (Batch 1) (d) injected mice are shown. The C indicates the capsule of the nanoparticles at the injection sites. On the right side section close-ups (magnification 40×) of the tissue reactions are displayed.

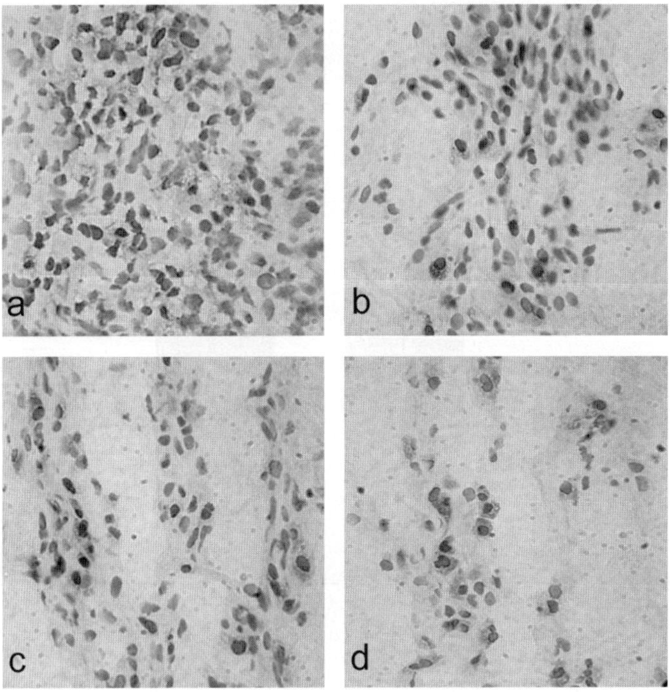

Figure 5. Immunohistochemical staining demonstrating CD11b-positive cells in skin tissue with subcutaneously-injected nanoparticles PLA (a), PS (b), HPC (c) (Batch 1) and PNIPAM (Batch 1) (d). Original magnification 400×.

elimination of surgery to remove an implanted delivery device after that the delivery system is exhausted.

Overall, we find that hydrogel hydrophilic nanoparticles possess better tissue compatibility than hydrophobic nanoparticles, at least PLA and PS nanoparticles. The real cause for differential tissue responses to hydrogel nanoparticles and to control nanoparticles is not clear. For example, the polymer volume fractions of the PS and PLA nanoparticles approach unity, while those of the PNIPAM and HPC are much lower. Although the detailed mechanism governing host responses to these nanoparticles is still yet to be determined, it is our belief that the hydrophobicity of both PLA and PS nanoparticles is, at least partially, responsible to strong foreign body reactions. It is well documented that, on hydrophobic surfaces, many proteins unfold and undergo conformational change [39, 40]. The conformation changes of plasma proteins, especially fibrinogen, trigger inflammatory cell accumulation and foreign body reactions [32–34, 41, 42]. Among hydrogel nanoparticles tested, PNIPAM nanoparticles elicit the least inflammatory cell accumulation and fibrotic tissue formation. This observation suggests that HPC and PNIPAM nanoparticles may serve as a good carrier for the applications of controlled delivery.

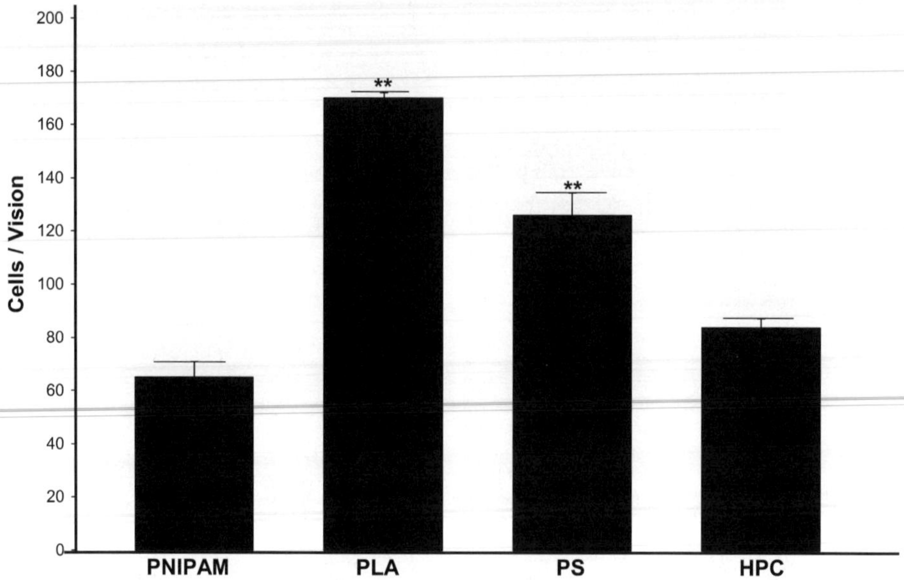

(a)

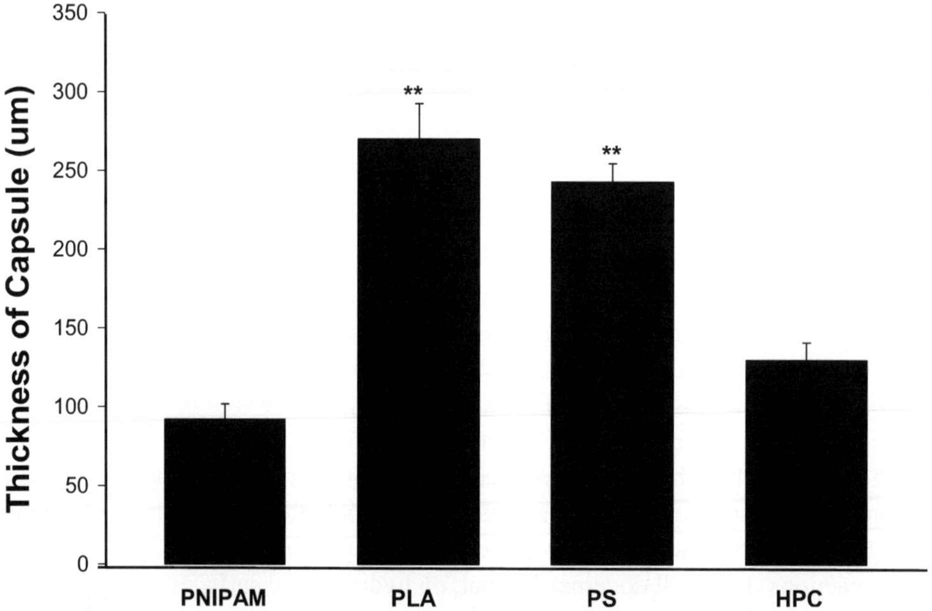

(b)

Figure 6. Histology of nanoparticles administered subcutaneously for 4 days. The cell numbers per view area (a), the thickness (b) and the cell numbers of CD11b-positive inflammatory cells (c), in capsule nearby nanoparticle implants were measured to assess the extent of tissue reactions to poly N-isopropylacrylamide (PNIPAM, Batch 1), PLA, PS and HPC (Batch 1) nanoparticles.

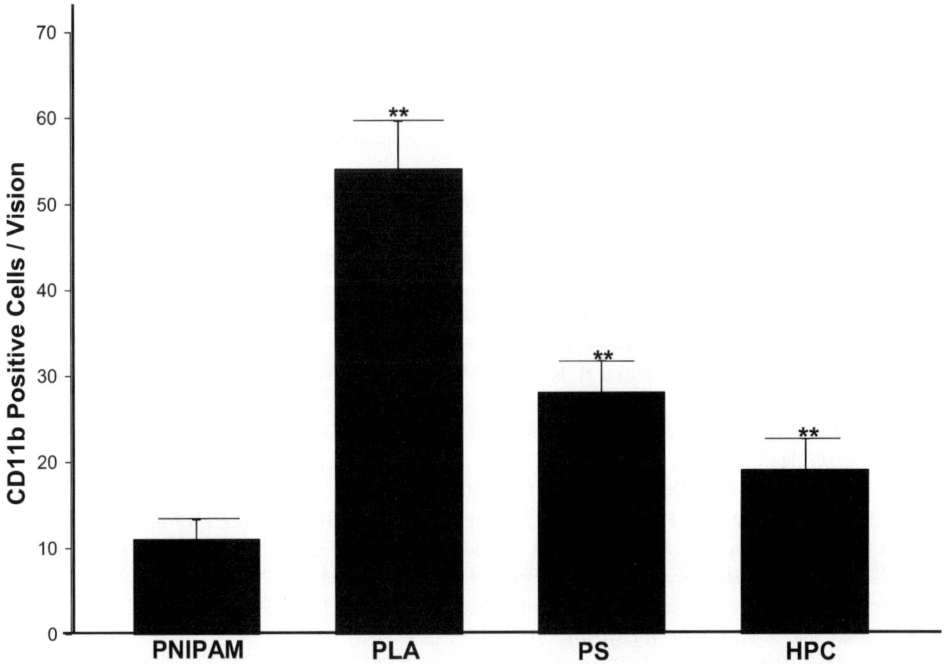

(c)

Figure 6. (Continued).

Acknowledgements

L. T. gratefully acknowledges the financial support from the American Heart Association and the National Institute of Health. Z. H. gratefully acknowledges the financial support from the US Army Research Office (Grant No. DAAD19–01-1-0596) and the Texas Advanced Technology Program.

REFERENCES

1. N. A. Peppas, *Hydrogels in Medicine and Pharmacy*. CRC Press, Boca Raton, FL (1987).
2. A. S. Hoffman, *Ann. N.Y. Acad. Sci.* **944**, 62 (2001).
3. T. Tanaka, *Phys. Rev. Lett.* **40**, 820 (1978).
4. S. Hirotsu, Y. Hirokawa and T. Tanaka, *J. Chem. Phys.* **87**, 1392 (**1987**).
5. Z. B. Hu, X. Zhang and Y. Li, *Science* **269**, 525 (1995).
6. Z. B. Hu, Y. Chen, C. Wang, Y. Zheng and Y. Li, *Nature* **393**, 149 (1998).
7. G. P. Misra and R. A. Siegel, *J. Control. Rel.* **81**, 1 (2002).
8. Y. Osada and A. Matsuda, *Nature* **376**, 219 (1995).
9. C. Wang, R. J. Stewart and J. Kopecek, *Nature* **397**, 417 (1999).
10. A. S. Hoffman, *J. Control. Rel.* **6**, 297 (1987).
11. A. S. Hoffman, A. Afrassiabi and L. C. Dong, *J. Control. Rel.* **4**, 213 (1986).
12. L. C. Dong and A. S. Hoffman, *J. Control. Rel.* **4**, 223 (1986).
13. T. Okano, Y. H. Bae, H. Jacobs and S. W. Kim, *J. Control. Rel.* **11**, 255 (1990).
14. A. Gutowska, Y. H. Bae, J. Feijen and S. W. Kim, *J. Control. Rel.* **22**, 95 (1992).

15. D. C. Harsh and S. H. Gehrke, *J. Control. Rel.* **17**, 175 (1991).
16. S. H. Gehrke, *Adv. Polym. Sci.* **110**, 81 (1993).
17. T. Cai, Z. B. Hu, B. Ponder, J. St. John and D. Moro, *Macromolecules* **36**, 6559 (2003).
18. Y. Qiu and K. Park, *Adv. Drug Deliv. Rev.* **53**, 321 (2001).
19. M. Berton, P. Turelli, D. Trono, C. A. Stein, E. Allemann and R. Gurny, *Pharm. Res.* **18**, 1096 (2001).
20. M. Michaelis, J. Matousek, J. U. Vogel, T. Slavik, K. Langer, J. Cinatl, J. Kreuter, D. Schwabe and J. Cinatl, *Anticancer Drugs* **11**, 369 (2000).
21. J. Panyam, S. K. Sahoo, S. Prabha, T. Bargar and V. Labhasetwar, *Int. J. Pharm.* **262**, 1 (2003).
22. U. Westedt, L. Barbu-Tudoran, A. K. Schaper, M. Kalinowski, H. Alfke and T. Kissel, *AAPS PharmSci.* **4**, 41 (2002).
23. N. Hussain, *Int. J. Pharm.* **214**, 55 (2001).
24. S. Sakuma, R. Sudo, N. Suzuki, H. Kikuchi, M. Akashi and M. Hayashi, *Int. J. Pharm.* **177**, 161 (1999).
25. X. H. Lu, Z. B. Hu and J. Gao, *Macromolecules* **33**, 8698 (2000).
26. J. Gao, G. Haidar, X. H. Lu and Z. B. Hu, *Macromolecules* **34**, 2242 (2001).
27. R. H. Pelton and P. Chibante, *Colloids Surf.* **20**, 247 (1986).
28. H. Fessi, F. Puisieux, J. P. Devissaguet, N. Ammoury and S. Benita, *Int. J. Pharm.* **55**, R1 (1989).
29. G. Odian, *Principles of Polymerization*, 3rd edn. Wiley, New York, NY (1991).
30. K. Li and D. H. Stöver, *J. Polym. Sci. Part A Polym. Chem.* **31**, 3257 (1993).
31. B. Chu, *Laser Light Scattering*. Academic Press, New York, NY (1991).
32. L. Tang and J. W. Eaton, *J. Exp. Med.* **178**, 2147 (1993).
33. L. Tang, T. P. Ugarova, E. F. Plow and J. W. Eaton, *J. Clin. Invest.* **97**, 1329 (1996).
34. L. Tang, T. A. Jennings and J. W. Eaton, *Proc. Natl. Acad. Sci. USA* **95**, 8841 (1998).
35. R. Diepold, J. Kreuter, P. Guggenbuhl and J. R. Robinson, *Int. J. Pharm.* **54**, 149 (1989).
36. V. Srinivasan, J. A. Pendergrass, K. S. Kumar, M. R. Landauer and T. M. Seed, *Int. J. Radiat. Biol.* **78**, 535 (2002).
37. Z. Cui and R. J. Mumper, *Int. J. Pharm.* **238**, 229 (2002).
38. A. Prokop, E. Kozlov, G. W. Newman and M. J. Newman, *Biotechnol. Bioeng.* **78**, 459 (2002).
39. T. J. Lenk, T. A. Horbett, B. D. Ratner and K. K. Chittur, *Langmuir* **7**, 1755 (1991).
40. D. R. Lu and K. Park, *J. Colloid Interface Sci.* **144**, 271 (1991).
41. W. J. Hu, J. W. Eaton and L. Tang, *Blood* **98**, 1231 (2001).
42. L. Tang, *J. Biomater. Sci. Polymer Edn* **9**, 1257 (1998).

Gels, Genes, Grafts and Giants, pp. 597–611 (2005)

Plasma protein adsorption to surfaces grafted with dense homopolymer and copolymer brushes containing poly(N-isopropylacrylamide)

JOHAN JANZEN [1], YEVGENIYA LE [1],
JAYACHANDRAN N. KIZHAKKEDATHU [1] and DONALD E. BROOKS [1,2,*]

[1] *Department of Pathology and Laboratory Medicine, 2211 Wesbrook Mall, University of British Columbia, Vancouver, BC, V6T 2B5, Canada*
[2] *Department of Chemistry, University of British Columbia, Vancouver, BC, V6T 2B5, Canada*

Received 24 October 2003; accepted 1 March 2004

Abstract—Growing polymer chains from surface initiators in principle allows much more dense polymer surface layers to be created than can be produced by grafting of whole (self-excluding) chains. We have utilized aqueous atom transfer radical polymerization to graft a series of cleavable hydrophilic poly(N-isopropylacrylamide) (PNIPAM) homopolymers and block copolymers of substituted acrylamides from polystyrene latex to give brushes of controlled MW and surface density. Average chain separations much less than their free solution radii of gyration have been achieved. Exposure to radiolabeled single proteins or to whole plasma and subsequent analysis by SDS-PAGE shows that PNIPAM brushes decrease protein adsorption relative to the latex surface or other substituted polyacrylamides. The PNIPAM brushes exhibit a second-order phase transition around 30°C as reflected by a decrease in the hydrodynamic thickness of the brush at higher temperatures. Total plasma protein adsorption is increased at 40°C compared to 20°C but there is significant differential adsorption behavior among the proteins detected by gel-electrophoresis analysis.

Key words: Plasma protein; adsorption; grafted latex; polymer brush; poly(N-isopropylacrylamide); block copolymer; SDS-PAGE; ^{125}I radiolabel.

INTRODUCTION

Allan Hoffman has been a prime mover in the application of polymer science to problems of biocompatibility [1–3]. He has been particularly active in studying and taking advantage of the environmental sensitivity of polymers containing N-isopropylacrylamide [4–8]. This polymer, due to the relatively hydrophobic isopropyl

*To whom correspondence should be addressed. Tel.: (1-604) 822-7081. Fax: (1-604) 822-7635. E-mail: don.brooks@ubc.ca

substituent on the hydrophilic acrylamide group, exhibits complex configurational properties as a function of temperature, pH and salt concentration. In aqueous solution the isolated polymer coil partially collapses with a decrease in radius of gyration and phase separation at temperatures above the well defined lower critical solution temperature (LCST) of 32°C due to an increase in inter- and intramolecular association as H-bonding with water weakens.

The early work of the Hoffman group on reduction of protein adsorption to surfaces to enhance biocompatibility [1–3] was an important part of the body of work that has shown that for many types of interfaces incorporation of neutral polymers such as poly(ethylene glycol) (PEG) [9–11], poly(N,N-dimethylacrylamide) (PDMA) [12] and poly(N-isopropylacrylamide) [13] reduces protein adsorption. At sufficiently high PEG concentrations protein tends to be excluded from the polymer coil. Protein adsorption is never eliminated, however; 5–10% or more of the control value typically remains [14–17]. Direct chemical grafting of preformed polymer to activated surfaces has been employed in virtually all these studies [14, 16, 18, 19]. However, coverage uniformity (detected by X-ray photoelectron spectroscopy, XPS) can be a problem [18] and high-density layers known as brushes, in which the average chain separation is less than twice the radius of gyration (R_g), are generally not achieved except for very-low-molecular-weight (MW) oligomers [10, 20, 21]. This is because in good solvents bound chains progressively exclude mobile chains as the layer density builds, reducing the probability of reaction between mobile chains and surface sites to negligible levels.

In the present work we have synthesized PNIPAM brushes from initiators on the surface of polystyrene (PS) latex or from active terminal groups on a first brush layer to form copolymers, using methods we have developed using aqueous atom transfer radical polymerization (ATRP) [22, 23]. The goal here was twofold: to study the effects of some substituted acrylamide polymer brushes on protein adsorption, and to specifically examine temperature effects on protein adsorption to PNIPAM brushes. We find that PNIPAM brushes are more effective than poly(methoxyethylacrylamide) (PMEA) brushes at reducing plasma protein adsorption to PS latex. PNIPAM brushes show a second order phase transition, detected as a decrease in hydrodynamic thickness of the polymer layer. Protein adsorption is higher above the transition temperature (approx. 40°C) than at 20°C.

MATERIALS AND METHODS

2-(2′-chloropropionato) ethyl acrylate (HEA-Cl) was synthesized as described in our earlier report [22]. Styrene (Aldrich, 99%; Sigma-Aldrich Canada, Oakville, ON, Canada) and N-isopropylacrylamide (NIPAM 97%; Sigma-Aldrich) were purified before use as described [22]. All other reagents were of highest purity from Sigma-Aldrich and used without further purification. Water was purified using a MilliQ Plus™ (Millipore, Bedford, MA, USA) water purification system and used in all experiments.

Synthesis of PS latex and surface polymerization

A narrowly dispersed batch of anionic PS seed latex was synthesized by surfactant-free emulsion polymerization of styrene using potassium persulfate as initiator [22]. A surfactant-free, random copolymer shell of styrene and HEA-Cl was next added as described [22] to provide a layer of ATRP initiator. The latex was then cleaned by exhaustive dialysis against water.

The surface polymerization reactions were performed in an argon-filled glove box due to the sensitivity of the Cu(I) complex to air [22, 23]. For a typical reaction, NIPAM (0.20 g, 1.74 mM) was stirred with HMTETA (8.6 mg, 37 μM), CuCl (1.5 mg, 15 μM), $CuCl_2$ (0.4 mg, 3.0 μM) and Cu powder (1.0 mg, 16 μM) for 3 min; 3.5 g (3%, w/w) of degassed Brij-35-stabilized PS shell latex was added to this under stirring at room temperature (22°C). The reaction was continued for 24 h and monomer conversion determined by reverse-phase high-pressure liquid chromatography (HPLC) of the supernatant solution. The grafted latex was cleaned by centrifugation and re-suspension alternately in water and 50 mM $NaHSO_3$ to remove adsorbed copper complexes. This was repeated for 8 to10 cycles until there were no detectable amounts of polymer, monomer or catalyst in the supernatant.

In ATRP reactions the terminal monomer retains an active chloride function and thus, it is possible to reinitiate polymerization simply by adding monomer and catalyst under the appropriate conditions. We have shown it is possible to form copolymers of a variety of monomers by this general route (data not shown). In the present case, PNIPAM was added to a first block of PMEA to form copolymers of the same surface concentration as the first block.

Analysis of the shell and polymer-grafted latex

All lattices were prepared as clean suspensions in water. The solid content was determined by freeze-drying. Latex surface charge was determined by conductometric titration [22].

Latex hydrodynamic size was measured by dynamic light scattering using a temperature controlled N4 Plus™ (Beckman Coulter, Miami, FL, USA) particle size analyzer. Aqueous dispersions of particles were thermally equilibrated for 5 min at 20°C unless otherwise noted. We define the difference between the grafted and shell latex radii to be the hydrodynamic thickness (HT) of the grafted layer.

Surface-grafted polymer was characterized after quantitative cleavage of the polymer chains from the latex. Cleavage was carried out on known amounts of grafted latex stirred with 2 M NaOH (final concentration approx. 0.35 M) at 4°C until no change in the amount released was detected (about 3 weeks). The supernatant and washings were collected and analyzed. It is known that N-substituted amides are very stable to alkali hydrolysis [24] and the pendent amide groups along the polymer chains are stable under the conditions of our experiments. Thus we were able to selectively hydrolyse the ester linkage between the chains and surface to obtain the grafted polymer. The supernatant and washings were collected

Table 1.
Grafted latex properties

Graft designation	Shell	Grafted latex		Chain			D/R_g
	Hydro-dynamic diameter (nm)	Hydro-dynamic thickness (HT, nm)	Chain separation (D, Å)	M_n (g/mol, $\times 10^{-3}$)	Polydispersity index (PDI, M_w/M_n)	Radius of gtyration (R_g, Å)	
PMEA33:43	651	17	43	33	1.24	66[a]	0.65
PMEA33-NIPAM13:43	651	72	43	46	1.21	84[a]	0.51
PMEA33-NIPAM36:43	651	153	43	68	1.37	102[a]	0.42
PMEA33-NIPAM7:43	651	40	43	39	1.4	76[a]	0.57
PNIPAM281:40	583	390	40	281	1.26	238	0.17

[a] R_g calculated from M_n and a (0.129 for PNIPAM) using the equation $R_g = a(M_n)^{3/5}$.

and analyzed. The hydrodynamic size of the grafted latex after hydrolysis was indistinguishable from that of the shell latex [22].

The polymer chain number average molecular weight (M_n), molecular weight distribution, mass per unit area of latex and radius of gyration (R_g) were determined by gel permeation chromatography (GPC) and multi-angle laser light scattering (MALLS) using a DAWN EOS (Wyatt Technology, Santa Barbara, CA, USA; laser wavelength $\lambda = 690$ nm) as described earlier [23]. GPC was in 0.1 M $NaNO_3$ on Ultrahydrogel™ (Waters, Milford, MA, USA) Linear and 120 columns in series. The refractive index increment (dn/dc) for PNIPAM in the mobile phase at 22°C was determined to be 0.164 ml/g at $\lambda = 620$ nm. A summary of the grafted latex characteristics are given in Table 1.

The notation used here to identify latex grafts is Polymer M_n:D, where Polymer identifies the monomer unit, M_n is dimensionless and D is the average grafted-chain separation on the latex surface in Å, e.g. PNIPAM 281:40 refers to a PNIPAM brush of $M_n = 281 \times 10^3$, $D = 40$ Å. Copolymer grafts are identified as Polymer1 M_{n1}-Polymer2 M_{n2}:D, so that PMEA 33-PNIPAM13:43 refers to a copolymer graft with an inner PMEA chain of $M_n = 33 \times 10^3$ to which a second PNIPAM chain of $M_n = 13 \times 10^3$ has been grafted, with an average copolymer chain separation of $D = 43$ Å.

Protein adsorption to latex

Adsorption of purified human serum albumin was determined by radiotracer methods after exhaustive washing of the latex by protein-free buffer. Adsorption to latex of human plasma was determined by polyacrylamide gel electrophoresis (PAGE) of protein desorbed from the latex by detergent after similar exhaustive washing. Thus the adsorption reported here measures 'irreversible' [25–27] binding.

Human serum albumin (HSA; 99%, globulin free; Sigma-Aldrich) was stored desiccated at –15°C and radiolabelled with carrier-free ^{125}I (Amersham Biosciences,

Baie D'Urfe, Quebec, Canada) using immobilized N-chloro-benzenesulfonamide (sodium salt) (Iodo-Beads™, Pierce Biotechnology, Rockford, IL, USA). Labelled protein was chromatographed on Sephadex G25™ (Amersham Biosciences, Amersham, UK) to remove free radiolabel. Protein concentration was determined from UV-Vis spectrophotometery using an OD increment for HSA of 0.541 at 278 nm [28]. Protein specific activities were between 10^4 and 10^6 dpm/mg and were corrected for free radiolabel. For the desorption recovery estimate, plasma was radiolabelled as a 10% solution. PAGE analysis of the product showed a normal plasma protein distribution. However, the specific activities of individual bands were not determined.

[^{125}I]HSA was equilibrated with latex suspensions for 40 min at room temperature. HSA adsorption to latex was at a final salt concentration of approximately 100 mM NaCl and 10 mM phosphate at pH 7.0. The volume of the adsorption suspensions was generally 800 μl. Latex-bound protein was separated from solution protein by repeated centrifugation at 13×10^3 rpm in an Eppendorf microcentrifuge, removal of a portion of the supernatant solution and re-suspension with an equal volume of protein-free buffer. The radiolabel activity of latex pellets was corrected for free-fluid activity. However, by the 5th or 6th wash cycle the free-fluid correction was generally less than 1% of the uncorrected value.

Latex adsorption area was calculated from latex mass, density, and the surface to volume ratio of spherical particles of known radius. The latex mass was determined from gravimetric addition of latex stock suspensions. Care was exercised to avoid loss of latex in the wash procedure. Gravimetric transfer of washed latex was required to avoid overestimation of binding due to adsorption to the tube used for equilibration. For all grafted lattices, the radius of the shell particle used to prepare the graft was used to calculate adsorption area.

For the plasma adsorption experiments, plasma was obtained from fresh, EDTA anti-coagulated human blood from a single donor and centrifuged (IEC Centra-8R) at room temperature and 3000 rpm for 15 min. A 400 μl aliquot of plasma was added to approx. 400 μl of water/latex suspension with a total latex area of 500 cm^2. Latex/plasma samples were equilibrated for 1 h at either room temperature or in an incubator at 40°C. Plasma adsorption by PAGE was found to be unchanged between 1 and 24 h.

Plasma-adsorbed latex was washed in a manner similar to that used in radiolabel experiments. After washing (typically 6 changes), latex pellets were subjected to a desorption 'wash' in 50 μl PAGE sample buffer concentrate (4xSB; 2% SDS solution, 0.0625 M Tris buffer, 10% glycerol, 0.01% bromophenol blue with or without 80 mM dithiothreitol) plus 150 μl water. The latex suspensions were thoroughly mixed and brought to boiling for 1 min. After cooling, the suspensions were centrifuged for 3 min and the supernatants were transferred into new, clean Eppendorf tubes. Radiolabelled HSA recoveries in the desorption wash were between 75 and 100%. Radiolabel plasma recovery was 77 ± 7%.

PAGE

PAGE was as described by Laemmli [29]. Desorbed protein and control solutions were loaded onto 4% 1-mm-thick gels. The following controls were always present on the gels: 1×SB for background OD measurements; size marker (SM) for molecular weight estimation (Bio-Rad prestained broad range Precision Protein Standards™ 250, 150, 100, 75, 50, 37 and 25 kDa); human plasma 1000 times diluted in 1×SB and protein desorbed from ungrafted shell latex. Gels were silver-stained following Morrissey [30] modified as follows: fix in 12% trichloroacetic acid for 30 min, stop development with 1 ml acetic acid in 100 ml water, background destain in 10% Rapidfix™ (Kodak) and store in water at 4°C for subsequent evaluation. It was estimated from OD calibration curves that 4 min 30 s was the optimal time for silver stain development.

Gels were photographed using a JVC color CCD camera fitted with a Hoya neutral filter for transmission illumination provided by an X-ray light box. Camera software white balance control of the red, blue and green (RGB) signals was set so that a nominal 1 optical density (OD) neutral density (ND) filter (5 × 5 cm, Newport) gave equal 8-bit R, G and B values. Line profiles (40 pixels wide) of the gel lanes were obtained using image analysis software (Northern Eclipse 6.0™, Empix Imaging, Mississauga, ON, Canada). The sample buffer lane was used for background correction. Lateral scans of ND filters showed that illumination by the light box was uniform on the scale of the gel holder and provided a calibration to convert RGB bit values to OD. The OD value of each ND filter was determined by measuring the average OD of the filter over 535–735 nm using a holmium oxide calibrated spectrophotometer (HP 8450A UV-Vis spectro- photometer, Hewlett Packard, Palo Alto, CA, USA).

Radiolabelled HSA was used to convert the R, G and B OD values from silver staining to protein amount. In the following, HSA monomer was used to calibrate the silver stain in the gels. The 8-bit to OD and OD to HSA monomer calibrations were applied independently for each color to provide protein mass estimates from each colour value.

Plasma protein molecular weight, quantity and identification in PAGE

Molecular weight calibration curves based on the protein standards were plotted for each gel as $\log_{10}$ of the molecular weight *vs.* relative mobility (R_f) and fit to a third order polynomial. Predictive 90% confidence limits for these data put the uncertainty of M_w estimates at between 5 and 6% for interpolated values. The uncertainty increased to about 7% at 400 kDa.

For each sample run, the line scan R, G and B bit values were converted to grams using the HSA-monomer calibration. The protein amount plotted against R_f was intergrated from $R_f = 0$ to $R_f = 1$ to determine the total protein. The total protein was sufficient to survey the adsorption behaviour of various grafted lattices. Total amounts in a given gel can be directly compared; however, amounts from different gels are subject to some variation in stain development. The amounts adsorbed in specific molecular weight ranges were determined with R_f intergration limits set using the molecular weight *vs.* R_f calibration.

Immunological identification followed the Amersham Pharmacia Biotech Western Blotting protocol RPN-2106. Polyclonal rabbit anti-human fibrinogen antibody (Behring, Marburg, France) at 1 in 5000 parts of skim milk/TBST buffer and mouse monoclonal anti-human albumin clone HSA-9 (Sigma-Aldrich) at $1:10^4$ dilution were used as primary antibodies. The secondary antibodies were respectively HRP donkey anti-rabbit conjugate and goat anti- mouse HRP both diluted to $1:10^4$. Development of blots was performed on Kodak X-OMAT™ scientific imaging film using Kodak developer and GBX™ fixer (VWR, Mississauga, ON, Canada).

N-terminal amino acid sequence analysis was performed by the University of British Columbia NAPS Protein Sequencing & Peptide Mapping Laboratory. For sequencing, proteins separated on 7% reduced or nonreduced PAGE gels were transferred to polyvinyldifluoride membranes and stained with Bio-Safe™ G-250 Coomassie Blue stain (Bio-Rad). Bands of interest were excised and sent for analysis.

RESULTS AND DISCUSSION

HSA adsorption

Polymer brushes are defined as grafts for which $D/R_g < 2$. For the work reported here (Table 1) it is seen that all of the grafts have values of $D/R_g < 1$, so that all of the surface polymer layers are within the dense brush regime. The polydispersities of the grafted chain populations are quite low, from 1.21 to 1.40, so these are relatively uniform systems with which to examine protein adsorption properties. For the copolymers it is difficult to calculate the molecular properties exactly, since their refractive index increment, required for the calculation of molecular weight properties, has not been measured independently due to a lack of sufficient material. Therefore, we have approximated this value as that of the dominant component.

In all cases the surface charge of the underlying latex is negative and is associated with the exposed sulfate groups that result from the potassium persulfate polymerization initiation used in shell synthesis. Although the latex surface is negative, the electrostatic potential near the exterior end of the brush, as estimated by the zeta potential, was zero for all the grafted latex (data not shown), due to the large HT values of the brushes (17–390 nm) compared to the electrical double layer thickness (approx. 1 nm). Hence, electrostatic effects on protein adsorption are likely negligible for all grafted surfaces. The effects of grafted brushes on the irreversible component of HSA adsorption is illustrated in Fig. 1. Duplicate measurements were not made for these samples due to the logistics of the experiments. However, error estimates based on counting, weight and volume data are 7 to 8% of the mean adsorption. For a similar experiment in which replicate samples were measured, the coefficient of variation for the adsorbed amount is 1.1% ($n = 4$), while error estimates computed from data as above are greater than 2%. Given the computed error estimates in Fig. 1, comparison of the adsorption differences and the estimated errors suggests that the adsorption difference between ShellXV and PMEA33:43 is significant (difference/sum of the errors = 6.2), between PMEA33:43 and PMEA33-NIPAM13:43 is likely significant (3.2) and between the three PMEA33-NIPAMxx:43 are likely not significant ($\leqslant 1.2$).

It is seen that adsorption to the bare shell latex, about 0.4 μg/cm^2, is as expected for a mixed hydrophobic/hydrophilic surface as it is close to the values reported for other mixed polymer surfaces exposed to protein [31]. The HEA-Cl esters and the sulfate initiators provide some hydrophilic character against the sea of styrene in the shell copolymer, providing a variety of types of binding sites for protein.

Addition of a relatively dilute brush of PMEA ($D/R_g = 0.65$) reduces the adsorbed amount of HSA from 0.39 to 0.14 μg/cm^2, in spite of the fact that the chains are approaching a mushroom regime. When a short chain (M_n approx. 7000) of a second monomer, NIPAM, is polymerized from the active ends of the PMEA grafted chains and the adsorption measurement again made, it is seen that the amount of HSA irreversibly bound is further decreased, from 0.14 to 0.08 μg/cm^2. This could be due to either the increase in molecular weight or to the different nature of the PNIPAM polymer compared to PMEA and the detailed manner in which it interact with HSA. When two higher molecular weight PNIPAM chains (M_n about 13×10^3 and 36×10^3) are added to the original PMEA brush, however, in both cases very little further decrease in adsorption is observed. This suggests that the nature of the interaction between PNIPAM and protein, not the total molecular weight, is the dominant factor in reducing adsorption.

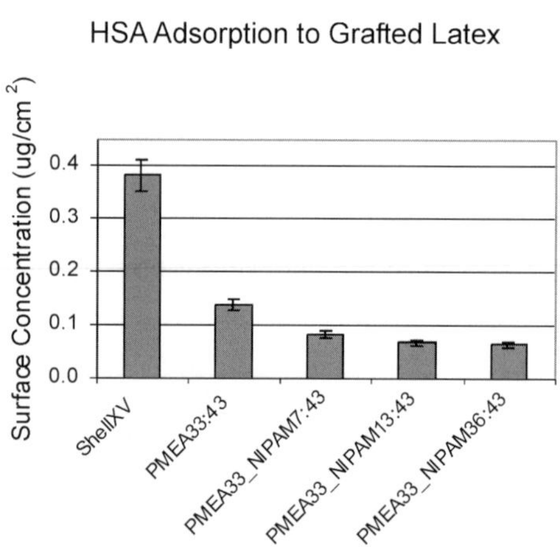

Figure 1. Radiolabelled human serum albumin adsorption to shell and grafted latex. Adsorption at 7.4 g/l protein for 40 min at 21°C in 0.1 M NaCl, 0.01 M $NaHPO_4$, pH 7 buffer. The latex area for the reported ads is 240 cm^2. Bound amount determined after six buffer washes to remove non-adsorbed and reversibly adsorbed protein. Error bars estimated from counting, weight and volume data. The errors are in the range of 7 to 8% of the means and due principally to uncertainty in the latex mass.

In considering interactions between proteins and neutral hydrophilic polymer layers one can anticipate a number of effects on the basis of thermodynamics. It is now well known that the change in entropy of the protein–polymer system plays a large role in determining the degree to which a protein is rejected from a polymer layer with which it comes in contact. This is the basis for the gel exclusion-like behavior that is observed when columns are packed of solid particles carrying surface grafted neutral polymer chains and protein partition coefficients calculated from the resulting chromatographic behavior [32–34]. It is found that protein molecular-weight-based separations are obtained, of the same magnitude as predicted by theories that ascribe the rejection to the unfavorable entropy of mixing caused by the reduced configurations and volume available to both protein and polymer when the species attempt to mix [33, 34]. Enthalpic interactions between protein and monomer can enhance or reduce rejection, although the effects are not predicted to be as large as the entropy changes. If the enthalpy is favourable for adsorption, higher molecular weight components could adsorb more strongly than lower M_w species as they do on solid surfaces.

In the present example, if it is accepted that much of the decrease in adsorption of HSA upon adding a short second PNIPAM block is due to the nature of the monomer, enthalpic effects might be indicated. In this interpretation PNIPAM would interact less favorably (weaker net attraction) with HSA than would PMEA. This is perhaps unexpected in that the side chain of MEA is more hydrophilic than that of NIPAM (Fig. 2) and generally it is expected that hydrophobic interactions are strong determinants of interactions of groups with proteins; the observed effect is in the opposite direction. One might also argue that the MEA side chain should increase rejection because it would increase main chain solubility more than NIPAM. This could allow PMEA to take up more configurations than PNIPAM. This in turn could produce larger entropy decreases when protein attempts to enter the brush and increase protein rejection. Again, this is opposite to the observed behaviour.

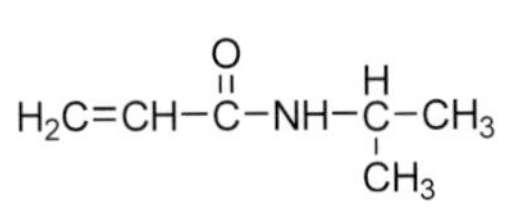

N-isopropylacrylamide

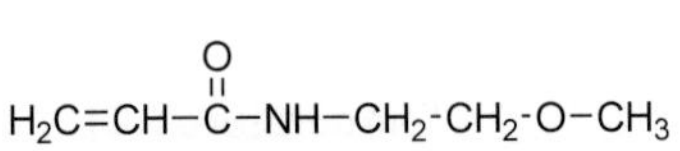

methoxyethylacrylamide

Figure 2. Structure of the N-isopropylacrylamide and methoxyethylacrylamide monomers.

Temperature-dependent adsorption of plasma proteins to latex carrying PNIPAM brushes

It is well known that PNIPAM in water exhibits an LCST at $T_c = 32°C$ and undergoes a phase transition from an open coil, soluble structure below T_c to a more globule, aggregated form above T_c, due to decreased solubility associated with reduced H-bonding with solvent. Well-defined grafted brushes of PNIPAM have not been examined for their configurational properties through this temperature range, however [23]. Fixing one end of the polymer on the latex surface and surrounding each chain with neighbors in an extended configuration in a brush in principle could affect the phase behavior. Figure 3 shows data illustrating that this is indeed the case. The thickness of a PNIPAM brush decreases between 20°C and 40°C. The transition is not sharp and, therefore, indicates a second-order phase transition, centered around 30°C. Interestingly, although above the LCST in aqueous solution PNIPAM molecules associate, the latex carrying the brushes did not aggregate at the higher temperatures, neither in the dilute solution used to measure HT, nor in concentrated suspensions on a microscope slide [23]. Presumably this is due to the limited access chains from one latex have to those on a neighboring particle when the chains collapse, at high surface density, above the critical temperature range.

We examined the effect of the brush collapse on irreversible protein adsorption by exposing the latex to plasma at 20°C and 40°C, washing in buffer, then removing the protein by boiling in SDS and running SDS-PAGE gels of the supernatants. From studies utilizing radiolabeled plasma we estimate this procedure releases 77 ± 7% of the adsorbed protein present after the buffer wash (data not shown). Figure 4 shows an example. Comparing lane 2, human plasma (non-reduced) with lane 5, protein eluted from the shell latex at 20°C, it is clear that the pattern of adsorbed protein is very different from that of plasma itself. Hence, there is considerable

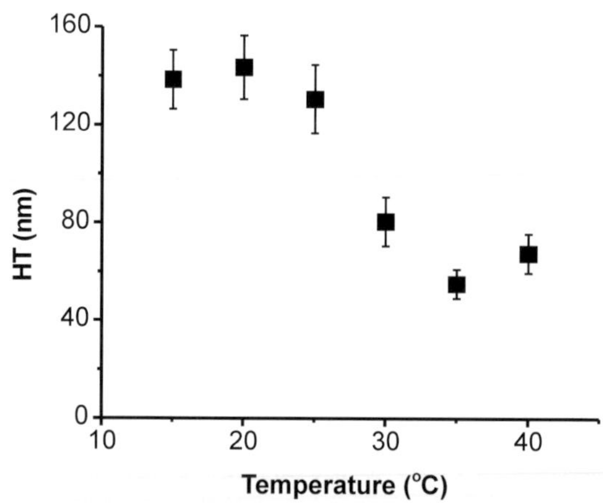

Figure 3. Hydrodynamic thickness (HT) of PNIPAM95:40 grafted brush as function of temperature. The temperature of was changed from 15 to 40°C with increments of 5°C and an equilibration time of 15 min. HT given as mean and standard deviation ($n = 3$).

selectivity in the protein adsorption/elution process, the pattern not simply reflecting the relative concentration of each component in plasma. Two bands in particular are evident in high concentration, one around 340 ± 20 kDa, which we showed by Western blotting was fibrinogen, and one at 117 ± 7 kDa which is not evident in the native plasma, lane 2. A similar band, eluted from hydroxylated PS latex, has been described by Luck *et al.* [35], who identified it as inter-α-trypsin inhibitor family heavy chain-related protein (IHRP). We excised this band, sequenced the first 11 amino acids and confirmed its identification as IHRP. This protein is known to be sensitive to digestion by plasma kallikrein but its function is unknown.

Introduction of a PNIPAM281:40 brush to the surface changed the pattern dramatically. At 20°C, where the brush is more hydrated and extended, fibrinogen adsorption/elution was strongly reduced (Fig. 4, lane 3). The IHRP band was not detectable in this experiment, although experiments with more dilute brushes (higher values of D/R_g) show the presence of this band (data not shown). There is a band at 80 kDa adsorbed/eluted at 20°C from PNIPAM281:40 that is not evident in the plasma lane nor in the material eluted from the shell latex, that we have not identified. Calibrated gel scans indicated that the total amount of

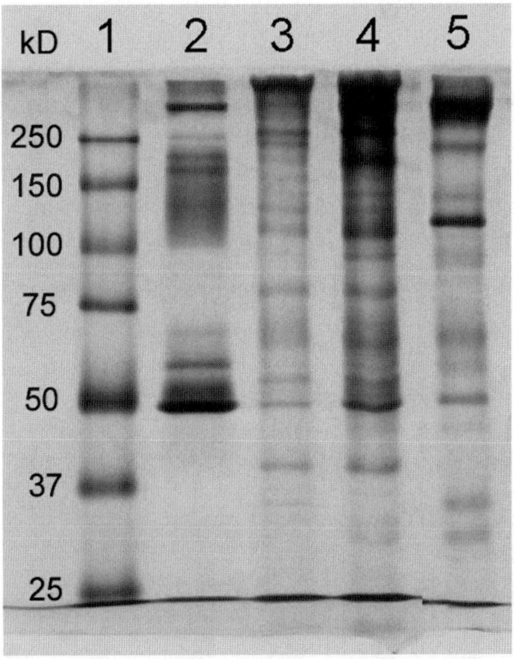

Figure 4. Silver-stained non-reduced SDS PAGE of standards and plasma proteins eluted from latex after adsorption with human plasma. The samples for each lane are: lane 1, M_w markers; lane 2, plasma; lane 3, PNIPAM281:40 latex at 20°C; lane 4, PNIPAM281:40 latex at 40°C; lane 5, Shell XXVI. The PAGE loads for lanes 3 and 4 were equal volumes of undiluted desorbed sample. The lane 5 sample was diluted 8-fold. The protein mass determined from gel staining for these three lanes are 0.33, 0.76 and 0.42 μg, respectively.

protein adsorbed/eluted from the shell was 210 ± 46 μg ($n = 3$ gels), while that associated with the PNIPAM-grafted latex at 20°C was 22 ± 5 μg ($n = 3$). Hence, as in the HSA case, addition of the brush strongly reduced the total amount of protein adsorbed. However specific proteins were both enriched and reduced in surface concentration on the PNIPAM latex relative to both native plasma and to protein adsorbed to the shell PS latex.

When the plasma was equilibrated with the PNIPAM latex at either 20°C or 40°C and the elution pattern examined (Fig. 4, lanes 3 and 4, respectively) there were again differences in adsorption/elution of specific proteins as well as in the total amount of protein associated with the latex. It is clear from Fig. 4 that much more total protein adsorbs and elutes at 40°C compared to 20°C, the integrals of the two curves, averaged over three gels each and corrected for dilutions, giving 62 ± 16 μg total protein at 40°C compared to 22 ± 5 μg at 20°C. In addition, specific proteins were found to be significantly enriched at 40°C compared to 20°C, particularly peaks near 350 kDa (fibrinogen), 200 kDa, 110 kDa and 50 kDa (HSA) [13]. The apparent 50 kDa value for HSA is due to the non-reduced nature of the sample and agrees with literature values. When reduced plasma samples are run the plasma band spans the range 69.5 to 61.4 kDa. This behaviour has been reported previously [36]. While we have not identified the peaks near 200 kDa and 110 kDa, the enhanced adsorption/elution shown for fibrinogen and HSA is not surprising, given that these molecules are often found enriched on hydrophobic surfaces. The collapsed PNIPAM brush behaves in this way due to the hydrophobic side chain on each monomer and reduced H-bonding at the higher temperature [5].

A coarser but more general comparison of the relative behavior of plasma proteins interacting with the three surfaces, compared to their abundance in native plasma, is given in Fig. 5. This shows the percentage of protein present in each molecular

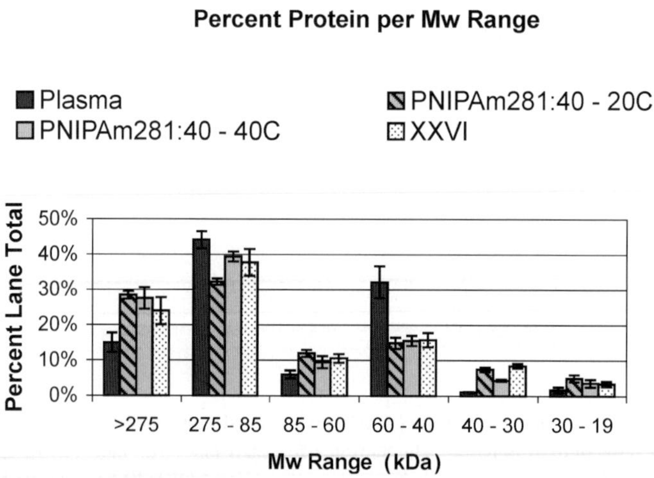

Figure 5. Percent of total eluted protein per molecular weight interval from analysis of silver stained protein by PAGE. The values are mean and sd for 4 samples of each of a PNIPAM281:40 adsorption at 20 and 40°C and 3 plasma and XXVI samples.

weight range for native plasma and plasma equilibrated with PNIPAM281:40 brushes at 20°C and 40°C, and the shell latex. Given that HSA runs in the 40–60 kDa Mw range under non-reducing conditions and that HSA, on the basis of concentration, is normally between 47 to 74% of the protein present in plasma [37], it appears that HSA is under-represented by about a factor of two. The possible causes of the discrepancy are differential staining of proteins by silver stain and nonlinearity in the protein-stain-OD relation which is not fully accounted for by our calibrations.

With this in mind, Fig. 5 illustrates the difference between plasma concentration and surface concentration over all molecular weight ranges and on all surfaces. Comparing the three surfaces as adsorbents, while total adsorption to the three surfaces differs significantly, the relative molecular weight selectivity is quite similar among the three, only specific proteins standing out as behaving differently, as noted above. This may again be due to the prevalence of hydrophobic associations between proteins and surface groups, moderated in the PNIPAM cases by general entropic repulsion accompanying protein entering the two forms of brush.

It is perhaps surprising that the PNIPAM281:40 brush did not reject protein more than was observed here. One might expect that hydrophilic brushes, with individual molecules packed quite tightly together ($D/R_g = 0.17$), would be poor substrates for attachment and that the entropic effect would dominate. Aside from the hydrophobic side chains, one other factor may mitigate against stronger rejection, however, namely the effect of radius of curvature of the surface and the extreme thickness of the brush (HT = 390 nm for PNIPAM281:40).

The radius of the shell latex was 292 nm, so the brush thickness is greater than the solid particle radius. Thus, if the area per PNIPAM molecule at the outer edge of the brush is compared to that at its base, the chain density is 5.3-times lower than the area per molecule at the latex surface, due just to the spherical geometry. Entropic repulsion increases monotonically with increase in polymer concentration [33, 34]. Hence, repulsion of protein will be much weaker at the outer edge of the spherical brush than it will be deeper in the surface polymer layer where the polymer concentration is higher. This effect could allow proteins with a weak attraction for NIPAM, likely due to hydrophobic interactions, to penetrate some ways into the brush before entropic repulsion becomes strong enough to discourage further incursion. Such protein would therefore be partitioned into the polymer layer and would be retained at an appropriate depth in the brush, depending on the balance between the enthalpic and entropic interactions with the protein. Models of this type have been discussed for flat surfaces carrying polymer [38] but have not, to our knowledge, been applied to spherical brushes previously.

The area per molecule would be increased by the above effect, relative to values calculated assuming all protein was adsorbed at the latex surface. This assumption was made for all adsorption values quoted in this paper. Hence, the true surface concentration for protein adsorbed within or to the brushes could be lower by a factor of up to approx. 5 than the values quoted, depending on how deeply into the brush the proteins were located. We have no information on this location at present.

CONCLUSIONS

PNIPAM brushes have been grown from functionalized latex surfaces by ATRP and examined for their ability to respond to temperature changes and affect protein adsorption. The brushes show a second-order temperature-dependent collapse at a lower temperature than free polymer. The presence of the brush reduces protein adsorption, both HSA and plasma, by about an order of magnitude compared to the bare latex surface when PNIPAM is below its phase transition temperature and the brush is fully extended. More protein adsorbs when the brush is collapsed, but the surface concentration is still lower than that bound to the bare latex. Plasma-protein interactions with the extended brush show significant specificity in both extended and collapsed configurations compared to the composition of native plasma. The extreme thickness of the PNIPAM brush, greater than the latex radius, likely plays a role in allowing protein to accumulate in the polymer layer due to the relatively low chain density at the outer edge of the brush associated with to the spherical geometry of the system.

Acknowledgements

We thank Suzanne C. Perry for performing the sequence analysis and Professor Evan Evans for the loan of neutral density filters. Financial support from the Canadian Institutes of Health Research, the Canadian Blood Services, the National Sciences and Engineering Research Council Canada and the Canada Foundation for Innovation is gratefully acknowledged.

REFERENCES

1. A. S. Hoffman, *J. Biomed. Mater Res. Biomed. Mater. Symp.* **5**, 77 (1974).
2. T. A. Horbett, P. K. Weathersby and A. S. Hoffman, *Thromb. Res.* **12**, 319 (1978).
3. A. S. Hoffman, *ACS. Symp. Ser.* **256**, 13 (1984).
4. A. S. Hoffman, *Clin. Mater.* **11**, 13 (1992).
5. T. G. Park and A. S. Hoffman, *Macromolecules* **26**, 5045 (1993).
6. G. Chen and A. S. Hoffman, *Bioconj. Chem.* **4**, 509 (1993).
7. Z. Ding, G. Chen and A. S. Hoffman, *Bioconj. Chem.* **7**, 121 (1996).
8. N. Malmstadt, P. Yager, A. S. Hoffman and P. S. Stayton, *Anal. Chem.* **75**, 2943 (2003).
9. K. L. Prime and G. M. Whitesides, *J. Am. Chem. Soc.* **115**, 10714 (1993).
10. T. McPherson, A. Kidane, I. Szleifer and K. Park, *Langmuir* **14**, 176 (1998).
11. Y. Mori, S. Nagaoka, H. Takiuchi, T. Kikuchi, N. Noguchi, H. Tanzawa and Y. Noishiki, *Trans. Am. Soc. Artif. Intern. Organs* **28**, 459 (1982).
12. H. Inoue, K. Fujimoto, Y. Uyama and Y. Ikada, *J. Biomed. Mater. Res.* **35**, 255 (1997).
13. A. Yamazaki, F. M. Winnik, R. M. Cornelius and J. L. Brash, *Biochim. Biophys. Acta* **1421**, 103 (1999).

14. C. Golander and E. Kiss, *J. Colloid Interface Sci.* **121**, 240 (1987).
15. W. R. Gombotz, PhD Thesis, Centre for Bioengineering, University of Washington, Seattle, WA (1988).
16. K. Bergstrom, K. Holmberg, A. Safranj, A. S. Hoffman, M. J. Edgell, A. Kozlowski, B. A. Hovanes and J. M. Harris, *J. Biomed. Mater. Res.* **26**, 779 (1992).
17. F. R. Llanos and M. V. Sefton, *J. Biomed. Mater. Res.* **27**, 1383 (1993).
18. J. M. Harris (Ed.), in: *Poly(ethylene glycol) Chemistry: Biotechnical and Biomedical Applications*, Chapter 15. Plenum Press, New York, NY (1992).
19. Y.-C. Tseng and K. Park, *J. Biomed. Mater. Res.* **26**, 373 (1992).
20. C. Pale-Grosdemange, E. S. Simon, K. L. Prime and G. M. Whitesides, *J. Am. Chem. Soc.* **113**, 12 (1991).
21. S. J. Sofia, V. Premnath and E. W. Merrill *Macromolecules* **31**, 5059 (1998).
22. K. N. Jayachandran, A. Takacs-Cox and D. E. Brooks *Macromolecules* **35**, 4247 (2002).
23. J. N. Kizhakkedathu, R. Norris-Jones and D. E. Brooks, *Macromolecules* **37**, 734 (2004).
24. D. Barton and W. D. Ollis, in: *Comprehensive Organic Chemistry* 2, p. 1004. Pergamon Press, Oxford (1979).
25. J. L. Brash and Q. M. Samak, *J. Colloid Interface Sci.* **65**, 495 (1978).
26. T. P. Burghardt and D. Axelrod, *Biophys. J.* **33**, 455 (1981).
27. R. Kurrat, J. E. Prenosil and J. J. Ramsden, *J. Colloid Interface Sci.* **185**, 1 (1997).
28. T. Peters, Jr., in: *The Plasma Proteins: Structure, Function and Genetic Control*, F. W. Putnam (Ed.), Vol. 1, 2nd edn, p. 147, Academic Press, London (1975).
29. U. K. Laemmli, *Nature* **227**, 680 (1970).
30. J. H. Morrissey, *Anal. Biochem.* **117**, 307 (1981).
31. C. F. Wertz and M. M. Santore, *Langmuir* **18**, 706 (2002).
32. D. Hritcu, W. Muller and D. E. Brooks *Macromolecules* **32**, 565 (1999).
33. D. E. Brooks, C. A. Haynes, D. Hritcu, B. M. Steels and W. Muller, *Proc. Natl. Acad. Sci. USA* **97**, 7064 (2000).
34. B. M. Steels, J. Koska and C. A. Haynes, *J. Chromatogr. B* **743**, 41 (2000).
35. M. Lück, B. R. Paulke, W. Schroder, T. Blunk and R. H. Muller, *J. Biomed. Mater. Res.* **39**, 478 (1997).
36. B. Kshirsagar, B. Wilson and R. C. Wiggins, *Clin. Chim. Acta* **143**, 265 (1984).
37. R. I. Handin, S. E. Lux and T. P. Stossel (Eds), in: *Blood: Principles & Practice of Hematology*, p. 2144. Lippincott, London (1995).
38. A. Halperin, *Langmuir* **15**, 2525 (1999).

Hydrogels

Gels, Genes, Grafts and Giants, pp. 615–624 (2005)

Preparation and characterization of pH-sensitive hydrogel of chitosan/poly(acrylic acid) co-polymer

LILI SHI, LIMING YANG, JIE CHEN *, YONG PEI, MIN CHEN, BAOJUN HUI and JIAN LI

Department of Chemical Engineering and Technology, School of Environmental and Chemical Engineering, Shanghai University (Jiading Campus), Jiading, Shanghai 201800, China

Received 26 September 2003; accepted 29 December 2003

Abstract—A pH-sensitive co-polymer hydrogel of chitosan/poly(acrylic acid) (CS/PAAc) was prepared by irradiating the aqueous solution mixture of chitosan and acrylic acid with ^{60}Co γ-ray irradiation. The effect of the composition of chitosan and AAc on the properties of the hydrogel, such as swelling ratio and pH-sensitivity, were determined. Fourier Transform Infrared (FT-IR) spectrometry was applied in the attenuated total reflectance (ATR) mode for analyzing the structure change of the hydrogels after the treatment in different pH buffer solutions.

Key words: Chitosan; poly(acrylic acid); hydrogel; pH-sensitivity.

INTRODUCTION

Chitin, or poly-β-(1-4)-N-acetyl-D-glucosamine, is one of the most abundant natural polymers, being second only to cellulose in the amount produced annually by biosynthesis [1]. Chitosan is a product obtained by alkaline deacetylation of chitin. It is insoluble in water, alkali and many organic solvents, but soluble in many acidic aqueous solutions. That limits the application of chitosan. Several methods were reported for the modification of chitosan, such as grafting hydrophilic monomers and oligomers onto chitosan backbone [2]. Chitosan contains $-OH$ and $-NH_2$ groups that can give rise to hydrogen bonding and the linear molecule express a sufficient chain flexibility, the conformation of which is highly dependent on ionic strength. These properties are considered essential for mucoadhesion [3]. Since it is a hydrophilic, biocompatible and biodegradable polymer of low toxicity, chitosan

*To whom correspondence should be addressed. E-mail: chenjjd@online.sh.cn or jchen@mail.shu.edu.cn

has been utilized in a variety of biomedical applications including wound dressings [4] and drug delivery systems, etc. [5, 6].

In recent years, many interesting results have been obtained in the development of transmucosal drug delivery (TMD) for drug administration [7]. Various synthetic and natural polymers have been investigated for their application in TMD system as mucoadhesive polymers, including poly(acrylic acid) (PAAc). Despite excellent mucoadhesive properties of PAAc, it has some drawbacks to be used in TMD system, such as high water solubility. In the previous study, a novel mucoadhesive polymer was prepared by template polymerization of acrylic acid in the presence of poly(ethylene glycol) (PEG) [8]. It was found that a polymer complex was formed between PAAc and PEG through hydrogen bonding. Now, much attention has been focused upon the polymer complex or semi-interpenetrating polymer networks (IPNs) of the system of CS and PAAc [9–13]. The IPNs, by their original definition, are composed of two or more chemically distinct components held together ideally and solely by their permanent mutual entanglements [14]. Ahn *et al.* used a LWUV lamp to irradiate the mixture of chitosan and AAc for preparing the CS/PAAc complex membranes. They assumed that the interpolymer complex between carboxylic group of PAAc and hydroxyl or amino one of CS may be formed and the dissolution rate of the PAAc would be retarded by the complex formation [15]. As CS/PAAc gel is pH-sensitive, it may potentially be used in specific regions such as the gastrointestinal (GI) drug delivery system.

In this paper, a CS/PAAc hydrogel was prepared by irradiating the aqueous solution mixture of CS and AAc with ^{60}Co γ-ray irradiation. Many polymers maybe be crosslinked by irradiation [16]. The structure changes of the hydrogels after the treatment in different pH buffer solutions were determined by FT-IR. The effects of pH, feed ratio and ionic strength on the swelling ratio of the gels were studied.

MATERIALS AND METHODS

Materials

Chitosan (CS) with a degree of deacetylation of 85% (supplied by Shanghai Chunlu Biochemistry) was used without any treatment. AAc was purchased from Shanghai Reagent. Other chemicals were analytical-grade from commercial sources made in China and used without any purification.

Preparation of CS/PAAc hydrogels

Solution mixtures with different CS/AAc compositions (ratio of 0.25–2 (w/w)) were prepared by dissolving 2 g CS in 40 ml AAc aqueous solution with a predetermined monomer concentration. The CS/PAAc co-polymer hydrogels with different feed ratio were obtained by exposing the solution mixture to ^{60}Co γ-ray irradiation (300 kCi, Shanghai Hexing Radiation Factory) with a dose of 25 kGy in moving condition. The samples were washed by immersing into distilled water for 24 h with stirring and water changing to remove AAc monomer and homopolymer. Samples were then treated with 2% acetic acid to remove the non-crosslinked CS and washed again with distilled water until the pH was 6–7. Afterwards, the samples were cut into small pieces, about $10 \times 5 \times 2$ mm in size.

FT-IR analysis

The CS/PAAc hydrogels were verified by Fourier transform infrared (FT-IR) spectrophotometry (Avatar 370, Nicolet) in the attenuated total reflectance (ATR) mode before and after the treatment with pH buffer.

Measurement of the swelling ratio

Different pH buffers with the same ionic strength ($I = 0.4$ M, adjusted with NaCl) were used. Gravimetric method was applied to obtain the swelling ratio. The dried samples were immersed into the buffer solution for the required period of time. After blotting with filter paper to remove the superabundant water on surface, the samples were then weighed immediately on an electronic balance.

The swelling reversibility of the hydrogels was determined by soaking a sample in two kinds of pH buffer alternately and the swelling ratio was measured as a function of time. The dried sample was first immersed into the pH 1 (or pH 6) buffer until swelling equilibrium. After washing with distilled water, it was transferred into pH 4 buffer, and was weighed every 10 min. Then it was put back into pH 1 buffer for the same treatment. The whole process was repeated for several times.

For studying the effect of ionic strength on swelling of the hydrogels, the samples were immersed in aqueous solutions of 2% (v/v) HCl with various NaCl concentrations (from 0.3% to 0.9%, w/v; $I = 0.05$–0.15 M) for 24 h, respectively. The swollen samples were weighed after the blotting to remove the surface water.

The swelling ratio of the hydrogel was expressed by the following formula:

$$\text{Swelling ratio (times)} = \frac{W_\text{s} - W_\text{d}}{W_\text{d}},$$

where W_s and W_d are the weights of swelled and dried samples, respectively.

RESULTS AND DISCUSSION

Effect of pH on the swelling ratio of CS/PAAc hydrogels

The effect of pH on the equilibrated swelling ratio of the hydrogels was determined in pH buffers in the range 1–10 and an ionic strength of 0.4 M. As shown in Fig. 1, most of the hydrogels appeared to swell well in both lower pH and higher pH conditions; however, they shrunk in the middle pH range. This interesting phenomenon is illustrated schematically in Fig. 2.

While in strong acidic solutions, most carboxylic groups of PAAc in the hydrogel were in the form of $-COOH$. The interaction between $-NH_3^+$ and $-COO^-$ might be disrupted by the acid, which led to chain stretch of CS and PAAc. So the CS/PAAc gels could be swelled quickly due to the hydration of ammonium ion. When the pH reached 3–4 (weak acid conditions), positively charged $-NH_3^+$ on CS and negatively charged $-COO^-$ on PAAc could form compact polyelectrolytes complex by ionic interaction, which limited the swelling ability, similar to the result reported by Hu *et al.* [17]. Furthermore, it may be assumed that hydrophobic side-chain aggregation and hydrogen bonds in the gels were relatively stronger, so it appeared the lowest swellability in this situation. However, with the increase of the pH (>5), CS almost stopped swelling and the ionized degree of PAAc increased (as shown in Fig. 2, equation), and the swelling ratio of the samples increased again due

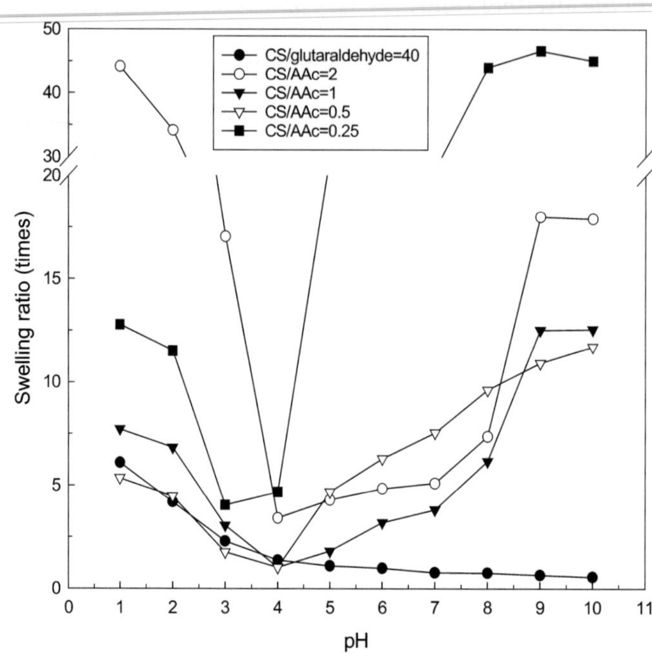

Figure 1. The effect of pH value on the swelling ratio of CS/PAAc hydrogels.

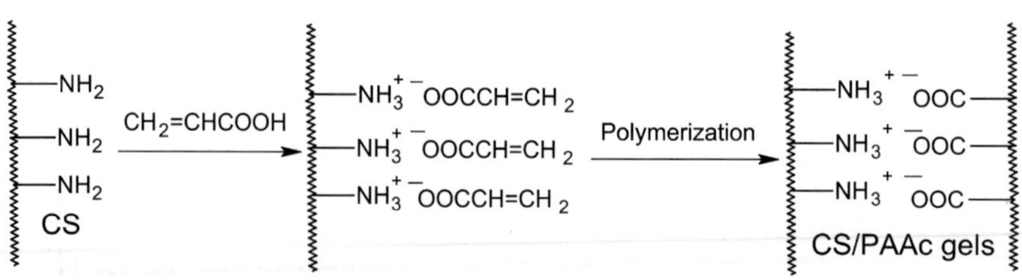

$$-NH_2 + -COO^- + H_2O \xleftarrow{OH^-} -NH_3^{+-}OOC- \xrightarrow{H^+} -NH_3^+ + -COOH$$

Figure 2. Schemas of the characteristics of CS/PAAc hydrogel in different pH solution.

to the increase of the osmotic pressure and charge repulsion. CS/PAAc gels with different feed ratio were all swollen well at high pH; oppositely, the swelling ratio of glutaraldehyde-crosslinked CS hydrogel, i.e. the single CS hydrogel, decreased with increasing pH over the complete pH range.

FT-IR analysis of CS/PAAc hydrogels

For comparing the difference of pure CS, pure PAAc and CS/PAAc, FT-IR spectrometer was applied in ATR mode and the spectra were obtained as shown in Fig. 3. The FT-IR spectrum of pure CS (Fig. 3, top panel) showed a strong amino characteristic peak at around 1587 cm^{-1}. The absorption bands at 1655 and 1325 cm^{-1} were characteristic of N-acetylated chitin and have been reported to be the amide-I and -III bands, respectively. Pure PAAc (Fig. 3, bottom panel) had a strong $-COOH$

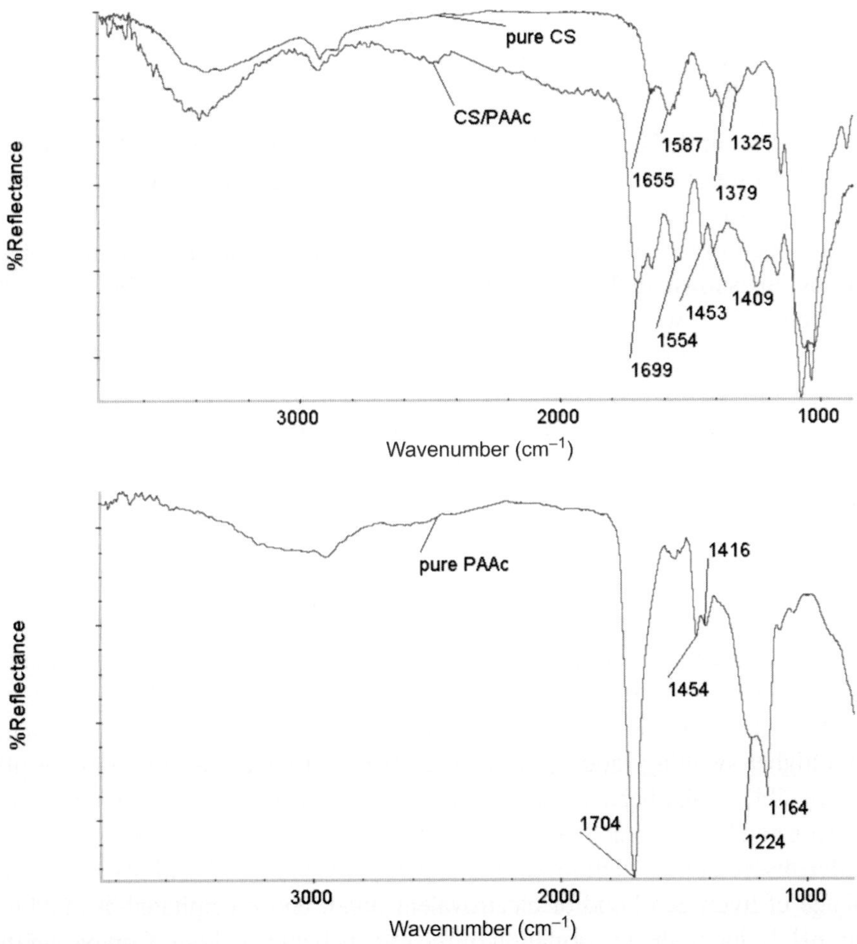

Figure 3. (Top) FT-IR spectra of pure CS and CS/PAAc hydrogel (CS/AAc = 0.25). (Bottom) FT-IR spectrum of pure PAAc.

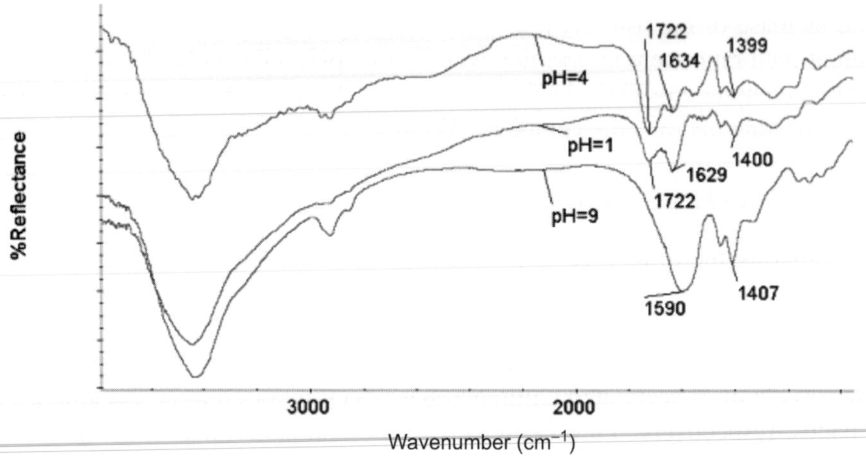

Figure 4. FT-IR spectra of CS/PAAc hydrogels (CS/AAc = 0.25) in pH buffer solutions with different pH.

peak at 1704 cm^{-1}. Compared to FT-IR spectrum of pure CS, CS/PAAc had a new peak appearing around 1699 cm^{-1}, which was the characteristic peak of the carboxylic group of PAAc.

FT-IR spectra of CS/PAAc hydrogels after the treatment in different pH buffer solutions are shown in Fig. 4. After the swelling in pH 1 buffer, the $-NH_3^+$ absorption of CS, which appeared at 1629 cm^{-1}, could be observed. The band at 1722 cm^{-1} was the peak of carboxylic group of PAAc. At pH 4, the absorption peak of $-NH_3^+$ shifted to 1634 cm^{-1} due to complex formation between CS and PAAc. When pH reached 9, two new absorption bands at 1590 and 1407 cm^{-1} could be assigned to asymmetrical and symmetric stretching vibrations of $-COO^-$ anion group. These results were similar to the scheme of Fig. 2.

Swelling reversibility in two kinds of pH buffers

The swelling behavior of CS/PAAc hydrogel appeared to be pH dependent, as shown in Fig. 5. The data were obtained from the determination of the sample with a CS/Aac ratio of 0.25 (w/w) in two kinds of pH buffer, alternately. The sample gave a higher swelling ratio at pH 1 (and pH 6) but lower swelling ratio at pH 4, and reversible. This behavior is similar to the result reported by Qu *et al.* [18]. Hydrogen bonding or ionic interaction made the two polymers interpenetrating. The complex dissociated at both lower and higher pH value (pH 1 and pH 6) due to the breakage of hydrogen bond or electrovalent bond, and precipitated at middle pH value (pH 4) due to the re-formation of the effects between them. Comparing these two figures (Fig. 5, top and bottom panel), the absorption and release rate of the hydrogel in pH 1 and 4 buffer solution was much faster than the one in pH 4 and 6.

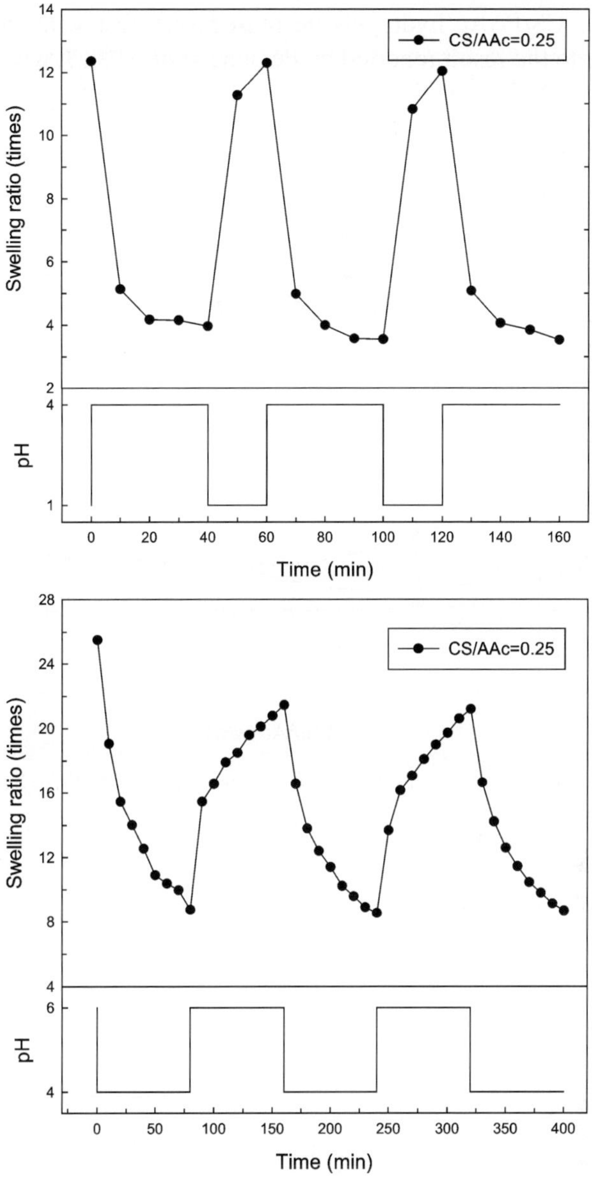

Figure 5. Swelling ratio of hydrogels as a function of time under repeated abrupt change of pH. (Top) Between pH 1 and 4; (bottom) between pH 4 and 6.

Effect of the feed ratio of CS/AAc on swelling ratio

The dependence of swelling ratio upon the feed ratio of CS and AAc was determined in distilled water and different pH buffer solutions, respectively. As shown in Fig. 6, in the buffers of pH 1 and pH 3, the swelling ratio of the CS/PAAc hydrogels increased with the increasing of CS ratio. In the buffer of pH 8, however, the swelling ratio of CS/PAAc hydrogels decreased very fast with the increasing of CS ratio, similar to the result reported by Peniche *et al.* [19]. It was obvious that an increase of CS would be beneficial for the swelling ratio of the CS/PAAc hydrogel

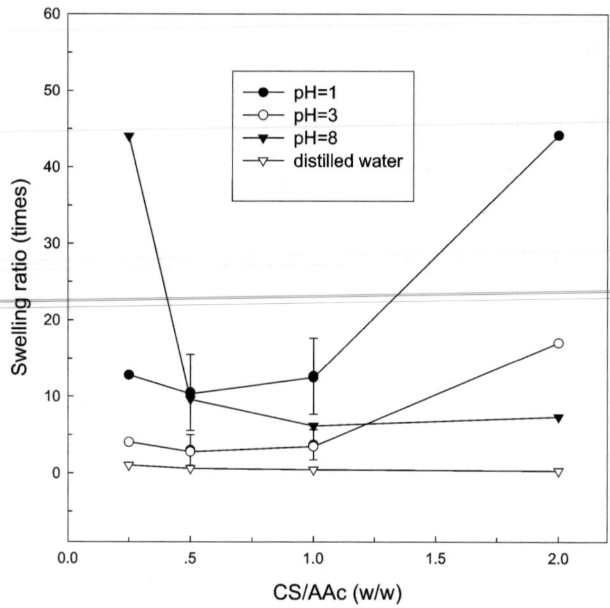

Figure 6. The effect of CS/AAc feed ratio on the swelling ratio of CS/PAAc hydrogels in distilled water and various pH buffers.

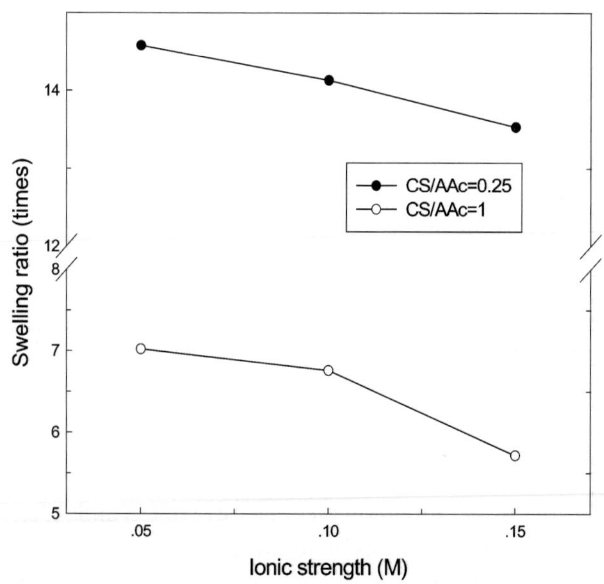

Figure 7. Effect of ionic strength on swelling ratio.

in acidic conditions. Oppositely, an increase of AAc would be good for increasing the swelling ratio in the basic conditions. In distilled water, on the other hand, the swelling ratio was very lower and the ratio of CS/AAc had almost no effect on the swelling.

Effect of ionic strength on swelling

Ionic strength sometimes affects the swelling ratio of hydrogels, when other factors are fixed. The effect of ionic strength on swelling ratio was determined at room temperature in aqueous solutions of various concentration of NaCl (I = 0.05–0.15 M), which contained 2% HCl. As shown in Fig. 7, the swelling ratio of the hydrogels decreased with the increase of ionic strength.

CONCLUSIONS

(1) Copolymer hydrogels of CS/PAAc were obtained by irradiating the aqueous solution mixture of chitosan and acrylic acid with ^{60}Co γ-ray.

(2) The CS/PAAc hydrogel obtained was pH sensitive, it swelled in lower and higher pH conditions and shrunk in the middle range of pH.

Acknowledgements

The authors thank the Shanghai Science and Technology Committee for financial support.

REFERENCES

1. S. G. Cao, Y. Q. Shi and G. W. Chen, *Polym. Bull.* **41**, 553 (1998).
2. H. Saito, X. Wu, J. M. Harris and A. S. Hoffman, *Macromol. Rapid Commun.* **18**, 547 (1997).
3. P. He, S. S. Davis and L. Illum, *Int. J. Pharm.* **166**, 75 (1998).
4. L. Shahabeddin, O. Damour, F. Berthod, P. Rousselle, G. Saintigny and C. Collombel, *J. Mater. Sci. Mater. Med.* **2**, 222 (1991).
5. M. L. Lorenzo-Lamosa, C. Remunan-Lopez, J. L. Vilajato and M. J. Alonso, *J. Control. Rel.* **52**, 109 (1998).
6. F. C. MacLaughlin, R. J. Mumper, J. Wang, J. M. Tagliafferi, L. Gill, M. Hinchliffe and A. P. Rolland, *J. Control. Rel.* **56**, 259 (1998).
7. D. Harris and J. R. Robinson, *J. Pharm. Sci.* **81**, 1 (1992).
8. H. K. Choi, O. J. Kim, C. K. Chung and C. S. Cho, *J. Appl. Polym. Sci.* **73**, 2749 (1999).
9. J. S. Ahn, H. K. Choi, M. K. Chun, J. M. Ryu, J. H. Jung, Y. U. Kim and C. S. Cho, *Biomaterials* **23**, 1411 (2002).
10. S. Y. Nam and Y. M. Lee, *J. Membr. Sci.* **135**, 161 (1997).
11. J. J. Jin, G. C. Chen, M. Y. Huang and Y. Y. Jiang, *React. Polym.* **23**, 95 (1994).
12. P. M. Torre, Y. Enobakhare, G. Torrado and S. Torrado, *Biomaterials* **24**, 1499 (2003).
13. H. F. Wang, W. J. Li and Y. H. Lu, *J. Appl. Polym. Sci.* **65**, 1445 (1997).
14. S. J. Kim, S. J. Park and S. I. Kim, *React. Funct. Polym.* **55**, 53 (2003).

15. J. S. Ahn, H. K. Choi and C. S. Cho, *Biomaterials* **22**, 923 (2001).
16. J. E. Wilson (Ed.), in: *Radiation Chemistry of Monomers, Polymers, and Plastics*, p. 369. Marcel Dekker, New York, NY (1974).
17. Y. Hu, X. Q. Jiang, Y. Ding, H. X. Ge, Y. Y. Yuan and C. Z. Yang, *Biomaterials* **23**, 3193 (2002).
18. X. Qu, A. Wirsén and A.-C. Albertsson, *Polymer* **41**, 4589 (2000).
19. C. Peniche, W. Argüelles-Monal, N. Davidenko, R. Sastre, A. Gallardo and J. S. Román, *Biomaterials* **20**, 1869 (1999).

Gels, Genes, Grafts and Giants, pp. 625–638 (2005)

Fibroin hydrogels for biomedical applications: preparation, characterization and *in vitro* cell culture studies

A. MOTTA [1,*], C. MIGLIARESI [1], F. FACCIONI [2], P. TORRICELLI [3], M. FINI [3] and R. GIARDINO [3]

[1] *Department of Materials Engineering and Industrial Technologies, University of Trento, via Mesiano 77, 38050 Trento, Italy*
[2] *Department of Biomedical Sciences, University of Verona, Policlinico GB Rossi, via delle Menegone 10, Verona, Italy*
[3] *Experimental Surgery Department, Research Institute Codivilla-Putti, IOR, via da Barbiano, Bologna, Italy*

Received 1 October 2003; accepted 29 January 2004

Abstract—Silk fibroin hydrogels prepared either by treating a 2% (w/v) silk fibroin aqeuous solution at 4°C (thermgel) or by adding 30% (v/v) of glycerol (glygel), were characterized by using Environmental Scanning Electron Microscopy (ESEM), Fourier Transform Infrared Spectroscopy (FT-IR), Differential Scanning Calorimetry (DSC), Thermogravimetrical Analysis (TGA) and molecular weight determination. The preparation procedure affected morphology and molecular weight of hydrogels, with no or negligible differences being displayed by FT-IR and DSC analyses. While thermgel presented a well uniform porous structure, the morphology of glygel appeared to be non-porous and heterogeneous. Glygel presented lower water content and lower degradation temperatures, associated with the presence of glycerol but likely also to less-organized protein structures. Cytoxicity tests with human osteoblast-like cells indicated that both gels were not cytoxic, while cell cultures pointed out a faster cell proliferation on glygel and a higher cell activation and differentiation on thermgel. These gels could be used as scaffolds able to promote *in situ* bone regeneration.

Key words: Silk; fibroin; hydrogels; cell cultures; biocompatibility.

INTRODUCTION

Silk is a natural fiber produced by spiders and some insects. Silkworms have been cultivated to produce silk in early Chinese civilizations, then the cultivation spread all over the world, since the middle age contributing to the present role of the silk

*To whom correspondence should be addressed. E-mail: antonella.motta@ing.unitn.it

as preferred materials for the fabrication of precious and fashionable fabrics and textiles.

Due to their excellent mechanical properties and flexibility, silk filaments have been used for years as suture threads. In some cases allergic and inflammatory reactions were observed [1, 2], these being attributed to sericins [3], a protein family that with fibroin composes the virgin silk filament. Compared to the whole silk filament, in fact, fibroin exhibited low inflammatory potential and adhesion of the immunocompetent cells [4], did not significantly adsorb the complement factors such as Bb and C1q [5] and did not show any appreciable fibrin polymerization when in contact with human plasma [6].

Recently, some authors have processed silk fibroin in order to obtain different kind of materials, such as films, nets, powder, re-spunned fibers and gels, which have been recommended for new biomedical and surgical applications. For example, regenerated fibroin films display a high oxygen permeability [7, 8], water vapour permeability [9], enzyme-immobilization ability [10], good compatibility to mammalian cells [11–14], and fibroin materials appear to stimulate osteoblast-based mineralization *in vitro* [15]. Pharmaceutical utility of silk fibroin as a possible material for an oral dosage for ederly patients was also investigated [16] and silk fibroin aggregates with different porosity were prepared as carriers for drug delivery [17]. Semi-interpenetrating network fibroin-based materials were studied as material for making an artificial muscle, due to their reversible swelling-shrinking behavior [7]. Silk fibroin fibers were also studied as materials for tissue engineering anterior cruciate ligaments because of their unique benefits in terms of mechanical properties as well as biocompatibility and slow degradation [18].

Hydrogels are of great interest in medicine, having a variety of applications for drug release, fabrication of prostheses for soft tissues, cells encapsulation and, most recently, becoming attractive in the field of tissue engineering as matrices for repairing and regenerating tissues and organs [19–22].

Aiming at applications in the field of tissue, in particular in bone regeneration, two different silk-fibroin-based hydrogels were prepared and their physical properties were investigated by Environmental Scanning Electron Microscopy (ESEM), Fourier Transform Infrared Spectroscopy (FT-IR–ATR), Differential Scanning Calorimetry (DSC), Thermal Gravimetric Analysis (TGA) and molecular weight determination. For some physical tests and for cell cultures a water solution cast fibroin film (FFC) has been used as a control.

Cytotoxicity, proliferation, differentiation and activation of osteoblast-like cells seeded on the fibroin hydrogels were evaluated by *in vitro* tests.

MATERIALS AND METHODS

Materials

Bombyx mori silk cocoons, bred and selected by Centro Sperimentale di Gelsibachicoltura (Como, Italy), were degummed by boiling twice, 1 h each, in aqueous solutions of Na_2CO_3 (1.1 g/l and 0.4 g/l, respectively), in order to remove the sericins. Degummed silk was then washed several times in distilled water, then dried at room temperature. The fibroin thus obtained was dissolved in 9.3 M LiBr (Fluka Chemical, Buchs, Switzerland) aqueous solution (10%, w/v) at 65°C for 3 h and filtered to eliminate impurities. LiBr was removed by dialysis against distilled water for three days at room temperature using a cellulose membrane with a MWCO: 3500 D (Pierce, Rockford, IL, USA). After dialysis the solution was filtered.

An aqueous solution of fibroin (2%, w/v) was treated following two different protocols, in order to obtain gels with different structures and properties.

In the following, Thermgel denotes a fibroin gel prepared by treating the fibroin water solution at 4°C for 3 days till gelation occurred, while Glygel denotes a gel produced by adding pure glycerol (99%, Janssen, Tilburg, The Netherlands) to the fibroin water solution in a ratio of 30% (v/v) and keeping the solution at room temperature until after 2 days the gel is formed; FFC is the control material made by a fibroin film cast from water solution at room temperature in a polystyrene dish, for the physical characterization, or directly in a 24-well polystyrene cell culture plate for *in vitro* tests. After casting FFC samples were made insoluble by treatment in 80% (v/v) methanol/water solution for 5 min at room temperature. Fibroin hydrogels were sterilized by γ-ray before cell cultures. FFC samples were disinfected by 70% ethanol overnight at 4°C and conditioned in Dulbecco's Modified PBS (pH 7.2) for 1 h at room temperature before *in vitro* tests.

Methods

The surface morphology of the fibroin hydrogels and film was examined with an environmental scanning electron microscope (E-SEM, Philips XL 30 ESEM, Philips, Eindhoven, The Netherlands), WD 10–15 μm, acc. V 20.0 kV. Samples were observed without any further treatment at vacuum pressure levels in the range of 0.9–4.0 Torr at a constant temperature of 5°C. As the pressure decreases from 4.0 Torr to 0.9 Torr, the water content in the gels decreases. The water content at the surface of the samples was evaluated by determining the equilibrium vapor pressure of water at 5°C and calculating a percent relative humidity at a given chamber pressure. Thus, the morphology of the surfaces was observed at increasing levels of dryness, up to near dry conditions.

The IR spectra of the samples were acquired at room temperature using a Fourier Transform Infrared Spectrophotometer (FT-IR–ATR, Bio-Rad FTS165, Bio-Rad, Hercules, CA, USA), equipped with a diamond probe, using 128 scans in reflectance mode at 8 cm^{-1} resolution.

Differential Scanning Calorimetry (DSC) measurements were performed with a Mettler Toledo DSC30 instrument with about 10 mg of each sample at a heating rate of 3°C/min. The experimental temperature range was between −100°C and +350°C for each scan. Experiments were conducted using a punched crucible to allow release of the vaporized material. The sample chambers were swept by N_2 gas during the analysis.

Thermogravimetric analysis (TGA) was conducted in a Mettler TG50 thermobalance, testing about 10 mg of material with nitrogen flowing at 100 ml/min and at a heating rate of 3°C/min in the range 30–600°C.

The weight average molecular weight (MW) of fibroin film and gels was determined by GPC, in order to detect potential molecular weight modifications induced by the process.

Fibroin samples were frozen in liquid nitrogen, lyophilized and then dissolved in 1,1,1,3,3,3-hexafluoro-2-propanol (Fluka, 5 mg/5 ml) with 0.01 N trifluoroacetate.

Molecular weight analysis was performed with a HPLC Waters system, equipped with a Spectra Physic P1500 isocratic pump, Jasco1570 UV/Vis injector and detector, and Shodex Protein KW-804 column, pre-column Shodex KW-G. The mobile phase, 20 mM Tris buffer (pH 8.0; composition: 0.15 M NaCl, 5 M urea), was filtered using a nylon membrane (0.2 μm) and degassed with helium in a ultrasound bath.

The calibration curve was obtained using protein standards (Pharmacia) and the samples were eluted at a flow rate of 1 ml/min. The injection volume was 50 μl, the samples and standard concentration was 0.1% (w/w), detection at 274 nm.

In vitro cell tests were performed with human osteoblast-like cells (MG63), cultivated in DMEM (10% FCS, 100 IU/ml penicillin and 100 μg/ml streptomycin solution) and incubated at 37°C in a humidified 95% air/5% CO_2 atmosphere. Cells were released at confluence with 0.05% (w/v) trypsin and 0.02% (w/v) EDTA, counted (Coulter Counter Z1, Beckman Coulter, Miami, FL, USA) and used at a concentration of 0.5×10^4 cells/ml for the experiment.

A cell suspension (0.5×10^4 cells in 100 μl) was directly seeded on every coated (samples of FFC, Thermgel and Glygel) well of 24-well plates. The same amount of osteoblast-like cells was plated in the same number of uncoated polystyrene wells as control. After a 2-h incubation to allow cell adhesion to the substrate, 900 μl DMEM (supplemented with 50 μg/ml ascorbic acid and 10^{-8} M β-glycerophosphate) was added to every well. Cultures were maintained in the same conditions as described above for 72 h, and the medium was replaced at 24 h.

At the end of the experiment, as indicated by the ISO10993 standards, supernatant from all wells was collected, in order to determine the expressed cell metabolic products through the following assays: Lactate dehydrogenase (LDH, Sigma kinetic method kit, Sigma, St. Louis, MO, USA), Interleukin 6 (IL-6, Quantikine human IL-6 immunoassay, R&D Systems, St. Paul, MN, USA), Alkaline Phosphatase activity (ALP, Sigma Kinetic method kit, Sigma), Type-I Collagen (CICP, Prolagen-C enzyme Immunoassay kit, Metra Biosystem, Palo Alto, CA, USA), Osteocalcin (OC, Novocalcin enzyme Immunoassay kit, Metra Biosystem) and Transforming Growth Factor-β 1 (TGF-β1, Quantikine human TGF-β1 Immunoassay, R&D Systems). The measured concentrations were normalized according to the cell number, derived by the WST-1 proliferation test. This is a colorimetric test based on a tetrazolium salt that is reduced to a soluble formazan salt by a reductase of the mitochondrial respiratory chain, active only in viable cells. The spectrophotometric absorbance of WST-1 directly correlates with the cell number.

Cell proliferation reagent WST-1 test was done to assess cell proliferation and viability: 100 μl of WST-1 solution and 900 μl of medium (1 : 10 final dilution) were added to the cell monolayers, and the multi-well plates were incubated at 37°C for 4 h. Supernatants were quantified spectrophotometrically at 450 nm with reference wavelength at 640 nm. Results were reported as optical density (OD).

Statistical evaluation of data was performed using the software package SPSS/PC$^+$ Statistics™ 7.5 (SPSS, Chicago, IL USA). Data are reported as mean ± standard deviation (SD) at a significance level of $P < 0.05$. After having verified normal distribution and homogeneity of variance, a one-way ANOVA was done for comparison among groups. Finally, Scheffè's post-hoc multiple comparison tests were done to detect significant differences between groups.

RESULTS AND DISCUSSION

Environmental Scanning Electron Microscopy (ESEM) was used to assess the effects of the gelation process on the surface morphology of fibroin hydrogels. Thermgel (Fig. 1A) had a highly porous network with pore sizes in the range of approx. 2–6 μm. To the eye, the hydrogel appears as a white, opaque material. ESEM micrographs of the surfaces of Glygel (Fig. 1B) showed for the most part that the morphology of the resulting gel is a non-porous, heterogeneous, diffuse array of protein material.

Conformational changes of the regenerated silk fibroin gels were also investigated using FT-IR. The spectrum of an untreated silk fibroin film (Fig. 2A) shows a shoulder at 1660 cm^{-1} and bands at 1540 and 1235 cm^{-1}, assigned, respectively, to amide-I, amide-II and amide-III bands of a random coil conformation [21]. Figure 2B illustrates the spectra of Thermgel and Glygel. A strong peak is observed for both samples at 1620 cm^{-1}, characteristic of β-structural frequencies [22]. A slight shoulder appearing at 1645 cm^{-1} in the spectrum of Thermgel is indicative of the presence of a random coil or intermediate structure. The peak for Thermgel at 1522 cm^{-1} is characteristic of a β-structure, and the shoulder at 1545 cm^{-1} is in the amide-II region and characteristic of random coil structure. These peaks are much less evident for Glygel. The two gel spectra are substantially different below 1450 cm^{-1}. The shoulder in the spectrum of Thermgel at 1256 cm^{-1} and the peak at 1235 cm^{-1} in the amide-III region is related to β-structure, while the Glygel spectrum shows only a relatively small peak at 1210 cm^{-1} and two other sharp peaks at 1110 cm^{-1} and 1039 cm^{-1}. These peaks are associated with the presence of glycerol. The formation of β-structures, that are evident in the IR spectra of thermgel and are likely masked by the glycerol in glygel is well known [23].

Figure 3 shows DSC curves of the samples carried out in an sealed punched crucible. The endothermic peak centered at −13°C in Glygel (Fig. 3A) corresponds

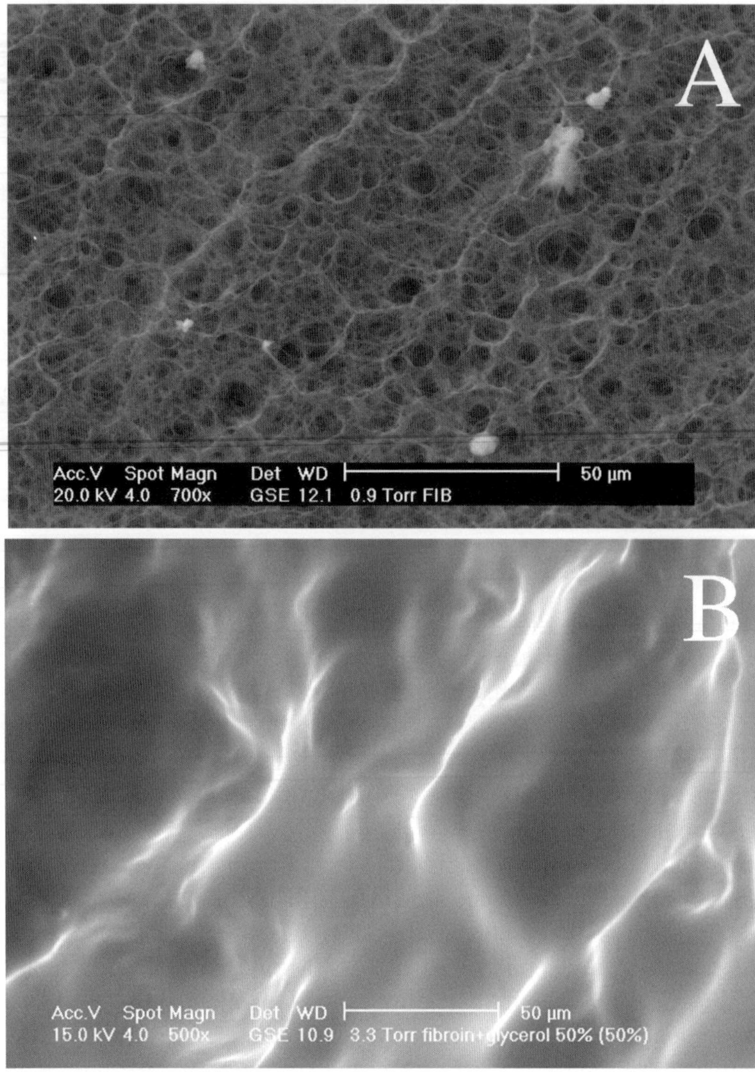

Figure 1. ESEM images of fibroin hydrogels. (A) Thermgel, (B) Glygel.

to the melting of the water-glycerol, the continuous downward slope up to +107°C is due to evaporation and escape of the solvent phase and the peak at 100°C represents the boiling of the remaining water-glycerol solution. The higher temperature endothermic peak is analogously due to the evaporation and escape of glycerol and to its boiling. Thermgel (Fig. 3B) shows the same overall behavior with two endothermic peaks attributable to the water melting and boiling and a high temperature (about 278°C) small peak that could be inferred to the fibroin degradation. This peak is likely masked in the Glygel curve.

The TGA of Thermgel is shown in Fig. 4A. Beyond about 200°C, the evaporation of water is complete and the measured weight loss is 96%. Beyond this temperature

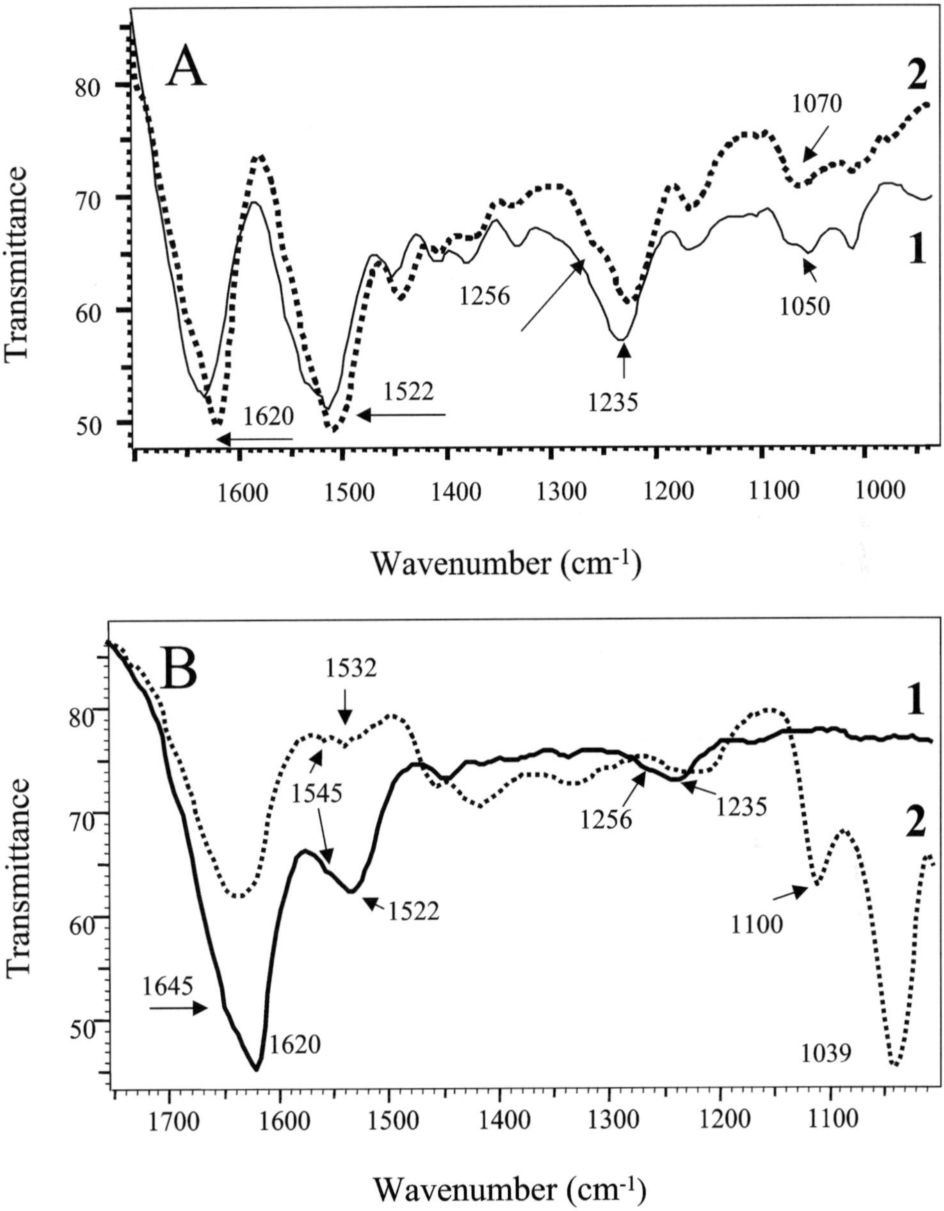

Figure 2. FT-IR–ATR analysis of fibroin samples. (A) fibroin film (FFC) before (1) and after (2) methanol treatment; (B) fibroin hydrogels: (1) Thermgel, (2) Glygel.

the remaining 4% of residue is mainly silk-protein-based structures, in agreement with the initial measured composition and the DSC data. At about 295°C the degradation of the SF protein occurs, leaving behind a residue of approx. 1.4%.

The TGA of the Glygel is shown in Fig. 4B. It indicates that the glycerol solution is evaporating over a broad temperature range that extends into the region of where

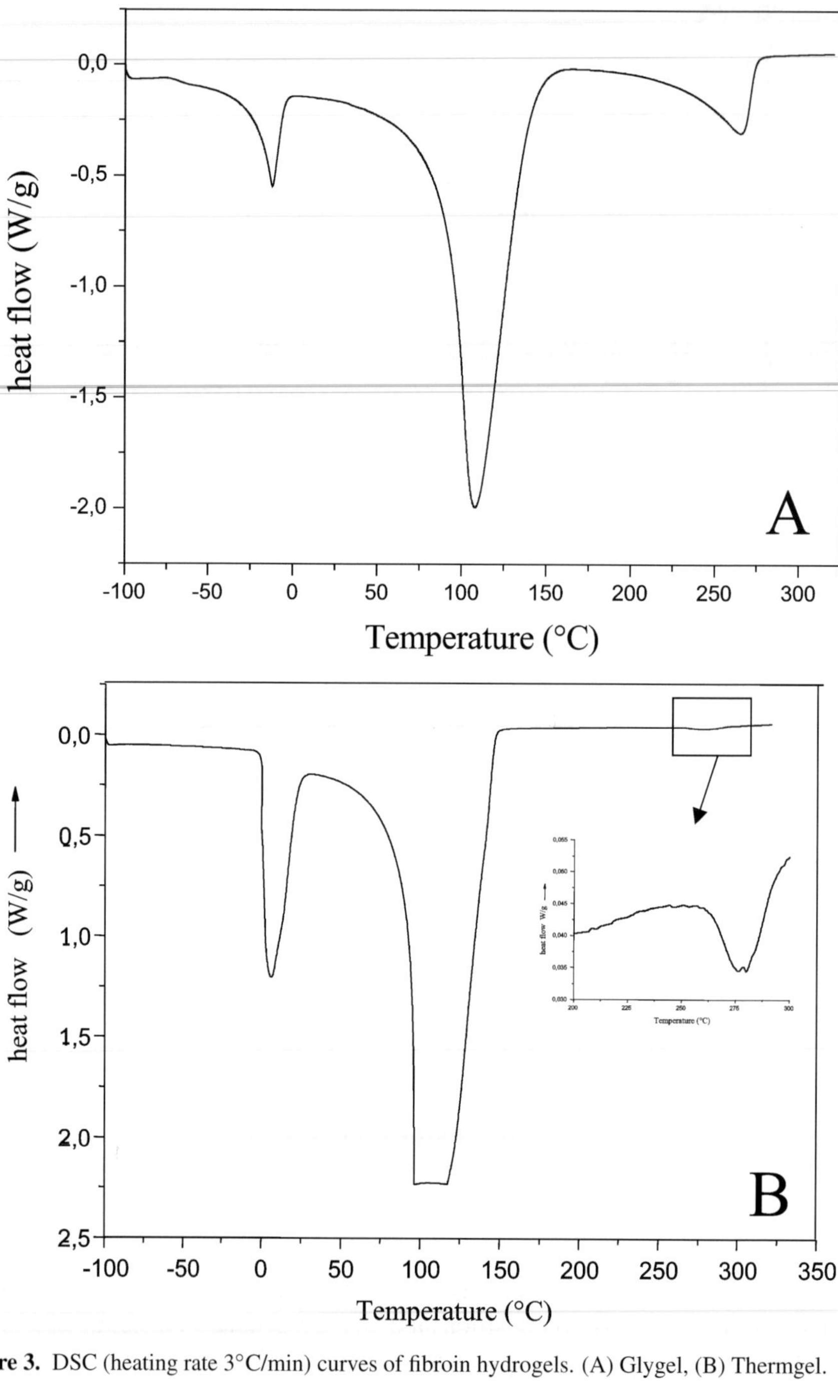

Figure 3. DSC (heating rate 3°C/min) curves of fibroin hydrogels. (A) Glygel, (B) Thermgel.

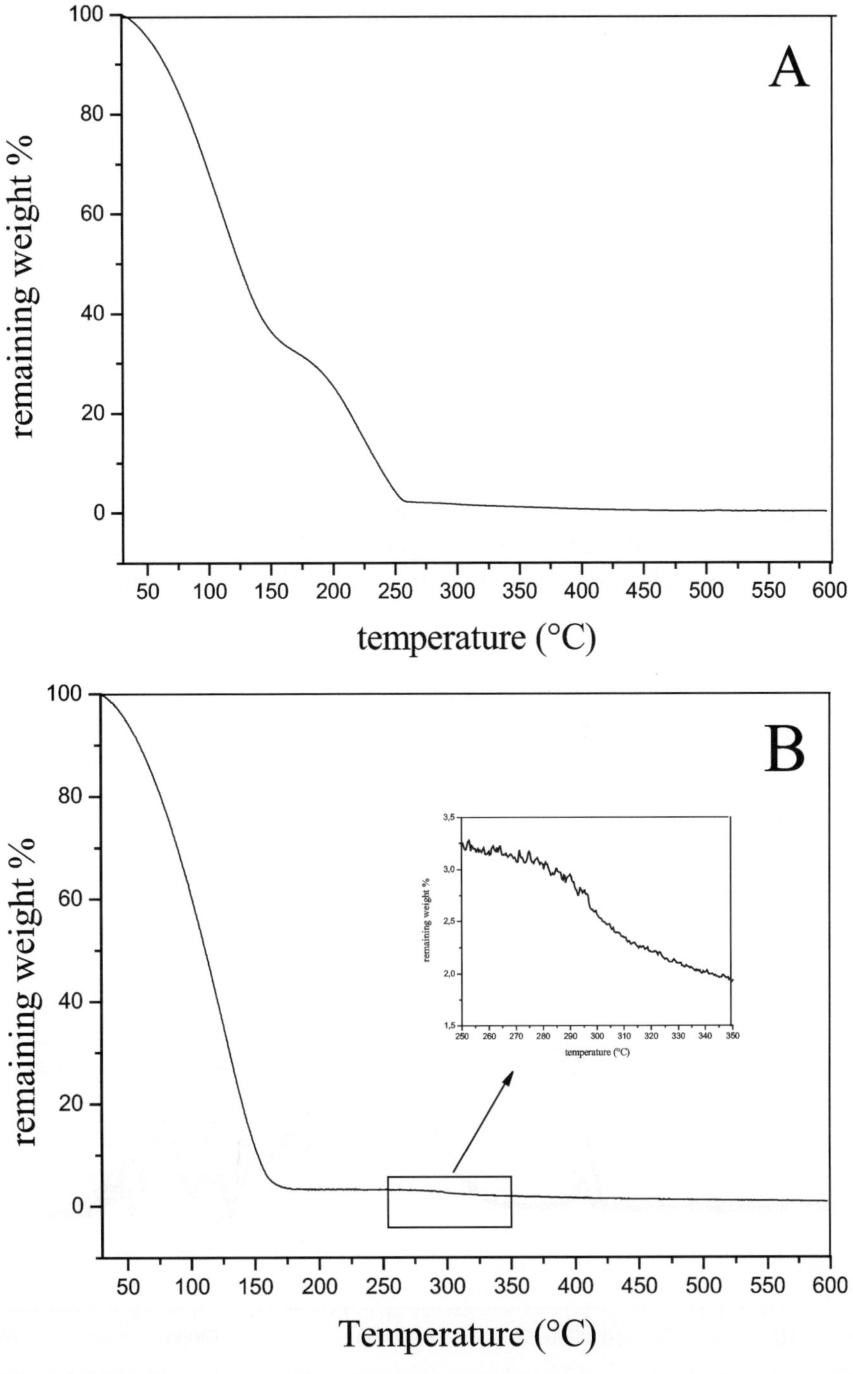

Figure 4. Remaining mass loss during TGA analysis of fibroin hydrogels. (A) Glygel, (B) Thermgel.

both the glycerol and SF protein are degrading. Beyond 200°C, the material consists mainly of glycerine and SF protein which then appears to slowly degrade in the region of 280–390°C, leaving a residue of approx. 1.2%.

Native fibroin molecule is composed of two subunits, heavy chain and light chain, the masses of which are reported in most papers to be about 350 and 25 kDa, respectively [24]. Comparison of the GPC curves in Fig. 5 indicates that both gelation procedures induced molecular mass degradation. While fibroin in FFC films is made by units weighing 310 and 20 kDa, the molecular mass of the heavier unit in Glygel and Thermgel dropped down to 226 and 101 kDa, respectively. In addition, in the thermgel fibroin well evident is the presence of peptides of 16 and 10 kDa. Such a degradation appears to control the thermodynamics and kinetics of the gel formation, with a cause–effect relation that has been observed on fibroin-water solutions stored for different times.

We have also observed degradation of fibroin in the random coil fibroin-water solutions. The lower the solution pH and the higher the temperature, the faster the degradation. The gelation process is accompanied by the formation of beta structures whose formation, in turn, is triggered by the molecular degradation. The organized gel is much more stable to hydrolytic degradation than the random coil fibroin.

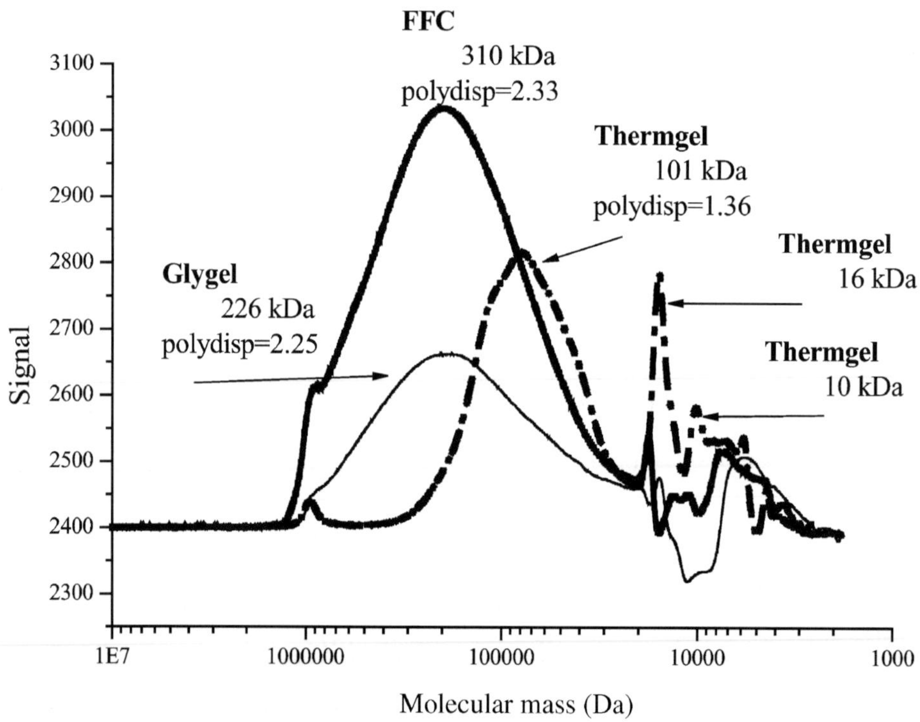

Figure 5. HPLC-GPC analysis of fibroin samples: FFC, Glygel, Thermgel.

In vitro *studies*

Biochemical and immunoenzymatic parameters of MG63 cells grown on the tested biomaterials and control were performed to evaluate cell proliferation and activity. Results of the WST-1 test, to assess cell proliferation, are summarized in Fig. 6. Data shown that the presence of gel influences cell proliferation, in fact cells grown on FFC and Thermgel showed proliferation values significantly lower when compared to control group (polystyrene of cell culture plates) and Glygel ($P < 0.0001$); Glygel values did not significantly differ from the control group.

Statistical evaluation of data was performed using SPSS/PC$^+$ Statistics™ 7.5 and data are reported as mean ± standard deviations (SD) at a significance level of $P < 0.05$. After having verified normal distribution and homogeneity of variance, a one-way ANOVA was done for comparison among groups. Finally, Scheffé's post-hoc multiple comparison tests were done to detect significant differences between groups.

LDH levels in the supernatant, whose concentration can be assumed as an index of cytotoxicity, for all materials did not significantly differ from control (Table 1). The production of Il-6, that plays a role as a mediator of inflammation, did not reveal significant differences from control either.

ALP, CICP and OC are products of osteoblast synthetic activity, their improvement indicating a positive cell response to the culture condition or stimulation applied.

Cells grown on FFC and Thermgel showed a significant enhancement of cell activity, demonstrated by the significantly higher level of ALP activity and TGFβ1 production, in comparison with control and Glygel groups. Moreover, CICP value was significantly higher in FFC group when compared to control and Glygel, while OC did not show differences among groups.

Fibroin gel appeared to have beneficial effects on cell activation, as indicated by ALP activity and collagen production increase, thus stimulating cell differentiation.

Table 1.

Biochemical and immuno-enzymatic evaluation of human MG63 osteoblast-like cells after 72 h of interaction with tested biomaterials and control ($n = 6$)

MG63	LDH (IU/1)	IL-6 (pg/ml)	ALP (IU/1)	CICP (ng/ml)	OC (ng/ml)	TGF-β1 (pg/ml)
Control	14.6 ± 2.2	2.3 ± 0.6	10.1 ± 1.6	12.6 ± 1.6	6.8 ± 0.3	432 ± 42
FFC	13.5 ± 1.7	3.1 ± 0.2	15.9 ± 1.3^a	17.9 ± 1.7^c	7.1 ± 0.4	603 ± 32^d
Thermgel	14.7 ± 2.0	3.0 ± 0.4	16.7 ± 1.8^b	16.4 ± 1.8	6.5 ± 0.9	487 ± 29^e
Glygel	18.0 ± 0.2	3.1 ± 0.2	8.8 ± 0.3	11.9 ± 1.3	6.3 ± 1.9	359 ± 34

Data were normalized according to the cell number. LDH, ns; IL-6, ns; OC, ns.

[a] FFC *vs.* control ($P < 0.01$) and *vs.* Glygel ($P < 0.005$).

[b] Thermgel *vs.* Glygel ($P < 0.005$).

[c] FFC *vs.* control ($P < 0.05$) and *vs.* Glygel ($P < 0.05$).

[d] FFC *vs.* control ($P < 0.005$), *vs.* Thermgel ($P < 0.01$) and *vs.* Glygel ($P < 0.0001$).

[e] Thermgel *vs.* Glygel ($P < 0.005$).

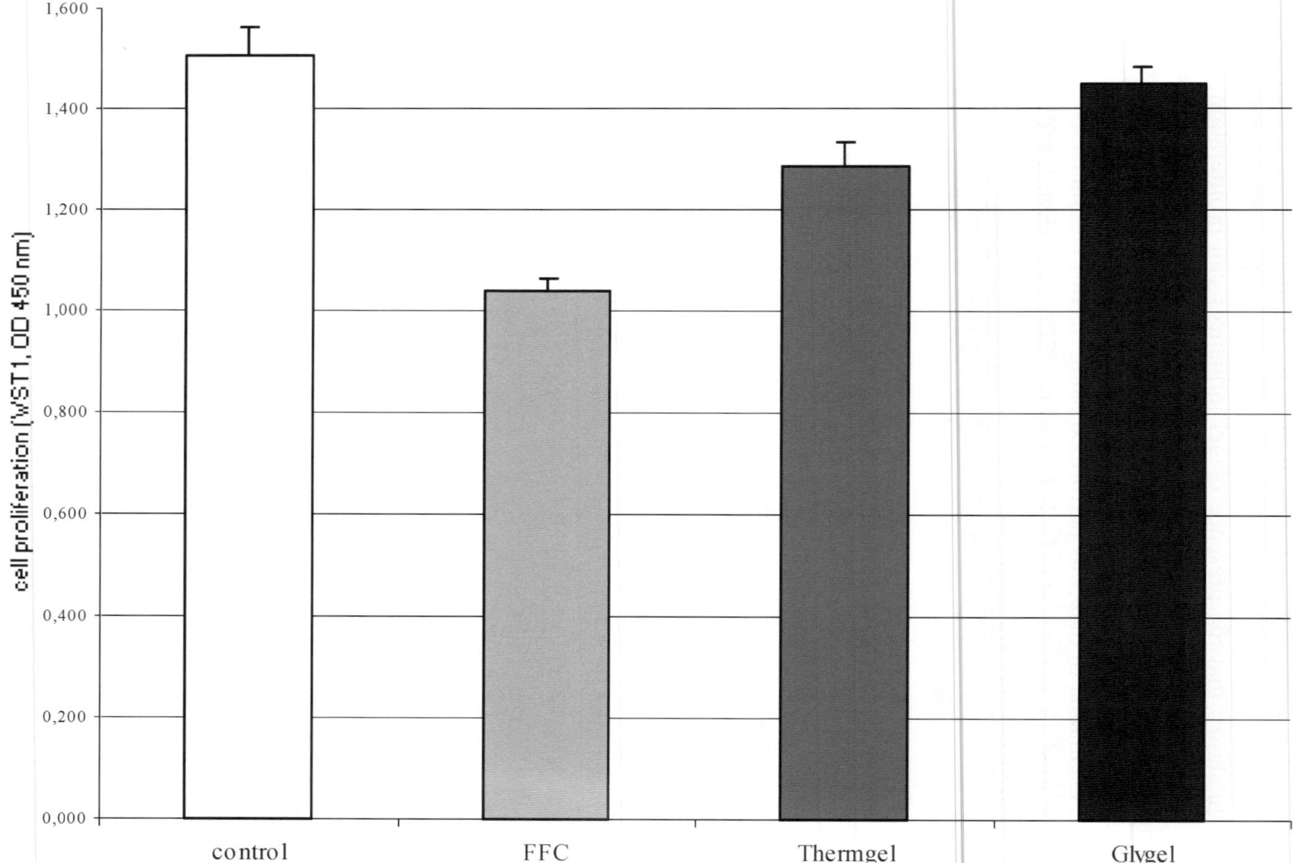

Figure 6. Human osteoblast-like cells (MG63) proliferation (WST1 test) after 72 h of interaction with the tested biomaterials and control ($n = 6$). FFC *vs.* Control, Glygel, Thermgel ($P < 0.0001$); Thermgel *vs.* Control, Glygel ($P < 0.0001$).

This finding was further supported by the increase in TGF-β1 production, that, as other cytokines and growth factors, is known to be an important mediator in the regulation of osteoblast differentiation.

CONCLUSIONS

Gels of fibroin have been prepared by treating a fibroin-water solution with glycerol (Glygel) or keeping the solution at 4°C (Thermgel). The two procedures produced gels with well distinct morphologies and fibroin molecular weights. In particular, while glygel was uniform and compact, Thermgel was porous with pores quite regular and homogeneously distributed. Molecular weight of fibroin in Glygel was higher than in Thermgel. Moreover, GPC analysis revealed for Thermgel the presence of low-molecular-mass peptides whose amino-acid composition is under investigation. *In vitro* cell culture indicated that both gels did not have any cytotoxic effect, as revealed by LDH level and IL-6 production.

While Glycel promoted osteoblast proliferation, significantly higher than the other gels and similar to the cell-culture PS plate control, Thermgel substrate mainly favored osteoblast activity and differentiation, as indicated by ALP activity, CICP production and TGF-β1 level. These favorable preliminary results have been also confirmed by *in vivo* tests that will be soon submitted for publication.

Acknowledgements

This work was carried out thank to the financial support of MIUR, PRIN2001. The authors thank Mrs. Claudia Gavazza for the HPLC-GPC tests and Mrs. Giovanna Salice for providing selected silkworm cocoons.

REFERENCES

1. S. Kurosaki, H. Otsuka, M. Kunitomo, M. Koyama, R. Pawankar and K. Matumoto, *J. Nippon Med. School* **66**, 41 (1999).
2. G. H. Altman, F. Diaz, C. Jakuba, T. Clabro, R. L. Horan, J. Chen, H. Lu, J. Richmond and D. L. Kaplan, *Biomaterials* **24**, 401 (2003).
3. H. K. Soong and K. R. Kenyon, *Ophtalmology* **91**, 479 (1984).
4. M. Santin, A. Motta, G. Freddi and M. Cannas, *J. Biomed. Mater. Res. Part A*. **46**, 382 (1999).
5. A. Motta, C. Migliaresi, A. W. Lloyd, S. P. Denyer and M. Santin, *J. Bioact. Compat. Polym.* **17**, 23 (2002).
6. M. Santin, S. P. Denyer, A.W. Lloyd and A. Motta, *J. Bioact. Compat. Polym.* **17**, 195 (2002).
7. X. Chen, W. Li, W. Zhong, Y. Lu and T. Yu, *J. Appl. Polym. Sci.* **65**, 2257 (1997).
8. N. Minoura, M. Tsukada and M. Nagura, *Polymer* **31**, 265 (1990).
9. N. Minoura, M. Tsukada and M. Nagura, *Biomaterials* **11**, 430 (1990).
10. M. Demura, T. Takekawa and T. Asakura, *Biomaterials* **13**, 276 (1992).
11. N. Minoura, S. Aiba, Y. Gotoh, M. Tsukada and Y. Imai, *J. Biomed. Mater. Res.* **29**, 1215 (1995).
12. K. Inouye, M. Kurokawa, S. Nishikawa and M. Tsukada, *J. Biochem. Biophys. Methods* **37**, 159 (1998).

13. K. Cai, K. Yao, Y. Cui, Z. Yg, X. Li, H. Xie, T. Qing and L. Gao, *Biomaterials* **23**, 1153 (2002).
14. K. Cai, K. Yao, Y. Cui, Z. Yg, X. Li, H. Xie, T. Qing and L. Gao, *Biomaterials* **23**, 1603 (2002).
15. S. Sofia, M. B. McCarthy, G. Gronowicz and D. L. Kaplan, *J. Biomed. Mater. Res.* **54**, 139 (2001).
16. T. Hanawa, A. Watanabe, T. Tsuchiya, R. Ikoma, M. Hidaka and M. Sugihara, *Chem. Pharm. Bull.* **43**, 284 (1995).
17. M. Tsukada, G. Freddi, N. Minoura and G. Allara, *J. Appl. Polym. Sci.* **54**, 507 (1994).
18. G. H. Altman, R. L. Horan, H. H. Lu, J. Moreau, I. Martin, J. C. Richmond and D. L. Kaplan, *Biomaterials* **23**, 4131 (2002).
19. A. S. Hoffman, *Adv. Drug Deliv. Rev.* **43**, 3 (2002).
20. T. C. Holmes, S. deLacalle, X. Su, G. Liu, A. Rich and S. Zhang, *Proc Natl. Acad. Sci. USA* **97**, 6728 (2000).
21. M. Kunioka and H. J. Choi, *Polym. Degrad. Stabil.* **59**, 33 (1998).
22. M. Yamamoto, Y. Tabata, L. Hong, S. Miyamoto, N. Hashimoto and Y. Ikada, *J. Control. Rel.* **64**, 133 (2000).
23. J. Nam and Y. H. Park, *J. Appl. Polym. Sci.* **81**, 3008 (2001).
24. T. Gamo, T. Inokuchi and H. Laufer, *Insect Biochem.* **7**, 285 (1977).

Gels, Genes, Grafts and Giants, pp. 639–654 (2005)

Synthesis of 3,4-dihydroxyphenylalanine (DOPA) containing monomers and their co-polymerization with PEG-diacrylate to form hydrogels

BRUCE P. LEE [1], KUI HUANG [1], F. NELSON NUNALEE [2], KENNETH R. SHULL [2] and PHILLIP B. MESSERSMITH [1,*]

[1] *Department of Biomedical Engineering, Northwestern University, 2145 N Sheridan Road, Evanston, IL 60208, USA*

[2] *Department of Materials Science and Engineering, Northwestern University, 2220 Campus Drive, Evanston, IL 60208, USA*

Received 1 October 2003; accepted 29 December 2003

Abstract—L-3,4-Dihydroxyphenylalanine (DOPA) is an unusual amino acid found in mussel adhesive proteins (MAPs) that is believed to lend adhesive characteristics to these proteins. Most previous efforts to incorporate DOPA into hydrogels have utilized oxidative cross-linking, which is hypothesized to reduce the adhesive properties of DOPA and requires reagents that are harmful to biological tissues. In this paper, we describe the synthesis of N-methacrylated DOPA monomers and their copolymerization with poly(ethylene glycol) diacrylate (PEG-DA) using either ultraviolet (UV) or visible light. The effect of DOPA containing monomers on gelation time, gel conversion and elastic modulus of the photocured hydrogels was investigated. Despite a retarding effect of DOPA on photopolymerization, DOPA was successfully incorporated into hydrogels with elastic moduli suitable for many biomedical applications. The incorporation of DOPA into hydrogels by photopolymerization may lead to new adhesive hydrogels for medical applications.

Key words: DOPA; poly(ethylene glycol); hydrogel; bioadhesive; photopolymerization; DMPA; contact mechanics.

INTRODUCTION

Bioadhesive hydrogels have many potential applications in medicine and dentistry, such as surgical adhesives and as vehicles for drug delivery to mucosal surfaces [1]. Clinically successful adhesive biomaterials should have the following primary attributes: (1) the ability to rapidly solidify or polymerize *in situ* from a fluid

*To whom correspondence should be addressed. Tel.: (1-847) 467-5273. Fax: (1-847) 491-4928. E-mail: philm@northwestern.edu

precursor; (2) the ability to form strong and durable bonds to tissue surfaces even in the biological environment; and (3) biocompatibility. However, the current generation of medical adhesives seldom meet all of these requirements due to lack of adhesion strength, possibility of viral contamination, poor adhesion in the wet environment and toxicity concerns [2]. Therefore, there exists a need for the development of new, safe and rapidly curable bioadhesives.

Successful water resistant adhesives have been found in marine organisms including mussels, barnacles, kelps and oysters, whose survival is dependent upon their ability to attach to wet surfaces [3]. Of particular interest are the mussel adhesive proteins (MAPs) secreted by marine mussel species such as *Mytilus edulis*. MAPs are remarkable underwater adhesive materials initially secreted as proteinaceous fluids that harden *in situ* to form an adhesive plaque that anchors marine and freshwater mussels to the substrates upon which they reside [3, 4]. One of the unique structural features of MAPs is the presence of L-3,4-dihydroxyphenylalanine (DOPA), an amino acid which is believed to be responsible for the adhesive characteristics of MAPs. Although the adhesive mechanisms of MAPs are not fully understood, oxidation of DOPA is believed to lead to intermolecular cross-link formation that results in the curing of the plaques [5].

In an effort to take advantage of the adhesive properties of DOPA, several groups have reported the synthesis and characterization of DOPA-containing peptides and polymers [4–13]. However, the curing of these DOPA-containing compounds was typically accomplished by oxidation of DOPA residues to form DOPA-quinone, which participated in intra- and/or intermolecular cross-linking reactions to form a gel network. The oxidized forms of DOPA that result from such reactions are believed to be less adhesive than unoxidized DOPA [5]. DOPA-containing proteins and polymers exhibit better adhesion to both metallic and mucosal surfaces when DOPA residues are not oxidized [11, 14]. Furthermore, complete characterization of the DOPA adhesion mechanism, as well as future *in vivo* applications, are complicated by the presence of oxidation reagents necessary for curing. Thus, we have been searching for alternative methods of hydrogel formation that can preserve the unoxidized adhesive form of DOPA [15].

The goal of this work was to incorporate DOPA into a gel network by photopolymerization. Synthesis of *N*-methacrylated DOPA monomers is described and these monomers were copolymerized with poly(ethylene glycol) diacrylate (PEG-DA) to form hydrogels by photopolymerization. Gel conversion and DOPA incorporation as a function of DOPA monomer concentration were determined for gels polymerized by ultraviolet and visible light initiating systems. Finally, a contact mechanical test was performed on the photocured gels and the elastic modulus of these gels was obtained by fitting the load-displacement data using a Hertzian relationship [16].

MATERIALS AND METHODS

Materials

PEG ($\overline{M}_w$ 8000), pentafluorophenol, 1,3-dicyclohexylcarbodiimide (DCC), 4,7,10-trioxa-1,13-tridecanediamine, fluorescein sodium salt (FNa_2) and ascorbic acid (AA) were purchased from Sigma (St. Louis, MO, USA). L-DOPA, sodium borate, methacryloyl chloride, *t*-butyldimethylsilyl chloride (TBDMS-Cl), di-*t*-butyl dicarbonate, methacrylic anhydride, sodium molybdate dihydrate, sodium nitrite, 2,2′-dimethoxy-2-phenyl-acetonephenone (DMPA), acryloyl chloride, 1,8-diazabicyclo[5.4.0]undec-7-ene (DBU), tetrabutylammonium fluoride (TBAF), 4-(dimethylamino)-benzoic acid (DMAB) and 1-vinyl-2-pyrrolidone (VP) were purchased from Aldrich (Milwaukee, WI, USA). Camphorquinone (CQ) was obtained from Polysciences (Warrington, PA, USA). Titanium oxide (TiO_2) surfaces were prepared by electron beam physical evaporation onto silicon (Si) wafer and cleaned in plasma chamber prior to testing. Unless otherwise noted, all chemical reagents were used as received.

Synthesis of N-methacryloyl 3,4-dihydroxyl-L-phenylalanine (**1**)

1.15 g (5.69 mmol) of $Na_2B_4O_7$ was dissolved in 30 ml of water. The solution was degassed with Ar for 30 min, after which 0.592 g (3.0 mmol) of L-DOPA was added and stirred for 15 min. 0.317 g (3.0 mmol) of Na_2CO_3 was then added, the solution was cooled to 0°C and 0.3 ml (3.0 mmol) of methacryloyl chloride was slowly added with stirring. The pH of the solution was maintained above 9 with Na_2CO_3 during the reaction. After stirring for 1 h at room temperature, the solution was acidified to pH 2 with concentrated HCl. The mixture was extracted with ethyl acetate three times. After washing with 0.1 M HCl and drying over anhydrous $MgSO_4$, the solvent was removed in vacuo to yield crude light brown solid. The product was further purified by elution from a silica gel column with dichloromethane (DCM) and methanol (95 : 5). After evaporating the solvent, a white, sticky solid was obtained with a product yield of 35%. ^{1}H-NMR (500 MHz, acetone-d_6): δ 7.1 *d* (1 H, $-NH-$); 6.6–6.8 (3 H, $C_6H_3(OH)_2-$); 5.68 *s* (1 H, C*H*H=); 5.632 s (unknown peak); 5.33 *s* (1 H, CH*H*=); 4.67 *m* (1 H, $-CH-$); 2.93–3.1 *m* (2 H, CH_2-); 1.877 *s* (3 H, $-CH_3$).

Synthesis of 3,4-bis(t-butyldimethylsilyloxy)-L-phenylalanine (**2**)

Compound **2** was prepared by modification of a literature procedure [17]. 3.60 g (24.0 mmol) of TBDMS-Cl was dissolved in 18 ml of anhydrous acetonitrile. 1.60 g (8.0 mmol) of L-DOPA was added to the solution, the suspension was stirred and cooled to 0°C and 3.6 ml of DBU (24.0 mmol) was added. The reaction mixture was then stirred for 24 h at room temperature. Addition of cold acetonitrile to the reaction solution resulted in a colorless precipitate. The precipitate was filtered and washed with cold acetonitrile several times followed by drying in vacuum. White powder was obtained with a yield of 78%. ^{1}H-NMR (500 MHz, methanol-*d*): δ 6.7–6.9 (3 H, $C_6H_3(O{-}Si{-})_2{-}$); 3.72 (m, 1 H, $-CH-$); 2.82–3.2 (m, 2 H, $-CH_2-$); 1.0 (d, 18 H, $-C(CH_3)$); 0.2 (d, 12 H, $Si{-}CH_3$).

*Synthesis of 3,4-bis(t-butyldimethylsilyloxy)-N-t-butyloxycarbonyl-*L-*phenylalanine* (**3**)

Compound **3** was prepared by modification of a literature procedure [17]. 1.60 g (3.77 mmol) of **2** was added to 10 ml of water containing 0.34 g (4.05 mmol) of $NaHCO_3$. 0.96 g (4.30 mmol) of di-*t*-butyl dicarbonate in 10 ml of tetrahydrofuran was added and the reaction mixture was stirred for 24 h at room temperature. After evaporation of tetrahydrofuran, 10 ml of water was added to the residue. The solution was acidified with dilute HCl to pH 5 and extracted three times with ethyl acetate. After drying over anhydrous $MgSO_4$, the solvent was removed *in vacuo*. The crude product was purified by column chromatography (silica gel; eluent, 10% methanol in DCM). A white solid was obtained with a yield of 70% after evaporating the eluding solvent. ^{1}H-NMR (500 MHz, methanol-*d*): δ 6.68–6.81(3 H, $C_6H_3(O{-}Si{-})_2{-}$); 4.28 (m, 1 H, $-CH-$); 2.78–3.08 (m, 2 H, $-CH_2-$); 1.4 (s, 9 H, $-O{-}C(CH_3)_3$); 1.0 (d, 18 H, $-Si{-}C(CH_3)_3$); 0.2 (d, 12 H, $Si{-}(CH_3)_2$).

*Synthesis of 3,4-bis(t-butyldimethylsilyloxy)-N-t-butyloxycarbonyl-*L-*phenylalanine pentafluorophenyl ester* (**4**)

1 g (1.90 mmol) of **3** and 0.351 g (1.90 mmol) of pentafluorophenol were dissolved in a solvent mixture of 24 ml of dioxane and 1 ml of DMF and 0.432 g (2.10 mmol) of DCC was added at 0°C. The solution was stirred for 1 h at 0°C and for 1 h at room temperature, after which the solution was filtered to remove dicyclohexylurea and evaporated *in vacuo*. The product **4** was purified by column chromatography (silica gel; eluent, hexane/ethyl acetate 11 : 2). After removing the eludent, a pure white, sticky solid was obtained with a yield of 66%. ^{1}H$-$NMR (500 MHz, $CDCl_3$): δ 6.65–6.81 (3 H, $C_6H_3(O{-}Si{-})_2{-}$); 4.85 (m, 1 H, $-CH-$); 3.05–3.2 (m, 2 H, $-CH_2-$); 1.41 (s, 9 H, $-O{-}C(CH_3)_3$); 1.0 (d, 18 H, $-Si{-}C(CH_3)_3$); 0.2 (d, 12 H, $-Si{-}(CH_3)_2$).

*Synthesis of N-(13′-amino-4′,7′,10′-trioxatridecanyl)-t-butyloxycarbonyl-3′,4′-bis(t-butyldimethylsilyloxy)-*L-*phenylalaninamide* (**5**)

0.869 g (1.26 mmol) of **4** in 10 ml of DCM was added drop wise to a mixture of 2.07 ml (9.44 mmol) of 4,7,10-trioxa-1,13-tridecanediamine and 1.32 ml (9.44 mmol) of Et_3N in 1 ml DMF over 30 min at 0°C. The solution was stirred at room temperature for another 2 h and then the solvent was removed under vacuum. The crude product was loaded onto silica gel and eluted with DCM, 5% methanol in DCM, 10% methanol in DCM and 15% methanol in DCM. The solvent was removed under vacuum to yield **5** as a white solid. The yield was 63%.

^{1}H-NMR (500 MHz, acetone-d_6): δ 7.38 (m, 1 H, $-$CON$H-$); 6.60–6.80 (3 H, C_6H_3(O$-$Si$-)_2-$); 5.26 (m, 1 H, $-$CON$H-$); 4.30 (m, 1 H, $-CH-$); 3.4–3.8 (m, 12 H, $-CH_2$O$-$); 3.03–3.4 (m, 4 H, $-CH_2-$NH$-$, $-CH_2-NH_2$); 2.78–3.02 (m, 2 H, $-CH_2-$); 2.0 (m, 2 H, $-CH_2-$); 1.7 (m, 2 H, $-CH_2-$); 1.39 (s, 9 H, $-$O$-$C($CH_3)_3$); 1.0 (d, 18 H, $-$Si$-$C($CH_3)_3$); 0.2 (d, 12 H, $-$Si$-(CH_3)_2$).

*Synthesis of N-(13-(N′-t-butyloxycarbonyl-*L*-3′,4′-bis(t-butyldimethylsilyloxy)-4,7-10-trio-xatridecanyl)-methacrylamide* (**6**)

0.572 g (0.79 mmol) of **5** and 0.166 ml (1.18 mmol) of Et_3N were dissolved in 5 ml of anhydrous chloroform, to which 0.176 ml (1.18 mmol) of methacrylic anhydride was added. The solution was stirred at room temperature for 3 h, then solvent was removed *in vacuo*. Pure **6** was obtained by column chromatography (silica gel; eluent, ethyl acetate) as a white, sticky solid with a yield of 61%. ^{1}H-NMR (500 MHz, $CDCl_3$): δ 6.60–6.80 (3 H, C_6H_3(O$-$Si$-)_2-$); 6.40 (m, 1 H, $-$CON$H-$); 5.71 s (1 H, CHH=); 5.30 *s* (1 H, CHH=); 5.096 (m, 1 H, $-$CON$H-$); 4.21 (m, 1 H, $-CH-$); 3.2–3.65 (m, 16 H, $-CH_2$O, $-CH_2-$NH$-$, $-CH_2-NH_2$); 2.80–2.99 (m, 2 H, $-CH_2-$); 1.96 s (3 H, $-CH_3$); 1.81 (m, 2 H, $-CH_2-$); 1.68 (m, 2 H, $-CH_2-$); 1.40 (s, 9 H, $-$O$-$C($CH_3)_3$); 1.0 (d, 18 H, $-$Si$-$C($CH_3)_3$); 0.2 (d, 12 H, $-$Si$-(CH_3)_2$).

*Synthesis of N-(13-(N′-t-Boc-*L*-3′,4′-dihydroxylphenylalaninamido)-4,7-10-trioxatridecanyl)-methacrylamide* (**7**)

To a 10 ml round-bottom flask were added 0.344 g (0.433 mmol) of **6**, 3 ml of THF and 0.137 g (0.433 mmol) of TBAF. The solution was stirred at room temperature for 5 min, then 3 ml of 0.1 M HCl was added. The solution was extracted three times with DCM, after which the solvent was evaporated in vacuum. **7** was obtained as a white solid by column chromatography (silica gel; eluent, 7% methanol in DCM) with a yield of 63%. ^{1}H-NMR (500 MHz, acetone-d_6): δ 7.90 (m, 1 H, $-$CON$H-$); 7.23–7.40 (d, 2 H, $C_6HH_2(OH)_2-$); 6.56–6.76 (3 H, $C_6HH_2(OH)_2-$); 5.930 (m, 1 H, $-$CON$H-$); 5.71 s (1 H, CHH=); 5.30 s (1 H, CHH=); 4.20 (m,1 H, $-CH-$); 3.1–3.60 (m, 16 H, $-CH_2$O, $-CH_2-$NH$-$, $-CH_2-NH_2$); 2.70–2.95 (m, 2 H, $-CH_2-$); 1.93 s (3 H, $-CH_3$); 1.78 (m, 2 H, $-CH_2-$); 1.65 (m, 2 H, $-CH_2-$); 1.39 (s, 9 H, $-$O$-$C($CH_3)_3$).

Synthesis of PEG-diacrylate (PEG-DA)

Acrylation of PEG was performed using a similar method as described by Hern and Hubbell [18]. 40 g (5 mmol) of PEG was dried by azeotropic evaporation in benzene and then dissolved in 150 ml of DCM. 4.18 ml (30 mmol) of Et_3N and 3.6 ml (40 mmol) of acryloyl chloride (Aldrich) was added to the polymer solution. The mixture was refluxed with stirring for 5 h and allowed to cool at room temperature overnight. Ether was added to the mixture to form a faint yellow precipitate. The crude product was then dissolved in saturated NaCl solution, which was heated to 60°C to form two layers. DCM was added to the top layer and $MgSO_4$ was added to remove moisture. After filtration of $MgSO_4$, the volume of the solvent was reduced in vacuum and the sample was precipitated in ether. The final product was dried in vacuum and stored at −15°C. The yield was 75%. ^{1}H-NMR (500 MHz, D_2O): δ 6.47 (d, 1 H, C*H*H=C−); 6.23 (m, 1 H, C=C*H*−C(=O)−O−); 6.02 (d, 1 H, CH*H*=C−); 4.35 (m, 2 H, −CH_2−O−C(=O)−C=C); 3.23–3.86 (PEG CH_2).

Photopolymerization

Precursor solutions of PEG-DA, **1**, **7** and photoinitiator were prepared and mixed immediately before photopolymerization. Stock solutions of PEG-DA (200 mg/ml) and **1** (40 mg/ml) were dissolved in N_2-purged phosphate buffered saline (PBS, pH 7.4), whereas **7** (60 mg/ml) was dissolved in 50 : 50 PBS/95% ethanol previously purged with N_2. To prepare the final polymerization mixture, solutions of **1** or **7** were combined with PEG-DA to achieve a final concentration of PEG-DA and DOPA derivatives of 150 mg/ml. 100 μl of the mixture was then added to a disc-shaped mold (100 μl, diameter 9 mm, depth 2.3 mm, Secure Seal® SA8R-2.0, Grace Bio-Lab, OR, USA) and irradiated for up to 20 min either with an UV lamp (Black Ray® Lamp, 365 nm, model UVL-56, UVP, Upland, CA, USA) or a blue light lamp (VIP®, 400–500 nm, Bisco, Schaumburg, IL, USA). For UV-initiated photocuring, DMPA (600 mg/ml in VP) was added to the polymeric solution to make a final concentration of 34 mM. Visible light-induced curing was performed using either CQ (100 mg/ml in VP, final concentration 150 mM) with DMAB (30 mg/ml in VP, final concentration 151 mM), or FNa_2 (188 mg/ml in PBS, final concentration 2 mM) with AA (100 mg/ml in PBS, final concentration 17 mM) as the photoinitiator. The final VP concentration was adjusted to be between 135 and 300 mM.

After irradiation, the gels were blotted with filter paper to remove the liquid surface layer and weighed. Percent gel conversion was then determined by dividing the weight of the gel by the weight of 100 μl of the precursor solution.

Determining DOPA incorporation

The amount of DOPA incorporated into the photopolymerized gel was determined using a modification of the colorimetric DOPA assay developed by Waite and Benetict [19]. Photo-cross-linked gels were stirred in 3 ml of 0.5 M HCl to extract DOPA monomers that were not incorporated into the gel network. 0.9 ml of the nitrite reagent (1.45 M sodium nitrite and 0.41 M sodium molybdate dihydrate) and 1.2 ml of 1 M NaOH were added to 0.9 ml of the extraction solution and the absorbance (500 nm) of the mixture was recorded using a Hitachi U-2010 UV-Vis spectrophotometer within 2 to 4 min of NaOH addition. Standard curves were constructed using known **1** and **7** concentrations.

Mechanical test

Hydrogels were formed in the shape of a hemisphere by loading 25 μl of the polymer mixture onto a glass slide treated with 1 H, 1 H, 2 H, 2 H-perfluorooctyltrichlorosilane. Gels were irradiated for 10 min, dialyzed in 0.15 M HCl for at least 24 h to extract unincorporated DOPA monomers and then equilibrated in PBS for greater than 15 min prior to testing. Figure 1 shows a detailed schematic diagram of the testing apparatus employed to characterize the mechanical properties of these gels. To determine the gel modulus, hemispherical gel caps were attached to one end of a steel cylinder (diameter 6 mm, length 30 mm) using superglue. The other end of the cylinder was attached to a piezoelectric stepping motor (IW-701-00, Burleigh Instruments, Fishers, NY, USA) aligned in series with a 50 g load transducer (FTD-G-50, Schaevitz Sensors, Hampton, VA, USA) with a resolution of approx. 0.1 mN. A fiber optic displacement sensor (RC100-GM2OV, Philtec, Annapolis, MD, USA) measured the axial movement of the steel rod. A TiO_2-coated Si wafer was positioned below the hydrogel and the TiO_2 surface was flooded with PBS in order to maintain the hydration of the gel. The indenter was advanced at 5 μm/s until a maximum compressive load of 4 mN was measured.

Elastic moduli were calculated by assuming Hertzian mechanics [16] for the specific case of non-adhesive contact between an incompressible elastic hemisphere and a rigid plane, in which case the Hertzian relationship between load (P_h) and

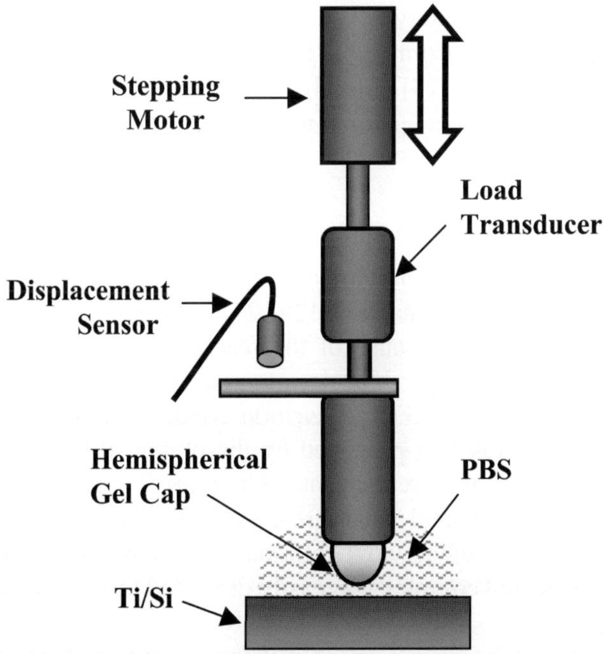

Figure 1. Testing apparatus for characterizing the mechanical properties of polymer hydrogels.

displacement (δ_h) becomes [20]:

$$P_h = \frac{16R^{1/2}E}{9}\delta_h^{3/2}, \quad (1)$$

where R and E are the radius of curvature and the elastic modulus of the hemispherical gel, respectively. The radius of curvature of the gels was determined from height and width measurements obtained from a photograph of the gel.

Statistical analysis

Statistical analysis of percent gel conversion, extent of DOPA incorporation and gel modulus was performed using analysis of variance (ANOVA) and Tukey post-hoc analysis using a significance level of $P = 0.05$.

RESULTS AND DISCUSSION

Novel photocurable DOPA-containing monomers were synthesized and copolymerized with poly(ethylene glycol) diacrylate (PEG-DA) to form DOPA-containing hydrogels. The new monomers **1** and **7** combine an adhesive moiety, DOPA and a polymerizable methacrylate group with or without an oligomeric ethylene oxide linker. To our knowledge, this is the first report of DOPA-containing hydrogels formed by photopolymerization. PEG was chosen as the structural gel-forming polymer because of its favorable properties of biocompatibility, non-toxicity and non-immunogenicity [21, 22]. Furthermore, photocuring of PEG-DA macromonomers to form biocompatible hydrogels have been demonstrated by numerous investigators [18, 23, 24]. Copolymerization of **1** and **7** with PEG-DA was hypothesized to be a convenient method to enhance the adhesive properties of PEG hydrogels, which are generally considered to be poorly adhesive.

Synthesis of polymerizable DOPA monomers **1** *and* **7**

Preparation of N-methacryloyl amino acids using methacryloyl chloride has been previously reported by Nishiyama *et al.* [25–27]. However, existing methods were not employed in this study because of the reactivity of the catechol side chain of DOPA toward methacryloyl chloride, as well as the known susceptibility of catechols to oxidation under alkaline reaction conditions. Instead, for synthesis of monomer **1** DOPA was first protected by dissolving in borate buffer at pH 9, forming a DOPA-borate complex stable at pH between 7 and 10 [19, 28] that would function to protect the catechol during reaction. As shown in Scheme 1, **1** was then prepared by adding methacryloyl chloride to the borate-protected DOPA, after which the borate-catechol complex was easily disrupted through acidification of the solution [19, 28].

Monomer **7** was synthesized with an oligomeric oxyethylene spacer between DOPA and methacrylate moieties, to enhance aqueous solubility of the monomer, as

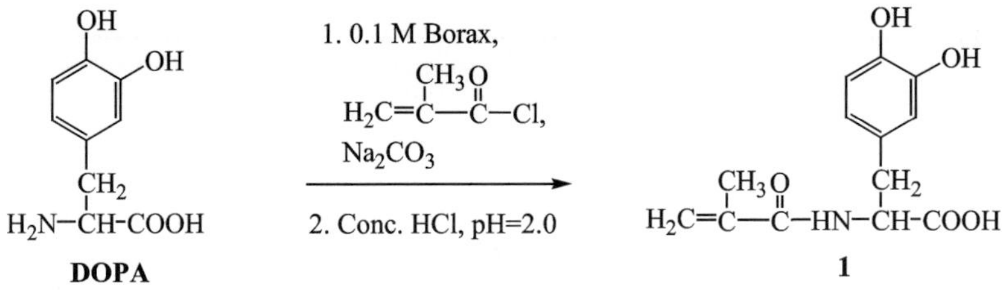

Scheme 1. Synthesis of **1**.

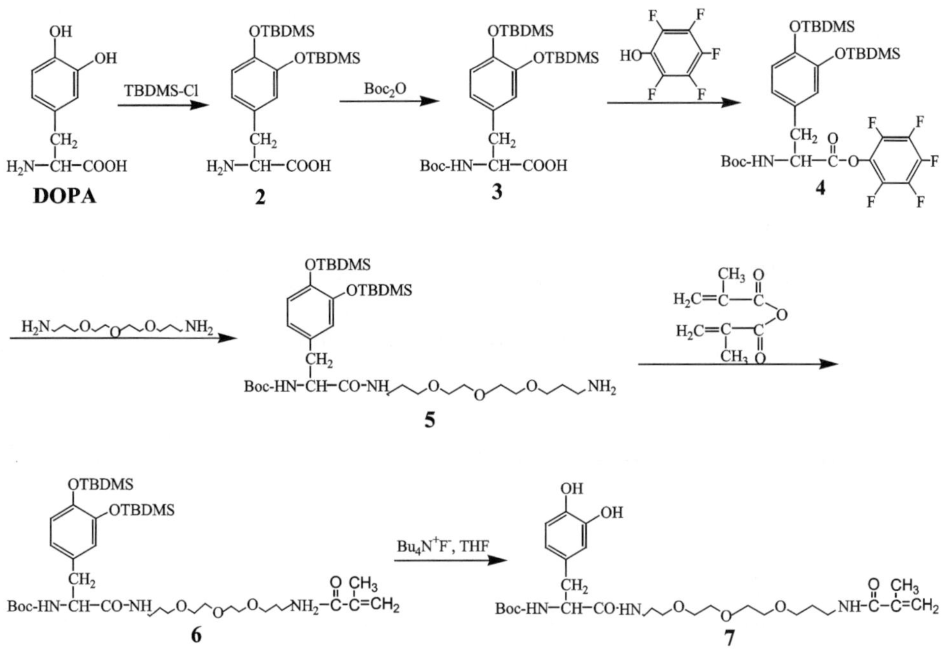

Scheme 2. Synthesis of **7**.

well as to facilitate the interaction of the pendant catechol with substrate surfaces. As shown in Scheme 2, in this case DOPA was protected with TBDMS and Boc groups, followed by activation of the DOPA carboxyl group with pentafluorophenol. The subsequent DOPA derivative was reacted with an excess of oxyethylene oligomer, 4,7,10-trioxa-1,13-tridecanediamine, after which the free amine terminus was reacted with methacryloyl chloride. TBAF, a mild deprotecting reagent [29], was then used to remove TBDMS.

Photopolymerization

The initiating systems used in this study (DMPA, CQ/DMAB, AA/FNa_2) have all been previously investigated for radical photopolymerization. Several investigators have used DMPA in the formulation of hydrogels from acrylated PEG macromers [18, 23, 24]. CQ in conjunction with an amine has been used exten-

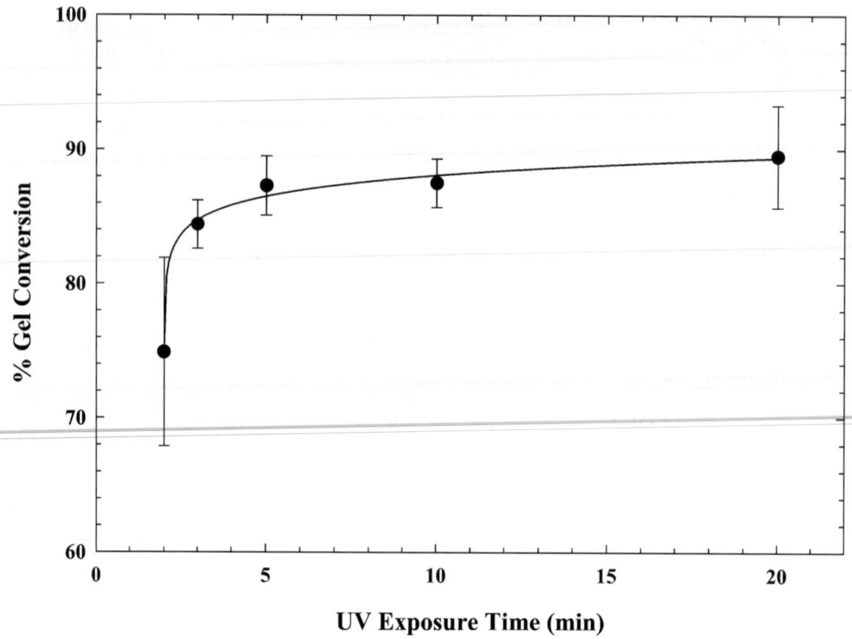

Figure 2. Percent gel conversion of PEG-DA (150 mg/ml) as a function of length of UV exposure.

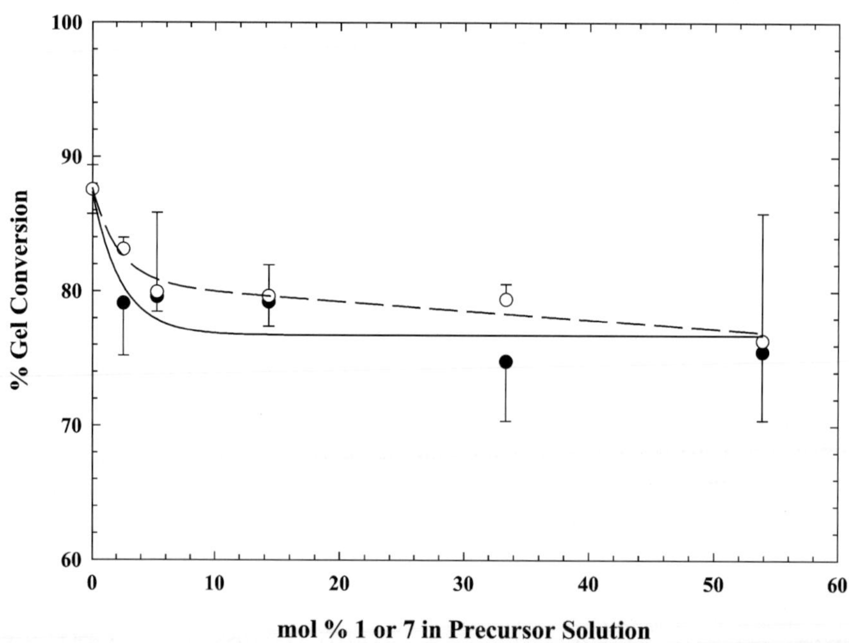

Figure 3. Percent gel conversion of DOPA-containing gels after 10 min UV exposure as function of mol% **1** or **7** in the precursor solution. The filled circles (●) and solid regression lines denote gels containing **1**, while open circles (○) and dashed regression lines denote gels containing **7**.

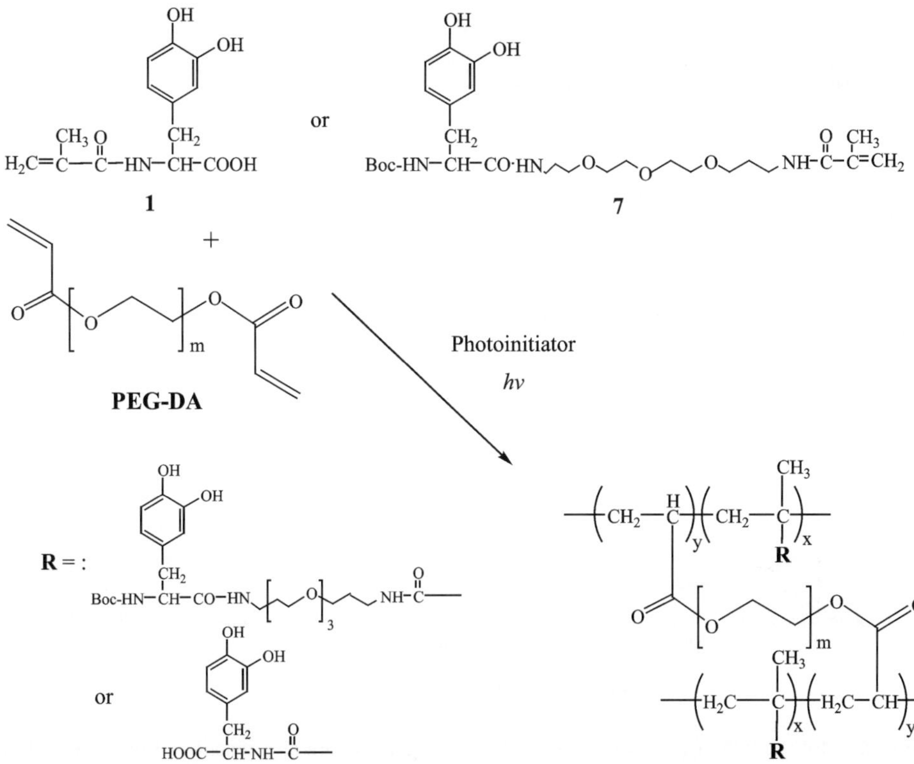

Scheme 3. Hydrogel formation via copolymerization of **1** or **7** with PEG-DA.

sively in the curing of dental resins [30, 31] and the usage of FNa_2 with a reducing agent for visible light photoinitiation have also been studied for biomedical purposes [32, 33].

After 2 min of UV irradiation, PEG-DA solutions containing DMPA as the initiator formed clear, colorless gels covered by a liquid surface layer, which was believed to be due to the inhibiting effect of molecular oxygen [34]. As shown in Fig. 2, gel conversion reached more than 75 wt% after 2 min of UV irradiation and increased to greater than 85 wt% upon irradiation for more than 5 min. Gelation of PEG-DA also occurred in 4 min or less when visible light initiators were used (4 min for CQ/DMAB and 3 min for FNa_2/AA).

Copolymerization of PEG-DA with **1** or **7** is illustrated in Scheme 3. Polymerization of these mixtures was qualitatively similar to pure PEG-DA, although addition of **1** or **7** to the PEG-DA precursor solution resulted in a decrease in gel conversion that was dependent on DOPA monomer concentration and initiating system. For example, in DMPA-initiated UV polymerization (Fig. 3), gel conversion was reduced to less than 85 wt% in the presence of 2.5 mol% or more **1** and **7**. However, the extent of gel conversion was not statistically different between **1** and **7** gels. Similar DOPA concentration-dependent inhibition was observed for the visible-light-induced initiators. For FNa_2/AA- and CQ/DMAB-initiated mixtures, addition of

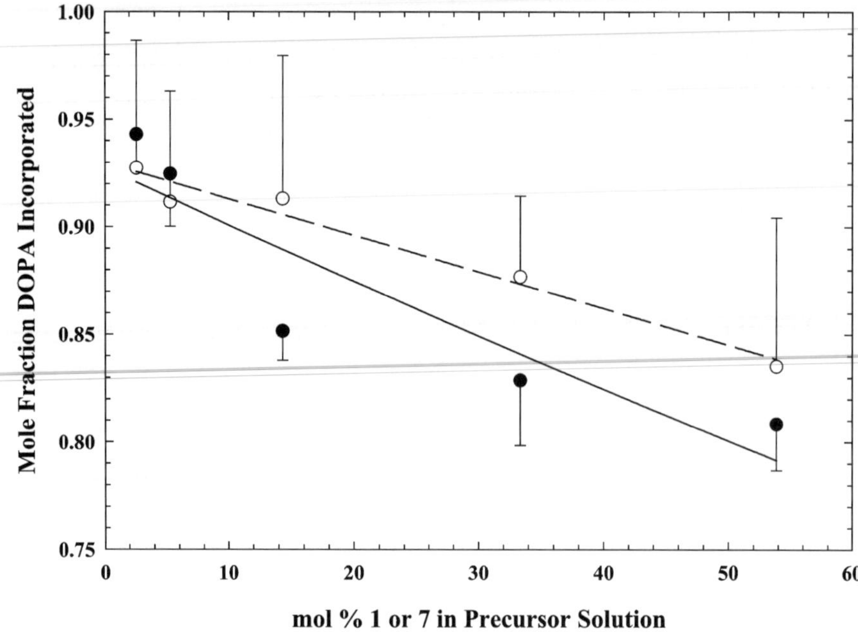

Figure 4. Mol fraction of DOPA incorporated into the gel network as a function of mol% **1** or **7** in the precursor solution. Symbols as in Fig. 2.

33.3 mol% of **1** increased the gelation time to more than 8 min. Nevertheless, solutions containing **1** and **7** were still capable of photocuring even at a relatively high mol% of DOPA.

The composition and elastic modulus of UV polymerized gels (DMPA initiator) were further analyzed in detail. After photocuring, DOPA-containing gels were dialyzed in 0.5 M HCl to extract unreacted DOPA monomer. To quantify the extent of DOPA incorporation, the dialysate was assayed using the DOPA colorimetric assay of Waite and Benedict [19] and the results used to calculate the amount of DOPA incorporated into the gel network. Figure 4 shows the mole fraction of DOPA incorporated into the gel network, as a function of mol% monomer **1** and **7** in the precursor solution. There was no significant difference in the mole fraction DOPA incorporated between samples containing **1** and **7**.

Direct evidence for the presence of DOPA in the gels was obtained by immersing the intact dialyzed hydrogels in nitrite reagent followed by NaOH [19]. The initially colorless gels turned bright yellow after the addition of the nitrite reagent and then red following the addition of excess base (Fig. 5). This color transition is typical of catechols [19], indicating that the unoxidized form of DOPA was incorporated into the hydrogels through photopolymerization. The intensity of the red color also reflected the concentration of DOPA incorporated into the photocured gels, as seen

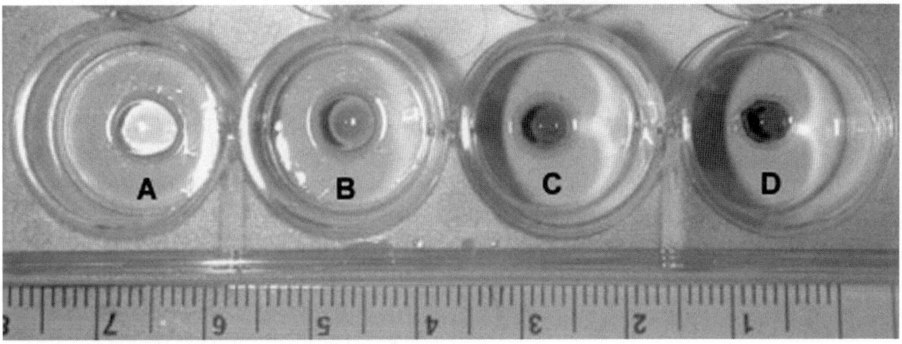

Figure 5. Photographs of UV-photocured PEG gels containing (A) 0, (B) 2.6, (C) 14.3 and (D) 33.3 mol% of **1** after dialysis with 0.5 M HCl and treatment with DOPA assay reagents.

in Fig. 5. PEG-DA gels that did not contain DOPA (Fig. 5A) remained colorless with the addition of nitrite reagent and NaOH solution.

Mechanical properties of DOPA-containing hydrogels

Load *versus* displacement data, obtained by indentation of hydrogel hemispheres onto a rigid flat surface, were recorded for selected gels to determine the elastic modulus. All of the loading data obtained from these experiments were fitted well ($R > 0.99$) by Hertzian curves (equation (1)). Equation (1) is valid when the average strain in the deformed region of the gel is small. For our geometry, the average strain can be approximated by $(\delta/R)^{1/2}$, which is indeed very small in our experiment. An example of a typical Hertzian fit is shown in Fig. 6 for a PEG-DA hydrogel. The non-linear relationship between P_{h} and δ_{h} is not an indication of a non-linear constitutive model for the material, but is instead due to the changing geometry as δ_{h} is increased. The load increases non-linearly with displacement because the contact area between the gel and the substrate increases with displacement. The curve-fitting procedure allowed elastic moduli to be calculated based on the proportionality factor of the curve fit.

Table 1 shows the average DOPA content in each gel type along with the mean values of E calculated by the methods described above. Addition of 33 mol% of monomers **1** and **7** into the precursor solutions lowered the average Young's modulus values of the photocured gel by 34% and 29%, respectively. The average Young's modulus values obtained for gels containing **1** and **7** were not statistically different. Despite the decrease in modulus compared to PEG-DA gels, DOPA-containing gels still exhibited moduli suitable for many biomedical applications.

While the presence of DOPA-containing monomers did not prevent polymerization, longer irradiation time was needed to achieve gelation when monomers **1** and **7** were present in the precursor solution. Gel conversion, extent of DOPA incorporation and elastic modulus also decreased with increasing DOPA concentration. These observations suggest that the DOPA-containing monomers exert a retarding effect on free radical polymerization. Although no literature references

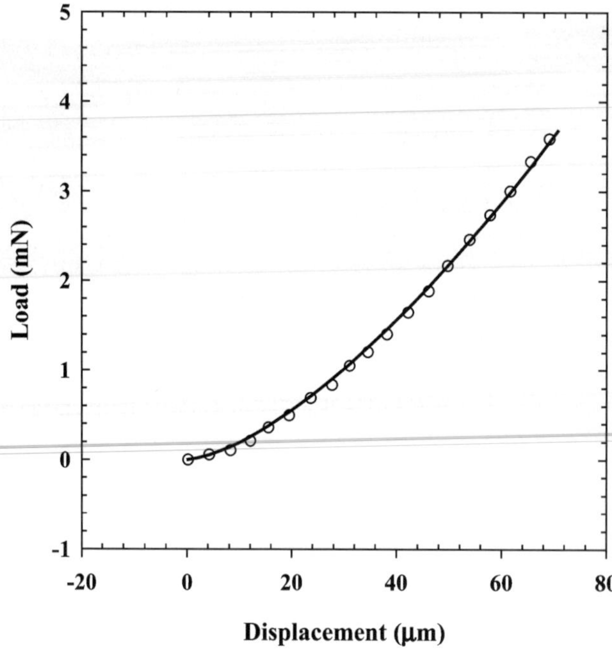

Figure 6. Load *vs.* displacement for a control gel pressed onto a PBS-wetted TiO_2 surface. The load data (○) is fitted with a Hertzian curve (equation (1), $R = 0.999$) as denoted by the solid line.

Table 1.
Young's moduli of DOPA-containing gels

	DOPA content[b] (µmol/g gel)		Young's modulus (kPa)	
	Average	S.D.	Average	S.D.
PEG-DA	—	—	72.3	7.69
PEG-DA + **1**[a]	15.0	0.25	47.4[c]	8.63
PEG-DA + **7**[a]	14.5	0.056	51.4[c]	9.91

[a] 33 mol% **1** or **7** in the precursor solution.
[b] Determined from DOPA assay.
[c] $P < 0.001$ relative to PEG-DA gels.

to inhibitory or retarding effects of catecholic compounds were found, related phenolic compounds such as hydroquinone and 4-methoxyphenol are known inhibitors of radical polymerization, acting to quench excited initiators and retard the initiation process [35–37]. In the presence of oxygen, these inhibitors can also interact with peroxyl radicals and terminate the propagating kinetic chain [38, 39]. Furthermore, melanin, a DOPA-containing polyphenol, has been found to act as a radical scavenger [40]. The presence of a liquid-like layer at the surface of the gels is consistent with an enhanced inhibition due to atmospheric oxygen and with the fact that the gels did not measurably adhere to the oxide substrates used for the mechanical measurements. Elucidation of the exact inhibition/retardation mechanism and its influence on the adhesive response of the gels will require further study.

In summary, *N*-methacrylated DOPA monomers were synthesized and copolymerized with PEG diacrylate to form hydrogels. Despite a retarding inhibitory effect of DOPA on photopolymerization, DOPA-containing monomers were successfully incorporated into PEG hydrogels. To our knowledge, this is the first report of a DOPA-containing hydrogel formed by photopolymerization. The resulting hydrogels do not require oxidizing reagents to gel and possess moduli sufficient for use in many biomedical applications. The incorporation of DOPA into hydrogels in the unoxidized state may prove to be an important tool in understanding the role of DOPA in mussel adhesive proteins and may lead to new adhesive hydrogels for biomedical applications. Experiments aimed at determining the adhesive properties of these hydrogels are underway.

Acknowledgements

This work was supported by NIH R01 DE14193 and by Bisco, Inc.

REFERENCES

1. E. Mathiowitz, D. E. Chickering and C.-M. Lehr (Eds), *Bioadhesive Drug Delivery Systems: Fundamentals, Novel Approaches and Development*. Marcel Dekker, New York, NY (1999).
2. D. Sierra and R. Saltz (Eds), *Surgical Adhesives and Sealants: Current Technology and Applications*. Technomic, Lancaster, PA (1996).
3. J. H. Waite, *Int. J. Adhesion Adhesives* **7**, 9 (1987).
4. H. Yamamoto, *Biotechnol. Genet. Eng. Rev.* **13**, 133 (1996).
5. M. Yu, J. Hwang and T. J. Deming, *J. Am. Chem. Soc.* **121**, 5825 (1999).
6. H. Yamamoto, *J. Adhesion Sci. Technol.* **1**, 177 (1987).
7. H. Yamamoto, *J. Chem. Soc., Perkin Trans.* **1**, 613 (1987).
8. H. Yamamoto, S. Yamauchi and S. Ohara, *Biomimetics* **1**, 219 (1992).
9. M. B. Berenbaum, J. I. Williams, H. R. Bhattacharjee, I. Goldberg, M. D. Swerdloff, A. J. Salerno and P. D. Unger, *Polym. Prepr.* **30**, 350 (1989).
10. M. D. Swerdloff, S. B. Anderson, R. D. Sedgwick, M. K. Gabriel, R. J. Brambilla, D. M. Hindenlang and J. I. Williams, *Int. J. Pept. Protein Res.* **33**, 318 (1989).
11. M. Yu and T. J. Deming, *Macromolecules* **31**, 4739 (1998).
12. K. Yamada, T. Chen, G. Kumar, O. Vesnovsky, L. D. T. Topoleski and G. F. Payne, *Biomacromolecules* **1**, 252 (2000).
13. B. P. Lee, J. L. Dalsin and P. B. Messersmith, *Biomacromolecules* **3**, 1038 (2002).
14. J. Schnurrer and C.-M. Lehr, *Int. J. Pharm.* **141**, 251 (1996).
15. K. Huang, B. P. Lee, D. Ingram and P. B. Messersmith, *Biomacromolecules* **3**, 397 (2002).
16. H. Hertz, *J. Reine Angew. Math.*, 156 (1881).
17. L. Nakonieczna, W. Przycgodzen and A. Chimiak, *Liebigs. Ann. Chem.*, 1055 (1994).
18. D. L. Hern and J. A. Hubbell, *J. Biomed. Mater. Res.* **39**, 266 (1998).
19. J. H. Waite and C. V. Benedict, *Methods. Enzymol.* **107**, 397 (1984).
20. K. L. Johnson (Ed.), *Contact Mechanics*. Cambridge University Press, New York, NY (1985).
21. S. Zalipsky and J. M. Harris, in: *Introduction to Chemistry and Biological Applications of Poly(Ethylene Glycol)*, p. 1. American Chemical Society, Washington, DC (1997).

22. J. M. Harris, *J. Macromol. Sci. Rev. Macromol.* **C25**, 325 (1985).
23. J. Elisseeff, W. McIntosh, K. Anseth, S. Riley, P. Ragan and R. Langer, *J. Biomed. Mater. Res.* **51**, 164 (2000).
24. M. B. Mellott, K. Searcy and M. V. Pishko, *Biomaterials* **22**, 929 (2001).
25. N. Nishiyama, T. Asakura, K. Suzuki, K. Horie and K. Nemoto, *J. Biomed. Mater. Res.* **29**, 107 (1995).
26. N. Nishiyama, T. Asakura, K. Suzuki, T. Sato and K. Nemoto, *J. Biomed. Mater. Res.* **40**, 458 (1998).
27. N. Nishiyama, K. Suzuki, T. Asakura, K. Komatsu and K. Nemoto, *J. Dent. Res.* **80**, 855 (2001).
28. Y. Yamamoto and Y. Miyahera, US Patent No. 5,098,999 (1992).
29. M. J. Sever and J. J. Wilker, *Tetrahedron* **57**, 6139 (2001).
30. M. Taira, H. Urabe, T. Hirose, K. Wakasa and M. Yamaki, *J. Dent. Res.* **67**, 24 (1988).
31. I. Watanabe and N. Nakabayashi, *Quintessence Int.* **24**, 335 (1993).
32. Y. Nakayama and T. Matsuda, *ASAIO J.* **41**, M374 (1995).
33. Y. Nakayama and T. Matsuda, *J. Biomed. Mater. Res.* **48**, 511 (1999).
34. E. Andrzejewska, L. A. Linden and J. F. Rabek, *Macromol. Chem. Phys.* **199**, 411 (1998).
35. I. V. Khudyakov, J. C. Legg, M. B. Purvis and B. J. Overton, *Ind. Eng. Chem. Res.* **38**, 3353 (1999).
36. M. Dossot, M. Sylla, X. Allonas, A. Merlin, P. Jacques and J.-P. Fouassier, *J. App. Polym. Sci.* **78**, 2061 (2000).
37. S. Fujisawa, M. Ishihara and Y. Kadoma, *Sar. Qsar. Environ. Res.* **13**, 617 (2002).
38. E. T. Denisov and V. V. Azatian (Eds), *Inhibition of Chain Reactions*. Institute of Problems of Chemical Physics University Press, London (2000).
39. J. J. Kurland, *J. Polym. Sci. Polym. Chem. Ed.* **18**, 1139 (1980).
40. G. Sichel, C. Corsaro, M. Scalia, S. Sciuto and E. Geremia, *Cell Biochem. Funct.* **5**, 123 (1987).

Gels, Genes, Grafts and Giants, pp. 655–668 (2005)

Hydrogen-bonding-driven spontaneous gelation of water-soluble phospholipid polymers in aqueous medium

MIZUNA KIMURA, KIKUKO FUKUMOTO, JUNJI WATANABE and KAZUHIKO ISHIHARA *

Department of Materials Engineering, School of Engineering, The University of Tokyo, 7-3-1 Hongo, Bunkyo-ku, Tokyo 113-8656, Japan

Received 1 October 2003; accepted 20 January 2004

Abstract—It has been found that mixing of two kinds of water-soluble phospholipids polymers, such as poly(2-methacryloyloxyethyl phosphorylcholine–*co*-methacrylic acid) (PMA) and poly(2-methacryloyloxyethyl phosphorylcholine-*co*-*n*−butyl methacrylate) (PMB), spontaneously forms a hydrogel in aqueous medium at room temperature without any chemical treatment. However, the mechanism of spontaneous gelation has not been clarified yet. The purpose of this study is to investigate the gelation mechanism of the hydrogel. Moreover, effects of ions on gelation and dissolution behavior were observed. We investigated the mechanism of the hydrogel formation by spectroscopic techniques and a rheological method with attention to the interactions between polymer chains. Both Raman spectroscopic analysis and FT-IR analysis revealed that carboxyl groups in methacrylic acid (MA) formed dimer when two polymer solutions were mixed, and the results of the rheological study showed dissociation of carboxyl groups caused dissolution of the hydrogel. Thus, the gelation occurred due to the formation of dimers by hydrogen bonding which acts as a physical cross-linking of polymer chains. The hydrogel dissolved in a large amount of aqueous medium. We also observed the addition of inorganic salts during the preparation of the hydrogel affected the gelation and dissolution behaviors by a rheological and a weight measuring method, respectively. The gelation period became longer in the presence of NaCl and $CaCl_2$ compared with that in the absence of these salts. NaCl and $CaCl_2$ disturbed the formation of hydrogen bonding between carboxyl groups by stabilization of carboxylate anion of the MA units. On the other hand, addition of $FeCl_3$ made the gelation period shorter and stabilized the hydrogel in the aqueous medium. This is because $FeCl_3$ can suppress dissociation of the carboxyl groups by acidic condition of $FeCl_3$ aqueous solution and cross-link the carboxylate anions in the PMA effectively.

Key words: Hydrogel; gelation; 2-methacryloyloxyethyl phosphorylcholine polymer; Raman spectroscopy; hydrogen bonding; carboxyl group; dissociation.

---

*To whom correspondence should be addressed. Tel.: (81-3) 5841-7124. Fax: (81-3) 5841-8647. E-mail: ishihara@bmw.t.u-tokyo.ac.jp

INTRODUCTION

Hydrogel structures can be found in living organisms, for example, in soft tissues like muscular tissues, arthrogenous and cornea. They perform impact adsorption, energy storage and signal transduction, and also play an important part in homeostasis. These facts indicate that the hydrogels have a good compatibility with living organisms. The first event of biological response against artificial materials after contact with living organisms, that is, protein adsorption and immunoresponse, hardly occurs on/in the hydrogel due to the high proportion of water in the hydrogel. Thus, the hydrogel is one of the interesting materials in the biomedical and pharmaceutical fields. Actually, many studies on the use of hydrogels for medical applications have been reported recently, including applications as adhesives [1], wound dressings [2, 3], scaffolds for tissue engineering [4, 5] and matrices for drug delivery [6–9]. Among these applications, *in situ* formation of hydrogels has attracted recent attention. A feature of these hydrogels is that they can be prepared in proper form and quantity for the treated area. There are three approaches for *in situ* preparation of hydrogels. One approach for gel formation is to initiate polymerization reactions of multifunctional monomers to form a three-dimensional polymer network. However, irradiation light, heat to initiate reactions or radicals generated causes injure the reaction site and remaining unreacted monomers show toxicity. A second approach is to initiate cross-linking reactions of reactive polymers or macromonomers. Photocurable gelatin hydrogel [10] can be categorized in this approach. This approach also involves a certain risk for toxicity of low-molecular weight and reactive compounds. A third approach is to use responsive 'smart' polymers to specific conditions of injected site. In general, smart polymers are responsive to change in temperature or pH and show gel–sol transition [11, 12]. There are few biocompatible ones among smart polymers; thermo-responsible smart polymers [13–15] are difficult to inject without gel–sol transition during injection.

In addition, these approaches are only available where reaction can be initiated easily. Studies performed up to now limit the applications to such as rectum or postoperative use.

We have been found that mixing of two kinds of water-soluble phospholipids polymers, such as poly(2-methacryloyloxyethyl phosphorylcholine-*co*-methacrylic acid) (PMA) and poly(2-methacryloyloxyethyl phosphorylcholine-*co*-*n*-butyl methacrylate) (PMB) spontaneously forms a hydrogel in aqueous medium at room temperature without any chemical treatment [9, 16, 17]. Since 2-methacryloyloxyethyl phosphorylcholine (MPC) polymers have excellent biocompatibility [18–21], the hydrogel is expected to be a material used in the biomedical field, for example as a drug reservoir or a wound dressing which can be prepared *in situ*. However, the mechanism of spontaneous gelation has not been clarified yet.

The main purpose of this study is to investigate the gelation mechanism of mixture of aqueous solutions containing PMA and PMB. Also the effect of ions on gelation and dissolution behaviors was observed. We investigated the mechanism of the hydrogel formation by spectroscopic techniques with attention to the interactions between these polymer chains. A rheological method was applied to study of gelation and dissolution process.

MATERIALS AND METHODS

Materials

MPC was synthesized by a reported method and purified by re-crystallization from acetonitorile [22]. BMA and MA were commercial reagents and used without further purification. PMA and PMB were synthesized by a conventional radical polymerization of corresponding monomers and purified by a membrane filtration technique. In this study, these polymers were supplied from NOF (Tokyo, Japan) as 5% aqueous solutions. The molecular weights of these MPC polymers were determined by gel-permeation chromatography with methanol/water as an elution solvent. Poly(ethylene oxide) standards were used for determining of the molecular weight. The number-average molecular weights of PMA and PMB were 2.7×10^5 and 1.5×10^5, respectively. To solidify these MPC polymers, the aqueous solutions were freeze-dried. The chemical structure of PMA and PMB is shown in Fig. 1. Sodium chloride (NaCl), calcium chloride ($CaCl_2$) and iron(III) chloride hexahydrate ($FeCl_3$) were purchased from Kanto (Tokyo, Japan) and used without further purification. Other reagents were extra-pure grade. A nylon mesh bag (for storage of pathological tissue specimens, 45 × 74 mm) was purchased from Eiken Kizai (Tokyo, Japan).

Preparation of hydrogel composed of PMA and PMB

An equal amount of PMA and PMB aqueous solutions (5 wt%) was added to a glass bottle. The glass bottle was sealed with cap and shaken vigorously with hand at room temperature for 60 s. Figure 2 shows the picture of PMA and PMB aqueous solutions and PMA/PMB hydrogel formed by shaking. After the gel formed, the bottle was opened and the gel was removed. The gel made from aqueous solutions of PMA and PMB was described as PMA/PMB hydrogel.

Raman spectroscopic measurement

Aqueous solutions (10 wt%) of PMA and PMB were prepared. The solution was poured in a glass cell for Raman spectrum measurement and the spectrum was

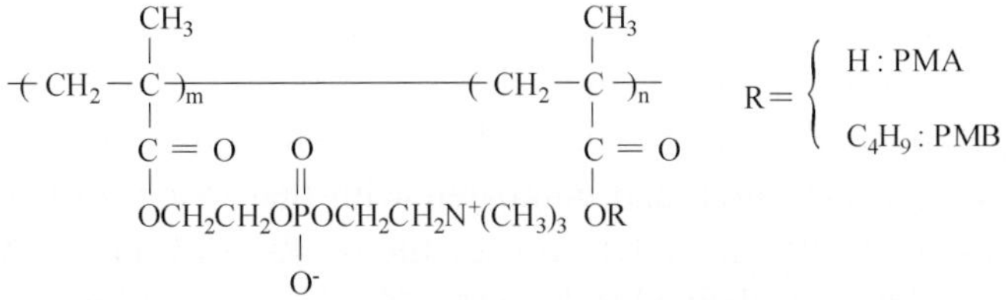

Figure 1. Chemical structure of water-soluble MPC polymers used in this study.

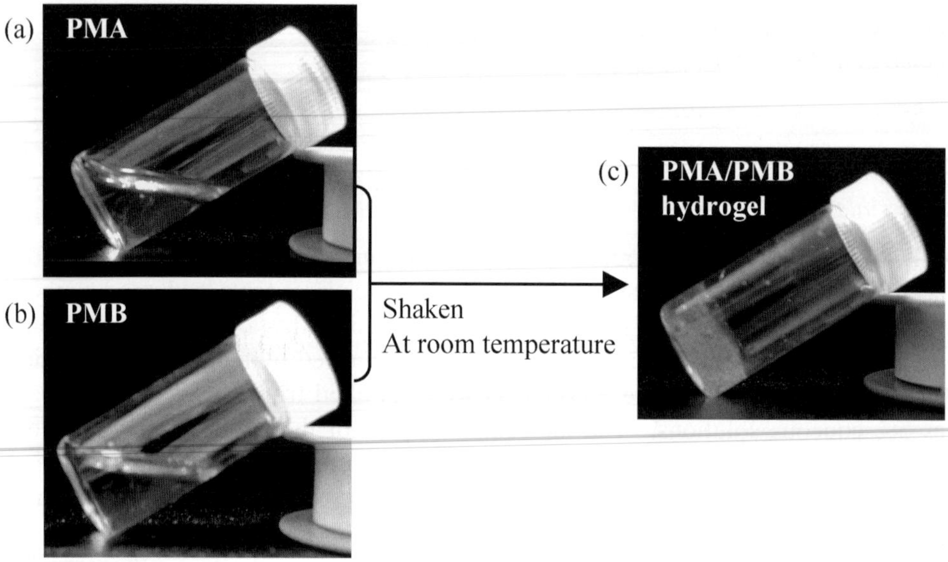

Figure 2. Pictures of PMA (a) and PMB (b) aqueous solutions and PMA/PMB hydrogel (c). Mixing of two polymer solutions and shaking at room temperature formed the hydrogel.

obtained using a Fourier transform infrared spectrometer (FT/IR-680, Jasco, Tokyo, Japan) equipped with a Raman attachment (RFT-600, Jasco). The measurement was carried out in the Raman shift range between 400 cm^{-1} and 4000 cm^{-1}. The limit of resolution was 8 cm^{-1}. Using the same procedure, Raman spectrum of PMA after addition of NaOH aqueous solution (2 M) to alkalinize was recorded. Then, the differential spectrum between the polymer aqueous solution before and after alkalization was obtained and analyzed using curve fitting and analysis software (GRAMS/386®, Galactic Industries, Salem, NH, USA).

In case of the PMA/PMB hydrogel, the gel was prepared beforehand and put in the cell for the measurement. Raman spectrum was recorded as the same conditions for the polymer solutions. Finally, the differential spectrum was obtained for the PMA/PMB hydrogel.

FT-IR spectrum measurement

After all polymers and gel were freeze-dried, FT-IR spectra were measured using a FT/IR-615 (Jasco) in the wavelength range of 400–4000 cm^{-1}. The mixture of PMA and PMB, that is, a mixture after PMA and PMB solutions were freeze-dried separately, was also measured and compared with that of the PMA/PMB hydrogel. To express transmittance intensities of the peaks between 1355–1000 cm^{-1} numerically, baseline was defined to be a straight line connecting spectrum points at 1355 cm^{-1} and 1000 cm^{-1}.

Measurement of viscoelastic property of the PMA/PMB hydrogel

Aqueous solutions of PMA (5 wt%, 0.75 ml) and PMB (5 wt%, 0.75 ml) were injected slowly into both sides of a vibration blade (1.71 cm^2) in a liquid cell. Just after injection, the these polymer solutions were mixed in the cell for preparation of the PMA/PMB hydrogel. Change in the elastic modulus (G') and the viscous modulus (G'') was recorded with a rheometer (Rheolograph-Micro, Toyoseiki, Tokyo, Japan). 1.5 h after injection, 0.2 ml distilled water or 2 M NaOH aqueous solution was added in the cell to the hydrogel and change in the G' and G'' was measured.

Dissolution behavior of the PMA/PMB hydrogel

PMA/PMB hydrogel was put into a nylon mesh bag and immersed in 100 ml Tris buffer solution (0.05 M, pH 7.4). The mesh bag with hydrogel was weighed at given time intervals and the weight of the PMA/PMB hydrogel remaining in the bag was determined. To investigate the effects of inorganic salts on the dissolution behavior of the PMA/PMB hydrogel, the hydrogel containing the desired salt was prepared as follows. One of the NaCl, $CaCl_2$ and $FeCl_3$ were dissolved in PMA aqueous solution just before the PMA solution with salt (0.5 ml) and PMB aqueous solution (0.5 ml) were mixed together for preparation of the PMA/PMB hydrogel. The final concentration of the salts was 0.14 M. In addition, we applied ethylenediamine-N,N,N′,N′-tetraacetic acid disodium salt dihydrate (EDTA, Kanto) to the immersion medium of the PMA/PMB hydrogel with $FeCl_3$.

Measurement of gelation time

By using the rheometer, gelation time after mixing the aqueous solutions of PMA and PMB was measured. The effect of additional salt was also investigated. In this experiment, NaCl and $CaCl_2$ were added to the PMA aqueous solution separately before preparation of the hydrogel with PMB. However, in the case of $FeCl_3$ addition, the salt was dissolved to PMB aqueous solution. Because when the $FeCl_3$ was added to the PMA aqueous solution, the possibility of spontaneous gelation of PMA alone was raised. The gelation time is defined as the time when the elastic modulus becomes greater than the viscous modulus [13].

RESULTS AND DISCUSSION

Interaction between MPC polymer chains

Raman scattering spectroscopy is an effective method to examine the state of a solute in the solution [23]. We used this method to study the state of carboxyl groups in MA unit of PMA in aqueous solution. In the aqueous solution of PMA, there are three kinds of carboxyl groups, that is, dissociated carboxylate anion, free carboxyl

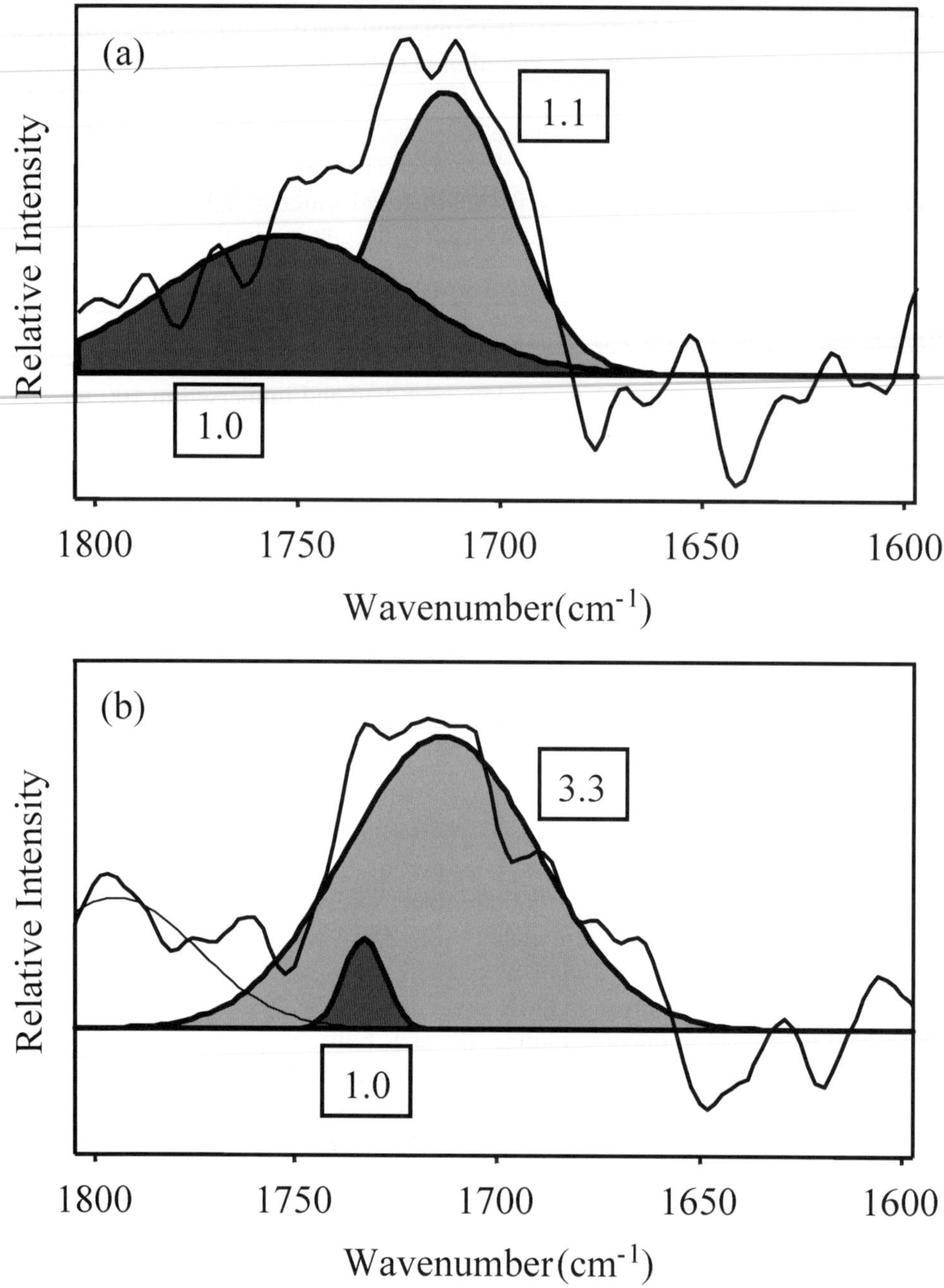

Figure 3. Raman spectra of PMA aqueous solution (a) and PMA/PMB hydrogel (b). In both cases, the differential spectra between the Raman spectrum before and after alkalization with NaOH were indicated.

group and dimer of the carboxyl groups. Raman shift of C=O stretching is observed at 1800–1700 cm^{-1} for free carboxyl groups and at 1750–1650 cm^{-1} for dimers of carboxyl groups. When the PMA aqueous solutions were alkalinized with NaOH aqueous solution, every carboxyl groups dissociate and turn into carboxylate anion, that is, neither free carboxyl groups nor dimers of carboxyl group can exist. Thus, differential spectrum between the Raman spectrum of PMA in aqueous solution and that of after alkalization indicated the specific carboxyl groups in MA unit of PMA. Raman spectroscopic analysis of the PMA/PMB hydrogel before and after alkalization was also carried out.

In Fig. 3, differential Raman spectra of PMA aqueous solution (Fig. 3a) and PMA/PMB hydrogel (Fig. 3b) are shown. The curve fitting and resolution into each state of carboxyl groups was carried out and well done. Boxed numbers demonstrate the ratio of the peak areas attributed to the dimer of carboxyl groups (bright part of the figure) against the respective peak area of free carboxyl groups (dark part of the figure). In case of PMA aqueous solution, it was 1.1. In contrast, in the PMA/PMB hydrogel the ratio increased to 3.3. That is, dimerization of carboxyl groups proceeded by gel formation. It is well known that the carboxyl groups can form dimer through hydrogen bonding between them [24]. This result indicated that dimerization of carboxyl groups in MA unit of PMA through hydrogen bonding acts as physical cross-linking points of hydrogel. In the aqueous solution, PMB chains associate themselves and form hydrophobic domains [25, 26]. When the aqueous solutions of PMA and PMB mixed together, the PMA chains penetrated into hydrophobic domains of PMB. Since the permittivity, or polarity is lower in the hydrophobic domains, the dissociation of carboxyl groups is suppressed. Thus, the carboxylate anion becomes the free carboxyl group and dimerization of carboxyl groups through hydrogen bond formation is likely to be induced.

Figure 4 shows the FT-IR spectra of PMA, PMB and the mixture in the dry state. When the FT-IR spectrum of PMA/PMB hydrogel is compared with that of mixture of PMA and PMB, we could observe an alteration in the transmittance around at 1240 cm^{-1} and 1170 cm^{-1}. The peak at 1240 cm^{-1} is identified as $O{=}P{-}O^-$ anti-symmetric stretching in MPC units, and the peak at 1170 cm^{-1} is identified as an overlap of C−O−C symmetric stretching from ester groups and C−O stretching from free carboxyl groups. Numbers beside the peaks demonstrate the ratio of transmittance intensity of the peak at 1170 cm^{-1} against that at 1240 cm^{-1}. In the case of the PMA and the PMA-PMB mixture, they were 0.96 and 0.93, respectively; that is, the peaks at 1170 cm^{-1} were almost the same with that at 1240 cm^{-1}. However, in the PMA/PMB hydrogel case it was 0.81 and the peak at 1240 cm^{-1} was greater than that at 1170 cm^{-1}. On the other hand, the peak at 1070 cm^{-1} ($O{=}P{-}O^-$) from MPC units was not changed, even when the PMA/PMB hydrogel was formed. Thus, the change in the spectrum is mainly due to the peak at 1170 cm^{-1} from free carboxyl groups in the MA unit. From these results of spectroscopic analysis, an increase in dimers of carboxyl groups means that hydrogen bonding

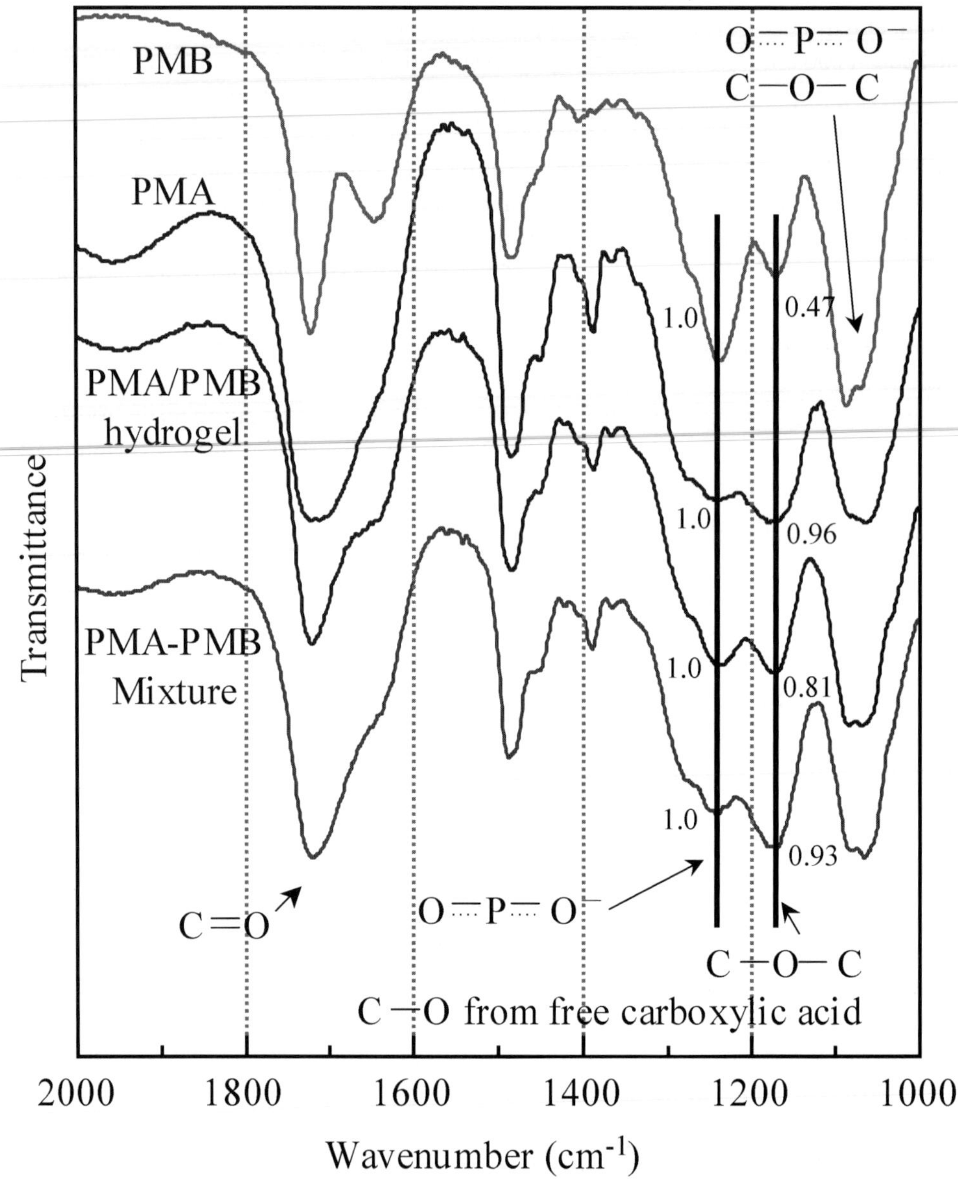

Figure 4. FT-IR spectra of freeze-dried PMA, PMB, their mixture and the PMA/PMB hydrogel.

between in MA units of PMA plays an important role as physical cross-linking for gelation in PMA/PMB hydrogel.

Viscoelastic properties of PMA/PMB hydrogel

Figure 5 illustrates the change in the viscoelastic curves of PMA/PMB hydrogels during formation and dissociation by addition of NaOH aqueous solution. When

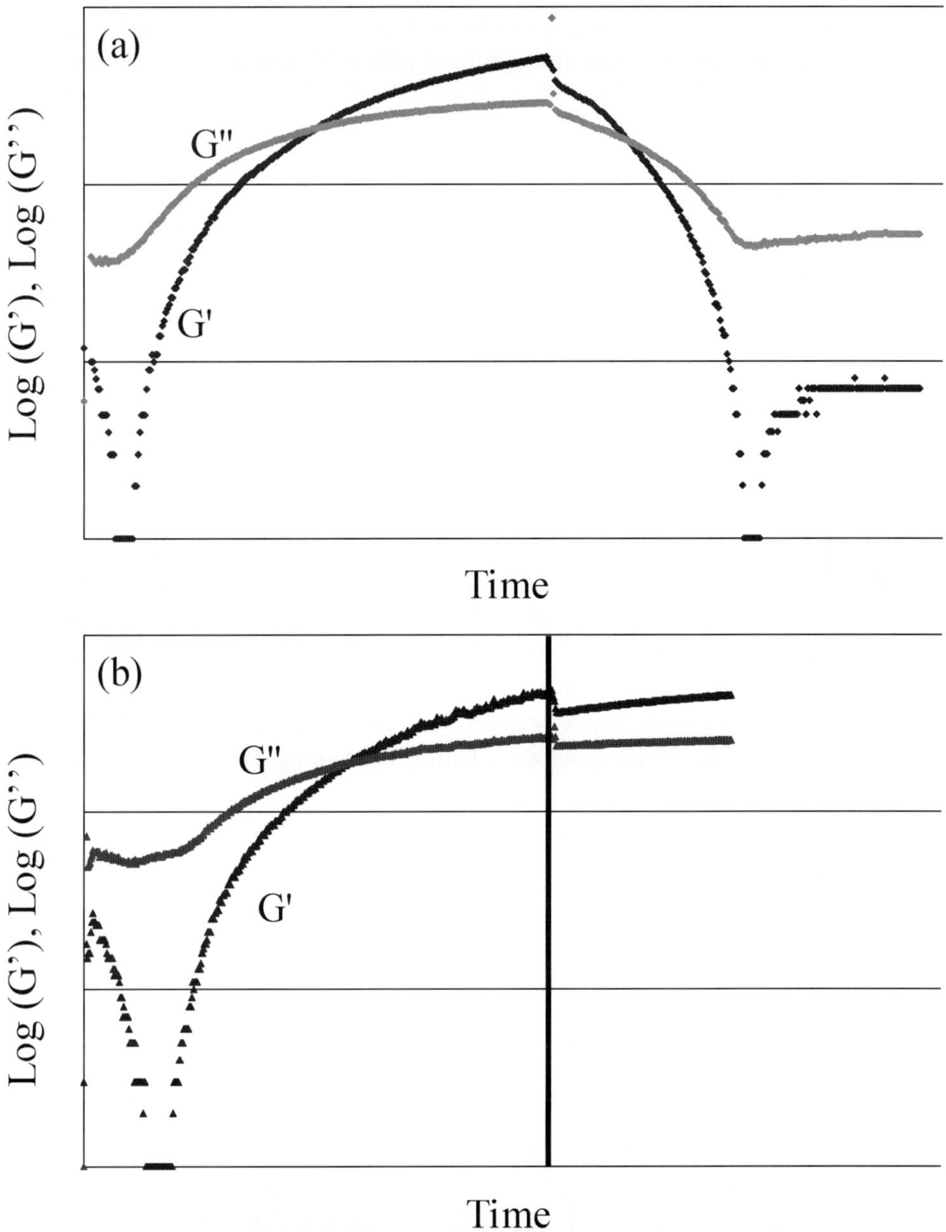

Figure 5. Change in the elastic modulus (G', darker line) and viscous modulus (G'', lighter line) during hydrogel forming process and after addition of NaOH aqueous solution (a) or distilled water (b).

aqueous solutions of PMA and PMB were mixed, elastic modulus (G') and viscous modulus (G'') changed with time. The G' value decreased at the initial stage of mixing, but it increased gradually and exceeded the G'' value. After gelation was completely finished, NaOH aqueous solution (Fig. 5a) or distilled water (Fig. 5b) was added to the hydrogel. Just after addition of NaOH aqueous solution, both the G' and G'' values decreased rapidly and the G' value became lower than the G'' value. That is, hydrogel was dissociated and the network structure did not exist any more. When distilled water was added to the hydrogel, both the G' and G'' values altered slightly but these values increased gradually again. In this case, the G' value maintained large compared with the G'' value. This behavior could be explained by swelling of hydrogel and decrease in the density of polymer networks by addition of distilled water. The network structure of hydrogel was kept during this process. The results also provide valuable evidence that hydrogen bonding between MA units acts as physical cross-linking in hydrogel.

We would like to understand the role of PMA chains in the formation of hydrogels. The PMA chains had the possibility of ionic interaction and cross-linking between them. Addition of inorganic salt with various metal cations may be suitable method. So, the effect of inorganic salt on gelation of PMA and PMB aqueous solutions was studied. We selected three kinds of inorganic salt, NaCl, $CaCl_2$ and $FeCl_3$.

Figure 6 shows the gelation time of PMA and PMB aqueous solutions with inorganic salt. Compared with gelation time in the absence of the salt, the presence of NaCl and $CaCl_2$ caused an increase in gelation time. On the other hand, $FeCl_3$ shortened the gelation time. One possible explanation is that Fe^{3+} interacted with carboxylate anions in PMA and formed cross-links through salt formation. In addition, since the PMB aqueous solution containing $FeCl_3$ was acidic, the

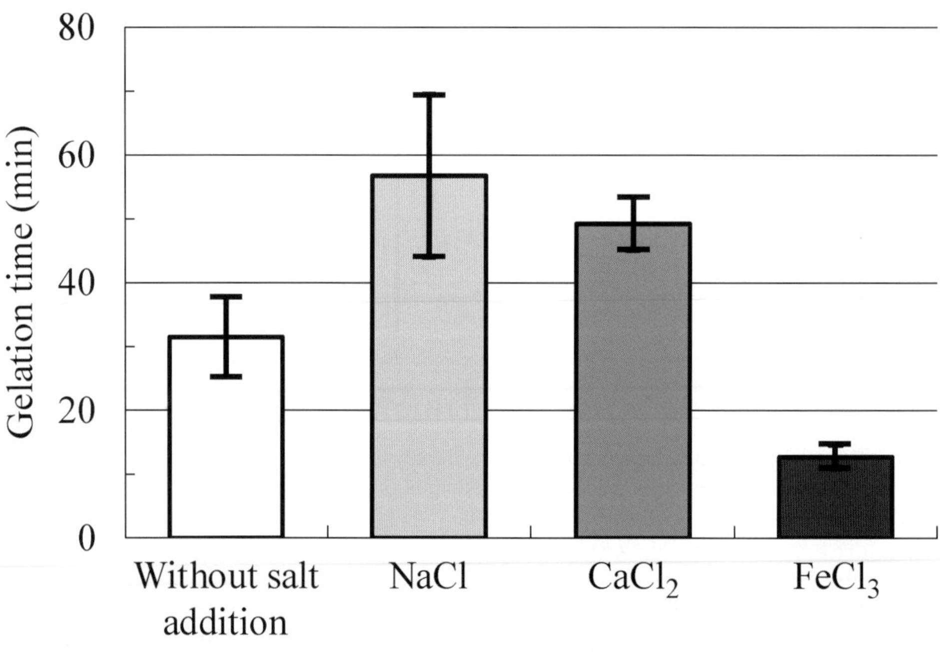

Figure 6. Effect of salts on the gelation times of PMA/PMB hydrogel.

dissociation of carboxyl groups in MA units of PMA was suppressed by a mix of these solutions. Both effects of Fe^{3+} may explain the shortening of gelation time.

Dissolution behavior of the PMA/PMB hydrogel

Considering the PMA/PMB hydrogel will be applied in the biomedical or pharmaceutical fields, it is necessary to control the dissolution behavior. We already found that the PMA/PMB hydrogel dissolves when the gel is immersed in a large amount of aqueous medium [9, 16, 17]. Moreover, this phenomenon depends on the pH condition of the medium. In the case of pH < 3, the hydrogel hardly dissolves, but at pH > 6, the dissolution starts immediately. Using this pH responsiveness, the PMA/PMB was considered to be applied in an oral drug delivery system [17]. Since polypeptide drugs, such as insulin, undergo enzymatic reaction in the stomach (low pH condition), it is necessary to protect the drug from this enzymatic reaction by encapsulation with an inert matrix. When the drugs subsequently go to the intestine (high pH condition), the drug will be released from the inert matrix and absorbed through the intestinal wall. The PMA/PMB hydrogel may have suitable potential for this system.

We considered that the inorganic salt also affects the dissolution behavior of the PMA/PMB hydrogel. Figure 7 shows the change in the relative weight of the hydrogel in the presence of various salts immersed in Tris-buffered solution. In the presence of NaCl and $CaCl_2$, the decrease in weight of the PMA/PMB hydrogel occurred with immersion time and it was much faster than in the absence of these salts. On the other hand, the weight of hydrogel slightly increased when the hydrogel was prepared in the presence of $FeCl_3$. This was due to the swelling of the hydrogel. Then the hydrogel dissolved gradually even in the presence of $FeCl_3$. These results indicated that the dissolution of the hydrogel depends on the ions in the hydrogel. We have already reported that the phosphorylcholine group in MPC unit is a hydrophilic and polar group, but it does not induce significant change in the water structure surrounding it [20, 27, 28]. Because the trimethylammonium cation and phosphate anion in phosphorylcholine group can form an innersalt between them, electrical charge of this group is nearly zero. Thus, we considered that the ions react with the co-monomer units of these MPC polymers, PMA and PMB. A reasonable explanation for the suppression of dissolution of the PMA/PMB hydrogel with $FeCl_3$ is that Fe^{3+} ions interact with carboxylate anions in MA units of PMA and form cross-links with PMA chains. In fact, after the addition of EDTA to the PMA/PMB hydrogel prepared in the presence of $FeCl_3$, the hydrogel swelled much more. Since the EDTA can deprive Fe^{3+} from the hydrogel and form coordination with Fe^{3+}, network structure became loose and resulted in swelling of the hydrogel. This phenomenon also implies that Fe^{3+} acted to form the network structure by cross-linking of PMA chains.

Dissolution of hydrogels prepared in the presence of NaCl and $CaCl_2$ was much faster than that in absence of salt. We considered two reasons for this phenomenon. One is that these cations (Na^+ and Ca^{2+}) stabilize carboxylate anions in MA unit

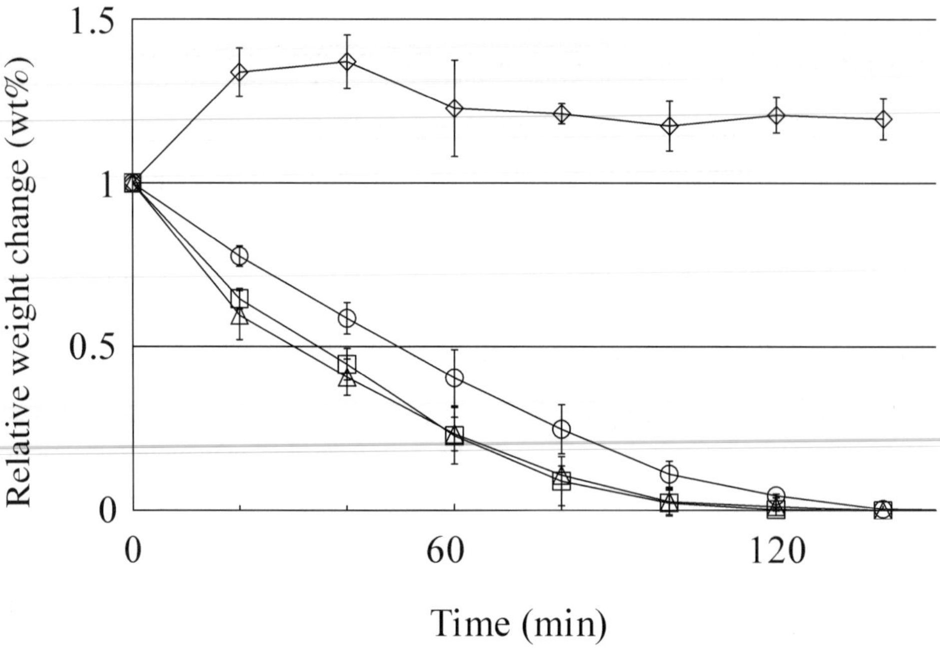

Figure 7. Change in weight of PMA/PMB hydrogel immersed in Tris-buffered solution. The various inorganic salts were added during preparation of the hydrogel. (○) in the absence of salt, (□) NaCl, (△) $CaCl_2$ and (◇) $FeCl_3$.

by salt formation. So, the number of free carboxyl groups, which can form dimers by hydrogen bonding between them and act as a cross-linker of the PMA chain, decreased. The other is related to water structure of the medium. It is known that Na^+ and Ca^{2+} induce formation of cluster structure of water molecules (structure forming ion), but Cl^- breaks that structure (structure breaking ion) [29]. The number of bonding water molecules per one ion is calculated to be 4.4 (Na^+ in NaCl, 0.1 M), 2.1 (Cl^- in NaCl, 0.1 M), 8.3 (Ca^{2+} in $CaCl_2$, 0.1 M) and 1.4 (Cl^- in $CaCl_2$, 0.1 M). Although the Ca^{2+} is stronger in structure formation ion than the Na^+, double concentration of the Cl^- in the PMA/PMB hydrogel system may disturb the clustering of water molecules. The hydrophobic interactions between BMA units in PMB are important to associate the PMB chains and to provide hydrophobic domain in aqueous medium [25, 26]. This is essentially due to clustering of water molecules. If the number or size of the hydrophobic domain decreased by presence of these salts, the network structure of the hydrogel became weaker compared with that in absence of the salt. We are now going to understand this phenomenon and will be reported elsewhere.

These findings mentioned above demonstrate that the addition of suitable inorganic salt during preparation of the PMA/PMB hydrogel makes it possible to control the dissolution time of hydrogel. In addition, a characteristic of spontaneous gelation of these polymers in aqueous medium indicates the PMA/PMB hydrogel system can be applied to *in situ* preparation of a drug reservoir or wound-dressing materials. Moreover, it is considered that the enzymes and cells will be immobilized without any adverse effects in the hydrogel easily. Considering the MPC polymers are well-known materials for their high biocompatibility, the PMA/PMB hydrogel has promise as a material to be used in the biomedical and pharmaceutical fields.

CONCLUSIONS

A hydrogel was spontaneously formed by mixing of aqueous solutions of PMA and that of PMB. The spectroscopic analysis of the PMA/PMB hydrogel indicated that the number of free carboxyl groups decreased and dimers of the carboxyl groups increased after gelation. Dissociation of carboxyl groups by alkalization destroyed the hydrogel structure. Thus, we concluded that the gelation was occurred due to formation by hydrogen bonding between free carboxyl groups in MA unit of PMA in hydrophobic PMB aggregates. Furthermore, we observed that both the gelation and dissolution time of the hydrogel could be controlled by addition of inorganic salt during preparation.

Acknowledgements

This study was financially supported by a Grant-in-Aid for the Development of Innovative Technology (No. 13202) and supported in part by a grant for 21st Century COE Program 'Human-Friendly Materials based on Chemistry' from the Ministry of Education, Culture, Sports, Science and Technology of Japan.

REFERENCES

1. N. A. Peppas and J. J. Sahlin, *Biomaterials* **17**, 1553 (1996).
2. J. B. Kane, R. G. Tompkins, M. L. Yarmush and J. F. Burke, in: *Biomaterials Science: An Introduction to Materials in Medicine*, B. D. Ratner, A. S. Hoffman, F. L. Schoen and J. E. Lemons (Eds), p. 360. Academic Press, San Diego, CA (1996).
3. M. Balasubramaniam, T. R. Kumar and M. Babu, *Burns* **27**, 534 (2001).
4. J. L. Drury and D. J. Mooney, *Biomaterials* **24**, 4337 (2003).
5. K. Y. Lee and D. J. Mooney, *Chem. Rev.* **101**, 1869 (2003).
6. I. McCulloch and S. W. Shalaby (Eds), *Tailored Polymeric Materials for Controlled Delivery Systems*. American Chemical Society, Washington, DC (1998).
7. S. M. Dinh, J. D. DeNuzzio and A. R. Comfort (Eds). *Intelligent Materials for Controlled Release*. American Chemical Society, Washington, DC (1999).
8. S. Mallapragada, M. Tracy, B. Narasimhan, E. Mathiowitz and R. Korsmeyer, *Biomaterials for Drug Delivery and Tissue Engineering.* Materials Research Society, Warrendale, PA (2001).
9. K. Ishihara, K. W. Nam, J. Watanabe, N. Sakamoto, K. Shuto and K. Suzuki, in: *Advances in Biomaterials and Drug Delivery Systems*, G. H. Hsiue, T. Okano, U. Y. Kim, H. W. Sung, N. Yui and K. D. Park (Eds), p. 203. Princeton International Publishing, Taipei (2002).
10. H. Okino, Y. Nakayama, M. Tanaka and T. Matsuda, *J. Biomed. Mater. Res.* **59**, 233 (2002).
11. A. Chenite, C. Chaput, D. Wang, C. Combes, M. D. Buschmann, C. D. Hoemann, J. C. Leroux, B. L. Atkinson, F. Binette and A. Selmani, *Biomaterials* **21**, 2155 (2000).

12. J. S. Park, Y. K. Oh, H. Yoon, J. M. Kim and C. K. Kim, *J. Biomed. Mater. Res.* **59**, 144 (2002).
13. T. H. Chen, H. D. Embree, E. M. Brown, M. M. Taylor and G. F. Payne, *Biomaterials* **24**, 2831 (2003).
14. Z. G. M. Wout, E. A. Pec, J. A. Maggiore, R. H. Williams, P. Palicharla and T. P. Johnston, *J. Parent. Sci. Technol.* **46**, 192 (1992).
15. G. Chandrashekar and N. Udupa, *J. Pharmacol.* **48**, 669 (1996).
16. K. W. Nam, J. Watanabe and K. Ishihara, *Biomacromolecules* **3**, 100 (2002).
17. K. W. Nam, J. Watanabe and K. Ishihara, *J. Biomater. Sci. Polymer Edn* **13**, 1259 (2002).
18. K. Ishihara, N. P. Ziats, B. P. Tierney, N. Nakabayashi and J. M. Anderson, *J. Biomed. Mater. Res.* **26**, 1397 (1991).
19. K. Ishihara, H. Oshida, Y. Endo, T. Ueda, A. Watanabe and N. Nakabayashi, *J. Biomed. Mater. Res.* **26**, 1543 (1992).
20. K. Ishihara, H. Nomura, T. Mihara, K. Kurita, Y. Iwasaki and N. Nakabayashi, *J. Biomed. Mater. Res.* **39**, 323 (1998).
21. K. Ishihara, E. Ishikawa, Y. Iwasaki and N. Nakabayashi, *J. Biomater. Sci. Polymer Edn* **10**, 1047 (1999).
22. K. Ishihara, T. Ueda and N. Nakabayashi, *Polym. J.* **22**, 355 (1990).
23. F. Lhert, F. Capelle, D. Blaudez, C. Heywang and J.-M. Turlet, *J. Phys. Chem. B.* **105**, 2672 (2001).
24. M. Meot-Ner, D. E. Elmore and S. Scheiner, *J. Am. Chem. Soc.* **121**, 7625 (1999).
25. K. Ishihara, Y. Iwasaki and N. Nakabayashi, *Polym. J.* **31**, 1231 (1999).
26. T. Konno, J. Watanabe and K. Ishihara, *J. Biomed. Mater. Res.* **65A**, 210 (2003).
27. H. Kitano, K. Sudo, K. Ichikawa, M. Ide and K. Ishihara, *J. Phys. Chem. B.* **104**, 11425 (2000).
28. H. Kitano, M. Imai, T. Mori, M. Gemmei-Ide, Y. Yokoyama and K. Ishihara, *Langmuir* **19**, 10260 (2003).
29. J. O'M. Bockris and P. P. S. Saluja, *J. Phys. Chem.* **76**, 2140 (1972).

Gels, Genes, Grafts and Giants, pp. 669–680 (2005)

Tethered protein/peptide-surface-modified hydrogels

JINGJING BI, J. CRAWFORD DOWNS and JEAN T. JACOB *

LSU Eye Center, Louisiana State University Health Sciences Center, 2020 Gravier Street, Suite B, New Orleans, LA 70112, USA

Received 7 October 2003; accepted 11 February 2004

Abstract—We investigated a wet chemistry method to covalently bond polyethylene glycol (PEG)-tethered extracellular matrix (ECM) proteins (laminin and fibronectin) or peptide (fibronectin-adhesion-peptide sequence) onto the surface of a poly(2-hydroxylethyl methacrylate-co-methylacrylic acid) (PHEMA/MAA) hydrogel that could potentially be used as a replacement for corneal tissue in the eye. An essential requirement for the success of such a surface in the biological environment is its ability to support the growth and attachment of corneal epithelial cells; ECM proteins are known to promote cellular attachment and growth. We hypothesized that the use of tethers or long hydrophilic chains would allow the attached ECM protein/peptide molecules to move two-dimensionally in space, thereby increasing their ability to bind with epithelial cell membranes. Additionally, the tethers would prevent the specifically added growth-enhancing factors from being obscured by any non-specific protein binding occurring on the hydrogel surface. In this surface-modification study, carbodiimidazole (CDI) was used to activate the carboxylic groups ($-$COOH) on the hydrogel surface in anhydrous dimethylsulfoxide (DMSO) before addition of the PEGylated proteins/peptide. The resulting tethered protein/peptide surface-modified hydrogels were analyzed in terms of percent grafting efficiency, biological activity and mechanical properties. X-ray photoelectron spectroscopy (XPS) and ^{125}I radioactive labeling demonstrated the successful covalent bonding between the moieties and the hydrogel surface. Radiolabeling and enzyme-linked immunosorbent assay (ELISA) studies indicated that the tethered proteins attached to the hydrogel surface at a concentration of approx. 0.1 μg/cm^2. ELISA testing further showed that tethered proteins remained biologically active. However, mechanical tensile testing indicated that the mechanical properties of these chemically modified hydrogels were somewhat altered in comparison with the unmodified hydrogel. Future studies will evaluate the cellular response to these surface-modified materials *in vitro* and *in vivo*.

Key words: PEGylation; extracellular matrix proteins; PHEMA/MAA hydrogel; surface modification; mechanical properties; XPS.

---

*To whom correspondence should be addressed. Tel.: (1-504) 412-1200, ext. 1326. Fax: (1-504) 412-1315. E-mail: jjacob@lsuhsc.edu

INTRODUCTION

The epithelial cell layer of the cornea plays an important role not only as a refracting surface and a barrier to pathogens, but also in the maintenance of tear film spreading and stability. Any device used to replace or augment the cornea must support a healthy anterior epithelial cell layer in order to maintain these precorneal functions. Treatment for end-stage corneal diseases, such as ocular pemphigoid, Stevens–Johnson syndrome and severe corneal burns, requires full-thickness corneal replacement in a small portion of the world population. However, corneal augmentation for refractive purposes could potentially affect the over 85 million contact lens wearers worldwide. Many current investigational devices designed to replace or augment the cornea have a central elastomeric or hydrogel segment that constitutes the optical portion of the device [1–3]. Hydrogels are used not only because their compliance allows them to respond to the pressure fluctuations caused during blink without inducing high shear forces at the material/tissue interface and damaging surrounding tissue, but also because they are able to maintain a hydrated environment similar to that of the natural tissues. However, corneal epithelial cell attachment to and growth over these hydrogel surfaces is severely limited [3].

A number of different surface modification techniques have been investigated in attempts to foster productive biological interactions between the hydrogel surface and the cell membrane of individual corneal epithelial cells. Franco *et al.* [4] evaluated the interactive effects of different surface functionalities with corneal epithelial cells using self-assembled monolayers of alkanethiols and alkylsiloxanes. Several investigators have used plasma modification of various surfaces including silicone rubber [5–7] and polyvinyl alcohol [8] to improve the corneal epithelial cell response. Adsorption and covalent attachment of extracellular matrix (ECM) proteins, including fibronectin and laminin, and collagens Type I and IV have also been investigated. Hydrogels preadsorbed with Type-I and -IV collagen showed a more rapid rate of cell growth than those coated with either fibronectin or laminin [9–12]. Adsorption of fibronectin to a collagen-modified surface resulted in further acceleration of corneal cell attachment [13, 14]. However, collagen-modified surfaces have been shown to be susceptible to collagenase activity over time. Cell adhesion peptides, including RGD, YIGSR and a novel collagen peptide sequence Gly-Pro-Nleu, have also been studied. However, all of these peptide modifications have produced only modest increases in the rate of epithelialization [15–19]. In general, coating and direct attachment of biological factors (both proteins and peptides) to these surfaces have not yielded the substantial increase in cell proliferation and growth necessary to maintain a functional corneal epithelial cell layer for any significant length of time (i.e. 3–6 months).

We have investigated an approach involving chemically tethering ECM proteins and peptides with cell-surface adhesion binding activities to a hydrogel surface to promote the attachment and growth of corneal epithelial cells. We hypothesized that the use of tethers or long hydrophilic chains would allow the attached molecules to move two-dimensionally in space, thereby increasing their ability to bind with the epithelial cell membranes. Additionally, the tethers would prevent the specifically added factors from being obscured by any non-specific protein binding occurring on the hydrogel surface after implantation.

In this study, we individually tethered two proteins and one peptide to polyethylene glycol (PEG) chains, including fibronectin and laminin, which are ECM proteins known to be involved in the wound-healing process or found in the corneal epithelial basement membrane, and fibronectin-adhesion-peptide sequence (FAP), a specific peptide sequence from the fibronectin molecule that has been shown to increase epithelial growth elsewhere in the body. The free end of the PEG chain was then specifically attached to pendant chemical groups on the hydrogel backbone chain. The resulting tethered protein-modified hydrogels were then characterized as to biological activity and physical properties.

MATERIALS AND METHODS

Materials

Ophthalmic grade 2-hydroxylethylmethacrylate (HEMA, Polysciences, Warrington, PA, USA) was used as received. Methacrylic acid (MAA, Polysciences) was purified by alumina (Polysciences) and stored at 4°C prior to use. Ethylene glycol dimethacrylate (EGDMA, Sigma-Aldrich, St. Louis, MO, USA), trimethylolpropane trimethacrylate (TMPTMA, Sigma-Aldrich) and Darocur® 1173 (Ciba Specialty Chemicals North America, Tarrytown, NY, USA) were used as received. Glucam E-20 was supplied by Vistakon (Jacksonville, FL, USA). Laminin (410 kDa, from the basement membrane of Engelbreth-Holm-Swarm Mouse sarcoma, Sigma-Aldrich), fibronectin adhesion-promoting peptide (FAP, 1023 Da, Bachem Biosciences, King of Prussia, PA, USA) and fibronectin (450 kDa, from human plasma, Sigma-Aldrich) were used as received unless specific treatment is mentioned. Custom synthesized polyethylene glycol (PEG, 3400 Da, Nektar Therapeutics (formerly Shearwater Polymers), Huntsville, AL, USA) was used as received. Two different types of activated PEGs with either N-hydroxylsuccinimide or azine as one end-group (HO-PEG_2-NHS or HO-PEG-NH_2-NH_2HCl) were used as the tethering moieties.

Base hydrogel polymerization

Poly(2-hydroxylethyl methacrylate-co-methacrylic acid) (PHEMA/MAA) was prepared by modifying the reported methods [20, 21]. A mixture of 93.84% (w/w) HEMA, 4.94% (w/w) MAA, 0.1% TMPTMA, 0.78% (w/w) EGDMA, 0.34% (w/w) Darocur® 1173 and 66.94% (w/w) Glucam E-20 was stirred under reduced pressure for 30 min at 25°C and then transferred to a custom-designed quartz mold with 0.5-mm spacers. The filled mold was exposed to UV irradiation at 254 nm for 40 min at room temperature. The polymer was then cured at 50°C for 2 h before removal from the mold. A clear and transparent hydrogel was obtained. The hydrogels were either placed in phosphate-buffered saline (PBS) at 70°C for 3 h to replace the solvent or stored in dimethyl sulfoxide (DMSO) for further grafting chemistry.

Tethering the proteins/peptide with PEG

The tethering chemistry between PEG and the ECM proteins/peptide was conducted in DMSO. PEG and the proteins/peptide were mixed in equal molar ratios in DMSO at room temperature. The mixture was stirred continuously in DMSO overnight. The reaction solution was then placed in a dialysis tube (MWCO 3500 Da; Pierce, Dallas, TX, USA) to remove all unreacted PEG and small molecular reagents. The dialysate solution of the PEGylated proteins/peptide was lyophilized and stored at 4°C prior to grafting.

Grafting the PEGylated ECM proteins/peptide to the PHEMA/MAA hydrogel surface

Carbodiimidazole (CDI, Sigma-Aldrich) was used to activate the PHEMA surface in the DMSO solution at room temperature for 2 h. The concentration of CDI in DMSO was 10 times that of the −COOH on the hydrogel surface in moles (for example, 0.141 g CDI in DMSO for 0.35 ml MAA). After the activation, the hydrogel was rinsed in pure DMSO before it was moved to a reactor that contained fresh DMSO solution. The PEGylated ECM proteins/peptide (1 mg/ml in DMSO) were added. The reaction proceeded at room temperature with constant stirring for 17 h. The tether-modified hydrogel was then moved through a solvent replacement system (PBS in DMSO at 50% (v/v), 25% (v/v), 5% (v/v) and 0% (v/v), respectively). All of the procedures and containers were maintained under sterile conditions to prevent contamination.

Characterization studies

Water content. Determination of the equilibrium water content (EWC) of the hydrogels before and after CDI chemistry was carried out according to the following method. A 0.5-cm diameter disc was cut from five samples of each hydrogel type. Each piece was cut, blotted, placed in a loosely capped, pre-weighed vial and weighed. The vials were placed in a vacuum oven (25 mmHg) at 35°C. Heating was discontinued after 3 days; the vacuum was maintained for another 2 days before measurements were taken. The hydrogel sample was determined to be dry when the weight did not change over three sequential weighings 24 h apart. The EWC (or percent water) was calculated with the following equation:

$$\text{Equilibrium water content} = \frac{W_{\text{wet}} - W_{\text{dry}}}{W_{\text{wet}}} \times 100. \quad (1)$$

Radiolabeling. Iodo Beads (Pierce) were activated with 0.5 μCi ^{125}I in PBS for 10 min. The PEGylated ECM proteins/peptide (0.25 mg) in 2 ml PBS solution were added to the activated beads and allowed to iodinate for 20 min. The iodinated protein/peptide solution was decanted, filtered through a Sephadex G-10 (Sigma-Aldrich) column and then dialyzed in a 0.01 M KI solution with a dialysis cassette (MWCO 3500 Da; Pierce) overnight at −20°C to remove any free ^{125}I. The dialyzed solution was lyophilized and stored dry. The radioactivity of the dry [^{125}I]PEGylated ECM proteins/peptide was determined by using a gamma counter (Gamma 5500, Beckman Instruments, Fullerton, CA, USA).

Grafting chemistry was carried out as described above with the radiolabeled and PEGylated ECM proteins/peptides. The surface-modified hydrogels obtained after solvent replacement were counted for radioactivity. Five samples of each hydrogel type were tested. The mean quantity of PEGylated protein/peptide grafted to each surface was determined.

The percent grafting efficiency was determined as the amount of tethered proteins/peptide bound per area of available carboxylic functional groups on the hydrogel surface (equation (7) below). The amount of the hydrogel surface area corresponding to carboxylic functional groups was determined by (1) calculating the percent of the surface area that consisted of the MAA part of the polymer (A_1 below) and (2) calculating the area between the carboxyl groups, assuming that a carboxyl group was attached to every other carbon along the PHEMA/MAA backbone (2.54 Å) and that the distance between the PHEMA/MAA backbone chains was the length of the hydrogen bond between the hydroxyl group and the carboxyl group (2.76 Å) (A_2 below; it was assumed that this amount was not affected by the cross-linking degree on the hydrogel surface). Specifically:

$$A_{\text{total}} = \text{total hydrogel surface area} \times (100 - EWC), \tag{2}$$

where A_{total} is the available PHEMA/MAA area on the hydrogel surface.

$$A_1 = A_{\text{total}} \times 5\%, \tag{3}$$

where A_1 is the available hydrogel area with MAA for grafting.

$$A_2 = l_1 \times l_2, \tag{4}$$

where l_1 is the C—C mono bond length and l_2 the hydrogen bond length between hydroxyl and carboxylic groups.

Then:

$$X = \text{Number of molecules of tethered protein/peptide on the surface}/A_1, \tag{5}$$

$$Y = \text{Number of COOH molecules}/A_2. \tag{6}$$

The molecular grafting efficiency is then defined as:

$$\%\ \text{Grafting efficiency} = 100 \times (X/Y). \tag{7}$$

Enzyme-linked immunosorbent assay (ELISA). The biological activity of the PEGylated proteins was determined by using a direct ELISA system. Briefly, wells of a 96-well plate were coated with PEGylated protein. Biotinylated primary monoclonal antibody to the adsorbed protein was then introduced into the wells. After incubation, the primary antibody was removed and streptavidin-conjugated alkaline phosphatase (Sigma) was added and further incubated for 2 h at 37°C. Wells were then washed with PBS and the substrate (p-nitrophenyl phosphate (1 mg/ml, Sigma) in 1 M diethanolamine buffer, pH 9.8, containing 0.5 mM $MgCl_2$) was added. The absorbancy at 405 nm was evaluated using a microplate reader and compared with the readings obtained for non-PEGylated protein. Using a similar ELISA method, the amount of tethered protein on the hydrogel surface was also determined [22]. Ten wells were measured for each free PEGylated protein and tethered protein-modified hydrogel surface and mean values calculated.

Surface analysis.

Fourier transform infrared spectroscopy-attenuated total reflectance (FT-IR–ATR). Surface characterization was carried out on a hydrogel sample in the hydrated state by means of FT-IR–ATR (Model 1620, Perkin Elmer, Wellesley, MA, USA). Incident angles of 30°, 45° and 60° were used on separate samples for penetration analysis at depths of 72, 41 and 32 nm, respectively, by means of the corresponding crystals.

X-ray photoelectron spectroscopy (XPS). Surfaces were characterized before and after grafting with the PEGylated ECM proteins/peptide using cold probe XPS. The experimental details are described completely elsewhere [23]. Briefly, a piece of hydrogel was frozen in liquid nitrogen (LN_2, −198°C) in a Dewar flask, then transferred to the introduction chamber where it was subjected to vacuum for roughly 1–2 min. The hydrogel was then transferred to the reaction chamber and onto the pre-cooled finger (−135°C). Typical base pressure in the reaction chamber was between 5×10^{-8} and 5×10^{-9} Torr. The sample was then transferred into the main chamber and analyzed. Incident angles of 30°, 45° and 90° were used to obtain an in-depth analysis of the surface structure over a range of 0–100 Å.

Mechanical testing. Mechanical tensile testing of the hydrogels before and after surface modification was performed using an Instron electromechanical test frame (Instron, Canton, MA, USA) fitted with an integrated MTS Renew computer-based controller and data acquisition system (MTS Systems, Eden Prairie, MN, USA). The machine was equipped with a custom-designed environmental chamber that maintained the test sample in a 37°C, pH 7.4, isotonic saline environment during testing. Dog-bone-shaped tensile specimens (20-mm gauge length, 5-mm gauge width, 20-mm grip end width and mean thickness of 1.46 mm) of each test sample were placed in custom-designed grips that prevented hydrogel specimen slippage during testing. After pre-tensioning to 0.001 N, each specimen was subjected to tensile displacement at a rate of 200 μm/s until failure. Load, stress and strain data were recorded at 20 Hz for each test. Young's modulus (stiffness) was calculated as the 0.5% offset of the slope of the linear regression of the stress-strain curve for each specimen. Five samples of each hydrogel type were tested.

Statistical analysis

An analysis of variance (ANOVA) was performed to determine the effects of surface chemistry modification on water content, bioavailability, maximum stress, maximum strain and Young's modulus. Confidence intervals were calculated using the standard error, which employs the pooled variance of all values in the ANOVA for each parameter [24].

RESULTS AND DISCUSSION

Water content

Table 1 lists the EWC of hydrogels before and after introduction of biologically active moieties to hydrogel surfaces. There was no significant difference between the EWCs of grafted and ungrafted hydrogels with the same MAA content.

Grafting efficiency as determined by radiolabeling

The conjugation between the PEGylated proteins or peptide and the hydrogel surface was consistent with previously published work using CDI chemistry for addition purposes [25]. Table 1 shows the mass/unit area of tethered protein or peptide on the modified hydrogel, as well as the determined grafting efficiency. Approximately 25% of all functional groups, specifically the $-$COOH group from MAA, were grafted with tethered laminin or tethered fibronectin according to ^{125}I characterization. The tethered FAP-modified hydrogel surface, however, showed twice as much grafting as the tethered protein-modified surfaces. This result may be explained by the smaller molecular weight and dimensional size of the PEGylated

Table 1.
Equilibrium water content, surface content and grafting efficiency of control and modified hydrogels

Hydrogel types	Equilibrium water content (%)	Surface content of tethered protein/peptide (μg/cm^2)	Grafting efficiency (%)
Base PHEMA/MAA hydrogel	70.86 ± 0.48	N/A	N/A
FAP-PEG tethered hydrogel	72.56 ± 0.48	0.161 ± 0.009	57 ± 5
Fibronectin-PEG tethered hydrogel	71.19 ± 0.48	0.072 ± 0.009	24 ± 5
Laminin-PEG tethered hydrogel	70.94 ± 0.48	0.084 ± 0.009	27 ± 5

All values are means ± standard error.

peptide, compared with the larger PEGylated proteins. It is likely that PEGylated-FAP is capable of penetrating farther into the hydrogel during the addition reaction than the two larger proteins and therefore binds to a considerably higher degree.

Biological activity

Table 2 shows the reactivity of the PEGylated protein and the tethered protein-modified hydrogels to their respective monoclonal antibodies as compared to free protein. The PEGylated proteins showed approx. 40% of the binding activity seen for free protein, whereas the tethered protein-modified hydrogel surfaces showed approx. 65% of the binding activity seen for the free protein. The differences between the PEGylated proteins and the tethered protein-modified hydrogel may be due to the methodology of the ELISAs performed. Measurement of the PEGylated protein binding activity required first coating a surface with the PEGylated protein. It is reasonable to assume that the PEGylated protein was laid down on the surface in a number of conformations, some with the protein attached to the surface and others with the PEG tail attached. Depending upon the attachment to the surface, the epitopes necessary for the antibody binding may or may not have been readily accessible. The tethered protein-modified hydrogel measurement used the hydrogel itself as the underlying attachment surface. Therefore, all of the tethered proteins would be in the most accessible position possible for antibody binding. Although the total amount of proteins/peptide tethered to the hydrogel surfaces, as determined by the radiolabeling analysis, was significantly less than a monolayer coating (0.6 to 1.0 $\mu g/cm^2$), the enhanced bioavailability of the protein through movement allowed by the tether should increase the cellular response to the surfaces.

Surface analysis

Using FT-IR–ATR with an incident angle of 30°, we observed the characteristic peaks of the carbonyl group at 1760 cm^{-1} and the hydrogen stretching of the hydroxyl, secondary and primary amines groups ($-OH$, $-NH_2$ and $-NH$) at 3500 to 3200 cm^{-1} for the hydrogel (data not shown). However, specific differentiation of the amine groups from the hydroxyl groups was not possible by FT-IR–ATR due to the overall contribution of the hydrogen stretching of water at 3600–3000 cm^{-1} and 2400–2200 cm^{-1}. Surface profiling with the 45° and 60° incident angles did not improve the band differentiation to any significant degree. Resolution of functional groups in low concentration on a sample surface by FT-IR–ATR requires

Table 2.
Mean ratio of bioactivity of free protein to PEGylated or tethered protein

Protein	PEGylated	Tethered to hydrogel
Laminin	1 : 0.41	1 : 0.68
Fibronectin	1 : 0.37	1 : 0.64

both a large number of internal reflections within the crystal at the incident angle to boost the peak resolution and a small quantity of overlaying functionalities. Drying of the hydrogel surface prior to analysis decreased the confounding influence of the water peaks somewhat; however, this was not enough to compensate for the severely diminished number of internal reflections at the higher incident angles. The number of internal reflections at the 30° incidence angle is 45, whereas at the 60° incidence angle there are only a third of that number.

XPS

Table 3 shows the results of XPS cold probe characterization of the hydrogel surfaces tethered with PEG alone and PEGylated laminin. Despite the fact that trace contaminants were found on some sample surfaces, marked elemental differences in the surfaces corresponding to the different surface chemistries were observed. C_{1s} and O_{1s} of the control sample showed similar results at all incident angles. The C/O ratio was 2.0 : 1 at 30° and 1.6 : 1 at 90°, which indicates that the surface chemistry on the control sample was uniform and consistent. Furthermore, comparing the C_{1s} values at a 30° incident angle for all the samples (Table 3) reveals that the concentration increased with the different surface chemistries, e.g. modified by PEG and PEGylated protein respectively. The same pattern was observed when O_{1s} was examined in these samples.

It was also found that the C_{1s} and O_{1s} concentrations varied with the incident angle. At the 90° angle, the C_{1s} and O_{1s} concentrations in the tethered protein-modified hydrogel sample were similar to those of the control sample. This depth phenomenon indicates that the surface grafting of PEGylated proteins took place on the very top surface of the hydrogel (<75 nm) [26].

An important determinant in XPS analysis of the presence of protein on the hydrogel surface is the N_{1s} concentration (Table 3). N_{1s} was observed in much

Table 3.
Cold probe and dried *in situ* XPS characterization of hydrogels: plain hydrogel, PEGylated surface-modified hydrogel and tethered laminin surface-modified hydrogel

Hydrogel samples	C_{1s}			O_{1s}			N_{1s}			F_{1s}			Si		
	30°	45°	90°	30°	45°	90°	30°	45 °	90°	30°	45°	90°	30°	45°	90°
Cold probe															
Base PHEMA/MAA	66.6	61.3	61.9	33.1	37.9	37.0	0.3	0.8	1.1	N/A	N/A	N/A	—	—	—
PEG tethered hydrogel	25.2	35	19.5	70.2	57.3	75.4	1.9	1.1	2.1	1.2	2.1	1.4	0.6	3.5	0.4
Laminin-PEG tethered hydrogel	41	61.4	61.7	53.5	37.0	36.5	3.5	1.0	1.2	2.0	0.6	0.5	—	—	—
Dried *in situ*															
Base PHEMA/MAA	56.5	63.7	67.9	33.4	30.4	31.1	2.3	1.2	0.7	7.8	4.7	0.3	—	—	—
PEG tethered hydrogel	49.3	62.2	64.9	32.5	31.9	32.8	3.2	1.1	0	5.3	2.1	1.1	2.7	1.8	0.6
Laminin-PEG tethered hydrogel	66.6	61.6	65.8	22.1	32.4	30.6	5.0	3.3	1.9	6.2	2.8	1.7	—	—	—

higher concentration on the tethered protein-modified hydrogel samples than on any of the control samples despite the some apparent contamination in the controls.

Table 3 also lists the XPS results of the same samples that were dried *in situ* after the cold-probe XPS characterization was carried out. This portion of the study was conducted 'concurrently', i.e. there was no further sample handling involved during this procedure. Samples were dried only by heating the sample stage. Comparison of the XPS analysis data for the dried *versus* frozen samples shows the lack of utility of the dried method for hydrogel surface structure analysis. Contamination by F and Si was increased significantly on these sample surfaces compared with the previous XPS analysis, most probably as a result of vacuum grease vapor released when the environment was heated. Since the PEG and proteins were no longer held extended from the surface by frozen water crystals, the elemental differences found with increasing depth into the hydrogel surface were no longer detectable with the *in situ* dried samples. In addition, any remaining isolets of ice crystals within the surface of the sample after *in situ* drying serve to confound the O_{1s} data. These phenomena imply that the *in situ* drying procedure changed the hydrogel's original surface structure, which was clearly identifiable under 'environmental' conditions.

Mechanical properties

The mechanical properties of the hydrogel are critical to its applications and the tensile material properties of the hydrogel before and after CDI chemistry modification are shown in Table 4.

Table 4 shows that soaking the hydrogel in DMSO led to significant decreases in both the peak stress and Young's modulus compared to the base hydrogel (Table 4) ($P < 0.05$, ANOVA). DMSO is an extremely polar solvent and we observed a high degree of swelling in the slightly crosslinked, control hydrogel sample. It has been reported that small molecular particles inside the PHEMA hydrogel could be washed out by storage medium [27], resulting in microvoids within the bulk

Table 4.
Tensile material properties of pHEMA/MAA hydrogels

Hydrogel type	Peak stress (MPa)	Peak strain	Young's modulus (MPa)
Base PHEMA/MAA	0.2385 ± 0.0275	0.4492 ± 0.0719	0.5340 ± 0.0249
Base PHEMA/MAA-DMSO	0.1615 ± 0.0275	0.4358 ± 0.0719	0.3862 ± 0.0249
Base PHEMA/MAA-CDI 30 min	0.2200 ± 0.0275	0.4007 ± 0.0719	0.5556 ± 0.0249
Base PHEMA/MAA-CDI 60 min	0.2220 ± 0.0318	0.4031 ± 0.0830	0.5487 ± 0.0287
Base PHEMA/MAA-CDI 90 min	0.2294 ± 0.0275	0.4264 ± 0.0719	0.5402 ± 0.0249
Base PHEMA/MAA-CDI 120 min	0.1789 ± 0.0275	0.3020 ± 0.0719	0.6015 ± 0.0249
FAP-PEG-tethered PHEMA/MAA	0.1757 ± 0.0390	0.3190 ± 0.1016	0.5599 ± 0.0352
Fibronectin-PEG-tethered PHEMA/MAA	0.1505 ± 0.0275	0.2586 ± 0.0719	0.5949 ± 0.0249
Laminin-PEG-tethered PHEMA/MAA	0.1934 ± 0.0318	0.3505 ± 0.0830	0.5526 ± 0.0287

Data are mean values ± 95% confidence intervals (ANOVA).

hydrogel. Microvoids can serve as the initiation point for crack propagation in a tensile test, which eventually results in mechanical failure. Hence, the marked decrease in mechanical strength of DMSO-soaked hydrogels may be attributed to microvoids induced by DMSO washout.

Treating the base hydrogels with CDI for 30, 60 or 90 min induced no significant changes in hydrogel mechanical properties. However, the samples treated for 120 min exhibited a significantly higher modulus but significantly lower peak stress and strain, compared with the untreated hydrogel (Table 4) ($P < 0.05$, ANOVA).

All further surface modifications were performed on hydrogels treated with CDI for 120 min and the mechanical properties of the final tethered protein/peptide-modified hydrogels were not significantly different from those of 120-min CDI-treated hydrogel ($P > 0.05$, ANOVA). This indicates that grafting of the PEGylated proteins/peptide onto the hydrogel had no significant effect on bulk mechanical properties. This is understandable, since the reaction between the CDI-activated hydrogel and PEGylated proteins/peptides takes place on the hydrogel surface, as indicated by environmental XPS and should, therefore, not affect the basic mechanical properties of the bulk hydrogel.

CONCLUSIONS

This paper describes the development and characterization of tethered protein/peptide-modified hydrogel surfaces. The proteins/peptide tethered to one end of a PEG chain remained biologically active. Tethering the free end of the PEGylated proteins/peptide to the surface of PHEMA/MAA hydrogels using CDI chemistry successfully introduced a bioactive surface onto the hydrogels. Environmental XPS was used to characterize the presence of biological moieties in low concentration on the tethered protein/peptide-modified hydrogel surfaces. Mechanical testing indicated that the hydrogel retains its bulk mechanical properties after the surface addition chemistry; however, prolonged exposure to CDI treatment can induce significant reductions in Young's modulus and, ultimately, in the tensile strength of the hydrogel. Further studies on the specific cellular response to these modified hydrogels will be reported elsewhere.

Acknowledgements

This work was supported in part by USPHS grants R01EY012367 (to J. T. J.) and P30EY02377 (departmental core grant) from the National Eye Institute. National Institutes of Health (Bethesda, MD, USA), and an unrestricted grant from Research to Prevent Blindness, Inc. (New York, NY, USA).

REFERENCES

1. J. T. Jacob-LaBarre and D. R. Caldwell, in: *Progress in Biomedical Polymers: Polymeric Materials Science and Engineering*, C. G. Geblein and R. L. Dunn (Eds), p. 27. Plenum Press, New York, NY (1990).

2. T. V. Chirila, *Biomaterials* **22**, 3311 (2001).
3. V. Trinkaus-Randall, X. Wu, R. Tablante and A. Tsuk, *Artif. Organs* **21**, 1185 (1997).
4. M. Franco, P. F. Nealey, S. Campbell, A. I. Teixeira and C. J. Murphy, *J. Biomed. Mater. Res.* **52**, 261 (2000).
5. G. H. Hsiue, S. D. Lee, C. C. Wang, M. H. Shiue and P. C. Chang, *Biomaterials* **14**, 591 (1993).
6. G. H. Hsiue, S.D. Lee, C. C. Wang, M. H. Shiue and P. C. Chang, *Biomaterials* **15**, 163 (1994).
7. S. D. Lee, G. H. Hsiue and P. C. Chang, *Biomaterials* **17**, 1599 (1996).
8. R. Latkany, A. Tsuk, M.-S. Sheu, I.-H. Loh and V. Trinkaus-Randall, *J. Biomed. Mater. Res.* **36**, 29 (1997).
9. V. Trinkaus-Randall, J. Capecchi, A. Newton, A. Vadasz, H. Leibowitz and C. Franzblau, *Invest. Ophthalmol. Vis. Sci.* **29**, 393 (1988).
10. H. Kobayashi and Y. Ikada, *Biomaterials* **12**, 747 (1991).
11. H. Kobayashi and Y. Ikada, *Curr. Eye Res.* **10**, 899 (1991).
12. D. F. Sweeney, R. Z. Xie, M. D. M. Evans, A. Vannas, S. D. Tout, H. J. Griesser, G. Johnson and J. G. Steele, *Invest. Ophthalmol. Vis. Sci.* **44**, 3301 (2003).
13. R. Z. Xie, D. F. Sweeny, G. J. Baumerm, G. Johnson, H. J. Griesser and J. G. Steele, *Aust. N. Zeal. J. Ophthalmol.* **25** (Suppl.), S46 (1997).
14. H. Thiessen, K. McLean, G. Johnson, J. G. Steele and H. J. Griesser, *Soc. Biomater. Meet. Transact.* **22**, 450 (1999).
15. J. R. Rochefort and J. T. Jacob, *Sixth World Biomater. Congr.* **1**, 10 (2000).
16. V. Trinkaus-Randall, M. Vural, J. Capecchi, C. Franzblau and H. M. Leibowitz, *Invest. Ophthalmol. Vis. Sci.* **32** (Suppl.), 1072 (1983).
17. K. Merritt, C. M. Griffith, Y. Deslandes, G. Pleizier and H. Sheardown, *J. Biomater. Sci. Polymer Edn* **12**, 647 (2001).
18. M. Goodman, M. Bhumralkar, E.A. Jefferson, J. Kwak and E. Locardi, *Biopolymers* **47**, 127 (1998).
19. G. Johnson, M. Jenkins, K. M. McLean, H. J. Griesser, J. Kwak, M. Goodman and J. G. Steele, *J. Biomed. Mater. Res.* **51**, 612 (2000).
20. I. M. Nunez, F. F. Molock, L. D. Elliot and J. D. Ford, US Patent No. 5,490,959 (1996).
21. M. J. Refojo, *J. Appl. Polym. Sci.* **9**, 3417 (1965).
22. M. Steinitz and L. Baraz, *J. Immunol. Methods* **238**, 143 (2000).
23. A. M. Hawkridge, G. R. Sagerman and J. A. Gardella, Jr., *J. Vacuum Sci. Technol, Part A* **18**, 567 (2000).
24. A. Agresti, *Categorical Data Analysis.* Wiley, New York, NY (1990).
25. C. H. Bamford, I. P. Middleton and K. G. Al-Lamee, *Polymer* **27**, 1981 (1986).
26. P. C. Schamberger, J. A. Gardella, G. L. Grobe and P. L. Valint, *Polym. Prepr.* **33**, 503 (1992).
27. S. H. Kim, A. Opdahl, C. Marmo and G. A. Somorjai, *Biomaterials* **23**, 1657 (2002).

Gels, Genes, Grafts and Giants, pp. 681–694 (2005)

Enhancement of stability of glucose oxidase by immobilization onto metal ion-chelated poly(N-vinyl imidazole) hydrogels

NURSEL PEKEL, BEKIR SALIH and OLGUN GÜVEN *

Department of Chemistry, Hacettepe University, Beytepe, 06532 Ankara, Turkey

Received 10 May 2004; accepted 25 August 2004

Abstract—Poly(N-vinylimidazole), PVIm, gels were prepared by γ-irradiation polymerization of N-vinylimidazole in aqueous solutions as an affinity gel for glucose oxidase (GOx). These affinity gels with a water swelling ratio of 1800% for plain polymeric gel and between 30–80% for Cu(II) and Co(II)-chelated gels at pH 6.0 in phosphate buffer were used in the GOx adsorption studies. Maximum metal ion adsorption capacity of these hydrogels was found to be 3.64 mmol/g dry gel for Cu(II) and 1.72 mmol/g dry gel for Co(II) leading to GOx adsorption capacities of 343 and 528 mg enzyme/g dry gel, respectively, as compared to 228 mg for the plain dry PVIm gel. Activity studies were carried out using plain and the metal ion-chelated form of this hydrogel to investigate the stability and retained activity of the GOx in different buffer solutions and at different temperatures. Activity of the enzyme, either in free or immobilized form on the gel, decreased dramatically in acetate buffer solutions. In phosphate buffer solution, however, stability of enzyme has been found to be significantly high reaching 90% retained activity at the end of a 40-day period at 4°C for Co(II) chelated systems. After immobilization of the enzyme onto metal-chelated hydrogel, the thermal stability of enzyme was enhanced significantly showing 23% activity, even at 75°C.

Key words: Glucose oxidase adsorption; affinity gels; poly(N-vinylimidazole); metal chelate adsorption; glucose oxidase activity.

INTRODUCTION

Purification of proteins and enzymes need several steps involving methods which select on the basis of molecular size (gel-permeation chromatography), electrical charge (ion-exchange chromatography), hydrophobicity (hydrophobic interaction chromatography) or biological recognition (bioaffinity chromatography). Metal

*To whom correspondence should be addressed. Tel./Fax: (90-312) 297-7977.
E-mail: Guven@hacettepe.edu.tr

chelate affinity chromatography introduces a new possibility for selectively interacting materials on the basis of their affinities for chelated transition metal ions. The separation is based on differential binding abilities of the proteins or enzymes to interact with chelated metal ions to a solid carrier [1, 2]. Metal chelate affinity chromatography of proteins, with metal chelate linked to Sepharose, was first introduced by Porath *et al.* [3]. They reported a model system using Zn(II) and Cu(II) columns in tandem for the fractionation of human serum proteins. Subsequent studies have shown the wide applicability of the technique and consistency of the methodology.

The specificity of enzymes and their ability to catalyse reactions make them attractive for applications in the biochemical, biomedical, industrial and analytical field. The major advantage of this specificity is that the yields are free of side products and, thus, cost-effective. The recovery and reusability of free enzymes and catalysts are quite limited and this has led to the development of a wide variety of immobilization techniques. Immobilization also offers operational advantages over free enzymes, such as choice of batch or continuous processes, rapid termination of reactions, controlled product formation, ease of removal from the reaction mixture and adaptability to various engineering designs [4]. Immobilization can be achieved through physical adsorption on solid supports, micro-encapsulation, and covalent bonding or matrix entrapment [5]. Among these, matrix entrapment seems to have advantages over the others, such as simplicity of the process, ease of transport of substrate and the products, and less possibility of inactivation during immobilization [6, 7]. The use of the open pore hydrogels either in anionic or cationic form gives some extra advantages for the separation of enzymes and especially for the activity studies of enzymes. Because of the open hydrophilic network of the gels, synthetic hydrogels have been gaining increasing popularity in various biomedical applications ranging from soft contact lenses to drug-delivery systems [8, 9]. On the other hand, the charged hydrogels are useful for the ion exchange and immobilization of enzymes [10–13]. In this case, adsorption and desorption of enzymes depend on certain external conditions, such as pH [14], temperature [15], ionic strength [16], pressure [17], ultrasound and electromagnetic radiation [18].

Glucose oxidase (GOx) has been mostly used in biological applications as biosensor, biocatalyst or drug-delivery systems. Some specific examples include the synthesis of glucose-sensitive hydrogels that could be used to deliver insulin to diabetic patients using an internal pH trigger [19]. The immobilization and stability of GOx on novel substrates have therefore continued to attract attention of scientists working in closely related fields.

The purpose of the present study is to prepare poly (N-vinylimidazole) based affinity gels containing Cu(II) and Co(II) ions (in chelate form) for metal-chelate affinity separation and enhanced stabilization of GOx enzyme activity. In this article, we present GOx adsorption properties of plain and Cu(II) and Co(II)-chelated hydrogels for the enhancement stability of enzyme in different buffer solutions and at different temperatures. GOx activity was measured as a function of

time in order to find the stability of GOx when it was adsorbed on/in the polymeric gels. The aim of using metal-chelated hydrogels is to see the additional effect of chelated form of metal ion on the stabilization of the enzyme if any.

MATERIALS AND METHODS

Materials

Glucose oxidase (lyophilized, Fraction V) was purchased from Sigma (St. Louis, MO, USA) and used as received. The information provided by the supplier indicates that it may contain up to 2% other enzymes as impurities. N-vinylimidazole (VIm) was obtained from Merck (Darmstadt, Germany) and distilled under reduced pressure and stored at 4°C until use. Copper chloride ($CuCl_2$) and Cobalt chloride ($CoCl_2$) were purchased from BDH (Poole, UK). All other chemicals were purchased from Merck, of reagent grade and used as received.

Preparation of PVIm gel

A VIm/water mixture (10 ml) containing 8 ml VIm and 2 ml water was placed in polyethylene straws of 3 mm diameter and irradiated in a ^{60}Co-γ source at a fixed dose rate of 4.5 kGy/h until gelation. The gel was taken out from straws and washed several times with distilled water to remove unreacted monomer and soluble polymer. Gels were dried in vacuum oven at 60°C for 3 days, and ground to obtain the desired size (800 μm) and then stored until use. Fine particles of hydrogels with 800 μm average size were used for the metal ion and GOx adsorption. The other details related to polymerization system and procedures are given elsewhere [20, 21].

Water content of PVIm gels

The swelling behavior of PVIm gels was investigated in distilled water. Dry gel pieces were placed in distilled water and kept at a constant temperature of 25 ± 0.5°C. Swollen gels were periodically removed and weighed. The swelling of gels was calculated by using the following expression:

$$\text{Swelling (\%)} = ((W_s - W_0)/W_0) \times 100,$$

where W_0 and W_s are weights of the gels before and after swelling, respectively.

Incorporation of Cu(II) and Co(II) ions into PVIm gels

Chelates of PVIm gels with Cu(II) and Co(II) ions were prepared as follows: PVIm gel particles were mixed with aqueous solutions containing 100–1200 ppm Cu(II) or Co(II) ions, at constant pH of 6.0 (adjusted with NaOH and HNO_3), which has been determined to be the optimum pH for Cu(II) and Co(II) chelate formation [22] at

room temperature. A very small amount of adsorption was observed at low pH (pH 2.5) due to the protonation of N atom at position 3 of imidazole ring. For the maximum metal ion adsorption, 1200 ppm solutions of $CuCl_2$ and $CoCl_2$ were used as the source of Cu(II) and Co(II) ions. The flasks were agitated magnetically at 1000 rpm for 2 days (sufficient to attain metal ion adsorption equilibrium). The concentration of the metal ion in the resulting solutions was determined with an UV-Vis Spectrophotometer (model PU 8715, Philips, UK).

Metal ion leakage from the PVIm gels was investigated in 0.02 M acetate and phosphate buffer in the pH range 4.0–8.5 and also in medium containing 1.0 M KSCN at pH 7.0. The gels suspended in the media mentioned above were stirred for 24 h at room temperature. After this period, the leakage of metal ions was determined in the supernatant using an atomic absorption spectrophotometer (GBC 932 AA, Australia).

GOx adsorption

GOx adsorption on plain, Cu(II) and Co(II)-chelated PVIm gels was studied at various pH values. The pH of the adsorption medium was changed between 4.0 and 8.5 by using two different buffer systems (0.02 M CH_3COONa/CH_3COOH for pH 4.0–6.0, 0.02 M K_2HPO_4/KH_2PO_4 for pH 7.0–8.5). The initial GOx concentration was changed between 0.5 and 10.0 mg/ml to determine the GOx adsorption capacity of the PVIm gels. In a typical adsorption experiment, GOx was dissolved in 20 ml 0.02 M phosphate buffer solution and a known amount of gel particles was added. The adsorption experiments were carried out for 2 h at 4°C, sufficient to reach equilibrium after which the gels were separated from the solution. The GOx adsorption capacity was determined by measuring the initial and final concentration of GOx in the adsorption medium spectrophotometrically at 276 nm, which is the maximum absorption wavelength of GOx used in this study.

GOx activity studies

The thermal stability of the immobilized enzyme onto plain and the chelated forms was determined by incubation at temperatures ranging from 25°C to 75°C for 15 min. After the incubation at the given temperature, enzyme activity was determined.

Activity studies for the free enzyme and enzyme adsorbed onto plain PVIm, Cu(II) and Co(II)-chelated PVIm were carried out using the TRINDER method [23]. Free enzyme or adsorbed enzyme was placed into a flask continuously oxygenated and then a known amount of glucose was added into the medium. After a standard waiting time for all systems, the reduced form of horseradish peroxidase dye was added into the reaction vessel. At the end of the reaction, the colored oxidase form of the dye was measured using UV-Vis spectrophotometry at 500 nm. Finally, GOx activity was calculated from the absorbance value of the oxidized form of the dye.

RESULTS AND DISCUSSION

Characteristics of PVIm and metal ion-chelated PVIm gels

Preparation and characterization details of PVIm gels have been given in our previous paper [24]. The gels prepared in this study are highly swellable (swelling ratio reaches 6000% for the protonated form of PVIm gel at pH 2.5). The swelling ratios of Cu(II) and Co(II)-chelated gels are rather low compared to the plain PVIm gel. This is due to interaction of imidazole nitrogen groups with metal ions and subsequent shrinking of gel structure. Despite considerable shrinking of metal ion-chelated gels, GOx adsorption has been found to be superior to plain gels.

Metal ion incorporation into the PVIm gels

Cu(II) and Co(II) incorporation into the PVIm gels at pH 6.0 (unprotonated) and at pH 2.5 (protonated) in 24 h has already been explained elsewhere [22]. Maximum metal ion adsorption of 3.64 and 1.72 mmol/g dry gel for Cu(II) and Co(II) has been achieved from aqueous solutions of 1200 ppm concentration onto unprotonated PVIm. The amount of metal ions adsorbed onto the hydrogels at low pH values is significantly lower, due to the protonation of active sites on hydrogels. This however, provides an easy way of GOx release from the hydrogels by simply decreasing the pH of the medium. Fresh enzyme can be loaded again onto hydrogels for repeated use at around neutral pH. The amount of Co(II) adsorbed onto hydrogel is nearly half of that for Cu(II). This is most probably due to the differences in the ionic radii and electronegativities of these two ions. Cu(II) and Co(II) have ionic radii of 71 and 88.5 pm, respectively [25], which makes the complexation of the larger radius ion, Co(II) in this case, sterically more difficult, hence the lower extent of chelation. The higher extent of chelation with Cu(II) may also arise from the difference between electronegativities of these two ions (Cu(II) 1.9, Co(II) 1.8), which is in favor of Cu(II).

GOx adsorption onto PVIm and metal ion-chelated PVIm

The maximum adsorption of GOx was observed at around pH 7.0. The adsorption behavior of GOx onto plain and metal-chelated PVIm gels at different pH values has already been reported in our previous papers [26, 27]. The maximum GOx adsorption capacities of PVIm, Cu(II)-PVIm and Co(II)-PVIm reached in this study are 228, 343 and 528 mg GOx/g dry gel, respectively. These are average values of triplicate measurements. Chelated Cu(II) and Co(II) ions significantly increased the GOx adsorption capacity of the gel (more than 100% in the case of Co(II)-chelation), possibly because of the specific favourable interactions between GOx and polymer associated metal ions to complete the coordination number of metal ions. In other words, metal ions promote the adsorption of GOx within the PVIm matrix. Considering the coordination number and geometries of metal ions

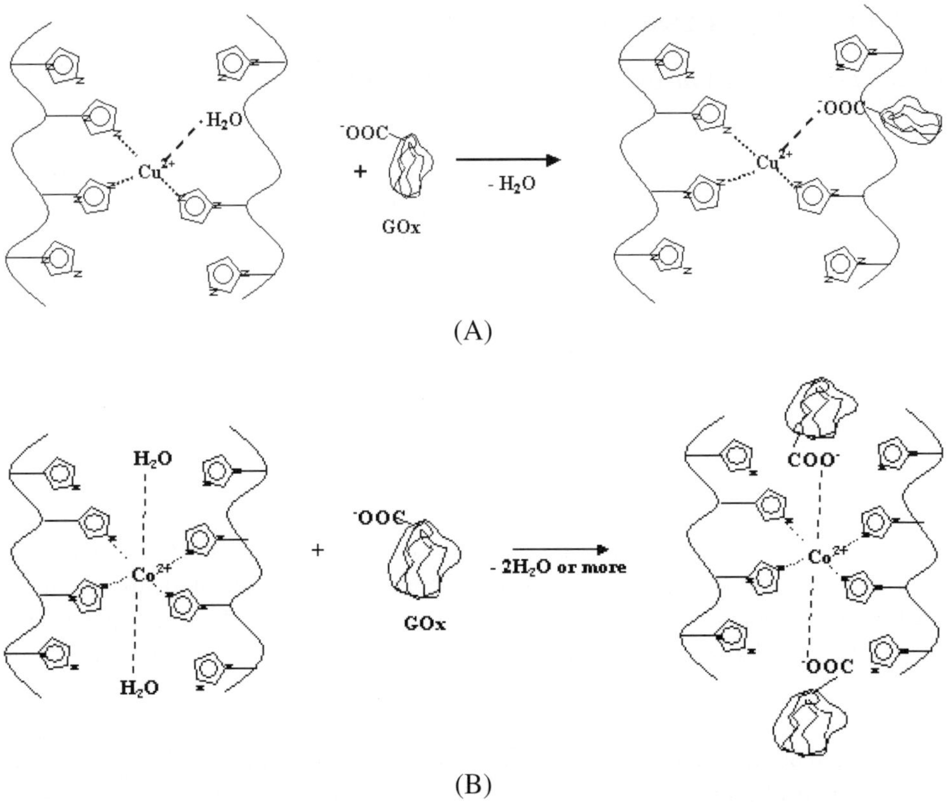

Figure 1. Proposed metal ion-chelated form of PVIm gels: (A) Cu(II) and (B) Co(II) chelation at pH 7.0.

associated with the imidazole ring of the polymer a schematic presentation of PVIm-metal ion-GOx is given in Fig. 1. The GOx adsorption capacity for the metal ion-chelated gels is significantly higher than in the plain gel, despite the more compact structure and very low swelling of metal ion-chelated gels. The relatively large difference in GOx adsorption by Co(II)-chelated gel compared to Cu(II)-chelation can be explained by referring to the differences in their coordination numbers. The coordination number of Co(II) in its complexes in aqueous media is usually 6. In the case of Cu(II), this is 4. Upon complexation of Co(II) and Cu(II) with VIm pendant groups the number of unoccupied coordination sites will be larger for Co(II), thus allowing more GOx molecules to be adsorbed.

In order to have a better understanding of the adsorption of enzyme onto metal ion-chelated hydrogels, competitive adsorption between metal ion and GOx was investigated. In this experiment, a mixture of metal ion and GOx was treated with hydrogels in phosphate buffer solutions and the decrease in metal ion and GOx concentration were followed using UV-Vis spectrophotometry at maximum absorption wavelengths of metal ions and GOx. The results obtained from these experiments are given in Fig. 2. As can be seen from Fig. 2, adsorbed metal ion

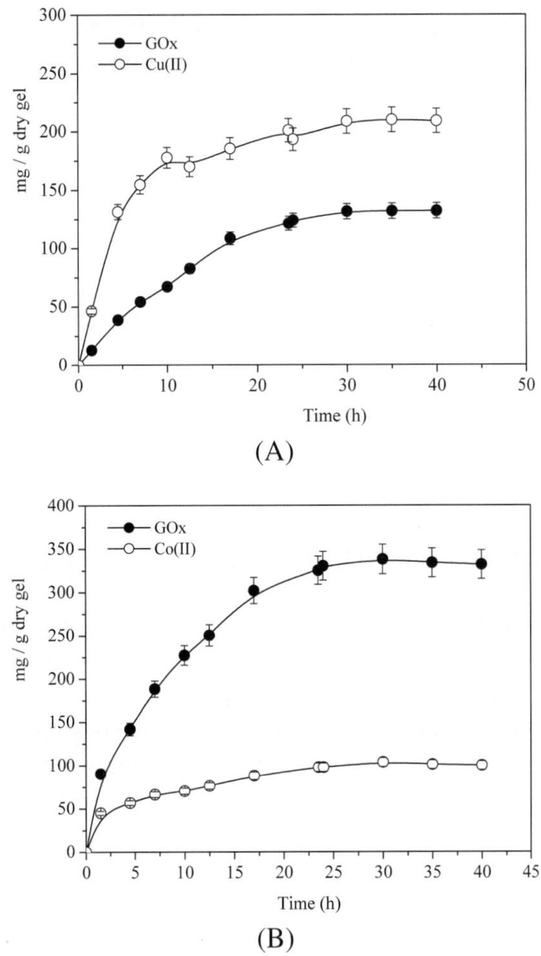

Figure 2. Adsorption of GOx and metal ions onto the PVIm hydrogels from the enzyme and metal ion mixture. (A) From GOx and Cu(II) ion mixture; (B) from GOx and Co(II) ion mixture. Adsorption conditions: initial concentration of enzyme, 10 mg of enzyme/ml; temperature 4°C. Initial concentration of the metal ions is 1200 ppm.

concentrations are approximately the same as in the case of non-competitive metal ion adsorption onto the hydrogels. However, the amount of adsorbed GOx in the case of competitive adsorption from the mixture of metal ion and GOx is low, 130 and 330 mg enzyme/g dry gel in the presence of Cu(II) and Co(II) ions, respectively. The equilibrium value of adsorption of both metal ions and GOx however follow the same trend as for GOx adsorbed on the metal ion chelated gels. Here, in the mixture of the metal ion and GOx, complexation between metal ion and enzyme takes place first. This restricts the adsorption of metal ions already attached to GOx on the active sites of hydrogels. In other words, the metal ion–enzyme complex in the solution contains lesser coordination vacancies to be coordinated with the imidazole nitrogen of hydrogels. As a result, adsorption tendency of GOx towards the gel was decreased when the GOx adsorption experiments performed simultaneously from the mixture of metal ion and enzyme. This shows us that, first, metal ion-chelated form of the hydrogels should be prepared and then enzyme should be loaded onto the metal ion chelated form of hydrogel, in order to increase the amount and activity of the enzyme immobilized on the hydrogels.

Stability of GOx immobilized onto PVIm gels

The stability of the enzymes in general depends on a number of factors, temperature and pH being the two most important. The stability of GOx is rather low at relatively high temperatures and even at room temperature. To investigate the stability of GOx at different temperatures, in different buffer solutions and in different immobilization forms, several experiments were conducted in this work. Activity loss of GOx with time at +4°C is shown in Fig. 3. As seen here, stability of the GOx immobilized onto PVIm, Cu(II)-PVIm and Co(II)-PVIm gels is considerably high as compared to the values reported in the literature (Table 1). In 10 days, the activity of the GOx remains to be reasonably high and not less than 90% of the original value for all gels at +4°C. At the end of the 40-day period, the activity of the GOx in PVIm, Cu(II)-PVIm and Co(II)-PVIm gels remained at 76%, 63% and 83% of original values, respectively. Plain and metal ion-chelated PVIm gels could be used safely for the removal of the GOx and also for biosensors applications within 10 days at +4°C without significant activity loss. Among the three different gel systems studied here, Co(II) chelated-PVIm gel has provided the highest stability for the immobilized enzyme and the activity loss is only about 22% at the end of the 70-day period. From our studies on specific/non-specific enzyme adsorption/release on these gels we have noted a leakage of about 5–10% GOx from the gels. The activity and stability studies take this behavior already into account.

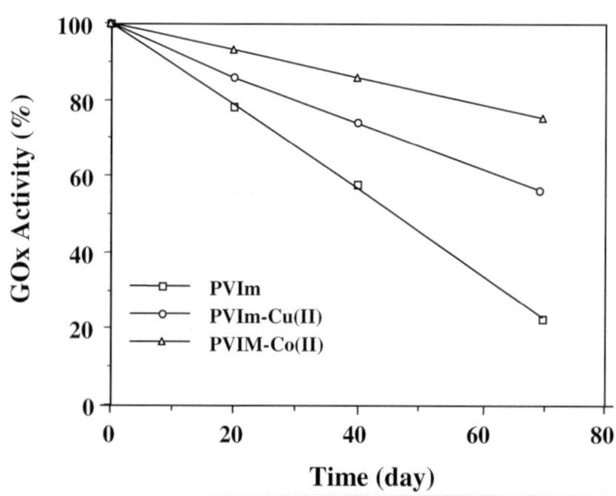

Figure 3. Stability of GOx immobilized in PVIm gels. Amount of GOx in PVIm gels: 228, 343 and 528 mg GOx/g dry gels for PVIm, Cu(II)- and Co(II)-chelated PVIm gels resp.; amount of metal ions in PVIm gels: 3.64 and 1.72 mmol for Cu(II)- and Co(II)-chelated PVIm gels; pH 7.0; temperature 4°C.

Table 1.
Results of activity studies for GOx immobilization

Polymer support	Support type	Immobilization method	GOx concentration range	GOx amount (mg/g support)	Activity (%)	Reference
Acrylic acid grafted magnetite	Particle	Chemical (covalent)	2.5 mg/ml	2.8	95 (after 9 months)	[28]
Electropolymerized o-Aminophenol	Film	Physical (adsorption)	2–40 mg/ml	0.4 mg/cm^2	50 (after 10 months)	[29]
Hyperbranched PAAc on Au-coated silica	Film	Physical (entrapment)	5.7 mg/ml	0.75 mg/cm^2	100 (on the 69th day)	[30]
GMA-co-AN	Membrane	Chemical (covalent)	0.1 wt%	0.036 mg/cm^2	94 (after 2 months)	[31]
VIm-co-AN	Membrane	Physical (adsorption)	0.1 wt%	0.023 mg/cm^2	100 (after 5 min)	[32]
AAc-g-PVDF	Membrane	Chemical (covalent)	4 mg/ml	0.27 mg/cm^2	55 (after 60 days)	[33]
Poly(2-hydroxyethyl methacrylate)	Gel	Physical (entrapment)	2–20 mg/ml	10	90 (after 60 days)	[34]
DEAEMA-g-PEG	Gel	Chemical (copolymerization)	2 mg/ml	0.33–0.66	—	[35]
PHEMA	Gel	Physical (entrapment)	1.5 mg/ml	1.5	50	[36]
Fe(III)-dye-PHEMA	Membrane	Physical (adsorption)	0.5–5 mg/ml	872 mg/m^2	—	[37]
Fe(III) chelated p(HEMA-co-MMA)	Microparticle	Physical (adsorption)	2 mg/ml	165	62	[38]
PVIm	Gel	Physical (adsorption)	0.5–10 mg/ml	228	25	[27]
Cu(II) chelated PVIm	Gel	Physical (adsorption)	0.5–10 mg/ml	343	55	[27]
Co(II) chelated PVIm	Gel	Physical (adsorption)	0.5–10 mg/ml	528	75 (after 70 days)	[27]

In order to see the effect of buffering medium on immobilized and free enzyme stabilities, tests were made in acetate and phosphate buffers at pH 7.0 and 25°C with time. In three different forms of the enzyme (free, PVIm hydrogel-Cu(II)-enzyme and PVIm hydrogel-Co(II)-enzyme systems), the stability of GOx was examined by measuring the activity for all enzyme forms at 25°C and the results are given in Fig. 4. In Fig. 4A, it could be seen that all forms of the enzyme in the acetate buffer were unstable and in the first hour, about 95% of original activity was lost. In phosphate buffer at 25°C, again free enzyme loses its activity almost completely in the first 5 h, but the other two forms, PVIm hydrogel-Cu(II)-enzyme and PVIm hydrogel-Co(II)-enzyme, were found to be more stable and not showing significant loss in their activities within 27 h (Fig. 4B). It can be concluded that in phosphate buffer, activity of the immobilized GOx enzyme in Co(II) and Cu(II) metal ion-chelated hydrogels could be retained without any significant loss in activity for a long period at room temperature.

The interesting result related to severe activity loss in acetate buffer needs further elaboration. This is probably due to the inhibitive interaction between acetate ion and the active sites of the enzyme. The active site of GOx is provided by a large deep pocket which is shaped like a funnel and is formed on one side by residues of the second molecule of the dimer. Part of the bottom of the catalytic cavity is formed by the mostly buried Asp-584, Tyr-515, His-559 and His-516, which are close to the front of isoalloxazine moiety, allowing them to be involved in the catalytic reaction or substrate binding [39]. The Asn amino-acid residue has an amino group site that can be easily protonated and further interact electrostatically with acetate anion and diminish the binding capacity of the enzyme with the substrate. In the case of phosphate, cavity size and the other additional interactions may not be effective to reduce the activity of enzyme. This is probably the reason for the activity of enzyme in phosphate buffer to be retained for a longer time without noticeable changes.

Effect of temperature and immobilization on the activity of enzyme

Temperature and the nature of buffer system have been observed to be very effective on the activity of GOx. At room temperature, free GOx loses its activity very fast, as seen in Fig. 4A and 4B, both in acetate and phosphate buffer solutions. Immobilization on metal ion-chelated PVIm hydrogels, however, seems to provide a solution, as shown in Fig. 4B. 95% of the original activity of GOx has been retained at room temperature for 1 day when it is immobilized onto Co(II)-chelated PVIm gels. When the amount of adsorbed GOx/g sorbent are compared with the literature values, a relatively higher amount of enzyme is seen to be immobilized onto plain and especially metal ion-chelated hydrogels in this study (Table 1). The activity of the enzyme was retained for a long time without any noticeable loss after its immobilization on the metal ion chelated hydrogels. In the physical adsorption of GOx on sorbents, the stability of enzyme is generally not retained for as long as in the covalent immobilization or entrapment processes. In our case, immobilization of GOx on the metal ion chelated hydrogels increased the activity and the stability

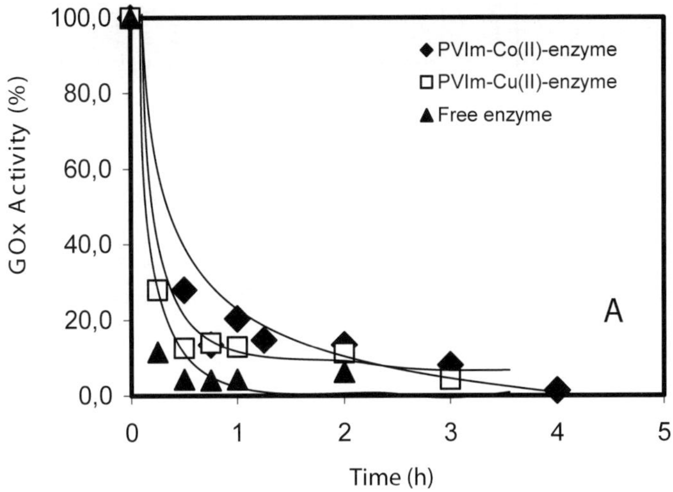

(A)

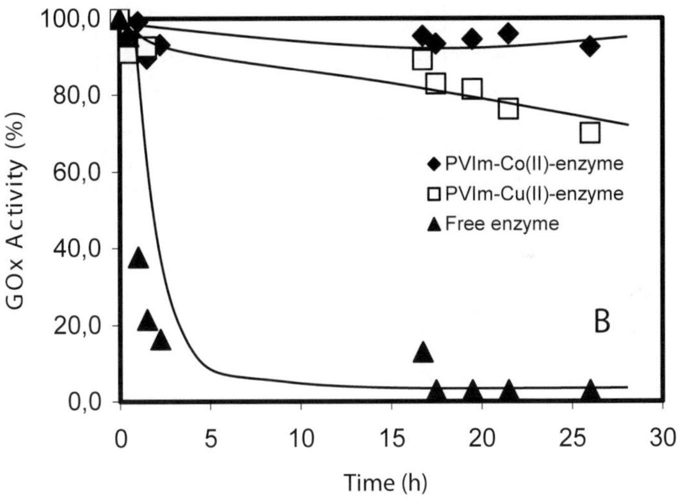

(B)

Figure 4. Effect of buffer type on GOx activity with time at 25°C: (A) in acetate buffer at pH 7.0; (B) in phosphate buffer at pH 7.0. Cu(II) and Co(II) loading, 3.64 and 1.72 mg/g dry gel, enzyme contents: 228 mg/g dry gel for plain PVIm, 343 mg/g dry Cu(II)-chelated PVIm, and 528 mg/g dry Co(II)-chelated PVIm gels.

of the enzyme compared to other physical adsorption modes, for a comprehensive comparison, see Table 1. Also, after loss of the activity of enzyme immobilized on the metal ion-chelated hydrogels, the hydrogels could be regenerated and reloaded with fresh enzyme for the repeated use of the same hydrogels.

Temperature is another very important parameter on the activity and the stability of enzymes either in the free or immobilized form. In this study, temperature stability of GOx was examined and compared with the literature results. We

Table 2.
Comparison of GOx activity at high temperatures

Polymer support	Support type	GOx amount (mg/g support)	Activity (%)			Reference
AAc-g-PVDF	Membrane	0.27 mg/cm^2	100 at 35°C	85 at 40°C	80 at 45°C	[33]
GMA-co-AN	Membrane	0.036 mg/cm^2	100 at 35°C	55 at 40°C	—	[31]
PHEMA	Gel	1.5	95 at 50°C	60 at 60°C	10 at 65°C	[36]
VIm-co-AN	Membrane	0.023 mg/cm^2	100 at 30°C	85 at 35°C	75 at 40°C	[32]
Acrylic acid grafted magnetite	Particle	2.8	50 at 40°C	40 at 60°C	—	[28]
Fe(III) chelated p(HEMA-co-MMA)	Microparticle	165	100 at 40°C	90 at 45°C	0 at 60°C	[40]
PVIm	Gel	228	70 at 37°C	12 at 50°C	7 at 75°C	This work
Cu(II) chelated PVIm	Gel	343	85 at 37°C	55 at 50°C	11 at 75°C	This work
Co(II) chelated PVIm	Gel	528	95 at 37°C	70 at 50°C	23 at 75°C	This work

observed that enzyme activity was significantly reduced with temperature both for free enzyme and enzyme immobilized on the plain hydrogels. Our results showed that GOx activity was retained considerably and activity loss with temperature was not too much adversely affected when the metal ion chelated form of the hydrogels were used in immobilization. This indicates that metal chelated form of the enzyme on PVIm hydrogels is getting more stable conformation and not being affected too much from temperature within the range studied here. At 75°C, the enzyme showed 7, 11 and 23% activity when immobilized onto plain, Cu(II)-chelated and Co(II)-chelated hydrogels, respectively. Especially in the case of Co(II)-chelated hydrogels, enzyme activity is still useful at 75°C for the determination of substrate in different media. These results show that the stability of GOx onto Co(II)-chelated hydrogel is significantly higher than the activities reported in other studies in the literature at 75°C (Table 2).

CONCLUSIONS

Poly(N-vinylimidazole) gels were prepared by irradiation of N-vinyl imidazole/water mixtures at ambient conditions by gamma rays. Metal ion adsorption capacities of the studied hydrogels are high enough to load glucose oxidase (GOx) on the metal ion-chelated hydrogels. The maximum chelated amounts of Cu(II) and Co(II) ions were 3.64 and 1.72 mmol metal ions/g dry gels, respectively. Metal-chelated form of hydrogels showed adsorption capacities of up to 528 and 343 mg GOx/g dry gels for Co(II)- and Cu(II)-chelated poly(N-vinylimidazole), respectively. Activities of the free enzyme and its metal ion-chelated form were found to be instable in acetate buffer at 25°C. Metal ion-chelated forms of enzyme, however, were found to be quite stable for a long time in phosphate buffer, even at 25°C. The stability of the immobilized enzyme is high for all gels and the activity is not less than 90% in the first 10 days for systems kept at +4°C. Co(II) chelated gel has higher stability and the enzyme activity loss in the 70 days is only about 22%. Metal ion chelation imparts considerable thermal stability to GOx when immobilized in the PVIm gels studied here.

Acknowledgements

This article is dedicated to Prof. Dr. Allan Hoffman on the occasion of his 70th birthday. O. G. appreciates support from TÜBA, Academy of Sciences of Turkey.

REFERENCES

1. M. Walenius, P. Flodin and J. Porath, *J. Chromatogr.* **604**, 5 (1992).
2. C. M. Zeng, J. L. Liao, K. Nakazato and S. Hjerten, *J. Chromatogr. A* **753**, 227 (1996).
3. J. Porath, J. Carlsson, I. Olsson and G. Belfrage, *Nature* **258**, 598 (1975).
4. L. Sottrup-Jensen, T. E. Petersen and S. A. Magnusson, *FEBS Lett.* **121**, 275 (1980).

5. J. G. Franchina, W. M. Lackowski, D. L. Dermody, R. M. Crooks and D. E. Bergbreiter, *Anal. Chem.* **71**, 3133 (1999).
6. R. Axen, in: *Insolubilized Enzymes*, M. Salmona (Ed.), p. 9. Raven Press, New York, NY (1974).
7. E. Katzir-Katchalski, in: *Enzyme Engineering*, H. H. Weetall and G. P. Royer (Eds), p. 12. Plenum Press, New York, NY (1980).
8. D. C. Rijken, G. Wijngaards, M. Z. Jong and J. Welbergen, *Biochim. Biophys Acta* **580**, 140 (1979).
9. D. C. Rijken and D. Collen, *J. Biol. Chem.* **256**, 7035 (1981).
10. I. Ohkubo, T. Kondo and N. Taniguchi, *Biochim. Biophys. Acta* **616**, 89 (1980).
11. K. Podual, F. J. Doyle III and N. A. Peppas, *Polymer* **41**, 3975 (2000).
12. S. Bharathi and M. Nogami, *Analyst* **126**, 1919 (2001).
13. S. Reiter, K. Habermüller and W. Schuhmann, *Sens. Actuat. B* **79**, 150 (2001).
14. M. Falamarzian, B. C. Moxley, B. A. Firestone and B. A. Siegel, *Proc. Int. Control Rel. Bioact. Mater.* **15**, 23 (1988).
15. A. S. Hoffman, in: *Polymers in Medicine III*, C. Migliaresi, E. Cheillini, P. Guisti and N. Luigi (Eds), p. 15. Elsevier, Amsterdam (1988).
16. R. A. Siegel and B. A. Firestone, *Macromolecules* **21**, 3254 (1988).
17. K. K. Lee, E. L. Cuslesler, M. Marchetti and M. A. McHugh, *Chem. Eng. Sci.* **45**, 766 (1990).
18. A. J. Grodinsky and A. M. Weiss, *Separ. Purif. Methods* **14**, 1 (1985).
19. C. M. Hassan, F. J. Doyle III and N. A. Peppas, *Macromolecules* **30**, 6166 (1997).
20. N. Pekel and O. Güven, *Colloid. Polym. Sci.* **277**, 570 (1999).
21. N. Pekel, Preparation and characterization of poly(N-vinylimidazole)/maleic acid hydrogels, and their use in the adsorption of metal ions, PhD Thesis, p. 110. Hacettepe University, Ankara (1998).
22. N. Pekel and O. Güven, *Polym. Bull.* **51**, 307 (2004).
23. A. J. Pesco and L. A. Kaplan, *Methods Clin. Chem.* **16**, 105 (1987).
24. N. Pekel and O. Güven, *Polym. Int.* **51**, 1404 (2002).
25. J. E. Huheey, in: *Inorganic Chemistry: Principles of Structure and Reactivity*, 4th edn, p. 114. Harper Collins, New York, NY (1993).
26. B. Salih, N. Pekel and O. Güven, *J. Appl. Polym. Sci.* **82**, 446 (2001).
27. N. Pekel, B. Salih and O. Güven, *J. Mol. Catal. B: Enzym.* **21**, 273 (2003).
28. E. M. Shimomura, H. Kikuchi, T. Yamauchi and S. Miyauchi, *J. Macromol. Sci. Pure Appl. Chem.* **A33**, 1687 (1996).
29. Z. Zhang, H. Liu and J. Deng, *Anal. Chem.* **68**, 1632 (1996).
30. J. G. Franchina, W. M. Lackowski, D. L. Dermody, R. M. Crooks, D. E. Bergbreiter, K. Sirkar, R. J. Russell and M. V. Pishko, *Anal. Chem.* **71**, 3133 (1999).
31. T. Godjevargova, V. Konsulov and A. Dimov, *J. Membr. Sci.* **152**, 235 (1999).
32. T. Godjevargova, V. Konsulov, A. Dimov and N. Vasileva, *J. Membr. Sci.* **172**, 279 (2000).
33. L. Ying, E. T. Kang and K. G. Neoh, *J. Membr. Sci.* **208**, 361 (2002).
34. Y. Arica and V. N. Hasirci, *Biomaterials* **8**, 489 (1987).
35. K. Podual, F. J. Doyle III and N. A. Peppas, *J. Control. Rel.* **67**, 9 (2000).
36. B. Schulz, A. Riedel and P. U. Abel, *J. Mol. Catal. B: Enzym.* **7**, 85 (1999).
37. M. Y. Arıca, H. N. Testereci and A. Denizli, *J. Chromatogr. A* **799**, 83 (1998).
38. A. Denizli, H. Yavuz and Y. Arıca, *Colloid. Surf. A: Phys.* **174**, 307 (2000).
39. J. Raba and H. A. Mottola, *Crit. Rev. Anal. Chem.* **25**, 1 (1995).
40. H. Yavuz, G. Bayramoğlu, Y. Kaçar, A. Denizli and M. Y. Arıca, *Biochem. Eng. J.* **10**, 1 (2002).

Immobilization of biomolecules

Gels, Genes, Grafts and Giants, pp. 697–716 (2005)

Design of highly specific ligands of fibrin for therapeutic applications

ANNA MARIA BIANUCCI [1], ILARIA MASSARELLI [2], FEDERICA CHIELLINI [2], CHAIM EIDELMAN [3] and EMO CHIELLINI [2,*]

[1] *Department of Pharmaceutical Science, University of Pisa, via Bonanno Pisano 6, 56126 Pisa, Italy*
[2] *Department of Chemistry and Industrial Chemistry, University of Pisa, via Risorgimento 35, 56126 Pisa, Italy*
[3] *Novetide Ltd., Deshanim Road, 26111 Haifa Bay, Israel*

Received 3 November 2003; accepted 8 March 2004

Abstract—The work presented here is aimed at designing high-affinity ligands for the fibrin γ(312–324) epitope. This epitope is specific for fibrin recognition, as it is exposed only on the fibrin surface, while in fibrinogen it is buried in the protein bulk. This property makes it very useful for therapeutic applications. In fact, it may be exploited in driving systems for targeted delivery of thrombolytic drugs toward the specific compartment where they are needed. It will then allow avoidance of serious unwanted side effects produced by a conventional systemic administration. The approach chosen for designing putative ligands is based on the known three-dimensional (3D) structure of the epitope. A wide virtual library made up of oligo-peptides and analogues designed by a combinatorial approach, on the basis of chemical complementarity criteria, has been screened by means of a docking/scoring approach (DOCK program). The peculiarity of the problem under study required a considerable effort in finding a method enabling the experimental validation of the design work results. In fact the selected biological target is absolutely new, so that neither a endogenous, nor a synthetic high-affinity ligand is known up to now. It does not allow for the validation of computational results by means of classical binding tests based on the use of known labeled high-affinity ligands. Preliminary binding essayes were so carried out by means of the Plasmon Surface Resonance (PSR) technique. The experimental results suggested that most of the molecules predicted to be good ligands by means of the selected computational tools, could carry the wanted affinity toward the selected target.

Key words: Computer modelling; targeted release; protein drugs; Biacore; fibrin ligands; fibrin epitope.

*To whom correspondence should be addressed. Tel.: (39-50) 221-9299/221-9445. Fax: (39-50) 28438. E-mail: emochie@dcci.unipi.it

INTRODUCTION

The aim of the present work is to design ligands characterized by highly specificity for polymerized fibrin, the main constituent of thrombus (fibrin clot). These ligands may be exploited in therapeutic applications. As an example they could be used in driving systems for targeted delivery of fibrinolytic drugs (anti-thrombotics), thus allowing avoidance of the unwanted severe side effects due to the systemic administration of such kind of drugs.

Basic rationale of the undertaken research

The coagulation system is a very delicate process and is regulated by a cascade of factors. Briefly, after vascular injury a series of protease activations culminates in the generation of thrombin, which mediates conversion of fibrinogen to fibrin. Fibrinogen is a covalent dimeric protein composed of three non-identical chains ($\alpha_2\beta_2\gamma_2$) that are interconnected by 29 disulfide bonds. Fibrinogen is cleaved by thrombin, resulting in the release of short peptides from the N termini of the α- and β-chains. The fibrin monomers polymerize spontaneously and non-covalently creating a fibrin gel network. The fibrin polymers are stabilized further by the action of factor XIIIa, a transglutaminase that introduces covalent crosslinks, into the network [1, 2].

Fibrin has only a temporary function; after it has fulfilled its role in tissue repair, it will be converted to soluble degradation products, with concomitant lysis of the blood clot. The crucial step in this fibrinolytic process is the activation of plasminogen (inactive proenzyme) to plasmin (active enzyme) by plasminogen activators such as tissue-type plasminogen activator (t-PA) and urochinase-type plasminogen activators (u-PA) [3, 4]. The plasmin formed converts fibrin to soluble degradation products.

In order to pursue our target in the task of therapeutic applications we need to design ligands that are specific for polymerized fibrin, but are unable to recognize fibrinogen, so that physiological processes cannot be perturbed. Taking into account the above consideration our starting point consisted of identifying structural differences between fibrinogen and fibrin.

More in detail, soluble fibrinogen (340 kDa) is a 450-Å-long disulfide-linked dimer of three nonidentical polypeptide chains (termed Aα, Bβ and γ). The N-termini of the six chains from the two halves of the molecule come together to form a small globular domain (the so-called disulfide knot) in the center. The C-termini of each of the three chains end in globular domains, those of the Bβ- and γ-chains being located at the ends of the molecule, generally interacting with each other adjacent to the central domain. Except for an extended flexible portion of the Aα chain immediately preceding the αC domain, the regions between the globular domains in each half-molecule are mainly α-helical and form irregular coiled coils bracketed by disulfide rings [5, 6].

The thrombin-catalyzed conversion of fibrinogen to fibrin centers upon the removal of two different pairs of peptides from the amino-terminal regions of the α- and β-chains. Removal of the fibrinopeptide A from the α-chain leads to the appearance of a new amino-terminal sequence beginning with the fragment Gly-Pro-Arg, whereas removal of the fibrinopeptide B from the β-chain unmasks the sequence Gly-His-Arg.

The cleavage of the two fibrinopeptides leads to the association of fibrin monomers into two-stranded half-staggered protofibrils [7, 8]; here the major contacts occur between the central and end domains on adjacent filaments and between the end domains within the same filament. The protofibrils then associate laterally to form thicker fibers.

The clots, which are strengthened by covalent cross-linking of the C-terminal end-to-end-bonded γ-domains (and at a slower rate of the αC domains), are later dissolved by plasmin primarily by proteolysis in the coiled-coil regions [9, 10].

Limited proteolysis of factor XIII-cross-linked fibrin gives rise to a set of fragments, the largest of which, fragment D, corresponds to the terminal globules with a residual stump of coiled coil.

Some modifications that occur after fibrinogen to fibrin conversion regard the exposure of critical sequences that were buried in the fibrinogen molecule. In particular, several sequences have been discovered and identified as fibrin-specific epitopes.

The idea of designing small specific ligands to be exploited for fibrin recognition arose once provided evidence that some amino-acid fragments, belonging both to fibrinogen and fibrin, are only exposed on the fibrin surface, while they are buried in fibrinogen. This observation was due to experiments based on antibody production. Such fragments belong to different chains. The Aα(148–160) epitope is located on the α chain, the Bβ(15–21) epitope belongs to the N-terminus of the β chain and the γ(312–324) epitope is located on the γ-chain near at the C-terminus.

The Aα(148–160) and the γ(312–324) epitopes are involved in accelerating the activation of plasminogen catalyzed by the tissue-type plasminogen activator. Detailed studies in the literature [11] showed that this rate-enhancing effect of fibrin is brought about by these two sites in the fibrin molecule. After the fibrinogen-fibrin conversion these sequences become accessible on the molecular surface. This has been experimentally proved by means two monoclonal antibodies respectively raised against synthetic Aα(148–160) and γ(312–324) fragments, which react with fibrin, while not giving any interaction with fibrinogen.

Some authors suggested that the polymerization, brought about in fibrin through factor XIIIa cross-linking, could be necessary for the exposure of these two 'fibrin-specific' (more correctly, 'polymerized fibrin-specific') epitopes [12].

Furthermore, fibrin specific monoclonal antibodies which recognize the N-terminus of the β-chain (Bβ(15–21) epitope) were obtained by using synthetic peptide antigens [13].

In a recent work it has been observed that the Aα(148–160) epitope interacts both with t-PA and plasminogen, while the γ(312–324) one interacts only with t-PA [14].

MATERIALS AND METHODS

Selection of the proper fibrin-specific targeting moiety

As it has already explained above, the main requirement which the fibrin fragment (epitope) needs to posses in order to be considered a proper biological target for our purposes, consists of its ability of allowing selective recognition of fibrin itself, being buried in fibrinogen. Moreover, it should be able to allow a selective recognition of polymerized fibrin, in order to avoid interference with monomeric fibrin, so that only fibrin which is already part of the clot is recognized. The epitope should not represent an important site for the interaction of fibrinolytic agents (such as t-PA and plasminogen) with fibrin in order to preferably minimize the eventual interference with the residual fibrinolytic process. Finally, the epitope must have high homology with the sequence of the corresponding epitope on rats or mice (which are usually employed in experimental assays).

As a final consideration it is necessary to point out that the above fibrin fragments constitutes non-usual biological targets. Neither endogenous ligands, nor previously developed interacting molecules are known at present. Because of that no approaches based on the properties of many already known ligands may be exploited in the search for molecules characterized by the wanted affinity and selectivity. This cut out all the approaches that exploit Quantitative Structure-Activity Relationships (QSAR) methods. Therefore the search could only be afforded by molecular design approaches based on the 3D structure of a proper biological target. The 3D structure of the Bβ(15–21) epitope is not completely available yet in the Protein Data Bank (PDB) [15] database (which contains protein and nucleic acids 3D structures) and the Aα(148–160) epitope is involved in the interaction with both t-PA and plasminogen. The γ(312–324) epitope instead is known to interact only with t-PA. On the basis of the above considerations, this latter epitope appeared to be the most suitable target for our design purposes.

Selection of a suitable computational tool

For structure-based drug design, many docking/scoring algorithms, implemented in different computer programs, are available either to screen very large databases or to optimize lead molecules. These programs use at least three family of scoring functions: knowledge-based, force field-based and empirical ones. In this work, we exploited a structure-based approach by screening many putative ligands belonging to a wide virtual library against the fibrin epitope. With regard to the computational tool, our choice felt on the DOCK program (version: 4.0), developed at the Molecular Design Institute (University of California at San Francisco) and freely distributed through Academic environment [16]. This program had previously been reported in the literature to perform at the best by well balancing accuracy and computational effort. The version currently available at the time this work started, while not allowing yet taking into account any potential flexibility of the macromolecular target, does allow to largely sample orientation and conformation of the selected ligands into the binding site. It allows also estimation of the binding energy corresponding to each screened configuration of the ligand–target complex, which is approximately calculated by means of empirical functions and force field contributions. Because of that, the DOCK interaction energies cannot be rigorously compared to experimental interaction energies, obtained from binding constant values, which are, instead, directly related to interaction free energies. The DOCK interaction energies are only capable of giving a qualitative estimate of the interaction energies by ranking them within analogous biological systems. More in detail, many configurations of the complex, corresponding to different orientations and conformations of the ligands in the binding site, are ranked on the basis of the estimated interaction energy values, so that optimal orientations and conformations of the ligands may be identified with good approximation. Sampling and ranking different conformations are accomplished through a genetic algorithm capable of performing a sort of conformational analysis of the ligand in the binding site. We would like to recall here the importance of sampling (even by rough procedures) many different conformations within the binding site of a target biomolecule instead of performing very accurate conformational analysis *in vacuo* or in simulated solvents. Only computational tools capable of exploring conformations of a ligand within the target macromolecules may give a reliable insight on the biologically active structure of the ligands under analysis.

Generation of a suitable virtual library

Since no ligands are known for the γ-epitope (312–324) we could not exploit any existing molecular template for our initial design work. It, thus, consisted of *de novo* designing molecules expected to possess the required binding properties. As a preliminary approach, we focused the design on the identification of peptidic derivatives able to recognize the γ-epitope. However, it must be taken into account that one of the most important consideration which limits the application of native peptides, is the rapid degradation of peptides by peptidase type enzymes. There are many specific and non-specific peptidases in biological systems which rapidly metabolize peptides.

In an effort to counteract these detrimental characteristics, some changes in native peptide structure have been considered. These modified structure are referred to as peptidomimetics: i.e. chemical structure derived from bioactive peptides which mimic natural molecules. It is widely believed that such modifications will enhance the desirable properties and avoid undesirable structural liability of native peptides [17]. The peptidomimetic compounds considered in this work were obtained, at the initial level, by introducing capping groups at the ending of each one of the peptide ligands considered.

The model of polymerized fibrin used for screening putative ligands by the DOCK program was obtained from PDB. The crystal structure of 'Fragment Double-D from human fibrin with two different bound ligands' (PDB entry 1FZC) was exploited as starting point. It consists of factor XIII-cross-linked fragment D (double-D) from human fibrin crystallized in the presence of two different peptide ligands. Its X-ray structure was determined at 2.3 Å resolution. The peptide Gly-Pro-Arg-Pro-amide, analogue of the knob fragment exposed by the thrombin-catalyzed removal of fibrinopeptide A, is considered a potent inhibitor of the polymerization and was found to bind to some amino acids, giving rise to a hole in the γ-chains. Unfortunately this hole cannot be exploited as target binding site for designing fibrin-specific ligands because it is accessible both in fibrinogen and fibrin structures, which enables Gly-Pro-Arg-Pro-amide to bind both proteins.

Starting from this structure a model of the domain containing the epitope of interest was built. It consists of a globular domain of about 150 amino acids surrounding the epitope γ(312–324). Due to the peculiarity presented by the symmetrical sub-fragments Asp-Asn-Asp-(Asn-Asp) our study started by attempting the design of ligands capable of binding them. The partial charge distribution on the 3D structure of the epitope was estimated. On the basis of its 3D structure the solvent accessible surface area was calculated by using the Homology module of the InsightII package [18]. The partial charge distribution was taken into account in designing the putative ligands that are required to show maximal chemical complementarity to the selected fibrin sequence. In most cases the ligands consist of charged peptidic molecules and the distribution of their charge must be complementary to the one found on the surface area of the epitope, which is expected to interact with them. On the basis of the above considerations, many ligands have been generated both in their free and in their mono or fully end-capped forms. The capping groups consisted of $-NH_2$ on the C-terminus and acetylic groups on the N-terminus. In the mono-capped forms only the $-NH_2$ terminus was included in the selected peptides. Eight tripeptidic ligands were designed against a central 3-amino-acid-long fragment made up of residues Asp-Asn-Asp. They were obtained by assembling in a combinatorial way three aminoacids at one time, chosen, as building blocks, among Arg, Asp, Glu and Lys, and taking into account the chemical and/or electrostatic complementarity. They were also considered in their free, mono-capped and fully-end-capped forms.

Then 40 pentapeptides were designed against the 5-amino-acid-long central fragment (Asp-Asn-Asp-Asn-Asp) of the γ-epitope. The same criterion used for generating tripeptides was followed (chemical complementarity with respect to the consisted fragment). Among the pentapeptides, only the ten best scoring derivatives, previously screened in their free form, were also analyzed in their mono- and fully-end-capped forms.

Molecular docking

The DOCK program (version 4.0) was used to screen our user-defined library for potential fibrin-specific ligands. Each compound was docked into the 150-amino-acid-long fibrin sub-domain containing the γ-(312–324)-epitope.

All crystallographic waters, ions and solvent were removed from the crystal structure of fragment double-D from human fibrin, as required by the DOCK program. Then the different modules constituting the whole DOCK package were applied. The SPHGEN module was used to create clusters of overlapping spheres in contact with fibrin solvent-accessible surface, from which the γ(312–324) was selected as binding site. The sphere positions in the dock site were selected using interactive graphical display of the surface curvature. At each binding site, the sphere cluster was used to create a negative image of the docking site. DOCK matched each compound in turn to the site negative image to search for favorable binding orientations. Finally, DOCK scored each orientation based on the ligand complementarity to the protein surface using an interaction force field score that included van der Waals forces and electrostatic terms. To assist the selection of potential ligands, the docked orientations were also evaluated with a scoring scheme based on the ligand surface area in contact with fibrin surface. All docking calculations were performed on a R10000 197 MHz processor (SGI).

Materials

The samples of molecules, suggested to possess good affinity by the DOCK screening, to be analyzed by SPR experiments, were prepared in house at Novetide. Fibrinogen from human plasma (F4883), human fibronectin (F2006), bovine thrombin (T4684) and GlyProArgPro-amide (G5779) were purchased from Sigma. The ready-to-use HBS-EP from Biacore (0.01 M HEPES pH 7.4, 0.15 M NaCl, 3 mM EDTA, 0.005% Surfactant P20) was used as running buffer during the binding experiments. Reagents for amino coupling, N-hydroxysuccinimide (NHS), N-ethyl-N′-dimethylaminopropyl)carbodiimide (EDC) and 1 M ethanolamine hydrocloride (pH 8.5), were supplied by Biacore as Amine Coupling Kit.

Selected approach for experimental validation

Due to the above mentioned characteristic of the molecular target, the assessment of suitable binding experiments turned out to be quite challenging.

Real-time Bio-molecular Interaction Analysis (BIA) from Biacore is an innovative technology for studying bio-molecular interactions in real time, without labelling any of the interacting moieties [19]. Biacore uses the Surface Plasmon Resonance (SPR) [20] optical phenomenon to monitor bio-molecular interactions as result of the changes in the mass concentration of macromolecules at the bio-specific interface. The critical device in the BIA facility consists of a bio-sensor (sensor-chip) made up of a grass slide covered with a carefully deposited layer of gold. A corollary of this feature is that all stages in an interaction sequence are monitored, in contrast to many conventional methods where detection is restricted to a single labelled component. Interactions are followed in real time, so that kinetic information is readily derived. In many cases, investigations can be performed without prior purification of components. SPR-based bio-sensors monitor interactions by measuring the mass concentration of bio-molecules close to a surface. The surface is specifically activated by anchoring on it one of the interacting partners. Sample containing the other partner(s) flows over the surface: when molecules from the sample bind to the interactant attached to the surface, the local concentration changes and an SPR response is measured. Plotting the response against time during the course of an interaction provides a quantitative measure of the progress of the interaction.

The kinetic analysis of sensor-grams from the interaction of various analytes with the immobilized protein is based on the following rate equation:

$$\mathrm{d}R/\mathrm{d}t = k_\mathrm{a} C R_\mathrm{max} - (k_\mathrm{a} C + k_\mathrm{d}) R_t,$$

where R_t is the plasmon-resonance response (RU) due to analyte interaction with immobilized protein at time t (s), k_a and k_d are the association ($\mathrm{M}^{-1}\,\mathrm{s}^{-1}$) and dissociation ($\mathrm{s}^{-1}$) rate constants, respectively, C is the concentration of the injected analyte (M) and R_max is the RU response at saturation of protein-binding sites. The rate constants are calculated from the analysis of sensor-grams using proper evaluation software and the dissociation equilibrium constant (K_D) is calculated as the ratio of $k_\mathrm{d}/k_\mathrm{a}$.

Calculations are carried out by numerical integration [21] and global curve fitting is done by non-linear least-squares analysis [22, 23]. The quality of the fitted data can be evaluated by comparison between the modeled and the experimental sensorgrams, as well as by statistical parameters such as the χ^2 given by the following equation:

$$\chi^2 = \frac{\sum_1^n (r_\mathrm{f} - r_x)^2}{n - p},$$

where r_f is the fitted value at a given point, r_x is the experimental value at the same point, n is the number of data points and p is the number of degrees of freedom. For good fitting to ideal data, χ^2 is of the same order of magnitude as the noise in RU (typically < 2). If the model fits the experimental data precisely, χ^2 represents the mean square of the signal noise (since this is the only deviation from the fitted curve). In practice, values of χ^2 below 10 are frequently acceptable. The form of the fitted curve in relation to the experimental data is often more valuable in assessing the goodness of fit than the absolute χ^2 value.

Hence, frequently a visual inspection of the experimental curves with overlaid fitted curves gives a good indication of the goodness of fit.

SPR experiments

In order to validate our predictive results, Surface Plasmon Resonance (SPR) experiments, based on Biacore X facility, were carried out on several compounds suggested to possess good affinity by the DOCK screening. CM5 type sensor-chip was used, characterized by a layer of carboxy-methylated dextran which provides a matrix for covalent attachment of proteins and forms a hydrophilic environment where the interaction will take place. Since fibrin could polymerize in the experimental conditions required for its coupling to the sensor-chip, before reaching the sensor-chip it-self, we started the experiments by binding fibrinogen to it instead of fibrin. Subsequent thrombin-driven digestion allowed us to obtain the wanted fibrin-coated sensor-chip. Briefly, fibrinogen was covalently coupled to the activated carboxy-methyldextran-coated sensor surface by the protocol described below. A mixture of N-hydroxysuccinimide (NHS) and N-ethyl-N′(dimethylaminopropyl)-carbodiimide (EDC) was injected into the dextran matrix of the sensor chip to activate it at the flow rate of 10 μl/min at 37°C and human fibrinogen in 10 mmol/l sodium acetate (pH 4.6) was immobilized on the matrix. The immobilization was carried out at pH 4.6 in order to control electrostatic pre-concentration of the protein in the surface matrix, which constitutes a critical factor. In the CM series sensor chips, the matrix is negatively charged at pH values higher than about 3, and positively charged proteins are electrostatically attracted to the surface. So that in order the electrostatic pre-concentration to take place, the fibrinogen molecule must have a net positive charge; therefore, the buffer pH must be below the isoelectric point (p*I*) of the fibrinogen molecule (5.6–5.8). The matrix exceeding active sites were blocked with ethanolamine-HCl (1 M). All materials were dissolved in HBS-EP at pH 7.4, which was also used as running buffer in the experiments. Functionality of the fibrinogen sensor chip was first assessed by measuring the specific binding of the known ligand H-GlyProArgPro-NH_2 (data not shown) capable to bind both fibrinogen and fibrin. 30 μl of the selected concentrations was injected at a flow rate of 30 μl/min. The fibrinogen surface was regenerated after each binding interaction with 50 mM HCl at a flow rate of 10 μl/min for 60 s followed by re-equilibration with running buffer. In order to convert immobilized fibrinogen into fibrin, we treated it with 60 μl of 100 nmol/l thrombin for 30 min at a flow of 2 μl/min. As digestion of fibrinogen immobilized on the sensor chip progressed, the resonance response gradually lowered. A marked decrease in the refractive index was observed as a function of the time, while thrombin injection proceeded, so indicating loss of material, due to on-chip digestion. This procedure has been previously proved to result in a fibrin monolayer that mimics a fibrin clot surface [24]. Fibrinogen to fibrin conversion, with loss of fibrinopeptides, was verified by increase in reactivity with fibronectin, a protein which binds fibrin but not fibrinogen. Hence, the sensor-chip surface was functionally characterized by measuring binding of fibronectin. Moreover the binding between fibrin and the small peptide H-GlyProArgPro-amide, was also measured, even if it is capable to bind both fibrinogen and fibrin indifferently.

The association between the immobilized fibrin and the added ligands was then monitored as the change in the SPR response. The fibrin surface was regenerated after each binding interaction with 50 mM HCl at a flow rate of 10 μl/min for 60 s followed by re-equilibration with running buffer.

RESULTS AND DISCUSSION

Generation of the virtual library

Before proceeding to the generation of a small initial virtual library, the solvent-accessible surface area of the target binding site was calculated, based on its 3D structure, by using the Homology module of the InsightII package [18]. The results are shown in Table 1. We can observe that residues Ser-313 and Thr-314 are not accessible to the solvent. Because of that we thought that such residues shouldn't be taken into account while generating the putative ligand library. The partial charge distribution was also calculated: it had been taken into account when designing the new putative ligands, so ensuring maximal chemical complementarity to the selected fibrin sequence. The partial charge distribution is shown in Fig. 1, where the positive residue Lys-321 is colored in blue and the negative residues Asp-316, Asp-318, Asp-320 and Glu-323 are colored in red.

DOCK screening

The eight tripeptides designed against the central three aminoacid long Asp-Asn-Asp fragments of the fibrin γ-epitope were analyzed in their free, mono-capped and fully-end-capped forms. The compound showing the most favorable interaction energy (best scoring compound) that was suggested by this first step of analysis is the derivative, mono-capped at carboxyl terminus, LysGluArg, which gave an interaction energy of −59.8 kcal/mol. Looking at its best conformation and orientation with regard to the biomolecular target (Fig. 2), we could observe that the tripeptide and the epitope form six H bonds which involve the pair of atoms shown in Table 2.

The group of derivatives designed against the central 5-amino-acid-long Asp-Asn-Asp-Asn-Asp fragment of the fibrin γ-epitope consists of pentapeptides. They

Table 1.
Solvent-accessible surface area for each residue of the γ(312–324) epitope

	Residue												
	Phe	Ser	Thr	Trp	Asp	Asn	Asp	Asn	Asp	Lys	Phe	Glu	Gly
Residue number	312	313	314	315	316	317	318	319	320	321	322	323	324
Surface area (Å)	0.193	0	0	112.1	61.29	30.09	64.51	35.03	16.58	119.5	57.18	155	31.2

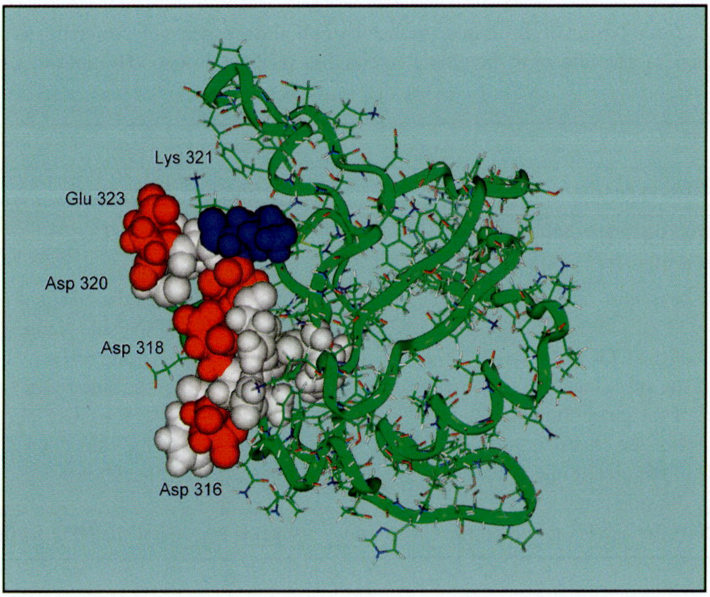

Figure 1. Charge distribution on the γ(312–324) eptiope. The positive residue Lys-321 is colored in blue and the negative residues Asp-316, Asp-318, Asp-320 and Glu-323 are colored in red.

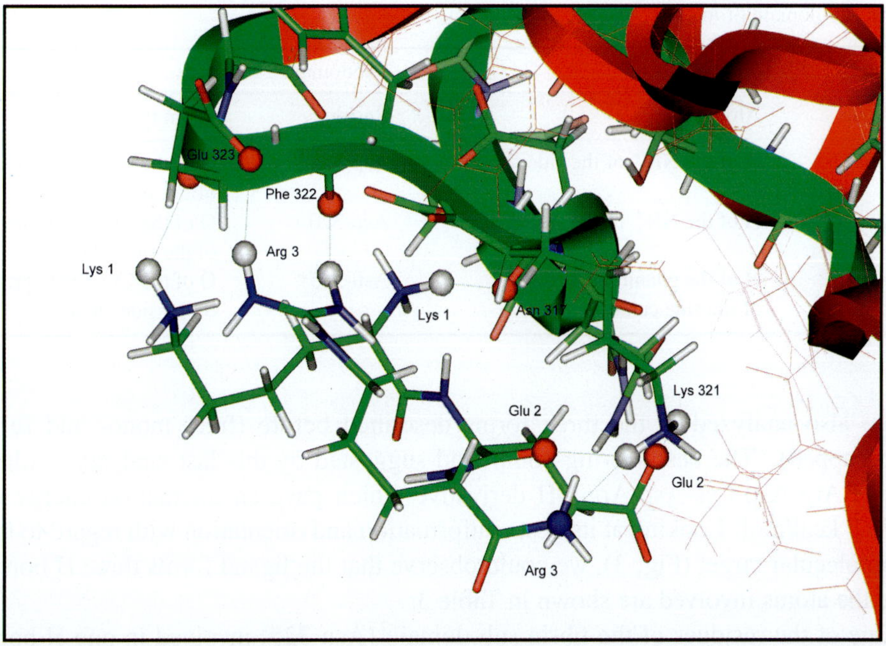

Figure 2. Zoom view on the interaction site between the mono-capped LysGluArg peptide derivative and the sub-domain of fibrin. The ball render is referred to the atoms involved in the formation of H bonds.

Table 2.
Residues and atom types involved in inter-molecular H bonds, respectively, referring to the tripeptide ligand predicted to give the most favorable interaction (H-LysGluArg-NH_2) and to the fibrin sub-domain of interest

Ligand		Subdomain	
Residue	Atom type	Residue	Atom type
Lys-1	H of the NH_3^+ of the side chain	Glu-323	O backbone
Lys-1	H of the NH_3^+ of the side chain	Asn-317	O backbone
Glu-2	O backbone	Lys-321	H of the NH_3^+ of the side chain
Glu-2	O of the COO^- group of the side chain	Lys-321	H of the NH_3^+ of the side chain
Arg-3	H of the guanidinic group of the side chain	Glu-323	O of the COO^- group of the side chain
Arg-3	H of the guanidinic group of the side chain	Phe-322	O backbone
Arg-3	N of the $-NH_2$ group on the C-terminus	Lys-321	H of the NH_3^+ of the side chain

Table 3.
Residues and atom types involved in inter-molecular H bonds, respectively, referring to the pentapeptide ligand predicted to give the most favorable interaction (H-ArgAspLysAspArg-OH) and to the fibrin sub-domain of interest

Ligand		Subdomain	
Residue	Atom type	Residue	Atom type
Lys-3	H of the NH_3^+ of the side chain	Asp-318	O of the COO^- group of the side chain
Lys-3	H of the NH_3^+ of the side chain	Asp-320	O of the COO^- group of the side chain
Arg-5	H of the guanidinic group of the side chain	Asn-325*	O of the CO$-NH_2$ group of the side chain

were also analyzed in the three forms described before (free, mono- and fully end-capped). The best scoring compound suggested by this last analysis resulted the H-Arg-Asp-Lys-Asp-Arg-OH derivative which gave an interaction energy of −75.7 kcal/mol. Looking at its best conformation and orientation with regard to the biomolecular target (Fig. 3), we could observe that the ligand forms three H bonds and the atoms involved are shown in Table 3.

One of the residues of the fibrin sub-domain (Asn-325) involved in one H bond with the ligand is located externally with respect to the epitope of interest. However, it is near to the binding site so that an interaction with such a residue could also play a role in selectively binding fibrin.

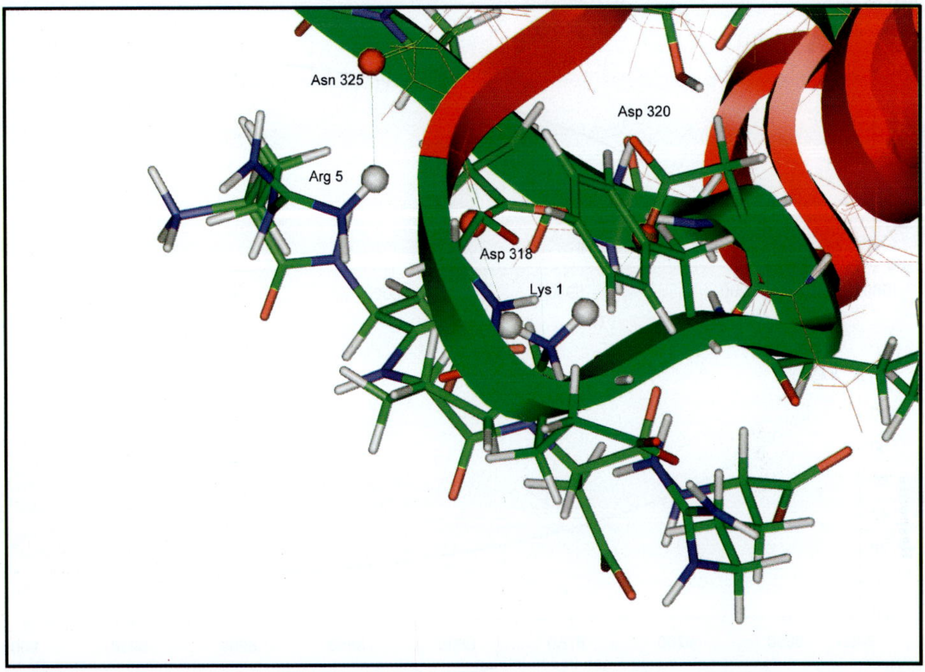

Figure 3. Zoom view on the interaction site between the free ArgAspLysAspArg peptide derivative and the sub-domain of fibrin. The ball render refers to the atoms involved in the formation of H bonds.

SPR essays

The preliminary SPR experiments were carried out according to the protocol described below. The ligands under study were flowed onto fibrin-immobilized sensor-chip at varying concentrations (ranging from 0.01 μM to 100 μM) and data were fitted to the Langmuir model assuming a 1 : 1 binding interaction. For each compound the BIA sensorgrams obtained were analyzed by means of the BIA evaluation 3.0.1. software.

Interaction of fibronectin with fibrin

This experiment was aimed at assessing the functionality of the fibrin-coated surface after the fibrinogen-fibrin conversion and to asses that fibrinogen was really converted since fibronectin is a protein capable to bind only fibrin. Only a significant curve was obtained, shown in Fig. 4A (association) and Fig. 4B (dissociation). The results referring to this interaction are reported in Table 4. The $K_{\rm D}$ measured for this interaction is about 0.6 nM. These findings are also quite in accordance with literature data reported for fibronectin binding to fibrin in which this constant could be about 7 nM [25]. Furthermore, this could mean that fibrin is perfectly functional once attached to the surface of the sensor-chip.

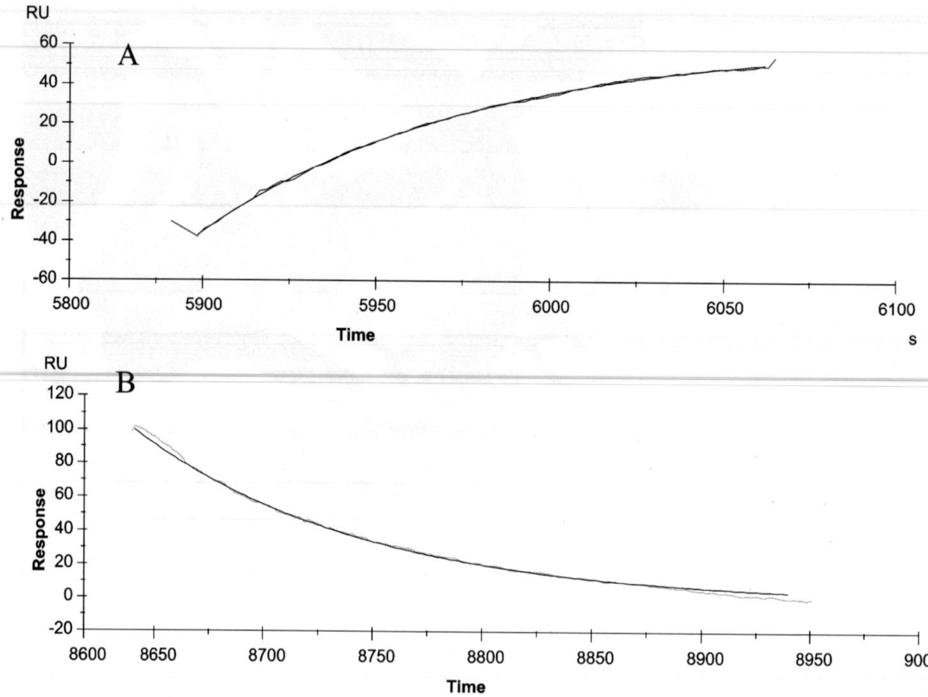

Figure 4. Analysis of the binding of fibronectin to immobilized fibrin. Fibronectin (1 nM) was added to immobilized fibrin, and its association (A) and dissociation (B) was monitored in real time while recording the resonance signal (Responce). Both experimental and corresponding fitted curves are shown.

Interaction of H-GlyProArgPro-amide with fibrin

This experiment was performed in addition to the previous one, even though it was not strictly needed for assessing the functionality of the fibrin-coated surface of the sensor-chip. In fact it was mainly aimed at assessing the validity of the following experiments involving small ligands instead of a quite big protein such as fibronectin. Figure 5 shows the experimental and fitted curves referring to the association (Fig. 5A) and dissociation phases (Fig. 5B) for each experiment carried out. The results referring to this interaction are also reported in Table 4. The K_D measured for this interaction is about 14 μM. These findings are in accordance with the ones found in literature where the H-GlyProArgPro-amide ligands binds to fibrin(ogen) (no selectivity) with a dissociation constant of 20 μM [26]. This ensured that the planned experiments involving the ligands previously designed were reasonably set up.

Interaction of new proposed ligands with fibrin. For each ligand several experiments were carried out varying the concentration of the samples (Table 4). As an example, the sensor-gram referring to ligand **4b** is reported in Fig. 6A and 6B.

Table 4.
Parameters derived from analysis of binding of fibronectin, H-GlyProArgPro-NH_2 and the ligands to immobilized fibrin

Compound	[Ligand]	k_a (M^{-1} s^{-1})	k_d (s^{-1})	K_A (M^{-1})	K_D (M)	χ^2
Fbronectin	1 nM	7.84×10^6	4.91×10^{-3}	1.6×10^9	6.27×10^{-10}	0.38
H-GlyProArgPro-NH_2	23 μM	3.54×10^3	0.137	2.59×10^4	3.87×10^{-5}	0.083
H-GlyProArgPro-NH_2	2.3 μM	7.38×10^3	0.117	6.29×10^4	1.59×10^{-5}	0.024
H-GlyProArgPro-NH_2	2.3 μM	1.01×10^4	0.228	4.44×10^4	2.25×10^{-5}	0.23
H-GlyProArgPro-NH_2	0.23 μM	2.52×10^4	0.16	1.57×10^5	6.36×10^{-6}	0.45
H-GlyProArgPro-NH_2	0.23 μM	3.43×10^4	0.297	1.15×10^5	8.67×10^{-6}	0.065
H-GlyProArgPro-NH_2	0.023 μM	8.93×10^4	0.184	4.85×10^5	2.06×10^{-6}	0.78
H-GlyProArgPro-NH_2	0.023 μM	9.46×10^4	0.238	3.98×10^5	2.52×10^{-6}	0.98
1c	100 μM	3.66	0.0238	154	6.5×10^{-3}	0.448
1c	100 μM	742	0.0123	6.03×10^4	1.66×10^{-5}	0.507
1c	100 μM	5.77	0.0286	202	4.96×10^{-3}	0.178
2a	100 μM	1.27×10^4	0.087	1.46×10^5	6.86×10^{-6}	0.83
2b	100 μM	4.56×10^3	0.0826	5.52×10^4	1.81×10^{-5}	0.656
2b	100 μM	5.41×10^3	0.0193	2.8×10^5	3.57×10^{-6}	0.226
2b	100 μM	719	0.0162	4.44×10^4	2.25×10^{-5}	0.183
2b	100 μM	111	0.0106	1.05×10^4	9.56×10^{-5}	0.741
2b–h	0.1 μM	1.09×10^4	0.0307	3.56×10^5	2.81×10^{-6}	1.04
2b–h	0.1 μM	3.06×10^3	0.0486	6.3×10^4	1.59×10^{-5}	0.553
2b–h	1 μM	3.71×10^3	0.0277	1.34×10^5	7.46×10^{-6}	0.501
2c	100 μM	2.11×10^3	0.0318	6.65×10^4	1.5×10^{-5}	0.385
2c	100 μM	270	0.0422	6.41×10^3	1.56×10^{-4}	1.32
2c	100 μM	73.8	0.0447	1.65×10^3	6.05×10^{-4}	0.261
2d	10 μM	1.15×10^5	0.0473	2.43×10^6	4.11×10^{-7}	1.54
2d	100 μM	1.11×10^4	0.107	1.03×10^5	9.68×10^{-6}	0.687
3b	10 μM	3.70×10^4	0.17	2.18×10^5	4.6×10^{-6}	0.664
3c	100 μM	4.40×10^3	0.0316	1.39×10^5	7.18×10^{-6}	0.328
4b	10 μM	2.97×10^4	0.0132	2.25×10^6	4.44×10^{-7}	0.491

The SPR experiments, although requiring further refinements, appear to provide valid encouraging insights. As it will be pointed out later, we observed a good correlation between our preliminary experimental data and the calculated affinities, within each one of the two sub-groups of designed ligands, i.e. tripeptides and pentapeptides.

Considerations on prediction reliability

In Table 5 the K_D values for the ligands in the two sub-groups (tripeptides and pentapeptides) are listed together with the relevant DOCK scores, and other information related to validation experiments. We have to point out here that the values listed in the last two columns of Table 5, which respectively refer to experimental K_D values and to DOCK interaction energies, should not be compared to each other within a strictly rigorous analysis. In fact, while K_D is related to a

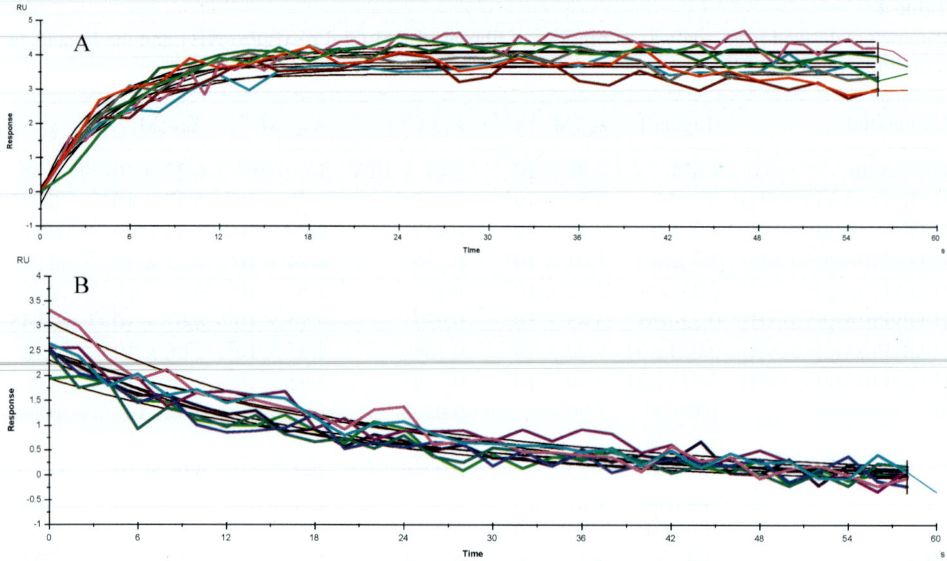

Figure 5. Analysis of the binding of H-GlyProArgPro-amide to immobilized fibrin. Increasing concentrations of the ligand were added to immobilized fibrin, and their association (A) and dissociation (B) was monitored in real time while recording the resonance signal (Response). The concentrations were 0.023, 0.23, 2.3 and 23 μM. Both experimental and corresponding fitted curves are shown.

thermodynamic quantity (the affinity of the ligand toward fibrin), the interaction energies are estimated by DOCK through empirical functions which do not account for all the entropy contributions. Nevertheless, both experimental and calculated values are expected to retain an analogous trend, being the entropy contributions due to the ligand flexibility, which is approximately the same for all the analyzed ligands. On this basis, a good agreement between the trends of the experimental and calculated values, reported in the two columns of Table 5, may be diagnostic, at a first approximation, for a good predictability of our design work, based on the DOCK screening.

In the pentapeptide series we can observe a K_D decrease from 3.83 mM for sample **1c** to 6.86 μM for sample **2a**, according to the decrease of the DOCK scores for the corresponding compounds. In the tripeptide series an analogous agreement can be detected. Only an inversion in the trend of the DOCK scores, referred to samples **3b** and **4b**, was found, in comparison with the trend shown by the K_D values obtained experimentally.

On the basis of these preliminary results a structure/affinity correlation could be highlighted. From the comparison between compounds **2a** and **2d** we can observe, for example, that the acetylation on the N-terminus leads to a decrease in affinity for the fibrin molecule. The same behavior is found in comparison between compounds **2b** and **2c**, and, within the tripeptide series, between compounds **3c** and **3b**.

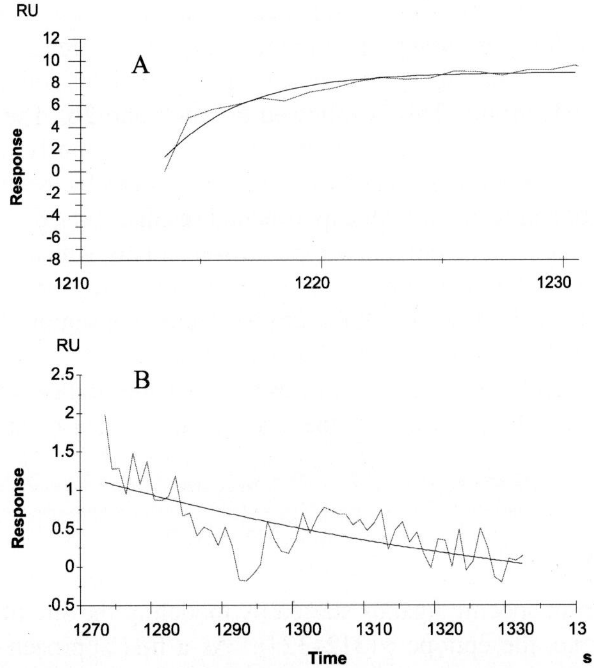

Figure 6. Analysis of the binding of ligand **4b** to immobilized fibrin. 10 μM of the ligand was added to immobilized fibrin, and its association (A) and dissociation (B) was monitored in real time while recording the resonance signal (Response). Both experimental and corresponding fitted curves are shown.

Table 5.
Pentapeptide and tripeptide ligands analyzed for experimental validation

Sample	ID	Sample purity	K_D	DOCK score
Petapeptides				
Ac-ArgGluArgAspArg-NH_2	**1c**	Low	3.83 mM	−47.74
Ac-ArgAspLysAspArg-NH_2	**2c**	Low	259 μM	−50.78
H-ArgAspLysAspArg-NH_2	**2b**	Low	34.9 μM	−62.25
H-ArgAspLysAspArg-NH_2	**2b–h**	High	8.72 μM	−62.25
Ac-ArgAspLysAspArg-OH	**2d**	High	5.05 μM	−64.01
H-ArgAspLysAspArg-OH	**2a**	High	6.86 μM	−75.70
Tripeptides				
Ac-LysGluArg-NH_2	**3c**	Low	7.18 μM	−47.06
H-LysGluArg-NH_2	**3b**	High	4.6 μM	−59.79
H-LysGluLys-NH_2	**4b**	Low	444 nM	−56.03

Compound **2d**, bearing the Ac capping group at its N-terminus, does not belong to the initial virtual library. Nevertheless, it was supplied by Novetide in addition to the one designed, as it is an intermediate of the synthesis of compound **2a**. Supplementary DOCK screening was carried out on it, being it available for measurements. Low purity means below 70%, while high purity means above 95%.

From the comparison among compounds **2a**, **2d**, **2b–h**, **2b** and **2c** we can observe that the highest affinity is shown by compound **2a**, that is free from end-capping group, the second highest affinity characterizes compound **2d**, with a single capped terminus at the NH_2 group. This is followed by **2b–h** and **2b**. The lowest affinity is shown by the fully-end-capped compound **2c**. As previously highlighted, the trend of interaction energies estimated for these compounds by means of the DOCK screening is in good agreement with experimental results.

Finally it has to be pointed out that samples **2b–h** and **2b**, which refer to the same molecule, the only difference between them being their purity, are correctly ranked by the SPR experiments. In fact, sample **2b**, less pure than sample **2b–h**, shows an observed K_D value higher than sample **2b–h**, as expected.

Whithin the tripeptide series, we can observe that the highest affinity between compounds **3c** and **3b** is shown by the one bearing at least one free terminus (compound **3b**).

CONCLUSIONS

The design of fibrin specific ligands started by modeling the site of interest in the fibrin molecule, i.e. the epitope γ(312–324). As a first approach a user-defined focused virtual library was screened by means of docking computational tools (DOCK, version 4.0) in order to predict correct orientations and conformations of the molecules in the site of interest and to rank these molecules according to empiric scoring functions. The screening revealed that most of the compounds, predicted by the DOCK program to possess high affinity toward fibrin (tripeptides and pentapeptides), show good binding properties.

The screening involved peptides in their free forms as well as in forms modified by insertion of suitable capping groups on their extremities, so that they could elude the action of peptidase enzymes. In particular it has been observed for the tripeptides derivatives, designed to interact with the central fragment of the epitope, that the mono-capped derivatives present binding characteristics similar to the corresponding analogs in their free form. This can suggest the use of the derivatives in the experimental screening rather then the parent ones. For what it concerns the pentapeptide derivatives the free form is the preferred ones with respect to the mono- and fully end-capped forms. From the structural analysis of the investigated complexes emerges that some groups within the epitope are very important for the interaction with the ligands. In particular, the most important residues of the epitope appear to be Asp-318, Asp-320, Lys-321, Phe-322 and Glu-323. Moreover, Asn-325, not comprised in the interval of interest, seems to play a significant role in enforcing the interaction of the ligands with the epitope. These residues most often act through their side chains as in the cases of Lys-321, Asp-318 and Asn-325, in which the H atoms of the $NH3^+$, the O atoms of the COO^- and the H atoms of the guanidine group of the side chains, respectively, are involved in the interactions with the ligands.

The experimental approach selected for validating the predictive work, based on the SPR technique and supported by the BIACORE facility, allowed the ligands to be tested against the whole fibrin molecule and was shown to be well suited for our purposes. The preliminary results obtained appear to support the computer-aided design of ligands afforded in this work. In spite of the need for further experiments to better evaluate the preliminary findings, it is worth noting that adopted while modeling system shows a good predictive power. Considering that some of the designed ligands show the wanted high affinity toward fibrin, further development of the design work presented here is necessary.

Acknowledgements

The present work was mostly performed within the framework of the EC-funded project G5RD-CT-2000-0294. Financial support from MIUR (Cofin 2002) was also exploited during the development of this research. We also thank the Dipartimento di Psichiatria, Neurobiologia, Farmacologia e Biotecnologie (DPNFB), University of Pisa, for the access to the BIACORE facility, and Prof. G. Giannaccini and Dr. G. Mascia (DPNFB) for helpful suggestion.

REFERENCES

1. S. J. Everse, G. Spraggon, L. Veerapandian, M. Riley and R. F. Doolittle, *Biochemistry* **37**, 8637 (1998).
2. K. P. Pratt, H. C. F. Cote, D. W. Chung, R. E. Stenkamp and E. W. Davie, *Proc. Natl. Acad. Sci. USA* **94**,7176 (1997).
3. D. Collen, *Thromb. Haemost.* **82**, 259 (1999).
4. J. I. Weitz, R. J. Stewart and J. C. Fredenburgh, *Thromb. Haemost.* **82**, 974 (1999).
5. J. H. Brown, N. Volkmans, G. Jun, A. H. Henschen-Edman and C. Cohen, *Proc. Natl. Acad. Sci. USA* **97**, 85 (2000).
6. G. Spraggon, S. Everse and R. Doolittle, *Nature* **384**, 455 (1997).
7. L. Stryer, C. Cohen and R. Langridge, *Nature* **197**, 793 (1963).
8. J. W. Weisel, G. N. Phillips and C. Cohen, *Ann. N.Y. Acad. Sci.* **408**, 367 (1983).
9. R. F. Doolittle, *Annu. Rev. Biochem.* **53**, 195 (1984).
10. R. R. Hantagan, C. W. Francis and V. J. Marder, in: *Hemostasis and Thrombosis: Basic Principles and Clinical Practice*, R. W. Colman, J. Hirsh, V. J. Marder and E. W. Salzman (Eds), p. 227. Lippincott, Philadelphia, PA (1994).
11. U. Haddeland, K. Sletten, A. Bennick, W. Nieuwenhuizen and F. Brosstad, *Thromb. Haemost.* **75**, 326 (1996).
12. M. W. Mosesson, K. R. Siebenlist, M. Voskuilen and W. Nieuwenhuizen, *Thromb. Haemost.* **79**, 796 (1998).
13. F. Chen, E. Haber and G. R. Matsueda, *Thromb. Haemost.* **67**, 335 (1992).
14. W. J. G. Schielen, H. P. H. M. Adams, K. van Leuven, M. Voskuilen, G. I. Tesser and W. Nieuwenhuizen, *Blood* **77**, 2169 (1991).
15. H. M. Berman, J. Westbrook, Z. Feng, G. Gilliland, T. N. Bhat, H. Weissig, I. N. Shindyalov and P. E. Bourne, *Nucleic Acids Res.* 28, 235 (2000).
16. T. J. A. Ewing and I. D. Kuntz, *J. Comput. Chem.* **18**, 1175 (1997).

17. M. Goodman and S. Ro, in: *Burger's Medicinal Chemistry and Drug Discovery, Volume 1: Principles and Pratcice*, M. E. Wolff (Ed.), p. 803. Wiley, New York, NY (1995).
18. InsightII, Biosym/MSI, San Diego, CA (1995).
19. L. G. Fagerstam, A. Frostell-Karlsson, R. Karlsson, B. Persson and I. Ronnberg, *J. Chromatogr.* **597**, 397 (1992).
20. J. H. Lakey and E. M. Ragget, *Curr. Opin. Struct. Biol.* **8**, 119 (1998).
21. T. A. Morton, D. G. Myszka and I. M. Chaiken, *Anal. Biochem.* **227**, 176 (1995).
22. D. J. O'Shaughnessy, M. Brigham-Burke, K. K. Soneson, P. Hensley and I. Brookes, *Anal. Biochem.* **212**, 457 (1993).
23. R. Karlsson and A. Falt, *J. Immunol. Methods* **200**, 121 (1997).
24. M. Dominguez, R. Montes, J. A. Paramo and E. Angles-Cano, *Biochim. Biophys. Acta* **1598**, 165 (2000).
25. E. Makogonenko, G. Tsurupa, K. Ingham and L. Medved, *Biochemistry* **41**, 7907 (2002).
26. E. Kaczmarek, M. H. Lee and J. McDonagh, *J. Biol. Chem.* **268**, 2474 (1993).

Gels, Genes, Grafts and Giants, pp. 717–734 (2005)

Synthesis of segmented poly(ether urethane)s and poly(ether urethane urea)s incorporating various side-chain or backbone functionalities

FELIX I. SIMONOVSKY [1], STEPHEN C. PORTER [2,*] and BUDDY D. RATNER [1,2,†]

[1] *Department of Bioengineering, University of Washington, P. O. Box 351720, Seattle, WA 98195-1720, USA*

[2] *Department of Chemical Engineering, University of Washington, P.O. Box 351750, Seattle, WA 98195-1750, USA*

Received 31 December 2003; accepted 15 September 2004

Abstract—A series of linear and branched functionalized high-molecular-weight segmented poly (ether urethane)s and poly(ether urethane urea)s were prepared by chain-extending isocyanate prepolymers based on poly(tetramethylene oxide) with molecular weight 1000 and 4,4′-diphenylmethane diisocyanate. Different functional groups were incorporated within the polymer backbone or on side-chains by using several chain extenders during the synthesis: glycerol, 2,2-bis (hydroxymethyl) propionic acid, 1H,1H,2H,3H,3H-perfluoroundecane-1,2-diol, 1H,1H,8H,8H-dodecanefluoro-1,8-octanediol, 1,3-diamino-2-hydroxypropane and 3,5-diaminobenzoic acid, using the method of gradual approach to stoichiometry. In some cases, pendant functional groups were used as reactive sites for the further attachment of side groups. Polymers were characterized using ^{1}H-NMR and FT-IR spectroscopies and GPC in conjunction with chemical structural confirmation by a model compound comparison study of 4,4′-diphenylmethane diisocyanate or trifluoro-*p*-tolyl isocyanate reacted with 1,3-diamino-2-hydroxypropane and 1,4-butanediol.

Key words: Synthesis; poly(ether urethane); poly(ether urethane urea); functionalized polymers.

INTRODUCTION

The potential for using poly(ether urethane)s (PEUs) in cardiovascular and other implant devices has stimulated considerable research interest in these polymers. The PEUs used in biomaterial applications are generally segmented block co-polymers

*Present address: Boston Scientific-Target, 47900 Bayside Parkway, Fremont, CA 94538, USA.

†To whom correspondence should be addressed. Tel.: (1-206) 685-1005. Fax: (1-206) 616-9763. E-mail: ratner@uweb.engr.washington.edu

comprised of a polyether soft segment and an aromatic isocyanate-diol or -diamine hard segment. The rather unique blend of material characteristics, inherent to this class of elastomers, is well known with the hard segment providing strength and the soft segment imparting an elastomeric quality. PEUs exhibit excellent mechanical properties, good processability and acceptable biointeractions. The chemistries that can be employed in the synthesis of PEUs are almost unlimited. Alterations of reactant ratios can be used to control mechanical properties. Similarly, a variety of chemical functionalities can be readily incorporated to facilitate alterations in their surface chemistries.

The surface chemistry of a material will direct the manner in which a biological system will interact with it; this is especially apparent in blood–material interactions and in *in vitro* applications. Consequently, biomaterials surface chemistry is a key factor in determining the ultimate fate of any medical device designed for biological contact. It is possible to design PEUs with a range of surface chemistries and resultant biological responses *via* the incorporation of different chemical functionalities into the polymer structure [1]. The presence of numerous types of chemical and biochemical functionalities at a biomaterial surface has been demonstrated to alter the biological interactions of those materials. However, there is currently no consensus on the identity of key surface chemical or biochemical modifications required to produce ideal biointeractions [2, 3]. Thus, there is a need for surface modified polymers and polymer supports representing a wide range of surface chemistries for use in biomaterial research.

Modification of PEUs to effect changes in bioreactivity has been performed by numerous researchers. However, methods for side-group modifications of PEUs have, for the most part, relied on the deprotonation of the urethane group to create a reactive nitrogen anion, or the high temperature attack of the urethane group by isocyanate to form allophanate linkages. While generally effective, these reactions are difficult to control. Also, they result in the consumption of hydrogen bonding sites and in some cases require reaction conditions that may degrade the polymer. By using different polyisocyanates and soft segment oligomers in conjunction with chain extenders having at least two active hydrogen atoms, a nearly infinite variety of linear as well as side-chain (or comb-shaped) PEU macromolecules can be produced. Linear or lightly branched PEUs may be produced using trifunctional chain extenders, if the third functionality is minimally- or non-reactive within the polymerization environment. Polymers constructed in this manner possess a residual functionality that may be further utilized in side-chain poly(ether urethane urea)s (PEUUs). Such side-chains can contain reactive hydroxyl and carboxyl groups as well as non-reactive fluorocarbon functionalities. Comb-shaped polymers were produced directly from non-linear diols (Fig. 1, Scheme A) or by reaction of an isocyanate with hydroxylated PEUs (Fig. 1, Scheme B).

Side-chain PEUs can be particularly effective in surface modification [4–6], provided that the surface energy of the side-chains is lower than both the hard and soft segments of the polymer. Surface energy minimization will cause spontaneous

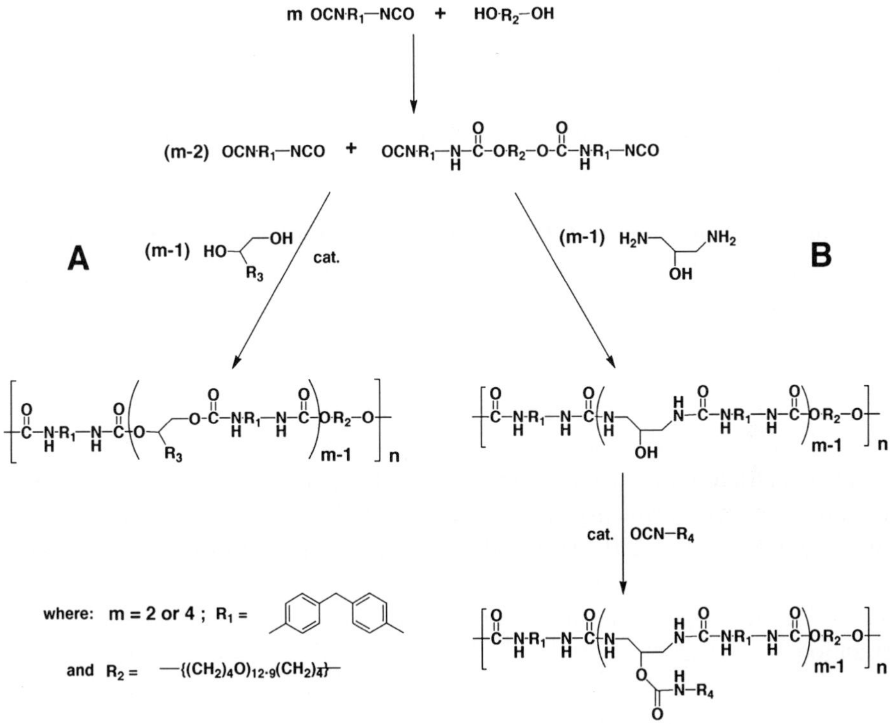

Figure 1. Generalized synthetic routes for the production of side-chain PEUs *via* (A) a di-functional chain extension or (B) tri-functional chain extension with subsequent reaction with an isocyanate.

migration and reorientation of the polymer side-chains to the polymer–air interface. Additionally, the added mobility of the side-chains relative to the backbone is thought to facilitate the formation of more uniform overlayers [7, 8]. While side-chains can be incorporated in relatively high and controllable densities, alteration of the phase separation behavior of the bulk polymer *via* disruption of hydrogen bonding can occur. Thus, side-chain homopolymers would not be suitable for systems requiring even moderate strength. However, in a low concentration physical blend with an unmodified polymer, most of the bulk physical properties should be retained and the side-chain polymers could act as surface-modifying additives.

Preliminary results on the synthesis of functionalized PEUs and PEUUs with linear and branched chains and the ability of these chemistry changes to modify surface properties were presented [9]. Here, we detail the synthesis and bulk characterization of these novel polymers.

MATERIALS AND METHODS

Materials

Poly(tetramethylene oxide) (PTMO, $M_n = 1000$) was obtained from E.I. du Pont de Nemours, dried 48 h at 60°C under vacuum before use and stored under dry nitrogen at room temperature. 4,4′-Diphenylmethane diisocyanate (MDI) was received from Bayer, stored under dry nitrogen at −12°C and distilled before use. Butane-1,4-diol (BD) was purchased from Aldrich, distilled under high vacuum and stored under dry nitrogen with activated molecular sieves at room temperature. Octadecyl isocyanate (ODI), α,α,α-trifluoro-p-tolyl isocyanate (TFTI) and 1,3-diamino-2-hydroxypropane (DAHP) were purchased from Aldrich and stored under dry nitrogen at 5°C. 3,5-Diaminobenzoic acid (DABA), ethylenediamine (EDA), 2,2-bis (hydroxymethyl) propionic acid (HMPA) and HPLC grade N,N-dimethylacetamide (DMAc) were purchased from Aldrich and stored under dry nitrogen at room temperature. 1H,1H,2H,3H,3H-perfluoroundecane-1,2-diol (PFUDD) and 1H,1H,8H,8H-dodecanefluoro-1,8-octanediol (DDFOD) were obtained from Oakwood Products and stored under dry nitrogen at 5°C. Lithium bromide (LiBr), dibutyltin dilaurate (DBTDL) and glycerol (GL) were purchased from Aldrich and stored under dry nitrogen at room temperature. Dimethyl-d_6 sulfoxide containing 1% (v/v) trimethylsilane (DMS-d_6/TMS) and dimethyl-d_7 formamide containing 1% (v/v) of TMS (DMF-d_7/TMS) were obtained from Cambridge Isotope Laboratories. All were used as received. The synthesis of the PEUs and PEUUs described in this paper were all performed *via* a multi-step solution addition polymerization.

Synthesis of isocyanate end-capped prepolymers

The scheme in Fig. 2 illustrates the first step in each polymer synthesis. PTMO was charged into a flamed and dry nitrogen purged 50-ml three-necked round-bottom flask and weighed. The flask was equipped with a stirrer and nitrogen inlet. A predetermined and weighed amount of MDI, corresponding to an isocyanate/hydroxyl (−NCO/−OH) ratio of 2 : 1 or 4 : 1, was added while stirring. The reaction mixture was then heated to about 80°C and stirred for 2 h.

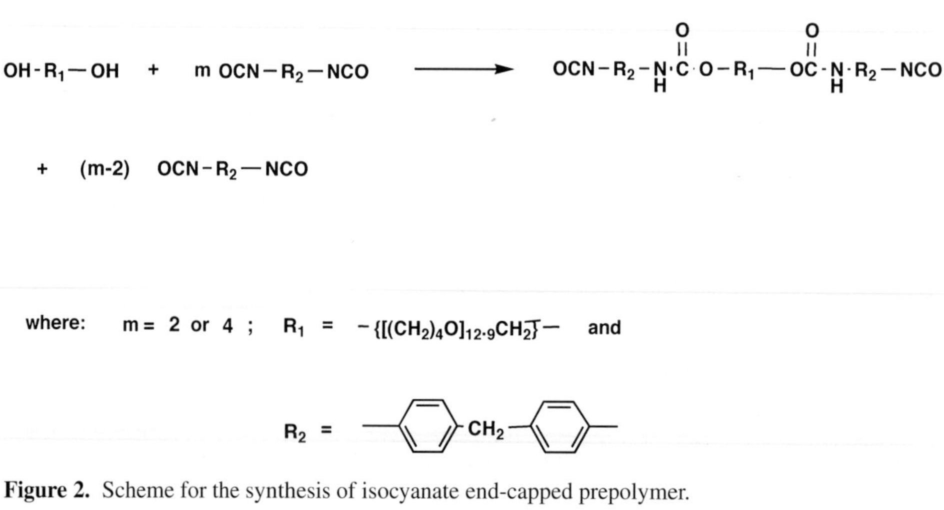

Figure 2. Scheme for the synthesis of isocyanate end-capped prepolymer.

Synthesis of linear PEUs

Diol chain-extension of the isocyanate end-capped prepolymer (IP) proceeded in accordance with the first reaction in the scheme in Fig. 3. The IP was dissolved in enough DMAc to yield a 20 wt% solution. A predetermined and weighed amount of GL, PFUDD, DDFOD or HMPA was dissolved in 5 ml DMAc, added to the IP solution at room temperature and stirred for 5–7 min. The OH/NCO ratio was either 1.05 : 1 or 1.20 : 1. The reaction mixture was homogenized by stirring for 10–15 min and then heated to about 80°C. The catalyst DBTDL (0.05 wt%) was added and stirring was continued for 10–15 min. The use of GL as a chain extender required special attention to the OH/NCO ratio in order to minimize the reactivity of the secondary hydroxyl groups. An essentially linear PEU possessing secondary hydroxyl functionality was formed.

The chain extension process was continued by the addition of small amounts of MDI or the concentrated solution of IP to the reaction mixture. Known quantities of melted MDI or IP solution and DBTDL were added dropwise to the reaction mixture to facilitate a gradual approach to stoichiometry described by the second reaction in the scheme in Fig. 3. This method of gradual approach to stoichiometry [10] was used to improve synthesis process control to produce reproducible, high-molecular-

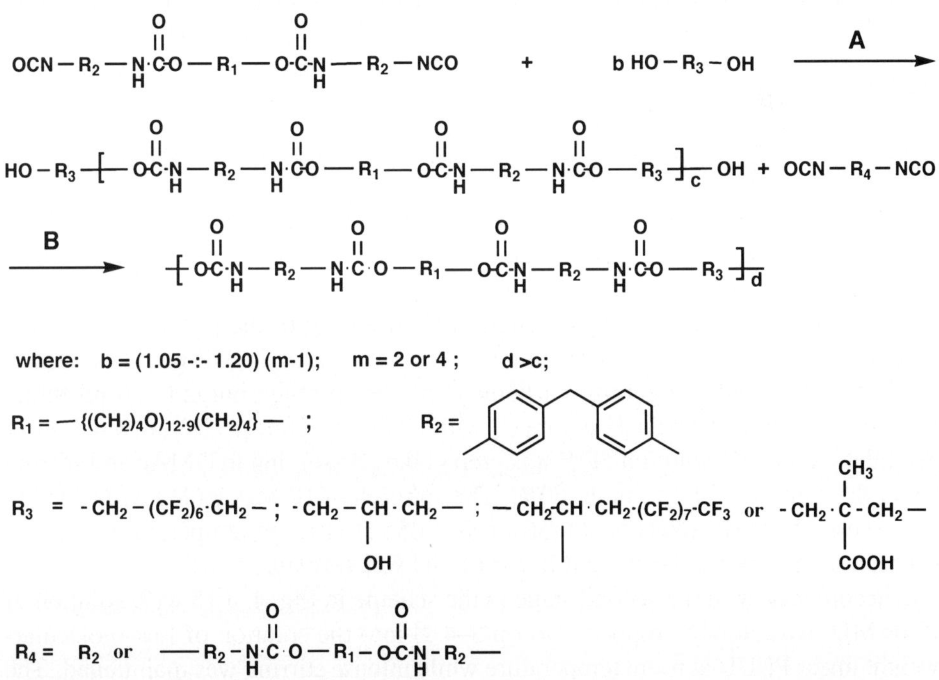

Figure 3. Scheme for the synthesis of (A) low-molecular-weight linear and side-chain PEUs based on IPs and a stoichiometric excess of polyol chain extenders and (B) the corresponding high-molecular-weight PEUs using low-molecular-weight polymer and a stoichiometric amount of diisocyanate-containing component. The chain extenders used were GL, HMPA, PFUDD and DDFOD.

weight PEUs and control the viscous properties of their concentrated solutions. The reaction mixture was monitored, until the complete disappearance of free NCO groups was noted, *via* a colorometric reaction with sodium nitrate in an acetone solution. Every time after the colorometric test the additional portions of MDI and DBTDL were added until viscous, concentrated solutions of the high-molecular-weight polymers (PEU-GL, PEU-PFUDD, PEU-DDFOD and PEU-HMPA) were formed. All the polymers were recovered *via* precipitation of a DMAc (10 wt%) solution in a large excess of deionized water.

^{1}H-NMR (δ, ppm, DMS-d_6/TMS) of PEU-GL: d = 1.57 (m, C$-$CH$_2-$C); 3.40 (t, O$-$CH$_2-$C); 3.88 (s, Ar$-$CH$_2-$Ar); 4.12 (m, C$-$C<u>H</u>(OH)$-$C); 5.45 (d, Ur$-$CH$_2-$C); 7.17–7.53 (two multiplets for each aromatic ring, Ar$-$H) and 9.50 (s, Ar$-$NH$-$C(O)O).

^{1}H-NMR (δ, ppm, DMF-d_7/TMS) of PEU-PFUDD: d = 1.57 (m, C$-$CH$_2-$C); 3.38 (t, O$-$CH$_2-$C); 3.86 (s, Ar$-$CH$_2-$Ar); 4.08 (m, C$-$CH(C)$-$O); 4.51 (m, CF$_2-$CH$_2-$C); 5.60 (d, Ur$-$CH$_2-$C); 7.14–7.55 (two multiplets for each aromatic ring, Ar$-$H) and 9.49 (s, Ar$-$NH$-$C(O)O).

^{1}H-NMR (δ, ppm, DMF-d_7/TMS) of PEU-DDFOD: d = 1.57 (m, C$-$CH$_2-$C); 3.38 (t, O$-$CH$_2-$C); 3.88 (s, Ar$-$CH$_2-$Ar); 4.87 (s, Ur$-$CH$_2-$CF$_2$); 7.16–7.55 (two multiplets for each aromatic ring, Ar$-$H) and 9.51 (s, Ar$-$NH$-$C(O)O).

^{1}H-NMR (δ, ppm, DMS-d_6/TMS) of PEU-HMPA: d = 1.21 (s, C$-$CH$_3$); 1.49 (m, C$-$CH$_2-$C); 3.32 (t, O$-$CH$_2-$C); 3.78 (s, Ar$-$CH$_2-$Ar); 4.22 (s, Ur$-$CH$_2-$C); 7.05–7.40 (two multiplets for each aromatic ring, Ar$-$H) and 9.49 (s, Ar$-$NH$-$C(O)O).

Synthesis of linear PEUUs based on DAHP

Another route to the formation of PEUs with hydroxyl side-groups used DAHP as the chain extender. Primary amines and secondary hydroxyls possess large differences in NCO-reactivity. Thus, reactions with the primary amines predominate in room temperature, non-catalyzed systems. Conforming to the first reaction in the scheme in Fig. 4, a predetermined and weighed amount of DAHP was dissolved in 5 ml of DMAc and charged into a flamed and dry nitrogen purged 100-ml three-necked round-bottomed flask, equipped with a stirrer, cylindrical funnel and nitrogen inlet. A 30 wt% solution of IP was prepared by dissolving in DMAc and stirring under dry nitrogen for 15 min at 60°C. The solution of IP was cooled and added to the excess of DAHP (NH$_2$/NCO ratio was 1.05 : 1) at room temperature. Thus, a solution of low-molecular-weight linear PEUU was formed.

In accordance with the second stage in the scheme in Fig. 4, a 15 wt% solution of IP or MDI was slowly dropped (1 drop/3–4 s) into the solution of low-molecular-weight linear PEUU at room temperature while intense stirring was maintained. The final NH$_2$/NCO ratio of the PEUU was nearly 1 : 1. Observance of the completion of the chain extension reaction was achieved as previously described. This was later confirmed by the lack of a NCO group absorption band (2260–2320 cm^{-1}) within the polymer infrared spectrum. The resultant high-molecular-weight PEUU-DAHP was recovered by precipitation. The method of gradual approach to stoichiometry was utilized by slow addition of dilute IP solution or monomeric diisocyanate in DMAc to the reaction mixture until a desirable PEUU molecular weight and viscosity of the concentrated solution was achieved. ^{1}H-NMR (δ, ppm, DMS-d_6/TMS) of PEUU-DAHP: $d = 1.48$ (m, C$-$CH$_2-$C); 3.32 (t, O$-$CH$_2-$C); 3.74 (s, Ar$-$CH$_2-$Ar); 4.05 (m, C$-$C<u>H</u>(OH)$-$C); 5.10–5.16 (d, N$-$CH$_2-$C); 6.15 (br.s, C(O)$-$NH$-$C); 7.06–7.37 (two multiplets for each aromatic ring, Ar$-$H); 8.56 (s, Ar$-$NH$-$C(O)N) and 9.50 (s, Ar$-$NH$-$C(O)O).

Synthesis of a linear PEUU based on DABA

Taking advantage of the large difference between aromatic-carboxyl and amine reaction rates with isocyanate, DABA chain extension was utilized in the formation of a PEUU containing aromatic COOH side-groups. Once more, following the scheme in Fig. 4, a predetermined and weighed amount of DABA was dissolved in 5 ml DMAc and charged into a similarly equipped and prepared flask as formerly mentioned. DMAc was added and the mixture was stirred at room temperature.

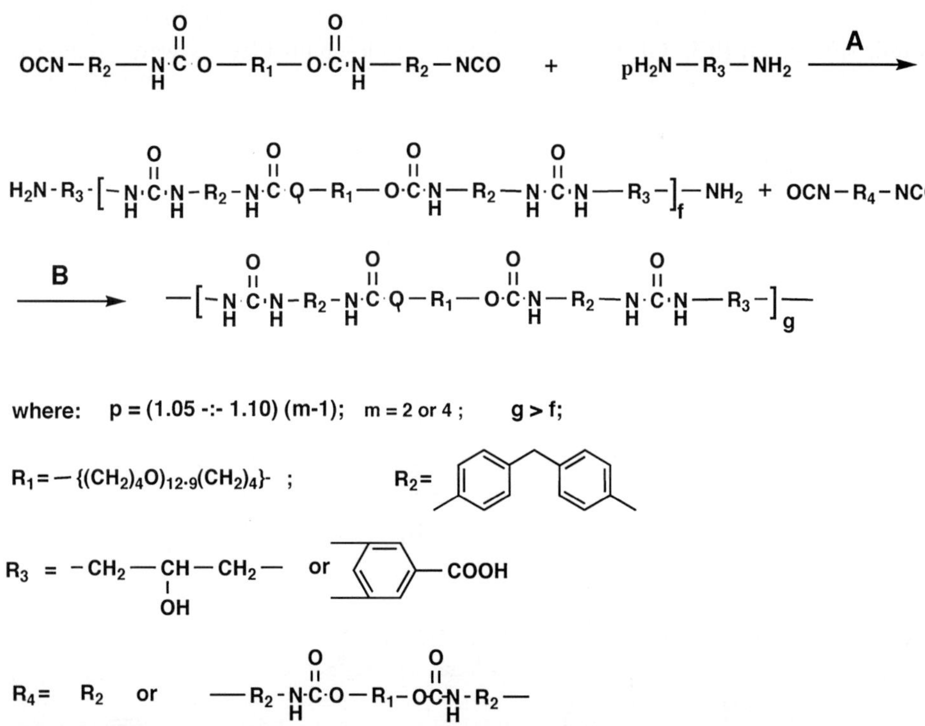

Figure 4. Scheme for the synthesis of (A) low-molecular-weight linear PEUUs based on IPs and a stoichiometric excess of diamine chain extenders and (B) the corresponding high-molecular-weight PEUUs containing OH or COOH side-groups using low-molecular-weight polymers and stoichiometric amounts of a diisocyanate-containing component. The chain extenders DAHP and DABA were used.

A cooled 40 wt% solution of IP in DMAc was added to the excess of DABA at room temperature to give a 1.10 : 1 NH_2/NCO ratio. Then a 20 wt% solution of IP was dripped (25–30 drops/min) into the low-molecular-weight linear PEUU solution with intensive stirring. The final NH_2/NCO ratio of the PEUU was 1 : 1. A viscous and homogeneous solution of high-molecular-weight PEUU-DABA was obtained and the polymer was recovered by precipitation. ^{1}H-NMR (δ, ppm, DMS-d_6/TMS) of PEUU-DABA: d = 1.50 (m, $C-CH_2-C$); 3.32 (t, $O-CH_2-C$); 3.82 (s, $Ar-CH_2-Ar$); 7.15–7.39 (two multiplets for each aromatic ring, Ar–H); 7.7–7.8 (s, N–Ar(COO)–N); 8.57 (s, Ar–NH–C(O)N); 8.89 (s, C(O)–NH–Ar) and 9.51 (s, Ar–NH–C(O)O).

Synthesis of side-chain PEUs and PEUUs based on ODI and TFTI

Octadecyl side-chains were added to the glycol chain extended PEU *via* the following procedure, described in the scheme in Fig. 5. A 20 wt% solution of PEU-GL in DMAc was charged into a flask equipped with a magnetic stirrer. An amount of ODI corresponding to 1 : 1 OH/NCO was added while stirring at room temperature under dry nitrogen. Then, 0.1 wt% DBTDL was added and stirring was continued at 80°C for 60 min. A viscous solution of the high-molecular-weight side-chain PEU-GL-ODI was formed. Side-chain PEUUs were created in a similar manner. A predetermined and weighed amount of the solid PEUU-DAHP was dissolved in DMAc in the flask utilizing a magnetic stirrer to yield a 15 wt% solution. ODI or TFTI was added while stirring under room temperature and dry nitrogen. The final ratio of NCO groups of monoisocyanate and OH groups of

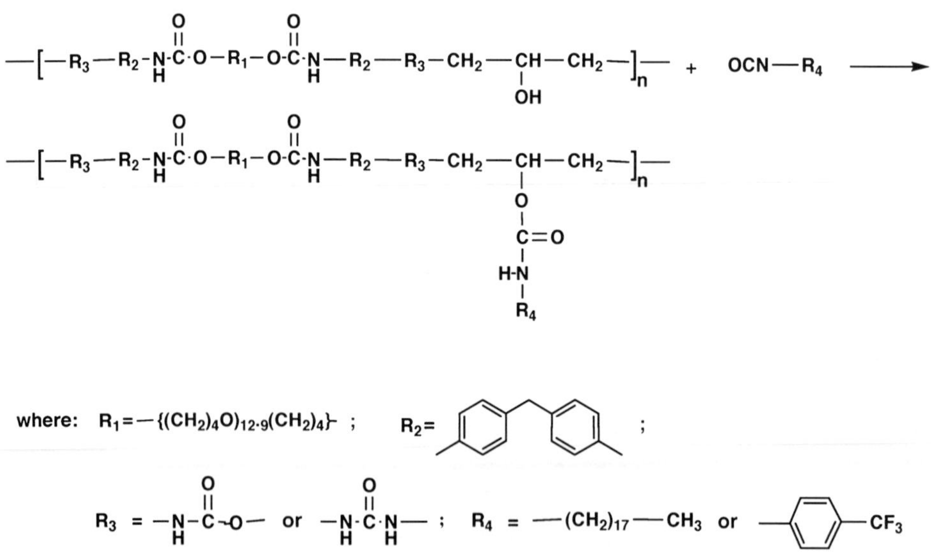

Figure 5. Scheme for the synthesis of side-chain high-molecular-weight polymers based on linear PEUs or PEUUs containing OH side-groups and monoisocyanates. The molar ratio of the OH side-groups to NCO groups of the monoisocyanate was 1 : 1.

DAHP was 1 : 1. DBTDL catalyst was added (0.2 wt%) and stirring was continued at 80°C for 60 min. As a result, DMAc solutions of side-chain PEUUs based on ODI (PEUU-DAHP-ODI) or TFTI (PEUU-DAHP-TFTI) were formed.

^{1}H-NMR (δ, ppm, DMS-d$_6$/TMS) of PEU-GL-ODI: d = 0.83 (t, C$-$CH$_3$); 1.21 (m, N$-$(CH$_2$)$_{17}$$-$C); 1.48 (m, C$-CH_2$$-$C); 3.30 (t, O$-CH_2$$-$C); 3.77 (s, Ar$-CH_2$$-$Ar); 4.07 (m, C$-$CH(OOCN)$-$C); 5.29 (d, Ur$-CH_2$$-$C); 7.03–7.43 (two multiplets for each aromatic ring, Ar$-$H); 8.52 (s, C$-$NH$-$C(O)O) and 9.50 (s, Ar$-$NH$-$C(O)O).

^{1}H-NMR (δ, ppm, DMS-d$_6$/TMS) of PEUU-DAHP-ODI: d = 0.83 (t, C$-$CH$_3$); 1.21 (m, N$-$(CH$_2$)$_{17}$$-$C); 1.49 (m, C$-CH_2$$-$C); 3.37 (t, O$-CH_2$$-$C); 3,75 (s, Ar$-CH_2$$-$Ar); 4.05 (m, C$-$CH(OOCN)$-$C); 5.14 (d, N$-CH_2$$-$C); 6.15 (t, C(O)$-NH-$C); 7.06–7.37 (two multiplets for each aromatic ring, Ar$-$H); 8.56 (s, Ar$-$NH$-$C(O)N) and 9.50 (s, Ar$-$NH$-$C(O)O).

^{1}H-NMR (δ, ppm, DMS-d$_6$/TMS) of PEUU-DAHP-TFTI: d = 1.47 (m, C$-$CH$_2$$-$C); 3.32 (t, O$-CH_2$$-$C); 3.75 (s, Ar$-CH_2$$-$Ar); 4.05 (m, C$-$CH(OOCN)$-$C); 5.12 (d, N$-CH_2$$-$C); 6.13 (t, C(O)$-NH-$C); 6.99–7.37 (two multiplets for each aromatic ring, Ar$-$H); 7.68 (m, F$_3$C–Ar$-$H); 8.56 (s, Ar$-$NH$-$C(O)N) and 9.50 (s, Ar$-$NH$-$C(O)O).

The structures of all these compounds, PEU-GL, PEU-GL-ODI, PEU-PFUDD, PEU-DDFOD, PEU-HMPA, PEUU-DAHP, PEUU-DAHP-ODI, PEUU-DAHP-TFTI and PEUU-DABA, are shown in Table 1.

Reference polymers

PEUs based on PTMO/MDI/BD = 1 : 2 : 1 (PEU-S$_1$) and PTMO/MDI/BD = 1 : 4 : 3 (PEU-S$_2$), as well as a PEUU based on PTMO/MDI/EDA = 1 : 2 : 1 (PEUU-S$_3$), were synthesized to serve as unmodified reference samples. The synthetic procedures were similar to the syntheses described above for PEUs and PEUUs, correspondingly.

^{1}H-NMR (δ, ppm, DMS-d$_6$/TMS) of PEU-S$_1$: d = 1.48 (m, C$-$CH$_2$$-$C); 3.32 (t, O$-CH_2$$-$C); 3.77 (s, Ar$-CH_2$$-$Ar); 7.03–7.45 (two multiplets for each aromatic ring, Ar$-$H) and 9.50 (s, Ar$-$NH$-$C(O)O).

^{1}H-NMR (δ, ppm, DMS-d$_6$/TMS) of PEU-S$_2$: d = 1.49 (m; C$-$CH$_2$$-$C), 3.31 (t; O$-CH_2$$-$C), 3.77 (s; Ar$-CH_2$$-$Ar), 7.06–7.37 (two multiplets for each aromatic ring, Ar$-$H) and 9.52 (s; Ar$-$NH$-$C(O)O).

^{1}H-NMR (δ, ppm, DMF-d$_7$/TMS) of PEUU-S$_3$: d = 1.57 (m; C$-$CH$_2$$-$C), 3.40 (t; O$-CH_2$$-$C), 3.83 (s; Ar$-CH_2$$-$Ar), 4.16 (t; N$-CH_2$$-$C), 6.32 (br.s; C(O)$-NH-$C), 7.06–7.58 (two multiplets for each aromatic ring, Ar$-$H), 8.57 (s; Ar$-$NH$-$C(O)N) and 9.54 (s; Ar$-$NH$-$C(O)O).

Synthesis of model urea compounds

In order to facilitate the identification of features within the ^{1}H-NMR spectra of the PEUUs, three model urea compounds were produced. These compounds, a diurea

Table 1.
Reactant ratios and corresponding hard segment repeat structures of synthesized polymers

Polymer and reactant ratio	Hard segment repeat structure
PEU-GL PTMO/MDI/GL = 1:4:3	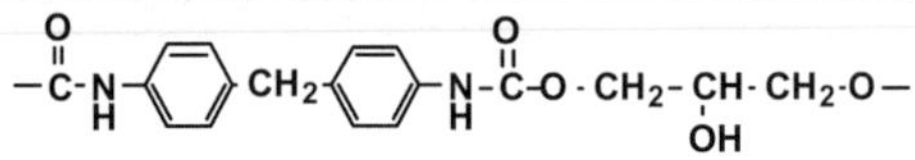
PEU-GL-ODI PTMO/MDI/GL/ODI = 1:4:3:3	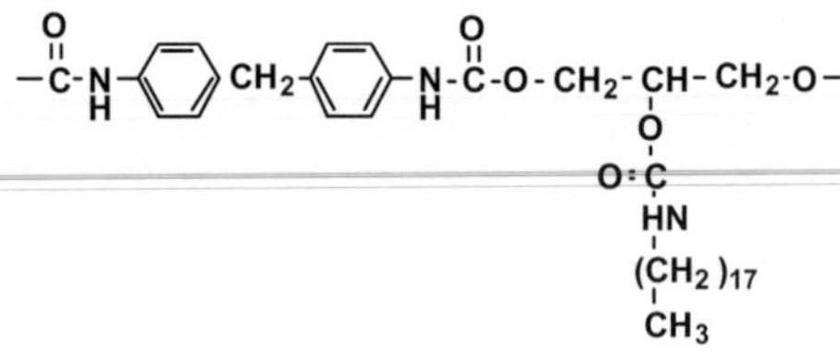
PEU-PFUDD PTMO/MDI/PFUDD = 1:2:1	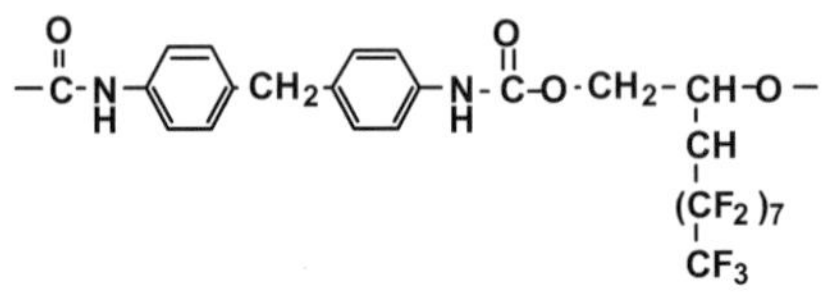
PEU-DDFOD PTMO/MDI/DDFOD = 1:2:1	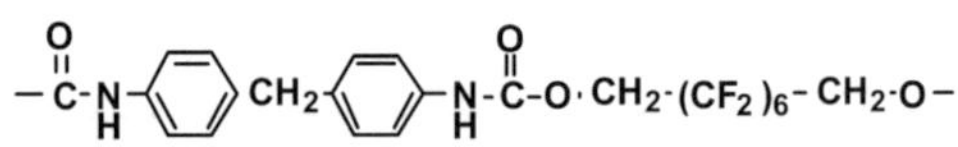
PEU-HMPA PTMO/MDI/HMPA = 1:4:3	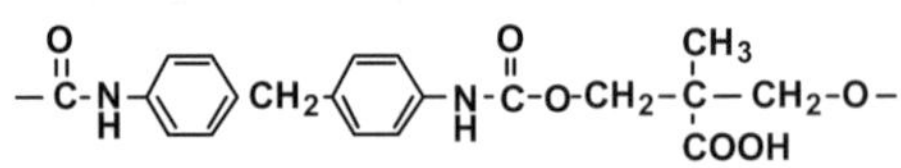
PEUU-DAHP PTMO/MDI/DAHP = 1:2:1	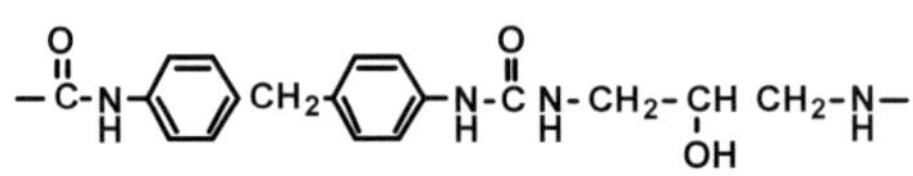
PEUU-DAHP-ODI PTMO/MDI/DAHP/ODI = 1:2:1:1	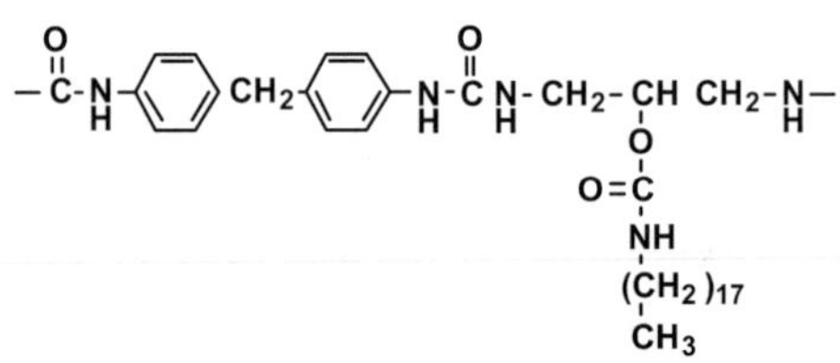
PEUU-DAHP-TFTI PTMO/MDI/DAHP/TFTI = 1:2:1:1	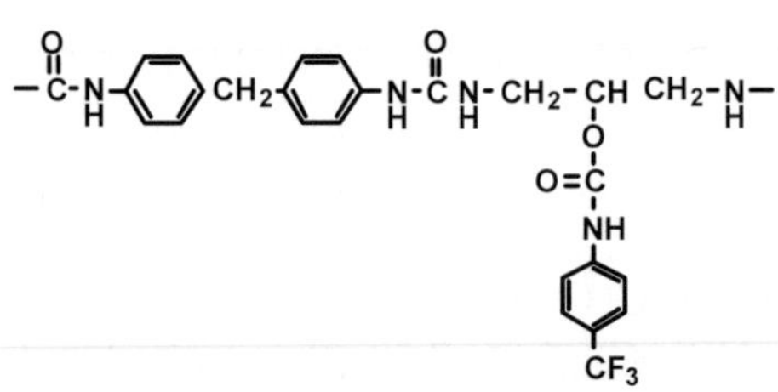
PEUU-DABA PTMO/MDI/DABA = 1:4:3	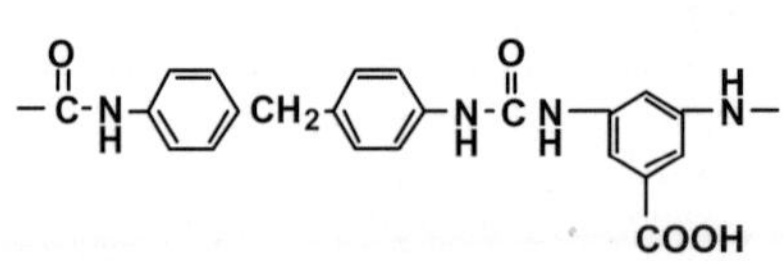

Table 2.
Reactant ratios and corresponding hard segment repeat structures of model compounds

Polymer and reactant ratio	Hard segment repeat structure
DUr DAHP/TFTI = 1 : 2	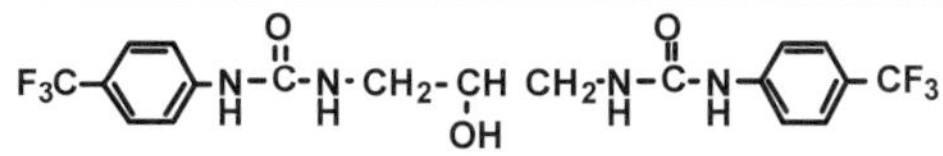
PUr DAHP/MDI = 1 : 1	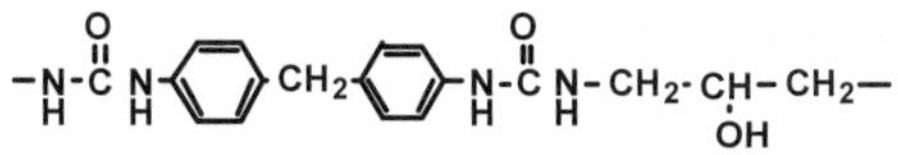
PUUr DAHP/MDI/BD = 0.3 : 1.3 : 1.0	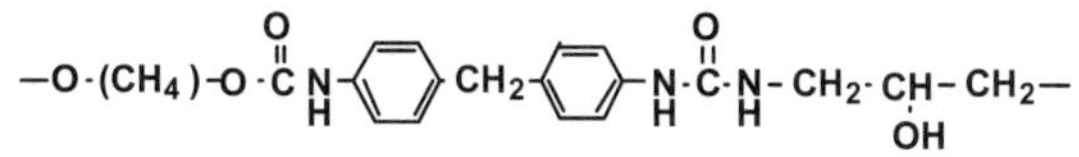

based on TFTI and DAHP (DUr), a polyurea based on MDI and DAHP (PUr) and a polyurethane urea based on BD, MDI and DAHP (PUUr), are shown in Table 2.

1.3-(α,α,α-trifluoro-p-tolylurea)-2-hydroxy propane (DUr)

A predetermined amount of DAHP was charged into a flamed and dry nitrogen purged 50-ml round-bottomed flask, equipped with a stirrer, cylindrical funnel and nitrogen inlet and weighed. DMAC was added and mixture was stirred at room temperature until a 10 wt% solution resulted. A calculated amount of TFTI was dissolved in DMAc resulting in a 10 wt% solution, charged into the cylindrical funnel and slowly dropped (1 drop/3–4 s) to the DAHP solution at about 0°C. The DAHP/TFTI ratio was 1 : 2. ^{1}H-NMR (δ, ppm, DMS-d$_6$/TMS) of DUr: $d = 3.64$ (m; C—C$\underline{\text{H}}$(OH)—C), 5,24 (d; N—CH$_2$—C), 6.40 (t; C(O)—NH—C), 7.02–7.73 (m; Ar—H) and 9.14 (s; Ar—NH—C(O)N).

Poly(2-hydroxypropane-1-diphenylmethane diurea) (PUr)

The synthesis as described above for DUr was used with MDI as the diisocyanate. The reaction was carried out at the room temperature. The DAHP/MDI ratio was 1 : 1. ^{1}H-NMR (δ, ppm, DMS-d$_6$/TMS) of PUr: $d = 3.53$ (m; C—C$\underline{\text{H}}$(OH)–C), 3.75 (s; Ar—CH$_2$—Ar), 5.13 (d; N—CH$_2$—C), 6.15 (t; C(O)—NH—C), 7.00–7.31 (two multiplets for each aromatic ring, Ar—H) and 8.54 (s; Ar—NH—C(O)N).

Poly(2-hydroxypropane-1-(diphenylmethane-4-urea-4′-n-butoxycarbamate)-3-(di-phenylmethane-4-urea-4′-carbamate)) (PUUr)

The method of synthesis as described above for the main samples was used. The synthesis of IP based on BD and MDI was carried out for 5 min at 80°C. Then IP was cooled and dissolved in DMAC resulting in a 15 wt% solution and then dropped into the 10 wt% DAHP solution at about 0°C. ^{1}H-NMR (δ, ppm, DMS-d$_6$/TMS) of PUUr: $d = 1.70$ (m; C—CH$_2$—C), 3.28 (t; O—CH$_2$—C), 3.79 (s; Ar—CH$_2$—Ar),

4.46 (m; C—C̲H̲(OH)—C), 5.13 (d; N—CH_2—C), 6.15 (t; C(O)—NH—C), 7.00–7.39 (two multiplets for each aromatic ring, Ar—H); 8.57 (s; Ar—NH—C(O)N) and 9.54 (s; Ar—NH—C(O)O).

Polymer recovery

All resultant PEUUs, PEUs and model components were isolated in a laboratory blender by precipitation from a 10 wt% DMAc solution into deionized water. The precipitated compounds were filtered, thoroughly washed with fresh deionized water at room temperature and finally dried under high vacuum for 5 h at 65–70°C.

Polymer characterization

GPC. Gel-permeation chromatography measurements were carried out using a Waters 590 HPLC pump connected in series with 100 Å and 1000 Å Nucleosil-C_{18} columns and a refractive index detector (Waters 410). A 0.05 M LiBr solution in DMAc was used as the mobile phase. The flow rate was 0.50 ml/min and the columns were heated to 80°C. Molecular weights were reported based upon the retention time of polystyrene standards ($M_n = 3 \times 10^3$–2×10^5). The apparent weight-average molecular weights of PEUUs and PEUs were in the range (5–12) × 10^4.

FT-IR. Transmission Fourier transform infrared spectra were acquired by using a BioRad FTS-60A/896 Fourier transform infrared spectrometer, equipped with a dry air purge system, between 3700 and 700 cm^{-1}. Typically, 256 scans at a resolution of 8 cm^{-1} were averaged for each sample. Each sample for infrared analysis was prepared by casting a thin film on a KBr window from a 7 wt% polymer solution in DMAc. The solvent was evaporated under vacuum at room temperature for 1 h and then the sample was further dried at 60°C for 2 h to more completely remove the solvent.

^{1}H-NMR. Proton nuclear magnetic resonance spectra were recorded on a Bruker ACF-200 (250 MHz) spectrometer using a TMS/solvent signal as an internal reference. The chemical shift was calibrated using DMS-d_6 (2.50 ppm) as an internal standard. While the PEUs chain-extended by the hydrogen carbon components are known to be soluble in DMS-d_6, the PEUs based on the fluoro-derivative diols are insoluble in this solvent; as such, DMF-d_7 was used for PEU-PFUDD and PEU-DDFOD. The identification of the basic NMR spectral features was aided by use of model compound spectra.

RESULTS AND DISCUSSION

GPC

Figure 6 shows the development of the GPC curve for PEU-GL at various points during the addition of MDI during the final stage of synthesis. The appearance of a

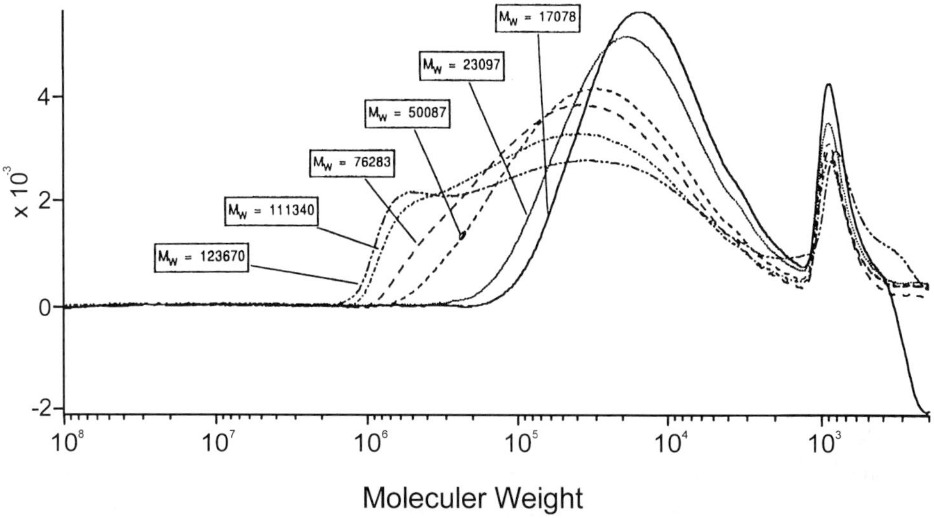

Figure 6. The changes in the PEU-GL GPC curve shape during the course of the synthesis.

high-molecular-weight shoulder, corresponding to an increasing branched polymer chain concentration, can be seen as the correlation between OH groups and NCO groups approaches stoichiometry. As a result, the polymer begins to gel. Due to the different reactivities with —NCO of the primary and secondary OH groups in GL, it was possible to obtain a soluble high-molecular-weight PEU ($M_w = 5 \times 10^4$) with OH groups attached to the backbone. However, the preferred method for the incorporation of hydroxyl side-groups utilized DAHP as the chain extender (PEUU-DAHP). The extreme difference in NCO reactivities between the primary amine groups and the secondary OH groups allowed nearly complete reaction of the NCO with the amines. No evidence of cross-linking was noted in the GPC spectra for PEUU-DAHP. Due to polymer/packing affinity, the GPC spectra for PEUU-DAHP and the side-chain variants, PEUU-DAHP-ODI and PEUU-DAHP-TFTI could not be compared quantitatively. However, a change in the polymer chemistry could be confirmed. The two side-chain PEUs displayed an increase in polymer retention time without any change in peak shape (Fig. 7), consistent with an average chemistry that is more similar to the hydrocarbon-modified chromatography packing material.

FT-IR

The IR survey spectra of PEU and PEUU samples are shown in Figs 8 and 9. The spectra are qualitatively similar. However, some deviations in peak position and/or relative intensity of certain bands can be observed between the functionalized polymers and their appropriate standards. A comparison of N—H absorption (3220–3420 cm^{-1}) for PEU-GL and PEU-S_2 demonstrates peak broadening in the PEU-GL spectrum. The appearance of a shoulder at approx. 3385 cm^{-1} can be attributed to

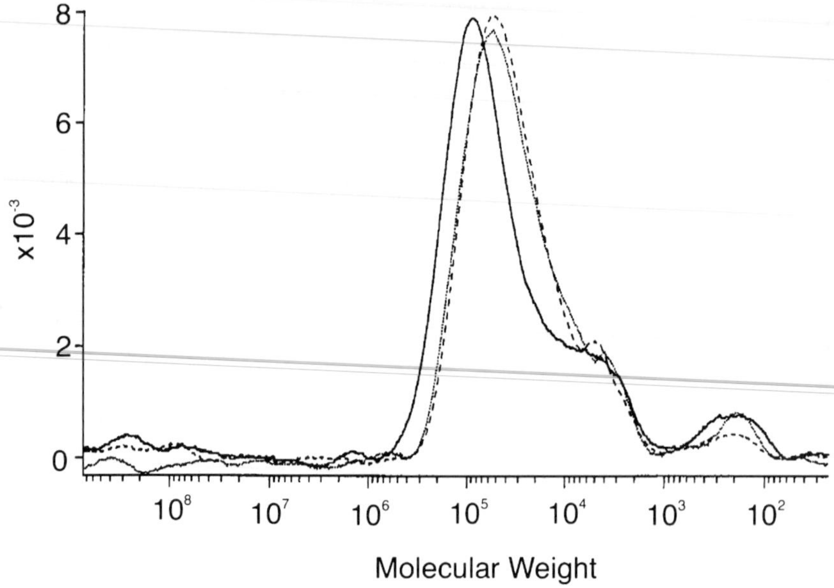

Figure 7. The GPC curves of the basic hydroxyl-functionalized PEUU-DAHP (solid line) and the two modified branched PEUUs: PEUU-DAHP-ODI with attached octadecylurethane (dotted line) and PEUU-DAHP-TFTI with attached trifluorotolylurethane (dashed line) side-chains.

the presence of hydroxyl side-groups. As expected, upon reaction with ODI to form PEU-GL-ODI this shoulder decreases. In the carbonyl stretching region (amide-I band), an increase in the intensity of the non-hydrogen-bonded C=O stretching (approx. 1733 cm^{-1}) relative to the hydrogen-bonded C=O stretching (approx. 1708 cm^{-1}) in the PEU-GL spectrum compared to the ratio A_{1733}/A_{1708} in the PEU-S_2 is noted. This may be an effect due to steric hindrance of some urethanes by neighboring hydroxyl side-groups in PEU-GL's backbone or from hydroxyl side-groups themselves. The ratio of the intensities for non-bonded to bonded carbonyl groups almost returns to that of the standard, PEU-S_2, after attachment of the octadecyl side-chain (PEU-GL-ODI). It might be related to the introduction of additional urethane groups, located at the base of the side-chains, which have the ability to form new hydrogen bonds. These changes are also correlated with the change in intensity of the amide III bands. Comparing PEU-S_1 and PEU-PFUDD, changes in the shape and intensity of all amide bands (I, II and III) are observed. In the amide-I band, a different splitting of the C=O absorption is observed. The incorporation of the fluorinated side-chains likely decreases the degree of the molecular association within the hard segment domains. This can be seen by an increase in the intensity of the non-hydrogen-bonded C=O stretching absorption (approx. 1733 cm^{-1}) relative to the bonded C=O stretching. The appearance of the shoulder in the amide-II region (at 1518 cm^{-1}) is assigned to the changing of the correlation of intensities of N–H bending and C−N stretching absorption in comparison with PEU-S_1. It is evident from the broadening of the amide-III

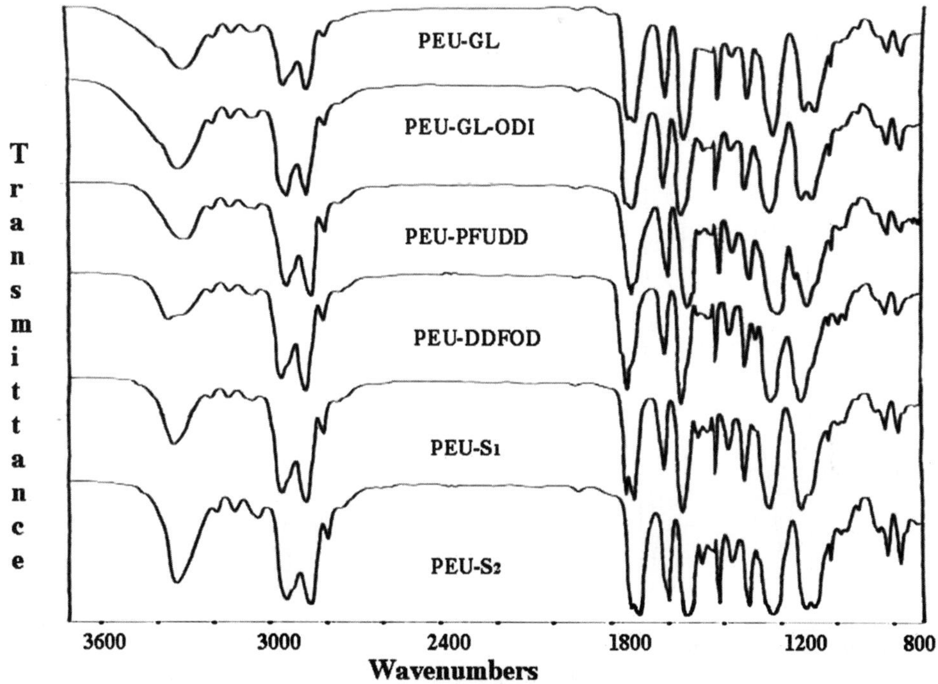

Figure 8. Superimposed FT-IR spectra of modified PEUs consisting of hydroxyl (PEU-GL), octadecyl (PEU-GL-ODI), fluorocarbon groups in the side (PEU-PFUDD) and in the backbone (PEU-DDFOD) chains and standards (PEU-S_1 and PEU-S_2) in the range 3700–700 cm^{-1}.

absorption band that there may be contributions from CF_2 groups in the side-chains of PEU-PFUDD. The existence of terminal CF_3 groups is demonstrated by the appearance of a peak at approx. 1152 cm^{-1}. The carbonyl-stretching region of linear DDFOD-extended PEU (PEU-DDFOD) demonstrates the higher association within the hard segment domains in comparison to PEU-PFUDD containing the fluorinated side-chains. The new peak at approx. 1280 cm^{-1} can be attributed to the CF_2 groups of the DDFOD-chain extender. The infrared spectrum of the PEUU-DAHP is different from the spectrum of the standard, PEUU-S_3. Despite the fact that the C=O stretching region for both PEUUs possesses two similar peaks attributed to free urethane and urea (approx. 1733 cm^{-1}) and hydrogen-bonded urethane (approx. 1708 cm^{-1}) groups, a difference in the intensity of the third peak attributed to hydrogen-bonded urea groups is observed. The presence of hydroxyl side-groups likely decreases the association within the hard segment domains. Thus, a decrease in the intensity of hydrogen bonded urea groups in the carbonyl stretching region can be attributed to steric hindrance of the hydroxyl groups of the diureahydroxypropane fragment of the DAHP hard segment. This corresponds to a decrease in the intensity of N−H-bending and C−H-stretching absorption in comparison with PEUU-S_3. For the case of CF_3-terminated groups in the side-chains (PEUU-DAHP-TFTI), a C−F-stretching band at approx. 1163 cm^{-1}

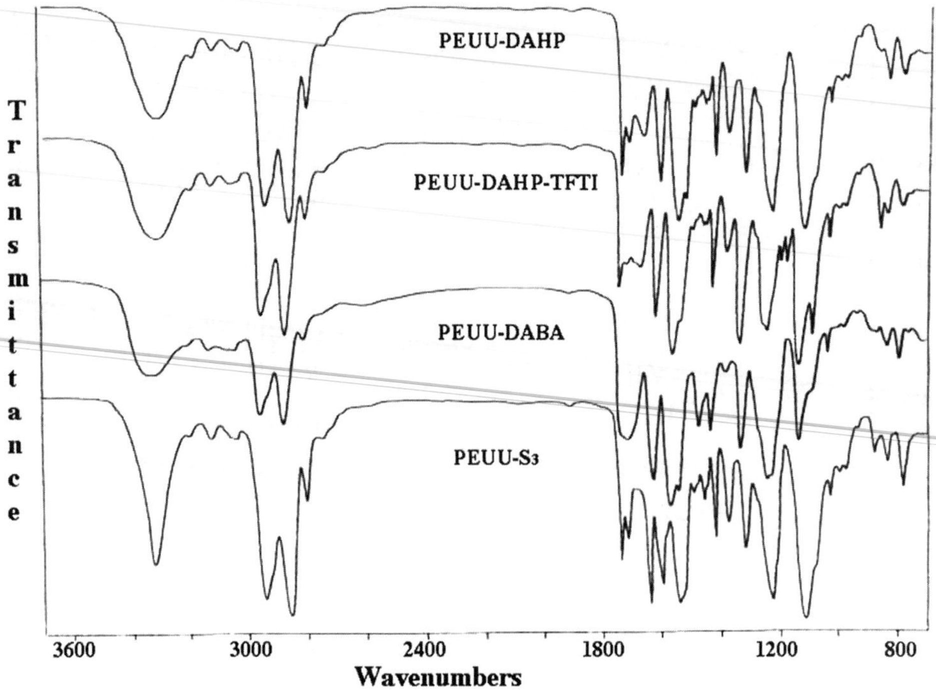

Figure 9. Superimposed FT-IR spectra of modified PEUUs consisting of hydroxyl (PEUU-DAHP), fluorocarbon (PEUU-DAHP-TFTI), carboxyl groups (PEUU-DABA) and standard (PEUU-S_3) in the range 3700–700 cm^{-1}.

characterizes the infrared spectrum. The carboxyl substitution in the aromatic diamine chain extender causes broading of the carbonyl-stretching band to such a degree that the free and bound carbonyl can no longer be individually distinguished. An additional new, unidentified, peak at approx. 1451 cm^{-1} occurs in the PEUU-DABA spectrum also and can be attributed to COOH-groups.

^{1}*H-NMR*

Poly(ether urethane)s. The signal intensities of urethane protons, phenyl ring protons, the outermost methylene protons of BD and PTMO, methylene bridge protons of MDI, internal methylene protons of BD and the butane fragment of the PTMO chain are dependent on the molar ratio of components present and display the same chemical shifts seen in both PEU-S_1 and PEU-S_2. In the case of PEU-GL-ODI, methylene and methyl protons of the octadecyl chain appear at 1.21 and 0.83 ppm, respectively. Also, the CH protons of the GL fragment of the hard segment indicate a multiplet at 4.07 ppm. The spectrum of side-chain fluoro-containing PEU-PFUDD shows protons signals at 9.49 ppm, belonging to the urethane groups that are formed based on the OH groups of PFUDD. CH-protons of PFUDD occurred at 4.08 ppm and the weak signal at 4.51 ppm can be assigned to protons of CH_2 groups connected with the CF_2-chain. The signal of the urethane protons of the linear fluoro-containing PEU-DDFOD is slightly shifted to the lower field. The CH_2 protons of the DDFOD fragment indicate a singlet at 4.87 ppm, characterizing the symmetric structure of the chain extender. The spectrum of PEU-HMPA is characterized by a signal at 9.49 ppm. This can be assigned to the urethane group protons formed by HMPA and the NCO group of MDI. In this polymer, methylene and methyl protons of the HMPA fragment appear at 4.22 and 1.21 ppm and the signals of carboxylic protons are not observed.

Poly(ether urethane urea)s. The basic differences in the ^{1}H-NMR spectral features of PEUU-S_3 in comparison to the two PEU-standards are the appearance of two signals attributed to nitrogen protons of the urea group: 8.57 ppm for Ar−NH−C(O)N protons and 6.32 ppm for C(O)−NH−C protons. This corresponds well to the data described by Nakano *et al.* [11]. PEUU-DAHP is the precursor to side-chain PEUUs that are formed via isocyanate addition reactions and it possesses signals associated with the urethane protons, NH protons of urea groups attached to an aromatic ring, phenyl ring protons, NH protons of urea connected with CH_2 groups of the chain extender, the multiplet of the CH protons of the DAHP fragment of the hard segment and the main signals of the methylene protons (same as PEU). The spectrum of PEUU-DAHP-ODI has acquired two new signals in comparison with PEUU-DAHP: the peaks at 1.21 ppm and 0.83 ppm are characteristic of the CH_2 and CH_3 protons from the octadecyl chain. This spectrum corresponds to the spectrum of PEU-GL-ODI, which possesses a similar octadecyl side-chain. In the other derivative of PEUU-DAHP based on TFTI, proton signals from the phenyl ring (7.68 ppm) are consistent with being located as a side-chain fluoro-containing PEUU-DAHP-TFTI. Thus, both of the above polymers (PEUU-DAHP-ODI and PEUU-DAHP-TFTI) show changes indicative of the formation of side-chains. Finally, PEUU-DABA is characterized by peaks in the range 7.7–7.8 ppm, which can be attributed to the phenyl ring of the DABA fragment of the hard segment.

CONCLUSIONS

An improved method was developed to synthesize functionalized linear and side-chain PEUs and PEUUs based on PTMO and MDI. The reactions occur in DMAc solution. The resultant polymers incorporate hydroxyl-, alkyl-, carboxyl- and fluorocarbon end-groups, attached to the backbone or as pendant units by using di- and trifunctional chain extenders containing functional groups with different reactivities. It was shown that by varying reaction conditions, the site of the reaction could be controlled with good accuracy and reproducibility permitting PEUs and PEUUs with reactive functionalities to be prepared. The functionalized PEUs and PEUUs prepared by the novel techniques described here were characterized using GPC, FT-IR and ^{1}H-NMR. They were found to be of high molecular weight, to possess the intended structures and, where appropriate, to contain active functional groups.

Acknowledgements

The authors wish to acknowledge the financial support of NIH (National Institutes of Health) grant HL25951 and UWEB, NSF EEC-952916, which has made these studies possible. The first author thanks Dr. Esmaeel Naeemi for useful discussion and his help in interpreting the ^{1}H-NMR data.

In the mid 1970s one of the authors (B. D. R.) was working with Professor Allan Hoffman on an NIH National Heart and Lung Institute contract to explore the blood compatibility of polyurethanes. Surprising observations were made on the relationship between surface chemistry and blood platelet reaction [12]. Those data continue to stimulate explorations on new compositions that can be used to elucidate the mechanisms of blood interaction. Thus, this paper, almost 30 years later, was inspired by ideas that can be traced back to Allan Hoffman's thinking on polyurethanes.

REFERENCES

1. N. M. K. Lamba, K. A. Woodhouse and S. L. Cooper, *Polyurethanes in Biomedical Applications*. CRC Press, Boca Raton, FL (1998).
2. B. D. Ratner, *J. Biomed. Mater. Res.* **27**, 837 (1993).
3. B. D. Ratner, *Biosens. Bioelectr.* **10**, 797 (1995).
4. R. Rahman and B. D. Ratner, *J. Polym. Sci. Part A Polym. Chem. Edn.* **27**, 2673 (1989).
5. T. Nakaya and Y. Li, *Designed Monomers Polymers* **6**, 309 (2003).
6. F. I. Simonovsky, Y. Wu, T. A. Horbett and B. D. Ratner, *Polym. Mater. Sci. Eng.* **88**, 104 (2003).
7. S. C. Yoon, B. D. Ratner, B. Ivan and J. D. Kennedy, *Macromolecules* **27**, 1548 (1994).
8. H. Tan, X. Xie, J. Li, Y. Zhong and Q. Fu, *Polymer* **45**, 1495 (2004).
9. F. I. Simonovsky, S. C. Porter and B. D. Ratner, *Trans. Soc. Biomater.* **20,** 461 (1997).
10. F. I. Simonovsky, A. A. Zaplatin, F. K. Samigullin and E. P. Khudyak, USSR Patent No. 1746688 (1992).
11. Y. Nakano, A. Nakashima, H. Hanahata and K. Okajima, *Polymer* **38**, 817 (1997).
12. S. R. Hanson, L. A. Harker, B. D. Ratner and A. S. Hoffman, *J. Lab. Clin. Med.* **95**, 289 (1980).

Gels, Genes, Grafts and Giants, pp. 735–741 (2005)

Immobilization of streptavidin-horseradish peroxidase onto a biotinylated poly(acrylic acid) backbone that had been radiation-grafted to a PTFE film

PATRICIA SADURNÍ [1], ALEJANDRO ALAGÓN [2], ROUSTAM ALIEV [1], GUILLERMINA BURILLO [1] and ALLAN S. HOFFMAN [3,*]

[1] *Instituto de Ciencias Nucleares, UNAM, Ciudad Universitaria, 04510 México, D.F., México*
[2] *Instituto de Biotecnología, UNAM, Morelos, México*
[3] *Department of Bioengineering, University of Washington, Seattle, WA 98195, USA*

Received 16 January 2004; accepted 10 May 2004

Abstract—Films of polytetrafluoroethylene (PTFE) were modified by radiation graft polymerization of acrylic acid (AAc). Optimal conditions for efficient AAc grafting were studied, including pre-irradiation dose in air, monomer concentration, temperature and time of the grafting process. Carboxylic groups of the grafted polyAAc were activated with carbodiimide (EDC) for biotinylation by reaction with 5-(biotinamido) pentylamine. Streptavidin-horseradish peroxidase (SA-HRP) was immobilized by affinity complexation of the SA with the biotin groups on the PTFE surface. The amount of active HRP immobilized on the PTFE films was determined as a function of the extent of polyAAc grafting. This study has demonstrated the utility of combining the processes of (a) radiation grafting of polymers with reactive groups onto inert polymers such as PTFE, (b) biotinylation of the graft polymer reactive groups, (c) immobilization of streptavidin on the biotinylated surface sites, followed by (d) immobilization of biotinylated, biologically active molecules *via* complexation of their conjugates with streptavidin. In this study, the last two steps were combined by immobilizing the complex of streptavidin and biotinylated HRP onto the biotinylated surface sites. The unique nature of this process is the ability to immobilize biotinylated molecules on an inert surface as PTFE.

Key words: Radiation graft polymerization; PTFE; acrylic acid; streptavidin-horseradish peroxidase.

INTRODUCTION

Immobilization of biomolecules on polymer supports has become important in medicine and biotechnology for a wide variety of applications, both *in vitro* and *in vivo*. The immobilization process is usually based on the reaction of

*To whom correspondence should be addressed. Tel.: (1-206) 543-9423. Fax: (1-206) 543-6124. E-mail: hoffman@u.washington.edu

polymer surface groups with reactive sites on the biomolecules. Radiation graft polymerization has been widely used as a promising method for modification of otherwise inert polymer surfaces, in order to create reactive surface groups (e.g., $-OH$, $-COOH$ or $-NH_2$) that can then be used for immobilization of bioactive molecules, such as peptides and proteins [1–9]. Biotin has been used to link proteins to the surfaces of polymers and membranes [10–12]. Radiation grafting of acrylic acid (AAc) onto fluorinated polymers has been carried out previously [13–18], but biotinylation of the AAc graft polymers for immobilization of biomolecules has not been previously investigated.

The objective of this work is to radiation graft AAc onto PTFE films, followed by biotinylation of the grafted polymer and immobilization of a streptavidin-peroxidase conjugate (SA-HRP) onto the biotinylated PTFE surfaces *via* the complexation of the SA to the biotin. HRP is a model for any number of biomolecules. This is a novel process that has not previously been investigated. PTFE is an important biomaterial that is widely used in vascular grafts. This process could be used to bind biomolecules such as heparin to the surface of PTFE.

MATERIALS AND METHODS

Materials

PTFE films of 0.1 mm thickness were kindly donated by CNRS (France); AAc was purchased from Materiales Plasticos (Mexico). 1-ethyl-3-(3-dimethylamino-propyl)carbodiimide hydrochloride (EDC) was from Sigma (Mexico). EZ-Link 5-(Biotinamido) penthylamine was obtained from Pierce (through Perstorp Biotec, Mexico). Hydrochloric acid was from Merck (Mexico). Sodium chloride was from Baker (Mexico). SA-HRP was purchased from Boehringer, 2, 2′-azino di(3 ethyl benzthiazoline sulfonic acid) (ABTS) was from Bio-Rad (Mexico).

Grafting

PTFE films were cut into 1×1.5 cm pieces, and they were irradiated in air with a ^{60}Co γ-source (Gammabeam 651 PT, Nordion International) at a dose rate of 8 kGy/h and doses of 10 to 50 kGy were delivered. The irradiated samples were placed in glass ampoules containing aqueous solutions of AAc at concentrations of 40 or 50%. The ampoules were de-aerated under vacuum by repeated freeze-thaw steps, sealed and placed in a water bath at temperatures from 35 to 50°C for 30–90 min. Care was taken to evenly expose both sides of the films to the AAc solutions. The grafted films were washed with hot water for 24–48 h to extract residual monomer as well as any free polyAAc homopolymer that had formed. This was followed by drying the samples under vacuum to a constant weight. This procedure was repeated until the sample weights did not change. Grafting of PAAc was confirmed by observation of the characteristic $-C=O$ bond using ATR–FT-IR.

Yields were determined gravimetrically by the equation:

$$\% \text{ Grafting} = 100 \times ((W - W_0)/W_0),$$

where W_0 and W are dry weights of the samples before and after grafting.

SA-HRP immobilization

Sets of 10 films of PTFE, each having different polyAAc graft contents, were swollen in distilled water for 30 min at room temperature. Then, 20 mg EDC in 15 ml distilled water was added gently to 50 ml Ependorff tubes. 18 mg 5-(biotinamido) pentylamine was then added at room temperature, and the reaction was allowed to proceed for 1 h at pH 5. The biotin-containing films were washed with 30 ml 0.5 M Tris/HCl and 0.15 M NaCl buffer solution at pH 8. 3 mm diameter circles of the washed biotinylated PTFE films were cut and placed into wells of polystyrene ELISA plates which contained 100 μl of SA-HRP at an HRP concentration of 10 μg/ml, and allowed to bind to the films at room temperature for 1 h. Then the films were washed five times with 200 μl of the same buffer and transferred to new, clean wells. After this procedure, 100 μl of ABTS substrate solution was added to the samples, and they were shaken for 5 min. ABTS (2,2′-Azinobis[3-ethylbenzothiazoline-6-sulfonic acid]-diammonium salt) is a water-soluble HRP dye that yields a green end product upon reaction with the atomic oxygen produced by HRP. The ABTS substrate solution that was used consists of a mixture of two freshly made solutions, one containing ABTS and the other containing sodium perborate and citric acid (50 mM adjusted at pH of 4 with 10 M NaOH). Sodium perborate decomposes under the acidic conditions to 500 mM hydrogen peroxide. The hydrogen peroxide is then converted to atomic oxygen by HRP, and the atomic oxygen reacts with ABTS to produce the green color. The reaction was stopped with the addition of 100 μl of 10% of SDS (sodium dodecylbenzene sulfonate) in water. The product was measured at 405 nm using a Beckman DV 650 spectrophotometer [19]. One unit of HRP activity corresponds to the decomposition of 1 μmol of hydrogen peroxide per min at 25°C under the specified conditions.

RESULTS AND DISCUSSION

Kinetics of the AAc grafting onto pre-irradiated PTFE depended on the key variables of pre-irradiation dose, concentration of monomer, temperature and reaction time (Fig. 1). It is known that the use of pure monomers in radiation graft polymerization is restricted by the undesirable homopolymerization that takes place, whether one is using the direct or pre-irradiation method [20]. Formation of homopolymer decreases significantly with decrease in pre-irradiation dose or monomer concentration. Furthermore, it is necessary to take into account the relatively low radiation stability of PTFE [21, 22] and therefore, using high radiation doses is generally undesirable for modification of this polymer by radiation grafting. Since PTFE is known to predominantly degrade by main chain scission under irradiation, it is likely that the grafting is initiated by a combination of trapped free radicals or radicals formed from decomposition of main chain peroxides. High doses could fragment the PTFE chains excessively, leading to increased yields of homopolymer as well as to PAAc-g-PTFE chains with short PTFE fragments that will be extracted during the wash step.

In our system, optimum AAc grafting was achieved by using a pre-irradiation dose of 10 kGy, with a monomer concentration of 40% and grafting temperatures of 40 and 50°C. When the pre-irradiation dose and monomer concentration were both low, homopolymerization was negligible and grafting was more gradual, allowing the monomer to penetrate into the film. That lead to the gradual and continual rise in grafting levels seen in Fig. 1 (diamonds, triangles and squares) as more and more monomer was able to add to the growing free radical chain ends in the viscous milieu within the film. The grafting rate was less at 35°C, and that can be explained by a combination of lower efficiency of thermal decomposition of the main chain peroxides along with lower efficiency of exposure of the frozen in free radicals. With increasing pre-irradiation doses and higher monomer concentrations above the optimum, the grafting process became complicated by AAc homopolymerization, as noted above. As the pre-irradiation dose increases, the initial rate of AAc grafting

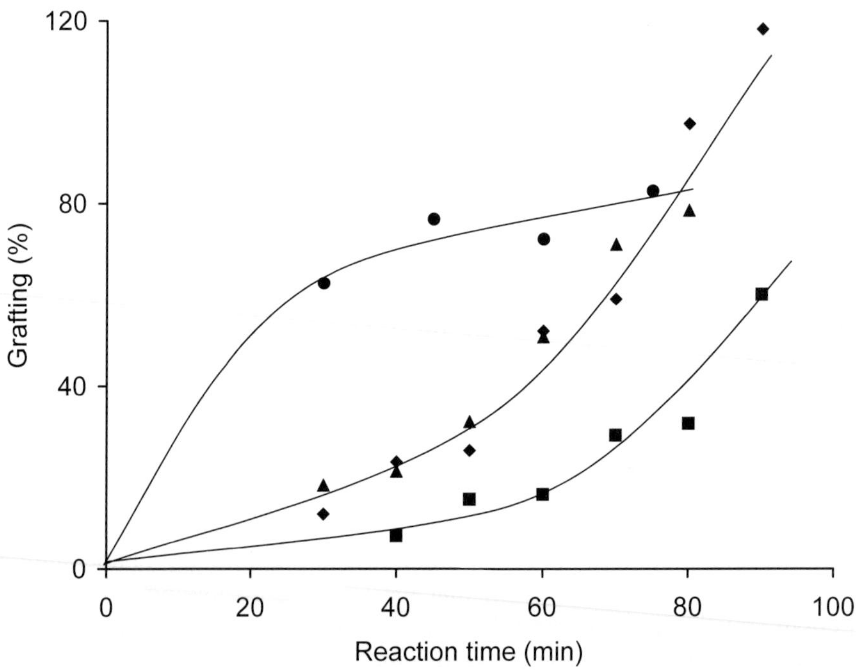

Figure 1. Kinetics of AAc grafting onto PTFE. Pre-irradiation dose, AAc concentration and grafting temperature: Circles: 50 kGy, 50% and 40°C; diamonds: 10 kGy, 40% and 40°C; triangles: 10 kGy, 40% and 50°C; squares: 10 kGy, 40% and 35°C.

will increase due to the higher radical concentration at the surface. However, radical fragments formed at the higher doses will also initiate hompolymerization, and monomer will be rapidly used up. This leads to a leveling out of the extent of grafting see in Fig. 1 (circles), along with a high yield of homopolyAAc, which is washed out during the wash steps.

Figure 2 illustrates the reaction scheme for biotinylation of the grafted polyAAc carboxyl groups. The carboxylic groups were activated with EDC. Stable amide bonds are formed between the primary amine of the biotin derivative and the PAAc carboxyl groups [23].

Figure 3 shows the amount of active HRP in the immobilized SA-HRP on the films as a function of grafting level. It can be seen that the amount of active HRP

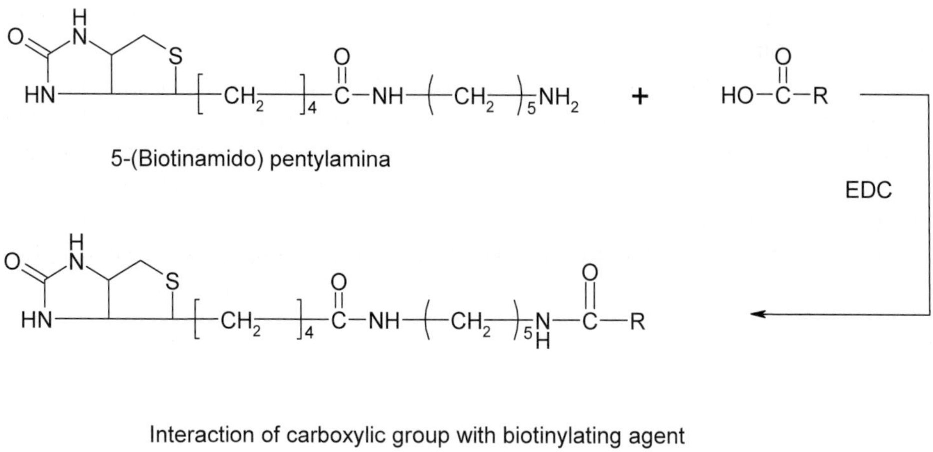

Figure 2. Conjugation of biotin to the carbodiimide-activated carboxylic group of PAAc.

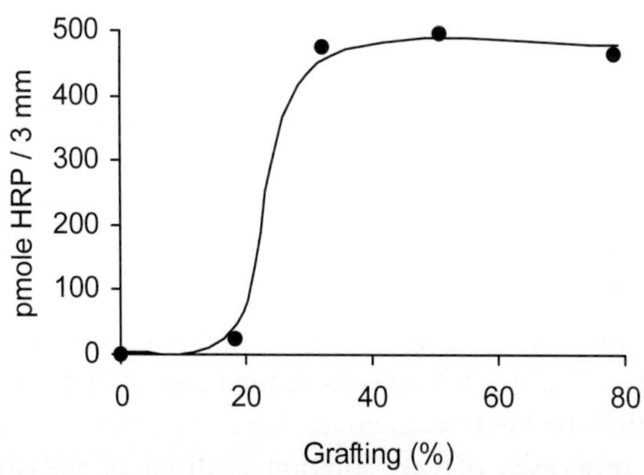

Figure 3. Amount of active HRP in the biotinylated films as a function of the extent of polyAAc grafting. Conditions of graft polymerization: Pre-irradiation dose 10 kGy; AAc concentration 40%; grafting temperature 50°C.

level per film increased rapidly between 20–30% PAAc grafting from essentially zero to about 400–500 pmol HRP/film. After that, the HRP content/film leveled out with further increases in grafting (Fig. 3).

These trends can be explained as follows: at low values of grafting, the grafted PAAc chains are most likely located only on the surface of the PTFE films because PTFE does not swell in the monomer solution. This would limit the immobilization of SA-HRP to an extremely thin layer on the surface, and that is probably undetectable by the method used to quantify HRP. But an increase in the extent of grafting promotes the continual diffusion of monomer into the depth of the film through the grafted PAAc layer and this would lead to continual penetration of the grafted layer into the film. It is important to note that the pre-irradiation dose must be below a critical level for there to be adequate penetration of free monomer into the film. This situation of radiation grafting onto polymers that do not swell in the monomer solutions has been investigated earlier, and similar observations were made of a rapid increase in grafting after a threshold level was reached [14, 24]. Between 20 and 30% grafting there is probably enough of an increase in the thickness of the grafted region to provide more and more COOH groups for reaction with the EDC and biotinylation agent. As grafting increases beyond 30%, the thickness of the grafted region should continue to increase. However, both the biotinylation reaction and the binding step of SA-HRP to the biotinylated groups were run at one time point of 1 h in each case. Thus, the concentration of biotinylated groups and the extent of binding of SA-HRP might be limited by the 1 h time period utilized for each reaction, and that could be the reason for the leveling out of the overall binding of SA-HRP to the films. If the times were increased for activation of COOH groups with EDC, for reaction of those activated groups with the biotinylation agent, and then for complexation of the SA-HRP with the biotin sites, one would expect to see a continual increase in binding with increase in grafting level. The goal of this study was to demonstrate the potential of combining (a) pre-irradiation grafting to inert materials such as PTFE plus (b) biotinylation of the grafted polymers, in order to permit surface immobilization on PTFE of biotinylated, bioactive biomolecules such as HRP *via* their complexes with streptavidin. We have successfully demonstrated this new surface modification process.

CONCLUSIONS

AAc was grafted onto pre-irradiated PTFE and biotin was conjugated to the carboxyl groups. The biotinylated sites were then used to bind SA-HRP by affinity, and it was shown that the HRP was active. This study has demonstrated the utility of combining the processes of (a) radiation grafting of polymers with reactive groups onto inert polymers such as PTFE, (b) biotinylation of the graft polymer reactive groups, (c) immobilization of streptavidin on the biotinylated surface sites, followed by (d) immobilization of biotinylated, biologically active molecules *via* complexation of their conjugates with streptavidin. In this study, the last two steps were combined by immobilizing the complex of streptavidin and biotinylated HRP onto the biotinylated surface sites. The unique nature of this process is the ability to immobilize biotinylated molecules on an inert surface as PTFE.

Acknowledgements

We thank F. García and S. Ham from ICN-UNAM and C. Gonzalez from IB, UNAM for their technical advice and assistance.

REFERENCES

1. A. S. Hoffman, G. Schmer, C. Harris and W. G. Kraft, *Trans. Am. Soc. Artif. Intern. Organ.* **18**, 10 (1972).
2. A. S. Hoffman, W. R. Gombotz, S. Uenoyama, L. C. Dong and G. Schmer, *Radiat. Phys. Chem.* **27**, 265 (1986).
3. L. C. Dong and A. S. Hoffman, *Radiat. Phys. Chem.* **28**, 177 (1986).
4. M. Alves da Silva, M. H. Gil, J. Guiomar, E. Lapa, E. Machado, M. Moreira, J. T. Guthrie and S. Kotov, *Radiat. Phys. Chem.* **36**, 589 (1990).
5. M. Carenza, *Radiat. Phys. Chem.* **39**, 485 (1992).
6. D. Müller-Schulte and W. Daschek, *Radiat. Phys. Chem.* **46**, 1043 (1995).
7. T. Godjevarova and A. Dimidov, *J. Appl. Polym. Sci.* **57**, 487 (1995).
8. L. M. Ferreira, J. M. S. Rocha, M. E. Andrade and M. H. Gil, *Radiat. Phys. Chem.* **52**, 207 (1998).
9. I.-K. Kang, B. K. Kwon, J. H. Lee and H. B. Lee, *Biomaterials* **14**, 787 (1993).
10. X. L. Huang, G. L. Catignani and H. E. Swaisgood, *J. Agr. Food Chem.* **43**, 895 (1995).
11. A. Breitwieser, S. Küpcü, S. Howorka, S. Weigert, C. Langer, K. Hoffman-Sommergruber, O. Scheiner, U. B. Sleytr and M. Sara, *BioTechniques* **21**, 918 (1996).
12. R. Graf and P. Friedl, *Anal. Biochem.* **273**, 291 (1999).
13. E.-S. A. Hegazy, *J. Appl. Polym. Sci.* **26**, 3117 (1981).
14. A. Chapiro and B. D. Gupta, *Eur. Polym. J.* **25**, 1145 (1989).
15. G. K. Kostov and A. N. Atanassov, *J. Appl. Polym. Sci.* **47**, 361 (1993).
16. A. Zouahri and A. Elmidaoui, *J. Polym. Sci.* **A34**, 1793 (1996).
17. M. S. Mohy Eldin, U. Bencivenga, M. Portaccio, S. Stellato, S. Rossi, M. Santucci, P. Canciglia, F. S. Gaeta and D. G. Mita, *J. Appl. Polym. Sci.* **68**, 625 (1998).
18. S. Turmanova, A. Trifonov, O. Kalaijiev and G. Kostov, *J. Membr. Sci.* **127**, 1 (1997).
19. B. Porstmann, T. Porstmann and E. Nugel, *J. Clin. Chem. Biochem.* **19**, 435 (1981).
20. V. S. Ivanov (Ed.), in: *Radiation Chemistry of Polymers*, p. 126. VSP, Utrecht (1992).
21. R. J. Woods and A. K. Pikaev (Eds), in: *Applied Radiation Chemistry: Radiation Processing*, p. 344. Wiley, New York, NY (1994).
22. T. R. Dargaville, G. A. George, D. J. T. Hill and K. Whittaker, *Prog. Polym. Sci.* **28**, 1355 (2003).
23. Y. Chen, E. T. Kang, K. G. Neoh and K. L. Tan, *Eur. Polym. J.* **36**, 2095 (2000).
24. B. L. Tsetlin, I. Yu. Babkin, V. Ya. Kabanov and A. N. Ponomarev, *High Energ. Chem.* **19**, 244 (1985).

Gels, Genes, Grafts and Giants, pp. 743–760 (2005)

Review

The relevance of the transfer of molecular information between natural and synthetic materials in the realisation of biomedical devices with enhanced properties

G. CIARDELLI [1,*], D. SILVESTRI [1], C. CRISTALLINI [2], N. BARBANI [1] and P. GIUSTI [1]

[1] *Department of Chemical Engineering, Industrial Chemistry and Materials Science, University of Pisa, Pisa, Italy*
[2] *Composite and Biomedical Materials Institute, CNR Section of Pisa, Pisa, Italy*

Received 24 January 2004; accepted 28 June 2004

Abstract—Past and recent attempts to introduce in synthetic polymers molecular information from natural substances through simple blending, template polymerization and molecular imprinting are reviewed. The most promising approaches that can open the way to the realisation of new materials with improved biocompatibility, antibody- or enzyme-like performances are analysed more deeply. The realisation of bioartificial blends from natural and synthetic polymers, molecularly imprinted nanospheres or membranes that can act as recognition element in (bio)sensing devices, as synthetic enzymes or as key constituents of body fluids purification tools is presented in order to make the reader aware of the fascinating possibilities that these techniques make available to the biomedical science and engineering in the close future. The last part of the paper describes recent attempts to insert recognition elements for large molecules as proteins, DNA segments, viruses or whole cells in synthetic polymer systems, in order to develop new systems in the treatments of diseases and for tissue-engineering applications.

Key words: Molecular information; molecular imprinting; biosensors; tissue engineering.

INTRODUCTION

The realisation of new materials or devices showing improved performances (e.g., cell and tissue compatibility, smart properties in drug delivery or tissue engineering, antibody mimics) is the challenge that biomaterials scientists operating in biomedical science and engineering are currently sharing. However, a complete review of the most promising efforts in this direction is outside the scope of this paper.

*To whom correspondence should be addressed. Tel.: (39-50) 511-271. Fax: (39-50) 511-266. E-mail: gianluca.ciardelli@ing.unipi.it

Therefore, we will focus on the most recent results that stem from the idea of anticipating the interactions between substances of biological or medical interest and synthetic systems at the molecular level in the realisation of new materials for biomedical applications. Particular attention will be devoted to the technique of molecular imprinting as a promising approach to generate molecular systems with superior recognition properties towards drugs, nucleic acids, proteins or cells.

BIOARTIFICIAL POLYMERIC MATERIALS

The combination of natural macromolecules with synthetic polymeric material has become of increasing interest in recent years to the scientific community for the potential that these materials have for applications in various fields, such as the biomedical and the environmental ones [1].

Bioartificial polymeric materials were initially conceived to produce new materials combining the characteristics of synthetic polymers, such as good mechanical properties, good processability, low production and transformation costs, with the specific properties of biodegradability and compatibility with cells and tissues associated with biopolymers [2]. The performance of these materials depends on the relationship that links the natural-synthetic interactions with the biocompatibility of the resultant material. The idea behind bioartificial polymeric materials is to reduce the interactions between the synthetic material and the living tissue, creating a material that can produce a better response from the organism with respect to a completely synthetic material since the interaction of the two components (the synthetic and the natural one) occurs before contact with living tissue. The first studies on bioartificial polymeric materials considered simple preparation techniques such as solution mixing. A mixture of a synthetic polymer and a biopolymer can be obtained, in which the miscibility between the two components depends on the various parameters linked to the nature of the polymers themselves, the functional groups present, the molecular weight, composition, temperature etc. To improve the degree of miscibility in these systems, a different method for the preparation of a bioartificial polymeric material is to perform polymerization onto a template (template polymerization) [3–6].

Template polymerization is a process involving the formation of a polymer (daughter polymer) in the presence of a macromolecule (template) added to the reaction system as preformed polymer. The reaction mechanism is determined by specific co-operative interactions (i.e., ionic interactions or hydrogen bonds) between the growing chain units and the polymer defined as template. The presence of the template can affect the reaction kinetics, the structural characteristics (i.e., average molecular weight, molecular weight distribution and stereoregularity) of the produced polymer and the reactivity ratios in the case of co-polymerizations. In general, after the template has directed the formation of the product, it is removed to yield the template-free product, although sometimes the template becomes inextricably locked into the structure it helps to create.

The polymerization of synthetic monomers onto biological templates produces a higher degree of miscibility between the two components than the simple mixing of preformed polymers [7]. Hydrogen bonding is one of the main interactions involved in the realisation of such materials. For example, while the presence of chitosan showed to poorly affect the rate of acrylic acid (AA) polymerization, the resulting material had different thermal behaviour and a more ordered structure with respect to the blends prepared from preformed polymers [8].

A clear template effect on AA polymerization in water was observed on the other hand due to the presence of dextran in the polymerization mixture. The polymerization rate is in fact maximised for an equimolar concentration of AA and dextran and doubled with respect to the rate of AA blank polymerization [7].

Results obtained in another study [9] showed that the presence of α-amylase (biological template) influences the polymerization of 2-hydroxyethylmethacrylate. In particular, an increase of the initial reaction rate with the increasing of the enzyme amount was observed.

The thermal stability of the polycomplexes obtained by template polymerization was higher with respect to those of the single components. The biological tests indicated that only a little amount of enzyme is released from the polycomplexes and that the retained enzyme maintains an adequate enzymatic activity. In addition, the presence of PVP has a protective action towards the enzyme immobilised inside the polycomplexes. This property can be used to retain the enzymatic functionality also in other immobilisation systems obtained by means of radical polymerization. The complex obtained by template polymerization of HEMA onto α-amylase and PVP seems to be an interesting system with potential application in enzyme immobilisation, since PVP exploits a stabilising action on the enzyme, while the formed PHEMA, interacting further with the other two components, leads to the formation of a stable polycomplex which precipitates from the reaction solution.

Ionic interactions can play also a role in the preparation of synthetic/natural polymers polycomplexes. The polymerization rate of sodium methacrylate is higher in the presence of gelatin at pH 4, where ionic interactions can be fully exploited, but not at pH 7 [7].

However, the complexity of these systems is higher than purely synthetic systems that, due to their relative simplicity, have been studied as models for the interactions between monomers and templates containing the same functional groups as those that would be present in a biological template. For example, studies have been made on the polymerization of sodium acrylate, methacrylate and vinyl sulfonate onto synthetic templates, such as polyallylamine and poly(vinyl alcohol), where the role of interactions between carboxylic, hydroxylic and amine groups is very important, the same groups involved when polymerizing onto protein-like templates [10–12].

An interesting evolution in bioartificial polymeric materials obtained from template polymerization is linked to the fact that the daughter polymer is capable of molecular recognition of the template, or of similar molecules. The information transfer from the template to the daughter polymer allows the daughter to acquire a memory for the selective recognition of the mother molecule. This attempt to imitate polymerization processes present in nature (as DNA replication, RNA and protein synthesis), has a few chances to succeed in producing macromolecular chains with such a precisely defined monomer sequence with a purely synthetic approach and no examples in this sense are present in the scientific literature up to now.

MOLECULARLY IMPRINTED POLYMERS (MIPs)

Polymerization onto a template is not the only method, however, by which information can be transferred from a template (either a low- or high-molecular-weight one) to a polymer that must subsequently recognise it. The alternative consists in a technology known as molecular imprinting [13–15]. This technology permits recognition sites for specific species to be inserted into a polymeric material through the polymerization of a monomer in the presence of the template, or through the dissolution of the preformed polymer in a solution containing the molecule to be recognised.

Polymeric materials showing selective recognition properties towards template molecules can, therefore, be realised in two different forms: cross-linked macromolecules as monoliths or micro/nanoparticles and phase inversion membranes.

In both cases, the spatial positioning is maintained by the polymer even after the template has been removed and a selective 'memory' of the template remains.

The suspension polymerization is a widely used method to obtain imprinted polymers, allowing the formation of microparticles. The water was considered to weaken or eliminate the non-covalent interactions (hydrogen bond, Van der Waals force, etc.) between functional monomer and template [16]. It is very difficult to recognize the template molecule in aqueous solution or polar solvent with MIPs prepared by aqueous suspension polymerization: numerous attempts (some with fair result) to overcome these recognition problems were carried out and are today in course because in nature, and also in biological environment, molecular recognition systems (such as enzyme and substrate, antigen and antibody, receptor and hormone) mostly performed in water. To improve recognition capability of synthetic materials in physiological conditions, the realization of polymers with a high content of cross-linking agent (which stabilizes the specific cavities) seems to partially solve the limitations of binding properties in polar medium.

In cross-linked polymers, the molecular recognition can be obtained through either the set up of covalent interactions between print molecules and monomers [17] or the development of reversible non-covalent interactions (mainly hydrogen bonding) between them [18].

In both cases, after the establishment of the interactions between functional monomers and templates, a polymerization process follows, usually in the presence of polyfunctional monomers (cross-linkers), which 'freezes' the macromolecule network in the configuration capable of interacting with the imprinting molecules (Fig. 1).

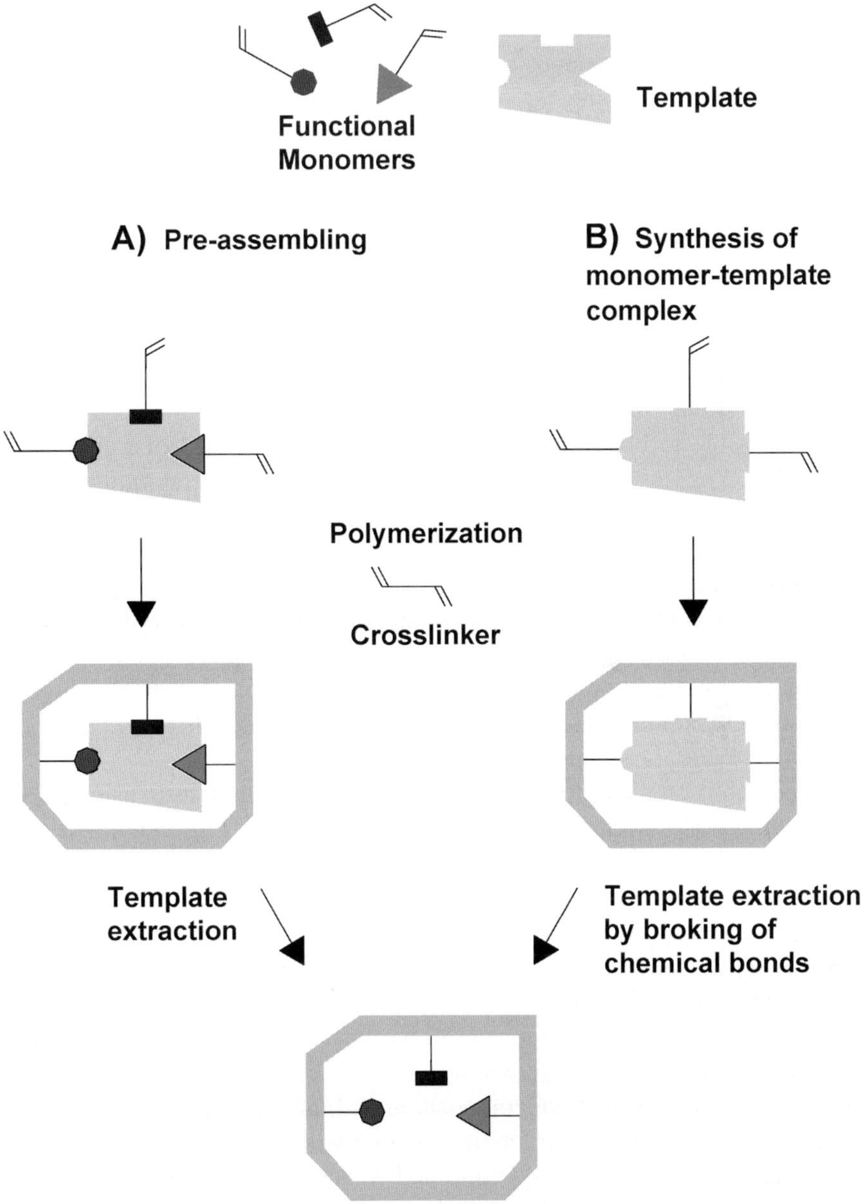

Figure 1. Schematic representation of the molecular imprinting process. Phase 1: (A) pre-assembly of functional monomers complementary with the template; (B) monomer(s)-template complex formation. Phase 2: freezing of molecular interactions between monomer(s) and template by polymerization in the presence of a cross-linking agent. Phase 3: extraction of the template allows to obtain free recognition cavities for the stamp molecule.

These materials are usually obtained in the form of a macroporous monolith, then ground and sieved to the required particle dimensions. The imprinted monolith is prepared through a suspension-polymerization (where each micelle acts as a single reaction site): the homogeneous solution containing the template, the functional monomer(s), the cross-linking agent and the initiator is dispersed or emulsioned in a pore-former solvent. The macroporous structure results from the phase separation between the polymer and the pore-former during polymerization. When the nuclei of the polymer chains coalesce, a cross-linked polymeric matrix is formed and a series of pores containing the pore-former solvent are created. The structure of the pores moves the extraction and rebinding of the template molecule easier. The porosity depends on the nature and amount of solvent and from the cross-linking agent concentration.

The dilution of the reacting phase influences the final dimensions of the particles; increasing the solvent amount, particles dimensions results to be smaller. In this way, it is possible to obtain a macrogel or a microgel instead of a macroporous monolith (Fig. 2). The characteristics of the solvent also have an influence on the morphology of the particles, while the presence of the template can cause variations in their dimensions [19]. In particular in this work [19] two different molar ratio between functional monomer and cross-linking agent were used, showing the possibility to control rebinding properties by varying the cross-linker amount. The microspheres obtained (regular in size and shape) are particularly suitable for the selective binding properties required to imprinted polymer: the high ratio surface/volume guaranteed an easy access of template molecule to recognition free cavities.

This approach is an alternative to conventional emulsion or suspension polymerization where usually surfactants or stabilizers are required to obtain MIPs in the form of micro or nanoparticles.

A recent work [20] had the objective to provide a general view of the behavior of different polymerization procedures on the final properties of the imprinted polymers, keeping the monomer feed composition and the amount of template molecule constant. The effect of the polymerization method on polymer morphology was reported, indicating the possibility to control the size and shape of MIP particles. Rebinding assays carried out in different media (toluene and aqueous buffer) showed a remarkable influence of material composition and morphology (resulting from the different polymerization method) on the specificity of template absorption.

The application of molecularly imprinted materials obtained by cross-linking polymerization for the realisation of devices for affinity separation, antibody binding and enzyme mimics and biomimetic sensors has been extensively reviewed [14, 21]. Recent developments (e.g., smart imprinted polymers), and perspective area evolution of MIP technology and applications were excellently examined by Mosbach and Ramström [22].

Significant effort has been devoted to developing rational approaches to design MIPs for the perspective application [23] and to screen their selectivity [24].

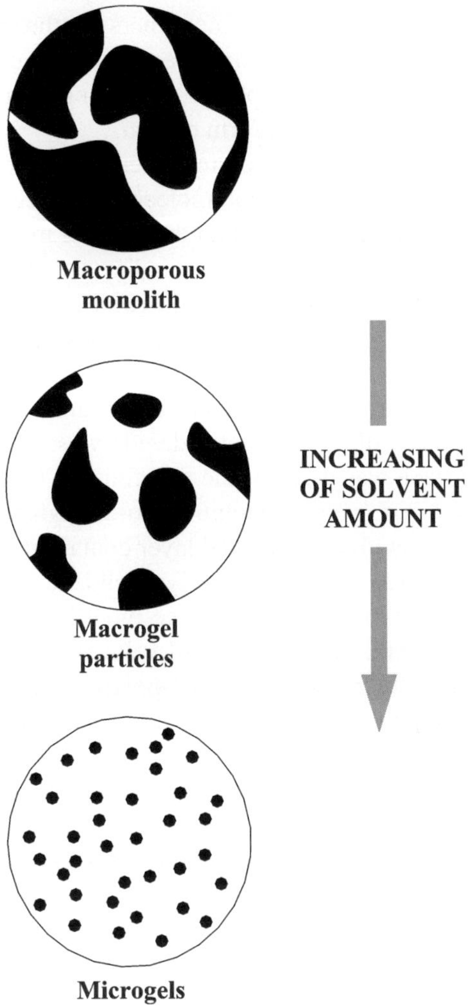

Figure 2. Influence of the solvent amount in the reaction mixture on the final particles dimension.

Lanza and Sellergren [25] tested six functional monomers (methyl methacrylate, 4-vinylpyridine, N-vinyl-α-pyrrolidone, methacrylic acid, hydroxyethyl methacrylate and (trifluoromethyl)acrylic acid) to rapidly screen the parameters of importance to reach a desired level of binding affinity and selectivity. It was demonstrated that a key role on the recognition properties of MIPs was played by the functional component, and the polymer showing the highest selectivity was prepared using methacrylic acid as functional monomer, also with respect to molecules of medical and biological interest too.

As previously anticipated, the phase inversion process allows insertion of molecular recognition cavities into polymeric membranes. Other approaches have been reported to produce molecular imprinted membranes starting from the first results obtained by Piletskii *et al.* [26] in producing a porous film through an *in situ* cross-linking polymerization of acrylic monomers in the presence of the template. The cross-linked diethylaminoethylmethacrylate showed selective permeability for individual nucleoside monophosphates, indicating the possibility to realize a recognition-able polymeric network in form of film.

An exhaustive outline of imprinted polymer membranes realization was carried out by Piletsky *et al.* [27]. The supramolecular devices that could perform highly selective functions may find potential application in membrane sensors and chromatography, but also in affinity separation and extracorporal fluids purification. Dry-phase separation [28], wet-phase inversion [29], surface imprinting [30] and *in situ* polymerization [31] were investigated as possible methods to prepare imprinted membranes.

Surface modification (using, for example, UV radiation) by grafting, in the presence of a template molecule, is a recent and widely used method to prepare MIP membranes. This approach was used for the first time by Kobayashi *et al.* in 1997 [30] with the realisation of a (poly-acrylonitrile-co-diethylamino-dithiocarbamoyl-methylstyrene) membrane with an imprinted layer containing theophylline.

Hilal and Kochkodan [32] investigated polyvinylidene fluoride and polyethersulfone microfiltration membranes, with surface modification with a thin layer of an imprinted polymer. This material resulted selective for adenosine 3,5-cyclic monophosphate (cAMP) *via* photoinitiated co-polymerization. The specific and non-specific template binding of MIP during filtration of aqueous solutions was studied for membranes with different degree of modification. MIP membranes showed the ability to bind cyclic adenosine monophosphate and this ability was related to the specific size and shape of the recognition sites and to the correct positioning of the functional groups involved in the template binding through ionic and hydrogen interactions [32].

This approach has been extended to selective rebinding of molecules of interest in the food industry [15], for the selective transport of target neutral molecules and for enantioselective separation of amino acids [33].

Phase separation from a polymeric solution in the presence of an active component (template), by solvent evaporation, was also described to prepare membranes with recognition properties [34]. The most applied method to realise polymeric imprinted membranes is, however, the already cited wet-phase inversion realised through the addition of a non-solvent to a homogeneous polymer solution containing an active component. Kobayashi and co-workers [35, 36] reported the preparation of membranes of polyacrylonitrile (PAN) and PAN-co-poly(acrylic acid) blends, prepared using the immersion precipitation method. These membranes, presenting an asymmetric porous structure, showed difference in sorption selectivity for the template theophylline in comparison to caffeine under ultrafiltration conditions. These results confirmed that during membrane formation, the complex between the template and the functional group of the polymer matrix are formed in the casting solution and then fixed during inversion precipitation. Recently Kobayashi *et al.* [37] realised imprinted membranes using three PAN co-polymers with vinylpyridine

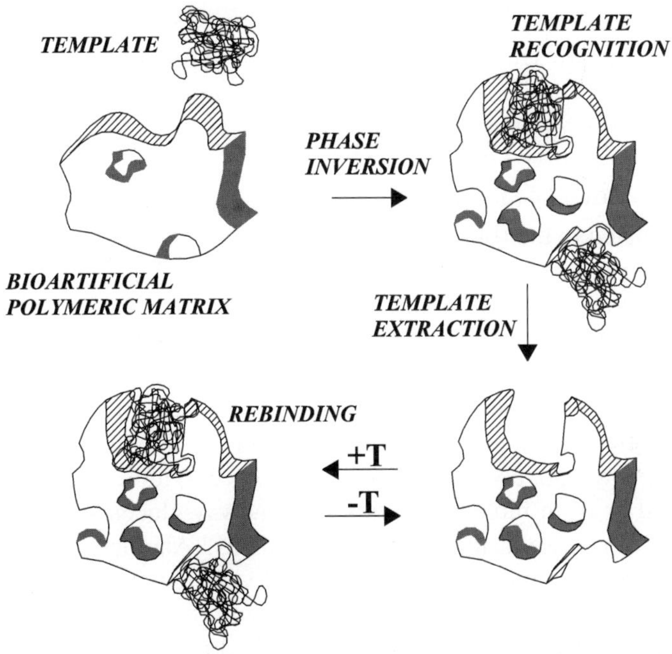

Figure 3. MIP membranes preparation: formation of selective sites during non-solvent induction phase separation and possibility of extraction and to tie up again of the template.

and styrene where caffeine was used as template. Results of rebinding experiments showed that caffeine-printed membranes recognise the template compared with caffeine analogues.

Membranes based on an acrylic acid–acrylonitrile co-polymer, realised through phase inversion method, have been modified introducing into their structure specific binding sites for uric acid (template). Materials prepared are intended to selectively remove uric acid from the blood in the case of increased serum uric acid values associated with different pathologies [38].

In Fig. 3 the mechanism of imprinted membranes preparation through non-solvent induction phase separation (NIPS) is illustrated.

More recently, our group presented the realisation of polymethylmethacrylate-co-acrylic acid membranes modified by the presence of molecularly imprinted acrylic nanospheres, either on the surface or in bulk, to achieve good drug release and selective rebinding in aqueous environments [39]. This work was focused on the possibility to realize a multi-component device showing good mechanical or functional properties of the film, and improved specific binding characteristics associated to imprinted nanoparticles loaded onto a polymeric membrane. This approach suggest interesting application in the development of composite membranes for filtration processes, where together with the typical specificity of membranes (relying on their diffusion properties regulated by pore sizes and/or chemical nature of the material), an additional feature with a significantly higher degree of selectivity is introduced with the presence of nanoparticles with imprinted cavities.

BIOMEDICAL APPLICATIONS OF MIPs

The technology of molecular imprinting has been more recently focused in finding applications related to the (bio)medical field.

Peppas and co-workers [40, 41] described recent developments of surface modifications and molecular imprinting of polymers for medical and pharmaceutical applications. Surface modification by chain addition to acrylic hydrogels can lead to materials with the ability to recognise biological and pharmaceutical compounds. By creation of additional recognition sites for determined molecules or functional groups on cell membranes, these systems can become able to realise a targeted release of drugs. Alternatively, toxic (or present in excess) molecules can be extracted from the blood or from biological fluids. In this way glucose-sensitive hydrogels were obtained to release insulin in diabetes patients. Another interesting research work for glucose recognition was carried out by Wizeman and Kofinas [42]; an higher binding capacities of glucose (compared with fructose) was observed, indicating an isomeric specificity of poly(allylamine hydrochloride) polymer toward the template employed.

An interesting work for potential application of MIPs in the biomedical field [43] involved the realisation of imprinted polymers for pinacolyl methylphosphonate, ethyl methylphosphonate and methane phosphonic acid, that are degradation products of nerve agents (when absorbed through the skin or *via* respiration, this lethal agents are highly toxic and have rapid effects). For their cross selectivity, the MIPs could recognise not only the print molecule, but also the degradation products of the other nerve agents, and the adsorbed degradation products could be quantitatively extracted using distilled water. The pinacolyl methylphosphonate-imprinted polymer was used as a solid phase extraction sorbent for all possible degradation products of nerve agents from human serum. Computational investigations have been applied to support experimental activities in designing selective MIPs (e.g., to selectively recognise creatine and creatinine [44]).

MIP membranes were used in biomimetic devices for controlled permeation applications, reporting as example a receptor-binding mediated membrane showing responsible behavior for controlled delivery of therapeutics [45]. Many practical applications in separations can be envisaged as continuous electrodialytic resolution of amino acid enantiomers [46].

(Bio)receptor mimics

Biosensor technology is widely developed and applied in the detection and quantitative determination of analytes. A selective chemical signal, generated when the sample molecule bind to the recognition site, is subsequently translated in an electric signal, amplified and converted in a readable form (Fig. 4). The main limitation

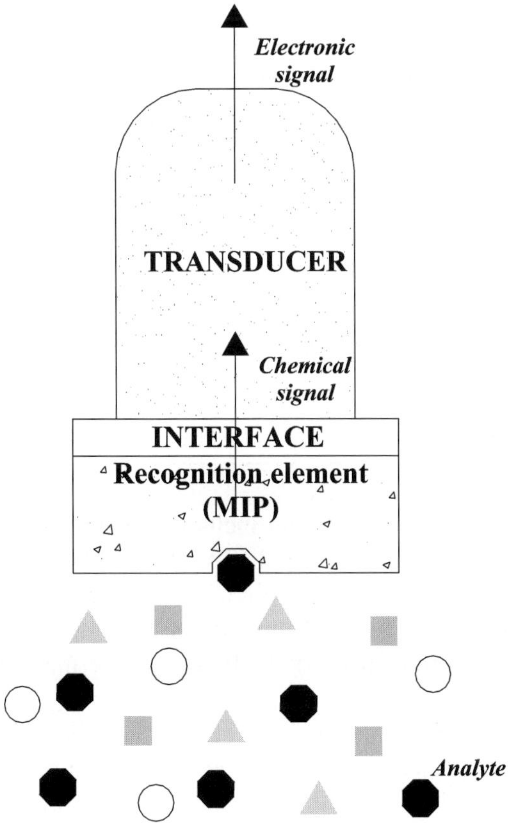

Figure 4. Schematic representation of a (bio)sensor.

of such an approach is the lack of suitable receptors for many compounds, their low stability and/or high cost. Therefore, the application of MIPs in biosensors, as recognition systems, instead of an enzyme or an antibody when immobilised at the interface between the sensor and the sample to be analysed represents a fascinating alternative. MIPs have the advantage of a greater stability even in severe condition, and can also widen the application palette to those cases where a suitable biomolecule is not available.

Recent studies have in fact shown that polymers prepared with the technology of molecular imprinting can lead to synthetic materials mimicking natural antibodies to determine drugs level in the human serum. Results obtained with the molecular imprinting technique have shown an accuracy comparable with that of the radio-immunological technique today in use, with advantages as the possibility to sterilise the polymers for their elevated chemical and physical resistance, the ability to maintain stable recognition sites, and the lower cost of synthetic polymers in comparison with the biomolecules. A further advantage of MIPs is that through suspension, seed or precipitation polymerization spherical beads can be prepared [47]. The printed microspheres display favourable binding characteristics, and were suc-

Table 1.
Summary of biosensor applications using molecularly imprinted polymers

Class	Typical analyte	Functional monomer	Detection range
Conductometry	Sialic acid	Allylamine + TVPhB	1–50 μM
	Morphine	MAA	Qualitative
	L-Phenylalanine	DEAEM	0.05–0.4 mM
Spectrometry	Chloramphenicol	DEAEM	1–1000 μg/ml
	Chloramphenicol	DEAEM	3–30 μg/ml
	Testosterone	MAA	0.10–1.25 mM
Potentiometry	Phenylalanine anilide	MAA	33–3300 μg/ml
Capacitance	Phenylalanine anilide	MAA	Qualitative
Amperometry	Morphine	MAA	0.1–10 μg/ml
pH	Glucose	STACNCu	0–25 mM
SPR	Theophyilline	MAA	0.4–6 mg/ml

cessfully used in various assay formats including radiolabeled immunoassay and enzyme-linked immunosorbent assay (ELISA). Since MIP microspheres, obtained by suspension polymerization, display binding characteristics similar to that of an affinity adsorbent, they may be used instead natural antibodies in various affinity separation processes. Applications of MIPs as bio- or medical sensors are summarised in Table 1.

A sensor for CAP (chloramphenicol, an effective antibiotic for both human and pets), using a CAP-imprinted polymer as the recognition element, was developed [48]; utilisation of diethylaminoethyl methacrylate as functional monomer has shown that the CAP template was much more strongly retained than its analogues. The molecular imprinting technology was used in the synthesis of polymers having a high affinity for testosterone, which is a biologically and clinically important steroid hormone [49, 50].

Several fundamental studies, which include radiolabelled ligand binding to corticosterone and cortisol [51], methyl-α-D-glucoside [52], morphine and leu-enkephalin [53] were presented. Whereas in several studies a binding was performed in organic solvent, one of them [54] explored the selective, high-affinity binding of the analyte in aqueous conditions.

Artificial enzymes

A new approach in the preparation of new MIPs with perspective application in the biological area followed the so-called anti-idiotypic imprinting ('imprinting the imprints'), whereby a mould of a molecular object (e.g., a protease inhibitor), is first allowed to be formed [55]. In the cavity which is formed in nanometer dimensions, not only the original compound can be obtained from its building stones, but also new structures with potentially superior properties. Alternatively, the enzyme itself can be used as a biological mould and the building stones of a chiral inhibitor assembled in its cavity, the active center of the enzyme [56].

PROTEIN RECOGNITION THROUGH MIPs

Usually, only relatively low-molecular-weight compounds (sugars, steroids, amino-acid derivatives, certain drugs and pesticides) are used as templates. The design of stable molecular recognition elements for proteins is, considering the complexity of these macromolecules, a difficult task [55]. Recent efforts in the realisation of systems able to recognise high size molecules are very promising. A polymer layer around silica beads, composed of acrylamide, acrylic acid and cross-linkers, with recognising properties towards glucose oxidase (GOD) in a pool of proteins, was realised by Hirayama *et al.* [57]. Hjertén and co-workers published two interesting works on protein recognition [58, 59]. In the first one [58], preliminary experiments aimed at the preparation of gel particles with the property to selectively recognise a particular protein (hemoglobin, cytochrome *c*, transferrin) were described, while in the second one [59] the same preparation method was utilised to specifically adsorb the human growth hormone, ribonuclease and horse myoglobin, indicating that the method may be universal or at least applicable to a great number of proteins. The gel with specific adsorption of the three model proteins was synthesized in order to demonstrate that the beads can be employed to remove (trace of) several proteins contaminating a sample (negative purification). Recently, a technique for coating microplate wells with molecularly imprinted polymers specific for proteins was presented by Bossi *et al.* [60]. 3-Aminophenylboronic acid was polymerised in the presence of the following templates: microperoxidase, horseradish peroxidase, lactoperoxidase and hemoglobin. Surface coating of polystyrene microtiter plates with a thin layer of a stable conjugated polymer obtained in the presence of the proteins, was finalised to overcome the major problem with the imprinting of large molecules: the restricted mobility of these molecules within highly cross-linked polymer networks and the poor reversibility and efficiency in binding. Another approach to solve these problems is the epitope approach [61, 62]. If a short peptide which represent only a small exposed fragment of a protein structure is used as template (see Fig. 5), the resultant macroporous MIP recognising the imprinted peptide is also able to recognise the whole protein molecule.

This system is present in nature where, in the recognition mechanism, an antibody interacts with only a small part of an antigen (the epitope, the antigenic site of a macromolecule). Successful recognition of oxytocin and oxytocin-related peptides was obtained in this way.

The realisation of a molecularly imprinted matrix using the phase inversion process can be an alternative solution to the epitope approach (cross-linked gels or grafting of a print layer) for the realisation of recognition systems toward species of larger size. Recently, our group prepared MIPs in the form of porous bioartificial ultrafiltration membranes (Fig. 6), imprinted with an enzyme to create systems able to recognise molecules as large as proteins in aqueous environment [63].

All these approaches presented are expected to face the intrinsic limitation of the non-covalent molecular imprinting approach to build up efficient systems for recognition in water environments. Since the recognition process relies on

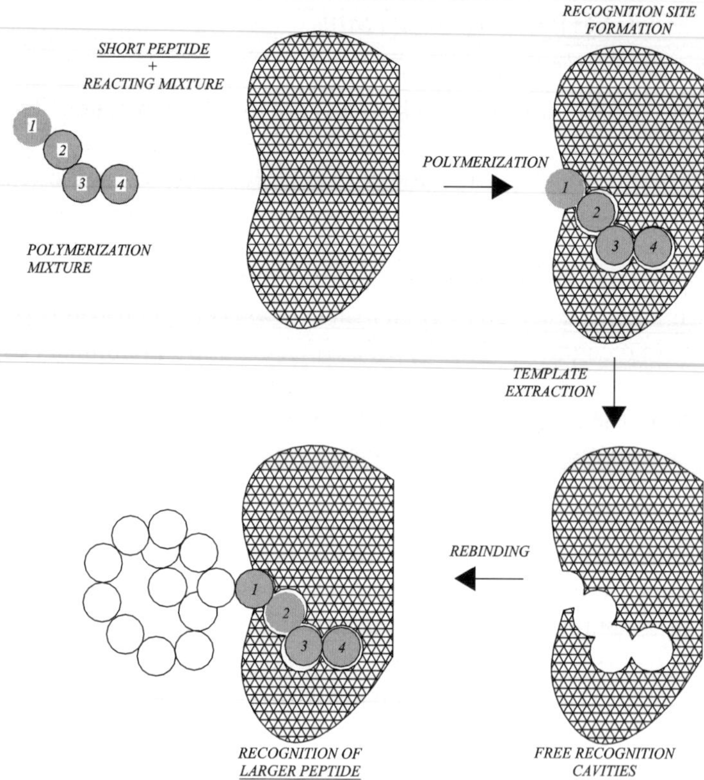

Figure 5. Epitope approach for the realisation of an imprinted polymeric matrix able to recognise large peptides.

hydrogen-bonding interactions to build up again selectively in the imprinting cavities, this mechanism is expected and showed up to be less efficient in protic media. Moreover, this drawback becomes more important when the analytes are strongly hydrophilic molecules which prefer to remain in the solvated state and do not enter the cavities to start the recognition process. This problem can be circumvented by using more hydrophilic cross-linking agents for the MIPs [64] in order to make the cavities more attractive for the water soluble molecules. Membrane systems are sometimes more efficient, since the molecules are forced to come in contact with the recognition cavities by the water flow during the filtration process [38]. The high sensitivity towards pH changes in developing efficient hydrogen bonding interactions makes the selective recognition of amino acids and peptides a relative tricky task.

However some pioneering attempts showed promising results. Rachkov *et al.* [65] reported the preparation of peptide-selective MIPs avoiding the use of organic solvents; a relatively large octapeptide (an analogue to peptide hormone angiotensin II) was used as template, and the innovative aspect is that the polymerization was carried out in an aqueous solution. A cross-linking agent miscible with

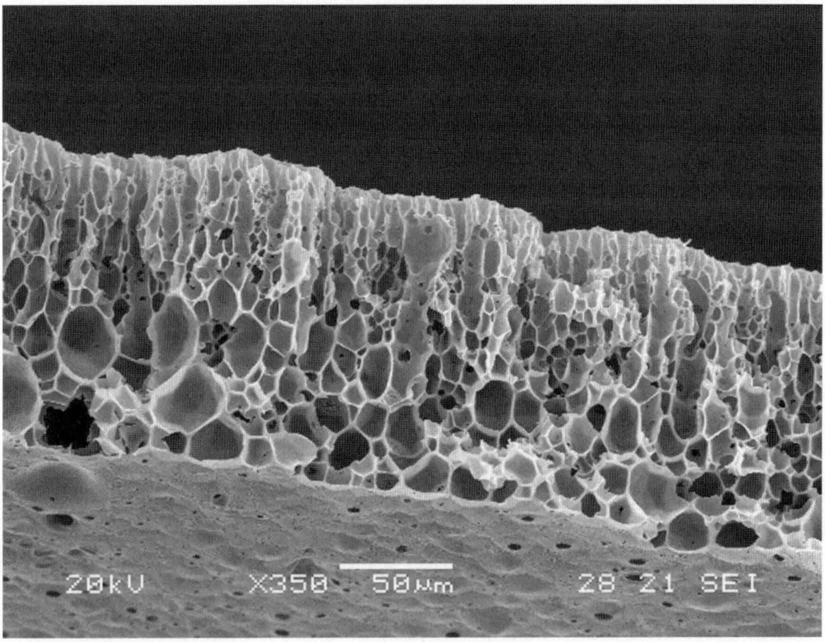

Figure 6. SEM photograph of dextran-ethylene-*co*-vinyl alcohol membrane imprinted with α-amylase.

water and compatible with the initiator was chosen. The selected azo-radical initiator (instead of a peroxide) showed an increased polymer chain growth with suppression of branch grafting and preliminary chain termination. MIPs realised with different template-monomer ratios showed stronger binding of template than their corresponding blank polymers.

Attempts to realise imprinted materials with binding behaviour towards *t*-butyloxycarbonyl phenylalanine were carried out using organic solvents as porogenic medium [66]; these imprinted porous microspheres showed moderate enantioselectivity and a sufficient separation capacity, but the performances in the binding assay still need to be improved. Ou *et al.* [67] and Parmpi *et al.* [68] were able to overcome the problems of recognition of large size molecules, also in aqueous environment, developing systems that exhibited good selective and binding behaviour. In particular Ou and co-workers, starting from previous attempts to rely on hydrogen bonding interaction for protein rebinding and recognition in aqueous environments, developed new systems based on polyacrylamides incorporating two different functional monomers (methacrylic acid and 2-(dimethylamino)ethyl methacrylate). In this way, protein recognition could be realised through the correct distribution of positive and negative charges in addition to moieties bearing acidic hydrogens. The electrostatic interactions, combined to hydrogen bonding, lead to superior imprinting efficiency in terms of selectivity.

The use of monomers which are able to exploit affinity interaction with the target molecule in addition to hydrogen bonding allowed for the development of more efficient recognition systems in some cases [69].

RECENT EFFORTS AND OUTLOOK

The successful transfer of molecular or structural information from a substance of biological or biomedical interest to a synthetic polymer by blending, template polymerization or molecular imprinting can render new smarter biomaterials accessible, which are able to recognise and selectively rebind toxins, cytokines, or growth factors opening the way to different approaches in the treatment of several protein-mediated kinds of diseases.

A further, even more stimulating step forward in this direction could be gained by creating polymeric systems that are able to selectively recognise more complex structures as double stranded DNAs, viruses, bacteria or cells.

Recently, attempts to develop sequence selective MIPs for double stranded (ds) DNA were presented. In order to construct the dsDNA recognizing MIPs, 4-vinyl-2,6-diamino-1,3,5-triazine (VDAT) was used as a functional monomer [70].

Polymer surfaces were imprinted through a bacteria-mediated lithographic technique. The resultant polymeric beads exhibit functionally anisotropic patches of the dimensions of the bacteria used as template [71]. Recently, MIP hydrogels for the biomimetic recognition of viruses for application in medicine (virus removal) and bioanalytics (virus sensing, identification or classification) were also obtained [72]. These stimulating approaches are however still at the pioneering level but nevertheless suggest interesting further developments towards practical applications.

Further improvement of this technique could lead to creating selective recognition sites in the MIPs for entire cells. These MIPs could then act as structures to support cell growth (scaffolds) to be used in tissue engineering, a leading research topic for the material scientist of the new millennium.

REFERENCES

1. P. Giusti, L. Lazzeri, S. De Petris, M. Palla and M. G. Cascone, *Biomaterials* **15**, 1229 (1994).
2. P. Giusti, L. Lazzeri and M. G. Cascone, in: *The Polymeric Materials Encyclopedia*, p. 538. CRC Press, Boca Raton, FL (1996).
3. V. A. Kabanov, *Makromol. Chem.* **3** (Suppl.), 41 (1979).
4. A. K. Srivastava, S. K. Nigam, A. Shukla, S. Saini, P. Kumar and N. Tewari, *J. Macromol. Sci. Rev. Macromol. Chem. Phys.* **C27**, 171 (1987).
5. C. H. Bamford, *Makromol. Chem.* **1**, 52 (1980).
6. G. Challa and Y. Y. Tan, *Pure Appl. Chem.* **53**, 627 (1981).
7. C. Cristallini, N. Barbani, P. Giusti, L. Lazzeri, M. G. Cascone and G. Ciardelli, *Macromol. Chem. Phys.* **202**, 2104 (2001).
8. P. Cerrai, G. D. Guerra, M. Tricoli, S. Maltinti, N. Barbani and L. Petarca, *Macromol. Chem. Phys.* **197**, 1 (1996).

9. G. Ciardelli, C. Cristallini, N. Barbani, G. Benedetti, A. Crociani, L. Travison and P. Giusti, *Macromol. Chem. Phys.* **203**, 1666 (2002).
10. C. Cristallini, G. Ciardelli, G. Polacco, A. Villani, L. Lazzeri and P. Giusti, *Polym. Int.* **48**, 1251 (1999).
11. I. Rainaldi, C. Cristallini, G. Ciardelli and P. Giusti, *Macromol. Chem. Phys.* **201**, 2424 (2000).
12. G. Ciardelli, N. Barbani, I. Rainaldi, P. Giusti and C. Cristallini, *Polym. Int.* **50**, 588 (2001).
13. G. Wulff, *Angew. Chem. Int. Ed. Engl.* **34**, 1812 (1995).
14. L. I. Andersson, I. A. Nicholls and K. Mosbach, *Adv. Mol. Cell Biol.* **15B**, 651 (1996).
15. P. A. G. Cormack and K. Mosbach, *React. Funct. Polym.* **41**, 115 (1999).
16. J. P. Lai, X. Y. Lu, C. Y. Lu, H. F. Ju and X. W. He, *Anal. Chim. Acta* **442**, 105 (2001).
17. G. Wulff and A. Sarhan, *Angew. Chem. Int. Ed. Engl.* **11**, 341 (1972).
18. K. Mosbach, *Trends Biochem.* **19**, 9 (1994).
19. L. Ye, R. Weiss and K. Mosbach, *Macromolecules* **33**, 8239 (2000).
20. N. Pérez-Moral and A. G. Mayes, *Anal. Chim. Acta* **504**, 15 (2004).
21. K. J. Shea, *Trends Polym. Sci.* **2**, 166 (1994).
22. K. Mosbach and O. Ramström, *Bio-Technology* **14**, 163 (1996).
23. J. Steinke, D. C. Sherrington and I. R. Dunkin, *Adv. Polym. Sci.* **123**, 81 (1995).
24. C. J. Allender, C. Richardson, B. Woodhouse, C. M. Heard and K. R. Brain, *Int. J. Pharm.* **195**, 39 (2000).
25. F. Lanza and B. Sellergren, *Anal. Chem.* **71**, 2092 (1999).
26. S. A. Piletskii, I. Y. Dubey, D. M. Fedoryak and V. P. Kukhar, *Biopolym. Kletka* **6**, 55 (1990).
27. S. A. Piletsky, T. L. Panasyuk, E. V. Piletskaya, I. A. Nicholls and M. Ulbricht, *J. Membr. Sci.* **157**, 263 (1999).
28. M. Yoshikawa, J. Izumi and T. Kitao, *Chem. Lett.*, 611 (1996).
29. T. Kobayashi, H. Y. Wang and N. Fujii, *Anal. Chim. Acta* **365**, 81 (1998).
30. H. Y. Wang, T. Kobayashi and N. Fujii, *J. Chem. Technol. Biotechnol.* **70**, 355 (1997).
31. J. Mathew-Krotz and K. J. Shea, *J. Am. Chem. Soc.* **118**, 8154 (1996).
32. N. Hilal and V. Kochkodan, *J. Membr. Sci.* **213**, 97 (2003).
33. I. A. Nicholls, K. Adbo, H. S. Andersson, P. O. Andersson, J. Ankarloo, J. Hedin-Dahlströ, P. Jokela, J. G. Karlsson, L. Olofsson, J. Rosengren, S. Shoravi, J. Svenson and S. Wikman, *Anal. Chim. Acta* **435**, 9 (2001).
34. M. Yoshikawa, J. Izumi, T. Kitao, S. Koya and S. Sakamoto, *J. Membr. Sci.* **108**, 171 (1995).
35. T. Kobayashi, H. Y. Wang and N. Fujii, *Chem. Lett.*, 927 (1995).
36. H. Y. Wang, T. Kobayashi and N. Fujii, *Langmuir* **12**, 4850 (1996).
37. T. Kobayashi, Y. Murawaki, P. S. Reddy, M. Abe and N. Fujii, *Anal. Ch. Acta* **435**, 141 (2001).
38. C. Cristallini, G. Ciardelli, P. Giusti and N. Barbani, *Macromol. Biosci.* **4**, 31 (2004).
39. G. Ciardelli, C. Cristallini, N. Barbani, B. Cioni, D. Silvestri and P. Giusti, *Mater. Res. Soc. Symp. Proc.* **787**, G 6.4.1. (2004).
40. P. Bures, Y. Huang, E. Oral and N. A. Peppas, *J. Control. Rel.* **72**, 25 (2001).
41. M. E. Byrne, K. Park and N. A. Peppas, *Adv. Drug Deliv. Rev.* **54**, 149 (2002).
42. W. J. Wizeman and P. Kofinas, *Biomaterials* **22**, 1485 (2001).
43. M. Zi-Hui and L. Qin, *Anal. Chim. Acta* **435**, 121 (2001).
44. S. Subrahmanyam, S. A. Piletsky, E. V. Piletska, B. Chen, K. Karim and A. P. F. Turner, *Biosens. Bioelectron.* **16**, 631 (2001).
45. J. Kost and R. Langer, *Trends Biothechnol.* **10**, 127 (1993).
46. M. Yoshikawa, J. Izumi, T. Kitao and S. Sakamoto, *Macromol. Rapid Commun.* **18**, 761 (1997).
47. L. Ye and K. Mosbach, *React. Funct. Polym.* **48**, 149 (2001).
48. R. Levi, S. McNiven, S. A. Piletsky, A. Rachkov, S.-H. Cheong, K. Yano and I. Karube, *Anal. Chem.* **69**, 2017 (1997).
49. S.-H. Cheong, A. Rachkov, J.-K. Park, K. Yano and I. Karube, *J. Polym. Chem.* **36**, 1725 (1998).

50. S.-H. Cheong, S. McNiven, A. Rachkov, R. Levi, K. Yano and I. Karube, *Macromolecules* **30**, 1317 (1997).
51. O. Ramström, L. Ye and K. Mosbach, *Chem. Biol.* **3**, 471 (1996).
52. A. G. Mayes, L. I. Andersson and K. Mosbach, *Anal. Biochem.* **222**, 483 (1994).
53. L. I. Andersson, R. Müller, G. Vlatakis and K. Mosbach, *Proc. Natl. Acad. Sci. USA* **92**, 4788 (1995).
54. K. Mosbach, Y. Yu, J. Andersch and I. Ye, *J. Am. Chem. Soc.* **123**, 12420 (2001).
55. B. Sellergren, *Angew. Chem. Int. Ed.* **39**, 1031 (2000).
56. Y. Yu, L. Ye, K. Haupt and K. Mosbach, *Angew. Chem. Int. Ed.* **41**, 4459 (2002).
57. K. Hirayama, M. Burow, Y. Morikawa and N. Minoura, *Chem. Lett.*, 731 (1998).
58. J.-L. Liao, Y. Wang and S. Hjertén, *Chromatographia* **42**, 259 (1996).
59. S. Hjertén, J.-L. Liao, K. Nakazato, Y. Wang, G. Zamaratskaia and H.-X. Zhang, *Chromatographia* **44**, 227 (1997).
60. A. Bossi, S. A. Piletsky, E. V. Piletska, P. G. Righetti and A. P. F. Turner, *Anal. Chem.* **73**, 5281 (2001).
61. A. Rachkov and N. Minoura, *Biochim. Biophys. Acta* **1544**, 255 (2001).
62. A. Rachkov and N. Minoura, *J. Chromatogr. A* **889**, 111 (2000).
63. D. Silvestri, C. Cristallini, G. Ciardelli, P. Giusti and N. Barbani, *J. Biomater. Sci. Polymer Edn.* **15**, 255 (2004).
64. P. Manesiotis, A. J. Hall and B. Sellergren, in: *Synthetic Receptors 2003*, Lisbon, Abstract P1.15 (2003).
65. A. Rachkov, M. Hu, E. Bulgarevich, T. Matsumoto and N. Minoura, *Anal. Chim. Acta* **504**, 191 (2004).
66. M. Glad, P. Reinholdsson and K. Mosbach, *React. Polym.* **25**, 47 (1995).
67. S. H. Ou, M. C. Wu, T. C. Chou and C. C. Liu, *Anal. Chim. Acta* **504**, 163 (2004).
68. P. Parmpi, L. D. V. Bolisay, P. Kofinas, *Mater. Res. Soc. Symp.* **787**, G6.2.1 (2004).
69. B. Sellergren, J. Wieschemeyer, K. S. Boos and D. Seidel, *Chem. Mater.* **10**, 4037 (1998).
70. O. Slinchenko, A. Rachkov, H. Miyachi, M. Ogiso and N. Minoura, in: *Synthetic Receptors 2003*, Lisbon, Abstract P2.73 (2003).
71. A. Aherne, C. Alexander, M. J. Payme, N. Perez and T. N. Vulfson, *J. Am. Chem. Soc.* **118**, 8771 (1996).
72. L. D. Bolisay, J. March, W. E. Bentley and P. Kofinas, in: *MRS Fall Meeting 2003*, Boston, MA, Abstract G3.1 (2003).

Tissue engineering scaffolds, entrapped cells, degradable polymers

Gels, Genes, Grafts and Giants, pp. 763–787 (2005)

Capillary zone electrophoresis as a tool to monitor the last stages of the degradation of water-sensitive polymers

CHRISTIAN BRAUD *

CRBA, CNRS, UMR 5473, Faculté de Pharmacie, Université Montpellier 1, 15 avenue Charles Flahault, BP 14491, 34093 Montpellier Cedex 5, France

Received 4 September 2003; accepted 25 November 2003

Abstract—In order to monitor the formation of the water-soluble by-products from chain-scission of degradable polymers used in the biomedical field, four capillary electrophoresis methods are discussed with the aim of giving the limits and performance for each. Three of them (electroosmotic flow reversal by dynamic adsorption of a polycation, multilayer polyelectrolyte coatings and physical binding of polyethylene oxide) are based on the use of dynamic coatings onto the inner surface of a fused silica capillary, a simple means to adapt performance to specific separations *via* modification and control of the electroosmotic flow of fused capillary. Using oligomers of lactic acid considered as standards the methods are compared. Other examples of ester-containing macromolecules (poly(hydroxybutyrate)), as well as degradable polyanions are described, namely N-acetylneuraminate polymer and poly(β-malic acid).

Key words: Capillary zone electrophoresis; degradable polymers; bioresorbable polymers; oligomers; lactic acid; poly(β-malic acid).

INTRODUCTION

Given the number of articles published every year, the field of degradable polymeric materials can be considered as one of the most attractive areas in present polymer science. Indeed, effective or potential applications are found in medicine, pharmacology and biology, as well as in the environment, regardless of their natural or synthetic origins.

Among these various domains of application, pharmacological ones, especially controlled drug delivery and drug carriers request polymeric systems that can be eliminated from the body after healing, i.e. bioresorbable systems [1–3]. Such systems can be based on the physical entrapment of a drug within a degradable

*E-mail: braud@univ-montp1.fr or crba@univ-montp1.fr

polymer matrix processed under the form of micro- or nano-particles/capsules, capsules, implants and even water-dispersed amphiphilic polymers.

Regulatory rules are progressively requesting better description of bioresorbable polymeric devices and this includes the identification and the assessment of degradation by-products generated by degradation at the manufacturing stage of the polymeric device itself (quality control), or during *in vivo* drug delivery, or after exhausting to eliminate the waste matrix.

Therefore, the search for analytical tools and methods aimed at monitoring degradation and degradation by-products is an attractive research area [4]. Bearing this in mind, methods based on liquid chromatography [5–10], gas chromatography and mass spectrometry [11–17] have been extensively considered.

To be degradable to low-molecular-weight molecules, the backbone of macromolecules has to include cleavable chemical functions such as amide, ester, urethane, carbonate, etc. Most of them are sensitive to water, a solvent medium common to all living systems, either by simple hydrolysis or with the help of more or less specific enzymes. The hydrolytic cleavage of these water-sensitive functions generates charged end-groups of the carboxylate or ammonium types ($-COO^-$ or $-NH_3^+$).

High-Performance Capillary Electrophoresis (HPCE), that was already used to characterize biopolymers [18], appears, in the Capillary Zone Electrophoresis (CZE) mode, as a technique of interest to monitor the formation of charged low molecular weight compounds going from water-soluble oligomers to unimers [19–25].

As CZE is a relatively new analytical technique in the case of polymer analysis [26], the related methods have rapidly progressed during the last decade. Moreover, this technique can be considered as the method of choice aside HPLC. It must be noted that CZE is a technique well suited for the monitoring of degradation by-products of degradable systems, since it only needs a few microliters of the supernatant solution to be performed.

In this paper, we want to revisit published data connected with CZE in order to emphasize the limits of the technique in the field of degradable polymers. Considering the case of the water-soluble oligomers of lactic acid, which is the simplest example of end-charged oligomers, four CZE techniques will be described. Their use will be discussed in the case of charged macromolecules (polyelectrolytes and proteins) that can be present in the degradation media. Last but not least, new data will be presented in order to further illustrate the performances of CZE.

CZE TECHNIQUES FOR THE MONITORING OF DEGRADATION BY-PRODUCTS OF MACROMOLECULES

To date, HPCE is performed in aqueous media where ionic charges are the driving force of the separation [27]. Nevertheless non-aqueous capillary electrophoresis is emerging [28], which allows separation of non water-soluble charged compounds.

For water-soluble polymers (and oligomers) bearing ionic charges, let us consider two aspects, i.e. end-charged macromolecules and charge-bearing macromolecules (polyelectrolytes and proteins). From the analytical point of view, these different aspects require different HPCE methods.

The case of end-charged macromolecules

End-charged macromolecules are molecules bearing generally one charge or, in more complex features, two charges at the maximum. Thus, the simplest HPCE technique for separation purposes of such species is CZE. The basis of CZE separation is related to the effective mobility of a charged analyte.

The instrumentation that allows CZE separation is relatively simple (Fig. 1). A narrow-bore capillary (20–100 μm i.d.) is filled up with a background electrolyte (generally a buffer). The two ends of the capillary are immersed into two vials filled up with the same background electrolyte. These vials are equipped with two platinum electrodes which are connected to a high voltage power supply. The analyte is introduced at the inlet side by pressure, siphoning or electrokinetic injection and is detected on line by UV (or fluorescence). Data are plotted to give an electropherogram of signal *versus* migration time.

The fused silica capillaries that are typically used for CZE separation have ionizable silanol groups at the silica/background electrolyte interface. The pK of

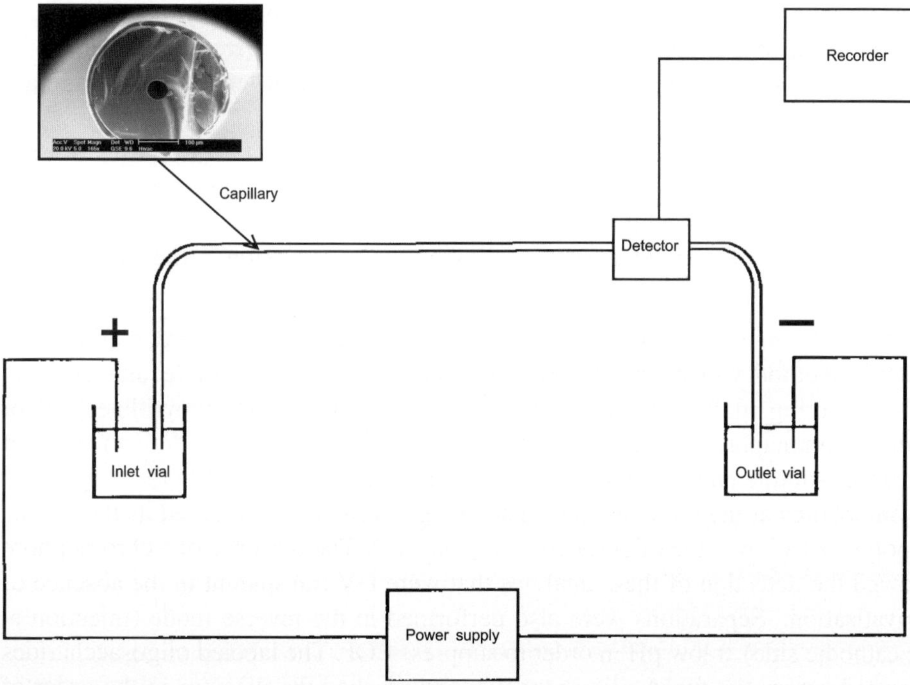

Figure 1. Scheme of a basic CZE instrument used according to the standard method.

fused silica is about 1.5; thus, the degree of ionization of silanol groups is controlled mainly by the pH of the background electrolyte. The stationary negative charges of the wall are compensated by mobile cations from the background electrolyte. When current is applied, these cations move towards the cathode creating a net flow of the background solution in the direction of the negative electrode. This phenomenon is called electroosmotis flow (EOF).

Currently, CZE principle is described using a fused silica capillary and a basic background electrolyte (standard method described later) [27]. Under these conditions, EOF is quite robust and most analytes migrate toward the cathode regardless of charge. For quantitative analysis, an EOF marker has to be introduced in the solution. The marker is an absorbing and neutral molecule that does not interact with the analyte. The effective velocity (v_i) and effective mobility (μ_i) can be deduced from the electropherogram by using equations (1) and (2):

$$v_\mathrm{i} = v_{\mathrm{i(net)}} - v_\mathrm{EOF} = \frac{L_\mathrm{D}}{t_\mathrm{i}} - \frac{L_\mathrm{D}}{t_\mathrm{EOF}}, \tag{1}$$

$$\mu_\mathrm{i} = \frac{L_\mathrm{D} \cdot L_\mathrm{T}}{V} \cdot \left[\frac{1}{t_\mathrm{i}} - \frac{1}{t_\mathrm{EOF}}\right], \tag{2}$$

where $v_{\mathrm{i(net)}}$ is the observed velocity (in cm/s) of ionic analyte i giving a peak at t_i (in s), v_EOF the observed velocity (in cm/s) of the EOF marker giving a peak at t_EOF (in s), μ_i the effective electrophoretic mobility (in cm^2/V per s) of ionic component i, V the applied voltage (in V), L_D the length from the inlet of the capillary to the detection window (in cm) and L_T the total length of the capillary (in cm).

The effective mobility of an analyte i is the result of two opposing forces: attraction of the electrode of opposite charge and friction of the charged molecule within the background solution.

Different methods can be used depending on the composition of the analytes.

CZE in fused silica capillary (standard method). The standard method was the first method which was reported for separation of macromolecular analytes by considering malto-oligosaccharides derivatives with 2-aminopyridine [29] or with 8-aminonaphthalene-1,3,6-trisulfonic acid (ANTS) [30]. The derivatization of the water-soluble and non-ionic oligosaccharidic chains brought both charges and chromophores at the reducing end. The charge/mass ratio decreased as the degree of polymerization of the oligosaccharide increased. The presence of a chromophore allowed the detection of these analytes that were UV transparent in the absence of derivatisation. Separations were also performed in the reverse mode (injection at the cathodic side) at low pH in order to suppress EOF. The labeled oligosaccharides migrated across the fused silica capillary and reached the detector in the order of increasing size, i.e. the most mobile comes first.

In the case of ANTS-labeled maltosaccharides, separations were also run using the standard method with phosphate solutions at pH 9.0 as background electrolyte [30]. Peaks reflecting the analytes appeared on electropherograms after the EOF marker, the higher the migration time the higher the effective mobility.

A similar behavior was observed for a mixture of alkylbenzene sulfonates with different alkyl lengths analyzed according to the normal mode with borate as background electrolyte [31].

Polyoxyethylene macromolecules, which were derivative at both ends with phthalic anhydride, were also successfully analyzed. Separations were performed in a background electrolyte composed of 57 mM boric acid + 35 mM 1,3-diaminopropane (pH 9.7)/acetonitrile (70 : 30, v/v) [32].

The first attempt to separate water-soluble oligomers of degradable polymers with the standard method was done by considering the lactic acid family [20, 21]. Such oligomers bear a carboxylic group at one of their terminus. This group is fully ionised when pH > 6.0. Samples of water-soluble (S)-lactic acid oligomers (OLA100, Scheme 1) were obtained by polycondensation of (S)-lactic acid syrup (85–90% (w/w) aqueous solution, from Janssen). Typically, 200 ml of the commercial solution was introduced in a 1000-ml round bottom flask and heated at 90°C under vacuum for 2 h. The reaction medium was maintained under argon during all the experiment. The mixture was then cooled at room temperature giving a waxy medium. An aliquot of approx. 10 mg of the resulting wax was mixed with 10 ml of 0.05 M phosphate buffer at pH 6.8. The solution (C approx. 0.1%) was filtered on 0.22 μm Millipore filter before being used as a standard solution.

These analytes were poor UV absorbers. Thus, only UV transparent background electrolytes at 200 nm can be used. In 0.1 M borate buffer (BB, pH 8.9, prepared by dissolving 6.184 g boric acid (Roth) in Milli-Q water, neutralizing the solution with 1 M NaOH up to pH 8.9 and diluting it with water up to a final volume of 1000 ml), poor separation was observed for the longer oligomers (Fig. 2a). Moreover, the basicity of the buffer did not guarantee the stability during the separation of these oligomers that were unstable at high pH. A 0.05 M phosphate buffer (PB, pH 6.8, prepared by mixing 2.00 g $NaH_2PO_4 \cdot 2H_2O$ (Prolabo) and 4.36 g $Na_2HPO_4 \cdot 12H_2O$ (Prolabo) in Milli-Q water up to a final volume of 500 ml, stored at 4–8°C and replaced every week) was then considered but the low value of the EOF did not allow separation in a short time, a critical request for HPCE techniques [23].

In conclusion, the standard method shows that in a series of related oligomers, the charge/mass ratio (c/m) which strongly varied from one oligomer to the other one is the key factor of the separation. However, the method is limited to oligomeric analytes that support basic background electrolytes where EOF is high as in the case of borate buffers. For this reason, we describe in the following other CZE methods based on modification and control of the EOF of a fused silica capillary *via* dynamic coating.

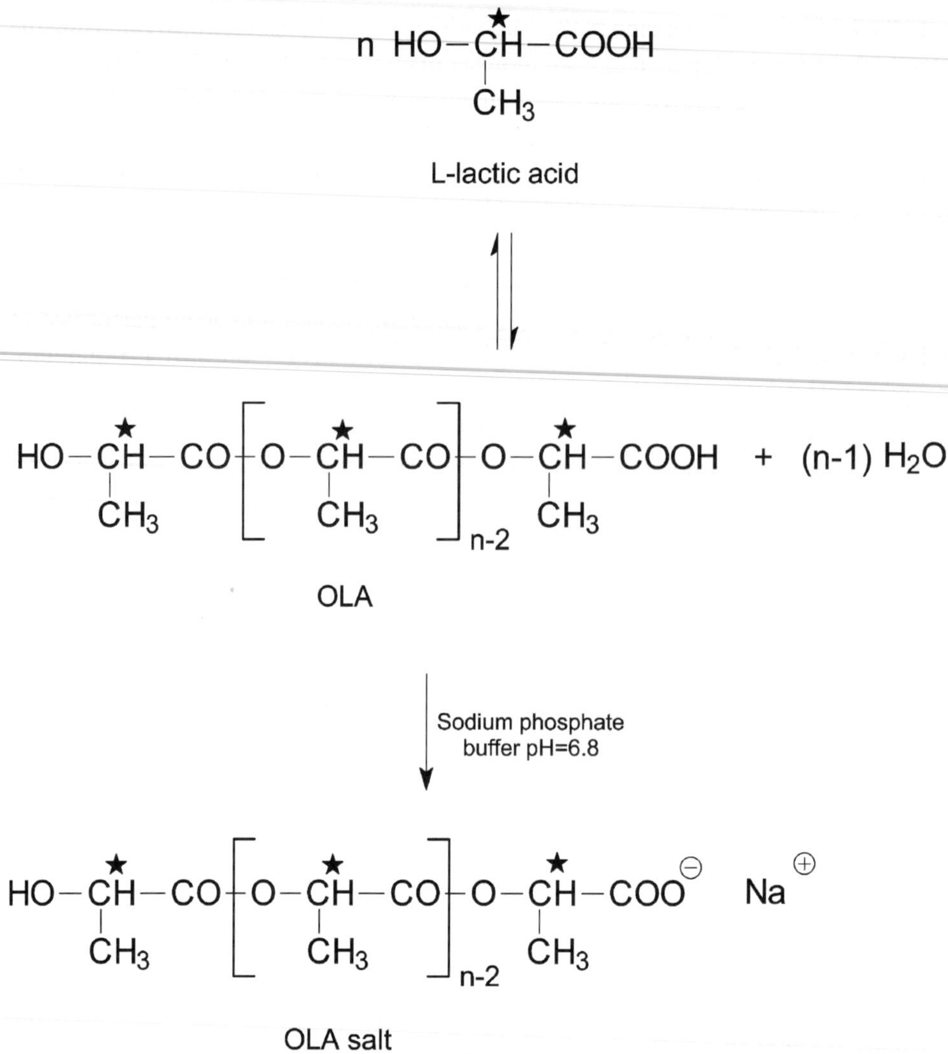

Scheme 1.

EOF reversal by dynamic adsorption of a polycation (DAP method). The DAP method is based on the reversal of EOF by physical adsorption of polycations onto the inner wall of a fused silica capillary [33]. At pH values where both the silanol groups on the inner wall of the capillary and the basic groups of the polycation are ionised, a simple rinse with a background electrolyte that contains a polycation, followed by a rinse with the same pure background electrolyte in order to remove unabsorbed polycations, allows a reversal of the EOF. Under these conditions, separation is performed in a background electrolyte that does not contain polycation molecules since the absorption process is strong enough to avoid any desorption during the run. Such a dynamic coating is fundamentally different from the one

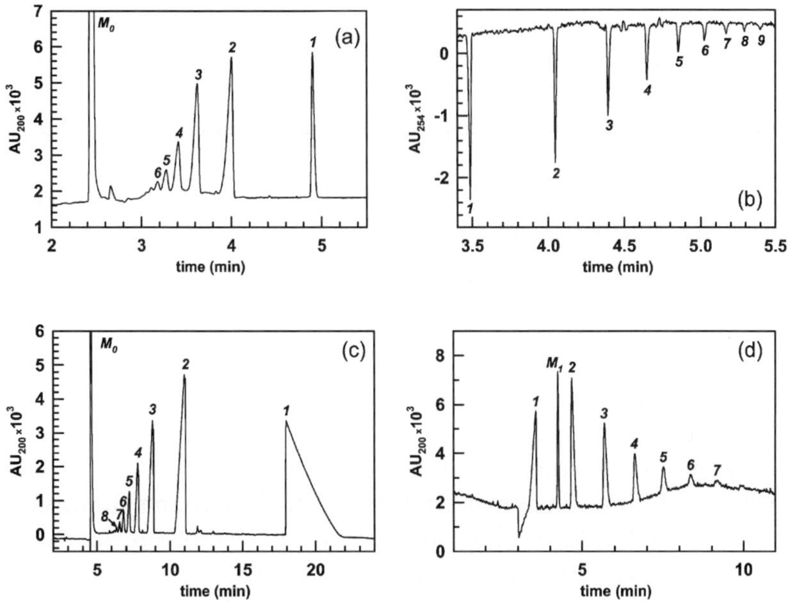

Figure 2. Electropherograms of OLA100 recorded in different separation modes. Data were collected using a P/ACE 5000 instrument (Beckman) equipped with UV detection. Fused silica capillaries were used (standard, DAP and SMIL methods: 50 μm i.d.; $L_T = 57.0$ cm; $L_D = 50.4$ cm; PBPEO method: 50 μm i.d.; $L_T = 47.0$ cm; $L_D = 40.4$ cm). (a) Standard method: the capillary was rinsed stepwise before each run with 0.1 M HCl for 2 min, then for 1 min with water, followed by 2 min with 0.1 M NaOH and, finally, 2 min with BB. The analyte was injected at the anodic side by the hydrostatic pressure method (10 s). Separation was performed at 25°C with an applied voltage of +25 kV. After the last run, the capillary was rinsed for 2 min with 0.1 M NaOH, then for 5 min with water and finally stored until the next analysis. UV detection was performed at 200 nm. (b) DAP method (running buffer: *p*ABE): indirect UV detection at 254 nm was used. The capillary was conditioned at the start by the following rinses: 1 M HCl for 30 min, water for 1 min, 1 M NaOH for 30 min and *p*ABE for 5 min. Furthermore, the capillary was rinsed before each run using the sequence 0.1% DEAE-Dextran in *p*ABE for 5 min, water for 1 min and *p*ABE for 3 min. Analyte was injected at the cathodic side by the hydrostatic pressure method (10 s). The analyte ($C \approx 1\%$ for model compounds) was diluted 50 times before analysis at 25°C with an applied voltage of −25 kV. After the last run, the capillary was rinsed for 30 min with 1 M NaOH, followed by 5 min with water before being stored wet. (c) SMIL method: a fused silica capillary was rinsed as followed: 1 M NaOH for 30 min, PB for 5 min, 5% PDDAC in PB for 5 min, PB for 5 min, 3% dextran sulfate (Sigma) in PB for 5 min and PB for 5 min. This series of rinses modified the capillary irreversibly. Furthermore, the capillary was rinsed before each run using the sequence 5% PDDAC in PB for 3 min, PB for 1 min, 3% dextran sulfate in PB for 3 min and PB for 5 min. The analyte was injected at the anodic side by the hydrostatic pressure method (10 s). Separation was performed at 25°C with an applied voltage of +25 kV. UV detection was performed at 200 nm. After the last run, the capillary was rinsed for 30 min with water before being stored wet. (d) PBPEO method: a fused silica capillary was rinsed as follows: 1 M HCl for 2.5 min, 0.1% PEO in 0.5 M HCl for 5 min and PB for 3 min. The analyte was injected at the cathodic side by the hydrostatic pressure method (10 s). Separation was performed at 25°C with an applied voltage of −25 kV. After the last run, the capillary was rinsed, first, for 30 min with 1 M NaOH, then for 5 min with water and kept as such before being stored wet. M_0, benzyl alcohol; M_1, sorbate anion, used as an internal mobility marker. Numbers on electropherograms are the degrees of polymerisation of the different oligomers.

obtained after rinsing with cationic surfactants. Indeed, the interaction of surfactants with the inner wall is weak and requires the presence of surfactant molecules in the background electrolyte to maintain the coating during the run [27]. It must be pointed out that surfactant molecules can physically interact with the analyte, thus modifying its effective mobility.

With the DAP method, polarity of the power supply is reversed (inlet side near the cathode) and the effective mobility is given by:

$$\mu_{\mathrm{i}} = \frac{L_{\mathrm{D}} \cdot L_{\mathrm{T}}}{V} \cdot \left[\frac{1}{t_{\mathrm{EOF}}} - \frac{1}{t_{\mathrm{i}}}\right]. \tag{3}$$

The major advantage of EOF reversal by polycation pre-adsorption was its versatility. Different polycations were used, namely medium-molecular-weight poly(diallyldimethylammonium chloride) (PDDAC), diethylaminoethyl dextran containing 40 mol% quaternised amino groups as side reactions [34] (DEAE Dextran), poly(diethylaminoethyl methacrylate) (P(DEAEMA)), poly(dimethylaminoethyl methacrylate) (P(DMAEMA)), poly(α-L-lysine), hydrobromide (P-α-lys, HBr, molecular mass 15–30 kDa) and free base form of poly(ε-L-lysine) (P-ε-lys). Prior to use, aqueous solutions of these polycations were purified by ultrafiltration using an Amicon cell equipped with a YM10 ultrafiltration membrane. EOF can be varied via the nature of the polycation [35] and the pH of the background electrolyte can go from basic to acid (Table 1).

The DAP method was successfully used to separate oligomers of lactic, 3-hydroxybutyric and 6-hydroxyhexanoic acids by using a 50 mM phosphate buffer (pH 6.8) as background electrolyte and DEAE-dextran as the polycation [21–23]. Based on reproducibility and separation performances, DEAE-dextran seems to be the more efficient in the series of the tested polycations, namely PDDAC, DEAE dextran, P(DEAE MA), P(DMEA MA), poly α-lys, HBr and poly ε-lys.

More recently, the combination of dynamic polycation coating (using DEAE Dextran) and indirect UV detection using p-anisate background electrolyte (pABE, 0.01 M), prepared by adding 0.7608 g p-anisic acid (Aldrich) and 1.2114 g Tris (tris(hydroxymethyl)aminomethane, Sigma) to Milli-Q water up to a final volume of 500 ml. The dissolution of p-anisic acid was accelerated by sonication. The pABE solution was kept at 4–8°C and replaced every week, which allowed the sep-

Table 1.
Migration time (min) of a neutral marker (benzyl alcohol) for a fused-silica capillary after a rinse with a PC solution for a fused capillary

PC	Phosphate buffer (0.05 M, pH 6.8)	Tris buffer (0.05 M, pH 7.4)	Phosphate buffer (0.05 M, pH 3.5)
DEAE Dextran	12.22	8.19	6.62
Poly(DEAE MA)	8.73	3.98	4.16
Poly(ε-lysine)	12.21	6.35	8.73

Capillary: 50 μm i.d., $L_{\mathrm{T}} = 57.0$ cm, $V = -25$ kV, temperature = 25°C.

aration of by-products of gluconic derivatives released from gluconic/glycolic/lactic copolymers [24]. Presently, this method seemed the most efficient method for the separation of oligomers of aliphatic hydroxyacids in simple degradation media, i.e. phosphate buffer, since the detection limit droped to C approx. 0.002%. Figure 2b exemplified such separation for lactic acid oligomers. Indirect UV detection could be considered for non-UV-absorbing analytes. It was first described to monitor the separation of free fatty acids [36]. The background electrolyte contained pABE as the unique anionic species and the detection wavelength was selected at 254 nm (or 280 nm). When the non-UV-absorbing analyte passed through the detection window, a displacement of pABE anions which was proportional to the negatively charged analyte created a negative absorption peak. In the case of lactic acid oligomers in dermatological formulations, this technique allowed one to increase the sensitivity of the detection of lactic acid up to less than 1 ppm [37], using the cationic surfactant: tetradecyltrimethylammonium bromide.

Multilayers polyelectrolyte coatings (SMIL method). The dynamic coating of a fused silica named SMIL (Successive Multiple Ionic-polymer Layer) leads to stable EOF, regardless of the pH of the background electrolyte [38]. The method is based on a rinse with a polycation, as in the case of the DAP method, followed by a second rinse with a polyanion. Under these conditions, the EOF is directed toward the cathode as for unmodified fused capillary. However, the value of the EOF is almost constant, regardless of the pH of the background electrolyte. It must be pointed out that SMIL coatings are very stable. The method is considered as profitable from an analytical viewpoint [38–40]. However, one must keep in mind that such dynamic coating is so strong that simple rinse cannot regenerate the genuine fused silica capillary.

This method was successful in separating the OLA oligomers under neutral pH condition, as illustrated in Fig. 2c.

The SMIL method can be selected when the standard method that requires basic background electrolyte cannot be used due to pH-instability of the analyte, or when the DAP method does not give satisfactory results.

Physical binding of PEO macromolecules (PBPEO method). The three preceding methods afford separation in simple media. By simple media we mean a medium such as phosphate-buffered saline at neutral pH, which is currently used to mimic body fluids (i.e. phosphate buffer at neutral pH). But in daily practice the ageing medium contains proteins. CZE of proteins with fused silica capillaries yields a complex problem linked to protein adsorption and non-reproducible results [41, 42]. As proteins are charged entities (except at their pI), successful CZE separation is a case by case procedure where both the structure of the protein and the charge of the inner wall of the capillary has to be carefully considered [43].

That is why the search for a 'neutral' capillary became attractive. Chemical modifications aimed at suppressing the negative charges of the fused silica capillary was the first tested route [27, 44]. The poor reproducibility of such chemical coatings led to consider physical adsorption of water-soluble polymers using rinsing procedures. Poly(vinyl alcool) was considered first [33]. Presently, the adsorption of polyethylene oxide (PEO) is preferred for analysis of complex media. Yeung *et al.* [45, 46] proposed a simple procedure to suppress the negative charges of a fused silica capillary. The method consists in rinsing the capillary first with a HCl solution and then with an acidic solution of PEO. Hydrogen interactions between −OH groups of silanol and oxygen atoms of PEO are, thus, maximized and the dynamic coating is stable at $pH < 7$.

This rinsing procedure was successfully used for the separation of OLA oligomers in PB (Fig. 2d). The M_1 peak was due to ascorbic acid, an anionic species used as an internal mobility marker. It must be pointed out that in the present separation mode, a neutral marker was useless because of the absence of EOF.

CZE of charge-bearing macromolecules

Whereas oligomers bearing one or two charges are easily separated using the above described methods, charge-bearing macromolecules can partly be separated according to molar masses because of unfavorable variation of c/m.

Under the term 'charge-bearing macromolecules', we have to consider, on the one hand, polyelectrolytes and, on the other hand, proteins. Polyelectrolytes are generally high-charge-density polymers with flexible chains and the same type of charge (polyanions or polycations). Proteins have to be distinguished from polyelectrolytes because of their low charge density, their polyampholytic structure and the lack of chain flexibility.

CZE of polyelectrolytes. Due to the fact that the c/m ratio was constant for high-MW molecules, there was no dependence of the effective mobility on MW for high-MW polyelectrolytes. Moreover, ion condensation reduced the effective charge of the polyion and was not MW dependent [47, 48]. For these reasons, Gel Capillary Electrophoresis (GCE) based on separations within sieving media gave the best results as far as MW discrimination was concerned [49, 50].

However, for low molar masses (i.e. in the range of oligoelectrolytes), the effective mobility of similar macromolecules became MW dependent. This was observed for synthetic poly(styrene sulfonate) (PSS) [51–54] and for by-products of hydrolytically or enzymatically degraded biopolyelectrolytes such as ss-DNA [52], N-acetylneuraminate polymers (NANA) [55–57], hyaluronate (HYAL) [58], galacturonate polymers [59] and poly(β-malate) (PMLA) [19]. Two major comments can be made from these examples.

First, polyions raised problems in CZE because of adsorption processes as previously described for DAP and SMIL methods. In fact, up to now, a polyanionic oligomer (oligoanion) was analysed with a silica capillary and a basic pH background electrolyte according to standard method only. For polycationic oligomer (oligocation), there was no data available yet.

Second, the variation of effective mobility *vs.* MW was complex. Considering PSS macromolecules — the unique source of commercially available polyanions with a narrow MW dispersity and a large range of MW — the effective mobility (in absolute value) of lower oligomers increased with MW when the degree of polymerization was less than about 10 [54]. A similar behavior was observed in the case of ss-DNA [52]. For NANA and HYAL also, mobility increased with MW. The only exception was for PMLA oligomers whose effective mobility decreased with MW [19]. This apparent contradiction can be explained if one put in parallel the variation of the c/m ratio as a function of the degree of polymerization (DP). If both chain-ends were neutral as in the case of PSS, ss-DNA, NANA or HYAL, $c/m = f(\mathrm{DP})$ obeyed the relation:

$$c/m = \frac{\mathrm{DP}}{\mathrm{DP} \cdot M_{\mathrm{RU}} + M_{\mathrm{ECG}}}, \tag{4}$$

where M_{RU} and M_{ECG} were the mass of the repeating unit of the polyelectrolyte and the mass of its chain-end groups, respectively.

For PMLA oligomers (Scheme 2), c/m *versus* DP varied differently (equation (5)). This was due to the presence of a negatively charged carboxylate group at one end.

$$c/m = \frac{\mathrm{DP} + 1}{\mathrm{DP} \cdot M_{\mathrm{RU}} + M_{\mathrm{ECG}}}. \tag{5}$$

In order to exemplify this feature, variation of c/m was represented as a function of DP for different polyelectrolyte families (Fig. 3). Thus, considering the high separation performance of CZE, direct correlation of effective mobility and the c/m value can be done in a defined series of oligomers.

CZE of proteins. Proteins are rigid objects whose charges are not randomly distributed at the protein surface. As a matter of fact, a protein has often part of its surface that can electrostatically interact with the surface of the capillary wall regardless of its charge.

However, when MW $< 50 \times 10^3$ successful separation was observed for acidic proteins using uncoated capillaries and for basic proteins using PC-modified fused silica capillaries [43].

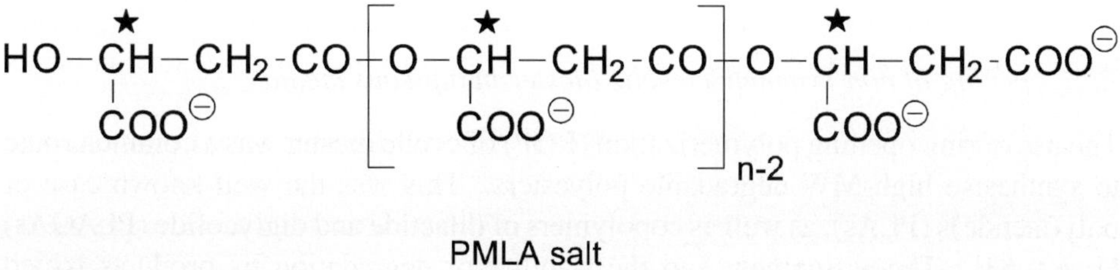

Scheme 2.

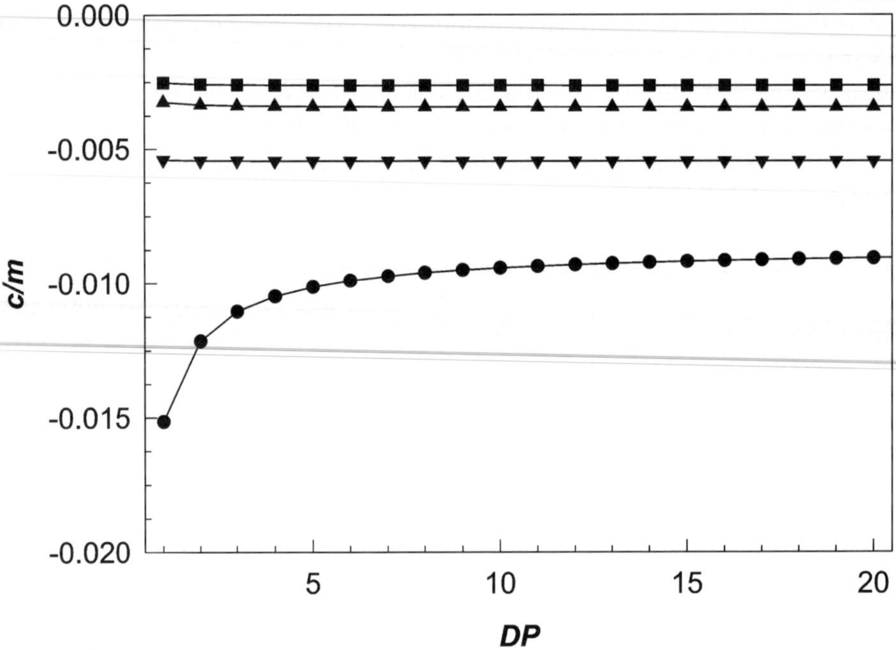

Figure 3. Calculated values of the charge/mass ratio (c/m) *versus* the degree of polymerization for PMLA (●), PSS (▼), NANA (▲) and HYAL (■). c/m is calculated using equation (4) or (5).

Another system to be commented on was the so-called 'protein charge ladder'. A protein was modified by partial blocking of its lysine residues with acylating agents [60, 61]. During the chemical coupling, the mass of the protein was practically unaffected. In contrast, its net positive charge strongly decreased. For a well-defined protein, the reaction generated modified proteins with different c/m ratio. The resulting CZE electropherograms showed peaks with different mobility, which constitute 'the ladder'.

APPLICATION TO DIFFERENT EXAMPLES

A selection of data dealing with ageing of degradable systems has been chosen in order to exemplify the possibilities which are offered by CZE to deeply analyze complex systems.

Ring-opening of non-symmetric cyclic diester in aqueous media

The use of ring opening polymerization (ROP) of cyclic diester was a common route to synthesise high-MW degradable polyesters. This was the well-known case of poly(lactide)s (PLAs), as well as copolymers of dilactide and diglycolide (PLAGAs) [1, 62, 63]. The assignment and the stability of degradation by-products issued during degradation of such polyester materials was easily done by considering parent oligomers generated by simple reactions such as mild polycondensation of hydroxyacids or ring-opening of cyclic dimers [20, 21].

In the research of new co-polyesters, other cyclic esters were considered to modulate the properties of the resulting material [64–72].

Using this strategy, the ageing of a copolymer of lactide and 3-(1,2,3,4-tetraoxo-butyldiisopropylidene) dioxane-2,5-dione (DIPAGYL) allowed us to monitor its fate in *in vitro* conditions [68]. DIPAGYL is a non-symmetric cycle constituted of a glycolic acid moiety and a protected gluconic acid moiety. The co-polymer generated up to 10 peaks assigned to dimeric and monomeric by-products. The formation of these by-products has been monitored by CZE using the DAP method and indirect UV detection with *p*-ABE [24]. Peak assignments were done by considering the ring-opening of DIPAGYL in PB and its *in vitro* degradation products.

The cleavage of the two different ester bonds (Scheme 3) produced two open dimers (I+II) depending on the cleaved ester bond (*1* or *2* in Fig. 4). The separation of corresponding peaks (Fig. 4a) exemplified the high separation performance of CZE because the two dimers had the same mass and only one negative charge. The difference in mobility was only due to difference in friction with the background electrolyte in relation with different configuration.

In the same vein, a similar double peak (1? and 2? in Fig. 5) was also observed for *in vitro* degradation products of $PLA_{26}GA_{48}$ copolymer (52 : 48% (w/w) D,L-lactide/glycolide in the monomer feed) by using the same CZE method as above. Many peaks appeared, as shown in Fig. 5. Some of them were easily assigned to defined oligomers. The two series of peaks corresponding to homo-oligomers of lactic acid (L, L_2, L_3 and L_4) and glycolic acid (G, G_2, G_3 and G_4) were easily

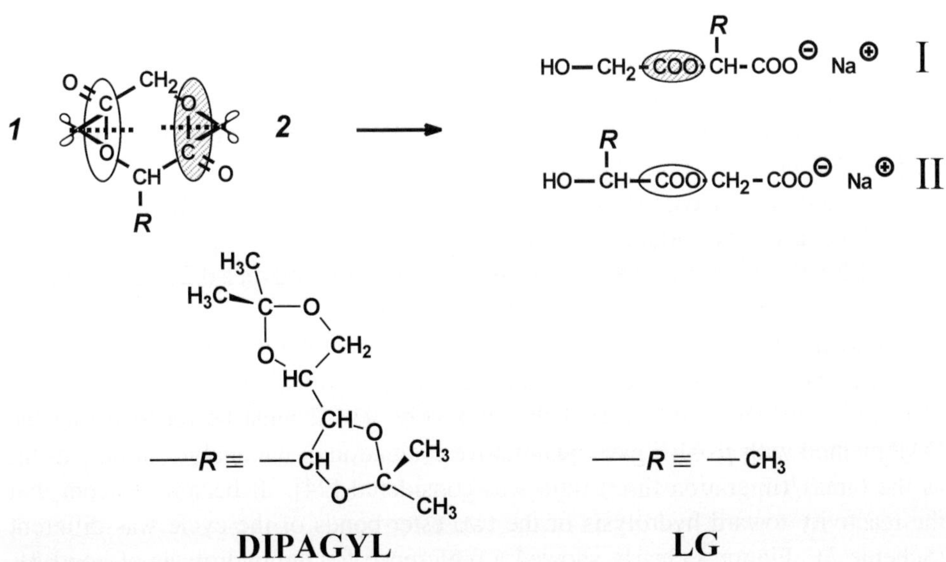

Scheme 3.

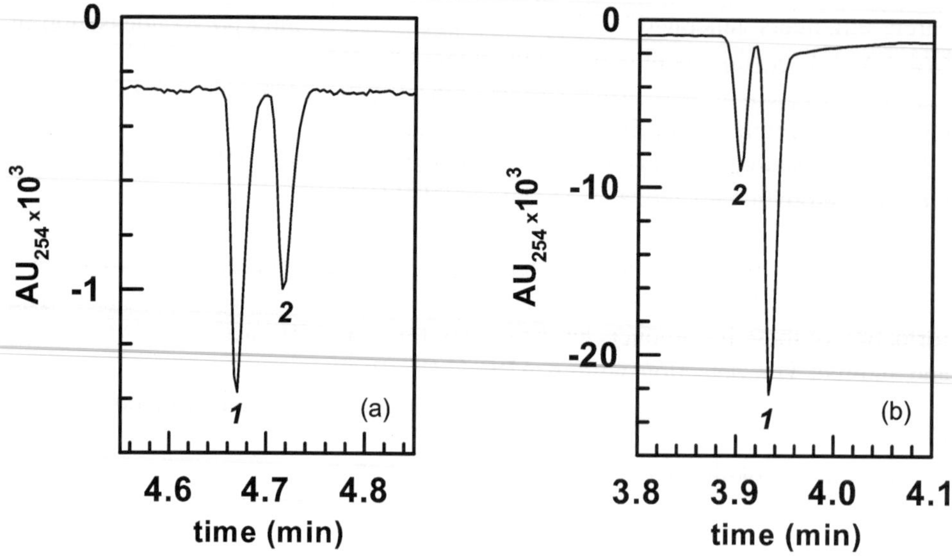

Figure 4. Electropherograms in the range of detection of the open dimers of DIPAGYL (a) and LG (b). Numbers *1* and *2* correspond to the peaks of structure I and II of the open dimmers, respectively (see Scheme 3). The electropherograms were recorded using the DAP method (running buffer, *p*ABE; PC, DEAE-Dextran).

assigned using our data on homo-oligomers. On the electropherogram, peak 1? and peak 2? were located between peak G_2 and peak L_2 and could be assigned to the two hetero-dimers. Similarly, the small peaks (noted 3? and 4?) observed between peaks L_2 and L_3, and between L_3 and L_4, respectively, could also be assigned to hetero-trimers and hetero-tetramers, based on the relation between charge and mass of such oligomers [24]. Whereas synthesizing trimers and tetramers was difficult, opening the ring of 3-methylglycolide (LG) could easily generate open dimers. For this purpose, a sample of LG was kindly supplied by Prof. K.-Y. Qiu (Peking University, Beijing, China).

LG (5 mg) was dispersed in 10 ml PB. As the reaction proceeded, the pH was maintained at 6.8 by adding 1 M NaOH. After 1 h at room temperature no crystal remained in the medium. The buffered solution was filtrated on a 0.22 μm Millipore filter. The resulting mixture was stored in solution.

The asymmetry of DIPAGYL and LG rings made a preference in the ring opening via pure hydrolysis in aqueous media. Peaks (identified by 1 and 2) were not identical, considering the area of the peak (Fig. 4). It must be recalled that the DAP method with *p*-ABE gave quantitative evaluation of the analyte amount as far as the (area)/(migration time) ratio was considered [24]. It became evident that the reactivity toward hydrolysis of the two ester-bonds of the cycle was different (Scheme 3). Figure 4 clearly showed a preference for the hydrolysis of bond #1, giving a larger amount of compound I compared with compound II.

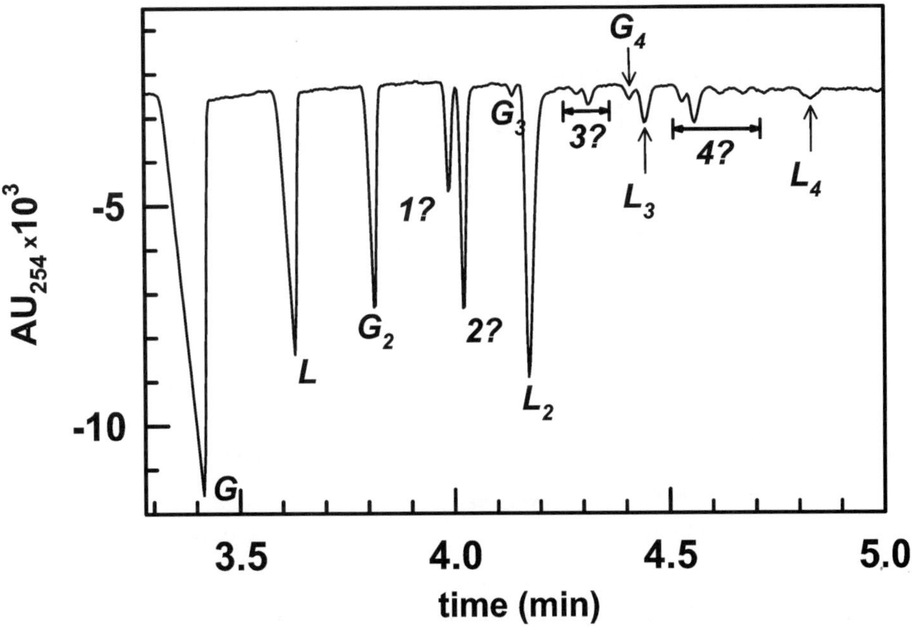

Figure 5. Electropherogram of $PLA_{26}GA_{48}$ copolymer after ageing for 32 days in PB at 37°C. The electropherogram was recorded using the DAP method (running buffer, *p*ABE).

The degradation of (R) 3-hydroxybutyrate oligomers

(R)-3-hydroxybutyrate polymer (P(3-HB)) is part of a larger family of polyhydroxyalkanoates (PHAs) which are polyesters of interest in the field of medical and pharmaceutical applications [73, 74]. While biocompatibility was the main property that was considered for PHAs, their degradability (or stability) also had to be considered.

The field of biodegradable polymers for environmental purposes has been well documented in the literature. PHAs seemed stable in *in vitro* conditions [75]. Their degradation proceeded *in vivo* when specific enzymes produced by bacteria present in the environment (soil, composts, activated sludges, etc.) were present [76]. Fortunately, these enzymes were not present in the body fluids, and the degradation of PHAs in the biomedical field was still a matter of debate.

During the synthesis routes used to transform P(3-HB) (Scheme 4) for specific applications, crotonic acid or derivatives were generally claimed [5, 13, 77, 78]. In our effort to synthesise (R)-3-hydroxybutyrate oligomers (O(3-HB)) by hydrolytic degradation of high-MW P(3-HB), crotonate contamination was observed and could be carefully monitored by CZE [22].

The degradation of water-soluble O(3-HB) in abiotic conditions revealed the presence, in small amounts, of a second series of oligomers which were characterized by a crotonate end-chain at one terminus (O′(3-HB)) [22]. The peak assignment corresponding to O′(3-HB) oligomers was made by comparison of electropherograms

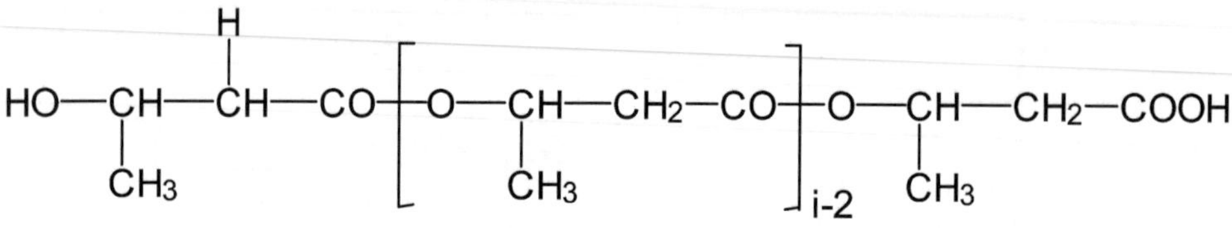

O(3-HB)

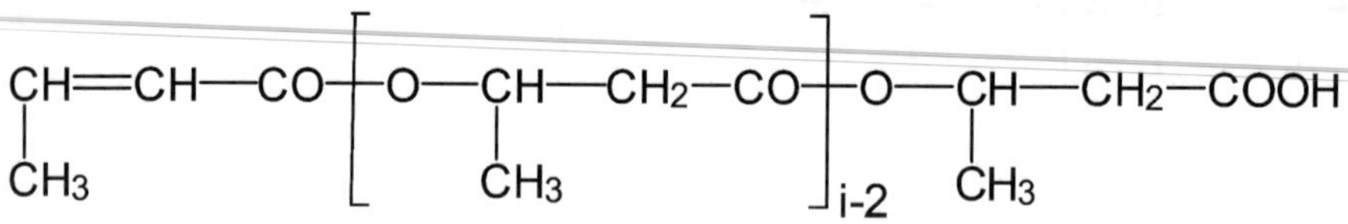

O′(3-HB)

Scheme 4.

recorded at 200 and 230 nm. At 230 nm, only O′(3-HB) oligomers absorb contrarily to 200 nm where both series of oligomers absorbed. At that time, the DAP method with indirect UV detection was not yet developed.

Here, a rapid comparison of two CZE techniques: the DAP method with a direct UV detection at 200 nm (Fig. 6a and 6b) as the one yet used, and the DAP method with *p*-ABE (Fig. 6c and 6d) was done for the analysis of water-soluble O(3-HB) during ageing. Initially, indirect UV detection (Fig. 6a), the two series of oligomers were detected. The crotonate oligomers series gave rise to minor peaks (noted 1′ through 4′). With the indirect UV detection (Fig. 6c), the crotonate series was not detected. Actually, the O′(3-HB) series was in small amount and could only be detected by direct UV detection because of the strong UV absorption at 200 nm of crotonate moieties. Upon ageing of the solution (50 days at 50°C), the crotonate oligomer peaks noted 1′ through 4′ became predominant (Fig. 6b), and were also detected through indirect UV detection (Fig. 6d). These electropherograms confirmed the generation of O′(3-HB) oligomers during ageing in the analyzed solution.

Some CZE data of degradation monitoring in complex media

From the analytical point of view of HPCE, the presence of enzymes in the analyte was a problem which was primarily connected to adsorption on the capillary wall of these proteins. We made some experiments on this absorption mechanism. Two features had to be commented. First, the absorption was so high that non-reproducible results were found; further rinses of the capillary with selected solutions (acidic, basic, urea ...) were ineffective with the consequence of putting away the capillary. Second, non-reproducible results were observed, but rinsing procedures can give back the initial CZE characteristics of the capillary. In both

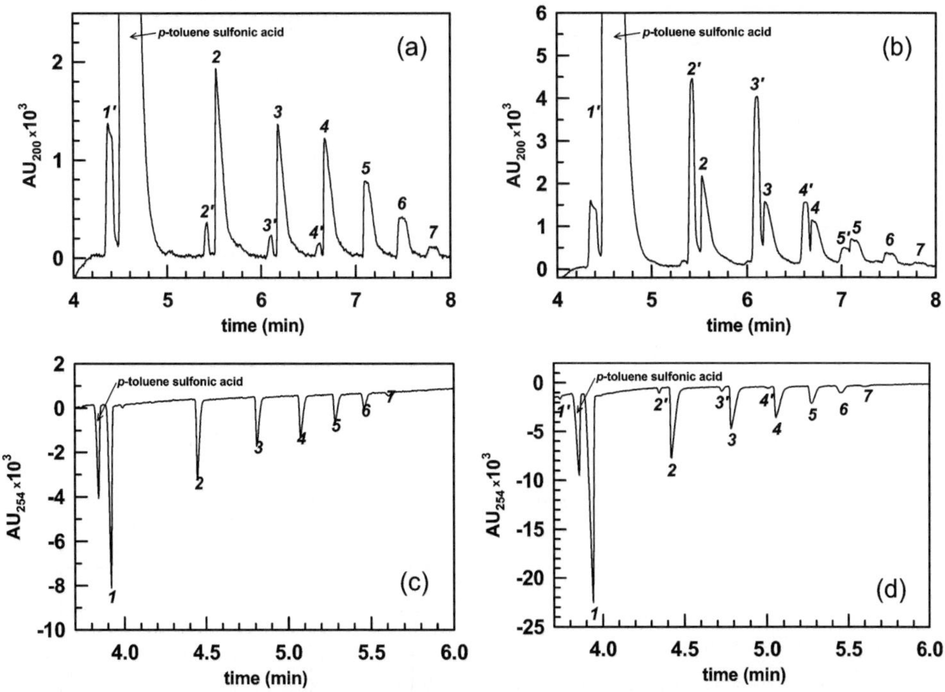

Figure 6. Comparison of direct and indirect UV detection for electropherograms of water-soluble 3-hydroxybutyrate oligomers (DAP method; PB (a) and (b); *p*ABE (c) and (d)). Numbered peaks from 1 to 7 correspond to O(3-HB) oligomers and the ones from 1′ to 5′ to O′(3-HB). DAP method (running buffer, PB): direct UV detection at 200 nm was considered. The capillary was rinsed everyday at starting as follows: 30 min with 1 M HCl, water for 1 min, 1 M NaOH for 30 min and PB for 5 min. Furthermore, the capillary was rinsed before each run using the sequence 0.1% DEAE-Dextran in PB for 5 min, water for 1 min and PB for 3 min. The analyte was injected at the cathodic side by the hydrostatic pressure method (10 s). Separation was performed at 25°C with an applied voltage of −25 kV.

cases, the stability of the EOF migration time by using an EOF marker was of great help to detect this adsorption mechanism. In order to exemplify this adsorption mechanism, we checked degradation of oligomers of lactic acid (OLA100) and of 6-hydroxyhexanoic acid (OCL) in the presence of enzymes: Proteinase K and *Pseudomonas* lipase which were currently used for degradation purposes in the field [25, 79]. We used the DAP method with indirect UV detection. When Proteinase K was present in the ageing solution, analysis could be performed, the EOF being stable. Unfortunately, ageing solutions containing *Pseudomonas* lipase could not be analyzed due to strong protein absorption. This feature has to be connected to the net charge of the protein at the pH of the background electrolyte (pH 8.1). No data on the p*I* of this protein are presently available.

Thus, we considered the fate of a set of water-soluble OLA oligomers in the presence of Proteinase K. The maximal activity of the enzyme was found at 37°C and pH 8.0 in Tris buffer [80]. With such conditions of pH and temperature, a

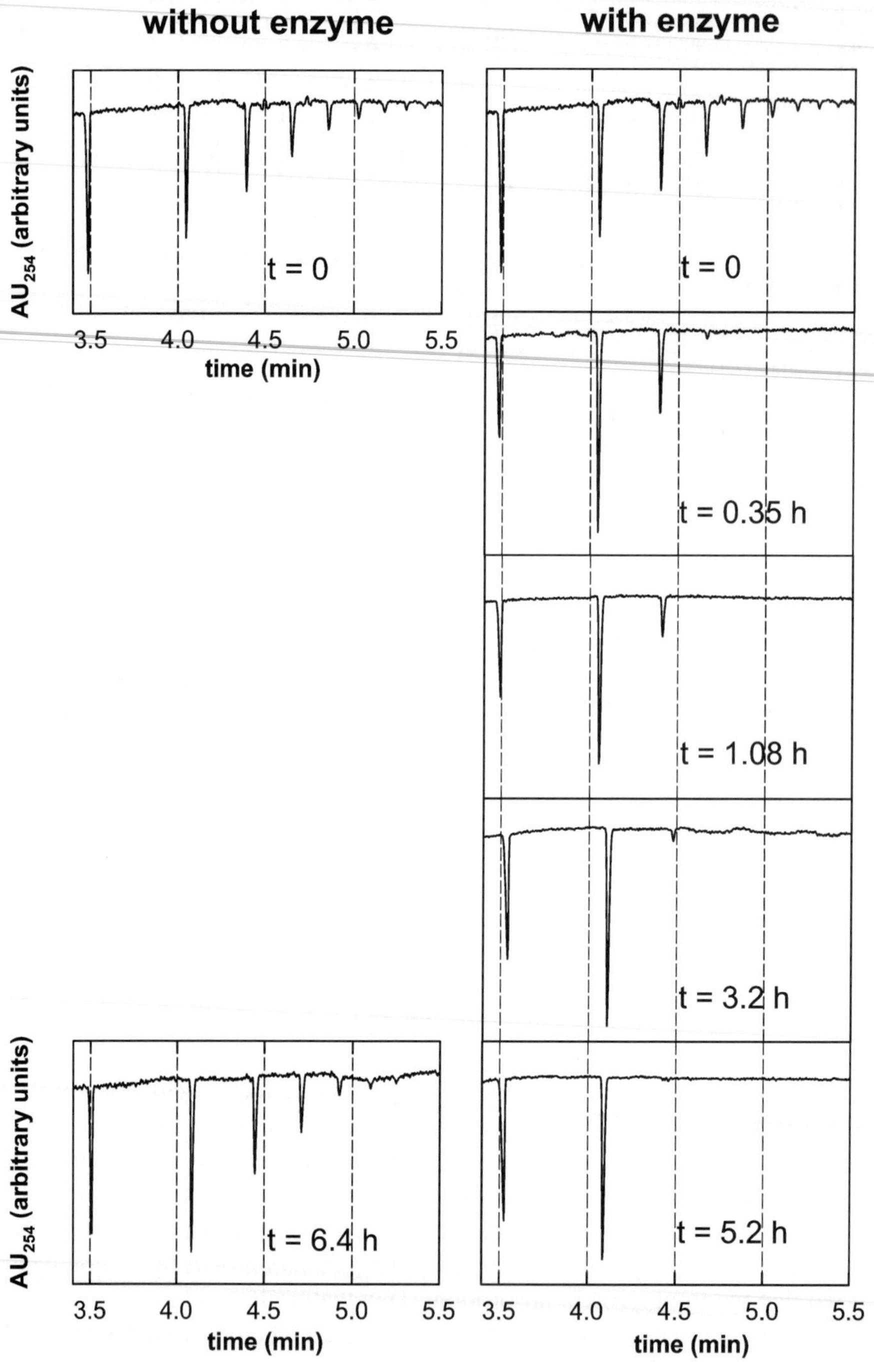

Figure 7. Electropherograms recorded during the ageing of OLA100 in the absence (left column) and in the presence (right column) of Proteinase K after different ageing time. The electropherograms were recorded using the DAP method (running buffer, *p*ABE; PC, DEAE-Dextran).

blank test consisting of the same OLA oligomer solution without enzyme was not significant, because pure hydrolytic degradation was observed. For this reason, we selected the following conditions: room temperature and phosphate buffer at pH 6.8, OLA oligomers being stable in the time window of the experiment. Under these experimental conditions where the activity of the enzyme was not maximum, very rapid degradation of the OLA oligomers in the presence of Proteinase K leading to dimer and monomer in less than 6 h took place (Fig. 7).

Up to now, the PBPEO method, which could be considered as the method of choice for such a separation, has not been validated.

PBPEO method for the monitoring of degradable polyanions

As discussed before, CZE of polyanions had been only performed by using the standard method (fused silica capillary filled with borate buffer). The use of a 'neutral capillary', as in the case of the PBPEO method, might be another possibility that will allow the use of a neutral background electrolyte. We checked this working hypothesis using PB as the background electrolyte and also as the degrading medium for two hydrolytically degradable polyanions: N-acetylneuraminate polymers (NANA, Scheme 5) [55–57] and the natural form of poly(β-malate) (nat-PMLA; kindly supplied by Prof. E. Holler, University of Regensburg, Regensburg, Germany) [81, 82].

Both polyanions (25 mg of dry sample under the sodium salt form) were dissolved in 10 ml of PB. Sodium azide (2 mg) was added to prevent bacterial development. Aliquots of these solutions were allowed to age at 37°C and 62°C.

Figure 8 shows the electropherograms recorded at the beginning of ageing. Peak 1 was due to the azide anion, a strongly UV-absorbing anion. In these ageing experiments, it can be profitably used as internal mobility marker. NANA and nat-PMLA showed two narrow peaks that can be assigned to non-dispersed polyanionic population. The respective migration time of 3.0 and 4.4 min agreed with the large difference of the c/m ratio between the two polyanions (−0.009 and −0.003, respectively; see Fig. 3).

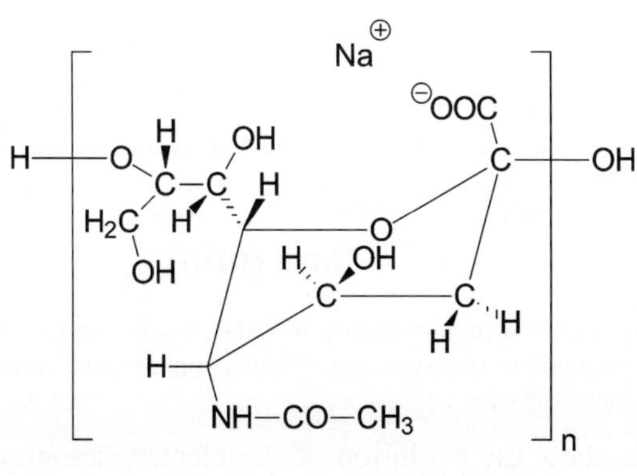

Scheme 5.

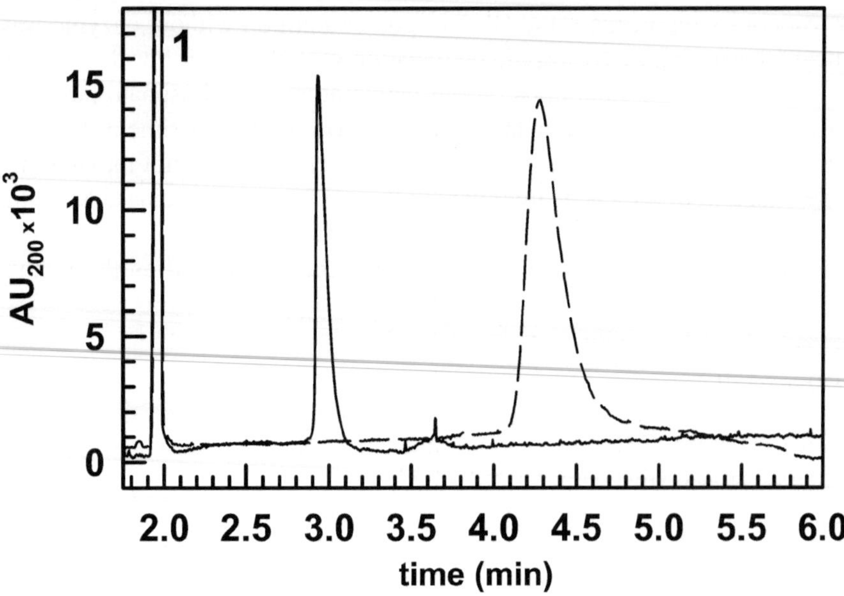

Figure 8. Electropherograms of PMLA (continuous line) and NANA (dashed line) at $t = 0$ recorded using the PBPEO method. Peak 1 is due to azide anion.

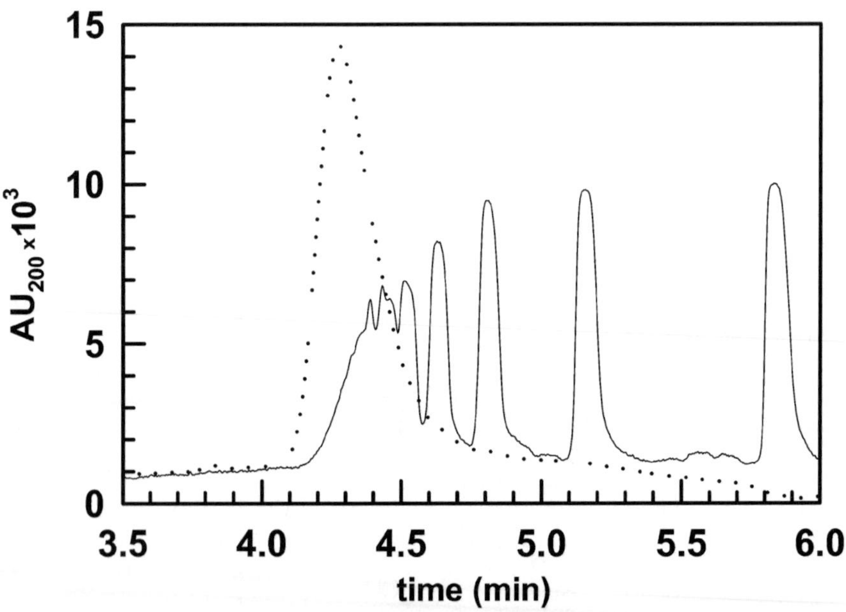

Figure 9. Electropherogram recorded after ageing of NANA solution in PB for 5 days at 62°C using the PBPEO method. The dotted line corresponds to the electropherogram recorded at $t = 0$.

Figures 9 and 10a show the evolution of the electropherograms after ageing of the PB solution at 62°C. As ageing time increased, discrete peaks corresponding to

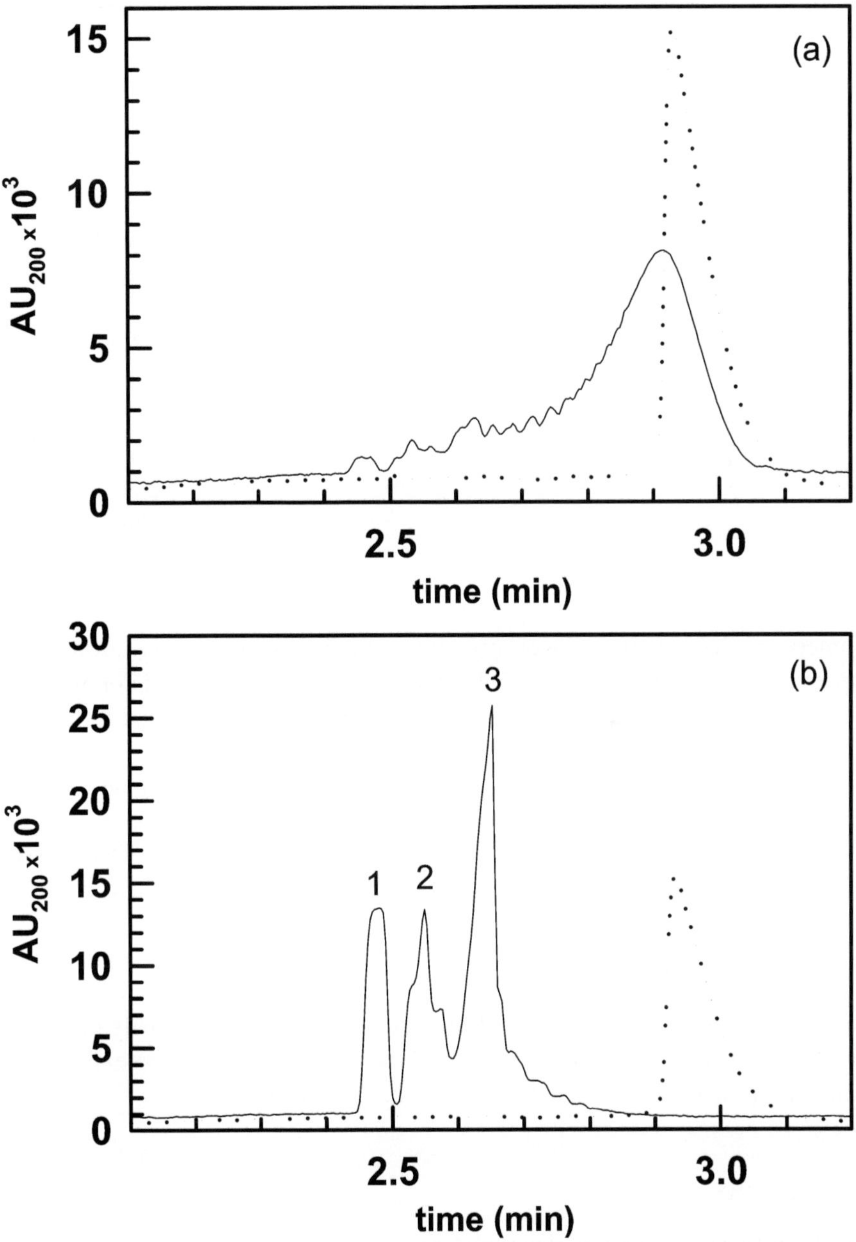

Figure 10. Electropherograms recorded after ageing of nat-PMLA solution in PB for 1 (a) and 9 days (b) at 62°C using the PBPEO method. The dotted line corresponds to the electropherogram recorded at $t = 0$.

oligomers were observed. The appearance of the peak and their intensity depended both on the ageing time and temperature. At 37°C similar electropherograms were recorded for a longer ageing time range (several months). As noted previously,

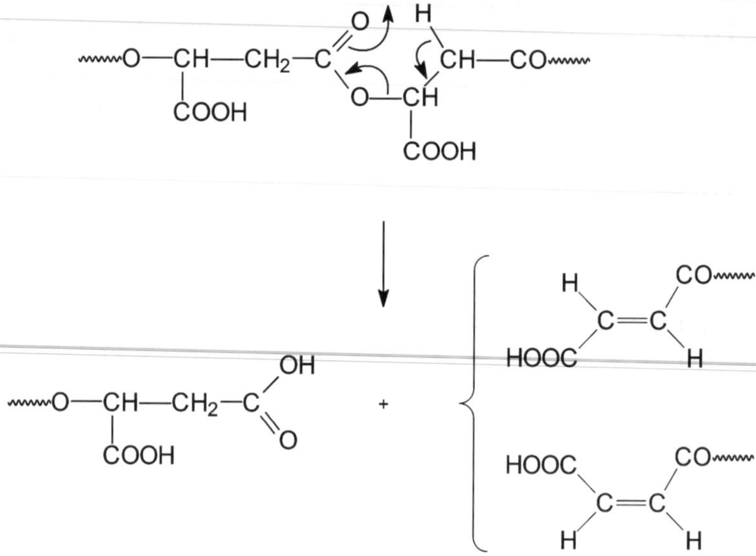

Scheme 6.

oligomers of NANA had lower effective mobility (higher migration time) as compared with the initial polyanion. The reverse was observed for nat-PMLA. Thus, the method using a 'neutral capillary' seemed to be effective to monitor oligoelectrolytic by-products.

After a long ageing time, the method allowed to detect the ultimate degradation products of nat-PMLA. In Fig. 10b three peaks can be observed. After spiking the aged solution with fumaric, maleic and malic acids, peaks 1, 2 and 3 were assigned to fumarate, maleate and malate anions, respectively. Since only dicarboxylic acids without any oligoelectrolyte remained in the final solution, the DAP method with UV detection allowed estimating the composition, which was 96% malate and 4% fumarate + maleate (the last diacids were not totally separated). Although they were only present in small amounts, fumarate and maleate were easily detected due their strong UV absorption at 200 nm, and their appearance can be explained by a β-elimination mechanism (Scheme 6) as in the case of PHB.

CONCLUSIONS

From the different examples described therein, it is evident that the charge/mass ratio plays a major role in the separation. Nevertheless, the open dimers from the ring-opening of LG and DIPAGYL show that separation of isomers can be done. This is also the case for oligoelectrolytes where the c/m ratio does not vary to a great extent in the range of oligomers.

When considering CZE as a tool for separating degradation by-products, the whole system has to be considered, including the potential by-products and the environment which causes the degradation.

Insofar as the four described CZE methods are concerned, each presents advantages and shortcomings connected with adsorption phenomena on the inner wall of the capillary. Finally, it must be pointed out that different and complementary information can be obtained from the same analyte just by selecting the relevant series of rinses in the same capillary.

Acknowledgements

Data presented in this review are a collection of results obtained from undergraduate students during academic projects. I would like to thank all of them and particularly R. Devarieux, A. Athlan and B. Saulnier. I would also thank Dr. M. Vert for all the discussions we had on the subject.

REFERENCES

1. M. Vert, in: *CRC Critical Reviews — Therapeutic Drug Carrier Systems*, S. D. Bruck (Ed.), p. 291. CRC Press, Boca Raton, FL (1986).
2. M. Vert, in: *Controlled Release of Drugs from Polymeric Particles and Macromolecules*, S. S. Davis and E. Illum (Eds), p. 117. Wright IOP, Bristol (1987).
3. L. Brannon-Pepas and M. Vert, in: *Handbook of Pharmaceutical Controlled Release Technology*, D. L. Wise, A. Klibanov, R. Langer, A. Mikos, L. Brannon-Pepas, N. A. Peppas, D. J. Trantalo, G. E. Wnek and M. J. Yaszemski (Eds), p. 99. Marcel Dekker, New York, NY (2000).
4. M. Hakkarainen (Ed.), in: *Degradable Aliphatic Polyesters*, p. 113. Springer, Berlin (2002).
5. D. W. McLellan and P. J. Halling, *J. Chromatogr. A* **445**, 251 (1988).
6. S. P. Sawan and J. J. Barry, *Polymer Prepr.* **29**, 299 (1988).
7. M. L. Maniar, D. S. Kalonia and A. P. Simonelli, *J. Pharm. Sci.* **78**, 858 (1989).
8. H. Abe, Y. Doi, H. Aoki, T. Akehata, Y. Hori and A. Yamaguchi, *Macromolecules* **28**, 7630 (1995).
9. S. Karjomaa, T. Suortti, R. Lempiainen, J. F. Selin and M. Itavaara, *Polym. Degrad. Stabil.* **59**, 333 (1998).
10. S. J. de Jong, E. R. Arias, D. T. S. Rijkers, C. F. van Nostrum, J. J. Kettenes van den Bosch and W. E. Hennink, *Polymer* **42**, 2795 (2001).
11. H. M. Burger, H. M. Muller, D. Seebach, K. O. Bornsen, M. Schar and H. M. Widmer, *Macromolecules* **26**, 4783 (1993).
12. S. Karlsson, M. Hakkarainen and A. C. Albertsson, *J. Chromatogr. A* **688**, 251 (1994).
13. S. Karlsson, C. Sares, R. Renstad and A. C. Albertsson, *J. Chromatogr. A* **669**, 97 (1994).
14. M. Scandola, M.-L. Focarete, G. Adamus, W. Sikorska, I. Baranowska, S. Swierczek, M. Gnatowski, M. Kowalczuk and Z. Jedlinski, *Macromolecules* **30**, 2568 (1997).
15. Z. Jedlindski, G. Adamus, M. Kowalczuk, R. Schubert, Z. Szewczuk and P. Stefanowicz, *Rapid Commun. Mass Spectrom.* **12**, 357 (1998).
16. J. X. Chen and J. A. Gardella, *Macromolecules* **32**, 7380 (1999).
17. G. Gallet, R. Lempiainen and S. Karlsson, *Polym. Degrad. Stabil.* **71**, 147 (2000).
18. B. L. Karger, A.S. Cohen and A. Guttman, *J. Chromatogr. A* **492**, 589 (1989).
19. C. Braud and M. Vert, *Polym. Bull.* **29**, 177 (1992).
20. C. Vidil, C. Braud, H. Garreau and M. Vert, *J. Chromatogr. A* **711**, 323 (1995).
21. C. Braud, R. Devarieux, H. Garreau and M. Vert, *J. Environ. Polym. Degrad.* **4**, 135 (1996).
22. A. Athlan, C. Braud and M. Vert, *J. Environ. Polym. Degrad.* **5**, 243 (1997).
23. C. Braud, R. Devarieux, A. Athlan, C. Ducos and M. Vert, *J. Chromatogr. B* **706**, 73 (1998).

24. J. Sélembarom, K. Marcincinova-Benabdillah, C. Braud and M. Vert, *Polym. Degrad. Stabil.* **68**, 281 (2000).
25. S. Li, M. Tenon, H. Garreau, C. Braud and M. Vert, *Polym. Degrad. Stabil.* **67**, 85 (2000).
26. J.A. Bullock, in: *Handbook of Capillary Electrophoresis Applications*, H. Shintani and J. Polonsky (Eds), p. 657. Blackie, London (1997).
27. R. Kuhn and S. Hoffstetter-Kuhn (Eds), *Capillary Electrophoresis: Principles and Practice*. Springer, Berlin (1993).
28. F. Steiner and M. Hassel, *Electrophoresis* **21**, 3994 (2000).
29. W. Nashabeh and Z. El Rassi, *J. Chromatogr. A* **514**, 57 (1990).
30. C. Chiesa and C. Horvath, *J. Chromatogr. A* **645**, 337 (1993).
31. P. L. Desbène, C. Rony, B. Desmazières and J. C. Jacquier, *J. Chromatogr. A* **608**, 375 (1992).
32. J. Bullock, *J. Chromatogr. A* **645**, 169 (1993).
33. G. Schomburg, D. Belder, M. Gilges and S. Motsch, *J. Capill. Electrophor.* **1**, 219 (1994).
34. Pharmacia Fine Chemicals, *Dextran fractions — Dextran sulphate — DEAE Dextran*. Upplands Grafiskka, Uppsala (1975).
35. M. Stéfani, C. Negrell, C. Braud and M. Vert, *J. Chromatogr. A* **864**, 335 (1999).
36. R. Roldan-Assad and P. Gareil, *J. Chromatogr. A* **708**, 339 (1995).
37. D. C. Sharman, *Analyst* **122**, 709 (1997).
38. H. Katayama, Y. Ishihama and N. Asakawa, *Anal. Chem.* **70**, 2254 (1998).
39. H. Katayama, Y. Ishihama and N. Asakawa, *Anal. Chem.* **70**, 5272 (1998).
40. T. W. Graul and J. B. Schlenoff, *Anal. Chem.* **71**, 4007 (1999).
41. T. Wehr, R. Rodriguez-Diaz and C. Liu, *Adv. Chromatogr.* **37**, 237 (1997).
42. F. T. A. Chen and R. A. Evangelista, in: *Handbook of capillary electrophoresis applications*, H. Shintani and J. Polonsky (Eds), p. 173. Chapman&Hall, London (1997).
43. E. Cordova, J. Gao and G. M. Whitesides, *Anal. Chem.* **69**, 1370 (1997).
44. F. Foret, L. Krivankova and P. Bocek (Eds), *Capillary Zone Electrophoresis*. VCH, Weinheim (1993).
45. J. Preisler and E. S. Yeung, *Anal. Chem.* **68**, 2885 (1996).
46. N. Iki and E. S. Yeung, *J. Chromatogr. A* **731**, 273 (1996).
47. F. Oosawa (Ed.), *Polyelectrolytes*. Marcel Dekker, New York, NY (1972).
48. G. S. Manning, in: *Polyelectrolytes*, E. Sélégny (Ed.), p. 9. Reidel, Dordrecht (1974).
49. H. Engelhardt and O. Grosche, in: *New Developments in Polymer Analytics I*, M. Schmidt (Ed.), p. 189. Springer, Berlin (2000).
50. M. E. Starkweather, D. A. Hoagland and M. Muthukumar, *Macromolecules* **33**, 1245 (2000).
51. J. B. Poli and M. R. Schure, *Anal. Chem.* **64**, 896 (1992).
52. D. A. Hoagland, E. Arvanitidou and C. Welch, *Macromolecules* **32**, 6180 (1999).
53. H. Cottet and P. Gareil, *Electrophoresis* **21**, 1493 (2000).
54. H. Cottet, P. Gareil, O. Theodoly and C. E. Williams, *Electrophoresis* **21**, 3529 (2000).
55. K. Kakehi, A. Hirose, T. Tamai, A. Taga and S. Honda, *Anal. Sci.* **12**, 171 (1996).
56. T. O. T. Terabayashi and Y. Kawanishi, *Carbohydr. Polym.* **29**, 35 (1996).
57. T. Ogawa, T. Terabayashi and Y. Kawanishi, *J. Chromatogr. A* **741**, 295 (1996).
58. Y. Park, S. Cho and R. J. Linhardt, *Biochim. Biophys. Acta* **1337**, 217 (1997).
59. J. T. Smith and Z. El Rassi, *J. High Resol. Chromatogr.* **15**, 573 (1992).
60. I. J. Colton, J. R. Anderson, J. Gao, R. G. Chapman, L. Isaacs and G. M. Whitesides, *J. Am. Chem. Soc.* **119**, 12701 (1997).
61. J. D. Carbeck, I. J. Colton, J. M. Gao and G. M. Whitesides, *Acc. Chem. Res.* **31**, 343 (1998).
62. M. Vert, in: *Biopolymers — Polyesters III: Applications and Commercial Products*, A. Steinbüchel and Y. Doi (Eds), p. 179. Wiley-VCH, Weinheim (2002).
63. P. Dobrzynski, J. Kasperczyk, H. Janeczek and M. Bero, *Polymer* **43**, 2595 (2002).
64. Y. Kimura, K. Shirotani, H. Yamane and T. Kitao, *Macromolecules* **21**, 3338 (1988).
65. A. Löfgren and A.-C. Albertsson, *J. Appl. Polym. Sci.* **52**, 1327 (1994).

66. A. Lofgren, A.-C. Albertsson, P. Dubois, R. Jerome and P. Teyssie, *Macromolecules* **27**, 5556 (1994).
67. T. Yamaoka, Y. Hotta, K. Kobayashi and Y. Kimura, *Int. J. Biol. Macromol.* **25**, 265 (1999).
68. K. Marcincinova-Benabdillah, M. Boustta, J. Coudane and M. Vert, *Macromolecules* **32**, 8774 (1999).
69. K. Marcincinova-Benabdillah, M. Boustta, J. Coudane and M. Vert, in: *Polymers from Renewable Resources: Biopolyesters and Biocatalysts*, C. Scholz and R. A. Gross (Eds), p. 200. American Chemical Society, Washington, DC (2000).
70. K. Marcincinova-Benabdillah, M. Boustta, J. Coudane and M. Vert, *Biomacromolecules* **2**, 1279 (2001).
71. A. Sodergard and M. Stolt, *Prog. Polym. Sci.* **27**, 1123 (2002).
72. S. Hiki, I. Taniguchi, M. Miyamoto and Y. Kimura, *Macromolecules* **35**, 2423 (2002).
73. M. Zinn, B. Witholt and T. Egli, *Adv. Drug. Deliv. Rev.* **53**, 5 (2001).
74. S. F. Williams and D. P. Martin, in: *Biopolymers — Polyesters III: Applications and Commercial Products*, A. Steinbüchel and Y. Doi (Eds), p. 91. Wiley-VCH, Weinheim (2002).
75. Y. Doi and K. Fukuda (Eds), *Biodegradable Plastics and Polymers*. Elsevier, Amsterdam (1994).
76. D. Jendrossek and R. Handrick, *Annu. Rev. Microbiol.* **56**, 403 (2002).
77. D. Seebach and M.G. Fritz, *Int. J. Biol. Macromol.* **25**, 217 (1999).
78. G. E. Yu and R. H. Marchessault, *Polymer* **41**, 1087 (1999).
79. L. Liu, S. Li, H. Garreau and M. Vert, *Biomacromolecules* **1**, 350 (2000).
80. D. F. Williams, *Eng. Med.* **10**, 5 (1981).
81. E. Holler, in: *Hanbook of Engineering Polymeric Materials*, N. P. Cheremisinoff (Ed.), p. 93. Marcel Dekker, New York, NY (1997).
82. B.-S. Lee, M. Vert and E. Holler, in: *Biopolymers — Polyesters I: Biological Systems and Biotechnological Production*, Y. Doi and A. Steinbüchel (Eds), p. 75. Wiley-VCH, Weinheim (2001).

Gels, Genes, Grafts and Giants, pp. 789–808 (2005)

Electrically conductive 2D-PAN-containing surfaces as a culturing substrate for neurons

R. OREN [1], R. SFEZ [2], N. KORBAKOV [2], K. SHABTAI [2], A. COHEN [1], H. EREZ [1], A. DORMANN [1], H. COHEN [4], J. SHAPPIR [3], M. E. SPIRA [1] and S. YITZCHAIK [2,*]

[1] *Department of Neurobiology, the Hebrew University of Jerusalem, Jerusalem 91904, Israel*
[2] *Department of Inorganic and Analytical Chemistry, the Hebrew University of Jerusalem, Jerusalem 91904, Israel*
[3] *School of Engineering, the Hebrew University of Jerusalem, Jerusalem 91904, Israel*
[4] *Department of Physical Chemistry, the Hebrew University of Jerusalem, Jerusalem 91904, Israel*

Received 17 October 2003; accepted 17 March 2004

Abstract—In the present contribution we report on a novel route to synthesize 2D-polyaniline (2D-PAN) on sulfonated-poly(styrene) (SPS) templates by allowing first monomer assembly followed by chemical oxidation to achieve polymerization. We show that *Aplysia* neurons grown on 2D-PAN exhibit an unusual growth pattern and adhesion to this conducting substrate that is manifested by the formation of giant lamellipodia. The lamellipodial domains are characterized by small gap between the plasma membrane and the 2D-PAN substrate (*ca.* 30 nm) and actin rich skeleton resembling the skeleton of growth cones. This behavior is characteristic to uniform substrates containing only 2D-PAN. However, in patterned substrates containing additionally poly(L-lysine) *Aplysia* neurons prefer to extend new neurites on the poly(L-lysine) domains.

Key words: Self-assembled monolayers; 2D-polyaniline (2D-PAN); surface modifications; micro-contact printing (μCp); cultured *Aplysia* neurons.

INTRODUCTION

The search for biocompatible surfaces is originating from the growing need to interface living cells, tissues and organs with 'devices' ranging from biosensors to transplanted organs. Thus, biomaterials surfaces are most often tailored with bioactive molecules for induction of optimal adhesion and cell properties. Among the large variety in surface modification techniques the most common ones are based on

*To whom correspondence should be addressed. Tel.: (972-2) 658-6971. Fax: (972-2) 658-5319. E-mail: sy@cc.huji.ac.il

chemical grafting [1], plasma activation, physisorption, Langmuir–Blodgett transfer [2], or chemisorbed self-assembled monolayers [3] and polyelectrolyte multilayers [4, 5]. Applications such as cell-based biosensors in general [6] and neuron–transistor conjugates in particular [7, 8] rely on extracellular matrix (ECM) components. Neurophilic layers that are most widely used include laminine, polylysine and fibronectin. The ECM-based surface modifications result in electronically insulating layers. An advantage for electrical coupling neurons to devices may be achieved by the development of a substrate that serves dual functions cell-adhesion and electrical conductivity. In the present study we have begun to examine the use of polyaniline (PAN) as such a substrate. Conducting polymers have been widely used in a large range of applications, varying from light emitting diodes (LED) [9] to biosensors [10]. PAN is one of the most investigated conducting polymers since its discovery about two decades ago [11]. The advantage of PAN is its ability to tune its electrical properties easily by proton doping [12] or simple versatile redox reactions [13]. These features made PAN an attractive building block in devices, e.g., as electrodes [14], charge-transfer layers [15] and electro-active layers [16]. Recently we have developed a novel route for the assembly of 2D-conducting polymers on oxide surfaces [17]. The method relies on the electrostatic assembly of a monolayer of monomers (anilinium ions) followed by their *in situ* polymerization (oxidation to PAN).

Layer-by-layer (LBL) assembly processes based on electrostatic or other intermolecular forces present an approach for formation of supramolecular architectures. This technique exploits the ionic interactions of oppositely charged molecules or macromolecules. A new approach to create chemically patterned surfaces utilizing polymers and copolymers was used. Chemical patterns were achieved by using the microcontact printing (μCp) [18].

μCp is a non-photolithographic method that routinely forms patterned SAMs containing regions terminated by different chemical functionalities with submicron lateral dimensions. An elastomeric PDMS stamp is used to transfer molecules of the 'ink' to the surface of the substrate by conformal contact. μCp was first demonstrated for SAMs of alkanethiolates on gold [19]. Patterned SAMs can be used as ultrathin resists in selective wet etching, or as templates to control the wetting, de-wetting, nucleation, growth and template chemical deposition [20] and form polymer microarrays [21]. μCp is attractive because it is simple and inexpensive technique that does not require routine access to clean rooms. Thus far, much of μCp has been directed toward metal, glass or semiconductor surfaces. More recently, other molecular systems such as polymers and ligands have been stamped onto surfaces of reactive alkanethiolate SAMs [22]. Similar approach has been used to stamp proteins onto a chemically activated polymer surface [23].

The ability to pattern polymeric thin films *in situ* through the use of chemically patterned surfaces as templates for ionic multilayer assembly was presented by Hammond *et al.* [24, 25]. They showed that patterned polymeric thin films with high selectivity can be obtained by adjusting ionic strength, pH and polyelectrolyte structure based on electrostatic, hydrogen bond and other secondary interactions. Thus, by creating patterned functional chemistry above the polyelectrolyte surface, alteration of any surface (metal, oxide and organic) covered with at least one layer of polyion is enabled. Here, we also demonstrate the ability to polymerize conducting polymers on predefined patterns and test them for guiding neural growth and development.

In the present report we show that *Aplysia* neurons grown on 2D-PAN exhibit an unusual growth and adhesion to this conducting substrate that is manifested in the formation of giant lamellipodia. The lamellipodial domains are characterized by small gap to the 2D-PAN substrate (*ca.* 30 nm) and actin rich skeleton resembling the skeleton of growth cones. This behavior is characteristic to uniform substrates containing 2D-PAN; however, in patterned substrates containing additionally poly(L-lysine) *Aplysia* neurons preferred the growth and development only on the poly(L-lysine) domains.

MATERIALS AND METHODS

2D-PAN assembly

Substrates and cleaning. Glass, quartz (Chemglass) and n-Si ⟨100⟩ (Virginia Semiconductors) substrates were cleaned and activated by the following cleaning procedure in order to increase the density of hydroxyl (silanol) groups on the surface. The substrates were ultrasonically cleaned in a 0.5% detergent solution at 60°C for 30 min, rinsed three times in triply-distilled water (TDW), followed by immersion in piranha solution (H_2SO_4/H_2O_2, 70 : 30 (v/v)) at 90°C for 1 h. Caution: the piranha solution should be handled with extreme care! After washing three times with TDW the RCA cleaning procedure [26] was employed. This involved sonicating the substrates in ammonium hydroxide solution ($H_2O/H_2O_2/NH_3$, 5 : 1 : 1 (v/v/v)) for 30 min followed by washing three times with TDW, rinsing with acetone and drying for 10 min at 130°C.

Chemicals. Aminopropyltrimethoxysilane (APT, Gelest) and aniline (Aldrich) were vacuum distilled before use. MeOH (Baker), sulfonated poly(styrene) SPS ($M_w = 50 \times 10^3$), Ammonium persulfate $(NH_4)_2S_2O_8$ (Aldrich) was used as received.

Instrumentation. Scanning force microscope (SFM) measurements were performed on a solver P47 (NT-MDT). The successive steps of substrate preparation were observed on dried samples in ambient conditions, using semi-contact mode with integrated Si probes NSC11 (Micromasch).

UV-Vis–NIR spectra were acquired with a Shimadzu spectrophotometer UV-3101PC on ITO, quartz or glass substrates. Contact angle measurements were performed using glass or Si with a Ramé–Hart (model 100) goniometer. XPS

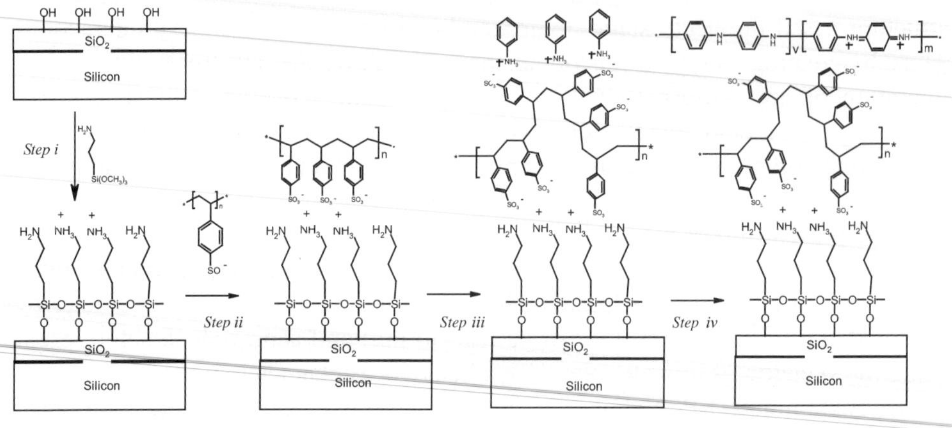

Scheme 1. The synthetic route for the assembly of 2D-PAN monolayer.

spectra were recorded using a monochromatized Al Kα X-ray source on a Kratos Axis-HS instrument. Ellipsometry measurements were carried out on a VB-200 variable angle spectroscopic ellipsometer (VASE, Woollam).

Procedure for 2D-PAN assembly. The consecutive assembly steps (Scheme 1) were carried out on clean and activated substrates. According to the first step (Scheme 1, step i) a clean and dried substrate of Si, quartz, glass or ITO was immersed in a 1 vol% solution of aminopropyltrimethoxysilane (APT) in methanol for 3–5 min followed by methanol rinse as described elsewhere [27]. The coupling layer containing the substrate was immediately placed in a 0.01 M solution of SPS in TDW (Scheme 1, step ii) followed by rinsing with TDW. The use of charged polymers as building blocks, known as electrostatic self-assembly (ESA) is entropically favorable [28]. The electrostatic adhesion of the positively charged anilinium monomers involved immersing the substrate in a 0.01 M anilinium solution (pH 1.5) (Scheme 1, step iii) followed by washing with HCl solution (pH 2). The anilinium-covered substrates were then put in a 5 mM persulfate solution (Scheme 1, step iv). The polymerization was completed after 1 h (Scheme 1, step iv): at this point the sample was rinsed and sonicated for 5 min in HCl solution (pH 2).

Surface patterning

Elastomer (PDMS) preparation. The masters for stamp fabrication were prepared by using conventional photolithography using s1828 photoresist to yield stripes with 10 μm (the space between the stripes is about 260 μm). Sylgard silicone elastomer 184 and Sylgard silicone elastomer curing agent (Dow Corning) were mixed at a ratio of 10 : 1 (w/w). Trapped air was removed by low vacuum. The master was placed in a cylindrically device, and the mixture was casted onto

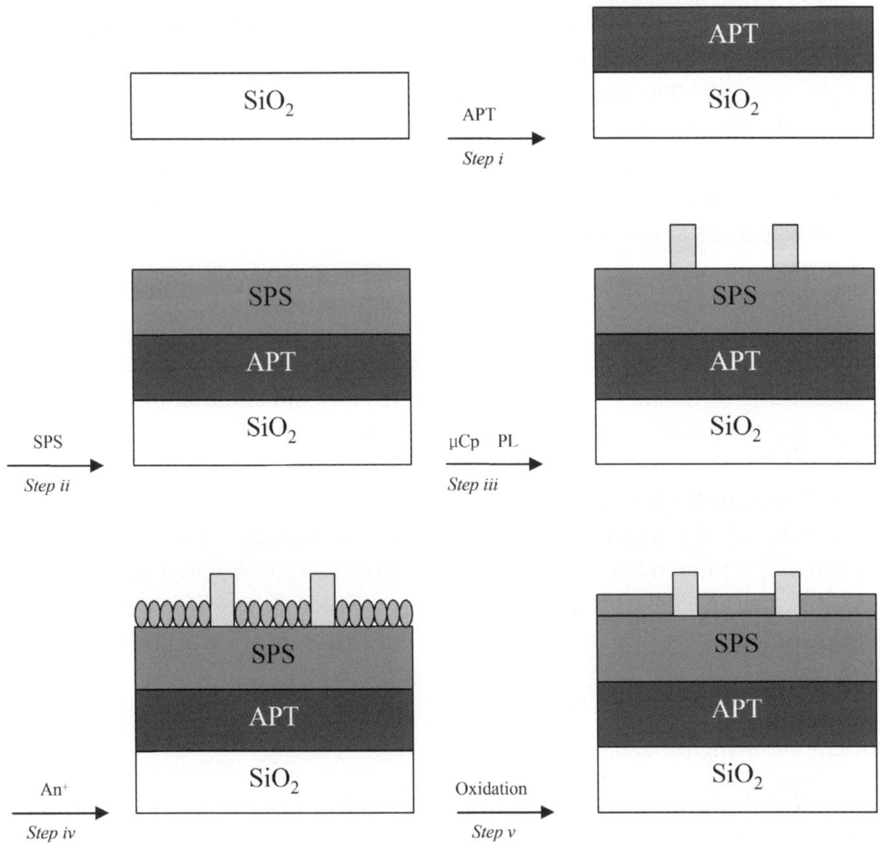

Scheme 2. μCP assembly of 2D-PAN next to poly(L-lysine) pattern.

the master. The polymer was allowed to cure for 2–3 days at room ambient in contact with air. The poly(dimethylsiloxane) (PDMS) was peeled carefully from the master and washed with acetone to remove photoresist residues followed by washing with EtOH and TDW. The stamps were dried using a spin-coater machine. The stamp was cut to about 1×1 cm pieces. The thickness of the stamp was about 0.5–0.7 cm.

Adsorption of template layer. Glass slides were cleaned in piranha solution followed by self-assembly with APT and SPS as described above (Scheme 2, steps i and ii). Poly(L-lysine) was patterned onto the adsorbed slides (according to the stamping procedure) and washed with TDW (Scheme 2, step iii). The patterned slides were soaked in 0.01 M aniline solution dissolved in 0.01 M HCl for 10 min, washed with 0.01 M HCl (Scheme 2, step iv) and polymerized using 0.5 mM ammonium persulfate dissolved in 0.01 M HCl solution for about 30 min, and washed with 0.01 M HCl (Scheme 2, step v). The slides were left to dry at room temperature.

Microcontact printing (μCp). The stamp was soaked with poly(L-lysine) inking solution for 0.5–1.0 h. The soaked stamp was spin-coated for 40 s at 1300 rpm to remove excess ink. The stamp was placed onto the SPS adsorbed slides with lines parallel to the 'length' dimension of the substrate. Light pressure was applied using 10 g weight to ensure contact between the stamp and the substrate. The stamp was removed after 3–4 min. The substrate and the stamp were washed with TDW. The substrate was dried at room temperature.

Experimental solutions

Leibovitz's L-15 Medium (Gibco-BRL) was supplemented for marine species (ms-L15) according to Schacher and Proshansky [29] by the addition of 12.5 g/l NaCl, 6.24 g/l D(+) dextrose, 3.15 g/l anhydrous $MgSO_4$, 344 mg/l KCl, 192 mg/l $NaHCO_3$, 5.7 g/l $MgCl_2 \cdot 6H_2O$ and 1.49 g/l $CaCl_2 \cdot 2H_2O$. Penicillin, streptomycin and amphotericin B (Biological Industries) were added up to final concentrations of 100 units/ml, 0.1 mg/ml and 0.25 μg/ml, respectively. The culture medium consisted of 5% filtered hemolymph obtained from *Aplysia fasciata* (collected along the Mediterranean coast) diluted in ms-L15. Artificial Sea Water (ASW): NaCl 460 mM, KCl 11 mM, $CaCl_2$ 10 mM, $MgCl_2$ 55 mM, HEPES 10 mM, adjusted to pH 7.6.

Neuronal culturing procedures

LUQ neurons from the abdominal ganglia of *Aplysia californica* were isolated and maintained in culture as described previously [30, 31]. Briefly, juvenile *Aplysia californica* (1–10 g) were anesthetized by injection of isotonic $MgCl_2$ solution (380 mM) into the animal's body cavity. Abdominal ganglia were dissected and incubated in ms L-15 containing 1% protease (type IX, Sigma), at 34°C for 1.5–2.5 h. Following the protease treatment the ganglia were desheated. The neurons were manually pulled out along with their original axon with the aid of a sharp glass microelectrode. The neurons were immediately plated in glass-bottom dishes coated either with poly(L-lysine) (Sigma), 2D-PAN, or patterned poly(L-lysine) on 2D-PAN (as described above) containing culture medium. The experiments were performed 2–6 days after plating, at room temperature (21–25°C) after replacing the culture medium with ASW.

Microscope imaging

The system used for confocal imaging consisted of an Olympus microscope IX70 equipped with a planApo 60 × 1.4 NA oil objective and a Bio-Rad Radiance 2000/AGR-3 confocal imaging system. Images of EGFP-tagged actin were acquired by excitation at 488 nm (argon laser) and the emitted fluorescence collected at 500–560 nm. Tetramethyl-rhodamine tubulin was excited by 543 nm (green HeNe laser) and the emitted fluorescence collected at 555–625 nm. Images were collected and processed using laserSharp and laserPix Bio-Rad software, respectively. The figures were prepared using LaserVox, Adobe Photoshop software.

Electrophysiological recordings

To examine the passive and active properties of the neurons a single glass micro-electrode was inserted into the cell body. The electrode was used for both current injection and voltage recording in the current clamp mode. The electrode were filed with 2 M KCl and had a resistance of 3–6 MΩ.

Electron microscopy

Electron microscopy was performed as described previously described [32]. The neurons were fixed in 3% glutaraldehyde (EMS) in ASW at pH 6.9. The cells were then washed in ASW and cacodilate buffer, pH 7.4 (EMS) and were post-fixed in 0.5% osmium tetraoxide (Next Chimica) and 0.8% $K_3Fe(CN)_6$. Dehydration was carried out through a series of ethanol solutions and finally the neurons were embedded in Agar 100 (Agar Scientific). The blocks were sectioned by a microtome (LKB). Thin sections of about 70 nm were stained by lead citrate, tannic acid and uranyl acetate and examined in a Jeol 100CX electron microscope.

RESULTS AND DISCUSSION

Electrostatic self-assembly of 2D-polyaniline

All reaction steps were monitored by UV-Vis–NIR absorption spectroscopy, advancing contact angle measurements, scanning force microscopy (SFM), X-ray photoelectron spectroscopy (XPS) and variable angle spectroscopic ellipsometry (VASE).

Substrate classification. The silicon substrates were characterized by SFM, contact angle, XPS and VASE. The contact angle of the silicon oxide was low (15°) as expected from a hydrophilic hydroxyl-terminated surface of Si's native oxide. The oxide's thickness was determined by VASE to be 20 ± 2 Å. SFM measurement showed a smooth surface with RMS roughness of 1.5 Å for image area of 200 × 200 nm.

Coupling layer deposition (Scheme 1, step i). The contact angle of the clean substrate increased from 15° to 60° upon coupling agent deposition, which agrees with earlier results [33] and probably indicates formation of more than a monolayer of APT. The thickness determined by VASE was 17 ± 2 Å for the coupling layer. The SFM measurements conducted after this treatment showed an inhomogeneous surface, with film height fluctuations of 3–5 Å. These results agreed with the ellipsometric measurements, which is an averaging method and indicates that the assembled monolayer is not perfectly ordered; nevertheless it apparently provides sufficient surface coverage to form a positively charged template for the following electrostatic assembly step (*vide infra*). XPS analysis showed an increase in the carbon peak (284.8 eV) relative to the substrate signal, and a complex N_{1s} line shape (400–402 eV), indicating different degrees of protonation of the amine group (about 40% protonated groups). The observed C/N ratio (5.5) compares favorably with that expected theoretically (5.0).

Surface sulfonation (Scheme 1, step ii). SPS was electrostatically self-assembled on the protonated, positively charged coupling layer. This step diminished dramatically the contact angle from 60° to 30° due to the hydrophilic character of this polyanion containing surface. Spectroscopic ellipsometry analysis gave a monolayer thickness of 10 ± 2 Å, which is the expected thickness for this monolayer. The SFM image obtained for this step (Fig. 1A) showed several different types of features: a web-like structure with features about 30 nm across and 2.5 nm high; a relatively continuous film with holes of 6 nm depth showing the film thickness; and patches of material as much as 100 nm in height. The web-like structure seems to be the dominant feature. The average height as estimated from bearing ratio and rms roughness is 10–12 Å, which corresponds well with the ellipsometric measurements.

XPS analysis of the substrate clearly showed the S_{2s} signal (232.1 eV) and Na_{1s} (1071.5 eV), both suggesting that the SPS with the sulfonate groups and its sodium counter-ions were indeed anchored to the surface. Moreover, the carbon signal was enhanced due to the polymer adhesion, and the N_{1s} signal shifted and showed one distinct peak (at 402.0 eV), suggesting a complete protonation of the APT layer. The S/N ratio and S/Na ratio were found to be 1.2 and 2.5, respectively. From the S/Na ratio it is possible to deduce the fraction of sulfonate groups pointing out from the surface (40%), as these groups are the only ones which did not exchange the Na counter ion and bind to N. From the S/N ratio, the fraction of sulfur that bound APT can be estimated (80%). Combining these two data, the APT-bound S/N ratio is 0.7. This suggests that only about 70% of the amine groups of the coupling layer were used for the polyanion anchoring to the surface. This is in agreement with the AFM and VASE data described above, which point to presence of more than a single monolayer of APT, such that about 30% of the amine groups are unavailable for ion exchange and bonding.

Anilinium monolayer and polymer formation (Scheme 1, step iii). In this stage a monolayer of anilinum-sulfonate is formed. The characteristic CT absorption of anilinium at 251 nm was not observed due to the mostly perpendicular alignment to the surface; however, two peaks at 200 (π–π^* transition) and 230 nm were observed. The spectrum obtained on the surface corresponds totally to what is observed in solution at the same pH (i.e., 3). The thickness was determined as 10±2 Å by VASE, indicating the monomer's adhesion to the rough polyanion template. The SFM

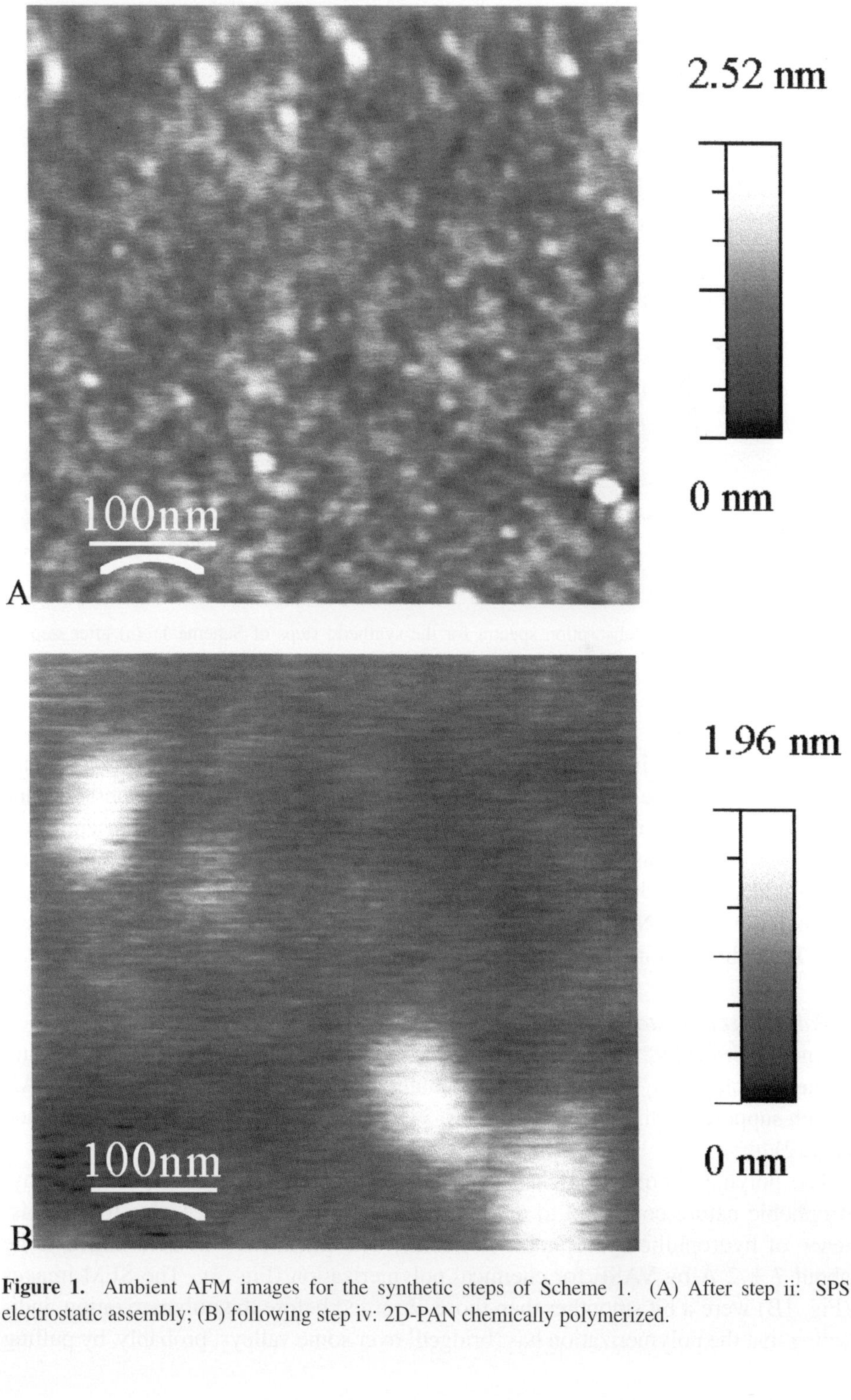

Figure 1. Ambient AFM images for the synthetic steps of Scheme 1. (A) After step ii: SPS electrostatic assembly; (B) following step iv: 2D-PAN chemically polymerized.

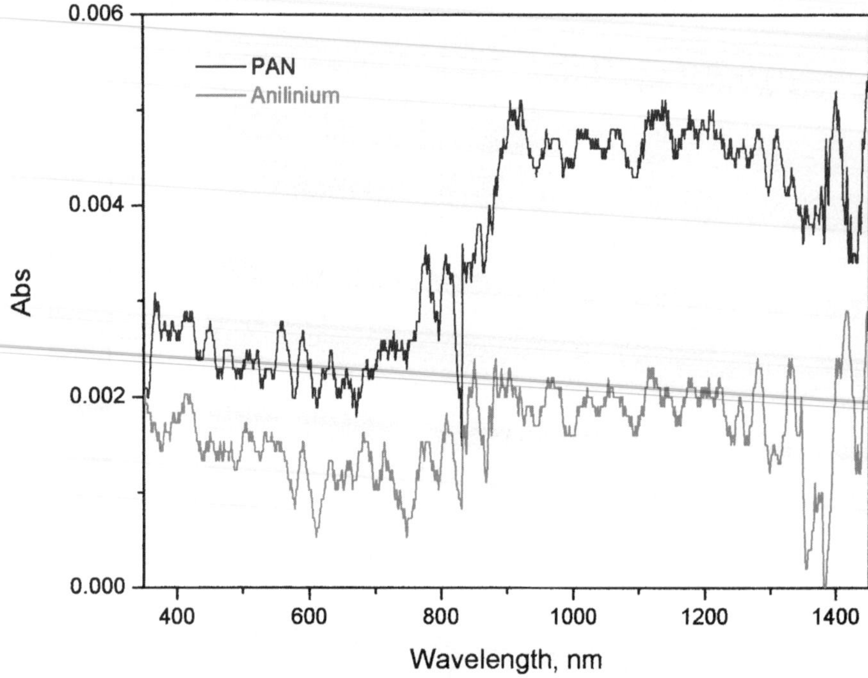

Figure 2. UV-Vis–NIR absorption spectra for the synthetic steps of Scheme 1: (a) after step iii: electrostatic assembly of anilinium on SPS; (b) following step iv: 2D-PAN chemically polymerized.

images were very similar to those obtained for the polyanion adhesion step. This fact suggests that the polyanion serves as a template for the anilinium monomers and no changes in the surface morphology were induced by the exchange of anilinium ions with sodium ions, indicating that the monolayer coverage is homogeneous and regular. XPS data showed a complete disappearance of the Na_{1s} signal and an increase in the N_{1s} and C_{1s} signals, suggesting the substitution of the sodium ions by anilinium ions. The C/N ratio agreed with the theoretical value of 9.5.

Polymerization step (Scheme 1, step iv). Polymerization (oxidation) was performed chemically. After sonication, the UV-Vis–NIR spectroscopy of the substrates displayed a broad spectrum containing the characteristic bands of PAN which supports anilinium oxidation [17, 34] and corresponds to polymer formation (Fig. 2).

The polymer's contact angle changed to about 60°, evidence of increased hydrophobic nature compared to the hydrophilic anilinium surface, due to the existence of hydrophilic/hydrophobic domains. The thickness was determined to be about 7 ± 2 Å by VASE for chemical polymerization (Fig. 3). The SFM images (Fig. 1B) were a bit smoother than those observed before the polymerization, indicating that the polymerization has 'bridged' over some valleys, probably, by pulling

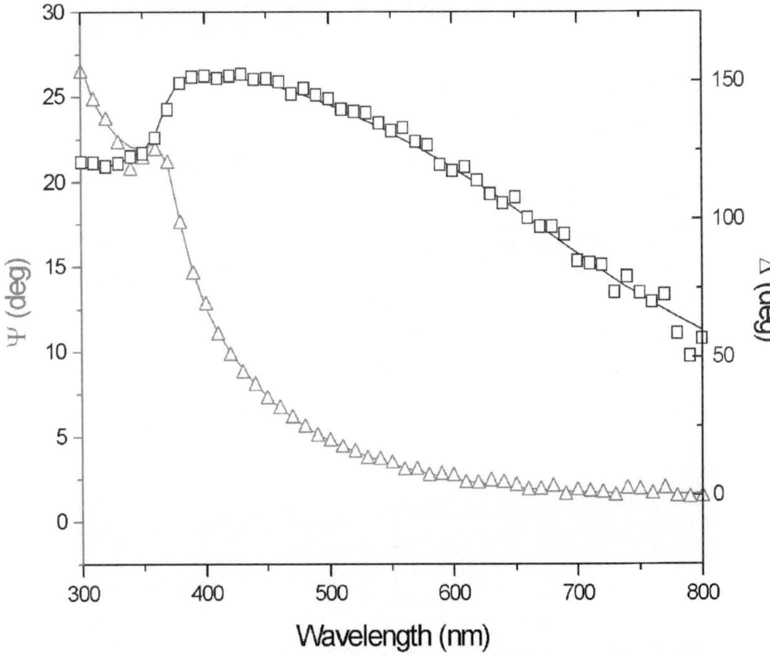

Figure 3. Ellipsometry spectra of 2D-PAN (Scheme 1, step iv).

the monomers from the low topographic zones into the upper ones. XPS analysis showed a significant increase in the N_{1s} signal at 400 eV, indicating a change in the oxidation state of the nitrogen. These results are consistent with polymer formation, in which two different oxidation states exist.

Biocompatibility of the modified surfaces

The effect of equally distributed 2D-PAN surfaces on the morphology and adhesion was studied using Aplysia's LUQ neurons plated on conventional poly(L-lysine) coated cover slides as previously described [29, 30] and on 2D-PAN. The morphology of the neurons was examined every day. Neurons grown on poly(L-lysine), in msL15 and 5% hemolymph maintained a cylindrically shaped of the main axon and its branches and survived for over three weeks (Fig. 1A). In contrast, the number of surviving neurons grown on 2D-PAN and msL15 with 5% hemolymph was lower (60%). The surviving neurons usually extended within a day or two giant lamellipodia (Figs 4B, 5A and 6). Examination of the growth pattern leading to the lamellipodia formation revealed that the giant lamellipodia were formed by expansion of the main cylindrical axon and neurites laterally to form flat structures (Figs 4–7). The giant lamellipodia spontaneously brake up leading to collapse of the structure and neuronal degeneration. The 2D-PAN-induced altered morphology suggests that PAN increases the adhesion of the plasma-membrane to the substrate.

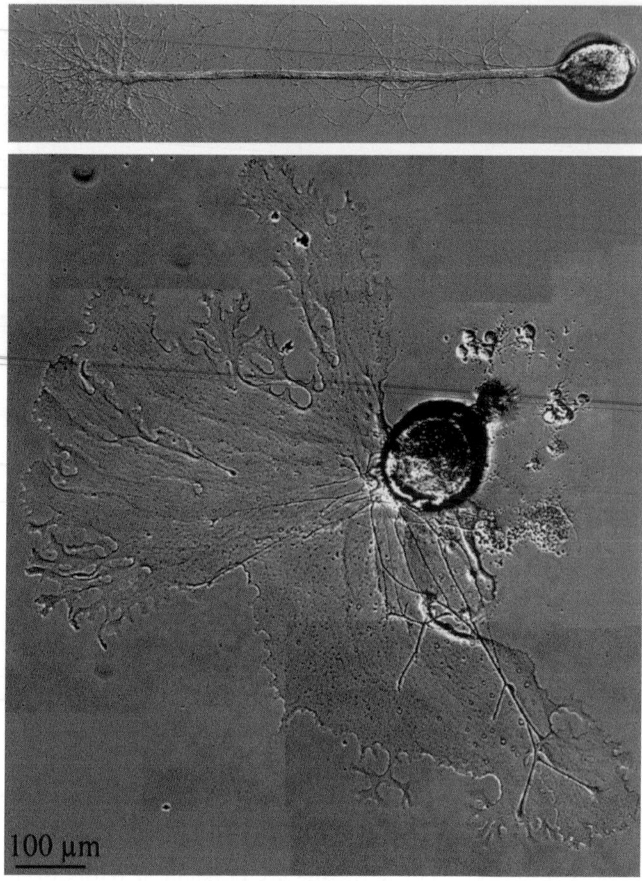

Figure 4. Comparison of the morphology of cultured LUQ neurons grown on (top) poly(L-lysine) and (bottom) 2D-PAN. Whereas the neurons grown on poly(L-lysine) extend neurites and spread out a thin lamellipodium only at the tip of the neurites (growth cones structures), neurons grown on 2D-PAN lose the cylindrical axonal structure and spread out as extensive flat lamellipodial-like structures.

Formation of giant lamellipodia is expected to involve altered organization of the cytoskeleton [35].

Structural modifications induced by the ECM may be associated with alterations in the passive and active membrane properties of the neurons. An example of an experiment in which we examined the excitable membrane properties of a neuron grown on 2D-PAN is shown in Fig. 5. Depolarization of the neuron by an intracellular microelectrode evoked action potentials of normal amplitude and shape. It is important to note that because of the particular shape of neurons cultured on 2D-PAN the neurons are most likely not isopotential [36]. Thus, we cannot state with certainty that the membrane forming the flat lamellipodia is excitable.

In view of the results suggesting that 2D-PAN increase the plasma membrane–substrate adhesion we turned to examined the dimensions of the 'extracellular

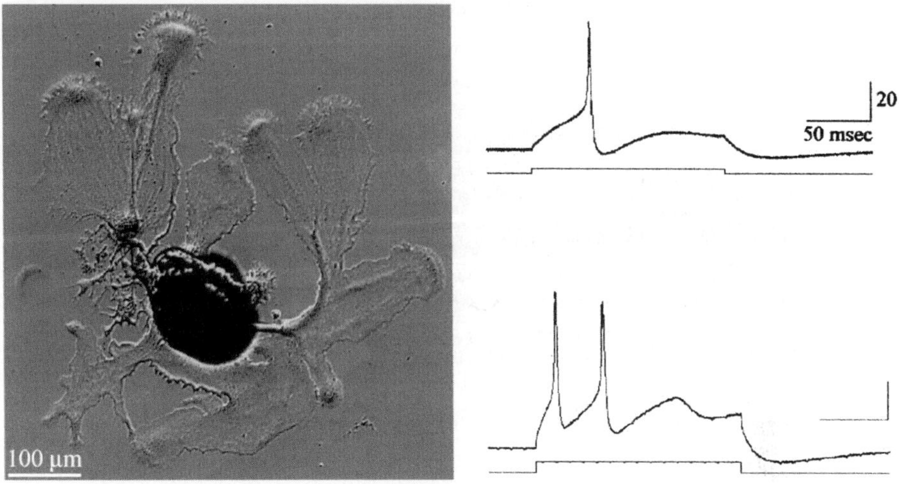

(A) (B)

Figure 5. The passive and active membrane properties of a neuron cultured on 2D-PAN substrate. Light microscope image of the neuron from which the intracellular recordings of the action potentials shown on the right-hand side were taken.

space' formed between the neuron's plasma membrane and the 2D-PAN-coated glass. To that end, the cultured neurons were fixed and embedded in agar-100 within the culturing dish. Embedded individual cells including the cover glass substrate were then cut out from the culture dish. The glass substrate was removed by its incubation in fluoric acid. Thereafter, the agar block (which is unaffected by the acid) is re-embedded in a flat mold (second embedding).

When cross sections are prepared a dark line is visualized at the interface between the agars cured during the first and the second embedding phases. Examination of the cross sections (Fig. 6) revealed that the gap between the plasma membrane and the substrate of neurons grown on 2D-PAN ranges between 20 and 100 nm. The space contains dark material which most likely represents secreted proteins. Interestingly, the extracellular gap between the plasma membranes of lamella grown on top of each other may be even smaller (Fig. 5A). Comparison of the gaps formed by neurons grown on 2D-PAN and poly(L-lysine) did not reveal significant differences. This observation implies that, while PAN increases the adhesion of the neuron and thereby alters its morphogenesis, it does not reduce the dimensions of the gap between the glass substrate and the neurons membrane.

The flat lamellipodia formed on 2D-PAN substrate contain the same repertoire of sub cellular organelles found in axons and growth cones of neurons grown on poly(L-lysine). Nevertheless, microtubules that serve as a major cytoskeletal component in axons dendrites and neurites (but less in growth cones lamellipodia) were only occasionally observed (Fig. 5A). To better understand the effects of 2D-PAN on the neurons cytoskeleton we used confocal microscope imaging of actin and microtubules.

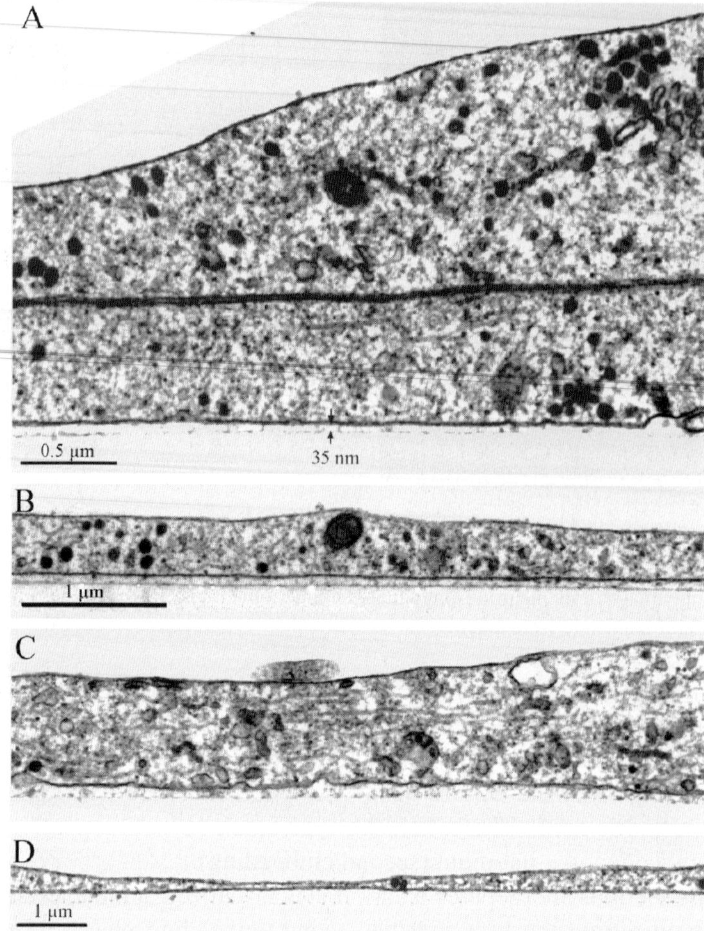

Figure 6. Electron micrographs of a lamellipodia extended on 2D-PAN substrate. (A–D) Enlargements of sections prepared from various regions. Note that the space between the substrate and the plasma membrane facings it is around 35 nm (arrows in A). Note that in some but not all sections the axoplasm contains microtubules.

Confocal microscope imaging of microtubules and actin

To study the distribution of microtubules and actin network in neurons grown on poly(L-lysine) and 2D-PAN we labeled the microtubules by intracellular microinjection of tetramethylrhodamin tubulin and actin by intracellular microinjection of EGFP-actin mRNA. 12–24 h after the injection of the fluorescent probes microtubules appeared as longitudinal bundles within the axoplasm. In neurons grown on poly(L-lysine), L-15 and 5% hemolymph, EGFP labeled actin was homogeneously distributed within the bulk of the axoplasm (Figs 7A′ and 8). Large numbers of EGFP-actin labeled spots were observed along the plasma membrane of the main axon facing the substrate (Figs 7A and 8). These actin spots most likely represent

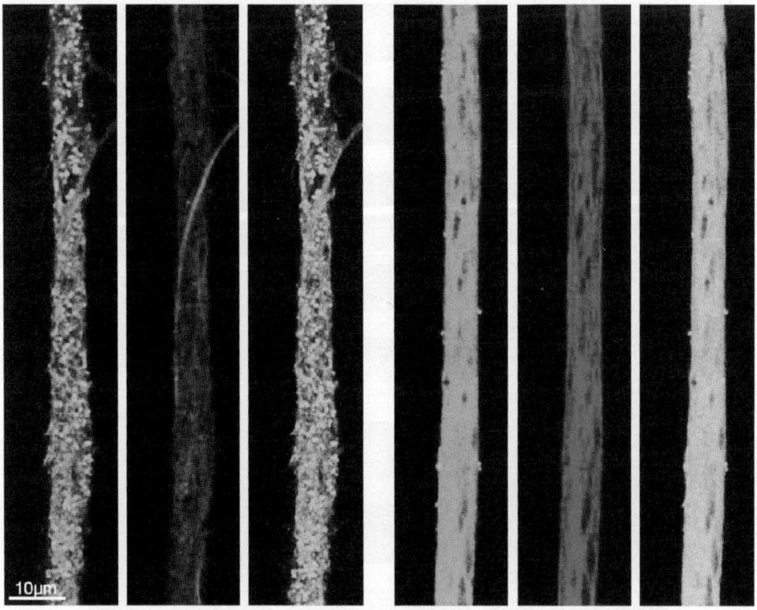

Figure 7. Confocal microscope images of an axon formed on poly(L-lysine) substrate. (A–C) at the level of the substrate; (A′–C′) 5 μm above the substrate. (A and A′) GFP-actin; (B and B′) rhodamine-labeled tubulin; (C and C′) merge image of actin and tubulin. Note the actin spots at the substrate level.

adhesion plaques. Labeled microtubules are distributed in bundles within the axoplasm facing the substrate as well as within the axons center (Fig. 7B and 7B′). The merged actin tubulin image of Fig. 7C′ clearly reveals that both components occupy the axoplasm. In contrast, in neuron grown on 2D-PAN L-15 for 24–48 h the actin and tubulin are separated forming a structure reminiscent of a growth cone (Figs 9 and 10). The center of the axon is occupied by tubulin (Figs 9 and 10) and the actin is localized to the edges of the flat lamellipodium (Figs 9 and 10). Few microtubules invade the axoplasm between the lamellipodial edge and the center. The complete sequence of events underlying the 2D-PAN-induced structural remodeling was not investigated yet, the present results however suggest that 2D-PAN induce local growth processes that transform a differentiated axonal structure into that of a growth cone. This process is also reflected by the flattening of the axon's diameter (compare the computer-reconstructed cross-sections of Figs 8 and 10).

In summary, the observations described above suggest that equally distributed 2D-PAN increase the adhesion of the plasma membrane to the substrate and that probably because of the mechanical tension produced by the abnormal growth pattern of the lamellipodia leads to injury and neuronal degeneration.

To compare the properties of poly(L-lysine) and 2D PAN as growth substrates for cultured *Aplysia* neurons we cultured the neurons on alternating strips of 2D PAN

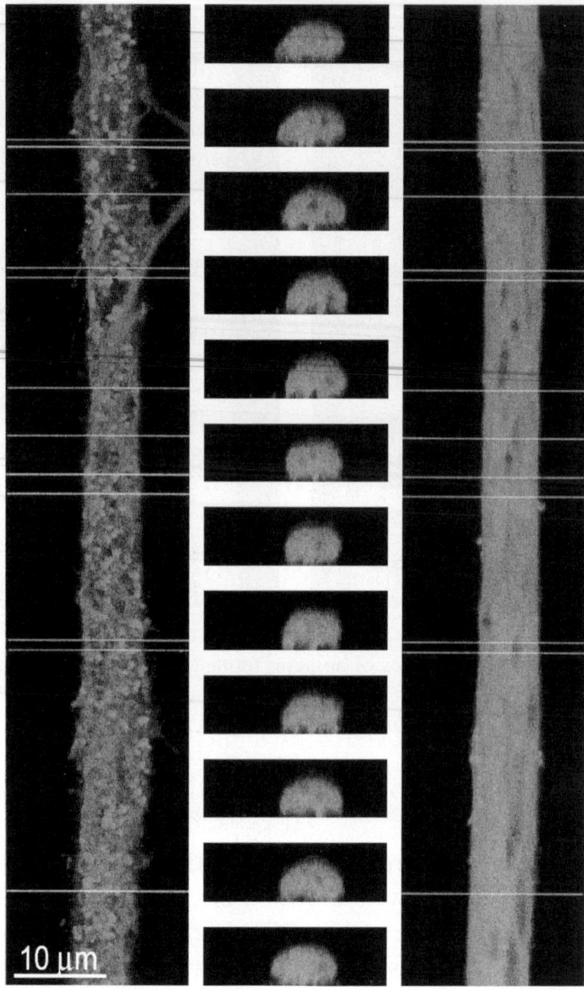

Figure 8. Computer reconstruction of cross-sections along the axon shown in Fig. 7. Note the cylindrical shape of the axon and the location of the actin spots.

and poly(L-lysine). For the experiment of Fig. 11 the glass substrate was patterned by alternating strips of 260 μm 2D PAN and 10 μm poly(L-lysine). The pattern was formed on SPS where first the poly(L-lysine) was stamped followed by filling the gap with anilinium monomers that were next chemically oxidized to give the 2D-PAN patterns (see Scheme 2, detailed characterization of these patterns will be given elsewhere (data not shown)). Clearly the neurites that extended after culturing grow along the poly(L-lysine) strips avoiding the 2D-PAN domains. Nevertheless, it should be noted that un-patterned growth is also observed around the neuronal cell bodies suggesting that neurites can grow on the PAN substrate.

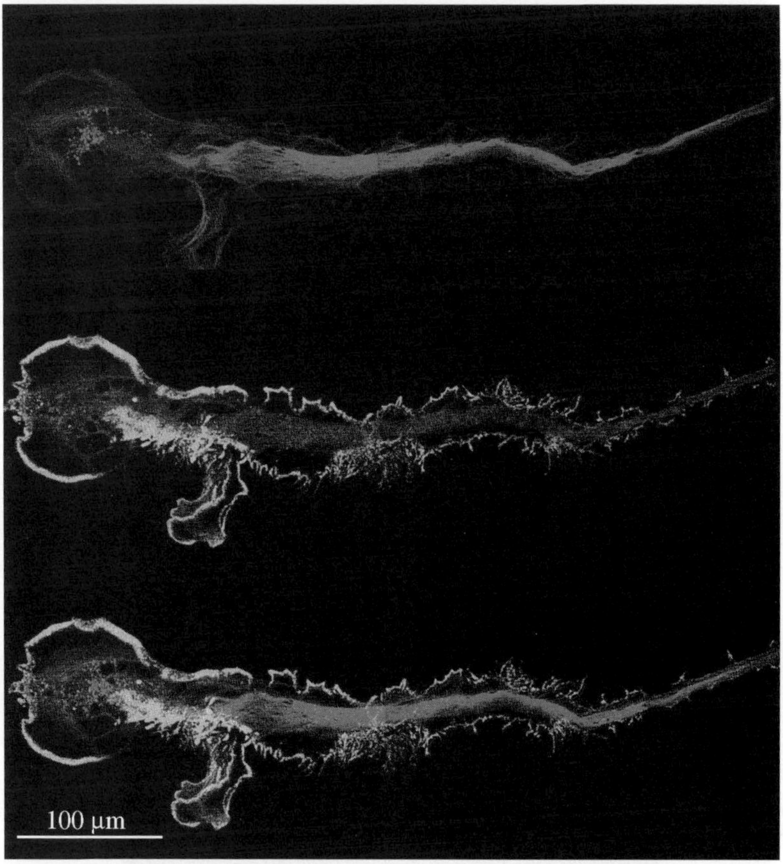

Figure 9. Reconstructed confocal image of a neuron grown on 2D-PAN for 24 h. The cell body is on the left hand side of the image. Note that in contrast to the axon grown on poly(L-lysine) (Fig. 7) the perimeters of the neuron grown on 2D-PAN reach within GFP-actin, while the center of the axon contains longitudinally-oriented microtubules.

CONCLUSIONS

The novel route to synthesize 2D-PAN on SPS templates by first allowing monomer assembly followed by chemical oxidation to achieve polymerization proved to be an easy and useful method for the assembly of nanometers thick layers on polyanion containing template surfaces. This method also proved to be compatible with the μCp method thus 2D-PAN can be integrated also into patterned surfaces. It was shown that *Aplysia* neurons grow on 2D-PAN exhibit an unusual growth and adhesion to this conducting substrate that is manifested in large area lamellipodium formation. The lamellipodia domains are characterized with small gap to the 2D-PAN substrate (*ca.* 30 nm) and actin-rich skeleton resembling the skeleton of growth cones. This behavior is characteristic to uniform substrates containing only 2D-PAN; however, in patterned substrates containing additionally poly(L-lysine)

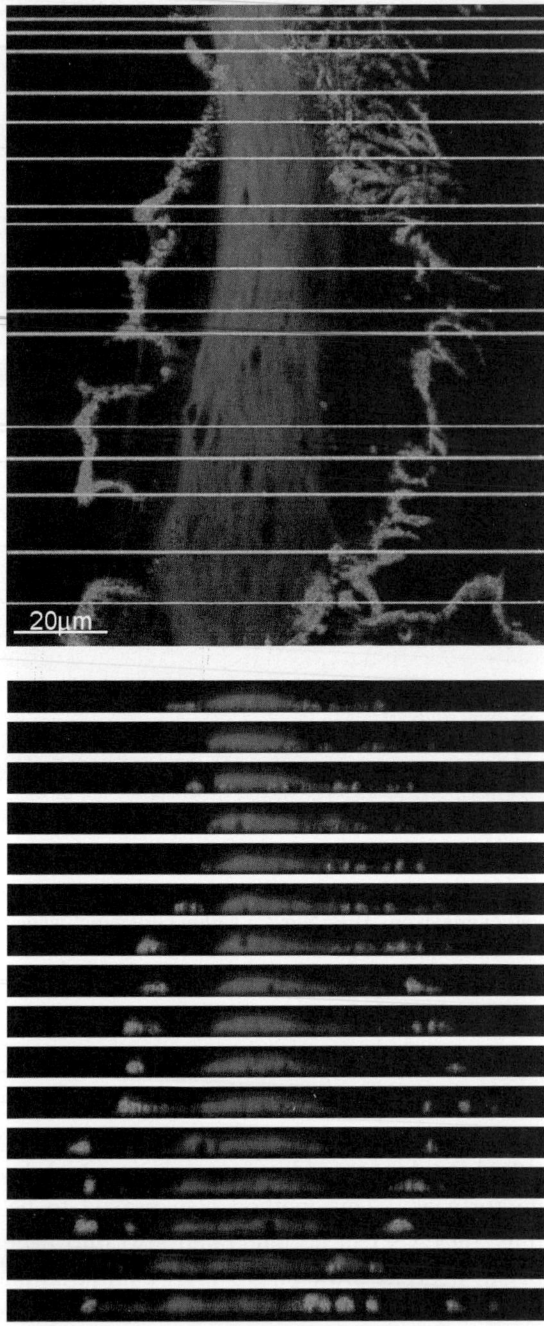

Figure 10. Computer reconstruction of cross-sections along the axon shown in Fig 9. Note the flat shape of the axon and the location of the actin to the perimeters of the axon that face the 2D-PAN substrate.

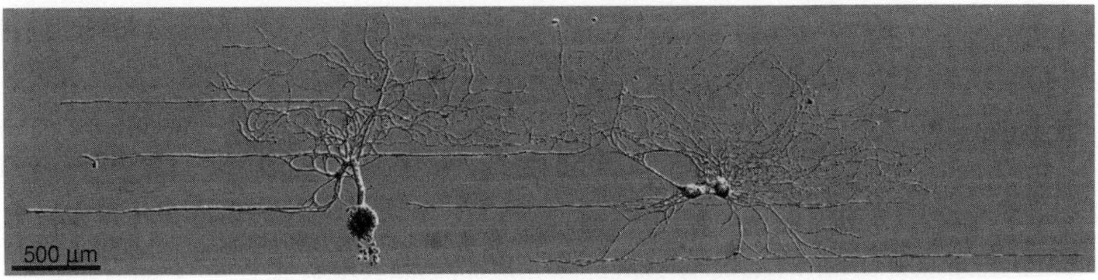

Figure 11. Preferential growth of new neurites on poly(L-lysine) substrate over 2D-PAN. A glass cover slide was patterned by alternating strips of 10 μm poly(L-lysine) and 260 μm 2D-PAN. When LUQ neurons were cultured on the matrix outgrowth of new neurites was observed mainly along the poly(L-lysine)-coated strip. Note, however, that short neurites also grow in areas with PAN stripe.

Aplysia neurons preferred the growth and development only on the poly(L-lysine) domains. On going experiments are designed to optimize PAN's oxidation state (doping level) to neuronal growth.

Acknowledgements

The project is funded by the European Union Fifth Program contract number: IST-1999-29091 (NEUMIC). R.S. thanks the Israeli ministry of science for an Eshkol post-doctoral fellowship. Part of the research was carried out in the Charles E. Smith Family Laboratory for Collaborative Research in Psychobiology. M.E.S. is the Levi DeViali Prof. in neurobiology. R.O. and R.S. contributed equally to the paper.

REFERENCES

1. J. R. Bain and A. S. Hoffman, *Biomaterials* **23**, 3347 (2002).
2. L. Li, J. Waida, L. Manna and A. P. Alivisatos, *Nano Lett.* **2**, 557 (2003).
3. V. Percec, M. Giodde, T. K. Bera, Y. Miura, I. Shiyanovskaya, K. D. Singer, V. S. K. Balagurusamy, P. A. Heiney, I. Schnell, A. Rapp, H.-W. Spiess, S. D. Hudson and H. Duan, *Nature* **419**, 384 (2002).
4. S.-W. Lee, C. Mao, C. E. Flynn and A. M. Belcher, *Science* **296**, 892 (2002).
5. N. Jessel, F. Atalar, P. Lavalle, J. Mutterer, G. Decher, P. Schaaf, J.-C. Voegel and J. Ogier, *Adv. Mater.* **15**, 692 (2003).
6. T. H. Park and M. L. Shuler, *Biotechnol. Prog.* **19**, 243 (2003).
7. P. Fromherz, *Ber. Bunsenges. Phys. Chem.* **100**, 1093 (1996).
8. A. Offenhausser, J. Ruhe and W. Knoll, *J. Vac. Soc. Technol.* **13**, 2606 (1995).
9. R. H. Friend, R. W. Gymer, A. B. Holmes, J. H. Burroughes, R. N. Marks, C. Taliani, D. D. C. Bradley, D. A. Dos Santos, J. L. Bréda, M. Lögdlund and W. R. Salaneck, *Nature* **397**, 121 (1999).
10. H. Sangodkar, S. Sukeerthi, R. S. Srinivasa, R. Lal and A. Q. Contractor, *Anal. Chem.* **68**, 779 (1996).
11. A. G. MacDiarmid, J. Chiang, M. Halpern, W. Huang, S. Mu, N. L. D. Somasiri, W. Wu and S. I. Yaniger, *Mol. Cryst. Liq. Cryst.* **121**, 173 (1985).
12. J. C. Chiang and A. G. MacDiarmid, *Synth. Met.* **13**, 193 (1986).
13. E. W. Paul, A. J. Rico and M. S. Wrighton, *J. Phys. Chem.* **89**, 1441 (1985).

14. G. Gustafsson, Y. Cao, G. M. Treacy, F. Klavetter, N. Colnari and A. J. Heeger, *Nature* **357**, 477 (1992).
15. P. K. H. Ho, M. Granstrom, R. H. Friend and N. C. Greenham, *Adv. Mater.* **10**, 769 (1998).
16. A. Talaie, J. Y. Lee, Y. K. Lee, J. Jang, J. A. Romagnoli, T. Taguchi and E. Mareder, *Thin Solid Films* **363**, 163 (2000).
17. R. Sfez, L. De-Zhong, I. Turyan, D. Mandler and S. Yitzchaik, *Langmuir* **17**, 2556 (2001).
18. J. L. Wilbur, A. Kumar, E. Kim and G. M. Whitesides, *Adv. Mater*. **6**, 600 (1994).
19. A. Kumar and G. M. Whitesides, *Appl. Phys. Lett.* **63**, 2002 (1993).
20. C. B. Gorman, H. A. Biebuyck and G. M. Whitesides, *Chem. Mater.* **7**, 526 (1995).
21. E. Kim, G. M. Whitesides, L. K. Lee, S. P. Smith and M. Prentiss, *Adv. Mater*. **8**, 139 (1996).
22. L. B. Goetting, T. Deng and G. M. Whitesides, *Langmuir* **15**, 1182 (1999).
23. Z. Yang and A. Chilkoti, *Adv. Mater*. **12**, 413 (2000).
24. S. L. Clark and P. T. Hammond, *Adv. Mater*. **10**, 1515 (1998).
25. S. L. Clark, E. S. Handy, M. F. Rubner and P. T. Hammond, *Adv. Mater*. **11**, 1031 (1999).
26. *RCA Engineering* **28** (1983).
27. M. Lahav, T. Gabriel, A. N. Shipway and I. Wilner, *J. Am. Chem. Soc.* **121**, 258 (1999).
28. G. Decher and J. D. Hong, *Macromol. Chem. Macromol. Symp.* **46**, 321 (1991).
29. S. Schacher and E. Proshansky, *J. Neurosci.* **3**, 2403 (1983).
30. M. E. Spira, D. Benbassat and A. Dormann, *J. Neurobiol.* **24**, 300 (1993).
31. M. E. Spira, A. Dormann, U. Ashery, M. Gabso, D. Gitler, D. Benbassat, R. Oren and N. E. Ziv, *J. Neurosci. Methods* **69**, 91 (1996).
32. M. E. Spira, R. Oren, A. Dormann and D. Gitler, *J. Comp. Neurol.* **457**, 293 (2003).
33. V. Burtman, A. Zelichenok and S. Yitzchaik, *Angew. Chem. Int. Ed.* **38**, 2041 (1999).
34. I. Turyan and D. Mandler, *J. Am. Chem. Soc.* **120**, 10773 (1998).
35. L. M. Grunebaum and T. J. Carew, *Learning Memory* **6**, 292 (1999).
36. U. Ashery, R. Penner and M. E. Spira, *Neuron* **16**, 641 (1996).

Gels, Genes, Grafts and Giants, pp. 809–826 (2005)

Controlled synthesis of biodegradable lactide polymers and copolymers using novel *in situ* generated or single-site stereoselective polymerization initiators

ZHIYUAN ZHONG, PIETER J. DIJKSTRA and JAN FEIJEN *

Department of Polymer Chemistry and Biomaterials, and Institute for Biomedical Technology, Faculty of Science and Technology, University of Twente, P.O. Box 217, 7500 AE Enschede, The Netherlands

Received 25 September 2003; accepted 18 February 2004

Abstract—Polylactides and their copolymers are key biodegradable polymers used widely in biomedical, pharmaceutical and ecological applications. The development of synthetic pathways and catalyst/initiator systems to produce pre-designed polylactides, as well as the fundamental understanding of the polymerization reactions, has continuously been an important topic. Here, we will address the recent advances in the ring-opening polymerization of lactides, with an emphasis on the highly versatile *in situ* generated initiator systems and single-site stereoselective initiators. The *in situ* generated initiators including *in situ* formed yttrium, calcium and zinc alkoxides all have been shown to bring about a rapid and living polymerization of lactides under mild conditions, which facilitated the preparation of a variety of advanced lactide-based biomaterials. For example, well-defined di- and tri-block copolymers consisting of hydrophilic poly(ethylene glycol) blocks and hydrophobic polyester blocks, which form novel biodegradable polymersomes or biodegradable thermosensitive hydrogels, have been prepared. In the past few years, significant progress has also been made in the area of stereoselective polymerization of lactides. This new generation of initiators has enabled the production of polylactide materials with novel microstructures and/or properties, such as heterotactic (−*RRSSRRSSRRSS*−) polylactide, crystalline syndiotactic (−*RSRSRSRSRSRS*−) polylactide and isotactic stereoblock ($-R_nS_nR_nS_n-$) polylactide, exhibiting a high melting temperature. The recently developed polymerizations using *in situ* generated initiators and stereoselective polymerizations have no doubt opened a brand-new avenue for the design and exploration of polylactides and their copolymers.

Key words: Polylactide; ring-opening polymerization; *in situ* polymerization; stereoselective polymerization; living polymerization; biodegradable polymers.

*To whom correspondence should be addressed. Tel.: (31-53) 489-2976. Fax: (31-53) 489-3823. E-mail: j.feijen@ct.utwente.nl

INTRODUCTION

Polylactides (PLAs), featuring biodegradability, biocompatibility and non-toxicity, are particularly useful for temporary medical devices, e.g. drug-delivery systems, resorbable sutures, medical implants and scaffolds for tissue engineering [1–3]. For example, in tissue engineering polylactides and copolymers of lactide and glycolide have appeared as one of the few ideal matrices [4]. These (co)polyesters can readily be processed into open, porous, three-dimensional scaffolds using different processing techniques, have sufficient mechanical strength to maintain the shape of the scaffold, thereby guiding tissue growth, are completely resorbable leaving only the newly formed natural tissue and do not induce a tissue response in the host. The surface of these polymeric scaffolds could be modified to provide optimal cell–scaffold interactions. Furthermore, these polymers could also be manufactured to provide controlled release of hormones and/or growth factors.

PLAs are a first family of polymers made entirely from annually renewable resources such as corn and sugar beets, are fully compostable and have performances comparable to conventional petro-chemical based polyolefins [5]. They have, therefore, also attracted much attention as sustainable environmentally friendly plastics used, e.g. as packaging materials, agricultural films and fibers [3, 5].

The PLA polymers can be prepared by direct polycondensation of lactic acid (Scheme 1a). However, the equilibrium nature of condensation reactions combined with the difficulties in removing trace amount of water during the late stage of

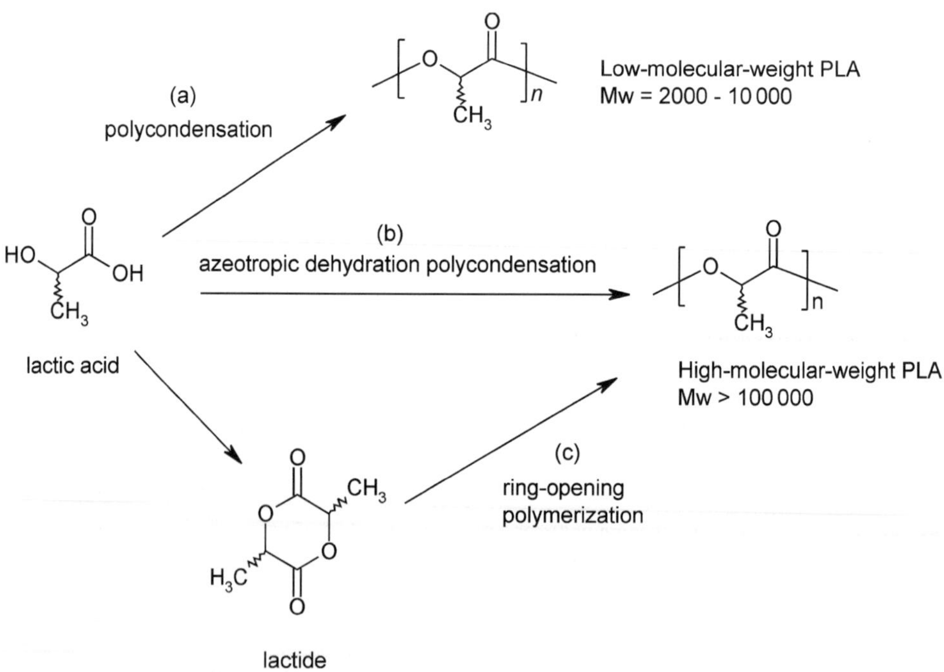

Scheme 1. Synthetic pathways to lactide polymers.

polymerization in general results in PLAs with a low molecular mass and thereby an insufficient strength for practical use. Recently, Mitsui Toatsu Chemicals invented an azeotropic distillation process for efficient dehydration, which enables the production of PLAs with molecular masses ranging from 100 up to 300 kDa employing direct polycondensation (Scheme 1b) [6, 7]. Nevertheless, this polymerization process suffers several drawbacks, among which are a need for a considerable amount of catalyst to achieve an acceptable polymerization rate, a need for a relatively large reactor as a result of low volume efficiency and a need for evaporation and recovery of the high boiling organic solvent.

In contrast, the ring-opening polymerization (ROP) of lactide (LA) in the presence of a catalyst and/or initiator has proven more versatile and efficient in producing PLA polymers (Scheme 1c) [8]. Stannous octoate has been the most often used catalyst because of its good solubility in the melt of LA monomers, high catalytic activity, low levels of racemization, as well as US FDA approval as a food additive. For example, the major PLA supplier Cargill-Dow LLC produces PLA high polymer using a stannous-octoate-catalyzed LA polymerization in the melt. The most probable polymerization mechanism involves tin(II) alkoxide growing species, presumably formed from stannous octoate and hydroxyl-containing impurities such as trace amounts of alcohol and water in the polymerization mixture [9, 10]. Frequently, hydroxyl-containing molecules, such as l-hexanol and l-octanol, are introduced on purpose to control the molecular weight as well as to accelerate the polymerization reaction. More recently, zinc lactate, which showed catalytic characteristics similar to stannous octoate, has been proposed as a less toxic catalyst substitute for PLA production [11, 12]. It should be noted, however, that for both stannous octoate and zinc lactate, high temperatures and long polymerization times are needed and significant transesterification reactions will take place. This has rendered these catalysts ideal for the preparation of random-type (usually amorphous) lactide copolymers such as poly(lactide-co-glycolide) and poly(ε-caprolactone-co-lactide), but made them less suitable for building advanced materials for instance functional polylactides and block copolymers of low polydispersities (M_w/M_n).

In the past two decades, a range of metal alkoxides, including alkali metal alkoxides [13], aluminum alkoxides [14], yttrium and lanthanide alkoxides [15], has been studied for the polymerization of lactides, motivated by a need for an efficient catalyst/initiator system to produce lactide-based materials with controlled parameters. The polymerization in general proceeds via nucleophilic attack of the alkoxide ligand to the carbonyl carbon followed by acyl-oxygen cleavage of the monomer, where the newly formed metal alkoxides continue to grow polymer chains (Scheme 2). The metal alkoxides, however, have a ubiquitous propensity to form aggregates. For example, yttrium isopropoxide and aluminum isopropoxide, both widely investigated for lactide polymerizations, exist as a pentanuclear complex with a formula of $Y_5(\mu\text{-O})(\text{O}i\text{Pr})_{13}$ [16] and a mixture of trimer ($(\text{Al}(\text{O}i\text{Pr})_3)_3$) and tetramer ($(\text{Al}(\text{O}i\text{Pr})_3)_4$) [17, 18], respectively. This aggregation behavior has

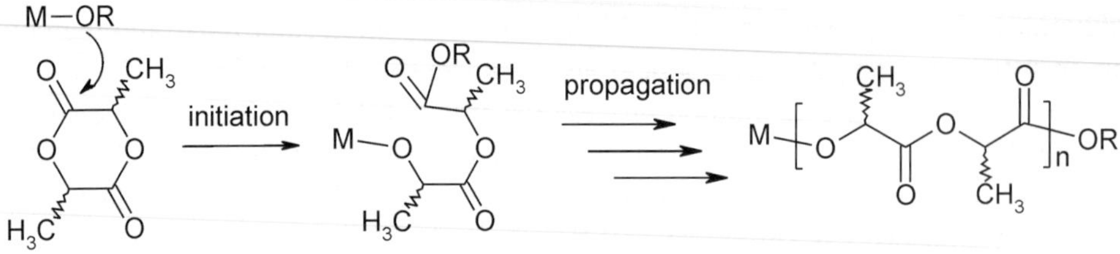

Scheme 2. General scheme for metal alkoxide (M-OR) initiated lactide polymerization.

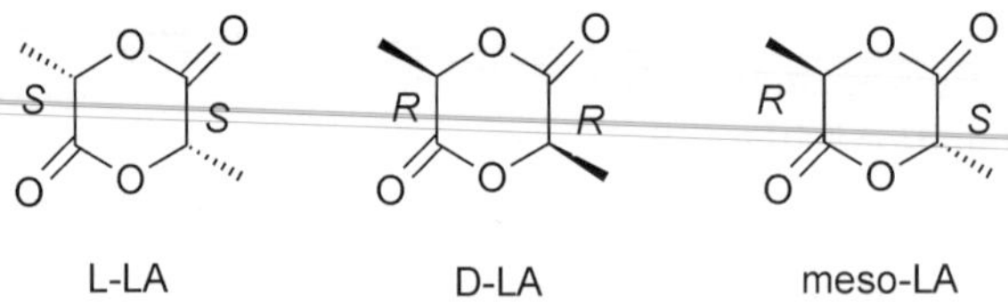

Scheme 3. LA stereoisomers. *rac*-LA is a 1 : 1 mixture of L-LA and D-LA.

resulted in a slow polymerization, a low initiating efficiency and/or complex polymerization kinetics.

A very unique property of PLA polymers is their main-chain chirality. The physical, mechanical as well as degradation properties of PLAs are intimately dependent on the chain stereochemistry [19, 20]. Isotactic poly(L-LA) and poly(D-LA), highly crystalline materials with a melting temperature (T_m) of 170°C, exhibit excellent mechanical properties and degrade rather slowly. By contrast, atactic PLAs are amorphous and subject to a comparatively fast degradation. Another interesting aspect of PLAs is that poly(L-LA) and poly(D-LA) enantiomeric chains readily co-crystallize in solution and in the melt to form a stereocomplex with a high melting temperature of 230°C [21] and enhanced tensile properties [22]. It was shown that even the stereo-diblock copolymer of L-LA and D-LA has a rather high melting point of 205°C [23]. It is, therefore, essential to control the stereochemical structure of PLA homopolymers and copolymers.

LA monomer, having two stereocenters, exists as three different stereoisomers: L-, D- and *meso*-LA (Scheme 3). It should be noted that using a non-selective catalyst, as often was the case, the polymerization of both *rac*-LA and *meso*-LA gives an atactic PLA. The stereochemistry of PLA chains can in principle be controlled by the feeding ratio of stereoisomers and/or stereoselection during PLA chain propagation. PLA is one of the few polymers in which the microstructure can easily be modified by polymerizing a controlled mixture of stereoisomers to afford high molecular weight amorphous or crystalline materials to meet specific applications. For example, PLA 70/30 (L-LA/*rac*-LA $= 70:30$), which is amorphous but with mechanical properties close to isotactic poly(L-LA), has currently been produced for several orthopedic implants [24–26]. On the other hand, stereoselection requires the use of specific catalysts and represents a challenging approach to control the stereochemical structure of PLAs [27].

Over the past few years, we have explored a range of *in situ* generated metal alkoxides based on soluble yttrium, calcium and zinc precursors. These *in situ* formed initiators, as a contrast to isolated metal alkoxides, induced an extremely fast living polymerization of lactides at room temperature. The resulting polylactides and copolymers have controlled molecular weight, low polydispersity and tailored end-groups and macromolecular architectures. Single-site metal alkoxide initiators with a general formula of L_n-M-OR where L_n is a bulky inert ligand stand for the other current strategy to gain good control over polymerization of lactides [28]. Of particular interest is that some of these single-site initiators were capable of effecting a stereoselective polymerization of lactides, in which polylactides of novel microstructures and properties have been prepared. These highly versatile *in situ* generated initiators and single-site stereoselective polymerization initiators represent the most exciting developments and have not been reviewed before. In this paper, we will describe state-of-the-art *in situ* polymerization and stereoselective polymerization of lactides. The preparative significance of these novel polymerizations will also be discussed.

HIGHLY EFFICIENT *IN SITU* GENERATED INITIATOR SYSTEMS

Yttrium alkoxides, formed *in situ* from sterically crowded tri(2,6-di-*tert*-butylphenoxy) yttrium and a less bulky alcohol (Scheme 4a), are the first reported *in situ* generated initiators that exhibit unprecedentedly high reactivity in initiating living ring-opening polymerization of LAs [29, 30]. The polymerization of L-LA in CH_2Cl_2 at room temperature was complete within a few minutes and furnished polymers of a low polydispersity ($M_w/M_n < 1.25$) and an expected M_n. This reaction was significantly faster than that of commercial yttrium isopropoxide cluster initiated L-LA polymerization where a period of 5 days was required in order to reach high monomer conversion under similar conditions. It should be noted that tri(2,6-di-*t*-butylphenoxy) yttrium, itself, was sluggish towards LA polymerization. The added alcohol molecule will determine the end-group functionalities, chain structure and molecular weight. The polymerization proceeded after instantaneous initiation and kinetic studies revealed a first-order polymerization in both monomer and initiator. In this study, *in situ* UV spectroscopy was established as a useful technique to follow the fast propagation of L-LA [30]. A comparative NMR study revealed that the species formed from $Y(OAr')_3$ and HO*i*Pr is different from $Y_5(\mu\text{-O})(O i\text{Pr})_{13}$. Lately, yttrium and neodymium alkoxides were also *in situ* generated by alcoholysis of yttrium and neodymium tris[bis(trimethylsilyl)amide], respectively. Both were reported highly active in the ring-opening polymerization of ε-caprolactone to give poly(ε-caprolactone)s with well-controlled molecular weight and low polydispersity [31, 32].

More recently, a calcium-based initiating system, generated *in situ* from bis(tetrahydrofuran)calcium bis[bis(trimethylsilyl)amide] ($Ca[N(SiMe_3)_2]_2(THF)_2$) and an adequate alcohol (Scheme 4b), was developed for the efficient polymerization of

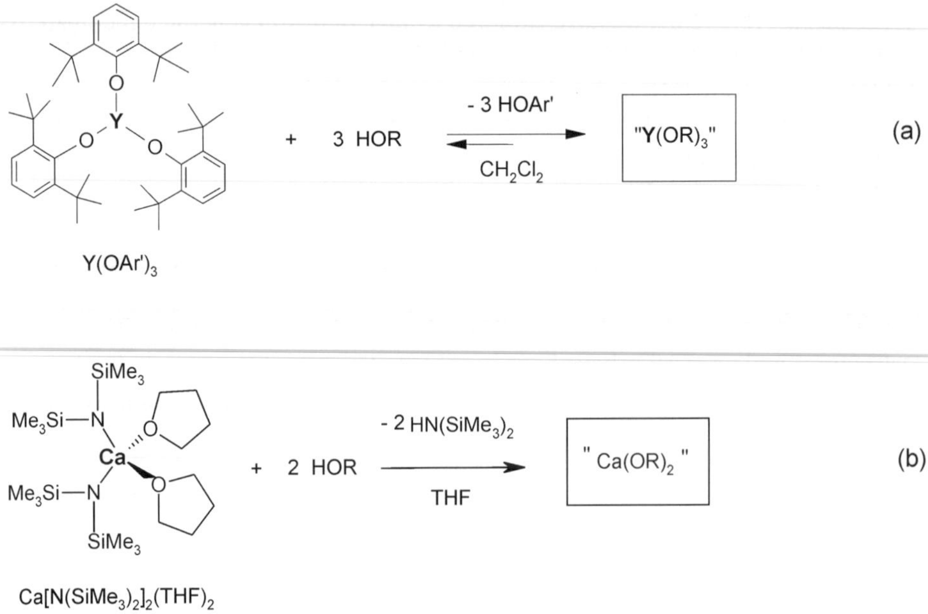

Scheme 4. *In situ* generation of yttrium alkoxides and calcium alkoxides.

lactides [33–35]. $Ca[N(SiMe_3)_2]_2(THF)_2$ was one of the few structurally well-characterized and easily prepared organocalcium compounds that can readily be dissolved in many common organic solvents such as THF and toluene. The solution polymerization of L-LA in THF using mild conditions was living, yielding PLAs of tailored macromolecular architectures and controlled molecular weights (Fig. 1). Both NMR and DSC analyses revealed the absence of racemization during L-LA polymerizations. The absence of an induction period suggested that both the alcoholysis of $Ca[N(SiMe_3)_2]_2(THF)_2$ and the subsequent initiation of L-LA polymerization are extremely fast. Polymerizations conducted at high 2-propanol/$Ca[N(SiMe_3)_2]_2(THF)_2$ ratios (e.g. 16 : 1) showed that all 2-propanol has initiated a chain polymerization of L-LA, indicating that the active species (coordinated alkoxides and/or growing polymer ends) transfer rapidly and reversibly with the dormant species (free alcohol molecules and/or hydroxyl-capped polymers). In addition, $Ca[N(SiMe_3)_2]_2(THF)_2$ was also useful for surface-initiated ROP of lactides or lactones from patterned hydroxyl group functionalized self-assembled monolayers (SAMs) or hydroxyl group functionalized monolayer-protected gold nanoparticles [36].

The ring-opening polymerization of D,L-3-methyl-glycolide (MG) in the presence of poly(ethylene glycol) (PEG) and $Ca[N(SiMe_3)_2]_2(THF)_2$ provided PMG-PEG-PMG triblock copolymers with alternating lactyl/glycolyl sequences of controlled molecular weight, low polydispersity index and uniform chain structure (Scheme 5) [37]. The site-specific cleavage of MG monomers is most probably due to the other site being more sterically hindered. A solubility test showed that

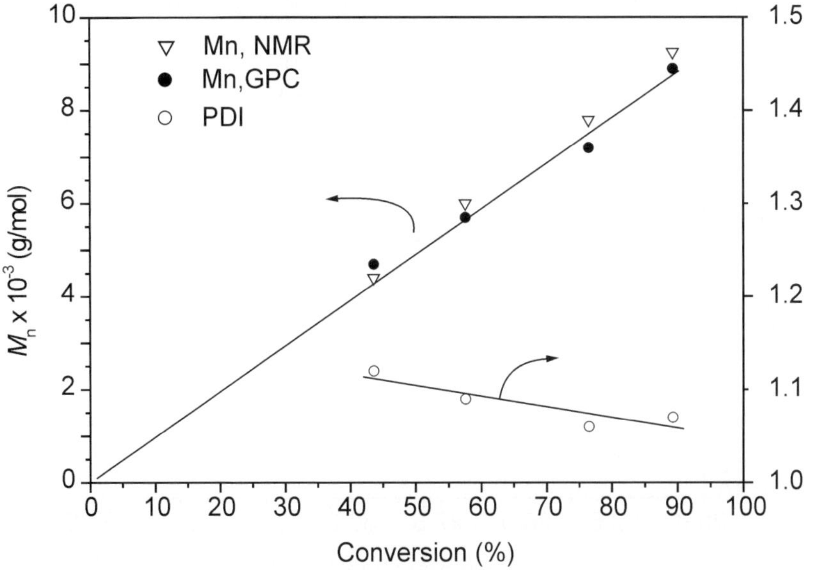

Figure 1. Poly(L-LA) molecular weight and polydispersity index (PDI) *versus* monomer conversion. $[M]_0/[2\text{-PrOH}]_0/[Ca]_0 = 150:2:1$, room temperature, THF, $[M]_0 = 0.8$ mol/l.

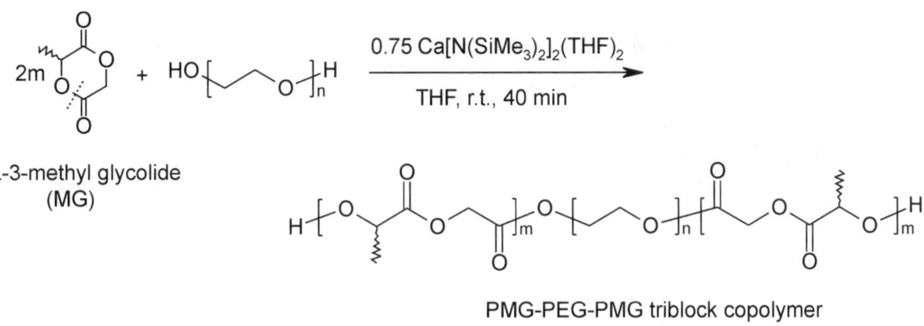

Scheme 5. Synthesis of PMG-PEG-PMG triblock copolymers by ROP of D,L-3-methyl glycolide (MG) using poly(ethylene glycol)/$Ca[N(SiMe_3)_2]_2(THF)_2$.

at relatively low temperatures (approx. 10°C) these low-molecular-weight copolymers form clear solutions in water up to high concentrations (50 wt%). Depending on molecular mass ratios of PMG and PEG blocks, a sol–gel transition (Fig. 2) or an increase in viscosity without gel formation was observed upon increasing the temperature of the aqueous solutions. These biodegradable thermosensitive hydrogels are interesting for many biomedical applications, e.g. as drug and cell carriers and as tissue-engineering matrices [38, 39].

In a further development, a single-site calcium catalyst containing chelating tmhd (H-tmhd = 2,2,6,6-tetramethylheptane-3,5-dione) ligands, $[(THF)Ca(tmhd)]_2[\mu\text{-}N(SiMe_3)_2](\mu\text{-tmhd})$, was synthesized (Scheme 6) [40, 41]. The polymerizations promoted by single-site catalysts are expected to have the advantage of

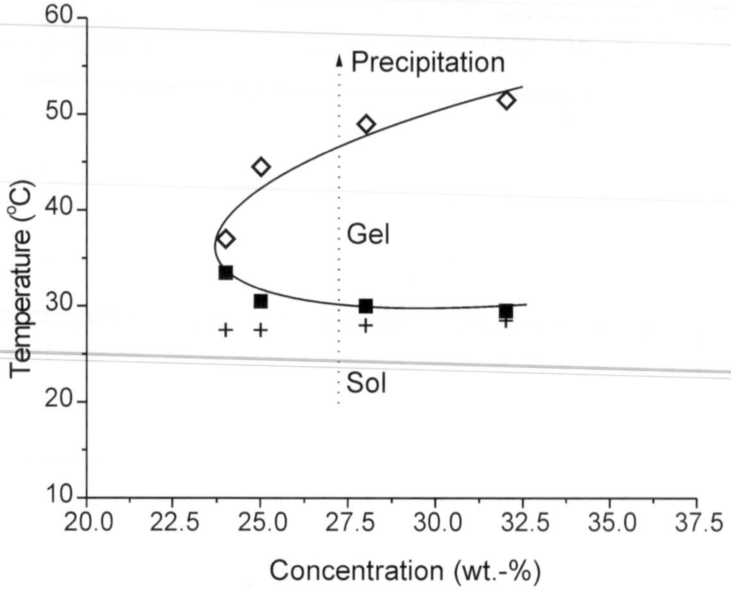

Figure 2. Phase diagram of PMG-PEG-PMG 1400-1450-1400 triblock copolymer in aqueous solution. (+) Cloud point, (■) sol–gel transition temperature, (◇) macrophase separation.

controlled kinetics and no gelation when using di- or multi-hydroxyl molecules. $[(THF)Ca(tmhd)]_2[\mu\text{-}N(SiMe_3)_2](\mu\text{-tmhd})$ exhibited a high catalytic activity and induced fast polymerization of L-LA in THF at room temperature, yielding PLAs with a high polydispersity and a higher molecular weight than calculated assuming that every catalyst molecule produces one polymer chain. On the contrary, in the presence of one equivalent of 2-propanol with respect to $[(THF)Ca(tmhd)]_2[\mu\text{-}N(SiMe_3)_2](\mu\text{-tmhd})$, L-LA was polymerized more rapidly up to high monomer conversions and the obtained PLAs had a low polydispersity, molecular weights as expected and evident end-groups, in accordance with the living nature of the active species [41]. In this way, also structurally defined poly(ε-caprolactone-b-L-lactide) block copolymers can be obtained by sequential polymerization. In order to compare the polymerization activity between *in situ* generated alkoxide compound and the isolated analogue, $[(THF)Ca(tmhd)]_2[\mu\text{-}N(SiMe_3)_2](\mu\text{-tmhd})$ was reacted in THF with isopropanol [40]. The recrystallization from toluene, however, gave the homoleptic dimutation product $[Ca(tmhd)_2]_3$ and a mixture of alkoxide-rich complexes of yet unknown composition. On the other hand, the structure of $[(THF)Ca(tmhd)]_2[\mu\text{-}N(SiMe_3)_2](\mu\text{-tmhd})$ was maintained upon alcoholysis with a sterically more demanding alcohol such as 1-phenylethanol and a compound with a formula $[(THF)Ca(tmhd)]_2[\mu\text{-OCH(Me)Ph}](\mu\text{-tmhd})$ was isolated (Scheme 6). Comparative kinetic experiments revealed that the polymerization of lactide in the presence of either initiating system, $[(THF)Ca(tmhd)]_2[\mu\text{-}N(SiMe_3)_2](\mu\text{-tmhd})$/2-propanol or $[(THF)Ca(tmhd)]_2[\mu\text{-OCH(Me)Ph}](\mu\text{-tmhd})$, was first-order in monomer up to high conversions. $[(THF)Ca(tmhd)]_2[\mu$-OCH

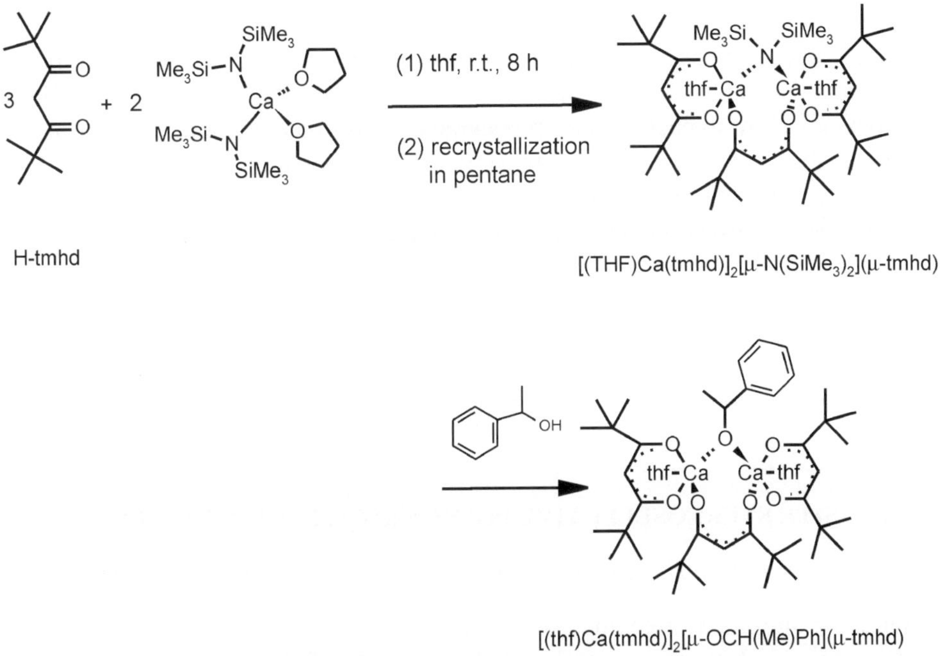

Scheme 6. Synthesis of $[(THF)Ca(tmhd)]_2[\mu\text{-}N(SiMe_3)_2](\mu\text{-tmhd})$ and $[(THF)Ca(tmhd)]_2[\mu\text{-}OCH(Me)Ph](\mu\text{-tmhd})$.

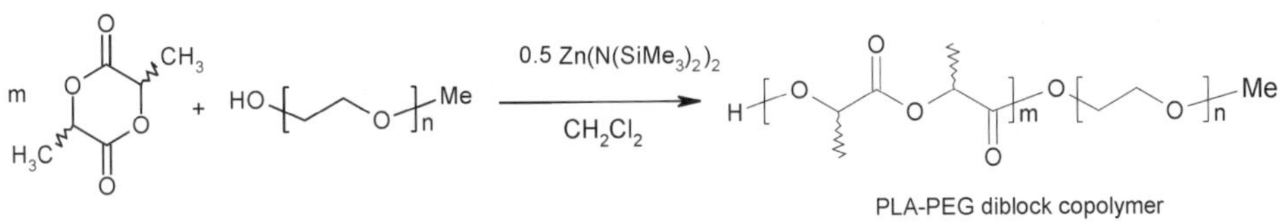

Scheme 7. Controlled synthesis of PLA-PEG diblock copolymers using monomethoxy PEG/$Zn(N(SiMe_3)_2)_2$ in CH_2Cl_2 at room temperature.

(Me)Ph](μ-tmhd) displayed a pronounced induction period before the polymerization was taking place, whereas the *in situ* initiating system $[(THF)Ca(tmhd)]_2[\mu\text{-}N(SiMe_3)_2](\mu\text{-tmhd})$/2-propanol initiated an immediate polymerization of lactide with a much higher polymerization rate as compared to isolated compound $[(THF)Ca(tmhd)]_2[\mu\text{-}OCH(Me)Ph](\mu\text{-tmhd})$ [41]. It is evident therefore that the *in situ* initiating systems have superior polymerization kinetics.

Very recently, zinc bis[bis(trimethylsilyl)amide], similar to $Ca[N(SiMe_3)_2]_2(THF)_2$, was also found to promote a fast controlled polymerization of lactides at room temperature in the presence of monomethoxy poly(ethylene glycol) (Scheme 7) [42, 43]. The resulting poly(*rac*-lactide)-b-poly(ethylene glycol) diblock copolymers had prescribed molecular weights with low polydispersities. Most interestingly, these diblock copolymers gave rise to polymeric vesicles, also referred to as polymersomes, by introducing an organic solution of the copolymer into an aqueous medium. By using different combinations of organic solvent and aqueous phase, the size of the polymersomes could be varied in a broad range from 50 nm up to 80 μm with narrow size distributions [42, 43]. These polymersomes are expected to be biodegradable and biocompatible. These novel biodegradable polymersomes have many potential applications for example in drug delivery systems, bioreactors and artificial cells. The catalyst used in this study, zinc bis[bis(trimethylsilyl)amide], is interesting due to its high activity, commercial availability and low toxicity of zinc residues.

These *in situ* initiating systems are remarkably active and versatile. PLA homopolymers and block copolymers with advanced architectures and functionalities can readily be prepared by taking advantage of the living polymerization character and the structure of the starting alcohol molecules.

SINGLE-SITE STEREOSELECTIVE POLYMERIZATION INITIATORS

Stereoselective polymerization of lactides via a chain-end control mechanism

A series of aluminum alkoxides based on achiral Schiff's base ligands has been explored for *rac*-LA polymerization (Scheme 8) [44–50]. The polymerization of *rac*-LA using **1** in toluene at 70°C showed a living character up to 60% conversion (M_w/M_n = 1.10–1.20), above which significant transesterification reactions took

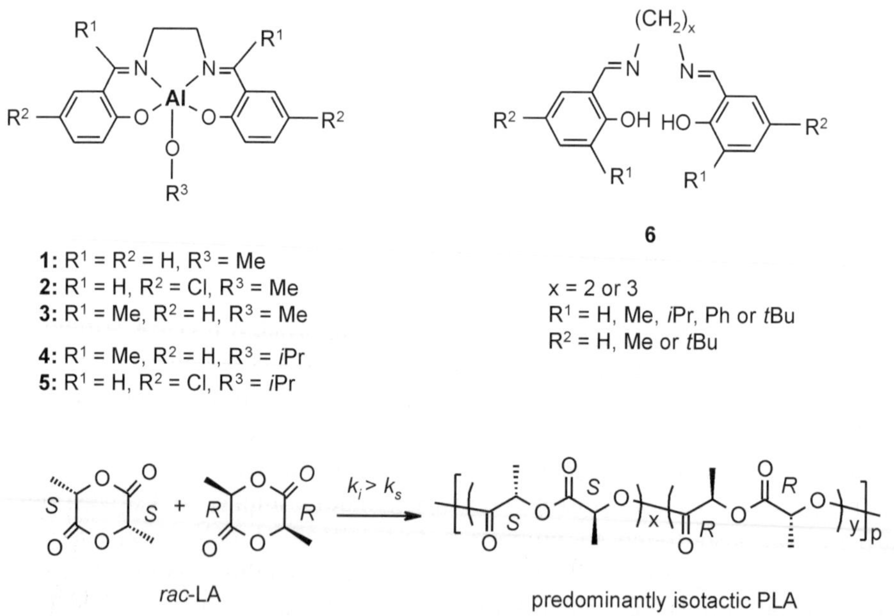

Scheme 8. Aluminum alkoxides based on achiral Schiff's base ligands for the stereoselective polymerization of *rac*-LA. k_i and k_s represent rates of isotactic and syndiotactic enchainment, respectively.

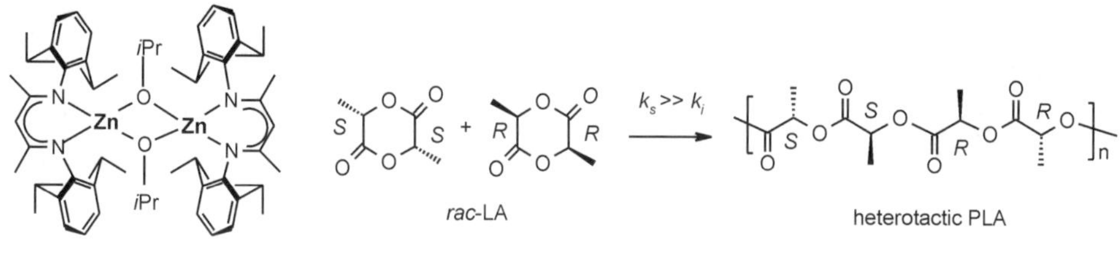

7

Scheme 9. Diiminate zinc alkoxide for heterotactic polymerization of *rac*-LA.

place [45, 46]. Most interestingly, the isolated polymers were crystalline. A ^{13}C-NMR spectroscopic analysis revealed long isotactic sequences of the obtained PLAs and a reactivity ratio (rates of homo/cross propagation, k_i/k_s) of 2.8. The preference for the isotactic enchainment was explained by a chain-end propagation mechanism. The modified versions, initiator **2** [48] and **3** [49], were capable of polymerizing *rac*-LA in CH_2Cl_2 at ambient temperature, yielding PLAs of low polydispersity ($M_w/M_n \leqslant$ 1.10 and 1.20 for **2** and **3**, respectively) and controlled M_n up to high conversions. An improved electrophilicity of the aluminum center in **2** due to the presence of electron withdrawing groups and an increased polarisability of the Al-OMe bond in **3** were proposed to account for the higher reactivity. The substitution of the methoxide by an isopropoxide group (initiator **4** and **5**) also led to an enhanced polymerization rate, although the polymerization was not controlled because of early occurring transesterifications [50]. More recently, achiral Schiff's base aluminum alkoxides *in situ* formed from ligand **6**, Et_3Al and benzyl alcohol were reported to display a preference for the formation of isotactic sequences for *rac*-LA polymerization in toluene at 70°C (Scheme 8) [51]. It is interesting to note that the catalytic reactivity of catalyst with a trimethylene backbone ($x = 3$) was much higher than that with an ethylene backbone ($x = 2$) and larger R^1 in the aromatic rings in general effected a higher isotacticity. The introduction of Ph-substituents into the aromatic rings enhanced both the polymerization rate as well as the selectivity. In contrast, the *t*-butyl substituents slowed the polymerization rate but gave the best selectivity.

Coates and co-workers have discovered that a diiminate ligated zinc complex, **7**, brought about the solution polymerization of *rac*-LA in a fast and living manner at a temperature ranging from 0 to 25°C, to yield a highly heterotactic (−*RRSSRRSSRRSS*−) PLA (Scheme 9) [52, 53]. Despite the chain stereoregularity, the polymer was amorphous with a T_g of 49°C. On the contrary, the magnesium analogue displayed no stereoselectivity under similar polymerization conditions.

Stereoselective polymerization of lactides via an enantiomorphic site control mechanism

The enantiopure (*R*)-binaphthyl Schiff base aluminum complex, (*R*)-**8**, was reported to initiate a faster and better-controlled polymerization of *rac*-LA than the

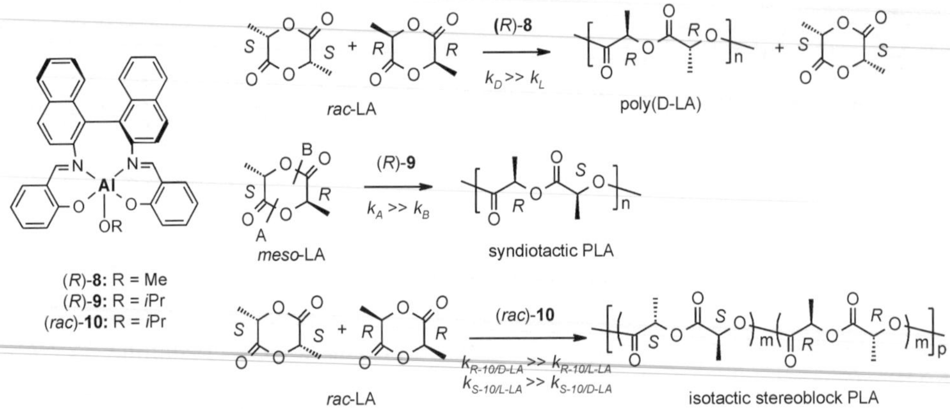

Scheme 10. Chiral binaphthyl Schiff's base aluminum alkoxides for the stereoselective polymerization of LAs.

parent achiral initiator **1** in toluene at 70°C and crystalline PLAs of controlled M_n and low polydispersity ($M_w/M_n = 1.05$–1.30) were obtained up to high conversions (Scheme 10) [54]. The optical rotation measurement of the PLAs at different conversions revealed that (*R*)-**8** has a preference for D-LA over L-LA. The optical purity of the polymer decreased with conversion because of the incorporation of L-LA, which was enriched in the monomer pool. The stereoselectivity originates from the stereogenic environment at the reactive center. Employing a site-control catalyst, (*R*)-**9**, syndiotactic (−*RSRSRSRS*−) PLA, which was crystalline with a $T_g = 34.1$°C and a $T_m = 152$°C, has been prepared for the first time by polymerizing *meso*-LA (Scheme 10) [55]. The polymerization of *rac*-LA using a racemic initiator, (*rac*)-**10**, gave PLAs of high melting temperatures ($T_m = 179$–191°C) as a result of stereocomplex formation between D-LA and L-LA enantiomeric segments produced by (*R*)-**10** and (*S*)-**10**, respectively [56–58]. Homonuclear decoupled ^{1}H-NMR spectroscopy revealed that the resulting PLA was not a mixture of enantiomerically enriched poly(L-LA) and poly(D-LA), but was an isotactic stereoblock copolymer of L-LA and D-LA for which a polymer exchange mechanism has been proposed [58].

Very recently, we discovered that cyclohexylsalen aluminum isopropoxide derived from Jacobsen ligand initiates a controlled and stereoselective polymerization of lactides. Jacobsen ligand is a commercial reagent used for many asymmetric reactions, e.g. asymmetric olefin epoxidation, asymmetric epoxide ring-opening reactions [59] and hydrolytic kinetic resolution of terminal epoxides [60]. The cyclohexylsalen aluminum isopropoxides were readily prepared with high yield by stoichiometric reaction of (*R,R*)- or (*rac*)-Jacobsen ligand with aluminum isopropoxide in toluene at 80°C (Scheme 11) [61–63]. The polymerizations of lactides in toluene at 70°C are living, affording in all cases PLAs with defined end-groups, prescribed molecular weight and low polydispersity ($M_w/M_n = 1.04$–1.09). The comparative polymerization kinetics (Fig. 3) showed that enantiopure

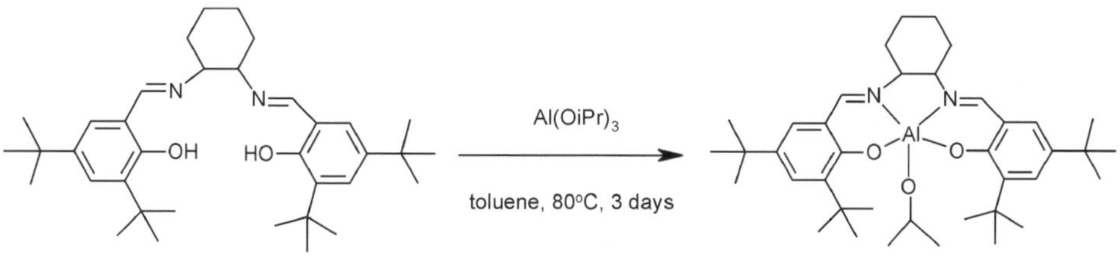

Jacobsen ligand

(1) (*R,R*)-**11**
(2) (*rac*)-**11**

Scheme 11. Synthesis of cyclohexylsalen aluminum isopropoxides.

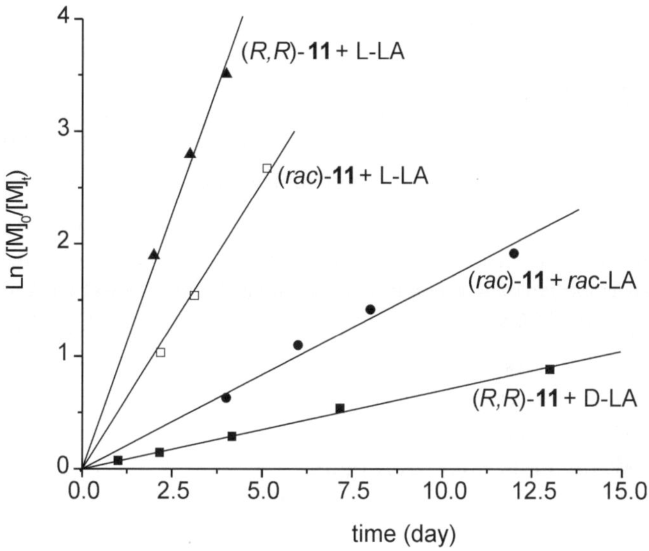

Figure 3. First-order kinetic plots for LA polymerizations in toluene at 70°C with $[M]_0/[Al]_0 = 62$ and $[M]_0 = 0.8$ M.

(*R,R*)-**11** polymerizes L-LA significantly faster than D-LA with a rate ratio k_L/k_D of about 14, indicating that (*R,R*)-**11** has a marked preference for L-LA over D-LA. This has been confirmed by the fact that PLAs obtained by polymerizing *rac*-LA using (*R,R*)-**11** at incomplete conversions have a pronounced levoratory optical rotation (Fig. 4) [63]. This is in contrast with binaphthyl Schiff's base aluminum alkoxides, in which the enantiopure (*R*)-initiator has a preference for D-LA. The equations useful for the determination of the stereoselectivity factor ($s = k_{fast}/k_{slow}$) for an asymmetric enantiomer-differentiating polymerization have recently been re-derived [64]. For example, the stereoselectivity can be calculated based on the specific rotation of the resulting polymer:

$$s = k_{rel} = \ln[1 - c(1 + ee)]/\ln[1 - c(1 - ee)],$$

where ee represents the enantiomeric excess of the monomer units in the polymer and c the monomer conversion. A selectivity factor ($s = k_L/k_D$) of 5.5 has thus been

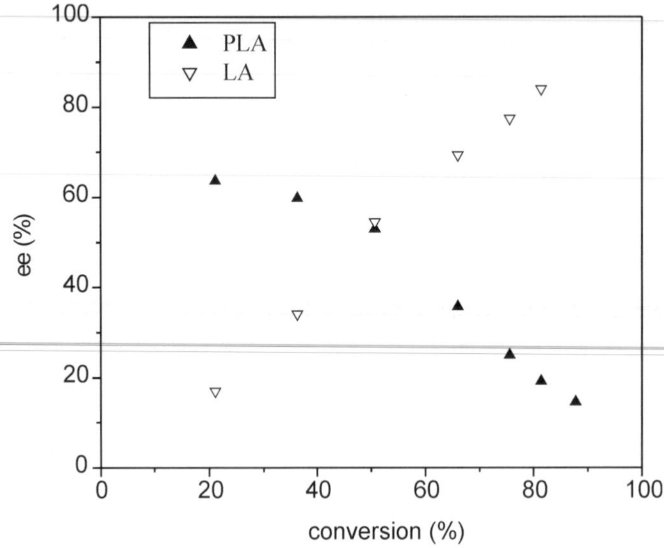

Figure 4. Plots for the enantiomeric excess (ee) of the monomers calculated from the optical purity of the polymer and monomer conversion (**7**) as well as the LA units in the recovered polymer (**4**) as a function of conversion for *rac*-LA polymerization initiated with (*R,R*)-**11** ($[LA]_0/[Al]_0 = 62:1$, $[LA]_0 = 0.8$ M, toluene, 70°C).

determined in (*R,R*)-**11** for *rac*-LA polymerization [63]. ^{13}C-NMR microstructural analysis (Fig. 5) revealed that PLAs resulting from *rac*-LA polymerization using (*R,R*)-**11** have comparably high isotacticity.

The polymerization of L-LA using (*rac*)-**11** has a rate constant slightly higher than half that of (*R,R*)-**11** for L-LA polymerization (Fig. 3). This is in line with our expectation, since only half of the (*rac*)-**11** has the (*R,R*)-configuration and the other half (*S,S*)-**11** can only slowly polymerize L-LA. (*R,R*)-**11** was also employed to polymerize a mixture of *meso*-LA and *rac*-LA (*meso*-LA/rac-LA: 83/17). The *meso*-LA, *rac*-LA and PLA display distinctive methyl resonances at δ1.70(d), 1.66 (d) and 1.55 (m), respectively, in the ^{1}H-NMR spectrum. The results showed that (*R,R*)-**11** also exhibited a preference for *meso*-LA polymerization over *rac*-LA with a relative rate k_{meso}/k_{rac} of 2.8 (diastereoselectivity). During the ring-opening of *meso*-LA, a syndiotactic selectivity was observed, which coupled with diastereoselectivity afforded a crystalline highly syndiotactic PLA at low conversion [36].

The polymerizaiton was first order in initiator where a k_p value of 9.02×10^{-3} $1\,mol^{-1}\,min^{-1}$ has been determined for (*rac*)-**11** initiated *rac*-LA polymerization in toluene at 70°C. This value is lower than that of *rac*-LA polymerization with $Al(O^iPr)_3$ ($k_p = 0.60\ 1\,mol^{-1}\,min^{-1}$) [65] or achiral Schiff's base aluminum methoxide ($0.061\ 1\,mol^{-1}\,min^{-1}$) [50] under similar conditions. This is most probably due to the rather rigid and bulky nature of the cyclohexylsalen ligand in cyclohexylsalen aluminum isopropoxide. It is worth mentioning herein that (*rac*)-

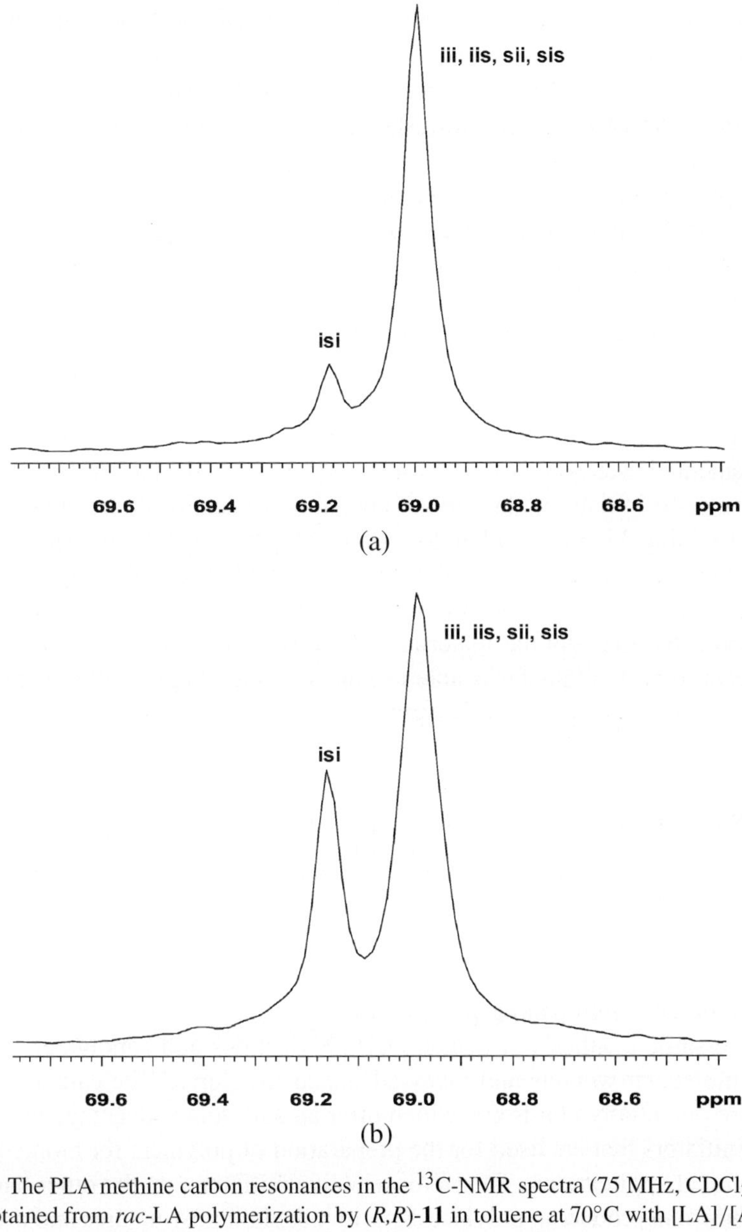

Figure 5. The PLA methine carbon resonances in the ^{13}C-NMR spectra (75 MHz, $CDCl_3$). (a) PLA sample obtained from *rac*-LA polymerization by (*R,R*)-**11** in toluene at 70°C with [LA]/[Al] = 62 : 1 and conversion = 87.8%. (b) PLA prepared by polymerizing *rac*-LA with the non-selective catalyst $Zn(OCH(Me)COO\textit{i}Pr)_2$.

10 brought about a much faster polymerization of *rac*-LA (1.05 l mol^{-1} min^{-1}) [58] than (*rac*)-**11**. The higher activity of (*rac*)-**10** as compared to (*rac*)-**11** might be due to the ability of binaphthyl moiety to delocalize electrons, the absence of bulky substituents in the aromatic rings in the ortho position, and a long diamino bridge (four carbons for (*rac*)-**10** *vs.* two carbons for (*rac*)-**11**). It has been shown that the replacement of ethylene diamino with trimethylene diamino in achiral Schiff's base aluminum alkoxide significantly increases the polymerization rate [51].

The polymerization of a L-lactide/D-lactide (molar ratio 80 : 20) mixture by enantiopure (*R,R*)-**11** in one pot furnishes an isotactic-atactic block copolylactide, which was highly crystalline with a T_m of about 155°C. Polymerization of *rac*-lactide applying (*rac*)-**11** yields isotactic stereoblock polylactides with a high T_m (approx. 185°C) and a high degree of crystallinity. Tapered stereoblock PLA with configurations varying from long L-sequences to long D-sequences throughout the polymer chain, could be prepared by the polymerization of *rac*-LA employing (*R,R*)-**11**. This stereoselective polymerization might also allow that optically pure D-LA, an expensive enantiomer, is kinetically separated from commodity *rac*-LA by selective polymerization of L-LA with (*R,R*)-**11**.

Most interestingly, excellent molar mass control as well as stereochemical control was implemented even when *rac*-LA was polymerized in the melt at 130°C [61, 62]. At a monomer-to-initiator molar ratio of 200, high conversions were obtained within 2 days, providing PLAs of rather low PDI ($M_w/M_n = 1.18$ for (*R,R*)-**11**, 1.37 for (*rac*)-**11**). The microstructural analysis showed that both polymers contain long isotactic sequences. (*R,R*)-**11** furnished an amorphous PLA, whereas (*rac*)-**11** afforded a hard crystalline material. These results are extraordinary, since no other catalyst reported thus far is able to impose such a high stereoselection in LA polymerization in the absence of solvents.

SUMMARY AND PERSPECTIVES

The past decade has witnessed several innovative PLA technologies, which have largely increased the scope of design and macromolecular engineering of biodegradable polymers. Particularly, the *in situ* generated initiating systems based on structurally characterized yttrium, calcium and zinc precursors have been shown to bring about an efficient ring-opening polymerization of lactides under mild conditions, allowing versatile synthesis of (functional) polylactides and copolymers of predetermined molecular weights and many advanced structures. The calcium and zinc systems are potentially non-toxic, which offer an additional advantage over current catalysts/initiators that are used for the preparation of polymers for biomedical and pharmaceutical applications. The stereoselective initiators, on the other hand, have shown promise in producing polylactides with novel microstructures and properties. It goes without saying that both polymerizations using *in situ* generated initiators and stereoselective polymerizations, as well as their synthetic potentials, warrant further investigation. Rational catalyst/initiator design especially assisted by, e.g. molecular modeling and combinatorial methods, may still afford more efficient initiator systems with improved stereoselectivity.

REFERENCES

1. K. E. Uhrich, S. M. Cannizzaro, R. S. Langer and K. M. Shakesheff, *Chem. Rev.* **99**, 3181 (1999).
2. R. P. Lanza, R. Langer and J. Vacanti, *Principles of Tissue Engineering*. Academic Press, San Diego, CA (2000).
3. Y. Ikada and H. Tsuji, *Macromol. Rapid Commun.* **21**, 117 (2000).
4. A. Atala and D. J. Mooney, in: *Tissue Engineering*, A. Atala (Ed.), p. 258. Birkhauser, Boston, MA (1997).
5. R. E. Drumright, P. R. Gruber and D. E. Henton, *Adv. Mater.* **12**, 1841 (2000).
6. M. Ajioka, K. Enomoto, K. Suzuki and A. Yamaguchi, *J. Environ. Polym. Degrad.* **3**, 225 (1995).
7. M. Ajioka, K. Enomoto, K. Suzuki and A. Yamaguchi, *Bull. Chem. Soc. Jpn.* **68**, 2125 (1995).
8. A. Lofgren, A. C. Albertsson, P. Dubois and R. Jerome, *J. Macromol. Sci. Rev. Macromol. Chem. Phys.* **C35**, 379 (1995).
9. A. Kowalski, A. Duda and S. Penczek, *Macromolecules* **33**, 7359 (2000).
10. H. R. Kricheldorf, I. Kreiser-Saunders and A. Stricker, *Macromolecules* **33**, 702 (2000).
11. G. Schwach, J. Coudane, R. Engel and M. Vert, *Polym. Int.* **46**, 177 (1998).
12. H. R. Kricheldorf and D. O. Damrau, *Macromol. Chem. Phys.* **198**, 1753 (1997).
13. B. T. Ko and C. C. Lin, *J. Am. Chem. Soc.* **123**, 7973 (2001).
14. D. Mecerreyes and R. Jerome, *Macromol. Chem. Phys.* **200**, 2581 (1999).
15. W. M. Stevels, P. J. Dijkstra and J. Feijen, *Trends Polym. Sci.* **5**, 300 (1997).
16. O. Poncelet, W. J. Sartain, L. G. Hubertpfalzgraf, K. Folting and K. G. Caulton, *Inorg. Chem.* **28**, 263 (1989).
17. V. J. Shiner, D. Whittaker and V. P. Fernandez, *J. Am. Chem. Soc.* **85**, 2318 (1963).
18. N. Ropson, P. Dubois, R. Jerome and P. Teyssie, *Macromolecules* **26**, 6378 (1993).
19. M. S. Reeve, S. P. McCarthy, M. J. Downey and R. A. Gross, *Macromolecules* **27**, 825 (1994).
20. J. R. Sarasua, R. E. Prud'homme, M. Wisniewski, A. Le Borgne and N. Spassky, *Macromolecules* **31**, 3895 (1998).
21. Y. Ikada, K. Jamshidi, H. Tsuji and S.-H. Hyon, *Macromolecules* **20**, 904 (1987).
22. H. Tsuji and Y. Ikada, *Polymer* **40**, 6699 (1999).
23. N. Yui, P. J. Dijkstra and J. Feijen, *Makromol. Chem.* **191**, 481 (1990).
24. J. M. Toth, M. Wang, J. L. Scifert, G. B. Cornwall, B. T. Estes, H. B. Seim and A. S. Turner, *Orthopedics* **25**, S1131 (2002).
25. M. J. Kaab, B. A. Rahn, A. Weiler, R. Curtis, S. M. Perren and E. Schneider, *Injury Int. J. Care Inj.* **33**, 37 (2002).
26. A. Joukainen, H. Pihlajamaki, E. A. Makela, N. Ashammakhi, J. Viljanen, H. Patiala, M. Kellomaki, P. Tormala and P. Rokkanen, *J. Biomater. Sci. Polymer Edn* **11**, 1411 (2000).
27. G. W. Coates, *J. Chem. Soc. Dalton Trans.*, 467 (2002).
28. B. J. O'Keefe, M. A. Hillmyer and W. B. Tolman, *J. Chem. Soc. Dalton Trans.*, 2215 (2001).
29. W. M. Stevels, M. J. K. Ankone, P. J. Dijkstra and J. Feijen, *Macromolecules* **29**, 3332 (1996).
30. W. M. Stevels, M. J. K. Ankone, P. J. Dijkstra and J. Feijen, *Macromolecules* **29**, 6132 (1996).
31. E. Martin, P. Dubois and R. Jerome, *Macromolecules* **33**, 1530 (2000).
32. K. Tortosa, T. Hamaide, C. Boisson and R. Spitz, *Macromol. Chem. Phys.* **202**, 1156 (2001).
33. Z. Y. Zhong, P. J. Dijkstra, C. Birg, M. Westerhausen and J. Feijen, *Macromolecules* **34**, 3863 (2001).
34. Z. Y. Zhong, M. J. K. Ankone, P. J. Dijkstra, C. Birg, M. Westerhausen and J. Feijen, *Polym. Bull.* **46**, 51 (2001).
35. Z. Y. Zhong, S. Schneiderbauer, P. J. Dijkstra, M. Westerhausen and J. Feijen, *J. Polym. Environ.* **9**, 31 (2001).
36. Z. Y. Zhong, in: *Novel Calcium and Aluminum-Based Initiators for the Controlled Ring-Opening Polymerization of Lactides and Lactones*, p. 121. University of Twente, Enschede (2002).

37. Z. Y. Zhong, P. J. Dijkstra, F. J. Jan, Y. M. Kwon, Y. H. Bae and S. W. Kim, *Macromol. Chem. Phys.* **203**, 1797 (2002).
38. A. S. Hoffman, *Artif. Organs* **19**, 458 (1995).
39. A. S. Hoffman, *Adv. Drug Deliv. Rev.* **54**, 3 (2002).
40. M. Westerhausen, S. Schneiderbauer, A. N. Kneifel, Y. Soltl, P. Mayer, H. Noth, Z. Y. Zhong, P. J. Dijkstra and J. Feijen, *Eur. J. Inorg. Chem.*, 3432 (2003).
41. Z. Y. Zhong, S. Schneiderbauer, P. J. Dijkstra, W. M. and J. Feijen, *Polym. Bull.* **51**, 175 (2003).
42. F. H. Meng, C. Hiemstra, G. H. M. Engbers and J. Feijen, *Macromolecules* **36**, 3004 (2003).
43. F. H. Meng, in: *Artificial Cells Based on Biodegradable Polymersomes*, p. 105. University of Twente, Enschede (2003).
44. A. Leborgne, V. Vincens, M. Jouglard and N. Spassky, *Makromol. Chem. Macromol. Symp.* **73**, 37 (1993).
45. A. Leborgne, M. Wisniewski and N. Spassky, *Abstr. Pap. Am. Chem. Soc.* **210**, 81 (1995).
46. G. Montaudo, M. S. Montaudo, C. Puglisi, F. Samperi, N. Spassky, A. LeBorgne and M. Wisniewski, *Macromolecules* **29**, 6461 (1996).
47. M. Wisniewski, A. LeBorgne and N. Spassky, *Macromol. Chem. Phys.* **198**, 1227 (1997).
48. P. A. Cameron, D. Jhurry, V. C. Gibson, A. J. P. White, D. J. Williams and S. Williams, *Macromol. Rapid Commun.* **20**, 616 (1999).
49. A. Bhaw-Luximon, D. Jhurry and N. Spassky, *Polym. Bull.* **44**, 31 (2000).
50. D. Jhurry, A. Bhaw-Luximon and N. Spassky, *Macromol. Symp.* **175**, 67 (2001).
51. N. Nomura, R. Ishii, M. Akakura and K. Aoi, *J. Am. Chem. Soc.* **124**, 5938 (2002).
52. B. M. Chamberlain, M. Cheng, D. R. Moore, T. M. Ovitt, E. B. Lobkovsky and G. W. Coates, *J. Am. Chem. Soc.* **123**, 3229 (2001).
53. M. Cheng, A. B. Attygalle, E. B. Lobkovsky and G. W. Coates, *J. Am. Chem. Soc.* **121**, 11583 (1999).
54. N. Spassky, M. Wisniewski, C. Pluta and A. LeBorgne, *Macromol. Chem. Phys.* **197**, 2627 (1996).
55. T. M. Ovitt and G. W. Coates, *J. Am. Chem. Soc.* **121**, 4072 (1999).
56. T. M. Ovitt and G. W. Coates, *J. Polym. Sci. Polym. Chem.* **38**, 4686 (2000).
57. C. P. Radano, G. L. Baker and M. R. Smith, *J. Am. Chem. Soc.* **122**, 1552 (2000).
58. T. M. Ovitt and G. W. Coates, *J. Am. Chem. Soc.* **124**, 1316 (2002).
59. E. N. Jacobsen, *Acc. Chem. Res.* **33**, 421 (2000).
60. M. Tokunaga, J. F. Larrow, F. Kakiuchi and E. N. Jacobsen, *Science* **277**, 936 (1997).
61. Z. Y. Zhong, P. J. Dijkstra and J. Feijen, *Angew. Chem. Int. Edn.* **41**, 4510 (2002).
62. Z. Y. Zhong, P. J. Dijkstra and J. Feijen, *Angew. Chem.* **114**, 4692 (2002).
63. Z. Y. Zhong, P. J. Dijkstra and J. Feijen, *J. Am. Chem. Soc.* **125**, 11291 (2003).
64. Z. Y. Zhong, P. J. Dijkstra and J. Feijen, *Macromolecules* **36**, 8198 (2003).
65. I. Barakat, P. Dubois, R. Jerome and P. Teyssie, *J. Polym. Sci. Polym. Chem.* **31**, 505 (1993).

Gels, Genes, Grafts and Giants, pp. 827–842 (2005)

Manufacture of elastic biodegradable PLCL scaffolds for mechano-active vascular tissue engineering

SUNG IN JEONG [2], SOO HYUN KIM [1], YOUNG HA KIM [1,*], YOUNGMEE JUNG [1], JAE HYUN KWON [2], BYUNG-SOO KIM [2] and YOUNG MOO LEE [2]

[1] *Biomaterials Research Center, Korea Institute of Science and Technology, P. O. Box 131, Cheongryang, Seoul 130-650, South Korea*
[2] *Department of Chemical Engineering, Hanyang University, 17 Haengdang-dong, Seongdong-gu, Seoul 133-791, South Korea*

Received 29 September 2003; accepted 16 January 2004

Abstract—A soft and very elastic poly(lactide-co-ε-caprolactone) (PLCL)(50 : 50, M_n 185 × 10^3) was synthesized. Tubular scaffolds were prepared by an extrusion-particulate leaching method for mechano-active vascular tissue engineering. The copolymer was very flexible but completely rubber-like elastic. Even the high porous PLCL scaffolds (90% salt wt) exhibited 200% elongation, but recovery over 85% in a tensile test. Moreover, the PLCL scaffolds maintained their high elasticity also in culture media under cyclic mechanical strain conditions. The highly porous scaffold (90% salt wt) withstood for an initial 1 week without any deformation and sustained for 2 weeks in culture media under cyclic stress of 10% amplitude and at 1 Hz frequency which are similar to the natural vascular conditions. Vascular smooth muscle cells (VSMCs) were seeded on to the PLCL scaffolds. The cell adhesion and proliferation on the scaffolds of various pore-size were increased with increasing pore size. For the pore sizes of 50–100 μm, 100–150 μm, 150–200 μm and 200–250 μm, the ratios of cell numbers were about 1 : 1.2 : 1.9 : 2.2, respectively, at both 12 h and 5 days. Similarly, the higher porous scaffolds exhibited more cell adhesion and proliferation compared to lower porous one, where the effect was more pronounced in the longer proliferation period. SMC-seeded scaffolds were implanted subcutaneously in athymic nude mice to confirm the biocompatibility. Such a high elastic property and proper biocompatibility to SMCs of PLCL scaffolds prepared in this study will be very useful to engineer SM-containing tissues such as blood vessels under mechanically dynamic environments (mechano-active tissue engineering).

Key words: Elastic PLCL; elastic scaffolds; smooth muscle cell; mechano-active tissue engineering.

*To whom correspondence should be addressed. Tel.: (82-2) 958-5340. Fax: (82-2) 958-5308. E-mail: yhakim@kist.re.kr

INTRODUCTION

Scaffolds for tissue engineering should regulate the organization and the subsequent proliferation of seeded cells to form new tissues. Therefore, they should have proper biocompatibility, mechanical stability, high surface area/volume ratios and interconnected structures [1–5]. Recently, several studies have shown that mechanical signals significantly regulate the phenotype of SMCs (smooth muscle cells) in two-dimensional culture systems [6–11]. Mechanical stimulation also affected the development of three-dimensional engineered SM or SM-containing tissues. Engineering blood vessels under pulsatile conditions resulted in enhanced mechanical strength, increased production of collagen and expression of contractile SMC phenotype [12–17]. The resultant blood vessels were also found to be patent for longer periods after implantation than those engineered without mechanical stimulation. It would be necessary to develop highly elastic scaffolds if one wishes to engineer SM tissues under cyclic mechanical strain conditions [17]. During the mechanical strain application, the scaffolds must maintain their mechanical integrity and deliver the mechanical signals to the SMCs adherent onto them. To achieve this, the scaffolds must be elastic and capable of withstanding cyclic mechanical strains without cracking or significant permanent deformations. In addition, the scaffolds must promote appropriate SMC adhesion and tissue formation.

Therefore, there have been several attempts to develop elastic scaffolds for vascular tissue engineering applications. Natural polymer scaffolds have been studied by cross-linking (e.g. collagen) with chemicals such as glutaldehyde, carbodiimide and divinylsulfone [10, 11]. Although such chemical cross-linking increases the elasticity of the scaffolds, these chemicals could be cytotoxic to cells [18, 19]. Manufacturing elastic synthetic biodegradable polymer scaffolds has been tried by bonding polyglycolic acid (PGA) fibers with poly(L-lactic acid)(PLA) [20]. However, these synthetic polymer scaffolds exhibited significant permanent deformation under cyclic mechanical strain conditions [20–24].

In a previous study, we attempted to develop a very elastic biodegradable poly(glycolide-co-caprolactone)(PGCL) (molar ratio 50 : 50) scaffolds [25]. Even the porous PGCL scaffold was a very flexible (elongation of 250%), but rubber-like elastic to maintain a complete recovery after 250% strain. In addition, the PGCL scaffolds in PBS at 37°C also displayed very little deformation (<5%) under cyclic loading for a couple of days [25]. In order to diversify the degradation behavior, a biodegradable poly(lactide-co-caprolactone) (PLCL) copolymer was synthesized in this study as lactide units instead of glycolide units may extend the degradation period. Generally, aliphatic polyesters such as PGA, PLA or PCL are degraded by hydrolysis and therefore their degradation rates are first of all decreased with their decrease in hydrophility (PGA > PLA > PCL) [26–28]. Then, tubular scaffolds were fabricated by an extrusion-particulate leaching method. The effects of the salt particle size and salt weight fraction on porosity, pore size, cell adhesion and proliferation of the extruded conduits were investigated. Elastic mechanical properties of the PLCL scaffolds were investigated with uniaxial tensile tests and long-term cyclic strain tests. SMCs were seeded on the scaffolds and subsequently implanted subcutaneously in athymic nude mice. The resulting explants were histochemically evaluated.

MATERIALS AND METHODS

Materials

L-Lactide (Purac Biochem, Gorinchem, Netherlands) was purified by recrystallization from dried ethyl acetate and thoroughly dried for 24 h under vacuum prior to use. ε-Caprolactone (Aldrich, Milwaukee, WI, USA) was dried over calcium hydride for 24 h at room temperature and then distilled under reduced pressure (*ca.* 0.3 mmHg) at 55°C. Stannous octoate (Sigma, St. Louis, MO, USA) was purified by vacuum distillation at 175°C (*ca.* 0.2 mmHg). 1,6-Hexanediol was distilled over calcium hydride, respectively. Toluene was thoroughly dried by distillation over sodium and benzophenone. All other chemicals and solvents were of analytical grade and used without further purification.

Synthesis of poly(lactide-co-ε-caprolactone) (PLCL)

The polymerization of PLCL was carried out in a 50 ml glass ampoule containing L-lactide (100 mmol), ε-caprolactone (100 mmol) and 1,6-hexanediol (0.5 mmol). Stannous octoate (1 mmol) was used as a catalyst. The ampoule was sealed under vacuum after purging three times with nitrogen at 90°C and heated to 170°C in an oil bath for 24 h with stirring. After reaction, the product was dissolved in chloroform and micro-filtered through a 0.45 μm pore membrane filter. The polymer was precipitated into an excess of methanol, filtered and dried under vacuum. The overall yield was 81%.

Poly(lactide-co-glycolide) (PLGA) copolymer was synthesized from L-lactide (0.077 mol) and glycolide (0.033 mol) in the presence of stannous octoate (1.5×10^{-5} mol) at 170°C for 20 h. The preparation of PLGA was identical to that of PLCL.

Preparation of tubular PLCL scaffolds

Tubular PLCL scaffolds were prepared by an extrusion-particulate leaching technique that consisted of the following steps; (1) The sieved NaCl particles were added into PLCL solutions (2%, w/w) in chloroform and the vortexed dispersions were extruded by using a home-made piston extrusion tool (Fig. 1). The extruded tubes had an inner diameter of 4 mm and an outer diameter of 6 mm. Five different polymer/salt compositions were applied in this study: 40 : 60, 30 : 70, 20 : 80, 10 : 90, 7 : 93 and 5 : 95 (all w/w), while the salt size was fixed to 100–200 μm. For the 10 : 90 weight ratio of polymer/salt study, four different salt sizes were used: 50–100 μm, 100–150 μm, 150–200 μm and 200–250 μm. (2) The solvent was

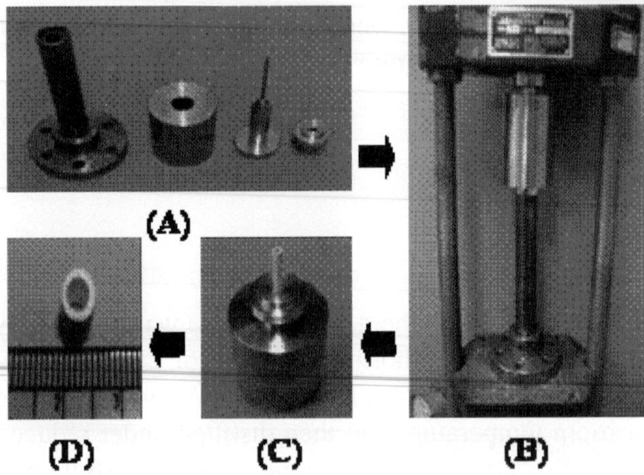

Figure 1. Extrusion tool for the manufacture of tubular PLCL scaffolds. (A) Extrusion cylinder, mold, piston; (B) extrusion process; (C) tubular scaffold extruded; (D) close view of a tubular scaffold.

allowed to evaporate from the covered Petri dish at room temperature over 48 h and then completely by vacuum drying at room temperature for 24 h. (3) The PLCL/salt composites were immersed in distilled deionized water with stirring for 48 h (the water was changed every 12 h) to leach out the salt. (4) The salt-free PLCL scaffolds were air-dried for 24 h and then vacuum-dried for 48 h. The scaffolds were stored in a desiccator under vacuum.

Scanning electron microscopy

The morphology of scaffolds was studied with a scanning electron microscope (SEM, Hitachi, Tokyo, Japan). The samples were coated with gold using a sputter-coater (Eiko IB3, Tokyo, Japan) and the microscope was operated at 15 kV to image samples. The pore diameter of PLCL scaffolds was measured from SEM photomicrographs of five different areas of the scaffolds.

Mercury intrusion porosimetry

A mercury porosimeter (AutoPore IV 9500 V1.03, Micromeritics, Norcross, GA, USA) was used for the measurement of porosity and pore size distribution of PLCL scaffolds. Each run required 2–4 samples to be loaded into the intrusion chamber, equivalent to a total sample weight of 0.034–0.234 g. The intrusion chamber was then filled with mercury at a pressure of 4.8 kPa and the samples were penetrated with mercury until the pressure reached 345 kPa. The intruded volumes and pressure were monitored during each run.

Mechanical characterization

The tensile properties of the scaffolds were measured using an Instron Model 5567 (Instron, Canton, MA, USA). A 10-N maximum load cell with a cross-head speed of 1 mm/min was used. Recovery was calculated using the following formula:

$$\text{Recovery } (\%) = ((L_2 - L_0)/(L_1 - L_0)) \times 100\%,$$

where L_0 is the original length, L_1 is the extended length and L_2 is the finial length after releasing stress. Dynamic mechanical tests ($n = 5$) of the scaffolds were performed using a custom-made mechanical strain apparatus that was previously described [25] (Fig. 2). The scaffolds were subjected to cyclic strain at amplitudes

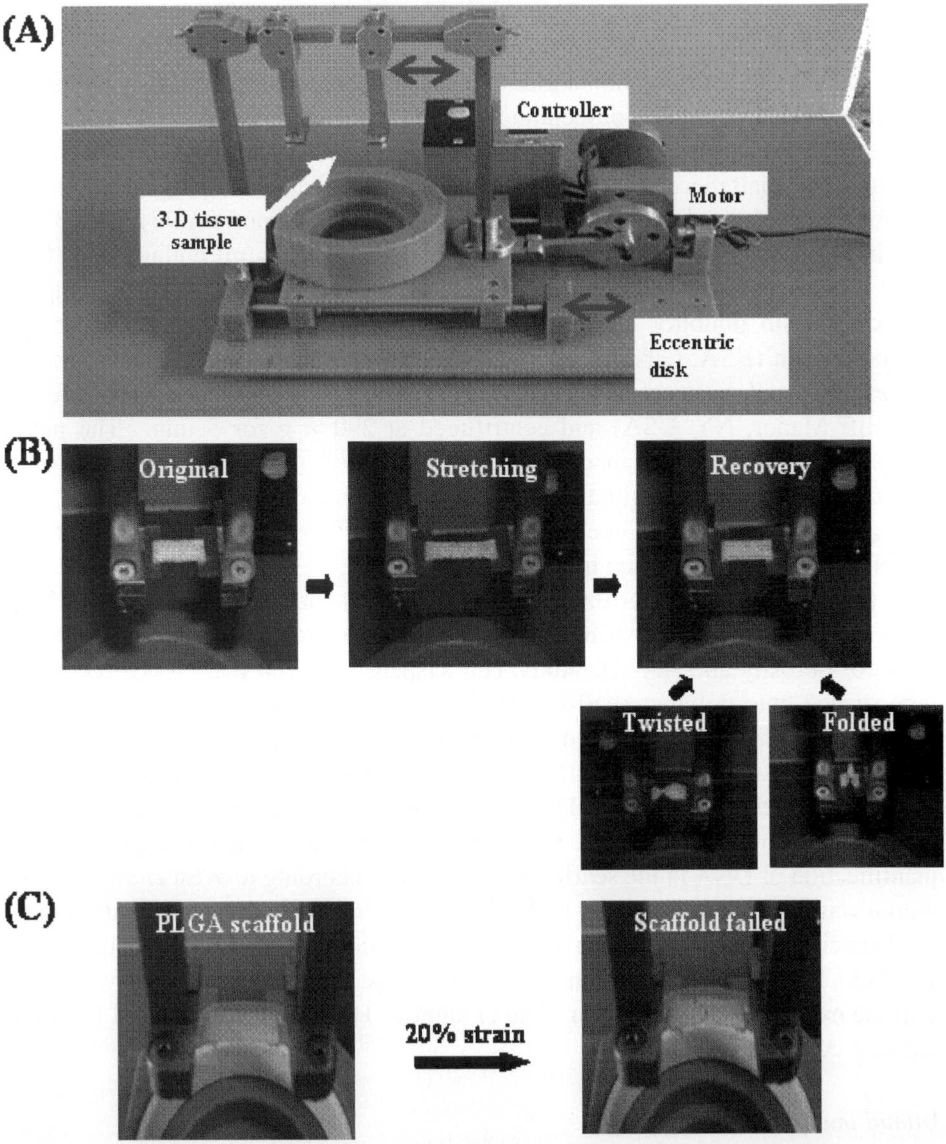

Figure 2. Cyclic strain apparatus (A) and the elastic behaviors of PLCL scaffolds (B) but broken PLGA scaffold (C).

of 10% and frequencies of 1 Hz (60 cycles/min) in Medium 199 (Gibco, Gaithersburg, MD, USA) supplemented with 10% (v/v) FBS (fetal bovine serum, Gibco), at 37°C over various time periods. The scaffolds were sterilized with ethanol before the cyclic strain test. We chose these ranges of cyclic strain amplitude and frequency because they are reported to be similar to those of vascular smooth muscle tissues *in vivo* [29, 30]. Elastic properties of the scaffolds were determined by measuring the permanent deformation of the scaffolds following cyclic strain applications.

Cell isolation and culture

SMCs were isolated from rabbit aortas using an adaptation of a previously published technique [31]. In brief, the descending aortas of 3.5 kg female New Zealand white rabbits, after removing endothelium, adventitia, fat and connective tissue, were cut into multiple small pieces and incubated in a sterile conical containing an enzymatic dissociation buffer under agitation on an orbital shaker (60 rpm) for 90 min at 37°C. This buffer contains 0.125 mg/ml elastase (Sigma), 1.0 mg/ml collagenase (CLS type I, 204 units/mg, Worthington Biochemical, Freehold, NJ, USA), 0.250 mg/ml soybean trypsin inhibitor (type 1-S, Sigma) and 2.0 mg/ml crystallized bovine serum albumin (BSA, Gibco). Following the complete dissolution of the matrix, the resultant cell suspension was filtered through a 100 μm Nitex filter (Tetko, Briarcliff Manor, NY, USA) and centrifuged at 200 $\times$ *g* for 5 min. The pellet was resuspended in growth medium consisting of Medium 199 supplemented with 10% (v/v) FBS, 2 mM L-glutamine (Gibco), 100 units/ml penicillin (Gibco) and 0.1 mg/ml streptomycin (Gibco) and cultured in tissue culture flasks (25 cm^2). SMCs of passage 2 were used in this study.

All the scaffolds were sterilized by UV irradiation for 5 min, immersed in 70% ethanol overnight and then washed by PBS three times. For the *in vitro* cell adhesion study for porosity and salt-size study, cell suspensions (100 μl, 1 $\times$ 10^6 cells/ml) were seeded onto polymer scaffolds (10 $\times$ 10 $\times$ 1 mm) located in 24-well tissue culture plates. After 3 h, 1 ml of culture medium was added to each well. Cell adhesion and proliferation on scaffolds were studied at 12 h and 5 days after seeding. The matrices were rinsed with PBS to remove unattached cells before harvesting the adherent cells and cell counting. Cell numbers were determined by a fluorometric quantification of DNA in the scaffolds by an assay according to West *et al.* [32]. An aliquot was read fluorometrically (DNA Fluorometer, Hoefer Scientific Instruments, San Francisco, CA, USA) using Hoechst dye number 33258 (Polyscience, Niles, IL, USA). Fluorescence emissions at 455 nm of the supernatant samples were read with the excitation set at 350 nm. The amount of DNA was extrapolated from the standard curve of SMC DNA.

Animal implantation study

20 μl of SMC suspension containing 1.0 $\times$ 10^7 cells was injected into the scaffolds (length 10 mm) in tissue culture dishes. After 1, 2 and 6 h, 0.5, 1 and 30 ml of

culture medium was added to each scaffold, respectively. The seeded scaffolds were implanted into subcutaneous pockets of female athymic mice (Jackson Laboratory, Bar Harbor, ME, USA), using a standard surgical procedure. Animals were killed 8 weeks post-implantation.

Histological analysis

Samples were fixed in 10% (v/v) buffered formalin, dehydrated with a graded ethanol series and embedded in paraffin. The specimens were cut into 4-μm-thick sections and stained with hematoxylin and eosin (H&E) for nucleus and cytoplasm or Masson's trichrome for collagen. For immunohistochemical analysis, they were stained using antibodies against SM α–actin (Dako, Carpenteria, CA, USA) or SM myosin heavy chain (Dako). The staining signal was developed with avidin-biotin complex immunoperoxidase kit (Vectastain® Elite ABC kit, Vector Laboratories, Burlingame, CA, USA) and DAB (3,3′-diaminobenzidine) substrate solution (Vector Laboratories). CM-DiI-labeled BMSCs in the retrieved specimens were detected using a fluorescence microscope (Eclipse E800, Nikon, Tokyo, Japan).

Statistical analysis

All the quantitative results were obtained from triplicate samples. Data were expressed as mean $\pm$ SD. Statistical analysis was carried out using the unpaired Student's t-test. A P value <0.05 was considered to be statistically significant.

RESULTS

A PLCL copolymer was synthesized by a ring opening polymerization of L-lactide and ε-caprolactone in the presence of stannous octoate. All the polymerizations were conducted in bulk with continuous stirring. Under these conditions, a high conversion in a homogeneous state is expected after a certain reaction time. The polymerizations were carried out at 170°C with a constant molar feed ratio (f_L/f_C) of 5.0 : 5.0. At temperatures lower than 150°C, caprolactone units were not completely incorporated into polymer chains due to the low reactivity of caprolactone. The overall yield of the PLCL copolymers was 81%. M_n and M_w/M_n measured by gel-permeation chromatography (GPC) were 185×10^3 and 1.58, respectively. The structure of PLCL was analyzed by ^{1}H-NMR (Table 1). These NMR results are consistent with those of NMR analysis of PLCL in the previous studies [33, 34]. The mole ratio of lactide to caprolactone in the copolymer was 5.1 : 4.9.

For the comparison, PLGA (7 : 3) hydrogel was also prepared. The M_n and M_w/M_n of PLGA copolymers measured by GPC were 132×10^3 and 2.17, respectively and the mole ratio of lactide to glycolide was 7.0 : 3.0.

Table 1.
Characterization of PLCL and PLGA

	f_L/f_C or f_L/f_G[a]	F_L/F_C or F_L/F_G[b]	M_n ($\times 10^{-3}$)	M_w ($\times 10^{-3}$)	MWD	T_g (°C)
PLCL	5:5	5.1:4.9	185	293	1.58	−6.3
PLGA	7:3	7.0:3.0	132	285	2.17	56.2

MWD = molecular weight distribution.
[a] Mole ratio in feed.
[b] Mole ratio in the copolymer.

Table 2.
Characterization of porous PLCL scaffolds prepared by extrusion

Salt particle size (μm)	Salt fraction (wt%)	Porosity (%)	Pore diameter (μm)
150–200	60	56.2	161.9
150–200	70	64.0	163.9
150–200	80	69.5	164.2
150–200	90	85.9	174.2
150–200	93	89.1	176.3
150–200	95	93.3	177.1
50–100	90	84.3	61.4
100–150	90	85.1	121.3
150–200	90	85.9	161.9
200–250	90	86.4	218.4

Porous biodegradable conduits were fabricated from PLCL copolymer using an extrusion-particle leaching technique (Fig. 1). SEM micrographs of the conduits revealed extensive pores which were roughly spherical in shape. The pores appeared connected to each other forming an interconnected network. The porosities and pore sizes of the scaffolds obtained can be varied independently by adjusting the salt fractions and the sizes, respectively. The effects of the weight fraction and particle size of NaCl on the porosity and pore diameter were investigated (Table 2). First of all, the matrix porosity increased with increasing the weight fraction of NaCl. For all the scaffolds, the measured porosities were very close to the calculated values. The pore diameter increased too as the salt particle size increased. Membranes prepared with salt weight fractions above 90 wt% had a uniform pore distribution, as observed by SEM, however those with 60, 70 and 80 wt% formed asymmetric structures with a dense skin at the outer surface of the scaffolds (Fig. 3).

Mechanical properties of PLCL scaffolds were measured and compared with those of PLGA scaffolds fabricated with the same method. PLCL scaffolds were flexible and highly elastic, while PLGA one was strong and brittle. In Fig. 4 the PLGA scaffold from 90 salt wt% exhibited only a few % of strain. On the contrary, the PLCL scaffolds were soft to show much larger elongation. The PLCL from 60 salt wt% exhibited a strain of 500% and even the highly porous one from 90 salt wt%

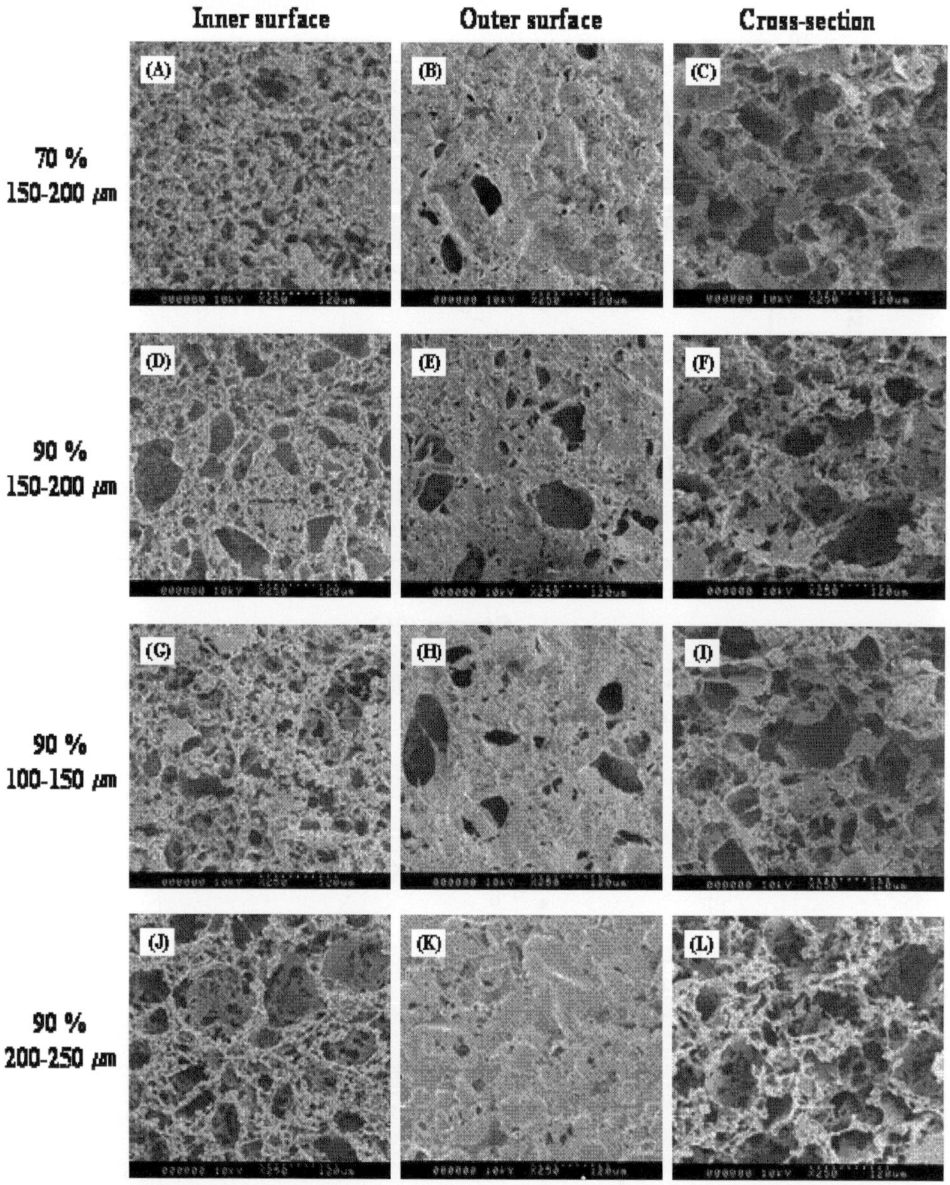

Figure 3. SEM micrographs of PLCL tubular scaffolds prepared by an extrusion-particulate leaching method. (A–C) were from 150–200 μm salt, 70 wt% salt fraction, (D–F) were from 150–200 μm salt, 90 wt% salt fraction, (G–I) were from 100–150 μm salt, 90 wt% salt fraction and (J–L) were from 200–250 μm salt, 90 wt% salt fraction. The original magnification was 250×.

was extended to 200%. The PLCL scaffolds could be easily twisted and bended, while PLGA was broken (Fig. 2). Furthermore, the PLCL scaffolds exhibited a complete rubber-like elasticity, which was not the case for PLGA scaffold. Elastic properties were evaluated by measuring recovery after stretching, as shown in Fig. 5.

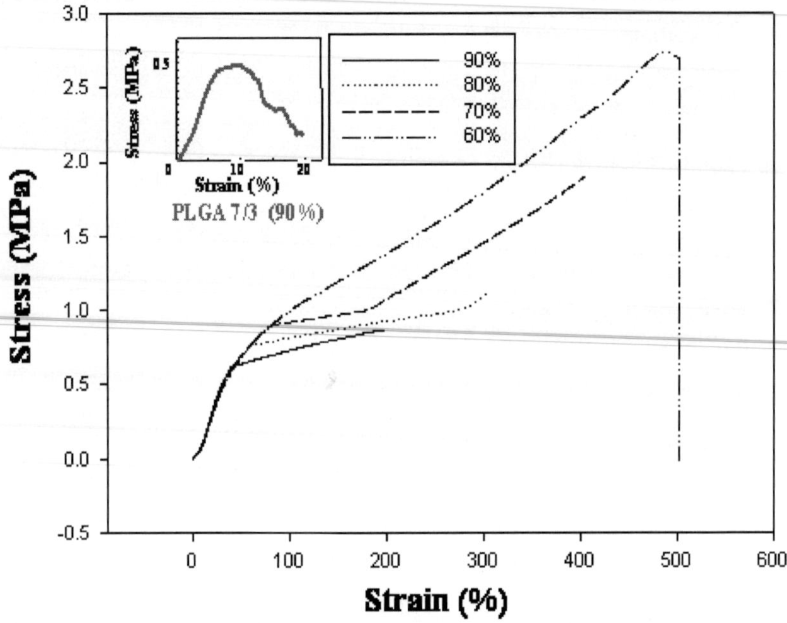

Figure 4. Representative tensile stress–strain curves of PLGA (90 wt% salt, on a different scale in the inset) and PLCL scaffolds of various salt fractions.

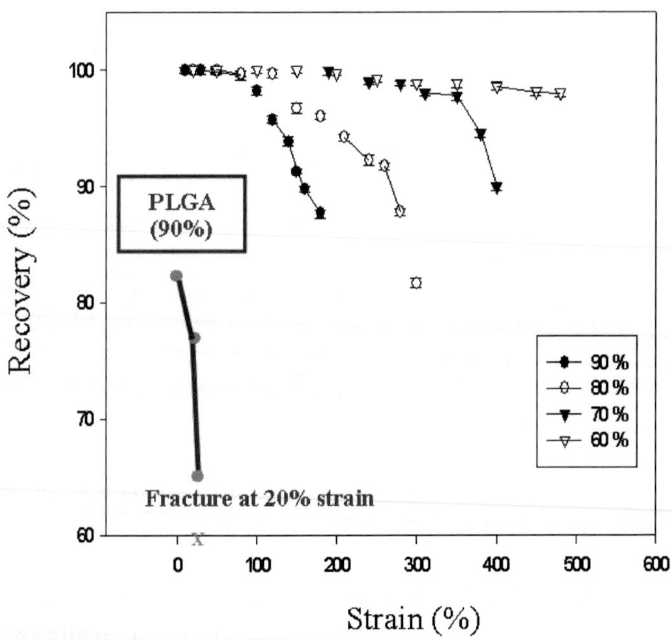

Figure 5. Recovery of PLGA and PLCL scaffolds for applied tensile strains.

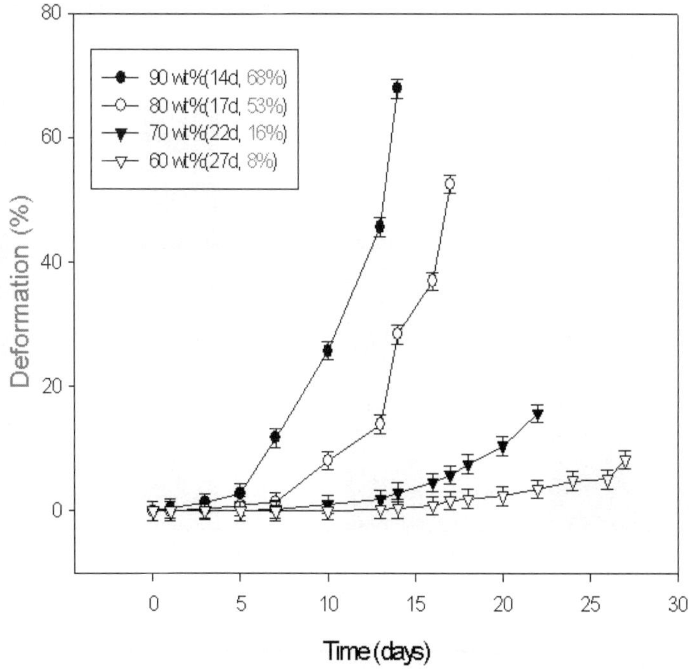

Figure 6. Deformations of PLCL scaffolds subjected to cyclic tensile loads with 10% amplitude and 1 Hz frequency for 27 days in the culture medium at 37°C ($n = 5$). The number of days and % in parentheses in the inset indicate the time and deformation at failure.

The PLCL scaffold from 60 wt% salt exhibited a complete recovery over 97% at strain applied of up to 500%. As the porosity increased, the recovery was decreased. But, the one from 90 wt% salt still showed 100% recovery at near 100% strain and 85% recovery at 200% strain. In contrast, PLGA scaffold had a large deformation and even broken at low strain as low as 20%. To study whether PLCL scaffolds would exhibit the elastic properties also under cyclic mechanical strain conditions, long-term dynamic mechanical tests were conducted. All the PLCL scaffolds were subjected to cyclic strain at 10% amplitudes and at 1 Hz frequencies for 27 days in the culture medium (M199) at 37°C (Fig. 6). All the scaffolds in M199 maintained the excellent elasticity even in the hydrolytic medium for a long time. The one from 60 salt wt% withstood for 4 weeks before breakage to leave 10% deformation. Even the higher porous one from 90 salt wt% held its elasticity for a week and failed after 2 weeks.

SMCs were seeded on each scaffold to study the optimal porosity and pore size for cell culture. First of all, small pore-sized scaffolds demonstrated low cell adhesion and proliferation, while large pore-sized ones increased cell adhesion and proliferation, as shown in Fig. 7A. The number of cells adhered and proliferated increased with increases in both pore size and porosity for 12 h and 5 days. For the pore sizes of 50–100 μm, 100–150 μm, 150–200 μm and 200–250 μm, the ratio of cell numbers adhered was about 1 : 1.2 : 1.9 : 2.2 at 12 h and 5 days, respectively.

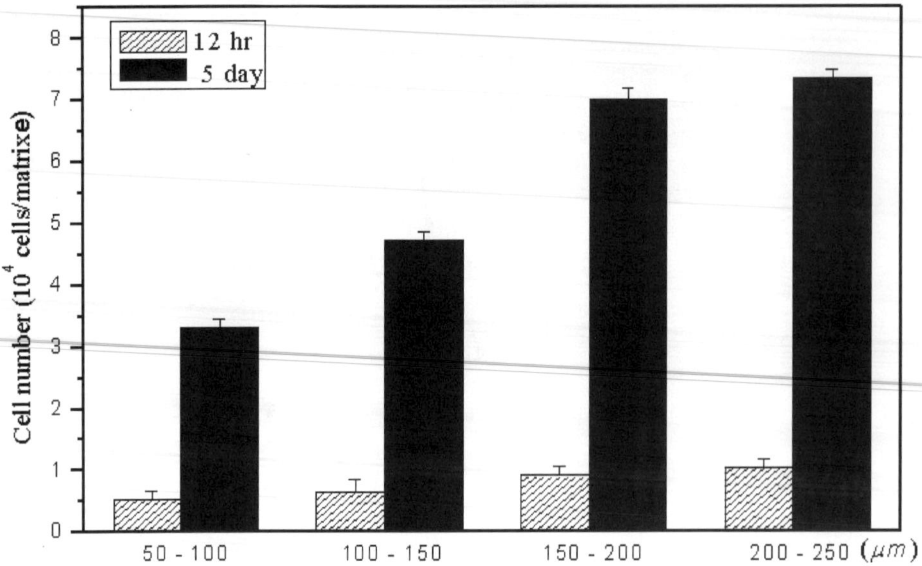

(A)

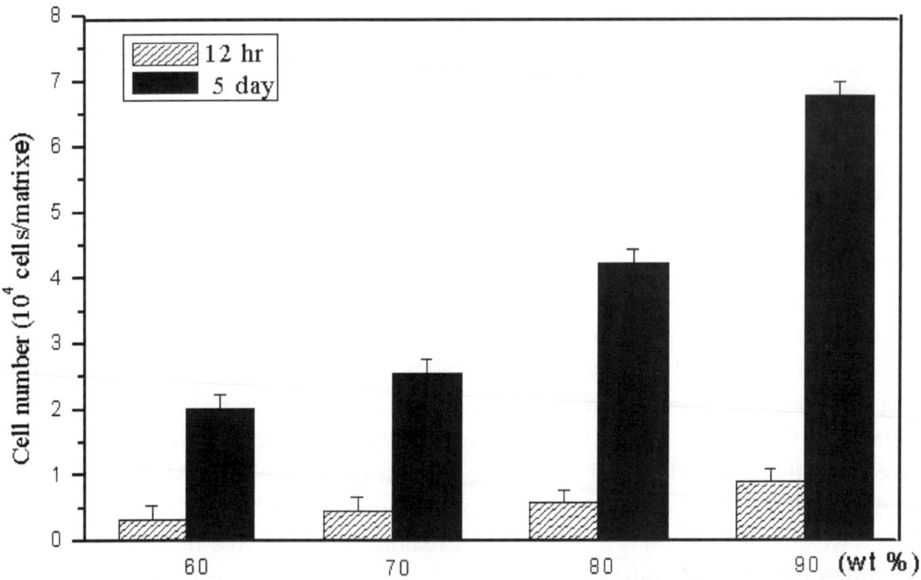

(B)

Figure 7. Number of SMCs cultured on PLCL scaffolds, (A) of various salt size at 90 wt% salt fractions, (B) of various salt fractions with 100–200 μm salt size; cell density was 1×10^6 cells/ml.

Similarly, the higher porous scaffolds exhibited more cell adhesion and proliferation compared to lower porous ones, where the effect was more pronounced with the longer proliferation period, as seen in Fig. 7B. This might due to the fact that the scaffold of large pore and high porosity yielded large interconnection and, therefore, provided an easy transport of media into the matrix inside.

In addition to the *in vitro* study above, the SMC-seeded scaffolds were implanted into subcutaneous spaces of nude mice for 8 weeks to investigate the bioactivity of the cells *in vivo*. The histological analysis of cross-sections of the cell–polymer constructs revealed that SMC were proliferated in large amounts and a tissue-like structure was formed *in vivo* after implantation, which confirmed the regular biofunctional activity of SMCs seeded (Fig. 8). The detailed result of the *in vivo* study together with the *in vivo* degradation behavior will be reported elsewhere [35].

DISCUSSION

A biodegradable PLCL 50 : 50 copolymer was synthesized from lactide and ε-caprolactone in the presence of stannous octoate as a catalyst. The random copolymer prepared here showed an very elastic characteristic. We have developed a new technique for manufacturing a very elastic tubular scaffold by an extrusion-particulate leaching method. The pore morphology of these conduits was mainly determined by the salt content and size. Increase in salt weight fraction resulted in increase in porosity. The pore size was very near to the salt size. The porosity, pore size and macroscopic dimension of scaffolds are the most important factors associated with cell adhesion and proliferation. For tissue engineering it is very necessary to obtain a maximal supply of nutrition by diffusion into tissue culture media *in vitro* or through newly formed blood vessels *in vivo*. It was shown in this study that the large pore size of 200–250 μm was more effective for SMCs culture than the smaller pore size.

Recently, several studies have demonstrated that mechanical signals play a critical role in engineering functional SM or SM-containing tissues (e.g. blood vessels) [12–16]. Mechanical stimulation increased collagen and elastin production, tensile mechanical strength and induced the phenotype of SMCs in engineered SM tissues to be more consistent with a contractile, differentiated phenotype [15, 16]. Blood vessels engineered under pulsatile conditions exhibited patency for longer periods after implantation than blood vessels engineered without mechanical stimulation [16]. In order to engineer any tissue under conditions of cyclic mechanical strain, it would be necessary to utilize elastic scaffolds. Here we have developed an elastic scaffold which can be used for such applications. PLCL scaffolds developed in this study exhibited a very appropriate mechanical property for cyclic mechanical strain applications. Unlike common PLGA scaffolds or PGA fibers, the PLCL scaffolds could keep its original dimension for an initial week in the culture media at 37°C, even under the cyclic stress conditions similar to natural vascular systems, and the scaffold was finally broken after 2 weeks. Therefore, such an elastic behavior of PLCL scaffolds will be very useful for tissue engineering of SM-tissue under cyclic mechanical stress.

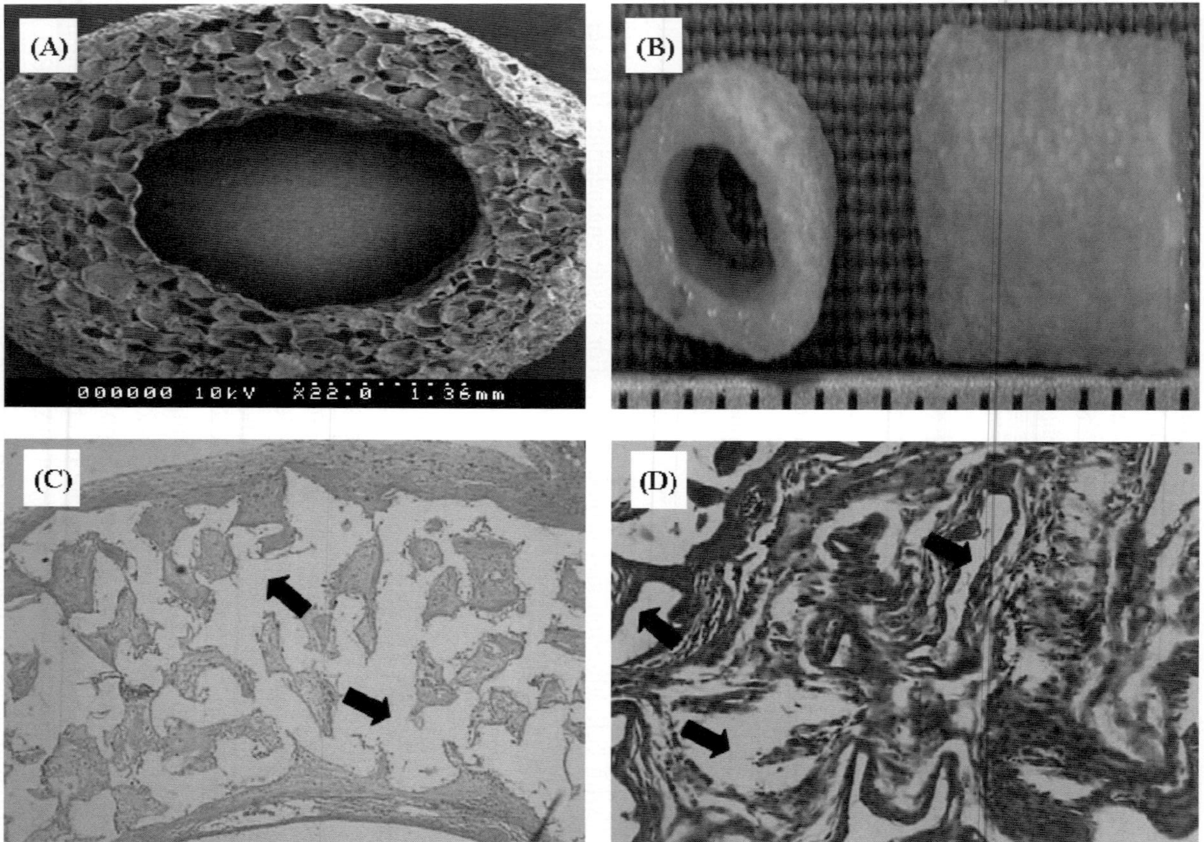

Figure 8. Photomicrographs of vascular SM tissue engineered with PLCL scaffolds *in vivo* for 8 weeks. (A) SEM of the fractured surface of a cylindrical PLCL scaffold (salt size 100–200 μm, salt fraction 90 wt%). (B) Specimen of SMC-seeded PLCL scaffolds explanted after 8 weeks. (C) H&E stain of explanted scaffold. (D) Masson's trichrome stain for collagen of explanted scaffold. Arrows indicate PLCL scaffold. The original magnification was 200×.

PLCL scaffolds allowed also appropriate interactions with seeded cells to induce tissue formation. The histological examinations indicated that SMCs seeded were well adhered on the scaffolds and proliferated (Fig. 8). It is well known that PLCL has appropriate biodegradability and biocompatibility. The scaffolds have proper physical characteristics including a pore size, interconnected pore structure and no skin layer on the external surfaces, which are important properties to affect tissue formation via nutritional diffusion and cell migration.

In summary, the PLCL scaffolds exhibited a unique elastic property and proper biocompatibility so that they will be very useful for vascular tissue engineering under cyclic mechanical stress (mechano-active tissue engineering) to obtain more native-like SM tissues. Actually, vascular tissue engineering with these elastic PCLC scaffolds demonstrated the effect of mechanical stress to enhance the proliferation of SMCs and formation of ECM in another study (data not shown).

Acknowledgements

This work was supported by Korean Ministry of Science and Technology, National Research Labs Program (Grants No. 2N23480 and 2N24620).

REFERENCES

1. D. W. Hutmacher, *Biomaterials* **21**, 2529 (2000).
2. C. S. Ranucci, A. Kumar, S. P. Batra and P. V. Moghe, *Biomaterials* **21**, 783 (2000).
3. P. X. Ma and J. W. Choi, *Tissue Eng.* **7**, 23 (2001).
4. D. W. Hutmacher, *J. Biomater. Sci. Polymer Edn* **12**, 107 (2001).
5. A. G. Mikos, G. Sarakinos, G. S. M. Leite, J. P. Vacanti and R. Langer, *Biomaterials* **14**, 323 (1993).
6. K. G. Birukov, V. P. Shirinsky, O. V. Stepanova, V. A. Tkachuk, A. W. A. Hahn, T. J. Resink and V. N. Smirnov, *Mol. Cell Biochem.* **144**, 131 (1995).
7. K. Kanda and T. Matsuda, *Cell Transplant.* **2**, 475 (1993).
8. P. Reusch, H. Wagdy, R. Reusch, E. Wilson and H. E. Ives, *Circ. Res.* **79**, 1046 (1996).
9. B. E. Sumpio, A. J. Banes, W. G. Link and G. J. Johnson, *Arch. Surg.* **123**, 1233 (1988).
10. M. E. Nimni, D. Cheung, B. Strates, M. Kodama and K. Skeikh, *J. Biomed. Mater. Res.* **21**, 741 (1987).
11. L. H. H. Olde Damink, P. J. Dijkstra, M. J. A. van Luyn, P. B. van Wachem, P. Nieuwenhuis and J. Feijen, *Biomaterials* **17**, 765 (1996).
12. F. Hess, *Microsurgery* **6**, 59 (1985).
13. L. E. Niklason, *Science* **286**, 1493 (1999).
14. L. E. Niklason and R. Langer, *Transplant. Immunol.* **5**, 303 (1997).
15. C. B. Weinberg and E. Bell, *Science* **231**, 394 (1986).
16. L. E. Niklason, J. Gao, W. M. Abbott, K. K. Hirschi, S. Houser, R. Marini and R. Langer, *Science* **284**, 489 (1999).
17. B. S. Kim, S. I. Jeong, S. W. Cho, J. Nikolovski, D. J. Mooney, S. H. Lee, O. J. Jeon, T. W. Kim, S. H. Lim, Y. S. Hong, C. Y. Choi, Y. M. Lee, S. H. Kim and Y. H. Kim, *J. Microbiol. Biotechnol.* **13**, 841 (2003).
18. L. L. H. Huang-Lee, D. T. Cheng and M. E. Mimni, *J. Biomed. Mater. Res.* **24**, 1185 (1990).

19. M. J. A. van Luyn, P. B. van Wachem, L. H. H. Olde Damink, P. J. Dijkstra, J. Feijen and P. Nieuwenhuis, *J. Biomed. Mater. Res.* **26**, 1091 (1992).
20. B. S. Kim and D. J. Mooney, *J. Biomech. Eng.* **122**, 210 (2000).
21. S. C. Woodward, P. S. Brewer, F. Montarned, A. Schindler and D. G. Pitt, *J. Biomed. Mater. Res.* **19**, 437 (1985).
22. D. G. Pitt, M. M. Gratzl, G. L. Kimmel, J. Surles and A. Schindler, *Biomaterials* **17**, 215 (1996).
23. P. van der Valk, A. W. J. van Pelt, H. J. Busscher, H. P. de Jong, Ch. R. H. Wildevuur and J. Arends, *J. Biomed. Mater. Res.* **17**, 807 (1983).
24. T. Nakamura, Y. Shimizu, Y. Takimo, T. Tsuda, Y. Li, T. Kiyotani, M. Teramachi, S. Hyon, Y. Ikada and K. Nishiya, *J. Biomed. Mater. Res.* **42**, 475 (1998).
25. S. H. Lee, B. S. Kim, S. H. Kim, S. W. Choi, S. I. Jeong, I. K. Kwon, S. W. Kang, J. Nikolovski, D. J. Mooney, Y. K. Han and Y. H. Kim, *J. Biomed. Mater. Res.* **66A**, 29 (2003).
26. B. T. Tighe, A. J. Amass and M. Yasin, *Macromol. Symp.* **37**, 133 (1997).
27. R. T. MacDonald, S. P. McCarthy and R. A. Gross, *Macromolecules* **29**, 7356 (1996).
28. A. Park and L.G. Cima, *J. Biomed. Mater. Res.* **31**, 117 (1996).
29. D. N. Ku and C. Zhu, in: *Hemodynamic Forces and Vascular Cell Biology*, B. E. Sumpio (Ed.), p. 1. RG Landes, Austin, TX (1993).
30. D. J. Patel, W. G. Greenfield, W. G. Austen, A. G. Morrow and D. L. Fry, *J. Appl. Phys.* **20**, 459 (1965).
31. A. Rothman, T. J. Kulik, M. B. Taubman, B. C. Berk, C. W. J. Smith and B. Nadal-Ginard, *Circulation* **86**, 1977 (1992).
32. D. C. West, A. Sattar and S. Kumar, *Anal. Biochem.* **147**, 289 (1985).
33. J. Kasperczyk, *Macromol. Chem. Phys.* **200**, 903 (1999).
34. J. W. Pack, S. H. Kim, I. W. Cho, S. Y. Park and Y. H. Kim, *J. Polym. Sci. Polym. Chem.* **40**, 544 (2002).
35. S. I. Jeong, B. S. Kim, S. W. Kang, J. H. Kwon, Y. M. Lee, S. H. Kim and Y. H. Kim, *Biomaterials* **25** (2004) (in press).

Gels, Genes, Grafts and Giants, pp. 843–862 (2005)

Review

Bioactive glasses for *in situ* tissue regeneration

LARRY L. HENCH [1,2,*], IONNIS D. XYNOS [2] and JULIA M. POLAK [2]

[1] *Department of Materials, Imperial College London, Prince Consort Road, London, SW7 2BP, UK*
[2] *Tissue Engineering and Regenerative Medicine Centre, Imperial College London, Prince Consort Road, London, SW7 2BP, UK*

Received 3 October 2003; accepted 9 December 2003

Abstract—Historically the function of biomaterials has been to replace diseased or damaged tissues. Recent findings show that controlled release of the ionic dissolution products of bioactive glasses results in regeneration of tissues. The mechanism for *in situ* tissue regeneration involves upregulation of seven families of genes that control the osteoblast cell cycle, mitosis and differentiation. In the presence of critical concentrations of Si and Ca ions, within 48 h osteoblasts that are capable of differentiating into a mature osteocyte phenotype begin to proliferate and regenerate new bone. Osteoblasts that are not in the correct phase of the cell cycle and unable to proceed towards differentiation are switched into apoptosis by the ionic dissolution products. A controlled release of soluble Ca and Si from bioactive glass — resorbable polymer composites leads to vascularised soft tissue regeneration. Gene activation by controlled ion release provides the conceptual basis for molecular design of a third generation of biomaterials optimised for *in situ* tissue regeneration.

Key words: Review; regeneration; second-generation biopolymers; third-generation biopolymers; molecular tailoring; bioactive glass.

FIRST-GENERATION BIOMATERIALS

The purpose of all early biomaterials was to replace diseased, damaged or ageing tissues. Thus, the materials were selected to match the physical properties of the replaced tissues with a minimal toxic response in the host; i.e. to be as 'bioinert' as possible [1–4]. More than 50 types of implants made from 40 different first-generation biomaterials are used annually to improve the quality of life of millions of people world-wide [1].

Professor Alan Hoffman has devoted nearly 50 years to understanding the mechanisms of biological response to polymeric materials for medical applications. The findings of many students and colleagues under Professor Hoffman's direction have greatly advanced the field and has led to second- and third-generation biopolymers

*To whom correspondence should be addressed. E-mail: l.hench@imperial.ac.uk

that are molecularly tailored for specific combinations of physical and biological properties. This article presents a description of the parallel development of second- and third-generation inorganic, ceramic biomaterials, specifically designed for tissue regeneration.

SECOND-GENERATION BIOMATERIALS

A characteristic of a 'bioinert' tissue response is formation of a non-adherent fibrous capsule at the materials–host interface. In the 1970s it was discovered that certain compositions of $Na_2O-CaO-P_2O_5-SiO_2$ glasses formed a strong, adhesive bond to bone and, under certain conditions, to soft connective tissues [5–7]. These biomaterials have become known as 'bioactive', with a controlled reaction in the physiological environment [3–18].

Poor survivability of first-generation biomaterials in some clinical uses stimulated development of bioactive prostheses. By the mid-1980s bioactive materials had reached clinical use in a variety of orthopaedic and dental applications, including various compositions of bioactive glasses, ceramics, glass-ceramics and composites. Synthetic hydroxyapatite (HA) ceramics began to be routinely used as porous implants, powders and coatings on metallic prostheses to provide bioactive fixation [3, 8–10]. The presence of sparingly soluble HA coatings led to a tissue response (termed osteoconduction) in which bone grew along the coating and formed a mechanically strong interface. Bioactive glasses and glass-ceramics were used as middle-ear prostheses to restore the ossicular chain and treat conductive hearing loss and as oral implants to preserve the alveolar ridge from the bone resorption that follows tooth extraction [19–26]. A mechanically strong and tough bioactive glass-ceramic, A-W glass-ceramic, was used for replacement of vertebrae in patients with spinal tumours [8, 14, 27–29].

Sufficient clinical data became available to conduct survivability analyses on first- and second-generation biomaterials for several of these medical and dental applications [30]. Bioactive glasses and glass-ceramics could be compared for the first time with bioactive ceramics, such as the synthetic calcium phosphates, and bioinert metals and polymers [30–33]. Many failures of bioinert middle ear prostheses [24] led to the development of bioactive glass devices [21–24]. But even more important, a new concept of combining bioactivity with the physical properties of biopolymers emerged in the form of bioactive composites. By the 1990s bioactive composites, such as HA particles in a polyethylene matrix, had become the standard for repair and replacement of bones in the middle ear [34].

This second generation of materials was developed specially for medical applications, in contrast to first generation biomaterials which were usually modified from 'off the shelf' compositions.

Another advance achieved in second-generation biomaterials was the development of resorbable biomaterials that exhibited controlled chemical breakdown and resorption when implanted in the body. In this manner, the interface problem is resolved, because the foreign material is ultimately replaced by regenerating tissues, and ultimately there is no discernible difference between the implant site and the host tissue [1]. An example of this is the biodegradable suture, in which the polymer composed of polylactic (PLA) and polyglycolic (PGA) acids hydrolytically decomposes into CO_2 and H_2O. By 1984 clinical use of resorbable polymers as sutures was routine [2]. Resorbable fracture fixation plates and screws in orthopaedics and controlled-release drug-delivery systems were in their infancy [4]. Resorbable calcium and phosphates were beginning to be used as bone graft substitutes [9, 31, 34].

THE CLINICAL CHALLENGE FOR THE FUTURE

The clinical success of bioinert, bioactive and resorbable implants has been a vital response to the medical needs of a rapidly ageing population. However, survival analyses of skeletal prostheses [30, 35] and artificial heart valves [36] show that a third to a half of prostheses fail within 10 to 25 years, and patients require revision surgery. Twenty years of research has had only small effects on failure rates [30]. Continuing this path of trial-and-error experiments that require use of many animals and human clinical trials is prohibitively expensive. Improvements of first- and second-generation biomaterials are limited in part because all man-made biomaterials used for repair or restoration of the body represent a compromise [1, 37, 38]. Living tissues can respond to changing physiological loads and biochemical stimuli, but synthetic materials cannot. This limits the lifetime of artificial body parts. It also signals that we have reached a limit to our current medical paradigm that emphasises replacement of tissues. It is time to emphasise a more biologically-based method for the repair and regeneration of tissues.

THIRD-GENERATION BIOMATERIALS

The separate concepts of bioactive materials and resorbable materials have converged into a new, third generation of biomaterials; bioactive materials are being made resorbable and resorbable polymers are being made bioactive [39]. Molecular modifications of resorbable polymers and bioactive composite systems elicit specific interactions with cell integrins and thereby direct cell proliferation, differentiation, and extracellular matrix production and organisation. Third-generation bioactive glasses and macroporous foams are being designed to activate genes that stimulate regeneration of living tissues.

Two alternative routes of repair are now available with the use of these third generation molecularly tailored biomaterials.

Tissue engineering

Progenitor cells are seeded onto biologically active resorbable scaffolds. The cells grow outside the body and become differentiated and mimic naturally-occurring tissues. These tissue-engineered constructs are then implanted into the patients to replace diseased or damaged tissues. With time the scaffolds are resorbed and replaced by host tissues that include a viable blood supply and nerves. The living tissue-engineered constructs adapt to the physiological environment and should provide long-lasting repair. Clinical applications include repair of articular cartilage, skin and the vascular system, although stability of the repaired tissues needs improvement.

In situ *tissue regeneration*

This approach involves the use of biomaterials in the form of powders, solutions, or doped microparticles to stimulate local tissue repair. Bioactive materials release chemicals in the form of ionic dissolution products, or growth factors such as bone morphogenic protein (BMP), at controlled rates, by diffusion or network breakdown, that activate the cells in contact with the stimuli. The cells produce additional growth factors, such as vascular endothelial growth factor (VEGF), that in turn stimulate multiple generations of growing cells to self-assemble into the vascularized tissues *in situ* along the biochemical and biomechanical gradients that are present.

CLASSES OF BIOACTIVITY

Bioactive materials used for either tissue engineering scaffolds or for *in situ* tissue regeneration must possess controlled release kinetics that synchronise with the sequence of cellular changes occurring in wound repair [37]. If dissolution rates are too rapid the ionic concentrations are too high to be effective. If the rates are too slow the concentrations are too low to stimulate cellular proliferation and differentiation. For example, numerous studies have shown that there is very much more bone formed in just one week in the presence of bioactive glass 45S5 than is formed when synthetic hydroxyapatite (HA) or other calcium-phosphate ceramic particulates are placed in the same type of defect [40–42]. After several weeks of repair there is almost twice as much new bone in the defect containing bioactive glass. The amount of bone that is regenerated matches that originally present in the site. The architecture of the trabecular bone is also equivalent to the original bone [42]. The mechanical properties have been restored by the regenerated bone [43]. These large differences in rates of *in vivo* bone regeneration and extent of bone repair indicate that there are two classes of bioactive materials [3].

Class-A bioactivity leads to both osteoconduction and osteoproduction [38, 40] as a consequence of rapid reactions on the bioactive glass surface [5, 6, 11, 17, 44–53]. The surface reactions involve ionic dissolution of critical concentrations of soluble Si, Ca, P and Na ions that give rise to both intracellular and extracellular responses at the interface of the glass with its physiological environment.

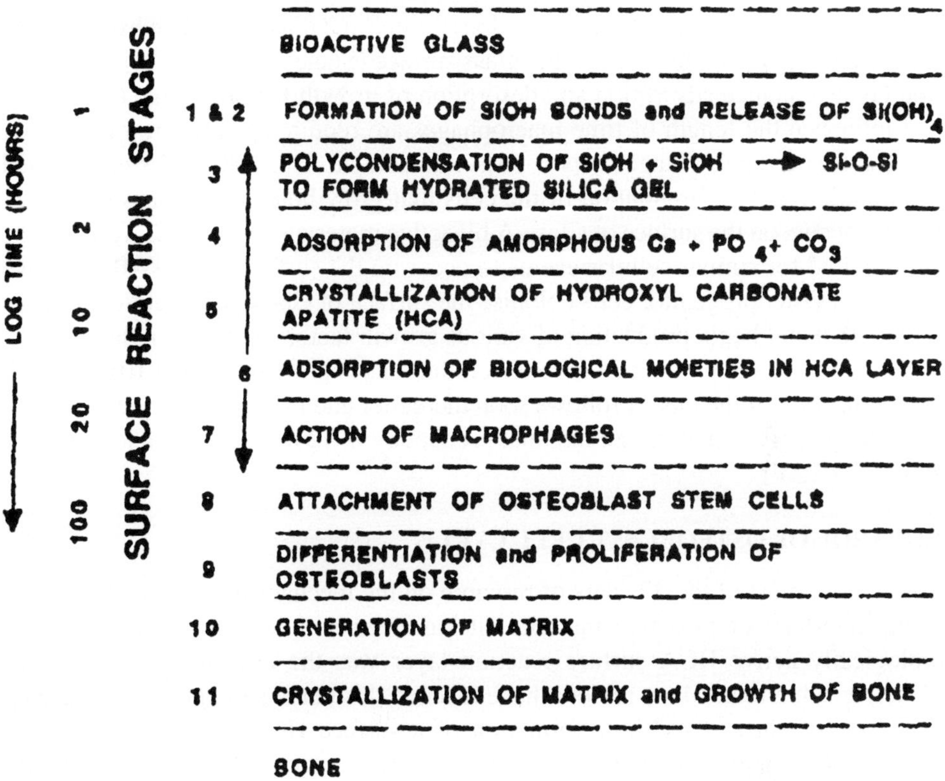

Figure 1. Time dependence (log scale) of the sequence of surface reaction stages (Stages 1–5) and biological reaction stages (Stages 6–11) of a Class-A bioactive glass (45S5 Biolglass®) in bone generation and repair.

Class-B bioactivity occurs when only osteoconduction is present; i.e. bone migration along an interface, due to slower surface reactions, minimal ionic release and only extracellular responses occur at the interface [37].

Details of evidence to support the distinction of Class-A and Class-B bioactive materials along with a description of the temporal sequence of material and cellular events involved in the regeneration of bone by Class-A bioactive materials are reviewed elsewhere [3, 10, 37, 54]. Many studies have led to the conclusion that there are 11 reaction stages involved in the regeneration of bone at the interface of a Class-A bioactive glass. The sequence and time required for these 11 reaction steps are summarised in Fig. 1.

GLASS SURFACE REACTIONS

The first five reaction stages shown in Fig. 1 occur at the surface of a Class-A bioactive glass very rapidly and go to completion within 2–24 h, depending upon localised concentration gradients. The effect is rapid release of soluble ionic species and formation of a high surface area hydrated silica and polycrystalline hydroxy carbonate apatite (HCA) bi-layer on the glass surface (Stages 1–5) [48–51]. The reaction layers enhance adsorption and desorption of growth factors (Stage 6) [55–57] and influence the length of time macrophages are required to prepare the implant site for tissue repair (Stage 7) [11, 58]. Attachment (Stage 8) and synchronised proliferation and differentiation of the cells that grow bone, osteoblasts (Stage 9), rapidly occurs on the surface of Class-A bioactive materials [59–64]. Several weeks are required for similar cellular events to occur on the surface of Class-B bioactive materials. Osteoprogenitor cells colonise the surface of Class-A bioactive materials within 48 h and begin production of various growth factors which stimulate cell division, mitosis, and production of extracellular matrix proteins, (Stage 10) [59–64]. Mineralization of the matrix follows soon thereafter and mature osteocytes, encased in a collagen-HCA matrix, are the final product by 6–12 days *in vitro* and *in vivo*.

CONTROL OF OSTEOBLAST CELL CYCLE

Throughout most of the 30 years of research on the mechanisms of bone formation at the interface of bioactive materials it has been assumed that formation of a biologically active HCA surface reaction layer was the critical requirement for bioactive behaviour. Many reviews have reached this conclusion [1–3, 11, 12, 37]. However, recent studies show formation of a surface HCA layer to be a useful but not the critical stage of reaction for bone regeneration. These studies greatly expand the understanding of Fig. 1, reaction Stages 8, 9 and 10, that are critical to regeneration of bone and tissue engineering of bone.

Xynos *et al.* compared the effect of Class-A bioactive glass (45S5 Bioglass®) *versus* a bioinert control (Thermanox® plastic) on the cell cycle of human osteoblasts (hOBs) [59–61]. The hOB cells were primary cultures obtained from excised femoral heads removed from patients aged 50–70 years, undergoing total hip replacements. Details of the experimental procedures are given elsewhere [59–61]. Various assays were used to quantify the percentages of cells in specific segments of the cell cycle.

In order for new bone to form it is essential for osteoprogenitor cells to undergo mitosis. There are very few cells in the bones of older people that are capable of dividing and forming new bone. The osteoprogenitor cells that are present must receive the correct chemical stimuli from their local environment that instruct them to enter the active segments of the cell cycle. Resting cells are in the G0 phase. Figure 2 summarises the sequence of cellular events that comprise a cell cycle. Every new cell cycle begins after a cell has completed a preceding mitosis.

During step 1 in Fig. 2, the G1 phase, the cell grows and carries out its normal metabolism [59, 60]. Osteoblasts are synthesising phenotype specific cellular products including alkaline phosphatase, an enzyme, which can be used as an osteoblast differentiation marker (Stage 9 in Fig. 1). A differentiated, functional osteoblast also produces tropocollagen macromolecules, which self assemble into

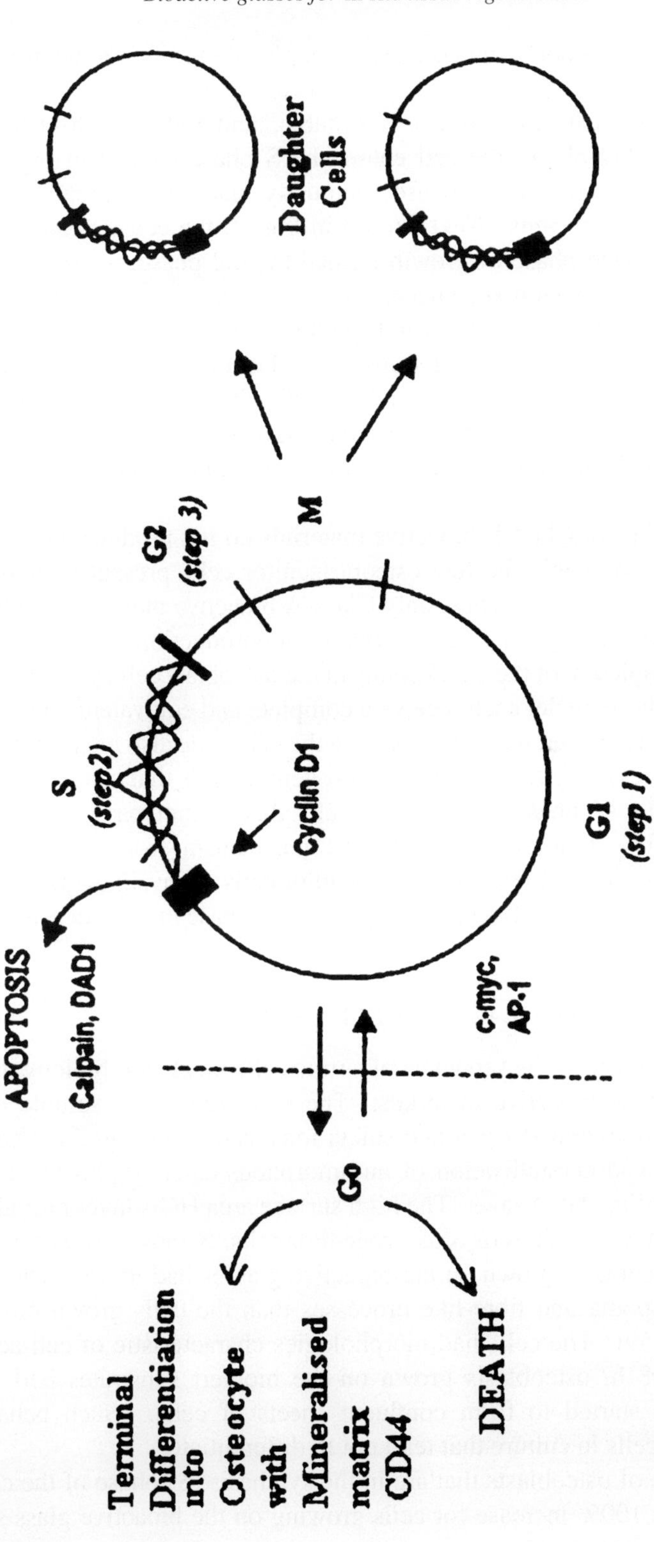

Figure 2. Cell cycle leading to terminal differentiation of an osteoprogenitor cell into osteocytes or apoptosis (left) or cell divisiion and mitosis (right) to create two daughter cells.

type-I collagen, the predominant collagenous molecule present in the mineralizable bone matrix.

If the local chemical environment is suitable, and following a critical period of growth in the G1 phase, the cell enters the S phase (step 2 in Fig. 2), when DNA synthesis begins. The S phase eventually leads to duplication of all the chromosomes in the nucleus. Next (step 3 in Fig. 2) the cell is ready to undergo mitosis with a second phase of growth termed the G2 phase. During G2 the cell prepares to undergo division and checks its replication accuracy using DNA repair enzymes. A critical increase in mass and synthesis and activation of various growth factors is necessary for the G2-M transition. Details of the feedback controls and cell cycle checkpoints are reviewed in Refs [59, 60]. If the local chemical environment does not lead to the full completion of either the G1 phase or the G2 phase, then the cell proceeds to programmed cell death, apoptosis, as shown in Fig. 2.

Bioinert materials or Class-B bioactive materials do not produce the local chemical environment to enable the few osteoprogenitor cells present to pass through these cell-cycle checkpoints. Thus, only Class-A bioactive materials produce rapid new bone formation *in vivo*, a process termed osteoproduction [40].

Successful completion of the checkpoints in the osteoblast cell cycle produces two daughter cells whose nuclei each receive a complete and equivalent complement of genetic material. However, the checkpoints in the cell cycle also mean that as we age we produce progressively fewer and fewer progenitor cells that can run the gauntlet and enter into the M phase unscathed. Cells that cannot pass the checkpoints become apoptotic. Our built-in protective mechanism from multiplication of damaged genes means that fewer osteoprogenitor cells are available to repair dying bone cells. The cumulative effect is a progressive decrease in bone density with age.

EFFECT OF BIOACTIVE GLASS ON CELL CYCLE

Figure 3 is a summary of the sequence of surface chemical, cell biology and tissue response of Class-A bioactive materials. The rapid release of soluble inorganic species, especially hydrated silica and calcia ions, shown in Fig. 3a, gives rise to rapid nucleation and crystallisation of an amorphous calcium-phosphate layer to form a polycrystalline HCA layer. The high surface area HCA layer provides many binding sites for the cytokines of osteoprogenitor cells. Xynos *et al.* showed that as early as 48 h osteoblasts grown on the bioactive glasses had more anchoring sites of numerous filopodia and fibre-like processes than the cells grown on the inert plastic controls [59]. The cells had morphologies characteristic of cell activation. In contrast, at 48 h, osteoblasts grown on the bioinert substrates had flattened and divided and started to form confluent sheets of cells. Such behaviour is characteristic of cells in culture that tend not to differentiate.

The percentage of osteoblasts that are in the synthesis (S) phase of the cell cycle at 48 h showed a 100% increase for cells growing on the bioactive glass substrate

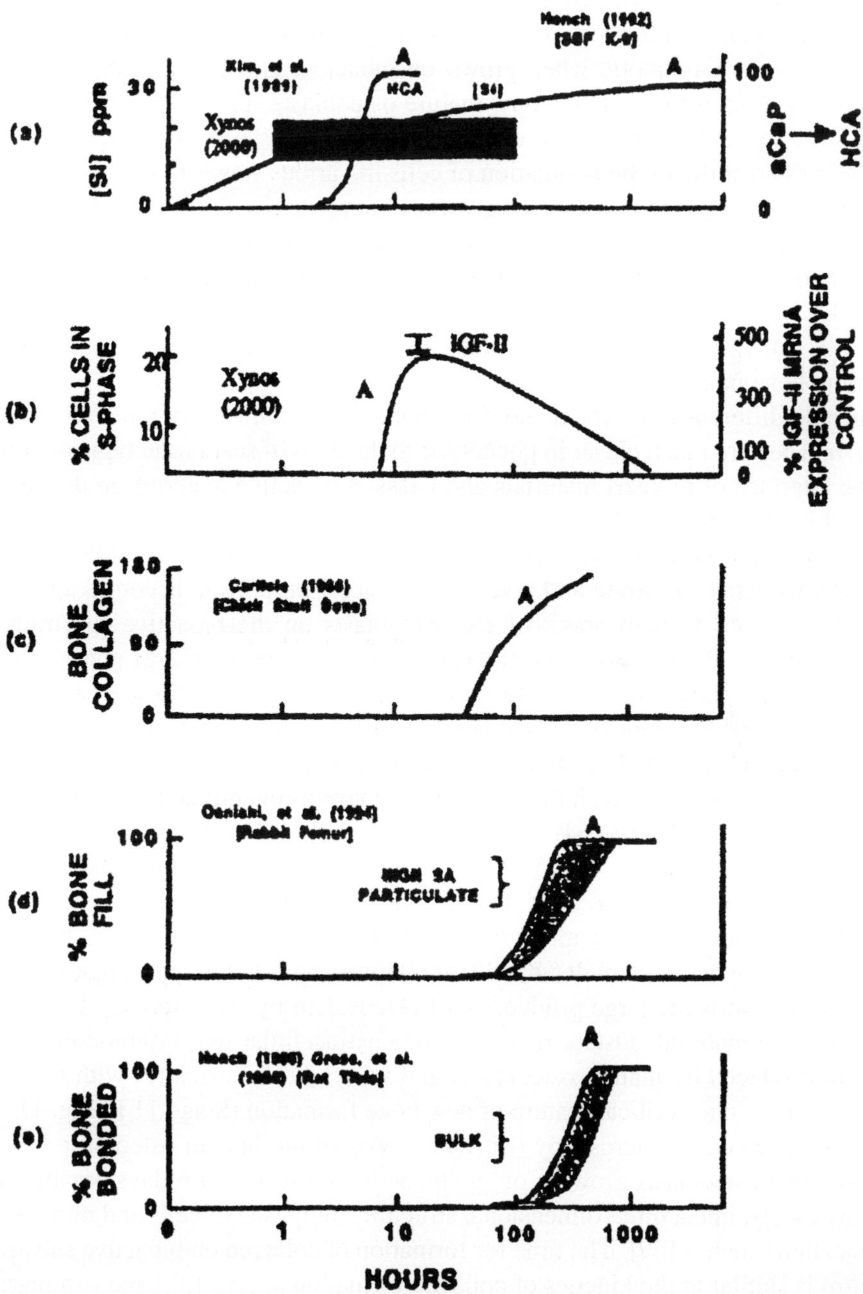

Figure 3. Effect of time on human osteoblast cultures and rabbit femoral bone repair in the presence of Class-A bioactive glass particles.

(Fig. 3b). Likewise, the percentage of osteoblasts that are in the G2-M phase on the bioactive substrate was over 100% higher than the control substrate [59]. The largest population shift occurred for cells in apoptosis. At 48 h, 5 times as many osteoblasts were apoptotic when grown on bioactive glasses. Primarily the cells capable of differentiating to become mature osteoblasts survived the first few days of growth on the bioactive glass substrates.

These rapid shifts in the population of cells in various stages of the cell cycle are important, as they lead to enhanced growth of new bone in the presence of Class-A bioactive glasses. Osteoblasts exposed to bioinert substrates quickly attach and enter into a cell cycle that leads to the M phase and formation of a confluent sheet of cells. More cells are alive and few of them are becoming apoptotic. However, the cells growing on the inert substrates are seldom capable of differentiating and forming bone matrix.

Such de-differentiated cells do not form bone nodules *in vitro* or mineralized bone *in vivo*. They are much closer in phenotype to the cells of scar tissue typically found at the interface of bioinert materials and Class-B bioactive materials at the earliest time of implantation.

The culture studies show that osteoblasts initially grow more slowly on the Class-A bioactive substrate and a very much larger proportion becomes apoptotic. However, by 48 h many more of the osteoblasts on the bioactive substrate are mitotically active; twice as many cells are in the S phase and G2-M phase. At this stage, alkaline phosphatase activity, characteristic of the osteoblastic phenotype, is expressed by both groups of cells at nearly equivalent levels [59–61].

The importance of the bioactive control of osteoblast cell cycle becomes apparent by 6 days in culture, just as it does at 6 days *in vivo* as quantified by several studies of Oonishi *et al.* [42, 63, 64]. By 6 days the proportion of osteoblasts in the S phase and especially the G2-M phase, is even greater for the cells growing on the bioactive material (Fig. 3b). The difference in proportion of cells eliminated by apoptosis becomes very much larger; the difference is greater than 700% and more of the bioactive stimulated cells are expressing alkaline phosphatase. Even more important is the large production of osteocalcin by the osteoblasts grown on the bioactive material. Osteocalcin is a bone extracellular matrix non-collagenous protein produced by mature osteoblasts and its synthesis correlates with the onset of mineralization, a critical feature of new bone formation (Stage 11 in Fig. 1).

Scanning electron microscopy (SEM) analysis of the human osteoblast cultures showed that osteoblasts growing on the bioactive substrate for 6 days organised by self-assembly, into a three-dimensional structure composed of cells and mineralised extracellular matrix [59]. The time for formation of collagen on bioactive substrates *in vitro* is similar to the kinetics of collagen formation *in vivo* [37] and summarised in Fig. 3c. The rate of forming mineralised bone nodules *in vitro* are also similar to the kinetics of bone growth *in vivo* summarised in Fig. 3d, e.

Additional confirmation of the 3D structure of the bone nodules was obtained by Xynos *et al.* using confocal scanning laser microscopy [48]. They mapped the 3D structure of the nodule showing the presence and organisation of the type-I collageneous matrix and calcium deposition within the bone nodules. The results confirm that human osteoblasts growing in culture in the presence of a bioactive glass self-assemble into a three-dimensional architecture and create a mineralised matrix.

This 3D architecture is created by the osteoblasts when the cells are exposed to release of critical concentrations of the soluble ionic constituents of the bioactive glass even if a substrate is not present. Approximately 17 to 20 ppm of soluble Si and 88 to 100 ppm of soluble Ca ions are required. The ions are provided by controlled dissolution of a bioactive glass substrate. It is also possible to react bioactive glass powders in tissue culture medium and create the critical concentrations of soluble inorganic ions in the culture medium. When osteoblasts are grown in this ionically conditioned medium they differentiate and form a mineralised extracellular matrix and create bone nodules. Thus, the role of the bioactive glass is primarily to release the critical concentrations of biologically active ions at the rate needed for cell differentiation [59, 60]. At 6 days osteoblasts grown on bioinert substrates show no evidence of formation of bone nodules [59].

By day 12 the differences in osteoblast cell cycle stimulated by the presence of the Class-A bioactive glass become solidified. More osteoblasts are present on the bioactive substrate and a significantly larger proportion exist in the S phase and the G2-M phase and are organising themselves into multilayers and 3D bone nodular structures [59, 60, 65]. Cells that are not capable of differentiating into the osteoblast phenotype are still being eliminated by apoptosis at a higher rate than on the bioinert control substrate. The number of bone nodules growing on the bioactive substrates continues to increase and the organisation of the nodules becomes increasingly more complex with large numbers of osteocytes within the nodules [59, 65, 66]. At day 12 there are still no bone nodules present on the bioinert substrates although the osteoblast-like cells are still healthy. The cellular markers suggest that the cells growing on the inert substrates are not capable of forming new mineralised bone, but are more similar to the fibroblast-like cells found in scar tissues.

GENETIC CONTROL OF THE OSTEOBLAST CELL CYCLE

Molecular biology studies by Xynos *et al.* have recently shown that the bioactive shift of the osteoblast cell cycle described above is under genetic control [61, 62]. Within a few hours exposure of human primary osteoblasts to the soluble chemical extracts of 45S5 Bioglass, several families of genes are activated, including genes encoding nuclear transcription factors and potent growth factors, especially IGF-II along with IGF-binding proteins and proteinases that cleave IGF-II from their binding proteins [61]. There is a 200 to 500% increase in the expression of these genes over those of the control cultures. Activation of several immediate early response genes and synthesis of growth factors is likely to modulate the cell-cycle response of osteoblasts to bioactive glasses. These findings indicate that

Class-A bioactive glasses enhance new bone formation (osteogenesis) through a direct control over genes that regulate cell-cycle induction and progression.

The entry of osteoblasts into the cell cycle (G0–G1 transition) and subsequent commencement of cell division is regulated by a family of transcription factors. These molecules do not solely trigger the initiation of cell division but provide the specific stimuli needed for the development of cells that bear the osteoblastic phenotype. These specific proteins must be present for a bone stem cell to become a differentiated bone-phenotype growing cell. The recent results by Xynos *et al.* [59–62] show that treating human osteoblast cultures with the ionic products of bioactive glass dissolution for 48 h activated expression of a large number of transcription factors and cell cycle regulators (Table 1). The transcription factors that were activated include c-*jun*, *fra-1* and c-*myc*, three well-characterised osteoblast transcription factors.

The importance of activating genes that express these transcription factors is summarised as follows: the c-*jun* protein product, Jun, forms a heterodimeric complex with Fos protein and interacts with the DNA element known as the AP1 binding site. The products of the *fos* (*FosB, fra-1, fra-2*) and *jun* (*c-Jun, JunB, JunD*) genes constitute transcription factor AP1. The AP-1 consensus binding sequence is present in the promoters of several osteoblast-specific genes, including alkaline phosphatase, osteocalcin, collagenase (MMP-1) and alpha1-collagen genes. The expression of AP-1 transcription factor by osteoblasts is correlated with osteogenesis *in vitro* and *in vivo* [67]. Transcriptional regulation by AP-1 expression precedes osteogenic differentiation of cartilage cells *in vitro* and AP-1 expression appears to play crucial roles in the early regulation of endochondral osteogenesis in both bone formation and fracture healing [67].

Transcription factor c-*myc* encodes the transcription factor c-Myc, which is a member of the Myc family of helix–loop–helix transcription factors [68]. c-*myc* is of great importance in controlling osteoblast growth and vitality. c-*myc* induction promotes cell proliferation by activating growth-promoting genes such as *rcl* [69], a downstream effector of c-Myc, which was also notably induced by the ionic products of bioactive glass dissolution (Table 1). The expression of c-*myc* is regulated by the c-*myc* transcriptional factor Puf [70], also moderately activated by the treatment, which binds to the c-*myc* promoter. It has been shown that c-*myc* expression is mediating parathyroid-hormone-induced differentiation [71] and post-fracture osteogenesis *in vivo* [72].

Osteoblast proliferation and phenotypic commitment is triggered by transcription factors c-Myc and AP-1, but depends on successful progression through the cell cycle, as described above. Certain cyclins are required for the progression from the G1 phase of the cell cycle to the synthesis (S) phase. These critical cyclins include cyclin D1 (G1/S-specific cyclin) which phosphorylates the product of the retinoblastoma gene [73] resulting in the release of transcription factors important for the initiation of DNA replication. Cyclin D1 is upregulated by 400% when osteoblasts are exposed to the ionic products of bioactive glass dissolution for 48 h

Table 1.
Transcription factors and cell cycle regulators

Factor	Activation (%)
RCL growth-related-c-*myc*-responsive gene	500
G1/S-specific cyclin D1 (CCND1)	400
26S proteinase regulatory subunit 6A	400
Cyclin-dependent kinase inhibitor 1 (CDKN1A)	350
cAMP-dependent transcription factor ATF-4	240
Cyclin K	200
c-*myc* purine-binding transcription factor *puf*	170
fos-related antigen (FRA1)	160
Proto-oncogene c-*jun*; transcription factor AP-1	150
c-*myc* oncogene	150

Table 2.
DNA synthesis, repair and recombination

DNA repair protein	Upregulation (%)
DNA exclusion repair protein ERCC1	300
mutL protein homolog	300
High-mobility-group protein (HMG-1)	230
Replication factor C 38-kDa subunit (RFC38)	200

(Table 1). This large increase in gene activation of cyclins demonstrates that the bioactive glass does not merely trigger the entry of osteoblasts into the cell cycle but also provides the vital stimulus needed for progression through the G1/S checkpoint, a crucial step for the successful completion of the cycle.

Two other important cell cycle regulators, CDKN1A [74] and cyclin K [75], were also activated by the ionic dissolution products of bioactive glass by 200% or more (Table 1). Both are involved in the regulation of the early stages of the mitotic cycle of the cells.

The cell cycle represents a highly synthetic phase in the life of an osteoblast. During this phase the cell produces proteins and nucleic acids that eventually result in the formation of two daughter cells, as illustrated in Fig. 2. Mistakes in the synthesis of proteins and nucleic acids are quite likely, especially in the mitosis of progenitor cells of older people. In order to avoid such mistakes being passed on during cell division the cell possesses an arsenal of mechanisms that can determine whether damage is present, evaluate its extent and correct it, if feasible.

The upregulation of DNA repair proteins by the ionic products of bioactive glass dissolution, listed in Table 2, indicates that these mechanisms are activated in human osteoblasts. At least four important genes involved in DNA synthesis, repair and recombination are differentially expressed at levels of >200% over control osteoblasts. When the damage is beyond repair the cell voluntarily exits the mitotic cell cycle through death by apoptosis. Apoptosis thereby prevents the creation of

Table 3.
Apoptosis regulators

Regulator	Upregulation (%)
Defender against cell death 1 (DAD-1)	450
Calcium-dependent proteinase small (regulatory) subunit; calpain	410
Deoxyribonuclease II (Dnase II)	160

abnormal cells and represents a means to regulate the selection and proliferation of functional osteoblasts; i.e. osteoblasts capable of synthesising the complex array of extracellular proteins and mucopolysaccharides required to form a mineralised matrix that is characteristic of mature osteocytes.

The treatment of the osteoblast cultures with the bioactive glass stimuli induced the expression of several important genes involved in apoptosis, as summarised in Table 3. The upregulated genes include calpain [76] and defender against cell death (DAD1) [77]. Through activation of the calpain system, a proteolytic mechanism is thought to mediate apoptotic cell death. On the other hand, DAD1, a regulator of N-linked glycosylation, is essential for cell survival, since DAD1 mutation was shown by Hong *et al.* [77] to induce embryonic apoptosis in mice.

As discussed earlier, the cell cycle does not merely provide the framework for cell proliferation but also determines to some extent cell commitment and differentiation. Bone cells cover a broad spectrum of phenotypes that include predominately the osteoblast, a cell capable of proliferating and synthesising bone cell-specific products such as Type-I collagen. However, a vital cellular population in bone consists of osteocytes that are terminally differentiated osteoblasts. Osteocytes are post-mitotic and not capable of cell division. They are capable of synthesising and maintaining the mineralised bone matrix wherein they reside. Thus, osteocytes represent the cell population responsible for extracellular matrix production and mineralisation, the final step in bone development and probably the most crucial one given the importance of collagen-hydroxyl carbonate apatite (HCA) bonding in determining the mechanical function of bone. Therefore, it is important to observe that the end result of the cell cycle activated by the ionic products of bioactive glass dissolution was the upregulation of numerous genes that express growth factors and cytokines (Table 4) and extracellular matrix components (Table 5). Also, there was 700% increase in the expression of CD44 (Table 6), a specific phenotypic marker of osteocytic differentiation, as shown by Hughes *et al.* [78] and a large increase in cell transduction molecules (Table 7).

cDNA microarray analysis by Xynos *et al.* showed that expression of a potent osteoblast mitogenic growth factor, insulin-like growth factor II (IGF-II) was increased to 300% by exposure of the osteoblasts to the bioactive glass stimuli [61]. This is an important finding because IGF-II is the most abundant growth factor in bone and is also a known inducer of osteoblast proliferation *in vitro*. It is produced locally by bone cells and is considered to exert mostly autocrine or paracrine effects.

Table 4.
Growth factors and cytokines

Factor	Activation (%)
Insulin-like growth factor II (IGF-II); somatomedin A	300
Macrophage-specific colony-stimulating factor (CSF1; MCSF)	260
Bone-derived growth factor	200
Vascular endothelial growth factor precursor (VEGF)	200

Table 5.
Extracellular matrix compounds

Factor	Activation (%)
Matrix metalloproteinase 14 precursor (MMP14)	370
Matrix metalloproteinase 2 (MMP2)	270
Metalloproteinase 1 inhibitor precursor (TIMP1)	220
TIMP2 (MI)	220
Bone proteoglycan II precursor; decorin	200

Table 6.
Cell surface antigens and receptors

Factor	Activation (%)
CD44 antigen hematopoetic form precursor	700
Fibronectin receptor beta subunit; integrin beta 1	600
N-sam; fibroblast growth factor receptor-1 precursor	300
Vascular cell adhesion protein-1 precursor (V-CAM1)	200

Table 7.
Signal transduction molecules

Factor	Activation (%)
MAP kinase-activated protein kinase 2 (MAPKAP kinase 2)	600
Dual specificity nitrogen-activated protein kinase 2	200
ADP-ribosylation factor 1	200

IGF-II expression is relatively high in developing bone periosteum and growth plate, healing fracture callus tissue and developing ectopic bone tissue, as reviewed by Xynos *et al.* [61].

Thus, it is important to determine whether the bioactive glass activation leads to a significant increase of the secretion of the IGF-II protein by the cells and whether it is present in a biologically active state in the cultures. Xynos *et al.* [61] confirmed the IGF-II mRNA upregulation using quantitative real-time PCR and also showed that the unbound IGF-II protein concentration was increased in the presence of the bioactive ions. The results indicate that the ionic dissolution products of Bioglass 45S5 may increase IGF-II availability in osteoblasts by inducing the transcription of the growth factor and its carrier protein (Table 4) and also by regulating the dissociation of this factor from its binding protein. The unbound IGF-II is likely to be responsible for the increase in cell proliferation observed in the cultures. Similar bioactive induction of the transcription of extracellular matrix components and their secretion and self-organisation into a mineralised matrix may be responsible for the rapid formation and growth of bone nodules and differentiation of the mature osteocyte phenotype.

BIOACTIVE FOAM SCAFFOLDS

The osteogenic response to the ionic dissolution products of bioactive glass described above has provided impetus to the design of several controlled release systems. The simplest way to control the rate of release of soluble calcium and silicon ions is to control the radius of curvature and surface area of particles. Oonishi *et al.* were the first to show *in vivo* the large effect that particle size had on bone regeneration in a critical size defect in a rabbit model [63, 64, 79].

The enhanced rate of bone formation in the presence of specific particle sizes of bioactive glass was examined by Gatti *et al.* [80], Maroothynaden and Hench [81] and Sepulveda and co-workers [82–86]. The results showed that the glass surface reactions (Stages 1–3 in Fig. 1) led to a rapid release of soluble Si, Ca and P in the first few hours. Subsequent growth and crystallisation of the HCA layer on the particles created a diffusion barrier and led to a greatly reduced rate of ion release. This understanding has led to development of 3D scaffold materials that provide controlled release of soluble ions from the mesoporous material of the foam, 1–20-nm pores in the gel-glass matrix and large interconnected macropores (100–300 μm diameter) for proliferation and differentiation of cells and maintenance of continuous vascularization. These third-generation biomaterials are termed foam scaffolds.

In sol–gel processing of compositions the $CaO-P_2O_5-SiO_2$ system is used to make the bioactive foam scaffolds [87–95]. There are several important characteristics of the gel-derived foam scaffolds. The pore interconnects are greater then 100 μm, which enables a 3D cellular structure to form, complete with vasculature. The pores are 200–300 μm in diameter, which enables self assembly of cells into 3D structures. The mesoporous network dissolves at a rate that releases the proper ionic concentration for osteogenesis, described above.

REGENERATION OF SOFT CONNECTIVE TISSUES

Obtaining and maintaining a blood supply in tissue-engineered constructs is necessary for their long-term stability following implantation. Recent studies have used third-generation bioactive, resorbable composites to enhance vascularization of a regenerated soft tissue construct in rats. A novel composites processing method was developed by Boccaccini *et al.* to impregnate a polyglycolic acid (PGA) mesh with controlled concentrations of sub-5 μm particles of 45S5 Bioglass® [96–100]. An optimal concentration of bioactive particles was established by Day *et al.*, using a rat fibroblast model with monitoring of cell proliferation and production of vascular endothelial growth factor (VEGF) [101]. Samples of the bioactive, resorbable composite mesh were implanted subcutaneously in adult rats. The composite meshes were infiltrated by fibroblasts and blood vessels. There were significantly greater number of blood vessels at 28 and 42 days post-implantation within the composite meshes compared with control PGA polymer meshes without the bioactive phase. The results for the *in vitro* studies indicate that the increased neovascularization occurring with meshes coated with 45S5 Bioglass® may be due to the bioactive phase stimulating release of VEGF, a highly pro-angiogenic factor that acts solely on endothelial cells.

IMPLICATIONS FOR THE FUTURE

A cellular and molecular basis for development of third-generation biomaterials provides the scientific foundation for molecular design of scaffolds for tissue engineering and for *in situ* tissue regeneration and repair, with minimally invasive surgery. The economic advantages of these new approaches may aid in solving the problems of caring for an ageing population. It should be feasible to design a new generation of gene-activating biomaterials tailored for specific patients and disease states. Tissue-engineered constructs based on a patient's own cells may be produced that can be used to select an optimal pharmaceutical treatment. The results suggest that bioactive stimuli may be used to activate genes in a preventive treatment to maintain the health of tissues as they age. Only a few years ago this concept would have seemed unimaginable. But we need to remember that only 37 years ago the concept of a material that would not be rejected by living tissues also seemed unimaginable.

Acknowledgements

The authors acknowledge the UK Engineering and Physics Research Council and the Medical Research Council for grants that have partially supported these studies.

REFERENCES

1. L. L. Hench, *Science* **208**, 826 (1980).
2. L. L. Hench and J. W. Wilson, *Science* **226**, 630 (1984).
3. L. L. Hench, *J. Am. Ceram. Soc.* **81**, 1705 (1998).

4. G. Hastings and P. Ducheyne (Eds), *Macromolcular Biomaterials*. CRC Press, Boca Raton, FL (1984).
5. L. L. Hench, R. J. Splinter, W. C. Allen and T. K. Greenlee, Jr., *J. Biomed. Mater. Res.* **2**, 117 (1971).
6. C. A. Beckham, T. K. Greenlee, Jr. and A. R. Crebo, *Calcif. Tissue Res.* **8**, 165 (1971).
7. J. Wilson, G. H. Pigott, F. J. Schoen and L. L. Hench, *J. Biomed. Mater. Res.* **15**, 805 (1981).
8. T. Yamamuro, L. L. Hench and J. Wilson (Eds), *Handbook on Bioactive Ceramics: Bioactive Glasses and Glass-Ceramics, Vol. 1*. CRC Press, Boca Raton, FL (1990).
9. S. F. Hulbert, J. C. Bokros, L. L. Hench, J. Wilson and G. Heimke, in: *High Tech Ceramics*, P. Vincenzini (Ed.), p. 189. Elsevier, Amsterdam (1987).
10. L. L. Hench and J. Wilson, *An Introduction to Bioceramics*. World Scientific, London (1993).
11. U. Gross, R. Kinne, H. J. Schmitz and V. Strunz, *Crit. Rev. Biocompatibil.* **4**, 155 (1988).
12. U. Gross and V. Strunz, *J. Biomed. Mater. Res.* **19**, 251 (1985).
13. T. Nakamura, T. Yamamuro, S. Higashi, T. Kokubo and S. Itoo, *J. Biomed. Mater. Res.* **19**, 685 (1985).
14. T. Kokubo, S. Ito, S. Sakka and T. Yamamuro, *J. Mater. Sci.* **21**, 536 (1986).
15. T. Kitsugi, T. Yamamuro and T. Kokubo, *J. Bone Jt. Surg.* **71A**, 264 (1989).
16. Ö. H. Andersson, G. Liu, K. H. Karlsson, L. Niemi, J. Miettinen and J. Juhanoja, *J. Mater. Sci. Mater. Med.* **1**, 219 (1990).
17. K. Kangasniemi and A. Yli-Urpo, in: *Handbook of Bioactive Ceramics, Vol. 1*, T. Yamamuro, L. L. Hench and J. Wilson (Eds), p. 97. CRC Press, Boca Raton, FL (1990).
18. J. Wilson and D. Nolletti, in: *Handbook of Bioactive Ceramics, Vol. 1*, T. Yamamuro, L. L. Hench and J. Wilson (Eds), p. 283. CRC Press, Boca Raton, FL (1990).
19. J. Wilson, A. E. Clark, E. Douek, J. Krieger, W. K. Smith and J. S. Zamet, *Bioceramics* **7**, 415 (1994).
20. C. A. Shapoff, D. C. Alexander and A. E. Clark, *Compend. Contin. Educ. Dent.* **18**, 352 (1997).
21. R. Reck, S. Storkel and A. Meyer, *Ann. NY Acad. Sci.* **523**, 100 (1988).
22. G. E. Merwin, in: *Handbook of Bioactive Ceramics, Vol. 1*, T. Yamamuro, L. L. Hench and J. Wilson (Eds), p. 323. CRC Press, Boca Raton, FL (1990).
23. E. Douek, in: *Proceedings of IVth International Symposium on Bioceramics in Medicine*, W. Bonfield (Ed.), London (1991).
24. K. Lobel, in: *Clinical Performance of Skeletal Prostheses*, L. Hench and J. Wilson (Eds), p. 214. Chapman & Hall, London (1996).
25. H. R. Stanley, M. B. Hall, A. E. Clark, J. C. King, L. L. Hench and J. J. Berte, *Int. J. Oral Maxillofac. Implant.* **12**, 95 (1997).
26. H. R. Stanley, A. E. Clark and L. L. Hench, in: *Clinical Performance of Skeletal Prostheses*, L. L. Hench and J. Wilson (Eds), p. 255. Chapman & Hall, London (1996).
27. T. Yamamuro, *Calcif. Tissue Res.* **8**, 89 (1971).
28. T. Yamamuro, J. Shikata, Y. Kakutani, S. Yoshii, T. Kitsugi and K. Ono, *Ann. NY Acad. Sci.* **523**, 107 (1988).
29. S. Yoshii, Y. Kakutani, T. Yamamuro, T. Nakamura, T. Kitsugi, M. Oka, T. Kokubo and M. Takagi, *J. Biomed. Mater. Res.* **22**, 327 (1988).
30. L. L. Hench and J. Wilson (Eds), *Clinical Performance of Skeletal Prostheses*. Chapman & Hall, London (1996).
31. C. P. A. T. Klein, J. G. C. Wolke and K. deGroot, in: *An Introduction to Bioceramics*, L. L. Hench and J. Wilson (Eds), p. 199. World Scientific, London (1993).
32. T. Yamamuro, L. L. Hench and J. Wilson (Eds), *CRC Handbook of Bioactive Ceramics, Vol. 2, Calcium Phosphate and Hydroxylapatite Ceramics*. CRC Press, Boca Raton, FL (1990).
33. T. Yamamuro, in: *CRC Handbook of Bioactive Ceramics, Vol. 2, Calcium Phosphate and Hydroxylapatite Ceramics*, T. Yamamuro, L. L. Hench and J. Wilson (Eds), p. 89. CRC Press, Boca Raton, FL (1990).

34. W. Bonfield, in: *CRC Handbook of Bioactive Ceramics, Vol. 2, Calcium Phosphate and Hydroxylapatite Ceramics*, T. Yamamuro, L. L. Hench and J. Wilson (Eds), Chapter 16. CRC Press, Boca Raton, FL (1990).
35. B. M. Wrobelewski, P. A. Fleming and P. D. Siney, *J. Bone Jt. Surg. Br.* **81**, 427 (1999).
36. F. J. Schoen, R. J. Levy and H. R. Piehler, *J. Soc. Cardiovasc. Pathol.* **1**, 29 (1992).
37. L. L. Hench and J. K. West, *Life Chem. Rep.* **13**, 187 (1996).
38. L. L. Hench, *Curr. Orthopaed.* **14**, 7 (2000).
39. L. L. Hench and J. M. Polak, *Science* **296**, 1014 (2001).
40. J. Wilson and S. B. Low, *J. Appl. Biomater.* **3**, 123 (1992).
41. A. E. Fetner, M. S. Hartigan and S. B. Low, *Compend. Contin. Educ. Dent.* **15**, 932 (1994).
42. H. Oonishi, L. L. Hench, J. Wilson, F. Sugihara, E. Tsuji, S. Kushitani and H. Iwaki, *J. Biomed. Mater. Res.* **44**, 31 (1999).
43. L. L. Hench, D. L. Wheeler and D. C. Greenspan, *J. Sol-Gel Sci. Technol.* **13**, 245 (1998).
44. L. L. Hench and H. A. Paschall, *J. Biomed. Mater. Res. Symp.* **4**, 25 (1973).
45. L. L. Hench and A. E. Clark, in: *Biocompatibility of Orthopedic Implants, Vol. 2*, D. F. Williams (Ed.), Chapter 6. CRC Press, Boca Raton, FL (1982).
46. U. M. Gross and V. Strunz, *J. Biomed. Mater. Res.* **14**, 607 (1980).
47. U. Gross, J. Brandes, V. Strunz, J. Bab and J. Sela, *J. Biomed. Mater. Res.* **15**, 291 (1981).
48. M. Ogino, F. Ohuchi and L. L. Hench, *J. Biomed. Mater. Res.* **14**, 55 (1980).
49. A. E. Clark, C. G. Pantano and L. L. Hench, *J. Am. Ceram. Soc.* **59**, 37 (1976).
50. L. L. Hench and H. A. Paschall, *J. Biomed. Mater. Res.* **5**, 49 (1974).
51. T. Kokubo, *J. Non-Cryst. Solids* **120**, 138 (1990).
52. I. Rehman, M. Karsh, L. L. Hench and W. Bonfield, *J. Biomed. Mater. Res.* **50**, 97 (2000).
53. L. L. Hench, H. A. Paschall, W. C. Allen and G. Piotrowski, *National Bureau of Standards Special Publication* **415**, 19 (1975).
54. L. L. Hench, *J. Biomed. Mater. Res.* **41**, 511 (1998).
55. K. D. Lobel and L. L. Hench, *J. Biomed. Mater. Res.* **39**, 575 (1998).
56. R. A. Latour, Jr., J. K. West, L. L. Hench and S. D. Trembley, *Bioceramics* **10**, 541 (1997).
57. J. K. West, R. A. Latour, Jr. and L. L. Hench, *J. Biomed. Mater. Res.* **37**, 585 (1997).
58. M. Bosetti, M. Cannas, J. Wilson and L. L. Hench, *Bioceramics* **11** (1998).
59. I. D. Xynos, M. V. J. Hukkanen, J. J. Batten, I. D. Buttery, L. L. Hench and J. M. Polak, *Calcif. Tiss. Int.* **67**, 321 (2000).
60. L. L. Hench, I. D. Xynos, L. D. Buttery and J. M. Polak, *J. Mater. Res. Innov.* **3**, 313 (2000).
61. I. D. Xynos, A. J. Edgar, L. D. Buttery, L. L. Hench and J. M. Polak, *Biochem. Biophys. Res. Commun.* **276**, 461 (2000).
62. I. D. Xynos, A. J. Edgar, L. D. K. Buttery, L. L. Hench and J. M. Polak, *J. Biomed. Mater. Res.* **55**, 151 (2001).
63. Y. Fujishiro, H. Oonishi and L. L. Hench, *Bioceramics* **10**, 283 (1997).
64. H. Oonishi, L. L. Hench, J. Wilson, F. Sugihara, E. Tsuji, M. Matsuwura, S. Kin, T. Yamamoto and S. Mizokawa, *J. Biomed. Mater. Res.* **51**, 37 (2000).
65. L. D. K. Buttery, S. Bourne, I. D. Xynos, H. Wood, F. J. Hughes, S. P. F. Hughes, V. Episkopou and J. M. Polak, *Tissue Eng.* **7**, 89 (2001).
66. R. Bielby, R. Pryce, L. L. Hench and J. M. Polak, *Tissue Eng.* **6**, 693 (2000).
67. A. E. Grigoriadis, Z. Q. Whang and E. F. Wagner, in: *Cellular and Molecular Biology of Bone*, M. Noda (Ed.), p. 15. Academic Press, San Diego, CA (1996).
68. M. Zorning and G. I. Evan, *Curr. Biol.* **1**, 1553 (1996).
69. B. C. Lewis, H. Shim, Q. Li, C. S. Lee, A. Maity and C. V. Dang, *Mol. Cell. Biol.* **17**, 4967 (1997).
70. M. A. Caligo, G. Cipollini, L. Fiore, S. Calvo, F. Basolo, P. Collechi, S. Pepe, M. Petrini and G. Bevilacqua, *Leukemia Res.* **20**, 161 (1996).
71. J. E. Onyia, J. Bidwell, J. Herring, J. Hulman and J. M. Hock, *Bone* **17**, 479 (1995).

72. O. O. Oni, *Injury* **31**, 363 (2000).
73. R. Donnelan and R. Chetty, *Mol. Pathol.* **51**, 1 (1998).
74. I. Sauroja, J. Smeds, T. Vlaykova, R. Kumar, L. Talve, M. Hahka-Kemppinen, K. Punnonen, C. T. Jansen, K. Hemminki and S. Pyrhonen, *Genes Chromosomes Cancer* **28**, 404 (2000).
75. T. J. Fu, J. Peng, G. Lee, D. H. Price and O. Flores, *J. Biol. Chem.* **274**, 34527 (1999).
76. M. K. Squier, A. C. Miller, A. M. Malkinson and J. J. Cohen, *J. Cell Physiol.* **159**, 229 (1994).
77. N. A. Hong, M. Flannery, S. N. Hsieh, D. Cado, R. Pedersen and A. Winoto, *Dev. Biol.* **220**, 76 (2000).
78. D. Hughes, D. M. Salter and R. Simpson, *J. Bone Miner. Res.* **9**, 39 (1994).
79. H. Oonishi, L. L. Hench, J. Wilson, F. Sugihara, E. Tsuji, S. Kushitani and H. Iwaki, *J. Biomed. Mater. Res.* **44**, 31 (1999).
80. A. M. Gatti, E. Monari, L. L. Hench and D. C. Greenspan, *Ceram. Trans.* **101**, 177 (2000).
81. J. Maroothynaden and L. L. Hench, *Key Eng. Mater.* **192–195**, 575 (2001).
82. P. Sepulveda, J. R. Jones and L. L. Hench, *Key Eng. Mater.* **192–195**, 629 (2001).
83. P. Sepulveda, J. R. Jones and L. L. Hench, *J. Biomed. Mater. Res. Appl. Biomater.* **58**, 734 (2001).
84. J. R. Jones, P. Sepulveda and L. L. Hench, *J. Biomed. Mater. Res. Appl. Biomater.* **58**, 720 (2001).
85. P. Sepulveda, J. R. Jones and L. L. Hench, *Key Eng. Mater.* **218–220**, 303 (2002).
86. P. Sepulveda, J. R. Jones and L. L. Hench, *J. Biomed. Mater. Res.* **61**, 301 (2002).
87. P. Saravanapavan and L. L. Hench, *J. Biomed. Mater. Res.* **54**, 608 (2000).
88. P. Sepulveda, J. R. Jones and L. L. Hench, *J. Biomed. Mater. Res.* **59**, 340 (2002).
89. P. Sepulveda, J. R. Jones and L. L. Hench, *Key Eng. Mater.* **218–220**, 287 (2002).
90. P. Sepulveda, J. R. Jones and L. L. Hench, *Key Eng. Mater.* **218–220**, 299 (2002).
91. R. F. S. Lenza, J. R. Jones, W. L. Vasconcelos and L. L. Hench, *J. Mater. Sci. Mater. Med.* **13**, 837 (2002).
92. J. R. Jones, R. F. S. Lenza, W. Vasconcelos and L. L. Hench, in: *Lung Transplantation and Pulmonary Disease*, N. Banner, J. Polak and M. Yacoub (Eds). Cambridge University Press, Cambridge (2002).
93. R. F. S. Lenza, J. R. Jones, W. L. Vasconcelos and L. L. Hench, *J. Biomed. Mater. Res.* **13** (2002).
94. P. Saravanapavan and L. L. Hench, *J. Non-Cryst. Solids* **318**, 1 (2003).
95. P. Saravanapavan and L. L. Hench, *J. Non-Cryst. Solids* **318**, 14 (2003).
96. A. Stamboulis, A. Boccaccini and L. L. Hench, *J. Mater. Sci. Mater. Med.* **13**, 843 (2002).
97. A. G. Stamboulis, A. R. Boccaccini and L. L. Hench, *Adv. Eng. Mater.* **4**, 105 (2002).
98. A. R. Boccaccini, J. A. Roether, L. L. Hench, V. Maquet and R. Jérôme, *J. Ceram. Eng. Sci. Proc.* **23**, 805 (2002).
99. J. A. Roether, A. R. Boccaccini, L. L. Hench, V. Maquet, S. Gautier and R. Jérôme, *Biomaterials* **23**, 3871 (2002).
100. J. A. Roether, J. Gough, A. R. Boccaccini, V. Maquet and L. L. Hench, *J. Mater. Sci. Mater. Med.* **13**, 1207 (2002).
101. R. Day, A. R. Boccaccini, J. A. Roether, S. Surey, A. Forbes, L. L. Hench and S. Gabe, *Gastroenterology* **122** (Suppl. 1), T875 (2002).

Gels, Genes, Grafts and Giants, pp. 863–875 (2005)

Xyloglucan as a synthetic extracellular matrix for hepatocyte attachment

SEOG-JIN SEO [1], IN-KYU PARK [1], MI-KYOUNG YOO [1], MAYUMI SHIRAKAWA [2], TOSHIHIRO AKAIKE [3] and CHONG-SU CHO [1,*]

[1] *School of Agricultural Biotechnology, Seoul National University, Seoul 151-742, South Korea*
[2] *Food, Food Additives and Chemicals Division, Dainippon Pharmaceutical Co., Suita, Osaka 564-0053, Japan*
[3] *Department of Biomolecular Engineering, Tokyo Institute of Technology, Yokohama 226-8501, Japan*

Received 23 September 2003; accepted 27 March 2004

Abstract—The possibility of employing naturally derived xyloglucan (XG) having galactose moieties in the side chain for the development of synthetic extracellular matrix in tissue engineering was studied. Hepatocyte adhesion to the XG-coated polystyrene (PS) dish was 73.9% after 30 min incubation, whereas that to the PS dish as a negative control was 59.1%. The hepatocyte adhesion to the XG-coated surface was dependent on the presence of Ca^{2+} ions, whereas that to the XG-coated surface could not be induced by Mg^{2+} ions alone, indicating specific interaction between galactose moieties of XG and asialoglycoprotein receptors of hepatocytes. From the results of fluorescence, confocal laser micrographs and flow cytometry, it was suggested that XG was internalized by hepatocytes through a receptor-mediated mechanism. The DNA synthesis of hepatocytes attached to the XG-coated surface was decreased with an increase of the coating concentration of XG and in the presence of epidermal growth factor (EGF). The spreading shapes of the hepatocytes attached to the surface in the presence of EGF at low concentration of XG (1 μg/ml) were enhanced. The hepatocytes attached to the surface at a high concentration of XG (200 μg/ml) showed round shapes with spheroids after 16 h in the presence of EGF.

Key words: Xyloglucan; extracellular matrix; hepatocyte; spheroid formation.

INTRODUCTION

Cellular growth and differentiation, in the two-dimensional cell culture as well as in the three-dimensional space of the developing organism, require an extracellular matrix (ECM) with which the cells can interact [1]. Natural ECMs such as col-

*To whom correspondence should be addressed. Tel.: (82-2) 880-4636. Fax: (82-2) 875-2494. E-mail: chocs@plaza.snu.ac.kr

lagen, laminin and fibronectin have been revealed to control spreading, migration, adhesion and proliferation of hepatocytes since they have several globular domains which are specialized for binding to the cell [2, 3]. Synthetic ECMs such as alginate [4] and chitosan derivatives [5, 6] substitute for many functions of the natural ECM, organizing cells into a three-dimensional architecture, providing mechanical integrity to the new tissue and providing a space for the diffusion of nutrients and metabolites to and from the cell [7, 8]. Development of new synthetic ECMs plays an important role in tissue engineering providing an alternative to organ and tissue transplantation, both of which suffer from a limitation of supply.

In previous studies, galactose-derivatized polystyrene (PS), poly(*N*-*p*-vinylbenzyl-4-*o*-β-D-galactopyranosyl-D-gluconamide) (PVLA), as a synthetic polymer, has been reported as an excellent synthetic ECM to guide hepatocyte adhesion through the unique asialoglycoprotein receptor (ASGPR)–galactose interaction, although ASGPR is a non-adhesion cell-surface receptor [9, 10].

Xyloglucan (XG), a polysaccharide derived from tamarind seeds, is composed of glucose units in the main chain and xylose and galactose units in the side chains. When XG is partially degraded by β-galactosidase, the resultant product exhibits thermally-reversible gelation in dilute aqueous solutions [11].

In this study, we aimed to study the possibility of employing XG as a synthetic ECM which control spreading, adhesion, proliferation and differentiation of hepatocytes in tissue engineering because XG, as a naturally-derived polymer, has galactose moieties in the side chain that bind with ASGPR of hepatocytes [12] and is expected to be a biocompatible synthetic ECM owing to usage of food additives such as jelly and ice cream.

In particular, the emphasis of this study was placed on a galactose moiety in the XG for the specific adhesive ligand to ASGPR in the two-dimensional cell culture compared with PVLA already reported previously as a synthetic ECM [9, 10].

The XG hydrogel will be expected to use hepatocyte encapsulation, hepatocyte transplantation and tissue-engineering applications owing to the thermoreversible gelation of XG.

MATERIALS AND METHODS

Materials

XG with a percentage of galactose removal of 48% by β-galactosidase [13] was kindly supplied by Dainippon Pharmaceutical (Osaka, Japan). The unit structures of XG are shown in Fig. 1. When 48% of galactose by β-galactosidase is removed from purified XG (purity 97%), the constituent ratio of hepta-, octa- and nonasaccharide is 48, 31 and 21 mol%, respectively. Galactose moieties are replaced with xylose ones after enzymatic degradation of original XG. The molecular mass of partially galactose-removed XG measured by light scattering varies from 108 to 827 kDa, indicating a broad polydispersion index (PDI). PVLA was prepared by the

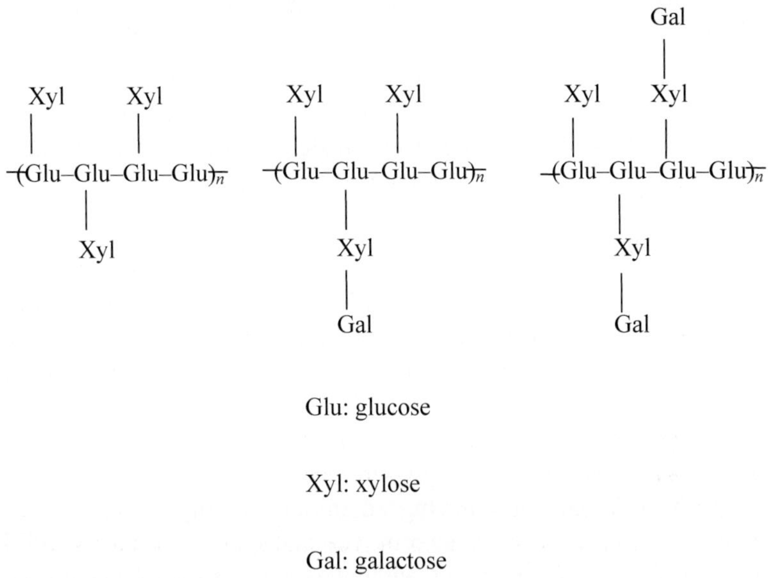

Figure 1. Chemical structure of XG.

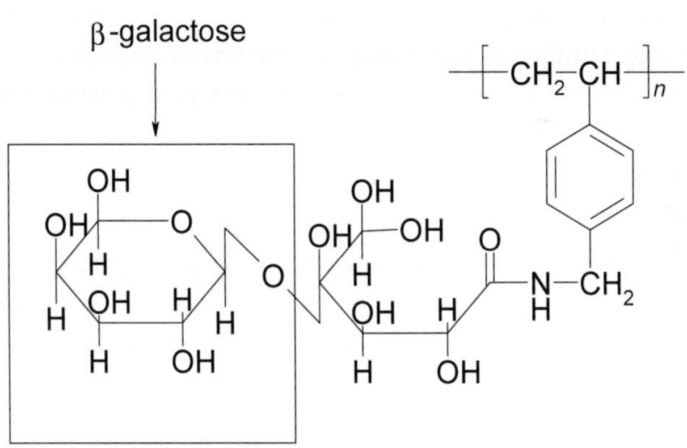

Figure 2. Chemical structure of PVLA.

same method as previously reported [14]. Degree of polymerization (DP), measured by light scattering, and PDI, measured by size-exclusion chromatography (SEC), of the prepared PVLA is 106 and 1.7, respectively. The structure of PVLA is shown in Fig. 2.

Coating of polystyrene (PS) dish

XG (1 μg/ml or 200 μg/ml) and PVLA (100 μg/ml) were dissolved in water. The polymer solutions were deposited onto non-treated PS dishes at 37°C for 24 h. Non-adsorbed polymer solutions were removed from the dishes and washed with water three times. The coated dishes were treated at 37°C for 2 h with PBS/0.5 wt% BSA to prevent non-specific cell adhesion and then washed with water three times.

Contact-angle measurement

The water contact angle of the XG- and PVLA-coated surfaces was measured at 25°C (or 37°C) and 65% relative humidity with the sessile drop method using distilled water. At least five readings were made on different parts of the surfaces after placing 5 μl of water droplets on them and averaged.

Hepatocyte attachment

Primary hepatocytes were isolated from ICR mice (5–7 weeks old, male) by the method of *in situ* collagenase perfusion method as described by Seglen [15]. Cell viability, measured by Trypan blue exclusion, was about 90%. The viable primary hepatocytes were suspended in Williams' E (WE) medium (Gibco-BRL), containing antibiotics (50 μg penicillin ml^{-1}, 50 μg streptomycin ml^{-1} and 100 μg neomycin ml^{-1}), HEPES (18 mM), epidermal growth factor (EGF, 20 ng ml^{-1}) and insulin (100 nM). Isolated hepatocytes having an initial viability of 90%, checked by Trypan blue exclusion, were suspended in XG- and PVLA-coated 24-well PS dish at 5×10^4 cells/ml in WE medium without serum. Isolated hepatocytes were incubated in a humidified air/CO_2 incubator (95 : 5, v/v). After a specific time period, the medium including free non-adherent cells was thoroughly washed with PBS solution. The number of collected non-adherent cells was counted using a Coulter counter. Hepatocyte adhesion was defined as a percentage of maximum adhesion, which was 100%.

Effect of divalent cations

The effect of divalent cations was investigated using Hanks' balanced salt solution (HBSS) in the presence and absence of 1.8 mM of Ca^{2+} and Mg^{2+} ions. The hepatocytes treated with medium in a humidified air/CO_2 incubator at 37°C for 60 min were plated and cultured on XG-coated (or PVLA-coated) dishes for 30 min. Non-adherent cells were collected and the number of cells was counted with a Coulter counter.

Observation of phase-contrast, fluorescence and confocal laser scanning micrographs

Isolated hepatocytes (1×10^5 cells/ml) were placed onto a cover glass in a plate at 37°C for 3 h. The plate was washed with PBS solution and FITC-XG (1 mg/ml) was added to the plate. After 30 min incubation, hepatocytes adhered onto the cover glass were transferred to a slide glass for phase-contrast, fluorescence and confocal laser scanning microscopy (Leica TCS NT).

Measurement of flow cytometery

Hepatocytes (1×10^6 cells/ml) were incubated with 1 mg/ml of FITC-XG or FITC-PVLA in WE containing 0.1% BSA and 0.1 μg/ml of propodium iodide (PI) for

60 min on ice with tapping every 5 min. All cells were resuspended in 1 ml of buffer (0.1% BSA and 1 mM $CaCl_2$ in HBSS) on ice to be measured by flow cytometry (FACSCalibur, Becton Dickinson) after washing with the buffer three times. Hepatocyte region was selected by adjusting the forward scatter (FS) and the side scatter (SS). Surface-labeled hepatocytes with FITC-XG or FITC-PVLA were selected from the region of PI-unlabeled hepatocytes to remove cytoplasm-labeled damaged hepatocytes.

DNA synthesis

8×10^3 cells/0.1 ml of isolated hepatocytes were placed into a 96-well PS dish precoated with XG and incubated at 37°C in a humidified air/CO_2 (95 : 5 v/v) incubator for 48 h. Then, 5-bromo-2′-deoxyuridine (BrdU) was added to the hepatocytes attached to the XG-coated PS dish for DNA synthesis [16] and the hepatocytes were incubated at 37°C for 24 h. The incorporation of BrdU was measured by the cell proliferation ELISA system version 2.

RESULTS AND DISCUSSION

Hepatocyte attachment

Figure 3 shows hepatocyte adhesion onto the XG- and PVLA-coated PS dishes against time at 37°C. The results indicated that hepatocyte adhesion to the XG-coated PS dish became as high as 73.9% after 30 min incubation, whereas that to the PS dish as a negative control was 59.1%. This result was almost similar to the PVLA-coated PS dish as a positive control, although there was a little difference of adhesion between them. It is thought that the slight difference of hepatocyte adhesion between XG and PVLA comes from more hydrophobic property of XG than PVLA although they have the same galactose moieties in the side chain. As a matter of fact, it was found that contact angles of water on the XG- and PVLA-coated surface at 25°C were 53.7 ± 2.0 (51.9 ± 3.8 at 37°C) and 40.6 ± 3.3° (43.6 ± 2.4° at 37°C) ($n = 5$), respectively, indicating that the XG-coated surface is more hydrophobic than the PVLA-coated surface. In previous studies, the hepatocyte adhesion to the PVLA-coated surface is inhibited by asialofetuin, a natural ligand for ASGPR and further hepatocytes require Ca^{2+} to adhere to the PVLA-coated surface. From the results, it is assumed that the hepatocyte adhesion onto the PVLA is mediated by the galactose-specific interactions between ASGPR of the hepatocytes and galactose residues of the PVLA [17–19]. Therefore, it is suggested that hepatocyte adhesion to the XG-coated surface is galactose-specific recognition between XG molecules and ASGPR of hepatocytes; a similar tendency was found for PVLA-coated surfaces.

Hepatocyte detachment at 4°C from the XG- or PVLA-coated surface *versus* time is shown in Fig. 4. It was found that hepatocyte adhesion onto the XG- or PVLA-coated surface at 37°C was not decreased after 2 h at 4°C, indicating that hepatocyte

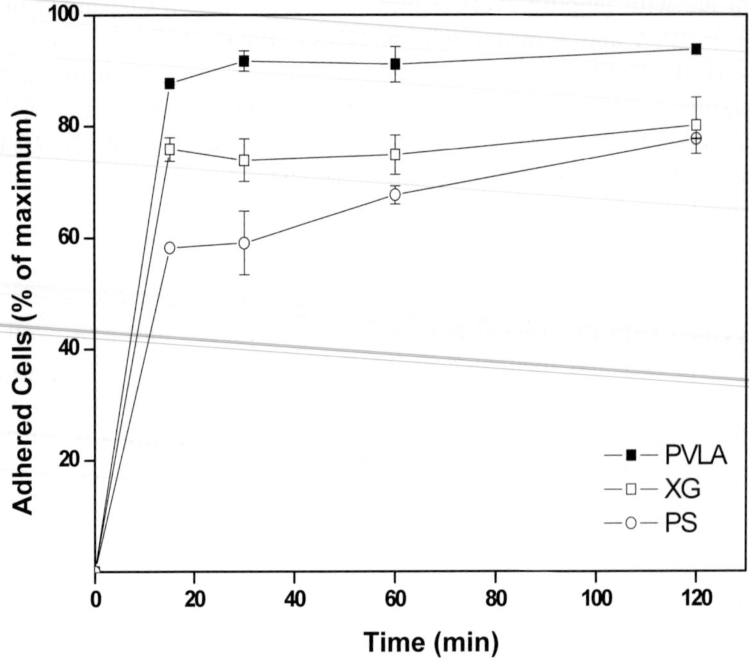

Figure 3. Hepatocyte adhesion onto the XG- and PVLA-coated PS surfaces against time at 37°C. Data are represented as mean ± S.D. ($n = 3$).

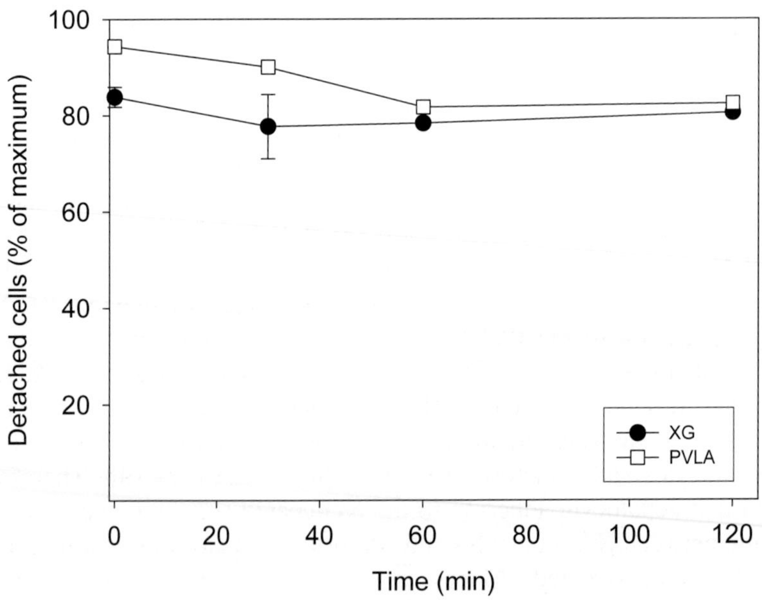

Figure 4. Hepatocyte detachment from the XG- and PVLA-coated PS surfaces against time at 4°C. Data are represented as mean ± S.D. ($n = 3$).

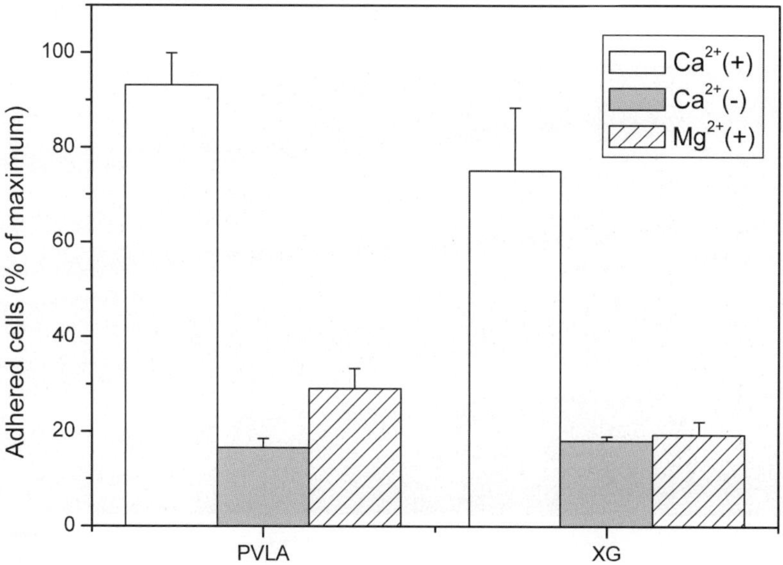

Figure 5. The effect of divalent cations (Ca^{2+} and Mg^{2+}) on hepatocyte adhesion to the surface. Data are represented as mean ± S.D. ($n = 3$).

adhesion occurs even at 4°C. Weigel reported that internalization of ASGPR by hepatocytes did not occur at 4°C, whereas recognition and binding of hepatocytes occurred through a receptor-mediated mechanism [20].

The interaction between ASGPR of hepatocytes and galactose ligands requires Ca^{2+} ions [21]. We checked the effect of divalent cations such as Ca^{2+} and Mg^{2+} on hepatocyte adhesion to the surfaces. The results are shown in Fig. 5. The results indicated that the hepatocyte adhesion to the XG-coated surface was dependent on the presence of Ca^{2+} ions, which was similar to the PVLA-coated surface. Also, hepatocyte adhesion onto the XG-coated surface could not be induced by Mg^{2+} ions alone. From the result of the effect of Ca^{2+} ions, it is suggested that hepatocyte adhesion to the XG-coated surface is mediated by ASGPR as the similar to the PVLA previously reported [17].

Figure 6 shows phase-contrast (Fig. 6a) and fluorescence (Fig. 6b) micrographs of hepatocytes incubated with FITC-XG according to time. From the fluorescence micrograph of the hepatocytes, it was found that the internalization of FITC-XG by hepatocytes was increased with time as shown in Fig. 6b.

Confocal laser scanning micrographs of hepatocytes incubated with FITC-XG for 2 h are shown in Fig. 7. Fluorescence was observed inside the cell membrane. Also, from the results of flow cytometry shown in Fig. 8, the fluorescence intensity of FITC-XG was stronger than that of control (or FITC-PVLA). These results indicate the internalization of XG by hepatocytes through a receptor-mediated mechanism. The affinity of ASGPR to natural and synthetic oligosaccharides having non-reducing galactose residues had been reported by Lee *et al.* [22, 23]. It is known that the binding efficiency of ASGPR is much enhanced with an increase of densities of

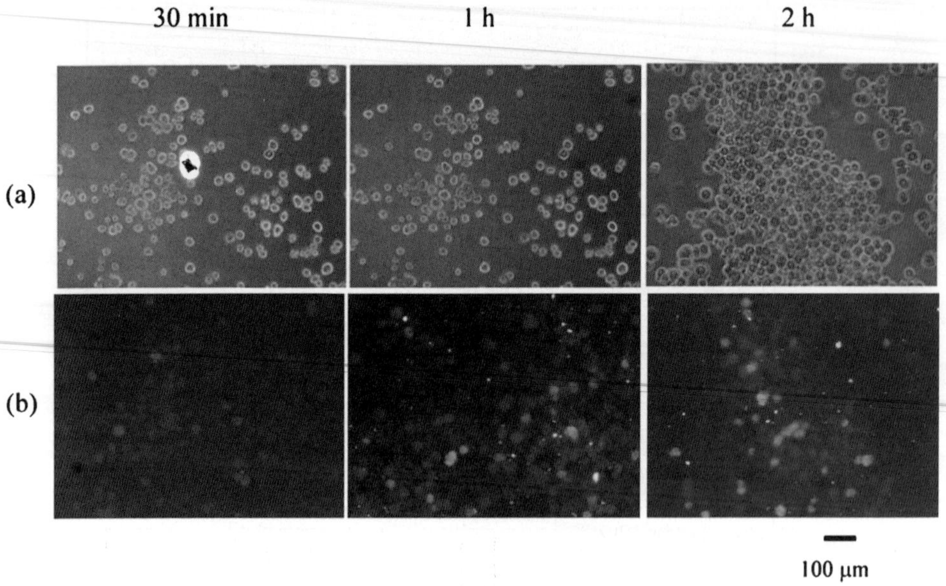

Figure 6. Phase-contrast (a) and fluorescence (b) microphotographs of hepatocytes incubated with FITC-XG (1 mg/ml) against time at 37°C.

the target sugars. Affinity is dramatically enhanced when the number of antennae of branching sugar chains is increased. Strong binding may be induced by the polymeric effect produced by highly concentrated galactose residues along XG molecules, in a similar fashion to PVLA. We assume that these adhesion responses on the XG substratum can be ascribed to galactose-specific interactions between XG and ASGPR on the surface of hepatocytes.

Proliferation of hepatocytes

DNA synthesis as an index of hepatocyte proliferation is shown in Fig. 9. The results indicate that the DNA synthesis of the hepatocytes attached to the XG-coated surface was decreased with an increase of the coating concentration of XG in the presence of EGF although DNA synthesis of the hepatocytes attached to the XG-coated surface was lower than that of the hepatocytes attached to the PS dish as a control. We assume that adhered hepatocytes at the lower galactose densities exhibited the higher DNA synthesis under the spreading morphology and the lower DNA synthesis under the round morphology. It is thought that DNA synthesis of the hepatocytes in the presence of EGF can be ascribed to the spheroid formation of adhered hepatocytes. The morphology of the adhered hepatocytes will be discussed below.

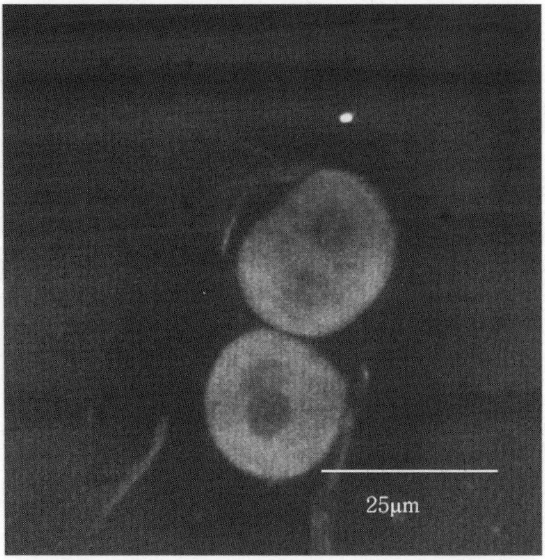

(a)

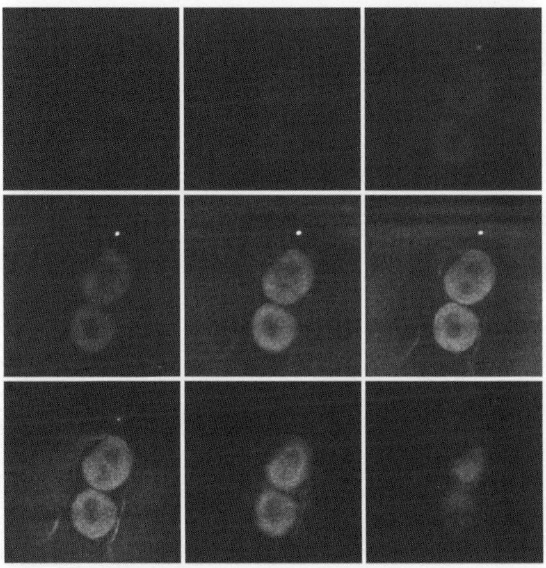

(b)

Figure 7. Confocal laser micrograph of hepatocytes incubated with FITC-XG (1 mg/ml) for 2 h. (a) Projection mode and (b) gallery mode.

Morphology of adhered hepatocytes

Figure 10 shows phase-contrast micrographs of adhered hepatocytes to the XG-coated surface at a low concentration of XG (1 μg/ml). Some cells attached to the surface showed spreading shapes after 16 h in the absence of EGF, whereas almost

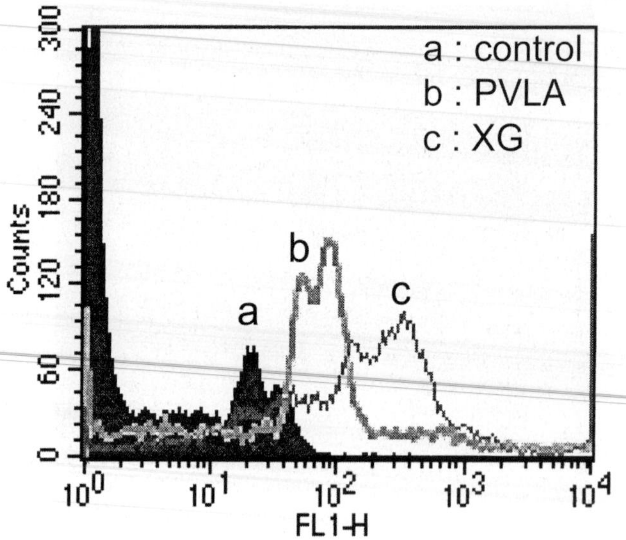

Figure 8. Flow cytometry of hepatocytes incubated with FITC-XG or FITC-PVLA for 1 h on ice.

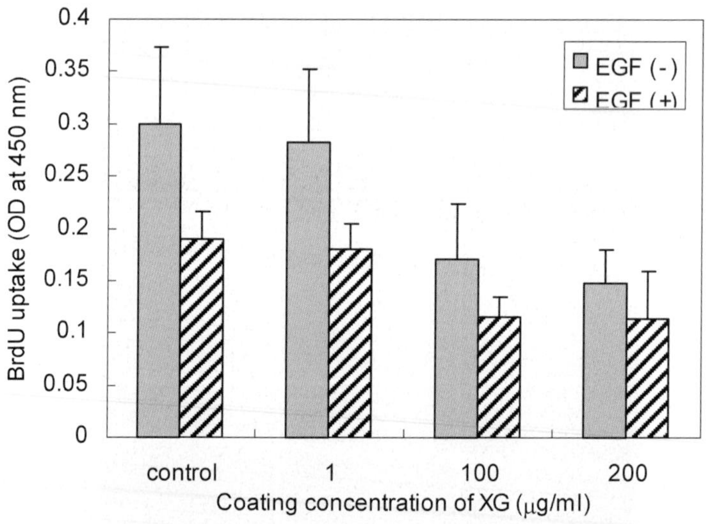

Figure 9. DNA synthesis of hepatocytes cultured on the XG-coated PS surface after 48 h. Concentration of EGF was 50 ng/ml. EGF (−) and EGF (+) represent without EGF and with EGF, respectively. Data are represented as mean ± S.D. ($n = 3$).

all cells attached to the surface showed more spreading shapes after 16 h in the presence of EGF, indicating that the spreading shapes of adhered hepatocytes in the presence of EGF were enhanced more than in the absence of EGF. Interestingly, spheroid formation of hepatocytes was observed after 48 h in the presence of EGF.

Figure 11 shows phase-contrast micrographs of adhered hepatocytes to the XG-coated surface at a high concentration of XG (200 μg/ml). Cells attached to

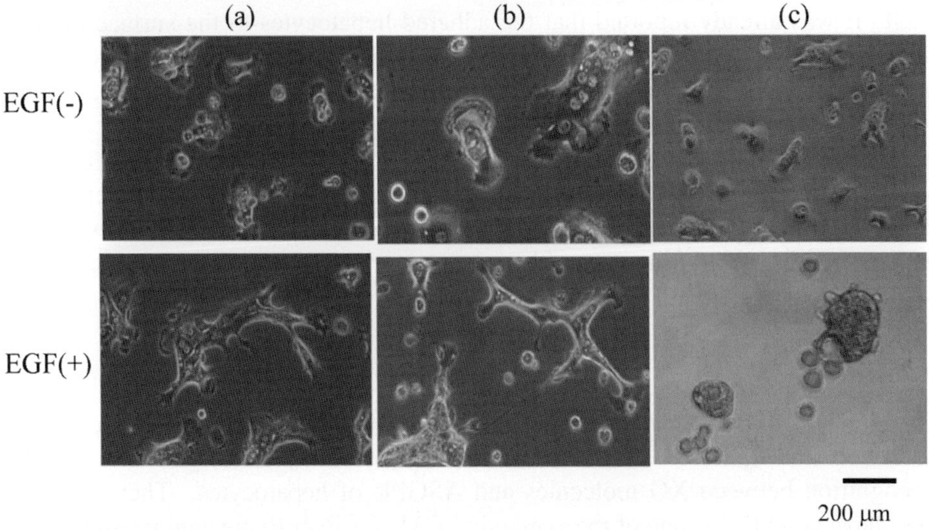

Figure 10. Phase-contrast micrographs of hepatocytes attached to the XG-coated surface at low concentration of XG (1 μg/ml) for (a) 16 h, (b) 24 h and (c) 48 h. EGF (−) and EGF (+) represent without EGF and with EGF, respectively.

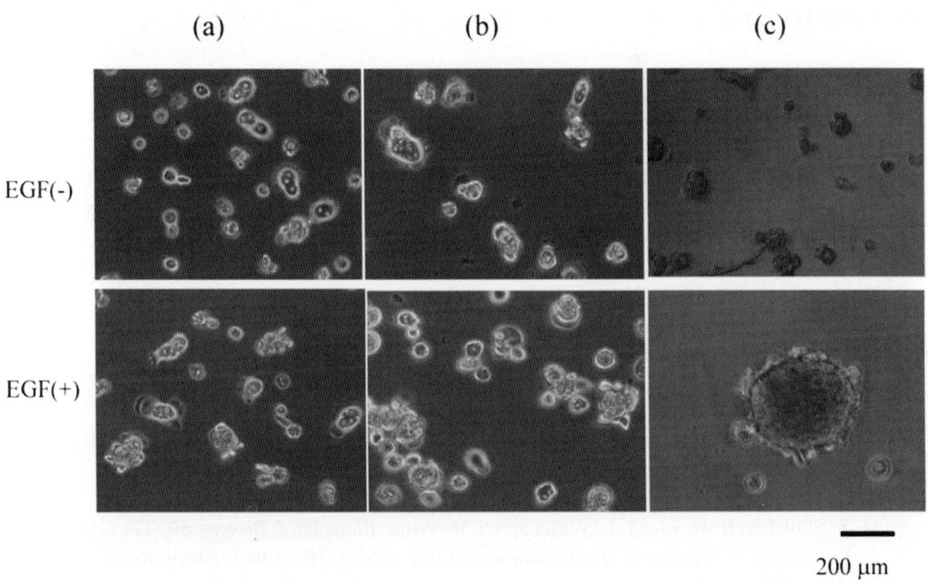

Figure 11. Phase-contrast micrographs of hepatocytes attached to the XG-coated surface at high concentration of XG (200 μg/ml) for (a) 16 h, (b) 24 h and (c) 48 h. EGF (−) and EGF (+) represent without EGF and with EGF, respectively.

the surface showed round shapes and exhibited spheroid formation after 48 h in the absence of EGF, whereas cells attached to the surface showed round shapes and exhibited many spheroid formations after 16 h in the presence of

EGF. It was already reported that the adhered hepatocytes to the surface through integrin-mediated recognition mechanism [3] had a round morphology at low coating densities, whereas the adhered hepatocytes to the surface through galactose-specific recognition mechanism had a spreading one [5]. It was also reported that differentiated functions of hepatocytes in alginate/galactosylated chitosan sponges were closely related to spheroid formation rather than to the hepatocyte adhesion [24]. Therefore, it has been proposed that galactose moieties of XG enhance the spheroid formation of hepatocytes.

CONCLUSIONS

Xyoglucan (XG) having galactose moieties in the side chain showed excellent adhesion and spheroid formation of hepatocytes due to the galactose-specific recognition between XG molecules and ASGPR of hepatocytes. Therefore, XG will have potential as one of the synthetic ECMs for liver tissue engineering.

Acknowledgements

This work was supported by the Korean Science and Engineering Foundation (F01-2003-000-00041-0).

REFERENCES

1. B. R. Olsen, in *Principles of Tissue Engineering*, R. Lanza, R. Langer, W. Chick and R. G. Landes (Eds), p. 47. Academic Press, Austin, TX (1997).
2. N. Sawada, A. Tomomura, C. A. Sattler, G. L. Sattler, H. K. Kleinman and H. C. Pitot, *Exp. Cell Res.* **167**, 458 (1986).
3. D. Mooney, L. Hansen, J. Vacanti, R. Langer, S. Farmer and D. Ingber, *J. Cell Physiol.* **151**, 497 (1992).
4. J. A. Rowley, G. Madlambayan and D. J. Mooney, *Biomaterials* **20**, 45 (1999).
5. I. K. Park, J. Yang, H. J. Jeong, H. S. Bom, I. Harada, T. Akaike, S. I. Kim and C. S. Cho, *Biomaterials* **24**, 2331 (2003).
6. K. Yagi, N. Michibayashi, N. Kurikawa, Y. Nakashima, T. Mizoguchi, A. Harada, S. Higashiyama, H. Muranaka and M. Kawase, *Biol. Pharm. Bull.* **20**, 1290 (1997).
7. A. J. Putnam and D. Mooney, *Nature Med.* **2**, 824 (1996).
8. M. S. Shoichet, R. H. Li, M. L. White and S. R. Winn, *Biotechnol. Bioeng.* **50**, 374 (1996).
9. A. Kobayashi, T. Akaike, K. Kobayashi and H. Sumitomo, *Macromol. Chem. Rapid Commun.* **7**, 645 (1986).
10. S. H. Kim, M. Goto, C. S. Cho and T. Akaike, *Biotechnol. Lett.* **22**, 1049 (2000).
11. S. Miyazaki, F. Suisha, N. Kawasaki, M. Shirakawa, K. Yamatoya and D. Attwood, *J. Control. Rel.* **56**, 75 (1998).
12. D. A. Wall, G. Wilson and A. L. Hubbard, *Cell* **21**, 29 (1980).
13. S. Yamanaka, Y. Yuguchi, H. Urakawa, K. Kajiwara, M. Shirakawa and K. Yamatoya, *SEN'I GAKKAISHI* **55**, 528 (1999).
14. A. Kobayashi, H. Sumitomo and Y. Ina, *Polym. J.* **17**, 567 (1985).
15. P. O. Seglen, *Methods Cell Biol.* **13**, 29 (1976).

16. D. Muir, S. Varon and M. Manthorpe, *Anal. Biochem.* **185**, 377 (1990).
17. A. Kobayashi, M. Goto, K. Kobayashi and T. Akaike, *J. Biomater. Sci. Polymer Edn* **6**, 325 (1994).
18. K. Kobayashi, A. Kobayashi, S. Tobe and T. Akaike, in *Neoglycoconjugates: Preparation and Applications*, Y. C. Lee and R. T. Lee (Eds), p. 261. Academic Press, San Diego, CA (1994).
19. K. Kobayashi, A. Kobayashi and T. Akaike, *Methods Enzymol.* **247**, 409 (1994).
20. P. C. Weigel, *J. Cell Biol.* **87**, 855 (1980).
21. R. Blomhoff, H. Tolleshaug and T. Berg, *J. Biol. Chem.* **257**, 7456 (1980).
22. Y. C. Lee, R. R. Townsend, M. R. Hardy, J. Lonngren, J. Arnarp, M. Haraldsson and H. Lonn, *J. Biol. Chem.* **258**, 199 (1983).
23. R. T. Lee, P. Lin and Y. C. Lee, *Biochemistry* **23**, 4255 (1984).
24. J. Yang, T. W. Chung, M. Nagaoka, M. Goto, C. S. Cho and T. Akaike, *Biotechnol. Lett.* **23**, 1385 (2001).

Gels, Genes, Grafts and Giants, pp. 877–894 (2005)

Development of perforated microthin poly(ε-caprolactone) films as matrices for membrane tissue engineering

A. S. HTAY [1], S. H. TEOH [1,*] and D. W. HUTMACHER [1,2]

[1] *Mechanical Engineering Department, Division of Bioengineering, National University of Singapore, Singapore*

[2] *Department of Orthopaedic Surgery, National University of Singapore, Singapore*

Received 15 October 2003; accepted 28 January 2004

Abstract—The design and fabrication of thin films based on bioresorbable polymers such as poly(ε-caprolactone) (PCL) has been the focus of a part of current biomedical research, especially as matrices for membrane tissue engineering. We have successfully developed perforated microthin PCL membrane for this purpose. Two critical issues are the control of moisture permeability and understanding the degradation of PCL microthin film. In order to increase the moisture permeability, PCL films were biaxially stretched to a thickness of 10 ± 3 μm and perforated with uniform array of holes (180–275 μm) using a Sony Robotic system. After perforation, the water vapour transmission rate was increased by 50% to a value of 47.6 ± 2.7 g/h per m^2. Accelerated hydrolytic degradations were performed in 5 M NaOH. The degraded samples were characterised for changes in weight, surface morphology, mechanical properties, crystallinity and molecular weight. Hydrolytic degradation commenced with random chain scission of backbone ester bonds on the film surface and followed by loss of material due to surface erosion. In general, the perforated films degraded faster than the unperforated microthin films. Scanning electron microscopic images showed that surface erosion led to extensive formation of micropores, microcracks and increased in surface roughness.

Key words: Poly(ε-caprolactone); biaxial stretching; water vapour transmission; tissue engineering; perforated film; accelerated degradation; sodium hydroxide.

INTRODUCTION

The design and fabrication of thin membrane-like matrices is currently investigated by a number of research groups in the field of biomedical sciences [1–6]. In the past decade, the use of films/membranes of both natural and synthetic origins in applications, such as wound dressings, drug-delivery devices and skin tissue engineering, has been widely studied [7]. In a study by Lu *et al.* [1], the degradation

*To whom correspondence should be addressed. Tel.: (65) 6874-6345. Fax: (65) 6777-3537. E-mail: mpetsh@nus.edu.sg.

properties of thin poly (DL-lactic-co-glycolic acid) (PLGA) films for applications in retinal pigment epithelium transplantation and guided tissue regeneration (GTR) were investigated. In several studies hyaluronic-acid-based matrices have been investigated for use in developing of skin equivalents [4–6]. In one such effort, by Hollander *et al.* [5], hyaluronan-based 20-μm-thick membranes with orderly arrays of laser-drilled micro perforations were used for cultivation of autologous keratinocytes and other cell types. Laserskin™ is a thin transparent membrane made of HYAFF (a hyaluronan derivative). In an *in vitro* study of the development of a bi-layered skin substitute, Zacchi *et al.* [6] reported that macropores of Laserskin membrane allowed the keratinocytes to migrate to the lower side at the interface between the natural polymer matrix and the plastic surface of the culture plate. However, the relatively weak mechanical properties of hyaluronan base materials render them difficult to handle.

Tissue-engineered epidermal graft based on a synthetic polymer membrane has been study intensively [7–9]. It was suggested that a thin film with good mechanical properties would be ideal for this application. A thin film would have a higher degree of flexibility but at the same time synthetic polymer membrane has good mechanical properties to allow user-friendly handling on the different anatomical wound sites. In addition, such a matrix should allow for better biochemical interaction between the graft and the underlying tissue. Another advantage of a thin matrix is related to its biodegradability. Such a film would invoke minimal foreign body reaction since it releases lesser degradation products. Micron-thin (in the range of 10 μm) bioresorbable films (that from here onwards will be referred to as microthin film) were successfully produced previously by biaxial stretching of poly(ε-caprolactone) (PCL) [10]. Cell-culture studies by our group showed that the films are highly biocompatible and has the potential to be used as a matrix material for so called 2D tissue engineering applications [9–12]. In a preliminary report, Ng *et al.* [13] evaluated the biocompatibility of these microthin films as a wound dressing in a rat model and results showed that the films did not induce excessive inflammatory reactions *in vivo*. Although no significant differences were detected in wound contraction measurement, tensile strength and histology results, a low level of fibrosis at the wound sites with microthin films was observed. However two issues remain to be investigated. One relates to the limited moisture permeability and the other on the slow degradation of PCL

It has been reported that the hydrolytic degradation of PCL is very slow, especially for those with high molecular weight. PCL is a hydrophobic material and its (ω-hydroxy acid) structure coupled with high initial molecular weight is estimated to take more than 3 years to degrade [14]. Previous literature has reported that the degradation mechanism followed the same principle as that of other aliphatic polyesters such as PLA and PGA [15]. An *in vitro* degradation study of PCL films in phosphate-buffered saline (PBS, pH 7.4) at 37°C showed no significant change in the weight of films, even after 12 months [16]. Observations under microscope also revealed no significant change in physical dimensions. Therefore, accelerated degradation of PCL films was considered. Acceleration of PCL degradation has been studied previously by four approaches:

1. Enzymatic degradation; introduction of enzymes into the degradation medium accelerates the degradation process.
2. Use of free radicals and Zn metal to initiate and catalyze the hydrolysis process.
3. Thermal degradation; i.e. increasing the degradation rate by elevating the temperature of degradation medium.
4. Base-promoted hydrolytic degradation (aqueous sodium hydroxide)

The role of enzymes in degradation of polyesters has been extensively studied. PCL is known to readily degrade in the presence of lipase-type enzymes [16–20]. The role of free radicals in the hydrolytic degradation of PDLA and PCL has also been investigated [21–23]. However, in the natural physiological environment, the occurrence of such enzymes and free radicals *in vivo* is unlikely unless under abnormal physiological conditions, hence these approaches may not be reflective of the *in vivo* degradation mechanism. Hydrolysis of PCL is reported to be faster in basic pH and at higher temperature [24–26]. However, the elevation of temperature has a drastic effect on the morphological structure and composition (e.g. crystallinity) of the microthin biaxially-stretched PCL film due to its low melting point (60°C). Therefore, it is not suitable to carry out PCL degradation at such elevated temperatures. Consequently, accelerated degradation in high basic pH medium is considered a preferred study.

This paper reports on the development of perforated microthin PCL films as matrices for membrane tissue engineering. The water vapour transmission rate, mechanical properties and accelerated degradation of these films in alkaline solution are reported.

MATERIALS AND METHODS

Film fabrication

Poly(ε-caprolactone) polymer was sourced from Sigma-Aldrich (Milwaukee, WI, USA) in pellet form ($M_n = 80 \times 10^3$, density $= 1.145$ g/cm^3). The PCL films were biaxially stretched to a draw ratio of 3×3, as described by Ng *et al.* [10]. Briefly describing the film fabrication process, the PCL pellets were first dissolved in methylene chloride (J. T. Baker, Phillipsburg, NJ, USA). This polymer solution was then cast over a glass mould (size 16 cm $\times$ 16 cm). The solvent was removed by slow evaporation (>6 h) and the film heat pressed to remove any defects such as surface pores before stretching the film using an in-house-built biaxial stretching machine. The solvent cast films were vacuum dried to further remove any remaining solvent for more than a day. Microthin films of thickness 10 ± 3 μm were obtained by stretching the heat pressed solvent cast films with initial sheet thickness of 0.1 ± 0.01 mm.

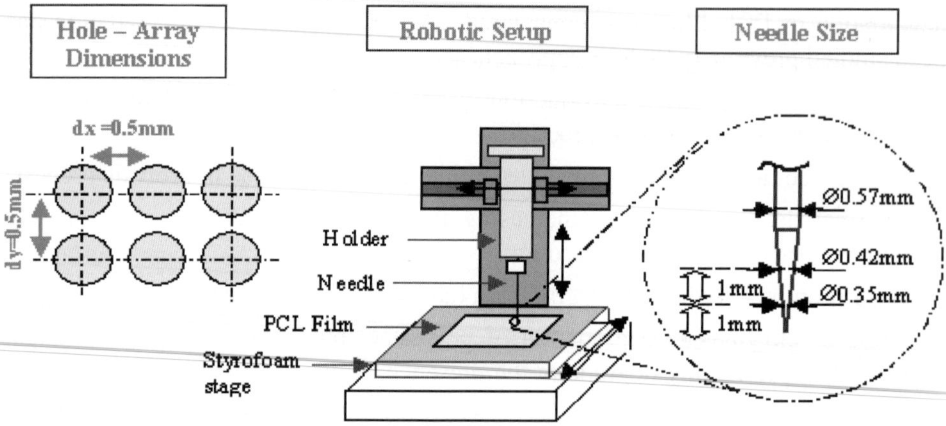

Figure 1. Diagram illustrating the robotic puncturing set-up for film perforation, needle size and hole-array dimensions.

Perforation of microthin films

The films were perforated after biaxial stretching. Perforations were made with the aid of a 3-axis Sony Desktop Robot (Model CAST_AU4/B2521E), a thin needle (0.57 mm diameter) and customized software to execute the perforation process. The films were mounted on a polystyrene foam stage and orderly arrays of holes were created by a vertical punching action using the needle. A diagram showing the dimensions of the needle, hole-array and the robotic puncturing set-up is depicted in Fig. 1. Adhesive tape was used to ensure that the film was flat and taut. The size of the perforation is a function of the diameter of needle, depth of puncture and speed of perforation. The software allowed for the selection of hole-centre to hole-centre distance, the depth and speed of perforation and the size of the hole-array. Through holes were obtained on the biaxially-stretched PCL films with the depth (z-height) of perforation set at 0.7 mm. The speed was set at 50 mm/s and the distance between two hole-centres was 0.5 mm.

Characterization of films

Physical properties. The thickness of the films was measured with a digital micrometer screw gauge (Mitutoyo 395-741-30, Singapore), which has an accuracy of 1 μm. The surface morphology of the films was captured under scanning electron microscope (SEM) (JSM-5600LV, Japan). Before scanning, the films were first gold sputtered using a current of 10 mA for 40 s with a fine gold coater (JFC-1200, Japan).

Surface wettability of films was determined by measuring water contact angle of the films with a VCA optima equipment (AST products™, Singapore). Five readings were taken using a 1-μl droplet of deionised water.

Mechanical properties. The tensile properties were determined using an Universal tensile tester (Instron®, Model: 4502, Singapore). Tensile force was applied to filmstrips of 50 mm × 10 mm at a cross-head speed of 5 mm/min using a 50-N load cell. An optical extensiometer with a lens of field-of-view 25 mm was used to measure the strain. The grip separation and gauge length were set at 30 mm and 5 mm, respectively. The films were fixed on paper window frames (size 20 mm × 30 mm, 10 mm outer border) made of thin paper cardboards to ensure that the films did not get pre-stressed during mounting onto the pneumatic grips. Thickness measurements were taken from five different regions along the strip and the average value was used. The experiment was carried out at $23 \pm 1°C$ and $80 \pm 5\%$ RH. The stress–strain curve was plotted using the data recorded by the Instron machine, and the ultimate tensile strength, elongation at break and modulus were calculated as defined by the equations below:

$$\text{Ultimate tensile strength (N/mm}^2) = \text{Breaking force (N)/Original cross-sectional area (mm}^2),$$

$$\text{Elongation at break (\%)} = (\text{Increase in length (mm)/Original length (mm)}) \times 100,$$

$$\text{Modulus} = \text{Initial slope of stress–strain curve.}$$

Crystallinity measurement. The thermal properties of PCL samples were studied by differential scanning calorimetry (DSC, Perkin-Elmer Pyris-6 DSC, Subambient). The instrument was calibrated with In and Zn standards. All samples were scanned from −20°C to 80°C at a heating rate of 5°C/min, using nitrogen as purge gas. The values of the degree of crystallinity were obtained from the first scan and the percentage of crystallinity was estimated using the enthalpy of fusion of 139.5 J/g for 100% crystalline PCL [27].

Molecular weight determination. The molecular weights of the undegraded and degraded films were determined using gel-permeation chromatography (GPC) equipped with a differential refractometer (Waters, Model 410, Milford, MA, USA) and an absorbance detector refractor (Waters 2690). The samples were dissolved in tetrahydrofuran (THF) (J. T. Baker) in a concentration of $\leqslant$0.1 ml% ($\leqslant$1 mg/ml) and were filled into 1.5-ml vials using 0.2-μm inorganic microfilters. These sample mixtures were eluted in a series of configurations through a Styragel column refractor at a flow rate of 1 ml/min, using THF as the mobile phase. Polystyrene standards (Polysciences, Warrington, PA, USA) were used to obtain a primary calibration curve. Average molecular weights (M_w and M_n) and polydispersity (M_w/M_n) were evaluated and each specimen was assessed three times.

Water vapour transmission rate (WVTR) studies

The moisture permeability of the perforated film was determined by measuring water vapour transmission rate across the films as stipulated by the water method

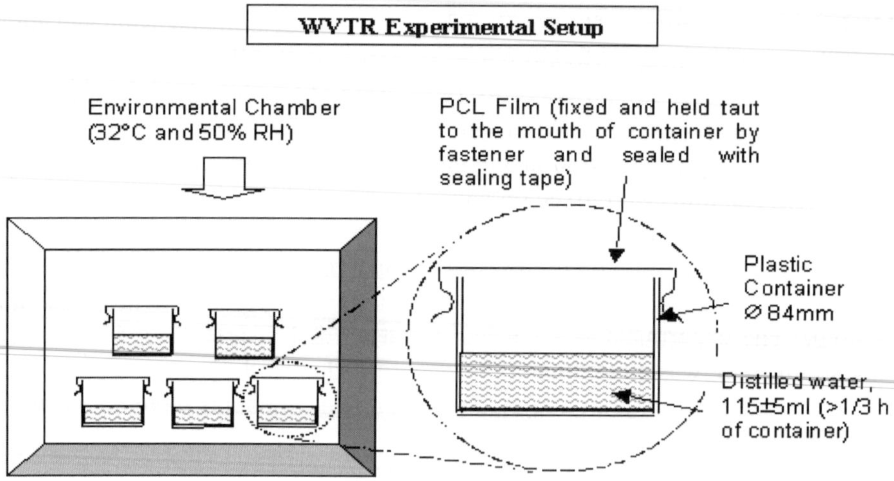

Figure 2. Experimental set-up for WVTR study.

procedure (ASTM Standard Method E96-00). Five specimen groups ($n = 5$) were tested. The test films were fixed onto the opening of round containers, which had an area of 5542 mm^2. The containers were filled with 115 ± 5 ml of deionised water. This was about a third of the container's height. An environmental chamber (Contherm®5200RHS, New Zealand) was used to ensure precise control of temperature and humidity, which were set at 32 ± 1°C and 50 ± 2% RH, respectively. Containers without films were used as control. The experimental set-up of the WVTR study is shown in Fig. 2. The loss of water mass was monitored over 15 days and WVTR was determined using the equation below:

$$\text{WVTR} = (G/t)/A,$$

where G = weight change (g), t = time during which G occurred (h), G/t = slope of the straight line (g/h) and A = test area (cup mouth area) (m^2).

Accelerated hydrolytic degradation

Accelerated hydrolytic degradation of the PCL films in 5 M NaOH was performed. PCL film strips of 50 mm × 10 mm were placed in 25-ml glass bottles and NaOH solution was filled to the brim to ensure films were completely immersed and the samples were incubated at 37°C. Films were retrieved for characterization every 4 h. Extreme care was taken during film retrieval and a flexible forceps was used to ensure only minimal force was exerted on the films. The films were thoroughly rinsed three times in deionised water. The films were then dried in a 40°C vacuum oven for 1 day before characterizing. Mass loss was determined using a microbalance (AG135, Mettler&Toledo) with an accuracy of 0.01 mg. DSC and GPC were used to determine changes in crystallinity and molecular weight. SEM was used to detect the surface morphological changes of the PCL films. A total of 11 samples were allocated for each time point.

RESULTS AND DISCUSSION

In our earlier study [10] the morphology, mechanical properties and gas permeability changes of the stretched and unstretched films were compared. Gas permeability of these PCL films with O_2 and N_2 were determined. It was shown that biaxially-stretched PCL films increased the gas permeability slightly but selectivity remained fairly constant. This higher gas permeability was attributed to the presence of microvoids and drawing defects. The gas transport mechanism was mainly due to Knudsen flow, indicating diffusion through voids and defects formed during biaxial stretching.

Thin PCL films fabricated by biaxially stretching have a number of significant advantages. They have a novel surface microtexture combined with relatively high tensile strength, and are also transparent, flexible and pliable. The results of a number of *in vitro* and *in vivo* studies indicate that biaxially stretched PCL films are bioresorbable and biocompatible [11–13]. As such they have shown to have great potential for use in tissue engineering applications especially in the areas of wound dressing, epidermal matrix and drug release.

Physical properties

A transparent and considerably flexible microthin film with an effective usage area of 10 cm^2, produced when stretched to a ratio of 3, is shown in Fig. 3. The film displays good flexibility and pliability. The cast film shows spherulite sizes in the range of 50–150 μm (Fig. 4). Defect-like structures were also seen in the microthin film (Fig. 5). These defects were formed due to some non-uniform stretching experienced by the film [10], as well as during solvent evaporation.

The perforated microthin film (which from here onwards will be referred to as perforated film) had flap-like structures, formed due to the puncturing and tearing action of the needle. They produced textured surfaces. A SEM image in Fig. 6 clearly shows that these flaps obstruct free flow and cause partial closure of holes.

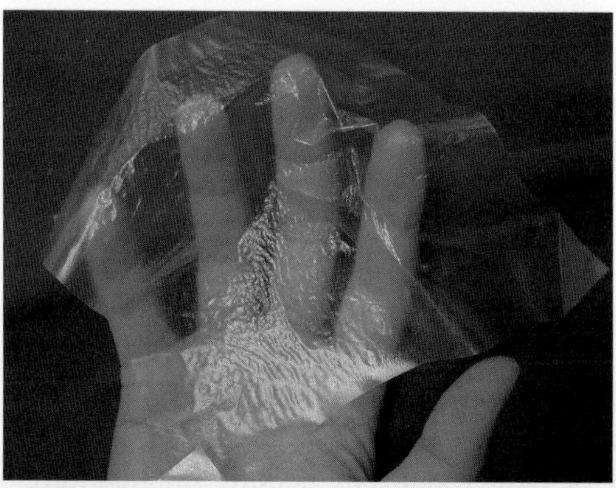

Figure 3. Picture of microthin film produced by biaxially stretching a cast PCL flim, showing transparency, high film flexibility and ease of handling.

Figure 4. Optical microscopy of solvent cast film.

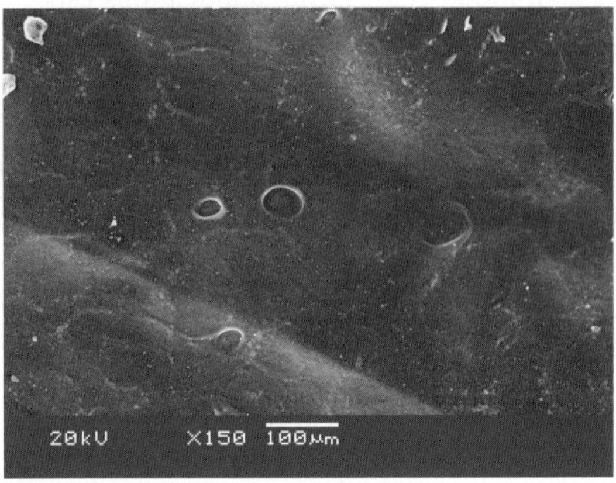

Figure 5. SEM scans of microthin film exhibiting defects.

An unobstructed hole had a diameter of 180 ± 5 μm, perforated film had a hole density of 400 holes/cm^2.

The microthin films were found to be relatively hydrophobic, with average water contact angles of 80.9 ± 2.0°. Treatment with 5 M NaOH solution improved the hydrophilicity of the NaOH-treated films, which recorded a contact angle of 59.6 ± 3.9°.

Mechanical properties

Tensile testing provides an indication of the strength and elasticity of the film, which can be reflected by tensile strength and elongation at break. It was suggested that ideal matrices for skin tissue engineering application should have suitable mechanical properties and a high degree of flexibility [13]. Tensile tests were carried out on all film groups ($n = 10$) and the average values for yield stress,

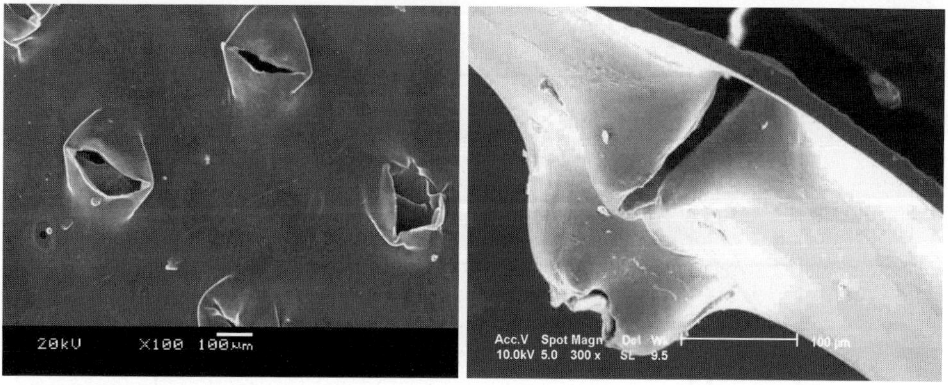

Figure 6. The SEM image on the left-hand side shows the surface morphology of the perforated films and the SEM image on the right-hand side shows the cross-sections of the perforated film, showing the presence of flaps on film causing partial closure of the hole.

Table 1.
Mechanical properties of microthin and perforated PCL film

Property	Microthin	Perforated
Thickness (μm)	10.4 ± 1.8	12.5 ± 1.4
Yield stress (MPa)	27.2 ± 4.6	17.9 ± 3.6
Yield strain (%)	9.6 ± 1.8	9.0 ± 1.6
Modulus (MPa)	279.5 ± 34.5	205.5 ± 29.9
Tensile strength (MPa)	44.0 ± 10.4	29.2 ± 3.9
Elongation at break (%)	81.3 ± 11.6	69.6 ± 19.7

yield strain, elastic modulus, elongation at break and film thickness are tabulated in Table 1. A typical stress–strain curve showing the typical behaviour of the PCL film under tensile load is shown in Fig. 7. Biaxially-stretched PCL films have fibrillar morphology as a result of the stretching [10], hence the yield and tensile strength increases. Results showed that microthin films had a tensile strength of 43.9 ± 10.4 MPa, elongation at break of $81.3 \pm 11.6\%$ and modulus of 279.49 ± 34.5 MPa, comparable to commercially available wound-dressing products such as Omiderm®, which has a tensile strength of 60.3 ± 4.5 MPa and maximum elongation at break of $56.3 \pm 3\%$ [22]. The mechanical properties were decreased after the perforation. For example, the tensile strength, elongation at break and modulus after perforation were 29.2 ± 3.9 MPa, $69.6 \pm 19.7\%$ and 205.46 ± 29.85 MPa, respectively. These values are expected as the holes are stress concentrators. The drop in modulus will mean that the perforated films are more pliable and this is an advantage for applications such as in wound dressing.

WVTR

Optimal moisture permeability through skin replacements, burn coverings and wound-dressings is important to minimize pain and help in the wound healing

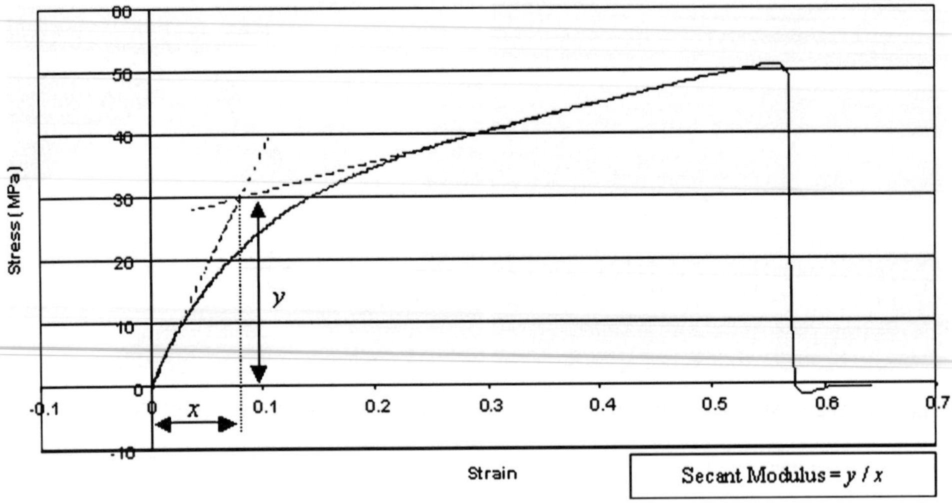

Figure 7. Stress–strain curve of a microthin film.

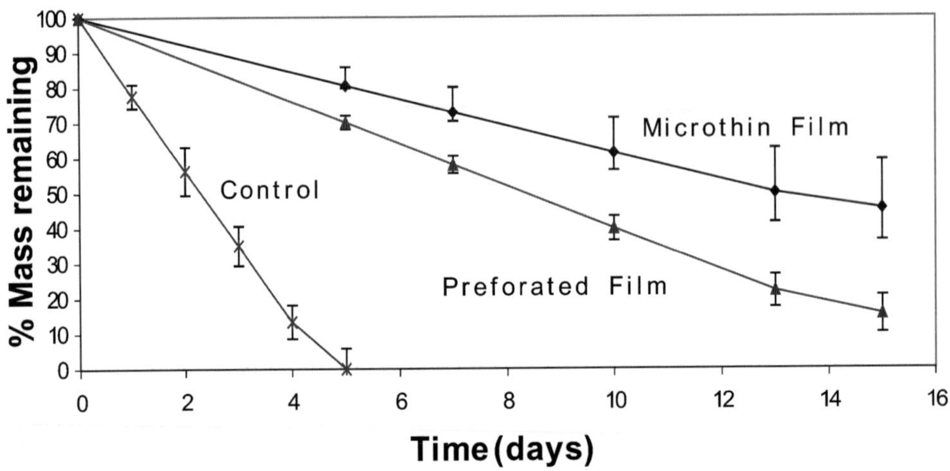

Figure 8. Graph showing trend in water mass loss over 15 days.

process [23]. In studies of wound-dressing materials WVTRs have been used to characterise the permeation properties [19–21, 28]. WVTR of free water surface (control; without film) was 163 g/h per m^2. As shown in Fig. 8, it was observed that the water mass loss rates for the two specimen groups were constant over the first 13 days. Hence, the total water mass loss during this period was used for the WVTR calculation. The WVTR for microthin film was 30.5 ± 3.5 g/h per m^2. Native skin has a WVTR value between 8.33 and 83.33 g/h per m^2 and polyurethane film has a WVTR value of 62.5–83.33 g/h per m^2 [29]. Loke *et al.* [30] reported chitosan based bilayer hydrogel and foam with WVTR values of 125.21 g/h per m^2 and 68.13 g/h per m^2, respectively. Hydroderm®, a polyurethane membrane,

Table 2.
WVTR values of microthin and perforated films compared to the native skin and other commercially available membranes

Specimen group	WVTR (g/h per m^2)
Without film (control)	135.5 ± 14.6
Microthin	30.5 ± 3.5
Perforated	47.6 ± 2.7
Skin [23]	8.33–83.33
Polyurethane [23]	62.5–83
Chitosan hydrogel [17]	125.21
Chitosan foam [17]	68.13

has varying water vapour permeability depending on the degree of hydration of the membrane [29]. Table 2 shows the results of the WVTR for the various commercial membranes as compared to the native skin. The WVTR of the perforated film was 50% more than that of microthin film, at 47.6 ± 2.7 g/h per m^2. The perforation of the microthin PCL films had improved the water vapour permeability of the original PCL microthin films significantly. Compared to the WVTR value that was obtained for a free water surface, 163 g/h per m^2, an microthin (unperforated) PCL film lost only 23% of water and a perforated film lost as much as 40% of water. Hence, perforated film could facilitate better drainage of excess fluid and prevention of fluid accumulation.

Accelerated hydrolytic degradation

Burkersroda *et al.* [31] demonstrated that it is possible to predict the erosion mechanism of water-insoluble biodegradable polymer matrices. They reported that all degradable polymers can undergo both surface erosion or bulk erosion and which way a polymer matrix erodes depends on three things: (1) the diffusivity of water inside the matrix, (2) the degradation rate of the polymer's functional groups and (3) the dimensions of the matrix.

According to Burkersroda's theoretical model and diffusion-reaction phenomena, matrices undergo surface erosion when the diffusivity of water into the polymer matrix is slow compared to rate of hydrolysis. Bulk degradation is considered to be a homogeneous event. This implied a constant water concentration gradient throughout the material. Surface degradation can be considered as a heterogeneous event because a gradient exists in the water concentration profile within the degrading material.

Mass loss. The dynamics of mass loss for the two films followed similar degradation profiles and are shown in Fig. 9. Both film types followed a steady decrease rate. At each time point, only intact film pieces were taken for characterization whereas irretrievable tiny fragments broken off from the main film strip were ig-

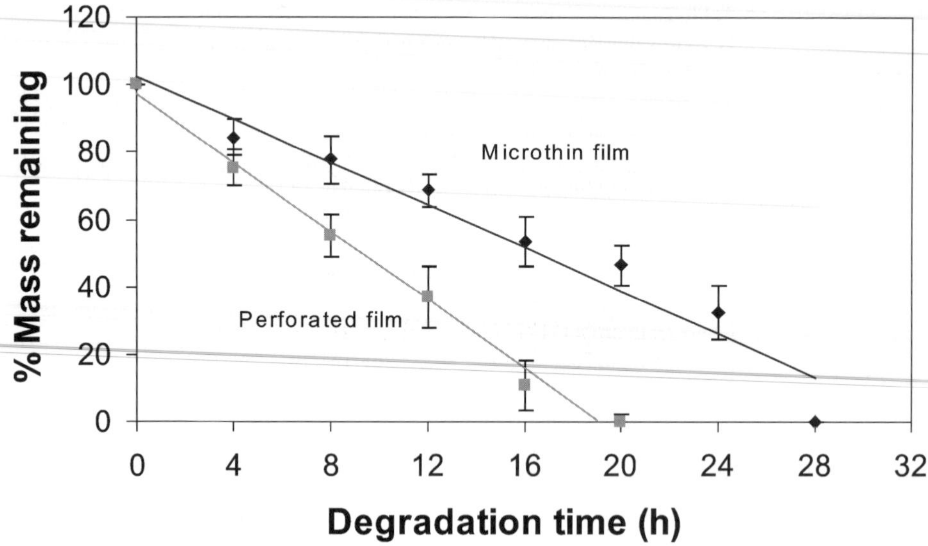

Figure 9. Graphical illustration of mass loss profiles of microthin and perforated films degraded in 5 M NaOH.

nored. Flexible forceps had been used to avoid exerting unnecessary force during the film retrieval, which might contribute in the fragmentation of films.

The microthin films degraded in 5 M NaOH solution showed considerable mass loss from the beginning of the experiment. The unperforated microthin films were dissolved completely around 28 h, perforated films dissolved completely after about 20 h. For both films, it was observed that towards their last time points, before they completely dissolved, they first broke up into large fragments (10–30% of the original film). Fragmentation of the microthin and perforated film was observed first at 20 h and 16 h, respectively. During the last few time points some of the film strips recovered for characterisation were smaller in size. In general, towards the last time points, the film fragments appeared fluffy and delicate. Erosion was so severe that a slight touch of the forceps during retrieval was enough to pulverise them. From Fig. 9, the degradation rate for the microthin film was about 1.5×10^{-2} g/h and that of the perforated film was 2.5×10^{-2} g/h, which is almost twice that of the microthin film. From observations it appeared that the PCL films degraded mainly by surface erosion. The advantage of surface eroding polymers is the predictability of the erosion process [32].

Degradation kinetics. From visual observations made, loss of film mass with degradation time can be described to follow two stages, (1) initial decrease in mass loss at a steady slow rate and (2) a secondary phase involving rapid loss of mass leading to fragmentation and then followed by complete dissolution.

As such it is hypothesized that initially the mass loss of the PCL films was due to the degradation of the amorphous regions. Once most of the amorphous regions

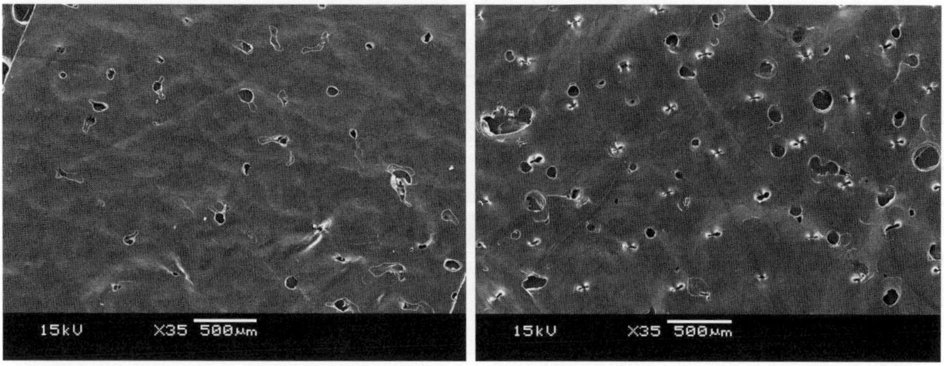

Figure 10. SEM images of degraded films in 5 M NaOH of microthin film at 12 h (right), showing formation of numerous micropores and perforated films, and (left) at 16 h, demonstrating the extent of micropores formation before the onset of fragmentation.

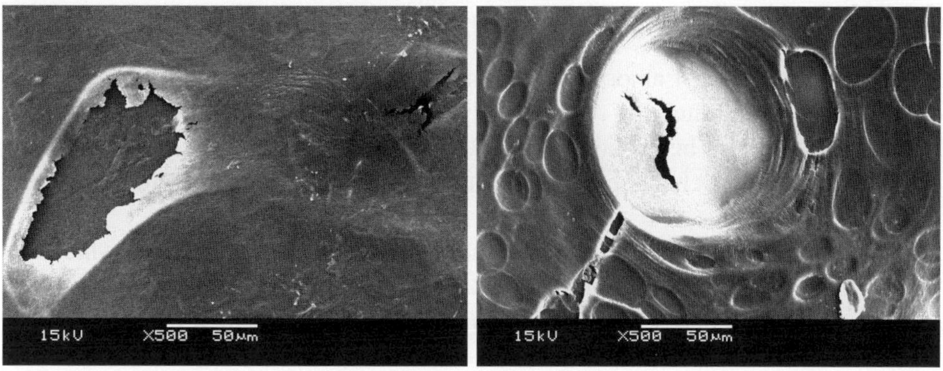

Figure 11. SEM images of 5 M NaOH degraded films showing (right) the erosion of top layers exposing secondary layers in unperforated film at 12 h and (left) image of perforated film at 12 h, showing that thinner regions were more readily attacked by NaOH.

were degraded away crystalline regions were attacked, leading to fragmentation, which corresponded to the rapid mass loss. This hypothesis is illustrated in the SEM pictures (Figs 10 and 11) where numerous holes and cracks were seen just before the last few data points. Thinner regions were more readily attacked by NaOH.

An initial increase in crystallinity of microthin and perforated films degraded in 5 M NaOH solution was recorded, i.e. from an initial of 60.8 ± 3.9% for both films to 74.1 ± 4.6% and 75.7 ± 6.7% by 8 h, respectively (Fig. 12). After 8 h the crystallinity remained constant until 20 h and 28 h, respectively, when they dissolved completely. Increase in crystallinity may indicate the preferential degradation in the amorphous domains of the PCL films. Tilstra *et al.* [33] showed that the amorphous part of PCL was degraded prior to the crystalline part in a biotic environment. Previously, uniaxially drawn (600%) PCL exposed to a *Fusarium* species exhibited a periodic surface structure perpendicular to the direction of orientation [34]. The authors suggested that the observed structure was due to the

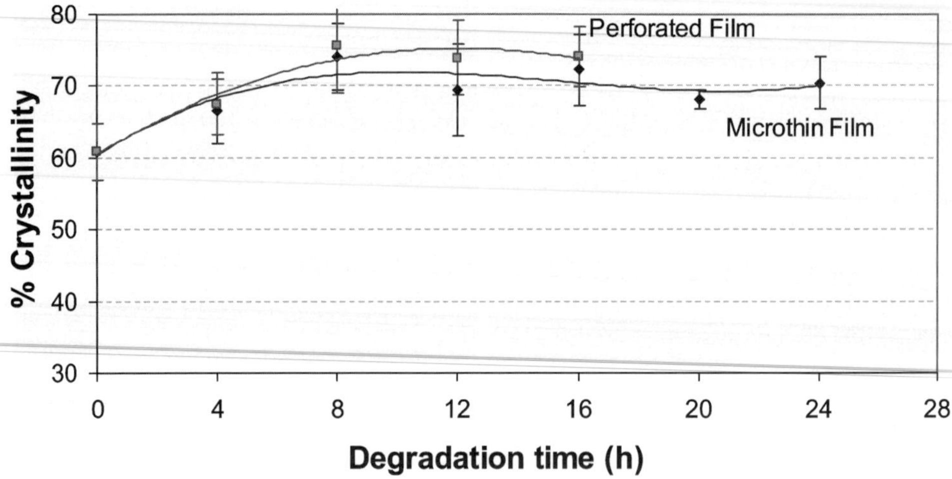

Figure 12. Graphical illustration of changes in crystallinity of films when degraded in 5 M NaOH.

selective degradation of amorphous regions between lamellae, lying perpendicular to the direction of stretching. Alternatively the increase in crystallinity of degraded PCL films could also be explained as due to degradation-induced crystallization of partially hydrolyzed chains present in amorphous micro-domains without loss of material [23].

Grizzi and co-workers [35] reported that the critical thickness above which the hydrolytic degradation of amorphous PLA/GA polymers becomes heterogeneous (surface erosion) is believed to be in the range of 250–300 μm. Burkersroda *et al.* [31] indicated that the PCL matrices must exceed 1.3 cm in thickness before they undergo surface erosion. However, our results appeared to indicate otherwise. This could have been due to ultrathin structures (10 ± 3 μm thick) giving rise to a high surface to volume ratio of the biaxially-stretched PCL film. Moreover, the PCL film was made of very high M_w PCL (80×10^3) and is highly hydrophobic with a water contact angle of 81°, hence the rate of diffusion of water into the bulk of the material will be very slow. The microstructure on the surface was also highly heterogeneous with highly drawn crystalline fibrillar structures and interspersed with amorphous phase. This can be seen in the atomic force microscope picture published earlier [10]. These conditions may explain why PCL films were more susceptible to surface erosion than bulk erosion.

Changes in film morphology. The gross appearances of all the original films are illustrated in Figs 10 and 11. Initially the films appeared transparent and due to the interaction with NaOH, the film surface became more opaque over time. On the whole, SEM images showed that erosion caused increase in surface roughness. In particular the perforated film showed significant surface erosion which has resulted in creation of dimple like structures. Extensive micropores formation was observed to have occurred earlier in perforated films than the unperforated microthin film.

Also perforations appeared to have assisted in the early disintegration of the film. Micropores were formed often at regions of 'weak structure', which include defects and thinner regions. Inherent defects are sites of high strain energy and hence more susceptible to hydrolytic attack. Erosion of layer by layer was also observed in both cases. Diffusion is one of the important factors affecting both surface and bulk erosions [31]. As seen from the SEM images, the surface becomes rougher with time.

One of the major factors affecting the overall transport (diffusion) in a polymer is the presence of defect structures such as voids and microcracks. The faster degradation of the perforated film was mainly due to its perforation and flap-like structures. Puncturing with the needle caused the material at the regions of the perforation to undergo large structural deformation due to the high tensile force. This resulted in film rupture, which created an opening as well as the flaps. Due to the shear forces experienced by the polymer during the perforation process, strain energy is induced at the perforation sites. Surfaces with higher energy state would be more susceptible to hydrolysis.

Molecular weight changes. Unlike the profile for mass loss there was no significant change in the weight-average molecular weight (M_w) and number-average molecular weight (M_n) for both films. They remain at 14×10^3 and 10×10^3, respectively, over the entire time-course of the experiment. Two deductions could be made: (1) little chain cleavage is needed to cause the surface erosion and mass loss seen during the accelerated degradation experiment or (2) chain scission is occurring mainly at the chain ends, leading to conservation of molecular weight.

Changes in tensile properties. All the films became stiffer and more brittle with degradation time. Figures 13 and 14 show that the decrease in tensile strength and elongation at break was in agreement with the findings of Pitt *et al.* [36] and Göpferich [37]. The changes in tensile strength can be described roughly as an exponential decrease for microthin film, whereas for perforated film a more linear trend was apparent. A rapid initial decrease was observed followed by a slower decrease at about 11 h and 10 h. Over a degradation time of 20 h elongation at break showed a similar decreasing trend.

CONCLUSIONS

Biaxially-stretched PCL membrane perforated with an array of holes using a robotic puncturing technique has been successfully achieved. This fabrication process produced thin membranes in the range of 10 μm thickness with a water vapour transmission rate of 47.6 $\pm$ 2.7 g/h per m^2, which was more than 50% greater than the unperforated film. This is within the range of the native skin. The mechanical properties such as the tensile strength, elongation at break and modulus

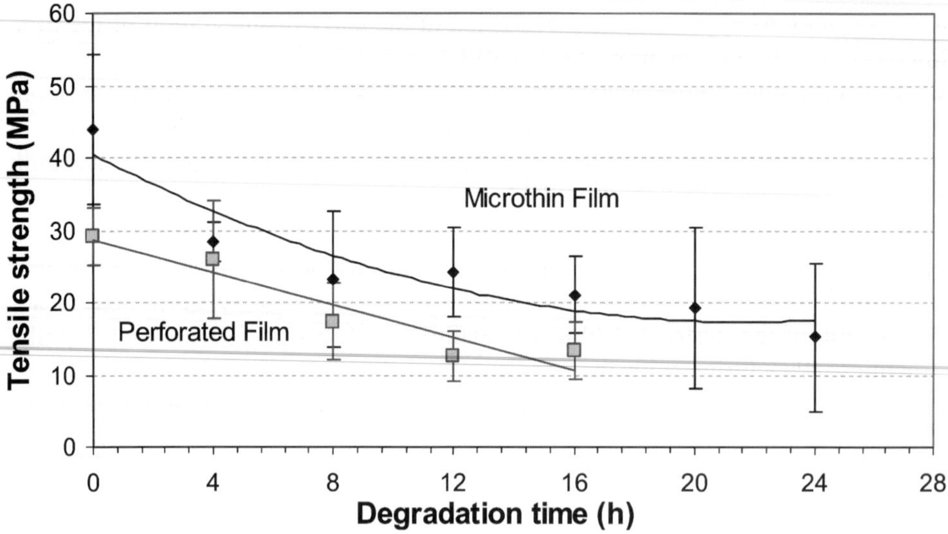

Figure 13. Variation of tensile strength for the microthin and perforated film subjected to 5 M NaOH accelerated degradation.

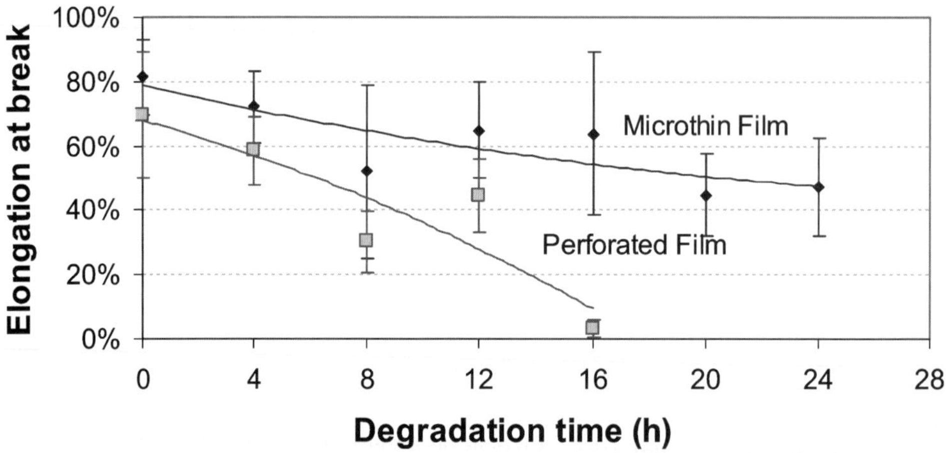

Figure 14. Variation of elongation at break for the microthin and perforated film subjected to 5 M NaOH accelerated degradation.

after perforation was 29.2 ± 3.9 MPa, 69.6 ± 19.7% and 205.46 ± 29.85 MPa, respectively, and these are deemed mechanically strong enough for handling and serve as good cell-template carrier. Accelerated degradation studies using 5 M NaOH revealed that the mechanism of hydrolytic degradation is predominately by surface erosion. SEM observation showed the erosion layer by layer. The SEM images also showed that surface erosion led to extensive formation of micropores, microcracks and increased in surface roughness. The degradation rate after perforation had increased almost by double. All the films became stiffer and more brittle with degradation time in 5 M NaOH.

REFERENCES

1. L. Lu, C. A. Garcia and A. G. Mikos, *J. Biomed. Mater. Res.* **46**, 236 (1999).
2. M. Maeda, K. Kadota, M. Kajihara, A. Sano and K. Fujioka, *J. Control. Rel.* **77**, 261 (2001).
3. L. J. Suggs, J. L. West and A. G. Mikos, *Biomaterials* **20**, 683 (1999).
4. P. K. Lam, E. S. Chan, E. W. To, C. H. Lau, S. C. Yen and W. W. King, *J. Trauma* **11**, 47 (1999).
5. D. Hollander, C. Soranzo, S. Falk and J. Windolf, *J. Trauma* **50**, 1125 (2001).
6. V. Zacchi, C. Soranzo, R. Cortivo, M. Radice, P. Brun and G. Abatangelo, *J. Biomed. Mater. Res.* **40**, 187 (1998).
7. D. W. Hutmacher and W. Vanscheidt, *Drugs Today* **38**, 113 (2002).
8. J. Teumer, J. Hardin-Young and N. L. Parenteau, in: *Frontiers in Tissue Engineering*, C. W. Patrick, Jr., A. G. Mikos and L.V. McIntire (Eds), p. 664. Elsevier, Oxford (1998).
9. J. F. Trent and R. S. Kirsner, *Int. J. Clin. Pract.* **5**, 408 (1998).
10. C. S. Ng, S. H. Teoh, T. S. Chung and D. W. Hutmacher, *Polymer* **41**, 5855 (2000).
11. H. L. Khor, K. W. Ng, A. S. Htay, J. T. Shantz, S. H. Teoh and D.W. Hutmacher, *J. Mater. Sci. Mater. Med.* **14**, 113 (2003).
12. K. W. Ng, D. W. Hutmacher, J. T. Schantz, C. S. Ng, H. P. Too, T. C. Lim, T. T. Phan and S. H. Teoh, *Tissue Eng.* **7**, 441 (2001).
13. K. W. Ng, H. L. Khor, H. N. Achuth, A. S. Htay, S. H. Teoh, S. Moochhala, T. C. Lim and D. W. Hutmacher, In vivo Evaluation of an ultra-thin substrate for epidermal regeneration, in: *Proceedings of the 11th International Conference on Biomedical Engineering*, Singapore, (2002).
14. W. P. Ye, S. D. Fu, H. J. Wei, Y. Y. Ji and X. Yong, *React. Funct. Polym.* **32**, 161 (1997).
15. S. Li and M. Vert, in: *Degradable Polymers: Principles and Applications*, G. Scott and D. Gilead (Eds), p. 43. Chapman&Hall, London (1995).
16. A. S. Htay, MEng Thesis, Mechanical Engineering Department, National University of Singapore, Singapore (2004).
17. M. Mochizuki and M. Hirami, *Polym. Adv. Tech.* **8**, 203 (1997).
18. Z. Gan, Q. Liang, J. Zhang and X. Jing, *Polym. Degrad. Stabil.* **56**, 209 (1997).
19. H. Tsuji and T. Ishizaka, *J. Appl. Polym. Sci.* **80**, 281 (2001).
20. S. Li, *J. Biomed. Mater. Res. (Appl Biomater.)* **48**, 342 (1999).
21. S. A. M. Ali, S. P. Zhong, P. J. Doherty and D. F. Williams, *Biomaterials* **14**, 648 (1993).
22. Y. Lanir, in: *Handbook of Bioengineering*, R. Skalak and S. Chien (Eds), p. 11.1. McGraw-Hill, New York, NY (1987).
23. S. Li., J. L. Espartero, P. M. Foch and M. Vert, *J. Biomater. Sci. Polymer Edn* **8**, 165 (1996).
24. N. S. J. A. Gerrits and P. J. Lemstra, *Polymer* **32**, 1770 (1991).
25. C. G. Pitt, in: *Biodegradable Polymers as Drug Delivery Systems*, M. Chasin and R. Langer (Eds), p. 71, Chapter 3. Marcel Dekker, New York, NY (1990).
26. J. McMurry, *Organic Chemistry*, 4th edn. Brooks/Cole, London (1996).
27. V. Crescenzi, G. Manzini, G. Calzolari and C. Borri, *Eur. Polym. J.* **8**, 449 (1972).
28. A. Widra, in: *Encyclopedia of Polymer Science and Engineering*, H. F. Mark, N. M. Bikales, C. G. Overberger and G. Menges (Eds), Volume 15, 2nd edn, p. 335. Wiley, New York, NY (1989).
29. H. O. Rennekampff, V. Kiessig and J. F. Hansbrough, *J. Surg. Res.* **62**, 288 (1996).
30. W. K. Loke, S. K. Lau, L. Y. Lim, E. Khor and K. S. Chow, *J. Biomed. Mater. Res. (Appl Biomater.)* **53**, 8 (2000).

31. F. Burkersroda, L. Schedl and A. Göpferich, *Biomaterials* **23**, 4221 (2002).
32. I. Grizzi, H. Garreau, S. Li and M. Vert, *Biomaterials* **16**, 305 (1995).
33. A. Göpferich and R. Langer, *Biopharmacology* **41**, 81 (1995).
34. L. Tilstra and D. Johnsonbaugh, *J. Environ. Polym. Degrad.* **1**, 257 (1993).
35. P. Jarrett, W. J. Cook, J. P. Bell, S. J. Huang and J. A. Cameron, *Polym. Prepr.* **22**, 351 (1981).
36. C. G. Pitt, F. I. Chasalow, Y. M. Hibionada, D. M. Klinas and A. Schindler, *J. Appl. Polym. Sci.* **26**, 3779 (1981).
37. A. Göpferich, *Biomaterials* **17**, 103 (1996).

Gels, Genes, Grafts and Giants, pp. 895–906 (2005)

Amido-modified polylactide for potential tissue engineering applications

NILMINI K. ABAYASINGHE [1], K. PRASANNA U. PERERA [1], CHUCK THOMAS [2], ANNIE DALY [3], S. SURESH [1], KAREN BURG [2], GRAHAM M. HARRISON [3] and DENNIS W. SMITH, Jr. [1,*]

[1] *Department of Chemistry, Clemson University, Clemson, SC 29634, USA*
[2] *Department of Bioengineering, Clemson University, Clemson, SC 29634, USA*
[3] *Department of Chemical Engineering, and the Center for Advanced Engineering Fibers and Films, Clemson University, Clemson, SC 29634, USA*

Received 24 October 2003; accepted 22 January 2004

Abstract—Poly(ester amide) copolymers based on L-lactide (**2**) and a new depsipeptide (**1**) were prepared by ring opening polymerization in the presence of $Sn(Oct)_2$ as the catalyst. Variable monomer feed ratios up to 2.3 mol% **1** afforded copolymers containing ester and amido functional groups in the backbone. Lower glass transition temperatures and reduced crystallization kinetics and crystallinity compared to homo-polylactide (PLA) was achieved with low levels of amido incorporation. A reactivity comparison between enchainment of **2** and **1** was determined using *in situ* infrared spectroscopy. An increase in shear viscosity was observed with the increase of **1** content as determined by rheology studies. Cellular compatibility of the co-polymers was investigated by seeding D1 mouse stem cells onto films and characterizing cell morphology by optical microscopy. Preliminary results indicate that these novel materials exhibit reduced cell attachment compared to PLA and, pending further exploration, may have potential use in biomedical applications.

Key words: Polylactide; depsipeptide; poly(ester amide); morpholinedione; tissue engineering.

INTRODUCTION

Over the past few decades, research efforts have focused on the design of thermoplastics using renewable resource agricultural materials rather than traditional petroleum derived plastics. Lactic acid, the precursor for commercial lactide monomer and polylactide (PLA), is obtained from the fermentation of corn or beet sugar. Primarily, PLA is used in food packaging, agricultural films and biomedical appli-

*To whom correspondence should be addressed. Tel.: (1-864) 656-5020. Fax: (1-864) 656-6613. E-mail: dwsmith@clemson.edu

cations such as sutures [1–9]. Large-scale production of PLA was recently initiated by Cargill Dow Polymers, suggesting potential high market demand in the near future [10–12].

Morpholine-2,5-dione derivatives (depsipeptides) are a class of monomers that have been used to obtain biodegradable copolymers with alternating amido and ester vertebrae. These materials degrade into nontoxic components *in vitro* and *in vivo*; therefore, they have attracted attention for specific biomedical applications. Ring opening polymerization of depsipeptides with lactide, glycolide, or caprolactone have been reported with variable properties compared to the corresponding homopolymers [13, 14].

We have recently prepared a novel depsipeptide monomer, 6,6′-dimethyl-2,5-morpholinedione (**1**) in good overall yield (40%), the synthesis of which has been reported elsewhere (Ref. [15] and data not shown). Copolymerization of the new monomer (**1**) with L-lactide (**2**) was accomplished in the presence of $Sn(Oct)_2$ catalyst. The thermal characterization of the novel copolymers by Differential Scanning Calorimetry (DSC) revealed that the glass transition temperature and the crystalline melting temperature of the copolymers decreases with the increment of the percentage of **1** in the polymer backbone. Further we have demonstrated that the crystallinity of the copolymers can be decreased by incorporating a small percentage of **1** into the polymer. It has been found that 8% of **1** co-monomer in the main chain resulted in a completely amorphous copolymer (Ref. [15] and data not shown).

Here we report our initial investigations on the relative reactivity of **2** and **1** with a ReactIR™ 1000 reaction analysis system, using attenuated total reflection (ATR) mid-infrared spectroscopy [16–19]. This system has the capability of recording an infrared spectrum of materials that are contacted with the chemically resistant probe, and it can be applied to a range of reaction conditions. Hillmyer and co-workers previously reported the kinetics of the ring opening polymerization of lactides using ReactIR™ spectroscopy for zinc and tin alcoxide catalyst systems [20, 21]. We present detailed FT-IR experimental results for the homo- and co-polymerization of **2** with **1** in this study. In addition, we report on the rheological behavior by measuring the complex viscosity as a function of depsipeptide content. Finally, we assessed the copolymer's potential use as a scaffold for tissue engineering applications by cell adhesion studies.

MATERIALS AND METHODS

Materials

L-Lactide was generously provided by Poly-Med (Pendleton, SC, USA) and Ortec (Easley, SC, USA), and was recrystallized from ethyl acetate and vacuum dried prior to use. New monomer 6,6′-dimethyl 2,5-morpholinedione (**1**) was synthesized as described elsewhere (Ref. [15] and data not shown). Stannous 2-ethylhexanoate (Aldrich) was stored under nitrogen prior to use. House nitrogen filtered through a Drierite filter/dryer was used for all synthetic operations. Other chemicals and reagents were purchased from Aldrich or Fisher Scientific and used as received unless otherwise stated.

Co-polymerization

A general procedure for copolymerization of **1** and **2** is described below. Monomer **1** (0.2 g, 1.4 mmol), **2** (0.8 g, 5.6 mmol) and $Sn(Oct)_2$ 70 μl of 0.01 mol l^{-1} solution in anhydrous toluene was transferred to an ampoule. After evaporation of the solvents, the tube was sealed in vacuum and kept at 130°C in an oil bath. After 24 h, the crude product was dissolved in chloroform and precipitated into methanol. The resulting polymer was dried *in vacuo*. Yield: 85%. ^{1}H-NMR (500 MHz, $CDCl_3$) δ (ppm): 6.60 (s, 1H), 5.1 (m, 2H), 4.6 (s, 2H), 1.6 (d, 12H). ^{13}C-NMR (500 MHz, $CDCl_3$) δ (ppm): 172.0, 168.0, 71.9, 69.0, 57.1, 24.8, 16.0. FT-IR (film) (cm^{-1}): 1174, 1181, 1208, 1522, 1685, 1760, 1852, 1900, 3400.

Gel-permeation chromotography (GPC)

GPC samples were prepared in HPLC grade $CHCl_3$ (6 mg in 1 ml of $CHCl_3$). GPC data were collected using a Waters 2690 Alliance System equipped with two consecutive Polymer Labs PLGel 5 mm Mixed-D and Mixed-E columns with refractive index detection at 35°C.

In situ *ReactIR spectroscopy (ReactIR™)*

In situ mid-FT-IR spectra for the polymerizations were collected using an ASI Applied Systems ReactIR™ 1000 reaction analysis system equipped with a light conduit and DiComp (diamond composite) insertion probe. The probe was inserted into the reaction flask containing the predetermined amount of the monomers and the initiator (monomer/initiator = 100 : 1), and it was sealed under nitrogen. The flask was heated to the desired temperature using heating tape and spectra were collected every 2 min (8 scans with 4 cm^{-1} resolution) for 1 h and every 5 min (128 scans with 4 cm^{-1} resolution) for the next 3 h. After collecting the spectra, the data was further analyzed with ReactIR™ (version 2.2) software. Due to overlapping of peaks of interest, integrating the area was difficult. Therefore, the peak height with a specified baseline was used for the calculations.

Rheology

The rheological properties of poly(**1**-*co*-**2**) were investigated using a Rheometric Scientific ARES Rotational Rheometer equipped with 8-mm parallel plates and a 0.3-mm gap between the plates. A nitrogen atmosphere was employed and a dynamic frequency sweep was used to measure the complex viscosity as a function of frequency. Dynamic strain sweeps were performed to establish the linear viscoelastic regime. All tests were repeated a minimum of three times to insure reproducibility.

Table 1.
Composition and molecular weight data for the co-polymers.

Entry	Poly(**1**-*co*-**2**)[a] (% monomer **1**)	M_n[b] ($\times 10^{-3}$)	M_w/M_n[b]
PLA-a	0	52.6	1.2
Poly(**1**-*co*-**2**)-b	0.1	33.0	1.6
Poly(**1**-*co*-**2**)-c	0.9	40.6	1.5
Poly(**1**-*co*-**2**)-d	1.3	37.0	1.6
Poly(**1**-*co*-**2**)-e	2.3	40.0	1.7

[a] Determined by ^{1}H-NMR.
[b] GPC in chloroform *vs.* polystyrene.

Cell seeding experiments

Three films of each material type were sized to fit a 12-well tissue culture polystyrene plate, using an arch punch. All films were cleaned with ethanol and rinsed with sterile medium prior to use. The materials included poly(**1**-*co*-**2**)-c, poly(**1**-*co*-**2**)-e and PLA-a (Table 1). The six surfaces were seeded to 20% confluence using D1 mouse stem cells. Teflon rings were used to immobilize each film, resulting in a cell seeding density of approx. 32 000 cells per well. Dulbecco's Modified Eagle's Medium, supplemented with 10% fetal bovine serum, 1% fungizone and 1% antibiotic-antimycotic, was used for the incubation period. The cultures on all polymer surfaces were maintained for 10 days. Representative photographs of all cellular surfaces were acquired using an inverted microscope.

RESULTS AND DISCUSSION

Synthesis

We have reported the synthesis of the novel depsipeptide monomer **1** and the structural and thermal characterization of its copolymer with **2** [15, 16]. The general synthetic route to copolymerize **1** and **2** to obtain poly(**1**-*co*-**2**) is shown in Scheme 1.

The ring-opening co-polymerization of **1** and **2** was performed in glass ampoules under high vacuum catalyzed by $Sn(Oct)_2$. A series of copolymers containing **1** and **2** was prepared at 1–10 g scale in 85–90% yield. The molecular weights of the polymers were measured using GPC relative to polystyrene standards. The molecular weight data collected for poly(**1**-*co*-**2**) with different percentages of **1** in the polymer backbone are summarized in Table 1.

The most commonly used catalyst for the ring-opening polymerization of lactide, $Sn(Oct)_2$, exhibited excellent activity in the co-polymerization. It is reported to be an efficient catalyst, which allows complete conversion of monomers even at a very high monomer/catalyst ratio (typically 10 000 : 1) with minimum racemiza-

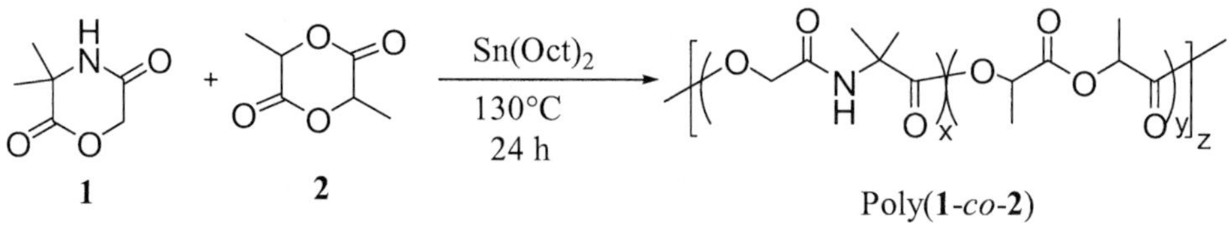

Scheme 1. Random co-polymerization to poly(**1**-*co*-**2**).

$$\sim\!\sim M_1^* + M_1 \xrightarrow{k_{11}} \sim\!\sim M_1 \qquad R_{11} = k_{11}[M_1^*][M_1]$$
$$\sim\!\sim M_1^* + M_2 \xrightarrow{k_{12}} \sim\!\sim M_2 \qquad R_{12} = k_{12}[M_1^*][M_2]$$
$$\sim\!\sim M_2^* + M_1 \xrightarrow{k_{21}} \sim\!\sim M_1 \qquad R_{21} = k_{21}[M_2^*][M_1]$$
$$\sim\!\sim M_2^* + M_2 \xrightarrow{k_{22}} \sim\!\sim M_2 \qquad R_{22} = k_{22}[M_2^*][M_2]$$

$[M_1^*]$, $[M_2^*]$ = concentration of the active propagating polymer having a **1** or **2** at the end. $[M_1]$, $[M_2]$ = **1** or **2** monomer concentration.

Scheme 2. Kinetic expressions and assumptions for the homopolymerization and copolymerization of **2** with **1**.

tion [16–18]. The number average molecular weights of the copolymers ranged from 33 000–52 000 with relatively narrow polydispersity.

In situ *FT-IR spectroscopy (ReactIR™)*

Homopolymerization of **2** as well as the copolymerization of **1** and **2** were monitored utilizing *in situ* FT-IR analysis. Kinetic assumptions utilized for these reactions are summarized in Scheme 2. The carbonyl absorption for pure **2** generally appears at 1759 cm^{-1} as a narrow strong peak [19]. The peaks at 2995 and 2945 cm^{-1} correspond to $-CH_3$ and $-CH$-groups of the lactide moieties. During homopolymerization of **2**, the peak at 1240 cm^{-1} from lactide ring stretching disappeared, and a new peak at 1181 cm^{-1} appeared due to the forming PLA. It was used for kinetic analysis using *in situ* FT-IR. The waterfall plot for this region is shown in Fig. 1a. The preliminary analysis for the homopolymerization of **2** exhibits a linear first-order kinetic plot ($k_{obs} = 0.08$ min^{-1}) (Fig. 2a).

For homopolymerization of **2**, the rate of polymerization is:

$$R_{22} = -\frac{d[\mathbf{2}]}{dt} = k_{22}[M_2^*][M_2],$$
$$R_{22} = \frac{d[-M_2]}{dt} = k_{22}[-M_2] = \frac{d[PLA]}{dt} = k_{22}[PLA],$$

where k_{22} = absolute propagation rate constant and $[-M_2]$ = [PLA] = PLA concentration.

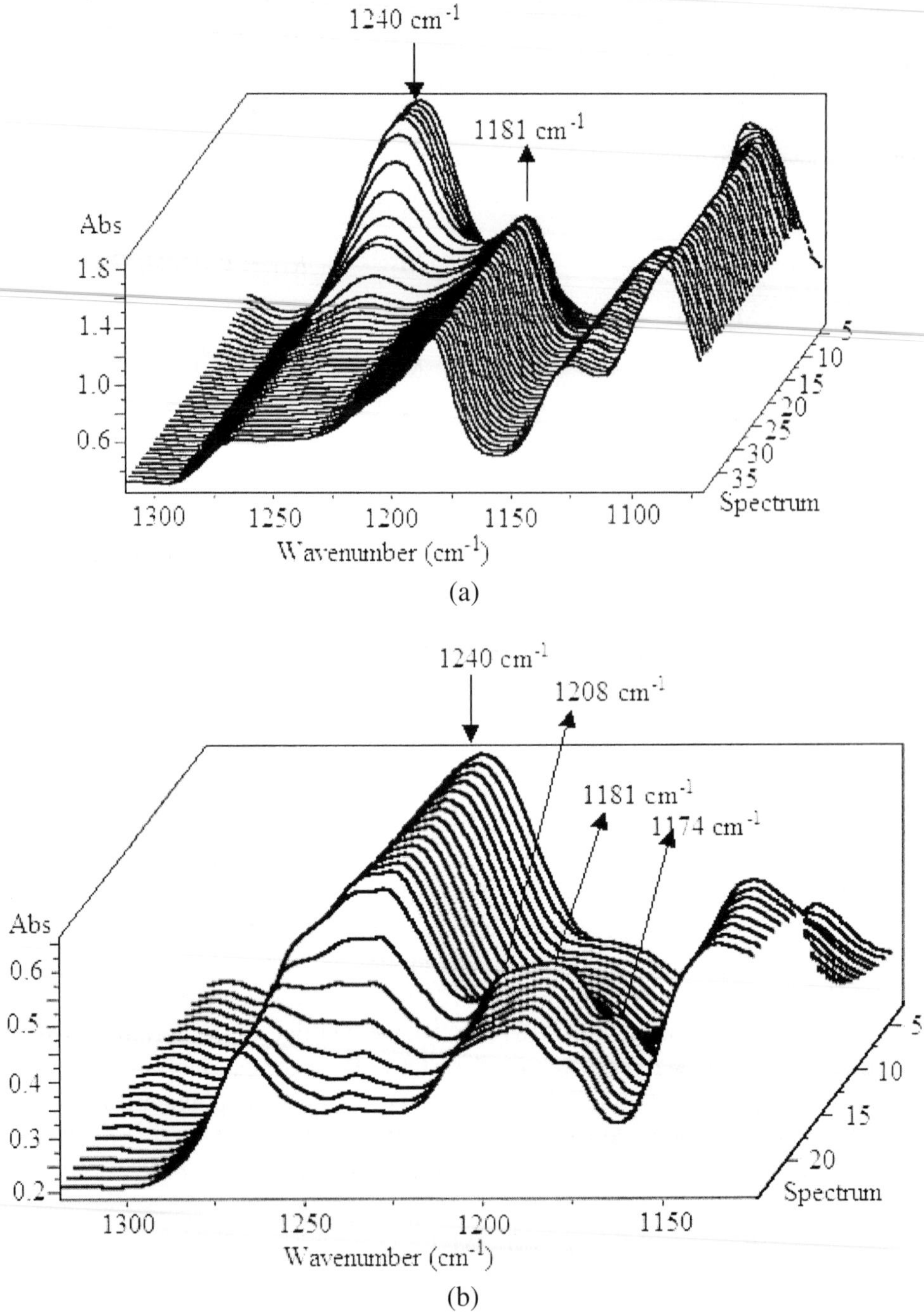

Figure 1. Ester region of the waterfall plot for homopolymerization of **2** (a) and copolymerization of **1** and **2** (b) by *in situ* FT-IR spectroscopy.

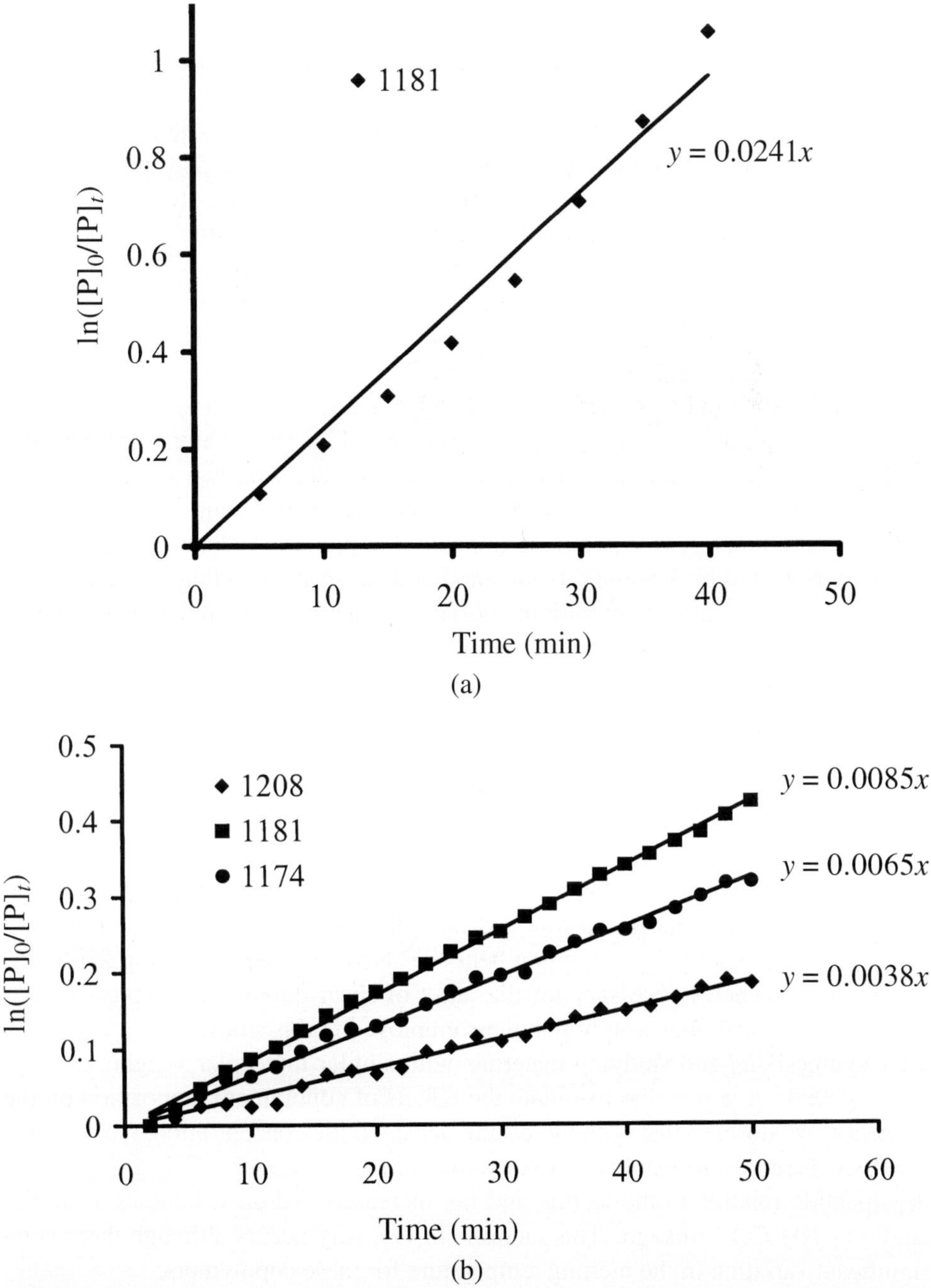

Figure 2. First-order kinetic plots for homopolymerization of **2** (a) and copolymerization of **1** and **2** (b) determined by FT-IR peak height at 1174, 1181 and 1208 cm^{-1}.

The first-order rate constant is expressed as:

$$k_{obs(2)} = k_{22},$$
$$k_{obs(2)} = (\ln[[PLA]_0/[PLA]_t])/t = \ln([P]_0/[P]_t)/t,$$

where $[P]$ = peak height at 1881 cm^{-1} for PLA determined by ReactIR. For the copolymerization of **1** and **2**, kinetic plots of $\ln([P]_0/[P]_t)$ vs time for the peaks at 1174, 1181 and 1208 cm^{-1} would correlate to the relative rates for the reactions between **1,2**; **2,2** and **1,1** and provide k_{obs} for $k_{21} + k_{12}$ (assuming $k_{21} = k_{12}$), k_{22} and k_{11}, respectively.

Homopolymerization and co-polymerization is performed using the same experimental conditions and monomer: initiator ratio (100 : 1) in melt at 130°C. In addition to the peak at 1181 cm^{-1}, two additional peaks were observed at 1174 and 1208 cm^{-1} that could be assigned to **1–2** and **1–1** ester linkages in the copolymer. The FT-IR waterfall plot for this region is shown in Fig. 1b and a first-order dependence for each peak was observed as shown in Fig. 2b. The k_{obs} values obtained using these respective plots were 0.0065, 0.0085 and 0.0035 min^{-1}. This observation illustrates that a growing chain terminated by **2** prefers to add another **2** unit and a chain terminated by **1** prefers to add another **2** rather than adding a **1** unit. These values are in good agreement with the polymer structure given in Scheme 1, which strongly suggests a copolymer with a random distribution of monomer **1** amido linkages.

Rheology

The complex viscosity, η^*, as a function of frequency for PLA-a, poly(**1**-*co*-**2**)-b and poly(**1**-*co*-**2**)-e is shown in Fig. 3. The dynamic frequency sweep was performed at 180°C over a frequency range from 1 rad/s to 500 rad/s. The samples exhibit a typical Newtonian plateau at low and intermediate shear rates. In addition, all of the samples exhibit shear thinning behavior at frequencies exceeding 200 rad/s. The high frequencies necessary for the onset of shear thinning are typical for the relatively low molecular weight polymers employed in this study.

By synthesizing and studying materials with similar molecular weights (M_n approx. 35 000), it is possible to isolate the effects of copolymer composition on the viscosity. We observe that with increasing depsipeptide concentration, the complex viscosity increases, probably due to the effect of the additional $-CH_3$ groups in the depsipeptide relative to the lactide and the increased hydrogen bonding from the amdio (—NH-CO-) linkage. This increase in viscosity occurs although there is no significant variation in the melting temperature for these copolymers.

To investigate polymer degradation in an air atmosphere, a dynamic time sweep was performed on poly(**1**-*co*-**2**)-b (Fig. 4). The test was performed at a temperature of 180°C and a frequency of 10 rad/s for 2 h. The complex viscosity decreases continually with time, indicating degradation of the polymer sample.

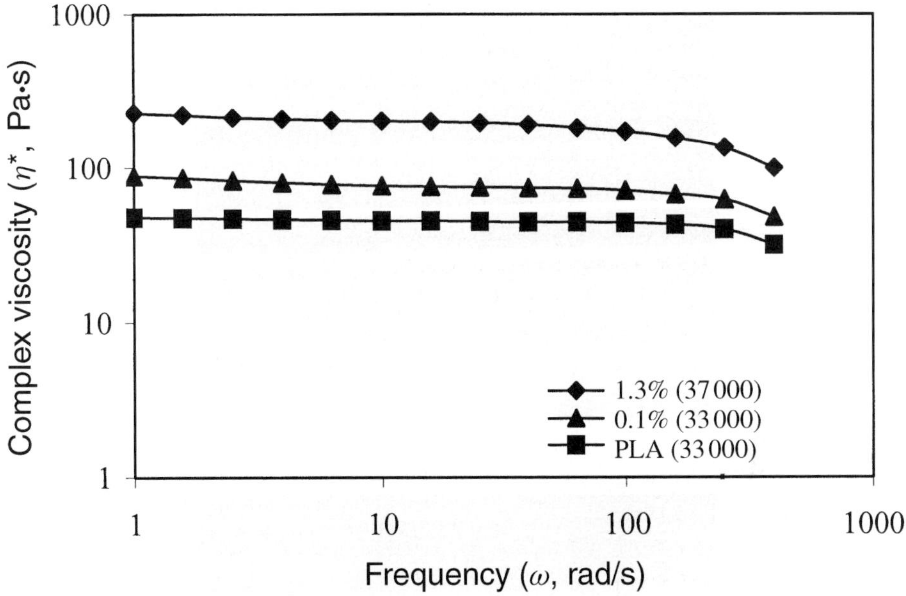

Figure 3. Complex viscosity as a function of copolymer composition.

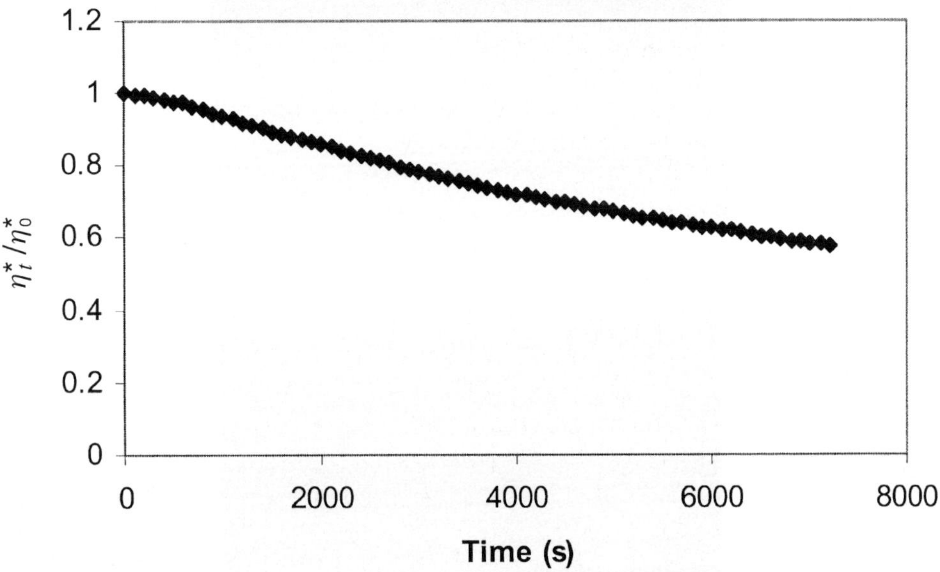

Figure 4. Isothermal complex viscosity as a function of time in air at 180°C.

Cell seeding experiments

Figure 5 illustrates our preliminary cell seeding results for the various films. The spindle-shaped cells on the PLA-a films were dense near the center, but only grew sporadically outward. Although the cells are healthy in appearance, they did not

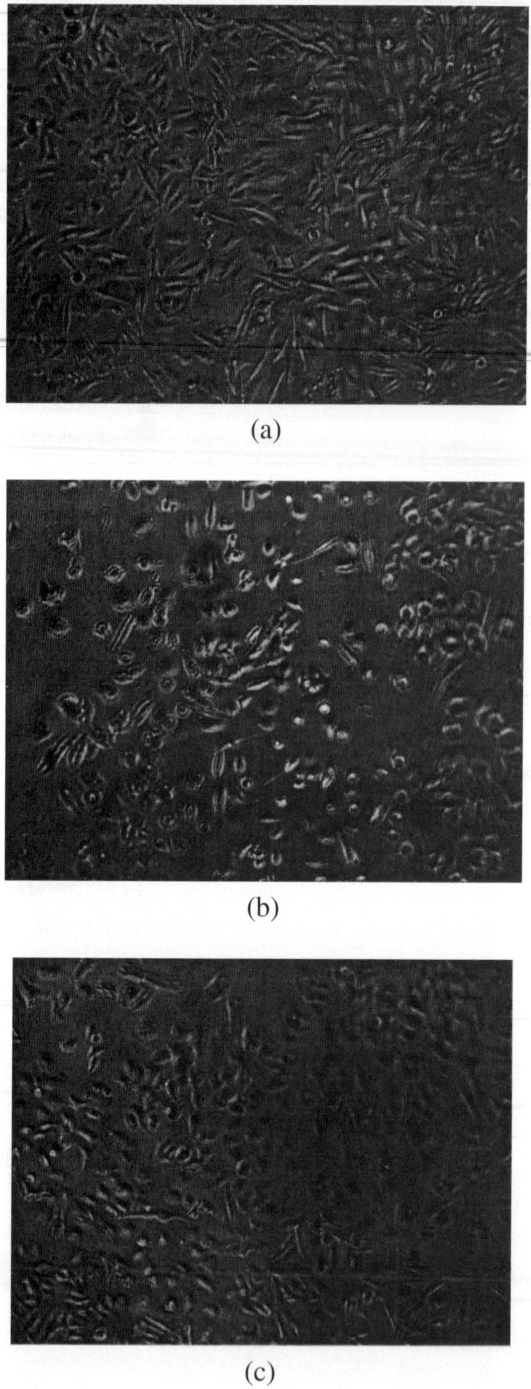

Figure 5. Optical micrographs ($\times 10$ magnification) of cells on polymer films for PLA-a (a), poly(**1**-*co*-**2**)-c (b) and poly(**1**-*co*-**2**)-e (c).

appear to proliferate or migrate on the surface during the 10-day study. The cells seeded on the poly(**1**-*co*-**2**)-e films were dense in a few widespread areas. They were spindle-shaped as well, but many rounded cells were also observed, indicating a lower state of viability with increased amounts of **1**. The cellular organization on these films appears to be random and little proliferation or migration occurred over the duration of the study.

The poly(**1**-*co*-**2**)-c films presented cells that formed dense bands in two of the three wells. Few cells were located outside of these bands. The heterogeneity in adhesion implies that there are slight variations in the consistency of the copolymers, either with respect to the bulk or the surface. All three replicates of the poly(**1**-*co*-**2**)-c films had more rounded cells than the other two surfaces. Polylactide is a hydrophobic material and thus does not attract high cell adhesion without surface modification or treatment, therefore, the low level of proliferation is not surprising [22–24]. The lower state of viability on the copolymers may be due in part to the increased polydispersity index associated with these systems, causing a more dynamic surface with which the cells must interact. It may be possible to increase the cell viability by extracting the monomeric and oligomeric species following synthesis and/or film processing [22–24].

CONCLUSIONS

Novel depsipeptide monomer dimethyl-2,5-morpholinedione was copolymerized with L-lactide to obtain amido functionalized copolymers by standard polymerization techniques. The relative reactivity of monomers **1** and **2** was determined by *in situ* ReactIR™ spectroscopy to be **2–2** ⋙ **1–2** > **1–1**. Shear viscosity was altered by varying the depsipeptide content in the copolymer. Cell adhesion studies showed that adhesion of D1 mouse stem cells on copolymer films of monomers **2** and **1** was attenuated compared to pure PLA films. Materials of this type, that appear to be nontoxic but do not appear to induce high cell proliferation, may be extremely useful in regenerating tissues that mandate high mechanical integrity and low cell volume (e.g. cartilage or spinal disc).

Acknowledgements

We thank the South Carolina Commission of Higher Education, Cryovac Sealed Air Corporation for financial support. This work was also supported by the Engineering Research Centers Program of the National Science Foundation under NSF Award Number EEC-9731680. We are also grateful to Poly-Med, Inc. (Pendleton, SC, USA) and Ortec Inc. (Easley, SC, USA) for generously donating lactide monomer. D.S. is a Cottrell Scholar of Research Corporation.

REFERENCES

1. W. Schnabel (Ed.), in: *Polymer Degradation, Principles and Practical Applications,* p. 154. Hanser, Munich (1981).
2. A. C. Albertsson and S. Karlsson, *Acta Polym.* **46**, 114 (1995).
3. H. R. Kricheldorf and I. Kreiser, *Makromol. Chem.* **188**, 1861 (1987).
4. G. S. Kumar (Ed.), in: *Biodegradable Polymers, Prospects and Progress*, p. 3. Marcel Dekker, New York, NY (1987).
5. F. H. Otey, R. P. Westhoff and W. M. Doane, *Ind. Eng. Chem. Prod. Res. Dev.* **19**, 502 (1980).
6. R. Langer and N. J. Peppas, *Macromol. Sci. Rev. Macromol. Chem.* **23**, 61 (1983).
7. H. Kobayashi, S. H. Hyon and Y. Ikada, *J. Biomed. Mater. Res.* **25**, 1481 (1991).
8. A. U. Daniels, M. K. D. Chang and K. P. Andriano, *J. Appl. Biomater.* **1**, 57 (1990).
9. M. Chasm and R. Langer, in: *Biodegradable Polymers, Prospects and Progress*, M. Chasm and R. Langer (Eds), p. 1. Marcel Dekker, New York, NY (1990).
10. R. E. Drumright, P. R. Gruber and D. E. Henton, *Adv. Mater.* **12**, 1841 (2000).
11. E. T. H. Vink, K. R. Rabago, D. A. Glassner and P. R. Gruber, *Polim. Degrad. Stabil.* **80**, 403 (2003).
12. R. Westervelt, *Chem. Week.* **162**, 9 (2000).
13. J. Felder, S. J. Lee, S. W. Kim and J. Feijen, *Makromol. Chem. Rapid Commun.* **7**, 193 (1986).
14. P. J. A. In't Veld, W. Ye, R. Klap, P. J. Dijkstra and J. Feijen, *Makromol. Chem.* **193**, 1927 (1992).
15. N. K. Abayasinghe, S. Suresh and D. W. Smith, Jr., *Polym. Prepr.* **44**, 847 (2003).
16. A. Kowalski, A. Duda and S. Penczek, *Macromolecules* **33**, 7359 (2000).
17. H. R. Kricheldorf and K. Saunders, *Macromolecules* **33**, 702 (2000).
18. J. Lunt, *Polym. Degrad. Stab.* **59**, 145 (1998).
19. H. R. Kricheldorf, C. Boettcher and K. Tonnes, *Polymer* **33**, 2817 (1992).
20. C. K. Williams, N. R. Brooks, M. A. Hillmyer and W. B. Tolman, *Chem. Commun.*, 2132 (2002).
21. K. B. Aubrecht, M. A. Hillmyer and W. B. Tolman, *Macromolecules* **35**, 644 (2002).
22. K. J. L. Burg, W. D. Holder, Jr., C. R. Culberson, R. J. Beiler, K. G. Greene, A. B. Loebsack, W. D. Roland, D. J. Mooney and C. R. Halberstadt, *J. Biomater. Sci. Polymer Edn* **10**, 147 (1999).
23. K. J. L. Burg and S. W. Shalaby, *J. Biomater. Sci. Polymer Edn* **9**, 15 (1997).
24. K. J. L. Burg, J. M. Allan, S. L. Roweton and S. W. Shalaby, *Ann. N. Y. Acad. Sci.* **831**, 217 (1997).

Gels, Genes, Grafts and Giants, pp. 907–920 (2005)

Hyaluronic-acid-based semi-interpenetrating materials

XUEJUN XIN, A. BORZACCHIELLO *, P. A. NETTI, L. AMBROSIO and L. NICOLAIS

Institute of Composite and Biomedical Materials, CNR and Interdisciplinary Research Center in Biomaterials (CRIB), University of Naples, 'Federico II', Piazzale Tecchio 80, 80125 Napoli, Italy

Received 30 October 2003; accepted 12 March 2004

Abstract—In order to enhance the mechanical performances of hyaluronic acid (HA) without compromising its biological activity, HA has been interpenetrating with a fibrillar collagen scaffold. The semi-interpenetrating materials were obtained by mixing HA with different molecular weight and a pepsin-solubilized collagen (atelocollagen) solution, and then by inducing collagen fibrillogenesis. Results indicate that molecular weight of HA significantly influences the mechanical properties of the semi-interpenetrating materials and more specifically stronger material results from the use of low-molecular-weight (LMW) HA. According to the dynamic mechanical data the composite collagen-LMW HA has a higher elastic modulus than collagen, whereas the opposite is true for the high-molecular-weight (HMW) HA. This result highlights the role of specific interactions that occur between collagen and HA during the gel formation in controlling the network mechanical stability. LMW HA may, probably, interact more strongly with collagen during the fibrillogenesis process than HMW HA due to the higher mobility of the chains and the weaker homologous interactions. Moreover, morphological observations showed that LMW HA is intimately interdispersed within the collagen network and completely coated the fibrils, which act as mechanical support.

Key words: Hyaluronic acid; collagen; semi-interpenetrating network; scaffold; gel.

INTRODUCTION

Hyaluronic acid (HA) is a high-molecular-weight polysaccharide that is a basic component of the extracellular matrix of the connective tissue and is present in large amounts in the vitreous humour of the eye, the synovial fluid of the joints, and in soft tissues such as umbilical cord and rooster comb [1]. Hyaluronic acid molecule has a set of unique chemical-physical and biological properties, which allow it to regulate and control several tissue physiological functions *in vivo*. Many of the biological and physical properties are a consequence of its extraordinary water-retaining capability [2]. In aqueous solutions HA shows a high excluded volume

*To whom correspondence should be addressed. E-mail: bassunta@unina.it

since it behaves as an expanded random coil occupying a large hydrodynamic volume [3–5]. For this reason it is generally postulated that it controls and regulate the trafficking of fluids and macromolecules within the interstitial space of tissues [6, 7]. Moreover, it also contributes to the mechanical properties of soft tissues in compression [8] and plays a major role in joint lubrication [9]. From a biological standpoint, HA molecule plays a major role in tissue growth and remodelling. Indeed, it has been shown that it interacts specifically with endogenous receptors such as CD 44, RHAMM, ICAM-1 regulating and guiding cellular migration, growth and adhesion [10].

For all the above reasons HA has long attracted research interest for its potential use as scaffold for tissue regeneration. However, the high water affinity strongly limits the use of the unmodified molecule since it leads to poor mechanical properties. Furthermore, the use of unmodified HA as scaffold material for tissue engineering is strongly limited by its poor processability and handling properties. To circumvent these limitations several approaches have been attempted to modify the HA molecule to reduce its water affinity by controlling the ratio of hydrophobic/hydrophilic groups on the polymer backbone [11, 12] or by inter-molecular cross linking of HA molecule [13]. For instance, HA has been chemically modified by coupling reactions, such as esterification of the carboxyl groups of HA with therapeutically inactive or active alcohol, obtaining a wide range of ester-derivatives which show enhanced processability and mechanical properties in the wet state compared to unmodified HA [14]. These derivatives have been engineered into a variety of medically useful configuration different products such as sponge, membranes and gels on which cells can be seeded [15]. HA has also been cross-linked by using a variety of cross-linking agents such as formaldehyde, vinyl sulphone, carbodiimides, bis carbodiimides and others [16–18]. However, although chemical modifications lead to an enhancement of mechanical and processability properties on one hand, they may bring about severe reduction of the biocompatibility. Indeed, several concerns have been expressed for the potential toxicity of some of the cross linking agents and of the degradation products [19]. Furthermore, the introduction of foreign group can hinder HA biological activity.

The aim of this study was to assess the feasibility to mechanically stabilize HA by interpenetrating it with a fibrillar scaffold. This approach has been inspired by the assembly of the extracellular matrix of mature soft tissue in which the HA is physically interconnected to the fibrous components to form a firm solid gel [20]. As a model of fibrillar scaffold we have chosen collagen for the following reasons: (a) it is the main fibrous component of the extracellular matrix of soft tissues; (b) it provides a major component of tissue strength; it is already used as scaffold material for soft tissue regeneration [21]. The collagen-HA semi-interpenetrating network obtained by letting the collagen to fibril in presence of HA molecules of different molecular weight, were mechanically and morphological characterized to asses the efficiency of the molecular interpenetration and its effect on the network stabilization.

MATERIALS AND METHODS

HA powder, obtained by roost-comb, with a molecular weight of 155×10^3 (LMW) and 1.2×10^6 (HMW) was kindly supplied by FAB (Fidia Advanced Biopolymers, Abano Terme, Italy) and used as received. Solutions of HA at different concentrations (from 1 to 60 mg/ml) were prepared using bi-distilled water.

Acid solution (pH 2) of pepsin-solubilized bovine dermal collagen (2.9 mg/ml) was purchased from Vitrogen® (Cohesion Technologies, Palo Alto, CA, USA) and used without any further treatment. Vitrogen collagen solution is made of the 95–98% type-I collagen with the remainder being comprised of type-III collagen.

A sterile 10× concentrate phosphate-buffered saline solution (PBS), containing 0.2 M Na_2HPO_4, 1.3 M NaCl, at pH 7.4, was used (Sigma).

Collagen gel

Collagen gels were prepared by diluting the Vitrogen collagen solution in PBS. In detail, 8 ml of collagen were mixed with 1 ml of PBS, (8 : 1 volume), subsequently 1 ml of 0.1 M NaOH solution was added and mixed. The pH of the solution was then adjusted to 7.4 ± 0.2 by the addition of a few drops of 0.1 M HCl or 0.1 M NaOH.

The solution was then incubated in a thermostatic bath at 37°C for 1 h to allow completion of the fibrillogenesis of collagen. During the incubation, the atelocollagen molecules assemble to each other to form collagen fibres [20], and after 1 h a firm collagen gel was obtained.

HA-collagen semi-interpenetrating gel

The semi-interpenetrating HA-collagen gel was obtained by promoting collagen fibrillogenesis in the presence of HA. HA powder (100 mg) was mixed with the collagen solution, at an initial concentration of 2.9 mg/ml. The resulting solution was diluted in PBS (8 : 1 volume), then 1 ml of 0.1 M NaOH solution was added and the final HA concentration was 10 mg/ml. The pH of the solution was then adjusted to 7.4 ± 0.2 by NaOH and HCl. The solution was incubated at 37°C for 1 h to allow collagen fibrillogenesis. After the incubation the collagen was completely fibrillated and the semi-interpenetrating material obtained appeared as a hydrated gel. The composite collagen-HA were prepared using LMW- and HMW HA.

Dynamic-mechanical analysis

The dynamic-mechanical properties of the gels were evaluated on a strain-controlled Bohlin VOR Rheometer (Bohlin Reologi, Lund, Sweden), in a parallel plate geometry (PP30 cell). The tests were carried out at temperature of 37 ± 0.01°C. To avoid water evaporation the cell containing the sample was enclosed in a Plexiglas chamber connected to a water reservoir.

The small-amplitude oscillatory shear experiments were performed to measure $G'(\omega)$ (shear storage modulus or elastic modulus) and $G''(\omega)$ (shear loss modulus or viscous modulus). Preliminary strain sweep tests at a fixed oscillation frequency (consisting in monitoring the viscoelastic properties while logarithmically varying the strain amplitude γ_0) were performed on the materials to determine the strain amplitude range at which linear viscoelasticity holds [22]. The oscillation frequency has been varied from 0.05 to 10 Hz. The tests were repeated at least four times.

Collagen fibrillogenesis was followed starting from the atelocollagen solution and evaluating the evolution with time of the elastic modulus at a fixed frequency (0.01 Hz). Specifically the atelocollagen solution, with and without HA, after the adjustment of its pH at 7.4, was placed between the rheometer plates at 37°C to let the fibrillogenesis to occur while registering the time evolution of the elastic modulus. To compare the rate of gelation of the different gels, the fibrillogenesis curves were expressed in terms of a normalized elastic modulus $G'_{\rm red}$ given by the ratio $(G'(t) - G'_0)/(G'_{\rm f} - G'_0)$, where G'_0 and $G'_{\rm f}$ are the elastic modulus of the starting solution and of the fibrillated gel, respectively.

Morphological properties

The structures of the semi-interpenetrating gels and collagen were evaluated by Scanning Electron Microscope (SEM) and Transmission Electron Microscope (TEM). For SEM observations, the sample was prepared according to well-established protocols [23]. Briefly, the materials were quickly frozen using liquid nitrogen, cut and then the water was eliminated by freeze-drying (K850 Emitech). The dried samples were then attached to sample stubs and sputter-coated with gold and examined under a Hitachi S-2300 Scanning Electron Microscope. Imaging were obtained at an accelerate voltage of 12–15 kV. TEM experiments were carried out on the materials after the staining that was performed with 2% (w/w) phosphotungstic acid as staining agent. The samples were stained onto a copper grid, which was covered by a Formavr (polyvinyl–formaldehyde) film [24]. The grids were air-dried and TEM measurements were obtained by means of a Philips CM 12 TEM apparatus.

RESULTS

Dynamic-mechanical properties

The dynamic mechanical properties of the semi-interpenetrating materials were evaluated and compared to that of collagen gels and HA. The HA solutions showed all range of behaviour spanning from dilute to highly entangled solutions, depending upon both molecular weight and concentration. Figures 1 and 2 show the storage and loss modulus as function of the frequency and concentrations for LMW and HMW, respectively. For LMW HA solutions, the loss modulus is always higher

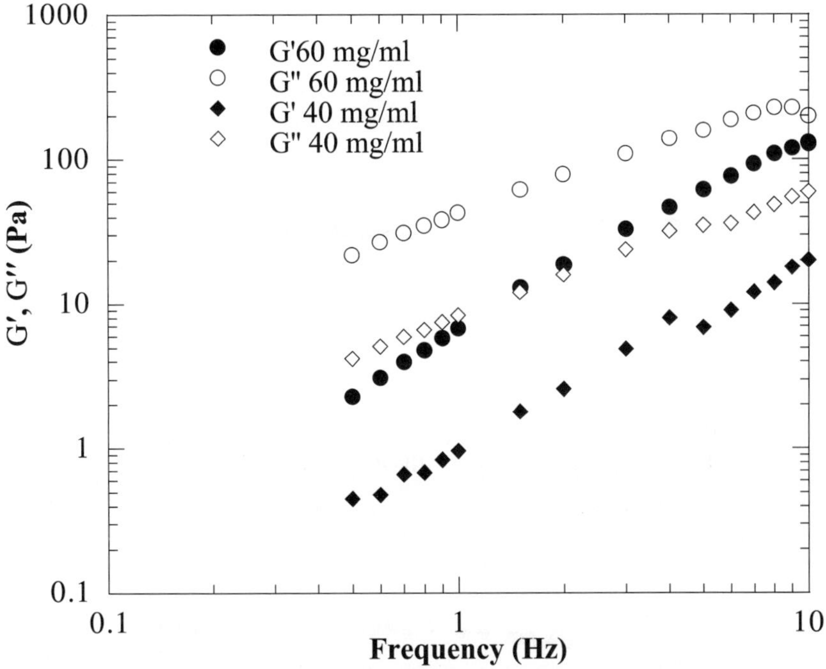

Figure 1. Mechanical spectra of LMW HA (MW 155×10^3) as function of concentration ($T = 37°C$).

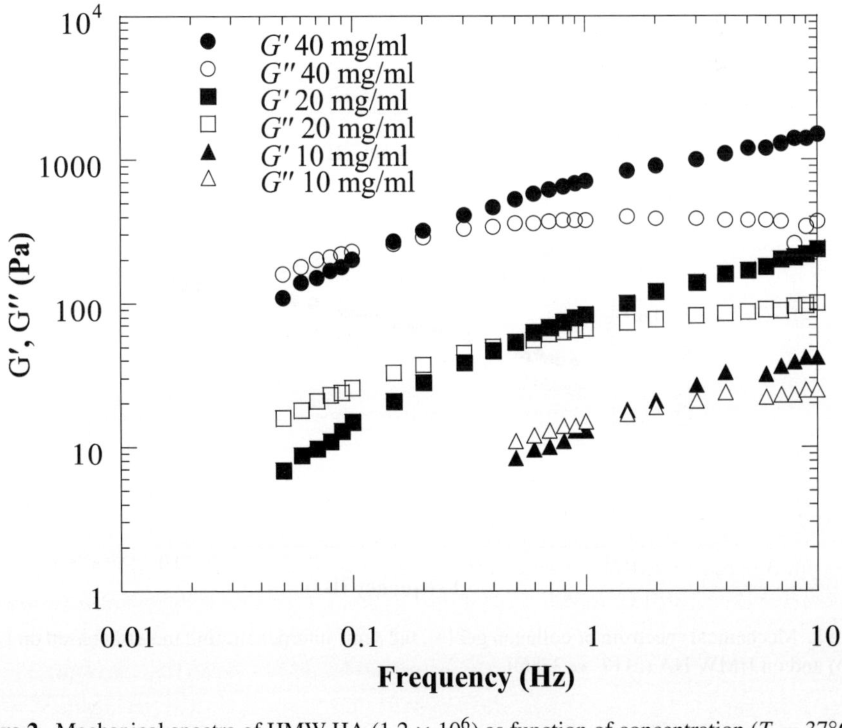

Figure 2. Mechanical spectra of HMW HA (1.2×10^6) as function of concentration ($T = 37°C$).

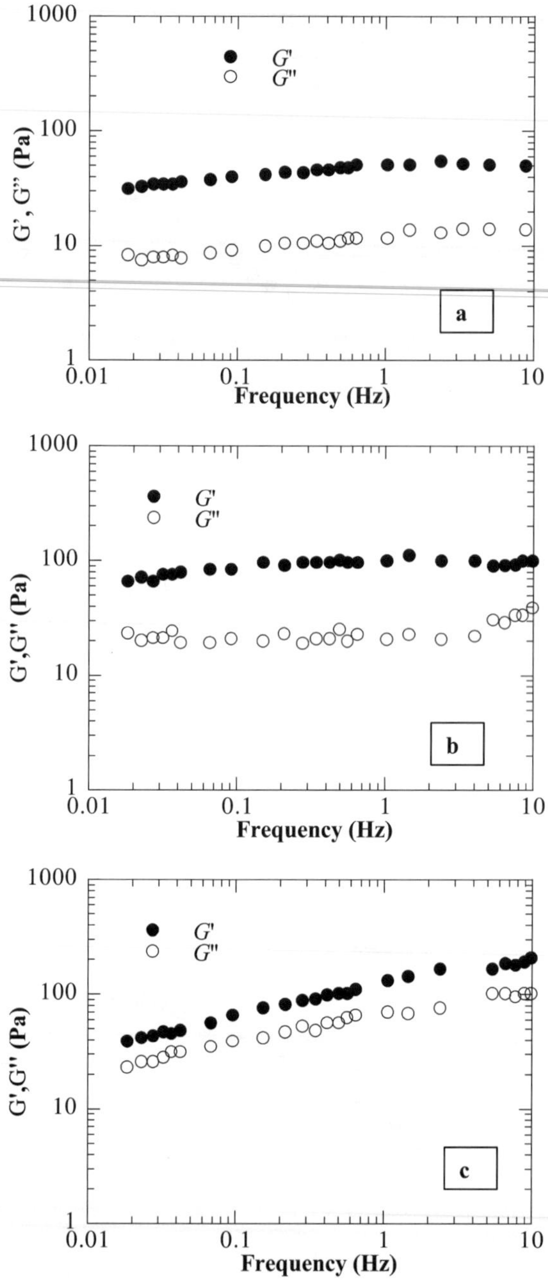

Figure 3. Mechanical spectrum of collagen gel (a), the semi-interpenetrating material based on LMW HA (b) and on HMW HA (c) ($T = 37°C$).

than the storage modulus. The relationship between G' and G'' with frequency is typical of that of dilute solutions, i.e. $G'' \propto \omega$ and $G' \propto \omega^2$. At higher molecular weight, G'' is higher than G' only at low frequency. As the frequency increases, G' approaches G'' and at high frequency G' exceeds G''. In this region both moduli depend weakly upon the frequency. The point of crossover, at which the G' curve crosses G'', depends strongly on the concentration, moving to the lower frequency as the concentration increases. Those features suggest the existence of an entangled network of HA chains. At low frequency where the viscous modulus is higher than the elastic one, the network is transient in the sense that the molecular rearrangements are faster than the period of oscillation and therefore the chains flow by continuously entangling and disentangling. At higher frequencies, the chain cannot disentangle and the material behaves as cross-linked network where the principal mode of accommodation of the applied stress is by network deformation [9, 25, 26]. The enhancement in the elastic properties of the HA solutions with MW is consistent with the change in the appearance of the materials from watery solutions to gel like. Collagen showed dynamic properties typical of a gel (Fig. 3a) [27, 28]. Indeed for the collagen, in agreement with literature data [29], the elastic modulus is higher than the viscous modulus and both moduli, within the frequency range analysed, were almost parallel to each other and not strongly frequency dependant. This behaviour has been observed for many biological gels in which the chains are physically cross-linked into a network. The nature of cross-links could involve disparate forces such as electrostatic, hydrophobic and hydrogen bonding interaction [25, 26]. These interactions arise in the collagen gel during the fibrillogenesis [20].

In Fig. 3b and 3c are shown the G' and G'' curves for the semi-interpenetrating materials with low and high molecular weight, respectively. Their dynamic mechanical behaviour was qualitatively similar to that of collagen gel but the value of the elastic modulus was significantly higher. The value of G' increases of about twofold when the collagen fibrillogenesis occurs in presence of LMW HA. This result suggests that the presence of LMW HA induces a reinforcement of collagen gel. Surprisingly the gel realised with HMW did not show the same kind of behaviour. Even if the HMW HA solutions showed enhanced viscoelastic properties than LMW HA, the final semi-interpenetrating material was stronger when using LMW HA. Indeed, the elastic modulus of HMW HA-collagen showed more marked frequency dependence and for frequency up to 1 Hz was lower than that of the LMW HA composite.

The rate of fibrillogenesis was also affected by the presence of HA molecules. Figure 4 shows the variation of normalised modulus with time during the fibrillogenesis for the collagen and the semi-interpenetrating materials. All the curves have the same shape consisting of a lag period during which the increase of the elastic modulus was not relevant, followed by a sigmoid portion or growth period characterised by a rapid increase of the modulus until equilibrium is reached that correspond to a maximum plateau of G_{red}. Curves of similar shape have been re-

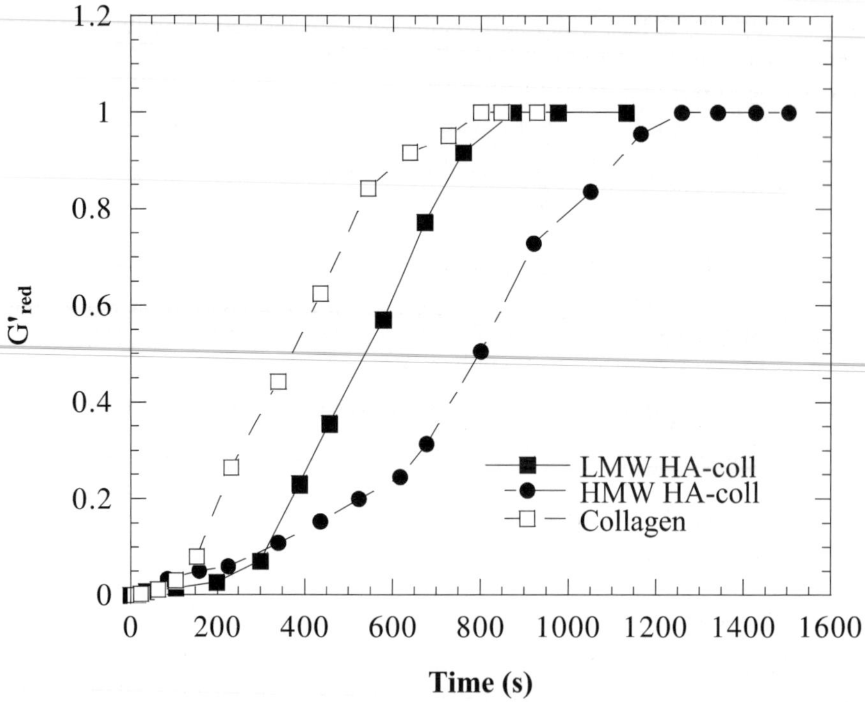

Figure 4. Evolution with time of the ratio $G_{\mathrm{red}} = (G'(t) - G'_0)/(G'_\mathrm{f} - G'_0)$ for semi-interpenetrating materials based on LMW HA (LMW HA-coll), HMW HA (HMW HA-coll) and collagen gel ($T = 37°\mathrm{C}$).

ported [20] when, to study the kinetic of aggregation, the evolution with time of other collagen properties, such as absorbance, were registered. It is evident from Fig. 4 that both the lag and the growth period increase for the presence of the HA molecules, being the fibrillogenesis delay, were dependent on the HA molecular weight. The semi-interpenetrating material containing HMW, indeed, fibrillated in about 1200 s, while about 900 and 800 s were necessary to complete the fibrillogenesis for the LMW HA-collagen material and the collagen gel, respectively.

Morphological properties

To investigate the morphological basis of the semi-interpenetrating materials mechanical properties, the electron microscopic analysis was performed. The SEM investigation showing the fibrillar organization of collagen is reported in Fig. 5a. An open interconnecting porous network consisting of collagen fibril [29] was clearly evident. The collagen fibrils was also visualised by TEM (Fig. 6a) and with staining showed the characteristic banding pattern along the fibre axis of about 67 nm, the D-period. The D-period results from the packing of individual collagen molecules, which are about 300 nm, into a staggered overlap arrangement. The molecules are packed side by side with adjacent molecules staggered along their axis by 67 nm, resulting in a gap of about $0.6D$ between the end of one molecule and the beginning

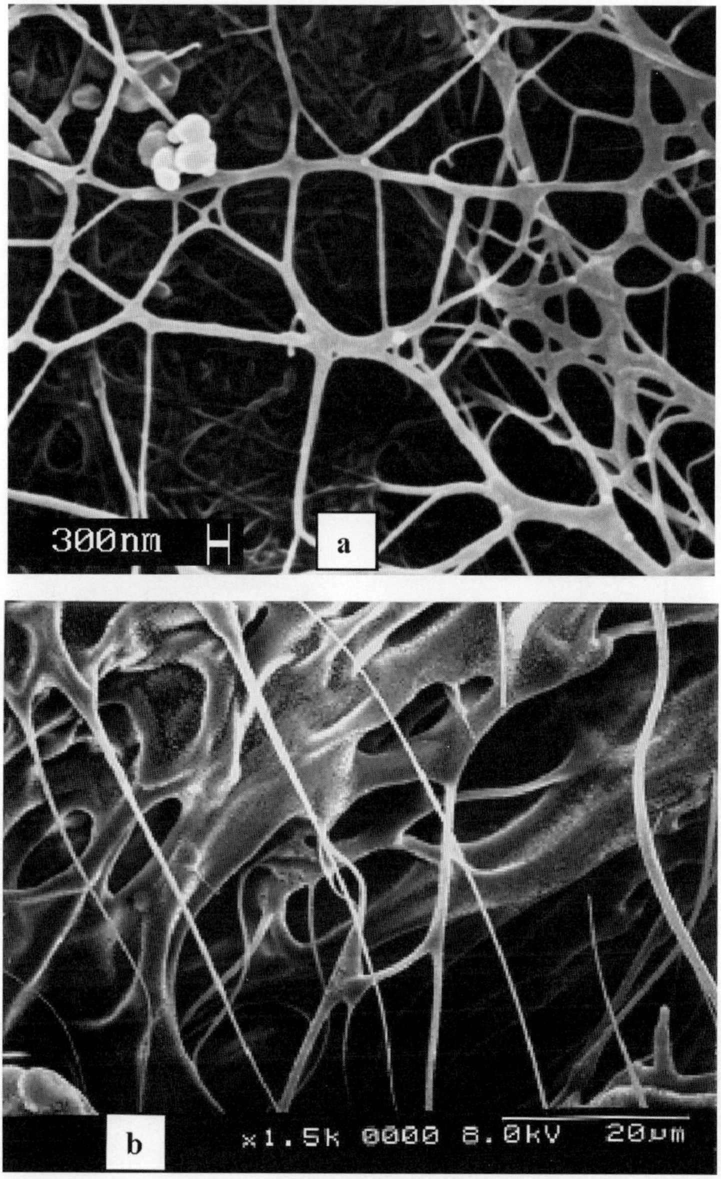

Figure 5. SEM micrograph of freeze-dried collagen gel (×1640) (a) and the semi-interpenetrating material based on LMW HA (×1500) (b).

of the next [20]. The morphological characteristics of the semi-interpenetrating gel based on LMW HA are shown in Fig. 5b. From the SEM micrograph it is evident that the amorphous HA coats the collagen fibrils. The collagen fibrils acts as bridging cross element within the HA matrix, in a kind of web-like structure. The amorphous HA took the fibrillar shape of the collagens scaffold that act as mechanical support. The binding and the adhesive interactions between HA and collagen

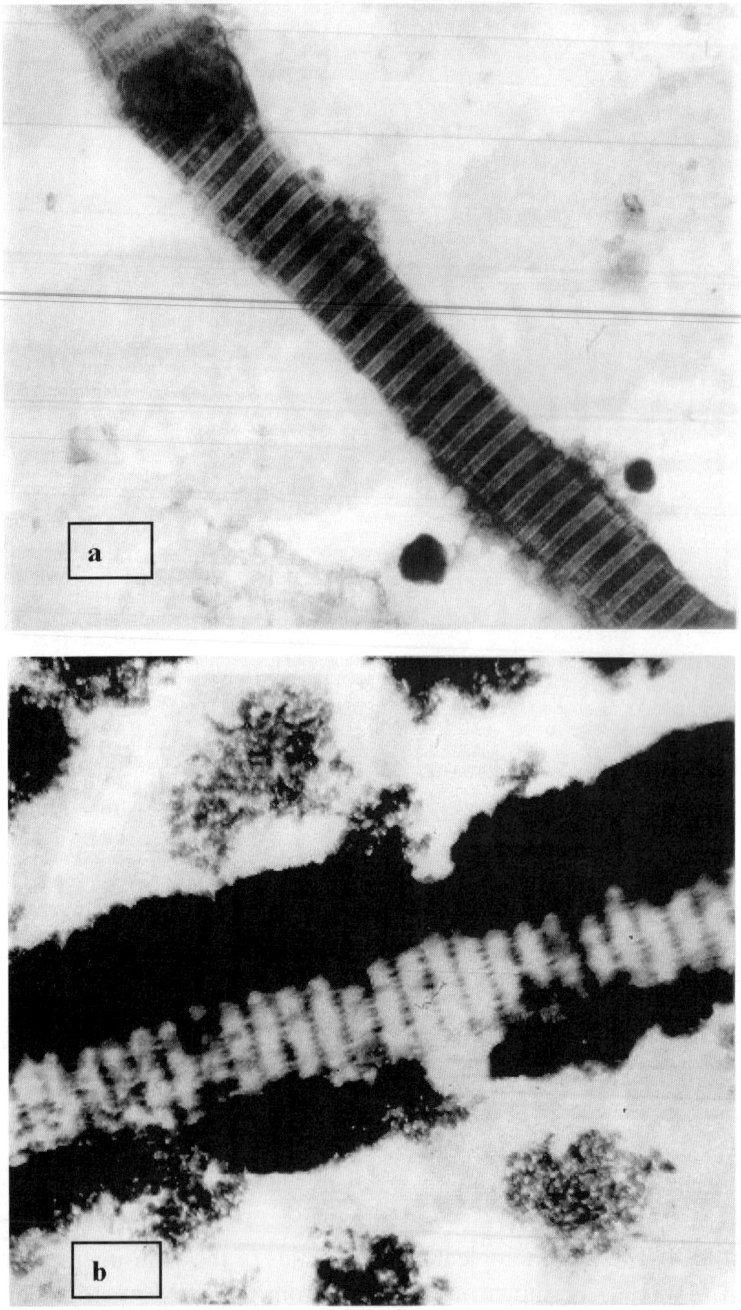

Figure 6. TEM micrograph of collagen gel (a) and the semi-interpenetrating material based on LMW HA (b).

are also clear from the TEM observation (Fig. 6b). The TEM micrograph of the LMW-HA-based material shows the typical banding pattern of the collagen fibrils, but in this case HA coated the fibril. The presence of LMW HA did not alter the normal banded fibril formation.

DISCUSSION

Most of the strategies that use HA as materials for soft tissue regeneration have been based on chemical modifications to enhance its mechanical properties. HA indeed has an extraordinary water retaining capability that allows it to regulate and control several important tissue functions *in vivo* but, on the other hand, limits the possibility of using the unmodified molecules because of the inconsistent mechanical properties. The chemical modification approach has led to HA derivatives with highly improved mechanical properties [25, 26], but unfortunately with impaired biological activity. The modified molecule cannot be properly recognised by the cells and the capabilities of regulating and guiding cellular migration, growth and adhesion, is therefore reduced. In this work, unmodified HA has been scaffolded onto collagen network and in this way it fully covers the collagen fibres, as shown from the morphological observations (Figs 5b and 6b). The HA matrix is intimately interdispersed in the fibrillar collagen. The collagen fibrils act as a binding cross element within the HA matrix, in a kind of a web-like structure.

The collagen-HA gel structure strongly resembles the organization of the connective tissues ECM. The ECM, indeed, is a composite structure consisting of a highly hydrated amorphous matrix comprised mainly of glycosamminoglycans and proteoglycans reinforced by collagen that is the major protein component of the animals ECM [8, 30]. The collagen content of tissue is typically 100-fold greater, on a weight/volume basis, than hyaluronan in the ECM composite material; consequently, the collagen network acts as a constrain for hyaluronan and the manner in which the collagen constrains the HA governs important physiological functions such as interstitial transport [31]. The semi-interpenetrating HA/collagen gels mimic the assembly of the soft tissue ECM. If the cells are seeded in these porous semi-interpenetrating materials they will experience, mostly, unmodified HA and this could be advantageous for cell activity. It has been reported that HA can stimulate cell migration indirectly by providing an extracellular or pericellular environment conducive to migration. HA-mediated cell-surface signalling events have been correlated with either a cell migration or a cell-proliferation response [32]. On the other hand this molecule seems not to favour cell adhesion. Contradictory results can be found in literature and HA has been shown to promote or inhibit cell adhesion depending on a series of variables such as the cell type, the length of the HA molecules and its physical form, soluble or immobilised [33].

It is also worth to point out that the collagen-HA materials exhibit a mechanical behaviour similar to that of collagen (Fig. 3a–c) but possess enhanced mechanical properties. Indeed the value of G' increases about twofold when the collagen fibrillogenesis occurs in presence of LMW HA. In this case the mechanical properties improvement is particularly unexpected if considered that LMW HA posses poor mechanical properties behaving as viscous solutions (Fig. 1). This result can be explained considering that during the collagen fibrillogenesis physical interactions among the fibrils arise, thus stabilising the collagen network [20, 29, 34]. When fibrillogenesis occurs in presence of polyanion molecules, such as HA, as in our case, further interactions could be promoted resulting in a strengthening of the network.

Again the improvement of the gel mechanical properties in presence of LMW HA could be considered a mimicking of what occurs *in vivo*. It has been widely reported that the interactions among collagen and HA can control the tissue mechanical properties [8, 31]. To the collagen network is ascribed the tensile stiffness and strength of the tissues but it has been widely accepted that the interactions with proteoglycans and HA add structural rigidity to the ECM [8, 34]. Indeed, in synovial fluids and in the articular cartilage HA is associated with proteins, forming complex known as HA protein complex (HAP), having a molecular weight in the order of 1 million, controlling their mechanical properties [8].

A similar mechanical improvement was not obtained by using HMW. The elastic modulus of the HMW IP material is even lower than that the LMW semi-interpenetrating materials, for frequency below 1 Hz. This result is also unexpected since HMW HA posses strong viscoelastic properties; indeed, at high frequency HMW behaves similarly to collagen gel ($G' > G''$) (Fig. 2). This apparent contradiction could be explained considering that collagen fibrillogenesis is strongly influenced by the mobility of the dispersing phase. It is known that fibrillogenesis process can be dissected in two steps (Fig. 4): a nucleation (or lag) phase and a growth phase [35]. Initially, the collagen fibril formation goes through the nucleation of primary aggregates, dimers and trimers of collagen molecules. During this phase the mechanical properties of the collagen solution do not change strongly, later these subunits start to aggregate laterally to give rise to microfibrils and then to the largest fibrils that form a three-dimensional network stabilised by physical and covalent cross-links. A rapid increase of the mechanical parameter is registered during this phase. Because the collagen molecules and the molecules clusters have to move in the suspending medium in order to aggregate, the fibrillogenesis is strongly influenced by the mobility of the dispersing phase.

In this study the dispersing phase contains HA molecules that behave as an expanded random coil occupying a large hydrodynamic volume that increases rapidly with the increase of molecular weight thus decreasing their mobility. Moreover, from the dynamic mechanical analysis we concluded that LMW HA macromolecules may be considered as individual flow units and no entanglement occurs among polymer chains. Due to their higher mobility and the weaker homologous interactions, LMW HA may interact more strongly with collagen during fibrillogenesis than HMW HA. The HMW chains, indeed, form topological interactions among them that reduce their and collagen molecules mobility. As consequence of the reduced mobility and of the topological hindrance, the collagen molecules could be segregated from the HA molecules and start to nucleate and to aggregate without interacting with the HMW molecules. As a consequence of a possible segregation the number of the nuclei formed could be reduced and of different shape, leading to a less dense network structure having weaker mechanical properties [36]. These considerations are in agreement with our result that the rate of fibrillogenesis is reduced in presence of HMW. It has been reported that a less dense network, i.e. less nuclei formed in the lag phase, also correspond to a retarded precipitation [34–36].

CONCLUSIONS

The feasibility of mechanically stabilising HA by interpenetrating it with a fibrillar scaffold has been assessed. Collagen has been fibrillated in presence of HA having different molecular weight. The semi-interpenetrating materials showed the same mechanical behaviour of the collagen gel. In particular the dynamic viscoelastic functions of the composite materials, obtained using LMW HA, were increased compared to that of collagen. A similar mechanical properties improvement was not obtained by interpenetrating collagen with HMW HA. This result highlights the importance of specific interactions occurring between HA and collagen during the gel formation in controlling the mechanical stability of the interpenetrating network. LMW HA, due to its higher mobility compared to HMW HA, presumably promotes more effective interaction with the collagen network and this leads to stronger mechanical properties. The morphological observation showed that HA coated the collagen fibrils. The collagen fibrils acting as bridging cross element within the HA matrix in a kind of web like structure. By interpenetrating HA with collagen it is possible to obtain firm gels and preserving the physical-chemical and biological properties of HA.

REFERENCES

1. T. C. Laurent, *The Chemistry, Biology and Medical Applications of Hyaluronan and its Derivatives*. Portland Press, London (1998).
2. T. C. Laurent and J. R. E. Fraser, *FASEB J.* **6**, 2397 (1992).
3. E. R. Morris, D. A. Rees and E. J. Welsh, *J. Mol. Biol.* **138**, 383 (1980).
4. M. Terbojevich, A. Cosani and M. Palumbo, *Carbohydr. Res.* **149**, 363 (1986).
5. E. Gibbs, W. Merril and K. A. Smith, *Biopolymers* **6**, 777 (1968).
6. J. Welsh, D. A. Rees, E. R Morris and J. K. Madden, *J. Mol. Biol.* **138**, 375 (1980).
7. T. C. Laurent, B. Preston, H. Petroft, B. Gustafsson and M. McCabe, *Eur. J. Biochem.* **53**, 129 (1975).
8. V. C. Mow, M. H. Holmes and W. M. Lai, *J. Biomech.* **17**, 377 (1984).
9. L. Ambrosio, A. Borzacchiello, P. A. Netti and L. Nicolais, *J. Macromol. Sci. Pure Appl. Chem.* **A36**, 991 (1999).

10. U. Agren, C. Tolg, F. Paiwand, S. A. Paiwand, S. A. Turkeley, R. Harrison and E. A. Turley, in: *New Frontiers in Medical Sciences: Redefining Hyaluronan*, G. Abatangelo and P. H. Weigel (Eds), p. 63. Elsevier Science, Amsterdam (2000).
11. F. Della Valle and A. Romeo, European Patent No. 216453 (1987).
12. S. P. Zhong, D. Campoccia, P. J. Doherty, R. L. Williams, L. Benedetti and D. F. Williams, *Biomaterials* **15**, 359 (1994).
13. M. Mensitieri, L. Ambrosio, L. Nicolais, D. Bellini and M. O'Regan, *J. Mater. Sci. Mater. Med.* **7**, 695 (1996).
14. L. Benedetti, R. Corvito, T. Berti, A. Berti, F. Pea, M. Mazzo, M. Moras and G. Abatangelo, *Biomaterials* **14**, 1154 (1993).
15. P. Brun, R. Cortivo, M. Radice and G. Abatangelo, in: *New Frontiers in Medical Sciences: Redefining Hyaluronan*, G. Abatangelo and P. H. Weiger (Eds), p. 269. Elsevier Science, Amsterdam (2000).
16. T. Malson and B. Lindqvist, PCT WO 90/09401 (1990).
17. A. N. De Belder and T. Malson, PCT WO 86/00912 (1986).
18. T. Matsuda, M. J. Moghaddam and K. Sakurai, US Patent No. 5,462,976 (1993).
19. D. Campoccia, P. Doherty, M. Radice, P. Brun, G. Abatangelo and D. F. Williams, *Biomaterials* **19**, 2101 (1998).
20. M. Djabourov, J. Lechire and F. Gaill, *Biorheology* **30**, 191 (1993).
21. C. H. Lee, A. Singla and Y. Lee, *Int. J. Pharm.* **221**, 1 (2001).
22. J. D. Ferry, *Viscoelastic Properties of Polymers*. Wiley, New York, NY (1970).
23. J. A. Buckwalter and L. C. Rosemberg, *J. Biol. Chem*, **257**, 9830 (1982).
24. A. C. Lin and M. C. Goh, *J. Microsc.* **205**, 205 (2002).
25. A. Borzacchiello and L. Ambrosio, *J. Biomater. Sci. Polymer Edn* **12**, 307 (2001).
26. A. Borzacchiello, P. A. Netti and L. Ambrosio, in: *New Frontiers in Medical Sciences: Redefining Hyaluronan*, G. Abatangelo and P. H. Weigel (Eds), p. 195. Elsevier Science, Amsterdam (2000).
27. A. H. Clark and S. B. Ross-Murphy, *Adv. Polym. Sci.* **83**, 61 (1987).
28. S. B. Ross-Murphy and K. P. Shatwell, *Biorheology* **30**, 217 (1993).
29. S. Hsu, A. M. Jamieson and J. Blackwell, *Biorheology* **31**, 21 (1994).
30. L. Soby, A. M. Jamieson, J. Blackwell, H. U. Choi and L. C. Rosemberg, *Biopolymers* **29**, 1587 (1990).
31. J. R. Levick, in: *Interstitial-Lymphatic Liquid and Solute Movement*, N. C. Staub, J. C. Hogg and A. R. Hargens (Eds), p. 124. Karger, Basel (1987).
32. J. Y. Lee and A. P. Spicer, *Curr. Opin. Cell Biol.* **12**, 581 (2000).
33. F. Boraldi, M. A. Croce, D. Quagliano, R. Sammarco, E. Carnevali, R. Tiozzo and I. Pasquali-Ronchetti, *Tissue Cell* **35**, 37 (2003).
34. G. C. Wood and M. K. Keech, *Biochem. J.* **75**, 588 (1960).
35. G. C. Wood, *Biochem. J.* **75**, 598 (1960).
36. G. C. Wood, *Biochem. J.* **75**, 605 (1960).